Teacher, Student, and Parent
One-Stop Internet Resources

Chemistry **online**

Log on to chemistryca.com

ONLINE STUDY TOOLS

- Section Self-Check Quizzes
- Chapter Tests
- Standardized Test Practice
- Vocabulary PuzzleMaker

ONLINE RESEARCH

- WebQuests
- Prescreened Web Links
- Safety Links
- Chemistry in the News
- Periodic Table Links
- Science Fair Ideas

INTERACTIVE ONLINE STUDENT EDITION

- Complete Interactive Student Edition
- Textbook Updates

FOR TEACHERS

- Teacher Bulletin Board
- Teaching Today—Professional Development

SAFETY SYMBOLS

SAFETY SYMBOLS	HAZARD	EXAMPLES	PRECAUTION	REMEDY
DISPOSAL	Special disposal procedures need to be followed.	certain chemicals, living organisms	Do not dispose of these materials in the sink or trash can.	Dispose of wastes as directed by your teacher.
BIOLOGICAL	Organisms or other biological materials that might be harmful to humans	bacteria, fungi, blood, unpreserved tissues, plant materials	Avoid skin contact with these materials. Wear mask or gloves.	Notify your teacher if you suspect contact with material. Wash hands thoroughly.
EXTREME TEMPERATURE	Objects that can burn skin by being too cold or too hot	boiling liquids, hot plates, dry ice, liquid nitrogen	Use proper protection when handling.	Go to your teacher for first aid.
SHARP OBJECT	Use of tools or glassware that can easily puncture or slice skin	razor blades, pins, scalpels, pointed tools, dissecting probes, broken glass	Practice common-sense behavior and follow guidelines for use of the tool.	Go to your teacher for first aid.
FUME	Possible danger to respiratory tract from fumes	ammonia, acetone, nail polish remover, heated sulfur, moth balls	Make sure there is good ventilation. Never smell fumes directly. Wear a mask.	Leave foul area and notify your teacher immediately.
ELECTRICAL	Possible danger from electrical shock or burn	improper grounding, liquid spills, short circuits, exposed wires	Double-check setup with teacher. Check condition of wires and apparatus.	Do not attempt to fix electrical problems. Notify your teacher immediately.
IRRITANT	Substances that can irritate the skin or mucous membranes of the respiratory tract	pollen, moth balls, steel wool, fiberglass, potassium permanganate	Wear dust mask and gloves. Practice extra care when handling these materials.	Go to your teacher for first aid.
CHEMICAL	Chemicals can react with and destroy tissue and other materials	bleaches such as hydrogen peroxide; acids such as sulfuric acid, hydrochloric acid; bases such as ammonia, sodium hydroxide	Wear goggles, gloves, and an apron.	Immediately flush the affected area with water and notify your teacher.
TOXIC	Substance may be poisonous if touched, inhaled, or swallowed.	mercury, many metal compounds, iodine, poinsettia plant parts	Follow your teacher's instructions.	Always wash hands thoroughly after use. Go to your teacher for first aid.
FLAMMABLE	Flammable chemicals may be ignited by open flame, spark, or exposed heat.	alcohol, kerosene, potassium permanganate	Avoid open flames and heat when using flammable chemicals.	Notify your teacher immediately. Use fire safety equipment if applicable.
OPEN FLAME	Open flame in use, may cause fire.	hair, clothing, paper, synthetic materials	Tie back hair and loose clothing. Follow teacher's instruction on lighting and extinguishing flames.	Notify your teacher immediately. Use fire safety equipment if applicable.

 Eye Safety Proper eye protection should be worn at all times by anyone performing or observing science activities.

 Clothing Protection This symbol appears when substances could stain or burn clothing.

 Animal Safety This symbol appears when safety of animals and students must be ensured.

 Handwashing After the lab, wash hands with soap and water before removing goggles.

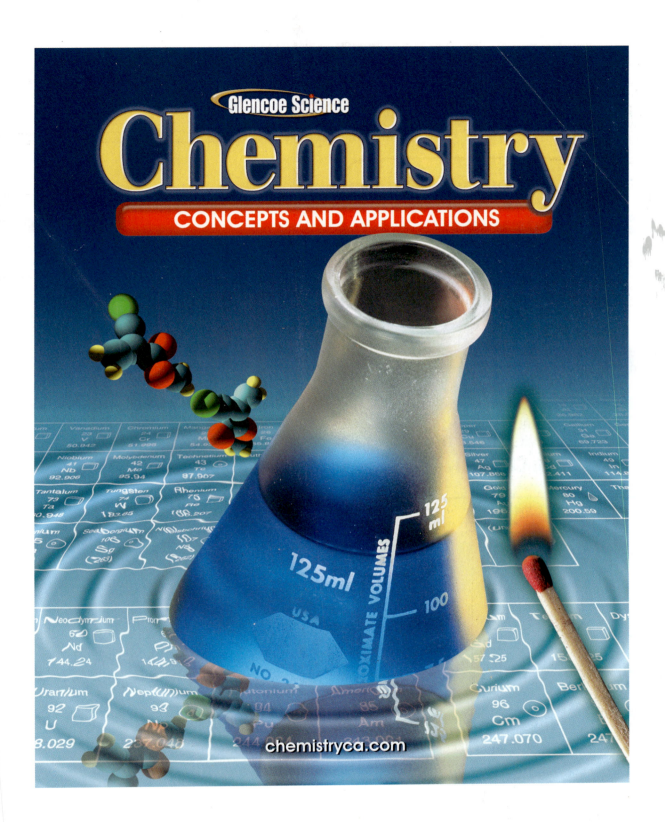

Glencoe Science

Chemistry

CONCEPTS AND APPLICATIONS

chemistryca.com

Mc Graw Hill **Glencoe**

New York, New York   Columbus, Ohio   Chicago, Illinois   Peoria, Illinois   Woodland Hills, California

Chemistry
Concepts and Applications

Visit the Chemistry Web site
chemistryca.com

You'll find:

Online Student Edition, Online Study Tools,
Interactive Tutor, Online Quizzes,
WebQuests, Teacher Forum, Problem of the
Week, Safety Links, Chemistry in the News,
Science Fair Ideas, Periodic Table Links,

and much more!

Glencoe/McGraw-Hill

A Division of The McGraw·Hill Companies

Send all inquiries to:
Glencoe/McGraw-Hill
8787 Orion Place
Columbus, Ohio 43240

ISBN 0-07-861798-7
Printed in the United States of America.

1 2 3 8 4 5 058/111 08 07 06 05 04

AUTHORS

John S. Phillips is a chemistry teacher at Forest Ridge School in Bellevue, Washington. He has been teaching chemistry at the high school and college levels for 16 years. Dr. Phillips has coordinated and led programs and workshops for teachers from kindergarten through college levels that encourage and support creative science teaching. He earned a B.A. degree in chemistry at Western Maryland College and a Ph.D. in chemistry from Purdue University. He is a member of the American Chemical Society, National Science Teachers Association, and Sigma Xi.

Victor S. Strozak is a science educator with 40 years teaching and administrative experience at both the high school and college levels. He holds a B.S. degree in chemistry from St. John's University, a M.S. in chemistry from New York University, and a Ph.D. in Science Education from New York University. Dr. Strozak taught chemistry and mathematics for six years at Xaverian High School in Brooklyn, New York and then moved on to New York City College of Technology, where he spent the next 31 years as a Professor of Chemistry, Dean of Science and Mathematics, and director of numerous science education projects. Dr. Strozak is currently the Senior Associate for Science Education at the Center for Advanced Study in Education at the Graduate Center of the City University of New York (CUNY).

Cheryl Wistrom is an associate professor of chemistry at Saint Joseph's College in Rensselaer, Indiana. She has taught chemistry and chemical education at the college level for six years. She earned her B.S. degree at Northern Michigan University and her M.S. and Ph.D. at the University of Michigan, where she carried out research on gene expression during aging of human cells. She has participated in summer institutes for educators at Pennsylvania State University and Miami University of Ohio. She is a member of the National Science Teachers Association and the American Chemical Society.

Contributing Writers

Helen Frensch, M.A.
Santa Barbara, CA

Nicholas Hainen, M.A.
Former Chemistry Teacher
Worthington High School
Worthington, OH

Zoe A. Godby Lightfoot, M.S.
Former Chemistry Teacher
Carbondale Community High School
Marion, IL

Mark V. Lorson, Ph.D.
Chemistry Teacher
Jonathan Alder High School
Plain City, OH

Robert Roth, M.S.
Pittsburgh, PA

Richard G. Smith, M.A.T.
Chemistry Teacher
Bexley High School
Bexley, OH

Patricia West
Oakland, CA

Safety Consultant

Douglas K. Mandt, M.S.
Science Education Consultant
Sumner, WA

High School Reviewers

Jon L. Allan, M.S.
University High School
Spokane, WA

William Allen, M.Ed.
Stevens Point Area Senior High School
Stevens Point, WI

Eddie Anderson
Oak Ridge High School
Oak Ridge, TN

Susan H. Brierley
Garfield High School
Seattle, WA

Robert A. Cooper, M.Ed
Pennsbury High School
Fairless Hills, PA

Sharon Doerr
Oswego High School
Oswego, NY

Jeffrey L. Engel, M.Ed., Ed.S.
Madison County High School
Danielsville, GA

Richard A. Garst
Ironwood High School
Glendale, AZ

Jo Marie Hansen
Twin Falls High School
Twin Falls, ID

Cynthia Harrison, M.S.A.
Parkway South High School
Manchester, MO

Vince Howard, M.Ed
Kentridge High School
Kent, WA

Stephen Hudson
Mission High School
San Francisco, CA

Michael Krein, M.S.
Coordinator of Chemistry
Stamford High School
Stamford, CT

**Sister John Ann Proach,
O.S.F., M.A., M.S.**
Science Curriculum Chairperson
Archdiocese of Philadelphia
Bishop McDevitt High School
Wyncote, PA

Eva M. Rambo, M.A.T.
Bloomington South High School
Bloomington, IN

Nancy Schulman, M.S.
Manalapan High School
Manalapan, NJ

Tim Watts, M.Ed.
Assistant Principal
Warren County Middle School
Front Royal, VA

Consultants

Larry B. Anderson, Ph.D.
Associate Professor
The Ohio State University
Columbus, OH

Ildiko V. Boer, M.A.
Assistant Professor
County College of Morris
Randolph, NJ

Marcia C. Bonneau, M.S.
Lecturer
State University of New York
Cortland, NY

James H. Burness, Ph.D.
Associate Professor
Penn State University
York, PA

Larry Cai
Graduate Teaching Associate
The Ohio State University
Columbus, OH

Sheila Cancella, Ph.D.
Department Chair, Science &
Engineering
Raritan Valley Community College
Somerville, NJ

James Cordray, M.S.
Berwyn, IL

Jeff Hoyle, Ph.D.
Associate Professor
Nova Scotia Agricultural College
Truro, Nova Scotia
Canada

Teresa Anne McCowen, M.S.
Senior Lecturer
Butler University
Indianapolis, IN

Lorraine Rellick, Ph.D.
Assistant Professor
Capital University
Columbus, OH

Marie C. Sherman, M.S.
Chemistry Teacher
Ursuline Academy
St. Louis, MO

Charles M. Wynn, Ph.D.
Chemistry Professor
Eastern Connecticut State University
Willimantic, CT

Chemistry: Concepts and Applications
CONTENTS IN BRIEF

Chapter 1 **Chemistry: The Science of Matter**
1.1 The Puzzle of Matter
1.2 Properties and Changes of Matter

Chapter 2 **Matter Is Made up of Atoms**
2.1 Atoms and Their Structure
2.2 Electrons in Atoms

Chapter 3 **Introduction to the Periodic Table**
3.1 Development of the Periodic Table
3.2 Using the Periodic Table

Chapter 4 **Formation of Compounds**
4.1 The Variety of Compounds
4.2 How Elements Form Compounds

Chapter 5 **Types of Compounds**
5.1 Ionic Compounds
5.2 Molecular Substances

Chapter 6 **Chemical Reactions and Equations**
6.1 Chemical Equations
6.2 Types of Reactions
6.3 Nature of Reactions

Chapter 7 **Completing the Model of the Atom**
7.1 Expanding the Theory of the Atom
7.2 The Periodic Table and Atomic Structure

Chapter 8 **Periodic Properties of the Elements**
8.1 Main Group Elements
8.2 Transition Elements

Chapter 9 **Chemical Bonding**
9.1 Bonding of Atoms
9.2 Molecular Shape and Polarity

Chapter 10 **The Kinetic Theory of Matter**
10.1 Physical Behavior of Matter
10.2 Kinetic Energy and Changes of State

Chapter 11 **Behavior of Gases**
11.1 Gas Pressure
11.2 The Gas Laws

Chapter 12 **Chemical Quantities**
12.1 Counting Particles of Matter
12.2 Using Moles

Chapter 13 **Water and Its Solutions**
13.1 Uniquely Water
13.2 Solutions and Their Properties

Chapter 14 **Acids, Bases, and pH**
14.1 Acids and Bases
14.2 Strengths of Acids and Bases

Chapter 15 **Acids and Bases React**
15.1 Acid and Base Reactions
15.2 Applications of Acid-Base Reactions

Chapter 16 **Oxidation-Reduction Reactions**
16.1 The Nature of Oxidation-Reduction Reactions
16.2 Applications of Oxidation-Reduction Reactions

Chapter 17 **Electrochemistry**
17.1 Electrolysis: Chemistry from Electricity
17.2 Galvanic Cells: Electricity from Chemistry

Chapter 18 **Organic Chemistry**
18.1 Hydrocarbons
18.2 Substituted Hydrocarbons
18.3 Plastics and Other Polymers

Chapter 19 **The Chemistry of Life**
19.1 Molecules of Life
19.2 Reactions of Life

Chapter 20 **Chemical Reactions and Energy**
20.1 Energy Changes in Chemical Reactions
20.2 Measuring Energy Changes
20.3 Photosynthesis

Chapter 21 **Nuclear Chemistry**
21.1 Types of Radioactivity
21.2 Nuclear Reactions and Energy
21.3 Nuclear Tools

APPENDICES
Appendix A Chemistry Skill Handbook
Appendix B Supplemental Practice Problems
Appendix C Safety Handbook
Appendix D Chemistry Data Handbook
Appendix E Answers to In-Chapter Practice Problems
Appendix F Try at Home Labs

Glossary/Glosario
Index

CONTENTS

CHAPTER 1   **Chemistry: The Science of Matter**   **3**

1.1   **The Puzzle of Matter**   **4**

A Picture of Matter
Using Models in Chemistry
Classifying Matter
Substances: Pure Matter
ChemLab 1.1   Observation of a Candle
ChemLab 1.2   Kitchen Chemicals
MiniLab 1.1   50 mL + 50 mL = ?
MiniLab 1.2   Paper Chromatography of Inks
MiniLab 1.3   Copper to Gold—The Alchemists' Dream
MiniLab 1.4   Waiter, what's this stuff doing in my cereal?

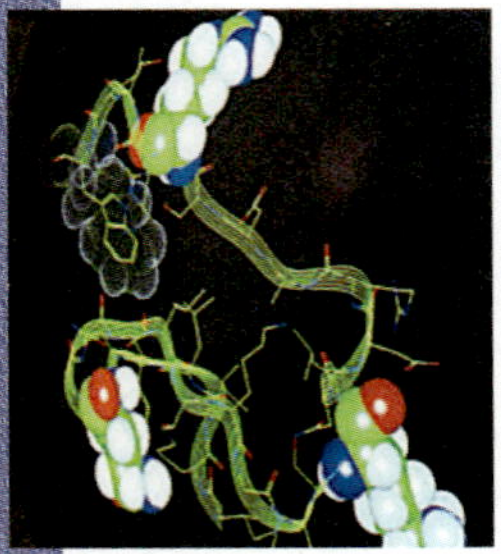

1.2   **Properties and Changes of Matter**   **34**

Identifying Matter by Its Properties
Chemical Properties and Changes
ChemLab 1.3   The Composition of Pennies
MiniLab 1.5   It's a Liquid, It's a Solid . . . It's Slime

CHAPTER 2   **Matter Is Made up of Atoms**   **51**

2.1   **Atoms and Their Structure**   **52**

Early Ideas About Matter
Development of the Modern Atomic Theory
The Discovery of Atomic Structure
Atomic Numbers and Masses
MiniLab 2.1   A Penny for Your Isotopes
ChemLab   Conservation of Matter

2.2   **Electrons in Atoms**   **69**

Electrons in Motion
The Electromagnetic Spectrum
Electrons and Light
The Electron Cloud Model
MiniLab 2.2   Line Emission Spectra
of Elements

CHAPTER 3 Introduction to the Periodic Table 85

3.1 **Development of the Periodic Table** 86
The Search for a Periodic Table
The Modern Periodic Table
MiniLab 3.1 Predicting the Properties of Mystery Elements

3.2 **Using the Periodic Table** 95
Relationship of the Periodic Table to Atomic Structure
Physical States and Classes of the Elements
Semiconductors and Their Uses
MiniLab 3.2 Trends in Reactivity Within Groups
ChemLab The Periodic Table of the Elements

CHAPTER 4 Formation of Compounds 119

4.1 **The Variety of Compounds** 120
Salt: A Familiar Compound
Carbon Dioxide: A Gas to Exhale
Water, Water Everywhere
MiniLab 4.1 Evidence of a Chemical Reaction: Iron Versus Rust

4.2 **How Elements Form Compounds** 130
When Atoms Collide
Ways to Achieve a Stable Outer Energy Level
How do ionic and covalent compounds compare?
MiniLab 4.2 The Formation of Ionic Compounds
ChemLab The Formation and Decomposition of Zinc Iodide

CHAPTER 5 Types of Compounds 153

5.1 **Ionic Compounds** 154
Formulas and Names of Ionic Compounds
Interpreting Formulas
MiniLab 5.1 A Chemical Weather Predictor

5.2 **Molecular Substances** 170
Properties of Molecular Substances
Molecular Elements
Formulas and Names of Molecular Compounds
MiniLab 5.2 Where's the calcium?
ChemLab Ionic or Covalent?

$7e^-$ $2e^-$ $8e^-$

CHAPTER 6   **Chemical Reactions and Equations**   **189**

6.1   **Chemical Equations**   **190**
     Recognizing Chemical Reactions
     Writing Chemical Equations
     Balancing Chemical Equations
   MiniLab 6.1 Energy Change

6.2   **Types of Reactions**   **202**
     Why Reactions Are Classified
     Major Classes of Reactions
   MiniLab 6.2 A Simple Exchange
   ChemLab Exploring Chemical Changes

6.3   **Nature of Reactions**   **210**
     Reversible Reactions
     Reaction Rate
   MiniLab 6.3 Starch-Iodine Clock Reaction

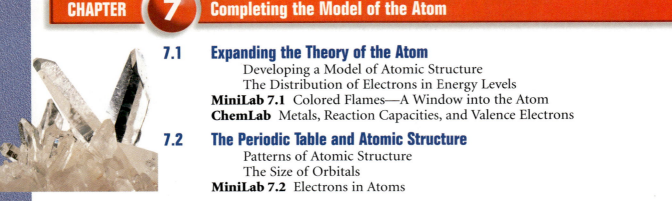

CHAPTER 7   **Completing the Model of the Atom**   **229**

7.1   **Expanding the Theory of the Atom**   **230**
     Developing a Model of Atomic Structure
     The Distribution of Electrons in Energy Levels
   MiniLab 7.1 Colored Flames—A Window into the Atom
   ChemLab Metals, Reaction Capacities, and Valence Electrons

7.2   **The Periodic Table and Atomic Structure**   **243**
     Patterns of Atomic Structure
     The Size of Orbitals
   MiniLab 7.2 Electrons in Atoms

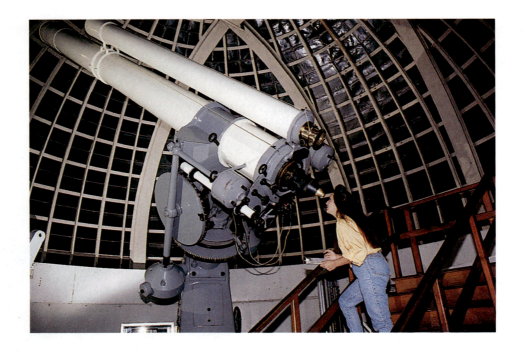

CHAPTER 8   Periodic Properties of the Elements   257

8.1   Main Group Elements   258
Patterns of Behavior of Main Group Elements
The Main Group Metals and Nonmetals
MiniLab 8.1 What's periodic about atomic radii?
ChemLab Reactions and Ion Charges of the Alkaline Earth Elements

8.2   Transition Elements   282
Properties of the Transition Elements
Other Transition Elements: A Variety of Uses
Lanthanides and Actinides: The Inner Transition Elements
MiniLab 8.2 The Ion Charges of a Transition Element

CHAPTER 9   Chemical Bonding   301

9.1   Bonding of Atoms   302
A Model of Bonding
Electronegativity: An Attraction for Electrons
The Ionic Extreme
The Covalent Extreme
Polar Covalent Bonds
Bonding in Metals
MiniLab 9.1 Coffee Filter Chromatography

9.2   Molecular Shape and Polarity   315
The Shapes of Molecules
How Polar Bonds and Geometry Affect Molecular Polarity
Ions, Polar Molecules, and Physical Properties
MiniLab 9.2 Modeling Molecules
ChemLab What colors are in your candy?

CHAPTER 10   The Kinetic Theory of Matter   339

10.1   Physical Behavior of Matter   340
States of Matter
The Kinetic Theory of Matter
Other Forms of Matter
MiniLab 10.1 Molecular Race

10.2   Kinetic Energy and Changes of State   348
Temperature and Kinetic Energy
Changing State
MiniLab 10.2 Vaporization Rates
ChemLab Molecules and Energy

CHAPTER 11   **Behavior of Gases**   **371**

11.1   **Gas Pressure**   **372**
  Defining Gas Pressure
  Devices to Measure Pressure
  Pressure Units
  MiniLab 11.1 Relating Mass and Volume of a Gas

11.2   **The Gas Laws**
  Boyle's Law: Pressure and Volume
  Kinetic Explanation of Boyle's Law
  Charles's Law: Temperature and Volume
  Kinetic Explanation of Charles's Law
  Combined Gas Law
  The Law of Combining Gas Volumes
  MiniLab 11.2 How Straws Function
  ChemLab Boyle's Law

CHAPTER 12   **Chemical Quantities**   **403**

12.1   **Counting Particles of Matter**   **404**
  Stoichiometry
  Molar Mass
  MiniLab 12.1 Determining Number Without Counting

12.2   **Using Moles**   **414**
  Using Molar Masses in Stoichiometric Problems
  Using Molar Volumes in Stoichiometric Problems
  Ideal Gas Law
  Theoretical Yield and Actual Yield
  Determining Mass Percents
  Determining Chemical Formulas
  MiniLab 12.2 Bagging the Gas
  ChemLab Analyzing a Mixture

CHAPTER 13   **Water and Its Solutions**   **435**

13.1   **Uniquely Water**   **436**
  Water: The Molecular View
  Intermolecular Forces in Water
  Water: Physical Properties Revisited
  More Evidence for Water's Intermolecular Forces
  MiniLab 13.1 How many drops can you put on a penny?

13.2   **Solutions and Their Properties**   **451**
  The Dissolving Process
  Solution Concentration
  Solution Properties and Applications
  Solutions of Gases in Water
  Colloids
  MiniLab 13.2 Hard and Soft Water
  ChemLab Solution Identification

CHAPTER 14 Acids, Bases, and pH — 479

14.1 Acids and Bases — 480
Macroscopic Properties of Acids and Bases
Defining Acids and Bases-A Submicroscopic Look
Acid Ionization
Submicroscopic Behavior of Bases
Other Acids and Bases: Anhydrides
The Macroscopic-Submicroscopic Acid-Base Connection
MiniLab 14.1 What do acids do?

14.2 Strengths of Acids and Bases — 497
Strong Acids and Bases
Weak Acids and Bases
The pH Scale
Using Indicators to Measure pH
MiniLab 14.2 Antacids
ChemLab Household Acids and Bases

CHAPTER 15 Acids and Bases React — 515

15.1 Acid and Base Reactions — 516
Types of Acid-Base Reactions
Strong Acid + Strong Base
Strong Acid + Weak Base
A Broader Definition of Acids and Bases
Weak Acid + Strong Base
MiniLab 15.1 Acidic, Basic, or Neutral?

15.2 Applications of Acid-Base Reactions — 531
Buffers to Regulate pH
The Acid-Base Chemistry of Antacids
Stoichiometry Revisited: Acid-Base Titrations
MiniLab 15.2 What does a buffer do?
ChemLab Titration of Vinegar

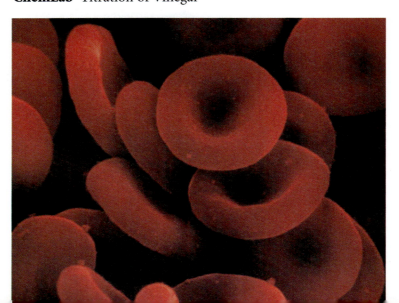

CHAPTER 16 Oxidation-Reduction Reactions 553

16.1 The Nature of Oxidation-Reduction Reactions 554
What is oxidation-reduction?
Identifying a Redox Reaction
Oxidizing and Reducing Agents
MiniLab 16.1 Corrosion of Iron
ChemLab Copper Atoms and Ions: Oxidation and Reduction

16.2 Applications of Oxidation-Reduction Reactions 563
Say Cheese: Redox in Photography
Having a Blast: Redox in a Blast Furnace
Redox in Bleaching Processes
Corrosion of Metals
Silver Tarnish: A Redox Reaction
Chemiluminescence: It's Cool
Biochemical Redox Processes
MiniLab 16.2 Testing for Alcohol by Redox

CHAPTER 17 Electrochemistry 583

17.1 Electrolysis: Chemistry from Electricity 584
Redox Revisited
Electrolysis
Applications of Electrolysis
MiniLab 17.1 Electrolysis

17.2 Galvanic Cells: Electricity from Chemistry 599
Electrochemical Cells
Better and Better Batteries
MiniLab 17.2 The Lemon with Potential
ChemLab Oxidation-Reduction and Electrochemical Cells

CHAPTER 18 Organic Chemistry 621

18.1 **Hydrocarbons** 622
Millions and Millions of Organic Compounds
Saturated Hydrocarbons
Unsaturated Hydrocarbons
Sources of Organic Compounds
MiniLab 18.1 How unsaturated is your oil?

18.2 **Substituted Hydrocarbons** 640
Functional Groups
The Sources of Functional Groups
MiniLab 18.2 A Synthetic Aroma

18.3 **Plastics and Other Polymers** 648
Monomers and Polymers
Polymerization Reactions
MiniLab 18.3 When Polymers Meet Water
ChemLab Identification of Textile Polymers

CHAPTER 19 The Chemistry of Life 667

19.1 **Molecules of Life** 668
Biochemistry
Proteins
Carbohydrates
Lipids
Nucleic Acids
Vitamins
MiniLab 19.1 DNA—The Thread of Life
ChemLab Catalytic Decomposition—It's in the Cells

19.2 **Reactions of Life** 692
Metabolism
Respiration
Fermentation
MiniLab 19.2 Yeast Plus Sugar—Let It Rise

CHAPTER 20 Chemical Reactions and Energy **707**

20.1 Energy Changes in Chemical Reactions **708**
Exothermic and Endothermic Reactions
Heat
Forces That Drive Chemical Reactions
MiniLab 20.1 Dissolving—Exothermic or Endothermic?

20.2 Measuring Energy Changes **719**
Calorimetry
Energy Value of Food
Energy Economics
MiniLab 20.2 Heat In, Heat Out
ChemLab Energy Content of Some Common Foods

20.3 Photosynthesis **733**
The Basis of Photosynthesis
The Chemistry of Photosynthesis
Energy and the Role of Photosynthesis

CHAPTER 21 Nuclear Chemistry **743**

21.1 Types of Radioactivity **744**
Discovery of Radioactivity
Nuclear Notation
Radioactive Decay
Detecting Radioactivity
Half–Life and Radioisotope Dating
ChemLab The Radioactive Decay of "Pennium"

21.2 Nuclear Reactions and Energy **761**
The Power of the Nucleus
Nuclear Fission
Nuclear Fusion
MiniLab 21.1 A Nuclear Fission Chain Reaction

21.3 Nuclear Tools **768**
Medical Uses of Radioisotopes
Nonmedical Uses of Radioisotopes
Sources of Radioisotopes
Problems Associated with Radioactivity
MiniLab 21.2 Radon—A problem in your home?

APPENDICES

Appendix A   **Chemistry Skill Handbook**   **785**

Measurement in Science
Making and Interpreting Measurements
Computations with the Calculator
Using the Factor Label Method
Organizing Information

Appendix B   **Supplemental Practice Problems**   **809**

Appendix C   **Safety Handbook**   **839**

Safety Guidelines in the Chemistry Laboratory
First Aid in the Laboratory
Safety Symbols

Appendix D   **Chemistry Data Handbook**   **841**

Table D.1   Symbols and Abbreviations
Table D.2   The Modern Periodic Table
Table D.3   Alphabetical Table of the Elements
Table D.4   Properties of Elements
Table D.5   Electron Configurations of the Elements
Table D.6   Useful Physical Constants
Table D.7   Names and Charges of Polyatomic Ions
Table D.8   Solubility Guidelines
Table D.9   Solubility Product Constants
Table D.10 Acid-Base Indicators

Appendix E   **Answers to In-Chapter Practice Problems**   **853**

Appendix F   **Try At Home Labs**   **863**

Glossary/Glosario   **000**

Index   **000**

Everyday Chemistry

How does your microwave work? Why do you hiccup? What makes a ruby red? These and other everyday questions are answered here as you explore their underlying chemistry.

Chapter	Title	Page
1	You Are What You Eat	19
2	Fireworks—Getting a Bang Out of Color	76
3	Metallic Money	110
4	Elemental Good Health	128
5	Hard Water	160
6	Whitening Whites	194
	Stove in a Sleeve	221
7	Colors of Gems	248
8	The Chemistry of Matches	275
9	Jiggling Molecules	320
10	Freeze Drying	353
11	Popping Corn	397
12	Air Bags	417
13	Soaps and Detergents	455
	Antifreeze	466
14	Balancing pH in Cosmetics	501
15	Hiccups	534
16	Lightning-Produced Fertilizer	571
17	Manufacturing a Hit CD	594
18	Chemistry and Permanent Waves	657
19	Clues to Sweetness	683
	Fake Fats and Designer Fats	685
20	Catalytic Converters	715
21	Radon—An Invisible Killer	777

ChemLabs

ChemLabs offer you the opportunity to discover how elements form compounds, why matter changes but is not created, what gives candies their colors, and many other chemical wonders. Develop your lab skills and become a practicing chemist.

Chapter	Title	Page
1.1	Observation of a Candle	8
1.2	Kitchen Chemicals	16
1.3	The Composition of Pennies	38
2	Conservation of Matter	56
3	The Periodic Table of the Elements	100
4	The Formation and Decomposition of Zinc Iodide	136
5	Ionic or Covalent?	172
6	Exploring Chemical Changes	206
7	Metals, Reaction Capacities, and Valence Electrons	236
8	Reactions and Ion Charges of the Alkaline Earth Elements	266
9	What colors are in your candy?	328
10	Molecules and Energy	362
11	Boyle's Law	384
12	Analyzing a Mixture	422
13	Solution Identification	456
14	Household Acids and Bases	504
15	Titration of Vinegar	542
16	Copper Atoms and Ions: Oxidation and Reduction	560
17	Oxidation-Reduction and Electrochemical Cells	606
18	Identification of Textile Polymers	650
19	Catalytic Decomposition—It's in the Cells	674
20	Energy Content of Some Common Foods	722
21	The Radioactive Decay of "Pennium"	752

miniLABS

With these short and easy activities, discover for yourself how chemistry is simple and exciting. Just as a picture is worth a thousand words, a MiniLab can take the place of many hours of study.

Title	Page
1.1 50 mL + 50 mL = ?	21
1.2 Paper Chromatography of Inks	22
1.3 Copper to Gold—The Alchemists' Dream	25
1.4 Waiter, what's this stuff doing in my cereal?	30
1.5 It's a Liquid, It's a Solid . . . It's Slime	40
2.1 A Penny for Your Isotopes	63
2.2 Line Emission Spectra of Elements	77
3.1 Predicting the Properties of Mystery Elements	89
3.2 Trends in Reactivity Within Groups	97
4.1 Evidence of a Chemical Reaction: Iron Versus Rust	122
4.2 The Formation of Ionic Compounds	135
5.1 A Chemical Weather Predictor	166
5.2 Where's the calcium?	171
6.1 Energy Change	196
6.2 A Simple Exchange	205
6.3 Starch-Iodine Clock Reaction	220
7.1 Colored Flames—A Window into the Atom	234
7.2 Electrons in Atoms	245
8.1 What's periodic about atomic radii?	262
8.2 The Ion Charges of a Transition Element	285
9.1 Coffee Filter Chromatography	312
9.2 Modeling Molecules	325

Title	Page
10.1 Molecular Race	343
10.2 Vaporization Rates	357
11.1 Relating Mass and Volume of a Gas	375
11.2 How Straws Function	388
12.1 Determining Number Without Counting	408
12.2 Bagging the Gas	420
13.1 How many drops can you put on a penny?	443
13.2 Hard and Soft Water	452
14.1 What do acids do?	482
14.2 Antacids	503
15.1 Acidic, Basic, or Neutral?	518
15.2 What does a buffer do?	532
16.1 Corrosion of Iron	557
16.2 Testing for Alcohol by Redox	568
17.1 Electrolysis	587
17.2 The Lemon with Potential	600
18.1 How unsaturated is your oil?	630
18.2 A Synthetic Aroma	646
18.3 When Polymers Meet Water	653
19.1 DNA—The Thread of Life	688
19.2 Yeast Plus Sugar—Let It Rise	699
20.1 Dissolving—Exothermic or Endothermic?	712
20.2 Heat In, Heat Out	726
21.1 A Nuclear Fission Chain Reaction	763
21.2 Radon—A problem in your home?	775

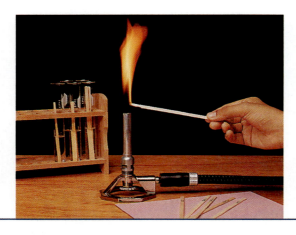

Launch Labs

With these short and easy activities, discover for yourself how chemistry is simple and exciting. Just as a picture is worth a thousand words, a Launch Lab can take the place of many hours of study.

Chapter	Title	Page
1	Where is It?	3
2	What's Inside?	51
3	Versatile Metals	85
4	Red Liquid to Clear Liquid	119
5	Elements, Compounds, and Mixtures	153
6	Observing a Chemical Reaction	189
7	Observing Electrical Charge	229
8	Magnetic Materials	257
9	Oil and Vinegar Dressing	301
10	Defying Density	339
11	More Than Just Hot Air	371
12	How Much is a Mole?	403
13	Solution Formation	435
14	Testing pH Using Natural Indicators	479
15	Physical or Chemical Change?	515
16	Observing an Oxidation-Reduction Reaction	553
17	A Lemon Battery?	583
18	Making Slime	621
19	Testing for Simple Sugars	667
20	Speeding Reactions	707
21	Chain Reactions	743

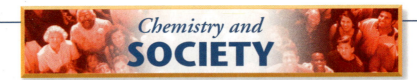

Chemistry and SOCIETY

How has chemistry helped solve some of the world's problems, such as finding cures for rare diseases, cleaning up pollution, and clothing the world's ever-increasing population? Take this opportunity to learn how chemistry is used to benefit society.

Chapter	Title	Page
1	Natural Versus Synthetic Chemicals	32
2	Recycling Glass	60
4	The Rain Forest Pharmacy	146
13	Water Treatment	447
14	Atmospheric Pollution	495
15	The Development of Artificial Blood	537
18	Recycling Plastics	659

CHEMISTRY & TECHNOLOGY

What new advances have been developed from the application of chemical principles? Explore how metals can retain memory, where to look for alternative energy sources, and how microscopes can now be used to see atoms in these amazing new stories of science in progress.

Chapter	Title	Page
3	Metals That Untwist	108
5	Carbon Allotropes: From Soot to Diamonds	176
6	Mining the Air	216
7	Hi-Tech Microscopes	240
8	Carbon and Alloy Steels	288
9	Chromatography	326
10	Fractionation of Air	354
11	Hyperbaric Oxygen Chambers	390
12	Improving Percent Yield in Chemical Synthesis	424
13	Versatile Colloids	470
14	Manufacturing Sulfuric Acid	484
16	Forensic Blood Detection	573
17	Copper Ore to Wire	590
20	Alternative Energy Sources	728
21	Archaeological Radiochemistry	754

How it Works

Look inside a tire gauge, an electric lightbulb, or a rechargeable battery to find out what makes them work. How does chemistry help you understand these simple everyday items?

Chapter	Title	Page
5	Cement	167
6	Emergency Light Sticks	197
8	Inert Gases in Lightbulbs	284
10	Pressure Cookers	359
11	Tire-Pressure Gauge	377
12	Electronic Balances	410
13	A Portable Reverse Osmosis Unit	468
15	Taste	519
15	Indicators	545
16	Breathalyzer Test	569
17	Nicad Rechargeable Batteries	612
17	Hydrogen-Oxygen Fuel Cell	614
20	Hot and Cold Packs	710
21	Smoke Detectors	748

PEOPLE in CHEMISTRY

Look inside a tire gauge, an electric lightbulb, or a rechargeable battery to find out what makes them work. How does chemistry help you understand these simple everyday items?

Chapter	Name	Page
5	**Forensic Scientist** Dr. John Thornton	167
6	**Plant-Care Specialist** Caroline Sutliff	197
8	**Chemist** Dr. William Skawinski	284
10	**Wastewater Operator** Alice Arellano	359
11	**Cosmetic Bench Chemist** Fe Tayag	377
12	**Metal Plater** Harvey Morser	410
13	**Pharmacist** John Garcia	468
15	**Biochemist** Dr. Lynda Jordan	519
13	**Pharmacist** John Garcia	468
15	**Biochemist** Dr. Lynda Jordan	519

Cross-Curricular Connections

It may not have occurred to you that chemistry is an intergral part of all your courses, not just the sciences. Learn in these features how chemistry is connected to literature, art, and history, as well as to the sciences of physics, biology, health, and earth science.

LITERATURE

Jules Verne and His Icebergs   26
The Language of a Chemist   96

ART

China's Porcelain   163
Glass Sculptures   346
Asante Brass Weights   411
Art Forger, van Meegeren—Villain or Hero?   759

HISTORY

Politics and Chemistry—Elemental Differences   58
Hydrogen's Ill-Fated Lifts   141
Lead Poisoning in Rome   271
Linus Pauling: An Advocate of Knowledge and Peace   307

PHYSICS

Aurora Borealis   73
Niels Bohr—Atomic Physicist and Humanitarian   232
Solid Rocket Booster Engines   566

BIOLOGY

Air in Space   203
Fluorides and Tooth Decay   280
Measurement of Blood Gases   487
Vision and Vitamin A   632
A Biological Mystery Solved with Tracers   772

HEALTH

Lithium Batteries in Pacemakers   610
Function of Hemoglobin   693

EARTH SCIENCE

Weather Balloons   387
Cave Formation   524
Bacterial Refining of Ores   727

Chemistry: The Science of Matter

Chapter Preview

Sections

1.1 The Puzzle of Matter

ChemLab 1.1 Observation of a Candle
ChemLab 1.2 Kitchen Chemicals
MiniLab 1.1 50 mL + 50 mL = ?
MiniLab 1.2 Paper Chromotography of Inks
MiniLab 1.3 Copper to Gold—The Alchemist's Dream
MiniLab 1.4 Waiter, What's This Stuff Doing in My Cereal?

1.2 Properties and Changes of Matter

ChemLab 1.3 The Composition of Pennies
MiniLab 1.5 It's a Liquid, It's a Solid… It's Slime

I Can't Believe It's Matter!

It is difficult to believe but everything in this picture has something in common. In fact, everything in the universe has something in common!

How is a baseball bat like cotton candy? How is the grass on the field related to your body? Those questions are answered by chemistry—chemistry is the what, how, and why of matter!

*L*aunch Lab

Where is it?

Matter is anything that has mass and volume, and the three most common forms matter can take on Earth are solids, liquids, and gases. How do the masses of these three states of matter compare?

Materials

- balloons (3)
- salt
- balance
- funnel
- graduated cylinder
- water
- scissors
- string

Safety Precautions

Do not eat or drink anything in the laboratory.

Procedure

1. Measure the mass of a balloon.
2. Insert the narrow end of a funnel into the opening of the balloon and fill it with water without stretching the bulb of the balloon. Tie off the balloon's end, measure the mass of the balloon and water, and calculate the mass of the water.
3. Repeat steps 1 and 2 using salt. The size of the salt-filled balloon should be approximately the same size as the water-filled balloon.
4. Repeat steps 1 and 2 using air. Blow just enough air into the balloon so that it is approximately the same size as the balloons filled with water and salt.

Analysis

Compare the masses of the solid, liquid, and gas. Hypothesize why the masses of the three states differ.

Reading Chemistry

Observe the Periodic Table of the Elements on pages 28–29. Write down a few chemical names of elements that seem familiar. Scan the chapter, looking for some of these elements. Note how the element symbols are used to represent different substances. Write down any questions that form.

Preview this chapter's content and activities at chemistryca.com

The Puzzle of Matter

SECTION PREVIEW

Objectives

✓ **Classify** matter according to its composition.

✓ **Distinguish** among elements, compounds, homogeneous mixtures, and heterogeneous mixtures.

✓ **Relate** the properties of matter to its structure.

Vocabulary

chemistry
matter
mass
property
scientific model
qualitative
quantitative
substance
mixture
physical change
physical property
solution
alloy
solute
solvent
aqueous solution
element
compound
formula

Have you ever worked a jigsaw puzzle? Imagine that you've just been hired by the Acme Puzzle Company to design the world's most difficult jigsaw puzzle. You'd probably want your puzzle to have a lot of pieces—thousands, or even tens of thousands. That way, no single piece would give away much information about the whole picture. Finally, you might pack the pieces in a plain box with no picture. That would be a tough puzzle.

A Picture of Matter

When you begin a study of chemistry, you start to work on a challenging puzzle—the puzzle of matter. Every chunk of matter is a puzzle piece. Your puzzle box is the universe, and the box contains many different kinds of pieces. Your job, like that of the puzzle solver at the top of the page, is to figure out how to connect all the different pieces. Keep in mind that, as with an ordinary jigsaw puzzle, your goal is not only to connect the pieces but also to see the complete picture that emerges.

Composition, Structure, and Behavior

Chemistry is the science that investigates and explains the structure and properties of matter. Matter is the *stuff* that's all around you: the metal and plastic of a telephone, the paper and ink of a book, the glass and liquid of a bottle of soda, the air you breathe, and the materials that make up your body. A more formal definition of **matter** is anything that takes up space and has mass. **Mass** is the measure of the amount of matter that an object contains. On Earth, mass is usually equated with weight. **Figure 1.1** compares two chunks of matter—one with a large mass and one with a small mass. On pages 785-791 of the *Skill Handbook* in the back of this book, you'll find a description of the metric system and the units used in chemistry. What isn't matter? The heat and light from a lamp are not matter; neither are thoughts, ideas, radio waves, or magnetic fields.

The structure of matter refers to its composition—what matter is made of—as well as how matter is organized. The **properties** of matter describe the characteristics and behavior of matter, including the changes that matter undergoes. **Figure 1.2** compares some different kinds of matter in terms of composition and behavior.

Figure 1.1
Comparison of Masses
Mass is a measure of the amount of matter in an object.

◄ The soda in the container has a mass of 1 kg. The kilogram is the unit of mass in the metric system.

The bus below has a large mass (about 14 000 kg or 15 tons), whereas the strand of hair at the right has a small mass (approximately 0.000 001 kg).

Figure 1.2
Comparing Composition and Behavior
Salt and water have different compositions, so it isn't surprising that they have different properties. Salt is made up of the elements sodium and chlorine, while water is made up of hydrogen and oxygen. You couldn't wash your hair with salt, just as you wouldn't sprinkle water on your popcorn. ▼

◄ Aspirin and sucrose (table sugar) are both composed of carbon, hydrogen, and oxygen, but you wouldn't use aspirin to sweeten your cereal or take a spoonful of sugar for a headache. Even though aspirin and sugar contain the same elements, differences in their structures determine their individual behaviors.

Aspartame and saccharin are substances with different compositions but similar tastes. Saccharin is made of carbon, hydrogen, nitrogen, oxygen, sodium, and sulfur. Aspartame also contains carbon, hydrogen, nitrogen, and oxygen, but it contains no sodium or sulfur. The way in which the components of aspartame and saccharin are organized must be a major factor in causing both of them to have a sweet taste. ▶

Figure 1.3

Some Properties of Iron

Many properties of iron are easy to observe.

Iron is strong, yet can be flattened and stretched.

Iron doesn't dissolve in water, but it rusts when exposed to air and water.

Iron turns to liquid at a high temperature.

Iron is a gray, shiny solid.

Iron is attracted by a magnet.

Iron conducts electricity.

You can determine some properties of a particular chunk of matter just by examining or manipulating it. What color is it? Is it a solid, liquid, or gas? If it's a solid, is it soft or hard? Does it burn? Does it dissolve in water? Does something happen when you mix it with another kind of matter? You determine all of these properties by examining and manipulating a chunk of stuff, as shown in **Figure 1.3.**

Although you can find out a lot about a piece of material just by looking at it and doing simple tests, you can't tell what something is made of only by looking at it, no matter whether it is a spoonful of sugar or a rock picked up by the *Spirit* or *Opportunity* Rovers on the surface of Mars. Measurements usually must be made or chemical changes observed. **Figure 1.4** shows, by a simple experiment, that sugar is composed of carbon, hydrogen, and oxygen. Most matter does not reveal its composition so easily.

Figure 1.4

The Composition of Sugar

When concentrated sulfuric acid is added to sucrose, an interesting reaction occurs. The sugar breaks down and forms water (composed of hydrogen and oxygen) **A**. The water is released as steam **B** (center), and black carbon is left behind **C** (right).

Figure 1.5

The Visible Results of Structure
When you look at the Sears Tower, you see its size and shape. These properties are the result of the building's structure, which is hidden from view under the exterior skin of the building. You cannot see the organization of the steel beams and bolts that hold the building together, or the system of pipes, wires, and ventilation ducts that thread through the building.

Examining Matter: The Macroscopic View of Matter

Observations of the composition and behavior of matter are based on a macroscopic view. Matter that is large enough to be seen is called macroscopic, so all of your observations in chemistry, and everywhere else, start from this perspective. The macroscopic world is the one you touch, feel, smell, taste, and see. The properties of iron shown in **Figure 1.3** are seen from a macroscopic perspective. But if you want to describe and understand the structure of iron, you must use a different perspective—one that allows you to see what can't be seen. What you can see of the Sears Tower, shown in **Figure 1.5,** is similar to a macroscopic perspective. The actual structure of the building is hidden from view.

In the same way, the appearance and properties of a piece of matter are the result of its structure. Although you may get hints of the actual structure from a macroscopic view, you must go to a submicroscopic perspective to understand how the hidden structure of matter influences its behavior.

The Submicroscopic View of Matter

The submicroscopic view gives you a glimpse into the world of atoms. It is a world so small that you cannot see it even with the most powerful microscope, hence the term *sub*microscopic. You learned in earlier science courses that matter is made up of atoms. Compared with the macroscopic world, atoms are so small that if the period at the end of this sentence were made up of carbon atoms, it would be composed of more than 100 000 000 000 000 000 000 (100 quintillion) carbon atoms. If you could count all of those atoms at a rate of three per second, it would take you a trillion years to finish. Fortunately, in chemistry, you will spend your time becoming acquainted with atoms rather than counting them.

ChemLab 1

Observation of a Candle

You have seen candles burn, perhaps on a birthday cake. But you probably have never considered the burning of a candle from a chemist's point of view. Michael Faraday, a 19th-century chemist, found much to observe as a candle burns. He wrote a book and gave talks on the subject. In this ChemLab, you will investigate the burning of a candle and the products of combustion.

Problem

What are the requirements for and characteristics of a candle flame? What are the products of the combustion of the candle?

Objectives

- **Observe** a candle flame and perform several tests.
- **Interpret** observations and the results of the tests.

PREPARATION

Materials

large birthday candles
matches
shallow metal dish
25 mL of limewater solution
250-mL beaker
500-mL Erlenmeyer flask
solid rubber stopper to fit the flask
wire gauze square
tongs

Safety Precautions

Keep all combustible materials, including clothing, away from the match and candle flames. Do not allow the limewater to splash into your eyes. If it does, immediately rinse your eyes for 15 minutes and notify your teacher.

PROCEDURE

1. Light a candle and allow a drop or two of liquid wax to fall into the center of the pan. Press the candle upright onto the melted wax before it can solidify. If the candle burns too low during the following procedures, repeat this step with a new candle.

2. Observe the flame of the burning candle for a few minutes. Try to observe what is burning and where the burning takes place. Observe the different regions of the flame. Make at least eight observations, and record them in a data table like the one shown.

3. Light a second candle and hold the flame about 2 cm to 4 cm to the side of the first candle flame. Gently blow out the first candle flame, then quickly move the flame of the second candle into the smoke from the first flame. Record your observations.

4. Relight the standing candle. With tongs, hold the wire gauze over the flame, perpendicular to the candle. Slowly lower the gauze onto the flame. Do not allow the gauze to touch the candle wax. If the flame goes out, quickly move the wire gauze off to the side. Record your observations.

5. Fill the 250-mL beaker with cold tap water, dry the outside of the beaker, and hold it about 3 cm to 5 cm above the candle flame. Record your observations.

6. Pour tap water into the pan or dish to a depth of about 1 cm.

7. Quickly lower an Erlenmeyer flask over the candle so that the mouth of the flask is below the surface of the water. Allow the flask to remain in place for approximately one minute. Record your observations.

8. Lift the flask out of the water, turn it upright, and add about 25 mL of limewater. Stopper the flask and swirl the solution for approximately one minute. Record your observations. If the solution becomes cloudy or chalky, calcium carbonate was formed, indicating the presence of carbon dioxide in the flask.

ANALYZE AND CONCLUDE

1. **Classifying** Which changes that you noted in step 2 were physical? Which were chemical?

2. **Making Inferences** Do your results in step 3 indicate that the candle wax burns as a solid, a liquid, or a vapor? Explain.

3. **Interpreting Data** One requirement for combustion is the presence of fuel. Interpret your results from steps 4 and 7 to determine the other requirements.

4. **Interpreting Data** Based upon your analysis of the observations from steps 5 and 8, what are two products of the combustion of the candle?

APPLY AND ASSESS

1. Sir Humphry Davy invented a safety lamp for miners in which a flame was surrounded by a wire gauze cylinder. Can you explain the reason why the lamp was constructed in this way?

2. What change in water level occurred in procedure 7? Propose an explanation for this change.

DATA AND OBSERVATIONS

Procedure step	Observations
2	
3	
4	
5	
7	
8	

A probe like this one moves up and down in response to the position of carbon atoms on the surface. A computer converts the probe's motion into a bumpy-looking image. ▶

This image of a crystal of graphite, a form of carbon, is obtained by scanning the surface with a fine, sensitive probe. ▼

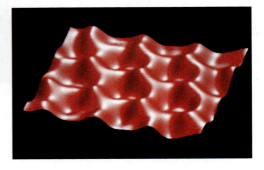

Figure 1.6
A Submicroscopic View of Matter
Although a scanning tunneling microscope provides a glimpse of the submicroscopic world, it does not truly produce pictures of atoms, at least not as we think of pictures in the macroscopic world.

Although individual atoms cannot be seen, the recently developed scanning tunneling microscope (STM), shown in **Figure 1.6,** is capable of producing images on a computer screen that show the locations of individual atoms. The STM can even be used to move individual atoms around on a surface.

Using Models in Chemistry

In your study of chemistry, you will use both macroscopic and submicroscopic perspectives. For example, sucrose and aspirin are both composed of carbon, hydrogen, and oxygen atoms, but they have different behaviors and functions. These differences must come about because of differences in the submicroscopic arrangement of their atoms. **Figure 1.7** shows models that reveal these submicroscopic differences.

Figure 1.7
Comparing the Structures of Aspirin and Sucrose
The different submicroscopic arrangements of the atoms in aspirin (left) and sucrose (right) cause the differences in their behavior. Don't worry about not understanding the complete meaning of these structures. Each ball represents an atom, and each bar between atoms represents a chemical connection between the atoms.

○ hydrogen

⬤ carbon

🔴 oxygen

Models Connect the Macroscopic and Submicroscopic Views

The drawings in **Figure 1.7** are tools that allow you to study what you cannot actually see—in this case, the arrangement of atoms. These drawings represent one type of model used in science. Sometimes, a model is something you can see and manipulate. These are the kinds of models you are already familiar with—model cars and airplanes or perhaps an architect's scale model of a proposed building. **Figure 1.8** shows that models are important tools in many fields. Models are used, tested, and revised constantly through new experiments. A model of the submicroscopic structure of a piece of matter must be able to explain the observed macroscopic behavior of that matter and predict behavior that has yet to be observed.

The model of aspirin (acetylsalicylic acid) in **Figure 1.8** is an example of a scientific model. A **scientific model** is a thinking device that helps you understand and explain macroscopic observations. Scientific models are built on experimentation. In Greece, a model of matter based on atoms was discussed about 2500 years ago, but this model was not a scientific model because it was never supported by experiments. It took until the 1800s before a scientific model of matter was proposed. This atomic model was developed and verified by experiments. It has withstood 200 years of prediction and experimentation with only slight modifications.

◄ A model of a prospective new airliner is made and studied before an airplane is actually built.

Figure 1.8

Using Models

Models help you see and understand structure, whether you're creating a new airplane or designing a drug that you hope will be an effective treatment for AIDS.

The model below is another, more informative way to represent the arrangement of atoms in the compound aspirin. ▼

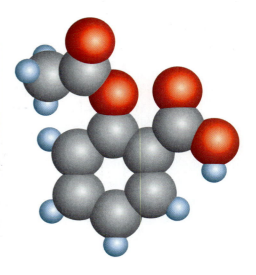

Chemists use computers to create models of new drugs in the search for new treatments for disease. ▶

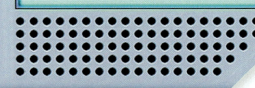

PEOPLE in CHEMISTRY

Meet Dr. John Thornton, Forensic Scientist

A poem by William Blake contains these words: "To see a world in a grain of sand and a heaven in a wildflower." In Dr. Thornton's work, that favorite line of poetry is literally true. To discover what Dr. Thornton means, read this interview, in which he shares his thoughts about the world of physical evidence.

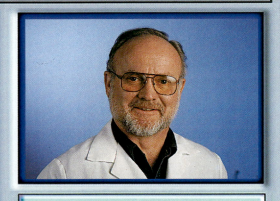

On the Job

 Dr. Thornton, could you tell us what you do at your lab, Forensic Analytical Specialties?

 We analyze physical evidence from the scene of a crime—hair, fiber, body fluids, bullets, paint, soil, glass, shoe impressions, fingerprints, drugs, and plant material. Any of this tangible, physical evidence can associate a person with a crime scene. That's half of it, the "whodunit"; the other half is the "howdunit."

Can you give some specific examples of each part of your analysis?

Fingerprints are the classic means of establishing that a person was at the scene of a crime. We develop fingerprints with chemicals: ninhydrin for prints on paper, and cyanoacrylate, which is actually superglue, on other items. As for the "how," here's an example of a case. A person claims he shot a person while defending himself against strangulation. If there's a gunpowder pattern on the victim's clothing, that

would tend to support the suspect's story. On the other hand, if there's no powder, indicating that the person was several feet away, that story doesn't hold up.

 It must be almost impossible to avoid leaving some kind of evidence behind, right?

 Yes. Frequently, the evidence that is most incriminating is so small that the perpetrator is oblivious to it, such as grains of pollen, sand, or tiny diatoms. Those would be a signature of a particular geographic location.

 Has your work ever helped locate a missing person?

 A few years ago, when a young woman was missing, a shoe was found near the freeway. Inside the shoe was about a thimbleful of fibers. I went through her sock drawer and found socks that matched every single type of fiber that was in the shoe. That clue indicated the direction of her disappearance, and we later found her.

Early Influences

 How did you get into this line of work?

 When I was in the seventh or eighth grade, I was just browsing through the local public library and stumbled upon a book called *Crime Investigation,* by Paul Kirk, who was probably the foremost forensic scientist at the time. From that point on, I knew what I wanted to do. What I *didn't* know was that I would later enroll in Kirk's university class. Even more improbably, I eventually took his place on the faculty after he retired.

 Were you a kid who liked things like decoder rings and puzzles?

 No, I was just a farm boy in the Central Valley. Everything there was flat, and I guess I was looking for some other horizons. I did that by reading.

Personal Insights

 What do you find most interesting about your work?

 Forensic science is really a study of how the world is put together. My work involves chemistry, physics, botany, and geology, so it's impossible to get bored.

 How do you go about tackling the problems posed by a set of clues at a crime scene?

 At any crime scene, you want to use physical evidence to tell a story. It's like looking at a tapestry from the back side: you can vaguely see that over here there's a unicorn, and over here there's a tree, and this might be a fence. It's difficult to get the clarity. I generally just sit down and think about things for a while. I let ideas wash over me from different directions. Then I play what-if: What if this happened at the scene, then what might have happened next? I try all of these ideas on for size, then I try to winnow the scientific wheat from the chaff.

 What kind of satisfaction do you derive from your work?

 I feel that the quality of justice is enhanced by a full and perhaps contentious airing of all the relevant issues and the physical evidence that we provide.

CAREER CONNECTION

Forensic chemists are assisted by people in these lines of work:

Crime Lab Technologist (Collects and analyzes physical evidence) BA from college with crime laboratory program

Fingerprint Classifier (Files and matches fingerprints) High school followed by study at police schools

Private Investigator (Collects facts pertaining to a crime) High school and detective-training program

Classifying Matter

Examples of matter range from grains of rice to the stars, from a drop of water to the rocks of the Grand Canyon, from a potato chip to a computer chip. It's all the stuff around us—all part of the same puzzle. You know that all these bits of matter connect in some way, just like the pieces of a jigsaw puzzle. But, there are so many different sizes and kinds of matter. How can you begin to make sense of the puzzle of matter?

Just as you do when you work a jigsaw puzzle, you first classify the pieces before trying to make connections. You might find pieces that share common properties, such as a flat edge or the same color, and place them in separate piles.

Classification by Composition

A powerful way to classify matter is by its composition. This is the broadest type of classification. When you examine an unknown piece of stuff, you first ask, "What is it made of?" Sucrose is composed of the elements carbon, hydrogen, and oxygen. This is a **qualitative** expression of composition. A qualitative observation is one that can be made without measurement.

After a qualitative analysis, the next question that you might ask is how much of each of the elements is present. For sucrose, the answer to that question is that 100 g of sucrose contains 42.1 g of carbon, 51.4 g of oxygen, and 6.5 g of hydrogen. This is a **quantitative** expression of composition. A quantitative observation is one that uses measurement. You make quantitative measurements every day when you answer such questions as What's the temperature? How long was the pass? How much do you weigh? **Figure 1.9** shows some quantitative measurements being made.

Figure 1.9

Measurement—A Quantitative Observation
In a medical examination, many of the doctor's observations are quantitative. Measured quantities, such as height and weight, can be compared with those of previous examinations. This boy has gained 5 pounds and grown 1 inch since his last examination.

Pure substance or a mixture?

The most general way to classify matter by composition is in terms of purity. There are only two categories here. A sample of matter is either pure—made up of only one kind of matter—or it is a mixture of different kinds of matter. What does it mean to say that a chunk of matter is pure?

Figure 1.10

Pure Products?
Products such as orange juice and soap are often advertised as pure. From a chemist's point of view, they are not pure but instead are complex mixtures containing many different substances.

The word *pure* is often used to describe common things, as in **Figure 1.10.** In chemistry, *pure* means that every bit of the matter being examined is the same substance. A **substance** is matter with the same fixed composition and properties. A substance can be either an element or a compound. Any sample of pure matter is a substance.

If the sugar in a bag from the supermarket is a substance, then it is pure sucrose. Every bit of matter in the bag must have the same properties and the same fixed composition as every other bit. Now, consider a bag of high-quality, dry, white sand. White sand is the common name for a substance called silicon dioxide. It is white and crystalline like sugar, and, when the sand is examined, every particle has the same fixed composition (53.2 percent oxygen, 46.8 percent silicon). Both sand and sucrose are substances, but they have different properties and compositions.

Mixed Matter

Suppose you mix the pure white sand with the pure white sucrose, as in **Figure 1.11.** You may not be able to notice any difference, but if you add this mixture to your cereal or tea, it's a different story. The sweet taste of the sugar is still there, but so is the grittiness of the sand. Every particle of this stuff is not the same, so the properties are not the same throughout. Some parts taste sweet, while others are gritty and tasteless. The composition is also not fixed, but instead depends on how much sand is mixed with the sugar.

Figure 1.11

A Mixture Retains the Properties of Its Components
Pure sugar is different from a mixture of sugar and sand. Each component of a mixture retains its behavior. Sugar dissolves and sweetens the tea, but sand is insoluble. It settles to the bottom of the cup.

Kitchen Chemicals

The chemical and physical properties of a substance make up a sort of fingerprint that characterizes the substance. In this ChemLab, you will test four unknown solids using three different liquids. The unknowns are common materials that you'd probably find in your kitchen. The results of your tests will give you the information you need to unravel the compositions of mixtures of two solids and three solids.

Problem

How can you identify a material by comparing its properties with those of known materials?

Objectives

- **Observe** the chemical and physical reactions of four common kitchen materials with three test reagents.
- **Compare and interpret** the reactions of the test reagents with five two-solid and three-solid mixtures of the common kitchen materials.
- **Infer** the composition of each of five unknown mixtures by comparing their reactions with those of the known materials.

PREPARATION

Materials

96-well microplate	3 thin-stemmed pipets
9 test tubes	masking tape
spatulas	marking pen

Safety Precautions

Do not touch or taste any of the solids or liquids, even though you may believe you know their identities.

PROCEDURE

1. Label four test tubes *A, B, C,* and *D.* Label five test tubes 1 through 5.

2. In the four lettered test tubes, place about 1 g of each of the labeled samples supplied by your teacher. These are the common kitchen materials.

3. In the numbered test tubes, place about 1 g of each of the numbered samples supplied by your teacher. These are the unknown mixtures. If you use the same spatula for each material, rinse and dry it before dipping into the next solid to avoid contaminating one material with another.

4. Label three long columns of wells on the microplate *I, II,* and *III.* Label nine rows of wells on the microplate *A, B, C,* and *D,* and 1 through 5 as shown in the photo.

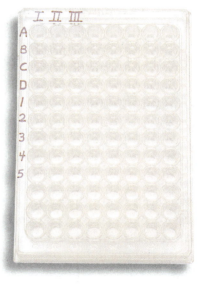

5. Place the microplate on a sheet of white paper.

6. Add a small amount of each material to the row of three wells that has the appropriate letter or number.

7. Observe and record the texture of each of the nine materials in a data table like the one shown.

8. Label the three pipets *I, II,* and *III.* From the containers of reagent liquids supplied by your teacher, draw into the bulb of each pipet the liquid corresponding to the label number on that pipet.

9. Add 3 drops of liquid *I* to each of the nine materials in column *I.*

10. Observe any changes that take place, and record them in the data table.

11. Repeat steps 9 and 10 using liquid *II* and then liquid *III.*

ANALYZE AND CONCLUDE

1. **Interpreting Data** What properties and reactions characterize each of the four kitchen solids?

2. **Drawing Conclusions** Can you positively identify the solids that are contained in any of the five mixtures? If so, identify the solids and explain your conclusions.

3. **Making Inferences** If you are unable to conclusively identify the solids in any of the mixtures, what are their likely identities? Explain.

APPLY AND ASSESS

1. Two of the four original solids, baking powder and baking soda, are often used in making baked goods. What characteristic probably makes them useful in baking? Which solids display this characteristic?

2. Baking powder is a mixture of two or more compounds, and it reacts with water or any other liquid that contains water. Baking soda is a single compound that reacts with acidic solutions but not with water. Which of the solids do you think is baking powder? Explain.

3. One of the solids is an organic compound you may have learned about in a biology course. It produces a characteristic color when combined with iodine. Which solid gave this reaction? What is the identity of this compound?

DATA AND OBSERVATIONS

Solid	Color	Texture	Reaction with liquid 1	Reaction with liquid 2	Reaction with liquid 3
Two-solid mixtures					
Three-solid mixtures					

Most of the matter you encounter every day is a mixture. A **mixture** is a combination of two or more substances in which the basic identity of each substance is not changed. In the sugar and sand mixture, the sand does not influence the properties of the sugar, and the sugar does not influence the properties of the sand. They are in contact with each other, but they do not interact with each other.

Unlike pure substances, mixtures do not have specific compositions. For example, a sand-and-sugar mixture can be any combination of sand and sugar. **Figure 1.12** shows some common mixtures.

Figure 1.12

Everyday Mixtures

This seascape shows two mixtures: seawater, a mixture of water with many salts and other soluble substances; and air, a mixture of nitrogen, oxygen, water vapor, argon, carbon dioxide, and other gases. ▶

▲ The exotic flavors of dishes from Asia and Africa arise from mixtures of spices such as cardamon, turmeric, cumin, allspice, and coriander.

▲ Solder is an alloy of tin and lead used by plumbers and electricians to seal a joint or connect one piece of metal to another.

▲ Blood is a complex mixture of many substances including water, proteins, glucose, fats, amino acids, and carbon dioxide.

◀ Sand, pulverized stone, minerals, salts, and substances from decayed plants and animals make up the rich mixture called soil.

Everyday Chemistry

You Are What You Eat

The morning is half over and you need some energy. So, you eat a piece of sour-apple candy. You have just consumed four chemicals—the sugar fructose for sweet taste and energy, citric acid for tartness, methyl butanoate for apple flavor, and red dye #2 for color. *Chemical* is just another name for "substance." Whenever you eat or drink, you consume chemicals that your body needs for energy, growth, and repair.

What chemicals are found in your body? The percentages of elements that make up the human body are shown in the graph. The elements found in your body are not free elements; they are in the form of compounds. For example, 67 percent of your body is water, a compound of hydrogen and oxygen.

Traces of iron, iodine, and others
Sulfur 0.3%
Sodium 0.2%
Magnesium 0.1%
Chlorine 0.1%
Potassium 0.4%
Phosphorus 1%
Calcium 1.5%
Nitrogen 3%
Hydrogen 10%
Carbon 18%
Oxygen 65%

Your body is a complex chemical factory. It checks to see if the right amount of each compound is present, decomposes the food you eat, and uses the new substances formed to make compounds needed for growth and repair. It breaks down other food components to obtain the energy it needs.

The chemistry of food You might start the day with a breakfast of melon, eggs, whole-wheat toast, milk, and tea. All of these foods are mixtures of many different compounds. If you decide to sweeten your tea with sugar, you will be adding a single compound—sucrose—to a mixture of water, caffeine, tannin, butyl alcohol, isoamyl alcohol, phenyl ethyl alcohol, benzyl alcohol, geraniol, hexyl alcohol, and essential oils. If you eat scrambled eggs, you are consuming a mixture of water, ovalbumin, coalbumin, ovomucoid, mucia, globulins, amino acids, lipovitellin, livetin, cholesterol, lecithin, lipids (fats), fatty acids, butyric acid, acetic acid, lutein, zeaxanthine, and vitamin A with a little sodium chloride (salt) sprinkled in. Doesn't that sound mouthwatering?

Exploring Further

1. **Analyzing** What color is burned food? What element is usually found in that color? What element do you think most foods have in common?

2. **Interpreting** Find out how organic and inorganic compounds differ. Which type makes up most of the human body? Why is the other type important to life?

The Separation of Mixtures into Pure Stuff

One characteristic of a mixture is that it can be separated into its components by physical processes. The word *physical* means that the process does not change the chemical identity of a substance. How could you separate a sand/sugar mixture into pure sand and pure sucrose? The simplest physical means would be to look at it with a microscope and separate the bits of sugar and sand with tweezers. You are right in thinking that there must be an easier way.

Separating mixtures by using physical changes is one easier way. A **physical change** is a change in matter that does not involve a change in the identity of individual substances. Examples of physical changes include boiling, freezing, melting, evaporating, dissolving, and crystallizing. Separation of a mixture by physical changes takes advantage of the different physical properties of the mixed substances. **Physical properties** are characteristics that a sample of matter exhibits without any change in its identity. Examples of the physical properties of a chunk of matter include its solubility, melting point, boiling point, color, density, electrical conductivity, and physical state (solid, liquid, or gas). Look at **Figure 1.13** to see one way to separate a mixture of sugar and sand using differences in the physical properties of the two substances.

Figure 1.13

Separating a Mixture by Physical Changes

Because sugar and sand have some different physical properties, a separation can be made by using a series of physical changes.

Step 1. Water is added and the mixture is stirred. The sugar dissolves in the water, but the sand does not.

Step 2. The mixture is passed through a filter that traps the sand but allows the sugar solution to pass through.

Step 3. The sugar solution is heated to evaporate the water.

Step 4. When all of the water has evaporated, pure sugar remains in the beaker. The separation is possible because of the difference in solubility of sugar and sand.

50 mL + 50 mL = ?

Nothing much happens when you make a mixture of sand and sugar. Some mixtures, such as water and alcohol, can give unexpected results that might contain clues to the structure of matter.

Procedure

1. Wear an apron and goggles.

2. Fill a 100-mL graduated cylinder exactly to the 50.0-mL mark with water that has been tinted with food coloring. Use a dropper to adjust the volume. Insert a thermometer and read the temperature of the water. Remove the thermometer.

3. Tilt the cylinder over almost as far as you can without spilling water and begin to add ethanol very slowly so that it does not mix with the water. Raise the cylinder slowly as you add more alcohol. Add the final alcohol with a dropper to adjust the level to exactly 100.0 mL. Assume the ethanol has the same temperature as the water.

4. Use a stirring rod to mix the contents of the cylinder as rapidly as possible. Immediately insert a thermometer. Read and record the temperature of the mixture.

5. Remove the thermometer and stirring rod, taking care that all liquid drips back into the cylinder. Read and record the volume to the nearest 0.1 mL.

Analysis

1. Describe what happened to the volume when the liquids mixed. Suggest a way to account for your observation.

2. Was heat absorbed or given off as the liquids mixed? How do you know?

Two Kinds of Mixtures

Sometimes, when you look at a sample of matter, it's easy to tell that the sample is a mixture. This kind of mixture is called a heterogeneous mixture. The prefix *hetero* means "different." A heterogeneous mixture is one with different compositions, depending upon where you look. The components of the mixture exist as distinct regions, often called phases. Orange juice and a piece of granite, as shown in **Figure 1.14,** are examples of heterogeneous mixtures.

Figure 1.14
Heterogeneous Mixtures
If you look closely at a glass of orange juice (left), you can see pieces of solid orange pulp floating in the liquid. If you look at a granite rock (right), you can see areas eof different color that indicate that the rock is composed of crystals of different substances.

Paper Chromatography of Inks

The inks in marking pens are often mixtures of dyes of several basic colors. In this MiniLab, you will use the technique of chromatography to analyze the ink from several pens.

Procedure

1. Wear an apron and goggles.

2. Obtain a clear plastic cup that is at least 6 cm high. From a coffee filter, cut a strip of paper about 2.5 cm wide and about 2.5 cm longer than the height of the cup.

3. Place the paper strip in the cup so that the bottom of the strip just rests on the bottom of the cup.

4. Push a pencil through the top of the paper in such a way that when the pencil rests on the top of the cup, the paper is suspended with its lower edge just touching the bottom.

5. Prepare several strips in the same way, one strip for each type of marker ink you will analyze.

6. Using one water-soluble ink marker for each strip, draw a narrow, horizontal line across the strip about 2 cm up from the bottom. If possible, include one black or brown marker.

7. Add about 1 cm of water to the cup and suspend the first

strip in the water. The marker line must be above the water level when the strip is suspended in the cup.

8. Cover the top of the cup loosely with clear plastic wrap to reduce evaporation. Observe and record the effect on the marker line as the water moves up the paper.

9. Remove the strip from the cup or beaker when the water level has risen to just below the pencil. Lay the strip on a paper towel to dry.

10. Repeat procedures 6 through 9 with each of your marker strips.

Analysis

1. Capillary action is the movement of a liquid upward through small pores that exist in some materials. Did you note any evidence of capillary action in this MiniLab?

2. Does your evidence indicate that any of the marker inks were composed of more than one pigment?

3. Which colors contained the greatest number of pigments?

The separation of sand and sugar took advantage of a difference in the physical properties of the two substances. Sugar dissolves in water but sand does not. When sugar dissolves in water, the two pure substances, sugar and water, combine physically to form a mixture that has a constant composition throughout. This means that no matter where you sample the mixture, you find the same combination of sugar and water. Even with a powerful light microscope, you could not pick out a bit of pure sugar or a drop of pure water. This type of mixture is called a homogeneous mixture. The prefix *homo* means "the same." Homogeneous mixtures are the same throughout. Another name for a homogeneous mixture is **solution.** Even

Table 1.1   Some Common Alloys

Name of Alloy	Percent Composition by Mass	Uses
Stainless steel	73-79% iron (Fe) 14-18% chromium (Cr) 7-9% nickel (Ni)	Kitchen utensils, knives, corrosion-resistant applications
Bronze	70-95% copper (Cu) 1-25% zinc (Zn) 1-18% tin (Sn)	Statues, castings
Brass	50-80% copper (Cu) 20-50% zinc (Zn)	Plating, ornaments
Sterling silver	92.5% silver (Ag) 7.5% copper (Cu)	Jewelry, tableware
14-karat gold	58% gold (Au) 14-28% silver (Ag) 14-28% copper (Cu)	Jewelry
18-karat white gold	75% gold (Au) 12.5% silver (Ag) 12.5% copper (Cu)	Jewelry
Solder (electronic)	63% tin (Sn) 37% lead (Pb)	Electrical connections

White gold alloy

though solutions may appear to be one pure substance, their compositions can vary. For example, you could make your tea very sweet by dissolving a lot of sugar in it or less sweet by dissolving only a little bit of sugar.

When you hear the word *solution*, something dissolved in water probably comes to mind. But liquid solutions do not have to contain water. Gasoline is a liquid solution of several substances, but it contains no water. Some solutions are gases. Air, for example, is a homogeneous mixture of several gases. Some solutions are solid. **Alloys** are solid solutions that contain different metals and sometimes nonmetallic substances. *Steel,* for example, is a general term for a range of homogeneous mixtures of iron and substances such as carbon, chromium, manganese, nickel, and molybdenum. Because pure gold is soft and bends easily, most gold jewelry is not made from pure gold, but rather from an alloy of gold with silver and copper. **Table 1.1** shows some common alloys and their compositions. The knife shown is stainless steel. The tableware is sterling silver, and the jewelry is white gold.

When you dissolve sugar in water, sugar is the **solute**—the substance being dissolved. The substance that dissolves the solute, in this case water, is the **solvent.** When the solvent is water, the solution is called an **aqueous solution.** Many of the solutions you encounter are aqueous solutions, for example, soda, tea, contact-lens cleaner, and other clear cleaning liquids. In addition, most of the processes of life occur in aqueous solutions.

WORD ORIGIN

heterogeneous
hetero (GK)
different, other
genea (GK)
origin, source
A heterogeneous mixture is different in different places.

Figure 1.15

Classifying Matter
The chart shows a way of classifying matter. Notice that mixtures can be either heterogeneous or homogeneous and can often be separated into pure substances by physical changes. You can also see that pure matter can be elements or compounds. You'll learn more about these classes of matter in the next section.

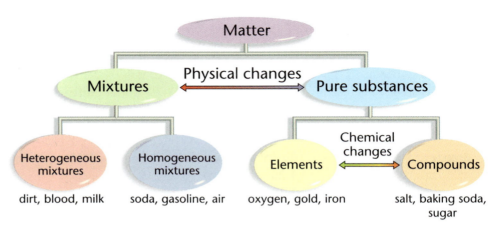

Matter

Mixtures ←— Physical changes —→ Pure substances

Heterogeneous mixtures
dirt, blood, milk

Homogeneous mixtures
soda, gasoline, air

Elements ←— Chemical changes —→ Compounds

Elements
oxygen, gold, iron

Compounds
salt, baking soda, sugar

The chemistry of the real world is mostly the chemistry of mixtures. Dig a hole, buy something at a grocery store, pick an apple from a tree, or take a deep breath. The stuff you dig, buy, pick, or inhale is a mixture. However, the behavior of mixtures is based on the composition, structure, and behavior of the pure substances that compose them. **Figure 1.15** summarizes the classification of matter from a chemical point of view.

Substances: Pure Matter

All matter is composed of substances. Although a pizza and a can of paint have different compositions, both are made of some combination of pure substances, and pure substances are chemicals. You, too, are made of chemicals. This book, your desk, your clothes, the air you breathe, the water you drink, the food you eat, and all the other stuff of the universe are all made of chemicals.

Elements: The Building Blocks

If you classify an unknown piece of matter as pure, it means that the stuff is made up of only one substance. But there are two types of substances. One type of pure substance can be broken down into simpler substances. This type of substance is called a compound. Another type of substance cannot be broken down into simpler substances. Such a substance is called an **element.** Elements are the simplest form of matter. They are the building blocks from which other forms of matter are made. All the substances of the universe are either elements, compounds formed from elements, or mixtures of elements and compounds. **Figure 1.16** shows some well-known elements.

WORD ORIGIN

homogeneous
homo (GK) same, alike
genea (GK) origin, source

A homogeneous mixture is the same throughout.

Diamonds are one form of pure carbon. ▶

Figure 1.16
Some Familiar Elements
Gold nuggets are valuable because they are almost pure gold. ▶

Copper to Gold: The Alchemists' Dream

An alchemist was a combination of magician and metallurgist who tried unsuccessfully to convert common metals to gold. The craft flourished from ancient times until the 18th century. Alchemists were not early chemists, as some people believe, but their practical knowledge about elements and compounds contributed to the work of the earliest true chemists. Like the alchemists, you will not turn copper into gold, but by allowing the copper in a penny to react with zinc under certain conditions, you may create an interesting alloy of the two metals.

Procedure

1. Wear an apron and goggles.
2. Clean a pre-1982 penny with steel wool or a pencil eraser.
3. Place 1 g of granular zinc in an evaporating dish. Add 20 mL of $1M$ zinc chloride solution ($ZnCl_2$). Use tongs to place the penny in the dish, and put the evaporating dish on a hot plate.
4. Heat the mixture until it just starts to boil. This should take about two minutes. Carefully stir the mixture with the tongs and turn the penny. Continue to heat and stir gently until the penny becomes covered with zinc and appears gray in color. This usually takes less than a minute.
5. Use the tongs to remove the penny from the liquid. Rinse the penny in a beaker of cold tap water, then pat it dry with a paper towel.
6. Using tongs to hold the penny, gently heat it in the cooler, outer portion of a Bunsen burner flame until it changes color. Record your observations.
7. Continue heating gently for two or three seconds longer, then immediately immerse the penny in a fresh beaker of cold water.
8. After the penny has cooled for about a minute, remove it from the water and pat it dry. Record your final observations.

Analysis

1. Does the evidence indicate that you created an alloy of copper and zinc? Explain.
2. What is the probable identity of this alloy?
3. What do you think you would see if you cut the penny in two and examined the cut edge with a powerful microscope?

◄ Mercury is the only metallic element that is a liquid at room temperature.

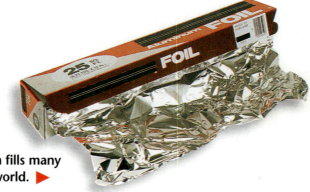

The element aluminum fills many needs in the modern world. ▶

Jules Verne and His Icebergs

You are underwater in a submarine in an Antarctic iceberg field. While sleeping in your bunk, you are awakened by a violent shock and thrown into the middle of your cabin. Surveying the situation, you realize that the submarine is on its side. This is what happens to P. Aronnax when he is aboard Captain Nemo's *Nautilus* in Jules Verne's science fiction novel *Twenty Thousand Leagues Under the Sea.*

How does Jules Verne explain the accident? In Verne's book, published in 1869, Captain Nemo explains that a mountain of floating ice, an iceberg, has turned over. When an iceberg is undermined at its base by warmer water or repeated shocks, its center of gravity rises, and the whole thing turns over. As the bottom of the iceberg turned upward, it struck the *Nautilus,* slid under its hull, and raised it onto an ice bed, where it lay on its side.

What concepts are involved? Density and center of gravity account for the accident. The iceberg was floating because the frozen fresh water (density 0.9 g/mL) composing it was less dense than the salt water (density 1.0 g/mL) in which it was floating. An iceberg floats in seawater because its density is less than that of the seawater. However, its density isn't a lot less than that of seawater so it floats with only about 15 percent of its mass above the water's surface.

The center of gravity of a mass is the point at which all the weight of an object seems to be located. The higher an object's center of gravity is above its support, the less stable the object is. If something happens to the bottom of an iceberg such that the center of gravity is shifted above the waterline, the iceberg will turn over.

An iceberg's center of gravity changes when large pieces of ice break off the berg. This can be caused by water freezing in cracks, or the vibration of the waves, or thunderous vibrations resulting from the breaking of other bergs. When an iceberg becomes unstable, small movements can cause it to topple over.

Connecting to Chemistry

1. **Acquiring Information** Find out what differences exist between icebergs in the Northern Hemisphere and in the Southern Hemisphere.

2. **Applying** If an iceberg were floating in fresh water instead of seawater, would more or less of its mass be above the water's surface? Explain.

Millions of substances are known to chemists, but only 112 are elements. These 112 elements combine with each other to form all the millions of known compounds. That's why the chemical elements are often referred to as the building blocks of matter.

Of the 112 known elements, only about 90 occur naturally on Earth. The remainder are synthesized, usually in barely detectable amounts, in high-energy nuclear experiments. Less than half of the 90 naturally occurring elements are abundant enough to play a significant role in the chemistry of everyday stuff.

The fact that most stuff is composed of a relatively small number of building blocks simplifies the puzzle of matter. On the other hand, the observation that such a small number of pieces creates such a variety of compounds means that elements must connect to each other in countless ways.

Organizing the Elements

Your classroom may have a large chart labeled *The Periodic Table of the Elements* hanging on the wall. This table will become the tool you will use most in your chemistry course. A similar table is printed on pages 92-93 of this book, as well as inside the back cover. The periodic table organizes elements in a way that provides a wealth of chemical information—much more than is evident to you now. **Figure 1.17** on the following two pages is a pictorial version of the periodic table. It shows the chemical symbols for the elements and photos of samples of many of the naturally occurring elements.

The symbols of the elements are part of the language of chemistry. As **Figure 1.18** indicates, the chemical symbols are a universal shorthand that is used to make chemistry communication understandable around the world. Just as it is easier and quicker for you to write *USA* instead of *United States of America,* it is easier to write *Al* instead of the word *aluminum.* As you can see, the symbol for aluminum is taken directly from the element's name, but some elements have symbols that don't correspond to their English names. Their symbols usually correspond to their names in Latin. Some examples are shown in **Table 1.2**.

Figure 1.18

Chemical Symbols Are Universal
These chemical research articles—written in journals of different countries—show that the symbols for the chemical elements are a universal language through which chemists can communicate worldwide.

Table 1.2	Some Historic Chemical Symbols		
Element	**Symbol**	**Origin**	**Language**
Antimony	Sb	stibium	Latin
Copper	Cu	cuprum	Latin
Gold	Au	aurum	Latin
Iron	Fe	ferrum	Latin
Lead	Pb	plumbum	Latin
Potassium	K	kalium	Latin
Silver	Ag	argentum	Latin
Sodium	Na	natrium	Latin
Tin	Sn	stannum	Latin
Tungsten	W	wolfram	German

Pewter—an alloy of tin and antimony

Figure 1.17

Pictorial Periodic Table
The periodic table shown here was designed to illustrate samples
of the elements. It differs somewhat from the most up-to-date
table that you will find on pages 92-93. Be sure to use this latter
table for reference throughout your chemistry course.

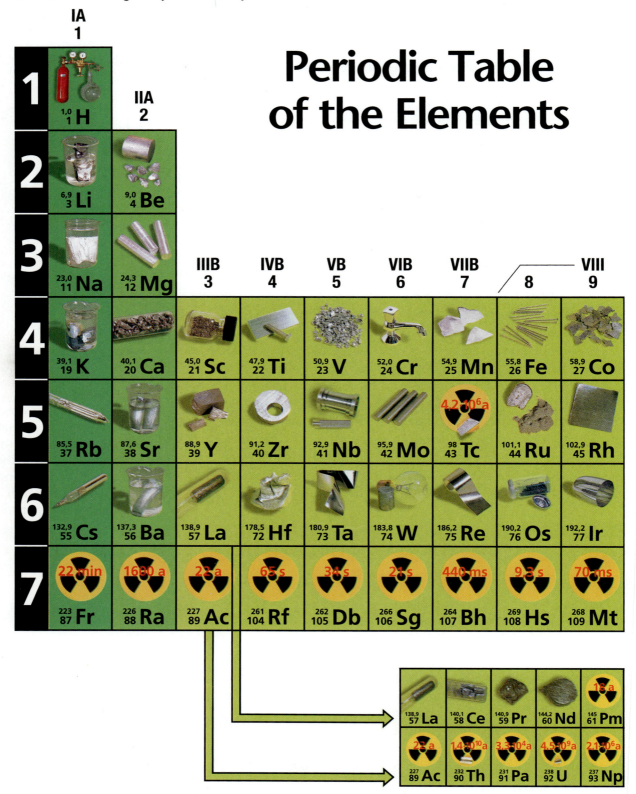

Periodic Table of the Elements

			IIIA 13	IVA 14	VA 15	VIA 16	VIIA 17	VIIIA 18

VIIIA
18

$^{4,0}_{2}$ He

IIIA
13

IVA
14

VA
15

VIA
16

VIIA
17

$^{10,8}_{5}$ B $^{12,0}_{6}$ C $^{14,0}_{7}$ N $^{16,0}_{8}$ O $^{19,0}_{9}$ F $^{20,2}_{10}$ Ne

$^{27,0}_{13}$ Al $^{28,1}_{14}$ Si $^{31,0}_{15}$ P $^{32,1}_{16}$ S $^{35,5}_{17}$ Cl $^{39,9}_{18}$ Ar

IB
11

IIB
12

10

$^{58,7}_{28}$ Ni $^{63,5}_{29}$ Cu $^{65,4}_{30}$ Zn $^{69,7}_{31}$ Ga $^{72,6}_{32}$ Ge $^{74,9}_{33}$ As $^{79,0}_{34}$ Se $^{79,9}_{35}$ Br $^{83,8}_{36}$ Kr

$^{106,4}_{46}$ Pd $^{107,9}_{47}$ Ag $^{112,4}_{48}$ Cd $^{114,8}_{49}$ In $^{118,7}_{50}$ Sn $^{121,8}_{51}$ Sb $^{127,6}_{52}$ Te $^{126,9}_{53}$ I $^{131,3}_{54}$ Xe

$^{195,1}_{78}$ Pt $^{197,0}_{79}$ Au $^{200,6}_{80}$ Hg $^{204,4}_{81}$ Tl $^{207,2}_{82}$ Pb $^{209,0}_{83}$ Bi

102 a
$^{209}_{84}$ Po   8,1 h
$^{210}_{85}$ At   3,8 h
$^{222}_{86}$ Rn

118 ms
$^{273}_{110}$ 1,5 ms
$^{272}_{111}$ 0,24 ms
$^{277}_{112}$

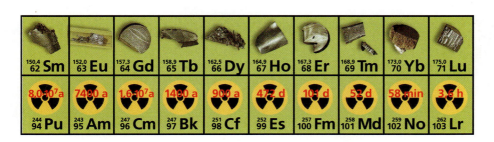

$^{150,4}_{62}$ Sm $^{152,0}_{63}$ Eu $^{157,3}_{64}$ Gd $^{158,9}_{65}$ Tb $^{162,5}_{66}$ Dy $^{164,9}_{67}$ Ho $^{167,3}_{68}$ Er $^{168,9}_{69}$ Tm $^{173,0}_{70}$ Yb $^{175,0}_{71}$ Lu

$8,0 \cdot 10^7$ a
$^{244}_{94}$ Pu   7480 a
$^{243}_{95}$ Am $1,6 \cdot 10^7$ a
$^{247}_{96}$ Cm   1480 a
$^{247}_{97}$ Bk   900 a
$^{251}_{98}$ Cf   472 d
$^{252}_{99}$ Es   101 d
$^{257}_{100}$ Fm   52 d
$^{258}_{101}$ Md   58 min
$^{259}_{102}$ No   3,6 h
$^{262}_{103}$ Lr

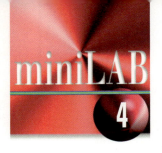

miniLAB 4

Get your...

Calcium
Copper
Iodine
Iron
Magnesium
Phosphorus
Potassium
Selenium
Zinc

Waiter, what's this stuff doing in my cereal?

Breakfast cereals are often fortified with elements and compounds to increase their nutritional value. In this MiniLab, you will test a common cereal for the presence of one of these additives.

Procedure

1. Tape a small, strong magnet to a pencil at the eraser end.

2. Place a sample of dry, fortified, cold cereal in a plastic bag.

3. Thoroughly crush the cereal with a rolling pin or other heavy object.

4. Pour the crushed cereal into a beaker and cover it with water.

5. Stir the cereal/water mixture for about ten minutes with your pencil-magnet stirrer. Stir slowly and easily for the last minute.

6. Remove the magnet from the cereal and examine it carefully. Record your observations.

Analysis

1. The substance attracted to your magnet is a common element. What is it?

2. Why do you think that this element is added to the cereal?

Compounds Are More Than One Element

You've learned that a compound is a pure substance that can be broken down into elements. A more complete definition is that a **compound** is a chemical combination of two or more different elements joined together in a fixed proportion. For example, if you were to collect and analyze samples of the compound water from a faucet, an iceberg, a river, and a rain puddle, you would find that every sample (barring impurities) is 11.2 percent hydrogen and 88.8 percent oxygen by mass. Every compound has its own fixed composition, and that composition results in a unique set of chemical and physical properties. The properties of the compound are different from the properties of the elements that compose the compound. You can see this in **Figure 1.19.**

Figure 1.19

Two Elements Combine to Form a New Substance

The elements that make up a compound are chemically combined to form a new substance with a unique set of properties. The element silver is a solid metal. The element bromine is a poisonous red liquid. Silver bromide, a compound of silver and bromine, is a crystalline solid that is used in photographic and print paper. Silver bromide has a unique set of physical and chemical properties and a fixed composition of 57.45 percent silver and 42.55 percent bromine.

Silver

+

Bromine

More than 10 million compounds are known and the number keeps growing. Some common compounds are listed in **Table 1.3** on page 33. New compounds are discovered and isolated from natural chemical sources such as plants and colonies of bacteria. Compounds are also synthesized in laboratories where they are tested for a variety of uses ranging from medicine to manufacturing.

Because the supply of useful chemicals from natural sources is often limited, chemists work to synthesize these compounds in the laboratory. The effort to synthesize taxol, an anticancer compound found in the bark of the Pacific yew tree, is an example of how nature often provides the lead in compound synthesis. If taxol can be made in the laboratory, then chemical engineers will try to find a way to produce it on an industrial scale.

Supplemental Problems

For more practice with solving problems, see Supplemental Practice Problems, Appendix B.

Formulas of Compounds

The second column in **Table 1.3** on page 33 gives the chemical formulas for the compounds listed. A **formula** is a combination of the chemical symbols that show what elements make up a compound and the number of atoms of each element. For example, the chemical formula of the compound sucrose is $C_{12}H_{22}O_{11}$. The formula tells you, in a compact way, that sucrose contains carbon, hydrogen, and oxygen. It also tells you that the smallest unit of sucrose, a molecule, contains 12 carbon atoms, 22 hydrogen atoms, and 11 oxygen atoms. Formulas provide a shorthand way of describing a submicroscopic view of a compound. You probably already use formulas like H_2O and CO_2 as a way of talking about water and carbon dioxide.

In later chapters, you'll learn more about compounds and why elements combine to form compounds. You'll also learn how to determine the formulas for many compounds.

Silver bromide

Natural Versus Synthetic Chemicals

If you look carefully at the vitamin display in a drugstore, you'll see some bottles marked *all natural*. Most of these vitamins have been extracted from plant or animal sources. Are natural vitamins, drugs, and other substances better than the purified or synthetic ones produced by drug companies? Maybe this will help you decide.

Aspirin: A common synthetic drug To get rid of a headache, would you rather drink a cup of willow bark tea or take two aspirin? The active ingredients in both remedies have similar chemical structures, and both are effective against headaches. However, the chemical from the willow bark, salicylic acid, has several harmful side effects including stomach pain. So does aspirin, but it is a more effective painkiller and can be taken in lower doses. Also, willow bark contains many other chemicals. After years of research, scientists made aspirin in the laboratory from salicylic acid and acetic anhydride. It contains only one active ingredient, acetylsalicylic acid. Not only is aspirin a pure substance, but also the difference in its structure eliminates the serious stomach pain caused by salicylic acid.

Development of new drugs What happens when a chemical discovered in nature is shown to be a potential treatment or cure for a disease? Scientists use the following procedures to make safe, effective drugs: (1) isolate and purify the drug, (2) determine its composition and structure, (3) search for a way to make it synthetically, (4) look for a cheap, easy way to produce it in large quantities, and (5) try changing the structure and composition of the original compound to improve on nature's model.

Taxol: A new cancer drug Scientists discovered that taxol, a chemical found in the bark of the Pacific yew tree, reduced the size of ovarian and breast cancer tumors in 30 percent of otherwise-untreatable patients. But scientists were concerned that demand for the drug might wipe out the population of yew trees, so they started to look for new sources of the drug. Chemists Andrea and Donald Stierle found a taxol-producing fungus growing on one yew. Other scientists discovered that needles from the European yew contain a chemical similar to taxol. Chemists experimented to figure out taxol's structure. In 1994, they succeeded in producing pure taxol in the lab. You might be wondering which taxol is better—the purified natural one or the synthetic one. Actually, the chemical structure and purity of both is identical. However, being able to make taxol synthetically is a real advantage; drug companies may be able to produce it more cheaply, and they can work on modifying its structure to make it more effective.

Trunk of Pacific yew

Analyzing the Issue

1. **Acquiring Information** Find out why scientists were concerned that the use of taxol could endanger the Pacific yew and whether this is still a concern.

2. **Modeling** In what ways are the structures of salicylic acid and acetylsalicylic acid (aspirin) alike? In what ways are they different?

3. **Debating** Debate the pros and cons of using natural herbal drugs, purified natural drugs, and synthetic drugs.

Table 1.3   Some Common Compounds

Compound Name	Formula	Uses
Acetaminophen	$C_8H_9NO_2$	pain reliever
Acetic acid	$C_2H_4O_2$	tart ingredient in vinegar
Ammonia	NH_3	fertilizer, household cleaner when dissolved in water
Ascorbic acid	$C_6H_8O_6$	vitamin C
Aspartame	$C_{14}H_{18}N_2O_5$	artificial sweetener
Aspirin	$C_9H_8O_4$	pain reliever
Baking soda	$NaHCO_3$	cooking
Butane	C_4H_{10}	lighter fuel
Caffeine	$C_8H_{10}N_4O_2$	stimulant in coffee, tea, some soda
Calcium carbonate	$CaCO_3$	antacid
Carbon dioxide	CO_2	carbonating agent in soda
Ethanol	C_2H_6O	disinfectant, alcoholic beverages
Ethylene glycol	$C_2H_6O_2$	antifreeze
Hydrochloric acid	HCl	called muriatic acid, cleans mortar from brick
Magnesium hydroxide	$Mg(OH)_2$	antacid
Methane	CH_4	natural gas, fuel
Phosphoric acid	H_3PO_4	flavoring in soda
Potassium tartrate	$K_2C_4H_4O_6$	cream of tartar, cooking
Propane	C_3H_8	fuel for cooking
Salt	$NaCl$	flavoring
Sodium carbonate	Na_2CO_3	washing soda
Sodium hydroxide	$NaOH$	drain cleaner
Sucrose	$C_{12}H_{22}O_{11}$	sweetener
Sulfuric acid	H_2SO_4	battery acid
Water	H_2O	washing, cooking, cleaning

SECTION REVIEW

Understanding Concepts

1. What three characteristics of matter does chemistry deal with?

2. List two ways in which a mixture differs from a pure substance.

3. How does a compound differ from a mixture?

Thinking Critically

4. **Applying Concepts** The element oxygen is a gas that makes up about 20 percent of Earth's atmosphere. Oxygen also is the most abundant element in Earth's crust, yet Earth's crust is not a gas. Explain this apparent conflict.

Applying Chemistry

5. **Everyday Materials** Matter may be subdivided into elements, compounds, and heterogeneous and homogeneous mixtures. Describe one material found in a household that belongs in each category.

Properties and Changes of Matter

SECTION PREVIEW

Objectives

✓ **Distinguish** between physical and chemical properties.

✓ **Contrast** chemical and physical changes.

✓ **Apply** the law of conservation of matter to chemical changes.

Review Vocabulary

Matter: anything that takes up space and volume.

New Vocabulary

volatile
density
chemical property
chemical change
chemical reaction
law of conservation
   of mass
energy
exothermic
endothermic

When the refuse truck pulls up at the landfill to empty its load, you might say, "What *is* all that stuff?" You know that it's all the throw-aways of modern life—paper, glass, metal, plastic, and more—but the question still remains: "What *is* it?" Chemists want to know what every bit of stuff is. What is it made of (composition)? How are its atoms arranged (structure)? What will the stuff do (behavior)? Any characteristic that can be used to describe or identify a piece of matter is a property of the matter. In fact, each different substance has its own set of properties, just as each person has unique fingerprints. By knowing the *fingerprint* of a substance, you can identify the substance.

Identifying Matter by Its Properties

Physical properties are those that don't involve changes in composition. Many physical properties are qualitative descriptions of matter, such as: *The solution is blue; The solid is hard;* or *The liquid boils at a low temperature.* Other physical properties are quantitative, which means that they can be measured with an instrument. Examples include: *An ice cube melts at 0°C; Iron has a density of 7.86 g/mL;* or *A mass of 35.7 g of sodium chloride dissolves in 100 mL of water.*

States of Matter

Most matter on Earth exists in one of three physical states: solid, liquid, or gas. A fourth state of matter, called plasma, is less familiar. You will learn about it in Chapter 10. The physical state of a substance depends on its temperature. You know that if you put liquid water into a freezer, it changes to solid water (ice), and if you heat liquid water to 100°C in a teakettle, it boils and becomes gaseous water (steam). But the physical state of a substance usually means its state at room temperature—about 20°C to 25°C. At room temperature, the physical state of water is liquid, salt is a solid, and oxygen is a gas.

Figure 1.20
The Freezing/Melting Temperature of Water
Because freezing and melting occur at the same temperature, both phases of water, solid and liquid, can be present when the temperature is exactly 0°C.

A physical property, closely related to the physical state of a substance, is the temperature at which the substance changes from one state to another. Water, for example, freezes (and melts) at 0°C. Salt (NaCl) melts (and freezes) at a much higher temperature, 804°C, and oxygen (O_2) freezes (and melts) at a much lower temperature, −218°C. The melting point and the freezing point of a substance are the same temperature, as you can see in **Figure 1.20,** which shows both water and ice at 0°C. Whether we use the word *freeze* or *melt* depends on how we usually encounter a substance. Water boils at 100°C, but it also condenses from a gas to a liquid at the same temperature. Therefore, boiling point and condensation point are the same temperature for each substance.

Changes in state are examples of physical changes because there is no change in the identity of the substance. Ice can melt back to liquid water, and steam will condense on a cool surface to liquid water. Some substances are described as **volatile,** which means that they change to a gas easily at room temperature. Alcohol and gasoline are more volatile than water. The substance naphthalene, used as mothballs, is an example of a solid substance that is volatile. You can readily smell alcohol, gasoline, and mothballs when open containers of these substances are present in a room because the liquid or solid has changed to gas and the molecules are present in the air.

Try at Home **Lab**

See page 863 in
Appendix F for
Comparing Frozen Liquids

◀ Matter in stone is packed much more densely than it is in Styrofoam. You could think of density in these photos as mass/box or g/box. The usual units are g/mL.

Figure 1.21

Density Compares the Masses of Equal Volumes

▲ The boxes have the same volume, but the Styrofoam balls that fit into the box have much less mass than the stones that fill the box.

Density

Density is another physical property of matter. Consider two identical boxes, one filled with Styrofoam balls and the other filled with stones. If you were to lift both boxes, as shown in **Figure 1.21,** you might say that the box of Styrofoam balls is light while the box of stones is heavy. The Styrofoam balls occupy a certain amount of space or volume (the box), but they have little mass because of the particular structure of Styrofoam. The stones occupy the same volume (the box), but they have a larger mass because of the particular structure of stone. The structure of stone packs much more mass into a given volume than the structure of Styrofoam does. When you compare the mass of the box of Styrofoam with the mass of an identical box of stones, you are observing the different densities of the two materials. **Density** is the amount of matter (mass) contained in a unit of volume. Styrofoam has a low density or small mass per unit of volume. Stones have a large density or a large mass per unit of volume.

In science, the density of solids and liquids is usually measured in units of grams (mass) per milliliter (volume) or g/mL. **Table 1.4** gives the densities of some common materials.

Table 1.4	Densities of Some Common Materials

Material	Density, g/mL
Water (4.0°C)	1.000
Ice (0°C)	0.917
Helium (25°C)	0.000164
Air (25°C)	0.00119
Aluminum	2.70
Lead	11.34
Gold	19.31
Cork	0.22-0.26
Sugar	1.59
Balsa wood	0.12

To find the density of a chunk of matter, it is necessary to measure both its mass and its volume. One technique for measuring these quantities is shown in **Figure 1.22.** It could be used for any object that is heavier than water and does not dissolve in water.

Figure 1.22

Determining Density

Here's one way to determine the density of a solid such as lead.

1. **Fill a graduated cylinder with water to the 30.0-mL mark.**

2. **Weigh the cylinder with the water (106.82 g).**

3. **Carefully add a mass of lead to the graduated cylinder.**

4. **Reweigh the cylinder with the water and the lead (155.83 g).**

5. **Measure the volume of all the material in the cylinder (34.5 mL).**

The mass of lead = 155.83 g − 106.82 g = 49.01 g.
The volume of lead by displacement = 34.5 mL − 30.0 mL = 4.5 mL.
Density = mass of lead/volume of lead = 49.01 g/4.5 mL = 11 g/mL.

ChemLab 3

The Composition of Pennies

U.S. pennies have been composed of copper and zinc since 1959, but the ratio of copper to zinc has changed over the years because of increases in the price of copper. Copper and zinc are both metallic elements and they share many physical properties, but they have different densities. Pure copper has a density of 9.0 g/mL, while pure zinc has a density of 7.1 g/mL. By measuring the density of pennies from different years, it's possible to track changes in the composition of the penny.

Problem

What are the approximate compositions of pennies having various mint dates?

Objectives

- **Measure** mass and volume and determine the density of pennies.
- **Interpret** class data to determine approximately when the composition of pennies changed.

PREPARATION

Materials

5 pennies of varying mint dates
balance sensitive to 0.01 g
50-mL graduated cylinder having 1-mL graduations

PROCEDURE

1. Record the mint date of each penny in a table like the one shown.

2. Weigh each penny and record the mass to the nearest 0.01 g.

3. Run tap water into the graduated cylinder until it is approximately one-half full. Read and record the volume to the nearest 0.1 mL.

4. Carefully add the five pennies to the cylinder so that no water splashes out. Jiggle the cylinder to dislodge any trapped air bubbles. Read and record the total volume of the water and the five pennies.

DATA AND OBSERVATIONS

Volume of water (mL)	
Volume of water + 5 pennies (mL)	
Volume of 5 pennies (mL)	
Average volume of a penny (mL)	

Mint Date	Mass (g)	Density (g/mL)

ANALYZE AND CONCLUDE

1. **Calculating** Subtract the initial volume of water from the volume of the water plus pennies to calculate the volume of the five pennies. Divide the volume by 5 to calculate the average volume of a penny. Record this volume in your data table.

2. **Calculating** Calculate the density of each penny by dividing its mass by the average volume. Post your results so they are available to the other members of your class.

3. **Observing and Inferring** Does your group data show any pattern that relates the densities of the pennies to the mint dates?

4. **Classifying** Classify the pennies of each year that your group examined as mostly copper (8.96 g/mL) or mostly zinc (7.13 g/mL).

5. **Making Inferences** Look at all the data from your class. In what year do you think the composition of the penny changed? Use data to support your conclusion.

APPLY AND ASSESS

1. What factors do you think might lead to error in your density measurements? Which of these factors could not be corrected by improved technique?

2. Explain how you might determine the identity of an irregularly shaped solid that is soluble in water.

3. In 1943, all pennies issued by the U.S. mint were struck from zinc-plated steel. Two of these pennies are shown in the photo. Do research to learn why steel pennies were struck in 1943. What was the purpose of the zinc coating?

4. Why would increases in the price of copper cause the mint to change the composition of pennies?

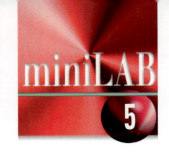

miniLAB 5

It's a Liquid, It's a Solid...It's Slime

In this MiniLab, you will work with a polymer, polyvinyl alcohol. A polymer is a large molecule that consists of a chain of smaller repeating units. By allowing polyvinyl alcohol to react with a borax solution, you will cross-link polymer molecules to form a gel not unlike the commercial green stuff you may have seen. You can then investigate some of the properties of the gel.

Procedure

1. Wear an apron, safety goggles, and disposable plastic gloves.

2. Place about 20 mL of polyvinyl alcohol solution in a plastic cup.

3. Use a craft stick to stir the solution vigorously as you add about 3 mL of borax solution. Add a drop of food coloring if you want colored slime.

4. Continue to stir the solutions together until they form a gel.

5. Remove the gel from the cup, shape it with your hands, and perform several tests on it.

Does it flow? Does it stretch or break? Can it be flattened? Record your observations.

6. Put your product in a self-sealing plastic bag, and dispose of it according to your teacher's instructions.

Analysis

1. What effect does cross-linking the polymer have on its properties? Can you explain this effect?

2. Can you name some polymers that occur in or are used to produce common materials?

Chemical Properties and Changes

Physical properties alone are not enough to describe a substance. For a complete description, you need to know about another set of properties called chemical properties. **Chemical properties** are those that can be observed only when there is a change in the composition of the substance. Chemical properties describe the ability of a substance to react with other substances or to decompose. A chemical property of iron, for example, is that it rusts at room temperature. Rusting is a chemical reaction in which iron combines with oxygen to form a new substance, iron oxide. Aluminum reacts with oxygen too, but the compound formed, aluminum oxide, coats the aluminum and protects it from further oxidation. Platinum does not react with oxygen at room temperature. Lack of reactivity is also a chemical property.

Have you ever noticed that hydrogen peroxide (H_2O_2) solution always comes in brown bottles? It's packaged in this way because in bright light, hydrogen peroxide breaks down into water and oxygen gas. The instability of a substance—its tendency to break down into different substances—is another chemical property. **Figure 1.23** illustrates other chemical properties of some substances. A chemical property always relates to a **chemical change,** the change of one or more substances into other substances. Another term for chemical change is **chemical reaction.**

Figure 1.23

Some Chemical Properties and Changes

The chemical properties of a substance describe how that substance changes to one or more new substances. It may change by reacting with another substance or by breaking down into simpler substances.

When vinegar is poured on baking soda (sodium hydrogen carbonate), bubbles of carbon dioxide gas form quickly. Water and sodium acetate are also formed.

Iron combines with oxygen to form iron oxide. The reaction shown is the same reaction as the rusting of iron, but it occurs much faster in pure oxygen in the flask.

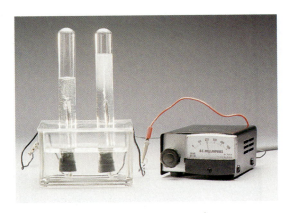

The stable compound water can be decomposed into hydrogen gas (left tube) and oxygen gas (right tube) by passing an electric current through it.

Hydrogen peroxide is another compound of hydrogen and oxygen. When the compound manganese dioxide is added to hydrogen peroxide, the peroxide breaks down rapidly into water and oxygen gas.

Atoms and Chemical Change

All matter is made of atoms, and any chemical change involves only a rearrangement of the atoms. Consider the chemical change shown in **Figure 1.23** in which water is broken down into hydrogen gas and oxygen gas. The reaction involves only hydrogen atoms and oxygen atoms. Whatever atoms are contained in the water that decomposes are found in the

Figure 1.24

Conservation of Mass (Atoms)

Two water molecules contain two oxygen atoms and four hydrogen atoms. When they decompose, they form one molecule of oxygen containing two oxygen atoms and two molecules of hydrogen containing four hydrogen atoms. Because all matter consists of atoms and the number of atoms is the same before and after the chemical change, you can say that matter is conserved.

Water molecules

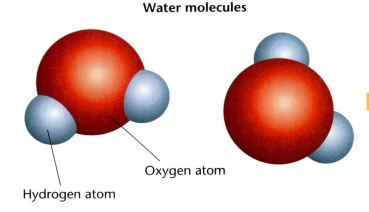

Oxygen atom

Hydrogen atom

Weight is different from mass. The weight of an object is a measure of the force of gravity on the object. Scientists generally say "weigh an object" when they really mean "measure the mass of an object on a balance."

hydrogen and oxygen molecules that are formed. **Figure 1.24** uses models of molecules of water, oxygen, and hydrogen to show that atoms do not just appear. Atoms do not just disappear. This is an example of the **law of conservation of mass,** which says that in a chemical change, matter is neither created nor destroyed. It would be equally correct to call this the law of conservation of matter.

Chemical Reactions and Energy

All chemical changes also involve some sort of energy change. Energy is either taken in or given off as the chemical change takes place. **Energy** is the capacity to do work. Work is done whenever something is moved. A carpenter does work whenever he or she picks up a hammer and drives a nail. The objects being moved can also be atoms and molecules. This is the kind of work that is encountered in chemistry.

Many reactions give off energy. For example, burning wood is a chemical change in which cellulose, and other substances in the wood, combine with oxygen from the air to produce mainly carbon dioxide and water. Energy is also produced and released in the form of heat and light. Chemical reactions that give off heat energy are called **exothermic** reactions. A dramatic example is the rapid decomposition of nitroglycerin shown in **Figure 1.25.**

Figure 1.25

An Exothermic Reaction

The energy released in the explosion of nitroglycerin breaks and moves rock. The complex nitroglycerin molecule is converted to four gaseous products: carbon dioxide, nitrogen, oxygen, and water vapor.

Hydrogen molecules   Oxygen molecule

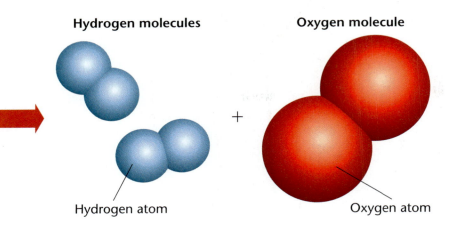

Hydrogen atom   +   Oxygen atom

Figure 1.26

An Endothermic Reaction
When the two compounds ammonium thiocyanate and barium hydroxide octahydrate are mixed, a reaction occurs in which heat is absorbed from the surroundings. The flask becomes so cold that if a film of water is on the bottom of the flask, it will freeze the flask to a wooden block.

    Some chemical changes absorb energy. Chemical reactions that absorb heat energy are called **endothermic** reactions. You can tell that the decomposition of water into oxygen and hydrogen is an endothermic reaction because it doesn't occur unless energy, in the form of an electric current, is passed through the water. When baking soda ($NaHCO_3$) is mixed into some kinds of cookie dough and the dough is baked in a hot oven, the baking soda absorbs energy and breaks down into carbon dioxide, water, and sodium carbonate (Na_2CO_3). The CO_2 and gaseous H_2O puff up the cookies. **Figure 1.26** shows another endothermic reaction.

    Photosynthesis is probably the most important endothermic process on Earth. Photosynthesis is a series of chemical reactions that absorb light energy from the sun and produce sugars from carbon dioxide and water. Oxygen is given off as a product of the reactions. Green plants, algae, and many kinds of bacteria carry out photosynthesis. When you eat sugars and starches, you are eating the molecules formed by endothermic photosynthesis reactions, as indicated in **Figure 1.27** on the next page. Your cells break down these molecules in an exothermic process that supplies you with energy.

Figure 1.27

From Sunlight to Food Energy

When you eat vegetables from a garden such as this, you are consuming substances that plants have synthesized using energy from sunlight.

Connecting Ideas

Chemistry makes connections among the composition, structure, and behavior of matter. As you study chemistry, you will learn how the chemical and physical properties of matter are important clues to its submicroscopic structure and behavior. You'll see how knowing the structure of an atom of an element can enable you to predict the chemical behavior of that element. You'll learn how the state of a substance at room temperature provides clues to the way its atoms are arranged. You'll find out why some things dissolve in water but others do not, how metals corrode, how batteries work, why compounds containing carbon are important to life, and how a nuclear reactor works.

Supplemental Problems

For more practice with solving problems, see Supplemental Practice Problems, Appendix B.

SECTION REVIEW

Understanding Concepts

1. What are the three states of matter?

2. Identify each of the following as either chemical or physical properties of the substance.

   a) Aluminum bends easily.
   b) Copper sulfate dissolves in water.
   c) Magnesium burns in air.
   d) Gold jewelry is unaffected by perspiration.
   e) Baking soda is a white powder.
   f) Fluorine is a highly reactive element.

3. Identify each of the following as either chemical or physical changes.

   a) A match lights when struck.
   b) Air is squeezed by a pump and forced into a tire.
   c) A lump of gold is pounded into a large, thin sheet.
   d) Baking powder bubbles and gives off CO_2 when it is moistened.
   e) A pan of water boils on the stove.
   f) Hydrogen sulfide gas causes silver to tarnish.

Thinking Critically

4. **Applying Concepts** A friend tells you that a newspaper is completely gone because it was burned up in a fire. Use the law of conservation of mass to write an explanation telling your friend what really happened to the newspaper.

Applying Chemistry

5. **Campfires** On cool evenings, a campfire warms the body and spirit. Is the burning of logs an exothermic or endothermic reaction?

---

chemistryca.com/self_check_quiz

REVIEWING MAIN IDEAS

1.1 The Puzzle of Matter

- Chemistry deals with the composition, structure, and behavior of matter.

- Macroscopic observations of matter reflect its submicroscopic structure.

- The quantity of matter in a sample can be measured by determining its mass.

- Descriptions of matter and its behavior are called properties.

- The composition of a sample of matter determines whether it is a substance or a mixture.

- Mixtures are heterogeneous or homogeneous (solutions).

- Mixtures may be separated by using physical changes that take advantage of the physical properties of the individual substances in the mixture.

- Substances are classified as elements or compounds.

- Elements are the building blocks of all matter.

- The composition of a compound is represented by its chemical formula.

1.2 Properties and Changes of Matter

- Every substance has a unique set of physical and chemical properties.

- The states of matter are solid, liquid, and gas. The state of a substance at room temperature is a physical property.

- The temperatures at which a substance changes state are physical properties of the substance.

- The density of a sample of matter is the amount of matter (mass in grams) in a unit volume (usually a milliliter).

- The chemical properties of a substance describe the chemical changes the substance undergoes.

- Chemical changes involve substances rearranging to form different substances.

- Another name for chemical change is *chemical reaction*.

- In a chemical reaction, atoms are never created or destroyed.

Vocabulary

For each of the following terms, write a sentence that shows your understanding of its meaning.

alloy	mass
aqueous solution	matter
chemical change	mixture
chemical property	physical change
chemical reaction	physical property
chemistry	property
compound	qualitative
density	quantitative
element	scientific model
endothermic	solute
energy	solution
exothermic	solvent
formula	substance
law of conservation	volatile
of mass	

UNDERSTANDING CONCEPTS

1. What is chemistry?

2. If you know the melting point of a pure substance, what does it tell you about its boiling point? Its freezing point?

3. What information does the chemical formula for a compound provide about the submicroscopic structure of the compound?

4. What is mass? How is the mass of an object determined?

5. If you say that sulfur is yellow, you are stating a property of sulfur. What is meant by *property*? Is color a physical or chemical property?

6. List three properties of iron.

7. What is the solvent in an aqueous solution of salt? Why are aqueous solutions important?

8. Could two objects with the same volume have different masses? Which, if either, would contain more matter?

9. What is meant by *pure matter*?

10. What is energy? What part does it play in chemistry?

11. Distinguish between exothermic and endothermic reactions, and give an example of each.

12. How is a qualitative observation different from a quantitative observation? Give an example of each.

13. Explain how a mixture is different from a compound.

14. Classify these as heterogeneous or homogeneous mixtures.
    a) salt water
    b) vegetable soup
    c) 14-k gold
    d) concrete

15. What is an element? A compound? Give an example of each.

16. Sucrose, $C_{12}H_{22}O_{11}$, is 51.5 percent oxygen by mass and only 6.4 percent hydrogen by mass, yet there are twice as many hydrogen atoms as oxygen atoms in the formula of sucrose. How can this be?

17. The formula of water is H_2O. The formula of hydrogen peroxide is H_2O_2. Which compound has the highest percentage of hydrogen by mass?

18. What is meant by *density*? Is there any way that a bag of Styrofoam balls could be heavier than a bag of stones?

19. Gold freezes at 1064°C. What is the melting point of gold?

20. Mercury freezes at −38.9°C; nitrogen boils at −195.8°C. How can a boiling point be lower than a freezing point?

21. Use **Table D.4** in the Appendix. Over what temperature range is iron a liquid? How does this liquid temperature range compare with that of krypton, Kr?

22. Is the diagram a molecule of nitroglycerin or a model of nitroglycerin? Explain.

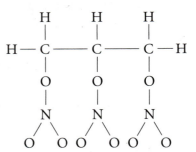

APPLYING CONCEPTS

Everyday Chemistry

23. Is the overall processing of food in your body an exothermic or endothermic process?

24. Air is a mixture of primarily nitrogen (78 percent) and oxygen (21 percent), with trace amounts of argon, neon, helium, and krypton. These pure elements have many applications. They are separated and purified by cooling air in a process known as fractionation. As air cools, the different elements liquefy based on their boiling points. In what order do the elements liquefy? Use **Table D.4.**

Chemistry and Society

25. A drug with a well-known brand name may also be made by another company and sold at a lower price. Both drugs are specified to have the same chemical formula and are approved by the Food and Drug Administration. Would it be safer to use the well-known brand? Explain.

26. A doorknob is coated with the alloy brass. What is an alloy? Give an example of another alloy.

27. Some iron filings accidentally dropped into a mixture of salt crystals and water. Describe two ways that you could separate the filings.

28. Ethanol, or ethyl alcohol, melts at $-114.1°C$ and boils at $78.5°C$. What is the physical state of ethanol at room temperature?

29. Using the information in question 28, what are the freezing points and the condensation points of ethanol?

30. Is the melting of candle wax an exothermic or endothermic change?

31. Which of these substances has the lowest boiling point: oxygen, water, or salt?

32. List five properties of a candle. Identify them as physical or chemical properties.

33. Is a tree homogeneous or heterogeneous? Explain.

34. Identify the following as an element, compound, homogeneous mixture, or heterogeneous mixture: hydrogen, polyethylene plastic, clear apple juice, cloudy apple cider, syrup, paint, and bronze.

● THINKING CRITICALLY

Observing and Inferring

35. **ChemLab 1** From macroscopic observations of a burning candle, what conclusions can you draw about events occurring on the submicroscopic level?

Observing and Inferring

36. **ChemLab 2** Why is starch not used in baking for the purpose of making cakes fluffy?

Measuring in SI

37. **ChemLab 3** Suppose you have a sample of an unknown mineral with a mass of 86 g. You place the sample in a graduated cylinder filled with water to the 55-mL mark. The sample sinks to the bottom of the cylinder and the water level rises to the 71-mL mark. What is the density of the sample?

Comparing and Contrasting

38. **MiniLab 1** Is the mixing of ethanol and water an exothermic or an endothermic process? Is it a chemical or physical change?

Designing an Experiment

39. **MiniLab 2** Marker inks of different colors, when separated by chromatography, may be found to contain some pigments of the same color. For example, a black marker and a blue marker might both contain a blue pigment. Design an experiment to show whether or not two pigments of the same color, found in different markers, are likely to be the same pigment.

Applying Concepts

40. **MiniLab 3** The appearance of a penny made of zinc and copper is different after heating the penny in zinc chloride solution. Is the new coating an element, a compound, or a mixture? Explain.

Observing and Inferring

41. **MiniLab 4** Are the chemicals in your food elements or compounds? Explain.

Observing and Inferring

42. MiniLab 5 Is a polymer an element or a compound? Explain.

43. Two solid compounds burn in the presence of oxygen. Can you conclude that they are the same compound?

Interpreting Chemical Structures

44. What does the formula C_2H_6O tell you about the substance it represents? Be as specific as possible.

CUMULATIVE REVIEW

In Chapters 2 through 21, this heading will be followed by review questions about skills and ideas you have learned in earlier chapters.

SKILL REVIEW

45. Making and Using a Table Make a table like the one shown below and fill in the columns. Identify each change or process as physical or chemical and justify your choice. When your table is complete, look at the two groups of processes. Within each group, list similarities shared by the processes. Are there differences within each group?

WRITING IN CHEMISTRY

46. Look at labels on items in your home or in the grocery store. Choose one item and gather information about it. List all of its ingredients. Describe it as a mixture or pure substance. List three of its properties. For information about the substance, look it up in a chemical handbook. If it is a compound, what is its formula? What are some of its physical and chemical properties? Find out why this substance is an ingredient in the item you chose. Find a similar item, made by a different company, and determine whether the two items have the same ingredients. Summarize your findings in a report.

PROBLEM SOLVING

47. A miner found a nugget that had a gold color. He realized that it could be precious gold metal or iron pyrite (FeS_2), which is a compound of iron and sulfur called fool's gold. The nugget had a mass of 16.5 g and displaced 3.3 mL of water. From this information and data from a chemical handbook, find out whether the miner had found gold.

Observation	Is it chemical or physical?	Explanation
1. Brewing tea or coffee		
2. Heating a pot of water to cook pasta		
3. Food scraps decomposing in a compost heap		
4. A seed germinating		
5. Sugar dissolving in water		
6. Vinegar and oil not mixing		
7. A termite eating wood and producing methane gas		
8. The density of water increasing as salt is added		

1. Matter is defined as anything that

  a) exists in nature.
  b) is solid to the touch.
  c) is found in the universe.
  d) has mass and takes up space.

2. Which of the following has no mass?

  a) air   **c)** atoms
  b) electricity   **d)** water

3. A model is best defined as

  a) a thinking device based on macroscopic observations.
  b) a thinking device based on microscopic observations.
  c) a thinking device based on constant experimentation.
  d) a thinking device based on a consensus of scientists.

4. When a spoonful of sugar is added to hot tea, the sugar is called a(n)

  a) solution.   **c)** solvent.
  b) alloy.   **d)** solute.

Elements	Density (g/mL)	Boiling Point (°C)	Melting Point (°C)
Argon	1.78	−186	−189
Bromine	3.12	59	−7
Gallium	5.91	2,403	30
Sodium	0.97	883	98
Tungsten	19.35	5,660	3,410

Use information from the data table above to answer questions 5–7.

5. Which element is a liquid at room temperature?

  a) argon
  b) bromine
  c) gallium
  d) sodium
  e) tungsten

6. Which element has the fewest atoms per cubic centimeter?

  a) argon   **d)** sodium
  b) bromine   **e)** tungsten
  c) gallium

7. Which element will exist as a gas in any location on planet earth?

  a) argon   **d)** sodium
  b) bromine   **e)** tungsten
  c) gallium

8. The chemical formula of chalk is $CaCO_3$. Identify the elements and calculate the number of atoms of each element in a chalk molecule.

  a) 1 calcium atom; 3 cobalt atoms
  b) 1 calcium atom; 1 carbon atom; 3 oxygen atoms
  c) 1 calcium atom; 1 chlorine atom; 3 oxygen atoms
  d) 1 calcium atom; 3 carbon atoms; 3 oxygen atoms

9. Which of the following substances is volatile?

  a) perfume   **c)** coal
  b) water   **d)** sugar

10. The law of conservation of mass states that

  a) matter cannot be created or destroyed.
  b) matter can be created but not destroyed.
  c) matter can be destroyed but not created.
  d) matter is always being created and destroyed.

Test Taking Tip

More Than One Graphic If a test question has more than one table, graph, diagram, or drawing with it, use them all. If you answer based on just one graphic, you've probaly missed an important piece of information. Make sure that you accurately analyze all graphics before answering the questions.

Chapter Preview

Sections

2.1 Atoms and Their Structure
MiniLab 2.1 A Penny for Your Isotopes
ChemLab Conservation of Matter

2.2 Electrons in Atoms
MiniLab 2.2 Line Emission Spectra
of Elements

What a View!

This molecule is odorless, tasteless, colorless and liquid at room temperature. Because of its unique properties, it makes life on Earth possible. Water contains two hydrogen atoms and one oxygen atom.

*L*aunch Lab

What's Inside?

It's your birthday, and there are many wrapped presents for you to open. Much of the fun is trying to figure out what's inside the package before you open it. In trying to determine the structure of the atom, chemists had a similar experience. How good are your skills of observation and deduction?

Safety Precautions

Materials

- a wrapped box from your instructor

Procedure

1. Obtain a wrapped box from your instructor.
2. Using as many observation methods as you can, and without unwrapping or opening the box, try to figure out what the object inside the box is.
3. Record the observations you make throughout this discovery process.

Analysis

How were you able to determine things such as size, shape, number, and composition of the object in the box? What senses did you use to make your observations? Why is it hard to figure out what type of object is in the box without actually seeing it?

What I Already Know

Review the following concepts before studying this chapter.
Chapter 1: the difference between an element and a compound; distinguishing between chemical and physical changes; the meaning of the law of conservation of mass

Reading Chemistry

Scan Chapter 2 for vocabulary words, and make two lists: one of familiar words, and one of new words. After defining the new words, create sentences using both the familiar and new vocabulary words.

Preview this chapter's content and activities at chemistryca.com

Atoms and Their Structure

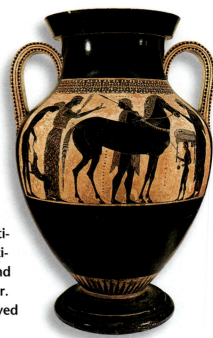

When did people first wonder about matter and its structure? Undoubtedly, early humans observed matter and questioned its behavior and composition. People in ancient Greece knew how to manipulate matter well enough to produce this beautiful vase. As years passed and methods of investigation improved, scientists from many times and cultures left written records of studies of matter. Current thoughts on the nature of matter evolved over a period of several thousand years.

SECTION PREVIEW

Objectives

✓ **Relate** historic experiments to the development of the modern model of the atom.

✓ **Illustrate** the modern model of an atom.

✓ **Interpret** the information available in an element block of the periodic table.

Review Vocabulary

Energy: the capacity to do work.

New Vocabulary

atom
atomic theory
law of definite
 proportions
hypothesis
theory
scientific law
electron
proton
isotope
neutron
nucleus
atomic number
mass number

Early Ideas About Matter

The current model of the composition of matter is based on hundreds of years of work that began when observers realized that different kinds of matter exist and that these kinds of matter have different properties and undergo different changes. About 2500 years ago, the Greek philosophers thought about the nature of matter and its composition. They proposed that matter was a combination of four fundamental elements—air, earth, fire, and water, as shown in **Figure 2.1.** These Greek philosophers also argued the question of whether matter could be divided endlessly into smaller and smaller pieces or whether there was an ultimate smallest particle of matter that could not be divided any further. The Greek philosophers were keen observers of nature but, unlike modern scientists, didn't test their hypotheses with experiments.

Figure 2.1

Greek Elements
The Greek philosophers thought that air, earth, fire, and water were the elements that formed all matter. Notice how the properties hot, dry, cold, and wet are associated with each element. These early ideas about the elements were not completely swept aside until the 19th century.

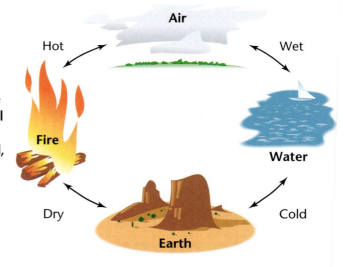

Democritus, 460-370 B.C., was a philosopher who proposed that the world is made up of empty space and tiny particles called atoms. Democritus thought that **atoms** are the smallest particles of matter and that different types of atoms exist for every type of matter. The idea that matter is made up of fundamental particles called atoms is known as the **atomic theory** of matter.

Development of the Modern Atomic Theory

In 1782, a French chemist, Antoine Lavoisier (1743-1794), made measurements of chemical change in a sealed container. He observed that the mass of reactants in the container before a chemical reaction was equal to the mass of the products after the reaction. For example, in a sealed container, 2.0 g of hydrogen always reacts with 16.0 g of oxygen to give 18.0 g of water. Lavoisier concluded that when a chemical reaction occurs, matter is neither created nor destroyed but only changed. Lavoisier's conclusion became known as the *law of conservation of matter*. This is another name for the law of conservation of mass that you learned in Chapter 1. This law is illustrated in **Figure 2.2.**

Proust's Contribution

In 1799, another French chemist, Joseph Proust, observed that the composition of water is always 11 percent hydrogen and 89 percent oxygen by mass. Regardless of the source of the water, it always contains these same

Figure 2.2

Law of Conservation of Matter
When magnesium burns in air, it gives off an intense white light and some white smoke. It is difficult to measure the amounts of magnesium and oxygen involved in the reaction under these open conditions. ▼

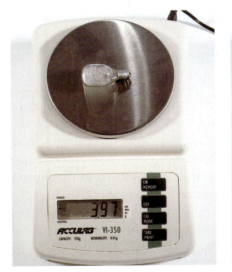
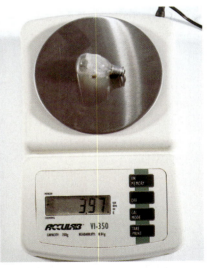
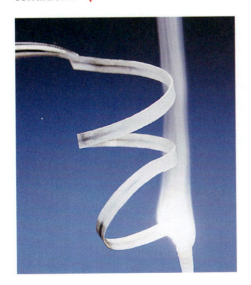

▲ The same reaction occurs in a flash-bulb, which contains fine magnesium wire and oxygen. Note the mass of the unused bulb on the left. The magnesium in the used bulb on the right has been burned, and the bulb now contains magnesium oxide. Notice that the mass did not change.

Two samples of natural FeS_2

46.5% Fe

53.5% S

Composition of Iron Pyrite

Figure 2.3

The Law of Definite Proportions

Iron pyrite, FeS_2, looks something like gold. It fooled many gullible people and got the name *fool's gold* as a result. The composition of pyrite is 46.5 percent iron and 53.5 percent sulfur by mass. These proportions are the same for every sample of pyrite. They are independent of the origin or amount of the substance.

percentages of hydrogen and oxygen. Proust studied many other compounds and observed that the elements that composed the compounds were always in a certain proportion by mass. This principle is now referred to as the **law of definite proportions. Figure 2.3** illustrates this law.

Dalton's Atomic Theory

John Dalton (1766-1844), an English schoolteacher and chemist, studied the results of experiments by Lavoisier, Proust, and many other scientists. He realized that an atomic theory of matter must explain the experimental evidence. For example, if matter were composed of indivisible atoms, then a chemical reaction would only rearrange those atoms, and no atoms would form or disappear. This idea would explain the law of conservation of mass. Also, if each element consisted of atoms of a specific type and mass, then a compound would always consist of a certain combination of atoms that never varied for that compound. Thus, Dalton's theory explained the law of definite proportions, as well. Dalton proposed his atomic theory of matter in 1803. Although his theory has been modified slightly to accommodate new discoveries, Dalton's theory was so insightful that it has remained essentially intact up to the present time.

The following statements are the main points of Dalton's atomic theory.
1. All matter is made up of atoms.

2. Atoms are indestructible and cannot be divided into smaller particles. (Atoms are indivisible.)

3. All atoms of one element are exactly alike, but they are different from atoms of other elements.

Dalton's atomic theory gave chemists a model of the particle nature of matter. However, it also raised new questions. If all elements are made up of atoms, why are there so many different elements? What makes one atom different from another atom? Experiments in the late 19th century began to suggest that atoms are made up of even smaller particles. Present-day chemistry explains the properties and behavior of substances in terms of three of these smaller particles. You will learn more about each of these particles in Section 2.2.

Atomic Theory, Conservation of Matter, and Recycling

Have you ever wondered what happens to the atoms in the stuff that you throw away? Where do the atoms go when waste is incinerated or buried in a landfill? As you've just learned, atoms are never created or destroyed in everyday chemical processes. So, in a sense, you can't throw anything away because there is no "away." When waste is burned or buried in a landfill, the atoms of the waste may combine with oxygen or other substances to form new compounds, but they don't go away. In natural processes, atoms are not destroyed; they are recycled. **Figure 2.4** shows how elemental nitrogen from the atmosphere is converted into compounds that are used on Earth, and then returns to the atmosphere.

In recent years, small towns, large cities, and entire states have discovered the benefits of recycling paper, plastic, aluminum, and glass. Labels on many supermarket bags, cardboard boxes, greeting cards, and other paper products say "Made from recycled paper." Scrap aluminum is easily recycled and made into new aluminum cans or other aluminum products. Have you noticed how newly paved roadways sparkle? The sparkle is a result of the addition of recycled glass to the paving material. Even ground-up tires can be added to asphalt for paving. By reusing the atoms in manufactured materials, we are imitating nature and conserving natural resources.

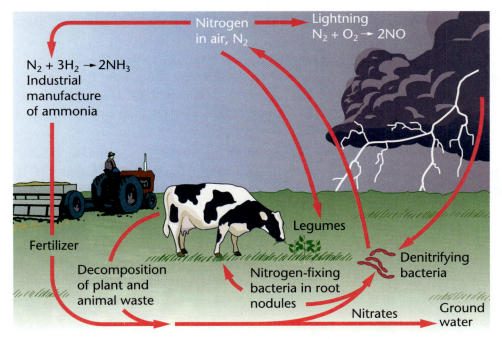

Figure 2.4
Recycling Nitrogen
Lightning, bacteria, industrial processes, and even lichens on tree branches convert nitrogen from the atmosphere into compounds. These compounds enter the plant and animal food chain. When plant and animal wastes decompose, bacteria in the soil produce free nitrogen, which returns to the atmosphere.

ChemLab

Conservation of Matter

Hydrochloric acid reacts with metals such as zinc, magnesium, and aluminum. A colorless gas escapes, and the metal seems to disappear. What happens to the atoms of the metals in these reactions? Are they destroyed? If not, where do they go?

Problem

What happens to the atoms of a metal when they react with an acid?

Objectives

- **Infer** what happens to atoms during a chemical change.
- **Compare** experimental results to the law of conservation of matter.

PREPARATION

Materials

granular zinc	hot plate
1M HCl	balance
125-mL Erlenmeyer flask	spatula
10-mL graduated cylinder	oven mitt

Safety Precautions

Use care when handling hydrochloric acid and hot objects. Wear goggles and laboratory aprons. Perform this ChemLab in a well-ventilated room.

PROCEDURE

1. Weigh a clean, dry, 125-mL Erlenmeyer flask to the nearest 0.01 g. Record its mass in a data table.
2. Obtain a sample of granular zinc, and add it to your flask. Weigh the flask with the zinc in it.

Record the total mass in the data table. Subtract to find the mass of zinc. The mass of zinc must be between 0.20 g and 0.28 g. If you have too much zinc, remove some with a clean spoon or spatula until the mass is in the range described.

3. Add 10 mL of 1M hydrochloric acid to the flask. Swirl the contents and look for signs of a chemical change. Record your observations in the data table.
4. Set the flask on a hot plate as shown in the photo.

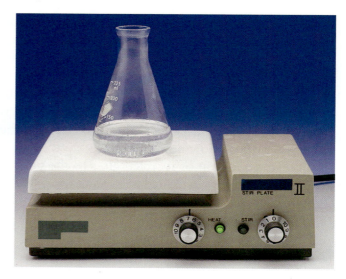

5. Set the hot plate on low to heat the flask gently. The liquid in the flask should not boil. Again look for signs that a chemical change is taking place, and record your observations in the data table.
6. Eventually, all the metallic zinc will disappear. It is important to heat the flask slowly. **CAUTION:** *Be careful not to inhale the fumes from the flask at this point.* When the liquid in the flask is almost gone, a white solid will appear. Stop heating as soon as you see this solid. Use an oven mitt to remove the flask from the hot plate The heat retained by the flask should be enough to completely evaporate the remaining liquid.

3. Observing and Inferring What happened to the zinc?

4. Making Inferences Why were the flask and its contents heated?

APPLY AND ASSESS

1. Chemists have determined that zinc chloride is 48 percent zinc. Use this information to compute the mass of zinc in your product. How does this mass compare with the mass of zinc you started with?

2. If the difference in the question above is greater than 0.04 g, how can you account for it?

3. How does this experiment support the law of conservation of matter?

4. How could you modify the procedure so that the law of conservation of matter is better demonstrated?

7. Let the flask cool. When all the liquid has evaporated, weigh the cool, dry flask and its contents. Record this mass in the data table. Calculate the mass of zinc chloride produced in the reaction.

ANALYZE AND CONCLUDE

1. **Interpreting Data** How did the mass of the product, zinc chloride, compare with the mass of the zinc?

2. **Interpreting Observations** How can you account for the difference in mass?

DATA AND OBSERVATIONS

Mass of empty Erlenmeyer flask	
Mass of Erlenmeyer flask with zinc sample	
Mass of zinc sample	
Mass of Erlenmeyer flask with reaction product (zinc chloride, $ZnCl_2$)	
Mass of zinc chloride produced	
What did you see when hydrochloric acid was first added to the flask?	
What did you see when you began to heat the flask?	
What did you see when all the liquid had evaporated?	

Politics and Chemistry–Elemental Differences

In the time of Antoine Lavoisier (1743-1794), many scientists were still trying to explain matter as combinations of the elements air, earth, fire, and water. Lavoisier's work changed the way chemistry was done, and today he is recognized as the first modern chemist. However, like other scientists of the 18th century, Lavoisier could not earn a living as a chemist, so he invested in a private firm that collected taxes for the king.

Remaking chemistry Lavoisier set out to reorganize chemistry. Lavoisier's habit of carefully weighing reactants and products in experiments led him to discover that the mass of materials before a chemical change equals the mass of the products after it, which is the basis of the law of conservation of matter. He also discovered that combustion is the result of reaction with oxygen.

International recognition Many scientists held Lavoisier in high esteem. Benjamin Franklin made a point to observe experiments by Lavoisier when he was in France soliciting support for the cause of the American Revolution. Lavoisier's experiments were also followed closely by Thomas Jefferson.

In 1774, the British chemist Joseph Priestley discussed one particular experiment with Lavoisier. Priestley explained that after heating "calx of mercury" (which we know today as mercury(II) oxide), metallic mercury remained and a gas was given off. When he placed a candle in the gas, it burned more brightly. He also found that if a mouse is placed in a closed jar with the gas, the mouse can breathe it and live. Priestley's gas was oxygen, but because he believed in an older theory of matter called the phlogiston theory, Priestley did not recognize it as an element. Lavoisier repeated Priestley's experiment and came to the history-making conclusion that air is not a simple substance but a mixture of two different gases. One of these gases, oxygen, supports combustion, promotes breathing, and rusts metals. Lavoisier gave oxygen its name.

Political price Lavoisier was not a member of the aristocracy. He belonged to the professional class from which many of the leaders of the French Revolution came. In spite of his class and the high regard for Lavoisier in the scientific world, his connection with the tax-collecting firm made him a target of suspicion. During the Reign of Terror that followed the French Revolution, Lavoisier was arrested and condemned to death in a trial that lasted less than a day. That same day, he was guillotined and his body thrown into a common grave—a victim of ignorance and mob rule.

Connecting to Chemistry

1. **Analyzing** Why was Lavoisier's role in the discovery of oxygen important even though he merely repeated Priestley's experiment?

2. **Applying** Lavoisier showed that a person uses more oxygen when working than when resting. Explain the reasoning behind Lavoisier's findings.

Hypotheses, Theories, and Laws

The first step to solving a problem, such as what makes up matter, is observation. Scientists use their senses to observe the behavior of matter at the macroscopic level. They then come up with a **hypothesis,** which is a testable prediction to explain their observations. For example, Lavoisier thought that matter was indestructible. He made a hypothesis that all the matter present before a chemical change would still be there after the change.

To find out whether a hypothesis is correct, it must be tested by repeated experiments. Lavoisier performed numerous careful experiments using different chemical changes. Scientists accept hypotheses that are verified by experiments and reject hypotheses that can't stand up to experimental testing.

A body of knowledge builds up as a result of these experiments. Scientists develop theories to organize their knowledge. A **theory** is an explanation based on many observations and supported by the results of many experiments. For example, Dalton's atomic theory was based on observations of matter performed again and again by many scientists. As scientists gather more information, a theory may have to be revised or replaced with another theory. In nonscientific speech and writing, the word *theory* is often used to mean an unsupported notion about something. As you can see, a scientific theory is heavily supported by evidence gained from experiments and continual observations. The methods of science just discussed are summarized in **Figure 2.5.**

A **scientific law** is simply a fact of nature that is observed so often that it becomes accepted as truth. That the sun rises in the east each morning is a law of nature because people see that it is true every day. A law can generally be used to make predictions but does not explain why something happens. In fact, theories explain laws. One part of Dalton's atomic theory explains why the law of conservation of matter is true.

Figure 2.5
The Methods of Science
Scientists make observations that lead to hypotheses. A hypothesis must be tested by multiple experiments. If the experimental results do not agree with the hypothesis, they become the observations that lead to a new hypothesis. A hypothesis that is refined and supported by many experiments becomes a theory that explains a fact or phenomenon in nature.

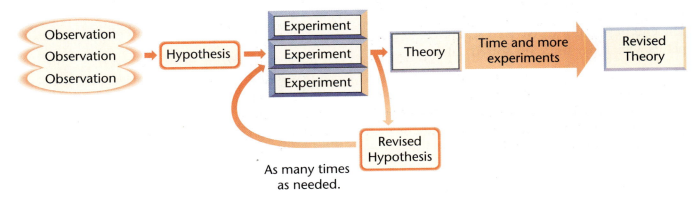

Recycling Glass

Imagine that you arrive home thirsty from a soccer game. You reach into the refrigerator and pull out an ice-cold bottle of apple juice. In seconds, the empty bottle is all that remains. But what about the law of conservation of matter? Because of this law, the bottle lingers on—as trash. Worse yet, unless the bottle is recycled, its matter will be conserved in a landfill.

Problems of recycling glass Glass manufacturers call broken glass that can be added to new material *glass cullet*. There is no difficulty in recycling uncolored glass or amber glass because the market for these kinds of cullet is strong. Only green glass, the kind found in some bottles, presents a problem.

Solutions in recycling methods Because of the color it adds, green glass cullet often can't be mixed with uncolored or amber glass cullet. For this reason, recycling centers often sort glass by color. The sorted glass is dropped down separate chutes and processed by separate glass crushers. The end products are delivered separately to glass plants, which pay more for sorted cullet.

Finding uses for recycled glass Many areas, such as the state of Washington, are searching for new markets for recycled glass. An example of one new use is adding ten percent green glass cullet or mixed glass cullet to asphalt for roads. This seems to be working well, but engineers will continue to study this surface for its durability. The glass particles make the roads sparkle, which is an added attraction.

Washington's Department of Trade and Economic Development has developed a list of more than 70 uses of glass cullet. These uses include manufacturing construction material such as fiberglass, foam glass, and rock wool insulation; using it in material for building roadbeds; and making pressed glass such as dinnerware and decorative glass. One other use of glass cullet is in removing pollutants in water runoff. In this extremely useful technology, glass cullet is given a static electrical charge, which makes it able to attract and hold small particles in runoff from areas such as airport runways.

Analyzing the Issue

1. **Acquiring Information** Do research to find out how new glass is manufactured. Learn whether cullet is routinely used in most glass manufacturing. Does the use of cullet reduce the cost of manufacturing glass? Does the use of cullet reduce the energy requirement for producing glass?

2. **Inferring** A company in Richmond, California, makes a business of washing wine bottles for reuse—9 million a year. If you were starting such a business, what marketing and distribution problems might you have to solve?

The Discovery of Atomic Structure

Dalton's atomic theory was *almost* true. Dalton had assumed that atoms are the ultimate particles of matter and can't be broken up into smaller particles and that all atoms of the same element are identical. However, his theory had to be modified as new discoveries were made in the late 19th and early 20th centuries. Today, we know that atoms are made up of smaller particles and that atoms of the same element can be nearly, but not exactly, the same. In this section, you'll examine the discoveries that led to the modern atomic theory.

The Electron

Because of Dalton's atomic theory, most scientists in the 1800s believed that the atom was like a tiny solid ball that could not be broken up into parts. In 1897, a British physicist, J.J. Thomson, discovered that this solid-ball model was not accurate.

Thomson's experiments used a vacuum tube such as that shown in **Figure 2.6.** A vacuum tube has had all gases pumped out of it. At each end of the tube is a metal piece called an electrode, which is connected through the glass to a metal terminal outside the tube. These electrodes become electrically charged when they are connected to a high-voltage electrical source. When the electrodes are charged, rays travel in the tube from the negative electrode, which is the cathode, to the positive electrode, the anode. Because these rays originate at the cathode, they are called cathode rays.

Thomson found that the rays bent toward a positively charged plate and away from a negatively charged plate. He knew that objects with like charges repel each other, and objects with unlike charges attract each other. Thomson concluded that cathode rays are made up of invisible, negatively charged particles referred to as **electrons.** These electrons had to come from the matter (atoms) of the negative electrode.

Modern cathode-ray tubes are the picture tubes in TV sets and computer monitors. These tubes use a varying magnetic field to cause the electron beam to move back and forth, painting a picture on a screen that's coated with luminescent chemicals.

The pole of a magnet bends the cathode rays in a direction at a right angle to the direction of the field. ▼

Figure 2.6

Cathode-Ray Tube
When a high voltage is applied to a cathode-ray tube, cathode rays form a beam that produces a green glow on the fluorescent plate. ▶

From Thomson's experiments, scientists had to conclude that atoms were not just neutral spheres, but somehow were composed of electrically charged particles. In other words, atoms were not indivisible but were composed of smaller particles, referred to as subatomic particles. Further experimentation showed that an electron has a mass equal to 1/1837 the mass of a hydrogen atom, the lightest atom.

Reason should tell you that there must be a lot more to the atom than electrons. Matter is not negatively charged, so atoms can't be negatively charged either. If atoms contained extremely light, negatively charged particles, then they must also contain positively charged particles—probably with a much greater mass than electrons. Scientists immediately worked to discover such particles.

Protons and Neutrons

In 1886, scientists discovered that a cathode-ray tube emitted rays not only from the cathode but also from the positively charged anode. These rays travel in a direction opposite to that of cathode rays. Like cathode rays, they are deflected by electrical and magnetic fields, but in directions opposite to the way cathode rays are deflected. Thomson was able to show that these rays had a positive electrical charge. Years later, scientists determined that the rays were composed of positively charged subatomic particles called **protons.** The amount of charge on an electron and on a proton is equal but opposite, but the mass of a proton is much greater than the mass of an electron. The mass of a proton was found to be only slightly less than the mass of a hydrogen atom.

At this point, it seemed that atoms were made up of equal numbers of electrons and protons. However, in 1910, Thomson discovered that neon consisted of atoms of two different masses, **Figure 2.7.** Atoms of an element that are chemically alike but differ in mass are called **isotopes** of the element. Today, chemists know that neon consists of three naturally occurring isotopes. The third was too scarce for Thomson to detect.

Because of the discovery of isotopes, scientists hypothesized that atoms contained still a third type of particle that explained these differences in mass. Calculations showed that such a particle should have a mass equal to that of a proton but no electrical charge. The existence of this neutral particle, called a **neutron,** was confirmed in the early 1930s.

Figure 2.7

Thomson's Isotopes of Neon
These diagrams represent the two isotopes of neon that Thomson observed. Both nuclei have ten protons, but the one on the left has ten neutrons and the one on the right has 12 neutrons.

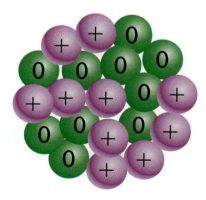

Neon-20 nucleus

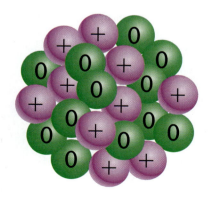

Neon-22 nucleus

A Penny for Your Isotopes

Many elements have several naturally occurring isotopes. Isotopes are atoms of the same element that are identical in every property except mass. When chemists refer to the mass of an atom of one of these elements, they really mean the average of the atomic masses of the isotopes of that element. You can use pennies to represent isotopes. Determine the mass of two penny isotopes and then the average mass of a penny.

Procedure

1. Get a bag of pennies from your teacher.

2. Sort the pennies by date. Group the pre-1982 pennies and the post-1982 pennies.

3. Weigh ten pennies from each group. Record the mass to the nearest 0.01 g. Divide the total mass by ten to get the average mass for one penny in that group.

4. Count the number of pennies in each group.

5. Using the data, determine the total mass of all of the pre-1982 pennies. In the same way, calculate the total mass of the post-1982 pennies.

Analysis

1. What is the mass of all the pennies?

2. Calculate the average mass of a penny. How does this mass compare with the average masses for each group?

3. Why were you directed to weigh ten pennies of a group and divide the mass by ten to get the mass of one penny? Why not just weigh one penny from each group?

Rutherford's Gold Foil Experiment

While all of these subatomic particles were being discovered, scientists were also trying to determine how the particles of an atom were arranged. After the discovery of the electron, scientists pictured an atom as tiny particles of negative electricity embedded in a ball of positive charge. You could compare this early atomic model to a ball of chocolate-chip cookie dough (except with microscopic chocolate chips). Almost at the same time, a Japanese physicist, Hantaro Nagaoka, proposed a different model in which electrons orbited a central, positively charged nucleus—something like Saturn and its rings. Both models are shown in **Figure 2.8.**

Nagaoka's model resembled a planet with moons orbiting in a flat plane. He believed the core had a positive charge, and the negative electrons were in orbit around it. ▼

Figure 2.8

Thomson's and Nagaoka's Atomic Models
Thomson pictured the atom as consisting of electrons embedded in a ball of positive charge. ▶

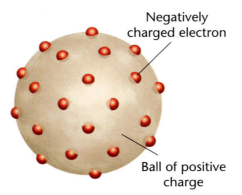

Negatively charged electron

Ball of positive charge

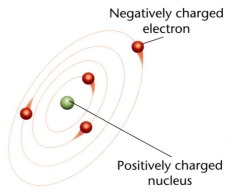

Negatively charged electron

Positively charged nucleus

In 1909, a team of scientists led by Ernest Rutherford in England carried out the first of several important experiments that revealed an arrangement far different from the cookie-dough model of the atom. Rutherford's experimental setup is shown in **Figure 2.9.**

The experimenters set up a lead-shielded box containing radioactive polonium, which emitted a beam of positively charged subatomic particles through a small hole. Today, we know that the particles of the beam consisted of clusters containing two protons and two neutrons and are called alpha particles. The sheet of gold foil was surrounded by a screen coated with zinc sulfide, which glows when struck by the positively charged particles of the beam.

Figure 2.9

The Gold Foil Experiment
Gold is a metal that can be pounded into a foil only a few atoms thick. Rutherford's group took advantage of this property in their experiment.

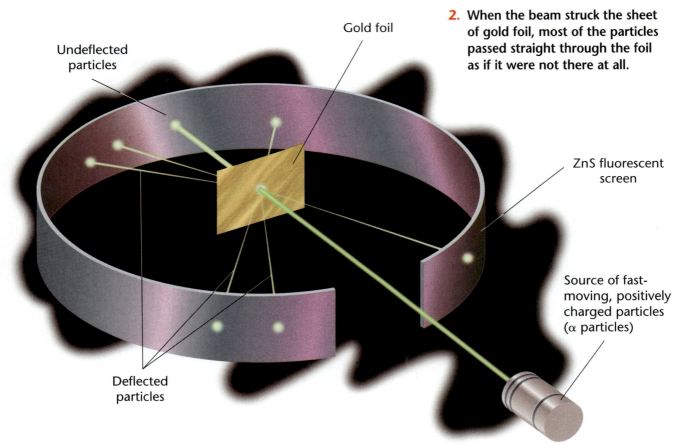

Undeflected particles

Gold foil

2. When the beam struck the sheet of gold foil, most of the particles passed straight through the foil as if it were not there at all.

ZnS fluorescent screen

Source of fast-moving, positively charged particles (α particles)

Deflected particles

3. However, a few of the beam's particles were deflected. Some were deflected slightly, but a few bounced almost straight back. Imagine throwing a baseball at a sheet of tissue paper and having the ball bounce back at you. Rutherford's scientists were just as surprised.

1. Rutherford's lead-shielded box contained radioactive polonium. As polonium decays, it emits helium nuclei, which consist of two protons and two neutrons. These nuclei are called alpha particles. Because they have no electrons, they are positively charged.

The Nuclear Model of the Atom

To explain the results of the experiment, Rutherford's team proposed a new model of the atom. Because most of the particles passed through the foil, they concluded that the atom is nearly all empty space. Because so few particles were deflected, they proposed that the atom has a small, dense, positively charged central core, called a **nucleus.** The new model of the atom as pictured by Rutherford's group in 1911 is shown in **Figure 2.10.**

Imagine how this model affected the view of matter when it was announced in 1911. When people looked at a strong steel beam or block of stone, they were asked to believe that this heavy, solid matter was practically all empty space. Even so, this atomic model has proven to be reasonably accurate. As an example of the emptiness of an atom, consider the simplest atom, hydrogen, which consists of an electron and a nucleus of one proton. If hydrogen's proton were increased to the size of a golf ball, the electron would be about a mile away and the atom would be about 2 miles in diameter. To get an idea of how small an atom is, consider that there are roughly 6 500 000 000 000 000 000 000 (6.5 sextillion or 6.5×10^{21}) atoms in a drop of water. If you need more help in expressing very large and very small quantities using scientific notation, study page 795 in the *Skill Handbook,* Appendix A.

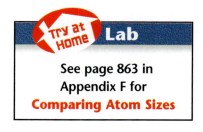

See page 863 in Appendix F for **Comparing Atom Sizes**

The team concluded that the nucleus is extremely small compared to the atom as a whole because so few particles were deflected. ▼

Figure 2.10

Discovery of the Nuclear Model

Remember that the beam's particles were positively charged. If the nucleus were negatively charged, the particles would attract each other. Because the beam was repelled, the nucleus must also be positively charged. ▶

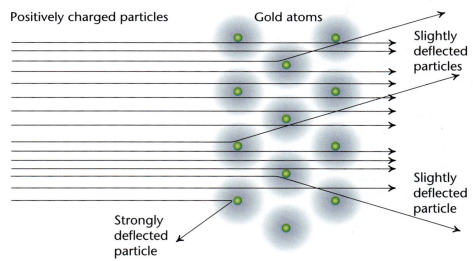

Positively charged particles   Gold atoms

Slightly deflected particles

Strongly deflected particle

Slightly deflected particle

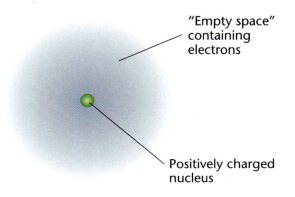

"Empty space" containing electrons

Positively charged nucleus

◀ Because electrons have little mass, the nucleus must contain almost all the mass of an atom. Therefore, the nucleus must be extremely dense and surrounded by empty space in which electrons were found.

Atomic Numbers and Masses

Supplemental Problems

For more practice with solving problems, see Supplemental Practice Problems, Appendix B.

Look at the periodic table hanging on the wall of your classroom or the one inside the back cover of your textbook. Notice that the elements are numbered from 1 for hydrogen, 2 for helium, through 8 for oxygen, and on to numbers above 100 for the newest elements created in the laboratory. At first glance, it may seem that these numbers are just a way of counting elements, but the numbers mean much more than that. Each number is the atomic number of that atom. The **atomic number** of an element is the number of protons in the nucleus of an atom of that element. It is the number of protons that determines the identity of an element, as well as many of its chemical and physical properties, as you will see later in the course.

Because atoms have no overall electrical charge, an atom must have as many electrons as there are protons in its nucleus. Therefore, the atomic number of an element also tells the number of electrons in a neutral atom of that element. You'll learn more about the arrangement of electrons in a later section.

Do you remember neutrons? These neutral particles are also found in the nucleus of all atoms except for the simplest isotope of hydrogen. The mass of a neutron is almost the same as the mass of a proton. The sum of the protons and neutrons in the nucleus is the **mass number** of that particular atom. Isotopes of an element have different mass numbers because they have different numbers of neutrons, but they all have the same atomic number. An isotope is identified by writing the name or symbol of the atom followed by its mass number. Recall that Thomson discovered that naturally occurring neon is a mixture of isotopes. The three neon isotopes are Ne-20 (90.5 percent), Ne-21 (0.2 percent), and Ne-22 (9.3 percent). All neon isotopes have ten protons and ten electrons. Ne-20 has ten neutrons; Ne-21 has 11 neutrons; and Ne-22 has 12 neutrons. Look at the isotopes of lithium in **Figure 2.11**. The properties of the common subatomic particles are summarized in **Table 2.1**.

Figure 2.11

The Particles of an Atom
The number of protons in an atom determines what element it is. For example, all atoms that contain three protons in the nucleus are atoms of lithium.

Naturally occurring lithium consists of two isotopes. The left isotope contains three protons, four neutrons, and three electrons and makes up 92.6 percent of lithium. ▼

The remaining 7.4 percent of lithium consists of atoms with three protons, three neutrons, and three electrons, shown at right. ▼

Isotopes are identified by placing the mass number after the name or symbol of the element. The two isotopes of lithium are Li-7 and Li-6. ▶

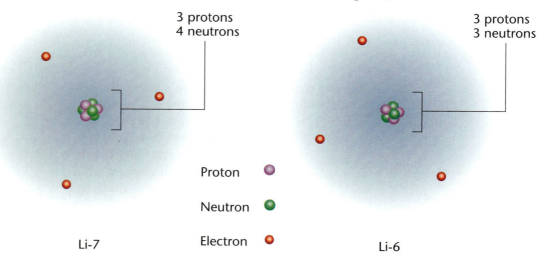

3 protons
4 neutrons

3 protons
3 neutrons

Proton

Neutron

Electron

Li-7

Li-6

Table 2.1   Particles of an Atom

Particle	Symbol	Charge	Mass Number	Mass in Grams	Mass in u
Proton	p^+	1+	1	1.67×10^{-24}	1.01
Neutron	n^0	0	1	1.67×10^{-24}	1.01
Electron	e^-	1–	0	9.11×10^{-28}	0.00055

Atomic Mass

Considering the number of atoms in just a drop of water, you can understand that atoms have extremely small masses, **Figure 2.12.** To state the mass of an atom in grams, you must deal with small numbers. **Table 2.1** shows you the mass in grams of protons, neutrons, and electrons. Notice the shorthand symbols for the particles. You will see these symbols elsewhere in the book.

As you can see, you must use some small numbers to state the mass of an atom in grams. In order to have a simpler way of comparing the masses of individual atoms, chemists have devised a different unit of mass called an atomic mass unit, which is given the symbol u. An atom of the carbon-12 isotope contains six protons and six neutrons and has a mass number of 12. Chemists have defined the carbon-12 atom as having a mass of 12 atomic mass units. Therefore, 1 u = 1/12 the mass of a carbon-12 atom. As you can see in **Table 2.1,** 1 u is approximately the mass of a single proton or neutron.

Refer again to the periodic table. Each box on the table contains several pieces of information about an element, as shown in **Figure 2.13.** The number at the bottom of each box is the average atomic mass of that element. This number is the weighted average mass of all the naturally occurring isotopes of that element. Scientists use an instrument called the mass spectrometer to determine the number and mass of isotopes of an element, as well as the abundance of each isotope. Today, scientists have measured the mass and abundance of the isotopes of all but the most unstable elements. These data have been used to calculate the average atomic masses of most elements. The isotopes of four common elements are illustrated in **Figure 2.14** on the next page.

Figure 2.12

Masses of Atoms
Recall that there are 6.5×10^{21} atoms in a drop of water. A kilogram or a gram would not be a convenient unit for objects so small.

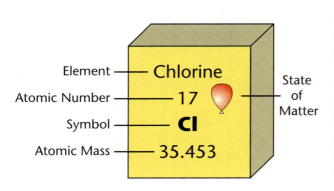

Element —— Chlorine
Atomic Number —— 17
Symbol —— **Cl**
Atomic Mass —— 35.453
State of Matter

Figure 2.13

Information in Each Block of the Periodic Table
Each block of the periodic table shows the atomic number of the element, its symbol and name, its physical state at room temperature, and the average mass of its atoms.

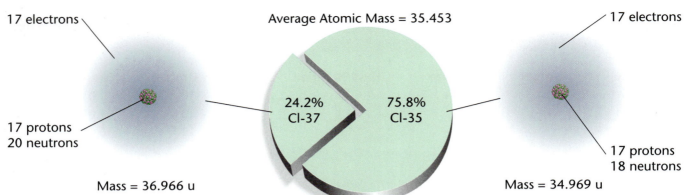

17 electrons

Average Atomic Mass = 35.453

17 electrons

17 protons
20 neutrons

24.2%
Cl-37

75.8%
Cl-35

17 protons
18 neutrons

Mass = 36.966 u

Mass = 34.969 u

Figure 2.14

Average Atomic Masses
Chlorine-35 makes up 75.8 percent of chlorine, and chlorine-37 makes up 24.2 percent. Suppose you have 1000 atoms of chlorine. On average, 758 of those atoms would have masses of 34.969 u for a total mass of $758 \times 34.969 = 26\,507$ u. Likewise, 242 atoms would have masses of 36.966 u for a total mass of 242×36.966 u $= 8946$ u. The total mass of the 1000 atoms is $26\,507$ u $+ 8946$ u $= 35\,453$ u, so the average mass of one chlorine atom is 35.5 u when rounded to the same number of digits as the percentages.

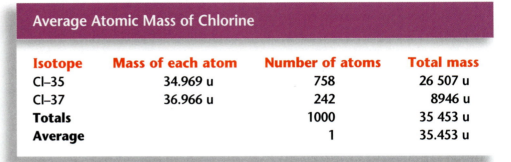

Average Atomic Mass of Chlorine

Isotope	Mass of each atom	Number of atoms	Total mass
Cl–35	34.969 u	758	26 507 u
Cl–37	36.966 u	242	8946 u
Totals		1000	35 453 u
Average		1	35.453 u

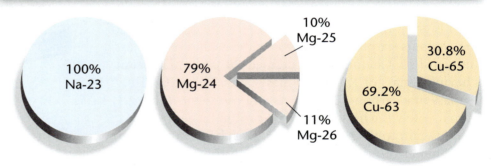

100%
Na-23

79%
Mg-24

10%
Mg-25

11%
Mg-26

30.8%
Cu-65

69.2%
Cu-63

▲ Several elements such as sodium consist of only one type of atom. Magnesium occurs as a mixture of three isotopes, and copper is a mixture of two.

SECTION REVIEW

Understanding Concepts

1. What conclusions about the nature of matter did scientists draw from the law of definite proportions?

2. How did scientists know that cathode rays had a negative electrical charge?

3. How do isotopes of an element differ from one another?

Thinking Critically

4. **Inferring** Why did Rutherford conclude that an atom's nucleus has a positive charge instead of a negative charge? Summarize the conclusions that Rutherford's team made about the structure of an atom.

Applying Chemistry

5. **Carbon Dating** The isotope of carbon that is used to date prehistoric fossils contains six protons and eight neutrons. What is the atomic number of this isotope? How many electrons does it have? What is its mass number?

Electrons in Atoms

Electrons are just one of the three main kinds of subatomic particles. Why do they have special importance in chemistry? Electrons are in motion in the outer part of an atom. When two atoms come near each other, it is these electrons that interact. These electrons and their arrangement in the atom greatly affect an element's properties and behavior.

Electrons in Motion

As you learned in Section 2.1, the atom is mostly empty space, except that the space isn't entirely empty. This space is occupied by the atom's electrons. Now, look more closely at how scientists have learned about the motion and arrangement of electrons.

Electron Motion and Energy

Considering that electrons are negative and that an atom's nucleus contains positively charged protons, why aren't electrons pulled into the nucleus and held there? Scientists in the early 20th century wondered the same thing.

Niels Bohr (1885-1962), a Danish scientist who worked with Rutherford, proposed that electrons must have enough energy to keep them in constant motion around the nucleus. He compared the motion of electrons to the motion of planets orbiting the sun. Although the planets are attracted to the sun by gravitational force, they move with enough energy to remain in stable orbits around the sun. In the same way, when we launch satellites, we use rockets to give them enough energy of motion so that the satellites stay in orbit around Earth, as shown in **Figure 2.15.** In a

Figure 2.15

Motion Versus Attraction
This satellite, as it travels around Earth, is subject to balanced forces. If its energy of motion is increased, its speed increases, and it moves to an orbit farther away from Earth.

similar way, electrons have energy of motion that enables them to overcome the attraction of the positive nucleus. This energy keeps the electrons moving around the nucleus. Bohr's view of the atom, which he proposed in 1913, was called the planetary model.

When a satellite is launched into orbit, the amount of energy determines how high above Earth it will orbit. Given a little more energy, the satellite will go into a slightly higher orbit; with a little less energy, it will have a lower orbit as shown in **Figure 2.16.** However, it seemed that electrons did not behave the same way. Instead, experiments showed that electrons occupied orbits of only certain amounts of energy. Bohr's model had to explain these observations.

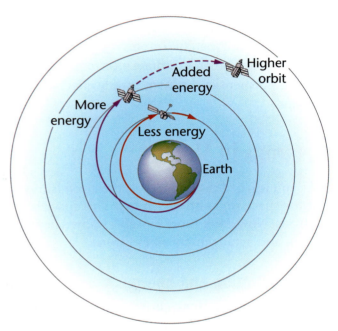

Figure 2.16

Satellites and Electrons

A satellite can orbit Earth at almost any altitude, depending on the amount of energy used to launch it. Electrons, on the other hand, can occupy orbits of only certain energies.

The Electromagnetic Spectrum

To boost a satellite into a higher orbit requires energy from a rocket motor. One way to increase the energy of an electron is to supply energy in the form of high-voltage electricity. Another way is to supply electromagnetic radiation, also called radiant energy. Radiant energy travels in the form of waves that have both electrical and magnetic properties. These electromagnetic waves can travel through empty space, as you know from the fact that radiant energy from the sun travels to Earth every day. As you may already have guessed, electromagnetic waves travel through space at the speed of light, which is approximately 300 million meters per second.

Waves Transfer Energy

If you've ever seen water waves breaking on a shoreline or heard objects in a room vibrate from the effects of loud sound waves, you already know that waves transfer energy from one place to another. Electromagnetic waves have the same characteristics as other waves, as you can see in **Figure 2.17.**

Figure 2.17
Properties of Waves
Waves transfer energy, as you can see from the damage being done by these huge waves during a hurricane. ▼

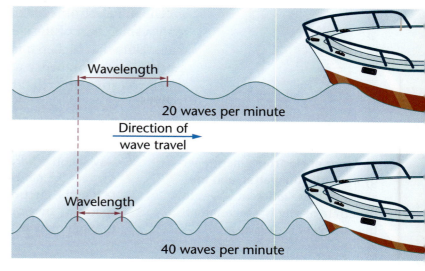

Wavelength

20 waves per minute

Direction of wave travel

Wavelength

40 waves per minute

▲ Waves are produced by something vibrating back and forth. Two properties of waves are frequency and wavelength. The number of vibrations per second is the frequency of the wave. The scientific unit for frequency is the Hertz (Hz).

▲ Wavelength is the distance between corresponding points on two consecutive waves. Because all electromagnetic waves travel at the same speed, the speed of light, wavelength is determined by frequency. A low frequency results in a long wavelength, and a high frequency results in a shorter wavelength.

Electromagnetic radiation includes radio waves that carry broadcasts to your radio and TV, microwave radiation used to heat food in a microwave oven, radiant heat used to toast bread, and the most familiar form, visible light. All of these forms of radiant energy are parts of a whole range of electromagnetic radiation called the **electromagnetic spectrum.** A portion of this spectrum is shown in **Figure 2.18.**

Figure 2.18
The Spectrum of White Light
White light is a mixture of all colors of visible light. Whenever white light passes through a prism or diffraction grating, it is broken into a range of colors called the visible light spectrum. When sunlight passes through raindrops, it is broken into the colors of the rainbow.

The full electromagnetic spectrum is shown in **Figure 2.19.** Notice that only a small part of the electromagnetic spectrum is made up of visible light. Note that higher-frequency electromagnetic waves have higher energy than lower-frequency waves. This is an important fact to remember when you study the relationship of light to atomic structure.

Figure 2.19

The Electromagnetic Spectrum
All forms of electromagnetic energy interact with matter, and the ability of these different waves to penetrate matter is a measure of the energy of the waves.

A. Gamma rays have the highest frequencies and shortest wavelengths. Because gamma rays are the most energetic rays in the electromagnetic spectrum, they can pass through most substances.

B. X rays have lower frequencies than gamma rays have but are still considered to be high-energy rays. X rays pass through soft body tissue but are stopped by harder tissue such as bone.

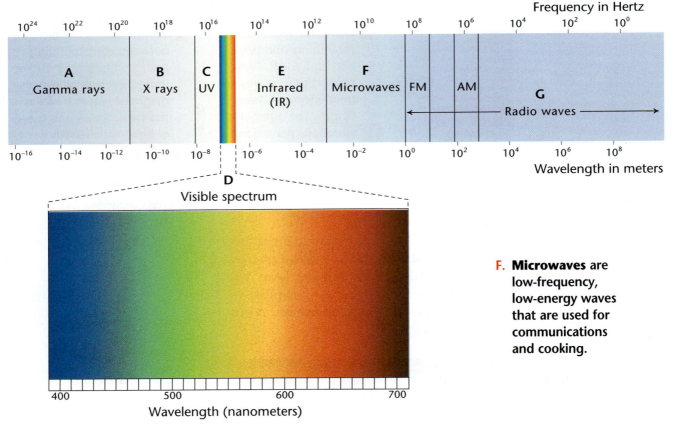

Visible spectrum

Wavelength (nanometers)

F. Microwaves are low-frequency, low-energy waves that are used for communications and cooking.

C. Ultraviolet waves are slightly more energetic than visible light waves. Ultraviolet radiation is the part of sunlight that causes sunburn. Ozone in Earth's upper atmosphere absorbs most of the sun's ultraviolet radiation.

D. Visible light waves are the part of the electromagnetic spectrum to which our eyes are sensitive. Our eyes and brain interpret different frequencies as different colors.

E. Infrared waves have less energy than visible light. Infrared radiation is given off by the human body and most other warm objects. We experience infrared rays as the radiant heat you feel near a fire or an electric heater.

G. Radio waves have the lowest frequencies on the electromagnetic spectrum. In the AM radio band, frequencies range from 550 kHz (kilohertz) to 1700 kHz and wavelengths from about 200 m to 600 m.

Aurora Borealis

In real life, you may never see colored lights such as those brightening the night sky in the photograph. You are looking at the *aurora borealis*—a fantastic light show seen only in high northern latitudes. The lights were once thought to be reflections from the polar ice fields. An aurora occurs from 100 to 1000 km above Earth.

Cause of auroras An aurora is attributed to solar wind, which is a continuous flow of electrons and protons from the sun. These high-energy, electrically charged particles become trapped by Earth's magnetic field, and they penetrate to the ionosphere. There, the particles collide with oxygen and nitrogen molecules and transfer energy to them. The energy causes electrons in these atoms and molecules to jump to higher energy levels. When the electrons return to lower energy levels, they release the absorbed energy as light.

Characteristics of the aurora When the frequencies of radiant energy released by the molecules are in the visible range, they can be seen as an aurora. Atomic oxygen, releasing energy at altitudes between 100 to 150 km, emits a whitish-green light. Molecular nitrogen gives off red light.

The aurora is most frequently seen in polar latitudes because the high-energy protons and electrons move along Earth's magnetic field lines. Because these lines emerge from Earth near the magnetic poles, it is there that the particles interact with oxygen and nitrogen to produce a fantastic display of light. Auroras also may be seen in extreme southern latitudes. These displays are the *aurora australis.*

WORD ORIGIN

aurora:
Aurora (L) the Roman goddess of dawn

The light of the aurora is a bit like the first light of dawn.

Connecting to Chemistry

1. **Applying** How does the *aurora borealis* relate to the structure of an atom?

2. **Inferring** What characteristic of an aurora indicates that it is caused by solar winds rather than a reflection of polar ice?

Electrons and Light

What does the electromagnetic spectrum have to do with electrons? It's all related to energy—the energy of motion of the electron and the energy of light. Scientists passed a high-voltage electric current through hydrogen, which absorbed some of that energy. These excited hydrogen atoms returned the absorbed energy in the form of light. Passing that light through a prism revealed that the light consisted of just a few specific frequencies—not a whole range of colors as with white light, as shown in **Figure 2.20.** The spectrum of light released from excited atoms of an element is called the **emission spectrum** of that element.

Figure 2.20

Emission Spectrum of Hydrogen
When hydrogen atoms are energized by electricity, they emit light. ▼

A prism will separate the light emitted by hydrogen into four visible lines of specific frequencies (colors). Hydrogen also emits light in the non-visible ultraviolet and infrared parts of the spectrum. ▼

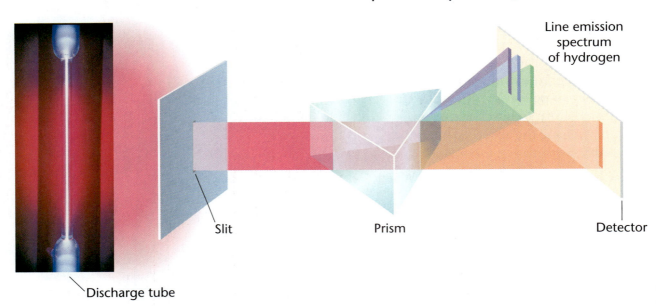

Line emission spectrum of hydrogen

Slit

Prism

Detector

Discharge tube

Evidence for Energy Levels

What could explain the fact that the emission spectrum of hydrogen consists of just a few lines? Bohr theorized that electrons absorbed energy and moved to higher energy states. Then, these excited electrons gave off that energy as light waves when they fell back to a lower energy state. But, why were only certain frequencies of light given off? To answer this question, Bohr suggested that electrons could have only certain amounts of energy. When they absorb energy, electrons absorb only the amount needed to move to a specific higher energy state. Then, when the electrons fall back to a lower energy state, they emit only certain amounts of energy, resulting in only specific colors of light. This relationship is illustrated in **Figure 2.21.**

Figure 2.21

Energy Levels in a Hydrogen Atom

This diagram shows how scientists interpret the emission spectrum. Notice that all the lines in the visible portion of the spectrum result from electrons losing energy as they fall from higher energy levels to the second energy level. The highest level represents the energy the electron has when it is completely removed from the atom.

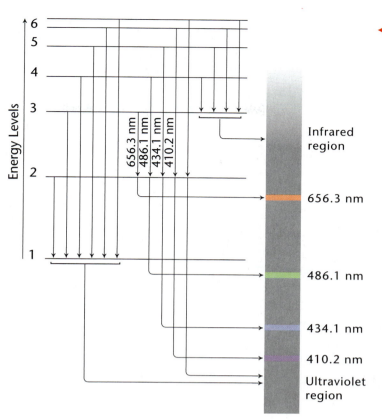

◄ Notice that drops of electrons to the lowest level emit ultraviolet frequencies, and shorter drops of electrons to the third level emit infrared frequencies.

Because electrons can have only certain amounts of energy, Bohr reasoned, they can move around the nucleus only at distances that correspond to those amounts of energy. These regions of space in which electrons can move about the nucleus of an atom are called **energy levels.** Energy levels in an atom are like rungs on a ladder. You can compare the movement of electrons between energy levels to climbing up and down that ladder, as shown in **Figure 2.22.**

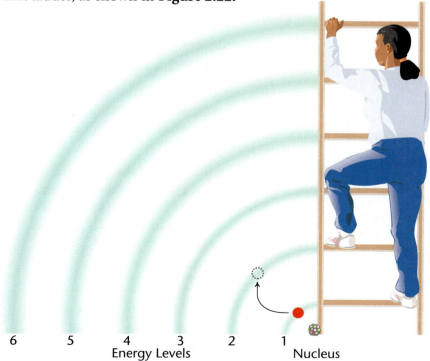

Figure 2.22

Energy Levels Are Like Rungs on a Ladder

When you climb up or down a ladder, you must step on a rung. You can't step between the rungs. The same principle applies to the movement of electrons between energy levels in an atom. Like your feet on a ladder, electrons can't exist between energy levels. Electrons must absorb just certain amounts of energy to move to higher levels. The amount is determined by the energy difference between the levels. When electrons drop back to lower levels, they give off the difference in energy in the form of light.

Everyday Chemistry

Fireworks–Getting a Bang Out of Color

When you see a colorful display of fireworks, you probably don't think about chemistry. However, creators of fireworks displays have to keep chemistry in mind. Packing what will provide "the rocket's red glare" for the 4th of July display is done by hand. Employees learn to follow proper precautions to avoid fires and explosions.

Chemistry of fireworks Typical fireworks contain an oxidizer, a fuel, a binder, and a color producer. The oxidizer is the main component, making up from 38 to 64 percent of the material. A common oxidizer is potassium perchlorate, $KClO_4$. The presence of chlorine in the oxidizer adds brightness to the colors by producing light-emitting chloride salts that make each color flame sparkle. When it oxidizes a fuel such as aluminum or sulfur, it produces an exothermic reaction with noise and flashes. Aluminum or magnesium makes the flash dazzling blue-white. The loud noise comes from the rapidly expanding gases. Other fuels not only raise the temperature but also bind the loosely assembled materials together.

Producing color A metallic salt, such as those shown in the table, is added to produce a specific color-emission spectrum. Care must be taken in selecting the ingredients so that the oxidizer is not able to

Flame Color	Color-Causing Salts
Red	Strontium salts
Yellow	Sodium salts
Green	Barium salts
Blue or Blue-Green	Copper salts

react with the metallic salt during storage and cause an explosion.

Exploring Further

1. **Applying** Based on your knowledge of the electromagnetic spectrum, which of the salts in the table has an emission spectrum with the shortest wavelength?

2. **Inferring** Oxidizers in fireworks are chemicals that break down rapidly to release oxygen, which then burns the fuel. Why is it necessary to provide an oxidizer in the mixture rather than relying on oxygen in the air?

For more information about the science involved in fireworks, visit the Chemistry Web site at **www.chemistryca.com**

Line Emission Spectra of Elements

Emission spectra of elements are the result of electron transitions within atoms and provide information about the arrangements of electrons in the atoms. Observe and compare a spectrum from white light with the emission spectra of several elements.

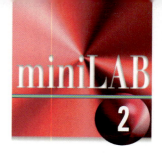

Procedure

1. Obtain a diffraction grating from the teacher. Hold it only by the cardboard edge and avoid touching the transparent material that encloses the diffraction grating.

2. Observe the emitted light from an incandescent bulb through the grating as you hold it close to your eye. Record your observations.

3. Next, observe the light produced by the spectrum tube containing hydrogen gas and record your observations. It may be necessary for you to move to within a few feet of the spectrum tube in order to effectively observe the emission spectrum. **CAUTION:** *The spectrum tube operates at a*

high voltage. Under no circumstances should you touch the spectrum tube or any part of the transformer.

4. Repeat procedure 3 with other spectrum tubes as your teacher designates.

Analysis

1. How do you explain why only certain colors appear in the emission spectra of the elements?

2. If each hydrogen atom contains only one electron, how are several emission spectral lines possible?

3. How do you interpet the fact that other elements emit many more spectral lines than hydrogen atoms?

Hydrogen spectrum tube

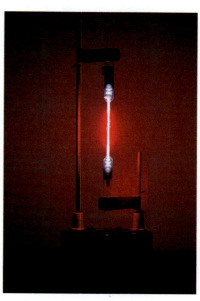

The Electron Cloud Model

As a result of continuing research throughout the 20th century, scientists today realize that energy levels are not neat, planetlike orbits around the nucleus of an atom. Instead, they are spherical regions of space around the nucleus in which electrons are most likely to be found. Examine **Figure 2.23.** Electrons themselves take up little space but travel rapidly through the space surrounding the nucleus. These spherical regions where electrons travel may be depicted as clouds around the nucleus. The space around the nucleus of an atom where the atom's electrons are found is called the **electron cloud.**

Figure 2.23

The Electron Cloud Model of the Atom
Pictured here is the electron cloud model of an atom.

Energy levels are concentric spherical regions of space around the nucleus. ▼

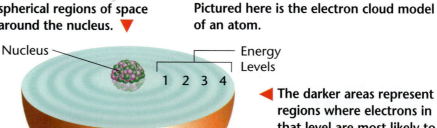

◄ The darker areas represent regions where electrons in that level are most likely to be found. Electrons are less likely to be found in the lighter regions of each level.

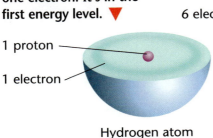

Figure 2.24
Energy Levels
A hydrogen atom has only one electron. It's in the first energy level. ▼

1 proton
1 electron

Hydrogen atom

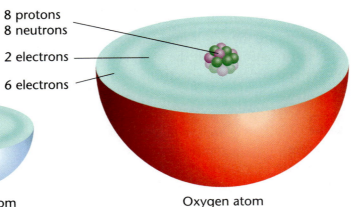

8 protons
8 neutrons
2 electrons
6 electrons

Oxygen atom

◀ An oxygen atom has eight electrons. Two of these fill the first energy level, and the remaining six are in the second energy level.

Electrons in Energy Levels

How are electrons arranged in energy levels? Each energy level can hold a limited number of electrons. The lowest energy level is the smallest and the closest to the nucleus. This first energy level holds a maximum of two electrons. The second energy level is larger because it is farther away from the nucleus. It holds a maximum of eight electrons. The third energy level is larger still and holds a maximum of 18 electrons. **Figure 2.24** shows the energy levels and electron clouds in hydrogen and oxygen atoms.

You will learn more about the details of electron arrangement in Chapter 7. For now, it's important to learn about the electrons in the outermost energy level of an atom. The electrons in the outermost energy level are called **valence electrons.** As you can see in **Figure 2.24,** hydrogen has one valence electron and oxygen has six valence electrons. You can also use the periodic table as a tool to predict the number of valence electrons in any atom in Groups 1, 2, 13, 14, 15, 16, 17, and 18. All atoms in Group 1, like hydrogen, have one valence electron. Likewise, atoms in Group 2 have two valence electrons. Atoms in Groups 13-18 have three through eight valence electrons, respectively. You may have seen nesting dolls like the ones shown in **Figure 2.25.** These dolls serve as a visual analogy for the energy levels in an atom.

Figure 2.25

Nesting Dolls Model Energy Levels
If the nested dolls represent an atom, the largest doll is the outermost energy level. In this analogy, the largest doll also represents the valence electrons. You can continue to peel away the dolls until you reach the nucleus. ▼

▲ If you open the upper half of the doll, you will discover an identical, smaller doll inside. This doll represents the inner core of the atom, which consists of the nucleus and all electrons except the valence electrons.

Why do you need to know how to determine the number of outer-level electrons that are in an atom? Recall that at the beginning of this section, it was stated that when atoms come near each other, it is the electrons that interact. In fact, it is the valence electrons that interact. Therefore, many of the chemical and physical properties of an element are directly related to the number and arrangement of valence electrons.

Lewis Dot Diagrams

Because valence electrons are so important to the behavior of an atom, it is useful to represent them with symbols. A **Lewis dot diagram** illustrates valence electrons as dots (or other small symbols) around the chemical symbol of an element. Each dot represents one valence electron. In the dot diagram, the element's symbol represents the core of the atom—the nucleus plus all the inner electrons. The Lewis dot diagrams for several elements are shown in **Figure 2.26.**

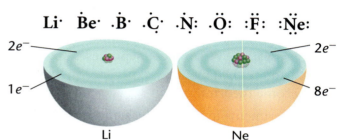

Figure 2.26

Lewis Dot Diagrams
The Lewis dot diagrams for these elements illustrate how valence electrons change from element to element across a row on the periodic table.

Connecting Ideas

A Lewis dot diagram is a convenient, shorthand method to represent an element and its valence electrons. You have used the periodic table already as a source of information about the symbols, names, atomic numbers, and average atomic masses of elements. In Chapter 3, you will learn that the arrangement of elements on the periodic table yields even more information about the electronic structures of atoms and how those structures can help you to predict many of the properties of elements.

Supplemental Problems

For more practice with solving problems, see Supplemental Practice Problems, Appendix B.

SECTION REVIEW

Understanding Concepts

1. For each of the following elements, tell how many electrons are in each energy level and then write the Lewis dot symbol for each atom.

   a) argon, which has 18 electrons
   b) magnesium, which has 12 electrons
   c) nitrogen, which has seven electrons
   d) aluminum, which has 13 electrons
   e) fluorine, which has nine electrons
   f) sulfur, which has 16 electrons

2. What change occurs within an atom when it emits light?

3. How does the modern electron cloud model of the atom differ from Bohr's original planetary model of the atom?

Thinking Critically

4. **Relating Cause and Effect** Describe how scientists concluded that electrons occupy specific energy levels.

Applying Chemistry

5. **Waves** Give an everyday example of how you can tell that light waves transfer energy.

REVIEWING MAIN IDEAS

2.1 Atoms and Their Structure

- Experiments in the 18th century showed that matter is not created or destroyed in ordinary chemical reactions and that a compound is always composed of the same proportion of elements.

- Dalton's atomic theory proposed that matter is made up of indestructible atoms and that the atoms of an element are alike but differ from atoms of other elements.

- Scientists make testable hypotheses based on observation. When these hypotheses are tested and refined by many experiments, they become theories.

- Experiments in the late 19th and early 20th centuries revealed that the atom has a tiny nucleus made up of protons and neutrons. Electrons move around the nucleus. The mass of the atom is concentrated in the nucleus.

- An atom's atomic number is the number of protons in its nucleus. The number of electrons in an atom equals the number of protons.

- Atoms of the same element always have the same number of protons and electrons but vary in the number of neutrons.

2.2 Electrons in Atoms

- Emission spectra led scientists to propose that electrons move around an atom's nucleus in orbits of certain energies.

- In the current model of the atom, electrons move around the nucleus in specific energy levels.

- Energy levels are spherical regions in which electrons are likely to be found. The greater the energy of the level, the farther from the nucleus the level is located.

- Electrons can absorb energy and move to a higher energy level. When they drop back to lower levels, they release specific amounts of energy in the form of electromagnetic radiation.

- Lewis dot diagrams can be used to represent the valence electrons in a given atom.

Vocabulary

For each of the following terms, write a sentence that shows your understanding of its meaning.

atom	law of definite
atomic number	proportions
atomic theory	Lewis dot diagram
electromagnetic	mass number
spectrum	neutron
electron	nucleus
electron cloud	proton
emission spectrum	scientific law
energy level	theory
hypothesis	valence electron
isotope	

UNDERSTANDING CONCEPTS

1. Why could the Greek philosophers not develop a real theory about atoms?

2. How does Dalton's atomic theory explain the law of conservation of mass?

3. What was Lavoisier's contribution to the development of the modern atomic theory?

4. How is the light of an emission spectrum of an element produced?

5. Draw the Lewis dot diagram for silicon, which contains 14 electrons.

6. Why is the region around the nucleus of an atom referred to as an *electron cloud*?

7. What are valence electrons? Why are they important to understanding the chemistry of an element?

APPLYING CONCEPTS

8. An atom's nucleus contains 28 protons, and the atom has a mass number of 60. How many neutrons does this atom contain? How many electrons? What is the identity of the atom?

History Connection

9. Lavoisier's discovery that air was not a substance but rather a mixture of nitrogen and oxygen is described as "history-making." Why do you think this description is accurate?

Physics Connection

10. How does solar wind differ from electromagnetic radiation from the sun?

Everyday Chemistry

11. Rocket boosters used to launch the space shuttle must carry both fuel and an oxidizer. Why would an oxidizer be needed in such rockets?

Chemistry and Society

12. Even though the law of conservation of mass applies to both, why is waste glass a more serious problem than leaves that have fallen from trees or food waste?

THINKING CRITICALLY

Relating Cause and Effect

13. How was Bohr's model of the atom consistent with the results of Rutherford's gold foil experiment?

Observing and Inferring

14. How did scientists account for the fact that the emission spectrum of hydrogen is not continuous but consists of only a few lines of certain colors?

Applying Concepts

15. Dalton's second statement of atomic theory on page 54 says that atoms are indestructible. This is still true as far as chemical changes are concerned. However, we now know that atoms can be broken apart in atomic fission reactions, and some mass is converted to energy.

a) Rewrite Dalton's second statement to make it true.

b) Examine the third statement of Dalton's theory. What evidence do we have that it is untrue? Restate this part to make it fit today's knowledge.

Comparing and Contrasting

16. What is the relationship between an atom that has 12 protons, 12 neutrons, and 12 electrons and one that has 12 protons, 13 neutrons, and 12 electrons?

Interpreting Data

17. MiniLab 1 Suppose you obtained the data in the following table for MiniLab 1. Complete the table and determine the average penny mass of the pennies in the mixture.

	Type of Penny	
	Pre-1982	Post-1982
Mass of 10	30.81 g	25.33 g
Mass of 1		
Number of Pennies in Mixture	34	55
Mass of Each Type		
Average Penny Mass		

Applying Concepts

18. MiniLab 2 Which of the lines of the visible emission spectrum of hydrogen represents the greatest energy drop?

Observing and Inferring

19. ChemLab Why do you think the amount of zinc that you could use in the experiment was limited to 0.28 g?

Making Comparisons

20. An electromagnetic wave has a frequency of 10^{21} Hz. Use **Figure 2.19** to determine which type of electromagnetic wave has this frequency. Compare the wavelength and energy of this wave to those of a second wave with a frequency of 10^{17} Hz.

Interpreting Data

21. Ferric chloride is a compound that occurs in nature as the mineral molysite. Ferric chloride is 34.4 percent iron and 65.6 percent chlorine by mass. A chemist analyzes two compounds that are composed of iron and chlorine. Her results are summarized in the following table. Which of the two compounds might be ferric chloride? Explain your answer.

Compound	Mass of Sample	Mass of Fe	Mass of Cl
A	12.00 g	5.29 g	6.71 g
B	14.50 g	4.99 g	9.51 g

CUMULATIVE REVIEW

22. Distinguish among a mixture, a solution, and a compound. (Chapter 1)

23. Suppose you stir some sugar in a glass until it dissolves completely. Describe this action and its result using the terms *solvent, solute, physical change,* and *aqueous solution.* (Chapter 1)

24. Suppose you are given two identical bottles filled completely with clear liquids. Describe a way you could determine whether the liquids have different densities. You cannot open the bottles. (Chapter 1)

SKILL REVIEW

25. Relating Cause and Effect A chemistry student decided to weigh a bag of microwave popcorn before and after popping it. Following the directions on the bag, she opened it imme-diately after microwaving it to let the steam escape and then weighed it. She noticed that the mass of the unpopped popcorn was 0.5 g more than the mass of the popped popcorn. Was matter destroyed when the popcorn popped? Explain your answer.

WRITING IN CHEMISTRY

26. Do research to learn about the phlogiston theory. How did people who believed the phlogiston theory explain chemical changes such as burning, oxidation of metals, and the smelting of ores to obtain the pure metal? Write a paragraph in which you attempt to answer the question "What was phlogiston?" using modern scientific terms.

PROBLEM SOLVING

27. A chemist recorded the following data in an experiment to determine the composition of three samples of a compound of copper and sulfur obtained from three different sources. Determine the ratio of mass of copper to mass of sulfur for each sample. How do these ratios compare? What law of chemistry does this experiment illustrate? Does the result mean that atoms of copper and sulfur occur in this compound in the same numerical ratio? Explain your answer.

Sample	Mass of Sample	Mass of Cu	Mass of S
1	5.02 g	3.35 g	1.67 g
2	10.05 g	6.71 g	3.34 g
3	99.6 g	66.4 g	33.2 g

1. Why would Greek philosophers not be considered scientists?

   a) They proposed inaccurate theories about matter and the structure of the universe.
   b) They did not have modern scientific instruments at their disposal.
   c) They did not test their hypotheses with extensive experimentation.
   d) Many of their observations were based on myths and superstitions.

Chemist's Notes about Statements Made During a Food Chemistry Lecture
A. The total number of atoms in food and drink taken into the body must equal the total number of atoms stored or expelled by the body.
B. A packaged food cake can sit on a shelf for years without growing mold.
C. Based on experimental evidence gathered on large numbers of test subjects, it is believed that the consumption of excessive amounts of soft drinks increases the chance of kidney disease.
D. In a study group of 34 people, the average person lost 4 pounds during a weeklong diet of whole grains, fresh fruit, and fresh vegetables.

Use the table above to answers questions 2–4.

2. Which statement is a hypothesis?

   a) A   c) C
   b) B   d) D

3. Which statement is a theory?

   a) A   c) C
   b) B   d) D

4. Which statement is a scientific law?

   a) A   c) C
   b) B   d) D

5. Positively charged, subatomic particles located in the nucleus of an atom are called

   a) electrons.   c) neutrons.
   b) isotopes.   d) protons.

6. Two different isotopes will differ in

   a) the number of electrons in their atoms.
   b) the number of neutrons in their atoms.
   c) the number of protons in their atoms.
   d) the amount of empty space in their atoms.

7. Why did Rutherford conclude from his gold foil experiment that an atom is mostly comprised of empty space?

   a) The positively charged particles shot into the foil were deflected by the nuclei of the gold atoms.
   b) The positively charged particles shot into the foil were attracted to electrons of the gold atoms.
   c) Most of the particles shot through the gold foil passed straight through the material.
   d) The radioactive particles shot into the gold foil caused the gold atoms to give off their own radiation.

8. What is the definition of an atomic mass unit?

   a) 1/12 of the mass of a carbon-12 atom
   b) a small unit used to measure subatomic particle masses
   c) the mass of one proton or one neutron
   d) the mass of one electron

9. What will determine the distance between an orbiting electron and the nucleus of an atom?

   a) the amount of energy in the electron
   b) the mass of the electron
   c) the energy level holding the electron
   d) the electromagnetic frequency of the electron

Test Taking Tip

Practice, Practice, Practice Practice to improve your performance on standardized tests. Don't compare yourself to anyone else.

3 Introduction to the Periodic Table

Chapter Preview

Sections

3.1 Development of the Periodic Table
MiniLab 3.1 Predicting the Properties of Mystery Elements

3.2 Using the Periodic Table
MiniLab 3.2 Trends in Reactivity Within Groups
ChemLab The Periodic Table of the Elements

Artistic Patterns in Nature

As seasons change from year to year, they form regular patterns. The periodic table organizes the chemical elements to show how they repeat their properties in regular patterns. These patterns are the key to differences, as well as similarities, in behavior and properties in all elements.

Launch Lab

Versatile Metals

A variety of processes can be used to shape metals into different forms. Because of their physical properties, metals are used in a wide range of applications.

Safety Precautions

Be careful when bending the copper samples, as they may have sharp edges.

Materials

- tape
- samples of copper
- light socket with bulb, wires, and battery

Procedure

1. Observe the different types of copper metal that your teacher gives you. Write down as many observations as you can about each of the copper samples.
2. Try gently bending each copper sample (do not break the samples). Record your observations.
3. Connect each copper sample to the circuit. Record your observations.

Analysis

What properties of copper are similar in all of the samples? How do the samples of copper differ? List several common applications of copper. What properties make metals such as copper so versatile?

What I Already Know

Review the following concepts before studying this chapter.
Chapter 1: physical and chemical properties of elements
Chapter 2: relationship of atomic number to number of protons and electrons in an atom

Reading Chemistry

Scan the chapter headings. Write down a few ways that the periodic table groups elements together. Write down any questions that form about new vocabulary concerning the periodic table. Try to answer the questions while reading the chapter.

Preview this chapter's content and activities at chemistryca.com

Development of the Periodic Table

Objectives
✓ **Outline** the steps in the historical development of the periodic table.

✓ **Predict** similarities in properties of the elements by using the periodic table.

Review Vocabulary
Electron cloud: space around the nucleus of an atom where the atom's electrons are found.

New Vocabulary
periodicity
periodic law

Changes that occur in regular, predictable cycles are natural and comfortable. The seasons change from spring to summer to fall to winter. Farmers know they can plant crops in spring and harvest them in summer or autumn. You probably look forward to summer as a time to relax, enjoy water sports, and get a break from school, but you'll be anticipating school again in the fall. Similarly, even the early chemists looked for regular, dependable changes in the properties and behavior of the elements that would allow them to make predictions. They were amassing many chemical facts. Any patterns or regularities that could organize all their data would be useful. That's why chemists began more than 100 years ago to search for an arrangement of the elements that would show patterns. It's also why the periodic table is on the walls of most chemistry labs today.

The Search for a Periodic Table

By 1860, scientists had already discovered 60 elements and determined their atomic masses. They noticed that some elements had similar properties. They gave each group of similar elements a name. Copper, silver, and gold were called the coinage metals; lithium, sodium, and potassium were known as the alkali metals; chlorine, bromine, and iodine were called the halogens. Chemists also saw differences among the groups of elements and between individual elements. They wanted to organize the elements into a system that would show similarities while acknowledging differences. It was logical to use atomic mass as the basis for these early attempts.

Döbereiner's Triads

In 1829, the German chemist J.W. Döbereiner classified some elements into groups of three, which he called triads. The elements in a triad had similar chemical properties, and their physical properties varied in an orderly way according to their atomic masses. **Table 3.1** shows the atomic mass, density, boiling point, and melting point for each of three elements in the halogen triad—chlorine, bromine, and iodine. **Figure 3.1** shows these three elements at room temperature.

Figure 3.1

Physical States and Colors of the Halogens
As atomic mass increases, the state of the elements in this triad changes. At room temperature, chlorine (left) is a gas, bromine (center) is a liquid, and iodine (right) is a solid. Colors change from greenish yellow to reddish orange to violet.

Table 3.1 shows that the atomic mass of the three elements increases from 35.5 to 79.9 to 127 from chlorine to bromine to iodine. More importantly, the atomic mass of bromine—the middle element—is 79.9 about the average of the atomic masses of chlorine and iodine:

$$\frac{35.5 + 127}{2} = 81.3$$

The fact that the atomic mass of the middle element lies about midway between the other two members of the triad is an important characteristic of triads. The table also shows that density, melting point, and boiling point all increase as atomic mass increases. The values for bromine are between those of chlorine and iodine. Iodine, with the highest atomic mass, has the highest density, boiling point, and melting point.

Table 3.1   The Halogen Triad

Element	Atomic Mass (u)	Density (g/mL)	Melting Point (°C)	Boiling Point (°C)
Chlorine	35.5	0.003 21	−101	−34
Bromine	79.9	3.12	−7	59
Iodine	127	4.93	114	185

The triad in **Table 3.2** shows a relationship among the densities of three metals that is true for many triads. The density of strontium (2.60 g/mL) is near the average of the densities of calcium (1.55 g/mL) and barium (3.62 g/mL):

$$\frac{1.55 + 3.62}{2} = 2.58 \text{ g/mL}$$

Density increases with increasing atomic mass, as it does in the chlorine, bromine, and iodine triad. But in the case of the triad in **Table 3.2,** this increase in densities is more evident than it is for the halogens because the halogens exist in three different states—solid, liquid, and gas.

Table 3.2 Metal Triad

Element	Atomic Mass (u)	Density (g/mL)	Melting Point (°C)	Boiling Point (°C)
Calcium	40	1.55	842	1500
Strontium	88	2.60	777	1412
Barium	137	3.62	727	1845

Melting points for calcium, strontium, and barium show a similar trend. The average of the atomic masses of calcium and barium is about 88.5, which is close to 88, the actual atomic mass of strontium. However, the pattern of the boiling points in this triad is irregular. The irregular pattern of boiling points is typical of triads involving metals.

Döbereiner's triads were useful because they grouped elements with similar properties and revealed an orderly pattern in some of their physical and chemical properties. The concept of triads suggested that the properties of an element are related to its atomic mass.

Mendeleev's Periodic Table

The Russian chemist, Dmitri Mendeleev, was a professor of chemistry at the University of St. Petersburg when he developed a periodic table of the elements. Mendeleev was studying the properties of the elements and realized that the chemical and physical properties of the elements repeated in an orderly way when he organized the elements according to increasing atomic mass. For example, beryllium resembled magnesium, and boron resembled aluminum. Patterns of repeated properties began to appear.

In 1869, Mendeleev published a table of the elements organized by increasing atomic mass. He listed the elements in a

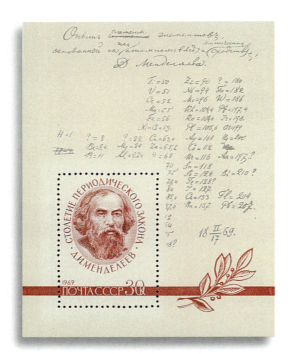

Figure 3.2

Mendeleev's Table of 1869
Elements in horizontal rows of Mendeleev's first table displayed similar properties. Mendeleev wrote question marks in the table in places where elements, then unknown, would eventually be placed.

Predicting the Properties of Mystery Elements

When Mendeleev arranged the elements according to their atomic masses, some elements didn't fit. He resolved this problem by predicting the existence and properties of elements that were unknown at the time. In this MiniLab, you'll predict some properties of two unknown elements.

miniLAB
1

Procedure

1. Unknown element A is in Group 14 below silicon and above tin. Unknown element B is in Group 16 below sulfur and above tellurium.

Group 13	Group 14	Group 15	Group 16	Group 17
	Si		S	
Ga	element A	As	element B	Br
	Sn		Te	

2. The following information is given for the surrounding elements and in this sequence: symbol, density in g/mL, melting point in K, atomic radius in pm. Si, 2.4, 1680, 118; As, 5.72, 1087, 121; Sn, 7.3, 505, 141; Ga, 5.89, 303, 134; S, 2.03, 392, 103; Br, 3.1, 266, 119; Te, 6.24, 723, 138.

3. Refer to the locations of these elements on the portion of the periodic table shown here, and average the values of the four surrounding elements to predict the densities, melting points, and atomic radii of elements A and B.

Analysis

1. What are the identities of elements A and B?

2. How do your predicted values for the three properties of element A compare with the actual values? Look up the actual values in a chemical handbook or obtain them from your teacher.

3. How do your predicted values for the three properties of element B compare with the actual values?

4. If your predicted values are fairly close to the actual values, how do you explain the approximate correlation?

vertical column starting with the lightest. When he reached an element that had properties similar to another element already in the column, he began a second column. In this way, elements with similar properties were placed in horizontal rows, as shown in **Figure 3.2.** Notice the question mark at atomic mass 180 and its position next to Zr = 90. Then refer to **Figure 3.3.**

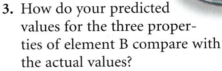

Figure 3.3

Similar Elements Are Neighbors

This zircon crystal contains zirconium (Zr = 90 on Mendeleev's table). It also contains hafnium, Mendeleev's unknown element (? = 180). The chemical and physical properties of zirconium and hafnium are so similar that the two elements always occur together in nature and are difficult to separate.

Mendeleev later developed an improved version of his table with the elements arranged in horizontal rows. This arrangement, shown in **Table 3.3**, was the forerunner of today's periodic table. Patterns of changing properties repeated for the elements across horizontal rows. Elements in vertical columns showed similar properties. An analogy can be made to the changes in the monthly calendar shown in **Figure 3.4.**

Table 3.3 Mendeleev's Table of 1871

Group	I	II	III	IV	V	VI	VII	VIII
Formula of Oxide	R_2O	RO	R_2O_3	RO_2	R_2O_5	RO_3	R_2O_7	RO_4
	H							
	Li	Be	B	C	N	O	F	
	Na	Mg	Al	Si	P	S	Cl	
	K	Ca	eka-	Ti	V	Cr	Mn	Fe, Co, Ni
	Cu	Zn	eka-	eka-	As	Se	Br	
	Rb	Sr	Yt	Zr	Nb	Mo	—	Ru, Rh, Pd
	Ag	Cd	In	Sn	Sb	Te	I	
	Cs	Ba	Di	Ce	—	—	—	
	—	—	—	—	—	—	—	
	—	—	Er	La	Ta	W	—	Os, Ir, Pt
	Au	Hg	Tl	Pb	Bi	—	—	
	—	—	—	Th	—	U	—	

Mendeleev's insight was a significant contribution to the development of chemistry. He showed that the properties of the elements repeat in an orderly way from row to row of the table. This repeated pattern is an example of periodicity in the properties of elements. **Periodicity** is the tendency to recur at regular intervals—like the appearance of Halley's comet every 76 years.

One of the tests of a scientific theory is the ability to use it to make successful predictions. Mendeleev correctly predicted the properties of several undiscovered elements. In order to group elements with similar properties in the same columns, Mendeleev had to leave some blank spaces in his table. He suggested that these spaces represented undiscovered elements.

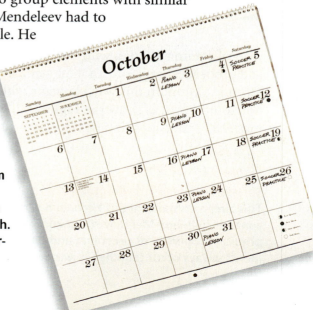

Figure 3.4
Monthly Calendar
A calendar is a table of days, with the days organized into weeks. From left to right in a given week, the days change from Sunday to Saturday and then repeat the next week. All of the days with the same name fall into vertical columns, and people generally do similar things on those days throughout the month. For example, you might have piano lessons each Thursday afternoon and soccer practice every Saturday morning.

II	III	IV
Mg	Al	Si
Ca	**eka-**	Ti
Zn	**eka-**	**eka-**

◀ **Figure 3.5**

Prediction of Eka-Aluminum and Eka-Silicon

Mendeleev left two blank spaces between zinc and arsenic and named the elements eka-aluminum and eka-silicon. *Eka* is from the Sanskrit meaning "one." In Mendeleev's earlier arrangement of elements, the unknown elements were one place from aluminum and silicon.

Figure 3.5 shows a part of Mendeleev's table in which there are two spaces for unknown elements to the right of zinc (Zn). He called these spaces eka-aluminum and eka-silicon, respectively. Based on their locations, Mendeleev predicted several of the properties of these undiscovered elements. Both elements were discovered during his lifetime. French chemists discovered eka-aluminum in 1875 and named it gallium (Ga). Eka-silicon was discovered in Germany in 1886 and was named germanium (Ge). **Table 3.4** shows the remarkable agreement between the observed properties of germanium and the properties predicted 17 years earlier by Mendeleev. The striking resemblance between Mendeleev's predicted properties and the actual properties of gallium and germanium was one of the factors that led chemists to accept his theory of periodicity among the elements and his organization of the elements into a periodic table.

▲ Eka-aluminum was discovered in 1875 and named gallium. Gallium's melting point is so low that the metal melts from the heat of the human hand.

Table 3.4 Properties of Germanium

Property	Predicted (1869)	Actual (1886)
Atomic mass	72 u	72.6 u
Color	dark gray	gray-white
Density	5.5 g/mL	5.32 g/mL
Melting point	very high	937°C
Formula of oxide	EsO_2*	GeO_2
Density of oxide	4.7 g/mL	4.70 g/mL
Oxide solubility in HCl	slightly dissolved by HCl	not dissolved by HCl
Formula of chloride	$EsCl_4$*	$GeCl_4$

* Es stands for eka-silicon

Mendeleev was so confident of the periodicity of the elements that he placed some elements in groups with others of similar properties even though arranging them strictly by atomic mass would have resulted in a different arrangement. An example is tellurium. The accepted atomic mass of tellurium (Te) was 128, so it should have been placed after iodine (I), which had an accepted atomic mass of 127. But the properties of tellurium logically placed it in the group with oxygen and sulfur, ahead of iodine, and the properties of iodine matched those of chlorine and bromine. Mendeleev placed tellurium with oxygen and sulfur and assumed that the atomic mass of 128 was incorrect. Mendeleev's placement of tellurium turned out to be correct even though his assumption about its atomic mass was not.

Supplemental Problems

For more practice with solving problems, see Supplemental Practice Problems, Appendix B.

PERIODIC TABLE OF THE ELEMENTS

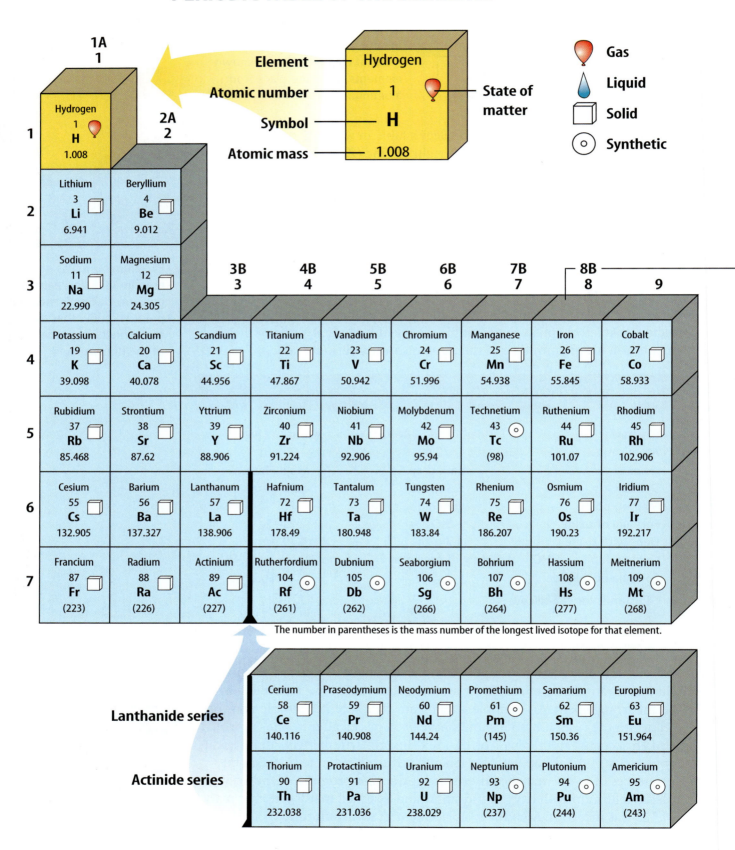

Element — Hydrogen
Atomic number — 1
Symbol — H
Atomic mass — 1.008
State of matter

🎈 Gas
💧 Liquid
⬜ Solid
⊙ Synthetic

1A 1	2A 2	3B 3	4B 4	5B 5	6B 6	7B 7	8B 8	9
Hydrogen 1 **H** 1.008								
Lithium 3 **Li** 6.941	Beryllium 4 **Be** 9.012							
Sodium 11 **Na** 22.990	Magnesium 12 **Mg** 24.305							
Potassium 19 **K** 39.098	Calcium 20 **Ca** 40.078	Scandium 21 **Sc** 44.956	Titanium 22 **Ti** 47.867	Vanadium 23 **V** 50.942	Chromium 24 **Cr** 51.996	Manganese 25 **Mn** 54.938	Iron 26 **Fe** 55.845	Cobalt 27 **Co** 58.933
Rubidium 37 **Rb** 85.468	Strontium 38 **Sr** 87.62	Yttrium 39 **Y** 88.906	Zirconium 40 **Zr** 91.224	Niobium 41 **Nb** 92.906	Molybdenum 42 **Mo** 95.94	Technetium 43 ⊙ **Tc** (98)	Ruthenium 44 **Ru** 101.07	Rhodium 45 **Rh** 102.906
Cesium 55 **Cs** 132.905	Barium 56 **Ba** 137.327	Lanthanum 57 **La** 138.906	Hafnium 72 **Hf** 178.49	Tantalum 73 **Ta** 180.948	Tungsten 74 **W** 183.84	Rhenium 75 **Re** 186.207	Osmium 76 **Os** 190.23	Iridium 77 **Ir** 192.217
Francium 87 **Fr** (223)	Radium 88 **Ra** (226)	Actinium 89 **Ac** (227)	Rutherfordium 104 ⊙ **Rf** (261)	Dubnium 105 ⊙ **Db** (262)	Seaborgium 106 ⊙ **Sg** (266)	Bohrium 107 ⊙ **Bh** (264)	Hassium 108 ⊙ **Hs** (277)	Meitnerium 109 ⊙ **Mt** (268)

The number in parentheses is the mass number of the longest lived isotope for that element.

Lanthanide series	Cerium 58 **Ce** 140.116	Praseodymium 59 **Pr** 140.908	Neodymium 60 **Nd** 144.24	Promethium 61 ⊙ **Pm** (145)	Samarium 62 **Sm** 150.36	Europium 63 **Eu** 151.964
Actinide series	Thorium 90 **Th** 232.038	Protactinium 91 **Pa** 231.036	Uranium 92 **U** 238.029	Neptunium 93 ⊙ **Np** (237)	Plutonium 94 ⊙ **Pu** (244)	Americium 95 ⊙ **Am** (243)

Metal

Metalloid

Nonmetal

Recently discovered

			3A **13**	**4A** **14**	**5A** **15**	**6A** **16**	**7A** **17**	**8A** **18**
								Helium 2 **He** 4.003
			Boron 5 **B** 10.811	Carbon 6 **C** 12.011	Nitrogen 7 **N** 14.007	Oxygen 8 **O** 15.999	Fluorine 9 **F** 18.998	Neon 10 **Ne** 20.180
1B **11**	**2B** **12**		Aluminum 13 **Al** 26.982	Silicon 14 **Si** 28.086	Phosphorus 15 **P** 30.974	Sulfur 16 **S** 32.065	Chlorine 17 **Cl** 35.453	Argon 18 **Ar** 39.948

10

Nickel 28 **Ni** 58.693	Copper 29 **Cu** 63.546	Zinc 30 **Zn** 65.39	Gallium 31 **Ga** 69.723	Germanium 32 **Ge** 72.64	Arsenic 33 **As** 74.922	Selenium 34 **Se** 78.96	Bromine 35 **Br** 79.904	Krypton 36 **Kr** 83.80
Palladium 46 **Pd** 106.42	Silver 47 **Ag** 107.868	Cadmium 48 **Cd** 112.411	Indium 49 **In** 114.818	Tin 50 **Sn** 118.710	Antimony 51 **Sb** 121.760	Tellurium 52 **Te** 127.60	Iodine 53 **I** 126.904	Xenon 54 **Xe** 131.293
Platinum 78 **Pt** 195.078	Gold 79 **Au** 196.967	Mercury 80 **Hg** 200.59	Thallium 81 **Tl** 204.383	Lead 82 **Pb** 207.2	Bismuth 83 **Bi** 208.980	Polonium 84 **Po** (209)	Astatine 85 **At** (210)	Radon 86 **Rn** (222)
Darmstadtium 110 **Ds** (281)	Unununium * 111 **Uuu** (272)	Ununbium * 112 **Uub** (285)		Ununquadium * 114 **Uuq** (289)				

* Names not officially assigned. Discovery of element 114 recently reported. Further information not yet available.

Gadolinium 64 **Gd** 157.25	Terbium 65 **Tb** 158.925	Dysprosium 66 **Dy** 162.50	Holmium 67 **Ho** 164.930	Erbium 68 **Er** 167.259	Thulium 69 **Tm** 168.934	Ytterbium 70 **Yb** 173.04	Lutetium 71 **Lu** 174.967
Curium 96 **Cm** (247)	Berkelium 97 **Bk** (247)	Californium 98 **Cf** (251)	Einsteinium 99 **Es** (252)	Fermium 100 **Fm** (257)	Mendelevium 101 **Md** (258)	Nobelium 102 **No** (259)	Lawrencium 103 **Lr** (262)

The Modern Periodic Table

Döbereiner and Mendeleev both observed similarities and differences in the properties of elements and tried to relate them to atomic mass. Look at the modern periodic table shown on pages 92 and 93 and notice that as in Mendeleev's table, elements with similar chemical properties appear in the same group. For example, tellurium is in the same group as oxygen and sulfur, where Mendeleev placed it.

Mendeleev based his periodic table on 60 or so elements. At present, elements up to atomic number 112 have been discovered or synthesized. Many of the elements now known as the transition elements, lanthanides, and actinides were unknown in 1869 but today occupy the center of the table. The noble gases, such as the helium shown in **Figure 3.6,** were also unknown in Mendeleev's time, but now fill column 18 of the table.

There are several places in the modern table where an element of higher atomic mass comes before one of lower atomic mass. This is because the basis for ordering the elements in the table is atomic number, not atomic mass. The atomic number of an element is equal to the number of protons in the nucleus. Atomic number increases by one as you move from element to element across a row. Each row (except the first) begins with a metal and ends with a noble gas. In between, the properties of the elements change in an orderly progression from left to right. The pattern in properties repeats after column 18. This regular cycle illustrates periodicity in the properties of the elements. The statement that the physical and chemical properties of the elements repeat in a regular pattern when they are arranged in order of increasing atomic number is known as the **periodic law.**

Figure 3.6
One Use for Helium
The lightest of the elements in column 18, helium (He), is a regular at parties and celebrations. Because helium and the other noble gases are unreactive, they were not easily discovered.

SECTION REVIEW

Understanding Concepts

1. What is the modern periodic law? How does it differ from Mendeleev's periodic law?
2. What are two factors that contributed to the widespread acceptance of Mendeleev's periodic table?
3. Which of the Döbereiner triads shown are still listed in the same column of the modern periodic table?

Triad 1	Triad 2	Triad 3
Li	Mn	S
Na	Cr	Se
K	Fe	Te

Thinking Critically

4. **Using a Table** Use the periodic table to separate these 12 elements into six pairs of elements having similar properties.
   Ca, K, Ga, P, Si, Rb, B, Sr, Sn, Cl, Bi, Br

Applying Chemistry

5. **Locating Elements** Use the periodic table to identify by name and symbol the elements that have the following atomic masses.

   a) 30.974
   b) 137.327
   c) 18.998
   d) 118.710

Using the Periodic Table

Have you ever tried to look up information in the tables in the sports section or the financial section of the newspaper? Or have you used a train or bus schedule to plan a trip? Tables give useful information, but they're sometimes hard to read. Some tables cram a lot of information into a small space and use many abbreviations and symbols. If you're not familiar with the codes, it's hard to obtain any useful information. But once you figure out how the data are organized and how to translate the symbols, it isn't difficult at all. It's the same with using the periodic table. Once you are familiar with the setup and symbols used in the table, you'll be able to obtain information about an element just by looking at its position in the periodic table.

Relationship of the Periodic Table to Atomic Structure

Chemists invariably have a periodic table on the walls of their offices and laboratories. It's a ready reference to a wealth of information about the elements, and it helps them think about their work, make predictions, and plan experiments based on those predictions. When you learn how to use the periodic table, you'll also find yourself referring to the periodic table to help you organize all the information you're learning about the elements.

In the modern periodic table, elements are arranged according to atomic number. Recall from Chapter 2 that the atomic number of an atom tells the number of electrons it has. If elements are ordered in the periodic table by atomic number, then they are also ordered according to the number of electrons they have. The lineup starts with hydrogen, which has one electron. Helium comes next in the first horizontal row because helium

The Language of a Chemist

. . . I have enjoyed looking at the world from unusual angles, inverting, so to speak, the instrumentation; examining matters of technique with the eye of a literary man, and literature with the eye of a technician.

In these words, Primo Levi describes the central paradox of his life. Born in Turin, Italy, in 1919, Primo Levi was trained as a chemist. In 1944, he was arrested as a member of the Italian Anti-Fascist resistance movement and deported to a concentration camp at Auschwitz, Poland. There, his knowledge of chemistry played a pivotal role in keeping his body as well as his spirit intact, as described in his memoirs, *Survival in Auschwitz* and *The Reawakening*. After the war, Levi continued to write until his death in Turin in April 1987.

Elements of a life Each essay in his memoirs, *The Periodic Table*, carries the name of an element. Some elements recall autobiographical events; others cause the author to reflect on human nature and the natural world. "Hydrogen" is a recollection of his days as a chemistry student; Levi explores the explosive properties of the element and of youth. In his final essay "Carbon," Levi finds an element that unites all living things. He follows an atom of carbon on its journey through rocks, leaves, wine, milk, blood, and finally muscle, where the atom allows the author to end his tale.

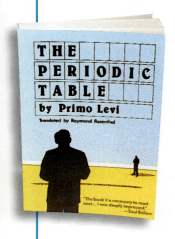

Chemist and writer In an essay, "The Language of Chemists (I)," Levi reflects on the many ways in which chemists represent reality. He traces the history of benzene from an ancient resin to the discovery of its structural formula. As he does, Levi makes the reader aware of how chemists use both words and symbols to describe a material.

Connecting to Chemistry

1. **Interpreting** Explain this quote by Levi: *"Chemistry is the art of separating, weighing and distinguishing: these are useful exercises also for the person who sets out to describe events or give body to his own imagination."*

2. **Inferring** Why do you think Levi chose the title *The Periodic Table*?

---

has two electrons. Lithium has three. Notice on the periodic table that lithium starts a new **period,** or horizontal row, in the table. Why does this happen? Why does the first period have only two elements? You learned in Chapter 2 that electrons in atoms occupy discrete energy levels. Only two electrons can occupy the first energy level in an atom. The third electron in lithium must be at a higher energy level. Lithium starts a new period at the far left in the table and becomes the first element in a group. A **group,** sometimes also called a family, consists of the elements in a vertical column. Groups are numbered from left to right. Lithium is the first element in Group 1 and in Period 2. Check this location on the periodic table.

Trends in Reactivity Within Groups

You've discovered that trends in the physical and chemical properties of elements occur both horizontally and vertically on the periodic table. In this MiniLab, you'll compare the reactivities of two elements in Group 2—magnesium and calcium—and three elements in Group 17—chlorine, bromine, and iodine.

miniLAB

2

Procedure

1. Wear an apron and goggles.

2. To compare the reactivities of magnesium and calcium, use forceps to drop a small piece of each element into two small beakers containing water to a depth of about 1 cm. Observe how rapidly each element reacts with water to produce bubbles of hydrogen.

3. Pour 1 mL of NaBr solution into a small test tube, and add three drops of chlorine water. Stir the solution with the tip of a micro-tip pipet. Add 1 mL of lighter fluid. Draw the entire mixture into the microtip pipet. Squeeze the pipet bulb, expelling the mixture back into the test tube. Repeat this step several times so that the liquids are thoroughly mixed. (**CAUTION:** *Use care when handling chlorine water.*)

4. Draw the liquids back into the pipet, invert it, and cover the tip with a cap made by cutting the bulb of another microtip pipet. Set the inverted pipet in a small beaker, and allow time for the upper layer to separate from the lower layer. If no color is evident in the upper layer, expel the liquids back into the test tube, and add five more drops of chlorine water. Then repeat the remainder of procedures 3 and 4 until a color is detected in the upper layer.

5. Using another test tube and microtip pipet, repeat procedures 3 and 4 with 1 mL of an NaI solution, chlorine water, and lighter fluid.

Analysis

1. Which of the two metallic elements, magnesium or calcium, is the more reactive? If this trend holds true for other groups of metallic elements, are the more reactive metals located toward the top or the bottom of the periodic table?

2. Interpret your results from procedures 3 and 4 in this way. If chlorine is more reactive than bromine, it will replace the bromine in NaBr and produce an orange color due to bromine in the upper layer. If chlorine is more reactive than iodine, it will replace iodine in NaI and produce a violet color characteristic of iodine in the upper layer. The ease with which the replacement reactions occur is an indication of the reactivities of the two elements. Given this information, how do the reactivities of chlorine, bromine, and iodine compare? If this trend holds true for other groups, are the more reactive nonmetallic elements toward the top or the bottom of the periodic table?

Elements with atomic numbers 4 through 10 follow lithium and fill the second period. Each has one more electron than the element that preceded it. Neon, with atomic number 10, is at the end of the period. Eight electrons are added in Period 2 from lithium to neon, so eight electrons must be the number that can occupy the second energy level. The next element, sodium, atomic number 11, begins Period 3. Sodium's 11th electron is in the third energy level. The third period repeats the pattern of the second period. Each element has one more electron than its neighbor to the left, and these electrons are in the third energy level.

Atomic Structure of Elements Within a Period

The first period is complete with two elements, hydrogen and helium. Hydrogen has one electron in its outermost energy level, so it has one valence electron. Can you see that helium must have two valence electrons? Every period after the first starts with a Group 1 element. These elements have one electron at a higher energy level than the noble gas of the preceding period. Therefore, Group 1 elements have one valence electron. As you move from one element to the next across Periods 2 and 3, the number of valence electrons increases by one. Group 18 elements have the maximum number of eight valence electrons in their outermost energy level.

Group 18 elements are called the noble gases. The **noble gases,** with a full complement of valence electrons, are generally unreactive. The period number of an element is the same as the number of its outermost energy level, so the valence electrons of an element in the second period, for example, are in the second energy level. A Period 3 element such as aluminum, **Figure 3.7,** has its valence electrons in the third energy level.

Atomic Structure of Elements Within a Group

The number of valence electrons changes from one to eight as you move from left to right across a period; when you reach Group 18, the pattern repeats. For the main group elements, the group number is related to the number of valence electrons. The main group elements are those in Groups 1, 2, 13, 14, 15, 16, 17, and 18. For elements in Groups 1 and 2, the group number equals the number of valence electrons. For elements in Groups 13, 14, 15, 16, 17, and 18, the second digit in the group number is equal to the number of valence electrons. **Figure 3.8** illustrates the electron dot structures for the main group elements.

Figure 3.7

Aluminum—Group 13, Period 3

This sculpture is made of aluminum, the Period 3 element in Group 13. Aluminum has two electrons in the first energy level, eight electrons in the second energy level, and three valence electrons in the third or outermost energy level.

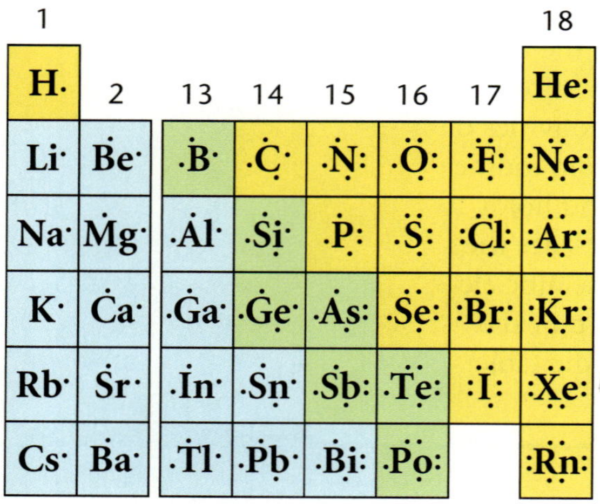

Figure 3.8

Lewis Dot Structures for Some Elements
The number of valence electrons is the same for all members of a group. Verify the relationship between group number and the number of valence electrons.

Because elements in the same group have the same number of valence electrons, they have similar properties. Sodium is in Group 1 because it has one valence electron. Because the other elements in Group 1 also have one valence electron, they have similar chemical properties.

Chlorine is in Group 17 and has seven valence electrons. All the other elements in Group 17 also have seven valence electrons and, as a result, they have similar chemical properties. Throughout the periodic table, elements in the same group have similar chemical properties because they have the same number of valence electrons.

Four groups have commonly used names: the alkali metals in Group 1, the alkaline earth metals in Group 2, the halogens in Group 17, and the noble gases in Group 18. The word *halogen* is from the Greek words for "salt former" so named because the compounds that halogens form with metals are saltlike. The elements in Group 18 are called noble gases because they are much less reactive than most of the other elements.

Because the periodic table relates group and period numbers to valence electrons, it's useful in predicting atomic structure and, therefore, chemical properties. For example, oxygen, in Group 16 and Period 2, has six valence electrons (the same as the second digit in the group number), and these electrons are in the second energy level (because oxygen is in the second period). Oxygen has the same number of valence electrons as all the other elements in Group 16 and, therefore, similar chemical properties. **Figure 3.9** shows representations of the distribution of electrons in energy levels in the first three elements of Group 16.

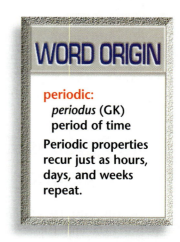

WORD ORIGIN

periodic:
periodus (GK)
period of time
Periodic properties recur just as hours, days, and weeks repeat.

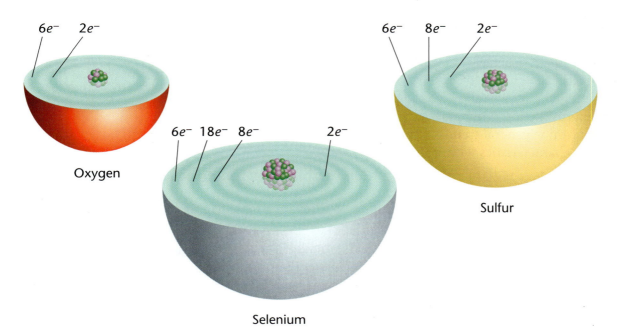

Figure 3.9

Electrons in Energy Levels—Group 16
Oxygen, sulfur, and selenium have six electrons in their outermost energy levels as predicted by their group number. Note in the diagram of selenium that the third energy level can hold 18 e^-. You will learn more about energy levels in Chapter 7. Predict what a representation of tellurium, another Group 16 element, might look like.

ChemLab

SMALL SCALE

The Periodic Table of the Elements

In the mid-1800s, scientists found that when all the known elements were arranged in a sequence of increasing atomic mass and according to their properties, the properties of the elements repeated in a regular, or periodic, manner. In this lab, you will investigate some representative elements in several of the vertical groups or families of the periodic table and classify the elements as metals, nonmetals, or metalloids. In general, metals are solids at room temperature. They have a metallic luster and are malleable. Metals conduct electricity, and many react with acids. On the other hand, nonmetals can be either solid, liquid, or gas at room temperature. If a nonmetal is a solid, it's likely to be brittle rather than malleable. Nonmetals do not conduct electricity and do not react with acids. Metalloids combine some of the characteristics of both metals and nonmetals.

Your experimental data will allow you to classify some elements as metals, nonmetals, or metalloids and to determine general trends in metallic and nonmetallic characteristics within the periodic table.

Problem

What is the pattern of metallic and nonmetallic properties of the elements in the periodic table?

Objectives

- **Observe** the properties of samples of the elements, including metals, nonmetals, and metalloids.
- **Classify** the elements as metals, nonmetals, or metalloids.
- **Analyze** your results to discover trends in the properties of the elements in the periodic table.

PREPARATION

Materials

stopperred test tubes containing small samples of carbon, nitrogen, oxygen, magnesium, aluminum, silicon, red phosphorus, sulfur, chlorine, calcium, selenium, tin, iodine, and lead
plastic dishes containing samples of carbon, magnesium, aluminum, silicon, sulfur, and tin
micro-conductivity apparatus
$1M$ HCl
test tubes (6)
test-tube rack
10-mL graduated cylinder
spatula
small hammer
glass marking pencil

Safety Precautions

Be cautious when using $1M$ HCl. If any of the acid touches your skin or eyes, immediately rinse with water and notify your teacher. Never test chemicals by tasting.

PROCEDURE

1. Prepare a table like the one shown for your data and observations.

2. Observe and record the appearance of each of the elements. Your description should include physical state, color, and any other observable characteristics such as luster. Don't open any of the test tubes.

3. Remove a small sample of each of the six elements in the dishes. Place the samples on a hard surface designated by your teacher. Gently tap each of the elements with a small hammer.

An element is malleable if it flattens when tapped. It is brittle if it shatters when tapped. Record your observations in the data table.

4. Test the conductivity of each of the six elements in the dishes by touching the electrodes of the micro-conductivity apparatus to a piece of the element. If the bulb lights, you have evidence of conductivity. Record your observations in the data table. Wash the electrodes with water, and dry between testing each element.

5. Use a graduated cylinder to measure 5 mL of water into each of the six test tubes.

6. Label each test tube with the symbol of one of the elements.

7. Using a spatula, put a small sample of each of the six elements (a 1-cm length of ribbon or 0.1-0.2 g of solid) into a test tube labeled with the symbol of the element.

8. Add approximately 5 mL of $1M$ HCl to each of the test tubes and observe the elements for at least one minute. Evidence of reaction is the formation of bubbles of hydrogen on the element. Record your observations in the data table.

ANALYZE AND CONCLUDE

1. **Interpreting Data** Which elements displayed the general characteristics of metals?

2. **Interpreting Data** Which elements displayed the general characteristics of nonmetals?

3. **Interpreting Data** Which elements displayed a mixture of metallic and nonmetallic characteristics?

APPLY AND ASSESS

1. Construct an abbreviated periodic table with seven 1-inch squares across and five squares down. Label the squares across the top from left to right as Groups 1 and 2 and 13-17. Label the squares down the side as Periods 2-6. Write the appropriate atomic number and element symbol in each of the squares. Based upon your answers to the Analyze and Conclude questions, write your classification of metal, nonmetal, or metalloid for each of the elements you observed and/or tested.

2. Do the metallic characteristics of the elements across a period seem to increase from left to right or from right to left?

3. Do the metallic characteristics of the elements in a group seem to increase from top to bottom or from bottom to top?

4. The metalloids indicate the approximate border between metals and nonmetals on the periodic table. Based upon your observations, draw a dark line along this border.

DATA AND OBSERVATIONS

Element	Appearance	Malleable or Brittle	Electrical Conductivity	Reaction with HCl
carbon				

Physical States and Classes of the Elements

Other practical information can be obtained from the periodic table. The arrangement of the table helps you determine the physical state of an element; whether the element is synthetic or natural; and whether the element is a metal, nonmetal, or metalloid.

Physical States of the Elements

The periodic table on pages 92 and 93 shows the states of the elements at room temperature and normal atmospheric pressure. Most of the elements are solid. Only two elements are liquids, and the gaseous elements, except for hydrogen, are located in the upper-right corner of the table.

Some elements are not found in nature but are produced artificially in particle accelerators like the one shown in **Figure 3.10.** These are known as synthetic elements. The synthetic elements, made by means of nuclear reactions, are marked on the periodic table. They include technetium, element 43, and all the elements after uranium, element 92. Although small amounts of neptunium and plutonium, elements 93 and 94, have been found in uranium ores, it is likely that they are the products of nuclear bombardment by radiation from uranium atoms.

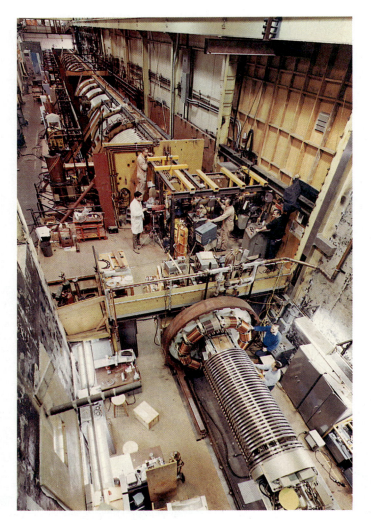

Classifying Elements

The color coding in the periodic table on pages 92 and 93 identifies which elements are metals (blue), nonmetals (yellow), and metalloids (green). The majority of the elements are metals. They occupy the entire left side and center of the periodic table. Nonmetals occupy the upper-right-hand corner. Metalloids are located along the boundary between metals and nonmetals. Each of these classes has characteristic chemical and physical properties, so by knowing whether an element is a metal, nonmetal, or metalloid, you can make predictions about its behavior. Elements are classified as metals, metalloids, or nonmetals on the basis of their physical and chemical properties.

Figure 3.10

The Super HILAC Accelerator at the University of California, Berkeley
Element 106, seaborgium, was synthesized in this linear accelerator. To change a nucleus and create a new element, a high-energy collision must occur between two particles. In the case of seaborgium, highly accelerated oxygen nuclei were smashed into a target of the heavy element californium.

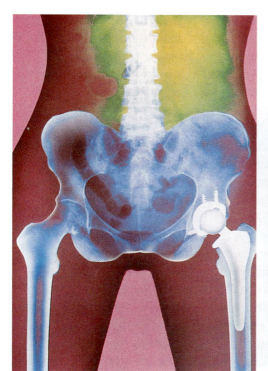

Figure 3.11

Structural Uses of Metals
Alloys of metals can be custom designed to meet a wide range of uses, such as hip replacements (left) and scaffolding for sky-scrapers (right).

Metals

Metals are almost everywhere. They make up many of the things you use every day—cars and bikes, jewelry, coins, electrical wires, household appliances, and computers. Because of their strength and durability, they're used in buildings and bridges, and for bone replacement as shown in **Figure 3.11. Metals** are elements that have luster, conduct heat and electricity, and usually bend without breaking. With the exception of tin, lead, and bismuth, metals have one, two, or three valence electrons. All metals except mercury are solids at room temperature; in fact, most have extremely high melting points.

The periodic table shows that most of the metals (coded blue) are not main group elements. A large number are located in Groups 3 through 12. Notice the elements in the fourth period beginning with scandium (Sc), atomic number 21, and ending with zinc (Zn), atomic number 30. These ten elements mark the first appearance of elements in Groups 3 through 12. From the fourth period to the bottom of the table, each period has elements in these groups. The elements in Groups 3 through 12 of the periodic table are called the **transition elements.** All transition elements are metals. Many are commonplace, including chromium (Cr), iron (Fe), nickel (Ni), copper (Cu), zinc (Zn), silver (Ag), and gold (Au). Some are less common metals but still important, such as titanium (Ti), manganese (Mn), and platinum (Pt). Some period 7 transition elements are synthetic and radioactive.

While the chemistry of the main group metals is highly predictable, that of the transition elements is less so. The unpredictability in the behavior and properties of the transition metals is due to the more complicated atomic structure of these elements.

A fascinating feature of chromium chemistry is the many colorful compounds it produces, such as bright yellow potassium chromate and brilliant orange potassium dichromate. These and other colored compounds are responsible for the element's name. Chromium comes from the Greek word *khroma*, which means "color." Trace amounts of chromium in otherwise-colorless mineral crystals produce the brilliant colors of rubies and emeralds.

In the periodic table, two series of elements, atomic numbers 58-71 and 90-103, are placed below the main body of the table. These elements are separated from the main table because putting them in their proper position would make the table very wide, as shown in **Figure 3.13.** The elements in these two series are known as the inner transition elements. Many of these elements were unknown in Mendeleev's time, but he did know of some of them and suspected that more would be discovered.

The first series of inner transition elements is called the **lanthanides** because they follow element number 57, lanthanum. The lanthanides consist of the 14 elements from number 58 (cerium, Ce) to number 71 (lutetium, Lu). Because their natural abundance on Earth is less than 0.01 percent, the lanthanides are sometimes called the rare earth elements. All of the lanthanides have similar properties.

The second series of inner transition elements, the **actinides,** have atomic numbers ranging from 90 (thorium, Th) to 103 (lawrencium, Lr). All of the actinides are radioactive, and none beyond uranium (92) occur in nature. Like the transition elements, the chemistry of the lanthanides and actinides is unpredictable because of their complex atomic structures. What could be happening at the subatomic level to explain the properties of the inner transition elements? In Chapter 7, you'll study an expanded theory of the atom to answer this question.

Figure 3.12

Nitrogen-Containing Explosives
Dynamite and TNT (trinitrotoluene) are nitrogen-containing explosives used to blast away rock in road building or efficiently demolish unwanted buildings.

Nonmetals

Although the majority of the elements in the periodic table are metals, many nonmetals are abundant in nature. The nonmetals oxygen and nitrogen make up 99 percent of Earth's atmosphere. Carbon, another nonmetal, is found in more compounds than all the other elements combined. The many compounds of carbon, nitrogen, and oxygen are important in a wide variety of applications like the one shown in **Figure 3.12.** Most **nonmetals** don't conduct electricity, are much poorer conductors of heat than metals, and are brittle when solid. Many are gases at room temperature; those that are solids lack the luster of the metals. Their melting points tend to be lower than those of metals. With the exception of carbon, nonmetals have five, six, seven, or eight valence electrons. **Table 3.5** summarizes the properties of metals and nonmetals.

Figure 3.13

The Long Version of the Periodic Table
This version of the periodic table makes clear the positions of the main group elements (pink), the transition elements (yellow), and the inner transition elements (gray).

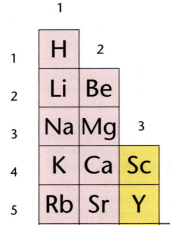

Metalloids

Metalloids have some chemical and physical properties of metals and other properties of nonmetals. In the periodic table, the metalloids lie along the border between metals and nonmetals. Silicon (Si) is probably the most well-known metalloid. Some metalloids such as silicon, germanium (Ge), and arsenic (As) are semiconductors. A **semiconductor** is an element that does not conduct electricity as well as a metal, but does conduct slightly better than a nonmetal. The ability of a semiconductor to conduct an electrical current can be increased by adding a small amount of certain other elements. Silicon's semiconducting properties made the computer revolution possible.

Table 3.5 Properties of Metals and Nonmetals	
Metals	**Nonmetals**
Bright metallic luster	Non-lustrous, various colors
Solids are easily deformed	Solids may be hard or soft, usually brittle
Good conductors of heat and electricity	Poor conductors of heat and electricity
Loosely held valence electrons	Tightly held valence electrons

Atomic Structure of Metals, Metalloids, and Nonmetals

The differences in the properties of the three classes of elements occur because of the different ways the electrons are arranged in the atoms. The number and arrangement of valence electrons and the tightness with which the valence electrons are held in an atom are important factors in determining the behavior of an element. In general, the valence electrons in a metal are loosely bound to the positive nucleus. They are free to move in the solid metal and are easily lost. This freedom of motion accounts for the ability of metals to conduct electricity. On the other hand, the valence electrons in atoms of nonmetals and metalloids are tightly held and are not easily lost. When undergoing chemical reactions, metals tend to lose valence electrons, whereas nonmetals tend to share electrons or gain electrons from other atoms.

Try at Home Lab

See page 864 in Appendix F for **Element Hunt**

General Properties of Metals, Nonmetals, and Metalloids

Most properties of metals, nonmetals, and metalloids are determined by their valence electron configurations. The number of valence electrons that a metal has varies with its position in the periodic table. Valence electrons in metal atoms tend to be loosely held. Nonmetals have four or more tightly held electrons, and metalloids have three to seven valence electrons.

① Familiar Metals

Polished silverware and copper jewelry are valued for their beautiful metallic luster. The ability of copper wire to conduct electricity makes it useful in electrical circuits. Some metals can be molded into objects, like this bronze figurine. Bronze is an alloy of two metals—copper and tin.

Silver and copper

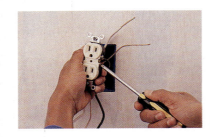
Copper wire

Bronze figurine

Many transition elements are important as structural materials. Iron is made into steel by mixing it with carbon. Sometimes, other metals are added to produce special properties. Iron mixed with manganese produces a steel hard enough to be used to make the jaws on rock-crushing machines. The combination of iron with vanadium produces a tough alloy used, among other things, to make the crankshafts in automobile engines.

Exposure to air and moisture causes the iron in steel to rust. Plating a steel surface with chromium protects it from corrosion. The appearance of some consumer products is enhanced by plating with bright, shiny chromium.

Steel rock crusher

Chrome plating

Most metals are solids at room temperature; mercury is the only metal that is a liquid at room temperature. Mercury is poisonous and should never be handled.

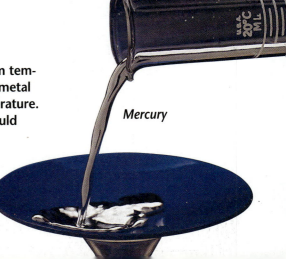
Mercury

2 Some Lanthanides and Actinides

Compounds of europium and ytterbium are used in the picture tubes of color televisions. Neodymium is used in some high-power lasers.

Neodymium in a laser

Europium and ytterbium in color TVs

Carbon in coal

3 Carbon and Some Other Nonmetals

Coal is nearly pure carbon. It is mined, often from strip mines, and burned as a fuel. Natural gas and oil are also carbon-rich fuels. Even though their appearance and physical properties are different, graphite and diamond are both naturally occurring forms of carbon. Bromine and iodine are used in high-intensity halogen lamps. Liquid nitrogen is used to maintain low temperatures; it can freeze the moisture in air, seen here as a white cloud.

Graphite and diamond

Halogen lamp

Liquid nitrogen

4 Metalloids

Silicon looks like a metal, but it's brittle and doesn't conduct heat and electricity well. Its melting point, 1410°C, is close to that of many metals. Elemental silicon (left) is melted, formed into a single crystal of pure silicon, and purified (back center). The crystal is sliced into thin wafers (right), and these are used to produce electronic devices (front center).

Silicon

Metals That Untwist

Those coated-wire ties for closing garbage bags are a great invention. The plastic coating makes it easy to twist the tie without cutting your hands or slicing through the bag. The wire is easy to twist and, once twisted, stays that way. Wouldn't you be surprised if a twisted tie began to untwist spontaneously? Wires can't do that, can they?

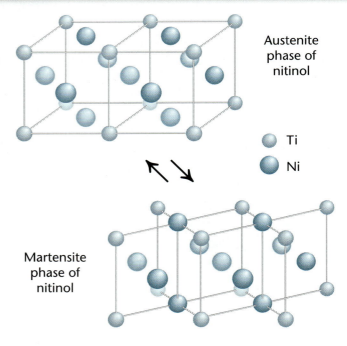

Austenite phase of nitinol

Ti
Ni

Martensite phase of nitinol

Shape-Memory Metals

Some alloys have a remarkable property. They revert to their previous shape when heated or when the stress that caused their shape is removed. These alloys are called shape-memory alloys. Above is a piece of shape-memory alloy wire that was bent and then heated by a small electric current. Notice that the wire reverted to its original shape as it was heated.

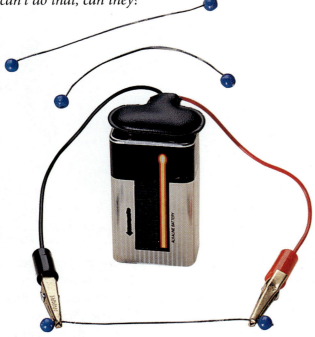

Different Solid Phases

Melting is the transition of a material from a solid to a liquid. Transitions from one phase to another can also take place within a solid. A solid can have two phases if it has two possible crystalline structures. It is the ability to undergo these changes in crystalline structure that gives shape-memory metals their properties. For example, an alloy having equal amounts of two metals may have one possible crystalline structure that is called the austenite phase. If the austenite phase is cooled under controlled conditions, the material takes on the martensite phase. The new crystal structure doesn't change the positions of the atoms throughout the material. However, the new internal organization gives the alloy new properties.

Nitinol

Nitinol is an alloy of nickel and titanium that has the austenite phase structure. Both the nickel and titanium atoms are arranged in cubes. As you can see above, each nickel atom is at the center of a cube of titanium atoms, and a titanium atom is at the center of each cube of nickel atoms. If nitinol is shaped into a straight wire (a), heated (b), and cooled past its transition temperature (c), then it takes on the martensite phase. Notice that its external shape doesn't change; it's still straight (d). But its structure in (d) allows it to be bent by an external stress in (e). Now, if the wire is heated, the stress is released and it reverts back to its initial shape in the martensite phase (f).

people who are both blind and deaf can feel and interpret its movements as it signs to them in American Sign Language. To do this, an optical character scanner reads texts and converts the characters into input signals, which then drive the fingers to form appropriate symbols.

Clot Captor

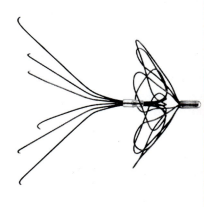

This device is used to capture blood clots in the vena cavae (main veins) before they reach the lungs. Folded inside its sheath, the clot captor is only 3 mm in diameter. But once inserted into the vein, the wire unfolds to a diameter of 28 mm. The clot captor is made of nitinol wire that has a transition temperature just below body temperature. The wire is folded and placed in a sheath. As the sheath is inserted into the vein, the wire is bathed in a cold saline solution. Slowly, the sheath is withdrawn. As the wire warms to body temperature, it unfolds into its opened, umbrella-like shape.

TEMPERATURE

(b) austenite

Transition temperature

(c) martensite

(f) martensite

(a) austenite   (d) martensite   (e) martensite

Robotic Arm

The Oaktree automatic arm contains Flexinol wires that act as muscles that can move the fingers in the robotic hand. The hand movements are so precise that

DISCUSSING THE TECHNOLOGY

1. **Inferring** Twisted nitinol-wire eyeglass frames unbend spontaneously at room temperature. Is the transition temperature of the frames above or below room temperature?

2. **Applying** Design a simple lever that could be raised and lowered smoothly using nitinol wires.

Everyday Chemistry

Metallic Money

The concept that a certain amount of a metallic substance could stand as a measure of goods and services dates back to the ancient Greeks. They were the first to use metal coins as measures of wealth.

Coinage metals Historically, copper, silver, and gold were logical choices for use in coins. These metals are not abundant in Earth's crust and are considered rare. Most elements occur combined in compounds, but copper, silver, and gold occur nearly pure in rocks fairly close to Earth's surface, making them easy to mine. People revere them because of their beauty and rarity. Finally, their properties allow them to be easily shaped, stamped, and marked for value.

Contemporary coinage metals Rarity makes metals more expensive. Gold and silver have become so expensive that they have vanished from most of the coins in your pocket. The United

States eliminated gold from its coins in 1934. Silver was eliminated in 1971. As a result, a Kennedy half-dollar minted in 1972 has a coating of 75 percent copper and 25 percent nickel on a pure copper core, called a platen. A Kennedy half-dollar minted in 1970 has a plating of 80 percent silver and 20 percent copper on a platen of 21 percent silver and 79 percent copper. Dimes and quarters minted after 1964 are made up of 75 percent copper and 25 percent nickel coated on a platen of 100 percent copper. If you hold a dime or quarter on edge, you can see the silvery-colored coating sandwiching the copper.

Cu 100%
1851

Cu 88%
Ni 12%
1859

Cu 95%
Sn and Zn 5%
1864

Cu 95%
Zn 5%
1962

Zn 97.5%
Cu 2.5%
1982

Exploring Further

1. **Applying** Give one chemical and one physical property of copper, silver, and gold that made them the metals of choice for use as coinage metals.

2. **Inferring** What metals make up today's U.S. coinage metals?

3. **Acquiring Information** What are some other uses of copper, silver, and gold?

Semiconductors and Their Uses

Your television, computer, handheld electronic games, and calculator are electronic devices that depend on silicon semiconductors. All have miniature electrical circuits that use silicon's properties as a semiconductor. You learned that metals generally are good conductors of electricity, nonmetals are poor conductors, and semiconductors fall in between the two extremes, but how do semiconductors work?

Electrons and Electricity

An electric current is a flow of electrons. Most metals conduct an electrical current because their valence electrons are not held tightly by the positive nucleus and are free to move. A copper wire is an example of a good conductor of electric current. **Figure 3.14** illustrates the flow of electrons in copper.

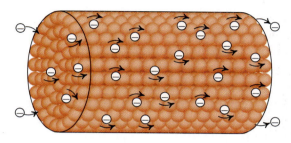

Direction of current flow

Figure 3.14

Electric Current in a Copper Wire

In a conductor such as a copper wire, valence electrons are free to move to produce an electric current.

At room temperature, pure silicon is not a good conductor of electricity. Silicon has four valence electrons, but they are held tightly between neighboring atoms in the crystal structure. You can see this clearly when you look at the structure of silicon in **Figure 3.15**.

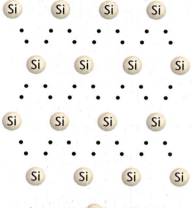

• = electron   Si = Silicon atom

Figure 3.15

Valence Electrons in Silicon

Valence electrons in silicon are localized between neighboring silicon atoms. These electrons hold the atoms together in the crystal, but no electrons are free to move beyond their positions. Therefore, no electrons are available to carry an electric current.

Electrical Conduction by a Semiconductor

The electrical conductivity of a semiconductor such as silicon can be increased by a process known as doping. Doping is the addition of a small amount of another element to a crystal of a semiconductor. If a small amount of phosphorus, which has five valence electrons, is added to a crystal of silicon having only four valence electrons, each phosphorus atom provides an extra electron to the crystal structure. These extra electrons are free to move throughout the crystal to form an electric current. A phosphorus-doped silicon crystal is shown in **Figure 3.16**.

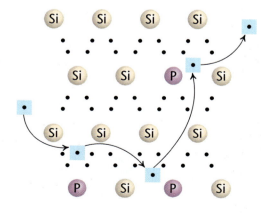

• = 1 electron

Figure 3.16

Silicon Doped with Phosphorus

In phosphorus-doped silicon, the extra electrons from the phosphorus atoms are not needed to hold the crystal together. They are free to move and carry an electric current. Arsenic and antimony are also used to dope silicon. Both have five valence electrons.

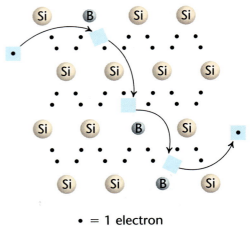

Figure 3.17

Silicon Doped with Boron
In boron-doped silicon, electrons move into and out of "holes"—spaces that are lacking an electron. This movement is an electric current.

• = 1 electron

WORD ORIGIN

semiconductor:
semi (L) half
conductus (L) to escort or guide
Semiconductors do not conduct electricity as well as metals do.

A semiconductor such as phosphorus-doped silicon is called an *n*-type semiconductor because extra electrons (*negatively* charged) are present in the crystal structure.

Silicon can also be doped with an element such as boron that has three valence electrons. Boron has fewer valence electrons than silicon, so when boron is added to a crystal of silicon, a shortage of electrons results. "Holes" are created in the silicon crystal structure. These are locations where there should be an electron but there is not because boron has one less electron than silicon. The movement of electrons into and out of these holes produces an electric current. Boron-doped silicon is an example of a *p*-type semiconductor because the holes act as if they are positive charges moving throughout the crystal. **Figure 3.17** illustrates boron-doped silicon.

Diodes

Many semiconductor devices are made by combining *n*- and *p*-type semiconductors to form a diode. This combination of semiconductors permits electrical current to flow in only one direction, from the negative terminal to the positive terminal.

Transistors like those shown in **Figure 3.18** are the key components in electrical circuits in devices such as computers, calculators, hearing aids, and televisions. They are used to amplify (increase the strength of) electrical signals. Transistors are exceedingly small, and their compact size and efficient operation have allowed the miniaturization of many electronic devices, such as laptop computers, heart pacemakers, and hearing aids. Transistors may be constructed by placing a *p*-type semiconductor between two *n*-type semiconductors, called an *npn*-junction, or by placing an *n*-type semiconductor between two *p*-type semiconductors, called a *pnp*-junction.

Supplemental Problems

For more practice with solving problems, see Supplemental Practice Problems, Appendix B.

Figure 3.18

Importance of Transistors
Transistors are used to increase the strength of electrical signals in such devices as television remote controls, telephones, and hearing aids.

Transistors, diodes, and other semiconductor devices are incorporated onto thin slices of silicon to form integrated circuits. Integrated circuits like the one shown in **Figure 3.19** may contain hundreds of thousands of devices on a slice of silicon called a chip. The small size of a chip—only a few millimeters in width—has made possible the amazing growth of computer technology.

Connecting Ideas

You already know that elements combine chemically to form compounds. The ways that elements combine depend entirely on their valence electrons. With your knowledge of the periodic table, you will be able to explain why elements combine and predict what compounds they will form.

Figure 3.19

Silicon and the Computer Revolution
Miniaturization of integrated circuits on silicon chips (top) allowed the size of computers to shrink from room-size (left) to laptop (right).

SECTION REVIEW

Understanding Concepts

1. Where are metals usually found on the periodic table? Where are nonmetals found? The metalloids?
2. How does the arrangement of elements in periods reflect electron structure?
3. What are the major differences in the physical properties of metals, nonmetals, and metalloids?

Thinking Critically

4. **Observing and Inferring** What can you deduce from the periodic table about the properties of the element barium?

Applying Chemistry

5. **Semiconductors** Germanium has the same type of structure and semiconducting properties as silicon. What type of semiconductor would you expect arsenic-doped germanium to be?

CHAPTER 3 ASSESSMENT

REVIEWING MAIN IDEAS

3.1 Development of the Periodic Table

- In his periodic table, Mendeleev organized the elements according to increasing atomic mass and placed elements with similar properties into groups.

- The modern periodic law states that the physical and chemical properties of the elements repeat in a regular pattern when they are arranged in order of increasing atomic number.

3.2 Using the Periodic Table

- Atomic structure and the number of valence electrons an element has can be related to an element's position on the periodic table.

- Elements may be classified as metals, nonmetals, or metalloids.

- The number of valence electrons and how tightly they are held determine the chemical properties of an element.

- The conductivity of semiconductors can be increased by adding small amounts of other elements.

Vocabulary

For each of the following terms, write a sentence that shows your understanding of its meaning.

actinide	nonmetal
group	period
lanthanide	periodic law
metal	periodicity
metalloid	semiconductor
noble gas	transition element

● UNDERSTANDING CONCEPTS

1. Describe element number 18 in terms of its period and group number, family name, and closest neighboring elements.

2. The modern periodic law says that the properties of the elements repeat periodically when the elements are arranged according to a certain trend. What is that trend?

3. Select the symbol of the element that fits the following descriptions.

   a) the Group 13 metal in the third period
   b) the Group 15 metalloid in the fourth period
   c) the lightest of the noble gases
   d) the halogen that exists as a liquid at room temperature
   e) the only metal that is a liquid at room temperature

4. Classify each of the elements in question 6 as a metal, nonmetal, or metalloid.

5. Write the electron dot structure for each of the following elements. What is the group number of each element?

   a) Cl   e) Kr
   b) Mg   f) Cs
   c) C   g) O
   d) Bi   h) P

6. An incomplete set of atomic mass, density, and melting point data is given for three elements in a triad. Following the patterns of Döbereiner's triads, predict a likely number for each missing value.

Element	Atomic Mass (u)	Density (g/mL)	Melting Point (K)
K	39.1		336
Rb		1.53	313
Cs	133	1.87	

chemistryca.com/vocabulary_puzzlemaker

APPLYING CONCEPTS

7. Which elements among the following have similar chemical properties?

a) Be   c) Cs   e) Ar
b) Sr   d) F   f) I

8. The symbols of some elements are written on the periodic table shown below. Answer the following questions for each element whose symbol is shown.

a) What is the element's row and column number?
b) Is the element a metal or nonmetal?
c) Is the element a solid, liquid, or gas at room temperature?
d) Draw the electron dot structure for each of the main group elements.

9. Explain why electrical conductivity increases in the elements in Group 14 as atomic number increases.

10. The noble gases share a unique chemical property. Describe this property and explain why they have it.

Chemistry and Technology

11. How is the transformation from one crystalline phase to another different from melting or boiling?

Everyday Chemistry

12. What properties of nickel and zinc make them good substitutes for copper and silver in coins?

Literature Connection

13. What two fields did Primo Levi combine in his life's work?

THINKING CRITICALLY

Identifying Patterns

14. In the system of numbering groups of elements from 1 to 18 on the periodic table, explain the significance of the group numbers 1, 2, 13, 14, 15, 16, 17, and 18.

Making Comparisons

15. Compare the atomic structure of metals, metalloids, and nonmetals. Explain how atomic structure accounts for differences in electrical conductivity.

Interpreting Data

16. **MiniLab 2** Graph density versus atomic mass for the elements in the table above. Describe the relationship between atomic mass and density for the elements. Use your periodic table to locate the elements. Based on these data, what can you say about the trend in density as you move down a column of elements?

Using a Table

17. **MiniLab 1** Mendeleev predicted the existence of eka-boron, which was unknown in his day. Eka-boron was located between calcium and titanium on the periodic table. What is the name of eka-boron now?

Relating Cause and Effect

18. **ChemLab** Explain how the metallic properties of the elements change as you move across a period from left to right. Why do the properties change in this manner?

Making Predictions

19. **MiniLab 2** Of the alkali metals lithium, potassium, and cesium, predict which is the most reactive and which is the least reactive.

20. The formulas of the chlorides of lithium, beryllium, boron, and carbon are $LiCl$, $BeCl_2$, BCl_3, and CCl_4, respectively. Use the periodic table to predict the formulas of the chlorides of potassium, magnesium, aluminum, and silicon.

Element	Helium	Neon	Argon	Krypton
Atomic Mass (u)	4.00	20.2	39.9	83.8
Density (g/L)	0.179	0.901	1.78	3.74

21. Argon has a melting point of −189.4°C and an atomic radius of 191 pm. Xenon has a melting point of −111.8°C and an atomic radius of 218 pm. Use this information to predict the melting point and atomic radius of krypton. Explain how you made your predictions.

Drawing Conclusions

22. Elements A and B are both in the same group. One element is in the second period and the other is in the fifth period. The density of element B is greater than the density of element A. Which element is in the second period? Explain your answer.

Making Inferences

23. The elements in Group 1 have one valence electron and they react vigorously with water at room temperature. The elements in Group 2 have two valence electrons and they react slowly with water at room temperature. The elements in Group 13 have three valence electrons and do not react with water at room temperature. What does this suggest about the reactivity of metals?

CUMULATIVE REVIEW

24. Explain how the law of definite proportions is illustrated by the formulas of the compounds given in question 30. (Chapter 2)

25. A neutral atom of argon has 22 neutrons and 18 electrons. What are the mass number and atomic number of argon? (Chapter 2)

26. Argon's 18 electrons are arranged in energy levels. How many energy levels are needed to accommodate argon's electrons and how many electrons are in each energy level? (Chapter 2)

27. What would you expect to see in the emission spectrum of argon to give evidence of the existence of energy levels? (Chapter 2)

28. An atom with a mass number of 196 has 40 fewer protons than neutrons. Find the atomic number and the identity of this atom. (Chapter 2)

29. How do the elements differ from most of the matter you see around you? (Chapter 1)

SKILL REVIEW

30. Look at the periodic table on pages 92 and 93. Find pairs of elements that would reverse order if the table were arranged according to increasing atomic mass as Mendeleev's periodic table was.

31. Describe the relationship that exists between the number of a period and the electron structure of atoms of elements in the period. Describe the change in electron structure as you move from left to right in a period.

WRITING IN CHEMISTRY

32. Mendeleev made predictions about germanium and several other elements. Three of these were gallium (Ga), scandium (Sc), and polonium (Po). Write a paragraph describing the accuracy of his predictions.

PROBLEM SOLVING

33. The density of aluminum is 2.7 g/mL and that of iron is 7.9 g/mL. Assume that manufacturers of soda cans could use the same volume of each metal to make soda cans. Compare the mass of an aluminum soda can with that of a comparable can made from iron. Explain your answer.

34. The chemical formula for zinc sulfide is ZnS. Use the periodic table to predict the formulas of the following similar compounds.

a) zinc oxide   **c)** mercury sulfide

b) cadmium oxide   **d)** zinc selenide

Standardized Test Practice

Element Symbol	Atomic Mass (u)	Density (g/mL)	Melting Point (°C)	Boiling Point (°C)
Cl	35.5	0.00321	−101	−34
Br	79.9	3.12	−7	59
I	127	4.93	114	185

Use the Halogen Triad Table above to answer Questions 1 and 2.

1. In which state does chlorine exist at room temperature?

   a) gas
   b) plasma
   c) liquid
   d) solid

2. As the atomic mass of a halogen increases, the point at which it turns from a liquid to a gas

   a) decreases.
   b) increases.
   c) fluctuates.
   d) remains constant.

3. Why was Mendeleev's periodic table a powerful tool for the science of chemistry in the nineteenth century?

   a) Mendeleev's periodic table organized element data into columns and rows.
   b) Mendeleev's periodic table allowed chemists to measure the densities of elements.
   c) Mendeleev's table allowed chemists to measure boiling and melting points of elements.
   d) Mendeleev's table allowed him to predict the properties of undiscovered elements.

4. Which of the following elements is a metalloid?

   a) aluminum
   b) argon
   c) arsenic
   d) calcium

5. Periodic law states that elements show a

   a) repetition of their physical properties when arranged by increasing atomic radius.
   b) repetition of their chemical properties when arranged by increasing atomic mass.
   c) periodic repetition of their properties when arranged by increasing atomic number.
   d) periodic repetition of their properties when arranged by increasing atomic mass.

6. It can be predicted that element 118 would have properties similar to a(n)

   a) alkali earth metal.
   b) halogen.
   c) metalloid.
   d) noble gas.

7. Which of the following statements is true about the number of valence electrons and the Lewis Dot structures of elements in the same group?

   a) Elements in the same group have the same number of valence electrons but a different Lewis dot structure.
   b) Elements in the same group have the same number of valence electrons and the same Lewis dot structure.
   c) Elements in the same group have a different number of valence electrons but the same Lewis dot structure.
   d) Elements in the same group have a different number of valence electrons and a different Lewis dot structure.

8. A semiconductor is

   a) a substance that conducts electricity well.
   b) a substance that conducts electricity poorly.
   c) a substance that conducts electricity better than metals but not as well as nonmetals.
   d) a substance that conducts electricity better than nonmetals but not as well as metals.

Test Taking Tip

Calculators Are Only Machines If your test allows you to use a calculator, use it wisely. The calculator can't figure out what the question is asking. That is still your job. Figure out which numbers are relevant, and determine the best way to solve the problem before you start punching keys.

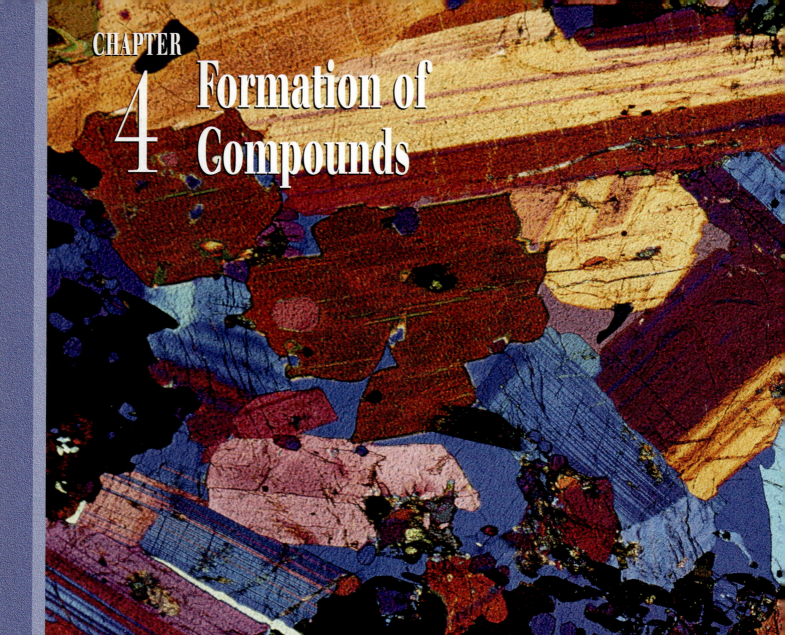

CHAPTER
4 Formation of Compounds

Chapter Preview

Sections

4.1 The Variety of Compounds

MiniLab 4.1 Evidence of a Chemical Reaction: Iron Versus Rust

4.2 How Elements Form Compounds

MiniLab 4.2 The Formation of Ionic Compounds

ChemLab The Formation and Decomposition of Zinc Iodide

Compounds—Submicroscopic to Macroscopic

Every compound has its own unique set of physical and chemical properties. Each macroscopic property you observe has a submicroscopic explanation. The beautiful mosaic of color in this volcanic mineral, known as eucrite, is the result of submicroscopic combinations of matter.

Start-up Activities

Launch Lab

Wine to Water

When substances go through a chemical reaction to form new compounds, the new substances often have different physical properties from the original reactants. How can a chemical reaction change a red liquid to a clear liquid?

Materials

- 500-mL beaker
- 25-mL graduated cylinder
- water
- red food coloring
- laundry bleach
- stirring rod

Safety Precautions

- Perform this experiment in a well-ventilated area or under a hood.
- Bleach can damage skin or clothing.
- Immediately notify your teacher of any spills.

Procedure

1. Pour 300 mL of water into a 500-mL beaker.
2. Add three drops of red food coloring and stir the mixture until the water turns uniformly red.
3. Measure 15 mL of bleach and pour the bleach into the beaker.
4. Stir the mixture and observe any changes in the solution.

Analysis

Research the compound in bleach that could account for the color change of the mixture. What new compounds might be formed during this reaction to account for the color change?

What I Already Know

Review the following concepts before studying this chapter.
Chapter 1: how elements and compounds differ
Chapter 2: the structure of atoms; the arrangement of electrons in an atom

Reading Chemistry

Before scanning the chapter, look at the definition of the word "compound." Then, scan the chapter, noticing the different illustrations of compounds. Pick one and try to draw it in a different way, using other shapes, spacing, or perspective. Be imaginative with your drawing.

Preview this chapter's content and activities at **chemistryca.com**

The Variety of Compounds

Chemistry is often like detective work. To figure out the submicroscopic structure of a compound, you first have to examine some of its macroscopic properties. Just as no two human fingerprints are the same, no two substances have exactly the same combination of physical and chemical properties. The structure of a substance at the atomic level determines its macroscopic properties. Therefore, looking at properties of compounds provides important clues about their submicroscopic structure and how they form from atoms. That is powerful detective work. You can begin this detective work by examining the properties of three familiar compounds: table salt, carbon dioxide, and water.

Salt: A Familiar Compound

What is the most popular food additive? In most kitchens, the answer is salt. It is used in cooking and at the table to enhance the flavor of food. Chemists refer to it as sodium chloride. The chemical name tells you what elements make up the compound; sodium chloride contains the elements sodium and chlorine.

Even though it's possible to make sodium chloride from its elements, salt is so abundant on Earth that it is used to manufacture the elements sodium and chlorine. Sodium chloride occurs naturally in large, solid, underground deposits throughout the world and is dissolved in the world's oceans. Salt can be obtained by mining these solid deposits, **Figure 4.1,** and by the evaporation of seawater.

Besides enhancing food's flavor, sodium chloride is an essential nutrient that plays crucial roles in living things. If you live in an area that gets snow and ice in the winter, salt is sometimes used to melt ice on roads, as shown in **Figure 4.1.**

Figure 4.1

Getting and Using Salt

Underground salt mining accounts for about 90 percent of the world's salt production. These salt deposits formed when ancient seas evaporated millions of years ago. ▼

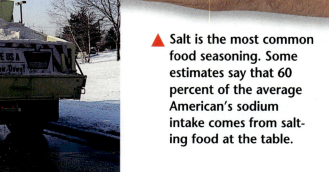

In some regions, trucks spread salt crystals on icy roads to melt the ice. The salt lowers the melting point of ice to about 15°F. If the air temperature is below 15°F, the salt won't do much good. ▶

▲ Salt is the most common food seasoning. Some estimates say that 60 percent of the average American's sodium intake comes from salting food at the table.

Physical Properties of Salt

You already know some of the physical properties of table salt. It is a white solid at room temperature. If you look at table salt under a magnifier, you'll notice that the grains of table salt are little crystals shaped like cubes. These crystals are hard, but when you press down on them with the back of a spoon, the crystals shatter. This shattering shows that the crystals are brittle. If sodium chloride is heated to a temperature of about 800°C, it melts and forms liquid salt. Solid sodium chloride does not conduct electricity, but melted sodium chloride does. Salt also dissolves easily in water. The resulting solution is an excellent conductor of electricity, as shown in **Figure 4.2.**

Figure 4.2

Conductivity of Salt

In order to light the bulb, electric current must flow between the two electrodes. As you can see, no current flows through the dry salt crystals (left). When the salt is dissolved in water, the solution conducts electricity and the bulb lights (right). Pure water alone does not conduct electricity.

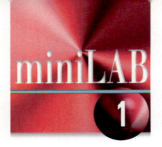

miniLAB 1

Evidence of a Chemical Reaction: Iron Versus Rust

Here's a chemical reaction whose result everyone has seen—the rusting of iron. Iron metal combines with oxygen in the air to form rust, Fe_2O_3, iron(III) oxide. This reaction is an example of a familiar compound forming from its component elements. One property of iron that everyone is familiar with is that it is attracted to a magnet. Attraction to a magnet is a simple property to measure—all you need is a magnet.

Procedure

1. Obtain a small wad of fresh steel wool in one small paper cup and another small wad of rusty steel wool in another small paper cup.

2. Obtain a 3" × 5" index card and a magnet wrapped in a plastic bag.

3. Test the fresh steel wool with the magnet. Record your observations.

4. Hold the rusty steel wool over the white card, and lightly rub the rusty steel wool between your thumb and forefinger. Some fine rust powder should fall onto the white card.

5. Next, hold the card up and slowly move the magnet under the card. Record your observations.

Analysis

1. What effect did the magnet have on the fresh steel wool?

2. What did you observe when the magnet was moved under the card with the rust powder?

3. Was your pile of rust powder a pure substance? How does your experimental evidence support your answer?

4. What evidence do you have that the rust is a different substance from iron itself?

Chemical Properties of Salt

The chemical properties of sodium chloride are not so useful in our detective work to determine its submicroscopic structure. Salt does not react readily with other substances. It could sit in a salt shaker for hundreds or even thousands of years and still remain salt. It does not have to be handled in any special way or be stored in a special container. Compounds with these chemical properties are referred to as stable or unreactive. You can get more clues about salt's submicroscopic structure by answering the question: How do the properties of salt compare with the properties of its component elements, sodium and chlorine?

Properties of Sodium

Sodium is a shiny, silvery-white, soft, solid element as you can see in **Figure 4.3.** From its location on the left side of the periodic table, you know that it is a metallic element. Sodium melts to form a liquid when it is heated above 98°C. Sodium must be stored under oil because it reacts with oxygen and water vapor in the air. In fact, it is one of the most reactive of the common elements. When a piece of sodium is dropped into water, it reacts so violently that it catches fire and sometimes explodes. Because of its high reactivity, the free element sodium is never found in the environment. Instead, sodium is always found combined with other elements.

Properties of Chlorine

The element chlorine, shown in **Figure 4.3,** is a pale green, poisonous gas with a choking odor. Because chlorine kills living cells and is slightly soluble in water, it is an excellent disinfectant for water supplies and swimming pools. You can tell that chlorine is a nonmetal by its position in the upper-right portion of the periodic table. Chlorine gas must be cooled to −34°C before it turns to a liquid. Like sodium, it is among the most reactive of the elements and must be handled with extreme care. Chlorine is needed for many industrial processes, such as the manufacture of bleaches and plastics. Because of its industrial importance, large quantities of chlorine must be transported in railroad tank cars, tanker trucks, and river barges. If a train that is carrying chlorine derails, entire communities are evacuated until the danger of a chlorine leak passes.

Look again at the photo on the opening page of this chapter. When sodium and chlorine react to form sodium chloride, two dangerous elements combine to form a stable, safe substance that we consume every day. What could be happening in such a change? You will find the details in Section 4.2, but first look at two other common compounds whose properties are different from those of sodium chloride.

Figure 4.3

A Comparison of Sodium and Chlorine

Sodium is a metal, but it is soft enough to be cut with a knife. Where it has been freshly cut, you can see that sodium has a silvery luster that is typical of many metals. ▶

Chlorine, in the flask on the right, is a greenish, poisonous gas at room temperature. If you've ever used liquid chlorine bleach, you've probably smelled chlorine. ▶

Carbon Dioxide: A Gas to Exhale

Carbon dioxide is a colorless gas. Take a deep breath and hold it for a few seconds. What you have inhaled is air, a colorless mixture of nitrogen and oxygen gases with small amounts of argon, water vapor, and carbon dioxide. Now, exhale. The mixture of gases that you exhale contains more than 100 times the amount of carbon dioxide that was in the air that you inhaled, as you can see in **Table 4.1.** In contrast, the quantity of oxygen is reduced by five percent. While the air was in your lungs, chemical and physical processes exchanged some of the oxygen for carbon dioxide. Carbon dioxide is an important chemical link between the plant and animal world. Green plants and other plantlike organisms take in carbon dioxide and give off oxygen during photosynthesis. Both plants and animals, including humans, use oxygen and give off carbon dioxide during respiration.

Table 4.1   Composition of Inhaled and Exhaled Air

Substance	Percent in Inhaled Air	Percent in Exhaled Air
Nitrogen	78	75
Oxygen	21	16
Argon	0.9	0.9
Carbon dioxide	0.03	4
Water vapor	variable (0 to 4)	increased

Physical Properties of Carbon Dioxide

Carbon dioxide, like sodium chloride, is also a compound, but its properties differ from those of sodium chloride. For example, salt is a solid at room temperature, but carbon dioxide is a colorless, odorless, and tasteless gas. When carbon dioxide is cooled below −80°C, the gas changes directly to white, solid carbon dioxide without first becoming a liquid. Because the solid form of carbon dioxide does not melt to a liquid, it is called dry ice, as shown in **Figure 4.4.** Carbon dioxide is soluble in water, as anyone who has ever opened a carbonated beverage knows. A water solution of carbon dioxide is a weak conductor of electricity. You can make carbon dioxide from its elements by burning carbon in air. Coal and charcoal are mostly carbon.

Chemical Properties of Carbon Dioxide

Like sodium chloride, carbon dioxide is relatively stable. Carbon dioxide is used in some types of fire extinguishers because it does not support burning, **Figure 4.4.** Photosynthesis is probably the most significant chemical reaction of carbon dioxide. In photosynthesis, plants use energy from the sun to combine carbon dioxide and water chemically to make simple sugars. Plants use these sugars as raw materials to make many

▲ Carbon dioxide does not support burning. In fact, it is often used to put out fires. Fire extinguishers like this one are filled with compressed carbon dioxide. Because CO_2 is denser than air, it displaces air and deprives the fire of a supply of oxygen.

Figure 4.4

Properties of Carbon Dioxide

◄ Solid carbon dioxide is called dry ice. Under ordinary conditions, it does not melt into a liquid; instead, it changes directly into a gas. The dry ice in this cylinder is immersed in water and is producing bubbles of carbon dioxide gas. The white vapors you see here are not carbon dioxide, but condensed water vapor carried along with the cold CO_2 gas. You can tell that carbon dioxide is denser than air.

other kinds of compounds, from cellulose in wood and cotton to oils such as corn oil and olive oil. Photosynthesis is only one part of a natural cycle of chemical reactions known as the carbon cycle.

The Properties of Carbon

As with sodium chloride, the properties of carbon dioxide differ from the properties of its elements. Carbon is a nonmetal and is fairly unreactive at room temperature. However, at higher temperatures, it reacts with many other elements. Charcoal is approximately 90 percent carbon. As anyone with a charcoal grill knows, carbon burns fairly easily and is an excellent source of heat, as shown in **Figure 4.5.** Carbon forms a huge variety of compounds. In fact, the majority of compounds that make up living things contain carbon. Carbon compounds are so significant that an entire branch of chemistry, called organic chemistry, is dedicated to their study.

Figure 4.5

A Chemical Property of CO_2
When you burn charcoal to cook chicken, carbon and oxygen in the air combine to produce carbon dioxide. When elements combine to form a substance that is more stable, the reaction often gives off energy in the form of heat.

The Properties of Oxygen

Like carbon, oxygen is another nonmetal. It is a colorless, odorless, and tasteless gas that makes up about 21 percent of the air you breathe. When materials such as paper burn in air, they react with oxygen, which is why people commonly say that oxygen supports burning.

Oxygen gas becomes a liquid when it is cooled to −183°C, and it is slightly soluble in water. The gills of fishes absorb dissolved oxygen from their water environment. Oxygen is more reactive than carbon and combines with many other elements. A prime example of its reactivity is the process of rusting, in which the element iron combines with oxygen from air. Many of the compounds that make up Earth's crust contain oxygen. Oxygen is the most abundant element in Earth's crust, as shown in **Figure 4.6.**

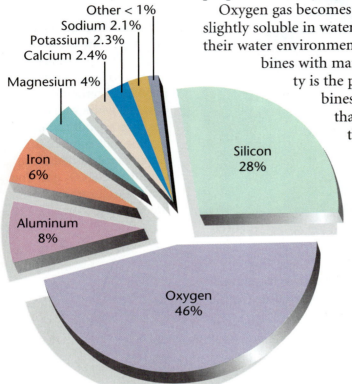

Other < 1%
Sodium 2.1%
Potassium 2.3%
Calcium 2.4%
Magnesium 4%
Iron 6%
Aluminum 8%
Silicon 28%
Oxygen 46%

Figure 4.6

Elemental Composition of Earth's Crust
As you can see from the graph, oxygen makes up about 46 percent of Earth's crust. Nearly all of this oxygen occurs in compounds with other elements. Silicon is the next most abundant element in Earth's crust. Therefore, you won't be surprised to learn that much of the oxygen is tied up in silicon dioxide, commonly known as sand.

Water, Water Everywhere

Water is the third familiar compound you will compare in your detective work on the submicroscopic structure of compounds. The formal chemical name of water is dihydrogen monoxide, but nobody calls it that. Water covers approximately 70 percent of Earth's surface and also makes up about 70 percent of the mass of the average human body.

Physical Properties of Water

The properties of water are different from those of sodium chloride and carbon dioxide. Water is the only one of the three compounds that occurs in Earth's environment in all three states of matter, as shown in **Figure 4.7.** At sea level, liquid water boils into gaseous water (steam) at 100°C and freezes to solid water (ice) at 0°C. Pure water does not conduct electricity in any of its states. Water is also excellent at dissolving other substances. It is often called the universal solvent in recognition of this valuable property. Water plays a vital role in the transport of dissolved materials, whether the aqueous solution is flowing down a river; up the xylem in a tree; or through the veins, capillaries, and arteries of your circulatory system.

Figure 4.7

Three States of Water
The iceberg shows water in its solid state, ice. In the ocean, water is in the liquid state. The atmosphere contains water vapor. Clouds consist of water vapor that has condensed back into small droplets of liquid water.

Chemical Properties of Water

Water is a stable compound; it doesn't break down under normal conditions and does not react with many other substances. Perhaps the most interesting property of water is its ability to act as a medium in which chemical reactions occur. Nearly all of the chemical reactions in the human body and many important reactions on Earth occur in an aqueous environment. Without water, these reactions could not occur or would occur extremely slowly. In addition, water and carbon dioxide are the starting materials for photosynthesis, the process that makes life on Earth possible. Now, compare the properties of water with those of its component elements, hydrogen and oxygen. The properties of oxygen were described on page 126.

The Properties of Hydrogen

Hydrogen is the lightest and most abundant element in the universe. Hydrogen is usually classified as a nonmetal. Like oxygen, the element hydrogen is an odorless, tasteless, and colorless gas. Hydrogen is a reactive element. Because of its reactivity, it is seldom found as a free element on Earth. Instead, it occurs in a variety of compounds, particularly water. Hydrogen reacts vigorously with many elements, including oxygen, as shown in **Figure 4.8.** This reaction forms water. The temperature of hydrogen gas must be lowered to a frigid −253°C before it turns to liquid. Hydrogen does not conduct electricity and is only slightly soluble in water.

Figure 4.8

The Reaction of Hydrogen and Oxygen
This welder is using a torch that burns hydrogen in oxygen. The heat produced is so intense that the torch can even be used underwater. The product of the reaction is water.

Everyday Chemistry

Elemental Good Health

Zinc and calcium

Potassium

Chromium

Selenium

Do you know that 60 elements are commonly found in the human body? Slightly fewer than half of them are essential for life, although scientists think most of the others play some role in life processes. The elements currently known to be essential are listed in the table below.

Although hydrogen, carbon, oxygen, and nitrogen make up almost 96 percent of the mass of the human body, minerals are also essential for life processes.

Different roles Because calcium compounds make up the hard parts of bones and teeth, calcium is needed for their growth and maintenance. Iron is an important element because it is the active part of the blood's hemoglobin molecule, which carries oxygen to the cells. Fluorine helps in the formation and maintenance of teeth and may prevent osteoporosis, which is the disintegration of bone. You may not know that magnesium is necessary for the functioning of nerves and muscles. So is potassium. Zinc and selenium are necessary for the activity of enzymes needed for cell division and growth and for the functioning of the immune system.

Elements Necessary for Life

Minerals	Non-minerals
F, Na, Mg, Si, P, S, Cl, K, Ca, V, Cr, Mn, Fe, Co, Ni, Cu, Zn, As, Se, Sn, I	H, C, O, N

Different amounts Maintaining the proper level of each mineral in your body is important for health. Nutritionists have established amounts of these elements that you should have in your daily diet. The amounts are described as the Recommended Dietary

Allowance (RDA) and Estimated Safe and Adequate Dietary Intakes (ESAI). For example, the RDA for calcium for teenagers is 1200 milligrams, while that of iodine is 150 micrograms (0.000 150 g). You may think an amount of 150 micrograms can't be too important, but it is crucial to the function of your thyroid gland, which helps control your metabolism and growth. If you use iodized salt, which contains a little potassium iodide, you probably are getting the proper RDA. Likewise, eating a well-balanced diet of the five food groups will maintain the proper levels of all the minerals that are elemental to good health.

Exploring Further

1. **Inferring** Why might cooking foods in boiling water reduce their mineral content?

2. **Acquiring Information** Why might consuming more than the Recommended Dietary Allowance of minerals be harmful?

Chemistry Online

To find out more about the role of minerals in healthy eating, visit the Chemistry Web site at **chemistryca.com**

Using Clues to Make a Case

You've seen that elements combine to form compounds whose properties differ greatly from those of the elements themselves. **Figure 4.9** shows another example. On the submicroscopic level, this clue indicates that atoms of elements react chemically to form some combinations that are much different from the original atoms. Also, if atoms of elements always combined in the same way, it's likely that all compounds would be similar. However, you've just studied three compounds that have greatly differing properties. On the submicroscopic level, this clue indicates that atoms must be able to combine in different ways to form different kinds of products. With the knowledge of the structure of atoms that you learned in Chapters 2 and 3, you can now examine the different ways that atoms can combine.

Figure 4.9

Iron Reacting with Chlorine
Here is another example of two elements reacting to form a compound whose properties are different from either element. Heated steel (iron) wool reacts with chlorine gas to form iron(III) chloride, the brown cloud you see in the flask.

Supplemental Problems

For more practice with solving problems, see Supplemental Practice Problems, Appendix B.

SECTION REVIEW

Understanding Concepts

1. Use water as an example to contrast the properties of a compound with the elements from which it is composed.

2. Classify the following substances as elements or compounds.

   a) table salt   **d)** chlorine gas
   b) water   **e)** carbon dioxide gas
   c) sulfur   **f)** dry ice

3. Give an example of a clue that indicates that atoms can combine chemically with each other in more than one way.

Thinking Critically

4. **Comparing and Contrasting** Using sodium chloride, carbon dioxide, and water as examples, what can you say about the chemical reactivity of compounds compared to the elements of which they are composed?

Applying Chemistry

5. **Using the Periodic Table** Find the elements that make up the compounds discussed in this section. Which elements are metals? Which elements are nonmetals? Compare the properties of the three compounds in terms of whether their component elements are two nonmetals or a metal and a nonmetal.

How Elements Form Compounds

It's time to return to the submicroscopic world of atoms in order to explain how and why atoms combine to form such a variety of compounds. In the 19th century, before electrons were discovered, chemists tried to visualize the way atoms combine to form compounds. Some early models pictured atoms with hooks that could attach to the hooks of other atoms. Today, chemists know a great deal more about the structure of atoms, so they have a much clearer model of chemical combination.

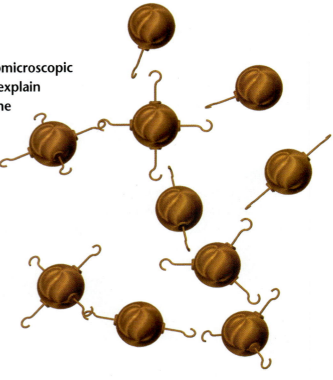

When Atoms Collide

When elements react, atoms of the elements must collide. It is what happens during that collision that determines what kind of a compound forms. How does the reaction of sodium and chlorine atoms to form salt differ from the reaction of hydrogen and oxygen atoms to form water?

When atoms collide with each other, what really comes into contact? As you learned in Chapter 2, the nucleus is tiny compared to the size of the atom's electron cloud. Also, the nucleus of an atom is buried deep in the center of the electron cloud. Therefore, it is highly unlikely that atomic nuclei would ever collide during a chemical reaction. In fact, reactions between atoms involve only their electron clouds.

When you studied the periodic table in Chapter 3, you learned that the properties of elements repeat because the pattern of the outermost (valence) electrons repeats in each period. It is this arrangement of valence electrons of an atom that is primarily responsible for the atom's chemical properties. You will not be surprised, then, to learn that it is the valence electrons of colliding atoms that interact. But what kinds of interactions between valence electrons are possible? For some additional clues to what takes place when atoms combine to form compounds, look at a group of elements with unusual properties—the noble gases.

Noble Chemical Stability

Of all of the elements, the elements of Group 18 are a curious bunch. They are notorious for their almost complete lack of chemical reactivity. In fact, they have some practical uses just because they are not reactive, as you can see in **Figure 4.10.** Despite the fact that all of these elements occur naturally in the environment, not a single *compound* of any of these elements has ever been found naturally in the environment.

This group of unreactive elements used to be called the inert gases because chemists thought these elements could never react to form compounds. However, in the 1960s, chemists were able to react fluorine, under conditions of high temperature and pressure, with krypton and with xenon to form compounds. Since that time, a few additional compounds of xenon and krypton have been produced. Still, no one has been successful at synthesizing compounds of helium, neon, and argon. Now that chemists know that these elements aren't completely inert, they are called noble gases.

Figure 4.10

Uses for Noble Gases

◄ Incandescent lightbulbs are filled with noble gases, usually argon and krypton, to protect the filament. The tungsten filament gets so hot that it will react with all but the most inert elements.

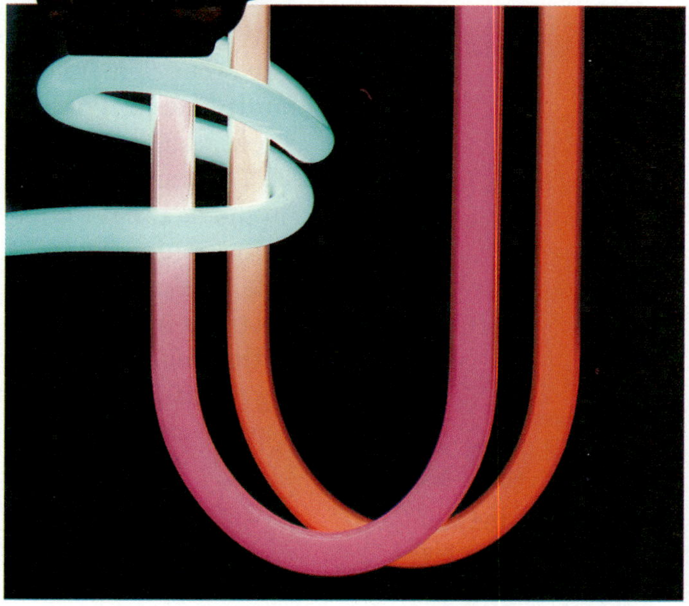

Noble gases and mixtures of noble gases are used in eye-catching light displays often referred to as neon lights. These gases give off different colors of light when a high-voltage electric current passes through them. Neon produces a bright orange color, argon produces blue, and helium gives off yellow-white. Lighting designers obtain different colors by adding mercury or other substances to the gas mixture. Sometimes, colored glass is used. ►

The Octet Rule

The lack of reactivity of noble gases indicates that atoms of these elements must be stable. Noble gases are unlike any other group of elements on the periodic table because of their extreme stability. As you know, the elements of a vertical group on the periodic table have similar arrangements of valence electrons. Each noble gas has eight valence electrons, except for helium, which has two. Because electron arrangement determines chemical properties, the electron arrangements of the noble gases must be the cause of their lack of reactivity with other elements.

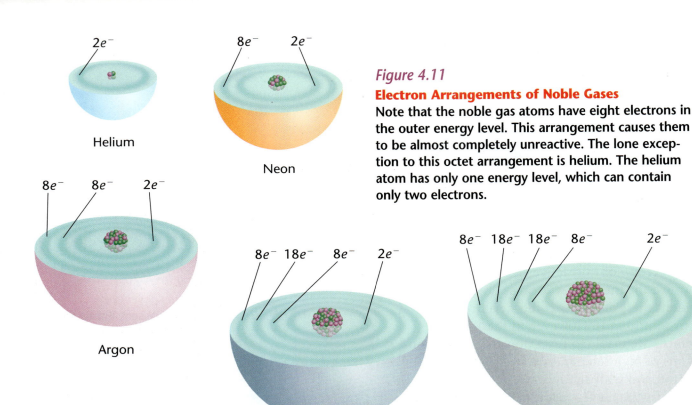

Helium

Neon

Argon

Krypton

Xenon

Figure 4.11

Electron Arrangements of Noble Gases
Note that the noble gas atoms have eight electrons in the outer energy level. This arrangement causes them to be almost completely unreactive. The lone exception to this octet arrangement is helium. The helium atom has only one energy level, which can contain only two electrons.

What does the electron arrangement of noble gases have to do with the way other elements react? Compare the electron arrangements of the noble gases in **Figure 4.11.** Today, scientists know that atoms don't have hooks. Rather, atoms combine because they become more stable by doing so. The modern model of how atoms react to form compounds is based on the fact that the stability of a noble gas results from the arrangement of its valence electrons. This model of chemical stability is called the **octet rule.** The octet rule says that atoms can become stable by having eight electrons in their outer energy level (or two electrons in the case of some of the smallest atoms). In other words, elements become stable by achieving the same configuration of valence electrons as one of the noble gases, a **noble gas configuration.**

Ways to Achieve a Stable Outer Energy Level

If atoms collide with enough energy, their outer electrons may rearrange to achieve a stable octet of valence electrons—a noble gas configuration—and the atoms will form a compound. Remember that electrons are particles of matter, so the total number of electrons cannot change during chemical reactions. Next, think about how valence electrons might rearrange among colliding atoms so that each atom has a stable octet. There are only two possibilities to consider. The first is a transfer of valence electrons between atoms. The second possibility is a sharing of valence electrons between atoms. The reactions discussed in Section 4.1 provide good examples of both of these possibilities.

Fact of the **MATTER**

Radon, Rn, is the last member of the noble gas group. It is a radioactive element that is formed by the radioactive decay of radium. Although it is found naturally on Earth, its presence is fleeting because it decays rapidly to other elements.

Electrons Can Be Transferred

You know that when sodium and chlorine are mixed, a reaction occurs and a compound, sodium chloride, forms. The photograph at the beginning of this chapter showed the macroscopic view of this reaction. What can be happening at the atomic level? Begin by picturing a collision between a sodium atom and a chlorine atom. Locate these atoms on the periodic table. Sodium is in Group 1, so it has one valence electron. Chlorine is in Group 17 and has seven valence electrons.

In previous chapters, you have used Lewis dot structures to represent an atom and its valence electrons. You will now be able to use these models to show what happens when atoms combine. The electron dot structures of the atoms are shown below.

$$Na^{\cdot} \qquad \cdot \ddot{\underset{\cdot\cdot}{Cl}} :$$

How can the valence electrons of atoms rearrange to give each atom a stable configuration of valence electrons? If the one valence electron of sodium is transferred to the chlorine atom, chlorine becomes stable with an octet of electrons. Because the chlorine atom now has an extra electron, it has a negative charge. Also, because sodium lost an electron, it now has an unbalanced proton in the nucleus and therefore has a positive charge.

$$Na^{\cdot} + \cdot \ddot{\underset{\cdot\cdot}{Cl}} : \rightarrow [Na]^{+} + [: \ddot{\underset{\cdot\cdot}{Cl}} :]^{-}$$

It's easy to see how chlorine has achieved a stable octet of electrons, but how does sodium become stable by losing an electron? Look at the position of sodium on the periodic table. By losing its lone valence electron, sodium will have the outer electron arrangement of neon. Sodium's stable octet consists of the eight electrons in the energy level below the level of the lost electron. **Figure 4.12** summarizes the way that sodium and chlorine react.

Figure 4.12
The Reaction of Sodium and Chlorine Atoms
The transfer of an electron from a sodium atom to a chlorine atom forms sodium and chloride ions. Examine the diagram carefully to see how this transfer gives both ions a stable octet.

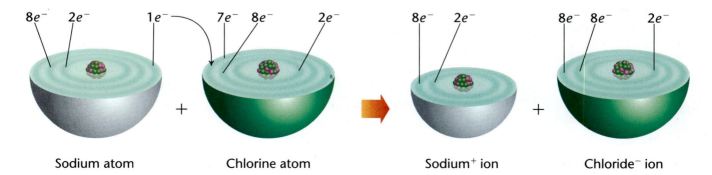

| Sodium atom | Chlorine atom | Sodium⁺ ion | Chloride⁻ ion |

Table 4.2 Reaction of Sodium and Chlorine

| | Sodium atom | + | Chlorine atom | → | Sodium ion | + | Chloride ion |
	Na	+	Cl	→	Na$^+$	+	Cl$^-$
Number of protons	11		17		11		17
Number of electrons	11		17		10		18
Number of outer-level electrons	1		7		8		8

Now that each atom has an octet of outer-level electrons, they are no longer neutral atoms; they are charged particles called ions. An **ion** is an atom or group of combined atoms that has a charge because of the loss or gain of electrons. Ions always form when valence electrons rearrange by electron transfer between atoms. A compound that is composed of ions is called an **ionic compound.** The transfer of a single electron changed a reactive metal, sodium, and a poisonous gas, chlorine, into the stable and safe compound, sodium chloride. Note that only the arrangement of electrons has changed. Nothing about the atom's nucleus has changed. This result is clear when you compare the atoms and ions in **Table 4.2.**

Ions Attract Each Other

Remember that objects with opposite charges attract each other. Once they have formed, the positive sodium ion and negative chloride ion are strongly attracted to each other. The strong attractive force between ions of opposite charge is called an **ionic bond.** The force of the ionic bond holds ions together in an ionic compound.

Even the smallest visible grain of salt contains several quintillion sodium and chloride ions. Every positively charged sodium ion attracts all nearby negatively charged chloride ions and vice versa. Therefore, these ions do not arrange themselves into isolated sodium ion/chloride ion pairs. Instead, the ions organize themselves into a definite cube-shaped arrangement, as shown in **Figure 4.13.** This well-organized structure is a crystal. Solid substances are composed of crystals. A **crystal** is a regular, repeating arrangement of atoms, ions, or molecules.

Figure 4.13

Crystal Structure of NaCl
Positive sodium ions attract negative chloride ions to form a cube-shaped arrangement in sodium chloride. In this arrangement, six chloride ions surround every sodium ion, and six sodium ions surround every chloride ion. The forces holding each ion in place are ionic bonds. ▼

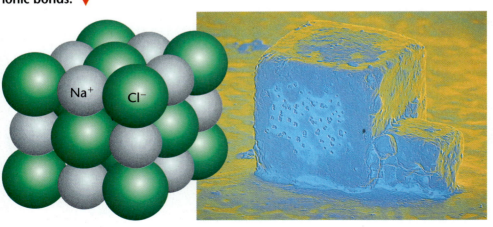

Na$^+$ Cl$^-$

◄ The macroscopic result of this submicroscopic arrangement is a cube-shaped salt crystal.

The Formation of Ionic Compounds

A sodium atom reacts by losing an electron to form a sodium 1+ ion. A chlorine atom gains one electron to form a chloride 1− ion. In this MiniLab, you will consider other combinations of atoms.

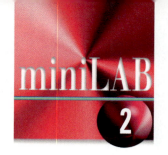
Procedure

1. Cut three paper disks about 7 cm in diameter for each of the elements: Li, S, Mg, O, Ca, N, Al, and I. Use a different color of paper for each element. Write the symbol of each element on the appropriate disks.

2. Select atoms of lithium and sulfur, and lay the circles side-by-side on a piece of corrugated cardboard.

3. Using thumbtacks of one color for lithium and another color for sulfur, place one tack for each valence electron on the disks, spacing the tacks evenly around the perimeters.

4. Transfer tacks from the metallic atoms to the nonmetallic atoms so that both elements achieve noble gas electron arrangements. Add more atoms if needed.

5. Once you have created a stable compound, write the ion symbols and charges and the formula and name for the resulting compound on the cardboard.

6. Repeat steps 2 through 5 for the remaining combinations of atoms.

Analysis

1. Why did you have to use more than one atom in some cases? Why couldn't you just take more electrons from one metal atom or add extra ones to a nonmetal atom?

2. Identify the noble gas elements that have the same electron structures as the ions produced.

The Results of Ionic Attraction

How does ionic bonding affect the macroscopic properties of a substance? The cubic shape of a salt crystal is a result of the cubic arrangement of sodium and chloride ions. Given the strong attractive force between sodium and chloride ions and the degree of organization among them, it's not surprising that sodium chloride is a solid at room temperature. All particles of matter are in constant motion. Raising the temperature of matter causes its particles to move faster. In order for a solid to melt, its temperature must be raised until the motion of the particles overcomes the attractive forces and the crystal organization breaks down. Breaking the strong crystal structure of sodium chloride requires a lot of energy, which is the reason that sodium chloride must be heated to more than 800°C before it melts.

When you press salt crystals, they don't bend or flatten out. When enough force is applied, the salt crystals suddenly shatter. This hardness and brittleness provide macroscopic evidence for the strength and rigidity of the submicroscopic structure of the salt crystal. Trying to break an ionic crystal is like trying to break a well-laid brick wall, **Figure 4.14**.

Figure 4.14

Ionic Crystals Are Strong
The crystal structure of an ionic compound is a lot like a well-laid brick wall. It takes a great deal of force to break a brick wall.

135

ChemLab

The Formation and Decomposition of Zinc Iodide

Compounds are chemical combinations of elements. Many chemical reactions of elements to form compounds are spectacular but must be run under special laboratory conditions because they are dangerous. The reaction of sodium and chlorine to form sodium chloride pictured at the beginning of Chapter 4 is a good example. If elements react spontaneously to form compounds, that is a good indication that the compound state is more stable than the free element state. To break a stable compound down into its component elements, energy must be put into the compound. Electricity is often used as this energy source. The process of decomposing a compound into its component elements by electricity is called electrolysis.

Problem

Can a compound be synthesized from its elements and then decomposed back into its original elements?

Objectives

- **Compare** a compound with its component elements.
- **Observe** and monitor a chemical reaction.
- **Observe** the decomposition of the compound back to its elements.

PREPARATION

Materials

10 × 150-mm test tube
test-tube rack
test-tube holder
100-mL beaker
spatula
plastic stirring rod
zinc
iodine crystals
distilled water
9-V battery with terminal clip and leads
two 20-cm insulated copper wires stripped at
    least 1 cm on each end

Safety Precautions

CAUTION: *Iodine crystals are toxic and can stain the skin. Use care when using solid iodine. The reaction of zinc and iodine releases heat. Always use the test-tube holder to handle the reaction test tube.*

PROCEDURE

1. Obtain a test tube and a small beaker. Place the test tube upright in a test-tube rack.

2. Carefully add approximately 1 g of zinc dust and about 10 mL of distilled water to the test tube.

3. Carefully add about 1 g of iodine to the test tube. Record your observations in a table like the one shown.

4. Stir the contents of the test tube thoroughly with a plastic stirring rod until there is no more evidence of a reaction. Record any observations of physical or chemical changes.

5. Allow the reaction mixture to settle. Using a test-tube holder, carefully pick up the test tube and pour off the solution phase into the small beaker.

6. Add water to the beaker to bring the volume up to about 25 mL.

7. Obtain a 9-V battery with wire leads and two pieces of copper wire. Attach the copper wires

to the wire leads from the battery. Make sure that the wires are not touching each other.

8. Dip the wires into the solution and observe what takes place. Record your observations.

9. After two minutes, remove the wires from the solution and examine the wires. Again, record your observations.

ANALYZE AND CONCLUDE

1. **Observing and Inferring** What evidence was there that a chemical reaction occurred?

2. **Comparing and Contrasting** How did you know the reaction was complete?

3. **Making Inferences** What term is used to describe a reaction in which heat is given off? How can you account for the heat given off in this reaction?

4. **Checking Your Hypothesis** What evidence do you have that the compound was decomposed by electrolysis?

5. **Drawing Conclusions** Why do you think the reaction between zinc and iodine stopped?

DATA AND OBSERVATIONS

APPLY AND ASSESS

1. What role did the water play in this reaction?

2. Do you think zinc iodide is an ionic or covalent compound? What evidence do you have to support your conclusion?

3. The formula of zinc iodide is ZnI_2. Use Lewis dot structures to show how it forms from its elements. Hint: Zinc atoms have two valence electrons.

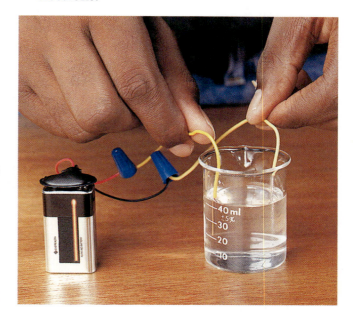

Step	Observations
3. Addition of iodine to zinc	
4. Reaction of iodine and zinc	
8. Electrolysis of solution	
9. Examination of wires	

Representing Compounds with Formulas

Rather than writing out the name *sodium chloride* every time you refer to it, you can use a much simpler system. You can write its chemical formula, NaCl. The formula of a compound tells what elements make up the compound and how many atoms of each element are present in one unit of the compound. Water is written as H_2O. This formula means that water consists of two hydrogen atoms combined with one oxygen atom. When sodium and chlorine atoms react, they form ions, which then arrange themselves into a crystal. The ratios of the elements that make up salt crystals do not change, and so the formula NaCl does not change.

You have seen how the electrons are transferred from sodium to chlorine to form a strong crystal arrangement. There are many other ionic compounds, as you will see later in Chapter 5. Now, look at another example from Section 4.1 to see a different way that atoms can combine to achieve a stable outer level of electrons.

Electrons Can Be Shared

In Section 4.1, you learned that the reaction of hydrogen and oxygen forms water. What happens when hydrogen and oxygen atoms collide? Hydrogen has only one valence electron. Oxygen, a Group 16 element, has six valence electrons. Could these atoms achieve the stable electron configuration of a noble gas by transfer of electrons? If the oxygen atom could pick up two more valence electrons, it would have a stable octet—the noble gas configuration of neon.

What about hydrogen? Could a hydrogen lose its single valence electron? Your first inclination might be to treat hydrogen just like sodium, but be careful. If hydrogen loses an electron, it is left with no electrons, and that isn't the electron structure of a noble gas. Maybe hydrogen could gain an electron so that its electron arrangement is like that of helium. But, both atoms cannot gain electrons.

Colliding atoms transfer electrons only when one atom has a stronger attraction for valence electrons than has the other atom. In the case of

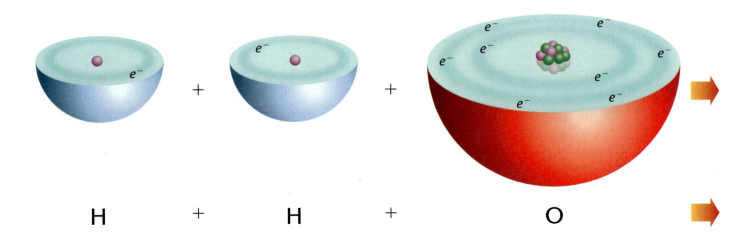

H   +   H   +   O

sodium and chlorine, chlorine attracts sodium's valence electron strongly, whereas sodium holds its electron weakly. Therefore, the electron moves from sodium to chlorine, forming positive and negative ions in the process. You will learn more about this process and the factors that influence it in Chapter 9. In the case of hydrogen and oxygen, neither atom attracts electrons strongly enough to take electrons from the other atom. When atoms collide with enough energy to react, but neither attracts electrons strongly enough to take electrons from the other atom, the atoms combine by sharing valence electrons.

To understand how water forms, start by looking at the electron dot structures of hydrogen and oxygen.

$$H\cdot \qquad \cdot \ddot{O}:$$

Hydrogen requires one more electron to have the same electron arrangement as helium, while oxygen requires two more electrons to have neon's arrangement. Hydrogen and oxygen can share one electron from each atom. This sharing is shown by placing two dots representing electrons between the atoms.

$$H\!:\!\ddot{\underset{\cdot\cdot}{O}}\!:$$

This arrangement makes hydrogen stable by giving it two valence electrons, but it leaves oxygen with only seven valence electrons. Oxygen's octet can be completed by sharing an electron with another hydrogen atom. (This explains why water has the formula H_2O.)

$$H\!:\!\ddot{O}\!: \quad + \quad H\cdot \quad \rightarrow \quad \underset{H}{H\!:\!\ddot{O}\!:}$$

Just as in the case of the formation of sodium chloride by ionic bonding, all the parts present before the reaction are still there after the reaction. What has changed in the combining of hydrogen and oxygen atoms? The valence electrons no longer reside in the same positions. This is what always happens in a chemical reaction; electrons rearrange. **Figure 4.15** summarizes the reaction between hydrogen and oxygen.

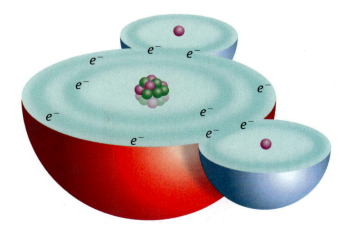

H_2O

Figure 4.15
Formation of Water by Electron Sharing
The stability of the atoms in a water molecule results from a cooperative arrangement in which the eight valence electrons (six from oxygen and one each from two hydrogens) are distributed among the three atoms. By sharing an electron pair with the oxygen, each hydrogen claims two electrons in its outer level. The oxygen, by sharing two electrons with two hydrogens, claims a stable octet in its outer level. By this method, each atom achieves a stable noble gas configuration.

Electron Sharing Produces Molecules

The attraction of two atoms for a shared pair of electrons is called a **covalent bond.** Notice that in a covalent bond, atoms share electrons and neither atom has an ionic charge. A compound whose atoms are held together by covalent bonds is a **covalent compound.** Water is a covalent compound. Although water is made up of hydrogen and oxygen atoms, these have combined into water molecules, each having two hydrogen atoms bonded to one oxygen atom. A **molecule** is an uncharged group of two or more atoms held together by covalent bonds. Sometimes chemists refer to covalent compounds as molecular compounds. The terms mean the same thing. **Figure 4.16** compares a compound with ionic bonds to a compound having covalent bonds.

Figure 4.16

Comparing an Ionic and a Covalent Compound
Iron(III) chloride is a typical ionic compound. It is crystalline at room temperature, melts at a high temperature (300°C), and dissolves in water. ▼

▲ Ethanol, also known as ethyl alcohol, is a typical covalent compound. It is a liquid at room temperature but evaporates readily into the air. Ethanol boils at 78°C and freezes at −114°C. Unlike many covalent compounds, ethanol dissolves in water. In fact, ethanol sold in stores as rubbing alcohol contains water.

More Than Two Electrons Can Be Shared

When charcoal burns, carbon atoms collide with oxygen to form CO_2. Carbon is in Group 14 and has four valence electrons. Oxygen is in Group 16 and has six valence electrons.

Carbon and oxygen are like hydrogen and oxygen when it comes to the question of sharing or transferring electrons. Neither atom is able to attract electrons away from the other atom. In fact, two nonmetallic elements usually achieve stability by sharing electrons to form a covalent compound. On the other hand, if the reacting atoms are a metal and a nonmetal, they are much more likely to transfer electrons and form an ionic compound.

Hydrogen's Ill-fated Lifts

Imagine the surprise of the citizens of Paris on that morning of December 1, 1783. There was Jacques Charles and his assistant gliding above their rooftops in a basket suspended from a large balloon. Charles and his assistant had filled the balloon with hydrogen and became the first humans to ride in such a lighter-than-air vehicle. By World War I, hydrogen-filled balloons were being used to carry military personnel aloft to observe troop movements.

In 1936, Germany launched the *Hindenburg,* a rigid airship originally designed to use helium for lift. Helium is slightly less buoyant than hydrogen, but it is a noble gas. Thus, it does not react with anything, whereas hydrogen burns explosively in air. However, the Germans had to continue the use of hydrogen in the airship because they had no source of helium. In the following year, the *Hindenburg* carried more than 1300 passengers in transatlantic flights. While attempting to dock in Lakehurst, New Jersey, on May 1, 1937, the airship exploded and burned when the hydrogen ignited. Thirty-six people were killed. The accident was the final chapter in the use of hydrogen in lighter-than-air vehicles.

The chemical reaction The reaction that destroyed the *Hindenburg* was the burning of hydrogen gas.

$$2H_2(g) + O_2(g) \rightarrow 2H_2O(g)$$

Once ignited, this reaction occurs spontaneously. However, the reaction can be controlled, such as in the main engine of the space shuttle.

Space shuttle fuel
The main engine of the space shuttle

booster uses liquid hydrogen and oxygen as fuel. These materials are stored in separate sections of a huge external fuel tank attached beneath the shuttle. The energy released during the reaction thrusts the shuttle into orbit.

During the launch of the space shuttle *Challenger* on January 28, 1986, the reaction became uncontrolled. One minute and 13 seconds after takeoff, the external tank and shuttle exploded, killing the shuttle's seven crew members. The accident was caused by defects in the design of O-rings that joined sections of the solid-fuel booster engines, which are attached to the sides of the shuttle.

Connecting to Chemistry

1. **Inferring** Much research has been done on developing hydrogen-fueled cars and trucks. Why would such a car be almost entirely nonpolluting? What factors do you think might affect the public's acceptance of such a vehicle?

2. **Hypothesizing** What do you think is the reason that so much energy is produced when hydrogen reacts with oxygen? Recall what happens when these atoms bond.

Challenger *explosion*

Consider the reaction of carbon and oxygen to form CO_2. Look at electron dot structures for the participating atoms below.

$$:\overset{\cdot}{\underset{}{\text{O}}}\cdot \qquad \cdot\overset{\times}{\underset{\times}{\text{C}}}\times \qquad \cdot\overset{\cdot}{\underset{}{\text{O}}}:$$

Can you arrange the 16 valence electrons from these three atoms to produce a molecule in which all three atoms have a stable configuration? You know that at least one bond must exist between the carbon and each oxygen, so start there. Here's an approach to the puzzle. Have each oxygen share an electron with carbon as in the following dot structures.

$$:\overset{\cdot}{\underset{\cdot}{\text{O}}}:\overset{\times}{\underset{\times}{\text{C}}}:\overset{\cdot}{\underset{\cdot}{\text{O}}}:$$

This arrangement gives carbon six electrons and each oxygen seven, but still no atom has an octet. What else can be done? There's no law in chemistry that says atoms must bond by sharing only one pair of electrons. What happens if they share two pairs? You now have double covalent bonds, as shown in this dot structure.

$$:\overset{\cdot}{\underset{\cdot}{\text{O}}}::\text{C}::\overset{\cdot}{\underset{\cdot}{\text{O}}}:$$

Now, count all the electrons around each atom, including the ones that are shared. You'll see that each atom has a stable octet. Study the result of this electron sharing in **Figure 4.17.** By sharing electrons, the three atoms achieve a more stable arrangement than they had as three separate atoms. A molecule of carbon dioxide is more stable than one carbon atom and two oxygen atoms. As with water, the molecule of carbon dioxide is different from the sum of its parts. The macroscopic properties of carbon dioxide are a result of the unique properties of carbon dioxide molecules, not the properties of carbon or oxygen atoms.

Figure 4.17

Electron Sharing in CO_2
When one carbon atom and two oxygen atoms react, the carbon atom shares two pairs of electrons with each oxygen. This arrangement gives all atoms a stable octet.

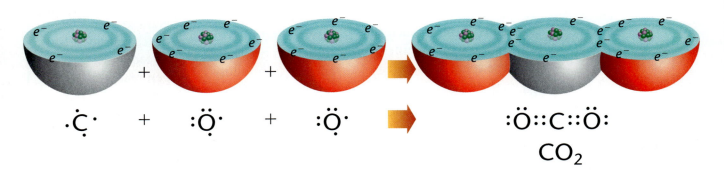

How do ionic and covalent compounds compare?

Now you can relate the submicroscopic models of the formation of NaCl, H_2O, and CO_2 to their macroscopic properties mentioned in Section 4.1. When elements combine, they form either ions or molecules. No other possibilities exist. The particles change dramatically, whether they change from sodium atoms to sodium ions or hydrogen and oxygen atoms to water molecules. This change explains why compounds have different properties from the elements that make them up.

Explaining the Properties of Ionic Compounds

The physical properties of ionic compounds are directly related to the fact that ionic compounds are composed of well-organized, tightly bound ions. These ions form a strong, three-dimensional crystal structure. This model of the submicroscopic structure explains the general observation that ionic compounds are crystalline solids at room temperature. Just like NaCl, these solids are generally hard, rough, and brittle. Ionic compounds usually have to be heated to high temperatures in order to melt them because the attractions between ions of opposite charge are strong. It takes a lot of energy to break the well-organized network of bound ions. Compare their appearance with the properties described above. Some typical ionic compounds are shown in **Figure 4.18**.

Potassium chloride (KCl) is used as a salt substitute because it has a taste similar to that of salt. ▼

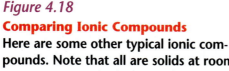

Figure 4.18
Comparing Ionic Compounds
Here are some other typical ionic compounds. Note that all are solids at room temperature and are soluble in water. Not all ionic solids are soluble in water, though.

◀ Copper(II) sulfate ($CuSO_4$) is sometimes used to treat the growth of algae in swimming pools and in water-treatment plants.

◀ Sodium hydrogen carbonate ($NaHCO_3$), commonly known as baking soda, is also called sodium bicarbonate.

Another physical property of ionic compounds is their tendency to dissolve in water. When they dissolve in water, the solution conducts electricity, as you saw in **Figure 4.2.** Ionic compounds also conduct electricity in the liquid (melted) state. Any compound that conducts electricity when melted or dissolved in water is an **electrolyte.** Therefore, ionic compounds are electrolytes. In order to conduct electricity, ions must be free to move because they must take on or give up electrons. Ionic compounds in the solid state do not conduct electricity because the ions are locked into position. Ionic compounds become good conductors when they melt. This is evidence that the ions are less bound and free to move in the liquid state.

Explaining the Properties of Covalent Compounds

As with ionic compounds, the submicroscopic model of the formation of covalent compounds explains many of the properties of these compounds. In particular, you can use this model to explain why typical covalent compounds such as H_2O and CO_2 have properties so different from ionic compounds.

In order to explain these differences, consider the submicroscopic organization of covalent compounds. Covalent compounds are composed of molecules. As you learned in this chapter, the atoms that compose molecules are held together by strong forces—covalent bonds—that make the molecule a stable unit. The molecules themselves have no ionic charge, so the attractive forces between molecules are usually weak.

The forces between particles that make up a substance are called **interparticle forces.** These forces are illustrated in **Figure 4.19.** It is the great difference in the strength of interparticle forces in covalent compounds compared to those of ionic compounds that explains many of the differences in their physical properties.

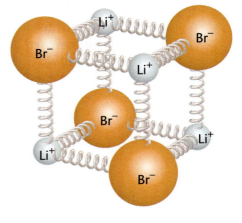

Figure 4.19

Comparing Interparticle Forces in Ionic and Covalent Compounds
In an ionic compound such as lithium bromide (LiBr), interparticle forces are strong because of the attraction between ions of opposite charges. ▶

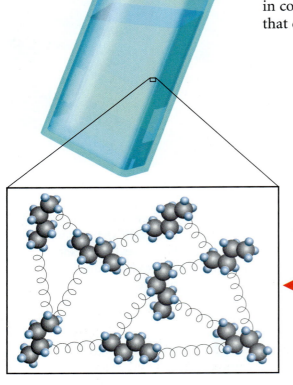

◀ Butane (C_4H_{10}) is the fuel commonly found in plastic, disposable lighters. Butane is a covalent compound, but the molecules have no electrical charge, so the attraction between them is weak. In fact, if the butane were not held under pressure in the lighter, it would immediately boil away to a gas.

Whereas all ionic compounds are solids at room temperature, many covalent compounds are liquids or gases at room temperature. Note, however, that many covalent compounds—sugar, for example—will form crystals if there is enough attractive force between molecules. In Chapter 9, you will learn why some molecules attract each other. Many of the covalent compounds that are solids at room temperature will melt at low temperatures. Examples are sugar and the compounds that make up candle wax and fat. Compare the properties of some covalent substances in **Figure 4.20.** Molecular (covalent) compounds do not conduct electricity in the pure state. Many covalent compounds, such as those in gasoline and vegetable oil, do not dissolve in water, although others such as sugar will dissolve. The solubility of covalent compounds in water varies, but in general, covalent compounds are usually less soluble in water than ionic compounds. What accounts for these differences?

Try at Home | **Lab**

See page 864 in Appendix F for
Mixing Ionic and Covalent Liquids

Figure 4.20

Comparing Covalent Compounds

Covalent compounds are composed of molecules in which atoms are bonded by electron sharing. Because of weak interparticle forces between molecules, covalent compounds tend to be gases or liquids at room temperature. They tend to be insoluble in water, although many are extremely soluble.

Table sugar ($C_{12}H_{22}O_{11}$) is called sucrose. It is an example of a covalent compound that is a crystalline solid soluble in water.

Gasoline and oil are mixtures of covalent compounds. Spilled oil does not dissolve in water, but instead floats on the water in thin layers.

Candle wax and butter are mixtures of covalent compounds. Because their molecules are larger and heavier, they form solids but melt at low temperatures.

In places where natural gas is not available, many people use propane (C_3H_8) to heat their homes and cook food. It is delivered to businesses and homes in pressurized tank trucks.

The Rain Forest Pharmacy

Long ago, Samoan healers dispensed a tea brewed from the bark of a native rain forest tree, *Homalanthus nutans,* to help victims of yellow fever, a viral disease. Researchers have since identified the bark's active ingredient, prostratin. Prostratin is now being investigated for possible use as a drug for other viral diseases.

Using plants and products from plants for medicinal purposes has a long history. Such common drugs as aspirin, codeine, and quinine were originally derived from plants. However, only about 0.5 percent of all plant species have been studied intensively for chemical makeup and medicinal benefit. Because most of the world's 250 000 species of flowering plants live in rain forests, researchers are now combing these areas of dense vegetation for new substances to fight diseases.

Screening plants One method of collecting plants for drug research is to select many different species of plants from one locale and test them for possible medicinal uses. Because of the great diversity of plants, researchers hope to find an even greater diversity of chemical substances. Agencies such as the National Cancer Institute routinely use this screening method. Another method of screening plants for research is the phylogenetic survey, in which researchers study close relatives of plants already known to produce beneficial medicinal substances. Researchers hypothesize that similarities in the evolution of these plants may have produced similar properties in the biochemical substances these plants produce.

Ethnobotanical approach Interest is growing in another method of screening plants for drug research. This method is based on the work of ethnobotanists. Ethnobotanists study the medicinal uses of plants by native peoples. Ethnobotanists have learned that the sophisticated knowledge concerning the use of medicinal plants possessed by these people can be used to identify plants that may be important for future research. The discovery of prostratin is an example of the ethnobotanical approach to screening plants for potential use in drug research.

The future Isolating and testing of substances from plants may take years. Within this time, rain forests may be significantly altered. Most rain forests are found in developing countries where the citizens want to work for better lives and grow enough food to be self-sufficient. Because of deforestation to obtain lumber as well as space for open-field agriculture, rain forest environments are disappearing. Some researchers and environmental scientists are setting up preserves within the rain forests to maintain the forests' biodiversity. Others are attempting to grow rapidly disappearing plant species in rain forest nurseries. With these methods, scientists are trying to maintain the complex rain forest ecosystems for future research.

Analyzing the Issue

1. **Inferring** Why might drug researchers investigate a plant that has few pests as a potentially useful plant?

2. **Acquiring Information** How might research into the pharmaceutical value of rain forest substances have an impact on the present and future uses of rain forests?

Interparticle Forces Make the Difference

Interparticle forces are the key to determining the state of matter of a substance at room temperature. You already know that ions are held rigidly in the solid state by the strong forces between them. Because there are much lower interparticle forces in covalent compounds, their molecules are less tightly held to one another. Therefore, they are more likely to be gases or liquids at room temperature.

Because there are no ions in covalent compounds, you do not expect them to be electrical conductors. Ionic compounds tend to be soluble in water while molecular compounds do not. This difference is also explained by interparticle forces. Ions are attracted by water molecules, but many covalent molecules are not and, therefore, do not dissolve. Solubility in water and the nature of water solutions is a major topic in chemistry. You will learn more about solutions in Chapter 13.

Connecting Ideas

Now that you know that two major types of compounds exist, you may wonder where else you come into contact with these compounds on a day-to-day basis. In Chapter 5, you'll look at more examples of ionic and covalent compounds, including compounds more complex than the simple ones used as examples in this chapter. You'll also learn the important practical skill of naming and writing the formulas of compounds, as well as how to identify a few special categories of compounds such as acids, bases, and organic compounds.

Supplemental Problems

For more practice with solving problems, see Supplemental Practice Problems, Appendix B.

SECTION REVIEW

Understanding Concepts

1. Explain why sodium chloride is a neutral compound even though it is made up of ions that have positive and negative charges.

2. Describe two ways in which atoms become stable by combining with each other. Compare and contrast the compounds that result from two kinds of combinations.

3. Why do you think that sodium chloride has to be heated to 800°C before melting, but candle wax will start to melt at 50°C?

Thinking Critically

4. **Making Predictions** Look at the following diagram and explain how you can determine whether the compound being formed is ionic or covalent. Do you think it is more likely to be a solid or a gas at room temperature? Explain.

$$Ca{\cdot} + {\cdot}\ddot{\underset{\cdot\cdot}{B}r}{:} + {\cdot}\ddot{\underset{\cdot\cdot}{B}r}{:} \rightarrow [Ca]^{2+} + [{:}\ddot{\underset{\cdot\cdot}{B}r}{:}]^- + [{:}\ddot{\underset{\cdot\cdot}{B}r}{:}]^-$$

Applying Chemistry

5. **Using Lewis Structures** Potassium metal will react with sulfur to form an ionic compound. Use the periodic table to determine the number of valence electrons for each element. Draw a Lewis dot structure to show how they would combine to form ions. How would you write the formula for the resulting compound?

REVIEWING MAIN IDEAS

4.1 The Variety of Compounds

- When compounds form, they have properties that differ greatly from the properties of the elements of which they are made.

- Sodium, a dangerously reactive metal, reacts with chlorine, a poisonous gas, to form the stable compound sodium chloride, table salt.

- Carbon, usually found as a black solid, reacts with oxygen in the air to form carbon dioxide, an unreactive gas that produces the fizz in soda pop.

- Hydrogen, the lightest gaseous element, reacts explosively with oxygen in the air to form water, a stable compound on which all life depends.

- The properties of compounds differ widely because of differences in what happens to their constituent atoms when they form.

4.2 How Elements Form Compounds

- Atoms become stable by reacting to achieve the outer-level electron structure of a noble gas (Group 18).

- One way to achieve a stable electron structure is to transfer electrons from one atom to another, thus forming charged ions of opposite charge. The ions attract each other to form a crystal. A compound formed in this way is an ionic compound.

- Another way that atoms can achieve stability is by sharing electrons with other atoms to form molecules. A compound formed by electron sharing is a covalent compound, also called a molecular compound.

- Two atoms can share more than one pair of electrons.

- In ionic compounds, the interparticle forces are attractions between ions of opposite electrical charge. These forces are much stronger than the interparticle forces between molecules of covalent compounds.

- It is the differences in strength of interparticle forces that account for many of the differences in physical properties between ionic and covalent compounds.

Vocabulary

For each of the following terms, write a sentence that shows your understanding of its meaning.

covalent bond	ionic bond
covalent compound	ionic compound
crystal	molecule
electrolyte	noble gas configuration
interparticle force	octet rule
ion	

UNDERSTANDING CONCEPTS

1. How is the compound sodium chloride different from the elements of which it is composed?

2. If you tried to breathe CO_2, you would suffocate. Why, then, is CO_2 essential to all life on Earth?

3. Why is water essential to life on Earth?

4. Why are Group 18 elements no longer called *inert*?

5. Describe the electron arrangement that makes an atom stable. Why is helium stable with a different arrangement?

6. Describe two processes by which elements can combine to form stable compounds. Name the type of bonding that results from each process.

7. Compare the formation of NaCl from sodium and chlorine to the formation of CO_2 from carbon and oxygen. In what ways are they similar? In what ways do they differ?

8. Sodium reacts with fluorine to form sodium fluoride, NaF, a common additive in toothpaste that prevents tooth decay. The electron exchange in this reaction is the same as that for sodium and chlorine. Use the format shown in **Table 4.2** to analyze this reaction.

9. What are three general properties of ionic compounds? How are these properties related to the submicroscopic structure of the compounds?

10. Why would you never expect to find Na_2Cl as a stable compound?

11. How is a sodium ion different from a sodium atom? From a neon atom?

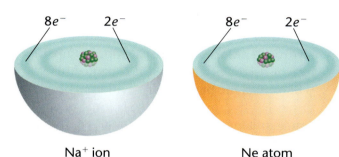

Na⁺ ion   Ne atom

12. The term *isoelectronic* is used to describe atoms and ions that have the same number of electrons. Which of the following are *isoelectronic*? Na^+, Ca^{2+}, Ne, K, O^{2-}, P^{3-}

13. What are three general properties of covalent compounds? How are these properties related to the submicroscopic structure of the compounds?

14. How do the basic particles of ionic and covalent compounds differ?

APPLYING CONCEPTS

15. A bag of pretzels lists the sodium content of the pretzels. Considering that sodium reacts violently with water, why don't you explode when you eat these sodium-containing pretzels? What does the label really mean to a chemist?

16. Hydrazine is a compound with the formula N_2H_4. What kind of compound is hydrazine?

Describe the formation of hydrazine from nitrogen and hydrogen atoms.

17. What does it mean to say that a chemical reaction is a rearrangement of matter?

18. When two atoms collide, what determines whether they will react by transferring electrons or by sharing electrons?

Everyday Chemistry

19. Carbon monoxide gas (CO) binds strongly to the iron atom of hemoglobin in blood. How does this action cause harm to the body?

History Connection

20. Do you think a rocket powered by hydrogen and oxygen causes a great deal of atmospheric pollution? Explain.

Chemistry and Society

21. List two economic factors that contribute to the loss of rain forests.

THINKING CRITICALLY

Applying Concepts

22. Several times in this chapter, it was stated that elements combine to form compounds. Considering what you learned in Section 4.2, what does the word *combine* mean in chemistry?

Interpreting Chemical Structures

23. **MiniLab 2** Which of these compounds could the following "tack model" represent: magnesium chloride, potassium sulfide, calcium oxide, or aluminum bromide?

 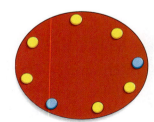

Observing and Inferring

24. MiniLab 1 Brass is an alloy of copper and zinc. Neither metal is magnetic. People who buy brass antiques at auctions, shows, or shops often carry a small magnet with them. What do you think they learn with this magnet?

Making Predictions

25. ChemLab In the electrolysis of zinc iodide, what limits your ability to recover all of the original zinc and iodine?

Observing and Inferring

26. ChemLab Why does the reaction between zinc and iodine speed up when the iodine is dissolved in water?

Interpreting Data

27. ChemLab In the synthesis of zinc iodide, the zinc was added in excess. What does *in excess* mean? What two experimental observations did you make to confirm this?

● CUMULATIVE REVIEW

28. When carbon in charcoal burns in air to form CO_2, is the process endothermic or exothermic? How do you know? (Chapter 1)

29. Describe the particle structure of an atom of potassium. Assume an atomic mass of 39. When a potassium atom forms an ion by reacting with chlorine, how will its structure change? (Chapter 2)

30. How does the electron arrangement of Group 18 elements relate to their chemical properties? (Chapter 3)

● SKILL REVIEW

31. Interpreting a Graph Look at the solubility graph and answer the following questions.

a) Which of the compounds on the graph are ionic compounds? How do you know?

b) How many grams of KBr will dissolve in 100 g of water at 60°C?

c) People think that salt is very soluble in water. Is that thinking accurate? Defend your answer.

● WRITING IN CHEMISTRY

32. Write an article about salt mining in the United States. Find out where the major salt deposits are located and how they came to be there. Describe what happens to the salt after it is removed from the mine. Tell how table salt is made, and find out what else it contains besides sodium chloride.

● PROBLEM SOLVING

33. When hydrogen and chlorine combine chemically, they form hydrogen chloride, HCl. Hydrogen chloride is a gas at room temperature; it becomes a liquid if it is cooled to −85°C. On the basis of this evidence, do you think that hydrogen chloride is ionic or covalent? Explain.

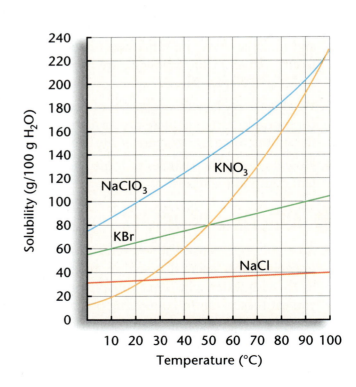

1. How do the properties of the compound magnesium oxide (MgO) compare with the properties of the elements magnesium and oxygen?

   a) The compound has completely different properties from the two elements.
   b) The compound has the same properties as the two elements.
   c) The compound has similar properties as the two elements.
   d) The properties of compounds and elements cannot be compared.

2. Why does oxygen make up such a large percentage of Earth's crust?

   a) Large quantities of oxygen are trapped in soil and rock deep beneath Earth's surface.
   b) Large quantities of water, which is made of oxygen, are trapped beneath Earth's surface.
   c) Low temperatures far beneath Earth's surface freeze atmospheric oxygen into solid form.
   d) Oxygen is highly reactive and bonds with other substances forming solid compounds.

Water Density	
State of Water	**Density (g/mL)**
Solid	0.917
Liquid	1.000
Gaseous (25°C)	0.0008

3. According to the chart above, how does the density differences of water in different states ensure the survival of aquatic life in Canada?

   a) In the winter, ice floats on lakes and ponds, insulating aquatic creatures from cold temperatures.
   b) In the winter, ice sinks to the bottom of lakes and ponds, causing organisms to hibernate.
   c) In the summer, water vapor sinks to the bottom of ponds and lakes, carrying needed oxygen.
   d) In the winter, water vapor sinks to the bottom of ponds and lakes, carrying needed oxygen.

4. Why is the gas hydrogen rarely found in elemental form on earth?

   a) Hydrogen is a rare element.
   b) Hydrogen is highly reactive.
   c) Hydrogen easily forms compounds.
   d) Hydrogen is an unreactive atmospheric gas.

5. What would be the result if nitrogen gas replaced noble gases inside incandescent light bulbs?

   a) The tungsten filament in the bulb would burn with a dimmer light.
   b) The tungsten filament in the bulb would combust.
   c) Chemical reactions would make the light bulb last longer.
   d) Chemical reactions would reduce the lifespan of the bulb.

6. The octet rule says

   a) atoms become less stable with eight electrons in their outer energy level.
   b) atoms become more stable with eight electrons in their outer energy level.
   c) atoms will change their configurations and become a noble gas.
   d) atoms will become more reactive and chemically bond with noble gases.

7. A compound composed of electrically charged atoms is called a(n)

   a) octet compound.
   b) chemically stable compound.
   c) ionic compound.
   d) covalent compound.

Test Taking Tip

Practice Under Test-Like Conditions
Ask your teacher to set a time limit. Then do all of the questions in the time provided without referring to your book. Did you complete the test? Could you have made better use of your time? What topics do you need to review? Show your test to your teacher for an objective assessment of your performance.

CHAPTER

5 Types of Compounds

Chapter Preview

Sections

5.1 Ionic Compounds
MiniLab 5.1 A Chemical Weather
Predictor

5.2 Molecular Substances
MiniLab 5.2 Where's the Calcium?
ChemLab Ionic or Covalent?

Covalent Crops

Farmers often spray chemical compounds on their crops as fertilizers, providing added nutrients, or as pesticides, to reduce crop destruction by insects. Most compounds, including those in living organisms, are covalently bonded. Fertilizers and pesticides used on crops are covalently bonded.

*L*aunch Lab

Elements, Compounds, and Mixtures

What are some differences among elements, compounds, and mixtures?

Safety Precautions

Materials

- plastic freezer bag containing the following labeled items:

copper wire	chalk (calcium carbonate)
small package of salt	piece of granite
sugar water in a vial	pencil

Procedure

1. Construct a data table and use it to record your observations.
2. Obtain a bag of objects. Identify each object and classsify it as an element, compound, heterogeneous mixture, or homogeneous mixture. The elements appear in the periodic table.

Analysis

1. If you know the name of a substance, how can you find out whether or not it is an element?
2. Examine the contents of your refrigerator at home. Classify what you find as elements, compounds, or mixtures.

What I Already Know

Review the following concepts before studying this chapter.
Chapter 2: arrangement of electrons in an atom
Chapter 3: periodicity of electron arrangements in atoms; importance of valence electrons
Chapter 4: formation of ionic compounds, formation of covalent compounds

Reading Chemistry

Carefully read the steps about how to write a chemical formula on page 158. Next, scan the different tables throughout the chapter. Write down the formulas for substances that appear often. Try to determine the elements in their makeup.

Preview this chapter's content and activities at chemistryca.com

Ionic Compounds

Seawater contains many dissolved substances, mostly dissolved sodium chloride. In Chapter 4, you learned that sodium chloride is an ionic compound. Another ionic compound found dissolved in seawater is magnesium chloride. Some common ionic compounds used in everyday life are potassium chloride, a salt substitute used by people avoiding sodium for health reasons; potassium iodide, added to table salt to prevent iodine deficiency; and sodium fluoride, added to many toothpastes to strengthen tooth enamel. You will learn how to use the language of chemistry to name and write the formulas of ionic compounds.

SECTION PREVIEW

Objectives

✓ **Apply** ionic charge to writing formulas for ionic compounds.

✓ **Apply** formulas to name ionic compounds.

✓ **Interpret** the information in a chemical formula.

Review Vocabulary

Ion: an atom or group of combined atoms that has a charge because of the loss or gain of electrons.

New Vocabulary

binary compound
formula unit
oxidation number
polyatomic ion
hydrate
hygroscopic
deliquescent
anhydrous

Formulas and Names of Ionic Compounds

Recall from Chapter 4 that the submicroscopic structure of ionic compounds helps explain why they share certain macroscopic properties such as high melting points, brittleness, and the ability to conduct electricity when molten or when dissolved in water. What is it about the structure of these compounds that gives them properties such as the one shown in **Figure 5.1?** The answer involves the ions of which they are made.

You have learned that ionic compounds are made up of oppositely charged ions held together strongly in well-organized units. Because of their structure, they usually are hard solids at room temperature and are difficult

Figure 5.1

Humpty Dumpty's Downfall
Eggshells are made mostly of ionic compounds such as calcium phosphate, $Ca_3(PO_4)_2$, which makes them brittle. When broken, eggshells shatter into many pieces that can't be put together again.

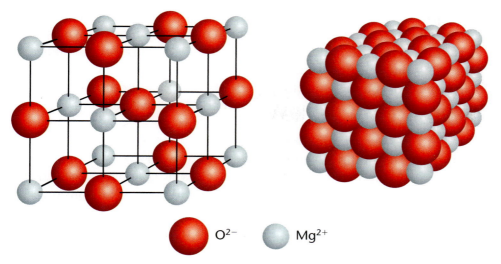

O^{2-} Mg^{2+}

Figure 5.2
Repeating Units
The structure of magnesium oxide is a repeating pattern of magnesium and oxide ions. Each Mg^{2+} ion is surrounded by six O^{2-} ions, which, in turn, are each surrounded by six Mg^{2+} ions. The structure as a whole is neutral. In the diagram on the left, the structure is expanded so you can better see the geometric arrangement.

to melt. Look at the structure of magnesium oxide in **Figure 5.2.** When ionic compounds melt or dissolve in water, their three-dimensional structure breaks apart, and the ions are released from the structure. These charged ions are now free to move and can conduct an electrical current.

Binary Ionic Compounds

Formulas are part of the language that is used to communicate information about substances. As a first step in studying this new language, you will learn how to name and write formulas for ionic compounds.

Sodium chloride (NaCl) contains only sodium and chlorine, and potassium iodide (KI) contains only potassium and iodine. Each is an example of a **binary compound,** which is a compound that contains only two elements. Binary ionic compounds can contain more than one ion of each element, as in CaF$_2$, but they are not composed of three or more *different* elements, as are more complex compounds.

To name a binary ionic compound, first write the name of the positively charged ion, usually a metal, and then add the name of the nonmetal or negatively charged ion, whose name has been modified to end in *-ide.* The compound formed from potassium and chlorine is called potassium chloride. Magnesium combines with oxygen to form a compound called magnesium oxide.

You are already familiar with one formula for an ionic compound—NaCl. Sodium chloride contains sodium ions that have a 1+ charge and chloride ions that have a 1− charge. You have learned that compounds are electrically neutral. This means that the sum of the charges in an ionic compound must always equal zero. Thus, one Na$^+$ balances one Cl$^-$ in sodium chloride. When you write a formula, you add subscripts to the symbols for the ions until the algebraic sum of the ions' charges is zero. The smallest subscript to both ions that results in a total charge of zero is 1. However, no subscript needs to be written because it is understood that only one ion or atom of an element is present if there is no subscript. The formula NaCl indicates that sodium chloride contains sodium and chloride ions, that there is one sodium ion present for every chloride ion in the compound, and that the compound has no overall charge.

WORD ORIGIN

binary:
bini (L) two by two
Binary compounds contain two, and only two, elements.

Figure 5.3

The Formula for Calcium Fluoride

When calcium fluoride forms from Ca and F, the two valence electrons from calcium are transferred to two fluorine atoms, leaving the Ca with a 2+ charge and each F with a 1− charge. ▶

If more than one ion of a given element is present in a compound, the subscript indicates how many ions are present. The mineral known as fluorite is calcium fluoride, which has the formula CaF_2. This formula indicates that there is one calcium ion for every two fluoride ions in the compound. In an ionic compound, a formula represents the smallest ratio of atoms or ions in the compound. In a covalent compound, the smallest unit of the compound is a molecule, so a formula represents a single molecule of a compound. However, ionic compounds do not form molecules. Their structures are repeating patterns of ions, as **Figure 5.3** shows. Should the formula of calcium fluoride be written as CaF_2, Ca_2F_4, or even Ca_3F_6? A properly written formula has the simplest possible ratio of the ions present. This simplest ratio of ions in a compound is called a **formula unit.** Each formula unit of calcium fluoride consists of one calcium ion and two fluoride ions. Each of the three ions has a stable octet configuration of electrons, and the formula unit has no overall charge. Although the sum of the ionic charges in both CaF_2 and Ca_2F_4 is zero, only CaF_2 is a correct formula. One formula unit of calcium fluoride has the formula CaF_2.

Predicting Charge on Ions

You have studied ionic compounds in which sodium becomes a positive ion with a single positive charge and calcium becomes a positive ion with two positive charges. Examine the periodic table to see if there is a way to predict the charge that different elements will have when they become ions. Which elements will lose electrons and which will gain electrons?

The noble gases each have eight electrons in their outer-energy levels. Metals have few outer-level electrons so they tend to lose them and become positive ions. Sodium must lose just one electron, becoming an Na^+ ion. Calcium must lose two electrons, becoming a Ca^{2+} ion. Most nonmetals, on the other hand, have outer-energy levels that contain four to seven electrons, so they tend to gain electrons and become negative ions. Trace the gain and loss of electrons in the example shown in **Figure 5.4.**

$$Ca\!:+\;\overset{\times\times}{\underset{\times\times}{:\!\overset{.}{F}\!\times}}+\;\overset{\times\times}{\underset{\times\times}{:\!\overset{.}{F}\!\times}}\rightarrow [Ca]^{2+}+\;\overset{\times\times}{\underset{\times\times}{:\!\overset{.}{F}\!:}}^{-}+\;\overset{\times\times}{\underset{\times\times}{:\!\overset{.}{F}\!:}}^{-}$$

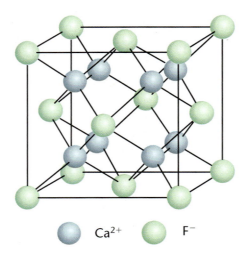

Ca²⁺   F⁻

◀ An ionic bond forms between the positive Ca^{2+} ion and each negative F^- ion. Although there are many Ca^{2+} and F^- ions in a crystal of CaF_2, one formula unit of CaF_2 contains one Ca^{2+} ion and two F^- ions.

Figure 5.4

Lime Is Calcium Oxide
The compound commonly called lime is calcium oxide. It is used to make steel and cement and is added to acidic lakes and soil to neutralize the effects of acidity. ▶

Do you recognize a periodic trend in ionic charges? For the elements in the main groups of the periodic table—Groups 1, 2, and 13 through 18—group numbers can be used to predict these charges. Because all elements in a given group have the same number of electrons in their outer-energy level, they must lose or gain the same number of electrons to achieve a noble-gas electron configuration. Metals always lose electrons and nonmetals always gain electrons when they form ions. The charge on the ion is known as the **oxidation number** of the atom. The oxidation numbers for many elements in the main groups are arranged by group number in **Table 5.1.** Oxidation numbers for elements in Groups 3 through 12, the transition elements, cannot be predicted by group number.

Alumina is the common name for aluminum oxide. It is used to produce aluminum metal, to make sandpaper and other abrasives, and to separate mixtures of chemicals by a technique called chromatography. Aluminum is in Group 13, so it loses its three outer electrons to become an Al^{3+} ion; oxygen is in Group 16 and has six valence electrons, so it gains two electrons to become an O^{2-} ion.

$$\dot{A}\dot{l}\cdot + :\!\ddot{O}\!\overset{\times}{\cdot} \rightarrow Al^{3+} + :\!\ddot{\underset{..}{O}}\!:^{2-}$$

Notice that one of aluminum's three electrons has not been taken up by the oxygen atom. Because all the electrons must be accounted for, more than one oxygen atom must be involved in the reaction. But, oxygen cannot gain only one electron, so a second aluminum atom must be present to contribute a second electron to oxygen. In all, two Al^{3+} ions must combine with three O^{2-} ions to form Al_2O_3. Remember that the charges in the formula for aluminum oxide must add up to zero.

$$2Al^{3+} + 3O^{2-} \rightarrow Al_2O_3$$

$$\dot{A}\dot{l}\cdot + \dot{A}\dot{l}\cdot + :\!\ddot{O}\!\overset{\times}{\cdot} + :\!\ddot{O}\!\overset{\times}{\cdot} + :\!\ddot{O}\!\overset{\times}{\cdot} \rightarrow Al^{3+} + Al^{3+} + :\!\ddot{\underset{..}{O}}\!:^{2-} + :\!\ddot{\underset{..}{O}}\!:^{2-} + :\!\ddot{\underset{..}{O}}\!:^{2-}$$

$$Ca\!:\, \rightarrow Ca^{2+} \quad \ddot{O}\!:\, \rightarrow :\!\ddot{O}\!:^{2-}$$
$$Ca^{2+} + :\!\ddot{O}\!:^{2-} \rightarrow CaO$$

▲ Calcium is a metal that loses two electrons to become a Ca^{2+} ion; oxygen is a nonmetal that must gain two electrons to achieve the stable octet of the noble gas neon, so it becomes an O^{2-} ion. Because a formula unit must be neutral, one Ca^{2+} ion can combine with only one O^{2-} ion. The formula for calcium oxide is CaO.

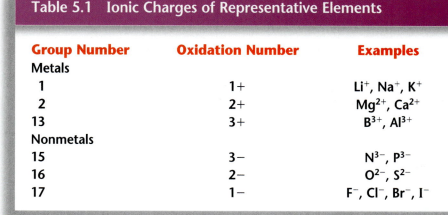

Table 5.1	Ionic Charges of Representative Elements	
Group Number	**Oxidation Number**	**Examples**
Metals		
1	1+	Li^+, Na^+, K^+
2	2+	Mg^{2+}, Ca^{2+}
13	3+	B^{3+}, Al^{3+}
Nonmetals		
15	3−	N^{3-}, P^{3-}
16	2−	O^{2-}, S^{2-}
17	1−	F^-, Cl^-, Br^-, I^-

Write the formula for an ionic compound containing sodium and sulfur.

Analyze • Sodium is in Group 1, so it has an oxidation number of 1+. Sulfur is in Group 16 and has an oxidation number of 2−.

Set Up • Write the symbols for sodium and sulfur ions in formula form, placing the positive ion first.

$$Na^+S^{2-}$$

Solve • The formula as written has one positive charge and two negative charges. To maintain neutrality, one more positive charge is needed to balance the 2− charge. This is accomplished by adding a second sodium ion and is indicated by placing the subscript *2* after the symbol for sodium in the formula. The correct formula is then written as Na_2S.

Check • Check to be sure that you have not changed the charges of the ions and that the overall charge of the formula is zero.

$$2(1+) + (2-) = 0 \quad \text{The formula as written is correct.}$$

PRACTICE PROBLEMS

Supplemental Problems

For more practice with solving problems, see Supplemental Practice Problems, Appendix B.

1. Write the formula for each of the following compounds.

   a) lithium oxide
   b) calcium bromide
   c) sodium oxide
   d) aluminum sulfide

2. Write the formula for the compound formed from each of the following pairs of elements.

   a) barium and oxygen
   b) strontium and iodine
   c) lithium and chlorine
   d) radium and chlorine

Compounds Containing Polyatomic Ions

The ions you have studied thus far have contained only one element. However, some ions contain more than one element. An ion that has two or more different elements is called a **polyatomic ion.** In a polyatomic ion, a group of atoms is covalently bonded together when the atoms share electrons. Although the individual atoms have no charge, the group as a whole has an overall charge. The formulas and names of some common polyatomic ions are shown in **Table 5.2.** Although the charge is shown to the right of the formula, it is the whole ion, rather than just the last atom listed, that is charged. **Figure 5.5** shows models of three common polyatomic ions.

Table 5.2   Common Polyatomic Ions

Name of Ion	Formula	Charge
ammonium	NH_4^+	1+
hydronium	H_3O^+	1+
hydrogen carbonate	HCO_3^-	1−
hydrogen sulfate	HSO_4^-	1−
acetate	$C_2H_3O_2^-$	1−
nitrite	NO_2^-	1−
nitrate	NO_3^-	1−
cyanide	CN^-	1−
hydroxide	OH^-	1−
dihydrogen phosphate	$H_2PO_4^-$	1−
permanganate	MnO_4^-	1−
carbonate	CO_3^{2-}	2−
sulfate	SO_4^{2-}	2−
sulfite	SO_3^{2-}	2−
oxalate	$C_2O_4^{2-}$	2−
monohydrogen phosphate	HPO_4^{2-}	2−
dichromate	$Cr_2O_7^{2-}$	2−
phosphate	PO_4^{3-}	3−

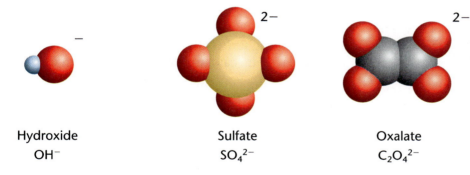

Hydroxide
OH^-

Sulfate
SO_4^{2-}

Oxalate
$C_2O_4^{2-}$

Figure 5.5

Polyatomic Ions
Polyatomic ions, such as hydroxide (left), sulfate (center), and oxalate (right), are composed of more than one atom. Electrons are shared between the atoms within the ion, forming covalent bonds, but the ion as a whole has a charge. Thus, polyatomic ions form ionic bonds with other ions to produce ionic compounds.

Ionic compounds may contain positive metal ions bonded to negative polyatomic ions, such as in NaOH; negative nonmetal ions bonded to positive polyatomic ions, such as in NH_4I; or positive polyatomic ions bonded to negative polyatomic ions, such as in NH_4NO_3. To write the formula for an ionic compound containing one or more polyatomic ions, simply treat the polyatomic ion as if it were a single-element ion by keeping it together as a unit. Remember that the sum of the positive and negative charges must equal zero.

Multiples of a polyatomic ion in a formula can be indicated by placing the entire polyatomic ion, without the charge, in parentheses. Write a subscript outside the parentheses to show the number of polyatomic ions in the compound. Never change the subscripts within the polyatomic ion. To do so would change the composition of the ion. The formula for the compound that contains one magnesium ion and two nitrate ions is $Mg(NO_3)_2$.

Everyday Chemistry

Hard Water

The term *hard water* doesn't describe water's physical state. It describes water in which calcium, magnesium, and hydrogen carbonate ions are dissolved. It is difficult to lather soap in hard water.

One of the compounds in soap that helps produce lather is sodium stearate, $NaC_{18}H_{35}O_2$, which dissolves in water. In hard water, calcium ions react with the stearate ions to form calcium stearate, $Ca(C_{18}H_{35}O_2)_2$. This material is insoluble and forms soap scum. Scum is often seen as a ring around sinks or tubs. If calcium ions are removed from the water by water softeners, soap will lather and no more scum will form.

Hard water can cause problems. When hard water is heated, calcium carbonate is formed from the reaction of calcium ions and hydrogen carbonate ions in the water. Because calcium carbonate is not soluble in water, it forms thick scales inside water heaters and water pipes. These scales often clog pipes and keep the heater from properly heating the water.

Ion Exchangers A common way of softening water—that is, reducing the number of calcium and magnesium ions—is by an ion exchanger. The ion exchanger usually contains a material, called a resin, made up of carbon, hydrogen, and sodium ions. As hard water passes through the resin, a cal-

cium ion or magnesium ion in the water is exchanged for two sodium ions. Thus, the water leaving the ion exchanger has fewer calcium and magnesium ions but many more sodium ions.

Exploring Further

1. **Interpreting** What is the charge of the stearate ion?

2. **Thinking Critically** Why do two sodium ions replace one calcium or one magnesium ion?

3. **Acquiring Information** Why are detergents more effective than soaps in hard water?

Chemistry Online

For more details about how water can be harder or softer, visit the Chemistry Web site at chemistryca.com

---

To name a compound containing a polyatomic ion, follow the same rules as used in naming binary compounds. Name the positive ion first, followed by the negative ion. However, do not change the ending of the negative polyatomic ion name. The name of the compound composed of calcium and the carbonate ion is calcium carbonate. Acids in groundwater can dissolve rocks made of calcium carbonate, such as limestone. Large, underground caverns are formed when the limestone is dissolved away slowly. Stalactites hanging from the ceiling and stalagmites rising from the ground are made when calcium carbonate precipitates from a water solution dripping through cracks in the cavern ceiling.

What is the formula for calcium carbonate? Calcium is in Group 2, so its ion has a 2+ charge. The carbonate ion has a 2− charge, as shown in **Table 5.2.** To form a neutral compound, one Ca^{2+} ion must combine with one CO_3^{2-} ion to give the formula $CaCO_3$.

SAMPLE PROBLEM 2 Writing a Formula Containing a Polyatomic Ion

Write the formula for the compound that contains lithium and carbonate ions.

Analyze
- Lithium is in Group 1, so its ion has a 1+ charge. According to **Table 5.2,** the carbonate ion has a 2− charge, and its structure is CO_3^{2-}.

Set Up
- Write the symbols for lithium carbonate in formula form.

$$Li^+CO_3^{2-}$$

Solve
- Determine the correct ratio of lithium ions to carbonate ions by examining their charges. In this case, the sum of the positive and negative charges does not equal zero. Two lithium ions are needed to balance the carbonate ion. Because you cannot change the charges of the ions, you must add a subscript of *2* to Li^+. The correct formula for lithium carbonate is Li_2CO_3.

Check
- Check to be sure that the overall charge of the formula is zero.

$$2(1+) + (2-) = 0 \quad \text{The formula as written is correct.}$$

SAMPLE PROBLEM 3 Writing a More Complex Formula

Write the formula for the compound that contains aluminum and sulfate ions.

Analyze
- Aluminum is in Group 13 and has an oxidation number of 3+. According to **Table 5.2,** the sulfate ion has a 2− charge.

Set Up
- Write the symbols for aluminum sulfate in formula form.

$$Al^{3+}SO_4^{2-}$$

Solve
- Determine the correct ratio of aluminum ions to sulfate ions by examining their charges. In this case, the sum of the positive and negative charges does not equal zero. To achieve neutrality, you must find the least common multiple of 3 and 2. The least common multiple is 6. How many Al^{3+} ions will be needed to make a charge of 6+, and how many SO_4^{2-} ions will be needed to make a charge of 6−? It will be necessary to have two Al^{3+} ions in the compound to balance three SO_4^{2-} ions. You should add a subscript of *2* to the aluminum ion and a subscript of *3* to the sulfate ion. The entire polyatomic ion must be placed in parentheses to indicate that three sulfate ions are present. Thus, the correct formula for aluminum sulfate is $Al_2(SO_4)_3$.

Check
- Check to be sure that the overall charge of the formula is zero.

$$2(3+) + 3(2-) = 0 \quad \text{The formula as written is correct.}$$

Supplemental Problems

For more practice with solving problems, see Supplemental Practice Problems, Appendix B.

3. Write the formula for the compound made from each of the following ions.

   a) ammonium and sulfite ions
   b) calcium and monohydrogen phosphate ions
   c) ammonium and dichromate ions
   d) barium and nitrate ions

4. Write the formula for each of the following compounds.

   a) sodium phosphate
   b) magnesium hydroxide
   c) ammonium phosphate
   d) potassium dichromate

Compounds of Transition Elements

In Chapter 3, you learned that the elements known as transition elements are located in Groups 3 through 12 in the periodic table. Transition elements form positive ions just as other metals do, but most transition elements can form more than one type of positive ion. In other words, transition elements can have more than one oxidation number. For example, copper can form both Cu^+ and Cu^{2+} ions, and iron can form both Fe^{2+} and Fe^{3+} ions. **Figure 5.6** shows the two compounds that iron forms with the sulfate ion. Zinc and silver are two exceptions to the variability of other transition elements; each forms one type of ion. The zinc ion is Zn^{2+} and the silver ion is Ag^+.

Figure 5.6

Two Compounds of Iron and Sulfate

Iron forms both Fe^{2+} and Fe^{3+} ions, each of which can combine with the sulfate ion. Some people use the older spelling, *sulphate.*

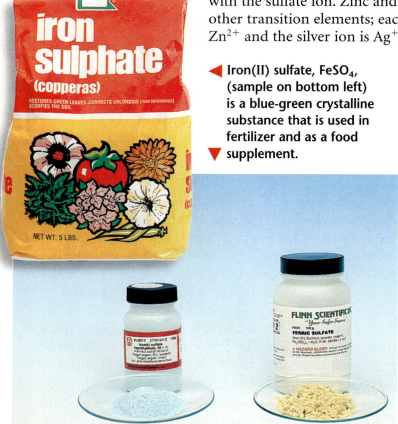

◀ Iron(II) sulfate, $FeSO_4$, (sample on bottom left) is a blue-green crystalline substance that is used in fertilizer and as a food supplement. ▼

Iron(III) sulfate, $Fe_2(SO_4)_3$, (sample on right) is a yellow crystalline substance that is used as a coagulant in water-purification and sewage-treatment plants. After the sewage is coagulated, it is filtered out, as shown here. ▼

China's Porcelain

See how proudly the dragon, an ancient symbol of Chinese culture, prances around the vase pictured here. Proud it should be, because this vase represents one of the greatest achievements of Chinese technology and art—glazed porcelain.

Clay, glaze, and fire By the third to sixth century A.D., the Chinese had invented glazed porcelain. They found that if a clay vessel, such as a bowl, is covered with a transparent glaze and then heated to a high temperature, a translucent ceramic material forms. This material is glazed porcelain. Unlike a fired-clay vessel, which remains slightly porous and opaque, the translucent vessel is sealed by a glasslike covering. By changing the chemical composition of the glaze, Chinese artisans were able to change the quality and color of the glaze. For example, when they added materials that reacted with each other to form tiny gas bubbles in the glaze, the porcelain appeared brighter because the surface of the bubbles reflected light.

Colored glazes One of the most important steps in glazing pottery was the addition of materials to the glaze to produce colored porcelains. These materials were solutions of transition element ions, such as iron, manganese, chromium, cobalt, copper, and titanium. During the firing of the glaze, these metals formed oxides. Because the metal ions in the oxides reflected only certain wavelengths of light, the glazes colored the porcelain. By varying the concentration and charges of the metal ions in the glaze, the Chinese were able to produce subtly colored porcelains. For example, cobalt produces a blue glaze, chromium a pink or green glaze depending on charge, and manganese a purple glaze. These beautiful colors have remained vivid over the course of thousands of years, and the techniques are still being used today.

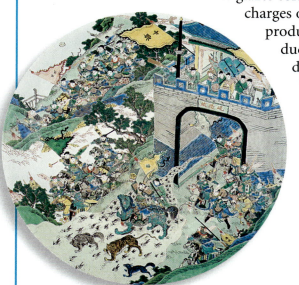

Connecting to Chemistry

1. **Applying** Why are porcelain dishes superior to wooden dishes?

2. **Thinking Critically** What properties of metallic compounds make them useful as colored glazes?

Chemists must have a way to distinguish the names of compounds formed from the different ions of a transition element. They do this by using a Roman numeral to indicate the oxidation number of a transition element ion. This Roman numeral is placed in parentheses after the name of the element. No additional naming system is needed for zinc and silver compounds because their formulas are not ambiguous. **Table 5.3** shows the naming of the two different ionic compounds formed when chloride ions combine with each of the two copper ions.

Table 5.3 Compounds of Copper and Chlorine			
Copper Ion	**Chloride Ion(s)**	**Formula**	**Name**
Cu^+	Cl^-	$CuCl$	copper(I) chloride
Cu^{2+}	$2Cl^-$	$CuCl_2$	copper(II) chloride

Table 5.4 shows the chemical names of some transition element ions. When you do Practice Problems 5 and 6, you will become familiar with these names. Note in the photos accompanying the table that the different ions of a transition element often form compounds of different colors. For example, CrO is black, Cr_2O_3 is green, and CrO_3 is red. Determine the oxidation number for chromium in each of these compounds.

Table 5.4 Names of Common Ions of Selected Transition Elements		
Element	**Ion**	**Chemical Name**
Chromium	Cr^{2+}	chromium(II)
	Cr^{3+}	chromium(III)
	Cr^{6+}	chromium(VI)
Cobalt	Co^{2+}	cobalt(II)
	Co^{3+}	cobalt(III)
Copper	Cu^+	copper(I)
	Cu^{2+}	copper(II)
Gold	Au^+	gold(I)
	Au^{3+}	gold(III)
Iron	Fe^{2+}	iron(II)
	Fe^{3+}	iron(III)
Manganese	Mn^{2+}	manganese(II)
	Mn^{3+}	manganese(III)
	Mn^{7+}	manganese(VII)
Mercury	Hg^+	mercury(I)
	Hg^{2+}	mercury(II)
Nickel	Ni^{2+}	nickel(II)
	Ni^{3+}	nickel(III)
	Ni^{4+}	nickel(IV)

CrO

Cr_2O_3

CrO_3

Suppose you wanted to write the formula for a compound containing a transition element. Look back at Sample Problem 1, where you learned to write the formula for a compound containing sodium and sulfur. How would you write the formula if it were iron(II) rather than sodium that combined with sulfur? Iron(II) has an oxidation number of 2+, and its ion can be written as Fe^{2+}. You know that the sulfide ion has a charge of 2− and can be written as S^{2-}. The charges balance in this case, and the formula for iron(II) sulfide is written as FeS, **Figure 5.7**.

You can write the formula for iron(III) sulfide in the same way. Just follow the steps in Sample Problem 3. The correct formula for iron(III) sulfide is Fe_2S_3. Note that the Roman numeral refers to the oxidation number of the iron and not to how many ions are in the formula.

How can you name a compound of a transition element if you are given the formula? Determining the charge of the transition element ion gives the clue needed to name the compound. In the formula $Cr(NO_3)_3$, you must determine the charge of the chromium ion in order to name the compound. Look first at the negative ion. Knowing that the nitrate ion has a charge of 1− and that there are three nitrate ions with a total charge of 3−, you can see that the chromium ion must have a charge of 3+ to maintain neutrality. Thus, this compound is named chromium(III) nitrate.

Try at Home Lab

See page 865 in Appendix F for **Iron Ink**

Figure 5.7
Fool's Gold
Iron disulfide (FeS_2), is commonly called fools gold. In this ionic compound, sulfur exists in the unusual form S_2^{2-}, and iron has an oxidation number of 2+.

PRACTICE PROBLEMS

Supplemental Problems

For more practice with solving problems, see Supplemental Practice Problems, Appendix B.

5. Write the formula for the compound made from each of the following pairs of ions.

   a) copper(I) and sulfite
   b) tin(IV) and fluoride
   c) gold(III) and cyanide
   d) lead(II) and sulfide

6. Write the names of the following compounds.

   a) $Pb(NO_3)_2$
   b) Mn_2O_3
   c) $Ni(C_2H_3O_2)_2$
   d) HgF_2

Hydrates

Many ionic compounds are prepared by crystallization from a water solution, and water molecules become a part of the crystal. A compound in which there is a specific ratio of water to ionic compound is called a **hydrate.** In a hydrate, the water molecules are chemically bonded to the ionic compound.

miniLAB 1

A Chemical Weather Predictor

Adding water to an anhydrous compound to form a hydrate often changes the physical properties—such as color—of the compound. Cobalt(II) chloride is such a compound. If you find that the color of the compound changes in accordance with the weather, perhaps cobalt(II) chloride can serve as a weather predictor.

Procedure

1. Place 5 mL of 95 percent ethanol in a small beaker.

2. Use a spatula to add a small amount of cobalt(II) chloride to the beaker. Stir until the compound dissolves.

3. Dip a cotton swab into the pink solution and use it to write the chemical formula of cobalt(II) chloride on a piece of white paper.

4. Dry the paper by holding it over a hot plate set on low or by putting it in a sunny location. What color is the formula now?

5. Keep your weather predictor in a convenient location, and check its color each morning and afternoon. Keep a three-week log of the time, the current weather, and the color of the treated paper.

Analysis

1. What is the formula of cobalt(II) chloride?

2. The hydrate of cobalt(II) chloride has six water molecules bonded to it. What is its formula?

3. From your observations, are you able to conclude that the cobalt(II) chloride test paper is a reliable weather predictor? Justify your answer.

Does your chemistry instructor often remind students to make sure that the lids on jars of chemicals are tightly closed? There is a good reason for sealing the jars tightly; some ionic compounds can easily become hydrates by absorbing water molecules from water vapor in the air. These compounds are called **hygroscopic** substances, and one example is sodium carbonate (Na_2CO_3). Substances that are so hygroscopic that they take up enough water from the air to dissolve completely and form a liquid solution are called **deliquescent**. A deliquescent substance is shown in **Figure 5.8.**

Figure 5.8

Deliquescent Substances

Sodium hydroxide (left) is an example of a deliquescent substance because it has a strong attraction for water molecules. Sodium hydroxide will absorb water molecules from the surrounding air and begin to dissolve (right). Eventually, it will absorb enough water to dissolve completely.

Cement

People have been using cementing materials for thousands of years. The stones in the Egyptian pyramids are held together by a mixture of sand and the mineral compound gypsum, which is calcium sulfate dihydrate. When this dihydrate is heated, water evaporates, forming a compound with one water molecule per two calcium sulfate formula units. Today, we know this binding material as plaster of paris.

1. Cement is made from a mixture of limestone and clay. The most important minerals in clay are the combinations of aluminum, oxygen, and silicon known as aluminum silicates.

2. Before this limestone-clay mixture can be used, it must be heated. Heating drives off carbon dioxide and forms new ionic compounds. This new mixture of calcium silicates, calcium aluminates, and calcium aluminum ferrates forms in clumps called clinker.

3. The clinker is ground and mixed with small amounts of calcium sulfate. This mixture is called portland cement.

4. Cement can be used alone to form a smooth, hard surface for roads or buildings, or it can be combined with sand and gravel to form a rougher material called concrete.

5. When concrete is mixed with water, silicate compounds hydrate and form gelatinous materials called gels.

6. The hardening process takes several days. During this time, some water is removed from the gels that formed around the sand and gravel, and calcium hydroxide absorbs carbon dioxide from the air to re-form calcium carbonates. Fibers that form from the cement materials interlock and strengthen the concrete.

Thinking Critically

1. What is the formula for the calcium sulfate dihydrate that makes up gypsum?

2. Tricalcium aluminate also becomes hydrated during solidification of cement to form $Ca_3Al_2O_6 \cdot 6H_2O$. What is the name of this hydrate?

Figure 5.9
Desiccants
Boxes of electronic equipment such as cameras and CD players usually contain small packets of a desiccant. The desiccant absorbs water vapor from the air, protecting the delicate metal parts against corrosion and preventing condensation of water vapor in the wiring of the equipment. Even a tiny quantity of water on a circuit board can create a short circuit. Desiccants are especially useful for packaging electronic equipment that is to be shipped overseas because ocean air in the holds of ships contains so much moisture.

Many of these compounds become hydrates by absorbing water from the air. As shown in **Figure 5.9,** compounds that form hydrates often are used as drying agents, or desiccants, because they absorb so much water from the air when they become hydrated.

To write the formula for a hydrate, write the formula for the compound and then place a dot followed by the number of water molecules per formula unit of compound. The dot in the formula represents a ratio of compound formula units to water molecules. For example, $CaSO_4 \cdot 2H_2O$ is the formula for a hydrate of calcium sulfate that contains two molecules of water for each formula unit of calcium sulfate. This hydrate is used to make portland cement and plaster of paris. To name hydrates, follow the regular name for the compound with the word *hydrate,* to which a prefix has been added to indicate the number of water molecules present. Use **Table 5.5** to find the correct prefix to use. The name of the compound with the formula $CaSO_4 \cdot 2H_2O$ is calcium sulfate dihydrate.

Heating hydrates can drive off the water. This results in the formation of an **anhydrous** compound—one in which all of the water has been removed. In some cases, an anhydrous compound may have a different color from that of its hydrate, as shown in **Figure 5.10.**

Table 5.5	Prefixes to Use in Naming Hydrates	
Molecules of Water		**Prefix**
1		mono-
2		di-
3		tri-
4		tetra-
5		penta-
6		hexa-
7		hepta-
8		octa-
9		nona-
10		deca-

Interpreting Formulas

You have learned how to write a formula to represent a formula unit of an ionic compound. Sometimes, it may be necessary to represent more than one formula unit of a compound. To do this, place a coefficient before the formula. Two formula units of NaCl are represented by 2NaCl, three formula units by 3NaCl, and so on.

A formula summarizes how many atoms of each element are present. Each formula unit of sodium chloride contains one sodium ion and one chloride ion. How many oxygen atoms are present in $3HNO_3$? Each formula unit contains three oxygen atoms. Because there are three formula units, a total of nine atoms of oxygen are present. As another example, consider how many atoms of hydrogen are in one formula unit of ammonium sulfate. The formula for ammonium sulfate is $(NH_4)_2SO_4$. Each ammonium ion contains four atoms of hydrogen. Because two ammonium ions are present, there are eight atoms of hydrogen in a formula unit of ammonium sulfate. How many hydrogen atoms are in $3(NH_4)_2SO_4$? To find out, simply multiply the eight hydrogen atoms in one formula unit by three formula units; 24 hydrogen atoms are present.

Figure 5.10

Forming an Anhydrous Compound
When blue copper(II) sulfate pentahydrate ($CuSO_4 \cdot 5H_2O$) (left) is heated, the water is driven off (center). The anhydrous compound, $CuSO_4$ (right), is white. Hydrated copper sulfate is used as a fungicide in water reservoirs.

SECTION REVIEW

Understanding Concepts

1. Explain why ionic compounds cannot conduct electricity when they are in the solid state.

2. Write formulas for each of the following ionic compounds.

   a) manganese(II) carbonate
   b) barium iodide dihydrate
   c) aluminum oxide
   d) magnesium sulfite
   e) ammonium nitrate
   f) sodium cyanide

3. Name the ionic compound represented by each formula.

   a) Na_2SO_4
   b) CaF_2
   c) $MgBr_2 \cdot 6H_2O$
   d) Na_2CO_3
   e) $KMnO_4$
   f) $Ni(OH)_2$
   g) $NaC_2H_3O_2$

Thinking Critically

4. **Interpreting Chemical Formulas** What information does the formula $3Ni(HCO_3)_2$ tell you about the number of atoms of each element that are present?

Applying Chemistry

5. **Toothpaste Ingredients** Examine the ingredient label on a tube of toothpaste. Write formulas for as many of the chemical names listed as you can. List whether each ingredient is an ionic or a covalent compound.

Molecular Substances

How many compounds can you name that are liquids or gases at normal room temperature? Water, carbon dioxide, and ammonia are just a few examples. Because most ionic compounds are solids at room temperature, the odds are pretty good that any compounds you thought of are members of the other major class of compounds described in Chapter 4—the covalent compounds. However, not all of them are liquids or gases. Some covalent compounds are solids at room temperature, for example sugar, mothballs, silica (sand), and the fats that make up butter and margarine. Most of the time, it is difficult to tell whether a solid compound is ionic or covalent by visual examination alone. Compare the crystals of sugar and salt shown here.

Properties of Molecular Substances

You know that ionic compounds share many properties. The properties of a **molecular substance**—a substance that has atoms held together by covalent rather than ionic bonds—are more variable than the properties of ionic compounds. Some molecular substances, such as polyethylene plastic and the fats in butter, are soft; rubber is elastic; and diamond and quartz are hard.

Although molecular substances have varied properties, some generalities can be made to distinguish them from ionic compounds. Molecular substances usually have lower melting points, and most are not as hard as ionic compounds, **Figure 5.11.** In addition, most molecular substances are less soluble in water than ionic compounds and are not electrolytes. The properties of most ionic and molecular substances are different enough that their differences can be used to classify and separate them from one

Figure 5.11

Crayons—Covalent or Ionic?
Crayons are made of covalent compounds. They are soft and are insoluble in water. If you have ever left crayons out in the sun, you know that they also have a low melting point.

Where's the calcium?

Calcium is an important part of the structure of bones and eggshells. If a bone is soaked in vinegar for several days, the structure of the bone will change. Vinegar contains acetic acid, which reacts with the calcium compounds in the bone to form calcium acetate.

Procedure

1. Pick most of the meat from a small, uncooked chicken bone.

2. Place the bone in a beaker, cover it with vinegar, and cover the beaker with a watch glass.

3. Label the beaker with your name and leave it for two days in the area indicated by your teacher.

4. Use forceps to remove the bone from the beaker, and blot it on a paper towel. Examine the bone and observe how it has changed.

5. Replace the bone in the vinegar and let it soak for two more days. Repeat step 4.

6. Straighten a paper clip. Holding the clip with forceps, dip it into the vinegar solution, then hold it in the blue flame of a Bunsen burner. If calcium is present in the vinegar, it will give an orange-red flame test.

Analysis

1. Describe the change in the properties of the bone after two days and after four days.

2. If the flame test verified the presence of calcium ions in the vinegar, what was the probable source of the calcium?

3. What do you conclude regarding the effect of ionic calcium compounds on the properties of bone? Do their properties in bone seem to correlate with those of typical ionic compounds?

4. How do the properties of the bone after soaking reflect the presence mainly of covalent compounds in the bone?

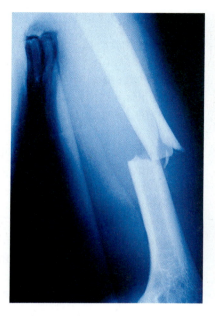

Calcium compounds make bones hard but brittle. Applying stress to brittle materials can cause them to break.

---

another. The separation of water from salt by distillation is one example that makes use of these property differences. **Distillation** is the method of separating substances in a mixture by evaporation of a liquid and subsequent condensation of its vapor. As you learned, solar stills make use of this method. A simple lab-distillation apparatus is shown in **Figure 5.12.**

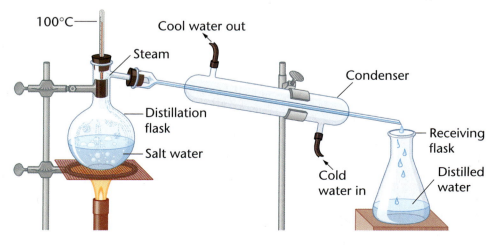

100°C
Cool water out
Steam
Condenser
Distillation flask
Salt water
Receiving flask
Cold water in
Distilled water

Figure 5.12

Distillation in the Lab
A soluble ionic compound such as NaCl can be separated from water using a distillation apparatus like this one. As the salt water is boiled in the distillation flask, the water turns to steam and the salt is left behind. The steam passes through a water-cooled condenser, where it condenses into pure, distilled water. The distilled water is collected in the receiving flask.

ChemLab

SMALL SCALE

Ionic or Covalent?

Compounds can be classified by the types of bonds that hold their atoms together. Ions are held together by ionic bonds in ionic compounds; atoms are held together by covalent bonds in molecular compounds.

You cannot tell whether a compound is ionic or molecular simply by looking at a sample of it because both types of compounds can look similar. However, simple tests can be done to classify compounds by type because each type has a set of characteristic properties shared by most members. Ionic compounds are usually hard, brittle, water-soluble, have high melting points, and can conduct electricity when dissolved in water. Molecular compounds can be soft, hard, or flexible; are usually less water-soluble; have lower melting points; and cannot conduct electricity when dissolved in water.

Problem

How can you identify ionic and molecular compounds by their properties?

Objectives

- **Examine** the properties of several common substances.
- **Interpret** the property data to classify each substance as ionic or molecular.

PREPARATION

Materials

glass microscope slide
grease pencil or crayon
hot plate
spatula
4 small beakers (50- or 100-mL)
stirring rod
balance
conductivity tester

graduated cylinder, small
thermometer (must read up to 150°C)
1- to 2-g samples of any 4 of the following:
   salt substitute (KCl), fructose, aspirin, paraffin,
   urea, table salt, table sugar, Epsom salt

Safety Precautions

Use care when handling hot objects.

PROCEDURE

1. Use a grease pencil or crayon to draw lines dividing a glass slide into four parts. Label the parts A, B, C, and D.

2. Make a data table similar to the one shown in Data and Observations.

3. Use a spatula to place about one-tenth (about 0.1 to 0.2 g) of the first of your four substances on section A of the slide.

4. Repeat step 3 with your other three substances on sections B, C, and D. Be sure to use a clean spatula for each sample. Record in your data table which substance was put on each section.

5. Place the slide on a hot plate. Turn the heat setting to medium and begin to heat the slide.

6. Gently hold a thermometer so that the bulb just rests on the slide. Be careful not to disturb your compounds.

Melting point determination

7. Continue heating until the temperature reaches 135°C. Observe each section on the slide and record which substances have melted. Turn off the hot plate.

8. Label four beakers with the names of your four substances.

9. Weigh equal amounts of the four substances (1-2 g of each), and place the weighed samples in their labeled beakers.

10. Add 10 mL of distilled water to each beaker.

11. Stir each substance, using a clean stirring rod for each sample. Note on your table whether or not the sample dissolved completely.

12. Test each substance for the presence of electrolytes by using a conductivity tester. Record whether or not each acts as a conductor.

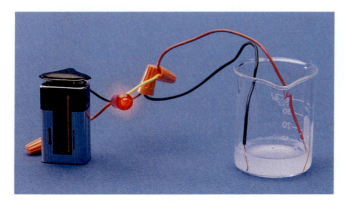

ANALYZE AND CONCLUDE

1. **Interpreting Observations** What happened to the bonds between the molecules when a substance melted?

2. **Comparing and Contrasting** Did all compounds melt at the same temperature?

3. **Classifying** Complete your data table by classifying each of the substances you tested as ionic or molecular compounds based on your observations.

APPLY AND ASSESS

1. What are the differences in properties between ionic and molecular compounds?

2. How did the melting points of the ionic compounds and the molecular compounds compare? What factors affect melting point?

3. The solutions of some molecular compounds are good conductors of electricity. Explain how this can be true when ions are required to conduct electricity.

4. Consider a mixture of sand, salt, and water. How can you make use of the differences in properties of these materials to separate them?

DATA AND OBSERVATIONS

Substance	Did it melt?	Did it dissolve in water?	Did the solution conduct electricity?	Classification
A				
B				
C				
D				

How does the submicroscopic structure of molecular substances contribute to their macroscopic properties? Because there are no ions, strong networks held together by the attractions of opposite charges do not form. The interparticle forces between molecules are often weak and easy to break. These weak forces explain the softness and low melting points of most molecular substances. Most molecular substances are not electrolytes because they do not easily form ions.

Molecular Elements

Molecules vary greatly in size. They can contain from just two to thousands or millions of atoms, as **Figure 5.13** shows. Most elements usually occur naturally in a combined form with another element; that is, they occur as compounds. However, in some cases, two or more atoms of the same element can bond together to form a molecule. A molecule that forms when atoms of the same element bond together is called a **molecular element.** Note that molecular elements are not compounds—they contain atoms of only one element. Why do atoms of these elements bond so readily to identical atoms? When they bond together, each atom achieves the stability of a noble-gas electron configuration.

Figure 5.13

Size of Molecules

Molecular substances can be as simple as two iodine atoms linked together as I_2 (bottom) or as complex as this protein, cytochrome *c* (top), which contains many thousands of atoms of carbon, hydrogen, oxygen, nitrogen, and sulfur linked together by covalent bonds. ▼

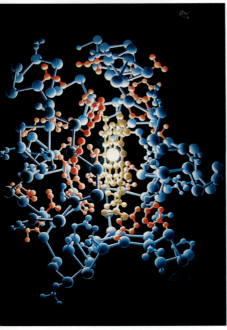

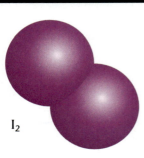

I_2

▲ Cytochrome *c* is found in all living cells that derive energy by breaking down food molecules in the presence of oxygen, and it is found in especially large quantities in hard-working muscle tissue.

Diatomic Elements

Seven nonmetal elements are found naturally as molecular elements of two identical atoms. The elements whose natural state is diatomic are hydrogen, nitrogen, oxygen, fluorine, chlorine, bromine, and iodine. Their formulas can be written as H_2, N_2, O_2, F_2, Cl_2, Br_2, and I_2, respectively. These molecules are referred to as diatomic elements. All except bromine and iodine are gases at room temperature; Br_2 is a liquid, and I_2 is a solid.

What can you learn by examining the structures of the diatomic elements? Electron dot diagrams offer clues. As an example, the chlorine atom has seven valence electrons and needs one more to achieve the configuration of the noble gas argon. If two chlorine atoms combine, they share a single pair of electrons, and each atom attains a stable octet configuration.

$$:\ddot{\text{C}}\text{l}\cdot \; + \; \cdot\ddot{\text{C}}\text{l}: \;\rightarrow\; :\ddot{\text{C}}\text{l}:\ddot{\text{C}}\text{l}:$$

Hydrogen, fluorine, bromine, and iodine molecules also are formed by the sharing of a single pair of electrons. Two oxygen atoms share two pairs of electrons to form O_2, and two nitrogen atoms share three pairs of electrons to form N_2.

$$:\ddot{\text{O}}::\ddot{\text{O}}: \qquad :\text{N}::\text{N}:$$

WORD ORIGIN

allotrope:
allos (GK) other
tropos (GK) way, manner
Allotropic elements have other ways of being arranged.

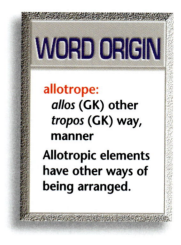

Allotropes

Although the diatomic form of oxygen, O_2, is most common in our atmosphere, oxygen also exists as O_3—ozone. The structure of ozone is different from that of diatomic oxygen. It consists of three atoms of oxygen rather than the two atoms in diatomic oxygen.

$$\dot{\text{O}}::\ddot{\text{O}}:\dot{\text{O}}:$$

Molecules of a single element that differ in crystalline or molecular structure are called **allotropes.** The properties of allotropes are usually different even though they contain the same element. This is because structure can be more important than composition in determining properties of molecules.

Oxygen and ozone are allotropes. The oxygen we breathe is O_2; it is found in the atmosphere. Ozone also occurs naturally and is formed from diatomic oxygen by lightning or ultraviolet light. You may have smelled the sharp odor of ozone during an electrical storm as it formed in the atmosphere by the action of lightning. Small amounts of ozone also are formed in TV sets or computer monitors when an electrical discharge passes through the oxygen in the air. Have you ever smelled it when sitting close to the screen? Because ozone is harmful to living things, it is advisable that you not sit close to your TV set or computer.

Although ozone formed near the surface of Earth is an undesirable component of smog, ozone also has many uses, such as that shown in **Figure 5.14** and in purifying water. The layer of ozone found high in Earth's atmosphere is helpful because it shields living things from harmful ultraviolet radiation from the sun.

Figure 5.14

Uses of Ozone
Ozone produced in machines like this is used to treat clothing, carpeting, and other materials that have been damaged by smoke and soot from a fire. Hotels also use ozone machines to remove the odor of cigarette smoke from rooms. The reactive ozone oxidizes large, smelly compounds in the smoke and soot into smaller, odorless compounds.

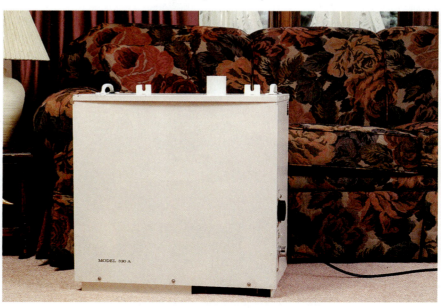

Carbon Allotropes: From Soot to Diamonds

Carbon is the most versatile element in forming allotropes. Organized or unorganized, atoms of carbon can take on an incredible number of arrangements, each different from the other and each forming a different allotrope. With all their diversity, these substances have one thing in common: they are made up solely of covalently bonded carbon atoms.

Graphite

Graphite

The most familiar form of carbon is graphite. Mixed with a little clay and formed into a rod, it becomes the lead in a pencil. Look at the structure of graphite. As you can see, the carbon atoms are linked to each other in a continuous sheet of hexagons (six-sided figures). Note that each carbon atom connects three different hexagons. It's clear that the structure of graphite is well organized. The arrays of hexagons are arranged in layers that are loosely held together. The looseness between layers is why graphite is useful in pencil lead. As you write, the surface of the paper pulls off the loosely held layers of carbon atoms.

Carbon Blacks

Carbon blacks make up most of the soot that collects in chimneys and becomes a fire hazard. They are formed by the incomplete burning of hydrocarbon compounds, as shown here. Each microscopic chunk of a carbon black is made of millions of jumbled chunks of layered carbon atoms. However, the layers lack the organization of graphite, giving carbon black its haphazard structure. Carbon blacks are used in the production of printing inks and rubber products.

Carbon black

Diamond

Another allotrope of elemental carbon is diamond. Besides being blinded by the brilliance of a cut diamond, you should know that diamond is the hardest natural substance. It's often used on the tips of cutting tools and drills. Can the structure of diamond explain its hardness? Look at the model of diamond. Every carbon atom is attached to four other carbon atoms which, in turn, are each attached to four more carbon atoms. Diamond is one of the most organized of all substances. In fact, every diamond is one huge molecule of carbon atoms. This organization of covalently bonded carbons throughout diamond accounts for its hardness. If you tried to write with a diamond, you'd only tear your paper because layers of carbon atoms do not slip off as they do in graphite. The organization of carbon atoms into diamond occurs under extreme pressure and temperature, often at depths of 200 km and over a long period of time. Diamonds range in age from 600 million to 3 billion years old.

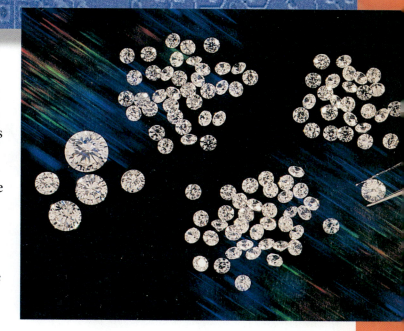

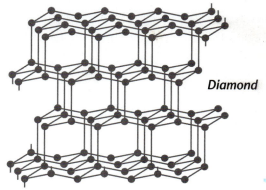

Diamond

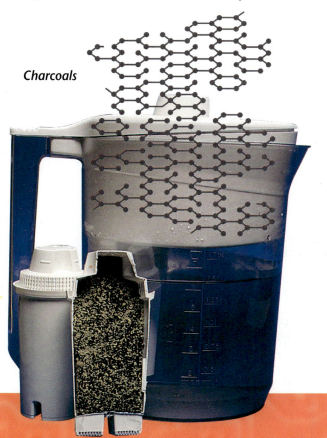

Charcoals

Charcoals

Charcoals—the kind you draw with or cook with—are another type of poorly organized carbon molecules. Charcoals are produced from the burning of organic matter. If you look closely at a chunk of charcoal, you can see that it's extremely porous. All these pores, pock marks, and holes give charcoal a large surface area. Some charcoal, called activated charcoal, has as much as 1000 m^2 of surface area per gram. This property makes activated charcoal useful in filtering water. Molecules, atoms, and ions responsible for unwanted odors and tastes in water are attracted to and held by the surface of the activated charcoal as water passes through it in this water-filtering pitcher.

Fullerenes

This is a model of buckminsterfullerene, C_{60}, which was named after the engineer and architect Buckminster Fuller, who invented the geodesic dome shown here. Both the dome and the molecule are unusually stable. The molecule is one of a group of highly organized allotropes of carbon called fullerenes. Buckminsterfullerene was discovered in soot in 1985, and its soccer-ball shape was confirmed in 1991. Since then, other naturally occurring and artificially produced fullerenes have been identified. Fullerenes have even-numbered molecular formulas such as C_{70} and C_{78}. The molecules of some fullerenes are hollow spheres, whereas molecules of others are hollow tubes. The cagelike structures of fullerenes are very flexible. After crashing into steel plates at speeds of 7000 m/s (about 16 000 miles/hour), C_{60} molecules rebound with their original shapes intact.

Fullerenes

Linear Acetylenic Carbon

This threadlike allotrope of carbon is organized into long spirals of bonded carbon atoms. Each spiral contains 300 to 500 carbon atoms. It's produced by using a laser to zap a graphite rod in a glass container filled with argon gas. The allotrope splatters on the glass walls and is then removed. Because they conduct electricity, these carbon filaments may have uses in microelectronics. Some linear acetylenic carbons may eventually form fullerenes, whereas others form soot.

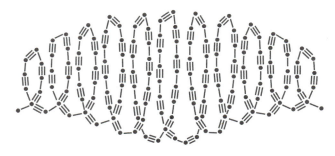

Linear acetylenic carbon

DISCUSSING THE TECHNOLOGY

1. **Applying** From their structures, predict how buckminsterfullerene, diamond, and graphite rank in increasing order of mass density. Explain.

2. **Thinking Critically** How would you describe a molecule of buckminsterfullerene? A molecule of linear acetylenic carbon? A molecule of diamond?

3. **Acquiring Information** What might be some possible uses of fullerenes and linear acetylenic carbons?

Phosphorus has three common allotropes: white, red, and black. All are formed from P_4 molecules that are joined in different ways, giving each allotrope a unique structure and properties, as shown in **Figure 5.15.**

Carbon has several important allotropes with different properties. Diamond is a crystal in which the atoms of carbon are held rigidly in place in a three-dimensional network. In graphite, the carbon atoms are held together closely in flat layers that can slide over each other. This property makes graphite soft and greasy-feeling and useful as a dry lubricant in locks.

Another set of carbon allotropes, the fullerenes, consist of carbon atom clusters. These molecules are unusually stable and are an exciting area of research for chemists because of their potential use as superconductors.

Figure 5.15

Phosphorus Allotropes
The white, red, and black allotropes of phosphorus have different properties. Note the differences in their structures.

Black phosphorus is a semiconductor, whereas the other two forms are not. ▼

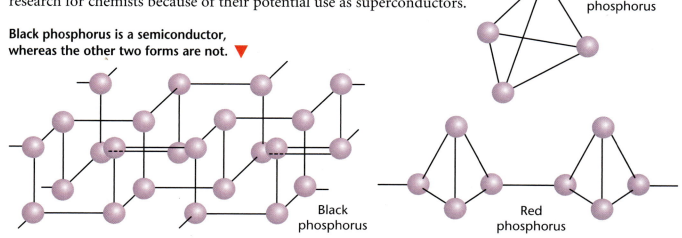

White phosphorus

Black phosphorus

Red phosphorus

◀ White phosphorus will ignite spontaneously in air, whereas the red form won't ignite unless it contacts a flame. For these reasons, white phosphorus (left) must be stored under water, and red phosphorus (right) is used in the strike pad of safety matches.

Formulas and Names of Molecular Compounds

Molecular compounds make up a large group; millions of molecular compounds are already known, and scientists are likely to discover or create many others. How can you possibly begin to study so many compounds? Before you can study their structures and properties and learn how these properties determine their usefulness, you should be able to name the compounds and write their formulas. Fortunately, chemists have devised a naming system for molecular compounds that is based on a much smaller number of rules than there are compounds.

Naming Binary Inorganic Compounds

Substances are either organic or inorganic. Compounds that contain carbon, with a few exceptions, are classified as **organic compounds.** Compounds that do not contain carbon are called **inorganic compounds.** How are inorganic compounds held together? If inorganic compounds contain only two nonmetal elements, they are bonded covalently and are referred to as molecular binary compounds.

To name these compounds, write out the name of the first nonmetal and follow it by the name of the second nonmetal with its ending changed to *-ide.* How do you know which element to write first? You write first the element that is farther to the left in the periodic table, with the exceptions of a few compounds that contain hydrogen. If both elements are in the same group, name first the element that is closer to the bottom of the periodic table. For example, sulfur dioxide is a compound containing sulfur and oxygen. The sulfur is named first because it is closer to the bottom of the periodic table than oxygen is.

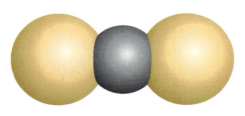

When naming a binary ionic compound, this is the last step. However, because nonmetal atoms can share different numbers of electron pairs, several different compounds can be formed from the same two nonmetal elements. Thus, an additional step is necessary to give an unambiguous name to a molecule. To name the compound correctly, add a prefix to the name of each element to indicate how many atoms of each element are present in the compound. The same prefixes that were used to indicate the number of water molecules in hydrates are used here. For example, CS_2 is named carbon disulfide, **Figure 5.16.** Refer to **Table 5.5** to review these prefixes.

A few other rules are helpful when naming molecular compounds. If only one atom of the *first* element is listed, the prefix *mono* is usually omitted. Also, if the vowel combinations *o-o* or *a-o* appear next to each other in the name, the first of the pair is omitted to simplify pronunciation. Thus, mononitrogen monooxide, NO, becomes nitrogen monoxide.

Now you are ready to practice naming molecular compounds. Several different molecules can be formed when different numbers of nitrogen and oxygen atoms combine. Look at their formulas in the first column of **Table 5.6,** and try to name them without looking at the names listed in the second column. The brown gas pictured is NO_2.

Figure 5.16

Carbon Disulfide

The compound represented by the formula CS_2 is named carbon disulfide because two sulfur atoms are bonded to one carbon.

Table 5.6   Formulas and Names of Some Molecular Compounds

Formula	Name
NO	nitrogen monoxide
NO_2	nitrogen dioxide
N_2O	dinitrogen monoxide
N_2O_5	dinitrogen pentoxide

Consider the two compounds that contain carbon and oxygen. The carbon contained in wood is converted to carbon dioxide when wood burns completely. The formula for this product is CO_2. If the carbon in wood burns incompletely, the highly toxic gas carbon monoxide is formed. What is the formula for carbon monoxide?

To write the formula of a molecular compound for which you are given the name, first write the symbols of each element in the order given in the name. Then add the appropriate subscript after each element that has two or more atoms present. Remember that the prefixes in the name tell how many atoms of each element are present. For example, the compound sulfur hexafluoride contains the elements sulfur and fluorine. Because the word *sulfur* has no prefix, it is understood that there is only one sulfur atom; thus, the symbol S does not require a subscript. The prefix *hexa* tells you that six fluorine atoms are in the compound, so the subscript 6 must be added to the F. The formula for sulfur hexafluoride is SF_6. Follow the rules for writing a formula for a molecular compound as you examine the formula shown in **Figure 5.17**.

N_2O_3

dinitrogen trioxide

Figure 5.17
Formulas of Binary Molecules
The formula for dinitrogen trioxide is written N_2O_3. Analyze the name of this compound to determine how its formula is written.

PRACTICE PROBLEMS

Supplemental Problems

For more practice with solving problems, see Supplemental Practice Problems, Appendix B.

7. Name the following molecular compounds.

   a) S_2Cl_2
   b) CS_2
   c) SO_3
   d) P_4O_{10}

8. Write the formulas for the following molecular compounds.

   a) carbon tetrachloride
   b) iodine heptafluoride
   c) dinitrogen monoxide
   d) sulfur dioxide

Common Names

A few inorganic molecular compounds have common names that all scientists use in place of formal names. Two of these compounds are water and ammonia. The chemical name for water is dihydrogen monoxide because each molecule contains two hydrogen atoms and one oxygen atom. If you wanted to get a glass of water at a restaurant, would you ask for dihydrogen monoxide? Probably not, at least not if you were really thirsty. Most people would not understand you because you used a name that even chemists never use for water. Although the formal names of both ionic and molecular compounds are simple to write once you learn the rules for the language of chemistry, there are good reasons for sometimes using common names. Which name you use will depend on your audience.

Fact of the **MATTER**

Vast deposits of methane trapped under high pressure in the pores of ice have been located deep under the ocean floor. This ice is as cold as the ice in your freezer, but it burns. If the methane in the ice can be harvested, it may replace Earth's dwindling supplies of fossil fuels because the deposits contain more than twice the amount of energy as in all fossil fuels combined.

The common acids are other examples of inorganic compounds that are known by common rather than formal names. Some names of common acids and bases that you will use frequently in chemistry laboratory experiments are listed in **Table 5.7.** Although they often do not follow the rules you have been learning, they will soon become so familiar that their formulas and names will be easy to remember.

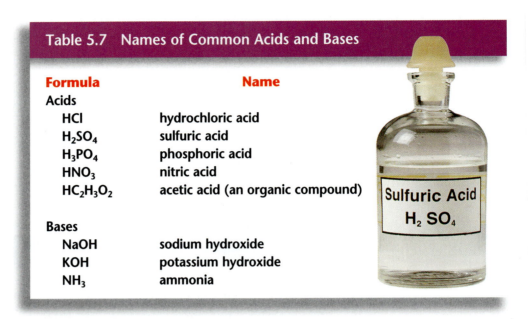

Table 5.7   Names of Common Acids and Bases

Formula	Name
Acids	
HCl	hydrochloric acid
H_2SO_4	sulfuric acid
H_3PO_4	phosphoric acid
HNO_3	nitric acid
$HC_2H_3O_2$	acetic acid (an organic compound)
Bases	
NaOH	sodium hydroxide
KOH	potassium hydroxide
NH_3	ammonia

Figure 5.18

Uses of Hydrocarbons
The structures of methane (left) and propane (right) are shown here. Count the number of carbon atoms in each. Many hydrocarbons are used for fuel. Methane is the main component in natural gas, and propane is used in gas grills.

Naming Organic Compounds

You have learned that most compounds that contain carbon are organic compounds. Organic compounds make up the largest class of molecular compounds known. This is because carbon is able to bond to other carbon atoms in rings and chains of many sizes.

Methane CH_4

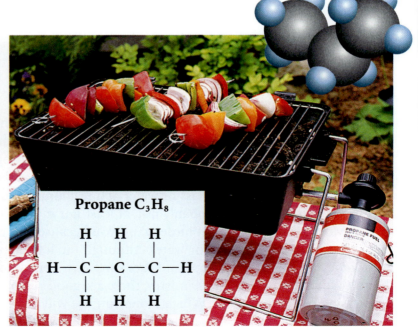

Propane C_3H_8

The name of even the most complex organic compound is based on the name of a **hydrocarbon,** an organic compound that contains only the elements hydrogen and carbon. Hydrocarbons occur naturally in fossil fuels such as natural gas and petroleum and are used mainly as fuels and the raw materials for making other organic compounds.

A carbon atom can form four covalent bonds. In the simplest hydrocarbon, methane, a single carbon is bonded to four hydrogen atoms. Methane is the main component of the natural gas that you burn when you light a Bunsen burner. The next simplest hydrocarbon, ethane, is formed when two carbon atoms bond to each other as well as to three hydrogen atoms apiece. The formulas and names of the first ten hydrocarbon chains are shown in **Table 5.8.** Note that the names of hydrocarbons are derived from the number of carbon atoms in the molecules. Do you recognize any of these hydrocarbons? What is propane used for? **Figure 5.18** shows its structure and one common use.

Table 5.8 Hydrocarbons	
Formula	**Name**
CH_4	methane
C_2H_6	ethane
C_3H_8	propane
C_4H_{10}	butane
C_5H_{12}	pentane
C_6H_{14}	hexane
C_7H_{16}	heptane
C_8H_{18}	octane
C_9H_{20}	nonane
$C_{10}H_{22}$	decane

Connecting Ideas

Formulas represent the known composition of real substances, but just because a formula can be written doesn't mean the compound actually exists. For example, you could easily write the formula HeP_2, but no such compound has ever been isolated. Compounds containing the noble gases helium, neon, and argon have never been found. In the next chapter, you will study the chemical changes that elements and compounds undergo and learn how to represent these changes in the language of chemistry.

SECTION REVIEW

Understanding Concepts

1. Write the formula for each of the following molecular compounds.

   a) carbon monoxide
   b) phosphorus pentachloride
   c) sulfur hexafluoride
   d) dinitrogen pentoxide
   e) iodine trichloride
   f) heptane

2. Write the name of the molecular compound represented by each formula.

   a) BF_3 d) IF_7
   b) PBr_5 e) NO
   c) C_2H_6 f) SiO_2

3. Explain what allotropes are and give two examples.

Thinking Critically

4. **Applying Concepts** Explain, in terms of electron structure, why carbon usually forms four bonds.

Applying Chemistry

5. **Tank of Gas** A tank of a substance delivered to a factory is labeled C_4H_{10}. What is the name of the substance in the tank? What is its most likely use?

REVIEWING MAIN IDEAS

5.1 Ionic Compounds

- Binary ionic compounds are named by first naming the metal element and then the non-metal element, with its ending changed to *-ide*. Subscripts are used in formulas to indicate how many atoms of each element are present in the compound.

- The position of an element in the periodic table indicates what charge its ions will have.

- Polyatomic ions have fixed charges and can combine with ions of opposite charge to form ionic compounds. These compounds are named by writing the name of the positive ion first and then the name of the negative ion.

- Most transition elements can form two or more positively charged ions. When naming a compound that contains a transition element, the oxidation number of the transition element is indicated by a Roman numeral in parentheses.

- Hydrates are ionic compounds bonded to water molecules. They are named by following the name of the compound with a prefix attached to the word *hydrate* to indicate how many water molecules are bound.

5.2 Molecular Substances

- Molecular substances have a greater variety of properties than do ionic compounds, but generally they have low melting points, low water solubility, and little or no ability to act as electrolytes.

- Seven elements occur naturally as diatomic molecules. They are hydrogen, nitrogen, oxygen, fluorine, chlorine, bromine, and iodine.

- Some elements exist in different structural forms called allotropes. Allotropes of an element have different properties.

- Binary molecular compounds are named by writing the two elements in the order they are found in the formula, changing the ending of the second element to *-ide*, and adding Greek prefixes to the element names to indicate how many atoms of each are present.

- It is important to know both the formal and common names of chemicals because both are part of the language of chemistry.

- The naming system for organic compounds is based on the names of hydrocarbons.

Vocabulary

For each of the following terms, write a sentence that shows your understanding of its meaning.

allotrope	hygroscopic
anhydrous	inorganic compound
binary compound	molecular element
deliquescent	molecular substance
distillation	organic compound
formula unit	oxidation number
hydrate	polyatomic ion
hydrocarbon	

UNDERSTANDING CONCEPTS

1. Which of the following substances are ionic and which are molecular?

   a) magnesium sulfate   d) ozone
   b) hexane   e) cesium chloride
   c) carbon monoxide   f) cobalt(II) chloride

2. Write the name for each of the following compounds containing polyatomic ions.

   a) $Ca(C_2H_3O_2)_2$ d) $MgSO_4$
   b) $NaOH$ e) $NaNO_2$
   c) $(NH_4)_2SO_3 \cdot H_2O$ f) $Ca(OH)_2$

3. Make a table comparing the properties of ionic and molecular compounds.

4. The metals in the following compounds can have various oxidation numbers. Predict the charge on each metal ion, and write the name for each compound.

   a) $FeCl_3$ **d)** $SnBr_4$
   b) CuF_2 **e)** FeS
   c) $AuBr_3$ **f)** $Pb(C_2H_3O_2)_2$

5. How can you tell if a compound is ionic or molecular by examining its formula?

6. How is Na_2HPO_4 a substance with two different types of bonding?

7. Write formulas for a bromine atom, ion, and molecule.

8. What happens to the composition of a hydrate when it is heated?

● APPLYING CONCEPTS

9. In Samuel Coleridge's poem *The Rime of the Ancient Mariner,* the mariner cried the following while on his ship far from shore. "Water, water everywhere, and all the boards did shrink/Water, water everywhere, nor any drop to drink." What did he mean?

10. Predict the effect of increasing acidity of rain on the rate of formation of limestone caves.

11. How could you determine quantitatively whether sodium hydroxide or calcium chloride is more deliquescent?

12. Why don't the noble gases form compounds easily?

13. How could you determine quantitatively whether the ionic compound table salt or the molecular compound table sugar is more soluble in water?

14. Explain why most elements do not occur naturally in their pure state.

Everyday Chemistry

15. What is hard water and how is it treated?

Art Connection

16. Suppose an artisan wanted to coat a clay vessel with a faint pink glaze. What material should

he or she add to a transparent glaze to achieve this result?

Chemistry and Technology

17. Use the structural organization of graphite to explain why it is a good lubricant.

How It Works

18. How does concrete differ from cement?

● THINKING CRITICALLY

Designing an Experiment

19. **MiniLab 1** Design an experiment to determine the minimum amount of water required to change the color of the anhydrous cobalt compound weather predictor.

Relating Cause and Effect

20. **MiniLab 2** Why did vinegar soften the chicken bone?

Making Predictions

21. **ChemLab** Would you expect a warm or a cool saturated solution of KNO_3 in water to be a better electrolyte? Explain.

Using a Table

22. The following table lists melting points for a number of ionic compounds. Use a periodic table to help you answer the following questions. Do the melting points of the sodium and potassium compounds increase or decrease as you move down Group 17? What does this suggest about the strength of the ionic bonds between these metals and the Group 17 nonmetals?

Compound	Melting Point (°C)
NaCl	804
NaI	651
KCl	773
KBr	730
NaF	993
KI	680
NaBr	755

Applying Concepts

23. Mercury(I) is unusual in that it often forms an ion that links with another mercury(I) ion. Thus, two mercury(I) ions are linked together in a single unit. What is the charge on this double ion? Write the formula for the compound that is formed from this double ion and chlorine.

CUMULATIVE REVIEW

24. How are physical changes different from chemical changes? (Chapter 1)

25. How does the atomic number compare with the number of electrons in a neutral atom? (Chapter 2)

26. How can the periodic table be used to determine the number of valence electrons in an element? (Chapter 3)

SKILL REVIEW

27. **Making and Using Graphs** Using the following data, construct a graph of melting point versus number of carbons and a graph of water solubility versus number of carbons. What is the relationship between chain length (number of carbon atoms) and melting point? Can you explain why? How are chain length and water solubility related?

Number of Carbon Atoms	Melting Point (°C)	Water Solubility (g per 100 mL)
1 (methane)	−183	0.0024
2 (ethane)	−172	0.0059
3 (propane)	−188	0.012
4 (butane)	−138	0.037
5 (pentane)	−130	0.036
6 (hexane)	−95	0.0138
7 (heptane)	−91	0.0052
8 (octane)	−57	0.0015
9 (nonane)	−54	insoluble
10 (decane)	−30	insoluble

WRITING IN CHEMISTRY

28. Write a set of descriptions comparing the structures of a soccer ball, a geodesic dome, and buckminsterfullerene. Is their similarity a coincidence?

PROBLEM SOLVING

29. Different ions and elements that have the same electronic structure are said to be isoelectronic. Na^+ is isoelectronic with Ne; both have ten electrons, including eight valence electrons. Mg^{2+} also is isoelectronic with Na^+ and Ne. Write the symbols for two ions or elements

30. a) Write the formulas for phosphorus trioxide and phosphorus pentoxide.
    b) What percent of the atoms in phosphorus trioxide are phosphorus? What percent are oxygen?
    c) What percent of the atoms in phosphorus pentoxide are phosphorus? What percent are oxygen?

31. Vitamin C is a covalent compound with the formula $C_6H_8O_6$. If a one ounce serving of potato chips provides you with 30 percent of the recommended daily value of Vitamin C, how many servings of the potato chips will you have to eat to get 100 percent of the recommended daily value of this vitamin?

32. Rubbing alcohol contains the covalent compound isopropanol. The most common form of rubbing alcohol available in drugstores contains 70 percent isopropanol and 30 percent water by volume.
    a) What is the ratio of isopropanol to water in the common form of rubbing alcohol?
    b) What is the volume of isopropanol in 200 mL of rubbing alcohol? What is the volume of water?

33. Impurities weaken the interparticle forces in a molecular substance and create irregularities in the crystal structure. What effect do impurities have on the melting point of a molecular substance?

1. A binary compound contains

   a) two elements.
   b) two ions.
   c) two oxidized elements.
   d) two bonds.

2. An oxidation number is

   a) the number of electrons an atom will lose.
   b) the number of electrons an atom will gain.
   c) the overall charge of an atom.
   d) the overall charge of an ion.

3. Which of the following choices is an example of a polyatomic ion?

   a) CO_2 c) MnO_4
   b) Mg^{2+} d) $NaCl$

4. What is the correct chemical formula for the ionic compound formed by the calcium ion (Ca^{2+}) and the acetate ion ($C_2H_3O_2^-$)?

   a) $CaC_2H_3O_2$
   b) $CaC_4H_6O_8$
   c) $(Ca)_2C_2H_3O_2$
   d) $Ca(C_2H_3O_2)_2$

5. Yttrium, a metallic element with atomic number 39, will form

   a) positive ions.
   b) negative ions.
   c) both positive and negative ions.
   d) no ions at all.

6. Copper (II) sulfate has the chemical formula

   a) $CuSO_4$
   b) Cu_2SO_4
   c) $Cu_2(SO_4)_2$
   d) CuS_2O_8

7. Which statement is NOT true about allotropes?

   a) Allotropes contain only one element.
   b) Allotropes have different oxidation numbers.
   c) The properties of allotropes are different.
   d) Allotropes have different molecular structures.

Formula and Names of Common Compounds Containing Nitrogen		
Formula	Molecular Compound Name	Common Name
?	Nitrogen monoxide	Nitrogen monoxide
NH_3	?	Ammonia
?	Dinitrogen tetrahydride	Hydrazine
N_2O	?	Nitrous oxide (Laughing gas)
NO_2	?	Nitrogen dioxide

Using the table above to answer questions 8–9.

8. What is the molecular compound name for laughing gas?

   a) mononitrogen dioxide
   b) nitrogen dioxide
   c) dinitrogen monoxide
   d) dinitrogen oxide

9. What is the molecular formula of hydrazine?

   a) N_4H_2
   b) N_2H_4
   c) $N_2(OH)_4$
   d) $N_4(OH)_2$

10. Hydrocarbons are most commonly used as

   a) acids.
   b) bases.
   c) fuels.
   d) explosives.

Test Taking Tip

When Eliminating, Cross It Out Consider each answer choice individually and cross out choices you've eliminated. If you can't write in the test booklet, use the scratch paper. List the answer choice letters on the scratch paper and cross them out there. You'll save time and stop yourself from choosing an answer you've mentally eliminated.

6 Chemical Reactions and Equations

Chapter Preview

Sections

6.1 Chemical Equations
: *MiniLab 6.1* Energy Change

6.2 Types of Reactions
: *MiniLab 6.2* A Simple Exchange
: *ChemLab* Exploring Chemical Changes

6.3 Nature of Reactions
: *MiniLab 6.3* Starch-Iodine Clock Reaction

What a Cool Chemical Reaction!

You probably do not think about chemical reactions while watching fireworks, but you are seeing chemistry in action. The beautiful colors are a result of metal salts such as barium chloride and strontium chloride undergoing chemical reactions.

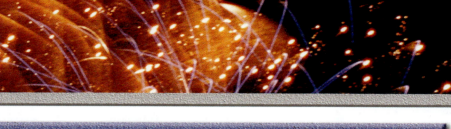

Start-up Activities

*L*aunch Lab

Observing a Chemical Reaction

Reactants are consumed in a chemical reaction as products are produced. What evidence can you observe that a reaction takes place?

Safety Precautions

Materials

- 10-mL graduated cylinder
- 100-mL beaker
- stirring rod
- 0.01M potassium permanganate ($KMnO_4$)
- 0.01M sodium hydrogen sulfite ($NaHSO_3$)

Procedure

1. Measure 5.0 mL of 0.01M potassium permanganate solution ($KMnO_4$) and pour it into a 100-mL beaker.
2. Add 5.0 mL of 0.01M sodium hydrogen sulfite solution ($NaHSO_3$) to the potassium permanganate solution while stirring. Record your observations.
3. Slowly add additional 5.0-mL portions of the $NaHSO_3$ solution until the $KMnO_4$ solution turns colorless. Record your observations.
4. Record the total volume of the $NaHSO_3$ solution you used to cause the beaker's contents to become colorless.

Analysis

What evidence do you have that a reaction occurred? Would anything more have happened if you continued to add $NaHSO_3$ solution to the beaker? Explain.

What I Already Know

Review the following concepts before studying this chapter.
Chapter 1: what a compound is; the meaning of a formula
Chapter 3: what periodicity is; how periodicity applies to elements
Chapter 4: reasons why atoms combine
Chapter 5: how to write chemical formulas

Reading Chemistry

Scan through Chapter 6. Use the headings you see to create a chapter outline. Make notes on your outline as you read the chapter.

Preview this chapter's content and activities at **chemistryca.com**

Chemical Equations

SECTION PREVIEW

Objectives

✓ **Relate** chemical changes and macroscopic properties.

✓ **Demonstrate** how chemical equations describe chemical reactions.

✓ **Illustrate** how to balance chemical reactions by changing coefficients.

Review Vocabulary

Allotrope: any of two or more molecules of a single element that have different crystalline or molecular structures.

New Vocabulary

reactant
product
coefficient

What do you remember about last summer's Fourth of July? Amazing bursts of color from fireworks shot over a lake? Mouthwatering aromas coming from a barbecue grill? Do the processes that result in those colors and smells have anything in common?

You learned in Chapter 1 that substances undergo both physical and chemical changes. A physical change does not change the substance itself, but a chemical change does. Did the chemicals making up the fireworks, charcoal, and barbecued food undergo chemical changes?

Recognizing Chemical Reactions

When a substance undergoes a chemical change, it takes part in a chemical reaction. After it reacts, it no longer has the same chemical identity. While it may seem amazing that a substance can undergo a change and become part of a different substance, chemical reactions occur around you all the time. Chemical reactions can be used to heat a home, power a car, manufacture fabrics for clothing, make medicines, and produce paints and dyes in your favorite colors. Reactions also provide energy for walking, running, working, and thinking.

Many important clues indicate when chemical reactions occur. None of them alone proves that such a change occurs because some physical changes involve one or more of these signs. Examine the photographs in **Figure 6.1** to see what clues to look for.

Figure 6.1

Signs of Chemical Reaction
When substances undergo chemical changes, observable differences usually occur. If you know what signs to look for, you can determine whether or not a chemical reaction has taken place.

Color changes often accompany chemical changes. If you place brownish-red iodine solution on a freshly cut potato, it reacts with white starch to produce a blue compound. ▶

◄ **Precipitation** of a solid from a solution can result from a chemical change. Using soap with hard water produces a precipitate called soap scum because the soap reacts chemically with ions in the water.

Energy changes occur during all chemical changes. Heat or light can be absorbed or released during a chemical reaction, such as when wood or another fuel rapidly combines with oxygen during burning. ►

◄ **Odor changes** can indicate that a substance has undergone a chemical change. When food spoils, odors change as a result of chemical changes within the food.

◄ **Gas release** sometimes occurs as a result of a chemical change. Automobile exhaust contains gases produced by the combustion of gasoline.

In order to completely understand a chemical reaction, you must be able to describe any changes that take place. Part of that description involves recognizing what substances react and what substances form. A substance that undergoes a reaction is called a **reactant.** When reactants undergo a chemical change, each new substance formed is called a **product.** For example, a familiar chemical reaction involves the reaction between iron and oxygen (the reactants) that produces rust, which is iron(III) oxide (the product). The simplest reactions involve a single reactant or a single product, but some reactions involve many reactants and many products. Examine the chemical reactions shown in **Figure 6.2.**

Complete the Description

Several possible observations help determine when a chemical reaction has taken place. But these observations don't completely describe what happens between reactants to form products. Have you ever seen what happens when baking soda and vinegar are mixed together? They react quickly, as you can tell by the bubbles that seem to explode out of the mixture, as shown in **Figure 6.3.** In describing this reaction, you could say that baking soda and vinegar turn into bubbles. But does that completely explain what happens? What are the bubbles made of? Do all of the atoms in vinegar and baking soda form bubbles? The reaction involves more than what can be determined by observation alone. Just as you can write a sentence to tell others what happened on your way to school today, chemists represent the changes taking place in a reaction by writing equations.

Figure 6.2

Parts of a Chemical Reaction

▲ Compounds in wood chemically combine with oxygen when they burn to form water and carbon dioxide.

Energy is needed by our bodies to perform daily activities. This energy is provided when glucose combines with oxygen in cells. This reaction forms the same two substances produced when compounds in wood combine with oxygen. What are the reactants in these reactions? What are the products? ▶

Figure 6.3

A Sample Word Equation
Vinegar and baking soda react vigorously, forming a bubbly product. This reaction was formerly used in fire extinguishers because the bubbles produced contain carbon dioxide, which is effective in putting out fires. This reaction can be described by the following word equation.

vinegar + baking soda →
sodium acetate + water + carbon dioxide

Word Equations

The simplest way to represent a reaction is by using words to describe all the reactants and products, with an arrow placed between them to represent change, as shown in **Figure 6.3.** As you can see in this word equation, reactants are placed to the left of the arrow, and products are placed to the right. Plus signs are used to separate reactants and also to separate products.

Vinegar and baking soda are common names. The compound in vinegar that is involved in the reaction is acetic acid, and baking soda is sodium hydrogen carbonate. These scientific names can also be used in a word equation.

acetic acid + sodium hydrogen carbonate →
sodium acetate + water + carbon dioxide

Chemical Equations

Word equations describe reactants and products, but they are long and awkward and do not adequately identify the substances involved. Word equations can be converted into chemical equations by substituting chemical formulas for the names of compounds and elements. Recall from Chapter 5 that these formulas can be written by using the oxidation numbers of the elements and the charges of the polyatomic ions. For example, the equation for the reaction of vinegar and baking soda can be written using the chemical formulas of the reactants and products.

$$HC_2H_3O_2 + NaHCO_3 \rightarrow NaC_2H_3O_2 + H_2O + CO_2$$

By examining a chemical equation, you can determine exactly what elements make up the substances that react and form.

Everyday Chemistry

Whitening Whites

Why might eating this meal be worrisome? There isn't a paper napkin in sight—only white linen ones! Not to worry. You'll be able to remove most evidence of sloppy etiquette with a six percent aqueous solution of sodium hypochlorite.

Household bleach Perhaps the most popular type of household bleach is an aqueous solution of sodium hypochlorite, NaClO. Sodium hypochlorite is made by reacting chlorine gas with an aqueous solution of sodium hydroxide, as shown in the following equation.

$$Cl_2(g) + 2NaOH(aq) \rightarrow$$
$$NaClO(aq) + NaCl(aq) + H_2O(l)$$

In an aqueous solution, NaClO does not exist as a complete unit, but as sodium ions, Na^+, and hypochlorite ions, ClO^-. The ingredient responsible for the bleaching action in this type of bleach is the hypochlorite ion. Many other solid and liquid bleaches contain hydrogen peroxide, H_2O_2, instead of sodium hypochlorite. In these bleaches, the active substance in solution is the perhydroxyl ion, HOO^-. What do the ClO^- and HOO^- ions have in common?

Bleaching reactions As you can see, each of these polyatomic ions carries a single negative charge. If each could react with a hydrogen ion, the following reactions would happen.

$$ClO^- + H^+ + \rightarrow HCl + O$$

$$HOO^- + H^+ + \rightarrow H_2O + O$$

Because the compounds HCl and H_2O are more stable than the hypochlorite and perhydroxyl ions, these reactions occur. The reactions result in a bleaching action because the released oxygen reacts with molecules of materials that cause the stain. The molecules of the compounds that cause color in a stain are structured in a way that gives them the physical property of producing color. In the reaction between these compounds and atomic oxygen, the compound or compounds formed have different structures. These structures do not have physical properties of producing color. So a bleach bleaches by rendering a colored compound colorless.

Exploring Further

1. **Applying** Liquid bleaches containing sodium hypochlorite are often sold in opaque, plastic containers because sunlight causes the compound to decompose to produce oxygen gas and sodium chloride. Write the balanced chemical equation for this reaction.

2. **Inferring** Why do you think bleaches containing sodium hypochlorite tend to damage finer fabrics more than bleaches containing hydrogen peroxide?

To find out more about bleaching, visit the Chemistry Web site at **chemistryca.com**

It may also be important to know the physical state of each reactant and product. How can we indicate that the bubbles we see during this reaction are CO_2? Symbols in parentheses are put after formulas to indicate the state of the substance. Solids, liquids, gases, and water (aqueous) solutions are indicated by the symbols (s), (l), (g), and (aq). The following equation shows these symbols added to the equation for the reaction of vinegar and baking soda.

$$HC_2H_3O_2(aq) + NaHCO_3(s) \rightarrow NaC_2H_3O_2(aq) + H_2O(l) + CO_2(g)$$

Now the equation tells us that mixing an aqueous solution of acetic acid (vinegar) with solid sodium hydrogen carbonate (baking soda) results in the formation of an aqueous solution of sodium acetate, liquid water, and carbon dioxide gas. If you had examined this equation before you mixed vinegar and baking soda, you could have predicted that bubbles would form.

Energy and Chemical Equations

Noticeable amounts of energy are often released or absorbed during a chemical reaction. Some reactions absorb energy. If energy is absorbed, the reaction is known as an endothermic reaction. **Figure 6.4** shows an example of an endothermic reaction. For a reaction that absorbs energy, the word *energy* is sometimes written along with the reactants in the chemical equation. For example, the equation for the reaction in which water breaks down into hydrogen and oxygen gases shows that energy must be added to the reaction.

$$2H_2O(l) + energy \rightarrow 2H_2(g) + O_2(g)$$

Figure 6.4

An Endothermic Reaction

▲ The reaction of ammonium chloride and barium hydroxide octahydrate is endothermic.

If these reactants are mixed at room temperature, which is about 20°C, the temperature in the mixture drops as energy is absorbed by the reaction. ▶

miniLAB 1

Energy Change

All chemical reactions involve an energy change. This change may be so slight that it can be detected only with sensitive instruments, or it may be quite noticeable.

Procedure

1. Place 25 g of iron powder and 1 g of NaCl in a resealable plastic bag.

2. Add 30 g of vermiculite to the bag, seal the zipper, and shake the bag to mix the contents.

3. Add 5 mL of water to the bag, reseal the zipper, and gently squeeze and shake the contents to mix them.

4. Hold the bag between your hands and note any changes in temperature.

Analysis

1. What did you observe? What type of reaction produces this kind of change?

2. Using the photo, what practical application can you think of for this reaction?

All reactions that occur in a Bunsen burner or gas grill or those used to power an automobile release energy. The How It Works feature shows another example of such a reaction. As you will recall from Chapter 1, reactions that release heat energy are called exothermic reactions. When writing a chemical equation for a reaction that produces energy, the word *energy* is sometimes written along with the products. For example, the equation for the reaction that occurs when you light methane in a Bunsen burner shows that energy is released. Some of this energy is in the form of light.

$$CH_4(g) + 2O_2(g) \rightarrow CO_2(g) + 2H_2O(g) + energy$$

You may have noticed that the word *energy* is not always written in an equation. It is used only if it is important to know whether energy is released or absorbed. For the burning of methane, energy would be written in the equation because the release of heat is an important part of burning a fuel. Energy would also be written in the equation that describes the reaction when water is broken down into hydrogen and oxygen because the reaction would not occur without the addition of energy. In many reactions, such as the formation of rust, energy might be released or absorbed but it is not included in the equation because it is not important to know about the energy involved in that particular reaction.

Emergency Light Sticks

The light given off by an emergency light stick is energy released from a chemical reaction. Light sticks are a good source of light when no electricity is available. Reactions in which light is given off are called chemiluminescent reactions. Different chemicals used in these reactions give off different colors, so light sticks come in many colors. The light from a light stick is only temporary; when the reactants are used up, light is no longer produced.

1. Light sticks are plastic rods that contain two solutions of chemicals. Enclosed inside a thin glass ampoule is one solution that is an oxidizing agent, and surrounding it in the rod is a second solution that contains a fluorescent dye.

2. When the light stick is bent, the glass ampoule is broken and the two solutions mix. The reaction begins. Energy is given off when the two solutions react.

3. The energy raises the energy level of the electrons of the dye molecules.

4. When the electrons drop back down to their original energy levels, the extra energy is given off as light. This light is sometimes referred to as cool light because no noticeable heat is given off in the reaction.

Thinking Critically

1. Explain why the chemical reaction in the light stick is not exothermic, even though it produces energy.

2. What are some advantages of light sticks over conventional light sources?

Balancing Chemical Equations

What do you think happens to the atoms in reactants when they are converted into products? Some products, such as the CO_2 produced from baking powder when a cake is baked, seem to disappear into the air. What really happens to them?

The Law of Conservation of Mass

Recall from Chapter 2 that you can test a reaction to determine whether the same amount of matter is contained in the products and the reactants. That type of experiment was first carried out by the French scientist Antoine Lavoisier (1743-1794), **Figure 6.5.** His results indicated that the mass of the products is always the same as the mass of the reactants that react to form them. The law of conservation of mass summarizes these findings. Matter is neither created nor destroyed during a chemical reaction.

Conservation of Atoms

Remember that atoms don't change in a chemical reaction; they just rearrange. The number and kinds of atoms present in the reactants of a chemical reaction are the same as those present in the products. When stated this way, it becomes the law of conservation of atoms. For a chemical equation to accurately represent a reaction, the same number of each kind of atom must be on the left side of the arrow as are on the right side. If an equation follows the law of conservation of atoms, it is said to be balanced.

How can you count atoms in an equation? The easiest way to learn is to practice—first with a simple reaction and then with some that are more complex. For example, consider the equation that represents breaking down carbonic acid into water and carbon dioxide.

$$H_2CO_3(aq) \rightarrow H_2O(l) + CO_2(g)$$

Because a subscript after the symbol for an element represents how many atoms of that element are found in a compound, you can see that there are two hydrogen, one carbon, and three oxygen atoms on each side of the arrow. All of the atoms in the reactants are the same as those found in the products.

Figure 6.5
Lavoisier's Contribution
One of the experiments that Lavoisier used to discover the law of conservation of mass was the decomposition of the red oxide of mercury to form mercury metal and oxygen gas. He weighed the amount of HgO that decomposed and found it to be the same as the total weight of Hg and O_2 produced.

Examine the equation for the formation of sodium carbonate and water from the reaction between sodium hydroxide and carbon dioxide.

$$NaOH(aq) + CO_2(g) \rightarrow Na_2CO_3(s) + H_2O(l)$$

Do both sides of the equation have the same number of each type of atom? No. One carbon atom is on each side of the arrow, but the sodium, oxygen, and hydrogen atoms are not balanced. The equation, as written, does not truly represent the reaction because it does not show conservation of atoms.

A Balancing Act

To indicate more than one unit taking part or being formed in a reaction, a number called a **coefficient** is placed in front of it to indicate how many units are involved. Look at the previous equation with a coefficient of 2 in front of the sodium hydroxide formula.

$$2NaOH(aq) + CO_2(g) \rightarrow Na_2CO_3(s) + H_2O(l)$$

Is the equation balanced now? Two sodium atoms are on each side. How many oxygen atoms are on each side? You should be able to find four on each side. How about hydrogen atoms? Now two are on each side. Because one carbon atom is still on each side, the entire equation is balanced; it now represents what happens when sodium hydroxide and carbon dioxide react.

The balanced equation tells us that when sodium hydroxide and carbon dioxide react, two units of sodium hydroxide react with each molecule of carbon dioxide to form one unit of sodium carbonate and one molecule of water. Look at a different balanced equation in **Figure 6.6.**

Figure 6.6

How many atoms?
Examine the balanced equation that shows what happens when carbon reacts with oxygen to form carbon dioxide (left). If a piece of coal contains 10 billion C atoms, how many molecules of O_2 will it react with? How many molecules of CO_2 will be formed?

C   +   O₂   ⟶   CO₂

| 1 carbon atom | 1 oxygen molecule | 1 molecule of carbon dioxide |

Why can't subscripts be changed when balancing an equation? Changing a subscript changes the identity of that substance. Look at the equation in Sample Problem 1. Changing the subscript of the oxygen in water to 2 changes water, H_2O, to hydrogen peroxide, H_2O_2, a different compound. Changing a coefficient simply means that you are changing the amount of that substance compared to the other substances in the reaction. Changing the coefficient of water to 2 means there are two molecules of water, $2H_2O$. The identity of water stays the same.

Write word and chemical equations for the reaction of hydrogen and oxygen gases to form gaseous water and release energy. This reaction powers the main stage of the space shuttle.

Analyze
- To write a word equation for the reaction, write the names of the reactants, draw an arrow, then write the name of any product. If there is more than one reactant or product, plus signs should separate them.

$$hydrogen + oxygen \rightarrow water + energy$$

Set Up
- To write the chemical equation, use chemical formulas to replace the names of the reactants and products in the word equation you wrote. Then add symbols to represent the physical state of each compound. Remember that hydrogen and oxygen occur as diatomic gases.

$$H_2(g) + O_2(g) \rightarrow H_2O(g) + energy$$

Solve
- To balance the atoms on each side of the arrow, count the number of atoms of each type on each side of the arrow. On the left are two hydrogen atoms and two oxygen atoms. On the right are also two hydrogen atoms but only one oxygen atom. Change the coefficient of the water to 2 so that the number of oxygen atoms will be balanced. Because that puts four hydrogen atoms on the right side of the arrow, the coefficient of hydrogen gas also must be changed to 2.

Problem-Solving HINT
Be sure to change only coefficients, not subscripts, when balancing equations.

$$2H_2(g) + O_2(g) \rightarrow 2H_2O(g) + energy$$

Check
- Make a final check of all atoms to make sure they are balanced.

Write word and chemical equations for the reaction that takes place when an aqueous solution of magnesium chloride is added to a silver nitrate solution. Aqueous magnesium nitrate and solid silver chloride form.

Analyze
- To write a word equation for the reaction, write the names of the reactants, draw an arrow, then write the name of any product. If there is more than one reactant or product, plus signs should separate them.

magnesium chloride + silver nitrate →
magnesium nitrate + silver chloride

Set Up
- To write the chemical equation, use chemical formulas to replace the names of the reactants and products in the word equation you wrote. Remember to use the oxidation number of an element and the charge on a polyatomic ion to write a correct formula. Then add symbols to represent the physical state of each compound.

$$MgCl_2(aq) + AgNO_3(aq) \rightarrow Mg(NO_3)_2(aq) + AgCl(s)$$

Solve
• To balance the atoms on each side of the arrow, count the number of atoms of each type on each side of the arrow. On the left are one magnesium atom and two chlorine atoms. On the right, there is also one magnesium atom but only one chlorine atom. Change the coefficient of the AgCl to 2 so that the number of chlorine atoms will be balanced. Because that puts two silver atoms on the right side of the arrow, the coefficient of $AgNO_3$ also must be changed to 2. This balances the oxygen and nitrogen atoms, as well.

$$MgCl_2(aq) + 2AgNO_3(aq) \rightarrow Mg(NO_3)_2(aq) + 2AgCl(s)$$

Check
• Make a final check of all atoms on both sides of the equation to make sure they are balanced.

PRACTICE PROBLEMS

Supplemental Problems

For more practice with solving problems, see Supplemental Practice Problems, Appendix B.

Write word equations and chemical equations for the following reactions.

1. Magnesium metal and water combine to form solid magnesium hydroxide and hydrogen gas.

2. An aqueous solution of hydrogen peroxide (dihydrogen dioxide) and solid lead(II) sulfide combine to form solid lead(II) sulfate and liquid water.

3. When energy is added to solid manganese(II) sulfate heptahydrate crystals, they break down to form liquid water and solid manganese(II) sulfate monohydrate.

4. Solid potassium reacts with liquid water to produce aqueous potassium hydroxide and hydrogen gas.

SECTION REVIEW

Understanding Concepts

1. Explain how you can tell whether a chemical reaction has taken place.

2. Write balanced chemical equations for the reactions described.

   a) sodium metal + chlorine gas $\rightarrow$
      sodium chloride crystals

   b) propane gas + oxygen $\rightarrow$
      carbon dioxide + water vapor + energy

   c) zinc metal + hydrochloric acid $\rightarrow$
      zinc chloride solution + hydrogen

3. Why is it important to balance a chemical equation?

Thinking Critically

4. **Applying Concepts** Use the law of conservation of mass to determine the following:

   a) The number of grams of CO_2 that form from 4.00 g C and 10.67 g O_2.
      $$C + O_2 \rightarrow CO_2$$

   b) The number of grams of water formed if 7.75 g H_2CO_3 forms 5.50 g CO_2.
      $$H_2CO_3 \rightarrow H_2O + CO_2$$

Applying Chemistry

5. **Catalytic Converter** In the catalytic converter of a car, a reaction occurs when nitrogen monoxide gas, NO, reacts with hydrogen gas. Ammonia gas and water vapor are formed. Write a balanced equation for this reaction.

Types of Reactions

From the kitchen of: Kris
Recipe for: Teriyaki (Japanese Dish)
Ingredients:

From the kitchen of: Kris
Recipe for: Israeli Falafel
Ingredients:
2 cups
⅓
1 c
½ t
2 cl
2 tak
¼ tea

From the kitchen of: Kris
Recipe for: Pakoras (Indian Cauliflower Fritters)
Ingredients:
1 cup sifted all-purposed flour   1⅓ cups water, approximately
½ teaspoon baking powder   vegetable oil for frying
1½ teaspoon salt
2 teaspoons ground coriander
1½ teaspoons ground cumin
¼ teaspoon tumeric
½ teaspoon ground ginger
1 medium-size cauliflower, broken into flowerets

Chemistry has a lot in common with cooking. What do you have to do to become a good cook? A lot of study and probably even more practice are necessary. The same is true of chemistry. If you combine the right amounts of the right reactants in the correct order under certain conditions, you will get the right products.

As you study chemistry, you will begin to recognize what types of reactants and conditions lead to certain products. In this section, you will learn to recognize five major classes of reactions. Just as an experienced chef can tell a sauté from a soufflé, you will soon be able to distinguish between combustion and decomposition.

Why Reactions Are Classified

Why are reactions grouped into classes? Think about why scientists classify plants and animals. If you were hiking in the Rocky Mountains for the first time, you might see animals you have never seen before, such as the ones in **Figure 6.7.** How would you decide whether or not to be cautious around them? You would probably decide what familiar animals they most resemble. If the animal looks like a cat, it is most likely a predator that you should be careful around. If it looks like a goat, you will probably realize that it is more likely to run from you than eat you.

Figure 6.7

Classifying

Around which animal would you be most cautious? Did you mentally classify the animals into categories of some type before answering?

Major Classes of Reactions

Just as there are thousands of species of animals, there are many different types of chemical reactions. Five types are common. If you can classify a reaction into one of the five major categories by recognizing patterns that occur, you already know a lot about the reaction.

In one type of reaction, two substances—either elements or compounds—combine to form a compound. Whenever two or more substances combine to form a single product, the reaction is called a **synthesis** reaction.

Air in Space

The concentration of carbon dioxide in Earth's atmosphere is regulated through a complex interplay of human, biological, and geological mechanisms. In a space vehicle, these mechanisms aren't available. If not controlled, carbon dioxide from the respiration of astronauts could become toxic to them. So how is air quality maintained within a space vehicle?

A 66-m^3 atmosphere The volume of the crew compartment of the space shuttle Orbiter is about 66 m^3. The air is maintained at a pressure of 100 kPa, which is similar to Earth's atmospheric pressure at sea level. During a flight, the composition of the air in the crew compartment is maintained at 79 percent nitrogen and 21 percent oxygen, which is almost identical to that of Earth's atmosphere. Oxygen is carried in the Orbiter as a liquid stored in two cryogenic tanks in the mid-fuselage. The gaseous oxygen from the tanks passes through pressurizing and heating nozzles and moves into the crew compartment. A five-member crew will normally consume about 4 kg of oxygen every day. Nitrogen is supplied from two systems, each made of two storage tanks, also in the mid-fuselage of the Orbiter. The atmosphere of the compartment is recycled about every seven minutes.

Filtering the air During recycling, odors are removed by filters containing activated charcoal granules, which absorb from the air the chemical substances that cause the odor. Carbon dioxide gas is removed from the air by reacting it with solid lithium hydroxide.

$$CO_2(g) + 2LiOH(s) \rightarrow Li_2CO_3(s) + H_2O(g)$$

The lithium hydroxide is stored in canisters that are changed every 12 hours. The used canisters are then stored for disposal when the Orbiter returns to Earth.

Connecting to Chemistry

1. **Applying** Identify by name the products of the reaction between carbon dioxide and lithium hydroxide.

2. **Comparing and Contrasting** How does the chemical removal of carbon dioxide from the atmosphere of the Orbiter compare with the geochemical removal of carbon dioxide from Earth's atmosphere?

Figure 6.8

A Synthesis Reaction

When iron rusts, iron metal and oxygen gas combine to form one new substance, iron(III) oxide. The balanced equation for this synthesis reaction shows that there is more than one reactant but only one product.

$$4Fe(s) + 3O_2(g) \rightarrow 2Fe_2O_3(s)$$

An example of a synthesis reaction involving elements as reactants is shown in **Figure 6.8.** A synthesis reaction also occurs when two compounds combine, such as when rainwater combines with carbon dioxide in the air to form carbonic acid, or when an element and a compound combine, as when carbon monoxide combines with oxygen to form carbon dioxide.

In a **decomposition** reaction, a compound breaks down into two or more simpler substances. The compound may break down into individual elements, such as when mercury(II) oxide decomposes into mercury and oxygen. The products may be an element and a compound, such as when hydrogen peroxide decomposes into water and oxygen, or the compound may break down into simpler compounds, as shown in **Figure 6.9.**

Figure 6.9

A Decomposition Reaction

When ammonium nitrate is heated to a high temperature, it explosively breaks down into dinitrogen monoxide and water. The decomposition reaction taking place is represented by a balanced equation that shows one reactant and more than one product.

$$NH_4NO_3(s) \rightarrow N_2O(g) + 2H_2O(g)$$

A Simple Exchange

Single-displacement reactions occur when one element replaces another in a compound. However, such a reaction does not automatically occur just because an element and a compound are mixed together. Whether or not such a reaction occurs depends on how reactive the element is compared to the element it is to displace.

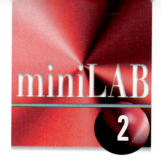

miniLAB
2

Procedure

1. Pour $0.1M$ AgNO$_3$ solution into a test tube until it is half full. Clean a piece of copper wire or copper foil with steel wool.
2. Drop the copper foil or copper wire into the solution.
3. Keeping the test tube absolutely still, observe what happens over a half-hour period of time.

Analysis

1. What changes do you observe in the wire? In the solution?
2. Write a balanced equation for the reaction.
3. Does copper displace silver in silver nitrate? Does silver displace copper in copper(II) nitrate? How do you know?

In a **single-displacement** reaction, one element takes the place of another in a compound, as shown in **Figure 6.10.** The element can replace the first part of a compound, or it can replace the last part of a compound.

Figure 6.10

Single Displacement

◀ If an iron nail is placed into an aqueous solution of copper(II) sulfate, the iron displaces the copper ions in solution, and copper metal forms on the nail.

$$Fe(s) + CuSO_4(aq) \rightarrow FeSO_4(aq) + Cu(s)$$

When the chlorine gas in the flask on the left is bubbled through an aqueous solution of sodium bromide, the chlorine replaces the bromine in the compound. The reddish-brown bromine can be seen in the solution. ▶

$$Cl_2(g) + 2NaBr(aq) \rightarrow 2NaCl(aq) + Br_2(l)$$

ChemLab

Exploring Chemical Changes

Most reactions can be classified into five major types. As you carry out this experiment, you'll observe examples of each of these types. In doing so, you will also learn to recognize many of the physical changes that accompany reactions.

Problem

What are some of the physical changes that indicate that a reaction has occurred?

Objectives

- **Observe** physical changes that take place during chemical reactions.
- **Compare** changes that take place during different types of chemical reactions.

PREPARATION

Materials

125-mL flasks (4)	large test tube and one-
balance	hole stopper with glass
hot plate	tube and rubber tubing
watch glass	attached
spatula	ring stand
stirring rod	test-tube clamp
lab burner	$0.1M$ $CuSO_4$
file	granular copper, Cu
new penny	powdered sulfur, S
250-mL flask	$CaCO_3$, finely ground
ice	saturated $Ca(OH)_2$
tongs	solution, limewater
100-mL graduated	$6M$ HCl
cylinder	$0.5M$ Na_2CO_3
	$0.5M$ $CuCl_2$

Safety Precautions

Wear an apron and goggles. Use care when handling hot objects. Dispose of the reaction mixture and products as instructed by your teacher.

PROCEDURE

For each of the following reactions, record in a data table all changes that you observe.

Synthesis Reaction

1. Place 50 mL $0.1M$ $CuSO_4$ in a 125-mL flask.
2. Place 1.6 g granular copper and 0.8 g powdered sulfur on a watch glass and mix together thoroughly with a spatula.
3. Heat the flask on a hot plate set at high until the solution begins to boil.
4. Stir the Cu/S mixture into the boiling $CuSO_4$ solution.
5. Continue boiling until a black solid forms.

Decomposition Reaction

1. Place 100 mL of saturated $Ca(OH)_2$ solution (limewater) in the 250-mL flask.
2. Add finely ground $CaCO_3$ to a large test tube until it is one-fourth full. Stopper the tube with the stopper/glass tube/rubber tubing assembly, and clamp the tube to the ring stand.
3. Light a laboratory burner, and begin to heat the test tube. Submerge the end of the rubber tubing into the limewater so that any gas produced in the tube will bubble through the limewater.
4. Continue heating the $CaCO_3$ until you observe a change in the limewater. The presence of CO_2 causes limewater to become cloudy.

Single-Displacement Reaction

1. Place 30 mL $6M$ HCl in a 125-mL flask.

2. Using a file, cut six 0.2-cm notches evenly spaced around the perimeter of a new penny.

3. Place the penny in the flask of acid and leave it in a fume hood overnight.

Double-Displacement Reaction

1. Add 25 mL $0.5M$ Na_2CO_3 and 25 mL $0.5M$ $CuCl_2$ to a 125-mL flask.

2. Swirl the flask gently until you observe the formation of a precipitate.

Combustion Reaction

1. Light a laboratory burner and adjust the air and gas supplies until the flame is blue. Observe what happens.

2. Using tongs, hold a flask or beaker with ice in it about 10 cm over the flame for approximately one minute. Move the flask away from the flame and observe the bottom of the flask.

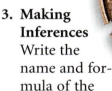

DATA AND OBSERVATIONS

Reaction	Observations
Synthesis	
Decomposition	
Single displacement	
Double displacement	
Combustion	

ANALYZE AND CONCLUDE

1. **Making Inferences** Which observations noted during each of the reactions indicated that a reaction had occurred?

2. **Comparing and Contrasting** What did all of the reactions have in common?

3. **Making Inferences** Write the name and formula of the

a) black solid formed in the synthesis reaction.

   b) gaseous product of the decomposition reaction.

   c) solid product of the decomposition reaction.

   d) pale blue precipitate in the double-displacement reaction.

   e) liquid product of the combustion reaction.

4. **Observing and Inferring** Explain how the penny changed during the single-displacement reaction. What would happen if a pre-1983 penny, which is solid copper, were used?

5. **Relating Concepts** Is energy a reactant or product of the combustion reaction?

APPLY AND ASSESS

1. Were there any physical changes that often occur during a reaction that you did not observe while doing this ChemLab? If so, what were they?

2. Write balanced chemical equations for all of the reactions carried out.

3. Why do you think pennies are no longer made from only copper metal?

In **double-displacement** reactions, the positive portions of two ionic compounds are interchanged. For a double-displacement reaction to take place, at least one of the products must be a precipitate or water. An example of a double-displacement reaction is shown in **Figure 6.11**.

Figure 6.11

Double Displacement

When clear aqueous solutions of lead(II) nitrate and potassium iodide are mixed, a double-displacement reaction takes place and a yellow solid appears in the mixture. This solid is lead(II) iodide, and it precipitates out because it is insoluble in water, unlike the two reactants and the other product.

$$Pb(NO_3)_2(aq) + 2KI(aq) \rightarrow PbI_2(s) + 2KNO_3(aq)$$

Table 6.1   Types of Reactions

Reaction Type	General Equation
Synthesis	element/compound + element/compound → compound
	Examples: $2Na(s) + Cl_2(g) \rightarrow 2NaCl(s)$
	$CaO(s) + SiO_2(l) \rightarrow CaSiO_3(l)$
Decomposition	compound → two or more elements/compounds
	Examples: $PCl_5(s) \rightarrow PCl_3(s) + Cl_2(g)$
	$2Ag_2O(s) \rightarrow 4Ag(s) + O_2(g)$
Single displacement	*element a + compound bc → element b + compound ac
	Example: $2Al(s) + Fe_2O_3(s) \rightarrow 2Fe(s) + Al_2O_3(s)$
	element d + compound bc → element c + compound bd
	Example: $Cl_2(aq) + 2KBr(aq) \rightarrow 2KCl(aq) + Br_2(aq)$
Double displacement	compound ac + compound bd → compound ad + compound bc
	Examples: $PbCl_2(s) + Li_2SO_4(aq) \rightarrow PbSO_4(s) + 2LiCl(aq)$
	$BaCl_2(aq) + H_2SO_4(aq) \rightarrow 2HCl(aq) + BaSO_4(s)$
Combustion	element/compound + oxygen → oxide(s)
	Examples: $CH_4(g) + 2O_2(g) \rightarrow CO_2(g) + 2H_2O(g)$
	$C_6H_{12}O_6(s) + 6O_2(g) \rightarrow 6CO_2(g) + 6H_2O(l)$

*The letters a, b, c, and d each represent different elements or parts of compounds. For example, in compound ac, *a* represents the positive part of the compound, and *c* represents the negative part.

Figure 6.12
Combustion

When welding is done with an acetylene torch, acetylene combines with oxygen to form carbon dioxide and water. This combustion reaction is exothermic, and enough energy is released to melt metal.

$$2C_2H_2(g) + 5O_2(g) \rightarrow$$
$$4CO_2(g) + 2H_2O(g) + \text{energy}$$

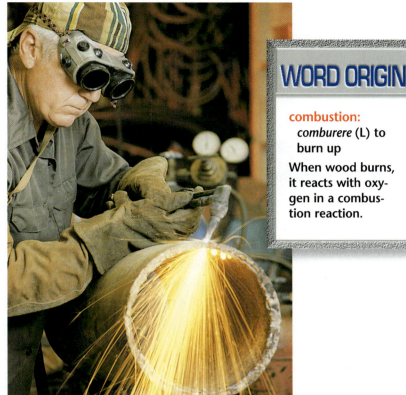

WORD ORIGIN

combustion:
comburere (L) to burn up
When wood burns, it reacts with oxygen in a combustion reaction.

The fifth common type of reaction is a combustion reaction. A **combustion** reaction is one in which a substance rapidly combines with oxygen to form one or more oxides, as shown in **Figure 6.12.**

Although there are exceptions, many thousands of specific reactions fit into these five classes. One example is in the Chemistry and Technology feature on pages 216-217. **Table 6.1** summarizes important information about each of these reaction types.

Supplemental Problems

For more practice with solving problems, see Supplemental Practice Problems, Appendix B.

SECTION REVIEW

Understanding Concepts

1. Explain why classifying reactions can be useful.

2. Using different symbols to represent different atoms, draw pictures to represent an example of each of the following kinds of reactions.

   a) synthesis
   b) decomposition
   c) single displacement
   d) double displacement
   e) combustion

3. Classify each of the following reactions.

   a) $N_2O_4(g) \rightarrow 2NO_2(g)$
   b) $2Fe(s) + O_2(g) \rightarrow 2FeO(s)$
   c) $2Al(s) + 3Cl_2(g) \rightarrow 2AlCl_3(s)$

   d) $BaCl_2(aq) + Na_2SO_4(aq) \rightarrow$
   $$BaSO_4(s) + 2NaCl(aq)$$
   e) $Mg(s) + CuSO_4(aq) \rightarrow Cu(s) + MgSO_4(aq)$

Thinking Critically

4. **Making Inferences** When a candle burns, wax undergoes a combustion reaction. Would a candle burn longer in the open or when covered with an inverted glass jar? Explain.

Applying Chemistry

5. **Decomposition** When a fungus breaks down the wood in a fallen tree, the biological process is called decomposition. What does that process have in common with the chemical decomposition reaction type?

Nature of Reactions

You know that changes constant-ly occur, but some changes are not permanent. For example, liquid water freezes into ice, but then ice melts and becomes liquid water again. In other words, the freezing process is reversed. Can chemical reactions be reversed? Can the product of a reaction become a reactant?

Reversible Reactions

Many reactions can change direc-tion. These reactions are called reversible reactions. Some reactions, such as the one shown in **Figure 6.13,** can be reversed based on the energy flow.

Not all chemical changes are reversible. Caves form, paint hardens, and fuel burns. These chemical changes result in new products, and the reac-tions are said to go to completion because at least one of the reactants is completely used up and the reaction stops. The reactions can't be reversed. But what happens when a reaction reverses?

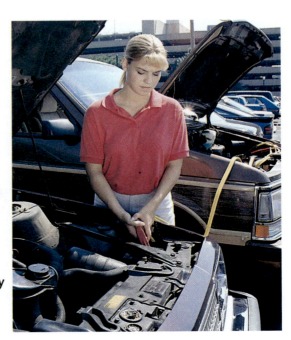

Figure 6.13
Reversible Recharge
When an automobile battery releases energy when the car isn't running, the reaction below moves to the right. If you leave the lights on and need to recharge the battery with a jump start, the reaction moves to the left while the car engine runs.

$$Pb(s) + PbO_2(s) + 2H_2SO_4(aq) \rightleftarrows$$
$$2PbSO_4(s) + 2H_2O(l) + energy$$

Picture what happens in a subway station when the door of a train opens for the first time that day. Passengers rush in, streaming across the platform and through the doors. No passengers get out because this is the first train of the day. When the train stops at the next station, more passengers rush in. A few probably leave the train at this point, as well. With each successive stop, more passengers will be getting on, but more and more will also be getting off. Look at **Figure 6.14.** To describe a similar situation in chemical reactions in which a reaction automatically reverses and there is no net (overall) change, we use the term *equilibrium.*

Equilibrium

When no net change occurs in the amount of reactants and products, a system is said to be in **equilibrium.** In most situations, chemical reactions exist in equilibrium when products and reactants form at the same rate. Such a system, in which opposite actions are taking place at the same rate, is said to be in **dynamic equilibrium.**

Reactions at equilibrium have reactants and products changing places, much like the passengers getting on and off a subway train. In equilibrium, reactants are never fully used up because they are constantly being formed from products. Eventually, reactants and products form at the same rate.

One example of a reversible reaction that reaches equilibrium occurs when lime, CaO, which is used to make soils less acidic, is formed by decomposing limestone, $CaCO_3$.

$$CaCO_3(s) \rightleftharpoons CaO(s) + CO_2(g)$$

Notice that the single arrow in the equation has been replaced with double arrows. Because an arrow shows what direction the reaction is going, the double arrow indicates that the reaction can go in either direction. In this case, $CaCO_3$ decomposes into CaO and CO_2. But, as those products form, CO_2 and CaO combine to form $CaCO_3$. The rate, or speed, of each reaction can be determined by how quickly a reactant disappears. Eventually, the rates of the two reactions are equal, and an equilibrium exists.

PEOPLE in CHEMISTRY

Meet Caroline Sutliff, Plant-care Specialist

An old gardening adage goes like this: A weed is just a flower that is growing where you don't want it. Landscape gardener Caroline Sutliff works to eradicate those hardy, invasive plants, such as *Taraxacum officinale* (better known as dandelions). In this interview, she tells about how she wages her weed battles with garden tools and chemicals.

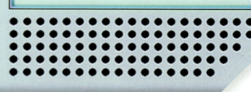

On the Job

 Ms. Sutliff, what do you do on the job?

 I take care of flowers, shrubs, and trees. The company for which I work maintains the grounds of business properties, such as real estate offices and banks, as well as private homes. My job involves working with chemicals such as herbicides, fungicides, and pesticides.

Are those chemicals hazardous?

They can be if you don't treat them properly. I have a spraying license that I earned through the Agricultural Service Extension, so I have learned all the safety precautions. Mixing the chemicals to get them ready for use is more hazardous than actually spraying them. For mixing, I wear a plastic-coated jacket and pants, heavy boots, rubber gloves, and a face mask.

 Do you and your coworkers know what to do in case of a mishap?

 We carry eyewash and first-aid kits in all of our trucks. We're trained to react quickly if we get a chemical on our clothes. In that case, we must remove the affected clothing and wash up quickly. When some chemicals come in contact with your skin, they are absorbed into the body and stored in fat cells. The cumulative effect could be harmful. Because I'm very careful and follow directions exactly, I haven't had any problems.

Why are chemicals necessary in your work?

On large properties, it's a matter of economy and time—two hours of hand weeding and cultivating versus a few minutes of spraying. However, chemical use is just part of our program. Before we spray chemicals, we cultivate, do some hand weeding, and fertilize and aerate beds of plants. Healthy plants are more likely to have a fighting chance against weeds and insects.

Do you use the information you learned in high school chemistry classes?

 Yes. It's important to be familiar with the properties of chemicals—to know what can be safely mixed and what can't. The container used for mixing is important chemically, too. For example, stainless steel containers can erode and change the makeup of chemicals so they won't have the same effects. Heavy plastic is used instead.

Early Influences

 As a child, were you interested in plants?

 I wasn't interested in planting things because I wanted immediate results. Plants don't work that way. Now I have more patience and love to tend to the flowers around my own home.

 Did you plan to enter the plant-care field later on?

 No. My intentions were to get a degree in psychology and then go to law school. While I was looking for a job as a paralegal, I began working in plant care and discovered I love working outside. Now I can't imagine being inside all day at a desk job.

Personal Insights

 Would you recommend that students investigate a plant-care job like yours?

 Only if they like hard, physical work. My workdays last ten or 12 hours, and I'm out in all kinds of weather. I certainly don't need to go to a gym to stay in shape! It's satisfying for me to see that more women are entering this field.

 How do you see the field of plant care changing in the years ahead?

 I think people will become even more aware of the value of plants, particularly trees. Trees can do so much to improve a neighborhood. They clean the air, shade the sidewalks, and beautify an area. Here in Iowa City, an organization called Project Green helps plant trees in public places. I hope to get involved with that group as a volunteer.

CAREER CONNECTION

The following careers are also associated with plant care.

Horticulturist Master's degree, research, and fieldwork
Landscape Architect Bachelor's degree, often followed by a licensing examination
Soil Conservation Technician Two-year college program

Did you ever have to make up your mind?

Saying that a reaction has reached equilibrium does not mean that equal amounts of reactants and products are present. Equilibrium just means that no net change is taking place—the amounts of reactants and products are not changing. Often, a reaction at equilibrium contains differing amounts of reactants and products.

Consider the following reaction in which phosphorus pentachloride decomposes into phosphorus trichloride and chlorine.

$$PCl_5(g) \rightleftarrows PCl_3(g) + Cl_2(g)$$

Actual measurements of the amounts of reactant and products show that the system is in equilibrium. But the measurements also show that there is more PCl_5 than PCl_3 and Cl_2 present, even though the rates of the reactions are equal. The reversible reaction will favor the direction that produces the most stable products, which are those that are least likely to change. In this example, PCl_5 is less likely to decompose than the two products, PCl_3 and Cl_2, are likely to combine. Reversible reactions at equilibrium are in a balance that favors stability.

Changing Direction

If a reaction reaches equilibrium, how can you obtain large quantities of a product? Won't the product constantly become a reactant? Keep in mind that reactions in equilibrium are stable. The French scientist Henri Louis Le Chatelier proposed in 1884 that disturbing an equilibrium will make a system readjust to reduce the disturbance and regain equilibrium, **Figure 6.15.** This principle regarding changes in equilibrium is called Le Chatelier's principle.

Chemical engineers can apply this tendency of reactions to stay at equilibrium to find ways to increase the yield of products in a reaction. For example, if products are removed from a reaction at equilibrium, more reactants will go on to form products so that balance is regained. If this removal continues, most of the reactants can be converted into products. For example, recall the reaction in which limestone decomposes into lime and carbon dioxide.

$$CaCO_3(s) \rightleftarrows CaO(s) + CO_2(g)$$

If the carbon dioxide is removed as it is produced, the reaction will favor the formation of more carbon dioxide to reestablish equilibrium. Thus, the reaction will shift in the direction that also produces more lime as a product.

Figure 6.15

A System in Stress
Before the dog started to drink, the water in this container reached a stable equilibrium level. As the dog drinks, this level is stressed and water comes from the bottle into the bowl to reestablish the equilibrium level.

When one product is a gas and other products and reactants are not, as in the previous example and in **Figure 6.16,** it is easy to see how the gas can be removed from the reaction. How can products that are not isolated gases be removed from a reaction? They usually can't be picked out manually because of the mix of reactants and products.

A product that does not dissolve in water can be removed if all other products and the reactants dissolve in water. A compound is **soluble** in a liquid if it dissolves in it; it is **insoluble** if it does not. An insoluble product will form a solid precipitate that sinks to the bottom of a liquid solution, as shown in **Figure 6.17.** The precipitate cannot react easily because it is somewhat isolated from the other substances in the reaction.

Adding Reactants or Energy

Adding more reactants has basically the same effect as removing products. For the reaction to reestablish equilibrium amounts of reactants and products, more products must be made if more reactants are added.

Adding or removing energy, usually in the form of heat, can also influence the direction of a reaction. Because energy is a part of any reaction, it can be thought of as a reactant or a product. Just as adding more reactants to a reaction pushes it to the right, so does adding more energy to an endothermic reaction. For example, the equation for a reaction that produces aluminum metal from bauxite, an aluminum ore, shows that energy must be added for products to form.

$$3C + 2Al_2O_3(s) + energy \rightleftarrows 4Al(l) + 3CO_2(g)$$

If more energy is added, the reaction goes to the right, forming more aluminum and carbon dioxide.

For an exothermic reaction, adding more energy pushes the reaction to the left. In the Haber process for producing ammonia from hydrogen and nitrogen, for example, energy is produced.

$$N_2(g) + 3H_2(g) \rightleftarrows 2NH_3(g) + energy$$

Adding energy favors the formation of nitrogen and hydrogen. Thus, temperature must be carefully controlled in the Haber process so that the desired product, ammonia, is produced in large quantities.

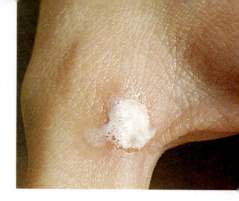

Figure 6.16

Removing a Gas

When hydrogen peroxide, H_2O_2, is poured onto a wound, the hydrogen peroxide decomposes to form water and oxygen. The gaseous oxygen bubbles away, preventing the re-formation of H_2O_2.

$$2H_2O_2(aq) \rightarrow 2H_2O(l) + O_2(g)$$

Notice that only a forward arrow is used because the removal of oxygen drives the reaction forward.

Figure 6.17

Forming a Precipitate

When potassium hydroxide is added to an aqueous solution of calcium chloride, calcium hydroxide and potassium chloride form. Calcium hydroxide is nearly insoluble in water, so it precipitates out as a solid.

Mining the Air

Diatomic nitrogen makes up about 79 percent of Earth's atmosphere. Nitrogen is an essential element for life, yet only a few organisms can use atmospheric nitrogen directly. A few species of soil bacteria can produce ammonia, NH_3, from atmospheric nitrogen. Other species of bacteria can then convert the ammonia into nitrite and nitrate ions, which can be absorbed and used by plants.

The Haber Process

Not all usable nitrogen compounds are produced naturally from atmospheric nitrogen. Ammonia can also be synthesized. The process of synthesizing large amounts of ammonia from nitrogen and hydrogen gases was invented by Fritz Haber, a German research chemist, and his English research assistant, Robert LeRossignol. The process was first demonstrated in 1909. Haber was granted the German patent for the process.

The Haber process involves a synthesis reaction. The two reactants are $H_2(g)$ and $N_2(g)$, and the products are $NH_3(g)$ and heat. The process produces high yields of ammonia by manipulating three factors that influence the reaction—pressure, temperature, and catalytic action.

Ammonia storage tanks

Fritz Haber

Pressure

In the ammonia synthesis reaction, four molecules of reactant, H_2 and N_2, produce two molecules of product, NH_3. According to Le Chatelier's principle, if pressure on the reaction or the system is increased, the forward reaction will speed up to lessen the stress because two molecules exert less pressure than four molecules. Increased pressure will also cause the reactants to collide more often, thus increasing the reaction rate. Haber's apparatus used a total pressure of 2×10^5 kPa, which was the highest pressure he could achieve in his laboratory.

Temperature

Two factors influence the temperature at which the process is carried out. A low temperature favors the forward reaction because this reduces the stress of the heat generated by the reaction. But high temperature increases the rate of reaction because of more collisions between the reactants. Haber carried out the process at a temperature of about 600°C.

operating ammonia plant in Oppau, Germany. The plant used nitrogen obtained from liquefied air and hydrogen obtained from methane and water. Throughout World War I, the Oppau plant and a later-built plant supplied Germany with ammonia for the production of explosives. Today, the Haber-Bosch process is the main source for producing commercial ammonia. Much of this ammonia is used in the production of agricultural fertilizers.

Catalyst

A catalyst is used to decrease the activation energy and thus increase the rate at which equilibrium is reached. Haber used osmium and uranium as catalysts in his apparatus.

The Process

Haber's process incorporated several operations that increased the yield of ammonia. The reactant gases entering the chamber were warmed by heat produced by the reaction. The reactant-product mixture was allowed to cool slowly after reacting over the catalyst. Ammonia was removed from the process by liquefaction, while unreacted nitrogen and hydrogen were recycled into the process.

The Haber process was gaining the attention of BASF, a leading German chemical manufacturer. Carl Bosch, a German industrial chemist employed by BASF, was given the task of making Haber's process commercially sound. He and his colleagues designed, constructed, and tested new reaction chambers; improved pressurizing pumps; and used inexpensive catalysts. By 1913, Bosch had built an

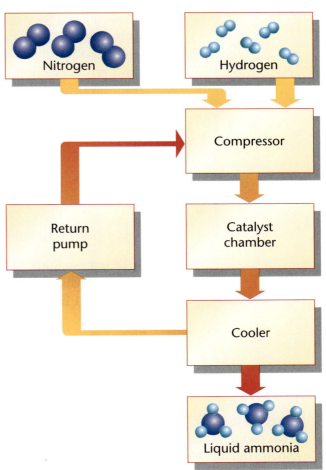

DISCUSSING THE TECHNOLOGY

1. **Applying** How would slow cooling of the reactant-product mixture affect the equilibrium of the reaction?

2. **Acquiring Information** How is ammonia used in the production of many agricultural fertilizers?

Reaction Rate

You have learned that many reactions reach equilibrium. At that time, reactants and products are being formed and broken down at the same rate. How fast is this rate? If you pour a solution of potassium iodide into a solution of lead nitrate, a yellow precipitate of lead iodide seems to form instantly. If you strike a match, you can see that it takes a little time before the match burns completely. Many reactions proceed even more slowly. To understand why reactions occur at different rates, look at what happens when reactions take place.

Activation Energy

For a reaction to occur between two substances, particles of those substances must collide. Not only do they have to collide, they have to hit each other with enough force to cause a change to take place. The amount of energy the particles must have when they collide is called the **activation energy** of the reaction.

Do exothermic reactions require activation energy? Consider the highly exothermic reaction between hydrogen and oxygen. Once it starts, this reaction produces enough energy to power the main stage of the space shuttle. However, hydrogen and oxygen molecules can exist in the same container for years without reacting. A spark can provide the activation energy needed to start the reaction. Once started, the reaction produces enough energy to keep itself going. Activation energy differs from reaction to reaction. Some reactions have such high activation energies that the reactions do not take place under normal conditions.

Speed of Reaction

Whether riding to school or completing math problems, the rate, or speed, at which the process takes place can be important. Look at an example of finding a rate in **Figure 6.18.** The rate at which a chemical reaction takes place is also important.

To determine how fast a reaction is taking place, you can measure how quickly one of the reactants disappears or how quickly one of the products

Figure 6.18
Reaction Rate
If ten green chameleons turn red per minute, the rate of this change can be expressed as ten green chameleons disappearing per minute or ten red chameleons appearing per minute.

appears. Either measurement will give an amount of substance changed per unit time, which is the rate of reaction.

Why is reaction rate important? The rate of a reaction is important to a chemical engineer designing a process to get a good yield of product. The faster the rate, the more product that can be made in a fixed amount of time. Rate of reaction is also important to food processors who hope to slow down reactions that cause food to spoil. Can the rate of a reaction be changed? Four major factors affect this rate.

Effect of Temperature

One factor that affects rate of reaction is temperature, **Figure 6.19.** Most reactions go faster at higher temperatures. Baking a cake speeds up the reactions that change the liquid batter into a spongy product. Lowering the temperature slows down most reactions. Photographic film and batteries stay useful longer if they are kept cool because the lower temperature slows the reactions that can ruin these products.

It's a Matter of Concentration

Changing the amounts of reactants present can also alter reaction rate. The amount of substance present in a certain volume is called the **concentration** of the substance. Raising the concentration of a reactant will speed up a reaction because there are more particles per volume. More particles result in more collisions, so the reaction rate increases, **Figure 6.20.** In most cases, concentration is increased by adding more reactant. If a fire is burning slowly, fanning the flames increases the amount of oxygen available and the fire burns faster. If a gas is involved, concentration may be increased by increasing pressure. Increasing pressure does not increase the number of particles, but it brings the particles closer together so collisions are more frequent. The Haber process, for example, uses high pressures to increase the rate of reaction of hydrogen and nitrogen to form ammonia. Lowering concentration decreases the rate of reaction. Many valuable historic documents are stored in sealed cases with most of the air removed to decrease the number of particles that might react with the paper.

Figure 6.19
Effect of Temperature on Reaction Rate
Adding heat to reactants helps break bonds and increases the speed at which molecules and atoms are moving. The faster they move, the more likely it is that they will collide and react. Removing heat slows down reactions. That's why freezing food might help keep it from spoiling as quickly.

Figure 6.20
Concentration and Reaction Rate
Adding more bumper boats increases the chance that a collision will occur. Taking some away decreases the odds of two boats meeting. When more particles are added to a reaction mixture, the chance that they will collide and react is increased.

miniLAB 3

Starch-Iodine Clock Reaction

In a clock reaction, an observable change indicating that a reaction has taken place occurs at some point in time after two or more different reactants are mixed. How soon this change occurs depends on the speed or rate of the reaction. Altering any factors that change reaction rate will change how quickly the observable change is seen.

Starch is an organic compound that reacts immediately with iodine to form a dark blue compound. The starch-iodine clock reaction involves timing the rate of formation of this blue compound after mixing two solutions. One solution reacts to produce iodine, and the other solution contains starch.

Procedure

1. Use a wax pencil to label five large test tubes with the numbers 1 through 5, and place them in a test-tube rack.

2. Place 10 mL of starch-containing solution in each of the five test tubes.

3. Place tube 4 in an ice bath and tube 5 in a water bath at 35°C. Leave them there for at least ten minutes.

4. Add the following amounts of iodine-producing solution to the indicated test tubes, stir carefully with a clean stirring rod or a wire stirrer, and record the time it takes for the blue color to appear.

   a) 10 mL to tube 1
   b) 5 mL to tube 2
   c) 20 mL to tube 3
   d) 10 mL to tube 4
   e) 10 mL to tube 5

5. Summarize your results in a table.

Analysis

1. What effect does changing the amount, and therefore the concentration, of a reactant have on the rate of a reaction? Why?

2. What effect does lowering the temperature have on the rate of a reaction? Why?

3. What effect does raising the temperature have on the rate of a reaction? Why?

Keep in mind that it doesn't matter how much the concentration of one reactant is increased if another reactant is depleted. Consider the situation that exists in **Figure 6.21.** Sometimes, when two reactants are present in a reaction, more of one than the other is available for reacting. The one that there is not enough of is called the **limiting reactant.** When it is completely used up, the reaction must stop. The reaction is limited because the other reactant alone cannot form any product.

Figure 6.21

Limiting Reactants
How many s'mores can you make if you have six graham cracker halves, three marshmallows, and two pieces of chocolate? Chocolate is the limiting reactant. When it is completely used up, only two s'mores have been made, and the s'more making must stop.

Everyday Chemistry

Stove in a Sleeve

That Napoleon stated "An army marches on its stomach" may be debated. However, few argue about the meaning of the statement. One of the necessities of military operations is supplying troops with meals, or *rations*. In the past, front-line troops often ate cold rations. Heating food by fire took too long, and the resulting smoke might have drawn enemy fire. What was needed was a portable, smokeless, self-contained method of quickly heating rations with readily available materials. The flameless heater was developed to meet these needs.

Flameless ration heaters

The flameless ration heater uses a heat-releasing chemical reaction similar to that of magnesium and water in forming magnesium hydroxide, $Mg(OH)_2$, and hydrogen gas. But pure magnesium can't be used in the heaters. Magnesium reacts with oxygen in the air to form a coating of magnesium oxide, MgO, that prevents the magnesium from reacting with other materials. Other materials are added to the magnesium to dissolve the MgO and promote the reaction between magnesium and water. A magnesium-iron alloy, composed of 95 percent magnesium and five percent iron, and a small amount of sodium chloride provide the desired effects. These materials are mixed with a powdered plastic to form a porous pad. Then the heat-producing reaction is started by adding water to the pad.

Hot MREs The heater is used to warm an individual portion of rations called a ready-to-eat meal, or MRE (*M*eal, *R*eady-to-*E*at). Each MRE comes in a pouch inside a cardboard sleeve. The MRE pouch slips into a small, plastic bag containing the Mg-Fe pad. After the water is added, the bag is placed inside the sleeve in which the MRE was packed.

The reaction transfers heat to the MRE pouch. The cardboard sleeve is an insulator to prevent heat loss. The reaction produces enough heat to raise the temperature of an 8-ounce ration serving by 60°C in about 12 minutes and keep the ration warm for about an hour. The heater and MRE can fit inside the pocket of a uniform, so it can be carried easily by soldiers. A hot meal-to-go might just be the morale booster a field soldier needs.

Exploring Further

1. **Classifying** Is the chemical reaction in the heater classified as an endothermic or an exothermic reaction?

2. **Applying** What other applications might flameless heating have?

What other chemicals produce heat? Visit the Chemistry Web site at chemistryca.com

Catalysts

Another way to change the rate of a reaction is to add or take away a catalyst. A **catalyst** is a substance that speeds up the rate of a reaction without being permanently changed or used up itself. Even though they increase the rate of reactions, catalysts do not change the position of equilibrium. Therefore, they do not affect how much product you can get from a reaction—just how fast a given amount of product will form.

How does a catalyst speed up a chemical reaction? You know that in chemical reactions, chemical bonds are broken and new ones are formed. The energy needed to break these bonds is the activation energy of the reaction. A catalyst speeds up the reaction by lowering the activation energy. This process can be compared to kicking a football over a goalpost. A player can kick the football over the goalpost at its regulation height, but if the height of the goalpost is lowered, the ball can be kicked over it using less energy.

Figure 6.22

Enzyme Action
Gelatin made with fresh pineapple will not harden properly, whereas gelatin made with canned pineapple will. Fresh pineapple contains active protease enzymes that break down protein molecules in the gelatin. Canned pineapple has been heated. Enzymes are heat-sensitive, so the proteases in the canned fruit are not active.

Many different compounds are able to act as catalysts. The most powerful catalysts are those found in nature. They are needed to speed up the reactions necessary for a cell to function efficiently. These biological catalysts are called **enzymes.** Enzymes help your body use food for fuel, build up your bones and muscles, and store extra energy as fat. Enzymes are involved in almost every process in a cell. For example, proteases are enzymes that break down proteins, as shown in **Figure 6.22.** These enzymes occur naturally in cells to help with recycling proteins so their parts can be used over and over. Proteases are also used in many common products, including contact lens-cleaning solution and meat tenderizer.

The importance of enzymes to our health can be seen when someone lacks a gene for a particular enzyme. For example, lactose intolerance results when an enzyme that breaks down lactose, the sugar in dairy products, is not produced in a person's digestive system, **Figure 6.23.** When lactose is not broken down, it accumulates in the intestine, causing bloating and diarrhea.

Slowing Down Reactions

Adding a catalyst speeds up a reaction. Would you ever want to slow down a reaction? Reactions that have undesirable products sometimes have to be slowed down. Food undergoes reactions that cause it to spoil. Medications decompose, destroying or limiting their effectiveness. Can any substances slow down these reactions?

A substance that slows down a reaction is called an **inhibitor.** Just as catalysts don't make reactions occur, inhibitors don't completely stop reactions. An inhibitor is placed in bottles of hydrogen peroxide to prevent it from decomposing too quickly into water and oxygen. If no inhibitor were added, the shelf life of hydrogen peroxide products would be much shorter because the molecules would decompose at a faster rate.

Try at Home Lab

See page 865 in Appendix F for **Preventing a Chemical Reaction**

Connecting Ideas

Now you are familiar with the types of chemical reactions that occur around you. You know the importance of equilibrium and are aware of what factors can affect the rate of reaction. But why do some substances react with each other while others don't? In the next chapter, you will find out how the structure of an atom affects the way it reacts with other atoms.

Supplemental Problems

For more practice with solving problems, see Supplemental Practice Problems, Appendix B.

SECTION REVIEW

Understanding Concepts

1. List and describe four factors that can influence the rate of a reaction.

2. Explain whether an exothermic reaction that is at equilibrium will shift to the left or to the right to readjust after each of the following procedures is followed.

   a) Products are removed.
   b) More reactants are added.
   c) More heat is added.
   d) Heat is removed.

3. Explain the difference between an inhibitor and a catalyst.

Thinking Critically

4. **Analyzing** Hydrogen gas can be produced by the reaction of magnesium and hydrochloric acid, as shown by this equation.

$$Mg(s) + 2HCl(aq) \rightarrow MgCl_2(aq) + H_2(g)$$

In a particular reaction, 6 billion molecules of HCl were mixed with 1 billion atoms of Mg.

   a) Which reactant is limiting?
   b) How many molecules of H_2 are formed when the reaction is complete?

Applying Chemistry

5. **Catalytic Converters** Catalytic converters on cars use the metals rhodium and platinum as catalysts to convert potentially dangerous exhaust gases to carbon dioxide, nitrogen, and water. Why don't cars need to have the rhodium and platinum replaced after they are used?

REVIEWING MAIN IDEAS

6.1 Chemical Equations

- Changes in color or odor, production or absorption of heat or light, gas release, and formation of a precipitate are all observable macroscopic changes that indicate that a chemical reaction may have occurred.

- Chemical equations are used to represent reactions. They are written using symbols and formulas for elements and compounds. Once the symbols and formulas are written, the equation can be balanced only by changing coefficients.

- Examining a chemical equation can tell you how elements and compounds change during a reaction. It may also tell you whether a reaction is endothermic or exothermic.

- Chemical equations must be balanced according to the law of conservation of mass, which states that matter cannot be created or destroyed in a chemical reaction.

6.2 Types of Reactions

- Although thousands of individual chemical reactions are known, most can be classified into five major classes that are based on patterns of behavior of reactants and products.

- The five general classes of reactions are synthesis, decomposition, single-displacement, double-displacement, and combustion.

6.3 Nature of Reactions

- Reversible reactions are those in which the products can react to form the reactants.

- Equilibrium occurs when no net change is taking place in a reaction.

- The direction in which a reaction shifts can be influenced by disturbing the equilibrium of reactants and products.

- How fast a reaction occurs can be influenced by temperature, concentration of reactants, and the presence of catalysts or inhibitors.

Vocabulary

For each of the following terms, write a sentence that shows your understanding of its meaning.

activation energy	equilibrium
catalyst	inhibitor
coefficient	insoluble
combustion	limiting reactant
concentration	product
decomposition	reactant
double displacement	single displacement
dynamic equilibrium	soluble
enzyme	synthesis

● UNDERSTANDING CONCEPTS

1. On which side of the arrow are the products of a reaction usually found? On which side are the reactants found?

2. Use the equation below to answer the following questions.

$$2Sr(s) + O_2(g) \rightarrow 2SrO(s)$$

   a) What is the physical state of strontium?
   b) What is the coefficient of strontium oxide?
   c) What is the subscript of oxygen?
   d) How many reactants take part in this reaction?

3. Use word equations to describe the chemical equations given.

   a) $AgNO_3(aq) + NaBr(aq) \rightarrow$
   $$AgBr(s) + NaNO_3(aq)$$
   b) $C_5H_{12}(l) + 8O_2(g) \rightarrow 5CO_2(g) + 6H_2O(g)$
   c) $CoCO_3(s) + energy \rightarrow CoO(s) + CO_2(g)$
   d) $BaCO_3(s) + C(s) + H_2O(g) \rightarrow$
   $$2CO(g) + Ba(OH)_2(s)$$

4. Explain why subscripts should never be changed when balancing an equation.

chemistryca.com/vocabulary_puzzlemaker

5. Balance the following equations.

  a) $Fe(s) + O_2(g) \rightarrow Fe_3O_4(s)$
  b) $NH_4NO_3(s) \rightarrow N_2O(g) + H_2O(g)$
  c) $COCl_2(g) + H_2O(l) \rightarrow HCl(aq) + CO_2(g)$
  d) $Sn(s) + NaOH(aq) \rightarrow Na_2SnO_2(aq) + H_2(g)$

6. Which of the five general types of reactions has only one reactant?

7. Hydrogen gas and iodine gas combine in a reversible exothermic reaction to form hydrogen iodide gas.

  a) Write a word equation for this reaction.
  b) Write a balanced chemical equation for this reaction.
  c) If more iodine gas is added after the reaction reaches equilibrium, will the reaction be shifted to the left or the right?
  d) If heat is added after the reaction reaches equilibrium, will the reaction be shifted to the left or the right?

● APPLYING CONCEPTS

8. List all the macroscopic changes that occur while a cake is made that indicate that chemical reactions have occurred.

9. One pollutant produced in automobile engines is nitrogen dioxide, NO_2. It changes to form nitrogen oxide, NO, and oxygen atoms when exposed to sunlight. Of what reaction type is this an example?

10. Catalytic converters help break down pollutants that result from gasoline combustion and are required in automobiles in the United States. Most catalytic converters contain a platinum-rhodium catalyst that coats the extensive surface of a honeycomb structure in the converter. Why would having the catalyst spread out over a large surface area be useful?

11. If you place a piece of a saltine cracker on your tongue for a few minutes, it begins to taste sweet. Do you think a chemical or physical change leads to this taste change? Why?

Everyday Chemistry

12. Explain why combining different cleaning products, such as those that contain ammonia and bleach, sometimes has deadly consequences. How could product manufacturers make the average person aware of hazards like this?

Chemistry and Technology

13. Why are high temperature and pressure needed for the Haber process?

Everyday Chemistry

14. What purposes might there be for a unit similar to an MRE that uses an endothermic reaction?

How It Works

15. When do you think it might be important that light sticks come in different colors?

Biology Connection

16. Why is it necessary that the gases breathed on the space shuttle are about the same composition as air on Earth?

● THINKING CRITICALLY

Relating Cause and Effect

17. Explain why it is best to store many chemicals in tightly sealed bottles in dark locations at cool or moderate temperatures.

Applying Concepts

18. **ChemLab** Why can't you find a reaction in the ChemLab that can be classified as two different reaction types?

Forming a Hypothesis

19. **MiniLab 1** Does the reaction taking place in MiniLab 1 go to completion, or does it reach an equilibrium? How do you know?

Making Predictions

20. **MiniLab 3** Predict the effect of each of the following on the speed of the starch-iodine clock reaction.

  a) adding an inhibitor to the reaction
  b) adding a catalyst to the reaction
  c) raising the temperature of the reaction

CUMULATIVE REVIEW

21. Compare the charges, masses, and atomic locations of electrons, protons, and neutrons. (Chapter 2)

22. Compare the number of electrons in the outer energy levels of metals, nonmetals, and metalloids. (Chapter 3)

23. Write a balanced equation for the formation of sodium chloride from sodium metal and chlorine gas. (Chapter 4)

24. Compare properties of ionic and covalent compounds. (Chapter 5)

SKILL REVIEW

25. **Relating Cause and Effect** Use what you have learned about factors that influence rates of chemical reactions to explain each of the following statements.

    a) Colors in the fabric of curtains in windows exposed to direct sunlight often fade.
    b) Meat is preserved longer when stored in a freezer rather than a refrigerator.
    c) Taking one or two aspirin will not harm most people, but taking a whole bottle at once can be fatal.
    d) BHA, or butylated hydroxyanisole, is an antioxidant often added to food, paints, plastics, and other products as a preservative.

26. **Modeling** Use toothpicks and Styrofoam balls of different colors to balance the following equations.

    a) $Cl_2O(g) + H_2O(l) \rightarrow HClO(aq)$
    b) $Fe_2O_3(s) + CO(g) \rightarrow Fe(s) + CO_2(g)$
    c) $H_2(g) + N_2(g) \rightarrow NH_3(g)$
    d) $ZnO(s) + HCl(aq) \rightarrow ZnCl_2(aq) + H_2O(l)$

27. **Interpreting Graphs** The graph shown represents the concentrations of two compounds, A and B, as they take part in a reaction that reaches equilibrium.

    a) Which compound represents the reactant in this reaction? The product?
    b) How long did it take for the reaction to reach equilibrium?
    c) Explain how the graph will change if more product is added one minute after the reaction reaches equilibrium.
    d) Summarize your answers in a report.

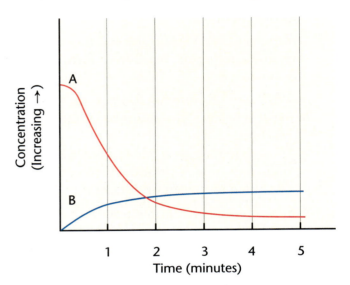

WRITING IN CHEMISTRY

28. Write a short paper describing five chemical reactions that commonly occur in your home, school, or neighborhood every day. Describe what you can see, smell, feel, hear, and taste as a result of these reactions.

PROBLEM SOLVING

29. Balance the following equations:

    a) $Xe(g) + F_2(g) \rightarrow XeF_6(s)$
    b) $Al(s) + H_2SO_4(aq) \rightarrow Al_2(SO_4)_3(aq) + H_2(g)$
    c) $CS_2(l) + O_2(g) \rightarrow CO_2(g) + SO_2(g)$
    d) $H_2SO_4(aq) + NaCN(s) \rightarrow$
       $HCN(g) + Na_2SO_4(aq)$
    e) $KClO_3(s) \rightarrow KCl(s) + O_2(g)$

30. Classify each reaction in question 29 as one of the five general reaction types.

1. When an iron nail is placed in water, a chemical reaction causes the nail to rust. What is (are) the reactant(s) in this reaction?

   a) iron
   b) water
   c) iron and rust
   d) iron and water

2. Which of the following chemical equations includes the formation of water?

   a) $2H_2O \rightarrow 2H_2 + O_2$
   b) $HC_2H_3O_2 + NaHCO_3 \rightarrow NaC_2H_3O_2 + H_2O + CO_2$
   c) $H_2 + O2 \rightarrow H_2O_2$
   d) H_2O (aq) $\rightarrow H_2O$ (s)

Free Energy of Chemical Reactions	
Chemical Reaction	**Free Energy (kJ/mol)**
a) $CH_4(g) + 2O_2(g) \rightarrow CO_2(g)$ $+ 2 H_2O(l)$	−890
b) $2H_2(g) + O_2(g) \rightarrow 2H_2O(g)$	−458
c) $2H_2(g) + O_2(g) \rightarrow 2H_2O(l)$	−572
d) $C_2H_4(g) + H_2(g) \rightarrow C_2H_6(g)$	+137
e) $6CO_2 + 6H_2O \rightarrow C_6H_{12}O_6$ $+ 6O_2 + 6H_2O$	+470

Use the table above to answer questions 3–4.

3. Which chemical reactions are endothermic?

   a) a, b, and c
   b) d and e
   c) all of the reactions
   d) none of the reactions

4. What is the amount of free energy associated with the conversion of liquid water into hydrogen and oxygen gas?

   a) −572 kJ/mol
   b) +572 kJ/mol
   c) −458 kJ/mol
   d) +458 kJ/mol

5. What is the product of this synthesis reaction?

   $$Cl_2(g) + 2NO(g) \rightarrow ?$$

   a) NCl_2
   b) $2NOCl$
   c) N_2O_2
   d) $2ClO$

6. What type of reaction is described by the following equation?

   $$Cs(s) + H2O(l) \rightarrow CsOH(aq) + H_2(g)$$

   a) synthesis
   b) decomposition
   c) single displacement
   d) double displacement

7. Lightning striking water to create hydrogen and oxygen gas is an example of

   a) synthesis.
   b) decomposition.
   c) single displacement.
   d) double displacement.

8. A system reaches chemical equilibrium when

   a) no new product is formed by the forward reaction.
   b) the reverse reaction no longer occurs in the system.
   c) the concentration of reactants in the system equals the concentration of products.
   d) the rate at which the forward reaction occurs equal the rate of the reverse reaction.

9. Which of the following can be used to speed up a reaction?

   a) increase temperature
   b) increase concentration
   c) add a catalyst
   d) all of the above

10. How can a reaction be slowed down?

   a) add an inhibitor
   b) increase the activation energy
   c) add more reactants
   d) all of the above

Test Taking Tip

Plan Your Work and Work Your Plan Plan your workload so that you do a little work each day rather than a lot of work all at once. The key to retaining information is repeated revieew and practice. You will retain more if you study one hour a night for five days in a row instead of cramming for five hours on a Sunday night.

7 Completing the Model of the Atom

Chapter Preview

Sections

7.1 Expanding the Theory of the Atom

MiniLab 7.1 Colored Flames—A Window Into the Atom

ChemLab Metals, Reaction Capacities, and Valence Electrons

7.2 The Periodic Table and Atomic Structure

MiniLab 7.2 Electrons in Atoms

Awesome Laser Show!

Lasers are a spectacular result of atoms absorbing energy. Energize electrons and these tiny particles respond by emitting pure colored light. Electrons drop from a higher energy level to a lower energy level and emit pure colored lights of specific frequencies.

*L*aunch Lab

Observing Electrical Charge

Electrical charge plays an important role in atomic structure and throughout chemistry. How can you observe the behavior of electrical charge using common objects?

Materials

- metric ruler
- plastic comb
- 10-cm long piece of clear plastic tape (4)
- hole punch
- paper

Procedure

1. Cut out small round pieces of paper using the hole punch and spread them out on a table. Run a plastic comb through your hair. Bring the comb close to the pieces of paper. Record your observations.
2. Fold a 1-cm-long portion of each piece of tape back on itself to form a handle. Stick two pieces of tape firmly to your desktop. Quickly pull both pieces of tape off of the desktop and bring them close together so that their non-sticky sides face each other. Record your observations.
3. Firmly stick one of the remaining pieces of tape to your desktop. Firmly stick the last piece of tape on top of the first. Quickly pull the pieces of tape as one from the desktop and then pull them apart. Bring the two tape pieces close together so that their non-sticky sides face each other. Record your observations.

Analysis

Use your knowledge of electrical charge to explain your observations. Which charges are similar? Which are different? How do you know?

What I Already Know

Review the following concepts before studying this chapter.
Chapter 2: the structure of the atom; electromagnetic radiation; the arrangement of electrons in energy levels
Chapter 3: the relationship of the periodic table to atomic structure

Reading Chemistry

Examine Figure 7.9 on page 244. Look closely at how the different elements are grouped, and try to determine what the elements in each group have in common. As you read, compare these reasons to the information given in the text.

Preview this chapter's content and activities at **chemistryca.com**

Expanding the Theory of the Atom

Electrons are strange. They don't behave like everyday objects. They're so small, they move so fast, and they seem to be in perpetual motion. They form clouds around the nucleus of the atom, but you can never be sure exactly where they are. Truly, electrons are strange. Their world is described in terms of probability, and that is different from the macroscopic world.

Scientists have discovered that electrons occupy a complex world of energy levels. They've described this world in terms of uncertainty, probability, and orbitals. In this section, you'll learn how to describe the energy levels in an atom and you'll see how electrons are arranged in these levels. The ways in which electrons are distributed in the energy levels of an atom account for many of the physical and chemical properties of the element.

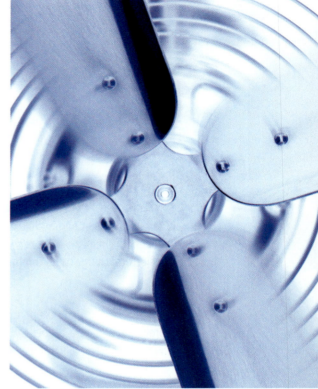

Developing a Model of Atomic Structure

You learned in Chapter 2 that in 1803, John Dalton proposed an atomic theory based on the law of conservation of matter. In his view, atoms were the smallest particles of matter. Then, in 1897, J.J. Thomson discovered electrons. The existence of electrons meant that the atom was made up of smaller particles. These particles include protons and neutrons, as well as electrons. At first, it wasn't clear how these subatomic particles were arranged. Scientists thought they were just mixed together like the ingredients in cookie dough. But in 1909, Ernest Rutherford performed an experiment in which he aimed atomic particles at a thin sheet of gold foil. He found that most particles went right through the foil, but some were deflected. These results suggested that most of the atom is empty space and that almost all of the mass of the atom is contained in a tiny nucleus. Rutherford proposed a nuclear model of the atom in which protons and neutrons make up the nucleus, and electrons move around in the space outside the nucleus.

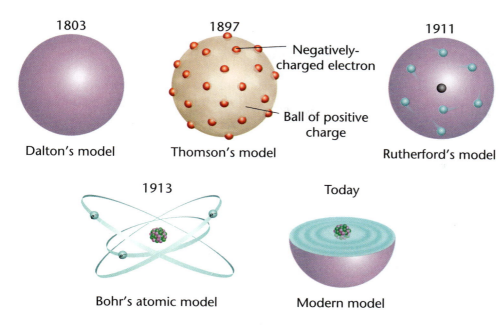

1803 Dalton's model

1897 Thomson's model
Negatively-charged electron
Ball of positive charge

1911 Rutherford's model

1913 Bohr's atomic model

Today Modern model

Figure 7.1

Building an Atomic Model The atomic theory evolved over a period of 2000 years. But it's the experimental evidence of the last 200 years that reveals the complex nature of the submicroscopic world. Because electrons are responsible for an element's chemical properties, chemists need an atomic model that describes the arrangement of electrons.

In 1913, the Danish physicist Niels Bohr suggested that electrons revolve around the nucleus just as planets revolve around the sun. Bohr's model was consistent with the emission spectrum produced by the hydrogen atom, but the model couldn't be extended to more complicated atoms. **Figure 7.1** illustrates the evolution of the atomic theory.

By 1935, the current model of the atom had evolved. This model explains electron behavior by interpreting the emission spectra of all the elements. It pictures energy levels as regions of space where there is a high probability of finding electrons. Before going on to the modern atomic theory, take another look at what you already know about atoms and electrons.

Building on What You Know

In the present-day model of the atom, neutrons and protons form a nucleus at the center of the atom. Negatively-charged electrons are distributed in the space around the nucleus. The electrons with the most energy are farthest from the nucleus and occupy the outermost energy level. Recall from Chapter 2 that evidence for the existence of energy levels came from the interpretation of the emission spectra of atoms. It's important to know about energy levels in atoms because it helps explain how atoms form chemical bonds and why they form particular kinds of compounds, for example ionic or covalent compounds.

Valence Electrons and the Periodic Table

The periodic table reflects each element's electron arrangement. In Chapter 3, you learned that the number of valence electrons is equal to the group number for elements in Groups 1 and 2 and is equal to the second digit of the group number for Groups 13 through 18. The period number represents the outermost energy level in which valence electrons are found. For example, lithium—in Group 1, Period 2—has one valence electron in the second energy level; sulfur—in Group 16, Period 3—has six valence electrons in the third level.

WORD ORIGIN

spectrum:
spectrum (L) appearance, specter

An emission spectrum reveals what cannot be seen directly.

Niels Bohr–Atomic Physicist and Humanitarian

Scientists are sometimes portrayed as strange and distant people. In fact, just like other people, they work, they interact with others, and they deal with everyday problems. They may differ from other people only in the way they use their intellects to be creative in areas of science. But they are like other people in that they may have strong ethical beliefs and the courage to fight for those beliefs. Niels Bohr's life exemplified these characteristics.

Bohr's atomic theory By the age of 28, Bohr had developed his atomic theory. Nine years later, in 1922, he received the Nobel Prize in Physics for this work. The basic ideas of Bohr's theory were that electrons move around the atom's nucleus in circular paths called orbits. These orbits are definite distances from the nucleus and represent energy levels that determine the energies of the electrons. Those electrons orbiting closest to the nucleus have the lowest energy; those farthest from the nucleus have the highest energy. If electrons absorb energy, they move to higher energy levels. When they drop down to lower energy levels, they release energy. Energy is absorbed and given off in definite amounts called quanta.

Theories change Parts of Bohr's theory are still accepted today, and parts are outdated. Electrons are thought to move around the nucleus, but not in definite paths. The idea of energy levels is correct, but now we know that they are regions of probability of finding electrons. An electron cannot be expected to be in any exact place. Electrons do jump to higher energy spaces as they gain energy and drop to lower ones when they lose energy.

Bohr helped develop the atomic bomb In 1939, Bohr attended a scientific conference in the United States, where he reported that Lise Meitner and Otto Hahn had discovered how to split uranium atoms—a process called fission. This dramatic announcement laid the groundwork for the development of the atomic bomb.

Bohr escaped from the Nazis In 1940, the Nazis invaded and occupied Niels Bohr's country, Denmark. Bohr was opposed to the Nazis, but he continued his position as director of the Copenhagen Institute for Theoretical Physics until 1943. When he learned that the Nazis planned to arrest him and force him to work in Germany on an atomic project, he and his family fled to Sweden under frightening conditions. In 1943, Bohr came to the United States and worked with scientists from all over the world on the Manhattan Project.

Connecting to Chemistry

1. **Interpreting** Bohr explained that the electrons in the outermost shell determine the chemical properties of an element. What did he mean?

2. **Thinking Critically** How important to chemistry is the physicist's work on atomic structure?

3. **Acquiring Information** Look up information on the Manhattan Project. Find out about the scientists involved, their nationalities, ages, genders, and their fields of expertise.

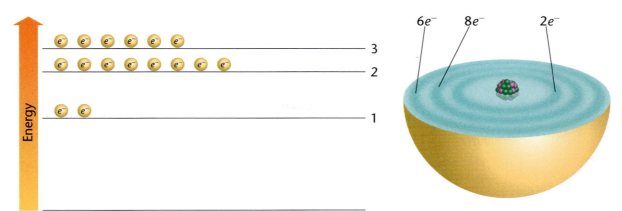

Energy Levels in Sulfur

6e⁻   8e⁻   2e⁻

You can use the periodic table to determine a complete energy-level diagram for any element. The atomic number of sulfur is 16. This means that sulfur has a total of 16 electrons. Sulfur is in Group 16 and Period 3, so it has six valence electrons in the third energy level. The remaining ten electrons must be distributed over the first two energy levels. **Figure 7.2** shows energy levels in sulfur.

Electromagnetic Radiation and Energy

Light, or electromagnetic radiation, can be described as waves having a range of frequencies and wavelengths. The higher the frequency of a wave and the shorter the wavelength, the greater the energy of the radiation. The lower the frequency and the longer the wavelength, the lower the energy. These relationships are used to calculate the exact amount of energy released by the electrons in atoms.

In the emission spectrum of hydrogen, shown on page 228, you can see the four different colors of visible light released by hydrogen as its electron moves from higher energy levels to a lower energy level. In Chapter 2, the analogy of energy levels as rungs on a ladder showed that electrons can move from one energy level to another, but they can't land between energy levels. By absorbing a specific amount of energy, an electron can jump to a higher energy level. Then, when it falls back to the lower energy level, the electron releases the same amount of energy in the form of radiation with a definite frequency. **Figure 7.3** shows how electron transitions between energy levels are related to amounts of energy.

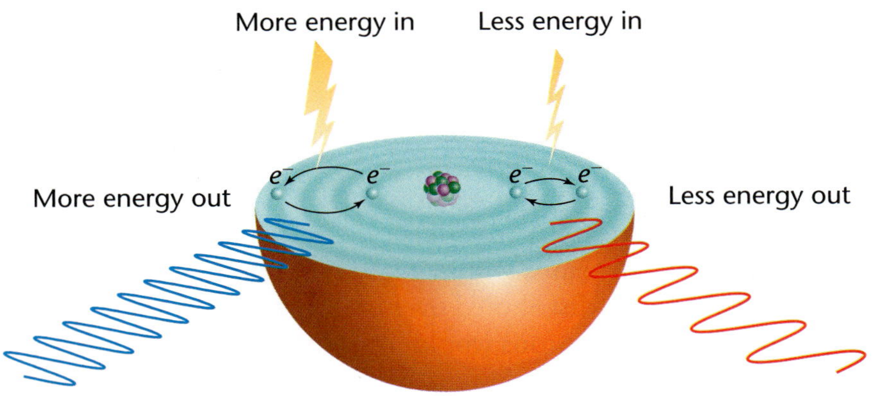

Figure 7.2
The Electron Distribution in Sulfur
The electrons in a sulfur atom, and in the atoms of all the other elements, are distributed over a range of energy levels shown in the diagram on the left. Relate the energies of the electrons in levels 1, 2, and 3 to the placement of the electrons in the model of the sulfur atom on the right.

More energy in   Less energy in

More energy out   Less energy out

Figure 7.3
Electron Transitions
The number of energy levels that an electron jumps depends on the amount of energy it absorbs. When an electron falls back to its original level, it emits energy in the form of light. The energy (color) of the light depends on how far the electron falls. The greater the energy given off, the more toward the violet end of the spectrum the color will be.

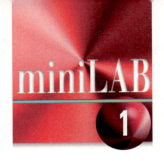

miniLAB 1

Colored Flames—A Window into the Atom

You've probably noticed that flames can be various colors, particularly in a fireworks display or when you burn logs treated with various salts in the fireplace. These colors are the result of electrons in metal atoms moving from higher energy levels to lower energy levels. The colors produced when compounds containing metals are heated in a flame can be used to identify the metals. The procedure is known as a flame test. In this MiniLab, you will study the flame tests of a few elements and identify an unknown element.

Procedure

1. Obtain from your teacher six wooden splints that are labeled and have been soaked in saturated solutions of the chlorides of lithium, sodium, potassium, calcium, strontium, and barium.

2. Light a laboratory burner or a propane torch held in a metal holder. Adjust the burner to give the hottest blue flame possible.

3. In turn, hold the soaked end of each of the splints in the flame for a short time. Observe and record the color of the flames. Extinguish the splint as soon as the flame is no longer colored.

4. Obtain a splint that has been soaked in a solution unknown to you. Perform the flame test and identify the unknown metallic element.

Analysis

1. What characteristic colors identify each of the metallic elements?

2. What is the identity of your unknown element?

3. Explain how you might test an unknown crystalline substance to determine whether it is table salt. A taste test is never recommended for an unknown compound.

Energy Levels and Sublevels

Just as the emission spectrum of hydrogen has four characteristic lines that identify it, so the emission spectrum for each element has a characteristic set of spectral lines. This means that the energy levels within the atom must also be characteristic of each element. But when scientists investigated multi-electron atoms, they found that their spectra were far more complex than would be anticipated by the simple set of energy levels predicted for hydrogen. **Figure 7.4** shows spectra for three elements.

Notice that these spectra have many more lines than the spectrum of hydrogen. Some lines are grouped close together, and there are big gaps between these groups of lines. The big gaps correspond to the energy released when an electron jumps from one energy level to another. The interpretation of the closely spaced lines is that they represent the movement of electrons from levels that are not very different in energy. This suggests that **sublevels**—divisions within a level—exist within a given energy level. If electrons are distributed over one or more sublevels within an energy level, then these electrons would have only slightly different energies. The energy sublevels are designated as *s, p, d,* or *f.*

Each energy level consists of sublevels that are close in energy. Each energy level has a specific number of sublevels, which is the same as the number of the energy level. For example, the first energy level has one sublevel. It's called the $1s$ sublevel. The second energy level has two sublevels, the $2s$ and $2p$ sublevels. The third energy level has three sublevels: the $3s$, $3p$, and $3d$ sublevels; and the fourth energy level has four sublevels: the $4s$, $4p$, $4d$, and $4f$ sublevels. Within a given energy level, the energies of the sublevels, from lowest to highest, are s, p, d, and f. **Figure 7.5** shows a diagram of the first three energy levels and an inside view of the sublevels within them. Notice how the sublevels within an energy level are close together. This explains the groups of fine lines in an element's emission spectrum. For example, you might expect three spectral lines with slightly different frequencies because electrons fall from the $3s$, $3p$, and $3d$ sublevels to the $2s$ sublevel. Because each of these electrons initially has slightly different energy within the third energy level, each emits slightly different radiation.

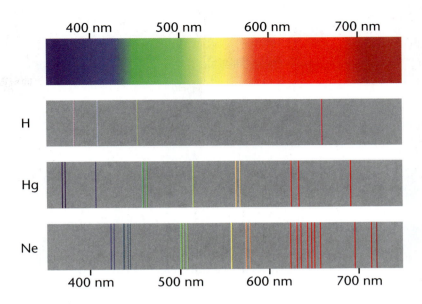

Figure 7.4

Comparison of Emission Spectra

The big gaps between spectral lines indicate that electrons are moving between energy levels that have a large difference in energy. The groups of fine lines indicate that electrons are moving between energy levels that are close in energy. The existence of sublevels within an energy level can explain the fine lines in the spectra of these elements.

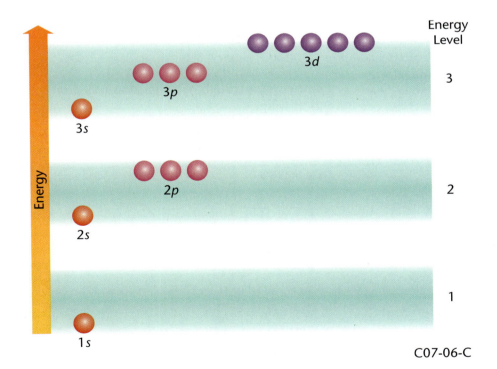

C07-06-C

Figure 7.5

Electron Distribution in an Atom

The diagram above shows the relative energies of the $1s$, $2s$, $2p$, $3s$, $3p$, and $3d$ sublevels. Electrons in the $1s$ sublevel are closest to the nucleus. Electrons in the $3s$, $3p$, and $3d$ sublevels are farthest from the nucleus.

ChemLab

SMALL SCALE

Metals, Reaction Capacities, and Valence Electrons

Many metals react with acids, producing hydrogen gas. If a metal reacts with acids in this manner, the amount of hydrogen produced is related to the number of valence electrons in the atoms of the metal. In this ChemLab, you will react equal numbers of atoms of magnesium and aluminum with hydrochloric acid and compare the reaction capacities of the two metals. Each reaction proceeds only as long as there are metal atoms to react.

Problem

How do the reaction capacities of magnesium and aluminum compare, and how are these capacities related to the valence electrons in the atoms of the two elements?

Objectives

- **Compare** the reaction capacities of magnesium and aluminum.
- **Interpret** the results of the experiment in terms of the numbers of valence electrons in the atoms of the two elements.

PREPARATION

Materials

thin-stemmed micropipets (2)	3M HCl
50-mL graduated cylinder	magnesium ribbon
water trough or plastic basin	aluminum foil
1M HCl	transparent, waterproof tape
	plastic wrap
	forceps (2)

Safety Precautions

Wear goggles and an apron. The bulb of the micropipet may become hot during the reaction. Hold the stem portion of the pipet with forceps.

PROCEDURE

1. Cut small slits in two micropipets as shown. Obtain a 0.020-g sample of magnesium ribbon from your teacher and insert the magnesium into the bulb of one pipet. Obtain a 0.022-g sample of aluminum foil and insert it into the bulb of the other pipet. Seal the slits with transparent, waterproof tape and label the two pipets *Mg* and *Al*.

2. Fill the water trough or plastic basin nearly full of water.

3. Fill the 50-mL graduated cylinder to the brim with water and cover the top with plastic wrap.

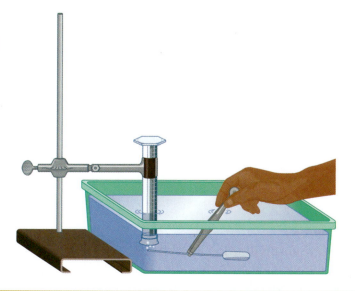

Hold the wrap tightly so that no water can escape and invert the cylinder, placing the top beneath the surface of the water in the trough. Remove the plastic wrap. No appreciable amount of air should be in the cylinder. If this is not the case, repeat the procedure. The inverted cylinder may be clamped in place or held by hand.

4. Using the pipet containing the magnesium, squeeze most of the air from the pipet bulb and draw 3 mL of 1M HCl into the pipet. (**CAUTION:** *Handle the hydrochloric acid solution with care. It is harmful to eyes, skin, and clothing. If any acid contacts your skin or eyes or any spillage occurs, rinse immediately with water and notify your teacher.*)

5. Hold the pipet with forceps as shown, and quickly immerse it in the water in the trough so that the pipet tip is inside the open end of the graduated cylinder.

6. Collect the hydrogen gas in the graduated cylinder. Allow the reaction to proceed until the magnesium ribbon is completely gone.

7. While the graduated cylinder is still in place, read the volume of hydrogen gas produced and record it in a table like the one shown.

8. Remove the tape from the pipet and rinse the pipet with water.

9. Repeat steps 4 through 8 using the micropipet containing aluminum and 3M HCl. Read and record the volume of hydrogen gas produced.

DATA AND OBSERVATIONS

Hydrogen from magnesium (mL)	
Hydrogen from aluminum (mL)	

ANALYZE AND CONCLUDE

1. **Interpreting Data** The volume of the hydrogen gas you collected is proportional to the number of molecules of hydrogen produced in the two reactions. Which element, aluminum or magnesium, produced more hydrogen molecules?

2. **Comparing and Contrasting** You used approximately equal numbers of atoms of magnesium and aluminum and enough HCl to react with all the atoms. Which of the two elements produced more hydrogen molecules per atom?

3. **Drawing Conclusions** Which element has the greater reaction capacity per atom? Use your volume data to express the relative reaction capacities of the two elements as a ratio of small, whole numbers.

4. **Relating Concepts** In this experiment, the atoms of both metals react by losing electrons to form positive ions. Relate the ratio of reaction capacities to the number of valence electrons each element has.

APPLY AND ASSESS

1. Write the balanced equations for the two chemical reactions you performed.

2. Had you performed this experiment with sodium, using approximately the same number of atoms as you used of magnesium and aluminum, predict the volume of hydrogen gas that would have been produced. (**CAUTION:** *Because sodium reacts explosively with water, it cannot be used safely in this experiment.*)

The Distribution of Electrons in Energy Levels

A specific number of electrons can go into each sublevel. An *s* sublevel can have a maximum of two electrons, a *p* sublevel can have six electrons, a *d* sublevel can have ten electrons, and an *f* sublevel can have 14 electrons. **Table 7.1** shows how electrons are distributed in the sublevels of the first four energy levels. Notice that the first energy level has one sublevel, the 1*s*. The maximum number of electrons in an *s* sublevel is two, so the first energy level is filled when it reaches two electrons. The second energy level has two sublevels, the 2*s* and 2*p*. The maximum number of electrons in a *p* sublevel is six, so the second energy level is filled when it reaches eight electrons—two in the 2*s* sublevel and six in the 2*p* sublevel. Look at the third and fourth energy levels and you can see that the third energy level can hold ten more electrons than the second because there is a *d* sublevel. The fourth can hold 14 more electrons than the third because there is an *f* sublevel.

Table 7.1   Distribution of Electrons in the First Four Energy Levels

Energy Level	Sublevel	Electrons in Sublevel	Electrons in Level
1	1*s*	2	2
2	2*s*	2	
2	2*p*	6	8
3	3*s*	2	
3	3*p*	6	
3	3*d*	10	18
4	4*s*	2	
4	4*p*	6	
4	4*d*	10	
4	4*f*	14	32

Energy level

4 xx xxxxxx xxxxxxxxxx xxxxxxxxxxxxxx

3 xx xxxxxx xxxxxxxxxx   *f*

2 xx xxxxxx   *d*
                 p

1 xx   x = 1 electron
   s

Orbitals

In the 1920s, Werner Heisenberg reached the conclusion that it's impossible to measure accurately both the position and energy of an electron at the same time. This principle is known as the **Heisenberg uncertainty principle.** In 1932, Heisenberg was awarded the Nobel Prize in Physics for this discovery, which led to the development of the electron cloud model to describe electrons in atoms.

The electron cloud model is based on the probability of finding an electron in a certain region of space at any given instant. Here's how it works. Pretend that you can photograph the single electron that is attracted to the nucleus of a hydrogen atom. Once every second, you click the shutter on your camera, forming images of the electron on the same piece of film. After a few hundred snapshots, you develop the film and find a scatter diagram.

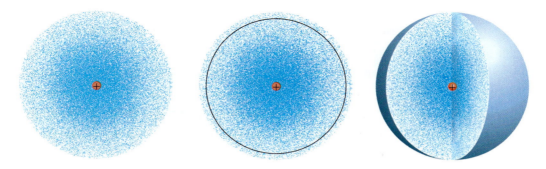

Sometimes, the electron is close to the nucleus. Other times, it's far away. Most of the time, it's in a small region of space that looks like a cloud. This electron cloud doesn't have a sharp boundary; its edges are fuzzy. If you ask someone to look at this picture, shown in **Figure 7.6,** and tell you where the electron is, he or she could say only that it's probably somewhere in the cloud. But this cloud of electron probability can be useful. If you draw a line around the outer edge enclosing about 95 percent of the cloud, within the region enclosed by the sphere, you can expect to find the electron about 95 percent of the time. The space in which there is this high probability of finding the electron is called an **orbital.**

Placing Electrons in Orbitals

Orbitals are regions of space located around the nucleus of an atom, each having the energy of the sublevel of which it is a part. Orbitals can have different sizes and shapes. There are four types of orbitals that accommodate the electrons for all the atoms of the known elements. Two simple rules apply to these four orbitals. First, an orbital can hold a maximum of two electrons. Second, an orbital has the same name as its sublevel. There is only one *s* orbital with, at most, 2 electrons. A *p* sublevel has, at most, 6 electrons, and, thus, there are three *p* orbitals, each with 2 electrons. **Figure 7.7** shows the shapes of *s* and *p* orbitals.

Figure 7.6

Hydrogen's Electron Cloud Most of the time, hydrogen's electron is within the fuzzy cloud in the two-dimensional drawing (left). A circle, with the nucleus at the center and enclosing 95 percent of the cloud, defines an orbital in two dimensions (center). The spherical model (right) represents hydrogen's 1*s* orbital in three dimensions.

Figure 7.7

Models of *s* and *p* Orbitals
Orbitals have characteristic shapes that depend on the number of electrons in the energy sublevels. An *s* orbital is spherically symmetrical about the nucleus; a *p* orbital has a dumbbell shape. Three *p* orbitals are aligned along the *x*, *y*, or *z* axis at each energy level.

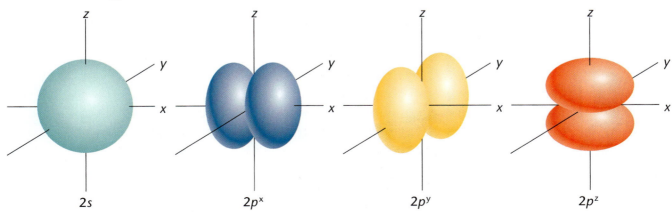

$2s$ $2p^x$ $2p^y$ $2p^z$

Hi-Tech Microscopes

If you were a chemistry student in the 1960s or earlier, you would have been told that no one could see an atom. Now, with the aid of computers and revolutionary microscopes, it's possible to generate two- and three-dimensional images of atoms. It's even possible to move atoms around and observe their electron clouds. What kinds of instruments can accomplish feats that were not even thought of a decade or two ago? They include three kinds of microscopes: scanning probe, scanning tunneling, and atomic force.

Scanning Probe Microscope (SPM)

SPMs use probes to sense surfaces and produce three-dimensional pictures of a substance's exterior, as shown below. The precise arrangement of atoms can be viewed in three dimensions. This is accomplished by measuring changes in the current as the probe passes over a surface. SPMs can pick up atoms one at a time, move them around, and form letters, such as IBM shown below. Accomplishments like this suggest that in the future, all the information in the Library of Congress could be stored on an 8-inch silicon wafer. Scientists may

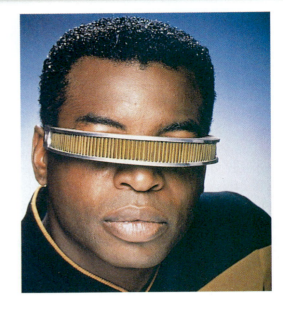

be able to use SPMs to learn why two surfaces stick together or how to shrink chip circuits to make computers work faster. They may even be able to put a thousand SPM tips on a 2-cm^2 silicon chip. Arrays of such chips could produce a working imitation of human sight. Could this mean that a small computer in the form of a visor similar to that worn by Geordi (above) on "Star Trek" will be available for the visually impaired in a few years?

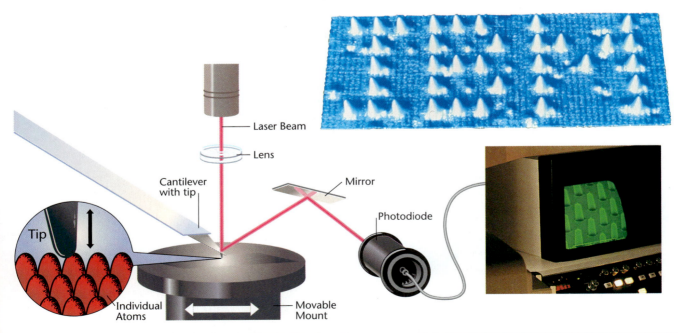

Laser Beam

Lens

Cantilever with tip

Mirror

Photodiode

Tip

Individual Atoms

Movable Mount

Scanning Tunneling Microscope (STM)

The scanning tunneling microscope also provides new technologies for chemists and physicists. The red areas in the photo below show the valence electrons of metal atoms that are free to move about in a metallic crystal. On the surface of the crystal, they can move in only two dimensions and behave like waves. Two imperfections on the surface of the crystal cause the electrons to produce concentric wave patterns.

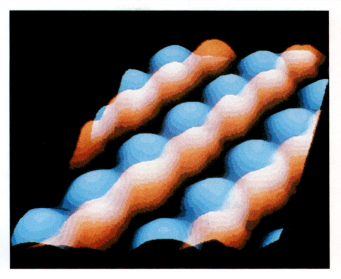

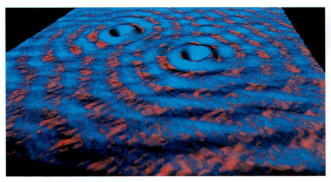

Wave patterns caused by electrons

Another current STM technology uses hydrogen to push aside surface atoms of semiconductors to reveal the atomic structure underneath. Using this method, atoms can be removed one layer at a time.

Recently, the STM has succeeded in forming images of electron clouds in atoms and molecules. The electron clouds of molecules can be calculated. An STM color-enhanced image of gallium arsenide is shown at right. The electron clouds of gallium appear blue. The electron clouds of arsenic appear red. Chemists have figured out the structure of atoms and molecules using indirect evidence from chemical reactions, physical experiments, and mathematical calculations. Now, they can view images of atoms while they experiment with them and base their work on both direct and indirect evidence.

Atomic Force Microscope (AFM)

The AFM, invented in 1985, uses repulsive forces between atoms on the probe's tip and those on the sample's surface to form images on the computer screen. An advantage of the AFM is that it does not have to work in a vacuum, so its samples require no special preparation. AFMs can provide images of molecules in living tissue and peel off membranes of living cells one layer at a time.

DISCUSSING THE TECHNOLOGY

1. **Comparing and Contrasting** Some scanning tunneling microscopes can remove layers of atoms. How does this capability compare with what the scanning probe microscope can do?

2. **Acquiring Information** Find information on the Near-Field Scanning Optical Microscope.

3. **Inferring** Compare the diagrams of the molecules with the computer-generated images of the molecules produced by the STM. How are the conclusions about the molecular structure obtained using conventional methods supported by the photos?

Electron Configurations

In any atom, electrons are distributed into sublevels and orbitals in the way that creates the most stable arrangement; that is, the one with lowest energy. This most stable arrangement of electrons in sublevels and orbitals is called an **electron configuration.** Electrons fill orbitals and sublevels in an orderly fashion beginning with the innermost sublevels and continuing to the outermost. At any sublevel, electrons fill the *s* orbital first, then the *p*. For example, the first energy level holds two electrons. These electrons pair up in the 1*s* orbital. The second energy level has four orbitals and can hold eight electrons. The first two electrons pair up in the 2*s* orbital, and the remaining six pair up in the three 2*p* orbitals. **Figure 7.8** shows that the overlap of the 2*s* and 2*p* orbitals results in a roughly spherical cloud. That's why the electrons in an atom can be represented as a series of fuzzy, concentric spheres. In the next section, you'll learn how the periodic table can be used to predict the electron configurations of the atoms.

Figure 7.8

Overlapping Orbitals
Because orbitals are regions of space, they can be placed one on top of another. Imagine a pair of electrons moving around in an orbital. At any instant, most of the orbital is empty space, and this space can be used by another pair of electrons. That's how orbitals can overlap.

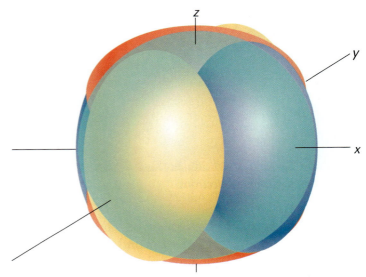

One 2*s* and three 2*p* orbitals

For more practice with solving problems, see Supplemental Practice Problems, Appendix B.

SECTION REVIEW

Understanding Concepts

1. How many *s* orbitals can the third energy level have? How many *p* orbitals? How many *d* orbitals?

2. How many electrons can each sublevel of the fourth energy level hold, and how many orbitals are required to accommodate them?

3. What is the shape of a *p* orbital? How do *p* orbitals at the same energy level differ from one another?

Thinking Critically

4. **Applying Concepts** What are two ways in which the elements with atomic numbers 12 and 15 are different?

Applying Chemistry

5. **Outdoor Lighting** Sodium vapor contained in a bulb or tube emits a brilliant yellow light when connected to a high-voltage source. Explain what is happening to the sodium atoms to produce this light.

The Periodic Table and Atomic Structure

Suppose you were about to play your first game of Chinese checkers. How would you know where to place the marbles on the game board to start the game? The game board has many indentations to accommodate the marbles, but the marbles can be placed in only one way at the start of the game. Without the rules of the game or maybe a diagram of the starting game board, you wouldn't know what to do.

The same is true of placing electrons in orbitals around a nucleus. Imagine yourself with the bare phosphorus nucleus and a bag of electrons. So many energy levels and sub-levels are available, each with one or more orbitals for electrons. How can you know where to put the electrons to produce the most stable electron configuration? You will learn some simple rules and discover that the periodic table provides a clear diagram for building electron configurations.

SECTION PREVIEW

Objectives
✓ **Distinguish** the *s, p, d,* and *f* blocks on the periodic table and relate them to an element's electron configuration.

✓ **Predict** the electron configurations of elements using the periodic table.

Review Vocabulary
Orbital: space in which there is a high probability of finding an electron.

New Vocabulary
inner transition element

Patterns of Atomic Structure

Electrons occupy energy levels by filling the lowest level first and continuing to higher energy levels in numerical order. Valence electrons of the main group elements occupy the *s* and *p* orbitals of the outermost energy level. The position of any element in the periodic table shows which orbitals—*s, p, d,* and *f*—the valence electrons occupy.

Orbitals and the Periodic Table

The shape of the modern periodic table is a direct result of the order in which electrons fill energy sublevels and orbitals. The periodic table in **Figure 7.9** is divided into blocks that show the sublevels and orbitals occupied by the electrons of the atoms. Notice that Groups 1 and 2 (the active metals) have valence electrons in *s* orbitals, and Groups 13 to 18 (metals, metalloids, and nonmetals) have valence electrons in both *s* and *p* orbitals. Therefore, all the main group elements have their valence electrons in *s* or *p* orbitals. Groups 1 and 2 are designated as the *s* region of

The Periodic Table

	1																	18
1s	1 H	2																2 He
2s	3 Li	4 Be											13	14	15	16	17	
3s	11 Na	12 Mg										2p	5 B	6 C	7 N	8 O	9 F	10 Ne
4s	19 K	20 Ca	3	4	5	6	7	8	9	10	11	12	3p	13 Al	14 Si	15 P	16 S	17 Cl	18 Ar

3d 21 Sc · 22 Ti · 23 V · 24 Cr · 25 Mn · 26 Fe · 27 Co · 28 Ni · 29 Cu · 30 Zn
4d 39 Y · 40 Zr · 41 Nb · 42 Mo · 43 Tc · 44 Ru · 45 Rh · 46 Pd · 47 Ag · 48 Cd
5d 57 La · 72 Hf · 73 Ta · 74 W · 75 Re · 76 Os · 77 Ir · 78 Pt · 79 Au · 80 Hg
6d 89 Ac · 104 Rf · 105 Db · 106 Sg · 107 Bh · 108 Hs · 109 Mt · 110 Ds · 111 Uuu · 112 Uub

4p 31 Ga · 32 Ge · 33 As · 34 Se · 35 Br · 36 Kr
5p 49 In · 50 Sn · 51 Sb · 52 Te · 53 I · 54 Xe
6p 81 Tl · 82 Pb · 83 Bi · 84 Po · 85 At · 86 Rn

5s 37 Rb · 38 Sr
6s 55 Cs · 56 Ba
7s 87 Fr · 88 Ra

114 Uuq

s region *d* region *p* region

4f 58 Ce · 59 Pr · 60 Nd · 61 Pm · 62 Sm · 63 Eu · 64 Gd · 65 Tb · 66 Dy · 67 Ho · 68 Er · 69 Tm · 70 Yb · 71 Lu
5f 90 Th · 91 Pa · 92 U · 93 Np · 94 Pu · 95 Am · 96 Cm · 97 Bk · 98 Cf · 99 Es · 100 Fm · 101 Md · 102 No · 103 Lr

f region

Figure 7.9

The Periodic Table: Key to Electron Configurations

It's not necessary to memorize electron configurations if you can interpret the *s, p, d,* and *f* blocks shown in this periodic table. When you write electron configurations, move from left to right through the periods, filling the orbitals that correspond to the *s, p, d,* and *f* blocks.

the periodic table, and Groups 13 to 18 are designated as the *p* region. Note that Groups 3 to 12 are designated as the *d* region and that each row of this region, except for that in Period 7, has ten elements. The block beneath the table is the *f* region, and each row in this region contains 14 elements.

Building Electron Configurations

Chemical properties repeat when elements are arranged by atomic number because electron configurations repeat in a certain pattern. As you move through the table, you'll notice how an element's position is related to its electron configuration. Hydrogen has a single electron in the first energy level. Its electron configuration is $1s^1$. This is standard notation for electron configurations. The number *1* refers to the energy level, the letter *s* refers to the sublevel, and the superscript refers to the number of electrons in the sublevel. Helium has two electrons in the $1s$ orbital. Its electron configuration is $1s^2$. Helium has a completely filled first energy level. When the first energy level is filled, additional electrons must go into the second energy level. Electrons enter the sublevel that will give the atom the most stable configuration, that with the lowest energy.

Lithium begins the second period. Its first two electrons fill the first energy level, so the third electron occupies the second level. Lithium's electron configuration is $1s^2 2s^1$. Beryllium has two electrons in the $2s$ orbital, so its electron configuration is $1s^2 2s^2$. As you continue to move across the second period, electrons begin to enter the *p* orbitals. Each successive element has one more electron in the $2p$ orbitals. Carbon, for example, has four electrons in the second energy level. Two of these are in the $2s$ orbital and two are in the $2p$ orbitals. The electron configuration for carbon is $1s^2 2s^2 2p^2$. At element number 10, neon, the *p* sublevel is filled with six electrons. The electron configuration for neon is $1s^2 2s^2 2p^6$. Neon has eight valence electrons; two are in an *s* orbital and six are in *p* orbitals.

Electrons in Atoms

The modern theory of the atom cannot tell you exactly where the electrons in atoms are placed. However, it does define regions in space called orbitals, where there is a 95 percent probability of finding an electron. The lowest energy orbital in any atom is called the 1s orbital. In this MiniLab, you will simulate the probability distribution of the 1s orbital by noting the distribution of impacts or hits around a central target point.

Procedure

1. Obtain two pieces of blank, white, $8\frac{1}{2}'' \times 11''$ paper and draw a small but visible mark in the center of each of the papers. Hold the papers together toward a light and align the center marks exactly.

2. Around the center dot of one of the papers, which you will call the target paper, draw concentric circles having radii of 1 cm, 3 cm, 5 cm, 7 cm, and 9 cm. Number the areas of the target 1, 2, 3, 4, and 5, starting with number 1 at the center.

3. Place a piece of poster board on the floor, and lay the target paper face up on top of it. Cover the target paper with a piece of carbon paper, carbon side down. Then place the second piece of white paper on top with the center mark facing up. Use tape to fasten the three layers of paper in place on the poster board and to secure the poster board to the floor.

4. Stand over the target paper and drop a dart 100 times from chest height, attempting to hit the center mark.

5. Remove the tape from the papers. Separate the white papers and the carbon paper. Tabulate and record the number of hits in each area of the target paper.

Analysis

1. How many hits did you record in each of the target areas? What does each hit represent in the model of the atom?

2. Make a graph plotting the number of hits on the vertical axis and the target area on the horizontal axis.

3. Which of the target areas has the highest probability of a hit? Relate your findings to the model of the atom.

Sodium, atomic number 11, begins the third period and has a single $3s$ electron beyond the configuration of neon. Sodium's electron configuration is $1s^2 2s^2 2p^6 3s^1$. If you compare this with the electron configuration of lithium, $1s^2 2s^1$, it's easy to see why sodium and lithium have similar chemical properties. Each has a single electron in the valence level.

Notice that neon's configuration has an inner core of electrons that is identical to the electron configuration in helium ($1s^2$). This insight simplifies the way electron configurations are written. Neon's electron configuration can be abbreviated $[He]2s^2 2p^6$. In the abbreviated form, neon's electron configuration is represented by an inner core of electrons from the noble gas in the preceding row (He), followed by the orbitals filled in the current period. The abbreviated electron configuration for sodium is [Ne]$3s^1$, where the neon core represents sodium's ten inner electrons. Its similarity to the [He]$2s^1$ configuration of lithium shows clearly that these Group 1 elements have the same number of valence electrons in the same type of orbital. **Table 7.2** shows electron configurations for all the elements in the second and third periods. Notice that elements in the same group have similar configurations. This is important because it shows that the periodic trends in properties, observed in the periodic table, are really the result of repeating patterns of electron configuration.

Table 7.2   Electron Configurations of Second and Third Period Elements

Second Period Elements	Configuration	Third Period Elements	Configuration
Lithium	[He] $2s^1$	Sodium	[Ne] $3s^1$
Beryllium	[He] $2s^2$	Magnesium	[Ne] $3s^2$
Boron	[He] $2s^22p^1$	Aluminum	[Ne] $3s^23p^1$
Carbon	[He] $2s^22p^2$	Silicon	[Ne] $3s^23p^2$
Nitrogen	[He] $2s^22p^3$	Phosphorus	[Ne] $3s^23p^3$
Oxygen	[He] $2s^22p^4$	Sulfur	[Ne] $3s^23p^4$
Fluorine	[He] $2s^22p^5$	Chlorine	[Ne] $3s^23p^5$
Neon	[He] $2s^22p^6$	Argon	[Ne] $3s^23p^6$

The Stable Configurations of the Noble Gases

Each period ends with a noble gas, so all the noble gases have filled energy levels and, therefore, stable electron configurations. The electron configurations for all the noble gases are shown in **Table 7.3.** All except helium have eight valence electrons. However, helium's two electrons fill its outermost energy level and are a stable configuration. These stable electron configurations explain the lack of reactivity of the noble gases. Noble gases don't need to form chemical bonds to acquire stability.

Table 7.3   The Electron Configurations of the Noble Gases

Noble Gas	Electron Configuration
Helium	$1s^2$
Neon	[He] $2s^22p^6$
Argon	[Ne] $3s^23p^6$
Krypton	[Ar] $4s^23d^{10}4p^6$
Xenon	[Kr] $5s^24d^{10}5p^6$
Radon	[Xe] $6s^24f^{14}5d^{10}6p^6$

Noble gases at work

What happens in the fourth period?

You might expect that after the $3p$ orbitals are filled in argon, the next electron would occupy a $3d$ orbital, but this is not the case. Potassium follows argon and begins the fourth period. Its configuration is $[Ar]4s^1$. Compare potassium's configuration to the configurations of lithium and sodium in **Table 7.2,** and recall that potassium is chemically similar to the Group 1 elements. Experimental evidence indicates that the $4s$ and $3d$ sublevels are close in energy, with the $4s$ sublevel having a slightly lower energy. Thus, the $4s$ sublevel fills first because that order produces an atom with lower energy. The next element after potassium is calcium. Calcium completes the filling of the $4s$ orbital. It has the electron configuration $[Ar]4s^2$.

Transition Elements

Notice in the periodic table that calcium is followed by a group of ten elements beginning with scandium and ending with zinc. These are transition elements. Now the $3d$ sublevel begins to fill, producing atoms with the lowest possible energy. Ten electrons are added across the row and fill the $3d$ sublevel. Just as the s block at each energy level adds two electrons and fills one s orbital, and the p block at each energy level adds six electrons and fills three p orbitals, so the d block adds ten electrons. There are five d orbitals. The f block adds 14 electrons, and they are accommodated in seven orbitals. The electron configuration of scandium, the first transition element, is $[Ar]4s^23d^1$. Zinc, the last of the series, has the configuration $[Ar]4s^23d^{10}$. **Figure 7.10** shows the configurations of the $3d$ transition elements. The six elements following the $3d$ elements, gallium to krypton, fill the $4p$ orbitals and complete the fourth period.

	3	4	5	6	7	8	9	10	11	12
$3d$	Scandium 21 **Sc** $[Ar]4s^23d^1$	Titanium 22 **Ti** $[Ar]4s^23d^2$	Vanadium 23 **V** $[Ar]4s^23d^3$	Chromium 24 **Cr** $[Ar]4s^13d^5$	Manganese 25 **Mn** $[Ar]4s^23d^5$	Iron 26 **Fe** $[Ar]4s^23d^6$	Cobalt 27 **Co** $[Ar]4s^23d^7$	Nickel 28 **Ni** $[Ar]4s^23d^8$	Copper 29 **Cu** $[Ar]4s^13d^{10}$	Zinc 30 **Zn** $[Ar]4s^23d^{10}$

Figure 7.10

The Electron Configurations of 3d Transition Elements
Ten electrons are added across the d block, filling the d orbitals. Notice that chromium and copper have only one electron in the $4s$ orbital. Such unpredictable exceptions show that the energies of the $4s$ and $3d$ sublevels are close.

Like most metals, the transition elements lose electrons to attain a more stable configuration. Most have multiple oxidation numbers because their s and d orbitals are so close in energy that electrons can be lost from both orbitals. For example, cobalt (atomic number 27) forms two fluorides—one with the formula CoF_2 and another with the formula CoF_3. In the first case, cobalt gives up two electrons to fluorine. In the second case, cobalt gives up three electrons.

Chemistry

Colors of Gems

Have you ever wondered what produces the gorgeous colors in a stained-glass window or in the rubies, emeralds, or sapphires mounted on a ring? Compounds of transition elements are responsible for creating the entire spectrum of colors.

Transition elements color gems and glass Transition elements have many important uses, but one that is often overlooked is their role in giving colors to gemstones and glass. Although not all compounds of transition elements are colored, most inorganic colored compounds contain a transition element such as chromium, iron, cobalt, copper, manganese, nickel, cadmium, titanium, gold, or vanadium. The color of a compound is determined by three factors: (1) the identity of the metal, (2) its oxidation number, and (3) the negative ion combined with it.

Impurities give gemstones their color Crystals have fascinating properties. A clear, colorless quartz crystal is pure silicon dioxide (SiO_2). But a crystal that is colorless in its pure form may exist as a variety of colored gemstones when tiny amounts of transition element compounds, usually oxides, are present. Amethyst (purple), citrine (yellow-brown), and rose quartz (pink) are quartz crystals with transition element impurities scattered throughout. Blue sapphires are composed of aluminum oxide (Al_2O_3) with the impurities iron(II) oxide (FeO) and titanium(IV) oxide (TiO_2). If trace amounts of chromium(III) oxide (Cr_2O_3) are present in the Al_2O_3, the resulting gem is a red ruby. A second kind of gemstone is one composed entirely of a colored compound. Most are transition element compounds, such as rose-red rhodochrosite ($MnCO_3$), black-grey hematite (Fe_2O_3), or green malachite ($CuCO_3 \cdot Cu(OH)_2$).

Citrine

Quartz

Amethyst

How metal ions interact with light to produce color Why does the presence of Cr_2O_3 in Al_2O_3 make a ruby red? The Cr^{3+} ion absorbs yellow-green colors from white light striking the ruby, and the remaining red-blue light is transmitted, resulting in a deep red color. This same process occurs in all gems. Trace impurities absorb certain colors of light from white light striking or passing through the stone. The remaining colors of light that are reflected or transmitted produce the color of the gem.

Adding transition elements to molten glass for color Glass is colored by adding transition element compounds to the glass while it is molten. This is true for stained glass, glass used in glass blowing, and even glass in the form of ceramic glazes. Most of the coloring agents are oxides. When oxides of copper

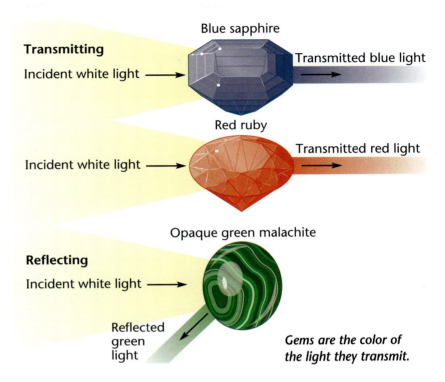

Transmitting
Incident white light → Blue sapphire → Transmitted blue light

Incident white light → Red ruby → Transmitted red light

Reflecting
Incident white light → Opaque green malachite
Reflected green light

Gems are the color of the light they transmit.

or cobalt are added to molten glass, the glass is blue; oxides of manganese produce purple glass; iron oxides, green; gold oxides, deep ruby red; copper or selenium oxides, red; and antimony oxides, yellow. Some coloring compounds are not oxides. Chromates, for example, produce green glass, and iron sulfide gives a brown color.

Exploring Further

1. **Applying** Explain why iron(III) sulfate is yellow, iron(II) thiocyanate is green, and iron(III) thiocyanate is red.

2. **Acquiring Information** Find out what impurities give amethyst, rose quartz, and citrine their colors.

To learn more about the differences between synthetic and natural gemstones, visit the Chemistry Web site at chemistryca.com

Rose quartz

The rusting of the transition element iron shows that iron can have more than one oxidation number. In the process of rusting, **Figure 7.11,** iron first forms the compound FeO. This compound continues to react with oxygen and water to form the familiar orange-brown compound called rust, Fe_2O_3. Because oxygen requires two electrons to achieve a noble-gas configuration, it takes the two 4s electrons from iron to form the FeO compound. The more complex Fe_2O_3 forms when two iron atoms give up a total of six electrons to three oxygen atoms. Each iron atom must surrender its two 4s electrons and one of its 3d electrons.

Figure 7.11

The Rusting of Iron
Once beautiful and sturdy, mighty iron can turn into a mass of rust as a result of the action of air and water. First, iron reacts with oxygen in the presence of water to form FeO. In FeO, iron's oxidation number is 2+ because it has lost its two 4s electrons. Then, FeO continues to combine with oxygen to form the familiar orange-brown compound, Fe_2O_3. In this oxide, iron's oxidation number is 3+ because it has lost two 4s electrons and one 3d electron.

Inner Transition Elements

The two rows beneath the main body of the periodic table are the lanthanides (atomic numbers 58 to 71) and the actinides (atomic numbers 90 to 103). These two series are called **inner transition elements** because their last electron occupies inner-level 4f orbitals in the sixth period and the 5f orbitals in the seventh period. As with the *d*-level transition elements, the energies of sublevels in the inner transition elements are so close that electrons can move back and forth between them. This results in variable oxidation numbers, but the most common oxidation number for all of these elements is 3+.

The Size of Orbitals

Hydrogen and the Group 1 elements each have a single valence electron in an *s* orbital. Hydrogen's configuration is $1s^1$; the valence electron configuration of lithium is $2s^1$. For sodium, it's $3s^1$; for potassium, $4s^1$. Continuing down the column, rubidium, cesium, and francium have valence configurations of $5s^1$, $6s^1$, and $7s^1$, respectively. How do these *s* orbitals differ from one another? As you move down the column, the energy of the outermost sublevel increases. The higher the energy, the farther the outermost electrons are from the nucleus. The *s* orbitals, occupied by the

Try at Home Lab

See page 866 in Appendix F for **Comparing Orbital Sizes**

valence electrons of hydrogen and the Group 1 elements, are described as spheres around the nucleus. As the valence electron gets farther from the nucleus, the *s* orbital it occupies gets larger and larger. **Figure 7.12** shows the relative sizes of the 1*s*, 2*s*, and 3*s* orbitals.

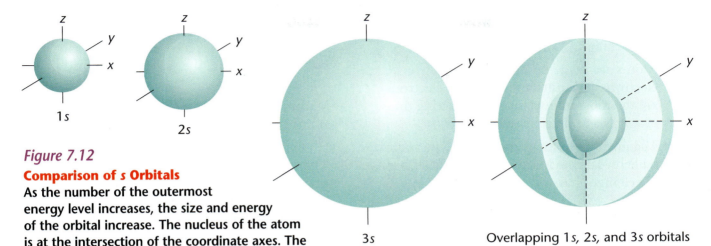

1*s*

2*s*

3*s*

Overlapping 1*s*, 2*s*, and 3*s* orbitals

Figure 7.12

Comparison of *s* Orbitals
As the number of the outermost energy level increases, the size and energy of the orbital increase. The nucleus of the atom is at the intersection of the coordinate axes. The model on the right shows the overlap of the 1*s*, 2*s*, and 3*s* orbitals. This model could represent sodium.

Connecting Ideas

An important skill that you learned in this chapter is how to use the periodic table to write electron configurations. It should be clear to you now that the organization of the table arises from the electron configurations of the elements. With this added insight, you are ready to learn in Chapter 8 about trends in properties and patterns of behavior of the elements. Knowing electron configurations and periodic trends will help you organize what may seem to be a vast amount of information.

Supplemental Problems

For more practice with solving problems, see Supplemental Practice Problems, Appendix B.

SECTION REVIEW

Understanding Concepts

1. Use the periodic table to help you write electron configurations for the following atoms. Use the appropriate noble gas, inner-core abbreviations.

   a) Ca   **d)** Cl
   b) Mg   **e)** Ne
   c) Si

2. Identify the elements that have the following electron configurations:

   a) $1s^2 2s^2$ **d)** $1s^2 2s^2 2p^2$
   b) $1s^2$ **e)** $[Ne]3s^2 3p^4$
   c) $1s^2 2s^2 2p^5$ **f)** $[Ar]4s^1$

3. What is the difference between a filled and an unfilled orbital?

Thinking Critically

4. **Applying Concepts** What is the lowest energy level that can have an *s* orbital? What region of the periodic table is designated as the *s* region? Are the elements in this region mostly metals, metalloids, or nonmetals?

Applying Chemistry

5. **Periodic Table** Why do the fourth and fifth periods of the periodic table contain 18 elements rather than eight elements as in the second and third periods?

CHAPTER 7 ASSESSMENT

REVIEWING MAIN IDEAS

7.1 Expanding the Theory of the Atom

- Atoms are made up of protons, neutrons, and electrons. The outermost valence electrons are the most important for predicting the properties of an element.

- The position of an element in the periodic table reveals the number of valence electrons the element has.

- Electromagnetic spectra provide information about energy levels and sublevels in an atom.

- Electrons are found only in levels of fixed energy in an atom. They cannot be located between energy levels.

- Energy levels have sublevels, which are partitioned into orbitals. An orbital is a region of space where there is a high probability of finding an electron. An orbital can hold a pair of electrons.

7.2 The Periodic Table and Atomic Structure

- The organization of the periodic table reflects the electron configurations of the elements.

- The active metals occupy the *s* region of the periodic table. Metals, metalloids, and non-metals fill the *p* region.

- Families of elements have similar electron configurations and the same number of valence electrons.

- Within a period of the periodic table, the number of valence electrons for main group elements increases from one to eight.

- The transition elements, Groups 3 to 12, occupy the *d* region of the periodic table. These elements can have valence electrons in both *s* and *d* orbitals, so they frequently have multiple oxidation numbers.

- The lanthanides and actinides, called the inner transition elements, occupy the *f* region of the periodic table. Their valence electrons are in *s* and *f* orbitals. Inner transition elements exhibit multiple oxidation numbers.

Vocabulary

For each of the following terms, write a sentence that shows your understanding of its meaning.

electron configuration
Heisenberg uncertainty principle
inner transition element
orbital
sublevel

UNDERSTANDING CONCEPTS

1. Explain what an electron cloud is.

2. Describe the shapes of the *s* and *p* orbitals.

3. What is an octet of electrons? What is the significance of an octet?

4. How many electrons fill an orbital?

5. What is the maximum number of electrons in each of the first four energy levels?

6. How many *p* orbitals can an energy level have? What is the lowest energy level that can have *p* orbitals?

7. What does each of the following symbols represent: 2*s*, 4*d*, 3*p*, 5*f*?

8. What is the maximum number of 2*s*, 3*p*, 4*d*, and 4*f* electrons in any sublevel?

9. Where is the *f* region of the periodic table? Name the two series of elements that occupy the *f* region.

10. Why does the first period contain only two elements?

11. Which of the following elements are inner transition elements? Explain your answer. V, Er, Hg, Cl, Po, Cm

APPLYING CONCEPTS

Physics Connection

12. What significance did Bohr's model of the atom have in the development of the modern theory of the atom?

Everyday Chemistry

13. Why do the transition metals impart color to gemstones such as emeralds and rubies?

Chemistry and Technology

14. Explain why the development of the scanning probe, the scanning tunneling, and the atomic force microscopes is important to modern chemists.

15. Identify the elements that have the following electron configurations:
    a) $[Kr]5s^2 4d^3$ c) $[Xe]6s^2$
    b) $1s^2 2s^2 2p^1$ d) $[Ar]4s^1 3d^{10}$

16. How is the light given off by fireworks similar to an element's emission spectrum?

17. What are valence electrons? Where are they located in an atom? Why are they important?

18. Sodium and oxygen combine to form sodium oxide, which has the formula Na_2O. Use the periodic table to predict the formulas of the oxides of potassium, rubidium, and cesium. What periodic property of the elements are you using?

19. Where are the valence electrons represented in this electron configuration of sulfur, $[Ne]3s^2 3p^4$?

THINKING CRITICALLY

Observing and Inferring

20. **MiniLab 2** Why are electrons best described as electron clouds?

Applying Concepts

21. **ChemLab** In an experiment to find out the reaction capacities of magnesium and aluminum, different amounts of the two metals were used, but the amount of hydrochloric acid was kept the same. Would it be possible to obtain correct results in this way? Explain.

Relating Cause and Effect

22. Explain the following observation. A metal is exposed to infrared light for 3 seconds and no electrons are emitted. When the same metal is exposed to ultraviolet light, thousands of electrons are emitted.

Using a Table

23. Each element emits light of a specific color in a flame test. The characteristic wavelengths emitted by some elements are shown in the following table. Using the electromagnetic spectrum shown on page 235, state the color associated with the wavelengths. Of the wavelengths listed, which represents the highest-energy light?

Element	Wavelength	Element	Wavelength
Ag	521 nm	Fe	492 nm
Au	461 nm	K	405 nm
Ba	553 nm	Mg	519 nm
Ca	397 nm	Na	589 nm

Interpreting Data

24. The emission spectrum of sodium shows a pair of lines close together near 500 nm. What might be the explanation for these lines?

Comparing and Contrasting

25. Refer to the periodic table and compare the similarities and differences in the electron configurations of the following pairs of elements: F and Cl, O and F, Cl and Ar.

Sequencing Events

26. **MiniLab 1** Explain what happens to the electrons of a metal atom when a splint soaked with a chloride of the metal is placed in a flame and a color is produced.

CUMULATIVE REVIEW

27. For elements that form positive ions, how does the charge on the ion relate to the number of valence electrons the element has? (Chapter 2)

28. Describe John Dalton's model of the atom, and compare and contrast it with the present-day atomic model. (Chapter 2)

29. What is the significance of an octet of electrons in forming both ionic and covalent compounds? (Chapter 4)

30. How many atoms of each element are present in five formula units of calcium permanganate? (Chapter 5)

WRITING IN CHEMISTRY

31. The word *laser* is an acronym for *light a*mplification by *s*timulated *e*mission of *r*adiation. Lasers have a number of applications besides light shows. Find out how lasers are produced, what substances are used in lasers, and the ways laser light is used. Write a paper describing your findings.

SKILL REVIEW

32. Using a Graph Using the graph of energy versus frequency, answer the following questions. What happens to energy when frequency is doubled? What happens to energy when frequency is halved?

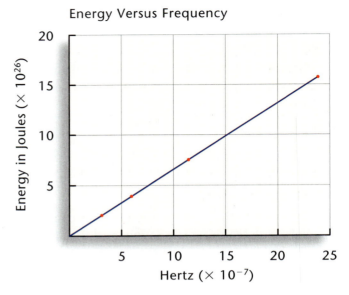

Energy Versus Frequency

PROBLEM SOLVING

33. Use the diagram of the electromagnetic spectrum below to list the following types of radiation in order of increasing wavelength: microwaves that cook food, ultraviolet radiation from the sun, X rays used by dentists and doctors, the red light in a calculator display, and gamma rays. Which type of radiation has the highest energy? The lowest energy?

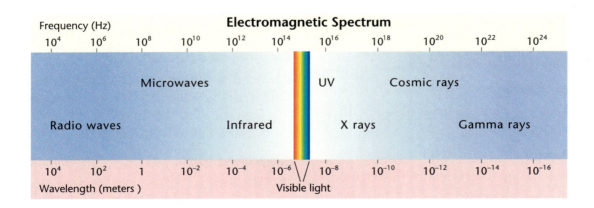

Standardized Test Practice

1. The modern model of the atom replaced
   a) Dalton's model.
   b) Thomson's model.
   c) Rutherford's model.
   d) Bohr's model.

2. When an electron falls back to its original energy level, it
   a) emits energy.
   b) absorbs energy.
   c) transfers energy.
   d) stores energy.

3. Which of the following sublevels has the greatest amount of energy?
   a) s
   b) p
   c) d
   d) f

4. Which of the following statements about the Heisenberg uncertainty principle is true?
   a) The principle states that the location of an electron in the electron cloud cannot be determined.
   b) The principle states that the position and energy of an electron cannot be known at the same time.
   c) The principle states that the exact location of an electron around an atom cannot be measured.
   d) The principle states that the exact energy of an electron around an atom cannot be accurately measured.

5. What is the electron configuration of xenon?
   a) 2,8,18,18,7,1
   b) 2,8,18,32,18,7,1
   c) 2,8,18,18,8
   d) 2,8,18,32,18,8

6. Why are noble gases unreactive?
   a) They have a stable electron configuration.
   b) They have filled energy levels.
   c) They have the maximum number of valence electrons.
   d) All of the above.

Halogen Electron Configurations	
Halogen	Electron Configuration
Fluorine	2,7
Chlorine	2,8,7
Bromine	2,8,18, 7
Iodine	2,8,18,18,7
Astatine	2,8,18,32,18,7

Use the table to answer questions 7–8.

7. Why do halogens have a −1 oxidation number?
   a) They have two electrons in their first orbital.
   b) They need one electron in the outer energy level.
   c) They lose one electron in one of their sublevels.
   d) They need one electron in their 2p orbitals.

8. What do all the halogens have in common?
   a) They have the same number of valence electrons.
   b) They have the same number of orbitals.
   c) They have the same number of sublevels.
   d) The have the same number of electrons in each orbital.

9. The d orbital holds a maximum of
   a) 2 electrons.
   b) 6 electrons.
   c) 10 electrons.
   d) 18 electrons.

10. Elements with d orbitals are called
    a) noble gases.
    b) transition elements.
    c) metals.
    d) nonmetals.

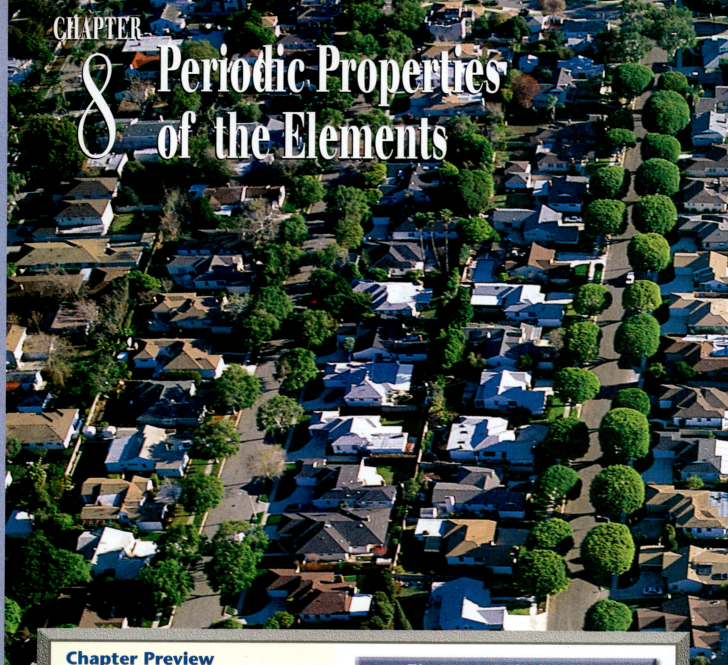

Chapter Preview

Sections

8.1 Main Group Elements
MiniLab 8.1 What's Periodic About
Atomic Radii?
ChemLab Reactions and Ion
Charges of the Alkaline
Earth Elements

8.2 Transition Elements
MiniLab 8.2 The Ion Charges of a
Transition Element

Elemental Similarities

Elements are arranged in the periodic table in a logical format, much like a well-designed neighborhood. Moving across a row, numbers of protons in atoms, like house numbers, change and indicate different elements. Elements in specific columns have similar characteristics that allow them to act in predictable ways, much like family groups.

Launch Lab

Magnetic Materials

You know that magnets can attract some materials. In this lab, you will classify materials based on their interaction with a magnet and look for a pattern in your data.

Safety Precautions

Materials

- bar magnet
- aluminum foil
- paper clips
- coins
- hair pin
- soda cans
- empty soup cans
- variety of other items

Procedure

1. Working with a partner, review the properties of magnets. Arrange the bar magnets so that they are attracted to one another. Then arrange them so that they repel one another.
2. Test each item with your magnet. Record your observations.
3. Test as many other items in the classroom as time allows. Predict the results before testing each item.

Analysis

Look at the group of items that were attracted to the magnet. What do these items have in common?

What I Already Know

Review the following concepts before studying this chapter.
Chapter 3: properties of metals, nonmetals, and metalloids and their positions on the periodic table

Reading Chemistry

Scan the chapter, and write down a list of several new vocabulary words. Look closely at the words, and try to derive their meanings. Next, go back to the text to find the words' definitions. Some vocabulary words also appear in graphs or diagrams.

Preview this chapter's content and activities at **chemistryca.com**

Main Group Elements

You have a chemical storehouse in your own home. Under the kitchen sink, you'll probably find dishwasher detergent, steel-wool soap pads, ammonia window cleaner, and a variety of other cleaning products. In your pantry, there could be vinegar, baking powder, and baking soda. Your medicine cabinet probably contains toothpaste, deodorants, and various medications. Read the labels on these products, and you'll find that they contain many simple compounds such as sodium chloride ($NaCl$), or table salt; sodium hydrogen carbonate ($NaHCO_3$), or baking soda; sodium hypochlorite ($NaOCl$), the active agent in bleach; or sodium hydroxide ($NaOH$), which is present in drain cleaners.

SECTION PREVIEW

Objectives

✓ **Relate** the position of any main group element in the periodic table to its electron configuration.

✓ **Predict** chemical behavior of the main group elements.

✓ **Relate** chemical behavior to electron configuration and atomic size.

Review Vocabulary

Inner transition element: one of the elements in the actinide or lanthanide series.

New Vocabulary

alkali metal
alkaline earth metal
halogen

Patterns of Behavior of Main Group Elements

Recall from Chapter 7 that elements in the same group (vertical column) of the periodic table have the same number of valence electrons, and because of this, they have similar properties. But elements in a period (horizontal row) have properties different from one another. This is because the number of valence electrons increases from one to eight as you move from left to right in any row of the periodic table except the first. As a result, the character of the elements changes. **Figure 8.1** illustrates the main group elements and shows that each period begins with two or more metallic elements, which are followed by one or two metalloids. The metalloids are followed by nonmetallic elements, and every period ends with a noble gas.

Patterns in Atomic Size

Recall that the size of an atom increases in any group of elements as you go down the column because the valence electrons are found in energy levels farther and farther from the nucleus. But how does atomic radius change across a period from left to right? Take Period 2 as an example. You might expect the size of the second-period atoms to increase across the period from lithium on the left to fluorine on the right because both the atomic number and, therefore, the number of electrons increases. However, the opposite is true. The lithium atom, with only three

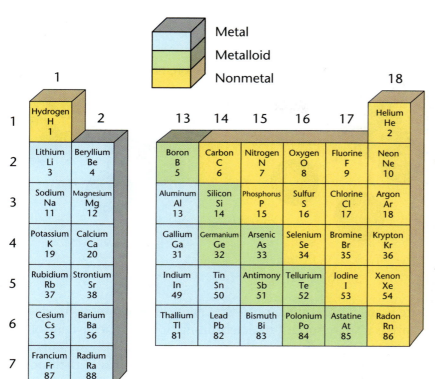

Metal
Metalloid
Nonmetal

Figure 8.1

Trends in Metallic Properties
The pattern metal-metalloid-nonmetal-noble gas is typical for the main group elements in each period. Period 2 begins with a metal, lithium, and ends with a noble gas, neon. In between are the metal beryllium; the metalloid boron; and the nonmetals carbon, nitrogen, oxygen, and fluorine. Remember that the most active metals, Groups 1 and 2, are in the *s* region of the periodic table. The metalloids, nonmetals, and less active metals are in the *p* region of the periodic table.

electrons, is actually larger than the fluorine atom, which has nine. **Figure 8.2** illustrates the relative sizes of atoms of the main group elements.

There's a simple explanation for the trend of decreasing atomic size across a period. Picture the valence electron on a lithium atom. The lithium nucleus has three protons, so there's an attractive force of 3+ acting on the electron. Lithium's valence electron is in the second energy level, and the attraction to the nucleus isn't too strong. Now, think about the valence electrons in a beryllium atom. Here, there is an attractive force of 4+ from the four protons in the beryllium nucleus. But the outer electrons are still in the second energy level, so the larger attractive force of the beryllium nucleus pulls these electrons a little closer to the nucleus, and the electron cloud gets a little smaller. With each increase in nuclear charge across the period, the outer electrons are attracted more strongly toward the nucleus, resulting in smaller size. In **Figure 8.2**, compare the size of fluorine, with a nuclear charge of 9+, to the size of lithium.

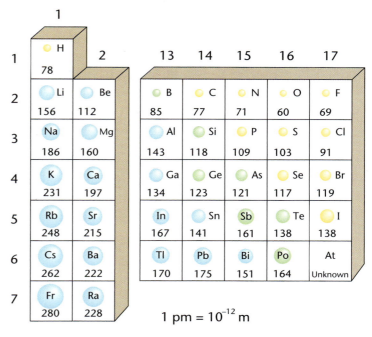

$1 \text{ pm} = 10^{-12} \text{ m}$

Figure 8.2

Atomic Radii of Main Group Elements
Atomic radius (plural, *radii*) is a measure of the size of an atom. The spheres in the table represent the relative sizes of the atoms as measured by X-ray diffraction studies. The actual atomic radius is given in picometers beneath each sphere. Atomic size is a periodic property of the elements. Can you see the pattern in every period?

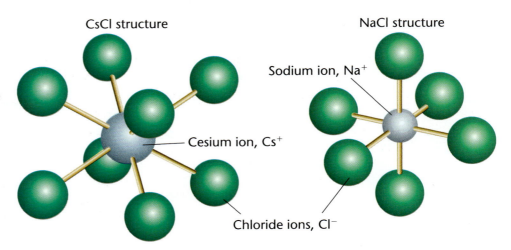

CsCl structure

NaCl structure

Sodium ion, Na⁺

Cesium ion, Cs⁺

Chloride ions, Cl⁻

Figure 8.3

Ionic Size and Crystal Structure

The cesium ion is larger than the sodium ion so it's possible for eight chloride ions to fit around a single cesium ion in the CsCl crystal lattice. The smaller sodium ion can accommodate only six chloride ions in the NaCl structure.

Ionic Size

Atomic size is an important factor in the chemical reactivity of an element. Ionic size is also important in determining how ions behave in solution and the structure of solid ionic compounds. **Figure 8.3** shows how the structures of two ionic compounds differ because of the sizes of their positive ions. How does the size of an atom change when it becomes an ion? When metallic atoms lose one or more electrons to become positive ions, they acquire the configuration of the noble gas in the preceding period. This means that the outermost electrons of the ion are in a lower energy level than the valence electrons of the neutral atom. The electrons that are not lost by the atom experience a greater attraction to the nucleus and pull together in a tighter bundle with a smaller radius. The result is that all positive ions have smaller radii than their corresponding atoms. **Figure 8.4** shows a comparison of lithium and sodium with their positive ions.

When an atom gains electrons to become a negative ion, the atom acquires the electron configuration of the noble gas at the end of its period. But the nuclear charge doesn't increase with the number of electrons. In the case of fluorine, a nuclear charge of 9+ must hold ten electrons in the F^- ion. The result is that all the electrons are held less tightly, and the radius of the ion is larger than the neutral atom. **Figure 8.4** shows how the sizes of the fluoride and chloride ions compare with the fluorine and chlorine atoms.

Figure 8.4

Sizes of Atoms and Their Ions

Lithium and sodium lose the single electron from their outermost energy level. The ions that form are smaller because the remaining electrons are at a lower energy level and are attracted more strongly to the nucleus. Fluorine and chlorine become negative ions by adding an electron. When electrons are added, the charge on the nucleus is not great enough to hold the increased number of electrons as closely as it holds the electrons in the neutral atom.

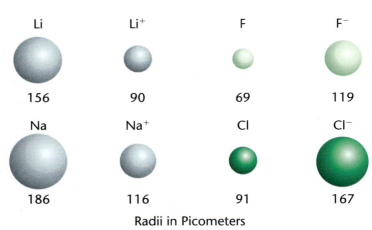

Li	Li⁺	F	F⁻
156	90	69	119
Na	Na⁺	Cl	Cl⁻
186	116	91	167

Radii in Picometers

Patterns in Ionic Radii

In **Figure 8.4,** you can see that the sodium ion is larger than the lithium ion. This trend in increasing ionic size continues as you go down the periodic table in Group 1, as shown in **Figure 8.5.** Notice the same trend for the positive ions in Groups 2 and 3 and for the negative ions in Groups 15, 16, and 17. **Figure 8.5** also shows how the sizes of both positive and negative ions change across a period.

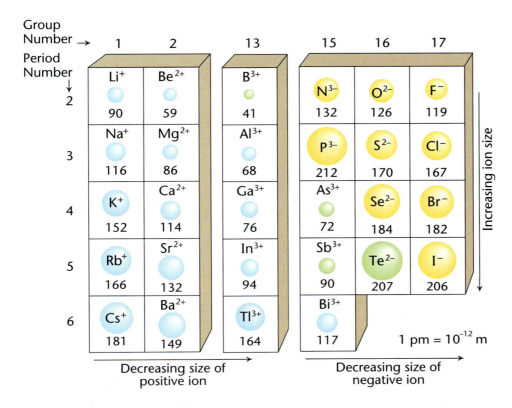

Figure 8.5

Trends in Ionic Radii
Ionic radii increase down the table in any group because of the increasing distance of the outermost electrons from the nuclear charge. Ions of atoms in the same period with 1+, 2+, and 3+ charges (Groups 1, 2, and 13) decrease in size from left to right. Although the ions have the same electron configuration, nuclear charge increases from left to right, resulting in a stronger attraction for electrons and smaller size. Negative ions in the same period with 3−, 2−, and 1− charges (Groups 15, 16, and 17) show the same trend in size. Ionic radii decrease because nuclear charge increases.

Patterns in Chemical Reactivity in Period 2

You've already noticed that the character of the Period 2 elements changes from metal to metalloid to nonmetal to noble gas as you move across the period. How are the electron configurations of these elements related to the tendency of the metals to lose electrons, the nonmetals to share or gain electrons, and the noble gases to be unreactive?

Lithium is the most active metal in the second period because it can attain the noble-gas configuration of helium by losing a single electron. If lithium loses one electron from its $2s$ sublevel, its electron configuration changes from $1s^2 2s^1$ to $1s^2$. The resulting lithium ion has a 1+ charge and the same electron configuration as a helium atom. While this is not an octet, it is a noble-gas configuration. Elements tend to react in ways that allow them to achieve the configuration of the nearest noble gas.

Beryllium, the next element in the second period, must lose a pair of $2s$ electrons to acquire the helium configuration. It's harder to lose two electrons than it is to lose one, so beryllium is slightly less reactive than lithium. Nevertheless, beryllium does react by losing both of its $2s$ electrons and forming a 2+ ion with the helium electron configuration.

What's periodic about atomic radii?

Atomic radius is approximately the distance from the nucleus of an atom to the outside of the electron cloud where the valence electrons are found. The reactivity of the atom depends on how easily the valence electrons can be removed, and that depends on their distance from the attractive force of the nucleus. In this MiniLab, you will study the periodic trends in the atomic radii of the first 36 main group elements from hydrogen through barium.

Procedure

1. Obtain a 96-well microplate, straws of a size to fit the wells in the plate, scissors, and a ruler. The well plate should be oriented to correlate with the periodic table of the elements in the following way: Row 1 of the plate represents the first period, H1 as hydrogen, A1 as helium; Row 2 of the plate represents the second period, from H2 (lithium) to A2 (neon). Rows 3 to 7 correlate with Periods 3 to 7; however, only the main group elements will be represented. Label the well plate *Atomic radius in pm* (picometers).

2. Use **Figures 8.1** and **8.2** to help you make your model. Look up the atomic radius of each of the elements in Table 4 in Appendix C.

3. Convert atomic radius in picometers to an enlarged scale in centimeters by multiplying the atomic radius in picometers by the conversion factor,

1 cm/40 pm. For example, the atomic radius of hydrogen in centimeters is calculated in this manner: 78 pm × 1 cm/40 pm = 1.95 cm or 2.0 cm. To represent the atomic radius of hydrogen, cut a piece of straw 2.0 cm long. Cut a piece of straw to scale for each element, and insert each piece into the appropriate well of the plate.

Analysis

1. How do the atomic radii change as you go from left to right across a period? Explain your observation on the basis of the electron configurations of the elements.

2. How do atomic radii change as you go from top to bottom within a group or family? Explain your observation on the basis of the electron configurations of the elements.

3. Why is the atomic radius of the elements described as a periodic property?

If this pattern continued, you would expect boron to lose three electrons to attain the helium configuration. Sometimes, boron does react by losing electrons, but often it reacts by sharing electrons. Boron is the only metalloid in the period. That means boron sometimes behaves like a metal and loses electrons like its neighboring metals, lithium and beryllium. When it loses electrons, boron achieves the noble-gas configuration of helium. But more often, boron acts like a nonmetal and shares electrons. Boron is unusual because it has only three electrons to share and cannot acquire an octet of electrons by just sharing. Later, you'll learn more about boron's chemistry.

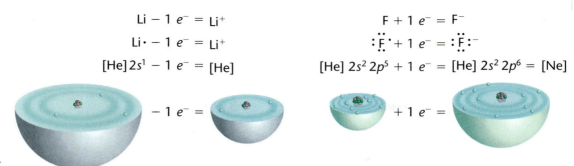

$$Li - 1\ e^- = Li^+$$
$$Li\cdot - 1\ e^- = Li^+$$
$$[He]\,2s^1 - 1\ e^- = [He]$$

$$F + 1\ e^- = F^-$$
$$:\overset{..}{\underset{..}{F}}\cdot + 1\ e^- = :\overset{..}{\underset{..}{F}}:^-$$
$$[He]\,2s^2\,2p^5 + 1\ e^- = [He]\,2s^2\,2p^6 = [Ne]$$

$$- 1\ e^- =$$

$$+ 1\ e^- =$$

Figure 8.6

Steps in the Formation of LiF

When an alkali metal such as lithium loses an electron, it attains the configuration of the noble gas in the preceding period. When a nonmetal such as fluorine gains an electron, it acquires the configuration of the noble gas at the end of its period. Ionic compounds such as LiF are combinations of a positive metal ion and a negative nonmetal ion, each having a noble-gas configuration.

Carbon, nitrogen, oxygen, and fluorine are nonmetals. Carbon, with the configuration $[He]2s^22p^2$, and nitrogen, with the configuration $[He]2s^22p^3$, share electrons to attain the noble-gas configuration of neon, $[He]2s^22p^6$. Oxygen, with the configuration $[He]2s^22p^4$, gains two electrons to form the oxide ion, O^{2-}. Fluorine, with the configuration $[He]2s^22p^5$, gains one electron to become the fluoride ion, F^-. **Figure 8.6** gives an example of the loss and gain of electrons by two elements in Period 2 and the ionic compound formed by these elements.

The Main Group Metals and Nonmetals

You can learn a lot about the elements in the products you use every day by studying the chemistry of the main group elements. To do so, recall that elements in a group are chemically similar to the first element in the group.

The Alkali Metals

The Group 1 elements—lithium (Li), sodium (Na), potassium (K), rubidium (Rb), cesium (Cs), and francium (Fr)—are called the **alkali metals.** The alkali elements are soft, silvery-white metals and good conductors of heat and electricity. Their chemistry is relatively uncomplicated; they lose their *s* valence electron and form a 1+ ion with the stable electron configuration of the noble gas in the preceding period.

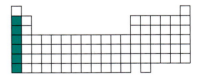

Because all of the alkali metals react by losing their single *s* valence electron, the most reactive alkali metal is the one that has the least attraction for this electron. Remember that the bigger the atom, the farther the valence electron is from the nucleus and the less tightly it's held. In the alkali metal family, francium is the largest atom and probably the most reactive, but francium has not been widely investigated because it is scarce and radioactive. Cesium (Cs) is usually considered the most active alkali metal—in fact, the most active of all the metals. Lithium, the smallest of the alkali metals, is the least reactive element in Group 1.

Reactions and Uses of the Alkali Metals

Because of their chemical reactivity, the alkali metals don't exist as free elements in nature. Sodium, for example, is found mostly combined with chlorine in sodium chloride. Metallic sodium is obtained from NaCl through a process called electrolysis in which an electric current is passed through the molten salt.

1 Spontaneous Reactivity

Oil protects sodium metal from spontaneous reaction with oxygen or moisture in the air. The metal is soft enough to be cut with a knife, and inside you can see the shiny metallic surface. Sodium and the other Group 1 elements are among the most active of all the metals. All Group 1 metals react vigorously with water. When they do, they replace hydrogen and form a hydroxide, as shown in the following equation.

$$2K + 2H_2O \rightarrow H_2 + 2KOH$$

2 Alkali Metals Form Hydroxides

So much heat is generated in the rapid reaction of potassium and water that the hydrogen gas produced in the reaction bursts into flames. The pink color of the water is due to the presence of the indicator phenolphthalein, which turns pink when the solution is alkaline. The pink color of the flame is characteristic of potassium. Potassium hydroxide (KOH) formed in the reaction makes the solution alkaline. Hydroxides are important household and industrial chemicals.

Sodium hydroxide is used in the digestion of pulp in the process of making paper (left). It's also used in making soap, in petroleum refining, in the reclaiming of rubber, and in the manufacture of rayon (right). In your household chemical storehouse, you'll find sodium hydroxide (lye) in oven cleaners and in the granular material you use to unclog drains. It is sodium hydroxide's ability to convert fats to soap that makes it effective as a kitchen drain cleaner.

Compounds of sodium and potassium are important to the human body because they supply the positive ions that play a key role in transmitting nerve impulses that control muscle functions. Potassium is also an essential nutrient for plants. It's one of the three major components of fertilizers; the other two are also main group elements—nitrogen and phosphorus.

The Alkaline Earth Metals

The Group 2 elements—beryllium (Be), magnesium (Mg), calcium (Ca), strontium (Sr), barium (Ba), and radium (Ra)—are called the **alkaline earth metals.** Their properties are

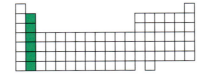

similar to those of the Group 1 elements. Like the alkali metals, they are too reactive to be found as free elements in nature. They lose both of their *s* valence electrons and form 2+ ions with the stable electron configuration of the noble gas in the preceding period. Because the Group 2 elements must lose two electrons rather than one, these metals are less reactive than the Group 1 elements. Each alkaline earth metal is denser and harder and has a higher melting point than the alkali metal that is its neighbor.

The most reactive element in the alkaline earth group is the one with the largest atomic radius and, therefore, the least attraction for its two valence electrons. Knowing this, you can predict that radium, the largest atom in the group, is the most reactive.

The trend to increasing reactivity with increasing size of atom for the alkaline earth metals is illustrated by the reaction of the elements with water, as shown in **Figure 8.7.** Beryllium does not react with water. Magnesium reacts with hot water. But calcium reacts with water to form calcium hydroxide [$Ca(OH)_2$], as shown by this equation.

$$Ca + 2H_2O \rightarrow H_2 + Ca(OH)_2$$

Figure 8.7

The Trend in Reactivity of the Alkaline Earth Metals
No reaction is visible when beryllium is placed in water, but bubbles of hydrogen are produced by the reaction of calcium with water. Strontium, barium, and radium react with water with increasing vigor.

ChemLab

SMALL SCALE

Reactions and Ion Charges of the Alkaline Earth Elements

The alkaline earth elements, Group 2 on the periodic table, are beryllium (Be), magnesium (Mg), calcium (Ca), strontium (Sr), barium (Ba), and radium (Ra). The positive ions of most of these elements react with negative oxalate ions ($C_2O_4^{2-}$) to form compounds that are insoluble in water. In this ChemLab, you will study the reactions of calcium, strontium, and barium with oxalate ions and determine the formulas for the insoluble products.

Problem

In what ratios do the positive ions of calcium, strontium, and barium react with negative oxalate ions to produce compounds?

Objectives

- **Observe** the reactions of the calcium, strontium, and barium ions with oxalate ions.
- **Determine** the formulas for the insoluble products and the charges on the ions of the alkaline earth elements.
- **Relate** the ion charges of the alkaline earth elements to their electron configurations.

PREPARATION

Materials

96-well microplates (3)
microtip pipets (4)
black paper
toothpicks (3)
marking pen
0.1 M calcium nitrate solution
0.1 M strontium nitrate solution
0.1 M barium nitrate solution
0.1 M sodium oxalate solution

Safety Precautions

Wear an apron and goggles. The solutions are toxic if ingested and may irritate the skin. Do not touch or ingest the solutions. Dispose of the products as instructed by your teacher.

PROCEDURE

1. Obtain three 96-well microplates and four labeled micropipets containing the four solutions: calcium nitrate solution, strontium nitrate solution, barium nitrate solution, and sodium oxalate solution. Label the microplates *calcium, strontium,* and *barium.*

2. Lay the calcium microplate near the edge of the lab bench with row H aligned with the edge of the bench. You will use only wells H1 through H9. You will see the product best if the microplate rests on a piece of black paper.

3. Add one drop of the calcium solution to well H1, two drops to well H2, and three drops to well H3. Continue in this way until you get to well H9, which receives nine drops.

4. Add one drop of the sodium oxalate solution to well H9, two drops to well H8, and three drops to well H7. Continue in this way until you get to H1, which receives nine drops.

5. Use a toothpick to stir the mixtures.

6. It may take several minutes for the reactions to be complete. When the insoluble product has settled in the bottoms of the wells, stoop so that your eyes are level with the microplate.

7. Determine and record the identity of the well (or wells) that contains the greatest depth, and therefore the greatest amount, of the insoluble product. Record your observations in a data table like the one shown.

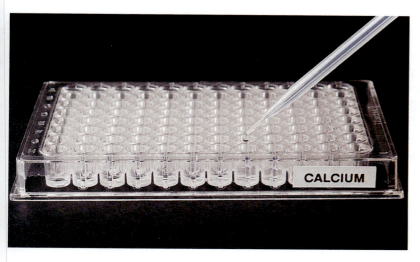

CALCIUM

8. Repeat procedures 2 through 7 with the strontium and oxalate solutions.

9. Repeat procedures 2 through 7 with the barium and oxalate solutions.

10. Dispose of the substances in all three well plates as directed by your teacher. Rinse the well plates with tap water and then with distilled water.

ANALYZE AND CONCLUDE

1. **Interpreting Data** What ratio of drops of the reactant solutions produced the maximum amount of insoluble product for each of the three reactions you performed?

DATA AND OBSERVATIONS

Sodium oxalate combined with	Number of well with maximum precipitate	Drops of oxalate solution	Drops of Group 2 ion solution
Calcium nitrate			
Strontium nitrate			
Barium nitrate			

2. **Interpreting Data** All the solutions you used were 0.1 molar, which means that they contained equal numbers of ions per drop. For example, if the maximum amount of product was produced from two drops of calcium solution and eight drops of oxalate solution, the insoluble product contained one calcium ion for each four oxalate ions. The formula for such a compound would be $Ca(C_2O_4)_4$. Interpret your results to determine the formulas for calcium oxalate, strontium oxalate, and barium oxalate.

3. **Drawing Conclusions** The oxalate ion has a charge of $2-$ and it combines with the positive alkaline earth ions in such a way that neutral compounds result. What do you think are the ion charges of calcium, strontium, and barium ions? Explain how you can infer these ion charges from your experimental results.

APPLY AND ASSESS

1. Write the electron configurations of calcium, strontium, and barium atoms. Relate the charges on calcium, strontium, and barium ions to the electron configurations of their respective atoms.

2. Use the same reasoning that you used to answer question 1 to predict the ion charges of potassium in Group 1 and gallium in Group 13. Explain your reasoning.

Reactions and Uses of the Alkaline Earth Metals

Beryllium has become a strategically important metal in the nuclear and weapons industries. Magnesium and beryllium are valued for their individual properties, but they are especially important when alloyed with other metals.

1 Important Properties of Magnesium

Alloys of magnesium are used where light weight and strength are important, as in this jet engine. Magnesium resists corrosion because it reacts with oxygen in the air to form a coating of magnesium oxide. The coating of MgO protects the metal underneath from further reaction with oxygen.

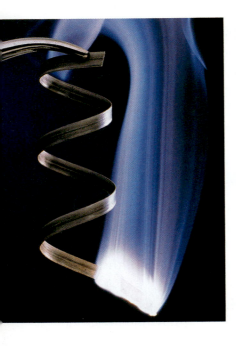

2 Reactions of Magnesium and Calcium

Magnesium oxide is also formed when magnesium is heated in air. It burns vigorously, producing a brilliant white light and magnesium oxide. In the process, magnesium loses two electrons to form the Mg^{2+} ion, and oxygen gains two electrons to form the O^{2-} ion. Together, they form the ionic compound MgO. The following equation shows what happens.

$$2Mg + O_2 \rightarrow 2MgO$$

Magnesium and calcium are essential elements for humans and plants. Plants need magnesium for photosynthesis because a magnesium atom is located at the center of every chlorophyll molecule.

Calcium ions are essential in your diet. They maintain heartbeat rate and help blood to clot. But the largest amount of dietary calcium ions is used to form and maintain bones and teeth. Bone is composed of protein fibers, water, and minerals, the most important of which is hydroxyapatite, $Ca_5(PO_4)_3OH$, a compound of calcium, phosphorus, oxygen, and hydrogen—all main group elements.

3 Strontium Reveals Its Presence

Strontium is a less well-known element of Group 2, but it's important, nevertheless. Because of its chemical similarity to calcium, strontium can replace calcium in the hydroxyapatite of bones and form $Sr_5(PO_4)_3OH$. This could be a problem only if the strontium atoms are the radioactive isotope strontium-90, which is hazardous if it is incorporated into a person's bones.

Strontium makes its presence known by the brilliant red color of a fireworks display. The red color also identifies strontium in laboratory flame tests.

Group 13 Elements

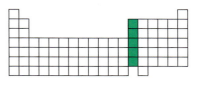

Only boron, the first element in Group 13, is a metalloid. The other Group 13 elements—aluminum (Al), gallium (Ga), indium (In), and thallium (Tl)—are metals. None of the metals are as active as the metals in Groups 1 and 2, but they're good conductors of heat and electricity. They are silvery in appearance and fairly soft. Group 13 metals tend to share electrons rather than form ionic compounds; in this respect, they resemble boron. Their valence configuration is s^2p^1, and they exhibit the 3+ oxidation number in most of their compounds.

Aluminum is the most abundant metallic element in Earth's crust and has so many desirable properties that it's becoming one of the world's most widely used metals. Aluminum's many uses result from its properties—low density, good electrical and thermal conductivity, malleability, ductility, and resistance to corrosion. Aluminum, like magnesium, is a self-protecting metal. On exposure to oxygen in the air, a protective layer of aluminum oxide (Al_2O_3) forms over the surface of the aluminum, preventing further reaction with oxygen.

$$4Al + 3O_2 \rightarrow 2Al_2O_3$$

Aluminum is obtained from its ore through a process that consumes 4.5 percent of the electricity produced in the United States. Recycling aluminum reduces costs by lowering the need for power.

The Uses of Group 13 Elements

Boron is a metalloid found in boric acid (H_3BO_3) and borax ($Na_2B_4O_7 \cdot 10H_2O$). Boric acid is one of the active ingredients in eyewash or contact lens-cleaning solution. Borax is the abrasive in some tough cleansing powders. It's also used as a water softener and is an important component in some types of glass.

1 The Importance of Aluminum

Think about how important aluminum is in your life. Aluminum foil and aluminum soda cans are everywhere. The antacid in your medicine cabinet may contain aluminum hydroxide, $Al(OH)_3$. Your antiperspirant or deodorant may contain an aluminum zirconium hydroxide or aluminum chlorohydrate.

Because aluminum is neither as hard nor as strong as steel, it is often alloyed with other metals to make structural materials. Aluminum alloys are used in automobile engines, airplanes, and truck bodies where high strength and light weight are important. The alloys used in the structure of airplanes are ten percent to 30 percent magnesium; the rest is aluminum. To save weight, some automobile engines use a magnesium alloy that is five percent to ten percent aluminum; the rest is magnesium.

Recycling aluminum from soda cans

2 Aluminum in Your Home

At home, you may find bicycles, outdoor furniture, ladders, and pots and pans that are made of aluminum or an aluminum alloy.

3 Aluminum as a Conductor

Even though aluminum doesn't conduct electricity as well as copper, it costs less to use aluminum than copper for transmission of electricity. Aluminum cables are much lighter than copper cables, so fewer support towers are needed to hold the miles and miles of cable that span the country. Fewer support towers means lower cost for all consumers of electricity.

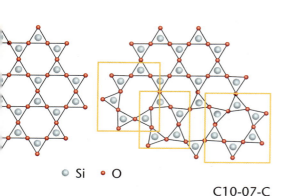

○ Si   • O

C10-07-C

4 Gallium's Low Melting Point

Gallium, indium, and thallium react much like aluminum. But gallium, shown here, has an unusually low melting point, 29.8°C. The heat of a hand is sufficient to liquefy the metal. For comparison, aluminum melts at 660°C.

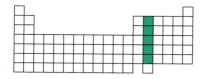

Group 14

The Group 14 elements—carbon (C), silicon (Si), germanium (Ge), tin (Sn), and lead (Pb)—exhibit a variety of properties. Carbon is a nonmetal, silicon and germanium are metalloids, and tin and lead are metals. Because the valence electron configuration for these elements is s^2p^2, a gain or loss of four electrons results in a noble-gas configuration. However, it's unusual for any element to gain or lose four electrons. Instead of gaining electrons to attain a noble-gas configuration, carbon, silicon, and germanium react by sharing electrons. But tin and lead, like the metals in the preceding groups, react by losing electrons. These larger elements at the bottom of the group lose electrons more easily than the smaller nonmetals at the top of the group because of the size of their atoms and the reduced attraction the nucleus has for the outermost electrons. As a group, the most common oxidation number is 4+.

Lead Poisoning in Rome

Could lead poisoning be partially responsible for the fall of the Roman Empire? Some researchers think so; they found that lead poisoning occurred in early history.

How lead poisoning happens Lead gets into the body when materials containing lead compounds are ingested with food and drink, or when lead-containing dust in the air is inhaled or absorbed through the skin. As lead intake increases, the body's ability to get rid of it decreases. Over a period of time, the accumulation of lead in the liver, kidneys, bones, and other body tissues becomes critical. Symptoms of abdominal pains, anemia, lethargy, and nerve paralysis of hands and feet develop.

An old and versatile metal Lead was highly valued in ancient times. The Egyptians refined and used lead as early as 3000 B.C. Deposits of galena, PbS, located near Athens, Greece, were processed and used by the Greeks in the sixth century B.C. But it was the Romans, in the first century B.C., who realized the full potential of lead. They processed it for a wide variety of purposes. Rome's famous system for supplying water to the populace was built with lead pipes. Beer and wine were stored in lead-glazed pottery and served in lead goblets. Cooking pots were made of lead.

Lead was everywhere The lead water pipes of the Roman plumbing system allowed lead to dissolve in the drinking water. Many types of food and drink were sweetened by a thick, sugary syrup called sapa. Sapa was made by boiling wine in a lead pot until much of the water and alcohol had evaporated. What remained was a tasty but poisonous confection. A small portion of sapa was lead(II) acetate, also known as sugar of lead. It was impossible for anyone living in ancient Rome to avoid ingesting lead. Because lead poisoning causes listlessness and mental failure, some researchers think that it contributed to the breakdown of the Roman ruling class and hastened the fall of the Empire.

Lead emissions When Rome was at its peak in lead production, it produced 80 000 metric tons every year. In studies of changes in atmospheric composition throughout history, researchers measured residues of lead from ancient Rome and Greece found in British peat bogs and in Swedish lake sediments. Lead emissions from Roman smelters at the height of Roman power were nearly as great as they were during the years of the Industrial Revolution in England from 1760 to 1840.

Connecting to Chemistry

1. **Acquiring Information** Find out how lead can get into water supplies today.

2. **Comparing and Contrasting** Use the library to find out about the problems the United States has with lead pollution and lead poisoning.

3. **Interpreting** The formula for the acetate ion is $C_2H_3O_2^-$. What is the formula for lead(II) acetate?

The Uses of Group 14 Elements

You'll find Group 14 elements in many of the products in your household chemical storehouse—carbon in charcoal briquettes, lead pencils, diamond jewelry, and almost every food item in your home. You'll learn more about carbon and carbon compounds in Chapters 9 and 18.

1 From Sand to Many Uses

Silicon, like boron, is a metalloid. It occurs in sand as silicon dioxide, SiO_2—sometimes called silica. About 59 percent of Earth's crust is made up of silica. In its elemental form, silicon is a hard, gray solid with a relatively high melting point, 1410°C.

Silicon is in window glass and in the chips that run computers. Compounds of silicon are found in lubricants, caulking, and sealants.

2 Special Glasses from Silicon

Silicon is important in semiconductors. It's also important in making alloys and in ceramics, glass, and cement. The glass-ceramic shown here doesn't expand when heated so it won't break when exposed to large temperature changes.

3 "Tin" Cans and Alloys

Tin (Sn) is best known for its use as a protective coating for steel cans used for food storage. The coating protects the steel from corrosion. Tin is also a principal component in the alloys bronze, solder, and pewter. Tin is a soft metal that can be rolled into thin sheets of foil.

4 The Lead-Acid Storage Battery

4 **The Lead-Acid Storage Battery**

Lead (Pb) has been known and used since ancient times. It's obtained from the ore galena (PbS). Lead is alloyed with tin in solder and cheaper grades of pewter. The most important use of lead is in the lead-acid storage batteries used in automobiles. The electrodes in this kind of battery are made of lead and lead(IV) oxide (PbO_2).

Group 15

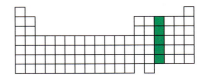

The trend in metallic properties is obvious as you go from the top of Group 15 to the bottom—from nitrogen (N) to phosphorus (P) to arsenic (As) to antimony (Sb) and bismuth (Bi). Nitrogen and phosphorus are nonmetals. They form covalent bonds to complete their outer-level configuration. Arsenic and antimony are metalloids and either gain or share electrons to complete their octets. Bismuth is more metallic and often loses electrons.

Group 15 elements have five valence electrons. Their valence-electron configuration is s^2p^3. They need only three electrons to attain the configuration of the noble gas at the end of their period. Nitrogen, phosphorus, and arsenic have an oxidation number of 3− in some of their compounds, but they can also have oxidation numbers of 3+ and 5+. Nitrogen is a component of proteins, deoxyribonucleic acid (DNA), and ribonucleic acid (RNA) so it's essential to life. Phosphorus is equally important because the phosphate group ($PO_4{}^{3-}$) is a repeating link in the DNA chain. The DNA molecule carries the genetic code that controls the activities of cells for many living organisms. Another important biological molecule, adenosine triphosphate, ATP, also contains phosphate groups that store and release energy in living organisms.

Nitrogen, as the chemically unreactive molecule N_2, makes up 78 percent by volume of Earth's atmosphere. Plants and animals can't use nitrogen in this form. Lichens, soil bacteria, and bacteria in the root nodules of beans, clover, and other similar plants convert nitrogen to ammonia and nitrate compounds. Lightning also converts atmospheric nitrogen to nitrogen monoxide (NO). Plants use these simple nitrogen compounds to make proteins and other complex nitrogen compounds that become part of the food chain.

The Uses of Group 15 Elements

Commercially, elemental nitrogen (N_2) is obtained from liquid air by fractional distillation. Much of it is converted to ammonia (NH_3), the familiar ingredient in some household cleaners.

1 Ammonia, the Essential Fertilizer

Ammonia is used as a liquid fertilizer applied directly to soil, or it can be converted to solid fertilizers such as ammonium nitrate, NH_4NO_3; ammonium sulfate, $(NH_4)_2SO_4$; or ammonium hydrogen phosphate, $(NH_4)_2HPO_4$. The bag of fertilizer shows the percentages of the essential main group elements—nitrogen, phosphorus, and potassium—that this fertilizer contains. Fertilizers are formulated in a variety of ways to provide the proper nutrients for different plant growth needs.

2 Two Allotropes of Phosphorus

White and red phosphorus are two common allotropes of phosphorus. Notice that the white phosphorus is photographed under a liquid because this form of phosphorus, which has the formula P_4, reacts spontaneously with oxygen in the air. Red phosphorus is used in making matches. You can read about it in *Everyday Chemistry*.

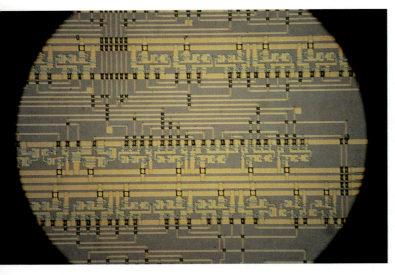

3 Gallium Arsenide Semiconductors

Arsenic is a metalloid found widely distributed in Earth's crust. An increasingly important use of the element is in the form of the binary compound gallium arsenide, GaAs. Because of its higher speed and performance, gallium arsenide is now replacing silicon in some of its semiconductor applications in electronic circuitry.

4 Antimony (Sb) is used primarily in alloys with other metals, particularly lead. Antimony improves the hardness and corrosion resistance of the metal.

Everyday Chemistry

The Chemistry of Matches

Making and lighting of friction and safety-matches involves a lot of chemistry.

Making friction matches Pinewood matchsticks are cut and dipped into a solution of borax (sodium tetraborate, $Na_2B_4O_7 \cdot 10H_2O$) or ammonium phosphate [$(NH_4)_3PO_4$] to make the matches safer. Next, the match head end is dipped into paraffin and then into a mixture of glue, coloring, a combustible material, and an oxidizing agent. Sulfur or diantimony trisulfide (Sb_2S_3) is used as the combustible material. Potassium chlorate ($KClO_3$) and manganese dioxide (MnO_2) are common oxidizing agents. Adding the tip—a mixture of tetraphosphorus trisulfide (P_4S_3), powdered glass, and a binder—is the final step.

Chemistry of friction matches The striking surface on a box of friction matches is powdered glass and glue. The P_4S_3 tip has a low kindling temperature. When the tip is rubbed on the striking surface, the heat from the friction causes it to ignite.

$$P_4S_3(s) + 6O_2(g) \rightarrow P_4O_6(g) + 3SO_2(g) + \text{heat}$$

The heat produced causes the potassium chlorate to decompose.

$$2KClO_3(s) \rightarrow 2KCl(s) + 3O_2(g)$$

The oxygen given off, combined with the heat from the first reaction, causes the sulfur to catch fire, which ignites the paraffin.

$$S(s) + O_2(g) \rightarrow SO_2(g) + \text{heat}$$

The burning paraffin carries the flame from the head to the wooden stem.

How safety matches work The wooden or paperboard sticks of safety matches are treated in a similar manner. Their heads contain diantimony trisulfide or sulfur, potassium chlorate or some other oxidizing agent, ground glass, and glue with paraffin underneath. These matches are called safety matches because they generally ignite only when they are rubbed across the striking surface on their box or packet. The striking surface serves the same function as the tip on the friction match; it ignites the head. The striking surface is a layer of red phosphorus, powdered glass, and glue. The friction of the match on the striking surface changes red phosphorus to white phosphorus.

$$P(\text{red}) + \text{heat} \rightarrow P(\text{white})$$

White phosphorus is formed; it ignites spontaneously in air and gives off enough heat to ignite the match head.

$$4P(\text{white})(s) + 5O_2(g) \rightarrow P_4O_{10}(s) + \text{heat}$$

Exploring Further

1. **Comparing and Contrasting** The first step in lighting a safety match is the conversion of red phosphorus to white phosphorus. Compare the chemical reactivities of these allotropes.

2. **Applying** Devise a method for making a match that would produce a colored flame.

Chemistry online

To find out more about the chemistry of matches, visit the Chemistry Web site at **chemistryca.com**

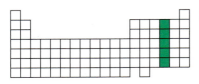

Group 16

The Group 16 elements—oxygen (O), sulfur (S), selenium (Se), and tellurium (Te)—are nonmetals, and polonium (Po) is a metalloid. Their valence-electron configuration is s^2p^4. With rare exceptions, oxygen gains two electrons and forms the oxide ion (O^{2-}) with the neon configuration. Oxygen reacts with both metals and nonmetals and, among the nonmetals, is second only to fluorine in chemical reactivity.

Oxygen is the most abundant element on Earth. It makes up 21 percent by volume of Earth's atmosphere and nearly 50 percent by mass of Earth's crust. Oxygen is present in the compound water and as oxides of other elements. Like nitrogen, oxygen gas (O_2) is obtained from fractional distillation of liquefied air.

There are two allotropes of oxygen—O_2, the most common, and O_3, called ozone. Ozone is a highly unstable and reactive gas that is considered a pollutant in the lower atmosphere. However, in the upper atmosphere, ozone protects Earth by absorbing harmful ultraviolet radiation from the sun. Ozone is responsible for the pungent odor you may notice during thunder and lightning storms or while operating your computer or other electronic equipment.

Both metals and nonmetals react directly with molecular oxygen to form oxides, as shown in the equations in **Table 8.1.**

Figure 8.8
Mining Sulfur
In the 1890s, the Frasch process (left) was invented by Herman Frasch. Hot water is pumped into an underground sulfur deposit where it melts the sulfur. Then, the liquid sulfur is brought to the surface by forcing air into the deposit. Liquid sulfur is shown solidifying after removal from a deposit (right).

Table 8.1 Reactions of Oxygen with Metals and Nonmetals	
Reaction of O_2 with Metals	**Reaction of O_2 with Nonmetals**
$4Na + O_2 \rightarrow 2Na_2O$	$C + O_2 \rightarrow CO_2$
$2Ca + O_2 \rightarrow 2CaO$	$S + O_2 \rightarrow SO_2$

Like oxygen, sulfur gains two electrons and forms the sulfide ion (S^{2-}) when it reacts with metals or with hydrogen. But in its reactions with nonmetals, sulfur can have other oxidation numbers. Much of the sulfur produced in the United States is taken from deposits of elemental sulfur by the Frasch process, shown in **Figure 8.8.**

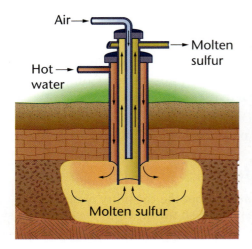

The Uses of Group 16 Elements

The largest industrial use of oxygen is in the production of steel, which is described later in Section 8.2. It's also used in the treatment of wastewater, as a part of H_2/O_2 rocket fuel, and in medicine to assist in respiration.

① Unstable Hydrogen Peroxide
In your household chemical storehouse, you'll find oxygen in solutions of hydrogen peroxide (H_2O_2), shown in a brown bottle. Peroxides are unstable compounds that decompose to produce molecular oxygen. The brown container helps to slow the decomposition of hydrogen peroxide by excluding light. The reaction is described by this equation.

$$2H_2O_2 \rightarrow 2H_2O + O_2$$

② Oxygen As an Antiseptic and Bleach
It's oxygen gas that produces the foam when you use hydrogen peroxide as an antiseptic to clean a cut or scrape. It's also oxygen that bleaches hair when a peroxide bleach is used. Some household cleansers use oxygen bleach rather than chlorine bleach.

③ One Use of Sulfuric Acid
Most elemental sulfur is converted to sulfuric acid (H_2SO_4), a key chemical in the production of a wide variety of products, such as fertilizers, automobile batteries, detergents, pigments, fibers, and synthetic rubber, as shown here.

④ An Application of Selenium's Photosensitivity
The chemistry of selenium and tellurium is similar to that of sulfur. Selenium has the property of increased electrical conductivity when exposed to light. This property has applications in security devices and mechanical opening and closing devices, where the interruption of a beam of light triggers an electrical response. But the most important application of selenium is in xerography, a process employed in modern photocopiers, as shown here.

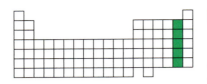

Group 17

The **halogens**—fluorine (F), chlorine (Cl), bromine (Br), iodine (I), and astatine (At)—are active nonmetals. Because of their chemical reactivity, they don't exist as free elements in nature. Their chemical behavior is characterized by a tendency to gain one electron to complete their s^2p^5 valence-electron configuration and form a 1− ion with a noble-gas configuration. Chlorine, for example, has the configuration $[Ne]3s^23p^5$. When it gains an electron, the chloride ion Cl⁻ is formed with the argon configuration $[Ne]3s^23p^6$. The halogens can also achieve a noble-gas configuration by sharing electrons.

Because the halogens react by gaining an electron, the most reactive element in the group is the one with the strongest attraction for an electron. Fluorine is the smallest of the halogens so it has the greatest ability to attract and hold an electron. As the size of the atoms increases down the group, the ability of the nucleus to attract and hold outer-level electrons decreases. Consequently, the reactivity of the halogens decreases as you go from fluorine to iodine. Iodine is the least active halogen because of its large atomic radius. Astatine, the largest of the halogens, is probably even less active than iodine, but it is scarce and radioactive.

The elemental halogens exist as diatomic molecules that are both highly reactive and toxic. Many household products contain chlorine compounds that can generate chlorine gas if not handled properly.

Try at Home Lab

See page 866 in Appendix F for **Periodic Properties of the Elements**

Uses of the Halogens

Fluorine and chlorine are both abundant in nature, and both are present in biologically essential compounds. Table salt (NaCl), used to flavor foods, provides all the chloride ions needed for a healthy diet. The iodide ion is also an essential trace element in the diet.

1 Biologically Important Iodine
The label on this box of table salt says that the salt is iodized. This means that the salt contains a small amount of the iodide ion, another biologically essential element. Iodine is absorbed by the thyroid gland, which regulates metabolism in the body. A swollen thyroid gland in the neck may indicate a lack of this trace element.

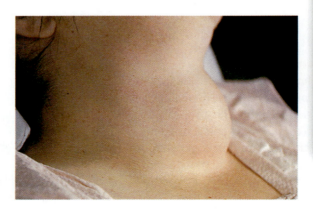

② Fluorides Prevent Tooth Decay

Many towns and cities add fluorides to their water supply, and sodium fluoride (NaF) or tin(II) fluoride (SnF_2) is often added to toothpastes to prevent tooth decay. You can read about the role of fluorides in preventing tooth decay in the *Biology Connection.*

③ Iodine As an Antibacterial

The halogens are important as antibacterial agents. Doctors use an iodine solution to sterilize the skin before surgery.

④ Chlorine Makes Water Safe

Chlorine is used in the water supply of most cities and towns and in swimming pools to kill bacteria. Chlorine added to swimming pools makes the water slightly acidic, so if your eyes are irritated after swimming in a pool, it's probably because of the acid.

⑤ Silver Bromide Coats Photographic Film

The compounds of the halogens are more important than the free elements. Compounds of chlorine with carbon, such as carbon tetrachloride and chloroform, are important solvents. Silver bromide (AgBr) is important in the light-sensitive coating on film.

Fluorides and Tooth Decay

Do you worry that you might get a cavity? If so, it's not surprising because total prevention of cavities is not yet possible. But since the 1950s, the problem of tooth decay has been greatly reduced.

Fluoridation and fluorides in toothpaste
Drinking water with minute amounts of fluorides helps prevent tooth decay. Studies by the U.S. Public Health Service show that people who drink water containing fluorides have lower rates of tooth decay. In the 1950s, cities and towns began to add fluorides to their water supplies. The results show that when children drink water containing 1 part per million (ppm) of sodium fluoride, NaF, or sodium silicofluoride, Na_2SiF_4, they have 70 percent fewer cavities than those who drink non-fluoridated water. Protection from tooth decay can also be obtained by using a toothpaste containing fluorides in the form of NaF, SnF_2, and Na_2PO_3F (MFP).

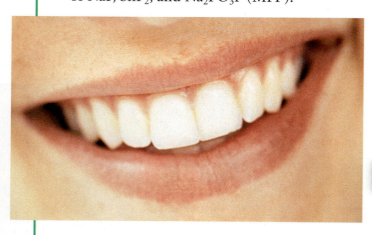

The decay process Tooth enamel is about 2 mm thick and 98 percent hydroxyapatite, $Ca_5(PO_4)_3OH$. Although hydroxyapatite is essentially insoluble in water, tiny amounts dissolve in the saliva in a process called demineralization.

$$Ca_5(PO_4)_3OH(s) \rightarrow$$
$$5Ca^{2+}(aq) + 3PO_4^{3-}(aq) + OH^-(aq)$$

The reverse process, remineralization, is the body's defense against bacterial acids.

$$5Ca^{2+}(aq) + 3PO_4^{3-}(aq) + OH^-(aq) \rightarrow$$
$$Ca_5(PO_4)_3OH(s)$$

The two equations show that this is a reversible reaction. In adults, the rates of demineralization and remineralization are equal, so equilibrium is established.

Bacteria and cavities Bacteria use sugar for energy and produce lactic acid. The acid causes the pH of saliva, which is normally 6.8, to drop below 6.0. When that happens, the rate of demineralization increases and tooth decay occurs.

Fluorides prevent cavities Fluoride compounds dissociate in water to form fluoride ions.

$$NaF(s) \rightarrow Na^+(aq) + F^-(aq)$$

$$SnF_2(s) \rightarrow Sn^{2+}(aq) + 2F^-(aq)$$

The fluoride ions replace the hydroxide ions in some of the $Ca_5(PO_4)_3OH$ and form fluorapatite, $Ca_5(PO_4)_3F$.

$$Ca_5(PO_4)_3OH(s) + F^-(aq) \rightarrow$$
$$Ca_5(PO_4)_3F(s) + OH^-(aq)$$

Fluorapatite is about 100 times less soluble than hydroxyapatite; it is also harder and denser, so tooth enamel is stronger and more resistant to bacterial attack.

Connecting to Chemistry

1. **Acquiring Information** Find out why only 50 percent of cities in the United States fluoridate their water supply.

2. **Applying** Check the ingredients on five different toothpastes to see whether they contain a fluoride compound, and if so, which one.

3. **Writing Equations** Write the reversible reaction of demineralization and remineralization of tooth enamel.

Group 18

Helium (He), neon (Ne), argon (Ar), krypton (Kr), xenon (Xe), and radon (Rn), the noble gases, were originally called the inert gases because chemists couldn't get them to react. Their lack of reactivity is understandable; all the noble gases have a full complement of valence electrons and, therefore, no tendency to gain or lose electrons. **Figure 8.9** shows one use of a noble gas that results from its lack of reactivity. In recent years, however, chemists have succeeded in making fluorine compounds of the heavier noble gases, krypton and xenon, but no reactions have been achieved for the lighter members of the group.

Trends in the chemical properties in each group of the main group elements are directly related to changes in atomic radii. For groups of elements that form compounds by losing electrons, the larger the atom, the more readily the atom gives up its electrons and the more reactive the atom is. For groups of elements that form compounds by gaining electrons, the larger the element, the less attraction it has for electrons and the less reactive the atom is. You will see similar trends among the transition elements.

Figure 8.9

Helium in Weather Balloons
Because helium does not burn, it is used rather than the lighter gas, hydrogen, to carry weather instruments into upper levels of the atmosphere. The instruments gather information on weather and atmospheric conditions.

Supplemental Problems

For more practice with solving problems, see Supplemental Practice Problems, Appendix B.

SECTION REVIEW

Understanding Concepts

1. Describe how the atomic radii of the main group elements change as you move across the third period. Give reasons for this trend.

2. Describe how the atomic radii of the elements in Group 2 change as you move down the group. Give reasons for this trend.

3. How does the size of a positive ion compare with the size of the neutral atom? How does the size of a negative ion compare with the size of the neutral atom? Give reasons for your answer.

Thinking Critically

4. **Comparing and Contrasting** Compare the way metals and nonmetals form ions and explain why they are different.

Applying Chemistry

5. **Hard Water** Soap scum forms when soap is used with hard water. This is because of the presence of magnesium and calcium ions. Which of the two elements, magnesium or calcium, reacts more easily with water?

Transition Elements

Some of the most intriguing elements are those that are beautiful, rare, and expensive. The might and power of kings and queens have been displayed through ornaments, jewelry, and works of art made of precious platinum, silver, and gold. But important as these metals are, they are only three of the transition elements—elements with such a variety of physical and chemical properties that they provide materials to fill a tremendous range of purposes. Iron, for example, is the world's most important structural material. Copper is known for its electrical conductivity. Chromium prevents the corrosion of other metals, and molybdenum can be alloyed with iron to give added hardness, corrosion resistance, and solidity at high temperatures. These are some of the metals that occupy the *d* block in the periodic table.

Properties of the Transition Elements

Each of the transition elements has its own properties that result from its atomic structure. For example, iron is strong. It's used for the structural framework of bridges and skyscrapers. But iron can also be reduced to a pile of reddish-brown rust if it is exposed to water and oxygen. Other transition metals may not be as strong as iron, but they also may not disintegrate in air like iron does. Fortunately, transition elements can be used together in alloys, as shown in **Figure 8.10.**

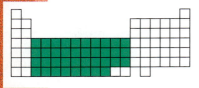

Figure 8.10
A Steel for All Purposes
When lightness, durability, and strength are needed for uses such as this racing wheelchair, a combination of transition elements, alloyed with iron, can provide the necessary properties. Here chromium, nickel, and magnesium (a main group element) are combined to produce a frame with the right properties.

Trends in Properties of the Transition Elements

With the exception of the Group 12 elements (zinc, cadmium, and mercury), the transition metals have higher melting points and boiling points than those of almost all of the main group elements. For example, in the fourth period (scandium to copper), the melting points range from 1083°C for copper (Cu) to 1890°C for vanadium (V). When you compare these melting temperatures to the melting temperatures of the main group metals, you find that only beryllium (Be) melts above 1000°C. Most of the other main group elements melt well below this temperature.

In any period, the melting and boiling points of the transition metals increase from Group 3 and reach a maximum in Group 5 or 6. Then they decrease across the remainder of the period. Tungsten (W) in Period 6, Group 6 has a melting point of 3410°C, the highest of any metal. It's because of its high melting point that tungsten is used as the filament in lightbulbs, as you'll see in *How it Works.* Mercury (Hg) in Period 6, Group 12 melts at −38°C, the lowest melting point of any metal. Mercury's liquid state at room temperature and its high density make it an important liquid for use in thermometers and barometers, as shown in **Figure 8.11.**

Figure 8.11

The Mercury Barometer

Any liquid could be used to make a barometer, but mercury is a good choice because a column of mercury only 76 cm high exerts a pressure approximately equal to the pressure of the atmosphere at sea level. This is because mercury has a high density for a liquid—13.6 g/mL. Other liquids, for example water with a density of 1.0 g/mL, require a column of liquid more than 30 feet high to equal the pressure of the atmosphere.

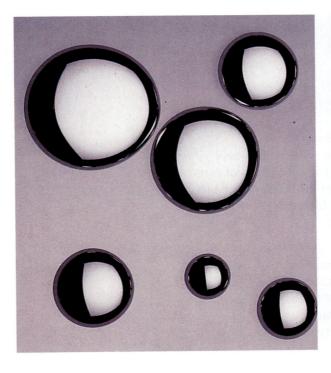

Multiple oxidation states are characteristic of the transition elements. Remember that iron gives up two electrons and forms the Fe^{2+} ion in its oxide, FeO. In another oxide, Fe_2O_3, iron gives up its two $4s$ electrons and one $3d$ electron to form the Fe^{3+} ion. Many of the transition elements can have multiple oxidation numbers ranging from 2+ to 7+. These oxidation numbers are due to involvement of the d electrons in chemical bonding. Recall that only some of the heavier main group elements such as tin, lead, and bismuth have multiple oxidation numbers. These elements also have d electrons that can be involved in bonding.

Inert Gases in Lightbulbs

Tungsten (W) is used as a filament in incandescent lightbulbs because it has a high melting point, 3420°C, and boiling point, 5850°C. But, nothing lasts forever. Because electricity continually passes through the filament, the metal eventually breaks as it vaporizes, and the bulb burns out.

1. If a lightbulb were filled with air, the filament would react with oxygen, burn, break, and lose its ability to provide light. If there were no gas inside the bulb, the filament would quickly vaporize and no electricity would flow.

2. If you've ever looked closely at a burned-out bulb, you may have seen a black coating of condensed metal on the inside of the bulb from the evaporated filament.

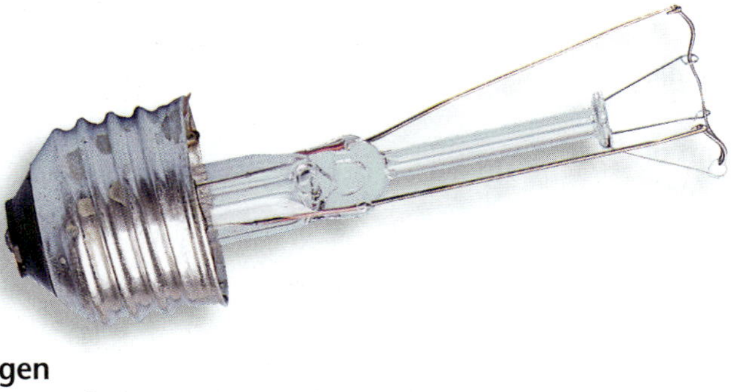

3. To prevent the filament from reacting and to slow its evaporation, lightbulbs traditionally have been filled with a mixture of nitrogen and argon. These gases carry heat away from the metal filament so it doesn't overheat and boil away. Nitrogen and argon don't react with the filament, and traditional bulbs can last as long as 750 hours.

In the search for longer-lasting lightbulbs, manufacturers are experimenting with various combinations of inert gases to replace the nitrogen/argon mixture. The most promising combination is a mixture of argon, krypton, and xenon, which produces a lightbulb that lasts 7500 to 10 000 hours—ten times longer than an ordinary lightbulb.

Thinking Critically

1. How would you go about finding a substance other than tungsten to use as the filament of a lightbulb?

2. What might be a disadvantage to the consumer of lightbulbs containing the relatively rare elements argon, krypton, and xenon?

The Ion Charges of a Transition Element

Iron, the most common transition element, has the electron configuration $[Ar]4s^2 3d^6$. The two electrons in the highest energy level, $4s^2$, are the ones most likely to be involved in the chemical reactions of iron. However, iron is a transition element and, like other transition elements, it has a partially filled d sublevel. Electrons in the d sublevel may also be involved in reactions. In this MiniLab, you will study some reactions of iron compounds and relate the results to the electron configuration of iron.

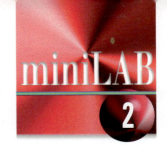

miniLAB 2

Procedure

1. Place about 20 mL of aqueous $FeCl_3$ solution into a 125-mL Erlenmeyer flask labeled *FeCl₃*. Add 2 drops of $1M$ NaOH solution. Describe the results in your data table.

2. Place about 20 mL of aqueous $FeCl_2$ solution into another 125-mL flask labeled *FeCl₂*. Add 2 drops of $1M$ NaOH solution. Describe the results in your data table.

3. Stopper the flask. Swirl and shake the mixture in the flask, labeled *FeCl₂*. Every 30 seconds, stop shaking the flask, and remove the stopper for a moment to admit more oxygen. Put the stopper back on the flask, and resume shaking until a change occurs.

4. Record your observations.

Analysis

1. Describe the colors of the two precipitates.

2. What are the charges on the iron ion in the two precipitates? Which electrons of the iron atom are probably lost to form the two ions?

3. Examine your results from procedure 3. Hypothesize what probably happened to the iron ions when oxygen entered the flask.

4. Iron(II) ions (Fe^{2+}) are useful as nutrients, whereas iron(III) ions (Fe^{3+}) are not. Using the results of this MiniLab, suggest a reason why elemental iron (Fe) is often added to breakfast cereals as a dietary supplement rather than an iron compound containing Fe^{2+} ions.

Trends in Atomic Size of Transition Elements

You learned that for the main group elements, atomic radius decreases from left to right in a period because of increasing nuclear charge. There is a similar trend for the transition elements, but the changes in atomic radii for the transition elements are not as great as the changes for the main group elements. You also learned that as you move from top to bottom in a main group, atomic radius increases. The same trend is seen among the transition elements. Atomic radius increases as you move from the fourth period to the fifth period in any group, but there is little change in atomic radius as you move from the fifth period to the sixth period. Because atomic size affects reactivity, you can expect the transition elements in Periods 5 and 6 to have similar chemical properties.

Iron: First Among the Transition Elements

Iron, and many steel alloys that are made from it, have been known and used since ancient times. Iron is the fourth most abundant element in Earth's crust and the second most abundant metal after aluminum.

1 Iron Is Essential in Human Life

Besides its importance as a structural metal, iron is an essential element in biological systems. It is the iron ion at the center of the heme molecule that binds oxygen. Iron in heme is the Fe^{2+} ion, and only the Fe^{2+} ion can bind oxygen. Heme is bound in the proteins hemoglobin and myoglobin. Hemoglobin supports life by transporting oxygen through the blood from the lungs to every cell of the body. Myoglobin stores oxygen for use in certain muscles.

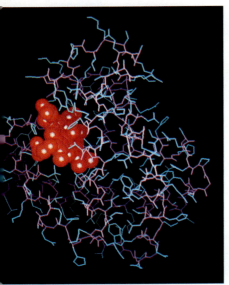

Hemoglobin molecule

Blast furnace

2 Separation of Iron from Its Ore

Iron is obtained from its ore (oxides) in a blast furnace. Iron oxide, Fe_2O_3, is mixed with carbon (coke) and limestone ($CaCO_3$) and fed continuously into the top of the furnace. Hot air is fed into the bottom of the furnace. Temperatures of 2000°C are reached as the mixture reacts while falling through the furnace. The process is complex, but the following equations are the important steps. First, coke is ignited in the presence of the hot air and converted to carbon dioxide.

$$C(s) + O_2(g) \rightarrow CO_2(g)$$

Then, CO_2 reacts with more coke to form carbon monoxide.

$$CO_2(g) + C(s) \rightarrow 2CO(g)$$

Carbon monoxide then converts iron ore (Fe_2O_3) to iron.

$$Fe_2O_3(s) + 3CO(g) \rightarrow 2Fe(l) + 3CO_2(g)$$

The molten iron, called pig iron, drops to the bottom of the furnace and is drawn off as a liquid. Impurities, called slag, form a layer on top of the iron and are drawn off separately. Pig iron, as it comes from the blast furnace, is a crude product containing impurities such as carbon, silicon, and manganese. The crude iron is refined, and then most of it is converted to steel.

Pig iron

3 Steelmaking

The first step in the production of steel is the removal of impurities from the pig iron. The second step is the addition of carbon, silicon, or any of a variety of transition metals in controlled amounts. Different elements give special properties to the final steel. Some steels are soft and pliable and are used to make things like fence wire. Others are harder and are used to make railroad tracks, girders, and beams. The hardest steels are used in surgical instruments, drills, and ordinary razor blades. These steels are made from iron mixed with small amounts of carbon.

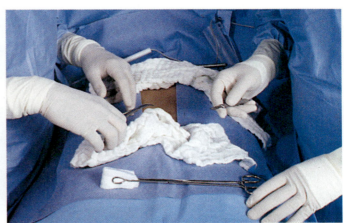

4 Heat-Treating

Heat-treating is a final step in the production of steel. When purified steel is heated to 500°C, the small amount of carbon contained in it combines with iron to form a carbide (Fe_3C), which dissolves in the steel. This makes the steel harder. Cooling the steel quickly in oil or water makes the hardness permanent.

Carbon and Alloy Steels

Steelmaking was already well advanced in ancient times. The earliest known steel objects were made between 1500 and 1200 B.C. About 1000 B.C., wootz steel was developed in India by heating a mixture of iron ore and wood in a sealed container.

Wootz steel later became known as Damascus steel because sword blades made from it had wavy surface patterns like Damask fabric. Damascus steel became famous because these swords kept their sharpness and strength after many battles. The knowledge of how to make Damascus steel was lost in the 1800s, but recently the process was redeveloped under the name superplastic steel. Collector hunting knives worth thousands of dollars are being made from superplastic steel.

What is steel and how is it classified?

Steel is an iron alloy containing small amounts of carbon (0.2 to 1.8 percent) and sometimes other elements such as chromium, manganese, nickel, tungsten, molybdenum, and vanadium. Steel cannot contain more than 1.8 percent carbon without becoming brittle, and the most common steels usually have closer to 0.2 percent carbon content. Commercial iron contains 2 to 4 percent carbon and is very brittle. Carbon steel contains only iron and carbon.

Damascus steel

Table 1 Carbon Steels

Name of Steel	Composition	Characteristics	Uses
Mild	Fe, less than 0.2% C	Malleable, ductile	Steel food cans, automobile bodies
Medium	Fe, 0.2–0.6% C	Less malleable/ductile	Structural purposes—beams, bridge supports
High	Fe, 0.7–1.5% C	Hard, brittle	Farm implements, drill bits, knives, springs, razor blades
UHCS, Superplastic*	Fe, 1.8% C	Corrosion- and wear-resistant, highly malleable	Engine components, earth movers, underside of tractors

*A few UHCSs are alloy steels.

Ninety Percent of Steel Is Carbon Steel

The properties of carbon steels depend upon the percent of carbon present. They are classed as mild, medium, and high on this basis. Because mild-carbon steels are ductile, sheets of it can be cold-formed to mold fenders and body parts for cars. Medium-carbon steels have more strength but are less ductile so they are used as structural materials. High-carbon steels are hard and brittle; they are used for wear-resistance purposes.

Superplastic Steel Is New Again

A new steel that can be formed into complex shapes called superplastic or ultrahigh-carbon steel (UHCS) has been developed. Damascus steels can be stretched up to 100 times their length without breaking. The limit for most steels is less than 1 1/2 times. At high temperatures, UHCS pulls like taffy. There is almost no waste from their use, thus conserving material and energy.

Table 2   Alloy Steels

Name of Steel	Composition in %*	Characteristics	Uses
Alnico I	Ni-20, Al-12, Co-5	Strongly magnetic	Loudspeakers, ammeters
Invar	Ni-36 to 50	Low coefficient of thermal expansion	Precision instruments, measuring tapes
Manganese steel	Mn-12 to 14	Holds hardness and strength	Safes
18-8 Stainless	Cr-18, Ni-8	Corrosion resistant	Surgical instruments, cooking utensils, jewelry
Tungsten steel	W-5	Stays hard when hot	High-speed cutting tools

* All alloys contain iron and 0.1-1.5% carbon.

Alloy Steels Contain Carbon and Other Elements

In alloy steels, iron is mixed with carbon and varying amounts of other elements, mainly metals. Added metals produce desired properties such as hardness and corrosion resistance (Cr), resistance to wear (Mn), toughness (Ni), heat resistance (W and Mo), and springiness (V). Stainless steel is a well-known, corrosion-resistant alloy steel. It contains ten to 30 percent chromium and sometimes nickel and/or silicon. Because of its outstanding magnetic properties, Alnico steel is used to make permanent magnets. Alnico magnets are used in voltmeters and ammeters to rotate the coil of wire connected to the pointer.

Working Steel

Two methods are used to shape steel into various shapes for specific purposes: hot-working and cold-working. In the hot-working process, steel is hammered, rolled, pressed, or extruded while it is very hot. Hammering and pressing are called forging. These processes were originally done by hand, in fact, hand forging continues as a craft in blacksmith shops. However, steam-powered hammers and hydraulic presses are used to forge most of the steel produced today. Extrusion involves forcing molten steel through a die that is cut to the desired

The development of methods of producing and working steel led to a revolution in construction in the 1880s. Concrete, reinforced by steel, became an important structural material. Steel beams made skyscrapers possible, and changed the shape of modern cities. Although the Sears tower is the highest building in the United States, it has so many tall neighbors that it no longer dominates the skyline of Chicago.

shape. Rolling is the most widely used method for shaping steel. The metal is passed between two rollers that move in opposite directions. The shape of the finished product determines the type of rollers used. Railroad tracks and I-beams are shaped in this way.

Cold-working includes rolling and extrusion, and in addition, drawing. Drawing involves pulling steel through a die rather than pushing it as in extrusion. Wires, tubes, sheets, and bars are shaped by drawing. Often cold-working processes follow hot-working to produce a more finished product.

DISCUSSING THE TECHNOLOGY

1. **Applying** The steel used in airplane motors is 43 percent Fe and 0.4 percent C; the other 56.6 percent is Cr, Ni, and Mo. What kind of steel is this? What are the functions of Cr, Ni, and Mo?

2. **Inferring** Using Table 2, how would you classify most of the metals used? (Hint: Look at their locations on the periodic table.) Draw a conclusion about most metals used in alloy steels.

3. **Acquiring Information** Find out how heat treatments, quenching, tempering, and annealing further alter the properties of different steels.

Other Transition Elements: A Variety of Uses

Fact of the MATTER

The principal ore of zirconium, a Group 4 element, is a silicate crystal called zircon. A zircon is a colorless crystal that can substitute for a diamond in jewelry. Often used in costume jewelry, jewelers refer to them as cubic zirconium.

You've learned that some transition elements are important in the production of steel because they impart particular properties to the steel. In addition, most of the transition elements, because of their individual properties, have a variety of uses in the production of the infrastructure and consumer products of the modern world.

The Iron Triad, Platinum Group, and Coinage Metals

Iron (Fe), cobalt (Co), and nickel (Ni) have nearly identical atomic radii, so it isn't surprising that these three elements have similar chemical properties. Like iron, both cobalt and nickel are naturally magnetic. Because of their similarities, the three elements are called the iron triad (group of three). Notice the positions of iron, cobalt, and nickel on the periodic table, as shown in **Figure 8.12.** They are in Period 4, Groups 8, 9, and 10. The elements below the iron triad in Periods 5 and 6—ruthenium (Ru), rhodium (Rh), palladium (Pd), osmium (Os), iridium (Ir), and platinum (Pt)—all resemble platinum in their chemical behavior and are called the platinum group. The platinum group elements are used as catalysts to speed up chemical reactions. Copper (Cu), silver (Ag), and gold (Au) in Group 11 are the traditional metals used for coins because they are malleable, relatively unreactive, and in the case of silver and gold, rare. You may have predicted that these elements would have similar chemical properties because they are in the same group. Notice the positions of the platinum group and the coinage metals in **Figure 8.12.**

Figure 8.12

The Iron Triad, Platinum Group, and Coinage Metals

The nearly identical atomic radii of the iron triad—iron, cobalt, and nickel —help explain the similar chemistry of these three elements. The similarities among the platinum group elements in Periods 5 and 6 emphasize the fact that there is little difference between the atomic radii of the elements in these periods in which inner *d* orbitals are being filled. The coinage metals show the expected similarity among elements in the same group.

Group Number → / Period Number ↓	3	4	5	6	7	8	9	10	11	12
4	Scandium Sc 21	Titanium Ti 22	Vanadium V 23	Chromium Cr 24	Manganese Mn 25	Iron Fe 26	Cobalt Co 27	Nickel Ni 28	Copper Cu 29	Zinc Zn 30
5	Yttrium Y 39	Zirconium Zr 40	Niobium Nb 41	Molybdenum Mo 42	Technetium Tc 43	Ruthenium Ru 44	Rhodium Rh 45	Palladium Pd 46	Silver Ag 47	Cadmium Cd 48
6	Lanthanum La 57	Hafnium Hf 72	Tantalum Ta 73	Tungsten W 74	Rhenium Re 75	Osmium Os 76	Iridium Ir 77	Platinum Pt 78	Gold Au 79	Mercury Hg 80

Iron triad   Platinum group   Coinage metals

Chromium

When chromium is alloyed with iron, tough, hard steels or steels that are corrosion-resistant are formed. Chromium is also alloyed with other transition metals to produce structural alloys for use in jet engines that must withstand high temperatures. A self-protective metal, chromium is often plated onto other materials to protect them from corrosion.

Chromium has the electron configuration $[Ar]4s^1 3d^5$ and exhibits oxidation numbers 2+, 3+, and 6+. When chromium loses two electrons, it forms the Cr^{2+} ion and has the configuration $[Ar]3d^4$. The Cr^{3+} ion results when chromium loses a second $3d$ electron. Chromium can lose six valence electrons and have an oxidation number of 6+. When it does, it loses all of its s and d electrons and assumes the electron configuration of argon. Potassium chromate (K_2CrO_4) and potassium dichromate ($K_2Cr_2O_7$) are two compounds in which chromium's oxidation number is 6+. **Figure 8.13** shows the brilliant colors that are typical of compounds of transition elements. Chromium gets its name from the Greek word for color, *chroma,* and many of its compounds are brightly colored—yellow, orange, blue, green, and violet.

Figure 8.13

Two Compounds of Chromium
The brilliant colors of these chromium compounds are typical of many transition metal compounds. In both the yellow potassium chromate (K_2CrO_4) and the orange potassium dichromate ($K_2Cr_2O_7$), chromium has a 6+ oxidation number.

Zinc

Like chromium, zinc is a corrosion-resistant metal. One of its principal uses is as a coating on iron and steel surfaces to prevent rusting. In the process called galvanizing, a surface coating of zinc is applied to iron by dipping the iron into molten zinc. Zinc is also important when alloyed with other metals. The most important of these alloys is the combination of zinc with copper in brass. Brass is used for making bright and useful objects like those shown in **Figure 8.14.**

Figure 8.14

The Importance of Brass
Brass can be worked into smooth shapes and drawn into the long, thin-walled tubes needed for musical instruments. Look around and see how many brass items you find that are both decorative and useful.

Lanthanides and Actinides: The Inner Transition Elements

The inner transition elements are found in the *f* block of the periodic table. In the lanthanides, electrons of highest energy are in the 4*f* sublevel. The lanthanides were once called *rare earth elements* because all of these elements occurred in Earth's crust as *earths,* an older term for oxides, and seemed to be relatively rare. The highest-energy electrons in the actinides are in the 5*f* sublevel. You probably won't find these elements among your household chemicals. Their names are unfamiliar except for uranium and plutonium, which are the elements associated with nuclear reactors and weapons. However, many of these elements, especially lanthanides, have important practical uses.

Cerium: The Most Abundant Lanthanide

Cerium is the principal metal in the alloy called misch metal. Misch metal is 50 percent cerium combined with lanthanum, neodymium, and a small amount of iron. Misch metal is used to make the flints for lighters. Cerium is often included in alloys of iron and other metals such as magnesium. A high-temperature alloy of three percent cerium with magnesium is used for jet engines. Some of cerium's compounds—for example, cerium(IV) oxide (CeO_2)—are used to polish lenses, mirrors, and television screens; in glass manufacturing to decolorize glass; and to make porcelain coatings opaque.

Other Lanthanides

Other lanthanides are used in the glass industry. Neodymium (Nd) is used not only to decolorize glass but to add color to glass. When added to the glass used for welders' goggles, neodymium and praseodymium (Pr) absorb the eye-damaging radiation from welding, as shown in **Figure 8.15.** They also decrease reflected glare when used in the glass of a television screen. A combination of the oxides of yttrium (Y), a transition element, and europium (Eu) produce a phosphor that glows a brilliant red when struck by a beam of electrons, such as in a TV picture tube. This phosphor is used with blue and green phosphors to produce realistic-looking television pictures. When rare earth phosphors are used in mercury-arc outdoor lighting, they change the bluish light of the mercury arc to a clear white light.

Figure 8.15
Lanthanides Provide Eye Protection
Intense light from a welding torch can harm eyes. Neodymium and praseodymium, incorporated into the lenses of goggles, absorb the damaging wavelengths.

Because europium, gadolinium (Gd), and dysprosium (Dy) are good absorbers of neutrons, they are used in control rods in nuclear reactors. Promethium (Pm) is the only synthetic element in the lanthanide series. It is obtained in small quantities from nuclear reactors and is used in specialized miniature batteries. Samarium (Sm) and gadolinium are used in electronics. Terbium (Tb) is used in solid-state devices and lasers.

Radioactivity and the Actinides

Uranium (U) is a naturally occurring, radioactive element used as a source of nuclear fuel and other radioactive elements. Plutonium (Pu) is one of the elements obtained from the use of uranium as a nuclear fuel. The isotope Pu-238 emits radiation that is easily absorbed by shielding. Pu-238 is used as a power source in heart pacemakers and navigation buoys. Other isotopes of plutonium are used as nuclear fuel and in nuclear weapons. Plutonium is the starting material for the synthetic production of the element americium, which is used in smoke detectors.

Some actinides have medical applications; for example, radioactive californium-252 (Cf) is used in cancer therapy. Better results in killing cancer cells have been achieved using this isotope of californium than by using the more traditional X-ray radiation.

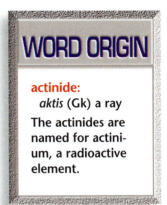

WORD ORIGIN

actinide:
aktis (Gk) a ray

The actinides are named for actinium, a radioactive element.

Connecting Ideas

If you can locate an element on the periodic table, you can predict its properties. Each element has unique characteristics because of its unique electron configuration. Together, the elements, their alloys, and compounds provide a wide variety of materials for countless applications. Compounds of the elements range from ionic to covalent, from polar to nonpolar. They have size and shape. In Chapter 9, you'll learn more about the formation of compounds and how to predict their shape and polarity.

SECTION REVIEW

Understanding Concepts

1. What are transition elements? What are inner transition elements? Describe where all the transition elements are found on the periodic table.

2. How do the electron configurations of the transition metals and inner transition elements differ from those of the main group metals?

3. Iron, aluminum, and magnesium all have important uses as structural materials, yet iron corrodes (rusts), whereas aluminum and magnesium usually do not corrode. Explain what causes iron to rust.

Thinking Critically

4. **Applying Concepts** Why do the transition metals have multiple oxidation numbers whereas the alkali metals and the alkaline earth metals have only one oxidation number, 1+ and 2+, respectively?

Applying Chemistry

5. **Steelmaking** Name three transition metals that are added to iron and steel to improve their properties and help prevent corrosion. Explain how each additive works.

REVIEWING MAIN IDEAS

8.1 Main Group Elements

- In a period of the periodic table, the number of valence electrons increases as atomic number increases. As a result, elements change from metal to metalloid to nonmetal to noble gas.

- Atomic size is a periodic property. As atomic number increases in a period, atomic radius decreases. As atomic number increases in a group, atomic radius increases.

- Positive ions have smaller atomic radii than the neutral atoms from which they derive. Negative ions have larger atomic radii than their neutral atoms.

- Positive ions in the same group increase in size down the group.

- In a group, each element has the same number of valence electrons. As a result, the elements in a group show similar chemical behavior.

- Metals react by losing electrons. The most reactive metals are those that give up electrons most easily. The metal with the biggest atom and smallest number of valence electrons is the most active metal. Cesium in the lower-left-hand corner of the periodic table is the most active metal.

- Nonmetals react by gaining or sharing electrons. The most reactive nonmetals are those that attract and hold electrons most strongly. The nonmetal with the smallest atom and greatest number of valence electrons is the most active nonmetal. Fluorine in the upper right-hand corner of the periodic table is the most active nonmetal.

8.2 Transition Elements

- Transition elements react by losing the valence electrons in their *s* orbitals. Many of the transition metals have more than one oxidation number because they also lose electrons from their *d* orbitals.

- The lanthanides and actinides react by losing the valence electrons in their *s* orbitals. Because these elements can also lose electrons from their *d* and *f* orbitals, they have multiple oxidation numbers.

Vocabulary

For each of the following terms, write a sentence that shows your understanding of its meaning.

alkali metal
alkaline earth metal
halogen

UNDERSTANDING CONCEPTS

1. From each of the following pairs of atoms, select the one with the larger atomic radius.

   a) K, Ca
   b) F, Na
   c) Mg, Ca
   d) Rb, Cs
   e) Ca, Sr
   f) S, C

2. What is the effect of atomic radius on the chemical reactivity of the halogens? Which is the most active halogen? Which is the least active?

3. Which of the following atoms increase in size when they become ions? Explain your answer. Cs, I, Zn, O, Sr, Al

4. Why do elements in a group have similar chemical properties?

5. What are the products of a reaction between an alkali metal and water? An alkaline earth metal and water?

6. Zinc is one of the few transition metals that has a single oxidation number. What is the oxidation number for zinc? How does zinc's electron configuration account for this oxidation number?

chemistryca.com/vocabulary_puzzlemaker

7. Using the periodic table as a guide, list the following ions in order of increasing ionic radius. Na^+, Cl^-, S^{2-}, F^-, Al^{3+}, Se^{2-}

8. Use the graph of atomic radius versus period number to answer these questions. How does the size of a lithium atom compare with that of a cesium atom? How does the size of a fluorine atom compare with that of an iodine atom? Which has the larger radius, lithium or fluorine? Based on their atomic radii, which is the most active alkali metal? Which is the most active halogen? Explain.

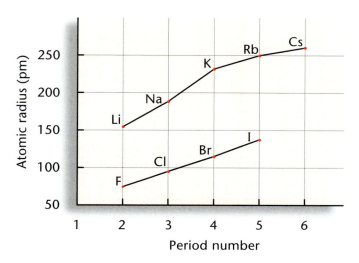

APPLYING CONCEPTS

How It Works

9. Why are noble gases used in lightbulbs? What advantages do they offer over a vacuum?

Chemistry and Technology

10. Explain the difference between carbon steel and alloy steel. Which type makes up the majority of steel produced?

History Connection

11. Use the periodic table to determine the oxidation numbers of lead and sulfur in galena, PbS, the common ore of lead.

Everyday Chemistry

12. Sodium silicofluoride, Na_2SiF_4, is added to water supplies to help prevent tooth decay. What is the oxidation number of silicon in Na_2SiF_4? Explain how you got your answer.

Biology Connection

13. Nitrogen, oxygen, and phosphorus are essential elements for maintaining life. What essential biological molecules contain these elements?

14. Describe one way in which nitrogen (N_2) is converted into a form that plants can use.

15. Magnesium and aluminum both react with oxygen to form oxides. Explain why they are considered corrosion-resistant metals.

THINKING CRITICALLY

Comparing and Contrasting

16. Compare the chemical behavior of oxygen when it combines with an active metal such as calcium to its behavior when it combines with a nonmetal such as sulfur.

Relating Concepts

17. Is carbon dioxide, CO_2, produced by sharing or transferring electrons? Draw the Lewis dot diagram for CO_2 and explain how each element achieves an octet of electrons.

Identifying Patterns

18. **ChemLab** Zinc exhibits the same oxidation number as the alkaline earth metals. What is the chemical formula of the compound formed when zinc combines with an oxalate ion?

Relating Cause and Effect

19. **MiniLab 1** What factor contributes to the difference in reactivity between potassium and cesium? Which element is more reactive?

Relating Concepts

20. **MiniLab 2** Write the electron configurations of the neutral iron atom and the ions Fe^{2+} and Fe^{3+}.

Making Predictions

21. Both sodium and cesium react with water. Predict the products of the reactions of sodium and cesium with water. Write balanced equations for both reactions.

Using a Table

22. The energy needed to remove the first electron from a gaseous atom to produce a gaseous ion is called the first ionization energy. First ionization energies for the main group elements are shown in the table. Examine the data and decide whether ionization energy is a periodic property. Describe how ionization energies change within a period and within a group.

First Ionization Energy kJ/mol

	1	2	13	14	15	16	17	18
1								He 2 2372
2	Li 3 520	Be 4 900	B 5 801	C 6 1087	N 7 1402	O 8 1314	F 9 1681	Ne 10 2081
3	Na 11 496	Mg 12 738	Al 13 578	Si 14 786	P 15 1012	S 16 1000	Cl 17 1256	Ar 18 1521
4	K 19 419	Ca 20 590						

CUMULATIVE REVIEW

23. Describe the physical properties of metals, metalloids, and nonmetals. (Chapter 3)

24. Write the electron configurations for each of the following atoms. Use the appropriate noble-gas inner core abbreviations. (Chapter 7)
  a) fluorine   **c)** titanium
  b) aluminum   **d)** argon

25. Explain why the gaseous nonmetals—hydrogen, nitrogen, oxygen, fluorine and chorine—exist as diatomic molecules, but other gaseous nonmetals—helium, neon, argon, krypton, xenon and radon—exist as single atoms. (Chapter 4)

SKILL REVIEW

26. Making and Using a Graph Use the data in the following figure to draw a graph of atomic radius versus atomic number for the second and third period elements in Groups 1, 2, 13, 14, 15, 16, and 17. Describe any patterns you observe.

	1	2	13	14	15	16	17
1	H 78						
2	Li 156	Be 112	B 85	C 77	N 71	O 60	F 69
3	Na 186	Mg 160	Al 143	Si 118	P 109	S 103	Cl 91
4	K 231	Ca 197	Ga 134	Ge 123	As 121	Se 117	Br 119
5	Rb 248	Sr 215	In 167	Sn 141	Sb 161	Te 138	I 138
6	Cs 262	Ba 222	Tl 170	Pb 175	Bi 151	Po 164	At Unknown
7	Fr 280	Ra 228					

$1 \text{ pm} = 10^{-12} \text{ m}$

27. Using the data in the above figure draw a graph of atomic radius versus atomic number for the elements in Groups 2 and 17. Describe any patterns you observe.

WRITING IN CHEMISTRY

28. The numbers 5-10-5 on a fertilizer package refer to the percentages of nitrogen (N), phosphorus (P), and potassium (K), respectively, in the fertilizer. Write a short paper describing the role of each of these elements in plants. Research other fertilizer compositions, and explain the purpose of varying the proportions of these nutrients.

PROBLEM SOLVING

29. Draw a sketch of the outline of a periodic table. Use the data from the diagram in question 43 of the Skill Review to help you draw arrows on the table showing changes in atomic radius and ionization energy in periods and groups as atomic number increases. Write a sentence describing the relationship between ionization energy and atomic radius in both periods and groups.

Standardized Test Practice

Element Characteristics			
Element	Atomic Number	Element Type	Atomic Radius (pm)
Boron	5	metalloid	85
Carbon	6	nonmetal	77
Nitrogen	7	nonmetal	71
Aluminum	13	metal	143
Silicon	14	metalloid	118
Phosphorus	15	nonmetal	109
Gallium	31	metal	134
Germanium	32	metalloid	123
Arsenic	33	metalloid	121

Use the table above to answer questions 1–4.

1. Across a period on the periodic table to the right, the atomic sizes of elements

a) increase.   c) vary.
b) decrease.   d) remain the same.

2. Down a group on the periodic table, the atomic sizes of elements

a) increase.   c) vary.
b) decrease.   d) remain the same.

3. With respect to atomic numbers, the atomic sizes of elements

a) increase.   c) vary.
b) decrease.   d) remain the same.

4. Why is the size of an aluminum atom larger than an atom of silicon?

a) The atomic radii of metal atoms are larger than the atomic radii of nonmetal atoms.
b) The positive charges in an aluminum atom's nucleus have a greater attraction on the atom's electron cloud.
c) The positive charges in a silicon atom's nucleus have a greater attraction on the atom's electron cloud.
d) The silicon atom contains more electrons.

5. Which of the following is true of the alkaline earth metals?

a) Alkaline earth metals have the same properties as the alkali metals.
b) Alkaline earth metals are the most reactive metals.
c) Alkaline earth metals have a greater density and hardness than Alkali metals.
d) Alkaline earth metals tend to lose three valence electrons and form positively charged ions.

6. Which of the following is not a member of group 15?

a) potassium   c) bismuth
b) arsenic   d) antimony

CHAPTER
9 Chemical Bonding

Chapter Preview

Sections

9.1 Bonding of Atoms
MiniLab 9.1 Coffee Filter
Chromotography

9.2 Molecular Shape and Parity
MiniLab 9.2 Modeling Molecules
ChemLab What Colors are in Your
Candy?

Hey, It's a Giant Buckyball!

You may be wondering what a buckyball is. A buckyball is an individual molecule of buckminsterfullerene—a carbon molecule with 60 carbon atoms bonded into a spherical ball. Graphite, diamond, and buckminster-fullerene are natural forms of carbon that have different arrangements.

Launch Lab

Oil and Vinegar Dressing

When preparing a meal, you combine different types of food. But when you mix different substances, do they always "mix"? How about making oil and vinegar dressing for tonight's salad?

Safety Precautions

Materials

- Beral-type pipette
- vinegar
- vegetable oil

Procedure

1. Fill the bulb of a Beral-type pipet about 1/3 full of vinegar and 1/3 full of vegetable oil. Shake the pipet and its contents. Record your observations.
2. Allow the contents to sit for about five minutes. Record your observations.

Analysis

Do oil and vinegar mix? What explanation can you give for your observations in this experiment? Why do the instructions on many types of salad dressings read "shake well before using"?

What I Already Know

Review the following concepts before studying this chapter.
Chapter 2: the use of electron dot diagrams to depict valence electrons
Chapter 4: properties of ionic compounds and covalent compounds
Chapter 8: trends in periodic properties of the elements

Reading Chemistry

Note the different figures used to illustrate the way molecules are structured, such as the gum drops pictured on pages 318–325. As you read the chapter, make drawings of several of the structures.

Preview this chapter's content and activities at chemistryca.com

Bonding of Atoms

You've probably used glue to repair a broken object or to make something new. The glue joins separate pieces together to form a stable whole. In Chapter 4, you learned that atoms bond with other atoms to form molecules and compounds. The glue that holds atoms together in a molecule is the sharing of electrons between the atoms.

SECTION PREVIEW

Objectives
✓ **Predict** the type of bond that forms between atoms by using electronegativity values.

✓ **Compare and contrast** characteristics of ionic, covalent, and polar covalent bonds.

✓ **Interpret** the sea of electrons model of metallic bonding.

Review Vocabulary
Alkali metal: any element from Group 1: lithium, sodium, potassium, rubidium, cesium, or francium.

New Vocabulary
electronegativity
shielding effect
polar covalent bond
malleable
ductile
conductivity
metallic bond

A Model of Bonding

Atoms either transfer electrons and then form ionic compounds or they share electrons to form covalent compounds. In both cases, the bond forms because of an increase in stability. By forming bonds, atoms acquire an octet of electrons and the stable electron configuration of a noble gas. Atoms are often more stable when they're bonded in compounds than when they're free atoms.

Dividing compounds into two bonding types, ionic and covalent, is convenient. If you know the type of bonds in a compound, you can predict many of its physical properties. **Table 9.1** summarizes the physical properties of ionic and covalent compounds. You can also reason in the opposite direction. If you know the physical properties of an unknown compound,

Table 9.1 Physical Properties of Ionic and Covalent Compounds		
Property	**Ionic Compound**	**Covalent Compound**
State at room temperature	crystalline solid	liquid, gas, solid
Melting point	high	low
Conductivity in liquid state	yes	no
Water solubility	high	low
Conductivity of aqueous solution	yes	no

you can predict its bond type. But predictions may not always be correct because there is no clear-cut division between ionic and covalent compounds. A compound may be partly covalent and partly ionic. A more realistic view of bonding is to consider that all chemical bonds involve a sharing of electrons. Electrons may be shared equally, but they also may be shared only slightly—almost not at all. The properties of any compound, particularly its physical properties, are related to how equally the electrons are shared. **Figure 9.1** shows a model of bonding as a sharing of electrons.

In Chapter 4, you explored the two extremes of this range of electron sharing: the ionic bond and the covalent bond. A purely ionic bond results when the sharing is so unequal that it is best described as a complete transfer of electrons from one bonding atom to another. A purely covalent bond results when electrons are shared equally. Most compounds fall somewhere in between these two extremes; they have some ionic characteristics and some covalent characteristics.

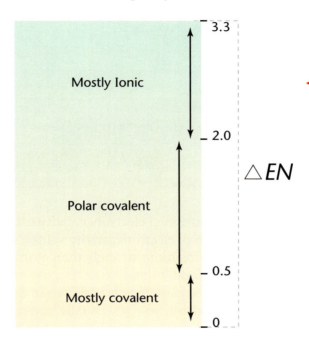

Figure 9.1

Electron Sharing Model of Bonding

◀ The bonding between atoms in compounds can be viewed as a range of electron sharing measured by electronegativity difference, ΔEN. This range contains three main classes of bonds—ionic, polar covalent, and covalent. Differences in electronegativity, ΔEN, are explained on the next two pages.

▲ Bonding can be thought of as a tug-of-war between two atoms for shared electrons.

Electronegativity: An Attraction for Electrons

You can think of bonding atoms as being in a tug-of-war for the shared valence electrons, as shown in **Figure 9.1.** To use this model of electron sharing, you need some way of determining how much tug each atom exerts on the shared electrons. The measure of the tug is electronegativity. **Electronegativity** is a measure of the ability of an atom in a bond to attract electrons. How each atom fares in a tug-of-war for shared electrons is determined by comparing the electronegativities of the two bonded atoms.

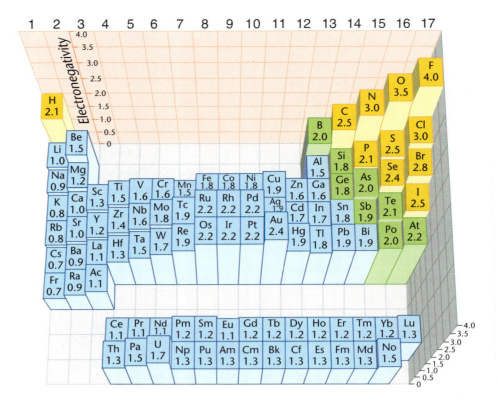

Figure 9.2

The Electronegativities of the Elements

Electronegativity is a periodic property. Notice how the heights of the columns change across each period and down each group. The heights are proportional to the electronegativities.

Atoms are assigned electronegativity values, as shown in **Figure 9.2.** Atoms with large electronegativity values, such as fluorine, attract shared valence electrons more strongly than atoms such as sodium that have small electronegativities.

The electronegativity values shown in **Figure 9.2** may not always be at your fingertips, but you will usually have a periodic table. Electronegativity is a periodic property. With only a few exceptions, electronegativity values increase as you move from left to right in any period of the periodic table. Within any group, electronegativity values decrease as you go down the group. That means that the most electronegative elements are in the upper-right corner of the table. Fluorine has the highest value of 4.0. It also follows that the elements having the lowest electronegativities are in the lower-left corner. The electronegativity of cesium is 0.7. The noble gases are considered to have electronegativity values of zero and do not follow the periodic trends.

The decrease in electronegativity as you move down a column occurs because the number of energy levels increases, and the valence electrons are farther from the positively charged nucleus. The nucleus has less attraction for its own valence electrons so they are held less tightly. Also, electrons in the inner energy levels tend to block the attraction of the nucleus for the valence electrons. This is known as the **shielding effect.** The shielding effect increases as you move down a column because the number of inner electrons increases. For example, magnesium has the electron configuration $1s^2 2s^2 2p^6 3s^2$ and an electronegativity of 1.2; calcium has the configuration $1s^2 2s^2 2p^6 3s^2 3p^6 4s^2$ and an electronegativity of 1.0. Although both have two valence electrons, calcium has eight more inner electrons than magnesium. These electrons shield the outer electrons from the attraction of the nucleus, as shown in **Figure 9.3.**

Figure 9.3

The Shielding Effect
The electrons in the first and second energy levels shield the two valence electrons in the magnesium atom from the full effect of 12 nuclear protons. ▼

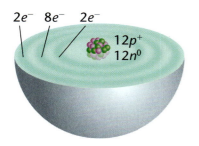

$2e^-$ $8e^-$ $2e^-$

$12p^+$
$12n^0$

Magnesium atom

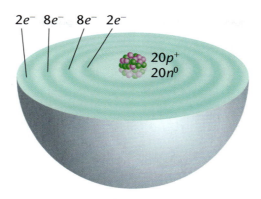

$2e^-$ $8e^-$ $8e^-$ $2e^-$

$20p^+$
$20n^0$

Calcium atom

◄ Calcium's two valence electrons are shielded from the full attraction of the 20+ nuclear charge by inner electrons in the first, second, and third energy levels. Because calcium has eight more inner electrons than magnesium, the valence electrons of calcium are held less tightly.

As you move across a row from left to right in the periodic table, the number of protons in the nucleus increases. With the increase in nuclear charge, the attraction of the nucleus for the valence electrons increases, and therefore electronegativity tends to increase across a row. In period 4, potassium in Group 1 has an electronegativity of 0.8, while bromine has a value of 2.8.

Because electronegativity varies in a periodic way, you can make predictions about differences in electronegativity by looking at the distance between bonding atoms on the table. In general, the farther the bonding atoms are from each other on the periodic table, the greater their electronegativity difference. **Figure 9.4** shows the trends in electronegativity.

The Ionic Extreme

The greater the difference between the electronegativities of the bonding atoms, the more unequally the electrons are shared. The electronegativity difference between two bonding atoms is often represented by the symbol ΔEN, where EN is an abbreviation for ElectroNegativity and Δ is the Greek letter *delta* meaning "difference." ΔEN is calculated by subtracting the smaller electronegativity from the larger, so ΔEN is always positive. For example, ΔEN for cesium and fluorine is $4.0 - 0.7 = 3.3$.

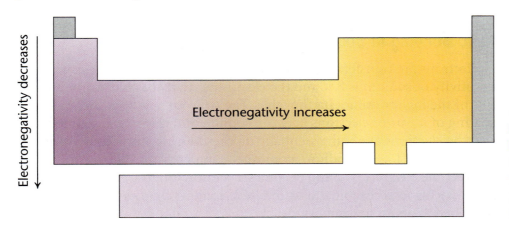

Electronegativity decreases

Electronegativity increases

Figure 9.4

Periodic Trends in Electronegativity
Electronegativity increases from left to right across a period and decreases from top to bottom down a group.

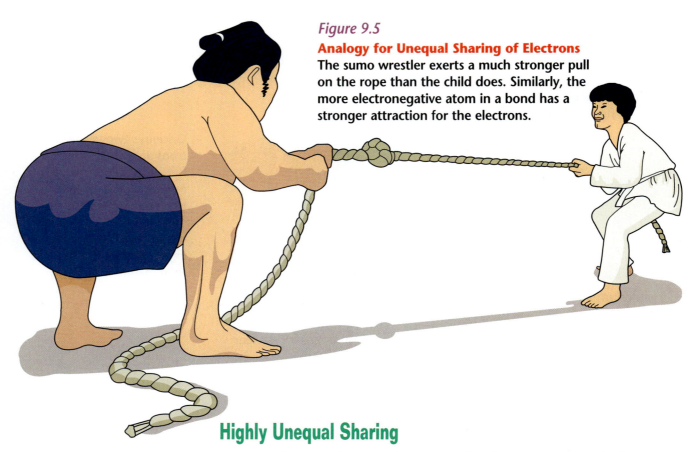

Figure 9.5

Analogy for Unequal Sharing of Electrons
The sumo wrestler exerts a much stronger pull
on the rope than the child does. Similarly, the
more electronegative atom in a bond has a
stronger attraction for the electrons.

Highly Unequal Sharing

Using the tug-of-war analogy, consider the situation in which a sumo
wrestler is at one end of a rope and a small child is at the other end, as
shown in **Figure 9.5.** There's no contest. The same is true in a chemical
bond when the ΔEN between bonding atoms is 2.0 or greater.

When the electronegativity difference in a bond is 2.0 or greater, the
sharing of electrons is so unequal that you can assume that the electron
on the less electronegative atom is transferred to the more electronegative
atom. This electron transfer results in the formation of one positive ion
and one negative ion. The bond formed by the two oppositely charged
ions is classified as a mostly ionic bond. Many bonds are classified as
ionic, but they have varying degrees of ionic character. The greater the
difference in the electronegativities of the two atoms, the more ionic the
bond. The greater the distance between the bonding atoms on the periodic table, the more ionic the bond between the atoms.

Ionic Bonding in Sodium Chloride

Electrons are transferred when sodium chloride forms. The electronegativity of sodium is 0.9, and the electronegativity of chlorine is 3.0,
one of the highest values on the table. The ΔEN for this bond is 2.1. The
sharing of a pair of electrons between sodium and chlorine is so unequal
that the electrons are both essentially on the chlorine atom, creating a
chloride ion, Cl^-. The sodium atom can't compete with chlorine for a
share in the electron pair and becomes a sodium ion, Na^+. Sodium ions
and chloride ions combine to form NaCl. Sodium chloride is best
described as an ionic compound.

Linus Pauling: An Advocate of Knowledge and Peace

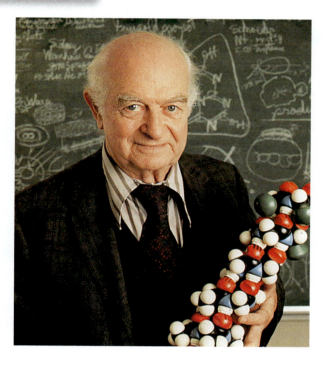

Some proclaim him one of the 20 greatest scientists of all time. But Linus C. Pauling claimed merely to have been well prepared and to have been in the right place at the right time. The time was the mid-1920s at the beginning of quantum physics.

Quantum theory of chemistry In 1925, Pauling was awarded a Ph.D. in chemistry from the California Institute of Technology, where he studied the crystal structure of materials. A year later, he was granted a Guggenheim Fellowship and traveled to Europe to study the quantum theory of the atom. Returning to CalTech, he merged his knowledge of material structures and quantum theory into the concept of the chemical bond. Pauling's book *The Nature of the Chemical Bond* was influential in providing a framework for researchers to study and predict the structures and properties of inorganic, organic, and biochemical compounds. To acknowledge the importance of his work in understanding the chemical bond, Pauling was awarded the 1954 Nobel Prize for Chemistry.

Antinuclear armaments Pauling was an outspoken critic of the atmospheric testing of nuclear bombs. He was convinced that the radioactive fallout from such testing would be hazardous to humans for many generations. Petitioning scientists worldwide, Pauling pleaded for an international ban on testing nuclear weapons. The announcement that Pauling had been awarded the 1962 Nobel Prize for Peace was made on the day that the world's first partial nuclear test ban went into effect.

Vitamin C In the early 1970s, Pauling became an advocate of the health benefits of taking megadoses (large doses) of vitamin C. His book *Vitamin C and the Common Cold* became a best-seller. Although his ideas are controversial, Pauling thought vitamin C might help eliminate minor ailments and be a possible cure for cancers.

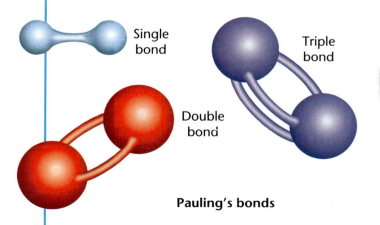

Single bond

Double bond

Triple bond

Pauling's bonds

Connecting to Chemistry

1. **Applying** Why is understanding the nature of chemical bonding important?

2. **Acquiring Information** Investigate the role Pauling played in the discovery of the structure of DNA.

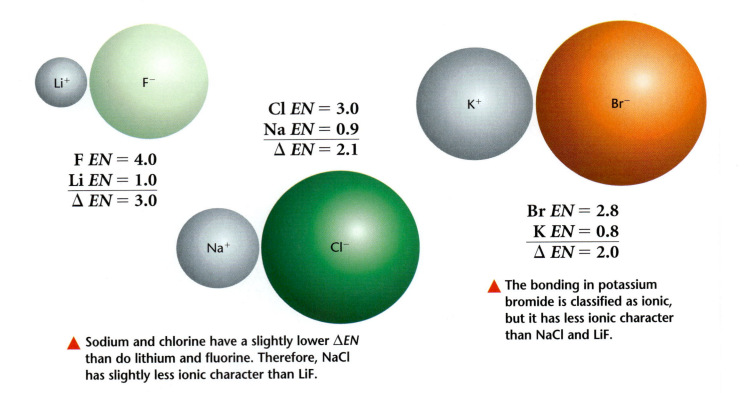

Cl $EN = 3.0$
Na $EN = 0.9$
$\overline{}$
$\Delta\ EN = 2.1$

F $EN = 4.0$
Li $EN = 1.0$
$\overline{}$
$\Delta\ EN = 3.0$

Br $EN = 2.8$
K $EN = 0.8$
$\overline{}$
$\Delta\ EN = 2.0$

▲ The bonding in potassium bromide is classified as ionic, but it has less ionic character than NaCl and LiF.

▲ Sodium and chlorine have a slightly lower ΔEN than do lithium and fluorine. Therefore, NaCl has slightly less ionic character than LiF.

Figure 9.6

Three Ionic Compounds
The electronegativity differences in lithium fluoride, sodium chloride, and potassium bromide show that they are best represented as ionic compounds.

Figure 9.6 compares the formation of sodium chloride with the formation of lithium fluoride and potassium bromide. For each of these salts, the ΔENs are equal to or greater than 2.0. Like sodium chloride, both lithium fluoride and potassium bromide are considered mostly ionic compounds. Notice that the two atoms in each bond are well separated from each other on the periodic table.

The Covalent Extreme

You have seen that when there is a large electronegativity difference between two atoms (2.0 or greater), the bond that forms between the two atoms is considered mostly ionic. What happens when there is no difference in electronegativity or if the electronegativity difference is small?

Equal Sharing

Now imagine a 300-pound sumo wrestler on one end of the rope, and another 300-pound sumo wrestler on the other end. This situation is the same as when two of the same atoms form a bond, for example, when two fluorine atoms form the fluorine molecule, F_2. Here is the electron dot structure for F_2.

$$:\ddot{F}:\ddot{F}:$$

Because two atoms of the same element are forming the bond, the difference in electronegativities is zero. In the fluorine molecule, a pair of valence electrons are shared equally. This type of bond is a pure covalent bond. All other diatomic elements (Cl_2, Br_2, I_2, O_2, N_2, and H_2) have pure covalent bonds. In all these molecules, the electrons are shared equally.

Sharing That's Close to Equal

Sometimes, the electronegativities of bonding atoms are close but not exactly the same. For example, carbon's electronegativity is 2.5 and hydrogen's is 2.1. A ΔEN of greater than zero always means unequal electron sharing. However, a bond in which the electronegativity difference is less than or equal to 0.5 is considered to be a covalent bond with electrons that are not shared exactly equally. All of the many compounds formed between carbon and hydrogen are considered covalent.

When the electronegativity difference is less than or equal to 0.5, the slightly unequal sharing of the electrons doesn't have a significant effect on the properties of the molecule. Low boiling points and melting points are typical of pure covalent compounds. Most of the elemental diatomic molecules are gases at room temperature—Cl_2, F_2, O_2, N_2, and H_2. Carbon disulfide, methane, and nitrogen dioxide, shown in **Figure 9.7**, are examples of covalent compounds in which the electron sharing is slightly unequal. These molecules are gases or low-boiling point liquids at room temperature.

Figure 9.7

Three Covalent Compounds
Carbon disulfide is an important solvent for waxes and greases. Methane is the principal component of natural gas. Nitrogen dioxide is used for making nitric acid and is also an atmospheric pollutant. All three compounds contain covalent bonds in which the sharing of electrons is more or less equal.

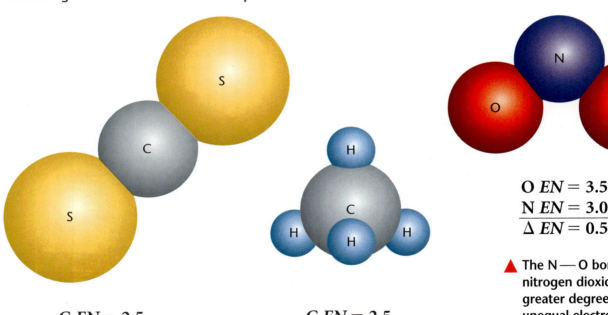

C EN = 2.5
S EN = 2.5
$\overline{\Delta EN = 0.0}$

▲ The C—S bonds in carbon disulfide are pure covalent bonds. The $\Delta EN = 0$, even though the atoms are different.

C EN = 2.5
H EN = 2.1
$\overline{\Delta EN = 0.4}$

▲ The ΔEN of 0.4 in the bonds of methane is not sufficient to significantly affect the properties of the compound.

O EN = 3.5
N EN = 3.0
$\overline{\Delta EN = 0.5}$

▲ The N—O bonds in nitrogen dioxide have a greater degree of unequal electron sharing than C—H bonds, but NO_2 is still considered a covalent compound.

Polar Covalent Bonds

Now consider a tug-of-war in which a 300-pound sumo wrestler is on one end of the rope and a 400-pound wrestler is on the other end. The unequal pull on the ends of the rope brings the middle of the rope closer to the 400-pound competitor. This is an analogy to a bond in which electrons are shared unequally. This type of bond falls between the two extremes of electron sharing: equal sharing (a covalent bond) and completely unequal sharing (an ionic bond). Bonds in which the pair of electrons is shared unequally have electronegativity differences between 0.5 and 2.0.

Unequal Sharing in Covalent Bonds

When the electronegativity difference between bonding atoms is between 0.5 and 2.0, the electron sharing is not so unequal that a complete transfer of electrons takes place. Instead, there is a partial transfer of the shared electrons to the more electronegative atom. The less electronegative atom still retains some attraction for the shared electrons. The bond that forms when electrons are shared unequally is called a **polar covalent bond.** A polar covalent bond has a significant degree of ionic character.

Polar covalent bonds are called *polar* because the unequal electron sharing creates two poles across the bond. Just as a car battery or a flashlight battery has separate positive and negative poles, so polar covalent bonds have poles, as shown in **Figure 9.8.** The negative pole is centered on the more electronegative atom in the bond. This atom has a share in an extra electron. The positive pole is centered on the less electronegative atom. This atom has lost a share in one of its electrons. Because there was

WORD ORIGIN

electronegative:
elektron (Gk)
amber (Rubbing
amber produces
electric charge.)
negare (L) to
deny

When highly
electronegative
atoms attract other
atoms' electrons,
they become more
negative.

Figure 9.8

Bonds with Positive and Negative Ends
When the sharing of electrons in a bond isn't equal, the bond is polar, as in the H — Cl bond. Just like this battery, the bond has two poles, one positive and one negative. The symbols δ^+ and δ^- (delta plus and delta minus) are used to show the distribution of partial charges in a polar covalent bond. The arrow points in the direction of the negative end of the bond.

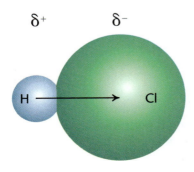

not a complete transfer of an electron, the charges on the poles are not 1+ and 1−, but δ^+ and δ^-. These symbols, delta plus and delta minus, represent a partial positive charge and a partial negative charge. This separation of charge, resulting in positively and negatively charged ends of the bond, gives the polar covalent bond a degree of ionic character.

Compounds with polar covalent bonds have different properties from compounds with pure covalent bonds. You saw that purely covalent compounds tend to have low melting points and boiling points. Carbon disulfide, CS_2, as shown in **Figure 9.7**, is a triatomic molecule, with a ΔEN equal to zero. Carbon disulfide boils at 46°C. Water is also a triatomic molecule, but the bonding in water is polar covalent. Even though water is a much lighter molecule than carbon disulfide, its boiling point is 100°C.

Water Has Polar Bonds

The ΔEN for the O—H bond is 1.4, so water molecules have polar bonds. When atoms of hydrogen and oxygen bond by sharing electrons, the shared pair of electrons is attracted toward the more electronegative oxygen. This unequal sharing causes an imbalance in the distribution of charge about the two atoms, as shown in **Figure 9.9**.

The effect of polar covalent bonds on boiling point is illustrated by comparing the boiling point of water with the other hydrides of Group 16 elements—sulfur, selenium, and tellurium. A hydride is a compound formed between any element and hydrogen. Water is the first hydride of Group 16. ΔEN for the O—H bond is 1.4. But ΔEN for both the S—H and the Se—H bonds is only 0.4, and ΔEN for the Te—H bond is 0.3. These electronegativity differences tell you that H_2O contains polar covalent bonds, but the bonds in H_2S, H_2Se, and H_2Te are essentially covalent. **Figure 9.10** shows that even though H_2O is the lightest of the four hydrides, its boiling point is significantly higher than the hydrides of the other members of Group 16.

Figure 9.9

Charge Distribution in an O—H Bond
Because oxygen is more electronegative than hydrogen, the electrons in an O—H bond spend more time near the oxygen atom than near the hydrogen atom. This distribution leads to a partial negative charge on oxygen and a partial positive charge on hydrogen.

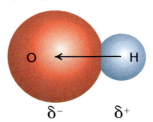

δ^- δ^+

Try at Home **Lab**

See page 867 in Appendix F for
Breaking Covalent Bonds

Figure 9.10

Boiling Points Reflect Bond Type
On the basis of the mass of the molecules, you might predict that H_2O would have the lowest boiling point of the four hydrides. Another factor, the polarity of the O—H bond, makes that prediction incorrect. Later, you'll learn how polar bonds often result in a polar molecule, as in the case of water.

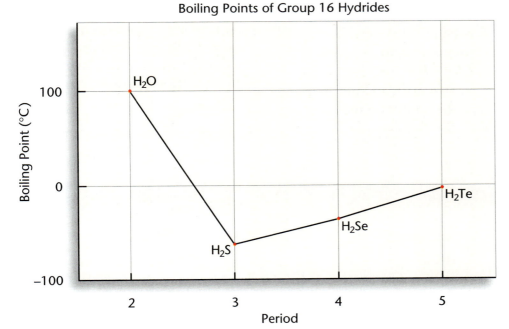

Boiling Points of Group 16 Hydrides

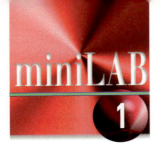

miniLAB

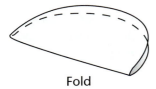

Fold

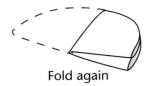

Fold again

Fold again

Coffee Filter Chromatography

Paper chromatography is a method of separating substances based upon their different attractions for the paper. The paper is called the stationary phase of the system, and the solvent is called the moving phase. In this experiment, the moving phase is water, a polar molecule. You will examine the composition of washable color marker inks by comparing the distances each component in the inks travels on a filter-paper chromatogram.

Procedure

1. Obtain a set of markers and two circular coffee filters.

2. Cover the bottom of a plastic cup with a small amount of water.

3. Lay the two coffee filters together on a dry surface. With the filters still together, fold the circles into eighths as shown.

4. Unfold the filters, but do not separate them. Find the center spot where all folds converge.

5. About 5 cm from the center, along a crease, make a dark mark with one of the pens. The mark should appear on both filters.

6. Continue for all eight creases, using a different color marker on each crease.

7. Separate the filters. Keep one filter unfolded as a control. Refold the other filter.

8. With a small amount of water in the bottom of the cup,

gently place the folded coffee filter, tip down, into the water.

9. When water reaches the top of the paper, gently remove the paper.

10. Gently open the paper. Compare each mark on the control filter with the corresponding mark on the chromatographed filter.

Analysis

1. Compare the marks on the control paper with the corresponding marks on the chromatographed filter. How are they different?

2. Which colors on the chromatographed filter are different from those on the control filter?

3. Polar substances tend to be attracted to other polar substances such as found in paper. Nonpolar substances move faster through the paper. Which of the inks contains the most polar substances?

PRACTICE PROBLEMS

Supplemental Problems

For more practice with solving problems, see Supplemental Practice Problems, Appendix B.

1. Calculate ΔEN for the pairs of atoms in the following bonds.

   a) Ca — S   c) C — Br   e) H — Br
   b) Ba — O   d) Ca — F

2. Use ΔEN to classify the bonds in question 1 as covalent, polar covalent, or ionic.

Bonding in Metals

Bonding in metals doesn't result in the formation of compounds, but it is an interaction that holds metal atoms together and accounts for some of the typical properties of metals and alloys. What are some of these properties?

Properties That Reflect Metallic Bonding

Metals and alloys are malleable and ductile, and they conduct electricity. When a metal can be pounded or rolled into thin sheets, it is called **malleable.** Gold is an example of a malleable metal, as shown in **Figure 9.11.** A chunk of gold can be flattened and shaped by hammering until it is a thin sheet. **Ductile** metals can be drawn into wires. For example, copper can be pulled into thin strands of wire and used in electric circuits, as illustrated in **Figure 9.11.** Electrical **conductivity** is a measure of how easily electrons can flow through a material to produce an electric current. Metals such as silver are excellent conductors because there is low resistance to the movement of electrons in the metal. These properties—malleability, ductility, and electrical conductivity—are the result of the way that metal atoms bond with each other.

Sea of Valence Electrons

The valence electrons of metal atoms are loosely held by the positively charged nucleus. Sometimes, metal atoms form ionic bonds with non-metals by losing one or more of their valence electrons and forming positive ions. However, in metallic bonding, metal atoms don't lose their

Figure 9.11

Malleability, Ductility, and Conductivity in Metals
Malleability, ductility, and electrical conductivity reflect the type of bonding in metals.

Copper is ductile and a good conductor of electricity. It is most commonly used in electrical circuits. ▼

▲ Gold is malleable. An artisan practices the ancient art of making and using gold leaf. Gold leaf is gold metal that has been flattened until a sheet of gold foil, only a few hundred atoms thick, is obtained.

valence electrons. Metal atoms release their valence electrons into a sea of electrons shared by all of the metal atoms. The bond that results from this shared pool of valence electrons is called a **metallic bond. Figure 9.12** illustrates a model of metallic bonding. The bonding interaction at the submicroscopic level explains what you observe at the macroscopic level. Although the metal atoms are bonded together in a large network, they are not bonded to any single atom. This explains why metals are often malleable and ductile.

The conductivity of metals can also be explained by the sea of electrons model of metallic bonding, as shown in **Figure 9.12.** Because the valence electrons of all the metal atoms are not attached to any one metal atom, they can move through the metal when an external force, such as that provided by a battery, is applied.

Figure 9.12

Atomic View of Metallic Bonding

Each atom in this model of a Group 2 metal releases its two valence electrons into a pool of electrons to be shared by all of the metal atoms. ▼

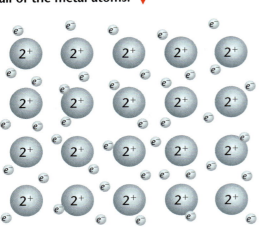

▲ Bonding in metals is not rigid. As a metal is struck by a hammer, the atoms slide through the electron sea to new positions while continuing to maintain their connections to each other. The same ability to reorganize explains why metals can be pulled into long, thin wires.

SECTION REVIEW

Understanding Concepts

1. Use ΔEN to classify the bonds in the following compounds as covalent, polar covalent, or ionic.

   a) H—S bond in H_2S
   b) S—O bond in SO_2
   c) Mg—Br bond in $MgBr_2$
   d) N—O bond in NO_2
   e) C—Cl bond in CCl_4

2. Using only a periodic table, rank these atoms from the least to most electronegative.

   Na, Br, I, F, Hg

3. Rank these bonds from the least to the most polar.

   a) C—F
   b) O—F
   c) Al—Br
   d) Cl—F
   e) O—H

Thinking Critically

4. **Properties of Metals** What aspect of metallic bonding is responsible for the malleability, ductility, and conductivity of metals?

Applying Chemistry

5. **Using Electronegativities** Do you expect LiF or LiCl to have a higher melting point? Explain.

Molecular Shape and Polarity

Y ou've probably seen small-scale models of structures such as buildings or bridges. Designers find it helpful to work with small models before producing a detailed plan for a full-scale structure. Models can also be enlarged versions of objects that are so small that they can't be seen or handled. This is the case for atoms and molecules. So far, you have only seen molecules represented on paper by formulas or electron dot diagrams. These representations tell you the kinds of atoms contained in a molecule and how many there are of each. Electron dot diagrams also tell you how the valence electrons are distributed around each atom. But these models do not tell you how the atoms are arranged in space. Because many properties of molecules are determined by the structure of the molecule, it's important to develop models of molecules in three dimensions. In this section, you'll learn how models are used to represent the bonding between atoms and the shapes of molecules.

SECTION PREVIEW

Objectives
✓ **Diagram** electron dot structures for molecules.

✓ **Formulate** three-dimensional geometry of molecules from electron dot structures.

✓ **Predict** molecular polarity from three-dimensional geometry and bond polarity.

Review Vocabulary
Electronegativity: a measure of the ability of an atom in a bond to attract electrons.

New Vocabulary
double bond
triple bond
polar molecule

The Shapes of Molecules

Models can help you visualize the three-dimensional structures of molecules. Consider the simplest molecule that exists—hydrogen, H_2. Two hydrogen atoms share a pair of electrons in a nonpolar covalent bond, as shown in the electron dot structure.

H:H

A model of this molecule might be made by connecting two gumdrops of the same color by a toothpick. The gumdrops represent the hydrogen atoms, and the toothpick represents the shared pair of electrons that makes up the covalent bond.

PEOPLE in CHEMISTRY

Meet Dr. William Skawinski, Chemist

An ordinary copy of *Mathematics for Physicists* is about two inches thick. However, the Braille version on Dr. Skawinski's bookshelf requires more than three feet of space. The chemist, who began losing his sight in childhood, consults this reference as he creates three-dimensional models of molecules. In the following interview, Dr. Skawinski talks about his innovative work and his love of chemistry.

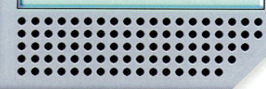

On the Job

 Dr. Skawinski, can you tell us how you go about building molecular models?

A I start by putting information about molecular structures into a CAD (computer-aided design) program. The program uses the information to calculate a structure for the molecule. Then stereolithography is used to make a physical model from the calculations. This is basically how it works. A table covered by a polymer film sits at the top of a tank containing liquid plastic. A laser traces a cross section of the bottom of the molecule in the film, and a slice of the model hardens. Then the table moves down a little and the laser traces another cross section, and so on to the top of the molecular design. After eight to 12 hours, I get one solid piece of plastic that replicates a molecule of a particular compound.

 By what factor are these models bigger than the real thing?

A A carbon atom, for example, has a radius of about 0.2 nanometers. The model is about an inch in diameter—many billions of times larger.

 Why are these three-dimensional models useful?

A Dealing with a physical model of a mathematical distribution can give anyone—blind or sighted—a better insight into what it represents.

 Do you have a favorite of the molecular models you've built?

 Beta-cyclodextrin has doughnut-shaped molecules that are interesting. People have remarked that many of the molecular models look like pieces of artwork.

Early Influences

 Do you remember specific incidents from your childhood that influenced your interest in science?

 I can remember being two years old and watching the flames of a gas water heater. The bright blue lights on a perfectly black background fascinated me. I also remember at about the age of five hammering red, yellow, and gray rocks in the backyard, breaking them up into powder, and mixing them with water to make colored suspensions.

 What were things like for you in high school?

 Even though I was having serious problems with my vision by that time, I was involved in an amateur rocket group. These rockets weren't the little four-inch rockets you buy through the mail and send up a couple of hundred feet. One rocket for which I mixed the chemical propellant reached an altitude of 42 000 feet. I loved the idea of propelling a vehicle into space—and I still do. In fact, if anyone offered me a ticket to another planet, I'd only pause long enough to pack a toothbrush!

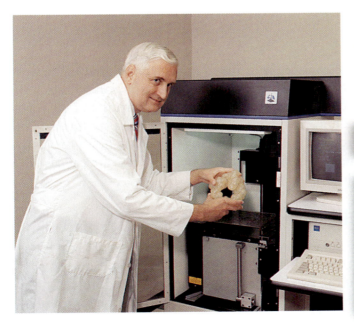

 Did your high school chemistry class interest you?

 My chemistry teacher made chemistry a part of the real world. Once he brought in a round, green glob about the size of a walnut. It was a flawed emerald. He explained that just a slight difference in chemical makeup differentiated that worthless stone from a priceless one.

Personal Insights

 Does the wonderment about science that you experienced as a child remain with you today?

 Of course. I think you really have to retain that trait to enter the field of science. Scientists are just a bunch of big kids who are fascinated with the world.

 How have you dealt with the challenge of your blindness?

I was able to see enough through college to learn a lot of critical information. But near the end of college, because my field of vision had narrowed so much, I could see just part of a word on a page at a time, and that was only with a strong magnifying glass. I managed to earn my master's degree and then my Ph.D. mostly through determination. My approach has always been to acknowledge obstacles and then concentrate on finding ways around them.

CAREER ▶ CONNECTION

The following job opportunities are also in the field of chemistry.

Chemical Engineer B.S. in chemical engineering
Chemical Laboratory Technician Two-year training program
Industrial Chemical Worker High school plus on-the-job training

Figure 9.13

The Geometry of Diatomic Molecules
Just as two gumdrops can be connected in only one way, all diatomic molecules are linear, whether they're composed of the same two atoms, as in H_2 and Cl_2, or different atoms, as in HCl.

What does a gumdrop model tell about the shape of a hydrogen molecule? As **Figure 9.13** shows, there's only one way to put this model together. When the two gumdrops are connected with a toothpick, they lie in a straight line. Therefore, a hydrogen molecule is linear. You could model other diatomic molecules, such as oxygen, nitrogen, chlorine, iodine, fluorine, and even hydrogen chloride, HCl. The model always predicts the same geometry: linear.

Modeling Water

How would you build a gumdrop model of a water molecule? First, you would draw the electron dot diagram. Remember that the dot diagram models the arrangement of valence electrons in the molecule in two dimensions. The electron dot diagram for a water molecule shows that each hydrogen shares a pair of electrons with the oxygen.

The eight valence electrons are distributed in such a way that the oxygen atom has an octet of electrons and the stable electron configuration of the noble gas neon. Each hydrogen has two valence electrons and the stable configuration of helium. Two pairs of valence electrons are involved in the bonding. These electrons are called bonding pairs. The other two pairs of valence electrons are not involved in the bonding. These are called nonbonding pairs, or lone pairs.

From Electron Dot Diagram to Model

To make a model of the water molecule, you need two colors of gumdrops, such as red for oxygen and yellow for hydrogen. You also need two toothpicks to represent the two covalent bonds, and two additional toothpicks to represent the lone pairs of electrons. Even though the lone pairs of electrons are not part of any bonds, they play a major role in determining the shape of a molecule. They are present and they occupy space.

How should the hydrogens be connected to the oxygen atom? Clearly, some rules are needed. Look at the dot diagram for water. Four pairs of electrons are on the oxygen. These electrons are all negatively charged, and because they all have the same charge, they repel each other. Therefore, they will form the three-dimensional arrangement around the oxygen atom that allows them to be as far from each other as possible. This arrangement is a geometric shape called a tetrahedron. In a tetrahedral arrangement, the repulsions between the electron pairs are minimized.

To create a model with tetrahedral geometry, place the four toothpicks in the red gumdrop in a three-dimensional arrangement with the largest possible angle between each adjacent toothpick. Two yellow gumdrops should then be placed at the ends of any two of the toothpicks. These represent the two O—H bonds. The other toothpicks represent the lone pairs. **Figure 9.14** shows the gumdrop model of a water molecule and visualizes the tetrahedral arrangement of four electron pairs.

From Model to Water Molecule

The models in **Figure 9.14** suggest that the three atoms in the water molecule are arranged in a bent structure. The angles in a perfect tetrahedron each measure 109.5°. The angle between the two bonds in the water molecule is 105°—a little less than the angle predicted by the gumdrop model. The difference between the predicted and experimental bond angle comes about because the lone pairs of electrons repel each other more than the shared pairs. In effect, the nonbonding electrons require more room, so they distort the tetrahedral arrangement by squeezing the bonding pairs closer together and decreasing the bond angle from 109.5° to 105°.

A space-filling model shows the electron clouds of each atom as spheres. The clouds overlap when two atoms form a bond. This model is a good representation of the water molecule. ▼

Figure 9.14

Visualizing a Water Molecule

In this gumdrop model of a water molecule, you can see that the three atoms form a bent structure. ▼

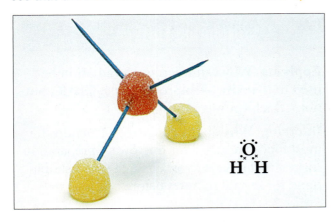

Four balloons, inflated equally and held together at a central point, arrange themselves in a tetrahedral shape. This is the most space-efficient arrangement for four things about a center point. In this model, the balloons represent the four electron pairs of the water molecule. ▶

Jiggling Molecules

How often have you warmed a snack by microwaves—that is, in a microwave oven? Maybe you were too hungry to notice that the food was a lot hotter than the dish or container. A clue to the cause of this sometimes-overlooked observation is the steam you blew away from the food as you waited for it to cool.

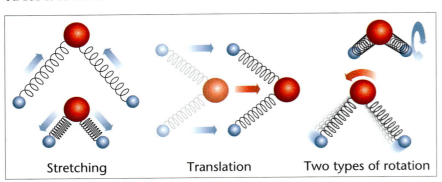

Stretching Translation Two types of rotation

Microwave heating Microwave radiation is a form of electromagnetic energy. The microwave radiation produced by a microwave oven has a wavelength of 11.8 cm and a frequency of 2.45 billion hertz. (A shorter way of writing a frequency of 2.45 billion hertz is 2.45 gigahertz, or 2.45 GHz.) The radiation travels through space and materials in the form of a moving electromagnetic field; that is, a field having an electric-field component and a magnetic-field component. The frequency of the wave is the number of oscillations or waves each second. The frequency also measures the energy of the waves. Microwaves have little effect on most molecules. However, the microwave's oscillating electromagnetic field interacts with positively and negatively charged polar molecules. As a result, the charged molecules oscillate (move back and forth). The increased motion means that the molecules have increased kinetic energy. (Kinetic energy is energy of motion and is directly related to temperature.) As microwaves are absorbed by substances containing polar water molecules, the temperature rises rapidly. Heat is transferred by conduction from the water molecules to other parts of the substance.

Microwave decomposition If you ever jiggled a garden rake to dislodge a leaf, you know that violent motion can be used to separate things. Researchers are applying the same concept to decompose molecules of some toxic substances by using microwaves. Molecules of compounds such as dihydrogen sulfide, sulfur dioxide, and nitrogen dioxide, which contribute to air pollution, are polar and can be decomposed into their nontoxic elements by the oscillations caused by microwave radiation. The oscillations become so violent that the attractions between the atoms in the molecule are no longer great enough to hold them together, and the molecule decomposes or falls apart. Here, the energy of the microwaves has overcome the energy of the chemical bond. Similar research is being done to use microwaves to initiate reactions in which toxic polar organic compounds are reacted with other substances, such as oxygen, to form nontoxic products. Applications of the research may produce cost-effective methods of using microwaves to control air pollution and clean up hazardous wastes.

Exploring Further

1. **Applying** Why can microwave radiation be used to detoxify trichloromethane, $CHCl_3$, but not tetrachloromethane, CCl_4?

2. **Inferring** Why do the cooking instructions for most foods packaged for microwave preparation suggest that after microwaving, the food should stand for several minutes before being served?

Chemistry Online

Hungry for more information about microwaves and how they cook? Visit the Chemistry Web site at **chemistryca.com**

Modeling Carbon Dioxide

Are all triatomic molecules bent like the water molecule? To find out, you can model carbon dioxide, CO_2, as you did water. Begin by drawing the electron dot diagrams for the two atoms. Carbon has four valence electrons, and each oxygen atom has six.

$$\cdot \ddot{C} \cdot \quad \cdot \ddot{O} \colon$$

To obtain a stable octet of electrons, the carbon atom needs four more electrons, and each oxygen atom needs two more electrons. Therefore, each oxygen atom must share two pairs of electrons with the carbon atom. A bond formed by sharing two pairs of electrons between two atoms is called a **double bond,** as illustrated by the electron dot structure for carbon dioxide.

$$\ddot{O} \colon\colon C \colon\colon \ddot{O}$$

If you count all the shared and unshared electrons around each of the three atoms, you'll see that each atom has an octet.

How can you determine the three-dimensional geometry of a carbon dioxide molecule? First, look at the arrangement of electrons around the central atom. Consider each double bond as one cloud of shared electrons. Two clouds are around the carbon. What geometry puts the two electron clouds as far apart as possible? Linear geometry, as illustrated in **Figure 9.15,** separates the clouds as much as possible. The model suggests that the three atoms in carbon dioxide are arranged in a straight line. This structure of CO_2 was proved correct by experiment.

Figure 9.15

Visualizing Carbon Dioxide

A gumdrop model of carbon dioxide predicts a linear structure with the C—O bonds pointed in opposite directions. ▼

The space-filling model of CO_2 is a good representation of the molecule. ▼

The balloon model represents the clouds of bonding electrons on either side of the carbon atom. A linear molecule results because of repulsion between the two clouds. ▶

A Model for Ammonia

Ammonia, NH_3, has three bonds from a central nitrogen to hydrogen atoms. Nitrogen has five valence electrons, and each hydrogen has one electron. Each hydrogen shares a pair of electrons with the nitrogen. Nitrogen's remaining two electrons form a nonbonding pair. This arrangement gives nitrogen a complete octet of electrons.

$$\text{H} \overset{..}{\underset{\underset{\text{H}}{\text{H}}}{\text{N}}} \text{H}$$

Count the number of electron pairs about the central nitrogen atom. There are four pairs—three bonding pairs and one lone pair. The four pairs avoid each other by arranging themselves in a tetrahedral arrangement, just as they did in the water molecule. But this time, three of the positions are N—H bonds, and the fourth position is the lone pair. The atoms in ammonia form a triangular pyramid structure. The three hydrogen atoms form the base of the pyramid with the nitrogen at the peak, as shown in **Figure 9.16.** Based on this tetrahedral arrangement, you would predict that the H—N—H bond angle is 109.5°. The experimentally determined structure is a triangular pyramid with a bond angle of 107°.

Figure 9.16

Looking at the Ammonia Molecule

Ammonia has one lone pair and three bonding electron pairs. When these are arranged around the central atom, the arrangement is tetrahedral, but the geometry of the four atoms is a triangular pyramid. The lone pair causes the H—N—H angles in ammonia to be 107°, slightly less than the predicted tetrahedral angle of 109.5°.

▲ The space-filling model of ammonia shows the overlap of the electron clouds of the nitrogen and hydrogen atoms.

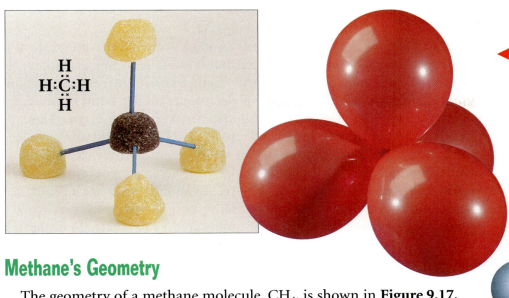

Figure 9.17
Modeling Methane

Each of the four electron pairs around the carbon atom is a C—H bonding pair, so the structure is symmetrical with all bond angles of 109.5°.

▲ The space-filling model of methane displays the symmetry of the molecule.

Methane's Geometry

The geometry of a methane molecule, CH₄, is shown in **Figure 9.17.** Methane is the simplest hydrocarbon compound. Hydrocarbons are organic compounds composed of only hydrogen and carbon. The electron dot structure for methane consists of a central carbon atom with four C—H single bonds, as shown in **Figure 9.17.**

Four pairs of electrons are positioned in a tetrahedral arrangement around the carbon atom, just as they are in the water and ammonia molecules. But in methane, all four pairs are shared between the carbon atom and the four hydrogen atoms. Because there are no lone pairs requiring extra space, the structure of methane is a perfect tetrahedron with bond angles of 109.5°.

Ethane in 3-D

Ethane, C₂H₆, is the second member of the hydrocarbon series known as the alkanes. Methane is the first and simplest. Alkanes are hydrocarbons that contain only carbon and hydrogen atoms with single bonds between all the atoms. An ethane molecule has two carbon atoms that form a bond to each other and to three hydrogen atoms. The electron dot diagram of ethane is shown in **Figure 9.18.** Each carbon atom has four single bonds. Just as in methane discussed above, a tetrahedral arrangement of bonding electron pairs around each carbon atom provides the most space for the electrons.

The space-filling model of ethane shows two connected tetrahedral arrangements around the two carbon atoms. ▼

Figure 9.18
Modeling Ethane
The geometric arrangement of atoms around each carbon atom in ethane is tetrahedral. All bond angles are 109.5° as predicted from the geometry. ▶

Ethene in 3-D

Ethene, C_2H_4, is the first member of another hydrocarbon series called the alkenes. Ethene is related to ethane, but it has four hydrogens rather than six. The common name for ethene is ethylene. In order for the carbon atoms in ethene to acquire an octet of electrons, a double bond must exist between the carbons. The electron dot diagram is shown in **Figure 9.19.**

Each carbon atom has three bonds: two C—H bonds and one C=C double bond. The most space-efficient arrangement of three electron clouds about a central atom is a flat, triangular arrangement, as shown in **Figure 9.19.** The H—C—H and H—C—C bond angles are all 120°. The geometry is rigid because of the double bond between the carbon atoms. In alkanes such as ethane, the atoms are free to rotate about the single C—C bond. No rotation can occur about a double bond, so the structure of ethene is fixed.

Figure 9.19

Modeling Ethene

The arrangement of hydrogen atoms around each carbon atom in ethene is a flat, triangular arrangement. All six atoms lie in the same plane. ▼

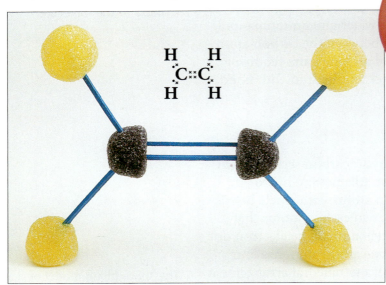

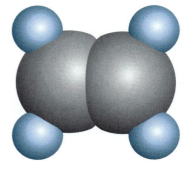

▲ Three equally inflated balloons, held at a central point, assume the triangular arrangement shown here. This is the most space-efficient arrangement of three things about a central point.

Like the hydrocarbon ethane, the space-filling model of ethene has two geometric centers. ▶

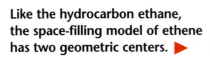

The Geometry of Ethyne

Ethyne, shown in **Figure 9.20,** is the first member of a hydrocarbon series called the alkynes. Its formula is C_2H_2. Notice that ethyne has two carbons like ethane and ethene, but only two hydrogens. Ethyne is more commonly known as acetylene. It is the fuel used in torches for cutting steel.

Modeling Molecules

The ability to build and interpret models is an important skill in chemistry. This exercise will give you practice working with models.

Procedure

1. Obtain a model kit from your teacher.

2. Draw the electron dot structure for one of these molecules.

   H_2, HCl, H_2O, CO_2, NH_3, CH_4, C_2H_6, C_2H_4, C_2H_2

3. Build a three-dimensional model of the molecule.

4. Draw a sketch of the geometric shape you predict for the molecule.

5. Repeat steps 2-4 for each of the molecules in the list.

6. Make a table like the one shown here and fill in the boxes with your data.

Analysis

1. How did the electron dot structure of each molecule help you predict its geometry?

2. Pick one of your assigned models. How many lone pairs does it have? How many bonding pairs are there?

Formula	Electron Dot Structure	Sketch of Predicted Geometry

You learned that in ethene, the two carbon atoms share two pairs of electrons in a double bond. In ethyne, the carbon atoms share *three* pairs of electrons to obtain a stable octet. A bond formed by sharing three pairs of electrons between two atoms is called a **triple bond.** The electron dot diagram for ethyne is shown in **Figure 9.20.**

Each carbon has two bonds, a C—H single bond and a C≡C triple bond, so two electron clouds are around each carbon atom. Linear geometry places two electron clouds as far as possible from each other. In ethyne, the four atoms are arranged in a straight line, so ethyne is a linear molecule.

The space-filling model of ethyne shows the arrangement of the electron clouds. ▼

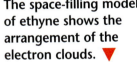

Figure 9.20

Looking at Ethyne

◄ Ethyne, when combined with excess oxygen, burns with a hot flame. It is used in welding torches.

H:C:::C:H

◄ The model shows that the geometry of ethyne is linear. The triple bond makes the molecule rigid.

Chromatography

Much of the stuff that's interesting in this world is a mixture. Blood, dirt, air, pizza—to name a few—are mixtures. These substances are complex, and realizing that they are complex may make them more interesting. Scientists have devised ways to analyze complex things. One way is chromatography.

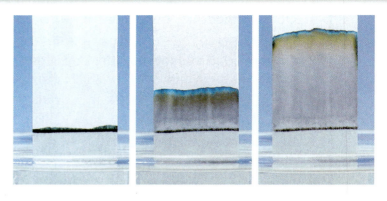

Paper Chromatography

One type of chromatography used to separate colored mixtures is paper chromatography. A porous paper is used as the stationary phase. Water or some other solvent is used as the mobile phase. A spot or a line of the mixture to be separated is placed on the paper. The solvent moves upward along the paper because of capillary action. As it reaches the spot, the mixture dissolves in the solvent. Now, the components of the mixture begin to migrate upward on the paper along with the solvent. Those components that have little attraction to the paper move almost as quickly as the solvent. Components of the mixture that have greater attraction for the paper migrate at a slower rate. The differences in the migration rates result in differences in the distances the separated components travel. Changing the solvent causes changes in the type of materials separated.

Going with the Flow

Believe it or not, this rained-upon poster is a good model of chromatography. Chromatography is a way of separating a mixture using differences in the abilities of the components to move through a material. All chromatography involves two phases—a stationary phase and a mobile phase. The movement of the mobile phase through the stationary phase allows separation to take place. Because the components of a mixture move at different rates, they eventually separate.

Thin-Layer Chromatography

In thin-layer chromatography, the stationary phase is a suspension made up of a material such as silica gel or cellulose in a solvent. The suspension is applied as a thin coating on a glass or metal plate, which is then dried. The

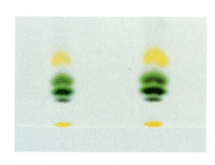

mixture being separated is applied to the bottom of the plate, and the plate is placed vertically in a solvent. Either the solvent vapor or the liquid itself acts as the mobile phase.

Gel Chromatography

Gel chromatography is often used to isolate the molecules of living, biological systems. The liquid phase carries the molecules through a stationary phase of a porous gel. The pore sizes can be manufactured precisely to affect the migration of molecules through the gel. Large molecules are blocked from moving through the gel. Because smaller molecules spend more time within the pores than midsize molecules, they take longer to migrate through the pores. As a result, gel chromatography can be used to isolate midsize molecules found in living materials.

Gas Chromatography

In gas chromatography, the mobile phase is a gas. The stationary phase is usually a liquid coating deposited on the interior of the tube through which the mixture, itself a mixture of gases, will migrate. Helium is often used as the mobile phase. The extent to which the components of a gaseous mixture interact with the coating of the tube determines their migration rates. The separated components of the mixture arrive at the end of the tube at different times, where they are analyzed and identified by a light spectrometer or mass spectrometer. Portable gas chromatographs, used in conjunction with mass spectrometers, can analyze trace amounts of gases that contribute to air pollution. Among other things, gas chromatography is used by researchers to analyze complex mixtures of compounds that constitute aromas and flavors.

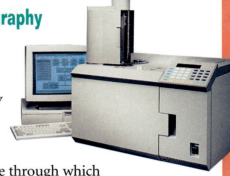

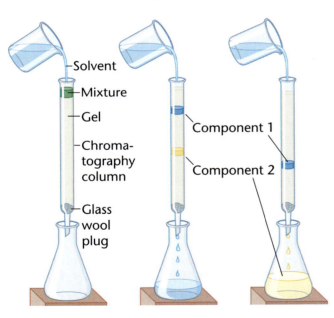

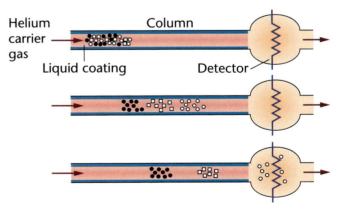

DISCUSSING THE TECHNOLOGY

1. **Analyzing** Identify the stationary and mobile phases of a rain-soaked poster.

2. **Designing** Design a laboratory method by which you could obtain pure samples of the components of a mixture that has been separated by using paper chromatography.

ChemLab

What colors are in your candy?

Yellow dye #5 is an artificial food coloring approved by the FDA, but some people are allergic to this dye. Many candies contain Yellow #5 as part of a mixture to color the candies. Dye mixtures can be extracted from the candy and separated into their component colors using paper chromatography. The yellow food coloring that you buy in the grocery store contains Yellow #5 and can be used as a reference standard.

Separations by paper chromatography are possible because different substances have different amounts of attraction for the paper. The greater the attraction the substance has for the paper, the slower it will move up the paper with the solvent.

Problem

Are there any colored candies that a person with an allergy to Yellow #5 can safely eat?

Objectives

- **Observe** separation of colors in dye mixtures.
- **Interpret** data to determine which candies contain Yellow #5.

PREPARATION

Materials

10 cm × 10 cm piece of Whatman #1 filter paper
large jar with lid   water
colored candy   salt
yellow food coloring   toothpicks
small plastic cup   ruler

PROCEDURE

1. Make a data table like the one shown.
2. Make a fine line with a pencil about 3 cm from one edge of the piece of filter paper.

3. Place a small amount of water in a plastic cup.
4. Dip the tip of a toothpick into the water.
5. Dab the moistened tip of the toothpick onto a piece of colored candy to dissolve some of the colored coating.
6. Place the tip of the toothpick with dye onto the filter paper to form a spot along the pencil line.

7. Remoisten the tip of the toothpick, and dab the same piece of candy to dissolve additional coating. Place the tip of the toothpick onto the filter paper on the same spot made in step 6. Repeat this step until a concentrated spot is obtained.
8. Using a new toothpick and fresh water, repeat steps 4 to 7 with a different colored piece of candy. Make a new spot for each piece of candy, and keep a record in your data table.
9. Dip a fresh toothpick into a drop of the yellow food coloring to be used as a reference standard. Make a spot along the pencil line and mark the location of the reference spot.
10. Carefully roll the paper into a cylinder. The spots should be at one end of the cylinder. Staple the edges. Avoid touching the paper.
11. Add water to the jar to a level of about 1.5 cm from the bottom. Sprinkle in a pinch of salt. Close the lid and shake.

12. Place the filter-paper cylinder into the jar so that the end with the spots is closest to the bottom of the jar. The water level must be at least 1 cm below the pencil line. Adjust the amount of water if necessary and close the lid.

13. Allow the water to rise to about 1 cm from the top of the filter paper.

14. Carefully remove the filter paper, open it flat, and mark the solvent edge (the farthest point the water traveled) gently with a pencil. Lay the filter paper on a paper towel to dry.

15. For each piece of candy spotted, measure the distance from the original pencil line to the center of each separated spot. Record these data in your data table. Some candies may have more than one spot.

16. Measure and record the distance from the original pencil line to the marked solvent edge.

17. Record the distance from the original pencil line to the center of each spot separated from the reference spot of Yellow #5.

ANALYZE AND CONCLUDE

1. **Interpreting Observations** Do any of the candies contain Yellow #5? How can you tell?
2. **Comparing and Contrasting** Do any of the candies contain the same dyes? Explain.
3. **Inferring** Which candies would be safe to eat if you were allergic to Yellow #5?
4. **Designing an Experiment** Devise a better way to remove the dye from the candy and place a spot on the paper. Carry out your experiment.

APPLY AND ASSESS

1. On what portion of the paper are the substances with the greater attraction for the paper? What conclusions can you draw about the molecular polarities of these dyes, given that the most polar component will have the greatest attraction for the paper?

2. Why was it important to use a pencil instead of a pen to mark the paper?

3. Why was it important to do the experiment in a closed jar?

4. What makes the water move up the paper?

5. How did the rate of water movement up the paper change as the water got higher on the paper? Suggest reasons why it changed.

DATA AND OBSERVATIONS

Solvent distance : _____ (distance from first pencil mark to solvent edge)

Original Spot	Distance (color 1)	Distance (color 2)	Distance (color 3)
Yellow #5 Reference			
Candy 1			

How Polar Bonds and Geometry Affect Molecular Polarity

Have you ever pulled clothing from the dryer and found that it was stuck together because of static cling? Static cling results from electrostatic attraction between positive and negative charges. Positive and negative charges arise in other ways besides the action of the clothes dryer, and some physical properties can be explained by them. For example, water molecules tend to bead up on smooth surfaces, and raindrops take a spherical shape, as shown in **Figure 9.21.** Why should water molecules stick together to form drops? The reason is that water molecules have positive and negative ends. Oppositely charged ends of the molecules attract one another, and the molecules stick together like the clothes in the dryer. Water is an example of how polar bonds and molecular geometry act together to affect the properties of compounds.

Figure 9.21

Water Molecules Attract Each Other

Water molecules attract one another because they have positive and negative ends. The water on this leaf beads up because water molecules on the surface of the drop are attracted to other water molecules below them. These attractions also explain the typical shape of a falling drop of water.

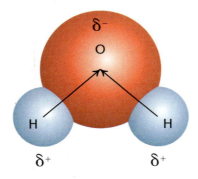

Water: A Polar Molecule

You saw earlier that the water molecule has a bent structure. The ΔEN of the O—H bond is 1.4 so the two O—H bonds in a water molecule are polar bonds. The oxygen end of the bond has a partial negative charge, while the hydrogen end of the bond has a partial positive charge. Because of its bent shape, the water molecule as a whole has a negative pole and a positive pole, as shown in **Figure 9.22.**

Figure 9.22

The Polar Water Molecule

The O—H bonds in a water molecule are polar. Because of water's bent shape, the hydrogen side of the water molecule has a positive charge, and the oxygen side has a negative charge. The arrows indicate the directions in which electrons are pulled.

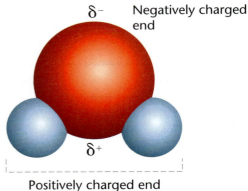

The water molecule is an example of how polar bonds, arranged in certain geometries, can result in a **polar molecule,** that is, a molecule that has a positive and a negative pole. A polar molecule is also called a dipole.

Ammonia: Another Polar Molecule

Ammonia, NH_3, is another example of a molecule with polar bonds. The N—H bond has a $\Delta EN = 0.9$. The geometry of an ammonia molecule is a triangular pyramid, as shown in **Figure 9.23.** When the three polar N—H bonds are arranged in this geometry, a polar molecule results. A center of net positive charge is located at the base of the pyramid, while a center of negative charge is on the nitrogen atom.

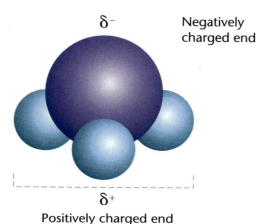

δ⁻   Negatively charged end

δ⁺
Positively charged end

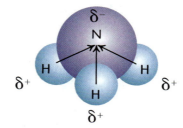

Figure 9.23

The Polar Ammonia Molecule
Like water, an ammonia molecule has two distinct sides. Because of polar bonds, the hydrogen side has a net positive charge and the nitrogen side has a net negative charge.

Carbon Dioxide: A Nonpolar Molecule

Carbon dioxide is another molecule with polar covalent bonds. The ΔEN for the C—O bond is 1.0, so the polarity of the C—O bond in carbon dioxide is similar to the polarity of the N—H bond in ammonia. But the geometry of CO_2 is different from the geometry of ammonia. The shape of a CO_2 molecule is linear. **Figure 9.24** shows that with the polar C═O bonds of a CO_2 molecule arranged in a straight line, the effects of the two polar bonds exactly cancel each other. As a result, there is no separation of positive and negative charge in the molecule. So, although carbon dioxide has relatively strong polar covalent bonds, CO_2 is a nonpolar molecule.

Figure 9.24

Nonpolar Carbon Dioxide
Both bonds in CO_2 have ΔEN equal to 1.0 and are polar, but the polar bonds oppose each other and therefore cancel each other's effects. Carbon dioxide is a nonpolar molecule. ▼

$$\delta^- \text{O} = \text{C} = \text{O} \ \delta^-$$

◄ Polar solvents such as water usually do not dissolve nonpolar substances. However, nonpolar CO_2 is slightly soluble in water; but under pressure, even more CO_2 dissolves. Bottling soda pop with CO_2 under pressure adds the fizz. When you release the pressure by opening the bottle, the CO_2 comes out of solution—sometimes too fast.

Figure 9.25

Dipole Interactions in Liquids and Solids

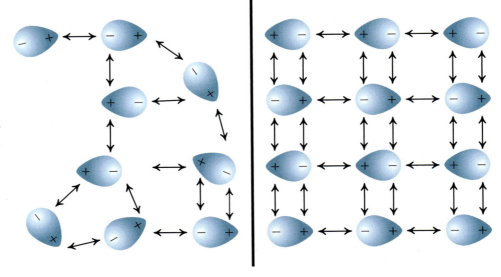

Figure 9.25

Dipole Interactions in Liquids and Solids
The force between dipole molecules is an attraction of the positive end of one dipole for the negative end of another dipole. Shown here are representations of dipole-dipole attractions in liquids (left) and solids (right).

Methane and Water Compared

Polar molecules attract each other because they have positive and negative ends. The diagram in **Figure 9.25** shows how dipoles interact. Because of this attraction, the properties of polar molecules differ from those of nonpolar molecules. For example, melting points and boiling points of polar substances tend to be higher than those of nonpolar molecules of the same size.

When you compare the physical properties of the polar molecule water with the nonpolar molecule methane, you see major differences. Although water and methane are approximately the same size and both are covalently bonded, water is a liquid at room temperature, whereas methane is a gas. **Table 9.2** shows a comparison of the melting points and boiling points of water and methane. Notice that the boiling point of water is 264°C higher than the boiling point of methane. This is macroscopic evidence for the submicroscopic attractions at work among water molecules.

Supplemental Problems

For more practice with solving problems, see Supplemental Practice Problems, Appendix B.

Table 9.2 Comparison of Melting and Boiling Temperatures		
Compound	**Melting Point**	**Boiling Point**
methane (nonpolar)	−182°C	−164°C
water (polar)	0°C	100°C

Ions, Polar Molecules, and Physical Properties

Recall from Chapter 4 that the submicroscopic interactions between the particles of a substance determine many of its macroscopic physical and chemical properties. For ionic compounds, the strong attractive force that binds positive and negative ions into well-ordered crystals keeps the physical properties of ionic substances to a relatively narrow range of variability.

Figure 9.26

Ionic and Covalent Compounds
Ionic compounds exhibit a narrower range of physical properties than covalent compounds.

Ionic compounds, such as copper(II) nitrate (left) and potassium chloride (below), tend to be brittle, solid substances with high melting points. ▼

The ionic solids potassium chloride and copper(II) nitrate are shown in **Figure 9.26.** Both have high melting points because of the strong attractions among their oppositely charged ions. On the other hand, sugar is covalently bonded, but polar bonds create relatively strong interactions that hold the molecules together in a solid crystal structure at room temperature. Ethanol, C_2H_6O, has covalent bonds and is a liquid at room temperature. Ethanol boils at a temperature 44° higher than dimethyl ether, which has the same formula but a different arrangement of atoms. The difference in boiling point occurs because ethanol is a polar molecule whereas dimethyl ether is less polar. Chlorine, Cl_2, is the most nonpolar of the substances shown in **Figure 9.26** because the bond between the two chlorines is completely covalent and, therefore, nonpolar. Interactions between chlorine molecules are minimal, so chlorine is a gas at room temperature.

Connecting Ideas

Whether you use gumdrops, electron dot diagrams, or supercomputers, the ability to model bonding between atoms is useful. By determining the shape and polarity of a molecule, you can predict its behavior and properties. In Chapter 10, you'll learn more about the forces between particles and the effects they have on the physical states of substances.

▲ Covalent substances may be solids, liquids, or gases at room temperature. Sugar (top, right) is a solid, ethanol (top, left) is a liquid, and chlorine (bottom) is a gas, yet all are covalently bonded.

SECTION REVIEW

Understanding Concepts

1. Draw the electron dot diagrams for each of the following molecules.

   a) PH_3 **c)** HBr
   b) CCl_4 **d)** OCl_2

2. Describe the shape of each molecule in question 1.

3. What is the difference between a single bond, a double bond, and a triple bond?

Thinking Critically

4. **Molecular Polarity** Chloroform, $CHCl_3$, is a molecule similar to methane in structure. Is chloroform a polar or nonpolar molecule?

Applying Chemistry

5. **Properties and Covalent Compounds** Sugar, water, and ammonia are all covalent compounds. How do these compounds demonstrate the variety of types of covalent compounds?

REVIEWING MAIN IDEAS

9.1 Bonding of Atoms

- Bonding can be viewed as a sharing of electrons.
- Electronegativity is a measure of the attraction that an atom has for shared electrons.
- Electronegativity difference, ΔEN, is a measure of the degree of ionic character in a bond.
- Electronegativity can be estimated from the periodic table.
- Metal atoms bond by sharing in a sea of valence electrons.

9.2 Molecular Shape and Polarity

- Models help you visualize what you can't see.
- The first step in building a model of molecular shape is to draw the electron dot diagram.
- Electron pairs about the central atom are either lone (nonbonding) pairs or bonding pairs.

- Electron pairs, placed as far from each other as possible, determine the geometry of a molecule.
- The polarity of the bonds and the shape of the molecule determine whether a molecule is polar or nonpolar.
- Interparticle forces determine many of the physical properties of substances. Interparticle forces between dipoles such as water are especially strong.

Vocabulary

For each of the following terms, write a sentence that shows your understanding of its meaning.

conductivity	metallic bond
double bond	polar covalent bond
ductile	polar molecule
electronegativity	shielding effect
malleable	triple bond

UNDERSTANDING CONCEPTS

1. Using only the periodic table, predict which atom in each pair has the higher electronegativity.

   a) Mg, Na   c) Ca, Sr
   b) Cl, F   d) C, O

2. How does the continuum model of bonding distinguish among ionic, covalent, and polar covalent bonding?

3. What is meant by a polar covalent bond? A nonpolar covalent bond?

4. Using only the periodic table, rank these bonds from the smallest to largest ΔEN: O—F, P—F, F—F, Al—F, Mg—F, N—F, K—F. Which do you predict to be ionic bonds? Check your answer by calculating the actual ΔEN using the data in **Figure 9.2.**

5. What is the shielding effect? Is shielding more important for carbon or for lead? Explain.

6. How does metallic bonding differ from ionic bonding? Covalent bonding?

7. What experimental evidence supports the model of metallic bonding?

8. What is a lone pair of electrons? What is a bonding pair?

9. CO_2 and H_2O are both triatomic molecules. How are their structures different?

10. Electronegativity versus atomic number is graphed for the fourth period elements in the diagram on the next page. Describe in words the general trend in electro-negativity in Period 4 as atomic number increases.

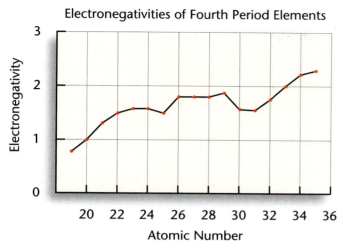

Electronegativities of Fourth Period Elements

APPLYING CONCEPTS

11. From the periodic table, select three sets of elements that would form ionic compounds. Write the formulas of the compounds.

12. Compounds containing the ammonium ion are often used as fertilizers. What is the geometry of an ammonium ion (NH_4^+)?

13. In a molecule with the formula $C_8H_{18}O_2$, what percentage of the total electrons participate in the bonding?

Chemistry and Technology
14. Which chromatographic technique would be appropriate for separating a mixture of gaseous hydrocarbons?

Everyday Chemistry
15. Explain how microwaves that cook food in a microwave oven might be used to decompose some pollutants in the atmosphere. What kinds of molecules could be decomposed in this way?

THINKING CRITICALLY

Comparing and Contrasting
16. Compare the molecules nitrogen trichloride (NCl_3) and carbon tetrachloride (CCl_4). How many pairs of electrons surround the central atom? How many pairs are bonding? Nonbonding? What are the shapes of the molecules?

Relating Cause and Effect
17. MiniLab 1 In separating a mixture by paper chromatography, why is it important to keep the spot of colored mixture that you apply to the filter paper as small but as concentrated as possible?

Making Predictions
18. Ethanol, an alcohol, and dimethyl ether both have the molecular formula C_2H_6O. Look up the structural formulas and draw electron dot diagrams for the two compounds. Use your diagrams to decide whether there is a difference in the polarity of the two molecules. Explain.

Designing an Experiment
19. ChemLab A green dye could be produced by mixing a yellow dye with a blue dye. How could you determine whether a dye was a mixture of a yellow dye and a blue dye or whether it was a pure green dye?

Making Predictions
20. MiniLab 2 Below is the electron dot structure for $AlBr_3$. What geometric shape do you predict for this molecule?

$$\ddot{\underset{\times\times}{Br}}\!:\!\overset{\times}{Al}\overset{\overset{\times}{Br}\overset{\times\times}{}}{\underset{\underset{\times\times}{Br}}{}}$$

Relating Concepts
21. Chemistry and Technology Why do you think helium is often used as the mobile phase in gas chromatography?

Comparing and Contrasting
22. Everyday Chemistry How are microwaves like visible light? How are they different from visible light?

CUMULATIVE REVIEW

23. Suppose you have a cube of gold measuring 1 cm on each side. You hammer it into a square measuring 15 cm on each side. What is the average thickness of the square? (Chapter 3)

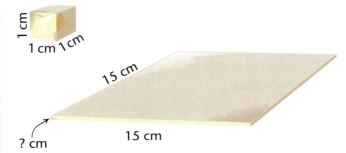

24. Carbon dioxide and carbon disulfide have identical Lewis dot structures. Why is this not surprising? (Chapter 7)

25. Would you expect K_3O to be a stable compound? Explain. (Chapter 4)

26. Does the mass number of an atom change when the atom forms a chemical bond? Explain. (Chapter 2)

SKILL REVIEW

27. Organizing Information The melting points for the sodium halide compounds are as follows.

NaF 992°C
NaCl 800°C
NaBr 755°C
NaI 651°C

Construct a graph of melting point versus ΔEN for these compounds. Explain the trend of the graph.

28. Organizing Information Complete the table below. Draw the best electron dot structure for each compound listed. In all cases, except H_2O, the first atom in the formula is the central atom. Predict the geometric arrangement of electron clouds around the central atom and use your prediction to determine the geometry of the molecule. From your predicted geometry, decide whether the molecule is polar or nonpolar. Water is given as an example.

PROBLEM SOLVING

29. There are three different structures for the hydrocarbon pentane, C_5H_{12}. These structures are called geometric isomers because each has a different shape. Draw electron dot diagrams for the three possibilities. You can substitute a dash, like the stick in the gumdrop models, to represent each pair of electrons.

WRITING IN CHEMISTRY

30. Pick one period in Linus Pauling's career and find out more about it. Locate a publication by Pauling from this period in his life and write a short article about his accomplishments.

Molecular Formula	Dot Structure	Geometry of Electron Pairs About the Central Atom	Molecular Geometry	Polar Bonds	Nonpolar Bonds	Polar Molecule
H_2O	H:Ö: H	tetrahedral	bent	O—H	none	yes
CCl_4						
$CHCl_3$						
CH_2Cl_2						
CH_3Cl						
CH_4						

Standardized Test Practice

1. Electronegativity is
   a) a measure of an atom's ability in a bond to attract electrons.
   b) a measure of an atom's ability to form negative charges.
   c) a measure of an atom's ability to chemically combine with other elements.
   d) all of the above.

Element	Electronegativity (EN)
Lithium	1.0
Iron	1.8
Sulfur	2.5
Nitrogen	3.0
Oxygen	3.5
Fluorine	4.0

Use the table above to answer questions 2–5.

2. What is the ΔEN of the bonds formed between oxygen and iron?
   a) 1.7
   b) 1.8
   c) 3.5
   d) 5.3

3. Which pair of elements will combine to form an ionic compound?
   a) lithium and iron
   b) oxygen and iron
   c) lithium and oxygen
   d) fluorine and oxygen

4. Which pair of elements will combine to form a covalent compound?
   a) lithium and iron
   b) fluorine and lithium
   c) iron and sulfur
   d) nitrogen and oxygen

5. Which pair of elements will combine to form a polar covalent compound?
   a) fluorine and sulfur
   b) fluorine and iron
   c) lithium and oxygen
   d) iron and sulfur

6. Which of the following is true about the shielding effect?
   a) The shielding effect increases as you move up a column of elements.
   b) Inner energy levels block the attraction of the nucleus on valence electrons.
   c) Added energy levels block the attraction of other ions on the nucleus of an atom.
   d) The shielding effect on atoms creates an increase in the atom's electronegativity.

7. Which compound will have a ΔEN equal to zero?
   a) CO
   b) H_2O
   c) O_2
   d) CH_4

8. Which of the following metallic properties does a sea of valence electrons explain?
   a) conductivity of electricity
   b) conductivity of heat
   c) malleability
   d) all of the above

9. What is the shape of a carbon dioxide molecule?
   a) bent structure
   b) straight line
   c) tetrahedral
   d) right angle

10. Why is water a polar molecule?
    a) The bent structure of a water molecule forms a negative oxygen end and a positive hydrogen end.
    b) The bent structure of a water molecule forms a negative hydrogen end and a positive oxygen end.
    c) The hydrogen and oxygen atoms in a water molecule form a tetrahedral structure.
    d) The hydrogen and oxygen atoms in a water molecule form a linear structure.

Test Taking Tip

Ask Questions If you've got a question about what will be on the test, the way the test is scored, the time limits placed on each section, or anything else. . .by all means ask! Will you be required to know the specific names of the gas laws, such as Boyle's law and Charles's law?

CHAPTER 10

The Kinetic Theory of Matter

Chapter Preview

Sections

10.1 Physical Behavior of Matter
MiniLab 10.1 Molecular Race

10.2 Kinetic Energy and Changes of State
MiniLab 10.2 Vaporization Rates
ChemLab Molecules and Energy

More Kinetic Energy—It's a Gas!

How is it possible for snow to be made of the same molecules as water in the lake and in the air? The positions and movements of atoms, ions, or molecules of a substance determine whether it is a solid, liquid, or gas.

Start-up Activities

Launch Lab

Defying Density

You know that an object sinks or floats in water based on its density. In this activity, you will explore an exception to this rule.

Safety Precautions

Be careful handling the pin, which has a sharp point.

Materials

- pin
- 600-mL beaker
- 400 mL of water
- detergent
- dropper

Procedure

1. Pour about 400 mL of water into a 600-mL beaker. Float the pin on the surface of the water.
2. Use a dropper to add one drop of water containing detergent to the beaker. Place the drop on the water surface near the wall of the beaker. Observe what happens.

Analysis

Is a metal pin likely to be more or less dense than water? How does the shape of the pin help it to float? Hypothesize about the reason for the pin's behavior before and after you added the detergent.

What I Already Know

Scan the chapter and make an outline using the headings, figures, and captions you notice. Make notes on your outline as you read the chapter.

Reading Chemistry

Scan the chapter and make an outline using the headings, figures, and captions you notice. Make notes on your outline as you read the chapter.

Preview this chapter's content and activities at chemistryca.com

Physical Behavior of Matter

Y ou know matter exists as gases, liquids, and solids because you can smell the perfume of flowers, pour fruit punch into a glass, and stack blocks of firewood. You know that air, water, and rocks all feel different, but can you describe their macroscopic properties? The properties that characterize these three states of matter—gas, liquid, and solid—reveal the organization of their submicroscopic particles.

SECTION PREVIEW

Objectives

✓ **Compare** characteristics of a solid, liquid, and gas.

✓ **Relate** the properties of a solid, liquid, and gas to the kinetic theory of matter.

✓ **Distinguish** among an amorphous material, liquid crystal, and plasma.

Review Vocabulary

Polar molecule: a molecule that has a positive and a negative pole.

New Vocabulary

solid
liquid
gas
Brownian motion
kinetic theory
ideal gas
pressure
crystal lattice
liquid crystal
amorphous material
plasma

States of Matter

Imagine trying to squeeze your textbook into a jelly jar. It can't be done. A **solid** is rigid with a definite shape, as shown in **Figure 10.1.** Solids are rigid because the atoms, ions, or molecules that make up a solid are fixed in place.

Figure 10.1

Properties of Solids
Whether its shape is natural, such as this crystal (left), or artificial, such as the silver coins (right) or platinum jewelry (bottom), a solid is rigid—it keeps its shape. Most elements are solids at room temperature.

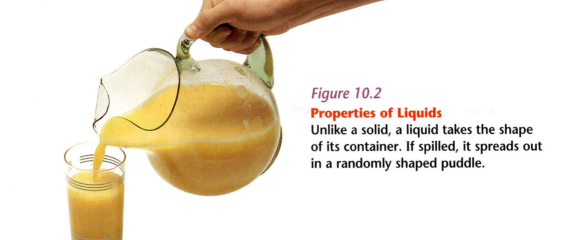

Figure 10.2
Properties of Liquids
Unlike a solid, a liquid takes the shape of its container. If spilled, it spreads out in a randomly shaped puddle.

The characteristics of a liquid differ dramatically from those of a solid. A **liquid** is flowing matter with a definite volume but an indefinite shape. Because its shape is indefinite, a liquid takes the shape of its container, as shown in **Figure 10.2.** If you have mopped up a spill, you have seen that the particles of a liquid can move and easily glide over each other. You can feel how a liquid flows by standing under a running shower.

Just as for a liquid, you can also feel the flow of gases that make up the atmosphere if you stand in a breeze. The flow of a gas feels different because the particles that make up a gas are farther apart than those of a liquid. If you blow up a balloon or a tire, you can observe some important properties of gases, as shown in **Figure 10.3.** From these observations, you see that a **gas** is flowing, compressible matter that has no definite volume or shape. The particles that make up a gas are much farther apart than they are in solids and liquids, and so, they can be easily pushed together.

Figure 10.3
Properties of Gases
The air completely fills the bag. Whatever the shape and volume of the bag, the air inside spreads out to fill it. If you squeeze the bag, you will observe that the air inside is compressible.

The Kinetic Theory of Matter

In 1827, Robert Brown, a Scottish botanist, was studying water samples with a microscope. He observed pollen grains suspended in the water moving continuously in irregular directions. Brown repeated his observations using dye particles in water, and noted that they too had random and erratic motions. The constant, random motion of tiny chunks of matter is called **Brownian motion** in honor of Robert Brown. In **Figure 10.4,** you can see why Brownian motion is cited as evidence of the random movement of particles of matter. Do only molecules of water display random motion? Is all matter in motion? The **kinetic theory** states that submicroscopic particles of all matter are in constant, random motion. The energy of moving objects is called kinetic energy.

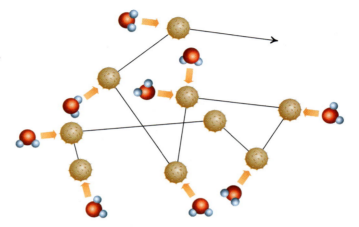

Figure 10.4

Brownian Motion

As viewed under a microscope, a pollen grain suspended in a water droplet traces an erratic path. Random movements of water molecules explain this path traced by the pollen grain. When a water molecule strikes the pollen grain, it pushes the grain in one direction. Then, another molecule strikes the grain and pushes it in a different direction.

Kinetic Model of Gases

In the kinetic theory, each particle of a gas moves like the air-hockey puck shown in **Figure 10.5.** The puck moves in a straight line until it strikes the side of the game board. Similarly, a gas particle can change direction only when it strikes the wall of its container or another gas particle.

Figure 10.5

Modeling the Motion of a Gas Particle

An air-hockey puck travels in a straight line until it strikes the side of the game board. Then, it rebounds in a straight line in a new direction. ▼

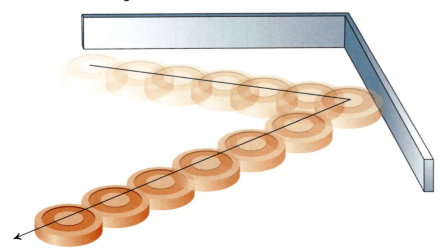

▲ Similarly, a gas particle moves through the space in its container in a straight line. The speed of the hockey puck is about 1 m per second, but the gas particle moves at a much faster rate, 10^2 to 10^3 m per second.

Molecular Race

Odors may be evidence of the diffusion of gases. What inferences can you make about two molecular gases by observing their diffusion through air?

Procedure

1. Make a tube by taping two transparent straws together end to end.

2. Tape the tube onto a black, horizontal surface. Label one end of the tube "NH_3" and the other "HCl."

3. Cut a cotton swab in half with scissors and wrap each cut end with masking tape thick enough to seal the tube.

4. Put on an apron and goggles.

5. Obtain containers of concentrated solutions of ammonia, $NH_3(aq)$, and hydrochloric acid, HCl, from your teacher. **CAUTION:** *Both solutions can damage eyes, skin, and clothing. Handle them with care. If any skin contact or spillage occurs, notify your teacher immediately.*

6. Dip one swab into the ammonia and the other into the hydrochloric acid. The cotton should be saturated but not dripping.

7. Simultaneously insert the swabs into the appropriate ends of the tube, pushing them far enough so that wrapped tape handles seal the ends of the tube as shown here.

8. Do not jostle or move the straws. After a few seconds, look closely for the white ring of ammonium chloride, the reaction product.

9. Measure and record the distances from the respective cotton swabs to the ammonium chloride ring.

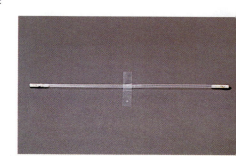

Analysis

1. Compare the diffusion rates of the two gases.

2. Propose an explanation for the difference in diffusion rates.

In air hockey, after each collision with the wall of the game board, the puck loses speed because some of its energy is transferred to the wall. When the puck has lost all its kinetic energy, it will stop gliding. But, unlike the puck, gas particles do not lose kinetic energy when they collide with the walls of their container or with another gas particle. Each gas particle rebounds without losing speed but in a new direction. Collisions of particles in a gas are called elastic collisions because no kinetic energy is lost.

A gas with particles that are in constant random motion but have no attraction for each other is called an ideal gas. The particles in an **ideal gas** undergo elastic collisions. Except at very low temperatures or very high pressures, nearly all real gases behave as ideal gases. The kinetic theory explains why gases fill their container and why they exert pressure on the walls of the container. The gas particles move randomly in all directions until they strike the walls or other particles and bounce back.

In the macroscopic world, blocking a volleyball is an example of how you can change the direction of motion. Remember the sting of the volleyball striking your hands? What you felt was pressure. **Pressure** is the force acting on a unit area of a surface; that is, for example, the force per square centimeter. Just as the volleyball exerted a force on a square centimeter of your skin, the particles in a gas exert a force on each square centimeter of the walls of the container when the walls deflect them. The outward pressure of the air inside a balloon is the force that keeps the balloon expanded. If the force is strong enough, the balloon will burst. You can feel this force by gently squeezing an inflated balloon or ball.

Earth's atmosphere, which is a mixture of gases, exerts a pressure, too. Atmospheric pressure is caused by the constant bombardment of the molecules and atoms in air. At sea level, this molecular bombardment exerts a pressure of about 14.7 pounds per square inch, as illustrated in **Figure 10.6.** Humans and other forms of life on Earth have adapted to atmospheric pressure. We are sensitive only to changes in pressure.

Figure 10.6
Experiencing the Pressure of Atmospheric Gases
The force of the bottle cap on the palm of the hand from three, 2-L bottles about equals the force that the gases of the atmosphere exert on the same area of skin. Of course, the pressure from these three bottles on this person's hand is in addition to normal atmospheric pressure.

Kinetic Model of Liquids

Just as a gliding air-hockey puck models a rebounding gas particle, **Figure 10.7** shows that marbles in a beaker model some behaviors of liquids. When liquids form puddles, interparticle forces maintain their volume but not their shape. The particles of a liquid can slide past each other, but they are so close together that they don't move as straight or as smoothly as an air-hockey puck. When you try to walk across a crowded sidewalk, you can't move quickly in a straight line, either.

Figure 10.7
Modeling Liquids
Magnetized marbles spread out evenly to take the shape of their container. The volume they occupy cannot be reduced. ▼

When the container is tipped, the magnetized marbles flow onto the table. ▼

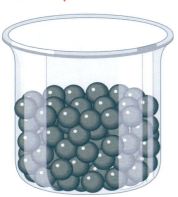

▲ When the container is swirled, the marbles flow with a swirling motion.

Kinetic Model of Solids

According to the kinetic theory, strong forces between particles explain the rigid structure of solids. Particles of a solid cannot move past each other, but they are in constant motion, bouncing between neighboring particles. In a solid, the particles occupy fixed positions in a well-defined, three-dimensional arrangement. The arrangement, which is repeated throughout the solid, is called a **crystal lattice,** as shown in **Figure 10.8.**

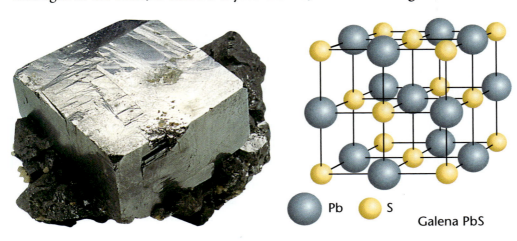

Pb S

Galena PbS

Figure 10.8
Crystal Lattices
Crystal lattices extend throughout solids. Shown here is the crystal lattice of lead(II) sulfide, which is found as the mineral galena.

Other Forms of Matter

Some forms of matter cannot readily be described as solid, liquid, or gas. Maybe they look like solids or gases but sometimes behave like a liquid. These forms of matter are liquid crystals, amorphous materials, and plasmas.

Liquid Crystals

When a solid melts, its crystal lattices disintegrate and its particles lose their three-dimensional pattern. However, when some materials called **liquid crystals** melt, they lose their rigid organization in only one or two dimensions. For example, the liquid crystal shown in **Figure 10.9** has rod-like molecules. The interparticle forces in a liquid crystal are relatively weak and their arrangement is easily disrupted. When the lattice is broken, the crystal can flow like a liquid. Liquid crystal displays (LCDs) are used in watches, thermometers, calculators, and laptop computers because liquid crystals change with varying electric charge.

In other liquid crystals, the parallel lines of molecules are arranged in layers. When these substances melt, the layers stay in place. ▼

Figure 10.9
Structures of Liquid Crystals
In some liquid crystals, the molecules are arranged in parallel lines. They keep the arrangement when the substances melt. ▶

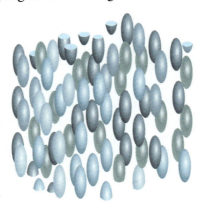

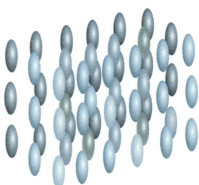

Glass Sculptures

In 1976, an automobile accident cost Dale Chihuly the sight in his left eye. You might think this accident would have ended the career of the artist who founded the Pilchuck Glass Center, a school for training glass artisans, in Stanwood, Washington, just five years before. Although Chihuly lost his depth perception and so cannot blow glass safely, he continues to train and inspire others in the art of blowing glass.

Making glass The glass used in making art objects is soda-lead glass. It is commonly called crystal or lead glass. Its composition is mostly silica (sand), sodium oxide, and lead oxide. Soda-lead glass has excellent optical qualities.

When glass is molten, it looks just like the sugar syrup you might cook on your stove to make hard candy. The ingredients are heated in a furnace to 1370°C, turning them into a syrupy mass. When it cools, it forms the amorphous material glass.

Turning glass into art Chihuly begins by making a design. One of his students blows the glass to the specifications of the design. The glassblower sticks an iron blowpipe into the molten glass and blows a large glass bubble. When Chihuly is satisfied that the bubble is the right size, the blower rolls it in tinted glass dust, which adheres to the surface. The different colored glass layers are fused by another firing in the furnace.

More blowing and careful shaping by pulling and pushing followed by further heating continues to transform the bubble. When cooled, it becomes an artwork—an astonishing marvel of human creativity.

Connecting to Chemistry

1. **Applying** What properties of glass make it ideal for repeated blowing and forming?

2. **Thinking Critically** Why is it scientifically incorrect to call a glass goblet a crystal goblet?

Amorphous Materials

Is the peanut butter you spread on your bread a solid? What about the wax in candles? Although such materials have a definite shape and fixed volume, they are not classified as solids but rather as amorphous materials. An **amorphous material** has a haphazard, disjointed, and incomplete crystal lattice. Candles and cotton candy are everyday examples of amorphous materials. **Figure 10.10** compares the structure of a solid with that of an amorphous material.

Figure 10.10

A Solid Becomes an Amorphous Material
The silicon dioxide crystal, SiO_2, has a regular honeycomb structure. ▶

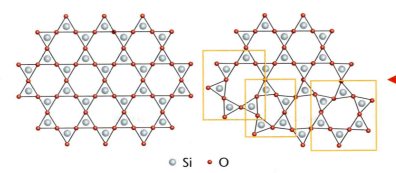

○ Si   • O

◀ If melted and then quickly cooled, silicon dioxide loses its regularity and becomes an amorphous material.

Plasmas

The most common form of matter in the universe but the least common on Earth is plasma. The sun and the other stars are composed of plasma. Plasmas can also be found in fluorescent lights, **Figure 10.11**. A **plasma** is an ionized gas. It can conduct electrical current but, like an ordinary conducting wire, it is electrically neutral because it contains equal numbers of free electrons and positive ions. Plasmas form at very high temperatures when matter absorbs energy and breaks apart into positive ions and electrons, or sometimes even into atomic nuclei and free electrons. In stars, the energy that ionizes the gases is produced by nuclear fusion reactions.

Supplemental Problems

For more practice with solving problems, see Supplemental Practice Problems, Appendix B.

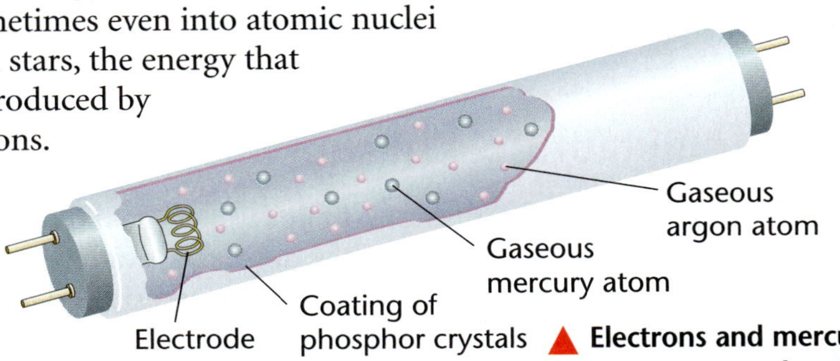

Gaseous argon atom

Gaseous mercury atom

Coating of phosphor crystals

Electrode

Figure 10.11

Fluorescent Light Tube
When a small electrical current heats the electrode, some electrons in the electrode material acquire enough energy to leave the surface and collide with molecules of argon gas, which become ionized. As more electrons are freed, they also ionize some of the mercury atoms, forming a plasma.

▲ Electrons and mercury ions collide with mercury atoms and excite their electrons to higher energy levels. When the excited electrons return to lower energy levels, they radiate energy in the form of invisible ultraviolet light. The fluorescent tube is coated with phosphor crystals, which absorb the ultraviolet light and radiate visible light.

SECTION REVIEW

Understanding Concepts

1. Analyze and compare the structure and shape of liquids and gases in terms of particle spacing and particle motion.

2. How do the particles of an ideal gas behave?

3. Why is a plasma called a high-energy state of matter?

Thinking Critically

4. **Hypothesizing** Instead of the three states of matter described in this chapter, matter is sometimes classified into only two states. In what properties would these two states differ?

Applying Chemistry

5. **Plant Care** How could fluorescent light tubes be altered to supply indoor lighting in the blue and red region of the visible spectrum for raising plants?

Kinetic Energy and Changes of State

After a big evening meal with your family, you settle down to watch your favorite TV show. In the meantime, the leftover spaghetti and tomato sauce has dried and set hard on the plates. It's your turn to do the dishes. Will you use hot or cold water? You will have to do more scrubbing if you use cold water to wash the greasy dishes than if you use hot water. Kinetic energy—provided by hot water or provided by your hard labor—does the work. What is kinetic energy, and how does kinetic energy change water molecules so that plates with dried-on food are easier to clean?

SECTION PREVIEW

Objectives

✓ **Interpret** changes in temperature and changes of state of a substance in terms of the kinetic theory of matter.

✓ **Relate** Kelvin-scale and Celsius-scale temperatures.

✓ **Analyze** the effects of temperature and pressure on changes of state.

Review Vocabulary

Brownian motion: constant, random motion of tiny chunks of matter.

New Vocabulary

temperature
absolute zero
Kelvin scale
kelvin (K)
diffusion
evaporation
sublimation
condensation
vapor pressure
boiling point
joule (J)
heat of vaporization
melting point
freezing point
heat of fusion

Temperature and Kinetic Energy

When you wash dishes with hot water, most of the water molecules are moving more rapidly than they do in cold water. They have more kinetic energy. Not all molecules of hot water in a sink have the same kinetic energy. They don't have the same speed. The same applies to a container of gas, such as an air-filled balloon. All the gas particles are moving randomly in the balloon at different rates.

Temperature and Particle Motion

The graph in **Figure 10.12** shows the distribution of speeds of the particles in a container of gas. The most common speed (kinetic energy) of the particles is represented by the peak near the center of the graph. According to the kinetic theory, the **temperature** of a material is a measure of the average kinetic energy of the particles in the material. For example,

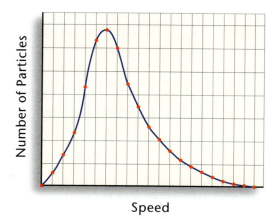

Number of Particles (vertical axis)

Speed (horizontal axis)

Figure 10.12

Distribution of Gas Particle Speeds
On a highway, some vehicles move slower and some faster than the flow of traffic. In a gas, some particles move slower or faster, but for most, the speed is near the average speed of the group. The peak of the graph represents the most common speed of the particles. The higher the temperature, the higher the average speed.

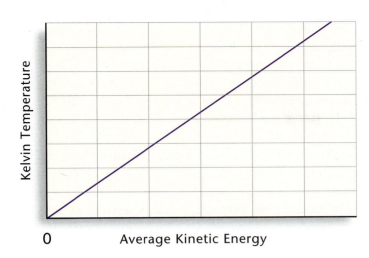

Kelvin Temperature

0   Average Kinetic Energy

Figure 10.13

Relating Average Kinetic Energy and Temperature
Because the graph is a straight line, the temperature of a gas is directly proportional to the average kinetic energy of its particles.

as a gas is heated, the average kinetic energy of its particles increases. This increase in the average kinetic energy of the particles of the gas can be measured as an increase in the temperature of the gas, **Figure 10.13.** As a gas is cooled, the average kinetic energy and speed of its particles decreases.

The Kelvin Scale

According to the graph in **Figure 10.13,** as the temperature of a gas increases, the average kinetic energy of its particles increases and as the temperature decreases, its kinetic energy decreases. Could the kinetic energy be as low as zero—no motion at all? You know that, according to the kinetic theory of matter, the submicroscopic particles are in constant motion.

As a substance is cooled, it loses more of its kinetic energy. The temperature at which a substance would have zero kinetic energy is called **absolute zero.** At this temperature, none of the particles would be moving at all. Their speed and their kinetic energy would both be zero. The temperature of a substance can be lowered to values near zero, but absolute zero has never actually been reached.

In **Figure 10.13,** the scale used for temperature, shown on the vertical axis, is the **Kelvin scale.** This scale is defined so that the temperature of a substance is directly proportional to the average kinetic energy of the particles and so the zero on the Kelvin scale corresponds to zero kinetic energy. Therefore, absolute zero corresponds to the zero on the Kelvin scale. **Figure 10.14** shows how the Kelvin scale is related to the Celsius scale, used throughout the world, and to the Fahrenheit scale, the scale used by weather reporters and on household ovens.

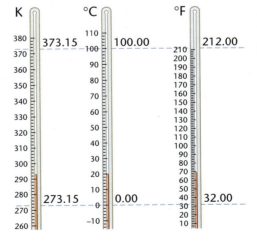

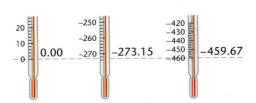

Figure 10.14

The Kelvin, Celsius, and Fahrenheit Scales
The Celsius scale is defined so that the temperature interval from the freezing point of water to the boiling point of water measures 100 degrees. Thus, on the Celsius scale, water freezes at 0°C and boils at 100°C. On the Fahrenheit scale, water freezes at 32°F and boils at 212°F. Note that on the Fahrenheit scale, the temperature interval between the freezing point and the boiling point of water is 180 degrees; Fahrenheit degrees are smaller than Celsius degrees.

If the Kelvin scale used in **Figure 10.13** had been the Celsius or Fahrenheit scale, the graph would be a straight line but the line would not pass through the origin. A zero reading on the Celsius and Fahrenheit scales does not correspond to zero kinetic energy. Only on the Kelvin scale is the temperature reading directly proportional to the kinetic energy. The Kelvin scale makes calculations for solving problems about gases simple, as you will appreciate in Chapter 11.

The divisions of the Fahrenheit and Celsius scales are called degrees, but the divisions of the Kelvin scale are called **kelvins (K).** The kelvin is the SI unit of temperature. Note that the degree symbol is not used with temperatures expressed in kelvins. For example, absolute zero is written as 0 K. On the Celsius scale, the temperature reading for absolute zero is $-273.15°C$, usually rounded to $-273°C$. A Celsius degree and a kelvin are the same size, so by shifting the Celsius scale up 273°, it will coincide with the Kelvin scale.

Temperature Conversions

Because the Celsius degree and the kelvin are the same size and Kelvin readings are 273 degrees higher than Celsius readings, any Celsius reading can be easily expressed as a Kelvin reading. Simply add 273 to the Celsius reading.

$$T_K = (T_C + 273) \text{ K}$$

For example, if the temperature of a room is 25°C, the Kelvin reading can be found as follows.

$$T_K = (25 + 273) \text{ K} = 298 \text{ K}$$

Similarly, a Kelvin reading can be expressed as a Celsius reading by subtracting 273.

$$T_C = (T_K - 273)°C$$

For example, human body temperature, 310 K, can be expressed on the Celsius scale.

$$T_C = (310 - 273) = 37°C$$

Use **Figure 10.14** to prove to yourself that 37°C is correct.

PRACTICE PROBLEMS

Supplemental Problems

For more practice with solving problems, see Supplemental Practice Problems, Appendix B.

Complete the following table.

Temperature	Celsius, °C	Kelvin, K
1. Melting point of iron		1808
2. Household oven	175-205	
3. Food freezer		255
4. Sublimation point of dry ice, $CO_2(s)$	−78.5	
5. Boiling point of nitrogen, N_2		77.4

Mass and Speed of Particles

When a rolling bowling ball knocks over pins, you can see an effect of the kinetic energy of moving objects. You know that kinetic energy depends on speed, so you would expect that if the ball were rolling faster, it would have the kinetic energy to knock over more pins. You also know it's harder to move heavier objects than lightweight ones. In other words, it takes more work and more kinetic energy to move heavier objects. **Figure 10.15** shows that the kinetic energy of a moving object such as a wagon or a gas particle depends on its mass and speed.

Figure 10.15

Mass, Speed, and Kinetic Energy
Two wagons with loads of different masses move at the same speed. Their kinetic energies are different because their masses are different. The wagon with the greater mass has the greater kinetic energy. Kinetic energy depends on both the mass and the speed of the moving body. ▶

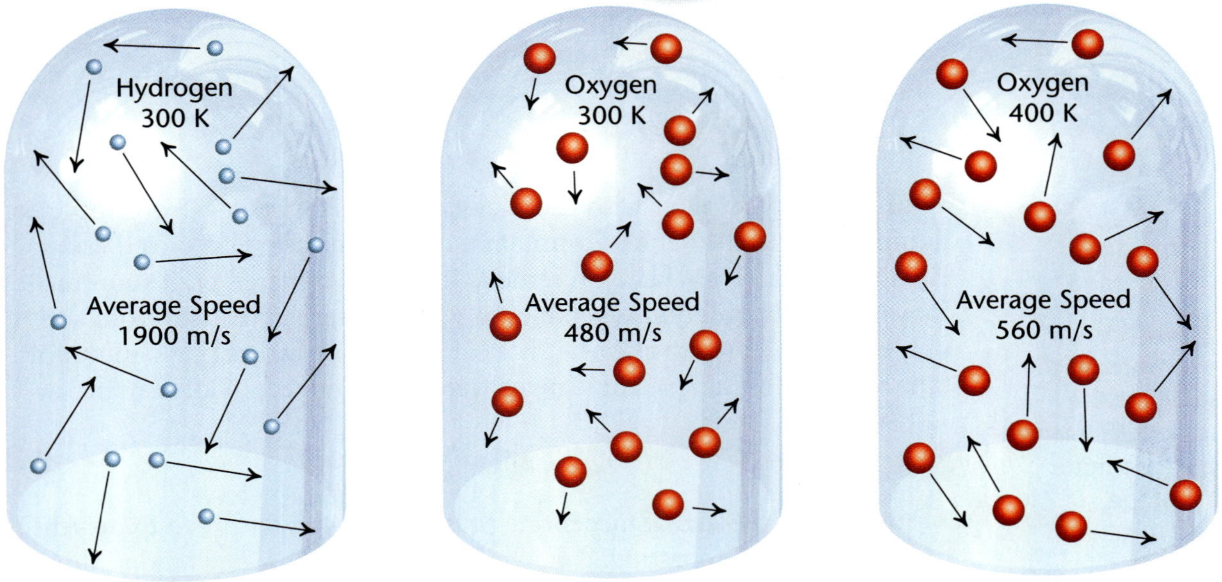

Hydrogen
300 K

Average Speed
1900 m/s

Oxygen
300 K

Average Speed
480 m/s

Oxygen
400 K

Average Speed
560 m/s

▲ The container of gas on the left contains hydrogen gas. Because the oxygen gas in the center and the hydrogen gas are at the same temperature, they have the same kinetic energy. Because the mass of an oxygen molecule is 32 u and the mass of a hydrogen molecule is only 2 u, the hydrogen molecules must have greater average speed.

▲ The two samples of oxygen on the right are at different temperatures. Because both gases are oxygen, the particles in the two containers have the same mass. The molecules of oxygen gas in the container at the higher temperature have greater kinetic energy because they are moving at a greater average speed.

After the wall separating the two chambers has been removed, gas flows between chambers until the number of molecules flowing to the left chamber is the same as the number flowing to the right. The gas diffuses between the two chambers until both have equal numbers of molecules. ▼

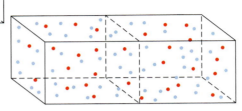

Diffusion

The motions of particles of a gas cause them to spread out to fill the container uniformly. **Diffusion** is the process by which particles of matter fill a space because of random motion, as shown in **Figure 10.16.** If you have seen dye such as food coloring spreading through a liquid, you have watched diffusion. Your sense of smell depends on diffusion and air currents for you to detect molecules of a gas that waft by your nose. Diffusion is slow, but in your lungs, oxygen reaches your blood rapidly enough by diffusion. Oxygen diffuses across the walls of tiny blood vessels called capillaries from the air sacs of your lungs that fill with air each time you inhale. The rate of diffusion of a gas depends upon its kinetic energy, that is, on the mass and speed of its molecules.

Changing State

You are very familiar with the changing states of water—from vapor to liquid and to ice. What environmental conditions are related to these changes of matter? When you remove ice from your freezer, it soon melts to water. When you boil vegetables in water, water vapor rises from the pot. From these observations, it is clear that temperature plays an important role in the changing of state of water, and indeed, of all matter.

Evaporation

You know that cans of turpentine must be kept tightly closed to prevent the liquid from evaporating. You may have noticed that wet laundry hung on a clothesline dries faster on a hot day and slower on a cold day. **Evaporation** is the process by which particles of a liquid form a gas by escaping from the surface. The area of the surface, as well as the temperature and humidity, affects the rate of evaporation, as shown in **Figure 10.17.**

Figure 10.17

Rate of Evaporation
The volume of water in the cup and puddle are the same. In the puddle, a larger surface area allows more molecules to escape.

Everyday Chemistry

Freeze Drying

Probably some of the foods you eat have been freeze-dried. In this process, foods are frozen and the ice is removed from the food by sublimation at low pressures in a vacuum chamber. Then, the water vapor produced by sublimation is removed from the chamber by pumps or water-vapor ejectors.

Advantages of freeze drying foods Food products that would deteriorate during heat processing are often freeze dried. For example, orange juice, which loses its taste in heat processing, is freeze concentrated before freeze drying. First, the pulp is separated from the juice. Then some of the water is removed from the juice by partially freezing it and removing the ice crystals. The pulp is returned to the juice before freeze drying.

Freeze-dried foods are lightweight and take up little room. Even foods such as chicken and potato salads have been freeze dried, packed in airtight cans, and eaten as field rations on military maneuvers.

Freeze drying biological specimens Freeze drying may be used to prepare specimens for scanning electron microscopy (SEM). Formerly, soft biological materials were chemically fixed and then air dried. Tissue structure collapsed and smaller appendages

became plastered against the body during drying. Freeze-dried biological tissue does not shrink or become otherwise distorted. Specimens of organisms as small as amoebas have been preserved by freeze drying.

Exploring Further

1. **Comparing** Since the time of the Incas, people living in the Altiplano of Peru and Bolivia have eaten *chuño*, potatoes that have been preserved by drying in cold, dry air at 4000 m above sea level. Compare this practice with freeze drying.

2. **Applying** Name some foods that are processed by freeze drying.

Chemistry online

For more about freeze drying, visit the Chemistry Web site at **chemistryca.com**

Liquids that evaporate quickly, such as perfume and paint, are volatile liquids. Applying perfumes with a spray bottle or splashing them on your skin increases their volatility and scent by increasing the size of the surface where evaporation takes place. Like the particles in a gas, the particles in a liquid have a distribution of kinetic energies. **Figure 10.18** applies the kinetic model to evaporation.

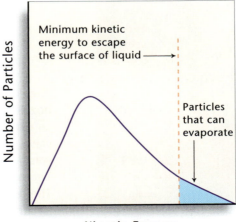

Figure 10.18
Evaporation of Liquids
Some particles in a liquid have enough kinetic energy to overcome interparticle forces. They leave the surface and become gas particles. Because the molecules that escape have higher kinetic energy, the average kinetic energy of the remaining particles decreases. Therefore, as liquids evaporate, they cool.

Fractionation of Air

In hospitals, at high altitudes, and on space walks, pure oxygen is used for life-support systems. Even greater supplies of oxygen are required by the steel and chemical manufacturing industries, where it is used in many reactions. Nitrogen, the most abundant gas in air, also has many industrial purposes. As a gas, it is used to insulate transformers and to exclude air from electric furnaces. Elemental nitrogen is used to manufacture fertilizers and explosives as well. Both oxygen and nitrogen are produced by the chemical industry by the fractional distillation of air.

Fractional Distillation

The components of dry air are nitrogen (78 percent), oxygen (21 percent), and smaller amounts of argon, carbon dioxide, neon, helium, krypton, hydrogen, xenon, and ozone (1 percent total). When a mixture such as air is fractionated, it is separated into its components. When the components are separated by differences in their boiling points, the method of separation is called fractional distillation.

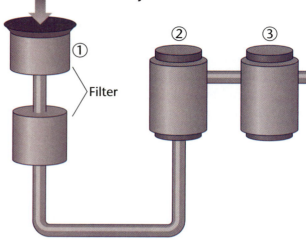

Air intake

1. **Soot and dirt are removed from air by filters.**

Filter

2. **Water vapor is removed from air either by cooling and removing the liquid water produced or by absorbing the moisture with a silica gel sponge. Then carbon dioxide reacts with lime.**

3. **The air is compressed to more than 100 times atmospheric pressure. Compression raises the temperature of the air.**

4. **In a heat exchanger, the air releases heat to the cooler surrounding fluid.**

5. **The cooled, compressed air passes through a nozzle into a chamber of larger diameter called an expansion valve. As the air passes through the valve, it expands and cools. (This cooling effect was first described by James Joule and William Thomson and is called the Joule-Thomson effect.) The temperature difference is so great that the air liquefies.**

6. As the liquefied air flows over several heated trays, it is warmed to the boiling point of nitrogen (−195.8°C). Most of the nitrogen, with a trace of oxygen, vaporizes.

7. The liquid oxygen and remaining liquid nitrogen are collected and passed into an upper chamber for another distillation at a higher temperature. Here the oxygen and nitrogen vaporize and separate because of their different densities.

8. After passing through expansion valves, the separated gases are liquefied again and bottled as liquid nitrogen and oxygen.

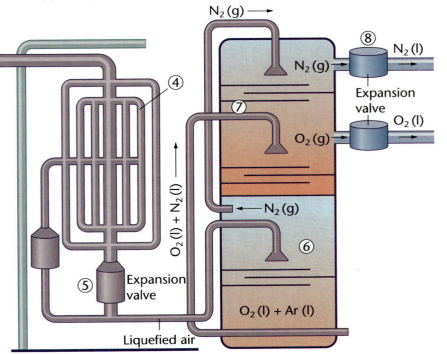

N_2 (g)

⑧ N_2 (l)

④

N_2 (g)

Expansion valve

⑦

O_2 (g)

O_2 (l)

O_2 (l) + N_2 (l)

N_2 (g)

⑥

⑤ Expansion valve

O_2 (l) + Ar (l)

Liquefied air

Liquid oxygen and liquid nitrogen are shipped in special insulated containers at temperatures slightly lower than their boiling points.

DISCUSSING THE TECHNOLOGY

1. Analyzing Why is it necessary to liquefy air to separate its components?

2. Inferring Why do you think liquid nitrogen is used to freeze food?

Sublimation

Some solid substances can change to the gaseous state directly, without melting first. The process by which particles of a solid escape from its surface and form a gas is called **sublimation.** For example, dry ice, solid carbon dioxide, does not melt but sublimes. Ice also sublimes. Some molecules in ice leave the surface and become water vapor. Sublimation of ice is the reason that food stored for a long time in a freezer becomes freezer burned.

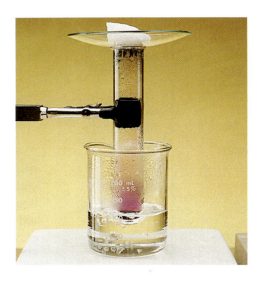

Figure 10.19
Sublimation and Condensation of Iodine
When solid iodine is heated in a beaker, it forms a purplish gas, which is free molecules of I₂. The iodine rises to the top of the tube, where it cools. There, it condenses and forms solid iodine again.

Condensation

When dew forms, water vapor in the air condenses into its liquid state. Vapor is used to describe the gaseous state of a substance that is a liquid at room temperature. Condensation is the reverse of evaporation. In **condensation,** gaseous particles come closer together—that is, they condense—and form a liquid or sometimes, as in **Figure 10.19,** a solid.

Vapor Pressure and Boiling

When a rain puddle dries up, the water molecules leave the liquid state and become water vapor. Finally, the puddle is gone. However, if liquid water is in a closed container, only some of the water evaporates. **Figure 10.20** describes how a liquid in a closed container comes to equilibrium with its vapor.

Figure 10.20
Evaporation of a Liquid in a Closed Container
Initially, some particles have sufficient kinetic energy to escape the liquid and form vapor. ▼

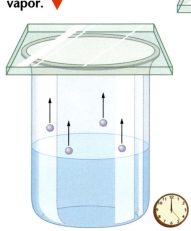

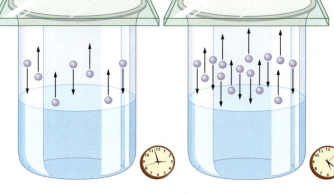

▲ As more vapor forms, collisions between particles knock some particles back into the liquid. However, the number of particles escaping the liquid is still far greater than the number of those returning to it.

◄ Eventually, the number of particles escaping from the liquid to the vapor will equal the number returning to the liquid from the vapor. Equilibrium has been reached. Although the liquid appears to have stopped evaporating, molecules are evaporating and condensing at equal rates.

Vaporization Rates

Not all molecules in a liquid move at the same speed. If the liquid is in an open container, the fastest molecules on the surface may have enough energy to escape from the liquid. What can you infer about liquids by observing their vaporization?

Procedure

1. Half fill a 250-mL beaker with water and set it on a hot plate. Put a thermometer in the water and monitor its temperature until it reaches about 60°C. Remove the beaker from the hot plate with beaker tongs and set it on the lab table.

2. Label three thin-stem pipets A, B, and C. Also, cut the bulbs of three other thin-stem pipets in half, creating three caps.

3. The teacher will provide small beakers of colored water, colored ethanol (C_2H_5OH), and hexane (C_6H_{14}). Draw water into the bulb of pipet A until it is about one-third full. Then invert it so that its stem remains filled with liquid and cap it as shown.

4. Grip the stem of the pipet with forceps, and immerse the bulb down in the hot water. Observe the liquid in the stem.

5. Measure and record the time it takes for the liquid to clear from the stem.

6. Repeat steps 3 to 5 for ethanol and hexane using pipets B and C, respectively.

Analysis

1. What can you infer about the rate of vaporization of these three liquids from the time they each take to clear from the pipet stem?

2. Rank the rate of vaporization of the three liquids from highest to lowest.

3. Considering how the molecules of the three substances vary in polarity, how do interparticle forces affect vaporization rates?

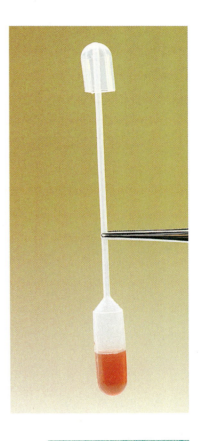

During evaporation of a liquid in a closed container, the amount of vapor and the pressure it exerts both increase. When equilibrium is reached, the pressure exerted by the vapor reaches its final, maximum value, and the volume of the liquid does not change. The pressure of a substance in equilibrium with its liquid is called its **vapor pressure.** The value of the vapor pressure of a substance indicates how easily the substance evaporates. When the pressure has reached this value, particles are evaporating and condensing at equal rates. For example, a volatile liquid such as ethanol has a high vapor pressure. The interparticle forces holding the ethanol molecules together as a liquid are weaker than those in water because ethanol molecules are less polar than water molecules. Therefore, it evaporates easily, and its vapor exerts more pressure in a closed cylinder. At the same temperature, a less volatile liquid such as water has a lower vapor pressure. Because the water molecule is polar, interparticle forces among water molecules are strong, and they tend to remain in the liquid state.

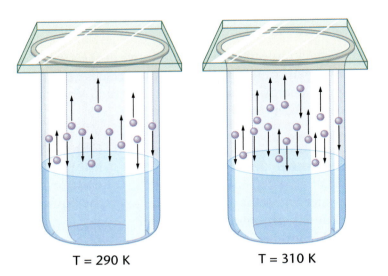

Figure 10.21
Vapor Pressure and Temperature
At the higher temperature, the vapor pressure is higher because more particles in the liquid have kinetic energy high enough to escape from the surface.

T = 290 K   T = 310 K

A liquid in a sealed container evaporates until its vapor pressure is high enough that rates of evaporation and condensation are equal. A liquid in an open container can never reach equilibrium with its vapor because the vapor is constantly escaping. But, exposed to the atmosphere, liquids having high vapor pressures will evaporate more quickly than those with low vapor pressures.

Temperature and vapor pressure are related, as shown in **Figure 10.21.** The rate of evaporation is higher at higher temperatures. If the temperature of a liquid such as isopropyl alcohol is raised high enough, not only will molecules of alcohol escape from its surface, but also, bubbles of the vapor will form below the surface, as shown in **Figure 10.22.** The **boiling point** of a substance is the temperature of the substance when its vapor pressure equals the pressure exerted on the surface of the liquid. For a liquid in an open container, the pressure exerted on its surface is atmospheric pressure. Atmospheric pressure is the force per unit area that the gases in the atmosphere exert on the surface of Earth. Normal boiling point is the temperature at which a liquid boils in an open container at normal atmospheric pressure. The normal boiling point of isopropyl alcohol is 82.3°C (355.5 K). The normal boiling point of mercury is 356.58°C (629.73 K) and that of ammonia is −33.35°C (239.80 K). Because mercury has such a low vapor pressure at room temperature, its temperature must be raised to a high level to boil under normal atmospheric pressure. On the other hand, the vapor pressure of ammonia at room temperature is so great that it boils well below room temperature at normal atmospheric pressure.

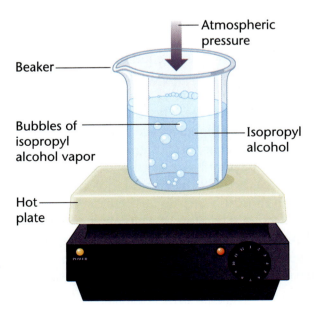

Atmospheric pressure

Beaker

Bubbles of isopropyl alcohol vapor

Isopropyl alcohol

Hot plate

POWER

Figure 10.22
Boiling
When a liquid is heated in an open container, the temperature rises until the vapor pressure equals the atmospheric pressure. Then the liquid boils. Small bubbles of vapor form below the surface and rise.

Pressure Cookers

A pressure cooker is a heavy, covered pot designed to trap steam formed from boiling water. Because the pot is tightly sealed, the pressure of the steam builds up inside the pot, raising the boiling point of the water. Because both the water and steam are at higher temperatures, the food cooks faster. Some fast-food restaurants fry chicken in pressure cookers as well as in open pans so that they can offer customers chicken of both textures and flavors. A commercial pressure cooker reaches an internal temperature of 140°C (284°F) and cooks chicken in 12 minutes—much shorter than the hour or so a conventional oven would require, but no faster than open-pan frying, which requires a much higher temperature and therefore has greater energy costs.

1. The sturdy construction of the pot allows pressure to build up safely inside the cooker.

2. The interlocking cover forms an airtight seal that keeps steam safely inside.

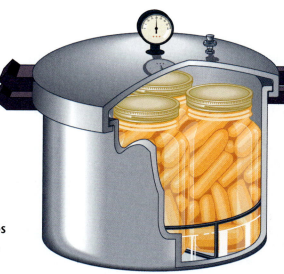

3. The rocker is a small weight that rests on top of a tiny valve built into the cover of the pressure cooker. This weight regulates the internal pressure and allows excess steam to escape safely. When the rocker vibrates gently, the cooker is heating at the proper rate.

4. Some pressure cookers, such as those used for canning food, have a pressure gauge that indicates the internal pressure in pounds per square inch (psi) so that you can regulate the internal pressure by adjusting the external heat.

5. The food rack supports food or jars above the water level. As steam fills the inside of the cooker, the pressure above the boiling water increases and the boiling temperature increases, as does the temperature of the steam. Food cooks more quickly because both steam and water are at a higher temperature than in an open pot.

Thinking Critically

1. At higher altitudes, how does a pressure cooker offset the effects of the lower atmospheric pressure on the boiling of water?

2. How does the rocker regulate the pressure inside?

When the pressure exerted on the surface of a liquid exceeds normal atmospheric pressure, vapor pressure must be higher than normal for the liquid to boil. To reach this higher vapor pressure, the liquid must be raised to a temperature higher than its normal boiling point. This is the principle behind the pressure cooker discussed in How It Works on page 559. Similarly, at pressures lower than normal atmospheric pressure, a liquid boils at a temperature below its normal boiling point. From these observations, you can conclude that the boiling point of a liquid increases when the pressure on the liquid increases and decreases when the pressure on the liquid decreases.

Heat of Vaporization

Probably, placing a pan of water on a burner, applying heat, and bringing the water to its boiling point is a familiar process to you. But you may be surprised to learn what the temperatures of steam and water are, as shown in **Figure 10.23.** You know that energy must be conserved. What happens to the energy supplied by the flame of the laboratory burner to the boiling water if it doesn't raise the temperature of the water? The rising bubbles of vapor as water changes state from liquid to gas within the boiling liquid are a clue.

The rising bubbles of steam are less dense than the liquid water because the molecules in the bubbles are farther apart than those in the liquid. To separate the particles and form a vapor requires overcoming the inter-particle forces holding them together in the liquid. The energy to overcome these forces comes from the heat of the flame.

Cloud physicists estimate that 2×10^{35} snowflakes have fallen on Earth in its 4.5-billion-year history. Water molecules bond in a hexagonal pattern when the liquid freezes. Because billions and billions of molecules make up a single snowflake and there are so many ways to arrange them, it is impossible to find two identical crystals.

Figure 10.23

Steam Is No Hotter Than Boiling Water

The setup on the left at zero time shows water beginning to boil. After more than 10 minutes of boiling (right), the thermometers still have the same readings. Because liquid water and water vapor have the same Kelvin temperature during boiling, the molecules of steam must have the same average kinetic energy as the molecules of liquid water.

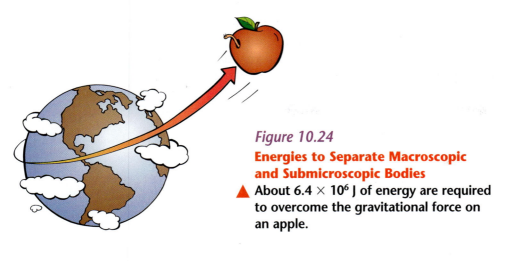

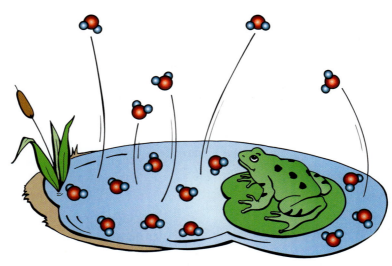

Figure 10.24

Energies to Separate Macroscopic and Submicroscopic Bodies

▲ About 6.4×10^6 J of energy are required to overcome the gravitational force on an apple.

About 2.26×10^6 J of energy are required to overcome the interparticle forces on the molecules in 1 kg of liquid water. ▶

To overcome the force of gravity that keeps an apple on the ground requires energy. The energy that does the work of separating the apple and the ground by 1 m is about 1 joule. The **joule** is the SI unit of energy required to lift a 1-kg mass 1 m against the force of gravity. To move the apple so far that it escapes Earth's gravitational force would require 6.4×10^6 J. About one-third of that amount, 2.26×10^6 J, is the energy needed to move the molecules in 1 kg of water far enough apart that they form water vapor, as shown in **Figure 10.24.**

The energy absorbed when 1 kg of a liquid vaporizes at its normal boiling point is called its **heat of vaporization.** The heat of vaporization of water is 2.26×10^6 joules per kilogram, written as 2.26×10^6 J/kg, which is about 500 times larger than the energy required to raise the temperature of 1 kg of water by 1 Celsius degree, 4180 J.

Because energy is always conserved, energy is released when vapor changes to a liquid. As the vapor condenses, the particles move closer together. For example, when 1 kg of steam condenses at water's normal boiling point, 100.0°C, it releases as heat the same energy it gained when it vaporized, 2.26×10^6 J. Perhaps you know that burns from steam are usually more severe than burns received from hot water. This occurs because the steam transfers a great deal of heat to the skin when it condenses.

ChemLab

Molecules and Energy

As you break from a saunter to a full gallop to get to your next class, your kinetic energy changes. Your energy was increased by the muscles of your legs propelling you down the hallway. In terms of energy transfer, the muscles of your legs transferred energy obtained from the foods you have eaten. Can you observe changes in a substance as energy is transferred to it?

Problem

How does energy transferred to or from a molecular substance affect the average kinetic energy of its molecules?

Objectives

- **Observe** the temperature changes and changes of state when a molecular substance is heated and cooled.
- **Make and use graphs** to analyze temperature changes.
- **Interpret** temperature changes in terms of the changes in the average kinetic energy of a substance's molecules.

PROCEDURE

1. Prepare two tables like those shown. Label one *Heating* and the other *Cooling*.

2. Pour 300 mL of tap water into a 400-mL beaker and place the beaker on a hot plate.

3. Place a thermometer in the beaker of water. Turn on the heat and monitor the water temperature until it reaches 90°C. Maintain the water temperature at 90°C by using the heat control of the hot plate or by adding cold water.

4. Half fill the test tube with stearic acid. Gently push the bulb of the second thermometer down into the substance. After the temperature of the thermometer has adjusted to the stearic acid, record this temperature in the first line of the *Heating Data* table.

5. Attach the clamp to the test tube and immerse the tube in the beaker of hot water as shown. Read and record the temperature and the physical state or states of the stearic acid every 30 seconds until all of the material has melted and its temperature is about 80°C.

6. Pour 300 mL of cold tap water into the second 400-mL beaker.

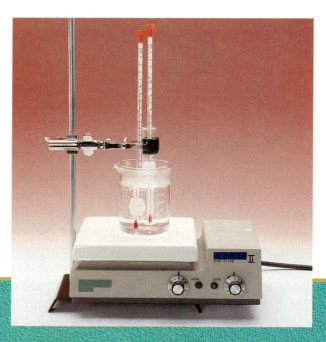

7. Remove the test tube and contents from the first beaker and immerse it in the cold water in the second beaker. Read and record in the *Cooling Data* table the temperature and physical state or states of the stearic acid every 30 seconds until the material has solidified.

ANALYZE AND CONCLUDE

1. **Making Graphs** Graph the heating data by plotting temperature readings on the vertical axis and time on the horizontal axis. Connect the data points with straight lines or smooth curves. Label the appropriate segments of the graph *solid, solid and liquid,* or *liquid.* Graph the cooling data in the same way.

2. **Interpreting Data** Divide each graph into three intervals by drawing two vertical lines at the points where the slope of the graph changes. Label the intervals of the first graph *A, B,* and *C* and those of the second graph *D, E,* and *F.*

3. **Drawing Conclusions** According to your data, what is the approximate melting point of stearic acid?

4. **Relating Concepts** Describe how the kinetic energy of the stearic acid molecules changed during each interval.

APPLY AND ASSESS

1. Describe how the molecular motion changed during each segment of the heating and cooling curves.

2. Suppose twice as much stearic acid were used. What would the graph look like? Make a sketch.

DATA AND OBSERVATIONS

Heating Data

Elapsed Time (s)	Temperature (°C)	Physical State
0		
30		
60		

Cooling Data

Elapsed Time (s)	Temperature (°C)	Physical State
0		
30		
60		

See page 867 in
Appendix F for
**Estimating Metric
Temperatures**

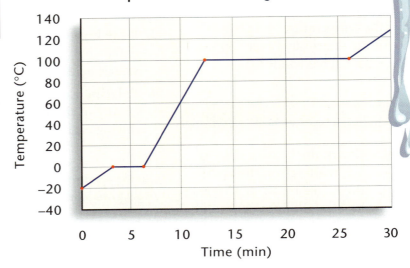

Figure 10.25

Changes in Temperature as a Solid Is Heated
A 0.100-kg sample of ice at −20°C is heated to
water vapor at 120°C. The rising line shows the
rise in temperature of the ice, water, and water
vapor. Notice how the graph is flat during
changes of state when heat is applied but that
the temperature does not change.

One definition of *energy*
is "the ability to do
work." The units of en-
ergy (joule, erg, foot-
pound) are also the
units of work.

Heat of Fusion

As with boiling and condensing, the kinetic energies of the particles of a substance do not change during melting or freezing. If
enough heat is applied to a solid, its crystal lattice disintegrates
and it becomes a liquid. The **melting point** is the temperature of
the solid when its crystal lattice begins to disintegrate. If more
heat is applied after the solid has reached its melting point, the
additional energy is used to overcome the interparticle forces
until the crystal lattice collapses and becomes a liquid. If a liquid substance is cooled, the temperature falls and the liquid
becomes a solid. The temperature of a liquid when it begins to
form a crystal lattice and becomes a solid is called its **freezing
point.** The energy released as 1 kg of a substance solidifies at its
freezing point is called its **heat of fusion.** The heat of fusion of
water, for example, is 3.34×10^5 J/kg. The same energy is absorbed
if 1 kg of a substance is heated until it melts.

As **Figures 10.25** and **10.26** show, the melting point and the
freezing point of a substance are the same temperature, provided
the pressure is the same. The rising lines and plateaus in the temperature graph as ice is heated to melting and then to boiling are
characteristic of all solid substances being heated from a solid to a liquid and then to a gas. The plateaus represent periods when heat is being
released or absorbed but the temperature is not changing. Similarly,
when a gas is cooled to liquid and then to solid, the graph of the temperature changes has the same shape.

Figure 10.26

Changes in Temperature as a Gas Is Cooled
As a 0.100-kg sample of water vapor at 120°C is cooled, it liquefies to water, then freezes to ice, then cools to −20°C. The graph falls as the steam is cooled. It levels off while the steam is condensing, falls again after all the steam has condensed, levels off again while the liquid water freezes, and then falls after all the water has frozen.

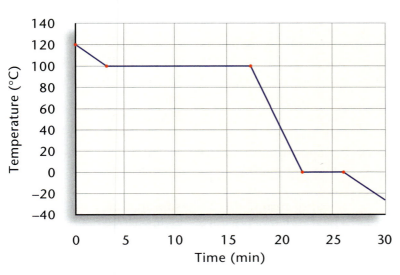

Connecting Ideas

The kinetic theory of matter explains the properties of solids, liquids, and gases, and explains changes of state in terms of interparticle forces and energy. It also quantitatively relates the pressure, volume, and temperature of gases. By studying how gases behave under different conditions, you will soon begin to understand how all matter behaves.

SECTION REVIEW

Understanding Concepts

1. Rank the following temperature readings in increasing order.

    32.0°F, 32.0°C, 32.0 K, 102.1°C, 102.1 K

2. Why will water in a flask begin to boil at room temperature as air is pumped out of the flask?

3. In terms of changes in total energy, how does the melting of 1 kg of water at 0°C differ from the freezing of 1 kg of water at 0°C?

Thinking Critically

4. **Analyzing** Why are most elements solids at room temperature?

Applying Chemistry

5. **Aerosol Sprays** Explain why volatile liquids are often used as propellants in aerosol spray cans of such things as paints and deodorants.

REVIEWING MAIN IDEAS

10.1 Physical Behavior of Matter

- The three ordinary states of matter are solid, liquid, and gas.
- The kinetic theory of matter postulates that particles of matter are in constant motion.
- Solids, liquids, and gases exist because of differences in interparticle forces.
- Amorphous materials lack the crystal lattice structures of solids.
- Plasma is the most common state of matter in the universe.

10.2 Kinetic Energy and Changes of State

- Changes in the temperature of a substance indicate changes in the average kinetic energy of its particles.
- The Kelvin-scale temperature of a substance is directly proportional to the average kinetic energy of its particles.
- Absolute zero, 0 K, is the temperature at which, theoretically, all particle motion stops.
- Changes of state occur without changing the temperature or kinetic energy of the particles.

Vocabulary

For each of the following terms, write a sentence that shows your understanding of its meaning.

absolute zero	joule (J)
amorphous material	kelvin (K)
boiling point	Kelvin scale
Brownian motion	kinetic theory
condensation	liquid
crystal lattice	liquid crystal
diffusion	melting point
evaporation	plasma
freezing point	pressure
gas	solid
heat of fusion	sublimation
heat of vaporization	temperature
ideal gas	vapor pressure

UNDERSTANDING CONCEPTS

1. Compare and contrast evaporation and boiling.
2. What is the difference between an amorphous material and a solid?
3. Why do most molecular solids have lower melting points than ionic solids?
4. Describe how vapor pressure is affected by increasing temperature.
5. Why are liquids less compressible than gases?
6. How do liquids and liquid crystals differ?
7. How is pressure exerted by a gas?
8. How does a real gas differ from an ideal gas?
9. Define a plasma and describe some of its properties. Where in the universe are plasmas commonly found? Where can plasmas be found on Earth?

APPLYING CONCEPTS

10. A bottle of vanilla extract was left uncovered in a cabinet overnight. When the cabinet door was opened, a strong odor of vanilla was detected. Explain this observation using the kinetic theory of matter.
11. What is the amount of energy absorbed when 5.00 g of water vaporize at 100°C?

Everyday Chemistry

12. Why is it necessary to place foods in low-pressure chambers during freeze drying?

Art Connection

13. How does glassblowing use the physical properties of gases?

Chemistry and Technology

14. Can fractional distillation be used to separate the substances in a compound? Explain.

THINKING CRITICALLY

Interpreting Data

15. The boiling points of five liquids are provided in either Celsius or Kelvin temperatures. List the liquids in order from the one with the lowest boiling point to the one with the highest boiling point.

acetone (C_3H_6O)	329 K
heptane (C_7H_{16})	98°C
nitromethane (CH_3NO_2)	374 K
benzene (C_6H_6)	80°C
sulfur trioxide (SO_3)	318 K

Comparing and Contrasting

16. A sample of $Cl_2(g)$ and $N_2(g)$ are both at 25°C. Compare the average kinetic energy and the most probable speed of the molecules of the gases.

Making Predictions

17. Predict whether the boiling point of water is greater or less than 100°C at the shoreline of the Dead Sea, which is 400 m below sea level. Explain your prediction.

Inferring

18. What shape would you expect an evaporating dish to have? Explain.

Comparing and Contrasting

19. Why does HCl have a higher boiling point than H_2?

Measuring in SI

20. How can both a Celsius-scale thermometer and a Kelvin-scale thermometer indicate the same temperature change but not the same final temperature reading?

Inferring

21. Explain why an amorphous material has a range of melting points rather than a fixed melting point.

Making Comparisons

22. The particles of which of the following gases have the highest average speed? The lowest average speed? Use the fact that NF_3 is the most massive and CH_4 is the least massive.
a) carbon monoxide (CO) at 90°C
b) nitrogen trifluoride (NF_3) at 30°C
c) methane (CH_4) at 90°C
d) carbon monoxide (CO) at 30°C

Relating Cause and Effect

23. How does the vapor of CO_2 affect its sublimation in a closed container?

Observing and Inferring

24. Why can you cool a cup of hot water by swirling the cup?

Using a Table

25. Examine the table and answer the following questions.
a) Which of the substances are gases at 50°C? At −50°C?
b) Which of the substances are liquids at 50°C? At −50°C?
c) Which of the substances are solids at 50°C? At −50°C?
d) Which substance has the smallest temperature range as a liquid?
e) Would it have been easier to answer part (d) if the Kelvin temperature scale had been used? Explain.

Substance	Freezing Point, °C	Boiling Point, °C
Bromine	–7	58
Mercury	–39	357
Propane	–188	–42
Radon	–71	–62
Silver	961	2195

Interpreting Graphs

26. The graph below shows the distribution of speeds of the particles of a gas at two different temperatures. Which curve, **A** or **B**, corresponds to the higher of the two temperatures? Explain your reasoning

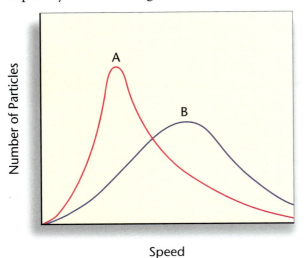

Applying Concepts

27. Name two ways to liquefy a gas.

Interpreting Graphs

28. ChemLab What can you say about the material with a heating curve such as the one shown below?

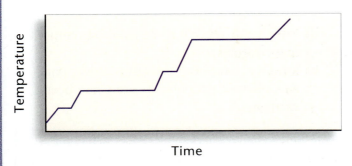

Making Predictions

29. MiniLab 1 How would the result of the experiment change if the NH₃ were first heated?

Inferring

30. MiniLab 2 How is the evaporation rate of nonpolar covalent molecules affected by molecular mass?

● CUMULATIVE REVIEW

31. What is the modern periodic law? How does it differ from the periodic law according to Mendeleev? (Chapter 3)

32. Explain how atomic radii influence the chemical reactivity of the alkaline earth metals. (Chapter 8)

● SKILL REVIEW

33. Ethanol boils at 79°C and melts at −114°C. A few grams of ethanol are heated from −130°C to +130°C. Graph the heating curve for ethanol. Show time on the horizontal axis and temperature on the vertical axis.

● WRITING IN CHEMISTRY

34. Microwave ovens cook most foods but have no effect on substances such as paper, microwave-safe glassware, and plastic. Write an article that explains how microwaves cook food and why they have no effect on certain other substances such as paper.

● PROBLEM SOLVING

35. A beaker of water is placed under a bell jar at room temperature. A vacuum pump removes all the air from the jar and the water begins to boil. Explain why the water boils at room temperature.

36. Ethanol melts at −114°C. Express the melting point of ethanol on the Kelvin scale.

Standardized Test Practice

Elemental Metals Boiling and Melting Points				
Metal Element	**Boiling Point (°C)**	**Boiling Point (K)**	**Melting Point (°C)**	**Melting Point (K)**
Uranium	1132	?	3818	?
Gold	?	1337	?	3080
Copper	1083	1356	2567	?
Silver	962	1235	962	?
Lead	?	600	?	2013

Use the table above for questions 1–3.

1. The boiling point of uranium is

   a) 859 K.   **c)** 1405 K.
   b) 1132K.   **d)** 4091 K.

2. The melting point of lead is

   a) 1740 °C.   **c)** 2286 °C.
   b) 1740 K.   **d)** 2286 K.

3. Which element has the highest melting point in kelvins?

   a) lead   **c)** gold
   b) copper   **d)** uranium

4. Which of the following has a definite volume but no definite shape?

   a) gas   **c)** solid
   b) liquid   **d)** plasma

5. Which statement explains the behavior of liquids?

   a) Liquid particles are attached to each other, and they can easily slide past each other.
   b) Liquid particles are attached and easily flow in straight lines when moved.
   c) Liquid particles are separated by spaces that allow them to move over each other.
   d) Liquid particles are held stiffly together until high temperatures loosen their bonds.

6. The most common form of matter in the universe is

   a) gas.   **c)** solid.
   b) liquid.   **d)** plasma.

7. Food dye spreading through a glass of water is an example of

   a) Brownian motion.
   b) distillation.
   c) sublimation.
   d) diffusion.

8. Snow turning directly into water vapor on a cold day is an example of

   a) sublimation.   **c)** vaporization.
   b) evaporation.   **d)** condensation.

9. An increase in the temperature of a substance results in a(n)

   a) increase in its boiling point.
   b) increase in its rate of evaporation.
   c) decrease in its boiling point.
   d) decrease in its rate of evaporation.

10. Heat of vaporization and heat of fusion are measurements of

   a) temperature.
   b) mass.
   c) energy.
   d) volume.

Test Taking Tip

Stock Up On Supplies Bring all your test-taking tools: number two pencils, black and blue pens, erasers, correction fluid, a sharpener, a ruler, a calculator, and a protractor. Bring munchies, too. You might not be able to eat them in the testing room, but they come in handy for a break, if you're allowed to go outside.

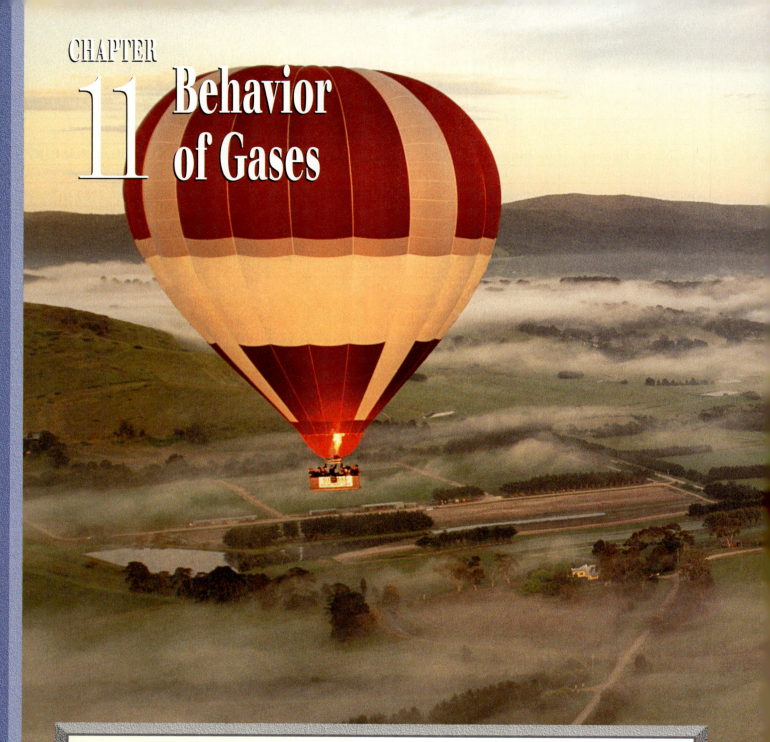

CHAPTER
11 Behavior of Gases

Chapter Preview
Sections
11.1 Gas Pressure
MiniLab 11.1 Relating Mass and Volume of Gas

11.2 The Gas Laws
MiniLab 11.2 How Straws Function
ChemLab Boyle's Law

Fly, Fly Away!

A hot air balloon is a great example of gases in action. As the burner heats up the air in the balloon, the air expands. The more the air expands, the higher the balloon rises.

Start-up Activities

Launch Lab

More Than Just Hot Air

How does a temperature change affect the air in a balloon?

Safety Precautions

Always wear goggles to protect eyes from broken balloons.

Materials

- 5-gal bucket
- round balloon
- ice
- string

Procedure

1. Inflate a round balloon and tie it closed.
2. Fill the bucket about half full of cold water and add ice.
3. Use a string to measure the circumference of the balloon.
4. Stir the water in the bucket to equalize the temperature. Submerge the balloon in the ice water for 15 minutes.
5. Remove the balloon from the water. Measure the circumference.

Analysis

What happens to the size of the balloon when its temperature is lowered? What might you expect to happen to its size if the temperature is raised?

What I Already Know

Review the following concepts before studying this chapter.
Chapter 10: properties of gases; the kinetic theory of gases

Reading Chemistry

Before reading the chapter, make a list of some ways we might use gases or gas pressure in everyday life. Next, scan the chapter to discover other ways we use gas. Note some of the chapter features, such as "Chemistry and Technology" on page 390.

Preview this chapter's content and activities at chemistryca.com

Gas Pressure

You might be surprised to see that an overturned tractor trailer can be uprighted by inflating several air bags placed beneath it. But, you're not at all surprised to see that an uprighted truck is supported by 18 tires inflated with air. Air can lift the tractor and support the truck because air is a mixture of gases, and gases exert pressure.

Defining Gas Pressure

Unless a ball has an obvious dent, you can't tell whether it is under-inflated by looking at it. You have to squeeze it. If it's soft, you know it needs to be pumped up with more air. The springiness of a fully inflated ball is the pressure of the air inside. **Figure 11.1** shows how the pressure of air inside a soccer ball rises as air is added. How can changes in gas pressure be explained by the kinetic theory?

Figure 11.1
Pumping up a Soccer Ball
Molecules in air are in constant motion and exert pressure when they strike the walls of the ball. The pressure they exert counterbalances two other pressures—atmospheric pressure on the ball plus the pressure exerted by the tough rubber of the ball itself.

Pumping more air into the soft ball (left) increases the number of molecules inside. As a result, molecules strike the inner wall of the ball more often and the pressure increases (right). The increase in pressure is counterbalanced by increased pressure from the tough wall. As a result, the ball becomes firmer and bouncier. ▼

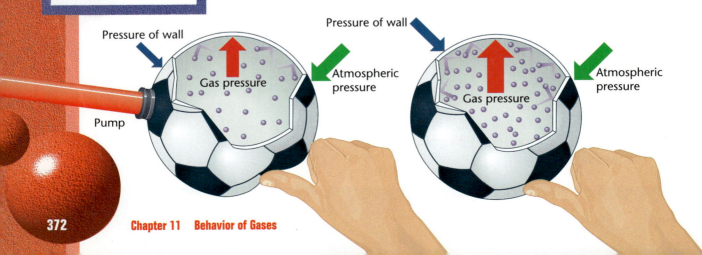

Pressure of wall
Pressure of wall
Gas pressure
Gas pressure
Atmospheric pressure
Atmospheric pressure
Pump

How are number of particles and gas pressure related?

Recall from Chapter 10 that the pressure of a gas is the force per unit area that the particles in the gas exert on the walls of their container. As you would expect, more air particles inside the ball mean more mass inside. In **Figure 11.2,** one basketball was not fully inflated. The other identical ball was pumped up with more air. The ball that contains more air has higher pressure and the greater mass.

From similar observations and measurements, scientists from as long ago as the 18th century learned that the pressure of a gas is directly proportional to its mass. According to the kinetic theory, all matter is composed of particles in constant motion, and pressure is caused by the force of gas particles striking the walls of their container. The more often gas particles collide with the walls of their container, the greater the pressure. Therefore, the pressure is directly proportional to the number of particles. For example, doubling the number of gas particles in a basketball doubles the pressure.

Figure 11.2

Two Basketballs at Unequal Pressures
The mass of the ball on the left is greater than the mass of the one on the right. The ball on the left has greater mass because it has more air inside and, therefore, is at a higher pressure than the ball on the right.

To demonstrate how a gas at constant temperature can be used to do work, let's examine the action of a piston as shown in **Figure 11.3.** The piston inside the cylinder is like the cover of a jar because it makes an airtight seal, but it also acts like a movable wall. When gas is added, as shown in the top cylinder, the gas particles push out the piston until the pressure inside balances the atmospheric pressure outside.

If more gas is pumped into the cylinder, the number of particles increases, and the number of collisions on the walls of the container increases. Because the force on the inside face of the piston is now greater than the force on the outside face, the piston moves outward. As the gas spreads out into the larger volume, the number of collisions per unit area on the inside face falls. Then, the pressure of the gas inside the container falls until it equals the pressure of the atmosphere outside and the piston comes to rest in its new position, farther out. The atmospheric pressure remains constant while the piston moves and the gas expands.

Figure 11.3

Expanding a Gas at Constant Pressure
The constant bombardment of molecules and atoms of the atmosphere exerts a constant pressure on the outside face of the piston. This pressure balances the pressure caused by the confined gas bombarding the inside face of the piston (top). ▶

When gas is added to the cylinder, the piston is pushed out (bottom). When the pressure of the new volume inside the cylinder balances atmospheric pressure, the piston stops moving out. ▶

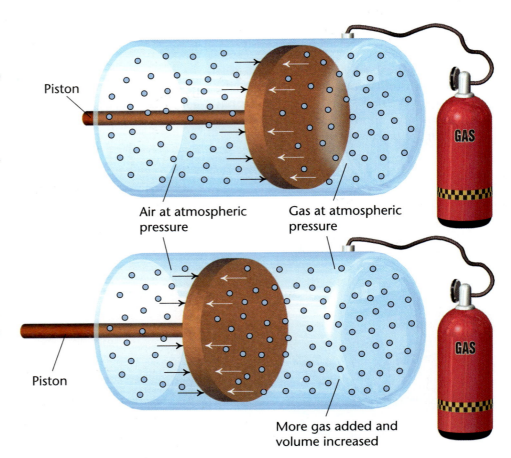

Piston

Air at atmospheric pressure

Gas at atmospheric pressure

GAS

Piston

More gas added and volume increased

GAS

How are temperature and gas pressure related?

How does temperature affect the behavior of a gas? You know from Chapter 10 that at higher temperatures, the particles in a gas have greater kinetic energy. They move faster and collide with the walls of the container more often and with greater force, so the pressure rises. If the volume of the container and the number of particles of gas are not changed, the pressure of a gas increases in direct proportion to the Kelvin temperature.

Relating Mass and Volume of a Gas

Recall from Chapter 4 that one property of carbon dioxide is that it changes directly from a solid (dry ice) to a gas. In other words, it sublimes. Can you determine some properties of gases from sublimed dry ice?

Procedure

1. Place a small, zipper-closure plastic bag on the pan of a zeroed balance. Wearing gloves and using tongs, insert a 20- to 30-g piece of dry ice into the bag.

2. Measure and record the mass of the bag and its contents. Quickly press the air out of the bag and zip it closed.

3. Immediately put the bag into a larger, clear plastic bag so that you can observe the sublimation process. After the inner bag is filled with gas, unzip the closure through the outer bag. Immediately remove the bag with dry ice from the larger bag. Press out the gas and zip it closed.

4. Measure and record the mass of the bag and its contents.

5. Use tongs and gloves to remove and dispose of the dry ice from the bag.

6. Determine the volume of the inner bag by filling it with water and pouring the water into a graduated cylinder. Record its volume.

7. Repeat steps 1-6 with a smaller zipper-closure bag and a smaller piece of dry ice.

Analysis

1. Calculate the mass of carbon dioxide that has sublimed in each trial.

2. Calculate the mass ratio by dividing the mass of sublimed CO_2 in Trial 1 by the mass that sublimed in Trial 2. Calculate the volume ratio of CO_2.

3. How do the volume and mass ratios compare?

4. What can you infer about the relationship between the volume and mass of a gas?

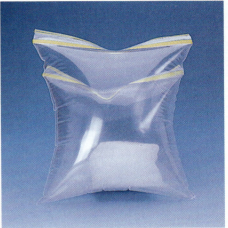

What if the volume, such as in the piston in **Figure 11.3,** is not held constant? How would a change in temperature affect the volume and pressure of a gas? If the temperature of the gas inside the piston chamber is raised, the pressure will momentarily rise. But, if the gas is permitted to expand as the temperature is raised, the pressure remains equal to the atmospheric pressure and the volume expands. Thus, the volume of a gas at constant pressure is directly proportional to the Kelvin temperature.

When the cylinder is in an automobile engine, the other end of the piston rod, shown in **Figure 11.3,** is attached to the crankshaft. When the mixture of gasoline and air in the cylinder ignites and burns, gases are produced. The gas inside the cylinder expands because the heat of burning raises the temperature. The pressure of expanding gases drives the piston outward. As the piston rod moves out and then back in, it turns the crankshaft, delivering power to the wheels that propel the vehicle forward.

Try at Home Lab

See page 868 in Appendix F for **Crushing Cans**

Devices to Measure Pressure

Although you can compare a property such as gas pressure by touching partially and fully inflated basketballs, this method does not give you an accurate measure of the two pressures. What is needed is a measuring device.

The Barometer

One of the first instruments used to measure gas pressure was designed by the Italian scientist Evangelista Torricelli (1608-1647). He invented the **barometer,** an instrument that measures the pressure exerted by the atmosphere. His barometer was so sensitive that it showed the difference in atmospheric pressure between the top and bottom of a flight of stairs. **Figure 11.4** explains how Torricelli's barometer worked. The height of the mercury column measures the pressure exerted by the atmosphere. We live at the bottom of an ocean of air. The highest pressures occur at the lowest altitudes. If you go up a mountain, atmospheric pressure decreases because the depth of air above you is less.

One unit used to measure pressure is defined by using Torricelli's barometer. The **standard atmosphere (atm)** is defined as the pressure that supports a 760-mm column of mercury. This definition can be represented by the following equation.

$$1.00 \text{ atm} = 760 \text{ mm Hg}$$

Because atmospheric pressure is measured with a barometer, it is often called barometric pressure.

Figure 11.4

How Torricelli's Barometer Works

A barometer consists of a tube of mercury that stands in a dish of mercury. Because the mercury stands in a column in the closed tube, you can conclude that the atmosphere exerts pressure on the open surface of the mercury in the dish. This pressure, transmitted through the liquid in the dish, supports the column of mercury. ▶

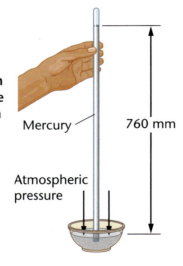

Mercury

760 mm

Atmospheric pressure

▲ The sphygmomanometer attached to the wall also uses a column of mercury to measure blood pressure.

The Pressure Gauge

Unfortunately, the barometer can measure only atmospheric pressure. It cannot measure the air pressure inside a bicycle tire or in an oxygen tank. You need a device that can be attached to the tire or tank. This pressure gauge must make some regular, observable response to pressure changes. If you have ever measured the pressure of an inflated bicycle tire, you're already familiar with such a device.

How it Works

Tire-Pressure Gauge

A tire-pressure gauge is a device that measures the pressure of the air inside an inflated tire or a basketball. Because an uninflated tire contains some air at atmospheric pressure, a tire-pressure gauge records the amount that the tire pressure exceeds atmospheric pressure.

The most familiar tire gauge is about the size and shape of a ballpoint pen. It is a convenient way to check tire pressure for proper inflation regularly. Proper inflation ensures tire maintenance and safety.

1. The pin in the head of the gauge pushes downward on the tire-valve inlet and allows air from the tire to flow into the gauge.

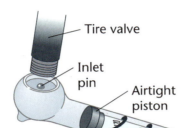

Tire valve

Inlet pin

Airtight piston

Spring

Calibrated scale

Pressure of tire

2. The air flowing into the gauge pushes against a movable piston that pushes a sliding calibrated scale.

3. The piston moves along the cylinder, compressing the spring until the force of air pressing on the surface of the piston is equal to the force of the compressed spring on the piston.

4. The pressure of the air in the cylinder, which is the same as the pressure of the air in the tire, is read from the scale.

Thinking Critically

1. What does the calibration of the scale indicate about the relationship between the compression of the spring and pressure?

2. Explain whether a tire-pressure gauge should be recalibrated for use at locations where atmospheric pressure is less than that at sea level.

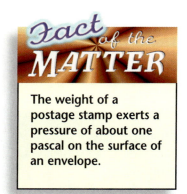
When you measure tire pressure, you are measuring pressure above atmospheric pressure. The recommended tire inflation pressures listed by manufacturers are gauge pressures; that is, pressures read from a gauge. A barometer measures absolute pressure; that is, the total pressures exerted by all gases, including the atmosphere. To determine the absolute pressure of an inflated tire, you must add the barometric pressure to the gauge pressure.

Pressure Units

You have learned that atmospheric pressure is measured in mm Hg. Recall from Chapter 10, atmospheric pressure is the force per unit area that the gases in the atmosphere exert on the surface of Earth. **Figure 11.5** shows two additional units that are used to measure pressure of one standard atmosphere.

The SI unit for measuring pressure is the **pascal (Pa),** named after the French physicist Blaise Pascal (1623-1662). Because the pascal is a small pressure unit, it is more convenient to use the kilopascal. As you recall from Chapter 1, the prefix *kilo-* means 1000; so, 1 **kilopascal (kPa)** is equivalent to 1000 pascals. One standard atmosphere is equivalent to 101.3 kilopascals.

Figure 11.5

Atmospheric Pressure

The weight of the air in each column pushing on the area beneath it exerts a pressure of one standard atmosphere. Each column of air extends to the outer limits of the atmosphere.

If the unit of area is the square inch and the unit of force is the pound, then the unit of pressure is the pound per square inch (psi). Expressed in these units, one standard atmosphere is 14.7 pounds per square inch or 14.7 psi. ▼

Expressed in SI units, one standard atmosphere is 101 300 pascals. Note that SI units are based on the square meter area and not the square inch. ▼

14.7 psi

1 in.　1 in.

101 300 Pa

1 m

1 m

Table 11.1 presents the standard atmosphere in equivalent units. Because there are so many different pressure units, the international community of scientists recommends that all pressure measurements be made using SI units, but pounds per square inch continues to be widely used in engineering and almost all nonscientific applications in the United States.

Table 11.1 Equivalent Pressures			
1.00 atm	760 mm Hg	14.7 psi	101.3 kPa

Pressure Conversions

You can use **Table 11.1** to convert pressure measurements to other units. For example, you can now find the absolute pressure of the air in a bicycle tire. Suppose the gauge pressure is 44 psi. To find the absolute pressure, add the atmospheric pressure to the gauge pressure. Because the gauge pressure is given in pounds per square inch, use the value of the standard atmosphere that is expressed in pounds per square inch. One standard atmosphere equals 14.7 psi.

$$44 \text{ psi} + 14.7 \text{ psi} = 59 \text{ psi}$$

The following Sample Problems show how to use the values in **Table 11.1** to express pressure in other units.

SAMPLE PROBLEM 1 **Converting Barometric Pressure Units**

In weather reports, barometric pressure is often expressed in inches of mercury. What is one standard atmosphere expressed in inches of mercury?

Analyze
- You know that one standard atmosphere is equivalent to 760 mm of Hg. What is that height expressed in inches? A length of 1.00 inch measures 25.4 mm on a meterstick.

Set Up
- Select the appropriate equivalent values and units given in **Table 11.1.** Multiply 760 mm by the number of inches in each millimeter to express the measurement in inches.

$$760 \text{ mm} \times \left(\frac{1.00 \text{ in.}}{25.4 \text{ mm}} \right)$$

The factor on the right of the expression above is the conversion factor.

Solve
- Notice that the units are arranged so that the unit *mm* will cancel properly and the answer will be in inches.

$$\frac{760 \text{ mm} \mid 1.00 \text{ in.}}{25.4 \text{ mm}} = 29.9 \text{ in.}$$

Check
- Because 1 mm is much shorter than 1 in., the number of mm, 760, should be much larger than the equivalent number of inches.

The method used in the Sample Problem to convert measurement to other units is called the **factor label method.** Study the Sample Problem again. The following equation gives the conversion factor because it contains both the given unit and the desired unit.

$$1.00 \text{ in.} = 25.4 \text{ mm}$$

You know that you can divide both sides of an equation by the same value and maintain equality. For example, you can divide both sides of the equation 1.00 in. = 25.4 mm by 25.4 mm, as shown below.

$$\frac{1.00 \text{ in.}}{25.4 \text{ mm}} = \frac{25.4 \text{ mm}}{25.4 \text{ mm}}$$

Now, simplify the right side. The conversion factor is on the left.

$$\frac{1.00 \text{ in.}}{25.4 \text{ mm}} = 1$$

In the Sample Problem, the height of mercury in millimeters was being multiplied by this conversion factor. Because multiplying any quantity by 1 doesn't affect its value, the height of the mercury column isn't changed—only the units are. The factor label method changes the units of the measurement without affecting its value. You will use the factor label method to solve problems in this chapter and in several of the following chapters. You will find more information on the factor label method on pages 801-803 of Appendix A, the Skill Handbook.

SAMPLE PROBLEM 2 — Converting Pressure Units

The reading of a tire-pressure gauge is 35 psi. What is the equivalent pressure in kilopascals?

Analyze • The given unit is pounds per square inch (psi), and the desired unit is kilopascals (kPa). According to **Table 11.1,** the relationship between these two units is 14.7 psi = 101.3 kPa.

Set Up • Write the conversion factor with kPa units as the numerator and psi units as the denominator. Note that the psi units will cancel and only the kPa units will appear in the final answer.

> **Problem-Solving HINT**
>
> In the factor label method, terms are arranged so that units will cancel out.

$$35 \text{ psi} \times \left(\frac{101.3 \text{ kPa}}{14.7 \text{ psi}} \right)$$

Solve • Multiply and divide the values and units.

$$\frac{35 \text{ psi} \mid 101.3 \text{ kPa}}{\mid 14.7 \text{ psi}} = \frac{35 \times 101.3 \text{ kPa}}{14.7} = 240 \text{ kPa}$$

Notice that the given units (psi) will cancel properly and the quantity will be expressed in the desired unit (kPa) in the answer.

Check • Because 1 psi is a much greater pressure than 1 kPa, the number of psi, 35, should be much smaller than the equivalent number of kilopascals.

Your skill in converting units will help you relate measurements of gas pressure made in different units.

Use Table 11.1 and the equation 1.00 in. = 25.4 mm to convert the following measurements.

1. 59.8 in. Hg to psi
2. 7.35 psi to mm Hg
3. 1140 mm Hg to kPa
4. 19.0 psi to kPa
5. 202 kPa to psi

Supplemental Problems

For more practice with solving problems, see Supplemental Practice Problems, Appendix B.

SECTION REVIEW

Understanding Concepts

1. Compare and contrast how a barometer and a tire-pressure gauge measure gas pressure.

2. At atmospheric pressure, a balloon contains 2.00 L of nitrogen gas. How would the volume change if the Kelvin temperature were only 75 percent of its original value?

3. A cylinder containing 32 g of oxygen gas is placed on a balance. The valve is opened and 16 g of gas are allowed to escape. How will the pressure change?

Thinking Critically

4. **Interpreting Data** Sketch graphs of pressure versus time, volume versus time, and number of particles versus time for a large, plastic garbage bag being inflated until it ruptures.

Applying Chemistry

5. **Overloaded** Why do tire manufacturers recommend that tire pressure be increased if the recommended number of car passengers or load size is exceeded?

The Gas Laws

An uninflated air mattress doesn't make a comfortable bed because comfort depends upon the pressure of the air inside it. So, you inflate it by huffing and puffing or by using the exhaust feature of a vacuum cleaner. Recall that when more air is added to the mattress, the pressure of the air inside increases. When the mattress is filled and its air valve is closed, it is far more comfortable. The air inside supports the walls of the mattress and pushes against the force exerted by atmospheric pressure on the outside surface of the mattress. As you add your weight to the mattress, do you know how the air inside is now acting upon the mattress walls, against the force of the atmosphere on its surface, and on you?

Boyle's Law: Pressure and Volume

As you know from squeezing a balloon, a confined gas can be compressed into a smaller volume. Robert Boyle (1627-1691), an English scientist, used a simple apparatus like the one pictured in **Figure 11.6** to

Figure 11.6

Relation Between Pressure and Volume

1. The pressure of the trapped air in the J tube balances the atmospheric pressure, 1 atm or 760 mm Hg.

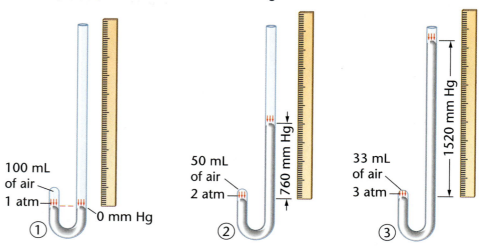

2. When mercury is added to a height of 760 mm above the height in the closed end, the volume of the trapped air is halved, and the pressure the air exerts is now 2 atm.

3. When an additional 760 mm of mercury is added to the column, the pressure of the trapped air is tripled.

SECTION PREVIEW

Objectives
✓ **Analyze** data that relate temperature, pressure, and volume of a gas.

✓ **Model** Boyle's law and Charles's law using kinetic theory.

✓ **Predict** the effect of changes in pressure and temperature on the volume of a gas.

✓ **Relate** how volumes of gases react in terms of the kinetic theory of gases.

Review Vocabulary
Pascal: SI unit for measuring pressure.

New Vocabulary
Boyle's law
Charles's law
combined gas law
standard temperature and pressure, STP
law of combining gas volumes
Avogadro's principle

compress gases. The weight of the mercury in the open end of the tube compresses air trapped in the closed end.

After performing many experiments with gases at constant temperatures, Boyle had four findings.

a) If the pressure of a gas increases, its volume decreases proportionately.
b) If the pressure of a gas decreases, its volume increases proportionately.
c) If the volume of a gas increases, its pressure decreases proportionately.
d) If the volume of a gas decreases, its pressure increases proportionately.

Because the changes in pressure and volume are always opposite and proportional, the relationship between pressure and volume is an inverse proportion. By using inverse proportions, all four findings can be included in one statement called Boyle's law. **Boyle's law** states that the pressure and volume of a gas at constant temperature are inversely proportional.

Recall from Chapter 10 that the kinetic energy of an ideal gas is *directly* proportional to its temperature and that the graph of a direct proportion is a straight line. The graph of an *inverse* proportion is a curve like that shown in **Figure 11.7.** Just as a road runs both ways, you can think of a gas following the curve A-B-C or C-B-A. The path A-B-C represents the gas being compressed, which forces its pressure to rise. As the gas is compressed by half, from 1.0 L to 0.5 L, the pressure doubles from 100 kPa to 200 kPa. As it is compressed by half again, from 0.5 L to 0.25 L, the pressure again doubles from 200 kPa to 400 kPa. Look at **Figure 11.9** on page 386 to see another example of this relationship. The reverse path C-B-A represents what happens when the pressure of the gas decreases and the volume increases accordingly.

Figure 11.7

Boyle's Law: An Inverse Relationship
As you follow the curve from left to right, the pressure increases and the volume decreases. Look at the change from points A to C. The volume is reduced from 1.0 L to 0.25 L; that is, the gas is compressed to one-fourth of its volume, and the pressure rises from 100 kPa to 400 kPa, which is four times as high. The volume/pressure relationship is a two-way system.

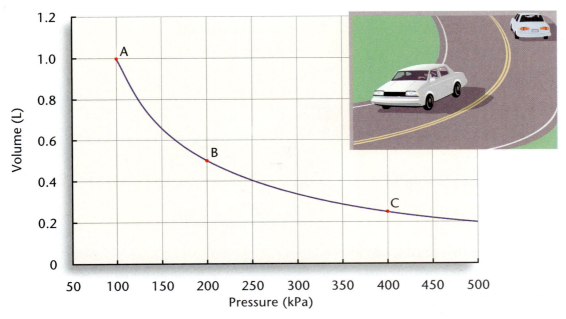

ChemLab

Boyle's Law

The quantitative relationship between the volume of a gas and the pressure of the gas, at constant temperature, is known as Boyle's Law. By measuring quantities directly related to the pressure and volume of a tiny amount of trapped air, you can deduce Boyle's law.

Problem

What is the relationship between the volume and the pressure of a gas at constant temperature?

Objectives

- **Observe** the length of a column of trapped air at different pressures.
- **Examine** the mathematical relationship between gas volume and gas pressure.

PREPARATION

Materials

thin-stem pipet	matches
double-post	metric ruler
screw clamp	scissors
fine-tip marker	small beaker
food coloring	water

Safety Precautions

Use care in lighting matches and melting the pipet stem.

PROCEDURE

1. Cut off the stepped portion of the stem of the pipet with the scissors.
2. Place about 20 mL of water in a beaker, add a few drops of food coloring, and swirl to mix.

3. Draw the water into the pipet, completely filling the bulb and allowing the water to extend about 5 mm into the stem of the pipet.

4. Heat the tip of the pipet gently above a flame until it is soft. (**CAUTION:** *If the stem accidentally begins to burn, blow it out.*) Use a metallic or glass object to flatten the tip against the countertop so that the water and air are sealed inside the pipet. Holding the pipet by the bulb, tap any water droplets in the stem down into the liquid. You should observe a cylindrical column of air trapped in the stem of the pipet.

5. Center the bulb in a double-post screw clamp, and tighten the clamp until the bulb is just held firmly. Mark the knob of the clamp with a fine-tip marker, and tighten the clamp three or four turns so that the length of the air column is 50 to 55 mm. Record this number of turns T under Trial 1 in a data table like the one shown. Measure and record the length L of the air column in millimeters for Trial 1.

6. Turn the clamp knob one complete turn, and record the trial number and the total number of turns. Measure and record the length of the air column.

7. Repeat step 6 until the air column is reduced to a length of 25 to 30 mm.

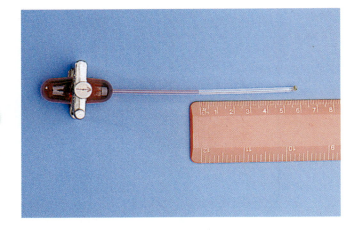

1. **Observing and Inferring** Explain whether the volume V of the air in the stem is directly proportional to the length L of the air column. What inferences can be made about the pressure P of the air in the column and the number of turns T of the clamp screw?

2. **Interpreting Data** Calculate the product LT and the quotient L/T for each trial. Which calculations are more consistent? If L and T are directly related, L/T will yield nearly constant values for each trial. On the other hand, if L and T are inversely related, LT will yield almost constant values for each trial. Are L and T directly or inversely related?

3. **Drawing Conclusions** Explain whether the data indicate that gas volume and gas pressure at constant temperature are directly related or inversely related.

1. For a mercury column barometer to measure atmospheric pressure as 760 mm Hg, the column containing the mercury must be completely evacuated. However, if the column is not completely evacuated, the barometer can still be used to correctly measure *changes* in barometric pressure. How is the second statement related to Boyle's law?

2. Using the kinetic theory, explain how a decrease in the volume of a gas causes an increase in the pressure of the gas.

From Robert Boyle's experiments, we know the relationship between the pressure and volume of a gas at constant temperature.

Pressure and Length Data				
Trial	Turns, T	Length of Air Column, L, (mm)	Numerical Value, LT	Numerical Value, L/T

The weather balloon in **Figure 11.8** illustrates Boyle's law. When helium gas is pumped into the balloon, it inflates, just as a soccer ball does, until the pressure of gas inside equals the pressure of the air outside. Because helium is less dense than air, the mass of helium gas is less than the mass of the same volume of air at the same temperature and pressure. As a result, the balloon rises. As it climbs to higher altitudes, atmospheric pressure becomes less. According to Boyle's law, because the pressure on the helium decreases, its volume increases. The balloon continues to rise until the pressures inside and outside are equal, when it hovers and records weather data.

Kinetic Explanation of Boyle's Law

You know that by compressing air in a tire pump, the air's pressure is increased and its volume is reduced. According to the kinetic theory, if the temperature of a gas is constant and the gas is compressed, its pressure must rise. Boyle's law, based on volume and pressure measurements made on confined gases at constant temperature, quantified this relationship by stating that the volume and pressure are inversely proportional. **Figure 11.9** shows how the kinetic theory explains Boyle's observations of gas behavior.

Figure 11.8

Weather Balloons and Boyle's Law
As the balloon rises, the weight of the column of air above it is shorter, and the pressure on the helium gas in the balloon decreases. At an altitude of 15 km, atmospheric pressure is only 1/10 as great as at sea level. When a balloon that was filled with helium at 1.00 atm reaches this altitude, its volume is about ten times as large as it was on the ground.

Figure 11.9

Modeling Boyle's Law

1. When the piston of the bicycle tire pump is pulled all the way out, the air pressure inside balances the air pressure outside.

3. When the piston is forced down farther and the air is compressed into one-fourth the volume of the pump, the frequency of collisions with the walls is four times as great. The air pressure inside the pump is 4 atm.

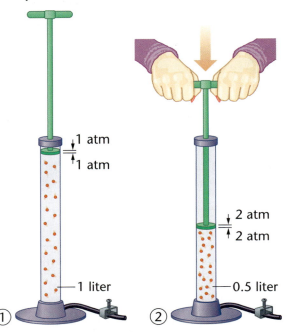

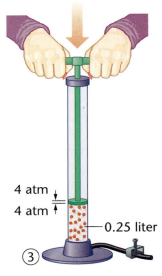

2. When the piston is forced halfway down, the average kinetic energy of the air particles is unchanged because the temperature is unchanged. They strike the wall with the same average force but, because they have been compressed into half the volume of the pump, the frequency of collisions with the walls doubles. The air pressure inside the pump is now 2 atm.

Weather Balloons

If you watch the sky any day at 1 P.M. EST, you may see a weather balloon like the one in the photograph. Every 12 hours, an instrument-packed helium or hydrogen balloon is launched at each of 70 sites around the United States. The balloons provide the data used to make the country's weather forecasts.

Weather balloon system Since World War II, meteorologists have used weather balloons to provide profiles of temperature, pressure, relative humidity, and wind velocity in the upper atmosphere. Besides the 70 launching sites in this country, there are more than 700 other sites around the world. The balloons are released at the same time in all countries—at 6:00 and 18:00 hours Greenwich Mean Time (GMT). Data from weather stations all over the world reach the United States via a central computer in Maryland. From there, the data are sent to regional weather service stations throughout the country for release to newspapers and television and radio networks.

Sending the balloons aloft The rubber weather balloons are inflated with either helium or hydrogen to a diameter of about 2 m. The inflated balloons become buoyant because the density of either gas is less than the density of air. An instrument package is attached to each balloon by a 15-m cord. As the atmospheric pressure decreases, the gas inside the balloon expands, carrying the balloon higher. When it reaches a height of 30 km, the volume of the expanded gas is so great that the balloon bursts. Then a parachute carries the package containing the radio transmitter and measuring instruments safely to the ground.

Gathering data While aloft, devices in the instrument box record and transmit temperature, relative humidity, and barometric pressure.

Relative humidity readings are obtained from a device containing a polymer that swells when it becomes moist. The swelling causes an increase in the electrical resistance of a carbon layer in the polymer. Electrical measurements of the resistance of the material indirectly measure changes in relative humidity.

Wind velocity is determined by tracking the balloon by radar or by one of the radio location systems, known as LORAN-C, OMEGA, or VLF.

The instrument package transmits barometric, relative humidity, and temperature data to the ground by radio transmission using a multiplexing system. The system transmits the different kinds of data in rotation, for example, temperature first, followed by humidity, and then pressure.

Connecting to Chemistry

1. **Thinking Critically** Some people thought that when weather satellites were deployed, weather balloons would become obsolete. Why do you think this did not happen?

2. **Inferring** Why would a meteorologist have to understand the gas laws?

miniLAB 2

How Straws Function

Each of us has used straws to sip soda, milk, or another liquid from a container. How do they function?

Procedure

1. Obtain a clean, empty food jar with a screw-on lid. Fill the jar halfway with tap water.

2. Put the ends of two soda straws into your mouth. Place the other end of one of the straws into the water in the jar and allow the end of the other straw to remain in the air. Try to draw water up the straw by sucking on both straws simultaneously. Record your observations.

3. Using a hammer and nail, punch a hole of about the same diameter as that of a straw in the lid of the jar. Push the straw 2 to 3 cm into the hole and seal the straw in position with wax or clay.

4. Fill the jar to the brim with water, and carefully screw the lid on so that no air enters the jar. Try to draw the water up the straw. Record your observations.

Analysis

1. Explain your observations in step 2.

2. Explain your observations in step 4.

3. Explain how a soda straw functions.

SAMPLE PROBLEM 3 Boyle's Law: Determining Volume

The maximum volume a weather balloon can reach without rupturing is 22 000 L. It is designed to reach an altitude of 30 km. At this altitude, the atmospheric pressure is 0.0125 atm. What maximum volume of helium gas should be used to inflate the balloon before it is launched?

Analyze
- At 30 km, the pressure exerted by the helium in the balloon equals the atmospheric pressure at that altitude, 0.0125 atm. What volume does this amount of helium occupy at 1 atm? The pressure is greater at high altitude by the factor 1.0 atm/0.0125 atm, which is greater than 1. By Boyle's law, the volume at launch is smaller than at high altitude by the factor 0.0125 atm/1.0 atm, which is less than 1.

Set Up
- Multiply the maximum volume by the factor given by Boyle's law.

$$V = 22\ 000\ \text{L} \times \left(\frac{0.0125\ \text{atm}}{1.0\ \text{atm}} \right)$$

Solve
- Multiply and divide the values and units.

$$V = \frac{22\ 000\ \text{L}}{} \frac{0.0125\ \cancel{\text{atm}}}{1.0\ \cancel{\text{atm}}} = \frac{22\ 000\ \text{L} \times 0.0125}{1.0} = 275\ \text{L}$$

Check
- As expected, the volume is less at sea level because the gas is compressed.

Two liters of air at atmospheric pressure are compressed into the 0.45-L canister of a warning horn. If its temperature remains constant, what is the pressure of the compressed air?

Analyze • The initial pressure of the air that was forced into the canister was, of course, 1 atm. Because the volume of air is reduced, its pressure increases. Multiply the pressure by the factor with a value greater than 1.

Set Up • $P = 1.00 \text{ atm} \times \left(\dfrac{2.0 \text{ L}}{0.45 \text{ L}} \right)$

Solve • $P = \dfrac{1.00 \text{ atm}}{} \left| \dfrac{2.0 \cancel{L}}{0.45 \cancel{L}} \right. = \dfrac{1.00 \text{ atm} \times 2.0}{0.45} = 4.4 \text{ atm}$

Check • Estimate and reason: Because the volume of the air was reduced from 2 L to about half a liter, the volume changed by the factor $\frac{0.5}{2}$ or about $\frac{1}{4}$. Thus, the pressure changed by a factor of about $\frac{4}{1}$. The final pressure, 4.4 atm, is about four times as large as the initial pressure, 1.00 atm.

PRACTICE PROBLEMS

For more practice with solving problems, see Supplemental Practice Problems, Appendix B.

Assume that the temperature remains constant in the following problems.

6. Bacteria produce methane gas in sewage-treatment plants. This gas is often captured or burned. If a bacterial culture produces 60.0 mL of methane gas at 700.0 mm Hg, what volume would be produced at 760.0 mm Hg?

7. At one sewage-treatment plant, bacteria cultures produce 1000 L of methane gas per day at 1.0 atm pressure. What volume tank would be needed to store one day's production at 5.0 atm?

8. Hospitals buy 400-L cylinders of oxygen gas compressed at 150 atm. They administer oxygen to patients at 3.0 atm in a hyperbaric oxygen chamber. What volume of oxygen can a cylinder supply at this pressure?

9. If the valve in a tire pump with a volume of 0.78 L fails at a pressure of 9.00 atm, what would be the volume of air in the cylinder just before the valve fails?

10. The volume of a scuba tank is 10.0 L. It contains a mixture of nitrogen and oxygen at 290.0 atm. What volume of this mixture could the tank supply to a diver at 2.40 atm?

11. A 1.00-L balloon is filled with helium at 1.20 atm. If the balloon is squeezed into a 0.500-L beaker and doesn't burst, what is the pressure of the helium?

Hyperbaric Oxygen Chambers

Hyperbaric oxygen (HBO) chambers have been used since the 1940s. They were used then by the Navy to treat divers with decompression sickness. In the last 20 years, researchers have shown that HBO treatment has many more medical applications. One of these applications is healing bone and muscle injuries.

Healing Sports Injuries

Several professional football teams have each acquired HBO units. The units have been dubbed *space capsules* because of their appearance. A player with a severe sprain, ordinarily dooming him to warm the bench for several weeks, may instead spend three one-hour sessions in the

HBO. The device compresses the air to 3 atm of pressure around the injured player, who is breathing pure oxygen through a mask. The higher air pressure in the HBO forces more oxygen to dissolve in the bloodstream, speeding up the flow of oxygen to 15 times the normal rate. The abnormal oxygen flow causes the blood vessels in the injured muscles to constrict, limiting swelling in that area. In two days, the injured player may walk without crutches. By the end of the week, the player could be back on the field.

After a Heart Attack

HBO therapy is also being used along with clot-dissolving drugs to save heart muscle and improve the quality of life for patients who have suffered heart attacks. In a study of heart-attack patients, half were given t-PA, a drug that prevents clots. The other half received the same drug, followed by two hours of HBO treatment. The patients who received the drug alone had lingering chest pains for an average of 10.75 hours. Those who had HBO therapy experienced pains for only 4.5 hours. In addition, normal electrical activity was restored sooner in the hearts of the patients who were in the hyperbaric chamber. Their electrocardiograms were normal in ten to 15 minutes—half the time it took for the other group to approach that goal.

Breath of Fresh Air for Premature Babies

One of the most important uses of HBO is to treat babies with hyaline-membrane disease. These babies suffer from respiratory distress soon after birth because the alveoli in their lungs fail to inflate. HBO therapy has increased the chances that the lungs will normalize so that these babies can survive.

DISCUSSING THE TECHNOLOGY

1. **Thinking Critically** What two factors cause the remarkable effects of HBO? Explain how.

2. **Hypothesizing** How might HBO be effective for treating second- and third-degree burns over large portions of the body?

Charles's Law: Temperature and Volume

You may have observed the beautiful patterns and graceful gliding of a hot-air balloon, but what happens to a balloon when it is cold? **Figure 11.10** shows some dramatic effects of cooling and warming a gas-filled balloon.

The French scientist Jacques Charles (1746-1823) didn't have liquid nitrogen, but he was a pioneer in hot-air ballooning. He investigated how changing the temperature of a fixed amount of gas at constant pressure affected its volume. The relationship Charles found can be demonstrated, as shown in **Figure 11.11.**

Figure 11.10

Cooling and Warming a Gas-Filled Balloon
At 77 K, the nitrogen in the beaker is so cold it is a liquid. It rapidly cools the air-filled balloons, shrinking them by reducing the volume of the balloons. When the balloons are removed from the beaker and the temperature of the air inside rises to room temperature, the balloons expand.

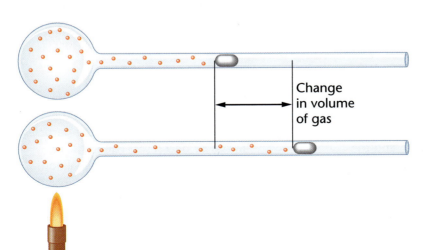

Figure 11.11

Demonstrating Charles's Law
The mercury plug is free to move back and forth in the horizontal tube because the end of the glass tube is open to the atmosphere. The pressure of the gas inside the bulb is always equal to the atmospheric pressure. The distance the mercury plug moves right or left measures the increase or decrease in the volume of gas as it is heated or cooled.

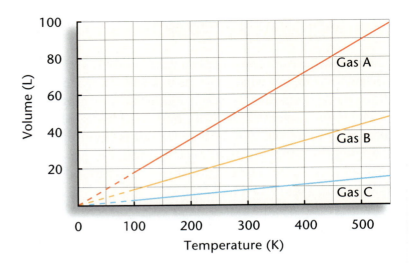

Figure 11.12

Volume and Temperature for Three Gases
The three straight lines show that the volume of each gas is directly proportional to its Kelvin temperature. The solid part of each line represents actual data. Part of each line is dashed because, as you recall from Chapter 10, when the temperature of a gas falls below its boiling point, a gas condenses to a liquid.

Charles's law states that at constant pressure, the volume of a gas is directly proportional to its Kelvin temperature, as shown in the graph in **Figure 11.12.** The straight lines for each gas indicate that volume and temperature are in direct proportions. For example, if the Kelvin temperature doubles, the volume doubles, and if the Kelvin temperature is halved, the volume is halved.

Kinetic Explanation of Charles's Law

Why did the air in the balloons expand when heated and contract when cooled? Study **Figure 11.13** to learn how the kinetic theory of matter explains Charles's law.

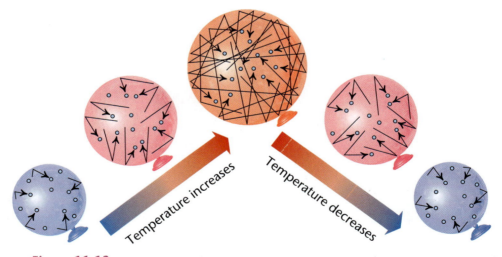

Figure 11.13

Modeling Charles's Law
▲ When the balloon is heated, the temperature of the air inside increases, and the average kinetic energy of the particles in the air also increases. They exert more force on the balloon, but the pressure inside does not rise above the original pressure because the balloon expands.

▲ When the balloon cools, the temperature of the air inside falls and the average kinetic energy of the particles in the air decreases. The particles move slower and strike the balloon less often and with less force. The balloon contracts and the pressure of the air inside the balloon continues to balance the pressure of the atmosphere.

A balloon is filled with 3.0 L of helium at 22°C and 760 mm Hg. It is then placed outdoors on a hot summer day when the temperature is 31°C. If the pressure remains constant, what will the volume of the balloon be?

Analyze

- Because the volume of a gas is proportional to its Kelvin temperature, you must first express the temperatures in this problem in kelvins. As in Chapter 10, add 273 to the Celsius temperature to obtain the Kelvin temperature.

> **Problem-Solving HINT**
>
> Remember that gas volumes and pressures are proportional to temperature only if the temperature is expressed in kelvins.

$$T_K = T_C + 273$$
$$T_K = 22 + 273 = 295 \text{ K}$$
$$T_K = 31 + 273 = 304 \text{ K}$$

Because the temperature of the helium increases from 295 K to 304 K, its volume increases in direct proportion. The temperature increases by the factor 304 K/295 K. Therefore, the volume increases by the same factor.

Set Up

- $V = 3.0 \text{ L} \times \left(\dfrac{304 \text{ K}}{295 \text{ K}} \right)$

Solve

- $V = \dfrac{3.0 \text{ L}}{} \dfrac{304 \text{ K}}{295 \text{ K}} = \dfrac{3.0 \text{ L} \times 304}{295} = 3.1 \text{ L}$

Check

- Does the answer have volume units? Does the volume increase as expected? Because the answer to both questions is yes, this solution is reasonable. Check your calculations to make sure your answer is correct.

Supplemental Problems

For more practice with solving problems, see Supplemental Practice Problems, Appendix B.

PRACTICE PROBLEMS

Assume that the pressure remains constant in the following problems.

12. A balloon is filled with 3.0 L of helium at 310 K and 1 atm. The balloon is placed in an oven where the temperature reaches 340 K. What is the new volume of the balloon?

13. A 4.0-L sample of methane gas is collected at 30.0°C. Predict the volume of the sample at 0°C.

14. A 25-L sample of nitrogen is heated from 110°C to 260°C. What volume will the sample occupy at the higher temperature?

15. The volume of a 16-g sample of oxygen is 11.2 L at 273 K and 1.00 atm. Predict the volume of the sample at 409 K.

16. The volume of a sample of argon is 8.5 mL at 15°C and 101 kPa. What will its volume be at 0.00°C and 101 kPa?

Combined Gas Law

You know that according to Boyle's law, if you double the volume of a gas while keeping the temperature constant, the pressure falls to half its initial value. You also know that according to Charles's law, if you double the Kelvin temperature of a gas while keeping the pressure constant, the volume doubles. What do you think would happen to the pressure if you doubled the volume and doubled the temperature? Suppose, like Robert Boyle and Jacques Charles, you were to investigate how gases behave. You would conduct experiments with gases just as they did. You would measure temperature, pressure, and volume for a sample of gas; expand the gas to twice the volume; raise the temperature to twice as high; and then measure the pressure. Following accepted principles of scientific methods, you would make many such experiments. You might choose to triple the volume and the temperature or change the volume and temperature by half or less to get a wide range of data. You might then use a computer to help make a graph of your data and look for relationships among your three variables.

Another approach might be to first double the volume while keeping the temperature constant or to double the volume and temperature and then measure the pressure. In what other ways could you compare relationships among temperature, pressure, and volume? Any one of the variables could be kept constant while varying another and measuring the effect on the third. You would find that doubling the volume first and the temperature second has exactly the same result as doubling the temperature first and the volume second. These results should not be surprising when you remember that you can read the curve in the graph for Boyle's law in either direction. Whether the gas was expanding or contracting, if you knew its volume you could determine its pressure. If you knew its pressure, you could determine its volume. One application of these gas laws can be seen at hot-air balloon events, **Figure 11.14.**

Figure 11.14

Flying High

Hot-air ballooning is a popular sport all around the world. The height of the balloon over the ground is controlled by varying the temperature of the air within the balloon.

Because Boyle's law of gases and Charles's law of gases are equally valid, it is possible to determine one of the three variables, regardless of the order in which the other two are changed. If you doubled the volume and the temperature at the same time, you would get exactly the same result as if you had doubled one first, then the other. For example, suppose you had 3 L of gas at 200 K and 1 atm. If you double the volume, according to Boyle's law, the pressure falls to 0.5 atm. If you then double the temperature, according to Charles's law, the pressure is increased to twice as high again, or 1 atm. Note that doubling the volume and then doubling the temperature brings the sample back to its initial pressure. The effect on the pressure of doubling the volume and doubling the temperature offset each other because pressure is inversely proportional to volume but directly proportional to temperature.

The combination of Boyle's law and Charles's law is called the **combined gas law.** The factors are the same as in the previous sample problems, but you may have more than one factor in a problem because more than one quantity may vary. The set of conditions 0.00°C and 1 atm is so often used that it is called **standard temperature and pressure** or **STP.**

SAMPLE PROBLEM 6 Determining Volumes at STP

A 154-mL sample of carbon dioxide gas is generated by burning graphite in pure oxygen. If the pressure of the generated gas is 121 kPa and its temperature is 117°C, what volume would the gas occupy at standard temperature and pressure, STP?

Analyze
- Reducing the pressure from 121 kPa to 101 kPa increases the volume of the carbon dioxide gas. Therefore, Boyle's law gives the factor 121 kPa/101 kPa. Because this factor is greater than 1, when it is multiplied by the volume, the volume increases. Cooling the gas from 117°C to 0.00°C reduces the volume of gas. To apply Charles's law, you must express both temperatures in kelvins.

$$T_K = T_C + 273 \qquad\qquad T_K = T_C + 273$$
$$= 117 + 273 \qquad\qquad = 0.00 + 273$$
$$= 390 \text{ K} \qquad\qquad = 273 \text{ K}$$

Because the temperature decreases, the volume decreases. Charles's law gives the factor 273 K/390 K, which is less than 1. Therefore, multiplying the volume by this factor decreases the volume.

Set Up
- Multiply the volume by these two factors.

$$V = 154 \text{ mL} \times \left(\frac{121 \text{ kPa}}{101 \text{ kPa}} \right) \times \left(\frac{273 \text{ K}}{390 \text{ K}} \right)$$

Solve
- The combined gas law equation is solved in the following steps.

$$V = \frac{154 \text{ mL} \mid 121 \text{ kPa} \mid 273 \text{ K}}{\mid 101 \text{ kPa} \mid 390 \text{ K}} =$$

$$\frac{154 \text{ mL} \times 121 \times 273}{101 \times 390} = 129 \text{ mL}$$

• Estimate to see whether the answer is reasonable. The reduction in pressure would expand the volume by a factor of about 12/10. Cooling the gas would contract it by a factor of about 7/10. Both factors together would alter the volume by a factor of about 84/100, which is less than 1. The final volume should be less than the initial volume. The final volume, 129 mL, is less than 154 mL, the initial volume.

PRACTICE PROBLEMS

Supplemental Problems

For more practice with solving problems, see Supplemental Practice Problems, Appendix B.

17. A 2.7-L sample of nitrogen is collected at 121 kPa and 288 K. If the pressure increases to 202 kPa and the temperature rises to 303 K, what volume will the nitrogen occupy?

18. A chunk of subliming carbon dioxide (dry ice) generates a 0.80-L sample of gaseous CO_2 at 22°C and 720 mm Hg. What volume will the carbon dioxide gas have at STP?

The Law of Combining Gas Volumes

When water decomposes into its elements—hydrogen and oxygen gas—the volume of hydrogen produced is always twice the volume of oxygen produced. Because matter is conserved, in the reverse synthesis reaction, the volume of hydrogen gas that reacts is always twice the volume of oxygen gas.

Experiments with many other gas reactions show that volumes of gases always react in ratios of small whole numbers. **Figure 11.15** presents the combining volumes for a synthesis reaction and a decomposition reaction. The observation that at the same temperature and pressure, volumes of gases combine or decompose in ratios of small whole numbers is called the **law of combining gas volumes.**

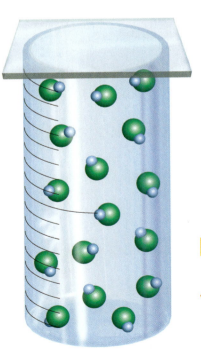

Figure 11.15

Comparing Volumes in Gas Reactions
When two liters of hydrogen chloride gas decompose to form hydrogen gas and chlorine gas, equal volumes of hydrogen gas and chlorine gas are formed—1 L of each. The ratio of hydrogen to chlorine is 1 to 1, and the ratio of volumes of hydrogen to hydrogen chloride is 1/2 to 1 or 1 to 2, the same as the ratio of chlorine to hydrogen chloride. Consider the reverse reaction—the composition of hydrogen gas with chlorine gas to form hydrogen chloride gas. What is the ratio of the reactants and the ratio of the product to each reactant?

2 liters HCl

Popping Corn

You recognize the smell, and that characteristic popping noise is a dead giveaway. Popcorn! What causes the kernels of popcorn to go through explosive changes?

History of popcorn The ancestor of the popcorn you eat was cultivated in the New World by Native Americans more than 7000 years ago. Popcorn was used as food, as decoration, and in religious ceremonies.

Popcorn kernels Popcorn kernels are extremely small and hard. They are wrapped in a tough, shell-like covering called the hull. The hull protects the embryo and its food supply. This food supply is starch located within the endosperm. Each kernel also contains a small amount of water.

Exploding the kernel When heated to about 204°C, the water in the kernel turns to steam. The expansion of the steam rips the remarkably tough hull with an explosive force, and the popcorn bursts open to 30 to 40 times its original size. The heat released by the steam bakes the starch into the fluffy product people like to eat.

Water content The amount of water in the kernel is an important aspect of popping. Food chemists have found that the kernel must contain about 13.5 percent water by mass to pop properly.

Exploring Further

1. **Hypothesizing** Suggest why too much or too little water present in the kernel greatly increases the number of unpopped kernels.

2. **Applying** Why is popcorn better stored in the freezer or refrigerator rather than on the shelf at room temperature?

3. **Acquiring Information** More than 1000 varieties of corn are grown in the world today. Write a report about as many uses for corn and corn products as you can.

Chemistry Online

To learn more about the mechanics of popcorn cooking, visit the Chemistry Web site at chemistryca.com

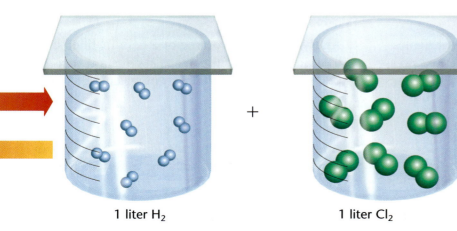

1 liter H_2 +   1 liter Cl_2

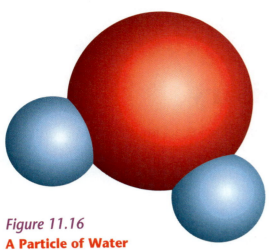

Figure 11.16

A Particle of Water
Avogadro determined that water is composed of particles.

In the synthesis of water, two volumes of hydrogen react with one of oxygen, but only two volumes of water vapor are formed. You might expect three. In the reverse reaction, which is the decomposition of water, two volumes of water vapor yield one volume of oxygen gas and two volumes of hydrogen gas. That mysterious third volume appears again. How can we account for it? Amedeo Avogadro (1776-1856), an Italian physicist, made the same observations and asked the same question. He first noticed that when water forms oxygen and hydrogen, two volumes of gas become three volumes of gas, and he hypothesized that water vapor consisted of particles, as shown in **Figure 11.16.** Each particle in the water vapor broke up into two hydrogen parts and one oxygen part. Two volumes of hydrogen and one volume of oxygen formed because there were twice as many hydrogen particles as oxygen particles. Today, we can show this by doing something that Avogadro could not do—write the chemical equation for the formation of water.

$$2H_2(g) + O_2(g) \rightarrow 2H_2O(g)$$

Avogadro was the first to interpret the law of combining volumes in terms of interacting particles. He reasoned that the volume of a gas at a given temperature and pressure must depend on the number of gas particles. **Avogadro's principle** states that equal volumes of gases at the same temperature and pressure contain equal numbers of particles.

Connecting Ideas

Knowing that equal volumes of gases at the same temperature and pressure have the same number of particles is the first step toward knowing how many particles are present in a sample of a substance. Knowing the number of particles in a reactant and the ratio in which the particles interact enables us to predict the amount of a product. This important number has been estimated.

> *Fact of the* **MATTER**
>
> Because of his observations of reactions involving oxygen, nitrogen, and hydrogen gases, Avogadro was the first to suggest that these gases were made up of diatomic molecules.

SECTION REVIEW

Understanding Concepts

1. The gas laws deal with the following variables—the number of gas particles, temperature, pressure, and volume. Which of these are held constant in Boyle's law, in Charles's law, and in the combined gas law?

2. Explain how an air mattress supports the weight of a person lying on it.

3. Why is it important to determine the volume of a gas at STP?

Thinking Critically

4. **Analyzing** If the volume of a helium-filled balloon increases by 20 percent at constant temperature, what will the percentage change in the pressure be?

Applying Chemistry

5. **Aerosol Cans** Use the kinetic theory to explain why pressurized cans carry the message "Do not incinerate."

---

chemistryca.com/self_check_quiz

REVIEWING MAIN IDEAS

11.1 Gas Pressure

- The pressure of a gas at constant temperature and volume is directly proportional to the number of gas particles.

- The volume of a gas at constant temperature and pressure is directly proportional to the number of gas particles.

- At sea level, the pressure exerted by gases of the atmosphere equals one standard atmosphere (1 atm).

- Gas pressure can be expressed in the following units: atmospheres (atm), millimeters of mercury (mm Hg), inches of mercury (in. Hg), pascals (Pa), and kilopascals (kPa).

- The factor label method is a simple tool for converting measurements from one unit to another.

11.2 The Gas Laws

- Boyle's law states that the pressure and volume of a confined gas are inversely proportional.

- Charles's law states that the volume of any sample of gas at constant pressure is directly proportional to its Kelvin temperature.

- According to the combined gas law, when pressure and temperature are both changing, both Boyle's and Charles's laws can be applied independently.

- The law of combining gas volumes states that in chemical reactions involving gases, the ratio of the gas volumes is a small whole number.

- Avogadro's principle states that equal volumes of gases at the same temperature and pressure contain equal numbers of particles.

Vocabulary

For each of the following terms, write a sentence that shows your understanding of its meaning.

Avogadro's principle
barometer
Boyle's law
Charles's law
combined gas law
factor label method
kilopascal (kPa)
law of combining gas volumes
pascal (Pa)
standard atmosphere (atm)
standard temperature and pressure, STP

UNDERSTANDING CONCEPTS

1. Name three ways to increase the pressure inside an oxygen tank.

2. What Celsius temperature corresponds to absolute zero?

3. What physical conditions are specified by the term STP?

4. What factors affect gas pressure?

5. Convert the pressure measurement 547 mm Hg to the following units.

   a) psi
   b) in. of Hg
   c) kPa
   d) atm

APPLYING CONCEPTS

6. The passenger cabin of an airplane is pressurized. Explain what this term means and why it is done.

7. Explain why balloons filled with helium rise in air but balloons filled with carbon dioxide sink.

8. At 1250 mm Hg and 75°C, the volume of a sample of ammonia gas is 6.28 L. What volume would the ammonia occupy at STP?

9. The volume of a gas is 550 mL at 760 mm Hg. If the pressure is reduced to 380 mm Hg and the temperature is constant, what will be the new volume?

10. A 375-mL sample of air at STP is heated at constant volume until its pressure increases to 980 mm Hg. What is the new temperature of the sample?

11. A sample of neon gas has a volume of 822 mL at 160°C. The sample is cooled at constant pressure to a volume of 586 mL. What is its new temperature?

12. The volume of a gas is 1.5 L at 27°C and 1 atm. What volume will the gas occupy if the temperature is raised to 127°C at constant pressure?

13. A 50-mL sample of krypton gas at STP is cooled to −73°C at constant pressure. What will be the volume?

14. How many liters of hydrogen will be needed to react completely with 10 L of oxygen in the composition reaction of hydrogen peroxide?

Chemistry and Technology

15. Compare the amount of oxygen you would receive in an HBO chamber with the amount you would receive breathing ordinary air.

Everyday Chemistry

16. Explain why the density of a popped kernel of popcorn is less than that of an unpopped kernel.

Earth Science Connection

17. Would chlorine gas be suitable for weather balloons? Explain.

THINKING CRITICALLY

Comparing and Contrasting

18. ChemLab How are inferences made about the relationship between volume and pressure in this chapter's ChemLab similar to inferences made from observing the apparatus described in **Figure 11.6**?

Observing and Inferring

19. MiniLab 1 How might you account for the possible popping of the bag containing the subliming dry ice?

Applying Concepts

20. MiniLab 2 Explain how air pressure helps you transfer a liquid with a pipet.

Making Predictions

21. Use Boyle's law to predict the change in density of a sample of air if its pressure is increased.

CUMULATIVE REVIEW

22. What are allotropes? Name three elements that occur as allotropes. (Chapter 5)

23. Can an atom of an element have six electrons in the $2p$ sublevel? (Chapter 7)

24. Classify the following bonds as ionic, covalent, or polar covalent using electronegativity values. (Chapter 9)

a) NH   c) BH

b) BaCl   d) NaI

SKILL REVIEW

25. Factor Label Method In a laboratory, 400 L of methane are stored at 600 K and 1.25 atm. If the methane is forced into a 200-L tank and cooled to 300 K, how does the pressure change?

WRITING IN CHEMISTRY

26. Modern manufacturing applies the properties of gases in many ways. List three examples of things you see and use every day that apply gas properties. Include at least one household product. Write a short explanation of the property used in each example.

PROBLEM SOLVING

27. A sample of argon gas occupies 2.00 L at −33.0°C and 1.50 atm. What is its volume at 207°C and 2.00 atm?

28. A welding torch requires 4500 L of acetylene gas at 2 atm. If the acetylene is supplied by a 150-L tank, what is the pressure of the acetylene?

chemistryca.com/chapter_test

Standardized Test Practice

1. Which of the following units is equivalent to 1.00 atm?

   a) 760 mm Hg
   b) 14.7 psi
   c) 101.3 kPa
   d) all of the above

2. 18.7 psi equals

   a) 0.36 kPa.
   b) 2.7 kPa.
   c) 128.9 kPa.
   d) 966.8 kPa.

Interpreting Graphs: Use the graph to answer questions 3–4.

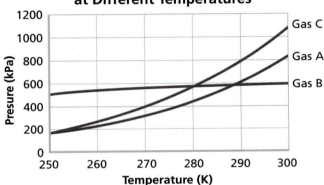

Pressures of Three Gases at Different Temperatures

3. The graph shows that

   a) as temperature increases, pressure decreases.
   b) as pressure decreases, volume decreases.
   c) as temperature decreases, moles decrease.
   d) as pressure decreases, temperature decreases.

4. The predicted pressure of Gas B at 310 K is

   a) 260 kPa.
   b) 620 kPa.
   c) 1000 kPa.
   d) 1200 kPa.

5. A sample of argon gas is compresses into a volume of 0.712 L by a piston exerting a pressure of 3.92 atm of pressure. The piston is slowly released until the pressure of the gas is 1.50 atm. The new volume of the gas is

   a) 0.272 L.
   b) 3.67 L.
   c) 1.8 L.
   d) 4.19 L.

6. Charles's Law states that the volume of a gas is

   a) inversely proportional to the temperature of the gas.
   b) directly proportional to the temperature of the gas.
   c) inversely proportional to the pressure of the gas.
   d) directly proportional to the pressure of the gas.

7. While it is on the ground, a blimp is filled with 5.66×10^6 L of He gas. The pressure inside the grounded blimp, where the temperature is 25°C, is 1.10 atm. Modern blimps are non-rigid, which means their volume is changeable. If the pressure inside the blimp remains the same, what will be the volume of the blimp at a height of 2300 m, where the temperature is 12°C?

   a) 5.66×10^6 L
   b) 2.72×10^6 L
   c) 5.40×10^6 L
   d) 5.92×10^6 L

8. When the temperature of the air inside a basketball is increased, the

   a) pressure of the air inside the ball increases.
   b) kinetic energy of the air particles decreases.
   c) the volume of the air increases.
   d) pressure of the air inside the ball decreases.

9. Which of the following conditions is STP?

   a) 0.00°C
   b) 760 mm Hg
   c) 273 K
   d) all of the above

10. A volume of gas has a pressure of 1.4 atm. What would the pressure be if you doubled the volume and reduced the temperature in kelvins by half?

    a) 0.35 atm
    b) 1.4 atm
    c) 2.8 atm
    d) 5.6 atm

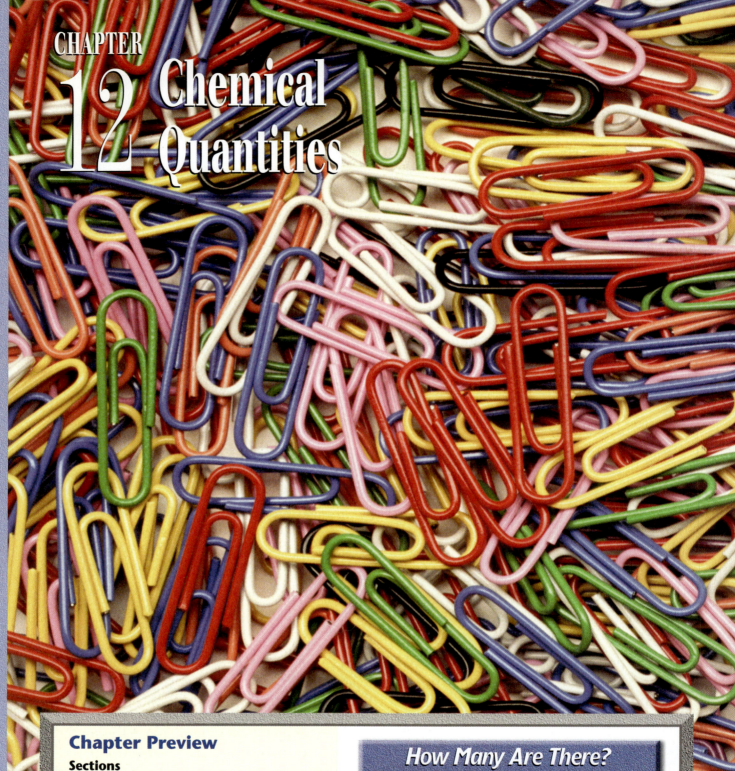

CHAPTER
12 Chemical Quantities

Chapter Preview

Sections

12.1   Counting Particles of Matter
  MiniLab 12.1 Determining Number
                 Without Counting

12.2   Using Moles
  MiniLab 12.2 Bagging the Gas
  ChemLab   Analyzing a Mixture

How Many Are There?

When chemical reactions occur, the products are very predictable. Determining the quantity of products produced is easier than determining the number of each color of the paper clips in this photo.

Start-up Activities

Launch Lab

How much is a mole?
Counting large numbers of items is easier when you use counting units like the dozen. Chemists use a counting unit called the mole.

Materials
- centimeter ruler
- paper clip

Procedure
1. Measure the length of a paper clip to the nearest 0.1 cm.
2. If a mole is 6.02×10^{23} items, how far will a mole of paper clips, placed end to end lengthwise, reach into space?

Analysis
How many light-years (ly) would the paper clips extend into space? (1 light-year = 9.46×10^{15} m). How does the distance you calculated compare with the following astronomical distances: nearest star (other than the sun) = 4.3 ly, center of our galaxy = 30,000 ly, nearest galaxy = 2×10^6 ly?

What I Already Know

Review the following concepts before studying this chapter.
Chapter 4: the types of compounds
Chapter 5: naming compounds; writing chemical formulas
Chapter 6: writing and balancing chemical reactions
Chapter 10: the factor label method; volume, pressure, and temperature relationships of gases

Reading Chemistry

Go through the chapter, and write down the sections heads. As you read, jot down several key points or vocabulary under each section head. Also, for each section, write down any questions you form about the topics.

Preview this chapter's content and activities at **chemistryca.com**

Counting Particles of Matter

Louis Staffilino of Dillonvale, Ohio, saved pennies for 65 years. When he deposited them in a bank in 1994, he had 40 large drums of pennies. His wealth represented an enormous counting job for the bank teller.

Just as a bank teller counts coins and bills, a chemist counts atoms, molecules, and formula units of substances. Unlike a bank teller, a chemist cannot count individual items because the particles of matter are too small and just too numerous. How do you determine the number of particles in a sample of matter without counting?

SECTION PREVIEW

Objectives

✓ **Compare and contrast** the mole as a number and the mole as a mass.

✓ **Relate** counting particles to weighing samples of substances.

✓ **Solve** stoichiometric problems using molar mass.

Review Vocabulary

Avogadro's principle: equal volumes of gases at the same temperature and pressure contain equal numbers of particles.

New Vocabulary

stoichiometry
mole
Avogadro constant
molar mass
molecular mass
formula mass

Stoichiometry

In Chapter 11, you learned that volumes of gases always combine in definite ratios. This observation, called the law of combining volumes, is based on measurements of the gas volumes. When Avogadro suggested that gases combine in fixed ratios because equal volumes of gases at the same temperature and pressure contain equal numbers of particles, he may have been thinking of particles rearranging themselves. Individual gas particles are so small that their rearranging cannot be observed, but the volumes of gases can be measured directly. Avogadro's principle is one of the earliest attempts to relate the number of particles in a sample of a substance to a direct measurement made on the sample.

Today, by using the methods of **stoichiometry,** we can measure the amounts of substances involved in chemical reactions and relate them to one another. For example, a sample's mass or volume can be converted to a count of the number of its particles, such as atoms, ions, or molecules.

Pennies are not small, but counting a drum full of pennies, like counting the number of particles in a gas, is a formidable task. Can we measure the pennies by grouping them in some conveniently large quantity and then counting the number of groups to find the total number of pennies in the drum? Can methods of stoichiometry help? **Figure 12.1** presents two ways to count large numbers of pennies. To count the pennies directly, you would have to handle every one—all 24 216 pennies. But you could also just make three measurements: the weight of the drum of pennies, the weight of the empty drum, and the weight of a group of 1000 pennies. Now, if you want to count atoms, what size group is most suitable? You will need a larger group than you used for the pennies—much larger than a thousand or even a million.

Atoms are so tiny that an ordinary-sized sample of a substance contains so many of these submicroscopic particles that counting them by grouping them in thousands would be unmanageable. Even grouping them by millions would not help. The group or unit of measure used to count numbers of atoms, molecules, or formula units of substances is the **mole** (abbreviated *mol*). The number of things in one mole is 6.02×10^{23}. This big number has a short name: the **Avogadro constant,** as illustrated in **Figure 12.2.**

Figure 12.1

Two Ways to Count Pennies
You could count the pennies directly. ▶

▲ You could calculate the number of pennies in a drum by using the mass of all the pennies, 70 125 g, and the mass of 1000 pennies. The mass of 1000 pennies is 2890.7 g. Use the ratio $\frac{1000 \text{ pennies}}{2890.7 \text{ g}}$ to find the total number of pennies.

$$70\ 125\ \cancel{g} \times \frac{1000 \text{ pennies}}{2890.7\ \cancel{g}} = 24\ 259 \text{ pennies}$$

The drum contains 24 259 pennies, worth $242.59.

Figure 12.2

One Mole Is a BIG Number.
If 6.02×10^{23} sheets of paper were placed in a stack, they would reach from Earth to the sun—more than a million times. The thickness of a sheet of paper is small, but an atom is much smaller. One mole of magnesium atoms is hardly a handful.

All kinds of submicroscopic particles can be conveniently counted using the Avogadro constant. There are 6.02×10^{23} carbon atoms in a mole of carbon, and there are 6.02×10^{23} carbon dioxide molecules in a mole of carbon dioxide. There are 6.02×10^{23} sodium ions in a mole of sodium ions. For that matter, there are 6.02×10^{23} eggs in a mole of eggs, but eggs are so large compared with the particles that interact in chemistry that you would have no need to estimate how many there are in a typical sample. Can you see a relationship here? For counting many small things, use a large unit of measure; for counting fewer things or larger things, use a small unit of measure, as shown in **Figure 12.3.**

Molar Mass

How many moles of methanol are in 500 g of methanol? Methanol is formed from CO_2 gas and hydrogen gas according to the balanced chemical equation below.

$$CO_2(g) + 3H_2(g) \rightarrow CH_3OH(g) + H_2O(g)$$

Suppose you wanted to produce 500 g of methanol. How many grams of CO_2 gas and H_2 gas would you need? How many grams of water would be produced as a by-product? Those are questions about the masses of reactants and products. But the balanced chemical equation shows that three molecules of hydrogen gas react with one molecule of carbon dioxide gas. The equation relates molecules, not masses, of reactants and products.

Like Avogadro, you need to relate the macroscopic measurements—the masses of carbon dioxide and hydrogen—to the number of molecules of methanol. To find the mass of carbon dioxide and the mass of hydrogen needed to produce 500 g of methanol, you first need to know how many molecules of methanol are in 500 g of methanol.

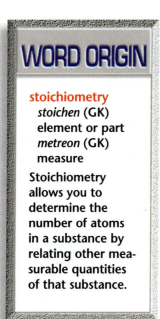

WORD ORIGIN

stoichiometry
stoichen (GK)
element or part
metreon (GK)
measure
Stoichiometry allows you to determine the number of atoms in a substance by relating other measurable quantities of that substance.

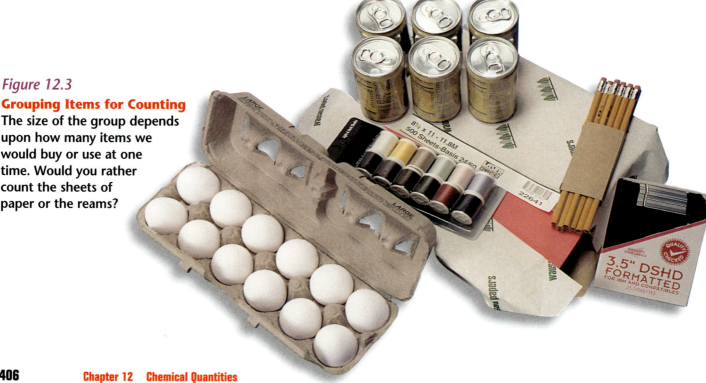

Figure 12.3
Grouping Items for Counting
The size of the group depends upon how many items we would buy or use at one time. Would you rather count the sheets of paper or the reams?

Remember the drum of pennies? By knowing the mass of one group of 1000 pennies and knowing the mass of all the pennies, you can find the number of groups of 1000 pennies. Then, finding the total number of pennies is an easy matter. You have a suitable unit of measure, the mole, and you know the mass of methanol you want to produce. But, you still need to know the mass of 1 mol of methanol molecules.

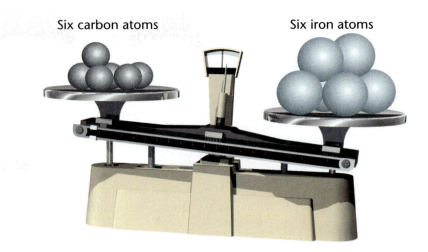

Six carbon atoms   Six iron atoms

Molar Mass of an Element

You know from Chapter 2 that average atomic masses of the elements are given on the periodic table. For example, the average mass of one iron atom is 55.8 u, where *u* means "atomic mass units." The atomic mass unit is defined so that the atomic mass of an atom of the most common carbon isotope is exactly 12 u, and the mass of 1 mol of the most common isotope of carbon atoms is exactly 12 g. The mass of 1 mol of a pure substance is called its **molar mass.** For example, the molar mass of iron is 55.847 g, and the molar mass of platinum is 195.08 g. Relative masses of elements are demonstrated in **Figure 12.4.** The molar mass is the mass in grams of the average atomic mass.

If an element exists as a molecule, remember that the particles in 1 mol of that element are themselves composed of atoms. For example, the element oxygen exists as molecules composed of two oxygen atoms, so a mole of oxygen molecules contains 2 mol of oxygen atoms. Therefore, the molar mass of oxygen molecules is twice the molar mass of oxygen atoms: $2 \times 16.00 \text{ g} = 32.00 \text{ g}$. **Figure 12.5** shows the molar masses of some elements.

Figure 12.4

Relative Masses of Iron and Carbon
An average iron atom is 4.65 times as heavy as an average carbon atom. Six iron atoms are 4.65 times as heavy as six carbon atoms. One mole of iron atoms is 4.65 times as heavy as 1 mol of carbon atoms.

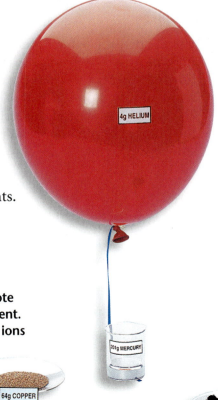

4g HELIUM

Figure 12.5

Molar Masses of Some Elements
Each sample contains 6.02×10^{23} atoms. Note that the masses of the samples are all different. Moles relate counts of atoms, molecules, or ions to mass because a molar mass always contains an Avogadro constant of particles.

32g SULFUR

64g COPPER

201g MERCURY

12g CARBON

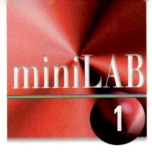

miniLAB

Determining Number Without Counting

Chemists and chemical engineers usually need to control the number of atoms, molecules, and ions in their reactions carefully. These particles are too small to be seen and too numerous to count, but their numbers can be determined by measuring their masses. Simulate this process by determining the approximate number of small, identical items in a large bag by weighing instead of counting.

Procedure

1. Count out ten items and weigh them. Record the mass.

2. Weigh and record the mass of an empty, self-sealing plastic storage bag.

3. Completely fill the bag with items and seal it.

4. Weigh and record its mass.

5. Develop and carry out an experiment asking yourself the following question, "How can I determine the number of buttons in the bag without opening it?"

Analysis

1. How many items are in the bag?

2. Explain how you determined that number.

SAMPLE PROBLEM **1** **Number of Atoms in a Sample of an Element**

The mass of an iron bar is 16.8 g. How many Fe atoms are in the sample?

Analyze
- Use the periodic table to find the molar mass of iron. The average mass of an iron atom is 55.8 u. Then the mass of 1 mol of iron atoms is 55.8 g.

Problem-Solving HINT

Remember that the units of molar mass are grams per mole, which can be used as a conversion factor.

Set Up
- To convert the mass of the iron bar to the number of moles of iron, use the mass of 1 mol of iron atoms as a conversion factor.

$$\frac{16.8 \text{ g Fe}}{} \left| \frac{1 \text{ mol Fe}}{55.8 \text{ g Fe}} \right.$$

Now, use the number of atoms in a mole to find the number of iron atoms in the bar.

$$\frac{16.8 \text{ g Fe}}{} \left| \frac{1 \text{ mol Fe}}{55.8 \text{ g Fe}} \right| \frac{6.02 \times 10^{23} \text{ Fe atoms}}{1 \text{ mol Fe}}$$

Solve
- Simplify the expression above.

$$\frac{16.8 \text{ g Fe}}{} \left| \frac{1 \text{ mol Fe}}{55.8 \text{ g Fe}} \right| \frac{6.02 \times 10^{23} \text{ Fe atoms}}{1 \text{ mol Fe}} =$$

$$\frac{16.8 \times 6.02 \times 10^{23} \text{ Fe atoms}}{55.8} = 1.81 \times 10^{23} \text{ Fe atoms}$$

Check
- Notice how the units of measure cancel to leave only Fe atoms.

Molar Mass of a Compound

As you learned in Chapter 4, covalent compounds are composed of molecules, and ionic compounds are composed of formula units. The **molecular mass** of a covalent compound is the mass in atomic mass units of one molecule. Its molar mass is the mass in grams of 1 mol of its molecules. The **formula mass** of an ionic compound is the mass in atomic mass units of one formula unit. Its molar mass is the mass in grams of 1 mol of its formula units. How to calculate the molar mass for ethanol, a covalent compound, and for calcium chloride, an ionic compound, is shown below.

Ethanol, C_2H_6O, a covalent compound

2 C atoms	2×12.0 u	=	24.0 u
6 H atoms	6×1.00 u	=	6.00 u
1 O atom	1×16.0 u	=	+16.0 u
molecular mass of C_2H_6O			46.0 u

mass of 1 mol C_2H_6O molecules	=	46.0 g
molar mass of ethanol		46.0 g/mol

Calcium chloride, $CaCl_2$, an ionic compound

1 Ca atom	1×40.1 u	=	40.1 u
2 Cl atoms	2×35.5 u	=	+71.0 u
formula mass of $CaCl_2$			111.1 u

mass of 1 mol $CaCl_2$ formula units	=	111.1 g
molar mass of calcium chloride		111.1 g/mol

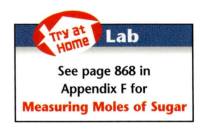

Try at Home **Lab**

See page 868 in Appendix F for **Measuring Moles of Sugar**

By making calculations with molar masses, you can find the number of molecules of methanol in your 500 g; the number of molecules of carbon dioxide and hydrogen that reacted; and the masses of carbon dioxide, hydrogen, methanol, and water in grams. **Figure 12.6** shows the molar masses of some compounds.

Figure 12.6

Molar Masses of Some Compounds
Each sample contains 6.02×10^{23} molecules of a covalent compound or 6.02×10^{23} formula units of an ionic compound. Each compound has its own molar mass.

Electronic Balances

Balances are devices used to measure mass. When a sample of a substance is placed on the balance pan, gravity on it pushes the pan down with a force proportional to the object's mass. A sensor uses an electromagnet to push the pan back to its original position. The electric current required to push the pan back to its original position is directly proportional to the gravitational force on the object on the pan. The mass of the object is displayed on a digital readout.

1. **The Null Position** The position of the balance is sensed by a light beam. When the beam is blocked, the balance is at the null position. The first step in using an electronic balance is to set the balance to zero so that the readout is zero when the pan is empty. This usually requires just the press of a button.

2. Most electronic balances allow the user to automatically tare the balance so that when an empty container is on the pan, the readout is zero.

3. **Weighing** When an object to be weighed is placed on the balance pan, the arm drops down and allows light to reach the detector, which sends an electrical signal to the control circuit.

4. The control circuit sends a correction current to an electromagnet under the balance pan. The electromagnet produces a magnetic force that pushes the pan up against gravity.

Null detector

Error signal

Servomotor

Correction current

Control circuit

Balance pan

Coil of electromagnet

5. The control circuit sends enough current to make the magnetic force exactly balance the gravitational force. The pan returns to the null position.

6. The current required to null the balance is proportional to the mass of the sample. The control circuit converts this current into a numerical value, a number of grams or kilograms, which can be seen in the readout on the balance.

Thinking Critically

1. What are some advantages of electronic balances over mechanical balances?

2. Why is it good practice to always re-zero an electronic balance before using it?

Asante Brass Weights

Imagine waking up in the morning after a rain and finding that gold has popped from the ground. Does it sound like a fairy tale? In the past, this was a common experience for people living in Ghana. Their land is an alluvial gold field. Besides picking up gold that has been bared by erosion and finding gold by panning streams, the Asante, one of about 35 language groups in Ghana, recovered gold by digging into sediment where smaller streams joined larger ones and by crushing gold-bearing quartz and rock and separating the gold by washing. Gold dust and small gold nuggets were their national currency.

Asante standard weights Although the Asante state was founded about 1700 A.D., as early as about 1400, the Asante created artistic, standardized weights. Their first weights were simple blocks, but later they made cubes, pyramids, triangular blocks, and rectangular blocks with carved designs. They cast natural objects like seeds, beetles, shells, and chicken feet. In the early 1700s, the carved designs on flattened shapes became more elaborate. While Europeans were colonizing America, the Asante were creating works of art with their weights.

Besides royal gold jewelry, they made artistic brass weights, spoons, and boxes. They also worked copper, zinc, tin, lead, and nickel. They cast objects in brass by the lost wax method. In this process, shapes were carved in wax, then encased in clay and baked. The hot wax was poured out, leaving a ceramic mold. Molten metal was poured into the mold. When the metal cooled, the mold was broken to remove the casting.

Today, Asante weights are collected as art forms—evidence of the advanced Asante culture of the past few centuries. Compared to the plain weights used in the United States, Asante weights add interest and beauty to everyday life. How enriching it must have been to use a 100-g porcupine or a 50-g headless fish to weigh gold dust for a purchase instead of paying with paper currency.

Connecting to Chemistry

1. **Hypothesizing** Do you think standard weights will soon be obsolete in scientific laboratories? Why?

2. **Applying** Why does gold form few compounds?

SAMPLE PROBLEM 2 Number of Formula Units in a Sample of a Compound

The mass of a quantity of iron(III) oxide is 16.8 g. How many formula units are in the sample?

Analyze

- Use the periodic table to calculate the mass of one formula unit of Fe_2O_3.

2 Fe atoms	2×55.8 u =	111.6 u
3 O atoms	3×16.0 u =	+ 48.0 u
formula mass of Fe_2O_3		159.6 u

Therefore, the molar mass of Fe_2O_3 (rounded off) is 160 g.

Set Up

- $$\frac{16.8\ \text{g}\ Fe_2O_3}{} \left| \frac{1\ \text{mol}\ Fe_2O_3}{160\ \text{g}\ Fe_2O_3} \right.$$

Now, multiply the number of moles of iron oxide by the number in a mole.

$$\frac{16.8\ \text{g}\ Fe_2O_3}{} \left| \frac{1\ \text{mol}\ Fe_2O_3}{160\ \text{g}\ Fe_2O_3} \right| \frac{6.02 \times 10^{23}\ Fe_2O_3\ \text{formula units}}{\text{mol}\ Fe_2O_3}$$

Solve

- $$\frac{16.8\ \cancel{\text{g}\ Fe_2O_3}}{} \left| \frac{1\ \cancel{\text{mol}\ Fe_2O_3}}{160\ \cancel{\text{g}\ Fe_2O_3}} \right| \frac{6.02 \times 10^{23}\ Fe_2O_3\ \text{formula units}}{1\ \cancel{\text{mol}\ Fe_2O_3}} =$$

$$\frac{16.8 \times 6.02 \times 10^{23}\ Fe_2O_3\ \text{formula units}}{160} =$$

$$6.32 \times 10^{22}\ Fe_2O_3\ \text{formula units}$$

Check

- The units of measure show that the calculation is set up correctly. The multiplication also checks.

PRACTICE PROBLEMS

1. Without calculating, decide whether 50.0 g of sulfur or 50.0 g of tin represents the greater number of atoms. Verify your answer by calculating.
2. Determine the number of atoms in each sample below.
   a) 98.3 g mercury, Hg
   b) 45.6 g gold, Au
   c) 10.7 g lithium, Li
   d) 144.6 g tungsten, W
3. Determine the number of moles in each sample below.
   a) 6.84 g sucrose, $C_{12}H_{22}O_{11}$
   b) 16.0 g sulfur dioxide, SO_2
   c) 68.0 g ammonia, NH_3
   d) 17.5 g copper(II) oxide, CuO

In the previous problems, you used the molar mass to convert a mass measurement to a number of moles. Now, you will learn to convert a number of moles to a mass measurement.

What mass of water must be weighed to obtain 7.50 mol of H_2O?

Analyze
- The molar mass of water is obtained from its molecular mass.

$$
\begin{array}{lll}
2 \text{ H atoms} & 2 \times 1.00 \text{ u} = & 2.00 \text{ u} \\
1 \text{ O atom} & 1 \times 16.0 \text{ u} = & \underline{16.0 \text{ u}} \\
\text{molecular mass of } H_2O & & 18.0 \text{ u}
\end{array}
$$

The molar mass of water is 18.0 g/mol.

Set Up
- Use the molar mass to convert the number of moles to a mass measurement.

$$
\frac{7.50 \text{ mol } H_2O}{} \Bigg| \frac{18.0 \text{ g } H_2O}{1 \text{ mol } H_2O}
$$

Solve

$$
\frac{7.5 \text{ mol } H_2O}{} \Bigg| \frac{18.0 \text{ g } H_2O}{1 \text{ mol } H_2O} = 7.50 \times 18.0 \text{ g } H_2O = 135 \text{ g } H_2O
$$

Check
- The final units are grams of H_2O, as expected. The multiplication also checks.

PRACTICE PROBLEMS

4. Determine the mass of the following molar quantities.

   a) 3.52 mol Si
   b) 1.25 mol aspirin, $C_9H_8O_4$
   c) 0.550 mol F_2
   d) 2.35 mol barium iodide, BaI_2

The concept of molar mass makes it easy to determine the number of particles in a sample of a substance by simply measuring the mass of the sample. The concept is also useful in relating masses of reactants and products in chemical reactions.

Supplemental Problems

For more practice with solving problems, see Supplemental Practice Problems, Appendix B.

SECTION REVIEW

Understanding Concepts

1. How is counting a truckload of pennies like counting the atoms in 10.0 g of aluminum? How is it different?

2. The average atomic mass of nitrogen is 14 times greater than the average atomic mass of hydrogen. Would you expect the molar mass of nitrogen gas, N_2, to be 14 times greater than the molar mass of hydrogen gas, H_2? Why?

3. Why is the following statement meaningless?

An industrial process requires 2.15 mol of a sugar-salt mixture.

Thinking Critically

4. **Analyzing** Determine the mass in grams of one average atomic mass unit.

Applying Chemistry

5. **Balances** If the mole is truly central to quantitative work in chemistry, why do balances in chemistry labs measure mass and not moles?

Using Moles

T he rose's distinctive scent is mostly the odor of one compound, geraniol. The balanced chemical equation for the formation of geraniol shows the ratios of carbon, hydrogen, and oxygen that interact. Although the particles are too small to be seen or weighed, a mole of carbon, hydrogen, or oxygen is large enough. Can you use the chemical equation to predict the masses of carbon, hydrogen, and oxygen that react and the mass of geraniol that is formed?

SECTION PREVIEW

Objectives

✓ **Predict** quantities of reactants and products in chemical reactions.

✓ **Determine** mole ratios from formulas for compounds.

✓ **Identify** formulas of compounds by using mass ratios.

Review Vocabulary

Mole: group or unit of measure used to count numbers of atoms, molecules, or formula units of a substance.

New Vocabulary

molar volume
ideal gas law
empirical formula

Using Molar Masses in Stoichiometric Problems

In Chapter 11, you saw how balanced chemical equations indicate the volume of gas required for a reaction or the volume of gas produced. Similarly, you can use balanced chemical equations and numbers of moles of each substance to predict the masses of reactants or products.

4 Predicting Mass of a Reactant

As you review this problem, note that nitrogen and hydrogen are related in terms of moles, not mass. In chemical reactions, when substances react, their particles react. Ammonia gas is synthesized from nitrogen gas and hydrogen gas according to the balanced chemical equation below.

$$N_2(g) + 3 H_2(g) \rightarrow 2 NH_3(g)$$

How many grams of hydrogen gas are required for 3.75 g of nitrogen gas to react completely?

Analyze
- The amount of hydrogen needed depends upon the number of nitrogen molecules present in 3.75 g and the mole ratio of hydrogen gas to nitrogen gas in the balanced chemical equation.

Set Up
- Find the number of moles of N_2 molecules by using the molar mass of nitrogen.

$$\frac{3.75 \text{ g } N_2}{} \left| \frac{1 \text{ mol } N_2}{28.0 \text{ g } N_2} \right.$$

To find the mass of hydrogen needed, first find the number of moles of H_2 molecules needed to react with all the moles of N_2 molecules. The balanced chemical equation shows that 3 mol of H_2 molecules react with 1 mol of N_2 molecules. Multiply the number of moles of N_2 molecules, as shown in *Set Up*, by this ratio.

$$\frac{3.75 \text{ g } N_2 \,|\, 1 \text{ mol } N_2 \,|\, 3 \text{ mol } H_2}{|\, 28.0 \text{ g } N_2 \,|\, 1 \text{ mol } N_2}$$

The units in the expression above simplify to moles of H_2 molecules. To find the mass of hydrogen, multiply the number of moles of hydrogen molecules by the mass of 1 mol of H_2 molecules, which is 2.00 g.

$$\frac{3.75 \text{ g } N_2 \,|\, 1 \text{ mol } N_2 \,|\, 3 \text{ mol } H_2 \,|\, 2.00 \text{ g } H_2}{|\, 28.0 \text{ g } N_2 \,|\, 1 \text{ mol } N_2 \,|\, 1 \text{ mol } H_2}$$

Solve

$$\frac{3.75 \text{ g } N_2 \,|\, 1 \text{ mol } N_2 \,|\, 3 \text{ mol } H_2 \,|\, 2.00 \text{ g } H_2}{|\, 28.0 \text{ g } N_2 \,|\, 1 \text{ mol } N_2 \,|\, 1 \text{ mol } H_2} =$$

$$\frac{3.75 \times 1 \times 3 \times 2.00 \text{ g } H_2}{28.0} = 0.804 \text{ g } H_2$$

Check • Check the multiplication to verify that the result is correct.

SAMPLE PROBLEM **5** **Predicting Mass of a Product**

What mass of ammonia is formed when 3.75 g of nitrogen gas react with hydrogen gas according to the balanced chemical equation below?

$$N_2(g) + 3 H_2(g) \rightarrow 2 NH_3(g)$$

Analyze • The amount of ammonia formed depends upon the number of nitrogen molecules present and the mole ratio of nitrogen and ammonia in the balanced chemical equation.

Set Up • As in Sample Problem 4, the number of moles of nitrogen molecules is given by the expression below.

> **Problem-Solving HINT**
>
> The balanced chemical equation is the source of conversion factors relating moles of one substance to moles of another substance.

$$\frac{3.75 \text{ g } N_2 \,|\, 1 \text{ mol } N_2}{|\, 28.0 \text{ g } N_2}$$

To find the mass of ammonia produced, first find the number of moles of ammonia molecules that form from 3.75 g of nitrogen. Use the mole ratio of ammonia molecules to nitrogen molecules to find the number of moles of ammonia formed.

$$\frac{3.75 \text{ g } N_2 \,|\, 1 \text{ mol } N_2 \,|\, 2 \text{ mol } NH_3}{|\, 28.0 \text{ g } N_2 \,|\, 1 \text{ mol } N_2}$$

Use the molar mass of ammonia, 17.0 g, to find the mass of ammonia formed.

$$\frac{3.75 \text{ g } N_2 \,|\, 1 \text{ mol } N_2 \,|\, 2 \text{ mol } NH_3 \,|\, 17.0 \text{ g } NH_3}{|\, 28.0 \text{ g } N_2 \,|\, 1 \text{ mol } N_2 \,|\, 1 \text{ mol } NH_3}$$

Solve

$$\frac{3.75 \text{ g N}_2}{28.0 \text{ g N}_2} \bigg| \frac{1 \text{ mol N}_2}{1 \text{ mol N}_2} \bigg| \frac{2 \text{ mol NH}_3}{1 \text{ mol N}_2} \bigg| \frac{17.0 \text{ g NH}_3}{1 \text{ mol NH}_3} =$$

$$\frac{3.75 \times 1 \times 2 \times 17.0 \text{ g NH}_3}{28.0} = 4.55 \text{ g NH}_3$$

Check
- Convince yourself that the factors are set up correctly and that the multiplication is correct.

PRACTICE PROBLEMS

Supplemental Problems

For more practice with solving problems, see Supplemental Practice Problems, Appendix B.

5. The combustion of propane, C_3H_8, a fuel used in backyard grills and camp stoves, produces carbon dioxide and water vapor.

$$C_3H_8(g) + 5O_2(g) \rightarrow 3CO_2(g) + 4H_2O(g)$$

What mass of carbon dioxide forms when 95.6 g of propane burns?

6. Solid xenon hexafluoride is prepared by allowing xenon gas and fluorine gas to react.

$$Xe(g) + 3F_2(g) \rightarrow XeF_6(s)$$

How many grams of fluorine are required to produce 10.0 g of XeF_6?

7. Using the reaction in Practice Problem 6, how many grams of xenon are required to produce 10.0 g of XeF_6?

Using Molar Volumes in Stoichiometric Problems

In Chapter 11, you used the law of combining volumes. Avogadro inferred from that law that equal volumes of gases contain equal numbers of particles. In terms of moles, Avogadro's principle states that equal volumes of gases at the same temperature and pressure contain equal numbers of moles of gases.

The **molar volume** of a gas is the volume that a mole of a gas occupies at a pressure of one atmosphere (equal to 101 kPa) and a temperature of 0.00°C. Under these conditions of STP, the volume of 1 mol of any gas is 22.4 L, as shown in **Figure 12.7**. Like the molar mass, the molar volume is used in stoichiometric calculations.

Figure 12.7

Molar Volumes of Gases
One mole of any gas at STP occupies 22.4 L. How large is that? It is the volume of a cube that is 28.2 cm on each edge. Each such volume contains 6.02×10^{23} atoms of a gaseous element or 6.02×10^{23} molecules of a molecular element or compound, but the mass of 1 mol is different for each element. One mole of helium floats because its mass is less than the mass of 22.4 L of air.

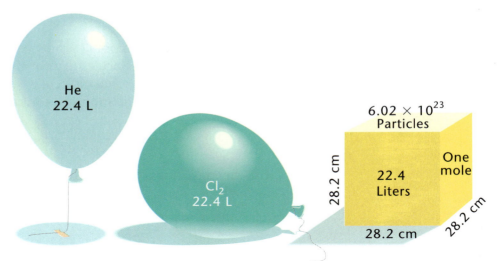

Everyday Chemistry

Air Bags

In 1990, on a Virginia hilltop, two cars collided in a head-on crash. Both drivers walked away with only minor injuries. A chemical reaction, along with seat belts, saved their lives. This was the first recorded head-on collision between two cars with air bags.

How air bags function Although air bags seem to inflate instantaneously, the process occurs in steps.

1. When a car collides with a rigid barrier at a speed of 12 mph or greater, two or three sensors on the front of the car send an electric current to fire the control unit 0.01 s after impact.

2. After 0.05 s, a chemical reaction in the stored air bag creates a gaseous product that inflates it and pushes open the cover on the steering wheel or on the passenger-side dash.

3. The driver or passenger strikes the inflated bag.

4. The bag deflates in 0.045 s as the gas escapes through holes at the base of the bag.

The chemical reactions Sodium azide (NaN_3) is the chemical that produces nitrogen gas to inflate the air bag. Sodium azide pellets, an igniter, inflator, and a tightly folded nylon air bag are stored under a breakaway cover in the steering wheel or dash. The igniter provides a current to decompose the sodium azide into nitrogen gas and sodium.

$$2NaN_3(s) \xrightarrow{\text{electricity}} 3N_2(g) + 2Na(s)$$

The sodium immediately reacts with iron(III) oxide in the pellet to form sodium oxide and iron.

$$6Na(s) + Fe_2O_3(s) \rightarrow 3Na_2O(s) + 2Fe(s)$$

The sodium oxide reacts with carbon dioxide and water vapor in the air to form sodium hydrogen carbonate.

$$Na_2O(s) + 2CO_2(g) + H_2O(g) \rightarrow 2NaHCO_3(s)$$

Reliability During a ten-year period, one car in a million *may* have an air-bag defect. Why are they so successful? There are no moving parts to wear out, all exposed components are tightly sealed, and the gold-plated electrical connectors corrode slowly.

Gas-volume relationships The driver's air bag requires 0.0650 m³ of nitrogen to inflate—no more, no less. The passenger's air bag needs 0.1340 m³. The pellet must have the exact amount of sodium azide needed to produce the correct amount of nitrogen. As with all expanding gases, pressure and temperature affect the amount of sodium azide needed. Because the nitrogen gas is formed in an explosion, it has to be cooled before it goes into the air bag.

Exploring Further

1. **Applying** If 130 g of sodium azide are needed for the driver's air bag, how much is needed for the passenger's bag? Explain.

2. **Inferring** What effect does the heat from the sodium azide reaction have on the pressure and volume of the nitrogen gas formed?

Chemistry Online

To find out more about air bags, visit the Chemistry Web site at chemistryca.com

In the space shuttle, exhaled carbon dioxide gas is removed from the air by passing it through canisters of lithium hydroxide. The following reaction takes place.

$$CO_2(g) + 2LiOH(s) \rightarrow Li_2CO_3(s) + H_2O(g)$$

How many grams of lithium hydroxide are required to remove 500.0 L of carbon dioxide gas at 101 kPa pressure and 25.0°C?

Analyze • In this problem, you use the molar volume to find the number of moles.

Set Up • The volume of gas at 25°C must be converted to a volume at STP.

$$V = 500.0 \text{ L CO}_2 \left(\frac{273K}{298K}\right) = 458 \text{ L CO}_2$$

Now, find the number of moles of CO_2 gas as below.

$$\frac{458 \text{ L CO}_2 \mid 1 \text{ mol CO}_2}{\mid 22.4 \text{ L CO}_2}$$

The chemical equation shows that the ratio of moles of LiOH to CO_2 is 2 to 1. Therefore, the number of moles of lithium hydroxide is given by the expression below.

$$\frac{458 \text{ L CO}_2 \mid 1 \text{ mol CO}_2 \mid 2 \text{ mol LiOH}}{\mid 22.4 \text{ L CO}_2 \mid 1 \text{ mol CO}_2}$$

To convert the number of moles of LiOH to mass, use its molar mass, 23.9 g/mol.

$$\frac{458 \text{ L CO}_2 \mid 1 \text{ mol CO}_2 \mid 2 \text{ mol LiOH} \mid 23.9 \text{ g LiOH}}{\mid 22.4 \text{ L CO}_2 \mid 1 \text{ mol CO}_2 \mid 1 \text{ mol LiOH}}$$

Solve •

$$\frac{458 \text{ L CO}_2 \mid 1 \text{ mol CO}_2 \mid 2 \text{ mol LiOH} \mid 23.9 \text{ g LiOH}}{\mid 22.4 \text{ L CO}_2 \mid 1 \text{ mol CO}_2 \mid 1 \text{ mol LiOH}} =$$

$$\frac{458 \times 2 \times 23.9 \text{ g LiOH}}{22.4} = 977 \text{ g LiOH}$$

Check • A mass of 1000 g of lithium hydroxide is about 40 mol of the compound. According to the chemical equation, about half as much, or 20 mol of CO_2, will be removed from the air. At STP, 20 mol of any gas occupy about 450 L, which will expand to about 500 L at 25°C.

PRACTICE PROBLEMS

8. What mass of sulfur must burn to produce 3.42 L of SO_2 at 273°C and 101 kPa? The reaction is $S(s) + O_2(g) \rightarrow SO_2(g)$.

9. What volume of hydrogen gas can be produced by reacting 4.20 g of sodium in excess water at 50.0°C and 106 kPa? The reaction is $2 \text{ Na} + 2 \text{ H}_2\text{O} \rightarrow 2 \text{ NaOH} + \text{H}_2$.

Gases are much less dense than solids, as illustrated by **Figure 12.8.**

Ideal Gas Law

Exactly how the pressure P, volume V, temperature T, and number of particles n of gas are related is given by the **ideal gas law** shown here.

$$PV = nRT$$

The value of the constant R can be determined using the definition of molar volume. At STP, 1 mol of gas occupies 22.4 L. Therefore, when $P = 101.3$ kPa, $V = 22.4$ L, $n = 1$ mol, and $T = 273.15$ K, the equation for the ideal gas law can be shown as follows.

$$101.3 \text{ kPa} \times 22.4 \text{ L} = 1 \text{ mol} \times R \times 273.15 \text{ K}$$

Now, we can solve for R.

$$R = \frac{101.3 \text{ kPa}}{} \left| \frac{22.4 \text{ L}}{1 \text{ mol}} \right| \frac{1}{273.15 \text{ K}} = \frac{8.31 \text{ kPa} \cdot \text{L}}{\text{mol} \cdot \text{K}}$$

Now, you can find the volume, pressure, temperature, and number of moles of a gas.

Figure 12.8
Explosives
Gases produced in powerful chemical reactions of explosives expand with great energy. One mole of a gas occupies much more space than 1 mol of any solid.

SAMPLE PROBLEM ❼ Using the Ideal Gas Law

How many moles are contained in a 2.44-L sample of gas at 25.0°C and 202 kPa?

Analyze • Solve the ideal gas law for n, the number of moles.

$$n = \frac{PV}{RT}$$

Set Up • $n = \dfrac{202 \text{ kPa} \times 2.44 \text{ L}}{\left(\dfrac{8.31 \text{ kPa} \cdot \text{L}}{\text{mol} \cdot \text{K}}\right) \times 298 \text{ K}}$

Solve • $n = \dfrac{202 \text{ kPa}}{} \left| \dfrac{2.44 \text{ L}}{} \right| \dfrac{1}{\dfrac{8.31 \text{ L} \cdot \text{kPa}}{\text{mol} \cdot \text{K}}} \left| \dfrac{1}{298 \text{ K}} \right. =$

$$\frac{202 \text{ kPa} \times 2.44 \text{ L} \times 1 \text{ mol} \cdot \text{K}}{8.31 \text{ L} \cdot \text{kPa} \times 298 \text{ K}} = \frac{202 \text{ mol} \times 2.44}{8.31 \times 298} = 0.199 \text{ mol}$$

Check • First, find the volume that 2.44 L of a gas would occupy at STP.

$$V = \frac{2.44 \text{ L}}{} \left| \frac{273 \text{ K}}{298 \text{ K}} \right| \frac{202 \text{ kPa}}{101 \text{ kPa}} = 4.47 \text{ L}$$

Then, find the number of moles in this volume.

$$\frac{4.47 \text{ L}}{} \left| \frac{1 \text{ mol}}{22.4 \text{ L}} \right. = 0.200 \text{ mol}$$

0.200 mol is close to the calculated value.

miniLAB 2

Bagging the Gas

Chemists and chemical engineers often need to determine amounts of reactants and products that will react efficiently and cost effectively. Use the molar volume to determine the amount of baking soda required to react with vinegar to yield just enough carbon dioxide to fill a one-quart, self-sealing plastic bag.

Procedure

1. Write the balanced equation for the reaction of baking soda (sodium hydrogen carbonate) and vinegar (acetic acid) that produces sodium acetate, water, and carbon dioxide.

2. Find the volume of a 1-quart, self-sealing plastic bag by filling it with water and then pouring the water into a graduated cylinder or measuring cup.

3. Calculate the mass of sodium hydrogen carbonate that will fill the bag with CO_2 gas when the compound reacts with excess acetic acid.

4. Put on an apron and goggles.

5. Weigh the calculated amount of sodium hydrogen carbonate and place it in a bottom corner of the bag. Use a plastic-coated twist tie to seal off this corner.

6. Pour about 60 mL of $1M$ acetic acid into the other bottom corner of the bag. Be careful not to allow the reactants to mix.

7. Squeeze the air from the bag and seal the zipper top.

8. Place the bag in a trash can or behind an explosion shield.

9. Release the twist tie, quickly mix the reactants, and allow the reaction to proceed.

Analysis

1. Show and explain the calculations that you used to determine the required mass of sodium hydrogen carbonate.

2. What mass of sodium hydrogen carbonate would be required to react with excess acetic acid to produce 20 000 L of carbon dioxide gas at STP for a water-treatment plant?

3. What would have happened in step 8 if the amount of acetic acid was insufficient to react with all of the baking soda?

4. Suppose the pressure of the gas in the bag was measured to be 101.5 kPa. Is this pressure consistent with the ideal gas law? Assume $T = 20$°C and $P = 101$ kPa.

PRACTICE PROBLEMS

Supplemental Problems

For more practice with solving problems, see Supplemental Practice Problems, Appendix B.

10. How many moles of helium are contained in a 5.00-L canister at 101 kPa and 30.0°C?

11. What is the volume of 0.020 mol Ne at 0.505 kPa and 27.0°C?

12. How much zinc must react in order to form 15.5 L of hydrogen, $H_2(g)$, at 32.0°C and 115 kPa?

$$Zn(s) + H_2SO_4 \rightarrow ZnSO_4(aq) + H_2(g)$$

Theoretical Yield and Actual Yield

The amount of product of a chemical reaction predicted by stoichiometry is called the theoretical yield. As shown earlier, if 3.75 g of nitrogen completely react, a theoretical yield of 4.55 g of ammonia would be produced. The actual yield of a chemical reaction is usually less than predicted. The collection techniques and apparatus used, time, and the skills of the chemist may affect the actual yield.

When actual yield is less than theoretical yield, you express the efficiency of the reaction as percent yield. The percent yield of a reaction is the ratio of the actual yield to the theoretical yield expressed as a percent. Suppose the actual yield was 3.86 g of ammonia. Then, the percent yield could be calculated by using the following equation.

$$\text{percent yield} = \frac{\text{actual yield}}{\text{theoretical yield}} \times 100\%$$

$$\frac{3.86 \text{ g NH}_3}{4.55 \text{ g NH}_3} \times 100\% = 84.8\%$$

This means that 84.8 percent of the possible yield of ammonia was obtained from the reaction. Calculating percentage yield is similar to calculating a baseball player's batting average, as shown in **Figure 12.9.**

A manufacturer is interested in producing chemicals as efficiently and inexpensively as possible. High yields make commercial manufacturing of substances possible. For example, in human experiments, taxol, a naturally occurring complex compound, is a strong agent against cancer. For ten years, chemists tried to synthesize this compound in the lab. In 1994, two independent academic research groups succeeded. However, the process is so complicated and time-consuming that the percent yield is probably not even one percent.

Figure 12.9

Batting Average and Percent Yield
In 1995, Alfredo Spinelli batted .352, which means:

$$\frac{\text{hits}}{\text{attempts}} = \frac{352}{1000}.$$

Just as batting averages measure a hitter's efficiency, percent yield measures a reaction's efficiency.

ChemLab

Analyzing a Mixture

Chemists often analyze mixtures to determine their compositions. Their analytical procedures may include gas chromatography, mass spectrometry, or infrared spectroscopy. In this ChemLab, you will use a double-displacement reaction between strontium chloride and sodium sulfate to analyze a mixture of sodium sulfate and sodium chloride.

Problem

What is the mass percent of sodium sulfate in a mixture of sodium sulfate and sodium chloride?

Objectives

- **Observe** the double-displacement reaction between strontium chloride and sodium sulfate.
- **Quantify** the amount of strontium sulfate produced.
- **Compare** the mass of strontium sulfate produced with the mass of sodium sulfate that reacted.

PREPARATION

Materials

funnel
wash bottle with distilled water
filter paper
250-mL beakers (2)
50-mL graduated cylinder
stirring rod
ring stand
iron ring
spatula
weighing dish or paper
balance

Safety Precautions

PROCEDURE

1. Weigh the weighing dish or weighing paper and record its mass in your data table where indicated.

2. Transfer a 0.500- to 0.600-g sample of the sodium sulfate-sodium chloride mixture to the dish or paper and weigh it. Record the total mass of the sample and container in your data table.

3. Transfer all of the mixture to a 250-mL beaker, add approximately 50 mL of distilled water, and stir slowly until the sample is completely dissolved.

4. Use a graduated cylinder to measure 15 mL of 0.500M strontium chloride solution, then pour it into the solution in the beaker. Stir slowly for 30 s to completely precipitate the strontium sulfate.

5. Obtain a piece of filter paper, fold it in half, and tear off one corner. Fold the paper in half again. Weigh the folded paper and record its mass in your data table where shown.

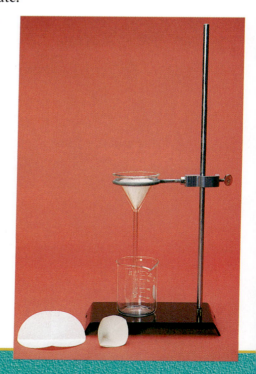

6. Clamp a funnel support or an iron ring to a ring stand, put the funnel in the support or ring, insert the folded filter paper in the funnel, and wet the paper with a small amount of distilled water from your wash bottle. Set an empty 250-mL beaker under the funnel to receive the filtrate.

7. Carefully transfer the mixture, including all of the precipitate, from the beaker to the filter paper, using a stirring rod and a wash bottle as demonstrated by your teacher.

8. Rinse the precipitate by pouring about 10 mL of distilled water into the filter paper.

9. When liquid no longer drips from the funnel, carefully remove the filter paper and residue from the funnel. Unfold the paper and spread it on a paper towel to dry overnight. Write your name on the paper towel.

10. Dispose of the filtrate according to your teacher's instructions.

11. Weigh the dry filter paper the following day, and record the mass of filter paper with residue in your data table.

DATA AND OBSERVATIONS

Mass of sample + dish	
Subtract mass of dish	
= Mass of sample	
Mass of $SrSO_4$ + filter paper	
Subtract mass of filter paper	
= Mass of $SrSO_4$	

ANALYZE AND CONCLUDE

1. **Interpreting Data** What is the balanced equation for the reaction between sodium sulfate and strontium chloride in aqueous solution?

2. **Making Inferences** After the reaction, does the sodium chloride produced and the unreacted sodium chloride from the original sample appear in the precipitate or the filtrate?

3. **Interpreting Data** What mass of strontium sulfate is produced in the reaction? Subtract the mass of filter paper from the mass of strontium sulfate plus filter paper.

4. **Interpreting Data** What is the mass and mass percent of sodium sulfate in the original sample?

APPLY AND ASSESS

1. Could you precipitate the sulfate from the sodium sulfate in the sample by adding an excess of aluminum chloride solution? Explain.

2. If your strontium sulfate precipitate was not completely dry when you weighed it, how would it affect your value of the mass percent of sodium sulfate in the sample?

3. How could you have tested the precipitate to be sure it contained strontium?

Improving Percent Yield in Chemical Synthesis

The single most important industrial chemical in the world is probably sulfuric acid. In the United States, its production exceeds 40 million tons per year. The strength of the U.S. economy can be gauged by the amount of sulfuric acid produced annually. Over the years, the manufacturing process of sulfuric acid has been improved to provide a higher and more economical yield.

Uses of Sulfuric Acid

Sixty percent of all manufactured sulfuric acid is used to make fertilizers. It is also used in manufacturing detergents, photographic film, synthetic fibers, pigments, paints, drugs, and other acids. It is the electrolyte in some batteries, acts as a catalyst and a dehydrating agent, and is a component in refining petroleum and metals.

The Lead-Chamber Process—A Low-Yield Process

The first industrial method, the lead-chamber process, is not commonly used now because its purity is low and its percent yield is only 60 to 80 percent. But it is much cheaper than the later and more productive contact process. The lead-chamber process is used for manufacturing sulfuric acid for applications that do not demand high purity.

The Contact Process—The High-Yield Process

The contact process is the most widely used commercial method. It is more expensive than the lead-chamber process, but it is simple and it produces high-purity sulfuric acid at a high percent yield—about 98 percent. In addition, it creates no by-products that pollute the atmosphere. The contact process has four steps.

Step 1

Sulfur is burned in air to produce sulfur dioxide, a stable compound. This reaction takes place quickly and readily.

$$S(s) + O_2(g) \rightarrow SO_2(g)$$

Impurities produced in combustion are removed from the sulfur dioxide so they will not react with and impede the catalyst in the next step.

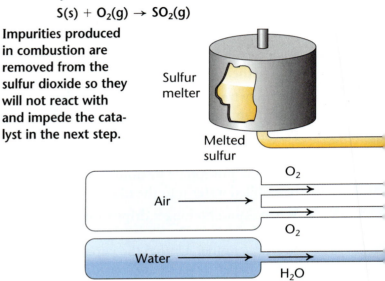

Sulfur melter

Melted sulfur

Air

O₂

O₂

Water

H₂O

Maximizing Percent Yield

Reaction yields in the contact process were increased in several ways—by selecting an optimum temperature, using an efficient catalyst, removing a product from a reaction that does not go to completion, and by controlling the rate of reaction of SO_3 with water. Keeping operating pressures at the correct values also increases yield.

Using Scientific Methods to Attain the Best Yields

To develop higher yields that are closer to the theoretical values, chemists adjust temperatures, pressures, or other conditions in their industrial processes. They look for better catalysts or new ways to deal with undesirable side reactions. Unwanted by-products are serious concerns to the chemist. Because they may cause environmental damage or be expensive to dispose of, by-products may raise production costs.

Step 3

Because sulfur trioxide reacts violently with water, it is bubbled through 98 percent concentrated sulfuric acid and forms pyrosulfuric acid, $H_2S_2O_7$.

$$SO_3(g) + H_2SO_4(l) \rightarrow H_2S_2O_7(l)$$

Step 4

When water is added to the pyrosulfuric acid, high-quality, 98 percent concentrated sulfuric acid is formed.

$$H_2S_2O_7(l) + H_2O(l) \rightarrow 2\ H_2SO_4(l)$$

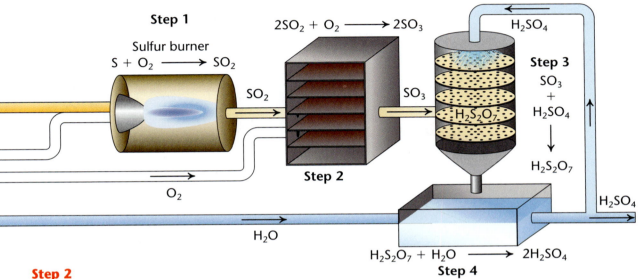

Step 1

Sulfur burner

$S + O_2 \longrightarrow SO_2$

$2SO_2 + O_2 \longrightarrow 2SO_3$

SO_2

SO_3

O_2

Step 2

H_2O

H_2SO_4

Step 3

SO_3
+
H_2SO_4

$H_2S_2O_7$

$H_2S_2O_7$

H_2SO_4

$H_2S_2O_7 + H_2O \longrightarrow 2H_2SO_4$

Step 4

Step 2

Because sulfur dioxide reacts slowly with excess oxygen, a catalyst, either vanadium pentoxide (V_2O_5) or finely divided platinum at a temperature of 400°C, is used. The reaction produces sulfur trioxide.

$$2SO_2(g) + O_2(g) \xrightarrow{catalyst} 2SO_3(g)$$

Sulfur trioxide is quickly removed from the contact chamber because it tends to form sulfur dioxide and oxygen again. Removing sulfur trioxide promotes the production of more sulfur trioxide.

Green Chemistry

The contact process is economically sound because the reactants are abundant and inexpensive and because it produces no unwanted by-products, which may be expensive to store or dispose of and which may pollute the environment. Ecologically sound chemical manufacturing is called green chemistry or green technology.

DISCUSSING THE TECHNOLOGY

1. **Hypothesizing** Roasting iron pyrite, FeS_2, can replace burning sulfur in Step 1 to produce sulfur dioxide. Write an equation to show what you think the reaction with pyrite is.

2. **Acquiring Information** Find out how sulfuric acid is used to manufacture hydrochloric acid and write the equation(s) of the reactions.

Determining Mass Percents

As you know, the chemical formula of a compound tells you the elements that comprise it. For example, the formula for geraniol (the main compound that gives a rose its scent) is $C_{10}H_{18}O$. The formula shows that geraniol is comprised of carbon, hydrogen, and oxygen. Because all these elements are nonmetals, geraniol is probably covalent and comprised of molecules.

In addition, the formula $C_{10}H_{18}O$ tells you that each molecule of geraniol contains ten carbon atoms, 18 hydrogen atoms, and one oxygen atom. In terms of numbers of atoms, hydrogen is the major element in geraniol. How can you tell whether it is the major element by mass? You can answer this question by determining the mass percents of each element in geraniol, which are shown in **Figure 12.10.**

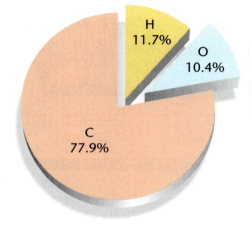

Figure 12.10

Mass Percents of Elements in Geraniol
This pie graph shows the composition of geraniol in terms of mass percents of the elements.

Suppose you have a mole of geraniol. Its molar mass is 154 g/mol. Of this mass, how many grams do the carbon atoms contribute? The formula shows that one molecule of geraniol includes ten atoms of carbon. Therefore, 1 mol of geraniol contains 10 mol of carbon. Multiply the mass of 1 mol of carbon by 10 to get the mass of carbon in 1 mol of geraniol.

$$\frac{10 \ \cancel{mol} \ | \ 12.0 \ \text{g C}}{| \ \cancel{mol}} = 120 \ \text{g C}$$

Now, use this mass of carbon to find the mass percent of carbon in geraniol.

$$\% \ C = \frac{120 \ \text{g C}}{154 \ \text{g} \ C_{10}H_{18}O} \times 100\% = 77.9\%$$

The mass percents of the other elements are calculated below in a similar fashion.

Mass of hydrogen in 1 mol geraniol:

$$18 \ \cancel{mol} \times \frac{1.00 \ \text{g H}}{\cancel{mol}} = 18.0 \ \text{g H}$$

$$\text{Mass percent of H} = \frac{\text{mass of H}}{\text{mass of geraniol}} \times 100\%$$

$$= \frac{18.0 \ \text{g H}}{154 \ \text{g} \ C_{10}H_{18}O} \times 100\% = 11.7\% \ \text{H}$$

Mass of oxygen in 1 mol geraniol:

$$\frac{1 \text{ mol}}{} \left| \frac{16.0 \text{ g O}}{\text{mol}} \right. = 16.0 \text{ g O}$$

$$\text{Mass percent of O} = \frac{\text{mass of O}}{\text{mass of geraniol}} \times 100\%$$

$$= \frac{16.0 \text{ g O}}{154 \text{ g C}_{10}\text{H}_{18}\text{O}} \times 100\% = 10.4\% \text{ O}$$

In the example above, you learned how to use the chemical formula and the molar masses to find the mass percent of a compound. You can also solve the reverse problem. You can use the mass percent to find the chemical formula of an unknown compound.

Determining Chemical Formulas

Suppose you analyzed an unknown compound and found that, by mass, it was 18.8 percent sodium, 29.0 percent chlorine, and 52.2 percent oxygen. Because this compound contains some metal and nonmetal elements, it may be an ionic compound. To determine its chemical formula, find the relative numbers of sodium, chlorine, and oxygen atoms in the formula unit of the compound.

Suppose you have 100.0 g of the unknown compound. Because you know the sample includes 18.8 g of sodium, 29.0 g of chlorine, and 52.2 g of oxygen, use the molar mass to find the number of moles of each element.

$$\frac{18.8 \text{ g Na}}{} \left| \frac{1 \text{ mol Na}}{23.0 \text{ g Na}} \right. = 0.817 \text{ mol Na}$$

$$\frac{29.0 \text{ g Cl}}{} \left| \frac{1 \text{ mol Cl}}{35.5 \text{ g Cl}} \right. = 0.817 \text{ mol Cl}$$

$$\frac{52.2 \text{ g O}}{} \left| \frac{1 \text{ mol O}}{16.0 \text{ g O}} \right. = 3.26 \text{ mol O}$$

You know that atoms combine in ratios of small whole numbers to form compounds. To find the whole-number ratios, divide the mole numbers by the smallest one.

$$\frac{0.817 \text{ mol Na}}{0.817} = 1.00 \text{ mol Na}$$

$$\frac{0.817 \text{ mol Cl}}{0.817} = 1.00 \text{ mol Cl}$$

$$\frac{3.26 \text{ mol O}}{0.817} = 3.99 \text{ mol O}$$

These values are whole numbers or very close to whole numbers. Therefore, the mole ratio for this compound is 1 mol Na :1 mol Cl :4 mol O. Similarly, the ratio of atoms in this compound is 1 Na :1 Cl :4 O.

The formula of a compound having the smallest whole-number ratio of atoms in the compound is called the **empirical formula.** The empirical formula of this unknown compound is $NaClO_4$.

What is the chemical formula for this compound? You have learned that the formula for an ionic compound represents the simplest possible ratio of the ions present and is called a formula unit. Chemical formulas for most ionic compounds are the same as their empirical formulas. Because the unknown compound is ionic, the chemical formula for a formula unit of the compound is the same as its empirical formula, $NaClO_4$. The compound is called sodium perchlorate.

As another example, suppose the mass percents of a compound are 40.0 percent carbon, 6.70 percent hydrogen, and 53.3 percent oxygen. Because all its elements are nonmetals, the compound is covalent. Imagine you have 100 g of the compound. Then you have 40.0 g of carbon, 6.70 g of hydrogen, and 53.3 g of oxygen. Use the molar masses of these elements to find the number of moles of each element.

$$\frac{40.0 \text{ g C} \mid 1 \text{ mol C}}{12.0 \text{ g C}} = 3.33 \text{ mol C}$$

$$\frac{6.70 \text{ g H} \mid 1 \text{ mol H}}{1.00 \text{ g H}} = 6.70 \text{ mol H}$$

$$\frac{53.3 \text{ g O} \mid 1 \text{ mol O}}{16.0 \text{ g O}} = 3.33 \text{ mol O}$$

Now, divide all the mole numbers by the smallest one.

$$\frac{3.33 \text{ mol C}}{3.33} = 1.00 \text{ mol C}$$

$$\frac{6.70 \text{ mol H}}{3.33} = 2.01 \text{ mol H}$$

$$\frac{3.33 \text{ mol O}}{3.33} = 1.00 \text{ mol O}$$

Because the ratio of moles is the same as the ratio of atoms, CH_2O is the empirical formula for this compound. But, the empirical formula is not always the chemical formula. Many different covalent compounds have the same empirical formula, as demonstrated in **Figure 12.11,** because atoms can share electrons in different ways.

Figure 12.11

Compounds Having the Empirical Formula CH₂O
For each of these compounds, the ratio of atoms is 1C:2H:1O. Because one molecule of each compound has a different number of each atom, it is a different compound with its own molecular formula.

To decide which multiple of the empirical formula is the correct molecular formula, you need the molar mass of the compound. Suppose a separate analysis shows that the molar mass of the compound is 90.0 g/mol. Therefore, the molecular mass of the compound is 90.0 u. The molecular mass of a CH_2O molecule is 30.0 u. By dividing the molecular mass of the compound by 30.0 u, you find the multiple.

$$\frac{90.0 \text{ u}}{30.0 \text{ u}} = 3$$

The molecular formula of the compound contains three empirical formula units. The molecular formula is $C_3H_6O_3$. The compound is lactic acid, the sour-tasting substance in spoiled milk. Lactic acid is also produced by active muscles, causing them to feel sore after strenuous activity.

Connecting Ideas

In this chapter, you learned to solve several kinds of stoichiometric problems. For each kind, you used the mole concept because when substances react, their particles interact. The number of particles at this submicroscopic level controls what happens macroscopically. In the next chapter, you will use the mole concept and the particle nature of matter to study the mixtures of substances called solutions.

SECTION REVIEW

Understanding Concepts

1. Octane, C_8H_{18}, is one of the many components of gasoline. Write the balanced chemical equation for the combustion of octane to form carbon dioxide gas and water vapor. Identify as many mole ratios among the substances in the combustion as you can.

2. If 25.0 g of octane burn as in Problem 1, how many grams of water will be produced? How many grams of carbon dioxide?

3. A manufacturer advertises a new synthesis reaction for methane with a percent yield of 110 percent. Comment on this claim.

Thinking Critically

4. **Predicting** The reaction of iron(III) oxide and aluminum is called the thermite reaction because of its intense heat. The iron produced is molten and was formerly used to weld railroad tracks in remote areas. Its balanced chemical equation is shown below.

$$Fe_2O_3 \text{ (s)} + 2Al\text{(s)} \rightarrow Al_2O_3\text{(s)} + 2Fe\text{(l)}$$

If 20.0 g of each reactant are used, which is the limiting reactant?

Applying Chemistry

5. **Blue Jeans** Indigo, the dye used to color blue jeans, is prepared using sodium amide. Sodium amide contains the following mass percents of elements: hydrogen, 5.17 percent; nitrogen, 35.9 percent; and sodium, 58.9 percent. Find an empirical formula for sodium amide.

CHAPTER 12 ASSESSMENT

REVIEWING MAIN IDEAS

12.1 Counting Particles of Matter

- The number of particles (atoms, molecules, or ions) in macroscopic matter controls the consumption and formation of substances in chemical reactions.
- One mole equals 6.02×10^{23}.
- Use the molar mass to convert mass to moles or moles to mass.

12.2 Using Moles

- A balanced chemical equation provides mole ratios of the substances in the reaction.
- The mole is a central concept in making chemical calculations.
- The ideal gas law is expressed in the following equation.

$$PV = nRT$$

- Percent yield measures the efficiency of a chemical reaction.

$$\text{Percent yield} = \frac{\text{actual yield}}{\text{theoretical yield}} \times 100\%$$

- Percent composition can be determined from the chemical formula of a compound.
- The empirical formula of a compound can be determined from its percent composition.
- The chemical formula of a compound can be determined if the molar mass and the empirical formula are known.

Vocabulary

For each of the following terms, write a sentence that shows your understanding of its meaning.

Avogadro constant
empirical formula
formula mass
ideal gas law
molar mass

molar volume
mole
molecular mass
stoichiometry

UNDERSTANDING CONCEPTS

1. A manufacturer must supply 20 000 connector units, each consisting of a bolt, two washers, and three nuts. How many of each part are needed?

2. Explain how you would use weighing to count 40 000 washers.

3. A reaction requires 0.498 mol of Cu_2SO_4. How many grams must you weigh to obtain this number of formula units?

4. What is the molecular mass of UF_6? What is the molar mass of UF_6?

5. What mass of copper contains the same number of atoms as 68.7 g of iron?

6. Explain why the mass of copper is not equal to the mass of iron in question 9.

7. Determine the molar mass of each of the following compounds.

   a) C_6H_5Br c) $(NH_4)_3PO_4$

   b) $K_2Cr_2O_7$ d) $Fe(NO_3)_3$

8. Calculate the mass of 0.345 mol of sodium nitrite, $NaNO_2$.

9. Calculate each of the following:

   a) the mass in grams of 0.254 mol of calcium sulfate

   b) the number of atoms in 2.0 g of helium

   c) the number of moles in 198 g of glucose, $C_6H_{12}O_6$

   d) the total number of ions in 10.8 g of magnesium bromide

10. Calculate the mass of silver chloride, AgCl, and dihydrogen monosulfide, H_2S, formed when 85.6 g of silver sulfide, Ag_2S, reacts with excess hydrochloric acid, HCl.

11. Aluminum nitrite and ammonium chloride react to form aluminum chloride, nitrogen, and water. What mass of aluminum chloride is present after 43.0 g of aluminum nitrite and

chemistryca.com/vocabulary_puzzlemaker

43.0 g of ammonium chloride have reacted completely?

12. A 10.0-g sample of magnesium reacted with excess hydrochloric acid to form magnesium chloride by the reaction below. When the reaction was complete, 30.8 g of magnesium chloride were recovered. What was the percent yield?

$$Mg(s) + 2HCl(aq) \rightarrow MgCl_2(aq) + H_2(g)$$

13. What is the molecular formula for each of the following compounds?

  a) empirical formula: CH_2;
    molar mass: 42 g/mol
  b) empirical formula: CH;
    molar mass: 78 g/mol
  c) empirical formula: NO_2;
    molar mass: 92 g/mol

14. An oxide of nitrogen is 26 percent nitrogen by mass. The molar mass of the oxide is approximately 105 g/mol. What is the formula of the compound?

APPLYING CONCEPTS

15. Hydrogen fuels are rated with respect to their hydrogen content. Determine the percent hydrogen for the following fuels.

  a) ethane, C_2H_6
  b) methane, CH_4
  c) whale oil, $C_{32}H_{64}O_2$

16. Oxidizers are generally compounds that contain relatively large mass percentages of oxygen. Hydrogen peroxide and sodium nitrate are good oxidizing agents. Compare the mass percent of oxygen in each of these compounds.

17. Serotonin is a compound that conducts nerve impulses in the brain. Serotonin is 68.2 percent C, 6.86 percent H, 15.9 percent N, and 9.08 percent O. The molar mass is 176 g/mol. What is the molecular formula of serotonin?

18. Suppose that the percentage yield of the reaction in the last problem is 80.0 percent. How many grams of KO_2 would need to be in the mask to insure enough oxygen?

Art Connection

19. How does the United States standardize measurements of mass and weight?

Everyday Chemistry

20. How many grams of sodium azide, NaN_3, are needed to fill a 0.0650-m^3 driver's air bag with nitrogen if the pressure required is 2.00 atmospheres and the gas temperature is 40.0°C? (Hint: 1 m^3 = 1000 L)

Chemistry and Technology

21. What are some uses of sulfuric acid?

THINKING CRITICALLY

Applying Concepts

22. What is the mass in grams of a single molecule of water?

23. Explain how a mole is used in chemistry as both a number and a mass.

Comparing and Contrasting

24. Distinguish between molar mass, formula mass, and molecular mass.

25. **ChemLab** Suggest a procedure to determine the mass percent of silver nitrate, $AgNO_3$, in a mixture of silver nitrate and sodium nitrate, $NaNO_3$.

Comparing and Contrasting

26. **MiniLab 1** Explain how your procedure in this activity is like using the molar mass to determine the number of particles in a sample of a known substance.

Relating Cause and Effect

27. **MiniLab 2** If you used calcium carbonate instead of baking soda, how would the amount of reactants needed change?

CUMULATIVE REVIEW

28. Distinguish between atomic number and mass number. How do these two numbers compare for isotopes of an element? (Chapter 2)

29. Why does the second period of the periodic table contain eight elements? (Chapter 7)

30. The volume of a gas at STP is 8.50 L. What is its volume if the pressure is 1250 mm Hg and the temperature is 75.0°C? (Chapter 11)

SKILL REVIEW

31. Using a Table Complete the table below. In the last column, rank the number of particles from smallest to largest. Did you use the number of moles or the number of individual particles for your ranking? Explain.

WRITING IN CHEMISTRY

32. A mole is such a large number that it is hard to envision just how big 6.02×10^{23} is. Write three of your own examples of the number of things in a mole. One example must use time, one must use distance, and the third is up to you.

PROBLEM SOLVING

33. A student carries out the following sequence of chemical reactions on a 0.635-g sample of pure copper.

$$Cu(s) + 2HNO_3(aq) \rightarrow Cu(NO_3)_2(aq) + H_2(g)$$
$$Cu(NO_3)_2(aq) + 2NaOH(aq) \rightarrow$$
$$Cu(OH)_2(s) + 2NaNO_3(aq)$$
$$Cu(OH)_2(s) \rightarrow CuO(s) + H_2O(l)$$
$$CuO(s) + H_2SO_4(aq) \rightarrow CuSO_4(aq) + H_2O(l)$$
$$CuSO_4(aq) + Mg(s) \rightarrow Cu(s) + MgSO_4(aq)$$

The student isolates and weighs the product at the end of each step before proceeding to the next. What is the theoretical yield of the copper product in each step?

Item	Formula	Molar Mass (g/mol)	Number of Particles	Rank
The formula units in 10.0 g of calcium fluoride				
The sodium ions in 10.0 g of sodium chloride				
The water molecules in 10.0 g of water				
The hydrogen atoms in 10.0 g of water				
The carbon dioxide molecules in 10.0 g of carbon dioxide				
The carbon monoxide molecules in 10.0 g of carbon monoxide				
The aspirin molecules, $C_9H_8O_4$, in 10.0 g of aspirin				
The carbon atoms contained within 10.0 g of aspirin				
The valence electrons in 10.0 g of aspirin				

Interpreting Graphs: Use the graph below to answer questions 1–4.

Supply of Various Chemicals in Dr. Raitano's Laboratory

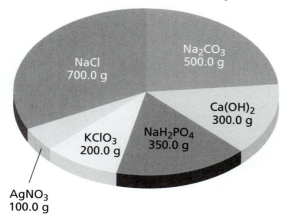

1. Pure silver metal can be made using the reaction shown below:

$$Cu(s) + 2AgNO_3(aq) \rightarrow 2Ag(s) + Cu(NO_3)_2(aq)$$

How many grams of copper metal will be needed to use up all of the $AgNO_3$ in Dr. Raitano's laboratory?

a) 18.70 g   c) 74.7 g
b) 37.3 g   d) 100 g

$$Na_2CO_3(aq) + Ca(OH)_2(aq) \rightarrow 2NaOH(aq) + CaCO_3(s)$$

2. The LeBlanc process, shown above, is the traditional method of manufacturing sodium hydroxide. Using the amounts of chemicals available in Dr. Raitano's lab, the maximum number of moles of NaOH that can be produced is

a) 4.05 mol.   c) 8.097 mol.
b) 4.72 mol.   d) 9.43 mol.

3. Sodium dihydrogen pyrophosphate ($Na_2H_2P_2O_7$), more commonly known as baking powder, is manufactured by heating NaH_2PO_4 at high temperatures:

$$2NaH_2PO_4(s) \rightarrow Na_2H_2P_2O_7\ (s) + H_2O\ (g)$$

If 444.0 g of $Na_2H_2P_2O_7$ are needed, how much more NaH_2PO_4 will Dr. Raitano have to buy to make enough $Na_2H_2P_2O_7$?

a) 94.0 g
b) 130.0 g
c) 480 g
d) none—the lab already has enough

4. Pure O_2 gas can be generated from the decomposition of potassium chlorate ($KClO_3$):

$$2KClO_3(s\ ) \rightarrow 2KCl(s) + 3O_2(g)$$

If half of the $KClO_3$ in the lab is used and 12.8 g of oxygen gas are produced, the percent yield of the reaction is

a) 12.8%.   c) 65.6%.
b) 32.7%.   d) 98.0%.

5. The variable P in the ideal gas law formula requires the unit

a) atm.   c) kPa.
b) mm Hg.   d) psi.

6. How many moles of helium occupy 26.4 L at a pressure of 103.0 kPa, and 10°C?

a) 0.03 mol   c) 1.16 mol
b) 0.86 mol   d) 32.72 mol

7. Red mercury(II) oxide decomposes at high temperatures to form mercury metal and oxygen gas:

$$2HgO(s) \rightarrow 2Hg(l) + O_2(g)$$

If 3.55 moles of HgO decompose to form 1.54 moles of O_2 and 618 g of Hg, what is the percent yield of this reaction?

a) 13.2%   c) 56.6%
b) 42.5%   d) 86.8%

8. How many grams of hydrogen make up 3.5 moles of glucose ($C_6H_6O_6$)?

a) 1.7 g   c) 21.0 g
b) 3.5 g   d) 101.5 g

Test Taking Tip

Watch the Little Words Underline words like *least*, *not*, and *except* when you see them in test questions. They change the meaning of the question!

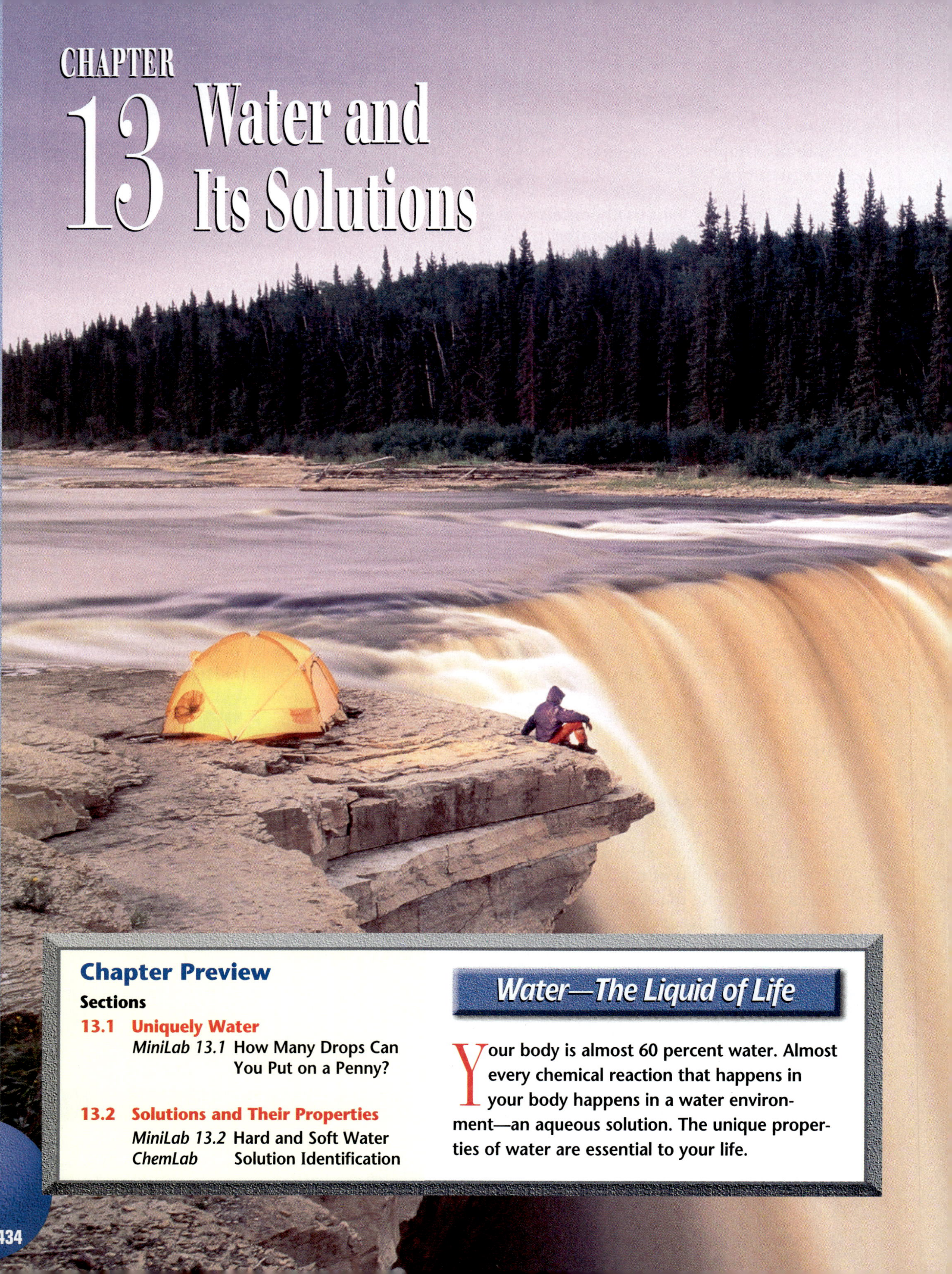

Chapter Preview

Sections

13.1 Uniquely Water
MiniLab 13.1 How Many Drops Can
You Put on a Penny?

13.2 Solutions and Their Properties
MiniLab 13.2 Hard and Soft Water
ChemLab Solution Identification

Water—The Liquid of Life

Your body is almost 60 percent water. Almost every chemical reaction that happens in your body happens in a water environment—an aqueous solution. The unique properties of water are essential to your life.

Launch Lab

Solution Formation

The intermolecular forces among dissolving particles and the attractive forces between solute and solvent particles result in an overall energy change. Can this change be observed?

Safety Precautions

Dispose of solutions by flushing them down a drain with excess water.

Materials

- balance
- 50-mL graduated cylinder
- 100-mL beakers (2)
- stirring rod
- ammonium chloride (NH_4Cl)
- calcium chloride ($CaCl_2$)
- water

Procedure

1. Measure 10 g of ammonium chloride (NH_4Cl) and place it in a 100-mL beaker.
2. Add 30 mL of water to the NH_4Cl, stirring with your stirring rod.
3. Feel the bottom of the beaker and record your observations.
4. Repeat the procedure with calcium chloride ($CaCl_2$).

Analysis

Which dissolving process is exothermic? Endothermic? Suggest some practical applications for dissolving processes that are exothermic and for those that are endothermic.

What I Already Know

Review the following concepts before studying this chapter.

Chapter 4: the meaning of electron dot structures

Chapter 9: how to determine molecular geometry; the meaning of electronegativity; the basis of polarity

Chapter 10: structure of solids, liquids, and gases; kinds of changes of state; energy in state changes

Reading Chemistry

As you look through the chapter, write down section headings. For each section, make a note of any questions or unfamiliar vocabulary words. While reading the chapter, write down key points for each section, and try to find answers to any questions in the text.

Preview this chapter's content and activities at **chemistryca.com**

Uniquely Water

Water may be a common substance on Earth, but its properties are anything but common. Those properties make water essential for life. You may well ask how it is that a substance like water can have such vital properties. A water molecule, after all, is just two atoms of hydrogen and one atom of oxygen linked by covalent bonds. Molecules don't get much simpler than H_2O. However, the simplicity of the composition of a water molecule is deceiving. As you will see, the electron distribution and the three-dimensional arrangement of those three atoms in the molecule are the source of the unique properties of water—those properties that cause it to play a major role in so many aspects of your world.

SECTION PREVIEW

Objectives

✓ **Demonstrate** the uniqueness of water as a chemical substance.

✓ **Model** the three-dimensional geometry of a water molecule.

✓ **Relate** the physical properties of water to the molecular model.

Review Vocabulary

Molar volume: the volume that a mole of gas occupies at a pressure of one atmosphere and a temperature of 0.00 °C.

New Vocabulary

hydrogen bonding
surface tension
capillarity
specific heat

Water: The Molecular View

Because water is so much a part of life, its properties are easy to take for granted. If you step back a bit and examine water scientifically, you will find that it is unusual among the compounds found on Earth. **Table 13.1** compares the physical properties of water against a group of molecules of similar mass. Look at the differences in physical state, melting point, and boiling point. Water definitely behaves differently from other small molecules. It is a unique substance.

When you use water for drinking or most other purposes, you use it in its liquid state. When you try to think of other common liquids, you probably think of water solutions. Water is most often thought of as a liquid. However, solid water, called ice, and gaseous water, called steam or water vapor, also exist in large quantities on Earth. Water is the only substance on Earth that exists in large quantities in all three states. Gaseous water is found around geysers and hot springs and in the atmosphere, where it plays a major role in determining weather. Ice occurs in glaciers and ice caps and acts as a huge reservoir of fresh water.

Table 13.1 The Uniqueness of Water

Substance	Formula	Molecular Mass (u)	State at Room Temperature	Melting Point (°C)	Boiling Point (°C)
Methane	CH_4	16	gas	–183	–161
Ammonia	NH_3	17	gas	–78	–33
Water	**H_2O**	**18**	**liquid**	**0**	**100**
Nitrogen	N_2	28	gas	–210	–196
Oxygen	O_2	32	gas	–218	–183

Most substances tend to be more dense as solids than they are as liquids. Water is an important exception. It is less dense as a solid than as a liquid. For that reason, ice floats in liquid water rather than sinking. If ice were like most solids and denser than the liquid state, lakes and ponds would freeze from the bottom up. The fact that the ice forms and remains on the top of lakes and ponds allows aquatic life in the liquid water below it to survive the winter cold. Also, if ice were like most solids, popular sports such as ice hockey, figure skating, and ice fishing, **Figure 13.1,** could never have developed.

Figure 13.1

Ice Has Low Density

The fact that ice is less dense than liquid water is the reason that ice collects at the top of the liquid rather than at the bottom. If it were not for this fact, ice-skating and ice fishing would not be possible.

What factors will explain the unique physical properties of water in its various states? **Figure 13.2** reviews some of the things you have already learned about the composition of the water molecule, its electron distribution, and its three-dimensional structure.

Figure 13.2

Geometry of the Water Molecule

The arrangement of electrons about the central oxygen in the water molecule relates to its three-dimensional geometry. There is a large electronegativity difference between the covalently bonded hydrogen and oxygen. Therefore, the electron pair is shared unequally. Because of the molecule's bent structure, the poles of positive and negative charge in the two bonds do not cancel, and the water molecule as a whole is polar.

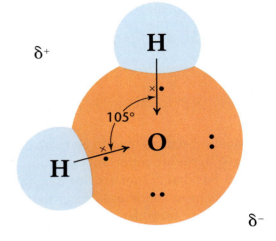

Intermolecular Forces in Water

Recall from Chapter 9 that molecules that are dipoles, such as water, have interparticle attractive forces. You can model the behavior of a group of water molecules by imagining that the molecules act like little bar magnets, as shown in **Figure 13.3.**

Notice that if you pulled one of the magnets in **Figure 13.3,** the position of all of the magnets would change. The same thing would happen if you could somehow reach into a group of water molecules and pull just one. The model demonstrates that attractive forces between objects do not create interactions between just two objects. The forces can also combine to give organization to groups of objects, whether they are bar magnets or water molecules.

Figure 13.3

Attractions and Order

A group of magnets will tend to orient themselves with the opposite poles of the different magnets pulled toward one another. Although dipoles aren't held by magnetic forces, a group of water molecules will orient in a similar way at the molecular scale because of electrical forces. The opposites attract and create order among molecules. This effect is especially great at low temperatures.

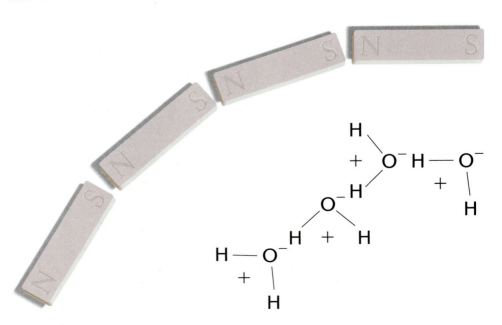

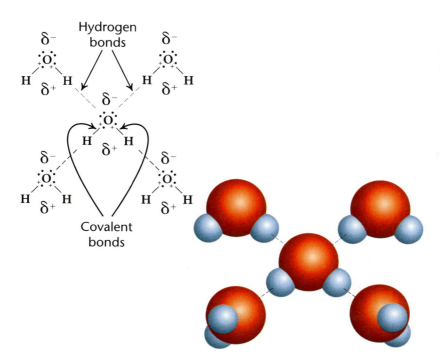

Figure 13.4

Hydrogen Bonding Versus Covalent Bonding in Water
Notice that water molecules have both intermolecular forces (hydrogen bonds, dashed lines) and intramolecular forces (covalent bonds, solid lines). The oxygens are highly electronegative and strongly attract hydrogen's electrons. As a result, the hydrogen nucleus, a proton, is partially exposed. The exposed hydrogen nucleus is attracted to the non-bonding pairs of electrons on the oxygen atom. The two hydrogens on the central water molecule are attracted at the same time to the oxygens of two other water molecules.

Modeling Water: Hydrogen Bonding

Although the intermolecular forces in water create order that involves large numbers of molecules, you can get a good picture of what's going on by modeling the interactions between just a few water molecules, as shown in **Figure 13.4.** Notice that the oxygen on one water molecule is attracted to the hydrogen atoms on other water molecules. The connections between the molecules are not full covalent bonds, but they are still fairly strong. The formation of such a connection between the hydrogen atoms on one molecule and a highly electronegative atom on another is called **hydrogen bonding.** Atoms that are electronegative enough to cause hydrogen bonding include oxygen, fluorine, and nitrogen.

Water: The Hydrogen Bonding Champion

Oxygen-hydrogen groups in molecules tend to promote intermolecular hydrogen bonding. In pure water, each water molecule may form hydrogen bonds with four other water molecules.

Any molecule that contains O—H bonds has the potential to form hydrogen bonds. Oxygen is sufficiently electronegative to attract hydrogen's single electron strongly so that the hydrogen atom almost becomes an exposed proton. Alcohols, organic compounds that contain O—H bonds, also form hydrogen bonds. Not surprisingly, some of the physical properties of alcohols are similar to those of water, as shown in **Figure 13.5.**

Because they have many O—H bonds, biological molecules such as proteins, nucleic acids, and carbohydrates also have the ability to form hydrogen-bond networks. These networks can be extensive, and their three-dimensional shape is important in determining their functions in living things.

Figure 13.5

Hydrogen Bonding in Methanol
Molecules other than water also form hydrogen bonds. This simplest alcohol, methanol, CH_3OH, has an O—H bond with which to form hydrogen bonds. This hydrogen bonding gives methanol some physical properties that are similar to those of water.

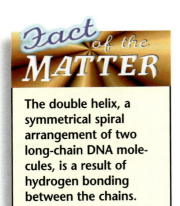
Water: Physical Properties Revisited

Look back at the properties in **Table 13.1.** You can see that compared to similar-sized molecules, water's physical properties are different. The most obvious difference is that water is a liquid at room temperature, whereas the others are gases. Water also has a high melting point and high boiling point for such a small molecule. Many of these unique physical properties are a result of the hydrogen bonding that occurs between water molecules. This is just another example of the relationship between submicroscopic structure and the macroscopic, directly observable behavior of substances. Let's examine a few of these properties in more detail.

States of Water

Water occurs primarily in the liquid and solid states on Earth, rather than as a gas. The intermolecular hydrogen bonds hold the water molecules together strongly enough that they cannot readily escape into the gaseous state at ordinary temperatures. That is why water has such a high boiling point for such a small molecule, 100°C. In order for water to boil, the temperature must be increased, adding enough energy to overcome the hydrogen bonds that are holding the water molecules together and separating them so that they can enter the gaseous state.

Ice Floats

You know that if you drop an ice cube into a glass of water, the ice floats. You also know this means that the density of the solid water is less than that of liquid water. **Figure 13.6** shows how the density of water changes as its temperature changes. Like most substances, liquid water shrinks and becomes denser when it is cooled. As water cools from 60°C, its volume decreases and its density increases. The water molecules move less rapidly, and they are able to be drawn closer together by dipole-dipole attractions. The volume of the water decreases because the molecules pull together. Meanwhile, the mass of water stays the same, so density increases.

However, when the water is cooled to about 4°C, something unusual happens. The volume stops decreasing. The density has reached a maximum. At this point, the water molecules have been pulled as close together as they can get. As the temperature is lowered below 4°C, the volume of the water begins to expand and its density decreases.

Figure 13.6

Density of Water at Various Temperatures
Water achieves its minimum volume, and therefore its maximum density, at 3.98°C.

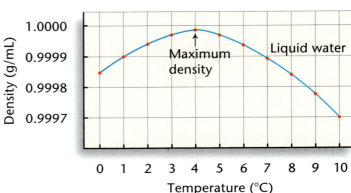

Figure 13.7

Freezing Water

Its Molecular Structure The six-pointed snowflakes reflect the open arrangement of bonds. In order to freeze, molecules of liquid water separate slightly from one another as a result of hydrogen bonding. The volume of cooling liquid increases as the molecules move apart and the density decreases. At 0°C, the liquid water freezes to solid water, ice. The volume rises and the density drops dramatically, from 1.000 g/mL to 0.99984 g/mL, as the water molecules assume the open arrangement. ▶

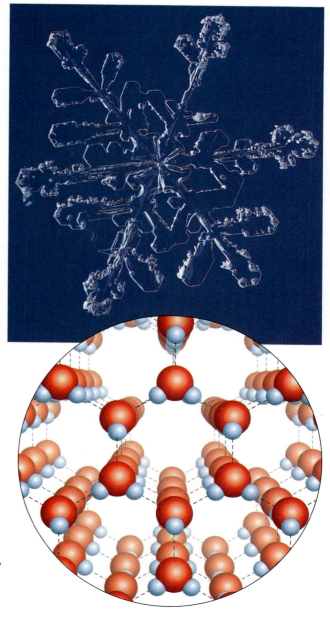

◀ **Its Expansion** The volume expansion that occurs when water freezes has some disadvantages, as you can see by looking at this burst jar.

You can account for this if you know what is happening to the molecular arrangement. Below 4°C, the water molecules are beginning to approach the solid state, which is highly organized. The water molecules begin to form the open arrangement shown in the model in **Figure 13.7.** The arrangement results from hydrogen bonding and is the most stable structure for the molecules in or near the solid state.

The fact that ice floats in liquid water has advantages for life that you read about earlier. However, it also has other consequences, as shown in **Figure 13.7.** Because liquid water expands significantly when it freezes, frozen water pipes can break and sidewalks can crack. The forces involved in the expansion of freezing water are surprisingly great—strong enough eventually to break boulders into small pieces. The processes of freezing and thawing contribute to soil formation and erosion and have been transforming the surface of Earth for many millions of years.

More Evidence for Water's Intermolecular Forces

The strong intermolecular forces that bind water molecules together show up in other properties of water. These observable properties provide more macroscopic evidence for the scientific model of the submicroscopic structure of water molecules. Keep in mind that the source of all of these properties is the unique combination of the water molecule's geometry and the electron distribution that makes the O—H bond highly polar.

Surface Tension

Have you ever watched water drip from a faucet? Each drop is composed of an enormous number of water molecules—roughly 2×10^{21}. The observation that this large number of molecules can hold together as a unit to form a single drop is more evidence for the presence of intermolecular forces.

A water molecule forms a drop because of **surface tension,** which is the force needed to overcome intermolecular attractions and break through the surface of a liquid or spread the liquid out. The higher the surface tension of a liquid is, the more resistant the liquid is to having its surface broken. **Figure 13.8** shows a molecular model of a water drop. The net inward force makes the surface of the drop contract and seem to toughen, behaving like a sort of skin.

Liquids other than water exhibit surface tension. Mercury is a good example of a liquid that has a high surface tension and strong interparticle attractive forces. When mercury is spilled, it forms droplets, much like the water beads that you see when it rains on a freshly waxed car.

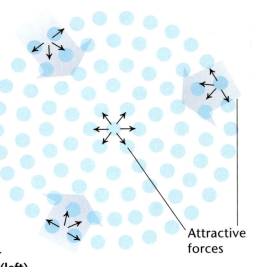

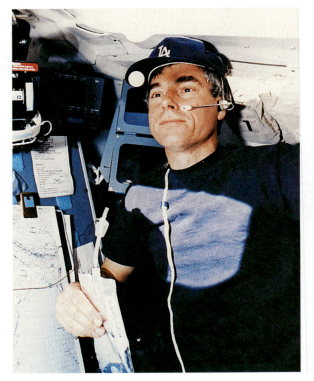

Figure 13.8

Surface Tension
Water drops in space, such as this one photographed on a space shuttle mission (right), show the perfectly spherical shape that is due to surface tension. In the drawing of a droplet of water (left), molecules in the middle of the drop are completely surrounded by other molecules and experience attractive forces (the small arrows) in all directions. Water molecules at the surface are not completely surrounded by other molecules. The forces acting on them are not equal in all directions, but they all pull inward with a net force (the large arrows). A water drop takes on its spherical shape because of this response to the inward force.

Attractive forces

How many drops can you put on a penny?

The surface tension of water allows it to bead up on many surfaces. In this MiniLab, you will compete to see who can deposit the most drops of water and the most drops of an aqueous detergent solution on a penny.

Procedure

1. Lay a penny flat on your lab table.

2. Fill a microtip pipet with tap water, and count the number of drops you can deposit on the penny before water spills over the edge. Record the number of drops.

3. Fill another microtip pipet with a detergent solution prepared by your teacher, and repeat the process. Record the number of drops.

Analysis

1. How is surface tension demonstrated in this experiment?

2. Which has the lower surface tension: the water or the aqueous detergent solution? What accounts for this fact?

Capillarity

Have you ever had a small sample of blood drawn from a finger during a routine blood test as shown in **Figure 13.9**? After your finger is pricked, the blood is drawn by touching a thin glass tube, called a capillary tube, to the blood drop. Even without any suction being applied, the blood rises up into the tube. You have seen another version of this effect if you have ever observed the curved meniscus surface that results when you place water in a narrow glass tube or graduated cylinder. The water level next to the walls of the glass tube is higher than that at the center, giving a concave surface at the top of the liquid.

WORD ORIGIN

capillarity:
capillus (L) hair
A liquid will rise by capillarity in a tube that is as narrow as a hair.

These examples of the rising of liquids in narrow tubes are illustrations of **capillarity,** or, as it is sometimes called, capillary action. Capillarity results from the competition between the interparticle attractive forces between the molecules of liquid and the attractive forces between the liquid and the tube that contains it.

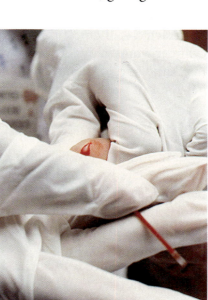

Figure 13.9
Capillarity
Capillarity draws blood into a thin tube after a finger is pricked.

Figure 13.10

Comparison of Water and Mercury in Glass Tubes

Water forms a concave meniscus, whereas mercury produces a convex meniscus. In the case of mercury, there is essentially no attractive force between the mercury atoms and the silicon dioxide to compete with the attractive forces between the mercury atoms themselves. The mercury forms a convex (high-centered) meniscus because the only force is the interparticle attractive force between mercury atoms, which produces surface tension.

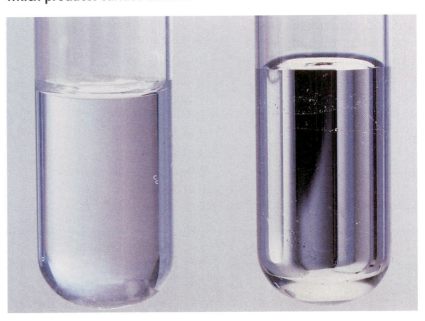

See page 869 in
Appendix F for
Measuring Capillarity

In this case of a glass tube, water molecules can form hydrogen bonds to the oxygen atoms in the silicon dioxide that makes up the glass. This attractive force between the water and the SiO_2 draws the water up the walls of the tube. Because the water molecules are also attracted to each other, more water rises upward. If the tube is narrow, like the capillary tubes used in blood tests, the liquid will be drawn high into the tube because nearly all the molecules are close to the walls and are thus strongly attracted to them. If the glass tube is of larger diameter, like a graduated cylinder, you see the upward capillary effect near the tube walls, causing the liquid surface to have a meniscus shape. **Figure 13.10** contrasts the behavior of water in a glass tube and mercury in a glass tube.

Water: Earth's Thermostat

Have you ever jumped into a pool of cool water on a hot day early in summer? Despite the fact that the air temperature is high, the water temperature lags behind and tends to stay lower. On the other hand, if you return and jump into that same pool in the evening when the air temperature has dropped, the pool water will be warmer than the outside air. The water once again has lagged behind its surroundings in changing temperature. Once it has heated up, it does not cool down quickly. You may have noticed the same kind of effect if you have ever put a pot of water on to boil and had to wait a long time for it to heat up.

What you were observing in these examples is the high specific heat of water. **Specific heat** measures the amount of heat, in joules, needed to raise the temperature of 1 g of substance by 1°C. The specific heats of water and some other common substances are shown in **Table 13.2.**

Table 13.2 Specific Heats of Some Substances	
Substance	**Specific Heat (J/g°C)**
Gold, Au	0.129
Copper, Cu	0.385
Iron, Fe	0.450
Glass	0.84
Cement	0.88
Wood	1.76
Ethanol, $C_2H_5OH(l)$	2.46
Water, $H_2O(l)$	4.18

Notice that water has the highest specific heat of any of the substances listed. This means that water must absorb or release more heat for its temperature to change by one Celsius degree than any of the other substances. The specific heat of water is 4.18 J/g°C. (The unit is read "joules per gram per degree Celsius.") In order to raise the temperature of 1 g of water by 1°C, you must add 4.18 J of heat. On the other hand, you must remove 4.18 J of heat from a 1-g sample of water to lower its temperature by 1°C. That is why it takes time for water in a swimming pool to warm up in the early part of the year and why the water, once warmed, cools off slowly when the outdoor temperature drops, **Figure 13.11.**

Figure 13.11

Warm Water Cools Slowly
The water in this swimming pool heats up during the day. Even though the outside temperature drops at night, the water tends to remain warm. Because of its high specific heat, the warm water has stored a great deal of energy.

Water serves as a great heat reservoir that moderates the temperature at Earth's surface. During the day, heat from the sun is absorbed by the oceans. Because water has a high specific heat, and because there's a great deal of water, the temperature of the oceans does not increase detectably despite all the heat the water has absorbed. In the evening, the stored heat in the world's oceans is released, helping to maintain the air temperature in the absence of the sun's energy. If there were no water on Earth's surface to serve as a temperature moderator, the daytime temperature would soar. Rocks have a much lower specific heat than water does. In the evening, Earth would turn frigid because rocks have so little capacity to store heat and to release it slowly. You have noticed this effect if you have ever experienced the extreme temperature difference between day and night in a desert during clear weather, **Figure 13.12.**

Figure 13.12
Low Heat Capacity of Dry Soil and Rock
A desert's temperature changes greatly at twilight. Without water, heat is given off quickly.

Water Evaporation/Condensation: No Sweat?

Recall that vaporization is the change in state of liquid to gas, and that the amount of heat required to vaporize a quantity of a liquid is called the heat of vaporization. During the vaporization of a liquid, interparticle forces must be overcome. Therefore, vaporization of a liquid is an endothermic (energy-absorbing) process. Condensation, the formation of a liquid from a gas, is an exothermic (energy-releasing) process. Because of the strong intermolecular forces, water absorbs a great deal of heat when it is vaporized. Thus, it has a high heat of vaporization. Water also loses a great deal of heat when it condenses—thus the effectiveness of steam in heating a room or in causing burns.

Evaporation is vaporization from the surface of a liquid. Evaporation of water is one of the mechanisms by which your body regulates its temperature. On a hot day, particularly if you are as active as the person in **Figure 13.13,** you perspire and cool down.

Figure 13.13
Perspiration and Cooling
Perspiration is a mechanism by which your body passes water to the surface of your skin via sweat glands. The vaporization of water is an endothermic (energy-absorbing) process, and as the water evaporates, it removes heat from the surface of your skin.

Water Treatment

With a twist of your wrist, you can make clean water flow from the faucet. Like most other people in industrialized countries, you may never stop to think how fortunate you are to have accessible, safe water. In many parts of the world, people are forced to carry water long distances. Then, after all their hard work, they have no assurance that the water does not carry disease. The water you use is probably safe because it is purified in a reliable water-treatment plant through the following series of steps.

1. Water is often conveyed over long distances through pipes from a lake or other freshwater source to the treatment plant. When the water to be treated enters the intake basin, it passes through bar screens that remove large, suspended solids and trash.

2. After the screening, pumps lift the water 6 m or more so that it can flow, by gravity, through the rest of the tanks.

3. Some particles present in water are so fine that they can be trapped only by coagulation. To do this, chemicals such as alum, chlorine, and lime are added by chemical applicators. The purpose of the alum—$Al_2(SO_4)_3 \cdot 18H_2O$—is to produce alum floc, a coagulant that causes fine, suspended solids to clump together. Chlorine kills bacteria. Lime causes precipitation of calcium carbonate and clears the water.

4. In the settling tank, bacteria, silt, and other impurities stick to the coagulants and settle to the bottom. The rest of the water moves on to a filtering tank.

5. The filtering tank allows the water to trickle through sand, gravel, and sometimes charcoal for a final cleansing.

6. When the water reaches the reservoir, it receives another treatment of chlorine. The purified water remains in the reservoir until it is pumped to homes, factories, and businesses.

Analyzing the Issue

1. **Acquiring Information** Find out and report on how biological treatment is used to clear water of organic waste from industries.

2. **Writing** In an essay, explain community activities that could be followed to ensure that local drinking water would not become polluted.

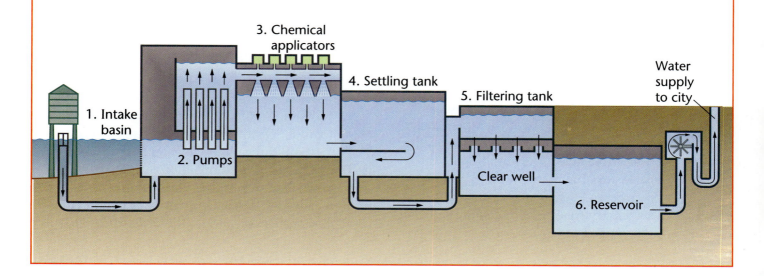

1. Intake basin
2. Pumps
3. Chemical applicators
4. Settling tank
5. Filtering tank
Clear well
6. Reservoir
Water supply to city

PEOPLE in CHEMISTRY

Meet Alice Arellano, Wastewater Operator

When alarm clocks buzz on weekday mornings, millions of people head for the bathroom to prepare for the day. Showers run, toilets flush, and teeth are brushed, generating a tremendous flood of wastewater that flows toward treatment plants. Wastewater operators like Alice Arellano are on the job, making sure this water is properly treated and protecting the health of people and the environment.

On the Job

 Ms. Arellano, can you explain your job responsibilities to us?

 I'm a wastewater control-room operator. Our plant is responsible for cleaning wastewater from this part of Austin, Texas.

 How does the cleaning process work?

First, the wastewater comes into our plant through a big line, about 54 inches in diameter. I've seen some pretty nasty stuff come through that pipe! The water goes through some bar screens, which remove the debris. The heavy solids, such as inorganic metals and sand, settle to the bottom of the chamber, where augers move them to a dump truck. The truck takes the inorganic waste to a landfill. The material that stays in suspension, the floatable stuff, goes into primary treatment. After primary treatment, then it goes to secondary treatment and some aeration tanks. Microorganisms break down the harmful bacteria. As I like to explain it in nontechnical language, the microorganisms start at the top of the tank and eat and party all the way down.

 After the water has been cleaned, where does it go?

 Into the creeks and eventually into the Colorado River and Lake Powell. The water is filtered and used again. There's a finite amount of water on Earth, so we have to use it over and over.

 **Does the weather have an effect on your work?**

When it rains, my work can get really tough. Ordinarily, the plant treats about 40 million gallons of water a day. Capacity is about 120 million gallons. But an unusually heavy storm can overwhelm the system with 400 million gallons. Then, all we can do is to add chlorine, open the valves, and let the water through.

How do people react when you talk about your profession?

 Some people are jealous because the job pays well and is steady work. Not everyone would like to do this job, though. The odor of ammonia can be strong. Once in a while, I get stomach viruses, even though I wear gloves and glasses and use all my safety equipment. However, I find the work fascinating. I'm trying to learn everything that goes on at the plant, so I ask everyone questions—the mechanics, the instrumentation people, and the electricians.

Early Influences

 Was this position your first one after you left school?

 I got my GED late because I became a mother quite young. I had a friend, a teammate on my softball team, who worked in the plant and encouraged me to get certified as an operator. I took classes in wastewater management because I really wanted to better myself.

 Do you find a use for the math and chemistry you took in school?

 I have to know the formulas for the amount of chlorine to add, as well as for the sulfur dioxide. That chemical removes the chlorine so the water discharge won't harm fish in the creeks. More than my knowledge of chemistry, I use all my senses. It's often a matter of experience.

Personal Insights

 Your job isn't a typical one for a woman, is it?

 Men are in the majority here, but that's changing. My daughter came to Take-Our-Daughters-to-Work Day and became interested in what I do.

 Do you have any advice for kids who might not like school?

 Just to try hard, and stay in school. I encourage my own daughter to leave the adult behavior until the time she's ready for adult responsibilities. There are always about a dozen kids who make my home their gathering place these days. Because I had problems when I was young, they trust me to help them with their problems.

CAREER ▸ CONNECTION

Other jobs vital for keeping water clean include the following.

Environmental Health Inspector College degree plus licensing

Environmental Technician Two-year accreditation program

Industrial Machinery Mechanic High school plus apprenticeship program

Water: The Super Solvent

Most of the water on Earth is not pure, but rather is present in solutions. Water is difficult to keep pure because it is an excellent solvent for a variety of solutes. Water is such a versatile solvent that it is sometimes called the universal solvent. Its ability to act as a solvent is one of its most important physical properties. As you will see, it is again the attraction of water molecules for other molecules, as well as for one another, that accounts for these solvent properties.

As **Figure 13.14** shows, nearly all of the water you consume is in aqueous solution. Aqueous solutions provide efficient means of transporting nutrients in plants, as well as the nutrients in your blood. Almost all of the life-supporting chemical reactions that occur take place in an aqueous environment. In the absence of water, these reactions would not occur.

Figure 13.14
Aqueous Beverages
Unless you buy distilled water or distill your own, everything you drink is an aqueous solution. Soft drinks, tea, coffee, spring water, and even tap water are all aqueous solutions.

Supplemental Problems

For more practice with solving problems, see Supplemental Practice Problems, Appendix B.

SECTION REVIEW

Understanding Concepts

1. List five physical properties that distinguish water from most other compounds of similar molecular size.

2. What is hydrogen bonding?

3. What is surface tension? Use the concept of surface tension to explain why a drop of water forms a sphere.

Thinking Critically

4. **Comparing** Ethylene glycol is a molecule with the formula $C_2H_4(OH)_2$. Each molecule has two O—H bonds. Ethanol, C_2H_5OH, has one O—H bond. How would you expect the boiling points of ethanol and ethylene glycol to compare?

Applying Chemistry

5. **Humidity and Cooling** Hot weather is more uncomfortable if the humidity is high. Explain why.

chemistryca.com/self_check_quiz

Solutions and Their Properties

The polar water molecule is capable of dissolving a range of compounds, from ionic compounds, such as sodium chloride, to covalent compounds, such as sugars. What properties of the resulting aqueous solutions make them different from pure water? Because most of the water with which you come into contact contains dissolved materials, these aqueous solution properties play an important role in your everyday life.

SECTION PREVIEW

Objectives

✓ **Compare and contrast** the ability of water to dissolve ionic and covalent compounds.

✓ **Distinguish** solutions from colloids.

✓ **Compare and contrast** colligative properties.

Review Vocabulary

Capillarity: the rising of a liquid in a narrow tube, sometimes called capillary action.

New Vocabulary

dissociation
unsaturated solution
saturated solution
supersaturated
 solution
heat of solution
osmosis
colloid
Tyndall effect

The Dissolving Process

Just as with the other physical properties of water, you can relate water's solvent properties to its molecular structure. The submicroscopic interactions that occur between water molecules and various solute particles determine the extent to which water is able to dissolve the solutes.

Water Dissolves Many Ionic Substances

You have probably added table salt, sodium chloride, to water and noticed that the salt dissolved. If you made a careful measurement, you'd discover that you could dissolve about 36 g of sodium chloride in 100 mL of water at room temperature. Salt, like a great many ionic compounds, is soluble in water. The salt solution is also an excellent conductor of electricity. This high level of electrical conductivity is always observed when ionic compounds dissolve to a significant extent in water.

What's going on in such a situation? Your model of water and its interactions explains why salt and many other ionic compounds dissolve in water and why the solutions conduct electricity. Remember that ionic solids are composed of a three-dimensional network of positive and negative ions, which form strong ionic bonds. Look at **Figure 13.15** to see how a crystal of sodium chloride dissolves in water.

miniLAB 2

Hard and Soft Water

Water is said to be hard if it has significant concentrations of certain ions, usually Ca^{2+} and/or Mg^{2+}. Water in which calcium and magnesium ions are present in low concentrations only or in which they have been substantially replaced with Na^+ ions is called soft water. Distilled water contains only dissolved gases and exceedingly few ions. Test hard water, soft water, and distilled water with a sodium oxalate ($Na_2C_2O_4$) solution. This solution will cause Ca^{2+} and Mg^{2+} to precipitate as calcium oxalate (CaC_2O_4) and magnesium oxalate (MgC_2O_4). How well do the three types of water produce suds when they are mixed with a soap solution?

Procedure

1. Wear an apron and safety goggles.

2. Add about 1 mL of hard water to a small test tube, about 1 mL of soft water to another small test tube, and about 1 mL of distilled water to a third small test tube.

3. Add to each of the three solutions 2 drops of $0.1M$ acetic acid and 2 drops of $0.1M$ sodium oxalate. Mix the contents of each of the tubes by tapping them with your finger. Observe the solutions against a black background and record your observations.

4. Add about 2 mL of hard water to a large test tube, about 2 mL of soft water to another large test tube, and about 2 mL of distilled water to a third large test tube.

5. Add to each of the three solutions about 1 mL of soap solution.

6. Shake each of the three solutions ten to 15 times by holding the test tube in your hand with your thumb over the top. Measure and record the height of the suds in each tube.

Analysis

1. If the oxalate hardness test is negative (that is, if no precipitate forms), is a complete absence of calcium and magnesium ions confirmed?

2. How does the hardness of water affect the ability to make soap suds?

Figure 13.15

A Model of the Dissolving of NaCl

When water is added, the polar water molecules surround the sodium and chloride ions, and the ionic compound dissociates. Because water molecules are polar and have a negative end and a positive end, they are attracted to both the positively charged sodium ions and the negatively charged chloride ions. Water molecules surround both types of ions. The opposite charges attract.

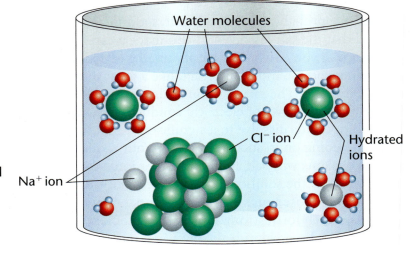

The process by which the charged particles in an ionic solid separate from one another is called **dissociation.** You can represent the process of dissolving and dissociation in shorthand fashion by the following equation.

$$NaCl(s) \xrightarrow{H_2O} Na^+(aq) + Cl^-(aq)$$

Water Dissolves Many Covalent Substances

Water is not only good at dissolving ionic substances. It also is a good solvent for many covalent compounds. Consider the covalent substance sucrose, commonly known as table sugar, as an example. You have probably observed that this substance, with the formula $C_{12}H_{22}O_{11}$, dissolves in water. In fact, it is highly soluble. It's possible to dissolve almost 200 g of sugar in 100 mL of water.

Take a look at **Figure 13.16.** It shows the molecular structure of a sucrose molecule. Notice that the structure has a number of O—H bonds. As you learned earlier, if a molecule contains O—H bonds, it will tend to be polar and it can form hydrogen bonds. One of the reasons that sugar is a solid rather than a liquid at room temperature is that sugar molecules have the ability to form hydrogen bonds with each other. The attractions hold the molecules in a crystal arrangement that ordinarily cannot be broken without an input of heat.

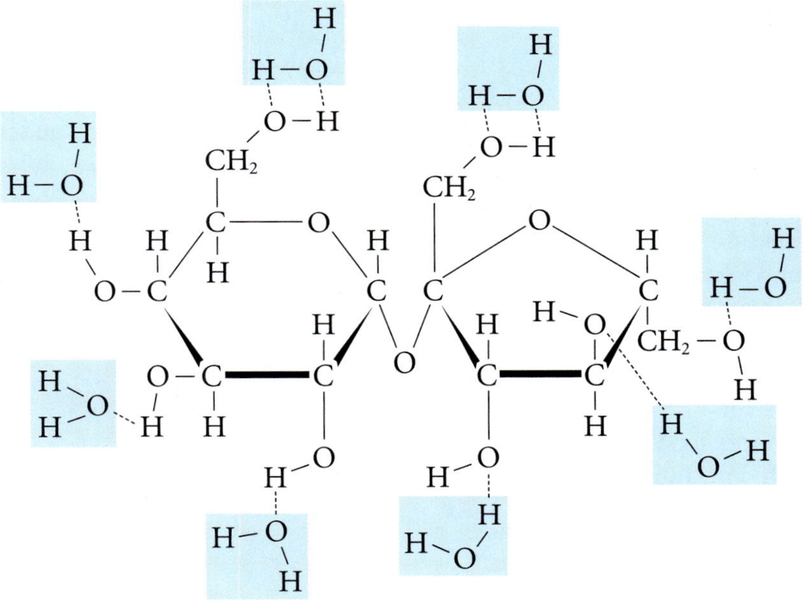

Figure 13.16

Sucrose

Sucrose, $C_{12}H_{22}O_{11}$, is a molecule that contains eight O—H bonds. When water is added to solid sucrose, each of the O—H bonds on the sucrose molecule is a possible site for hydrogen bonding with water. The intermolecular attractive forces between sucrose molecules are overcome and replaced by water-sucrose intermolecular attractive forces. This is why sugar is highly soluble.

This covalent sucrose model is similar to the ionic sodium chloride model in some ways. In both cases, interparticle attractive forces between the solvent and solute particles overcome attractive forces between solute particles. However, because sucrose is covalent, the sucrose molecules are simply separated from one another by water molecules. They do not dissociate into charged particles, but remain neutral molecules. Neutral molecules cannot conduct electricity because they have no charge. Therefore, an aqueous solution of sucrose is a nonconductor. The dissolving of sugar is represented by the following simple equation. Note that no ions are formed.

$$C_{12}H_{22}O_{11}(s) \xrightarrow{H_2O} C_{12}H_{22}O_{11}(aq)$$

Like Dissolves Like

Although water dissolves an enormous variety of substances, both ionic and covalent, it does not dissolve everything. The phrase that scientists often use when predicting solubility is "like dissolves like." The expression means that dissolving occurs when similarities exist between the solvent and the solute.

Consider the two examples of salt and sugar being dissolved in water. In the case of the ionic salt, water is "like" a charged ionic compound in the sense that it is polar, meaning that it has partially charged ends. The interactions that may occur between water molecules—dipole-dipole attractions and hydrogen bonds—are somewhat similar to the water-ion interactions that occur when salt dissolves. Thus, "like dissolves like." Water is polar and it tends to dissolve ionic substances.

In the case of sugar, water is like sucrose in that both compounds contain O—H bonds. More importantly, both substances are made up of polar molecules, with partially positive and negative ends. The molecular interactions between sucrose molecules are like the molecular interactions between water molecules. Again, like dissolves like. Water is polar and has hydrogen bonding, and it tends to dissolve substances that are polar or that form hydrogen bonds.

Oil and water are a classic example of two substances that do not mix; they do not form a solution. Oil is a mixture of nonpolar covalent compounds made up primarily of carbon and hydrogen. Given the composition of oil, you should not be surprised that oil does not dissolve in water. They are simply too *unlike*. When you try to place them in contact, there is little intermolecular force between solvent and solute that would allow them to mix. As shown in **Figure 13.17,** even after vigorous shaking, oil and water will rapidly separate into layers. However, different oils, which are nonpolar, are "like" enough to remain mixed.

Figure 13.17

Oil and Water Don't Mix, but Different Oils Do

When a bottle containing a mixture of baby oil (colorless) and water (dyed blue) is shaken, the oil and water seem at first to mix.

Upon sitting, the substances separate into two layers. ▶

In contrast, olive oil (greenish-gold) and safflower oil (nearly colorless) mix readily and remain mixed. ▶

Everyday Chemistry

Soaps and Detergents

Has a slippery piece of pizza ever made a greasy stain on your favorite jeans? You pour water on the telltale spot, but the water just runs off. You are learning the lesson that oil and water don't mix, unless you add soap. Oil is normally insoluble in water, but soap molecules achieve the seemingly impossible by causing oil to mix with water.

Making soap Though most people use soap for bathing and other detergents for washing clothes, there are important similarities between the two.

People may first have learned how to make soap when ashes from a fire fell into a pot of boiling fat. The people soon found uses for the smooth white gel that floated to the top of the mixture. The ashes supplied lye, which is a strong base such as sodium hydroxide or potassium hydroxide. When fats or oils react chemically with lye, the end products are soap—often sodium stearate—and glycerin. Examine the chemical formula of sodium stearate. A negative carboxyl group ion ($-COO^-$) is tied to a positive sodium ion (Na^+) and a long chain of $-CH_2-$ groups.

$$CH_3 - (CH_2)_{16} - COO^-Na^+$$

Like soaps, detergents have molecules with a polar end and a nonpolar end. Detergents typically contain sulfonates, which have SO_3 groups attached to a carbon chain or ring. The detergents are typically sodium salts of these sulfonates, and their molecules have an SO_3^- end rather than the COO^- end that is typical of soaps.

Soap forms a scumlike precipitate in the presence of the Mg^{2+} and Ca^{2+} ions that are present in hard water. Detergents, on the other hand, form soluble sulfonate salts in the presence of Mg^{2+} and Ca^{2+}. That makes detergents more effective than soaps in hard water.

How soaps and detergents work A soap molecule has two different parts: one end is hydrophilic, attracted to water, and the other end is hydrophobic, repelled by water. The hydrophobic part of the soap molecule consists of a long hydrocarbon chain that is structurally similar to oil and is therefore soluble in oil.

The hydrocarbon chains in soap are attracted to particles of oily grease and dirt. That part of the soap molecule forms a protective layer around the oily material. The hydrophilic end of the soap molecule (the $-COONa$ end) is attracted by polar water molecules. That causes the entire soap molecule, together with the oily material, to pull toward the wash water. The oil-soap complex is suspended in the water and rinsed away.

Exploring Further

1. **Inferring** How might rubbing and scrubbing with soap help to remove a greasy stain from clothing?

2. **Comparing** Some ancient peoples, such as the Egyptians and Romans, washed by rubbing themselves with oil, which was then scraped off. Compare the effectiveness of this method with the use of soap and water today.

For more about how soaps and detergents use chemistry to clean, visit the Chemistry Web site at chemistryca.com

ChemLab

Solution Identification

In an aqueous solution, ionic compounds are completely dissociated into ions. For example, an aqueous solution of barium nitrate, $Ba(NO_3)_2$, contains Ba^{2+} ions and NO_3^- ions. If aqueous solutions of ionic compounds are mixed, some ions may interact to form an insoluble product called a precipitate. For example, if aqueous solutions of barium nitrate and sodium sulfate are mixed, insoluble barium sulfate will precipitate. The complete formula equation for this reaction is written as follows.

$$Ba(NO_3)_2(aq) + Na_2SO_4(aq) \rightarrow$$
$$2NaNO_3(aq) + BaSO_4(s)$$

The barium sulfate is a white precipitate. Its formation as a result of mixing two unknown solutions could help you to identify the two reacting solutions. In this ChemLab, you will work with the aqueous solutions of six unknown ionic compounds. By observing the solutions and their interactions, you will be able to determine the identities of the compounds.

Problem

What are the identities of six unknown aqueous solutions?

Objectives

- **Observe** the interactions of the aqueous solutions of six compounds.
- **Interpret** the results of the interactions and use the results to identify the solutions.

PREPARATION

Materials

96-well microplate
10 mL each of 0.1M solutions of sodium carbonate, sodium iodide, copper(II) sulfate, copper(II) nitrate, lead(II) nitrate, barium nitrate—identified only as A through F
6 microtip pipets, labeled A through F
15 toothpicks
distilled water in a rinse bottle

Safety Precautions

Wear an apron and safety goggles. Wash your hands thoroughly after performing the lab. Some of the solutions are toxic.

PROCEDURE

1. Obtain a 96-well microplate. To produce all the combinations of solutions, you will be using only the 15 wells in the upper-right-hand corner.

2. Use the appropriately labeled microtip pipets to place three drops of solution A in each of five wells in the first horizontal row, three drops of B in each of four wells in the second row, three drops of C in each of three wells in the third row, three drops of D in each of two wells in the fourth row, and three drops of E in one well in the fifth row. The first letter in each box in the table in Data and Observations shows the arrangement.

3. For your second additions, use the appropriately labeled microtip pipets to add three drops of the solutions to the same wells to obtain the combinations shown in the Data and Observations table. Thus, you will place three drops of solution B in the single well in the first column, three drops of C in both of the wells in the second column, and so on up to the fifth column to which you will add three drops of solution F. You have now created all the possible combinations of pairs of solutions in the wells.

4. Stir the solutions in each well, using a clean toothpick for each.

5. Place the microplate on a white surface and look down through each well to detect the presence of a precipitate. Cloudiness is evidence of a suspended precipitate. Repeat the procedure on a black surface. Record in your data table the color of any precipitates. If no precipitate is formed, write *NR,* meaning "no reaction."

2. **Classifying** Use a handbook of chemistry to find out the colors of each of the products and whether they are soluble or insoluble in water.

3. **Drawing Conclusions** Identify the six solutions by relating your data and observations to the predicted colors and interactions. Assume that the color of an aqueous solution is generally the same as that of the corresponding solute compound, except that solutions of white compounds are colorless.

DATA AND OBSERVATIONS

Combinations				
A + B	A + C	A + D	A + E	A + F
	B + C	B + D	B + E	B + F
		C + D	C + E	C + F
			D + E	D + F
				E + F

ANALYZE AND CONCLUDE

1. **Formulating Models** Write balanced equations for all 15 possible double-replacement reactions, regardless of whether they actually occurred. Show both possible products in each case. If the two possible products are the same as the two reactants, you can rule out the possibility of reaction, so write *NR* (no reaction) after the equation.

APPLY AND ASSESS

1. Explain in general terms the reasoning you used in your identification process.

2. What are some other methods you might have used to identify the solutions?

Solution Concentration

Suppose someone handed you a bottle and said, "This is an aqueous ammonia solution." You'd know that it consists of ammonia dissolved in water, but you wouldn't know how much. In other words, you wouldn't know the concentration of the solution, which is the relative amount of solute and solvent.

Concentrated Versus Dilute

If you were making tea, you might choose the approximate concentration you desire based upon personal taste. If you like strong tea, you make what a chemist might call a concentrated solution of tea; a relatively large amount of tea is dissolved in the water, so the concentration is high. On the other hand, if you like weak tea, a chemist might call the resulting mixture a dilute solution; relatively little tea is dissolved in the water, so the concentration is low. **Figure 13.18** summarizes this terminology. Chemists never apply the terms *strong* and *weak* to solution concentrations. As you'll see in the next chapter, these terms are used in chemistry to describe the chemical behavior of acids and bases. Instead, use the terms *concentrated* and *dilute.*

Figure 13.18

Qualitative Expressions of Solution Concentration
When your tea is strong, it's concentrated. If you like weak tea, you prefer a dilute solution.

The Dead Sea in Israel receives water from several streams but has no outflow. Instead, water evaporates from it, which concentrates salt. The Dead Sea has a salinity of 25 percent—as compared with only about 3.5 percent for ocean water—and is saturated with salt. The salt makes the water uninhabitable for fishes but so buoyant that it is nearly impossible for swimmers to sink in it.

Unsaturated Versus Saturated

Another way of providing information about solution composition is to express how much solute is present relative to the maximum amount the solution could hold. If the amount of solute dissolved is less than the maximum that could be dissolved, the solution is called an **unsaturated solution.** The oceans of Earth are examples of unsaturated saltwater solutions. They could hold a higher concentration of salt than they do now. The maximum concentration of salt water is approximately 36 g of salt dissolved per every 100 g of water, or 36 percent by mass. Such a solution, which holds the maximum amount of solute per amount of the solution under the given conditions, is called a **saturated solution.**

Figure 13.19
Fudge from a Supersaturated Solution
In making fudge, you heat a highly concentrated mixture of sugar, chocolate, and a water-based solvent such as milk to a high enough temperature to make a sugar solution that is supersaturated. Next, you slowly cool the mixture to room temperature with a great deal of stirring. If you've done everything right, your fudge will be soft and creamy because the sugar will crystallize out as very small crystals.

An interesting third category of solution is called a **supersaturated solution.** Such solutions contain more solute than the usual maximum amount and are unstable. They cannot permanently hold the excess solute in solution and may release it suddenly. Supersaturated solutions, as you might imagine, have to be prepared carefully. Generally, this is done by dissolving a solute in the solution at an elevated temperature, at which solubility is higher than at room temperature, and then slowly cooling the solution. Fudge making involves preparation of a supersaturated solution, as shown in **Figure 13.19.**

Figure 13.20
Solubility Versus Temperature
The amount of solute required to achieve a saturated solution in water depends upon the temperature, as this graph shows. Most solutes increase in solubility as temperature increases.

Effect of Temperature on Solubility

As you realized in the fudge example above, the solubility of sugar increases as the temperature increases. More and more solute is able to dissolve at higher and higher temperatures. Temperature has a significant effect on solubility for most solutes. **Figure 13.20** shows how the solubilities of six different solutes change with temperature.

Notice that each solute behaves differently with temperature. The solubilities of some solutes, such as sodium nitrate and potassium nitrate, increase dramatically with increasing temperature. Notice how steeply the curves climb upward in the figure. Other solutes, like NaCl and KCl, show only slight increases in solubility with increasing temperatures. A few solutes, like cerium(III) sulfate, $Ce_2(SO_4)_3$, decrease in solubility as temperature increases.

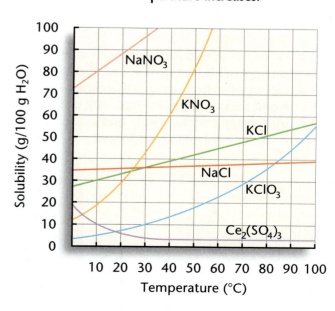

As you can see, heat plays an important role in determining solubility. The heat taken in or released in a dissolving process is called the **heat of solution.** For most solutes, the process of dissolving in a solvent is an endothermic process. The dissolving of ammonium nitrate, for example, is endothermic. You can write the equation for the process, incorporating the heat term, as follows.

$$NH_4NO_3(s) + heat \xrightarrow{H_2O} NH_4^+(aq) + NO_3^-(aq)$$

Notice that the heat in an endothermic process is written as if it were a reactant because it must be added to the substances that will form the products. To increase the solubility of NH_4NO_3, add more heat, which increases the temperature. This forces the process toward the production of more aqueous ions. When working with a solute that dissolves endothermically, you may notice that the mixture becomes cooler during the dissolving. The heat required for the process is taken from the solution, which therefore cools off.

The dissolving of some solutes is exothermic. For these solutes, the dissolving process releases heat. Calcium chloride is a good example of such a solute. Note that in the equation for the dissolving process, heat shows up on the product side.

$$CaCl_2(s) \xrightarrow{H_2O} Ca^{2+}(aq) + 2Cl^-(aq) + heat$$

Hot and cold packs for sports injuries, shown in **Figure 13.21,** use solutes whose dissolving is highly exothermic or endothermic. Therefore, the solutes have large heats of solution.

Molarity

Suppose you're a worker in a hospital pharmacy lab and you must prepare a salt solution that matches the salt concentration of a patient's blood. You need to measure things carefully. Blood is a dilute salt solution, but the term *dilute* gives only qualitative information. In this case, you need a quantitative concentration unit.

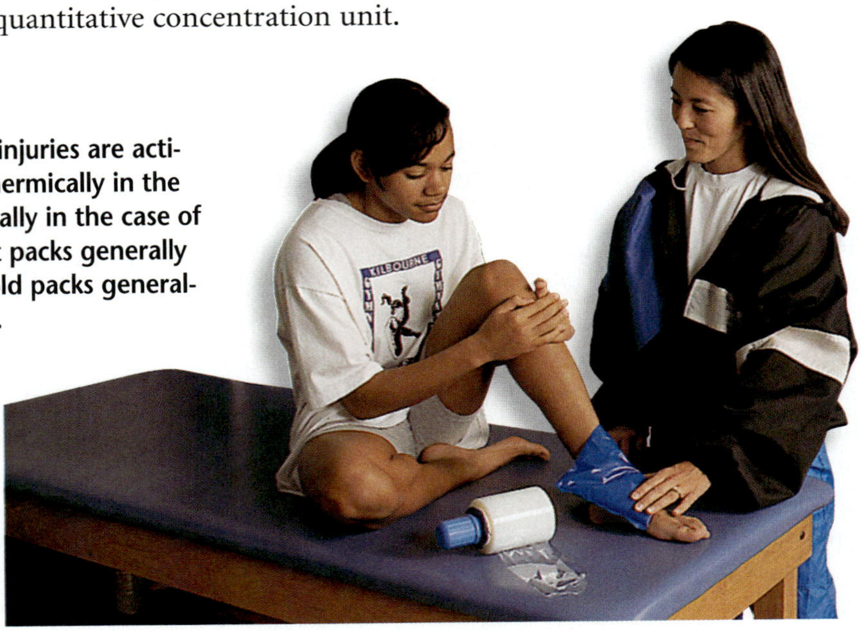

Figure 13.21

Using Heats of Solution
When hot or cold packs for sports injuries are activated, a solute that dissolves exothermically in the case of a hot pack, or endothermically in the case of a cold pack, dissolves in water. Hot packs generally use calcium chloride, $CaCl_2$, and cold packs generally use ammonium nitrate, NH_4NO_3.

Concentration units can vary greatly. They express a ratio that compares an amount of the solute with an amount of the solution or the solvent. For chemistry applications, the concentration term *molarity* is generally the most useful. Molarity is defined as the number of moles of solute per liter of solution.

Molarity = moles of solute/liter of solution

Note that the volume is the total solution volume that results, not the volume of solvent alone. Suppose you need 1.0 L of the salt solution mentioned above. In order to be at the same concentration as the salt in the patient's blood, it needs to have a concentration of 0.15 moles of sodium chloride per liter of solution. In other words, it must have a molarity of 0.15. To save space, you refer to the solution as 0.15*M* NaCl, where the *M* stands for "moles/liter" and represents the word *molar*. Thus, you need 1.0 L of a 0.15-molar solution of NaCl. How are you going to prepare it?

Assuming you're making an aqueous solution, you need to know only three things when working quantitatively: the concentration, the amount of solute, and the total volume of solution needed. The factor label approach to problem solving, as explained in Appendix A, is useful when you work problems such as the one on the next page. **Figure 13.22** shows the general solution-making steps in more detail using equipment that is available in labs where solutions are prepared quantitatively.

Step 1

Step 2

Step 3

Step 4

Figure 13.22

Quantitative Solution Preparation

Preparing solutions is important lab work. In step 1, the solute is weighed. In step 2, the solute is transferred to a volumetric flask, which holds a known volume. In step 3, enough water is added to dissolve the solute, and then more water is added to bring the solution volume up to the calibration mark on the flask. In step 4, the solution is shaken, stored in a stoppered container, and labeled.

How would you prepare 1.0 L of a 0.15M sodium chloride solution?

Analyze
- First, determine the mass of NaCl to add to a 1.0-L container. The 0.15M solution must contain 0.15 moles of NaCl per liter of solution.

Set Up
- You will need to use the molarity of the solution (0.15 mol NaCl/L solution) as a conversion factor to get from molarity to number of moles of NaCl. You will then use the molar mass of NaCl as a conversion factor to change moles of NaCl to grams of NaCl. To find the molar mass of NaCl (58.5 g/mol), add the atomic masses of Na and Cl, and apply the unit *grams/mole* to the sum.

> **Problem-Solving HINT**
>
> Remember that a conversion factor is any relationship that compares two different units.

Solve
- The proper setup, showing the conversion factors, is as follows.

$$\frac{1.0 \text{ L solution}}{} \left| \frac{0.15 \text{ mol NaCl}}{1 \text{ L solution}} \right| \frac{58.5 \text{ g NaCl}}{1 \text{ mol NaCl}}$$

Then carry out cancellations and calculate the answer.

$$\frac{1.0 \text{ L solution}}{} \left| \frac{0.15 \text{ mol NaCl}}{1 \text{ L solution}} \right| \frac{58.5 \text{ g NaCl}}{1 \text{ mol NaCl}} =$$

$$\frac{1.0}{} \left| \frac{0.15}{1} \right| \frac{58.5 \text{ g NaCl}}{1} = 8.8 \text{ g NaCl}$$

The result means you need to measure 8.8 g of NaCl, add some water to dissolve it, and then add enough additional water to bring the total volume of the solution to 1.0 L.

Check
- Check to be sure that all units were correctly handled and that the calculation can be repeated with the same result.

How would you prepare 5.0 L of a 1.5M solution of glucose, $C_6H_{12}O_6$?

Analyze
- You need to determine the number of grams of glucose to add to a 5.0-L container. The 1.5M solution must contain 1.5 mol of glucose per liter of solution.

Set Up
- Use the solution molarity as a conversion factor (1.5 mol glucose/L solution) to change from liters of solution to moles of glucose. Then, use the molar mass of glucose as another conversion factor to change from moles to grams of glucose. To find the molar mass of glucose (180 g/mol), add the atomic masses of 6 C, 12 H, and 6 O, and apply the unit *grams/mole* to the sum.

Solve
- The proper setup, showing the conversion factors, is as follows.

$$\frac{5.0 \text{ L solution}}{} \left| \frac{1.5 \text{ mol glucose}}{1 \text{ L solution}} \right| \frac{180 \text{ g glucose}}{1 \text{ mol glucose}}$$

Cancel units and carry out the calculation.

$$\frac{5.0 \; \cancel{\text{L solution}}}{} \; \Bigg| \; \frac{1.5 \; \cancel{\text{mol glucose}}}{1 \; \cancel{\text{L solution}}} \; \Bigg| \; \frac{180 \text{ g glucose}}{1 \; \cancel{\text{mol glucose}}} =$$

$$\frac{5.0}{} \; \Bigg| \; \frac{1.5}{1} \; \Bigg| \; \frac{180 \text{ g glucose}}{1} = 1400 \text{ g glucose}$$

The mass of glucose required is 1400 g. Weigh this mass, add it to a 5.0-L container, add enough water to dissolve the glucose, and fill with water to the 5.0-L mark.

Check • Check to be sure that all units were correctly handled and that the calculation can be repeated with the same result. All aspects of setup and calculation appear to be correct.

PRACTICE PROBLEMS

Supplemental Problems

For more practice with solving problems, see Supplemental Practice Problems, Appendix B.

1. How would you prepare 1.00 L of a 0.400M solution of copper(II) sulfate, $CuSO_4$?

2. How would you prepare 2.50 L of a 0.800M solution of potassium nitrate, KNO_3?

3. What mass of sucrose, $C_{12}H_{22}O_{11}$, must be dissolved to make 460 mL of a 1.10M solution?

4. What mass of lithium chloride, LiCl, must be dissolved to make a 0.194M solution that has a volume of 1.00 L?

SAMPLE PROBLEM 3 Calculating Molarity

You add 32.0 g of potassium chloride to a container and add enough water to bring the total solution volume to 955 mL. What is the molarity of this solution?

Analyze • You have all the information you need about the preparation of the solution. You know the solute, the mass, and the total volume of the solution in milliliters. What you need to know is the molarity (moles KCl/L) of the solution.

Set Up • You are given that there are 32.0 g of solute per 955 mL of solution, so this relationship can be expressed in fraction form with the volume in the denominator. Therefore, the initial part of the setup is as follows.

$$\frac{32.0 \text{ g KCl}}{955 \text{ mL solution}}$$

Determine that the molar mass of KCl is 74.6 g/mol by adding the atomic masses of K and Cl and applying the unit *grams/mole* to the sum. The conversion factor that must be used to convert from grams to moles of KCl is 1 mol KCl/74.6 g KCl.

$$\frac{32.0 \text{ g KCl}}{955 \text{ mL solution}} \; \Bigg| \; \frac{1 \text{ mol KCl}}{74.6 \text{ g KCl}} \cdots$$

Next, to convert milliliters to liters, given that there are 1000 mL solution/L solution, use that conversion factor in the setup.

$$\frac{32.0 \text{ g KCl}}{955 \text{ mL solution}} \left| \frac{1 \text{ mol KCl}}{74.6 \text{ g KCl}} \right| \frac{1000 \text{ mL solution}}{1 \text{ L solution}} \dots$$

Solve
- Cancel units and carry out the calculation, using the setup just developed.

$$\frac{32 \text{ g KCl}}{955 \text{ mL solution}} \left| \frac{1 \text{ mol KCl}}{74.6 \text{ g KCl}} \right| \frac{1000 \text{ mL solution}}{1 \text{ L solution}} =$$

$$\frac{32.0}{955} \left| \frac{1 \text{ mol KCl}}{74.6} \right| \frac{1000}{1 \text{ L solution}} =$$

0.449 mol KCl/L solution = 0.449M KCl

Check
- Check to be sure that all units were correctly handled and that the calculation can be repeated with the same result. All aspects of setup and calculation appear to be correct.

PRACTICE PROBLEMS

Supplemental Problems

For more practice with solving problems, see Supplemental Practice Problems, Appendix B.

5. What is the molarity of a solution that contains 14 g of sodium sulfate, Na_2SO_4, dissolved in 1.6 L of solution?

6. Calculate the molarity of a solution, given that its volume is 820 mL and that it contains 7.4 g of ammonium chloride, NH_4Cl.

Solution Properties and Applications

Throwing salt onto an icy sidewalk and adding coolant to a car radiator—what do these familiar activities have in common? They both relate to a set of interesting and useful solution properties that depend only on the number and concentration of solute particles. These properties are freezing-point depression and boiling-point elevation.

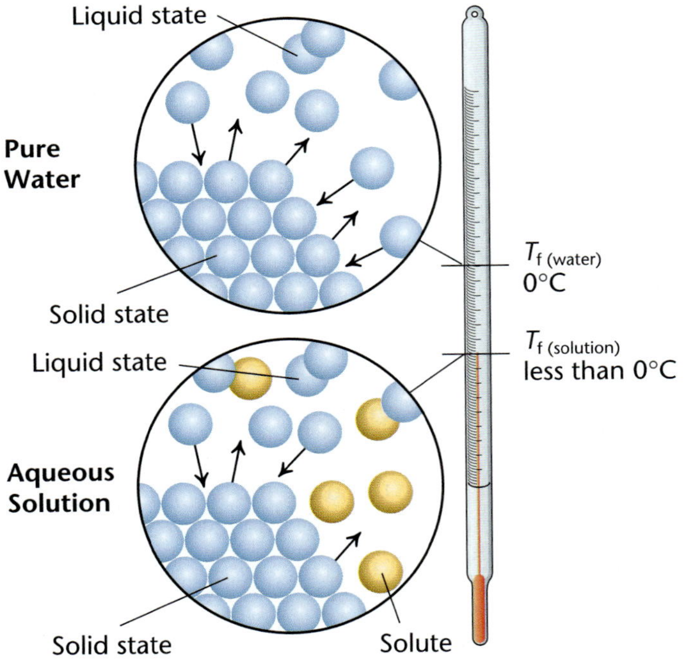

Liquid state

Pure Water

Solid state

Liquid state

Aqueous Solution

Solid state

$T_{f \text{ (water)}}$ 0°C

$T_{f \text{ (solution)}}$ less than 0°C

Solute

Figure 13.23

Disrupting Solvent Organization
A solute disrupts the high level of organization that is necessary if water is to be in the solid state. The freezing point of the solution is lower than the freezing point of the pure water. If the surrounding temperature is above the freezing point of the solution, the solution melts. The effect occurs because solute particles have replaced some of the solvent particles, and they interfere with the freezing process. The more solute particles there are, the more the freezing point is lowered.

Figure 13.24

Applying Freezing-Point Depression
Freezing-point depression has some practical uses, such as in deicing airplanes, a process that uses ethylene glycol as a solute.

Freezing-Point Depression

A solution always has a lower freezing point than the corresponding pure solvent. If you are interested only in aqueous solutions, this means that any aqueous solution will have a freezing point lower than 0°C. The amount that the freezing point is depressed relative to 0°C depends only upon the concentration of the solute. You have observed this property if you have carried out the winter salt-throwing you just read about. Salt placed on an icy sidewalk causes the ice to melt. This melting is a result of freezing-point depression. The salt dissolves in the water that makes up the ice and forms a solution that has a lower freezing point than pure water, as illustrated in **Figure 13.23.**

An ionic solute produces greater depression of freezing point than a covalent one because it dissociates into ions. One mole of ionic NaCl thus produces 2 moles of solute particles. It produces twice the interference with the freezing process and twice the desired effect as 1 mol of sucrose. Likewise, 1 mol of $CaCl_2$ has three times the effect of 1 mol of sucrose because each formula unit of $CaCl_2$ dissociates into three ions: one Ca^{2+} and two Cl^-. **Figure 13.24** gives some additional examples of the application of freezing-point depression.

Boiling-Point Elevation

You have just learned that the freezing point of a solution is lower than the freezing point of the pure solvent. It turns out that the boiling point of a solution is higher than the boiling point of the pure solvent. For aqueous solutions, this means that the solution boiling point will be greater than 100°C, assuming standard atmospheric pressure. The solute must also be nonvolatile; that is, not able to evaporate readily.

Solute particles affect boiling point because they take up space at the surfaces where the liquid and gas meet, which interferes with the ability of the solvent particles to escape the liquid state. The solute particles lower the vapor pressure of the solvent, so a higher temperature is required to bring the vapor pressure up to atmospheric pressure and cause boiling to occur. The higher the concentration of solute particles, the greater is the boiling-point elevation. Read about how antifreeze works as a coolant in an automobile engine in *Everyday Chemistry.*

Fact of the **MATTER**

The world's largest water-purification plant that uses reverse osmosis is in Jubail, Saudia Arabia. The plant provides 50 percent of the country's drinking water by using reverse osmosis to remove salt from seawater taken from the Persian Gulf.

Everyday Chemistry

Antifreeze

In 1885, Karl Benz of Germany invented and patented the first radiator for an automobile. This was a big change over simply cooling the engine by evaporation, which removed 4 L of water each hour and required constant replenishment. A radiator was able to recirculate the water used to cool the engine. After air-cooled water removed the heat created when the engine burned fuel, the water returned to the radiator to be recooled.

How chemical coolants work The first use of a chemical engine coolant, ethylene glycol, was tried in England in 1916 for high-performance military aircraft engines.

Ethylene glycol, $C_2H_4(OH)_2$, is still the major constituent of most automobile antifreeze solutions used in the United States. When it is added to water, the solution that is formed has a higher boiling point. This keeps the water in the radiator from boiling off, as it would without the coolant. Recall that the boiling point of a substance is the temperature at which the vapor pressure of the liquid phase equals the atmospheric pressure. When a solute is present, it crowds out some of the water molecules at the surface of the solution. This reduces the number of water molecules that can escape as water vapor. As a result, the solution has a lower vapor pressure than pure water. If the vapor pressure of the solution in the radiator is lowered, additional kinetic energy is needed to raise it to the same level as the atmospheric pressure. This makes the boiling point of the solution higher than that of water.

Ethylene glycol as an antifreeze The process of adding a solute to water also lowers the temperature at which water freezes. The number of degrees the freezing point is lowered is proportional to the number of solute particles dissolved in the water. Any solute added to water decreases its freezing point. A solution with a low freezing point, such as an ethylene glycol solution in a car radiator, is less likely to freeze in winter.

Properties of a good engine coolant Besides raising the boiling point and lowering the freezing point, a good coolant doesn't corrode the materials in the radiator and the pump. Over the years, material and design changes have caused problems of corrosion of metal parts in contact with coolants. Adding inhibitors to the coolant or changing the materials solves this problem.

A good coolant should also be easily disposed. Ethylene glycol is biodegradable. However, it is toxic to mammals if ingested. It has a sweet smell and taste, which appeals to animals and sometimes children. Care must be taken in disposing of these chemicals until a safer product is found.

Exploring Further

1. **Acquiring Information** Investigate how ice cream is made at home. Relate the information gathered to the use of an antifreeze in an automobile during winter.

2. **Applying** If a coolant becomes oily, murky, or rusty, what would this suggest?

To learn more about how solvents affect freezing and boiling points, visit the Chemistry Web site at **chemistryca.com**

Osmosis

Have you ever seen water being sprayed onto vegetables in a super-market? Some of the water is absorbed by the vegetables, which makes them plumper, crisper, and fresher looking. The water is able to move inside because cell membranes on the outside of the vegetables are selectively permeable—that is, they allow certain materials, such as water, to pass through them.

As you may recall from Chapter 10, gas molecules have a tendency to diffuse from an area of high concentration to an area of low concentration. Particles in a liquid have a similar tendency to move. In the case of the vegetables, the water that is naturally inside them contains solutes, such as sugars and salts. Because of the presence of these solutes, there is less water per unit volume than in pure water. If pure water is sprayed onto the outside of the vegetables, the water will tend to diffuse into the vegetables. It moves from the area of higher water concentration outside to the area of lower water concentration inside the vegetables. This flow of solvent molecules through a selectively permeable membrane, driven by concentration difference, is called **osmosis.**

Figure 13.25 shows a system that illustrates osmosis. Pure water is on the left side of a selectively permeable membrane, and an aqueous solution of sucrose is on the right side. The membrane allows water molecules to pass, but not sucrose molecules.

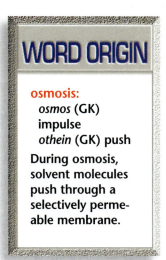

WORD ORIGIN

osmosis:
osmos (GK)
impulse
othein (GK) push

During osmosis, solvent molecules push through a selectively permeable membrane.

Figure 13.25

Osmosis
Note what takes place when pure water and a sucrose solution are separated by a selectively permeable membrane.

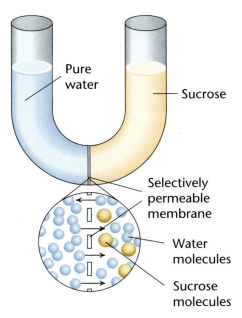

Osmosis

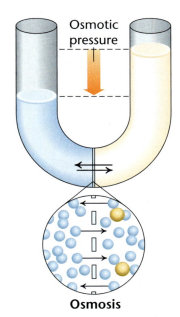

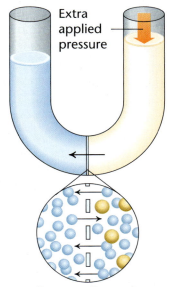

Reverse Osmosis

▲ As osmosis begins, water molecules diffuse more rapidly from the water into the sucrose solution than they diffuse from the sucrose solution into the water. As a result, the sucrose solution gains water, becomes more dilute, and the volume rises.

▲ The increasing height of the sucrose solution exerts a pressure that opposes the diffusion of water molecules from left to right. Eventually, the pressure becomes high enough that the rates of diffusion of water in both directions become the same.

▲ Adding extra pressure to the side with the sucrose solution can cause the water diffusion to move in the opposite direction, forcing water out of the solution and producing pure water. This process, called reverse osmosis, can be used to purify water, as discussed in *How It Works.*

A Portable Reverse Osmosis Unit

Reverse osmosis (RO) is the process by which water from a solution is forced through a selectively permeable membrane by the application of pressure to the solution side of the membrane. Portable RO units are commercially available that can purify seawater of its salts and make it drinkable. Approximately 27 atm of pressure needs to be applied to seawater in order to counteract the flow of solvent into the seawater through a selectively permeable membrane. In order to get a usable amount of water through the membrane, you need to apply about twice that pressure. An RO unit can apply these large pressures.

1. An operating handle is pushed down, and an attached piston puts pressure on seawater to drive it into a cylinder. Some of the water moves through a selectively permeable membrane, but the salts are left behind.

3. The remaining salty water that has not passed through the membrane is returned to the area behind the piston, where it provides some of the pressure needed to process the seawater.

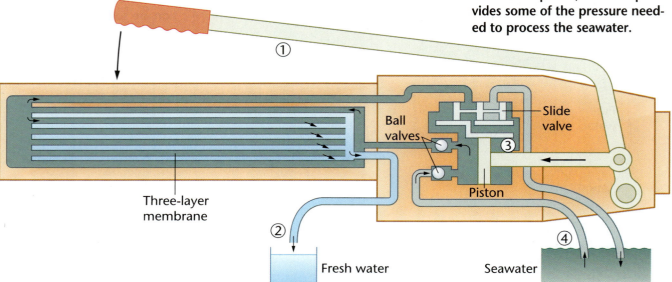

2. The pure water leaves the device and is collected.

4. The operating arm is lifted. The seawater behind the piston flows out as waste. New seawater enters and passes around movable ball joints, and the process is ready to begin again.

Thinking Critically

1. Why must the pressure required to operate a reverse osmosis unit be more than 27 atm?

2. Why isn't reverse osmosis more widely used as a purification method for water?

Solutions of Gases in Water

If you observe an unopened bottle of club soda, the liquid inside looks just like pure water. However, when you unscrew the cap, small bubbles of gas appear throughout the liquid and rise to the top. If the club soda is warm, the fizziness can be so intense that the liquid rises up and spurts out the top. **Figure 13.26** explains this sort of behavior.

The solubility of a gas in a liquid depends on the pressure of the gas pushing down on the liquid. The higher the pressure, the more soluble is the gas. For solutions of gases in liquids, gas solubility decreases as temperature increases. That's why club soda fizzes more vigorously when it is warm. The dependence of gas solubility on temperature and pressure is important in a number of areas, as shown in **Figure 13.26.**

Figure 13.26

Solutions of Gases in Water

The club soda (left) is a solution of carbon dioxide in water. In the unopened bottle, the gas is dissolved under pressure in the water. When the bottle is opened, the carbon dioxide gas trapped above the liquid escapes and the pressure drops. The solution is now supersaturated in carbon dioxide. Some of the dissolved carbon dioxide reenters the gaseous state and forms the bubbles that you see. ▼

▲ Deep under water, pressures are high. The nitrogen gas from the air in the lungs of divers gets dissolved at higher-than-normal concentrations in their blood. As they ascend from the depths, the pressure decreases and so does the blood solubility of the nitrogen. If divers come up too rapidly, the released nitrogen can form dangerous and painful bubbles in the blood vessels in a condition called the bends. To help prevent this, divers may use tanks of gas that contain helium mixed with oxygen instead of the usual nitrogen-oxygen mixture. The helium is less soluble in blood.

The fishes depend on dissolved oxygen in the water. If water temperature goes too high, the solubility of the oxygen may fall too low, and the fishes may die. That is one of the reasons thermal pollution from power plants can pose serious problems to aquatic life. ▶

Versatile Colloids

When you look at the photographs of the colloids on these pages, think about what these materials have in common. Recall that colloids are mixtures composed of tiny particles of one substance that are dispersed or evenly distributed in another substance. The particles of a colloid are intermediate in size between those of suspensions and true solutions. They range somewhere between the size of large molecules and a size great enough to be seen through a microscope.

Liquid Aerosols

Fog is an example of a liquid aerosol, which is formed when fine liquid droplets are dispersed in a gas. Fog appears when moist air near Earth's surface is cooled to the point at which the water vapor begins to condense. Other liquid aerosols with which you are familiar include spray deodorants and hair spray.

Solid Aerosols

At times, solid particles are dispersed in a gas. One example of a solid aerosol is the polluting soot particles that may be released into the air from an industrial smokestack. Such a problem could be remedied by placing a

Cottrell precipitator in the smokestack. The charged plates of that type of precipitator attract colloidal soot particles and remove them from the air. This is how air pollution has been reduced in many industrial cities.

Emulsions

Milk and mayonnaise are examples of emulsions. Emulsions are dispersions of fine droplets of liquid—generally a fat—in another liquid. Many emulsions are able to maintain their stability with the help of materials such as gums. Gums and other stabilizers thicken the liquid phase, making it less likely for the dispersed droplets to come together.

Sols

A sol is a fluid colloidal system in which fine solid particles are dispersed in a liquid medium. Most household paints are sols with finely ground pigments mixed with acrylic resins dissolved in water. Paint is applied to a surface as a liquid. With time, the water dries and the resins harden, leaving a thin solid film on the surface.

Gels

The food industry uses many colloids that have the ability to thicken or gel liquid foods. Gels are dispersions of giant macromolecules in liquid. Many gels are natural gums found in seaweeds and land plants. The natural substance pectin, found in fruits, is responsible for the gel structure in jelly.

Pastes

The basis for a paste is a concentrated dispersion of solids in a limited amount of liquid. To make a beautiful porcelain vase like the one in the photo, ground quartz and feldspar are mixed with a white clay called kaolin in a small portion of water. A paste is formed in which the water adheres to the surface of the clay, making the clay easy to work with.

Foam

You are familiar with the foam that forms when egg whites are beaten. Foam is a dispersion of gas bubbles in a liquid. Yeasts provide another kind of foam, as can be seen in bread dough. They do this by fermenting carbohydrates and giving off carbon dioxide gas, which produces tiny holes in the dough.

DISCUSSING THE TECHNOLOGY

1. **Classifying** Classify several health and beauty products according to the kinds of colloids described in this feature.

2. **Thinking Critically** Colloids do not pass through selectively permeable membranes. What can you conclude from this?

3. **Hypothesizing** Destruction of a colloid through a clumping process called coagulation can usually be achieved by heating. How would heat accomplish this?

WORD ORIGIN

colloid:
kolla (GK) glue
eidos (GK) form
A colloid often
contains particles
that are stuck
together in small
clumps.

Colloids

Sometimes, mixtures are partway between true solutions and heterogeneous mixtures. Such mixtures, called **colloids,** contain particles that are evenly distributed through a dispersing medium, and remain distributed over time rather than settling out. The major difference between a colloid and a solution is the size of the solute particles. Colloid particles are generally clumps that are ten to 100 times larger than typical ions or molecules dissolved in solutions. Because of their relatively large particle size, colloids play important roles in a variety of processes. Sometimes solid particles are dispersed in another solid, as shown in **Figure 13.27.** Some biological molecules, such as proteins, are large enough that their behavior is often best understood using a colloid model.

Because colloid particles are evenly dispersed, it is sometimes hard to tell colloids from true solutions. However, the larger particle size gives colloids some unique properties that help in identifying them. **Figure 13.28** illustrates one of these properties. Notice how a beam of light looks different when it passes through a solution and through a colloid. In the solution, the beam's path is hardly visible. As the light moves into the colloid, the light is partially scattered and reflected by the dispersed particles, and the beam becomes visible and broadens. You can observe the same phenomenon when you see the path of sunbeams through dusty air or the path of headlight beams on a foggy night. This light-scattering effect is called the **Tyndall effect.** It occurs because the dispersed colloid particles are about the same size as the wavelength of visible light (400 to 700 nanometers). The solute particles in a true solution are too small to produce this effect.

Figure 13.27
Stained Glass Windows
Brilliant colors can be given to glass, an amorphous solid, by adding certain solids. One part of nickel oxide particles is added to 50 000 parts of the ingredients normally used in making milk glass, which is frosty white, in order to produce a glass with a yellow tint. One part cobalt oxide particles added to 10 000 parts of the ordinary ingredients in glass produces blue glass. Gold, copper, or selenium oxide particles are used to produce red glass.

Figure 13.28

Tyndall Effect

Notice how the light beam becomes easily visible in the colloid because of light scattering. The headlight beam is visible in fog for the same reason.

Connecting Ideas

Now that you have learned about some of the properties of water, solutes, and colloids, you can understand why it is sometimes said that the chemistry of water is the chemistry of solutions. As you have discovered in this chapter, the physical properties of water make it a unique and important substance. Its ability to form solutions gives it an essential role in every aspect of life, including the reactions of acids and bases. As you will see, in these reactions, water is a subtle but important partner.

SECTION REVIEW

Understanding Concepts

1. Explain how a water molecule can be attracted to both a positive ion and a negative ion when dissolving an ionic compound.

2. Write equations for the dissociation of the following ionic compounds when they dissolve in water.

   a) Na_2SO_4 b) NaOH   c) $CaCl_2$

3. Characterize each of the following solutions as unsaturated, saturated, or supersaturated before the addition of solute.

   a) a solution that will produce a large amount of crystalline solid if only a small additional amount of solute is added
   b) a solution in which additional solute can be dissolved and remain in solution
   c) a solution in which, if additional solute is added, that solute will remain undissolved at the bottom of the container

Thinking Critically

4. **Comparing** Determine which solution has the highest concentration. Then rank the solutions from the one with the lowest to the one with the highest freezing point, and explain your answer.

   a) 0.10 mol of KBr in 100.0 mL of solution
   b) 1.1 mol of NaOH in 1.00 L of solution
   c) 1.6 mol of $KMnO_4$ in 2.00 L of solution

Applying Chemistry

5. **A Kitchen Colloid** Oil and aqueous solutions normally do not mix. However, if oil and vinegar or lemon juice are beaten together with egg, a stable mixture is formed. Explain how this is possible.

REVIEWING MAIN IDEAS

13.1 Uniquely Water

- Water has a number of unusual properties, such as high boiling point for its molecular size.

- On Earth, water exists primarily as liquid but also as solid and gas.

- The polarity of the water molecule is the source of many of water's unusual physical properties.

- Hydrogen bonds are formed by strong interactions between the hydrogen atoms on one molecule and a highly electronegative atom on another.

- Specific heat is the amount of heat needed to raise the temperature of 1 g of a substance by 1°C. Water has a high specific heat.

13.2 Solutions and Their Properties

- Like dissolves like. For example, polar solvents, such as water, tend to dissolve polar and ionic solutes, but not nonpolar ones.

- Interparticle forces between solvent and solute strongly influence solution formation.

- Ionic compounds dissociate when they dissolve in water.

- Hydrogen bonding plays an important role in the dissolving of many covalent compounds, such as sugar, in water.

- Solutions can be unsaturated, saturated, or supersaturated.

- Temperature affects solubility; generally, solubility of solids increases with increasing temperature.

- The molarity of a solution is equal to the moles of solute per liter of solution.

- Colligative properties of solutions, such as freezing-point depression and boiling-point elevation, are dependent only upon concentration of solute particles.

- Osmosis is movement of a solvent through a selectively permeable membrane under the influence of a concentration difference.

- Colloids are mixtures in which the dispersed particles are larger than those in solutions.

Vocabulary

For each of the following terms, write a sentence that shows your understanding of its meaning.

capillarity	saturated solution
colloid	specific heat
dissociation	supersaturated solution
heat of solution	surface tension
hydrogen bonding	Tyndall effect
osmosis	unsaturated solution

⬤ UNDERSTANDING CONCEPTS

1. List several ways in which water is unusual as a chemical substance.

2. Explain why water is polar.

3. Describe how the density of water relates to temperature, and explain why it does so.

4. What types of molecules form hydrogen bonds?

5. What is surface tension? Explain why it occurs.

6. Explain what capillarity is and give two examples of it.

7. In the figure to the right, which shows water molecules, identify the bonds as covalent or as hydrogen bonds.

8. Why does sucrose dissolve so well in water?

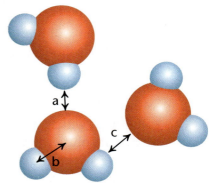

chemistryca.com/vocabulary_puzzlemaker

9. Explain how you would try to prepare a super-saturated solution.

10. Define *specific heat*. Why is the high specific heat of water important to Earth?

11. How does your body use perspiration to stay cool?

12. What is the molarity of a solution prepared by dissolving 0.217 mol of ethanol, C_2H_5OH, in enough water to make 100.0 mL of solution?

13. How many grams of sodium carbonate are needed to make 1.30 L of $0.890M$ Na_2CO_3?

APPLYING CONCEPTS

14. What mass of iron(III) chloride, $FeCl_3$, is needed to prepare 1.00 L of a $0.255M$ solution?

15. What is the molarity of a solution that contains 4.13 g of magnesium bromide, $MgBr_2$, in 0.845 L of solution?

16. What mass of potassium iodide, KI, is needed to prepare 5.60 L of a $1.13M$ solution?

17. To what volume in milliliters must 45.3 g of sodium benzoate, $NaC_7H_5O_2$, be diluted to give a final solution with a molarity of $0.250M$ $NaC_7H_5O_2$?

18. Explain how adding antifreeze to your car radiator protects it from freezing.

19. Explain why fresh water is a precious resource, despite the fact that Earth has plenty of water.

20. When you make pickles, the first step after washing the cucumbers is to soak them in a concentrated salt solution. What is the purpose of this step?

21. The specific heat of aluminum is 0.903 J/g°C; that of copper is 0.385 J/g°C. Suppose you have one cube each of Al and Cu; both cubes have a mass of 100 g and are at a temperature of 100°C. Which cube will release more heat when it cools to 20°C?

22. If you prepared a saturated aqueous solution of potassium chloride at 25°C, then heated the solution to 50°C, would you describe the resulting solution as saturated, unsaturated, or supersaturated? Explain.

23. 100.0 mL of an aqueous solution of $1.00M$ barium nitrate, $Ba(NO_3)_2$, is mixed with 100.0 mL of an aqueous solution of $1.00M$ sodium sulfate, Na_2SO_4. How many grams of barium sulfate will form?

$$Ba(NO_3)_2(aq) + Na_2SO_4(aq) \rightarrow$$
$$BaSO_4(s) + 2NaNO_3(aq)$$

Everyday Chemistry

24. Explain why soap is effective in removing greasy dirt.

25. How is a solute such as ethylene glycol able to help prevent the water in a car radiator from boiling?

How It Works

26. Why would it be more difficult to use a reverse osmosis unit to purify water from the Dead Sea than water from the ocean?

Chemistry and Technology

27. Contrast emulsions and sols and give an example of each.

Chemistry and Society

28. Explain why gravity settling alone is generally not sufficient as a water-purification treatment.

THINKING CRITICALLY

Relating Cause and Effect

29. **MiniLab 1** If you carefully place a steel razor blade flat on the surface of water, the razor blade can be made to float. Explain this result, given that the density of steel is much greater than that of water.

Drawing Conclusions

30. On a humid day, you may notice that water droplets condense on the outside of a glass of cold lemonade. What effect does this condensation have on the temperature of the glass of lemonade?

Interpreting Graphs

31. Use **Figure 13.20** to compare the solubilities in water of sodium chloride and potassium nitrate over the temperature range from 10°C to 40°C.

Drawing Conclusions

32. ChemLab Suppose you have spilled some blue copper(II) sulfate solution on a shirt. In washing the shirt, why should you avoid using washing soda (sodium carbonate)?

Making Predictions

33. MiniLab 2 Suppose that your tap water does not tend to produce suds when soap is mixed with it. What would you expect to happen if you added sodium oxalate test solution to a sample of your water? Why would you expect this to happen?

CUMULATIVE REVIEW

34. Explain how a potassium atom bonds to an atom of bromine. (Chapter 4)

35. Why does your skin feel cool if you dab it with rubbing alcohol? (Chapter 10)

36. How many liters of nitrogen are needed to react completely with 28 L of oxygen in the synthesis of nitrogen dioxide? (Chapter 11)

SKILL REVIEW

37 Written Summary of a Graph Examine the adjacent graph and write a summary of the information in it. In the course of your summary, answer the following questions.

   a) What is being graphed?
   b) What do the four different lines represent?
   c) Which compound has the lowest boiling point? Which has the highest?
   d) Moving down a group on the periodic table, what is the general trend in boiling point for a particular family of hydrogen compounds?
   e) Moving across a period, what is the general trend in boiling point for the hydrogen compounds shown?

   f) What does the dashed line show?
   g) Are boiling points of the second-period hydrogen compounds predicted from the behavior of the rest of the group? Explain any anomalies or unpredicted behavior.
   h) Which of the compounds are liquids at room temperature?

WRITING IN CHEMISTRY

38. Drinking water is a valuable resource. Write an article that traces the route and treatment of water from your faucet, down the drain, to your local municipal sewage-treatment plant, and back, ultimately, to your faucet. Your paper should give a general overview of municipal sewage treatment and drinking-water treatment.

PROBLEM SOLVING

39. Suppose you dipped a glass pipet tube into a beaker of water. Would the water level in the tube be higher or lower than the water level in the beaker? Explain. Suppose that you then immersed a clean, dry capillary tube into a beaker of molten wax and withdrew the tube in such a way that the inside walls of the tube became coated with a thin film of solid wax as the tube cooled. What would happen if you dipped this wax-coated tube into the beaker of water? Would the level of water inside the tube be higher or lower than it was for the non-wax-coated tube? Explain.

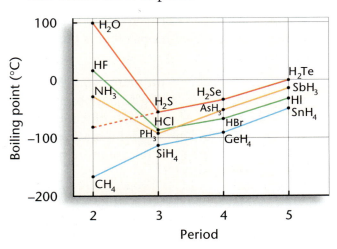

1. How is water a unique substance?

   a) Water is a liquid at room temperature even though it has a low molecular mass.
   b) In its solid state, water has a lower density than in its liquid state.
   c) Water has a high boiling point for a substance with such as a low molecular mass.
   d) All off the above.

2. Why does water have a boiling point of 100° C?

   a) The heavier mass of water molecules prevents them from detaching at lower temperatures.
   b) The lighter mass of water molecules causes them to detach at lower temperatures.
   c) The hydrogen bonds between water molecules prevent them from detaching at lower temperatures.
   d) The hydrogen bonds between water molecules cause them to detach at lower temperatures.

3. Water molecules form drops because of

   a) surface tension.   c) evaporation.
   b) capillarity.   d) sublimation.

4. Why do the ingredients in Italian salad dressing separate and not mix?

   a) The water is polar, and the oil is nonpolar.
   b) The oil is polar, and the water is nonpolar.
   c) Both water and oil are ionic substances.
   d) Both oil and water are covalent substances.

5. A supersaturated solution is a

   a) solution without a solute.
   b) solution with less than the usual maximum dissolved solute.
   c) solution with the maximum dissolved solute.
   d) solution with more than the usual maximum dissolved solute.

6. Molarity is defined as

   a) the number of liters of solution per molecules of solute.
   b) the number of moles of solute per liter of solution.

   c) the number of liters of solute per molecules of solution.
   d) the number of moles of solution per liter of solute.

7. What volume of a 0.125 M $NiCl_2$ solution contains 3.25 g $NiCl_2$?

   a) 406 mL   c) 38.5 mL
   b) 201 mL   d) 26.0 mL

Interpreting Graphs: Use the graph to answer questions 8–9.

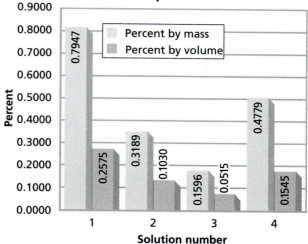

Bromine (Br_2) Concentration of Four Aqueous Solutions

8. The volume of bromine (Br_2) in 7.000 L of Solution 1 is

   a) 55.63 mL.   c) 18.03 mL.
   b) 8.808 mL.   d) 27.18 mL.

9. How many grams of Br_2 are in 55.00 g of Solution 4?

   a) 3.560 g   c) 1.151 g
   b) 0.084 g   d) 0.2628 g

Test Taking Tip

Take a Break! If you have a chance to take a break or get up from your desk during a test, take it! Getting up and moving around will give you extra energy and help you clear your mind. During your stretch break, think about something other than the test so you'll be able to get back to the test with a fresh start.

chemistryca.com/standardized_test

CHAPTER
14 Acids, Bases, and pH

Chapter Preview
Sections

14.1 Acids and Bases
Minilab 14.1 What Do Acids Do?

14.2 Strengths of Acids and Bases
MiniLab 14.2 Antacids
ChemLab Household Acids and Bases

Eat It or Clean With It?

Citrus fruits are a common source of acid. Because of their acidic composition, oranges and lemons are used in cleaning solutions. Acids also may chemically react with metal, and are a destructive force when air pollution comes in contact with marble and limestone structures. However, citrus fruits are ingested by humans on a daily basis as a common source of vitamin C.

Launch Lab

Testing pH Using Natural Indicators

How can you use red cabbage juice to determine the relative pH of several solutions?

Safety Precautions

Materials

- small test tubes (9)
- test-tube rack
- concentrated red cabbage juice in a dropper bottle
- labeled bottles containing: household ammonia, baking soda solution, 0.1 M hydrochloric acid solution, white vinegar, colorless carbonated soft drink, borax soap solution, distilled water
- grease pencil

Procedure

1. Construct a data table to record your observations.
2. Mark each test tube with the appropriate label.
3. Half-fill each test tube with the solution to be tested.
4. Add 10 drops of cabbage juice indicator to each test tube. Gently agitate each test tube to mix the solution.
5. Observe and record the color of each solution.

Analysis

Compare your observations with the table on page 504. Record in your data table the relative acid or base strength of each solution you tested.

What I Already Know

Review the following concepts before studying this chapter.
Chapter 5: names and formulas of common acids and bases
Chapter 13: dissociation of ionic compounds; hydrogen bonding

Reading Chemistry

Scan the key terms in the chapter for new words. Try to break down each word to find its origin, or recall a context in which you have heard the word used before. As you read, note the words' official definition in the text.

Preview this chapter's content and activities at chemistryca.com

Acids and Bases

A s you have discovered, classifying sub-stances into broad categories simplifies the study of chemistry. Consider some of the chemistry classification schemes you have used. In each scheme, the categories have generally been opposites, such as metal versus nonmetal, ionic versus covalent, and soluble versus insoluble. The materials in each category do not share exactly the same properties, but they share similar properties. Likewise, substances classified as acids or bases can be considered opposites. Sub-stances in each category share some general properties that make them dif-ferent from other substances. In this section, you will examine acids and bases from both a macroscopic and a submicroscopic level.

SECTION PREVIEW

Objectives

✓ **Distinguish** acids from bases by their properties.

✓ **Relate** acids and bases to their reac-tions in water.

✓ **Evaluate** the central role of water in the chemistry of acids and bases.

Review Vocabulary

Osmosis: the flow of solvent molecules through a selectively permeable membrane, driven by a concentra-tion difference.

New Vocabulary

acid
hydronium ion
acidic hydrogen
ionization
base
acidic anhydride
basic anhydride

Macroscopic Properties of Acids and Bases

Because they are present in so many everyday materials, acids and bases have been recognized as interesting substances since the time of alchemists. Simple, observable properties distinguish the two.

It's a Matter of Taste and Feel

Although taste is not a safe way to classify acids and bases, you proba-bly are familiar with the sour taste of acids. Lemon juice and vinegar, for example, are both aqueous solutions of acids. Bases, on the other hand, taste bitter.

Bases have a slippery feel. Like taste, feel is not a safe chemical test for bases, but you are familiar with the feel of soap, a base, on the skin. Bases, such as soap, react with protein in your skin, and skin cells are removed. This reaction is part of what gives soaps a slippery feel, as well as a cleans-ing action. **Figure 14.1** shows how this reaction makes some bases excel-lent drain cleaners.

Figure 14.1

Bases and Protein
Certain bases are excellent at dissolving hair, which is often the source of clogged drains. Hair is composed of protein.

Table 14.1   Top Ten Industrial Chemicals Produced in the United States in 1994

Chemical	Billions of Pounds	Acid/Base	Some Uses
Sulfuric acid, H_2SO_4	78.70	Acid	Car batteries; manufacture of chemicals, fertilizer, and paper
Nitrogen, N_2	67.54		
Oxygen, O_2	49.67		
Ethylene, C_2H_4	48.53		
Lime, CaO	38.35	Base	Neutralizes acidic soils
Ammonia, NH_3	37.93	Base	Fertilizer; cleaner; making rayon, nylon, and nitric acid
Propylene, C_3H_6	28.84		
Sodium hydroxide, NaOH	25.83	Base	Drain and oven cleaners; manufacture of soap and chemicals
Phosphoric acid, H_3PO_4	25.26	Acid	Making detergents and fertilizers; soft drinks
Chlorine, Cl_2	24.20		

Acids React with Bases

As you have learned, substances with opposite properties, such as acids and bases, tend to react with each other. You'll learn more about these acid-base reactions in Chapter 15.

The reactions of acids and bases are central to the chemistry of living systems, the environment, and many important industrial processes. **Table 14.1** shows the top ten industrial chemicals produced in the United States in 1994. Not surprisingly, acids and bases make up half of the top ten.

Litmus Test and Other Color Changes

Acids and bases cause certain colored dyes to change color. The most common of these dyes is litmus. When mixed with an acid, litmus is red. When added to a base, litmus is blue. Therefore, litmus is a reliable indicator of whether a substance is an acid or a base. **Figure 14.2** shows how vegetable dyes change color in the presence of an acid or a base. Dyes such as these are called acid-base indicators because they are often used to indicate whether substances are acids or bases.

Fact of the MATTER

In the chemistry lab, litmus test paper is used. Litmus paper is made by soaking paper in a solution of litmus and then drying it to remove the water. Litmus papers are usually available in a slightly basic form (blue) and a slightly acidic form (red).

Figure 14.2
Acid-Base Indicators
The ability of a substance to change the color of certain dyes is a good indication of whether the substance is an acid or a base. Common materials that act as acid-base indicators include litmus, red cabbage, radishes, tulips, and rose petals.

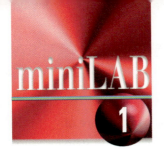

miniLAB 1

What do acids do?

Most acids tend to be reactive substances. Test the reactivities of three acids with several common substances, and develop an operational definition for acidic solutions.

Procedure

1. Wear an apron and goggles.

2. Use a labeled microtip pipet to add 10 drops of $3M$ hydrochloric acid, HCl, to wells D1-D6 of a clean, 24-well microplate. In the same manner, add 10 drops of $3M$ sulfuric acid, H_2SO_4, to wells C1-C6 and 10 drops of $3M$ acetic acid, $HC_2H_3O_2$, to wells B1-B6.

3. Dip blue litmus paper into the solutions in wells D1, C1, and B1. Record your observations.

4. Add 2 drops of bromothymol blue indicator solution to wells D2, C2, and B2. This indicator turns from blue to yellow as the solutions become more acidic. Record your observations.

5. In a similar manner, add marble chips (calcium carbonate) to wells D3, C3, and B3; pieces of zinc to wells D4, C4, and B4; pieces of aluminum to wells D5, C5, and B5; and a small amount of egg white to wells D6, C6, and B6. Record your observations.

6. Dispose of all materials as directed by your teacher. Rinse the microplate with tap water, then distilled water.

Analysis

1. Summarize the reactions of the three acids with the substances you tested. This summary constitutes an operational definition of an acid.

2. Which acid, although it had the same molar concentration as the other acids, reacted less noticeably? Explain this behavior.

Except for Group I carbonates, carbonate-containing compounds are almost completely insoluble in water. This makes naturally occurring substances such as marble and limestone stable materials for sculpting and building.

Reactions with Metals and Carbonates

Another characteristic property of an acid is that it reacts with metals that are more active than hydrogen. **Figure 14.3** shows how iron metal rapidly reacts with hydrochloric acid, HCl, to form iron(II) chloride, $FeCl_2$, and hydrogen gas. However, if you were to add a piece of copper metal to the acid, you could see that the acid will not react with copper metal. This property explains why acids corrode most metals. Bases do not commonly react with metals.

Another simple test that distinguishes acids from bases is the reaction of acids with ionic compounds that contain the carbonate ion, CO_3^{2-}, to form carbon dioxide gas, water, and another compound, as shown in **Figure 14.3.** A similar reaction, also shown in **Figure 14.3,** is the source of the destructive action of acidic pollution on marble and limestone sculptures. Bases do not react with carbonates.

Figure 14.3
Acids with Metals and Carbonates
Typical behavior in certain chemical reactions helps identify substances as acids.

lithium
potassium
calcium
sodium
magnesium
aluminum
zinc
chromium
iron
nickel
tin
lead
hydrogen
copper
mercury
silver
platinum
gold

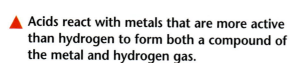
Decreasing activity

▲ Vinegar, a solution of acetic acid, reacts with egg shell, which is primarily calcium carbonate, to produce carbon dioxide, calcium acetate, and water.

$$2HC_2H_3O_2(aq) + CaCO_3(s) \rightarrow$$
$$CO_2(g) + Ca(C_2H_3O_2)_2(aq) + H_2O(l)$$

▲ Acids react with metals that are more active than hydrogen to form both a compound of the metal and hydrogen gas.

$$Fe(s) + 2HCl(aq) \rightarrow FeCl_2(aq) + H_2(g)$$

Calcium carbonate is the major component in limestone and marble. In the presence of acids in the environment, marble and limestone sculptures and buildings can be damaged or destroyed. ▶

Defining Acids and Bases– A Submicroscopic Look

The description of acids and bases in terms of their physical and chemical properties is useful for classification purposes. However, to understand these properties, you need to know about the behavior of acids and bases at the submicroscopic level.

Submicroscopic Behavior of Acids

The submicroscopic behavior of acids when they dissolve in water can be described in several ways. The simplest definition is that an **acid** is a substance that produces hydronium ions when it dissolves in water. A **hydronium ion,** H_3O^+, consists of a hydrogen ion attached to a water molecule.

Manufacturing Sulfuric Acid

You might not expect that a simple acid would acquire worldwide status, but sulfuric acid has done just that. Most industrialized nations produce significant quantities of the chemical. The United States alone produces 40 million tons of sulfuric acid every year.

With such quantities being produced, this product must have many uses. Ninety percent of the sulfuric acid made in the United States is used in the production of liquid fertilizers and other inorganic chemicals. The rest is used in refining petroleum, in steel production, and in producing organic chemicals. Sulfuric acid is also useful in removing unwanted materials from ores.

The Manufacturing Process

The production of sulfuric acid is fairly simple. It starts with burning sulfur to produce sulfur dioxide.

$$S(s) + O_2(g) \rightarrow SO_2(g)$$

The next step in the process is called the contact method because the sulfur dioxide and oxygen molecules are in contact with a catalyst, usually vanadium pentoxide, V_2O_5. When the sulfur dioxide and oxygen gases pass through a heated tube that contains layers of the pellet-size catalyst, the sulfur dioxide is converted to sulfur trioxide. To make sure the reaction is complete, contact with the catalyst takes place twice.

$$2SO_2(g) + O_2(g) \xrightarrow{\text{catalyst}} 2SO_3(g)$$

Then the sulfur trioxide is bubbled through a solution of sulfuric acid to produce pyrosulfuric acid, $H_2S_2O_7$. Pyrosulfuric acid is then added to water to produce sulfuric acid.

$$SO_3(g) + H_2SO_4(l) \rightarrow H_2S_2O_7(l)$$
$$H_2S_2O_7(l) + H_2O(l) \rightarrow 2H_2SO_4(l)$$

DISCUSSING THE TECHNOLOGY

1. **Hypothesizing** Why do you suppose sulfur trioxide is bubbled through a solution of sulfuric acid instead of through water to produce sulfuric acid? Isn't this inefficient?

2. **Inferring** Some people use the quantity of sulfuric acid produced by an industrialized nation as an economic indicator. Why is sulfuric acid production useful in this regard?

For example, hydrochloric acid is produced by dissolving hydrogen chloride gas, HCl, in water. Remember from Chapter 13 that water is a polar molecule that is able to form strong hydrogen bonds with solutes that also form hydrogen bonds. When HCl dissolves in water, it produces hydronium ions by the reaction shown below. HCl is definitely an acid; it produces H_3O^+ when dissolved in water.

$$HCl(g) + H_2O(l) \rightarrow H_3O^+(aq) + Cl^-(aq)$$

Acetic acid, $HC_2H_3O_2$, undergoes a similar reaction when it dissolves in water to form a vinegar solution.

$$HC_2H_3O_2(aq) + H_2O(l) \rightarrow H_3O^+(aq) + C_2H_3O_2^-(aq)$$

Notice the similarities in these two reactions. In both cases, the dissolved substance reacts with water to form hydronium ions and a negatively charged ion.

WORD ORIGIN

acid:
acidus (L) sour
One property of acids is that they taste sour.

Acidic Hydrogen Atoms

How and why are hydronium ions formed? At the submicroscopic level, the reaction of an acid with water is a transfer of a hydrogen ion, H^+, from an acid to a water molecule. This transfer forms the positively charged hydronium ion, H_3O^+, and a negatively charged ion. In an acid, any hydrogen atom that can be transferred to water is called an **acidic hydrogen.**

Take another look at the acetic acid example. Although a molecule of acetic acid, $HC_2H_3O_2$, contains four hydrogen atoms, only one is an acidic hydrogen that participates in the transfer. The other hydrogen atoms remain a part of the acetate ion. **Figure 14.4** shows that it is possible for acids to have more than one acidic hydrogen per molecule.

To help distinguish acids from other hydrogen-containing molecules, acidic hydrogens are written first in the formula. Any time hydrogen is the first element in a formula of a compound, the substance is an acid.

$HC_2H_3O_2$	HCl
Acetic acid	Hydrochloric acid
Monoprotic acids	

▲ Acids such as acetic acid, $HC_2H_3O_2$, and hydrochloric acid, HCl, are called monoprotic acids. Monoprotic acids contain only one acidic hydrogen.

H_2SO_4	$H_3C_6H_5O_7$
Sulfuric acid	Citric acid
a **di**protic acid	a **tri**protic acid

▲ All acids that have more than one acidic hydrogen per molecule are called polyprotic acids. Polyprotic acids with two acidic hydrogens are diprotic acids. Those with three acidic hydrogens are triprotic acids.

Figure 14.4
Acidic Hydrogen
If a hydrogen atom loses its electron, all that remains is a proton. Prefixes, used with the term *protic,* which refers to the remaining proton, indicate how many acidic hydrogens are present in an acid.

Figure 14.5

Steel-Making

Sulfuric acid, which is used to make steel, is an example of a diprotic acid.

Chemical Reaction Shorthand

You know that you can write an equation for the ionization of a specific acid. However, it is sometimes handy to represent the formation of hydronium ions when acids dissolve in water by a general equation. In this general equation, any monoprotic acid is represented by the general formula HA. Compare this general equation to the specific equation for the ionization of HCl.

$$HCl(g) + H_2O(l) \rightarrow H_3O^+(aq) + Cl^-(aq)$$

$$HA + H_2O(l) \rightarrow H_3O^+(aq) + A^-(aq)$$

Although the form of the general equation written above is the most complete, it is more convenient to use a shorthand form of the reaction. In the shorthand form, water is not shown as participating in the reaction, and the hydronium ion is represented as an aqueous hydrogen ion.

$$HA(aq) \rightarrow H^+(aq) + A^-(aq)$$

Similar equations apply to the transfer of hydrogen ions from polyprotic acids, such as sulfuric acid, which is used in **Figure 14.5.** Equations depicting the transfer are shown in **Figure 14.6.**

When using this convenient shorthand style, keep in mind that the water molecule is always an active participant in the reaction, even though it is not written in the equation.

Figure 14.6

Ionization of Polyprotic Acids

Polyprotic acids lose their acidic hydrogens one at a time. For a diprotic acid, there are two steps (A). For triprotic acids, there are three steps (B).

A General:

$$H_2A(aq) \rightarrow H^+(aq) + HA^-(aq)$$
$$HA^-(aq) \rightarrow H^+(aq) + A^{2-}(aq)$$

Example:

$$H_2SO_4(aq) \rightarrow H^+(aq) + HSO_4^-(aq)$$
$$HSO_4^-(aq) \rightarrow H^+(aq) + SO_4^{2-}$$

B

$$H_3A(aq) \rightarrow H^+(aq) + H_2A^-(aq)$$
$$H_2A^-(aq) \rightarrow H^+(aq) + HA^{2-}(aq)$$
$$HA^{2-}(aq) \rightarrow H^+(aq) + A^{3-}(aq)$$

$$H_3PO_4(aq) \rightarrow H^+(aq) + H_2PO_4^-(aq)$$
$$H_2PO_4^-(aq) \rightarrow H^+(aq) + HPO_4^{2-}(aq)$$
$$HPO_4^{2-}(aq) \rightarrow H^+(aq) + PO_4^{3-}(aq)$$

Measurement of Blood Gases

There are only four hydrogen ions in blood for every 100 000 000 other ions and molecules in blood. But enzyme reactions in the body are sensitive to small changes in the concentration of hydrogen ions, so it is crucial. Hydrogen ions affect acid-base relationships in body fluids.

Normal Ranges of Some Blood Components	
Plasma Component	**Normal Range**
HCO_3^-	23–29 m Eq/liter*
P_{CO_2}	35–45 mm Hg
P_{O_2}	75–100 mm Hg
pH	7.35–7.45
*expressed in molar equivalents per liter	

Interpreting acid-base status in blood
When a patient is ill, the doctor's role is to diagnose the patient's condition. Sometimes, this is difficult because different conditions may have similar symptoms. One helpful tool that the physician has is a blood test that provides information about the acid-base relationships in blood. This particular blood test will provide data on acidity (pH), pressure caused by dissolved carbon dioxide (P_{CO_2}), pressure caused by dissolved oxygen (P_{O_2}), and hydrogen carbonate concentration (HCO_3^-). The normal ranges of these components are shown in the table above.

Case histories To see how a physician uses acid-base relationships, it is interesting to consider case histories. In one case, after a pneumonia patient was put on a respirator, she failed to improve. Her blood test showed the following.

P_{CO_2}	17 mm Hg	HCO_3^-	18 m Eq/liter
P_{O_2}	75 mm Hg	pH	7.65

The low carbon dioxide level and the high pH while on the respirator were unexpected. These levels led the physician to check the settings on the respirator. The volume adjustment on the respirator had slipped. The patient was receiving twice the recommended quantity of air. This caused respiratory alkalosis, a condition of decreased acidity of the blood and tissues. When the respirator was adjusted, the blood levels returned to normal as the acid-base balance was reestablished, and the patient began to recover.

Connecting to Chemistry

1. **Hypothesizing** In a heart attack, blood flow to some parts of the heart may be stopped or greatly reduced. How might this affect the acid-base relationship in blood in the heart muscle?

2. **Applying** Blood gases may fail to show major abnormalities while the patient is at rest. Suggest a way to overcome this problem if the patient is not bedridden.

Acid Ionization

In the reaction of an acid with water, ions are formed from a covalent compound. When ions form from a covalent compound, the process is called **ionization.** Specifically, acids form ions in a process called acid ionization.

Acids Are Electrolytes

Because acids ionize to form ions in water, acidic solutions conduct electricity. As you learned in Chapter 4, substances that dissolve in water to form conducting solutions are called electrolytes. **Figure 14.7** compares the electrical conductivities of water, a solution of an ionic compound, and a solution of a weak acid.

Figure 14.7

Conductivity

The electrical conductivities of solutions are easy to compare by using a simple circuit. The brightness of the light indicates the relative electrical conductivity of the solutions.

▼ **Distilled water is nonconducting.**

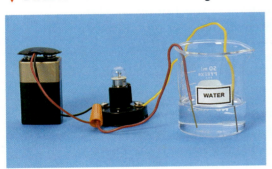

◄ **1*M* NaCl is an excellent conductor.**

▲ **1*M* HC$_2$H$_3$O$_2$ is a weak conductor.**

As opposed to solutions of ionic compounds (such as table salt), which are always excellent conductors of electricity, acidic solutions have electrical conductivities ranging from strong to weak. The range of electrical conductivities exhibited by different acidic solutions distinguishes acid ionization from ionic dissociation. The range also indicates that acids vary in their ability to produce ions.

Submicroscopic Behavior of Bases

The behavior of bases is also described at the molecular level by the interaction of the base with water. A **base** is a substance that produces hydroxide ions, OH$^-$, when it dissolves in water. There are two mechanisms by which bases produce hydroxide ions when they dissolve in water.

Simple Bases: Metal Hydroxides

The simplest kind of base is a water-soluble ionic compound, such as sodium hydroxide, that contains the hydroxide ion as the negative ion. When NaOH dissolves in water, for example, it dissociates into aqueous sodium ions and hydroxide ions, as shown below.

$$NaOH(s) \xrightarrow{H_2O} Na^+(aq) + OH^-(aq)$$

NaOH is definitely a base because it produces hydroxide ions when it dissolves in water. You can predict that any water-soluble or slightly water-soluble metal hydroxide will be a base when added to water.

Water plays a different role here than in the formation of hydronium ions when acids ionize in water. Water molecules do not chemically react with this type of base. The hydroxide ion is formed by simple ionic dissociation, and no transfer occurs between the base and the water molecules to form the hydroxide ions.

Just as a polyprotic acid in water produces more than one hydronium ion, it is possible for a formula unit of a metal hydroxide to produce more than one hydroxide ion. Calcium hydroxide, $Ca(OH)_2$, and aluminum hydroxide, $Al(OH)_3$, are examples of such bases, as shown in **Figure 14.8**.

Figure 14.8

Dissociation of Some Metal Hydroxides
All of these compounds are bases because they produce hydroxide ions when they dissolve in water.

Reversing the Transfer: Bases That Accept H⁺

A few bases are covalent compounds that produce hydroxide ions by an ionization process when dissolved in water. The ionization involves the transfer of a hydrogen ion from water to the base. The most common example of this type of base is ammonia, NH_3.

When ammonia gas dissolves in water, some of the aqueous ammonia molecules react with water molecules to form ammonium ions and hydroxide ions, as shown in this reaction.

$$NH_3(g) + H_2O(l) \rightleftharpoons NH_4^+(aq) + OH^-(aq)$$

Ammonia is a base. It produces hydroxide ions in water, but by a different mechanism than that in the NaOH example. In the reaction with ammonia, the water molecule is an active chemical reactant. Water molecules transfer hydrogen ions to ammonia molecules.

It is also helpful to have a general reaction for the ionization of a covalent base, which is represented by the letter *B*. Study the equation for the general reaction.

$$B + H_2O(l) \rightleftharpoons BH^+(aq) + OH^-(aq)$$

PEOPLE in CHEMISTRY

Meet Fe Tayag, Cosmetic Bench Chemist

It pays to be a careful reader of labels. Here's a hint from Ms. Tayag, who has formulated cosmetics for more than 20 years. Many cosmetic companies find it is a good selling point to add sunscreen to their products. But unless the container specifies an SPF (sun protection factor) as a number, there probably isn't enough sunscreen to do much good. In this interview, Ms. Tayag shares other cosmetics savvy that can make you a wiser shopper.

On the Job

 Ms. Tayag, what do you do on your job?

 I'm in the Research and Development Department of a cosmetic laboratory. I formulate cosmetics, such as shampoos, lotions, and bubble baths, for cosmetic companies.

 Can you give us an idea of a typical formula for a shampoo?

 A simple shampoo consists of a mixture of water, sodium lauryl sulfate, and an amide to make it foam. I heat the mixture and then I have to adjust the acidity. Most shampoos are neutral. If it's more basic than that, I adjust it with a citric acid solution. Then I cool it and check the viscosity, or how it flows. I don't want it to be either water-thin or molasses-thick, so I'll adjust the viscosity by using a 20 percent sodium chloride solution. Perfume and color are added to make it smell and look good.

 The label on a cosmetic usually lists water as the first ingredient. Does that mean it's mostly water?

 Yes. Face creams have the lowest amount of water, about 60 percent. The amount of water used depends on the skin type.

 What colorings do you add to the cosmetics you make?

 I use colors approved by the Food and Drug Administration. They come in basic colors, but I can combine them to produce other colors. Green is the most popular coloring for shampoo. That color seems to be associated with freshness and cleanliness.

 What trends do you see developing in cosmetics?

 There are more cosmetics created especially for different ethnic groups. Sunscreen is being added to more and more cosmetics as people become increasingly aware of the damage that ultraviolet light can do to the skin. And, because a large part of the population is getting older, ingredients that are supposed to delay the aging effects, such as antioxidants, are becoming more popular. In addition, there's lots of interest in antiallergenic, natural cosmetics that are both unscented and uncolored.

Early Influences

 How did you get interested in becoming a chemist?

 My father was a warehouse man for a cosmetics company in the Philippines, where I grew up. He was fascinated by the process of producing cosmetics. Because he thought the job of a chemist was a very dignified one, he encouraged me to study chemistry. Our family, which included eight children, was poor, but he somehow found a way to send me to college.

 Did you enjoy the study of chemistry in college?

 Not at first. I had a tough time in college and kept asking myself, "Why did I take this course?" I cried at exam time, but I couldn't bear to disappoint my father. I prayed a lot, and I thought of my father struggling to find money to pay my tuition. That helped me find the courage to continue. My third year of college marked a turning point, and my studies became easier for me.

 Were there people besides your father who influenced you in your career?

 A friend of my father helped me get a job in the cosmetics lab while I was still in school. I worked days and attended school at nights. That way, I got to see what chemistry was all about in the real world. I felt I was ahead of students who lacked practical experience. For instance, in my colloid chemistry class, I brought materials to school and demonstrated how to make a cleansing cream.

Personal Insights

 Do you consider chemistry to be a little like cooking?

 Yes. When I cook, I rely heavily on my senses of sight and smell. In my opinion, a keen sense of smell and a good ability to make observations are very important for a chemist, too.

 Some people think that cosmetics are frivolous. Do you agree?

 No. I think it's important for people's self-confidence to look nice. Because people want to remain attractive and young-looking, this is an industry that will never die out.

CAREER CONNECTION

These career opportunities are related to cosmetic chemistry.

Food and Drug Inspector College degree plus written examinations
Manufacturers' Sales Representative High school diploma
Cosmetologist State-administered exam

Bases as Electrolytes
The electrical conductivity of 1*M* NaOH is greater than that of 1*M* NH₃. These differences show that aqueous solutions of bases may be strong or weak electrolytes, based on whether many or few ions are in solution.

Bases Are Electrolytes

Because a base in water produces ions, you can predict that aqueous solutions of bases will conduct electricity. **Figure 14.9** compares the conductivity of a $1M$ NaOH solution with that of a $1M$ NH₃ solution. As with acids, the ability of basic solutions to conduct electricity varies, depending upon the base. This variability is evidence that differences exist in the ability of different bases to produce ions.

Why does water transfer H⁺ to bases?

Think back to why acids transfer hydrogen ions to water. The same model can be used to explain why water molecules lose hydrogen ions to covalent bases when they dissolve in water. Consider the example of ammonia. Ammonia is a polar molecule because it contains polar covalent N—H bonds. The nitrogen end of the molecule has a slight negative charge, and the hydrogen atoms each have a slight positive charge. A lone pair of electrons is also on the central nitrogen. Look at **Figure 14.10** to see what happens when polar ammonia molecules dissolve in polar water molecules.

Other Acids and Bases: Anhydrides

Two related classes of compounds do not fit the previous models of acids and bases, but they still act as acids or bases. These compounds are both oxides, which are compounds containing oxygen bonded to just one other element. These oxides are called anhydrides, which means that they contain no water.

Anhydrides differ, depending upon whether the oxygen is bonded to a metal or a nonmetal. Nonmetal oxides form acids when they react with water and are called **acidic anhydrides.** Metal oxides, on the other hand, react with water to form bases and are called **basic anhydrides.** In both of these reactions, water is an active reactant. Now, examine some examples of anhydrides.

Figure 14.10

Forming Ammonium and Hydroxide Ions

Because an aqueous solution of ammonia contains ammonium ions and hydroxide ions, such a solution is commonly referred to as ammonium hydroxide.

A Ammonia has a trigonal pyramid geometry and is a polar molecule.

C The H$^+$ bonds to the N in NH$_3$, using the lone pair of electrons on the N to form a fourth N—H bond and a stable ammonium ion, NH$_4{}^+$.

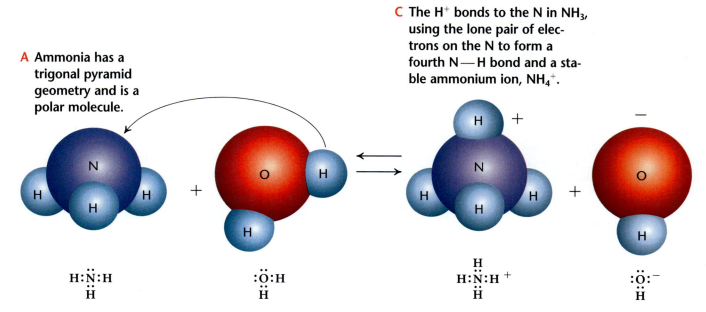

B The hydrogen bond that forms between the N end of NH$_3$ and the H end of H$_2$O is strong enough to pull an H$^+$ completely away from H$_2$O. The two electrons in the broken O—H bond remain as a lone pair on the O. The result is a stable hydroxide ion, OH$^-$.

D The electron dot structures show that these ions are stable. Each atom in the dot structures has a stable number of valence electrons.

Acidic Anhydrides and Acid Rain

Probably the most familiar acidic anhydride is carbon dioxide, CO$_2$. Water that has had carbon dioxide bubbled through it turns blue litmus to red, indicating that CO$_2$ and water form an acid, carbonic acid, H$_2$CO$_3$. A solution of CO$_2$ also has a slightly sour taste, which is one of the reasons that carbonated water is such a refreshing beverage.

Carbon dioxide is a minor component in Earth's atmosphere and an important component in the carbon cycle. Because CO$_2$ is always in the atmosphere, when it rains, CO$_2$ dissolves in rainwater, forming carbonic acid, H$_2$CO$_3$. The result is that rain is always slightly acidic. If rain is always acidic, why is increased acidity in rain such an environmental concern?

The acidity of normal rain does not damage the environment. However, other nonmetal oxides such as sulfur oxides are sometimes present in the atmosphere. Also, levels of carbon dioxide are sometimes higher than normal.

The major source of sulfur oxides in the atmosphere is the burning of sulfur-containing coal in power plants. As this type of coal burns in a furnace, sulfur dioxide gas, SO$_2$, is produced. The SO$_2$ escapes into the atmosphere, where it reacts with more oxygen to form sulfur trioxide, SO$_3$.

At room temperature, the reaction between nitrogen, N_2, and O_2 is slow and insignificant. At the high temperatures in an automobile engine, the reaction between N_2 and O_2 goes quickly, and large amounts of nitrogen oxides are produced in exhaust, **Figure 14.11.**

When sulfur oxides, nitrogen oxides, and increased amounts of carbon dioxide dissolve in rain, they undergo acid-forming reactions and produce what is commonly referred to as acid rain.

Examples:
$$SO_2(g) + H_2O(l) \rightarrow H_2SO_3(aq)$$
$$SO_3(g) + H_2O(l) \rightarrow H_2SO_4(aq)$$
$$2NO_2(g) + H_2O(l) \rightarrow HNO_3(aq) + HNO_2(aq)$$
$$CO_2(g) + H_2O(l) \rightarrow H_2CO_3(aq)$$

Acid rain has been significantly reduced over the past decade as new mechanisms for trapping nonmetal oxides before they get into the atmosphere have been developed. Read the Chemistry and Society feature for more information about this type of air pollution.

Figure 14.11

Nitrogen Oxides
The major source of nitrogen oxides in the atmosphere is automobiles.

Basic Anhydrides and Making Your Garden Grow

Unlike nonmetal oxides, which are covalent compounds, metal oxides are ionic compounds. When metal oxides react with water, they produce hydroxide ions.

Gardeners sometimes use lime to treat their soil, as shown in **Figure 14.12.** Lime is the common name for the chemical compound calcium oxide, CaO. When CaO is spread on soil, it reacts with water in the soil to form calcium hydroxide, $Ca(OH)_2$. This compound then forms calcium and hydroxide ions.

$$CaO(s) + H_2O(l) \rightarrow Ca(OH)_2(aq)$$
$$Ca(OH)_2(aq) \rightarrow Ca^{2+}(aq) + 2OH^-(aq)$$

A similar reaction was used historically to produce an important commodity, soap. Early soap makers used the basic properties of metal oxides. When wood burns, the metal atoms in the wood form solid metal oxides in the burning process. These metal oxides are predominantly those of sodium, potassium, and calcium. These metal oxides are ionic so they are solids, even at the high temperature of a roaring fire. They are the major component of the ash that is left when the fire burns out.

Figure 14.12

Making Soil More Basic
Adding lime to soil makes the soil less acidic and more favorable for growing many types of plants. Because $Ca(OH)_2$ is only slightly water soluble, it provides a longer-lasting source of base than provided by more soluble ionic hydroxides.

Atmospheric Pollution

The air you breathe is literally a matter of life and death. Atmospheric oxygen is taken into your body and, during respiration, reacts with glucose to produce the energy required for all the life processes that keep you going.

Unfortunately, the same air, at times, may contain materials that cause respiratory diseases and bring about other harmful effects. Air is often polluted with chemicals produced by human activity. Even Earth itself coughs up some of the same air pollutants during volcanic eruptions.

Introducing the major air pollutants The major chemicals that pollute the air are carbon monoxide, CO; carbon dioxide, CO_2; sulfur dioxide, SO_2; nitrogen monoxide, NO; nitrogen dioxide, NO_2; hydrocarbons; and suspended particles.

In addition, pollutants form under the influence of sunlight when oxygen, nitrogen oxides, and hydrocarbons react. These reactions produce ozone, O_3, and aldehydes such as formaldehyde, CH_2O. Why are pollutants a problem?

Acid rain What do the salmon and the pine trees in the photos have in common? Both have succumbed to the acid environment in which they live. They are two of the many victims of acid rain.

Unpolluted rain is not harmful. However, many industrial and power plants burn coal and oil. The smoke produced may contain large quantities of sulfur oxides, suspended particles, and nitrogen oxides. Automobiles also contribute to the problem by emitting similar oxides. These chemicals react with water in the air to form acids, such as sulfuric acid. These acids reach the surface of Earth in fog, rain, snow, and dew. Acid rain can have a disastrous effect when it reaches bodies of water and waterways. But if a lake has a high limestone content, it is able to somewhat neutralize the acid.

Smog Large cities with many automobiles may have another problem with airborne pollutants. It is called smog, which is a haze or fog that is made harmful by the chemical fumes and suspended particles it contains.

A type of smog known as photochemical smog frequently occurs in large cities in sunny, dry climates. When the pollutants from automobile exhaust enter the air and are exposed to sunlight, they interact to produce photochemical smog. This type of smog is generally worse on hot days and between 11 A.M. and 4 P.M. when exhaust has accumulated in the air.

Analyzing the Issue

1. **Acquiring Information** Research free trade in this country. How might the issue of free trade influence the problems of air pollution in this country?

2. **Thinking Critically** Older-model cars are responsible for the greatest amount of air pollutants being vented into the atmosphere. Hold a debate on whether these cars should be banished from the highways.

This early soap-making process is shown in **Figure 14.13.** The reaction of water and sodium oxide, Na_2O, one of the metal oxides in wood ash, is similar to that shown for lime and water.

$$Na_2O(s) + H_2O(l) \rightarrow 2NaOH(aq)$$

$$NaOH(aq) \rightarrow Na^+(aq) + OH^-(aq)$$

Figure 14.13

Making Soap

Lye was produced by collecting wood ashes and soaking them in water. After several days, the highly basic solution was separated from the undissolved ash and combined with animal fat. The lye reacted with the fat to make soap.

Supplemental Problems

For more practice with solving problems, see Supplemental Practice Problems, Appendix B.

The Macroscopic-Submicroscopic Acid-Base Connection

As you have discovered, the properties of acids and bases are determined by the submicroscopic interactions between the acid or base and the solvent water. For example, HCl and $HC_2H_3O_2$ both interact with water to cause a transfer of hydrogen ions from the acid to water molecules to form hydronium ions. Both solutions turn blue litmus red. Even though these properties are the same for both acids, remember that the conductivity of a $1M$ HCl solution is much greater than that of a $1M$ $HC_2H_3O_2$ solution.

A base interacts with water molecules to form hydroxide ions either by ionic dissociation or by the transfer of a hydrogen ion from a water molecule to the base. Consider a $1M$ NaOH solution and a $1M$ NH_3 solution. Both solutions are basic. They each turn red litmus to blue. But the NaOH solution shows strong electrical conductivity, while the NH_3 solution is only weakly conducting.

Why do different acids have some properties in common and yet differ in other properties? Why is the same thing true for bases? Section 14.2 will explain these differences.

SECTION REVIEW

Understanding Concepts

1. Make a table that compares and contrasts the properties of acids and bases.

2. Use a chemical equation to show how aqueous HNO_3 fits the definition of an acid.

3. Consider the oxides MgO and CO_2. For each oxide, tell whether it is a basic anhydride or an acidic anhydride. Write an equation for each to demonstrate its acid-base chemistry.

Thinking Critically

4. **Applying Concepts** Chemists often call a hydrogen ion a proton. Explain why an acid is sometimes called a proton donor, and a base is sometimes called a proton acceptor.

Applying Chemistry

5. **Using Soap** After using soap to wash dishes by hand, it is sometimes difficult to keep your hands from remaining slick. Explain why rinsing your hands in lemon juice would make them less slick.

Strengths of Acids and Bases

From reading Section 14.1, you know that all acids have a sour taste, but they may differ in how readily they react with another substance. You wouldn't hesitate to use the acetic acid in vinegar on a salad, but you certainly wouldn't use hydrochloric acid, which may be used to clean brick, on any type of food. All bases also share some properties but differ in others. You would readily use a dilute ammonia solution as a cleaner, but you certainly wouldn't let sodium hydroxide, which is used in drain cleaners, come in contact with your skin. These two bases differ greatly in how they react.

SECTION PREVIEW

Objectives
✓ **Relate** different electrical conductivities of acidic and basic solutions to their degree of dissociation or ionization.

✓ **Distinguish** strong and weak acids or bases by their degree of dissociation or ionization.

✓ **Compare and contrast** the composition of strong and weak solutions of acids or bases.

✓ **Relate** pH to the strengths of acids and bases.

Review Vocabulary
Ionization: the process in which ions form from a covalent compound.

New Vocabulary
strong base
strong acid
weak acid
weak base
pH

Strong Acids and Bases

What's going on? You know that when acids and bases are mixed with water, they form ions. Much of the behavior of acids and bases depends on how many ions are formed by a particular acid or base in water. The degree to which bases and acids produce ions depends on the nature of the acid or base.

Acids and bases are classified into one of two categories depending upon their strength, which is the degree to which they form ions. The strong category is reserved for those substances, such as NaOH and HCl, that completely dissociate or ionize and produce the maximum number of ions when dissolved in water. All other acids and bases are classified as weak because they produce few ions when dissolved in water.

Strong Bases

Sodium hydroxide, NaOH, is a **strong base** because when NaOH dissolves in water, all NaOH formula units dissociate into separate sodium and hydroxide ions. The dissociation of the base is complete. The strength of a base is based on the percent of units dissociated, not the number of OH^- ions produced. Some bases, such as $Mg(OH)_2$, are not very soluble in water, and they don't produce a large number of OH^- ions. However, they are still considered to be strong bases because all of the base that does dissolve completely dissociates.

Table 14.2   Common Strong Acids and Bases

Strong Acids
Perchloric acid, $HClO_4$
Sulfuric acid, H_2SO_4
Hydriodic acid, HI
Hydrobromic acid, HBr
Hydrochloric acid, HCl
Nitric acid, HNO_3

Strong Bases
Lithium hydroxide, LiOH
Sodium hydroxide, NaOH
Potassium hydroxide, KOH
Calcium hydroxide, $Ca(OH)_2$
Strontium hydroxide, $Sr(OH)_2$
Barium hydroxide, $Ba(OH)_2$
Magnesium hydroxide, $Mg(OH)_2$

The strong bases shown in **Table 14.2** are all ionic compounds that contain hydroxide ions. NaOH and KOH are the most common strong bases you will encounter. A $1M$ solution of NaOH and a $1M$ solution of KOH will each contain $1M$ OH^- because both compounds completely dissociate.

Strong Acids

HCl is a **strong acid** because no HCl molecules are in a water solution of HCl. Because of the strong attraction between the water molecules and HCl molecules, every HCl molecule ionizes. A $1M$ HCl solution contains $1M$ H_3O^+ and $1M$ Cl^-. Similarly, a $1M$ HNO_3 solution contains $1M$ H_3O^+ and $1M$ NO_3^-. As before, the labeling of a solution as $1M$ HNO_3 is not particularly descriptive of the submicroscopic composition of the solution.

Table 14.2 lists common strong acids and bases. Because they are used so often, it is helpful to memorize their names and formulas. They all completely dissociate or ionize into ions when they dissolve in water. If an acid or base is not listed in this group, it is considered to be a weak acid or base. However, the terms *strong* and *weak* are not absolute. Strength of acids and bases covers a wide range from extremely strong to extremely weak. Notice that the strongest bases are all hydroxides of Group I, the alkali metals, and Group II, the alkaline earth metals. *Alkali* is a term frequently used to refer to materials that have noticeably basic properties.

Figure 14.14

Some Common Weak Acids
Some common weak acids vary in their structures.

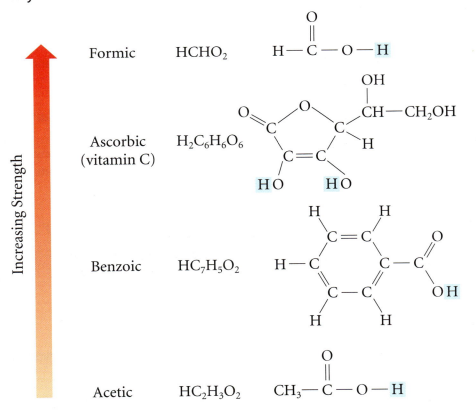

Increasing Strength

Formic $HCHO_2$

Ascorbic (vitamin C) $H_2C_6H_6O_6$

Benzoic $HC_7H_5O_2$

Acetic $HC_2H_3O_2$

Weak Acids and Bases

The weak category of acids and bases contains those with a wide range of strengths. This is the category into which most acids and bases fall. Instead of being completely ionized, weak acids and bases are only partially ionized.

Weak Acids

Acetic acid, $HC_2H_3O_2$, is a good example of a **weak acid.** In a $1M$ $HC_2H_3O_2$ solution, less than 0.5 percent of the acetic acid molecules ionize, and 99.5 percent of the acetic acid molecules remain as molecules. Another way to think of this is to consider 1000 acetic acid molecules in a water solution. On the average, only five of the 1000 molecules transfer their single hydrogen ion to a water molecule. The molarity of hydronium ion produced in a $1M$ $HC_2H_3O_2$ solution is much less than $1M$ due to this partial ionization. Other common weak acids include phosphoric acid, H_3PO_4, and carbonic acid, both of which are found in soft drinks.

The molecular structure of a weak acid determines the extent to which the acid ionizes in water. **Figure 14.14** shows the variety of structures of some common weak acids.

Figure 14.15 uses a graph format to show the dramatic difference in degree of ionization between a solution of a weak acid and one of a strong acid. A solution of weak acid contains a mixture of un-ionized acid molecules, hydronium ions, and the corresponding negative ions. The concentration of the un-ionized acid is always the greatest of the three concentrations.

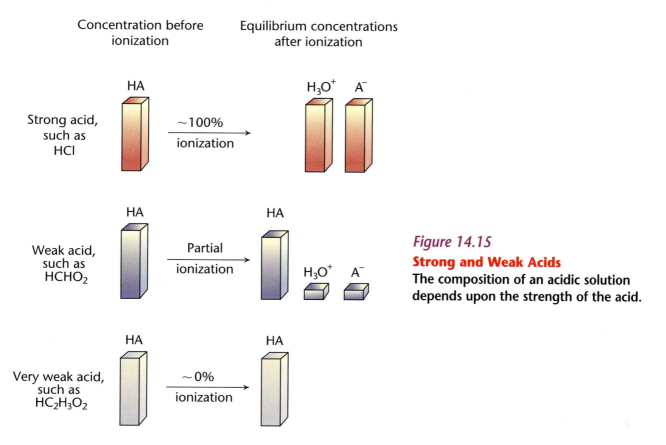

Figure 14.15
Strong and Weak Acids
The composition of an acidic solution depends upon the strength of the acid.

Weak Bases

Ammonia is a **weak base** because most of its molecules don't react with water to form ions. In a $1M$ aqueous solution of ammonia, only about 0.5 percent of the ammonia molecules react with water to form ammonium and hydroxide ions. About 99.5 percent of the ammonia molecules remain as intact molecules. The molarity of hydroxide ions in a $1M$ ammonia solution is much less than $1M$. The major dissolved component in a weak base solution is the un-ionized base. Other examples of bases that produce so few OH^- ions that they are considered to be weak bases are $Al(OH)_3$, and $Fe(OH)_3$.

Weak Is Not Insignificant

Although most acids and bases are classified as weak, their behavior is extremely significant. Most of the acid-base chemistry in living systems occurs through interactions between weak acids and bases. For example, amino acids, the small molecules that serve as the building blocks of proteins, have properties of both weak acids and weak bases. The amino portion of the molecule acts as a base when it comes into contact with a strong acid, and the acid part of the molecule acts as a weak acid when exposed to a base. The coiling of the DNA into a double helix is also due to the interactions between weak acids and bases. *Weak* does not mean "insignificant."

Strength Is Not Concentration

Although the terms *weak* and *strong* are used to compare the strengths of acids and bases, *dilute* and *concentrated* are terms used to describe the concentration of solutions. The combination of strength and concentration ultimately determines the behavior of the solution. For example, it is possible to have a concentrated solution of a weak acid or weak base or a dilute solution of a weak acid or weak base. Similarly, you can have a concentrated solution of a strong acid or strong base, as well as a dilute solution of a strong acid or strong base.

Figure 14.16

Developing the pH Scale
The Danish biochemist S.P.L. Sørenson developed the pH scale in 1909 while working on brewing beer. *pH* is an abbreviation in French for "pouvier d'hydrogene" or, in English, the "power of hydrogen."

The pH Scale

Because of the range of solution concentration, the range of possible concentrations of hydronium ions and hydroxide ions in solutions of acids or bases is huge. For example, a $6M$ solution of HCl has an H_3O^+ molarity of $6M$, but a $6M$ solution of $HC_2H_3O_2$ has an H_3O^+ molarity of $0.01M$.

In most applications, the observed range of possible hydronium or hydroxide ion concentrations spans $10^{-14}M$ to $1M$. This huge range of concentrations presents a problem when comparing different acids and bases. To make this range of possible concentrations easier to work with, the pH scale was developed by S.P.L. Sørenson, **Figure 14.16.**

Balancing pH in Cosmetics

With so many shampoos available, you may find it hard to know which type is best for your hair. Advertisements for each type tell you that their shampoo has more to offer than any other shampoo. How can you know which one to choose?

Cosmetic chemists have many tools for determining the effect of their products on different hair types. Using one technique, developed by NASA, they place a hair under a microscope connected to a TV screen that is hooked up to a computer. The computer evaluates the hair before and after treatment with the shampoo. Working with different types of hair allows these chemists to determine the best treatment for each type of hair.

Shampoos and pH balance The clear, outer layer of a strand of hair is the cuticle, which consists of the protein keratin. The cells of the cuticle are arranged like overlapping shingles. Shampoos that have a high pH make the entire hair shaft swell and push the cells of the cuticle away from the rest of the shaft. Harsh basic substances in solutions for permanents and hair coloring dissolve some of the cuticle, damaging the hair. Hair can also be damaged by the sun and excessive blow drying. Damaged hair is dull and dry.

In contrast, acidic substances in shampoos of low pH make the hair shaft tight and smooth by shrinking it and causing the cells of the cuticle to lie flat. Low-pH shampoos help restore damaged hair to its original condition and make it shine again. They also strengthen the keratin and increase the flexibility and elasticity of the hair. People with coarse, curly hair can benefit from using alkaline or high-pH shampoos. These products soften and relax the hair, making it softer and less curly.

Why balance pH in skin products? The outer layer of skin has a keratin structure just as hair does. Products aimed at making the skin look brighter and clearer have a higher pH. Their purpose is to

remove the top layer of keratin, which may consist of dead cells. The new cells underneath look fresh and vibrant. Occasional use of these products may be helpful, but regular use damages healthy skin by removing too many layers of cells.

Another problem with basic skin products is related to an acid mantle that bathes the top layer of the skin, the epidermis. This fluid—composed of oil, sweat, and other cell secretions—is a natural defense against bacterial infections. Strongly basic soaps can neutralize the protective acid mantle. People with acne or oily skin must be careful not to remove the acid mantle.

Exploring Further

1. **Inferring** If you live in an area where the water is hard, your hair may look dull. Why would rinsing with water to which lemon juice has been added help?

2. **Thinking Critically** Why would a person with acne use skin products that are pH-neutral or mildly acidic?

To learn more about the chemical reasoning behind commercial beauty tips, visit the Chemistry Web site at **chemistryca.com**

What is pH?

pH is a mathematical scale in which the concentration of hydronium ions in a solution is expressed as a number from 0 to 14. A scale of 0 to 14 is much easier to work with than a range from 1 to 10^{-14} (10^0 to 10^{-14}). The pH scale is a convenient way to describe the concentration of hydronium ions in acidic solutions, as well as the hydroxide ions in basic solutions.

Think about the pH numbers 0 to 14 and the hydronium ion concentration range. Notice that the pH value is the negative of the exponent of the hydronium ion concentration. For example, a solution with a hydronium ion concentration of $10^{-11}M$ has a pH of 11. A solution with a pH of 4 has a hydronium ion concentration of $10^{-4}M$.

How do these numbers relate to hydroxide ion concentrations? Experimental evidence shows that when the hydronium ion concentration and the hydroxide ion concentration in aqueous solution are multiplied together, the product is 10^{-14}. So, if the pH of a solution is 3, the hydronium ion concentration is $10^{-3}M$, and the hydroxide ion concentration is $10^{-14}/10^{-3}M$, which is $10^{-14-(-3)}M$, or $10^{-11}M$.

Supplemental
Problems

For more practice with solving problems, see Supplemental Practice Problems, Appendix B.

PRACTICE PROBLEMS

Find the pH of each of the following solutions.
1. The hydronium ion concentration equals:

    a) $10^{-5}M$ **b)** $10^{-12}M$ **c)** $10^{-2}M$

2. The hydroxide ion concentration equals:

    a) $10^{-4}M$ **b)** $10^{-11}M$ **c)** $10^{-8}M$

There are several easy ways to measure pH. Two common methods are shown in **Figure 14.17**.

▲ pH meters are instruments that measure the exact pH of a solution.

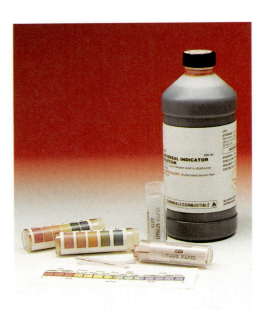

Figure 14.17

Measuring pH

pH is convenient because there are simple methods for measuring it in the lab or in the field.

Indicators register different colors at different pHs. A convenient way to store indicators is by soaking strips of paper in them and then drying the paper. Indicator paper is frequently sold because it is easy to use. ▶

Antacids

Many people experience a burning sensation known as heartburn after eating certain foods. Heartburn is caused by excess acid in the stomach and esophagus. Use your knowledge of acid chemistry to evaluate the effects of antacids that are commonly used to treat heartburn.

Procedure

1. Obtain four snack-size zipper-closure bags, and mark each with the name of an antacid to be tested.

2. To each of the four bags, add 5 mL of white vinegar, 10 mL of water, and enough cabbage juice indicator (probably 30 to 40 drops) to impart a distinct color.

3. Add the appropriate antacid tablet to each bag, squeeze out the excess air, and zip the bag closed. Be sure that the antacid tablet is immersed in the vinegar solution.

4. Squeeze the antacid tablets to break them into small pieces. Record your observations.

When the reactions have ceased or slowed markedly, note and record the colors and approximate pH values of the solutions. Consult the pH-color chart from the ChemLab on the next page.

Analysis

1. Describe the different ways in which the antacids reacted with the vinegar. Infer which of the antacids contain carbonates. Explain your answer.

2. Which of the antacids created the most basic final solution? Explain this answer in terms of how well the antacid works.

Interpreting the pH Scale

The pH scale is shown in **Figure 14.18.** The scale is divided into three areas. If a solution has a pH of exactly 7, the solution is said to be neutral. It is neither acidic nor basic.

Neutral
↓

```
0  1  2  3  4  5  6  7  8  9  10  11  12  13  14
←——— More acidic ———        ——— More basic ———→
```

Figure 14.18

The pH Scale

A pH of 7 is neutral. A pH less than 7 is acidic, and a pH greater than 7 is basic. As the pH drops from 7, the solution becomes more acidic. As pH increases from 7, the solution becomes more basic. The flowers shown are hydrangeas. Their blooms are blue when the plants are in acidic soil and pink when the soil is basic.

ChemLab

SMALL SCALE

Household Acids and Bases

Indicators often are used to determine the approximate pH of solutions. In this ChemLab, you will make an indicator from red cabbage and use the indicator to determine the approximate pH values of various household liquids. The cabbage juice indicator contains a molecule, anthocyanin, that accounts for the color changes.

Problem

What are the approximate pH values of various household liquids?

Objectives

• **Measure** and **compare** the pH values for various household liquids.
• **Compare** the functions of the liquids to their chemical makeup.

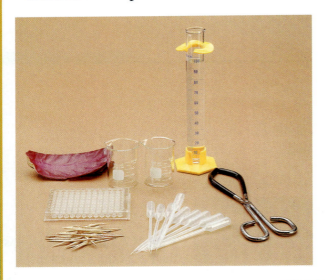

Indicator color	Relative pH
bright red	strong acid
red	medium acid
reddish purple	weak acid
purple	neutral
blue green	weak base
green	medium base
yellow	strong base

PREPARATION

Materials

red cabbage
hot plate
beaker tongs
100-mL beakers (2)
distilled water
microtip pipets (9)
96-well microplate
piece of white paper
100-mL graduated
  cylinder

toothpicks
solutions of:
  eyewash
  lemon juice
  white vinegar
  table salt
  soap
  baking soda
  borax
  drain cleaner

Safety Precautions

Use beaker tongs to handle hot beakers. Wear an apron and goggles. Some of the solutions to be tested are caustic, especially the drain cleaner. Avoid all contact with skin and eyes. If contact occurs, immediately wash with large amounts of water and notify the teacher.

PROCEDURE

1. Tear a red cabbage leaf into small pieces, and layer the pieces in a 100-mL beaker to a depth of about 2 cm. Add about 30 mL of distilled water.

2. Set the beaker on a hot plate, and heat until the water has boiled and become a deep purple color. Remove the beaker from the hot plate using beaker tongs, and allow it to cool. Pour off the cabbage juice indicator liquid into a clean beaker.

3. Set a clean microplate on a piece of white paper. Use the pipets to add 5 drops of eyewash to well H1, lemon juice to H2, white vinegar to H3, and solutions of table salt to H4, soap to H5, baking

soda to H6, borax to H7, and drain cleaner to H8. Use a clean pipet for each solution.

4. Draw the cabbage juice indicator solution into a clean pipet, and add 5 drops to each of the solutions in wells H1-H8. Stir the solution in each well with a clean toothpick.

5. Looking down through the wells, note and record the color of each solution in a data table such as the one shown. Using the color chart, record in the data table the approximate pH of each of the solutions.

ANALYZE AND CONCLUDE

1. **Interpreting Data** Are food items such as lemon juice or vinegar acidic or basic? These solutions are either tart or sour, so what ion probably accounts for this characteristic?

2. **Interpreting Data** Were the cleaning solutions acidic or basic? What ion is probably involved in the cleaning process?

3. **Observing and Inferring** How can you account for the great pH difference between lemon juice (citric acid solution) and eyewash (boric acid solution)?

4. **Using Variables, Constants, and Controls** Suppose that, in addition to the solutions, you tested a well containing pure distilled water. What purpose would this test serve?

APPLY AND ASSESS

1. Would your indicator work well to determine the pH of ketchup? Explain.

2. You may have noted that some shampoos are described as pH-balanced. What do the manufacturers mean by this phrase? Why would they do this to a soap or detergent?

3. Hypothesize about how other solutions at home would react with the cabbage juice indicator. Explain your predictions.

DATA AND OBSERVATIONS

Solution	Color	Approximate pH
Eyewash		
Lemon juice		
White vinegar		
Table salt		
Soap		
Baking soda		
Borax		
Drain cleaner		

Figure 14.19

Comparison of Concentrations
Look at the hydronium ion and hydroxide ion concentrations for three common solutions. For each solution, what is the product of the two concentrations?

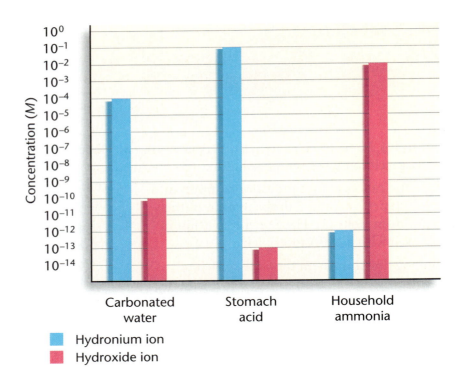

Legend:
■ Hydronium ion
■ Hydroxide ion

As the pH decreases, the concentration of hydronium ions increases, and the concentration of hydroxide ions decreases. Every one unit decrease in the pH means a factor of 10 increase in the hydronium ion concentration. For example, a solution with a pH of 4 and a solution with a pH of 3 are both acidic because their pHs are less than 7. The solution with a pH of 3 has ten times the concentration of H_3O^+ of the solution with a pH of 4. Small changes in pH can mean big changes in hydronium ion concentration.

Similarly, as the pH increases above 7, the concentration of hydroxide ions increases, and the concentration of hydronium ions decreases. For example, suppose you have a solution with a pH of 10 and another solution with a pH of 11. The solution with a pH of 11 has ten times more hydroxide ions than the solution with a pH of 10. The solution with a pH of 11 has one-tenth the concentration of hydronium ions of the solution that has a pH of 10.

In a neutral solution, the concentration of hydroxide ions and the concentration of hydronium ions are equal. **Figure 14.19** compares hydroxide and hydronium ion concentrations for several solutions with different pHs.

Figure 14.20

pH of Common Materials
The pH of an item may vary depending upon the solution concentration, the brand tested, and other variables.

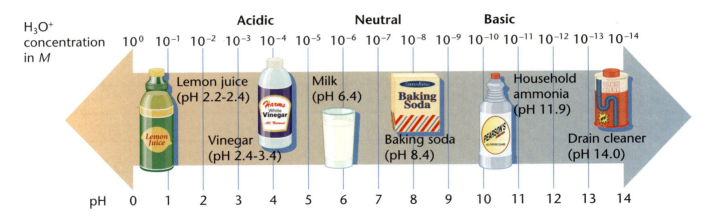

pH of Common Substances

Try at Home Lab

See page 869 in
Appendix F for
Testing for Acid Rain

Figure 14.20 gives the pHs of some common materials. Notice the range of pHs from the low pH of lemon juice to the high pH of drain cleaner.

Compare the pHs of vinegar and milk. Depending upon the brand and type of vinegar, the pH ranges from 2.4 to 3.4. If you have a bottle of vinegar with a pH of 3.4, the vinegar is definitely acidic. Milk, with a pH of 6.4, is also acidic, but much less so. The difference between 3.4 and 6.4 (3 pH units) may not seem like much, but remember that each unit of pH represents a power of 10. The hydronium ion concentration of the vinegar is 10^3, or 1000, times the hydronium concentration in the milk.

Using Indicators to Measure pH

Perhaps you've been to a swimming pool and have seen someone testing the water as in **Figure 14.21.** Among the tests done on swimming pool water is a test for pH by using indicators. The pH tells the condition of the water and its suitability for swimming.

The colored solutions that are used in this test are indicators that have different colors at different pHs. These dyes are not as precise as a pH meter, but they allow you to find approximate pH by comparing the color to a standard chart. **Figure 14.22** on the next page shows the colors of several indicators at different pHs. By choosing the right combination of these indicators, you can estimate pH across the entire pH range.

Figure 14.21
Checking Pool Water
A check of the pH of the water in a swimming pool is quick, easy, and inexpensive. Water that is not within a certain range of pH can harm skin or encourage the growth of bacteria.

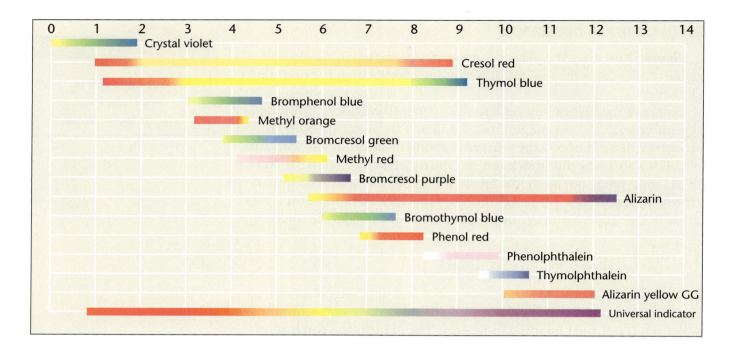

Figure 14.22
Indicators
Notice the colors of these different acid-base indicators over the pH scale. Some are better indicators at low pH, others at moderate pH, and yet others at high pH.

Connecting Ideas

You may have noticed that this chapter, which introduces the properties of acids and bases, follows a chapter that focuses on the properties of water. Although acids and bases have different properties, the common link between the two is the role that water plays in making the chemistry of acids and bases happen.

Except for those cases where ionic hydroxides dissolve in water to form free hydroxide ions, the behavior and strength of acids and bases are due to their ability to cause hydrogen ions to move to water molecules from an acid or from water molecules to a base. This movement of hydrogen ions between particles in solution can be used to demonstrate different types of acid-base reactions.

SECTION REVIEW

Understanding Concepts

1. Classify the following acids and bases as strong or weak: NH_3, KOH, HBr, $HCHO_2$, HNO_2, $Ca(OH)_2$.

2. What distinguishes a strong acid or base from a weak acid or base? Are there more strong acids and bases or weak acids and bases?

3. Other than water molecules, what particle has the highest concentration in an aqueous solution of ammonia? What has the lowest concentration? Why?

Thinking Critically

4. **Interpreting Data** The pH of normal rain is about 5.5 due to dissolved CO_2. Consider a sample of rainfall with a pH of 3.5. How does the hydronium ion concentration in these two rain samples compare?

Applying Chemistry

5. **Finding pH** A solution of unknown pH gives a pink color with phenolphthalein indicator and a blue-gray color with universal indicator. Use **Figure 14.22** to estimate the pH of the unknown solution.

REVIEWING MAIN IDEAS

14.1 Acids and Bases

- Acids have a sour taste, turn litmus red, and react with active metals, carbonates, and bases.

- Bases have a bitter taste, a slippery feel, turn litmus blue, and react with acids.

- Acids ionize by losing hydrogen ions to water molecules to form hydronium ions, H_3O^+.

- Bases form hydroxide ions, OH^-, when dissolved in water. Ionic hydroxides are bases because their ions dissociate. Covalent bases ionize by hydrogen ion transfer from water molecules to the base.

- Acidic anhydrides are nonmetallic oxides that react with water to form acids. Basic anhydrides are metallic oxides that react with water to form bases.

- Water plays a central role in the chemistry of acids and bases.

14.2 Strengths of Acids and Bases

- Strong acids and bases completely dissociate or ionize.

- Most acids and bases are weak. Weak acids and bases form few ions.

- The structure of acids and covalent bases determines their strength.

- The pH scale is a convenient way to compare the acidity and basicity of solutions.

Vocabulary

For each of the following terms, write a sentence that shows your understanding of its meaning.

acid	ionization
acidic anhydride	pH
acidic hydrogen	strong acid
base	strong base
basic anhydride	weak acid
hydronium ion	weak base

UNDERSTANDING CONCEPTS

1. Predict if the following substances will cause an increase, decrease, or no change in pH when added to water.

   a) NaOH   d) CO_2
   b) $H_2C_2O_4$ e) CH_4
   c) NH_3 f) Fe_2O_3

2. For each of the acid reactive metals in the last question, write a balanced equation that shows the reaction of each metal with aqueous hydrochloric acid, HCl(aq).

3. Calcium carbonate is the major component of limestone and marble. Sulfuric acid is one of the major components of acid rain. Write a balanced chemical reaction that shows how sulfuric acid reacts with calcium carbonate.

4. Write the chemical equation that shows the ionization of the acid HBr in water.

5. Write balanced equations that show what happens when NH_3 and KOH are placed separately in water. How do these reactions differ?

6. Explain the role of water in the ionization of an acid. Use the general equation to illustrate your explanation.

7. Explain the role of water in the ionization of a covalent base. Use the general equation for the reaction to illustrate your explanation.

8. Citric acid, $H_3C_6H_5O_7$, and vitamin C (ascorbic acid), $H_2C_6H_6O_6$, are polyprotic acids found in citrus fruits. Why are these acids classified as polyprotic acids? Write balanced equations that show the stepwise ionization of these acids in water.

9. You want to prepare magnesium acetate by an acid-base reaction. Write a reaction that will do this.

10. Give the formula of the product formed when potassium oxide reacts with water. Is the product an acid or a base?

11. What product forms when sulfur trioxide reacts with water? Is the product an acid or is it a base?

12. Write the formula of each of the following acids and identify each as a diprotic, a triprotic, or a monoprotic acid.

    a) sulfuric acid
    b) perchloric acid
    c) phosphoric acid
    d) hydrofluoric acid
    e) acetic acid

13. Dimethylamine is the strongest base of the three methyl derivatives of ammonia while trimethylamine is the weakest. How would the pH values of equal concentration solutions of these bases compare?

14. Why are soluble ionic hydroxides always strong bases?

15. What is the distinguishing factor that differentiates strong acids from weak acids?

16. What are the hydronium ion concentrations of each of the solutions in the previous question?

17. If the pH of a solution increases from 5.0 to 6.0, what has happened to the hydronium ion concentration?

APPLYING CONCEPTS

18. Natural rainfall has a pH of approximately 5.5. Write the balanced equation that shows why natural rainfall is acidic.

19. How many times greater is the hydronium ion concentration in natural rainfall (pH = 5.5) compared to pure water (pH = 7.0)?

20. A 0.10M solution of dihydrogen sulfide, H_2S, forms a solution with a pH of 4.0. What percent of the original H_2S molecules react with water to form hydronium ions?

21. Each of the following solutions was prepared by dissolving enough of the compound to make a solution that is 0.10M weak acid. The pH of each solution is measured. Rank the acids from weakest to strongest. Suppose a strong acid had been prepared like the others. What pH would you predict for a 0.10M solution of a monoprotic strong acid?

    a) Acetic acid
    b) Formic acid
    c) Benzoic acid

22. Blood has a pH of 7.4. Milk of magnesia, a common antacid for upset stomachs, has a pH of 10.4. What is the difference in pH between the two bases? Compare the hydroxide ion concentrations of the two solutions.

23. Lemon juice has a pH of about 2. Should you cut a lemon on a marble surface? Explain.

24. A substance that can act as an acid or a base is called amphoteric. Use the chemical equations to show how nitric acid and ammonia react with water to show that water is amphoteric.

25. An aqueous ammonia solution is sometimes called ammonium hydroxide. Why?

26. Why does acid rain dissolve marble statues? Write the reaction that occurs when rain containing sulfuric acid reacts with a statue made of marble, $CaCO_3$. Calcium sulfate is one of the products formed.

27. Finland, a country that has been hit hard by acid rain, has used lime, CaO, to try to return life to acidified lakes. Why was lime used?

28. A sour taste is related to acids. A bitter taste is related to bases. How is a salty taste related to acids and bases?

Everyday Chemistry

29. Some people rinse their freshly shampooed hair in diluted lemon juice or vinegar. Why might doing this be beneficial to your hair?

Biology Connection

30. After exercise, muscles sometimes become sore because of a buildup of lactic acid. How might this buildup temporarily affect the acid-base balance in the blood?

chemistryca.com/chapter_test

Chemistry and Society

31. Why might city dwellers with respiratory diseases such as asthma or emphysema be advised to stay inside on hot days?

Chemistry and Technology

32. From what you know about reactions that acids will undergo, infer how sulfuric acid might be used to remove unwanted materials from ores.

● THINKING CRITICALLY

Applying Concepts

33. What is the molarity of HCl, H_3O^+, and Cl^- in a 0.50M solution of HCl?

Relating Cause and Effect

34. Estimate the molarity of $HC_2H_3O_2$, H_3O^+, and $C_2H_3O_2^-$ in 0.50M $HC_2H_3O_2$.

Observing and Inferring

35. Examine the formulas of several polyprotic acids and compare them to the formulas of several monoprotic acids. What element is always present in monoprotic acids? What two elements are usually present in polyprotic acids?

Relating Cause and Effect

36. Baking powder, when mixed with water, produces bubbles that make a cake rise. From what you know about the properties of acids, infer what types of compounds are contained in baking powder and what gas is contained in the bubbles released by the reaction.

Applying Concepts

37. Hard water deposits around sinks may be composed of calcium carbonate and magnesium carbonate. You can buy commercial cleaners to remove these insoluble compounds, or you could use something from your kitchen. What might you try?

Interpreting Data

38. Identify the first compound in the following reactions as an acid or a base.

a) $C_5H_5N + H_2O \rightarrow C_5H_5NH^+ + OH^-$
b) $HClO_3 + H_2O \rightarrow H_3O^+ + ClO_3^-$
c) $HCHO_2 + H_2O \rightarrow H_3O^+ + CHO_2^-$
d) $C_6H_5SH + H_2O \rightarrow H_3O^+ + C_6H_5S^-$

39. One model of acid/base interactions focuses on the transfer of hydrogen ions in chemical reactions. In this model, acids are the H^+ donors and bases are the H^+ acceptors. Use the reactions in the previous question to classify the reactants as acids or bases.

Interpreting Chemical Structures

40. What is an acidic hydrogen? Draw the dot structure for acetic acid, and use it to explain why only one of the four hydrogens is acidic.

Observing and Inferring

41. ChemLab What would you expect the pH of an aqueous solution of the salt calcium chloride to be?

Forming a Hypothesis

42. MiniLab 1 Small pieces of zinc must be added to the acids instead of powdered zinc. Hypothesize why powdered zinc can't be used safely.

Observing and Inferring

43. MiniLab 2 Examine antacid ingredient lists. Some antacid tablets contain ingredients other than the antacid itself. What are some of these ingredients? Infer why the presence of these ingredients might affect the amount of antacid contained in an antacid tablet.

● CUMULATIVE REVIEW

44. Write a balanced equation for each of the following reactions involving acids and bases, and classify each reaction as one of the five general reaction types. (Chapter 6)

a) the reaction of water and lime to form calcium hydroxide
b) the reaction of sulfuric acid with zinc
c) the reaction of nitric acid with aqueous potassium hydroxide

45. What is the molar mass of each of the following substances? (Chapter 12)

   a) calcium carbonate
   b) chlorine gas
   c) ammonium sulfate

46. How many grams of sodium hydroxide are needed to prepare 2.50 L of 2.00M NaOH? (Chapter 13)

SKILL REVIEW

47. Making and Using Tables Complete each row of the table below using the information provided. For some cells, there is more than one possible correct answer. Assume that all substances are dissolved in pure water, and the solutions all have a molarity of 0.10M.

48. Designing an Experiment Design an experiment to determine the best material to use to store lemon slices. Explain your conclusions.

49. Sequencing Create a bar graph that shows solutions with the following pHs in order from the most acidic to the most basic: 7.6, 9.8, 4.5, 2.3, 4.0, 11.6.

WRITING IN CHEMISTRY

50. Find out about how sulfuric acid is made and its many commercial uses. Write an advertisement for sulfuric acid, including any research information you found out. Include graphics that illustrate things such as uses of the acid or a graph that indicates the amount of production by different countries.

PROBLEM SOLVING

51. A student buys some distilled water. She measures the pH of the water and finds it to be 6.0. She then boils the water, fills a container to the top with the hot water, and puts a lid on the container. When the water is at room temperature, the pH is 7.0. The student pours half the water into one container and stirs it for five minutes. She blows air through a straw into the other sample of water. She measures the pH of the samples of water and finds it to be 6.0 in both samples. Write an entry in your Science Journal that explains this series of pH measurements.

Name of Substance	Formula of Substance	Acid, Base, or Neutral	Strong or Weak Electrolyte	pH: >7, <7, or 7	Litmus Color	Major Particle(s) in Solution
Acetic acid						
	HCl					
		Acid	Strong			
Ammonia						
	LiOH					
		Base	Strong			
			Weak	<7		
		Neutral				Na$^+$, Cl$^-$
						H$^+$, NO$_3^-$
Phosphoric acid						

1. Acids are defined as substances that

   a) produce hydronium ions in water.
   b) have hydrogen in their compounds.
   c) produce hydroxide ions in water.
   d) neutralize bases.

2. A triprotic acid is

   a) an acid with three hydrogen atoms.
   b) an acid with three acidic hydrogen atoms.
   c) an acid with three hydronium ions.
   d) an acid with three extra protons.

3. Which of the following is the best description of the total conversion of the diprotic succinic acid ($H_2C_4H_4O_4$) into hydronium ions?

   a) $H_2C_4H_4O_4(aq) \rightarrow HC_4H_4O_4^-(aq) + H_3O(aq)$
   b) $H_2C_4H_4O_4(aq) \rightarrow H_2C_4O_4^{2-}(aq) + H_3O^+(aq)$
   c) $H_2C_4H_4O_4(aq) \rightarrow HC_4H_4O_4^-(aq) + H_3O(aq) \rightarrow C_4H_4O_4^{2-}(aq) + H_3O^+(aq)$
   d) $H_2C_4H_4O_4(aq) \rightarrow H_2C_4H_3O_4^-(aq) + H_3O^+(aq) \rightarrow H_2C_4H_2O_4^{2-}(aq) + H_3O^+(aq)$

4. Bases are defined as substances that

   a) produce hydronium ions in water.
   b) have hydrogen in their compounds.
   c) produce hydroxide ions in water.
   d) neutralize acids.

5. Which of the following is a strong acid?

   a) NaOH   **c)** H_3O^+
   b) $HC_2H_3O_2$ **d)** HCl

6. Which of the following compounds is a basic anhydride?

   a) CO_2 **c)** CaO
   b) SO_2 **d)** NO_2

Use the table in the next column to answer questions 7–9.

7. What is the pH of apples?

   a) −3   **c)** −10
   b) 3   **d)** 10

Substances	H_3O^+ concentration in M	pH
Eyewash	?	5
Apples	10^{-3}	?
Lemons	10^{-2}	?
Lye	10^{-14}	?
Household ammonia	?	12

8. The concentration of hydronium ions in eyewash is

   a) 3 times greater than the concentration of hydronium ions in lemons.
   b) 3 times less than the concentration of hydronium ions in lemons.
   c) 1,000 times greater than the concentration of hydronium ions in lemons.
   d) 1,000 times less than the concentration of hydronium ions in lemons.

9. The concentration of hydroxide ions in apples is

   a) $10^{-11}M$.
   b) $10^{11}M$.
   c) $10^{-3}M$.
   d) $10^{3}M$.

10. What is the H_3O^+ ion concentration of household ammonia?

   a) $10^{-2}M$.
   b) $10^{2}M$.
   c) $10^{-12}M$.
   d) $10^{12}M$.

Test Taking Tip

Don't Be Afraid To Ask For Help If you're practicing for a test and you find yourself stuck, unable to understand why you got a question wrong, or unable to answer it in the first place, ask someone for help. As long as you ask for help before the test, you'll do fine!

CHAPTER
15 Acids and Bases React

Chapter Preview
Sections

15.1 Acid and Base Reactions
  MiniLab15.1 Acidic, Basic, or Neutral?

15.2 Applications of Acid-Base Reactions
  MiniLab 15.2 What Does a Buffer Do?
  ChemLab Titration of Vinegar

What Happened Here?

When industrial plants burn fossil fuels, air pollutants are released into the atmosphere. Sulfur and nitrogen oxides often combine with moisture in the air, forming acid rain. This precipitation, in turn, may have a negative impact on otherwise healthy vegetation.

Launch Lab

Physical or Chemical Change?

Determine if a physcial or chemical change took place.

Safety Precautions

Materials

- 500-mL Erlenmeyer flask
- 1,000-mL graduated cylinder
- one-hole stopper with 15-cm length of glass tube
- 1,000-mL beaker
- 45-cm length of rubber (or plastic) tubing
- stopwatch or clock with second hand
- weighing dish   • balance
- baking soda   • vinegar

Procedure

1. Measure 300 mL of water. Pour water into the 500-mL Erlenmeyer flask.

2. Weight 15 g of baking soda. Carefully pour the baking soda into the flask. Swirl the flask until the solution is clear.

3. Insert the rubber stopper with the glass tubing into the flask.

4. Measure 600 mL of water and pour it into the 1,000-mL beaker.

5. Attach one end of the rubber tubing to the top of the glass tubing. Place the other end of the rubber tubing in the beaker. Be sure the rubber tubing remains under water.

6. Remove the stopper from the flask. Carefully add 250 mL of vinegar to the flask. Replace the stopper.

7. Count the number of bubbles coming into the beaker for 20 s. Repeat this two more times.

8. Record your data in a data table.

Conclude and Apply

1. Describe what you observed in the flask after the acid was added to the baking soda solution. Was it a physical or chemical change? How do you know? Was this process endothermic or exothermic? Calculate the average reaction rate based on the number of bubbles per second.

What I Already Know

Review the following concepts before studying this chapter.

Chapter 12: the mole concept; using the factor label method

Chapter 13: solution concentration using molarity

Chapter 14: properties and definitions of acids and bases

Reading Chemistry

Quickly skim the chapter, writing down the titles of the section heads. As you read, locate the main idea of each section and write it, in your own words, beneath each title.

Preview this chapter's content and activities at **chemistryca.com**

SECTION 15.1

Acid and Base Reactions

SECTION PREVIEW

Objectives

✓ **Distinguish** the overall, ionic, and net ionic equations for an acid-base reaction.

✓ **Classify** acids and bases using the hydrogen transfer definition.

✓ **Predict** and explain the final results of an acid-base reaction.

Review Vocabulary

pH: mathematical scale in which the concentration of hydronium ions in a solution is expressed as a number from 0 to 14.

New Vocabulary

neutralization
   reaction
salt
ionic equation
spectator ion
net ionic equation

As you learned in Chapter 14, acids and bases are opposites in most of their properties. But you also learned that there is a big difference between strong acids and weak acids and between strong bases and weak bases. Do all acids react with all bases? Do all acid-base reactions produce a neutral solution?

Types of Acid-Base Reactions

The reaction of an acid and a base is called a **neutralization reaction** because the properties of both the acid and base are diminished or neutralized when they react.

In most cases, the reaction of an acid with a base produces water and a salt. **Salt** is a general term used in chemistry to describe the ionic compound formed from the negative part of the acid and the positive part of the base. In the language of chemistry, sodium chloride, common table salt, is just one of a large number of ionic compounds that are called salts. KCl, NH_4NO_3, and $Fe_3(PO_4)_2$ are other examples of salts.

Consider the following neutralization reaction. Hydrochloric acid, HCl, is a common household and laboratory acid. Muriatic acid is the common household name of hydrochloric acid. It is often sold in hardware stores to be used in masonry work to remove excess mortar from brick. Sodium hydroxide, NaOH, is a common household and laboratory base. The common name of sodium hydroxide is lye. It is the primary component of many drain cleaners. **Figure 15.1** shows litmus tests before and after mixing these substances together.

Figure 15.1

Neutralization Reactions

A solution of hydrochloric acid, HCl, is added to exactly the amount of a solution of basic sodium hydroxide, NaOH, that will react with it. Litmus papers show that the resulting salt solution is neither acidic nor basic.

$$NaOH(aq) + HCl(aq) \rightarrow NaCl(aq) + H_2O(l)$$

Basic   Acidic   No color changes

After the reaction, the mixture contains only the salt sodium chloride, NaCl, dissolved in water. The litmus test shows no acid or base present in the reaction products.

Because both acids and bases may be either strong or weak, four possible combinations of acid-base reactions may occur. **Table 15.1** summarizes the possibilities. As you will find out later in this section, only three of the types are significant in everyday chemistry.

As long as one of the reactants is strong, the acid-base reaction goes to completion. As you learned in Chapter 6, a reaction goes to completion when the limiting reactant is completely consumed.

Although all of the reactions in **Table 15.1** are acid-base reactions, the submicroscopic interactions in each are different. Examine each possible type of acid-base reaction and see how they compare.

Table 15.1 Types of Acid-Base Reactions

Acid	Base
Strong	Strong
Strong	Weak
Weak	Strong
Weak	Weak

Strong Acid + Strong Base

A typical type of acid-base reaction is one in which both the acid and base are strong. The reaction of aqueous solutions of hydrochloric acid and sodium hydroxide shown in **Figure 15.1** is a good example of this type of reaction.

A Macroscopic View

It is easy to write and balance equations for strong acid-strong base reactions. In the previous example, HCl is the acid; NaOH is the base. The products are NaCl, which is a salt, and water. Now take a closer look at these reactants and their products.

The Submicroscopic View: Ionic Reactions

Recall from Chapter 14 that HCl, when dissolved in water, completely ionizes into hydronium ions and chloride ions because HCl is a strong acid. As you learned in Chapter 14, the hydronium ion is more conveniently written in shorthand as H^+.

$$HCl(aq) \rightarrow H^+(aq) + Cl^-(aq)$$

You also know that sodium hydroxide in water completely dissociates into sodium ions and hydroxide ions because NaOH is a strong base.

$$NaOH(aq) \rightarrow Na^+(aq) + OH^-(aq)$$

An overall equation for the reaction between NaOH and HCl shows each substance involved in the reaction. An overall equation does not indicate whether these substances exist as ions. The best way for you to model the submicroscopic behavior of an acid-base reaction is to show reactants and products as they actually exist in solution. Instead of an overall equation, an **ionic equation,** in which substances that primarily exist as ions in solution are shown as ions, can be written.

$$H^+(aq) + Cl^-(aq) + Na^+(aq) + OH^-(aq) \rightarrow$$
$$Na^+(aq) + Cl^-(aq) + H_2O(l)$$

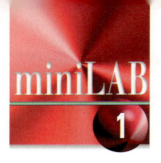

miniLAB
1

Acidic, Basic, or Neutral?

The salt potassium bromide forms in the acid-base neutralization reaction between hydrobromic acid and potassium hydroxide.

$$HBr(aq) + KOH(aq) \rightarrow H_2O(l) + KBr(aq)$$

Hydrobromic acid and potassium hydroxide are referred to as the parent acid and the parent base of potassium bromide. Test several aqueous salt solutions with bromothymol blue indicator to determine whether the solutions are acidic (yellow), basic (blue), or neutral (green).

Procedure

1. Wear laboratory aprons and safety goggles.

2. Use labeled microtip pipets to put six drops of sodium acetate solution in A1, potassium nitrate solution in A2, ammonium chloride solution in A3, sodium carbonate solution in A4, sodium chloride solution in A5, and aluminum sulfate solution in A6 of a 24-well microplate.

3. Add two drops of bromothymol blue indicator solution to each of the salt solutions. Stir each with a separate toothpick.

4. Set the microplate on a piece of white paper and look down through each well to determine the color of the solution. Record your results.

Analysis

1. According to the color of each solution that has bromothymol blue indicator added, is each of the solutions acidic, basic, or neutral?

2. Relate the relative strengths of the parent acids and bases to the results of using the indicator with the salt solutions.

3. What type of salt might be added to a product, such as shampoo, in order to make it slightly acidic so that it will not be harmful to skin and hair? Check the label of several brands of shampoo for the presence of such a salt.

Notice in the previous equation that in addition to showing the acid as completely ionized and the base as completely dissociated, the ionic compound NaCl is also dissociated. Water does not ionize much, so it is indicated as a molecule rather than H^+ and OH^- ions. The ionic aspects of this ionic equation are confirmed in **Figure 15.2.**

Figure 15.2
What ions are present?
When a conductivity apparatus is placed in solutions of each reactant and product of the reaction between a strong acid and a strong base, you can see that the acid, base, and salt exist as ions in solution.

Taste

Although the total flavor of a food comes from the complex combination of taste, smell, touch, texture or consistency, and temperature sensations, taste is a major factor. Three of the four fundamental tastes are directly linked to acids and bases. Your tongue has four different types of taste buds—sweet, salty, bitter, and sour—that are located at different places on your tongue. Only certain molecules and ions can react with these specific buds to produce a signal that is sent to a certain region of your brain. When these signals are received, your brain processes them, and you sense taste.

1. The taste buds that sense a bitter taste are located at the base of the tongue. Bases taste bitter. Many medications are basic, and pharmaceutical companies spend a lot of time and research trying to mask the bitter taste with other tastes.

Bitter

Sour

Salty

Salty and sweet

2. The taste buds that detect a sour taste are along the sides of the tongue. The sour taste comes from acids in your food. Sour-tasting foods include vinegar and citrus fruits.

Sweetness of Some Compounds

Compound	Relative Sweetness
Lactose	16
Glucose	74
Sucrose	100
Fructose	173
Aspartame	16 000
Saccharin	50 000

3. The taste buds that detect the salty and sweet tastes are located at the tip of your tongue. A salt is the product of an acid-base reaction. The sweet taste seems to depend a great deal on the properties of both acids and bases that are combined on a single molecule, but this taste is not as clear-cut as the other tastes.

As the table shows, aspartame, the popular artificial sweetener in soft drinks, is 160 times sweeter than sucrose, common table sugar. The sweetening ability of aspartame comes from a molecular structure that creates an impact 160 times that of sucrose on your taste buds that detect the sweet taste.

Thinking Critically

1. How are three of the four fundamental tastes linked to the properties of acids and bases?

2. What types of companies do you think would support research on taste?

3. What benefits might artificial sweeteners, such as aspartame, have over natural sweeteners?

Spectator Ions and the Net Ionic Reaction

Note that the ionic equation gives more information about how a strong acid-strong base reaction occurs. When you examine the two sides of the ionic equation on page 517, you see that Na^+ and Cl^- are present both as reactants and as products. Although they are important components of an overall equation, they do not directly participate in the chemical reaction. They are called **spectator ions** because they are present in the solution but do not participate in the reaction, **Figure 15.3.**

Figure 15.3

Spectators
The presence of spectators at a sporting event is important, but spectators do not actually participate in the game and do not determine the final outcome.

Why show the spectator ions in an equation if they aren't really involved in the reaction? An ionic equation can be simplified to take care of that problem. Just as in a mathematical equation, items common to both sides of the equation can be subtracted. This process simplifies the equation so that the reactants and products that actually change can be seen more clearly.

$$H^+(aq) + \cancel{Cl^-(aq)} + \cancel{Na^+(aq)} + OH^-(aq) \rightarrow \cancel{Na^+(aq)} + \cancel{Cl^-(aq)} + H_2O(l)$$

When ions common to both sides of the equation are removed from the equation, the result is called the **net ionic equation** for the reaction of HCl with NaOH.

$$H^+(aq) + OH^-(aq) \rightarrow H_2O(l)$$

The net ionic equation describes what is really happening at the submicroscopic level. Although solutions of HCl and NaOH are mixed, the net ionic equation is hydrogen ions reacting with hydroxide ions to form water.

Even though the strong acid and strong base in Sample Problem 1 are different from those in the HCl reaction with NaOH, the net ionic equation is the same. Hydrogen ions from the acid react with hydroxide ions from the base to form water. This equation is always the net ionic equation for a strong acid-strong base reaction.

SAMPLE PROBLEM 1 Equations—Strong Acid, Strong Base

Write the overall, ionic, and net ionic equations for the reaction of sulfuric acid with potassium hydroxide.

Analyze

- Decide whether the acid is a strong acid or a weak acid and whether the base is strong or weak. A list of strong acids and bases can be found in Chapter 14, **Table 14.2.** By looking at this table, you can see that sulfuric acid, H_2SO_4, is a strong acid. Potassium hydroxide, KOH, is a strong base.

Set Up

- Write an equation for the overall reaction. Because sulfuric acid is a diprotic acid, you need two moles of KOH for every one mole of H_2SO_4. Two moles of water and one mole of K_2SO_4 will be produced.

$$H_2SO_4(aq) + 2KOH(aq) \rightarrow K_2SO_4(aq) + 2H_2O(l)$$

- Write the ionic equation by showing H_2SO_4, KOH, and K_2SO_4 as ions. You must keep track of the coefficients from the overall equation and the formulas of the substances when writing the coefficients of the ions.

$$2H^+(aq) + SO_4^{2-}(aq) + 2K^+(aq) + 2OH^-(aq) \rightarrow 2K^+(aq) + SO_4^{2-}(aq) + 2H_2O(l)$$

Solve

- Look for spectator ions. In this reaction, K^+ and SO_4^{2-} are spectator ions. Subtract them from both sides of the equation to get the net ionic equation.

$$2H^+(aq) + \cancel{SO_4^{2-}(aq)} + \cancel{2K^+(aq)} + 2OH^-(aq) \rightarrow \cancel{2K^+(aq)} + \cancel{SO_4^{2-}(aq)} + 2H_2O(l)$$

$$2H^+(aq) + 2OH^-(aq) \rightarrow 2H_2O(l)$$

- Simplify the balanced net reaction by dividing coefficients on both sides of the equation by the common factor of 2.

Problem-Solving HINT

Coefficients should be in the smallest whole-number ratio possible.

$$\frac{2H^+(aq) + 2OH^-(aq) \rightarrow 2H_2O(l)}{2} \text{ results in } H^+(aq) + OH^-(aq) \rightarrow H_2O(l)$$

Check

- Take a final look at the net ionic equation to make sure no ions are common to both sides of the equation.

PRACTICE PROBLEMS

Supplemental Problems

For more practice with solving problems, see Supplemental Practice Problems, Appendix B.

Write overall, ionic, and net ionic equations for each of the following reactions.

1. hydroiodic acid, HI, and calcium hydroxide, $Ca(OH)_2$
2. hydrobromic acid, HBr, and lithium hydroxide, LiOH
3. sulfuric acid, H_2SO_4, and strontium hydroxide, $Sr(OH)_2$
4. perchloric acid, $HClO_4$, and barium hydroxide, $Ba(OH)_2$

The pH Perspective

The net ionic equation also shows why this reaction is called a neutralization reaction. The hydrogen ion from the acid reacts with the hydroxide ion from the base to form water, which is neutral.

Figure 15.4 shows the reaction of 50.0 mL of 0.100*M* HCl with 50.0 mL of 0.100*M* NaOH, which is exactly the right amount of base to react with all of the acid. As the NaOH solution is added to the HCl solution, the pH of the solution increases. After all of the NaOH solution is added, the pH of the final solution is 7.

What happens if one of the reactants is strong and the other is weak? A similar approach can be used.

Figure 15.4

Final pH

The reaction of a strong acid and a strong base is definitely a neutralization. The pH of a 0.100*M* HCl solution (top) is 1. The pH of a 0.100*M* NaOH solution (bottom left) is close to 13. When the reaction is complete, the pH is 7, which is the pH of a neutral solution (bottom right).

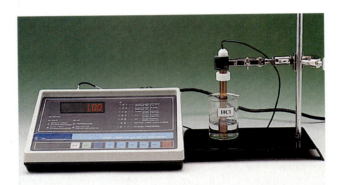

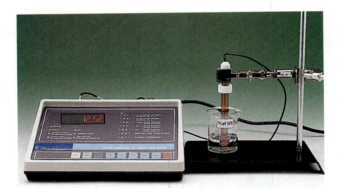

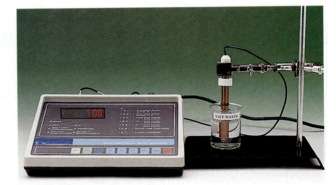

Strong Acid + Weak Base

Do the reactions change when the strength of an acid or base changes? Look at an example of what happens when a strong acid and a weak base are mixed together. Consider the reaction of hydrobromic acid and aluminum hydroxide. The overall equation shows the reactants and products.

$$3HBr(aq) + Al(OH)_3(s) \rightarrow AlBr_3(aq) + 3H_2O(l)$$

Hydrobromic acid, a strong acid, completely ionizes in water. All of the $Al(OH)_3$ that dissolves dissociates, so it is technically a strong base. However, because it is so insoluble, few OH^- ions are produced, and $Al(OH)_3$ acts as a weak base. Therefore, the ionic equation shows little dissociation of the base. The dissociated salt, $AlBr_3$, is also shown as ions.

$$3H^+(aq) + 3Br^-(aq) + Al(OH)_3(s) \rightarrow Al^{3+}(aq) + 3Br^- + 3H_2O(l)$$

The spectator ions in this equation are bromide ions. They are removed from both sides of the equation to produce the net ionic equation.

$$3H^+(aq) + Al(OH)_3(s) \rightarrow Al^{3+}(aq) + 3H_2O(l)$$

Compare this equation to the net ionic equation for a strong acid-strong base reaction.

A Strong Acid + NH_3

Recall from Chapter 14 that the most common weak base does not contain the hydroxide ion. Consider an equation for the reaction between hydrochloric acid and ammonia.

$$HCl(aq) + NH_3(aq) \rightarrow NH_4Cl(aq)$$

Notice that although the product is a salt, NH_4Cl, no water is produced in this overall reaction. As before, use the net ionic equation to understand the submicroscopic processes for this reaction of a weak base with a strong acid.

The Ionic Reaction: What's in Solution?

Recall that when ammonia dissolves in water, some of the ammonia molecules react with water to form ammonium ions and hydroxide ions. However, most of the ammonia molecules remain as molecules. Other than water, the major particle present in an aqueous solution of ammonia is ammonia molecules, NH_3.

A solution of ammonia is best represented by $NH_3(aq)$. A solution of HCl, as in the last case, is best represented as $H^+(aq)$ and $Cl^-(aq)$. The ionic reaction is written, as in Sample Problem 1, by representing what is actually in the reactant and product solutions.

$$H^+(aq) + Cl^-(aq) + NH_3(aq) \rightarrow NH_4^+(aq) + Cl^-(aq)$$

The ionic salt NH_4Cl is written as dissociated ions. **Figure 15.5** shows which of these particles are actually involved in the reaction.

Figure 15.5

What actually reacts?
A quick look at the ionic reaction shows that the chloride ion is a spectator ion because it appears on both sides of the reaction. You get the net ionic equation by subtracting the spectator Cl^- from both sides of the ionic equation.

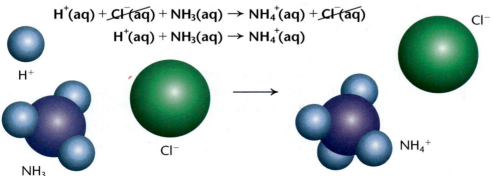

$$H^+(aq) + \cancel{Cl^-(aq)} + NH_3(aq) \rightarrow NH_4^+(aq) + \cancel{Cl^-(aq)}$$
$$H^+(aq) + NH_3(aq) \rightarrow NH_4^+(aq)$$

Cave Formation

When you think of a cave, what images come to mind? A massive, damp, cool, underground chamber running deep into Earth or a fantasy underground world with stone icicles rising from the floor and ornate columns seeming to support the ceiling? Whatever visual pictures you may imagine, caves are one of nature's wonders.

How caves are formed Caves form in limestone regions throughout the world. Limestone is calcium carbonate, which is only slightly soluble in water. The caves that form within these rocks are called solution caves.

What causes natural water to be acidic?
Most rainwater is slightly acidic because it contains carbon dioxide from the atmosphere. A small amount of the CO_2 dissolves in the water, but some of it reacts with the water to form carbonic acid.

$$CO_2(g) + H_2O(l) \rightarrow H_2CO_3(aq)$$

Carbonic acid forms a hydronium ion and a hydrogen carbonate ion.

$$H_2CO_3(aq) + H_2O(l) \rightarrow$$
$$H_3O^+(aq) + HCO_3^-(aq)$$

How does this acid form caves? The hydronium ions react with limestone to produce soluble ions.

$$H_3O^+(aq) + CaCO_3(s) \rightarrow$$
$$Ca^{2+}(aq) + HCO_3^-(aq) + H_2O(l)$$

The acidic water dissolves the limestone rocks, producing open spaces that contain water.

Stalactites and stalagmites During the second phase, clay, silt, sand, or gravel moves into the spaces. In the third phase, streams partially remove these materials and modify and enlarge the spaces. Stalactites and stalagmites now form by the reverse of the chemical and physical processes that formed the cave. The

water containing dissolved CO_2 and H_2CO_3 is saturated with $Ca(HCO_3)_2$. As it seeps through the roof of the cave, the water in each droplet slowly evaporates.

Some of the carbonic acid changes back into carbon dioxide and water. The pH of the water increases, and the solubility of $Ca(HCO_3)_2$ decreases. The $CaCO_3$ precipitates out slowly, forming stalactites over thousands of years. As the saturated water drops hit the floor, the same processes slowly form stalagmites. Sometimes, the two formations grow together, forming pillars.

Connecting to Chemistry

1. **Applying** Identify two physical changes and two chemical changes that occur in cave formation.

2. **Thinking Critically** Write the equations for cave formation from the $MgCO_3$ part of dolomite, $CaCO_3 \cdot MgCO_3$.

Write the overall, ionic, and net ionic equations for the reaction of nitric acid with ammonia.

Analyze
- Decide whether the acid is strong or weak and whether the base is strong or weak. Nitric acid, HNO_3, is a strong acid. Ammonia is a weak base.

Set Up
- Write an equation for the overall reaction.

$$HNO_3(aq) + NH_3(aq) \rightarrow NH_4NO_3(aq)$$

- Write the ionic equation. Because HNO_3 is a strong acid, you write it as completely ionized. NH_3 is a weak base, so you write it as NH_3. You dissociate the salt, ammonium nitrate, NH_4NO_3, into its component ions because it is an ionic compound.

$$H^+(aq) + NO_3^-(aq) + NH_3(aq) \rightarrow NH_4^+(aq) + NO_3^-(aq)$$

Solve
- Look for spectator ions. In this reaction, only the nitrate ion, NO_3^-, is a spectator ion. Subtract NO_3^- from both sides of the equation to get the net ionic equation.

$$H^+(aq) + \cancel{NO_3^-(aq)} + NH_3(aq) \rightarrow NH_4^+(aq) + \cancel{NO_3^-(aq)}$$
$$H^+(aq) + NH_3(aq) \rightarrow NH_4^+(aq)$$

- Note that this is the same net equation as in the HCl and NH_3 example shown in **Figure 15.5**.

Check
- Take a final look at the equation to make sure no ions are common to both sides of the equation.

PRACTICE PROBLEMS

Write overall, ionic, and net ionic equations for each of the following reactions.

5. perchloric acid, $HClO_4$, and ammonia, NH_3
6. hydrochloric acid, HCl, and aluminum hydroxide, $Al(OH)_3$
7. sulfuric acid, H_2SO_4, and iron(III) hydroxide, $Fe(OH)_3$

For all these examples of strong acid-weak base reactions, the net ionic equation differs from that for a strong acid and a strong base. The submicroscopic interactions in these strong acid-weak base reactions are between hydrogen ions and the bases.

Strong Acid-Weak Base and pH

Figure 15.6 shows that a solution of $0.100M$ ammonia is definitely a base. It has pH greater than 7. If you compare the pH of $0.100M$ NaOH with the pH of $0.100M$ NH_3, you see that ammonia is a weaker base because it has a lower pH.

Supplemental Problems

For more practice with solving problems, see Supplemental Practice Problems, Appendix B.

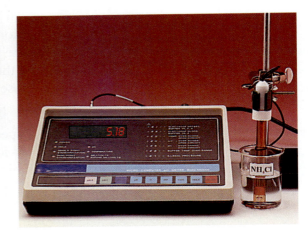

Figure 15.6
The Product Is Acidic
The reaction of a strong acid and weak base is not quite a neutralization. The pH of a 0.100*M* HCl solution is 1, and the pH of a 0.100*M* NH$_3$ solution is approximately 11. When equal volumes of the solutions are mixed and the reaction is complete, the pH is approximately 5.

Figure 15.6 also shows that when equal volumes of solutions of ammonia and hydrochloric acid of equal molarity are mixed, the pH of the final mixture is less than 7. A pH less than 7 means that the final reaction mixture is acidic. Therefore, more hydronium than hydroxide ions must be present in the final reaction mixture.

How can equal moles of base and acid react to produce a neutral solution in one reaction and an acidic solution in the second? The result must have something to do with the relative strengths of the acid and the base.

A Broader Definition of Acids and Bases

The reaction of a strong acid with a weak base demonstrates the need for a slightly broader definition of acids and bases. As you learned in the last chapter, much of the behavior of acids and bases in water can be explained by a model that focuses on the hydrogen ion transfer from the acid to the base. This model will also help explain why every acid-base reaction does not result in a neutral solution.

Hydrogen-Ion Donor or Acceptor

You can use the ability to exchange a hydrogen ion as the basis of a broader definition of an acid or a base. In this definition, called the Brønsted-Lowry definition of acids and bases, an acid is defined as a substance that donates, or gives up, a hydrogen ion in a chemical reaction. A base, not surprisingly, is just the opposite. A base is a substance that accepts a hydrogen ion in a chemical reaction.

Figure 15.7

Defining Acids and Bases by H⁺ Transfer

In a reaction between aqueous HCl and aqueous NH_3, several H^+ transfers occur.

A In the transfer of H^+ from HCl to a water molecule, HCl acts as an acid, and water acts as a base.

A. $HCl(aq) + H_2O(l) \rightarrow H_3O^+(aq) + Cl^-(aq)$
    acid   base

B. $H_3O^+(aq) + NH_3(aq) \rightarrow NH_4^+(aq) + H_2O(l)$
    acid   base

C. $H_2O(l) + NH_3(aq) \rightarrow NH_4^+(aq) + OH^-(aq)$
    acid   base

B In the transfer of H^+ from a hydronium ion to an ammonia molecule, the hydronium ion acts as an acid, and the ammonia molecule acts as a base.

C Water also reacts with ammonia molecules. Water acts as an acid, and ammonia acts as a base.

Take another look at the net ionic equation of HCl with NH_3 on page 523. If you adhere strictly to the definition of a base as a hydroxide-ion producer in water, none of these equations define ammonia as a base. Remember that hydroxide ions are produced, but the amount is so small that it is not shown in the equations.

$$H^+(aq) + NH_3(aq) \rightarrow NH_4^+(aq)$$

Using the new definition, you can definitely say that ammonia is acting as a base. It is accepting a hydrogen ion. **Figure 15.7** shows this reaction written in its most complete form and clearly shows the hydrogen ion transfer. The hydronium ion is acting as the acid because it donates the hydrogen ion.

Try at Home **Lab**

See page 870 in Appendix F for
Testing for Ammonia

It Takes Two to Transfer

Notice from **Figure 15.7** that the definitions of an acid as a hydronium-ion producer and a base as a hydroxide-ion producer are included in this H^+-transfer definition. When HCl reacts with water, it acts as an H^+ donor, so it is an acid. Water acts as an H^+ acceptor, so it is a base. When ammonia reacts with water, the ammonia molecule accepts H^+ from the water. Ammonia is the base and water is the acid. Remember that it takes two to transfer, so for every acid (an H^+ donor), there must be a base (an H^+ acceptor).

Although you generally think of water as neutral, a unique property of water can now be observed. Water can act as either an acid or a base, depending upon what else is in solution.

Water Not Required

Although most of the reactions that you will study occur in water, the H^+-transfer definition does not require water to be present. For example, the reaction between $HCl(g)$ and $NH_3(g)$ is shown in **Figure 15.8.** This reaction occurs in the gas phase. It involves the transfer of a hydrogen ion from a gaseous HCl molecule to a gaseous ammonia molecule to form a solid product. This gas reaction can now be classified as an acid-base reaction.

Figure 15.8

A Gas Phase Acid-Base Reaction
HCl(g) and NH₃(g) react to form NH₄Cl(s). Gases from the concentrated aqueous solutions react to form a smoke of solid ammonium chloride.

Weak Acid + Strong Base

Considering this new H^+-transfer acid-base definition, take a look at the type of acid-base reaction in which the acid is weak and the base is strong. An example is the reaction of acetic acid, $HC_2H_3O_2$, the weak acid present in vinegar, with sodium hydroxide. The equation of the overall reaction is similar to that of a strong acid-strong base reaction.

$$HC_2H_3O_2(aq) + NaOH(aq) \rightarrow NaC_2H_3O_2(aq) + H_2O(l)$$

The Ionic Reaction: What's in Solution?

As you know, acetic acid is a weak acid. In a solution of acetic acid, only a small fraction of the acetic acid molecules ionize. Other than water, the major particle present in an aqueous solution of acetic acid is $HC_2H_3O_2$. NaOH, as seen before, completely dissociates. The ionic equation shows what is present when the acid and base react. As in the previous cases, the salt is also written as dissociated ions.

$$HC_2H_3O_2(aq) + Na^+(aq) + OH^-(aq) \rightarrow$$
$$Na^+(aq) + C_2H_3O_2^-(aq) + H_2O(l)$$

Net Ionic Reaction: H^+ Transfer

The ionic equation contains the sodium ion as a spectator ion. Subtracting Na^+ from each side of the ionic equation gives the net ionic equation.

$$HC_2H_3O_2(aq) + \cancel{Na^+(aq)} + OH^-(aq) \rightarrow \cancel{Na^+(aq)} + C_2H_3O_2^-(aq) + H_2O(l)$$
$$HC_2H_3O_2(aq) + OH^-(aq) \rightarrow C_2H_3O_2^-(aq) + H_2O(l)$$

The net ionic equation shows that a weak acid and a strong base react by hydrogen ion transfer from the weak acid to the hydroxide ion. The acetic acid is the H^+ donor and serves as the acid. The hydroxide ion is the H^+ acceptor and serves as the base. Although this is an acid-base reaction, notice that H^+ is not involved as a reactant or product of the reaction. However, using the H^+-transfer definition, it is easy to include this reaction as an acid-base reaction.

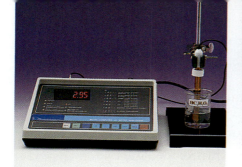  

The photos in **Figure 15.9** show what happens when 0.100M solutions of sodium hydroxide and acetic acid are mixed. Notice that the pH of the 0.100M HC$_2$H$_3$O$_2$ is greater than the pH of the 0.100M HCl solution in **Figure 15.6.**

As the sodium hydroxide solution is added to the acetic acid solution, the pH increases as the acidic hydrogens from acetic acid molecules react with the hydroxide ions. When equal volumes of the two are mixed, the final pH is greater than 7. The final reaction mixture is basic.

As in the case of the weak base-strong acid reaction, mixing equal moles of acid and base does not produce a neutral solution. Because the final pH in this case is basic, more hydroxide than hydronium ions must be present in the final reaction mixture.

Figure 15.9

Weak Acid, Strong Base The reaction of a weak acid and strong base does not result in a neutral solution. The pH of a 0.100M HC$_2$H$_3$O$_2$ solution is approximately 3, and the pH of a 0.100M NaOH solution is approximately 13. When the reaction is complete, the pH of the resulting solution is approximately 9.

SAMPLE PROBLEM 3   **Equations—Weak Acid, Strong Base**

Write the overall, ionic, and net ionic equations for the reaction of phosphoric acid with lithium hydroxide.

Analyze
- Identify the acid and base as strong or weak. Phosphoric acid, H$_3$PO$_4$, is a weak acid. Lithium hydroxide, LiOH, is a strong base.

Set Up
- Write a balanced equation for the overall reaction. The salt is lithium phosphate, Li$_3$PO$_4$. Water is also a product. You'll need three moles of LiOH for each mole of H$_3$PO$_4$. Three moles of water will be produced.

$$H_3PO_4(aq) + 3LiOH(aq) \rightarrow Li_3PO_4(aq) + 3H_2O(l)$$

Solve
- Now, write the ionic equation. Because H$_3$PO$_4$ is a weak acid, it is only partially ionized, and you write it in the ionic equation as H$_3$PO$_4$. LiOH is completely dissociated, as is the salt lithium phosphate. Be careful to keep the ionic equation balanced.

$$H_3PO_4(aq) + 3Li^+(aq) + 3OH^-(aq) \rightarrow$$
$$3Li^+(aq) + PO_4^{3-}(aq) + 3H_2O(l)$$

- Check for spectator ions. Li$^+$ is a spectator ion. Subtract Li$^+$ from both sides of the equation to get the net ionic equation.

$$H_3PO_4(aq) + \cancel{3Li^+(aq)} + 3OH^-(aq) \rightarrow \cancel{3Li^+(aq)} + PO_4^{3-}(aq) + 3H_2O(l)$$
$$H_3PO_4(aq) + 3OH^-(aq) \rightarrow PO_4^{3-}(aq) + 3H_2O(l)$$

Check
- Check to see that the reaction occurs through hydrogen ion transfer from phosphoric acid molecules to hydroxide ions. Every weak acid-strong base reaction occurs by this type of H$^+$ transfer.

Supplemental Problems

For more practice with solving problems, see Supplemental Practice Problems, Appendix B.

Write overall, ionic, and net ionic equations for the following reactions.

8. carbonic acid, H_2CO_3, and sodium hydroxide, NaOH

9. boric acid, H_3BO_3, and potassium hydroxide, KOH

10. acetic acid, $HC_2H_3O_2$, and calcium hydroxide, $Ca(OH)_2$

Weak and Weak: It's Uncertain

The strong-strong reaction plus the two types of weak-strong reactions are the favorable acid-base reactions. Looking at an acid-base reaction as occurring by H^+ transfer helps you to understand why the weak-weak reaction is not considered a favorable reaction.

Figure 15.10

Weak Acid-Weak Base
The weak acetic acid does not react with the weak base aluminum hydroxide.

Because neither a weak acid nor a weak base has a strong tendency to transfer a hydrogen ion, transfer between the two may occur, but it is uncommon. Reactions between a weak acid and a weak base generally do not play an important role in acid-base chemistry, as shown in **Figure 15.10**.

SECTION REVIEW

Understanding Concepts

1. Write the overall, ionic, and net ionic equations for the following reactions.

   a) perchloric acid and sodium hydroxide
   b) sulfuric acid and ammonia
   c) citric acid, $H_3C_6H_5O_7$, and potassium hydroxide

2. For each of the reactions in question 1, predict whether the pH of the product solution is acidic, basic, or neutral. Explain.

3. Identify the acid and the base in each of the following reactions.

   a) $HBr(aq) + H_2O(l) \rightarrow H_3O^+(aq) + Br^-(aq)$
   b) $NH_3(aq) + H_3PO_4(aq) \rightarrow$
   $\qquad\qquad NH_4^+(aq) + H_2PO_4^-(aq)$
   c) $HS^-(aq) + H_2O(l) \rightarrow H_2S(aq) + OH^-(aq)$

Thinking Critically

4. **Applying Concepts** Think about what acid and what base would react to form each of the following salts. When each of the salts is dissolved in water, will its solution be acidic, basic, or neutral? If not neutral, use an equation to explain why.

   a) NH_4Cl c) $LiC_2H_3O_2$
   b) NaCl   d) $NH_4C_2H_3O_2$

Applying Chemistry

5. **Lactic Acid Reaction** Lactic acid is produced in muscles during exercise. It is also produced in sour milk due to the action of lactic acid bacteria. The formula for lactic acid is $HC_3H_5O_3$. Write the overall, ionic, and net ionic equations for the reaction of lactic acid with sodium hydroxide. Will the pH of the product solution be greater than 7, exactly 7, or less than 7?

Applications of Acid-Base Reactions

Acid-base reactions play an important role in many chemical systems, whether you're interested in blood chemistry, the chemistry of acid rain and lakes, or the chemistry of your favorite shampoo. How is the acid-base balance maintained in each of these situations? Look at several applications of acid-base reactions and how balance is maintained.

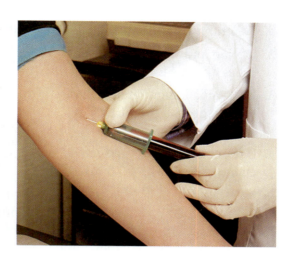

SECTION PREVIEW

Objectives
✓ **Evaluate** the importance of a buffer in controlling pH.

✓ **Design** strategies for doing acid-base titrations, and calculate results from titration data.

Review Vocabulary
Spectator ions: ions that are present in a solution but do not participate in a reaction.

New Vocabulary
buffer
titration
standard solution

Buffers to Regulate pH

Much blood chemistry depends on the acid-base balance in blood. The pH of your blood is slightly basic, about 7.4. If you're healthy, the pH of your blood does not vary by more than one-tenth of a pH unit. If you stop and think about all of the materials that go into and out of your blood, it is amazing that the pH remains so constant. Many of those materials are acidic or basic, which is why the constancy of the pH of blood is fascinating. Blood is an example of an effective buffer.

Buffers Defined

A **buffer** is a solution that resists changes in pH when moderate amounts of acids or bases are added. It contains ions or molecules that react with OH^- or H^+ if one of these ions is introduced into the solution.

Buffer solutions are prepared by using a weak acid with one of its salts or a weak base with one of its salts. For example, a buffer solution can be prepared by using the weak base ammonia, NH_3, and an ammonium salt, such as NH_4Cl. If an acid is added, NH_3 reacts with the H^+.

$$NH_3(aq) + H^+(aq) \rightarrow NH_4^+(aq)$$

If a base is added, the NH_4^+ ion from the salt reacts with the OH^-.

$$NH_4^+(aq) + OH^-(aq) \rightarrow NH_3(aq) + H_2O(l)$$

Look at another system that contains the weak acid acetic acid, $HC_2H_3O_2$, and the salt sodium acetate, $NaC_2H_3O_2$. If a strong base, OH^-, is added to the buffer system, the weak acid reacts to neutralize the addition.

$$HC_2H_3O_2(aq) + OH^-(aq) \rightarrow C_2H_3O_2^-(aq) + H_2O(l)$$

miniLAB 2

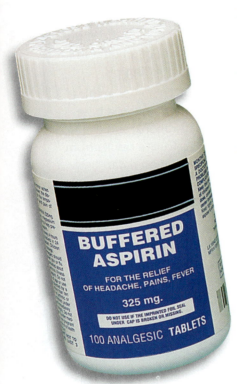

What does a buffer do?

Compare the amounts of acidic and basic solutions required to cause similar pH changes in two solutions: one a sodium chloride solution at a pH of 7 and the other a buffered solution with a pH of 7.

Procedure

1. Wear aprons and safety goggles.

2. Obtain and number four test tubes 1-4.

3. Use a small, graduated cylinder to measure 5.0 mL of NaCl solution into tubes 1 and 3.

4. Use another small, graduated cylinder to measure 5.0 mL of pH 7 buffer solution into tubes 2 and 4.

5. Add two drops of methyl orange indicator to tubes 1 and 2. The methyl orange is yellow if the pH is greater than 4.4 and will change to orange-red as the pH is lowered from 4.4 to 3.2.

6. Add 0.100M HCl solution drop by drop to tubes 1 and 2, stirring after each drop, just until the solution color changes to orange-red. Record the number of drops required for each solution.

7. Add two drops of phenolphthalein indicator to tubes 3 and 4. The phenolphthalein is colorless if the pH is lower than 8.2 and will change to pink or magenta as the pH is raised from 8.2 to 10.0.

8. Add 0.1M NaOH solution drop by drop to tubes 3 and 4, stirring after each drop, just until the solution color changes to pink or magenta. Record the number of drops required for each solution.

Analysis

1. Compare the number of drops of acid required to lower the pH of the two solutions sufficiently that the methyl orange changed color.

2. Compare the number of drops of base required to raise the pH of the two solutions sufficiently that the phenolphthalein changed color.

3. How would you describe the effect of the buffer on the pH of the solution?

This reaction takes care of the added OH^-. If H^+ is added, the acetate ion from the $NaC_2H_3O_2$ is available to neutralize the added H^+.

$$C_2H_3O_2^-(aq) + H^+(aq) \rightarrow HC_2H_3O_2(aq)$$

This system is shown in **Figure 15.11.**

Notice in **Figure 15.11** that the pH does not remain constant in the buffer solution. It changes slightly in the direction of the pH of the added acid or base. These pH changes are insignificant when you compare them to the changes that occur in the unbuffered solution.

These two buffer systems are common ones used in many laboratories. Take a closer look at the specific buffer chemistry of your blood, which operates just like these laboratory buffers.

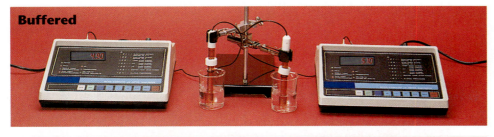

Buffered

Unbuffered

Figure 15.11

Buffers

A buffer maintains the pH of a solution at a fairly constant value. Compare what happens when acid and base are added to an acetic acid/sodium acetate buffer system at pH = 5 (top) to what happens when the same amount of acid and base are added to an unbuffered solution of pH = 5 (bottom).

Blood Buffer: Dissolved CO₂

The ability of blood to maintain a constant pH of 7.4 is due to several buffer systems. Dissolved carbon dioxide makes up one of the systems. Remember that when carbon dioxide dissolves in water, it produces carbonic acid, H_2CO_3.

$$CO_2(g) + H_2O(l) \rightarrow H_2CO_3(aq)$$

The other part of the blood buffer is the hydrogen carbonate ion, HCO_3^-. If something happens to increase OH^- in your blood, H_2CO_3 reacts to lower the OH^- concentration and keep the pH from increasing. If H^+ enters the blood, HCO_3^- reacts to keep the pH from decreasing.

added OH^-: $H_2CO_3(aq) + OH^-(aq) \rightarrow HCO_3^-(aq) + H_2O(l)$

added H^+: $HCO_3^-(aq) + H^+(aq) \rightarrow H_2CO_3(aq)$

Keep in mind that the level of CO_2 in the blood, and therefore the level of carbonic acid, is ultimately controlled by the lungs. If the amount of H^+ in the blood increases, a large amount of H_2CO_3 is produced, which lowers the H^+ concentration. In order to reduce the carbonic acid concentration produced, the lungs work to remove CO_2. Yawning is a mechanism that your body uses to get rid of extra CO_2.

On the other hand, rapid and deep breathing can cause a deficiency of carbon dioxide in the blood. This problem is called hyperventilation. It often occurs when a person is nervous or frightened. In this case, the air in the lungs is exchanged so rapidly that too much CO_2 is released. CO_2 blood levels drop, which causes the amount of carbonic acid in the blood to decrease. This causes the blood pH to increase and can be fatal if steps are not taken to stop this type of breathing.

When a person hyperventilates, he or she needs to become calm and breathe regularly. If the person breathes with a paper bag covering his or her nose and mouth, the concentration of CO_2 is increased in the air breathed. More CO_2 is forced into the blood through the lungs, and the pH of the blood drops to its normal level. **Figure 15.12** shows the narrow pH range of blood and summarizes the behavior of the buffer that controls the pH.

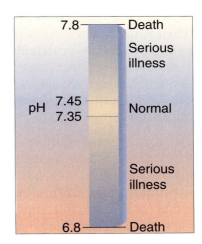

pH:
- 7.8 — Death
- Serious illness
- 7.45 / 7.35 — Normal
- Serious illness
- 6.8 — Death

Figure 15.12

Blood pH

The pH of human blood is maintained within a narrow range by a mixture of buffers. The H_2CO_3/HCO_3^- system is one of the important parts of the blood buffer.

Everyday Chemistry

Hiccups

Have you ever wondered what causes hiccups? Hiccups afflict almost everyone, but remain a scientific puzzle. They serve no useful purpose.

What are hiccups? Most of the time, hiccups are a harmless annoyance lasting a few minutes to several hours. They are repeated, involuntary spasms of your diaphragm, which is the dome-shaped breathing muscle separating the chest area from the abdomen. When hiccups start, the diaphragm jerks, and the air coming in is stopped when a small flap (the epiglottis) suddenly closes the opening to the windpipe (the glottis), resulting in the familiar "hic." No one is sure why this happens. It might start in a hiccup center in the brain stem by abnormal stimulation of nerves that control the diaphragm and the glottis.

How are they caused? Hiccups occur when your stomach is distended due to gas, overeating, or drinking carbonated beverages. Other causes are sudden temperature changes, such as drinking hot or cold beverages or taking a cold shower, and excitement or stress.

Possible cures A number of common cures stop the rhythmic reflex. These methods include massaging the area with a cotton swab, gargling with water, sipping ice water, eating a spoonful of dry sugar, and biting on a lemon. Interruptions of normal breathing such as sneezing, coughing, and sudden pain or fright may stop hiccups.

Most cures that seem to work are related to an increase of CO_2 in the blood, which slightly lowers the pH. The body has a buffer system to maintain a blood pH range of 7.35 to 7.45. Slight decreases in pH might turn off certain nervous controls that cause hiccups.

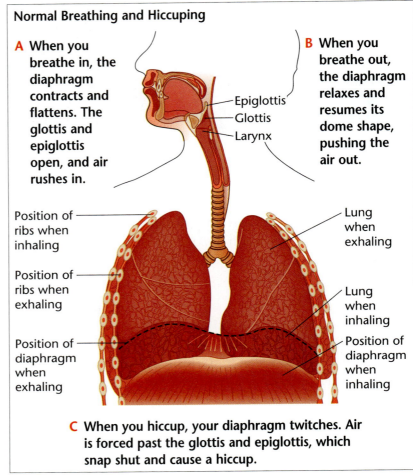

Normal Breathing and Hiccuping

A When you breathe in, the diaphragm contracts and flattens. The glottis and epiglottis open, and air rushes in.

B When you breathe out, the diaphragm relaxes and resumes its dome shape, pushing the air out.

Epiglottis
Glottis
Larynx

Position of ribs when inhaling

Position of ribs when exhaling

Position of diaphragm when exhaling

Lung when exhaling

Lung when inhaling

Position of diaphragm when inhaling

C When you hiccup, your diaphragm twitches. Air is forced past the glottis and epiglottis, which snap shut and cause a hiccup.

Exploring Further

1. **Comparing and Contrasting** Compare and contrast possible cures for hiccups.

2. **Hypothesizing** Hypothesize as to why breathing in and out of a paper bag or holding your breath might stop hiccups.

Chemistry online _____

For more information about the mystery of hiccups, visit the Chemistry Web site at **chemistryca.com**

Acid Rain Versus Acid Lakes

In Chapter 14, you learned about the sources of acid rain and about its impact on plant and animal life, as well as on human-made objects such as monuments and buildings. When acid rain falls on lakes and streams, it might be expected that the acidity of the water increases and the pH decreases.

This effect is true in some cases. Many lakes in the northeastern United States, southern Canada, northern Europe, and the Scandinavian countries have pHs as low as 4.0. The pH of a healthy lake is about 6.5.

Other lakes, many of which are in the midwestern United States, appear to get the same amount of acid rain as these low-pH lakes, but they do not show a dramatic lowering of pH. The key to the ability of these lakes, one of which is shown in **Figure 15.13,** to resist the pH-lowering effects of acid rain is their local geology.

Figure 15.13
Buffered Lakes
If the rock and soil that compose and surround a lake bed are rich in limestone, the lake can neutralize acid rain by acid-base reactions. These lakes have a capacity to absorb acid rain without an appreciable change in pH. The water in these lakes behaves as a buffer.

Limestone is primarily calcium carbonate, $CaCO_3$. Calcium carbonate reacts with carbon dioxide and water to form calcium hydrogen carbonate, $Ca(HCO_3)_2$, a water-soluble compound. Lakes in areas rich in limestone have significant concentrations of hydrogen carbonate ions.

$$CaCO_3(s) + CO_2(g) + H_2O(l) \rightarrow Ca(HCO_3)_2(aq)$$

Just as in the regulation of pH in the blood, hydrogen carbonate ions produced from calcium hydrogen carbonate form a base that can neutralize acid in lakes.

$$HCO_3^-(aq) + H^+(aq) \rightarrow CO_2(g) + H_2O(l)$$

The Acid-Base Chemistry of Antacids

The acid-base chemistry of stomach upset is big business. Even if you never use commercial antacids, a lot of other people do. The terms *acid indigestion* or *acid stomach* are a bit misleading. You need an acid stomach in order to be healthy. Remember that the pH of stomach acid, which is mostly hydrochloric acid, is about 2.5.

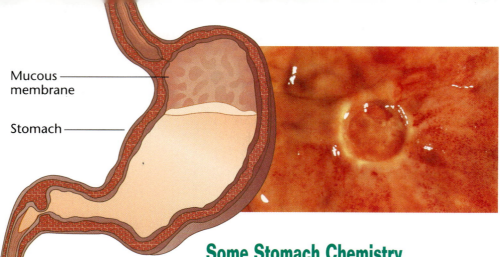

Mucous membrane

Stomach

Figure 15.14
Stomach Ulcer

A gastric ulcer forms when the membrane that protects the stomach lining breaks down and stomach acid attacks the stomach wall.

Some Stomach Chemistry

The combination of an acidic environment and enzymes in the stomach works together to break down the complex molecules that you eat into smaller molecules that can be transported through your blood and delivered to every cell in your body. These smaller molecules are the source of energy and structural material for the cells.

Your stomach is made of protein that is not much different from the protein in a hamburger. The inner walls of your stomach are coated by a basic mucous membrane that protects the stomach from the digestive power of the acid and enzymes. If the stomach contents become too acidic, this basic membrane breaks down by acid-base neutralization reactions. At the point where the barrier is neutralized, the gastric juices can begin to digest the protein that makes up the stomach wall, as shown in **Figure 15.14.** This causes the discomfort of acid indigestion and may cause more serious problems.

In most individuals, this breakdown of the mucous membrane corrects itself in a short time, and there is no long-term damage to the stomach. However, in some cases, the damage may be more long-term, and if not treated, a stomach ulcer may result. The correction process can be speeded up by using an antacid.

Antacids = Anti-Acids = Bases

Although a variety of over-the-counter and prescription antacids is available, they all share common acid-base chemistry. **Table 15.2** gives a list of the bases that are most widely used in antacids. The compounds can be divided into two categories: the hydroxide-containing bases and the carbonate-containing bases.

Fact of the Matter

An alternative approach to controlling acid indigestion is now available in over-the-counter medication. These drugs work by decreasing the secretion of acid in the stomach. These drugs were previously available only by prescription for persons who suffer severe acid indigestion or who have a tendency for gastric ulcers.

Table 15.2   Compounds Used in Antacids

Insoluble Hydroxides
Aluminum hydroxide, $Al(OH)_3$
Magnesium hydroxide, $Mg(OH)_2$

Carbonate-Based
Calcium carbonate, $CaCO_3$
Magnesium carbonate, $MgCO_3$
Sodium hydrogen carbonate, $NaHCO_3$
Potassium hydrogen carbonate, $KHCO_3$

The Development of Artificial Blood

You have a serious accident and need a blood transfusion immediately. There is no time to type your blood. Would you rather have someone else's blood or artificial blood? Right now, this choice is not available, but it may be in the near future. Under the described situation, artificial blood might be the better choice.

Why is artificial blood needed? The fear of blood transfusions is on the rise. Even though many safeguards are used, people are afraid of contracting diseases such as AIDS and hepatitis. Shortages of certain types of blood occur frequently, especially if a catastrophic disaster happens. Some people think the solution to these problems is artificial blood. Researchers have invested hundreds of millions of dollars to develop such a substance.

Types of artificial blood In real blood, hemoglobin in red blood cells carries oxygen. Many types of artificial blood focus on the hemoglobin aspect of the blood. One group is working on using the hemoglobin from outdated human blood from blood banks. Other scientists obtain it from genetically engineered bacteria, cattle, or pigs. Hemoglobin that has been removed or that was never in red blood cells causes problems. Its molecule falls apart. It does not carry oxygen well, and it sometimes clogs blood vessels. Scientists from different companies are working on these problems. Some are doing animal testing, some are already testing humans, and others have plans to do so in the near future.

Another approach to this problem is to use an oxygen-carrying perfluorocarbon emulsion that has been around since 1966. At that time, its oxygen-holding power was demonstrated by the fact that mice could breathe oxygen and survive when submerged in it. It is not considered a blood substitute by its manufacturer, but rather a drug-delivery system.

Artificial success? It will probably be several years before most oxygen-carrying blood substitutes will be approved by the FDA. However, one artificial blood component is currently in use. In the past, hemophiliacs have run a much higher risk of getting HIV because many units of blood were needed to get enough blood-clotting factor VIII. Now, the FDA has approved two genetically engineered factor VIIIs—Recombinate and Kogenate—so people who need factor VIII have a safe alternative.

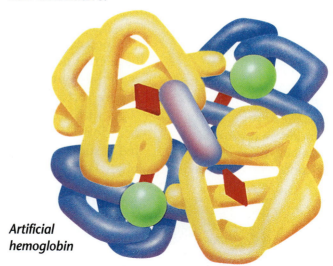

Artificial hemoglobin

Analyzing the Issue

1. **Acquiring Information** Research the structure of hemoglobin, and make a simplified model of hemoglobin.

2. **Inferring** Analyze why scientists working to develop artificial blood need an extensive knowledge of chemistry.

3. **Thinking Critically** What problems might arise with government control of artificial blood research? What advantages might there be?

Hydroxide Antacids

If you examine the ingredient list on an antacid, you probably recognize any hydroxide-containing compounds as bases. The hydroxides used in antacids have low solubility in water. Because the pH of saliva is neutral to slightly basic, these insoluble hydroxides do not dissolve and react until they get past your mouth and upper digestive tract and into the highly acidic environment of the stomach.

Milk of magnesia, a suspension of magnesium hydroxide in water, is a good example of this type of antacid. If you have ever used unflavored milk of magnesia, you may have noticed that it has a bitter taste, which is typical of a base.

Carbonate Antacids

You are familiar with the acid-neutralizing abilities of the carbonates and hydrogen carbonates from the discussion about blood buffering and acid rain. Carbonate and hydrogen carbonate antacids react with HCl to form carbonic acid, which decomposes into carbon dioxide gas and water.

$$CaCO_3(s) + 2HCl(aq) \rightarrow CaCl_2(aq) + H_2CO_3(aq)$$

$$NaHCO_3(aq) + HCl(aq) \rightarrow NaCl(aq) + H_2CO_3(aq)$$

$$H_2CO_3(aq) \rightarrow H_2O(l) + CO_2(g)$$

Many carbonates and hydrogen carbonates are insoluble in water and have great neutralizing power. They are the primary components of many over-the-counter antacid tablets. See an example of such an antacid in **Figure 15.15.**

As with any over-the-counter medication, antacids are designed for occasional use. Anyone who needs to use them frequently for attacks of indigestion may have a more serious health problem and should consult a physician.

Figure 15.15

Antacids

Antacids are hydroxide-containing bases or carbonate-containing compounds.

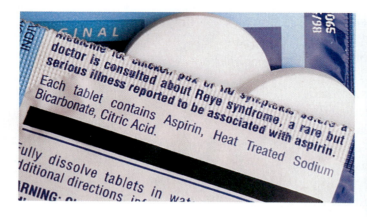

Some antacids contain a carbonate or hydrogen carbonate and a weak acid, such as citric acid. When added to water, these compounds react, producing a basic salt and carbonic acid, which decomposes into water and carbon dioxide. ▶

Stoichiometry Revisited: Acid-Base Titrations

Although acid-base reactions do not have to happen in aqueous solutions, many are water-based. In these reactions, it is frequently important to know the concentrations of the solutions involved.

The general process of determining the molarity of an acid or a base through the use of an acid-base reaction is called an acid-base **titration.** In a titration, the molarity of one of the reactants, acid or base, is known, but the other is unknown. The known reactant molarity is used to find the unknown molarity of the other solution. Solutions of known molarity that are used in this fashion are called **standard solutions.**

Titration of an Acid with a Base

Suppose you are in charge of cleaning up and doing inventory in the chemistry stockroom. Your first job is to catalog the variety of solutions that are on the shelf. One of the solutions you find is a tightly stoppered, well-labeled bottle of $0.100M$ NaOH.

On the next shelf, you find a tightly stoppered bottle that is not as well-labeled. The label has been damaged, and the only thing left of the label is M HCl. This tells that the solution contains HCl, but it does not tell the molarity. For the inventory form, you need to know the molarity of the solution. What can you do?

Do a Titration

You know that NaOH and HCl react completely.

$$HCl(aq) + NaOH(aq) \rightarrow NaCl(aq) + H_2O(l)$$

You know the concentration of the NaOH solution, so it is your standard solution. You can use the reaction, the volumes of acid and base used, plus the molarity of the base to determine the molarity of the unlabeled HCl. Follow the first part of this process in **Figure 15.16.**

Figure 15.16

Set Up the Titration
A titration requires mixing measured volumes of the standard solution and the unknown solution.

To add the NaOH to the HCl and to measure the amount of NaOH solution needed, a burette is used. A burette is a long, calibrated tube with a valve at the bottom. ▶

◀ You need to carefully measure a volume of the unknown HCl solution into a flask. For careful volume measurements, you can use a pipet. For this titration, add 20.0 mL of the unknown acid to the flask using a 20.0-mL pipet.

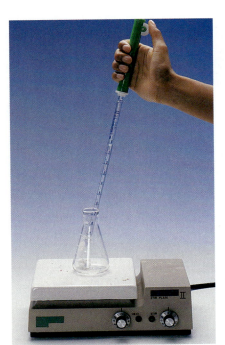

Using Indicators in a Titration

When the solution is neutral, you know that you have added exactly enough base to react with the amount of acid present, and you are at what is known as the endpoint of the titration. But how do you know when the endpoint is reached? Probably the best way to indicate the endpoint of a titration is to use an acid-base indicator. Different indicators change color at different pH values. Look at the indicators in **Figure 15.17.** For a titration of NaOH and HCl, the endpoint is reached when the solution reaches a pH of 7. Therefore, you need an indicator that changes color as close to pH = 7 as possible. The best indicator for this titration is bromothymol blue, which changes from yellow to blue close to pH = 7.

Figure 15.17

Indicators and Titration
The graphs show some of the more popular acid-base indicators used in the chemistry laboratory and the pH at which they change color.

Titration Process

Assume you have a burette containing an NaOH standard solution and a flask containing 20.0 mL of HCl of unknown concentration, as described in **Figure 15.16.** You know what indicator to use, according to **Figure 15.17.** To actually perform the NaOH-HCl titration, follow the process outlined in **Figure 15.18,** and record the data collected.

The reaction between a strong acid and a strong base results in a completely neutral solution. Bromothymol blue is an effective indicator for such reactions because it changes color at a pH of 7. ▶

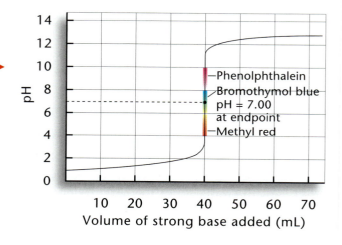

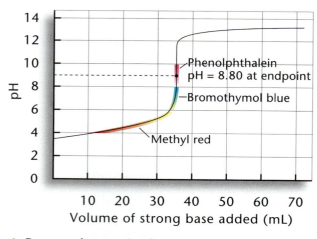

▲ Because the reaction between a weak acid and a strong base results in a slightly basic solution, the endpoint pH for a weak acid-strong base titration is greater than 7. For such a titration, phenolphthalein changes color at the endpoint.

The titration of a weak base with a strong acid has an endpoint pH that is less than 7. For this titration, methyl red is a good indicator because its pH range matches closely the endpoint pH. ▶

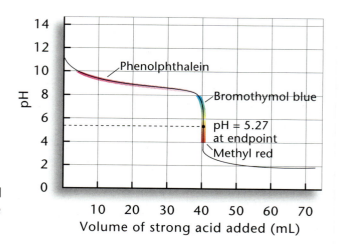

Figure 15.18

Titration Process
The following process shows how to perform and obtain data for an HCl-NaOH titration.

A Fill the burette with the NaOH solution. Be sure to record the exact reading on the burette. ▶

◀ **B** Add a few drops of bromothymol blue indicator to the HCl solution in the flask. The solution turns yellow.

C Titrate the HCl with the NaOH by slowly adding NaOH from the burette to the flask as the solution is constantly stirred. Continue to add the NaOH slowly until the color changes from yellow to blue. ▼

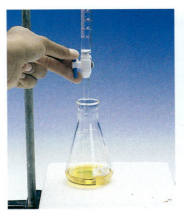 

D Measure the final volume of NaOH solution, which is 19.9 mL in this example. This reading means that the volume of NaOH used in this titration is 19.9 mL. ▼

Determining Concentration: Using Stoichiometry

You now have all your experimental data. How do you combine these experimental data into an experimental result—the molarity of the HCl solution?

First, summarize what you know. You know that 20 mL of the HCl solution reacts with 19.9 mL of 0.100M NaOH solution to reach the endpoint. From the balanced equation for the reaction, you know that one mole of HCl reacts with one mole of NaOH. Therefore, the number of moles of HCl in 20.0 mL of the HCl solution equals the number of moles of NaOH in 19.9 mL of 0.100M NaOH solution.

Now, use the factor label method to solve this solution stoichiometry problem, just as you used it to solve other stoichiometry problems. Because you know the concentration of the NaOH solution, first find the number of moles of NaOH involved in the reaction.

$$\frac{19.9 \text{ mL soln}}{} \left| \frac{1 \text{ L}}{10^3 \text{ mL}} \right| \frac{0.100 \text{ mol NaOH}}{1.00 \text{ L soln}} = 1.99 \times 10^{-3} \text{ mol NaOH}$$

Next, examine the balanced equation for the reaction and determine that, because their coefficients are the same, equal numbers of moles of NaOH and HCl react.

Titration of Vinegar

Vinegar is a solution of mostly acetic acid in water. It varies in concentration from about three percent to five percent by volume. The acetic acid in vinegar may be neutralized by adding sodium hydroxide.

$$HC_2H_3O_2(aq) + NaOH(aq) \rightarrow$$
$$H_2O(l) + NaC_2H_3O_2(aq)$$

You will titrate several brands of commercial vinegar with sodium hydroxide solution of a known concentration, and use your data to calculate the molarities and volume percentages of acetic acid in the vinegars.

Problem

What are the volume percentages of acetic acid in several brands of vinegar?

Objectives

- **Observe** acid-base titrations of several vinegars with a standard sodium hydroxide solution.
- **Calculate** the volume percentages of acetic acid in the vinegars.
- **Compare** the acetic acid concentrations of various brands of vinegar.

Brand A   Brand B   Brand C

PREPARATION

Materials

24-well microplate
several brands of vinegar
labeled microtip pipets
standard NaOH solution
phenolphthalein solution
toothpicks
distilled water in a wash bottle

Safety Precautions

Sodium hydroxide is caustic and can damage skin and eyes. If you come into contact with any of this solution, rinse the affected area with a large volume of water and notify the teacher. Wash hands thoroughly when you complete the lab.

PROCEDURE

1. Make a data table like the one shown.

2. Use a microtip pipet to add ten drops of the first type of vinegar to each of the wells—A1, B1, and C1—of the microplate. Record the brand of the first vinegar in your data table.

3. Use a clean microtip pipet to add one drop of phenolphthalein indicator solution to each of the three wells.

4. Set the microplate on a piece of white paper.

5. Use a clean pipet to carefully add a drop of the standard NaOH solution to the solution in well A1, and stir with a toothpick. Pause for about 30 seconds and look down through the well for evidence of a persistent, light pink, phenolphthalein color that indicates the endpoint of the titration. Repeat this process with each drop until the endpoint is reached. Record the num-

ber of drops of sodium hydroxide solution required to titrate the vinegar to the endpoint.

6. Repeat procedure 5 with the second sample of this vinegar, which is in well B1. If the result differs by more than one drop from that of the first titration, repeat again with the sample in well C1.

7. Repeat procedures 2 through 6 with the other brands of vinegar, using other columns of wells of the microplate. Record your data after each titration.

ANALYZE AND CONCLUDE

1. **Interpreting Data** From the two closest trials, find the average number of drops of NaOH required to titrate each vinegar. Use this average number of drops and the given molarity of the NaOH to calculate the molarity of acetic acid in each of the brands of vinegar. Assuming identical volumes for drops of vinegar and drops of NaOH solution, the ratio of reacting volumes in liters is the same as the ratio of reacting volumes in drops.

2. **Interpreting Data** Use your results to calculate the volume percentage of acetic acid in each brand of vinegar according to the formula:
$M_{acetic\ acid} \times (1.00$ percent acetic acid$/0.175M$ acetic acid$) =$ percent acetic acid by volume.

3. **Comparing and Contrasting** Which of the brands of vinegar you tested contained the highest volume percent of acetic acid?

APPLY AND ASSESS

1. How could you have changed the experimental procedure in order to more accurately determine the concentrations of the vinegars?

2. If the cost and volume of each brand of vinegar are available, calculate the cost per percent of acetic acid per unit volume for each. Which is the best buy based upon this criterion?

DATA AND OBSERVATIONS

Type of Vinegar	Trial 1—Drops of NaOH	Trial 2—Drops of NaOH	Trial 3—Drops of NaOH
Brand A			

$$NaOH(aq) + HCl(aq) \rightarrow NaCl(aq) + H_2O(l)$$

Because 1.99×10^{-3} mol NaOH react, 1.99×10^{-3} mol HCl present in solution also react.

Finally, use the volume to find the molarity of the acid.

$$\frac{1.99 \times 10^{-3} \text{ mol HCl}}{20.0 \text{ mL soln}} \left| \frac{10^3 \text{ mL}}{1 \text{ L}} \right. = \frac{0.0995 \text{ mol HCl}}{1 \text{ L soln}} = 0.0995M \text{ HCl}$$

Based on your single titration, the molarity of the HCl solution is 0.0995M. However, before you put this value on the label, you probably would repeat the titration for several additional trials in order to verify your analysis and be more confident of the value on the label. Study the following Sample Problem that gives another example of a titration.

SAMPLE PROBLEM **4** **Finding Molarity**

A 15.0-mL sample of a solution of H_2SO_4 with an unknown molarity is titrated with 32.4 mL of 0.145M NaOH to the bromothymol blue endpoint. Based upon this titration, what is the molarity of the sulfuric acid solution?

Analyze
- Because the molarity of the NaOH solution is known, the number of moles of NaOH involved in the titration can be calculated. The corresponding number of moles of H_2SO_4 can then be determined, and this figure can be used to calculate the molarity of the acid.

Set Up
- Write the balanced equation for the reaction. Remember that sulfuric acid is a diprotic acid.

$$H_2SO_4(aq) + 2NaOH(aq) \rightarrow 2H_2O(l) + Na_2SO_4(aq)$$

Solve
- Because the concentration of the NaOH solution is known, find the number of moles of NaOH used in the titration.

$$\frac{32.4 \text{ mL soln}}{} \left| \frac{1 \text{ L}}{10^3 \text{ mL}} \right| \frac{0.145 \text{ mol NaOH}}{1.00 \text{ L soln}} = 4.70 \times 10^{-3} \text{ mol NaOH}$$

- Using the balanced equation, find the number of moles of H_2SO_4 that react with 4.70×10^{-3} mol NaOH.

$$\frac{4.70 \times 10^{-3} \text{ mol NaOH}}{} \left| \frac{1 \text{ mol } H_2SO_4}{2 \text{ mol NaOH}} \right. = 2.35 \times 10^{-3} \text{ mol } H_2SO_4$$

- Find the concentration of the H_2SO_4.

$$\frac{2.35 \times 10^{-3} \text{ mol } H_2SO_4}{15.0 \text{ mL soln}} \left| \frac{10^3 \text{ mL}}{1 \text{ L}} \right. = \frac{0.157 \text{ mol } H_2SO_4}{1 \text{ L soln}} = 0.157M \text{ } H_2SO_4$$

Check
- Check to make sure that the final units are what they are supposed to be.

How it Works

Indicators

Indicators are acids and bases that have complicated structures and change color as they lose or gain hydrogen ions. The molecules must have acidic hydrogens, and they must have a relatively large number of carbon-carbon double bonds. The structure of the indicator thymol blue, $H_2C_{27}H_{28}O_5S$, satisfies both of these criteria.

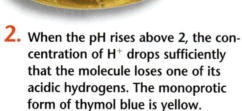

1. At low pH, thymol blue contains two acidic hydrogens. In this form, the molecule is red.

3. At pH of 8.9, the thymol blue molecule loses its second acidic hydrogen. With no acidic hydrogens, the molecule is blue.

2. When the pH rises above 2, the concentration of H^+ drops sufficiently that the molecule loses one of its acidic hydrogens. The monoprotic form of thymol blue is yellow.

However, a single indicator can give only a general indication of pH. For example, if a solution is blue with thymol blue, it does not reveal the actual pH. It indicates only that the pH is above 8.9. By combining indicators that span the range of pH from 1 to 14, it is possible to create color changes that allow more precise measurement of pH. pH paper is such a mixture. When the color of the paper is compared to the color chart, the pH can be determined to within about one pH unit.

Thinking Critically

1. What are two characteristics that molecules of indicators must have?

2. What do you think is the relationship between the number of acidic hydrogens in an indicator molecule and the number of color changes?

3. What might be some advantages of using indicator papers instead of a pH meter?

Supplemental Problems

For more practice with solving problems, see Supplemental Practice Problems, Appendix B.

11. A 0.100M LiOH solution was used to titrate an HBr solution of unknown concentration. At the endpoint, 21.0 mL of LiOH solution had neutralized 10.0 mL of HBr. What is the molarity of the HBr solution?

12. A 0.150M KOH solution fills a burette to the 0 mark. The solution was used to titrate 25.0 mL of an HNO_3 solution of unknown concentration. At the endpoint, the burette reading was 34.6 mL. What was the molarity of the HNO_3 solution?

13. A $Ca(OH)_2$ solution of unknown concentration was used to titrate 15.0 mL of a 0.125M H_3PO_4 solution. If 12.4 mL of $Ca(OH)_2$ are used to reach the endpoint, what is the concentration of the $Ca(OH)_2$ solution?

Connecting Ideas

Acid-base reactions are usually double-displacement reactions. If you examine the oxidation number of each element involved in a double-displacement reaction, you can see that the oxidation numbers of the elements do not change. Are there types of reactions in which the oxidation numbers do change?

Many important reactions, such as the rusting of an old car or the burning of a fuel, involve chemical reactions in which oxidation numbers do change. It's important to find out what causes these changes and investigate this important type of reaction.

SECTION REVIEW

Understanding Concepts

1. How does a solution that contains dissolved ammonia and ammonium chloride act as a buffer? Use net ionic equations to show how this buffer responds to added H^+ and OH^-.

2. Use chemical equations to show that the buffer system in blood works like the buffer in question 1.

3. Sequence the steps in an acid-base titration.

Thinking Critically

4. **Applying Concepts** The following are the endpoint pHs for three titrations. From the endpoint pH, indicate whether the titration involves a weak acid-strong base, a weak base-strong acid, or a strong acid-strong base reaction. Use **Figure 15.17** to select the best indicator for the reaction. What color will you see at the endpoint?

   a) pH = 4.65   b) pH = 8.43   c) pH = 7.00

Applying Chemistry

5. **Antacids** You wish to compare two different antacid tablets, brand X and brand Y. You crush each tablet and add each to 100 mL of 1.00M HCl. After stirring, you titrate the leftover HCl in the resulting solution with 1.00M NaOH. The brand X tablet requires less NaOH than the brand Y tablet requires. Which antacid neutralizes more acid?

REVIEWING MAIN IDEAS

15.1 Acid and Base Reactions

- Acid-base reactions are classified by the strength of the acid and base. Three reactions are of interest: strong acid-strong base, weak acid-strong base, and weak base-strong acid.

- Representing acid-base reactions by ionic and net ionic equations shows what is happening submicroscopically.

- Acids and bases in reactions can be identified using a hydrogen-ion transfer definition. An acid is an H^+ donor; a base is an H^+ acceptor.

- When acids and bases react, the pH of the final solution is dependent upon the nature of the reactants.

15.2 Applications of Acid-Base Reactions

- A buffer is a solution that maintains a relatively constant pH when H^+ ions or OH^- ions are added.

- The pH of blood is controlled in part by a buffer composed of carbonic acid, H_2CO_3, and the hydrogen carbonate ion, HCO_3^-.

- Antacids are bases that react with stomach acid.

- An acid-base titration uses an acid-base reaction to determine the molarity of an unknown acid or base.

Vocabulary

For each of the following terms, write a sentence that shows your understanding of its meaning.

buffer
ionic equation
net ionic equation
neutralization reaction
salt
spectator ion
standard solution
titration

UNDERSTANDING CONCEPTS

1. What are the three types of acid-base reactions that always go to completion? Give an example of each by writing an overall equation.

2. Describe what it means for a reaction to go to completion.

3. Write and balance the overall equation for each of the following reactions. Identify the type of acid-base reaction represented by the equation.

   a) potassium hydroxide + phosphoric acid
   b) formic acid, $HCHO_2$ + calcium hydroxide
   c) barium hydroxide + sulfuric acid

4. For each of the reactions in question 3, write the ionic and net ionic equations.

5. For each of the reactions in question 4, what ions are spectators? Will the final reaction mixture be acidic, basic, or neutral? Explain.

6. In words, not equations, explain the use and differences in the overall, ionic, and net ionic equations.

7. Define a *buffer solution*.

8. Write the pH control reactions for the carbon dioxide-based buffer in blood.

9. Why is it incorrect to define a buffer as a solution that maintains a constant pH?

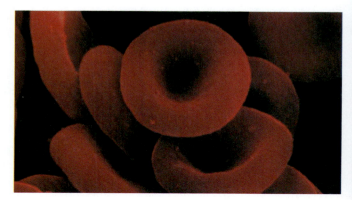

10. If the H^+ concentration in blood increases, what happens to the concentrations of H_2CO_3, HCO_3^-, and H^+?

11. What role do the lungs play in regulating blood pH?

12. Write two reactions that explain why lakes in limestone areas are capable of resisting pH decreases due to acid rain.

13. Antacids are classified into two types. What are they? Give an example of each type.

14. Consider an antacid that contains aluminum hydroxide. Write the overall equation that shows how this antacid reduces the acidity of stomach acid.

15. A student found that 53.2 mL of a $0.232M$ solution of NaOH was required to titrate 25.0 mL of an acetic acid solution of unknown molarity to the endpoint. What is the molarity of the acetic acid solution?

16. A student neutralizes 30.0 mL of a sample of sodium hydroxide with 28.9 mL of $0.150M$ HCl. What is the molarity of the sodium hydroxide?

17. A student finds that 23.1 mL of $0.200M$ potassium hydroxide are required to react completely with 25.0 mL of a phosphoric acid solution. What is the molarity of the H_3PO_4?

18. How does the endpoint pH of a strong acid-strong base titration compare with that of a weak acid-strong base titration?

19. How does the endpoint pH of a strong base-strong acid titration compare with that of a weak base-strong acid titration?

20. A 50.0-mL sample of an unknown monoprotic acid is titrated to the endpoint with 45.5 mL of $0.200M$ Ca(OH)$_2$. What is the molarity of the acid solution?

21. A 50.0-mL sample of aqueous ammonia is titrated to the endpoint with 36.3 mL of $0.100M$ H_2SO_4. What is the molarity of the ammonia solution?

22. What is hyperventilation? How does it change the pH of blood?

23. When 25.0 g of baking soda, $NaHCO_3$, and 25.0 mL of $1.00M$ HCl are mixed, is the final solution acidic, basic, or neutral?

APPLYING CONCEPTS

24. Methylamine, CH_3NH_2, is a weak base, as ammonia is. When methylamine completely reacts with hydrochloric acid, the final solution has a pH less than 7. Why are the products of this "neutralization" reaction not neutral? Use the net ionic equation to help in your explanation.

25. How many milliliters of $0.200M$ HCl are required to react with 25.0 mL of $0.100M$ methylamine, CH_3NH_2?

26. Vitamin C is also known as ascorbic acid, $HC_6H_7O_6$. A solution made from a vitamin C tablet is titrated to the endpoint with 12.3 mL of $0.225M$ NaOH. Assuming that vitamin C is the only acid present in the tablet, how many moles of vitamin C are in the tablet?

27. Why are magnesium hydroxide and aluminum hydroxide effective antacids, but sodium hydroxide is not?

28. How many milliliters of $0.100M$ NaOH are required to neutralize 25.0 mL of $0.150M$ HCl?

29. Concentrated HCl solutions are prepared by dissolving HCl(g) in water. Concentrated HCl is usually sold as a 12 M HCl solution. How many liters of HCl(g) at 25°C and 1 atmosphere pressure are required to make 10.0 liters of 12 M HCl?

30. Complete and balance the following overall equations.
    a) $KOH(aq) + HNO_3(aq) \rightarrow$
    b) $Ba(OH)_2(aq) + HCl(aq) \rightarrow$
    c) $NaOH(aq) + H_3PO_4(aq) \rightarrow$
    d) $Ca(OH)_2(aq) + H_3PO_4(aq) \rightarrow$

31. Dihydrogen phosphate and monohydrogen phosphate ions play an important role in maintaining the pH in intracellular fluid. Write equations that show how these ions maintain the pH.

chemistryca.com/chapter_test

32. The concentration of H_2CO_3 in blood is 1/20 of the concentration of HCO_3^-, yet the blood buffer is capable of buffering the pH against bases, as well as against acid. Explain.

33. Tartaric acid is often added to artificial fruit drinks to increase tartness. A sample of a certain beverage contains 1.00 g of tartaric acid, $H_2C_4H_4O_6$. The beverage is titrated with $0.100M$ NaOH. Assuming no other acids are present, how many milliliters of base are required to neutralize the tartaric acid?

34. How many grams of tartaric acid, $H_2C_4H_4O_6$, must be added to 150 mL of $0.245M$ NaOH to completely react?

35. Stomach acid is approximately $0.0200M$ HCl. What volume of stomach acid does an antacid tablet that contains 45.5 percent $Mg(OH)_2$ and weighs 355 mg neutralize?

36. Suppose the tablet in question 35 is composed of calcium carbonate. Is this tablet more effective than the one composed of $Mg(OH)_2$?

Chemistry and Society

37. What function of blood is most important when developing artificial blood?

Earth Science Connection

38. What acid most likely causes groundwater to be acidic? How does groundwater become acidic?

Everyday Chemistry

39. How is the CO_2 concentration in the blood related to hiccups?

How It Works

40. Explain how molecules and ions are related to taste.

How It Works

41. Can an indicator provide an exact pH? Explain.

THINKING CRITICALLY

Observing and Inferring

42. A sample of rainwater turns blue litmus red. Fresh portions of the rainwater turn thymol blue indicator yellow, bromophenol blue indicator green, and methyl red indicator red. Estimate the pH of the rainwater.

Interpreting Data

43. When formic acid, $HCHO_2$, reacts completely with NaOH, the resulting solution has a pH greater than 7. Why are the products of this neutralization reaction not neutral? Use the net ionic equation to help in your explanation.

Applying Concepts

44. Write an overall equation for the acid-base reaction that would be required to produce each of the following salts.

a) NaCl   **c)** $MgCl_2$ **e)** KBr
b) $CaSO_4$ **d)** $(NH_4)_2SO_4$

Observing and Inferring

45. ChemLab Explain why different bottles of the same brand of vinegar might contain solutions that have different pHs.

Making Predictions

46. MiniLab 1 Would a solution of iron(III) bromide, $FeBr_3$, be acidic, basic, or neutral?

Relating Cause and Effect

47. MiniLab 2 Explain why phenolphthalein and methyl orange are used as indicators in Mini-Lab 2.

CUMULATIVE REVIEW

48. Give the name of the compound represented by the formula $Mn(NO_3)_2 \cdot 4H_2O$, and determine how many atoms of each element are present in three formula units of the compound. (Chapter 5)

49. Terephthalic acid is an organic compound used in the formation of polyesters. It contains 57.8 percent C, 3.64 percent H, and 38.5 percent O. The molar mass is known to be approximately 166 g/mol. What is the molecular formula of terephthalic acid? (Chapter 12)

50. Write equations for the dissociation of the following ionic compounds when they dissolve in water. (Chapter 13)
  a) $CuSO_4$
  b) $Ca(NO_3)_2$
  c) Na_2CO_3

51. What is a monoprotic acid? A triprotic acid? Give an example of each. (Chapter 14)

SKILL REVIEW

52. Data Table Solutions of five different monoprotic acids are all $0.100M$. The pH of each solution is given. Rank the acids in the following table from weakest to strongest. For each solution, use the indicator table in **Figure 14.22** to predict the color that each solution would produce with the given indicator.

WRITING IN CHEMISTRY

53. Write an article about the effect of acid rain on a specific aspect of a local environment such as a lake or a forest. Give some history of the problem and indicate when local residents first realized a problem exists. What, if any, corrective measures have been taken to correct the problem? Is the environmental damage reversible?

PROBLEM SOLVING

54. Because antacids are frequently insoluble in water, they are often analyzed by dissolving them in a known volume of HCl with a known molarity. After the antacid has completely reacted, there is still HCl left in the solution. This excess HCl is then titrated with a standard NaOH solution. A 165-mg sample of an antacid tablet containing calcium carbonate is dissolved in 50.0 mL of $0.100M$ HCl. After complete reaction, the excess HCl is titrated with 15.8 mL of $0.150M$ NaOH. Sketch a flowchart that shows the steps in the analysis.

Solution	pH
A	5.45
B	1.00
C	3.45
D	4.50
E	2.36

Solution (weakest to strongest acid)	pH	Color in Bromphenol Blue	Color in Methyl Red	Color in Thymol Blue

1. What is the symbol for a hydronium ion?

a) H^+

b) H

c) OH

d) OH^-

Use the chemical equation to answer questions 2 and 3.

$$Mg(OH)_2(aq) + 2HCl \rightarrow MgCl_2(aq) + 2H_2O(l)$$

2. Which of the compounds in the equation is considered a base?

a) $Mg(OH)_2$

b) HCl

c) $MgCl_2$

d) H_2O

3. Which of the compounds in the equation is considered a salt?

a) $Mg(OH)_2$

b) HCl

c) $MgCl_2$

d) H_2O

4. Which ions are excluded from a net ionic equation?

a) weak acids or bases

b) negative ions

c) positive ions

d) spectator ions

5. Which of the following is true about a solution with a pH lower than 7?

a) The solution is a strong acid.

b) The quantity of hydronium ions is greater than the quantity of hydroxide ions.

c) The quantity of hydroxide ions is greater than the quantity of hydronium ions.

d) The solution is an equal mixture of moles of acid and moles of base.

6. In the equation, HBr is

$$HBr(aq) + H_2O(l) \rightarrow H_3O+ (aq) + Br^-(aq)$$

a) considered a strong base because it completely ionizes in water.

b) considered a weak base because it loses a positive charge to a water molecule.

c) considered a strong acid because it completely ionizes in water.

d) considered a weak acid because it loses a positive charge to a water molecule.

Weak Acid Ionization Constants	
Weak Acid	**Ionization Constant**
Hydrofluoric acid	6.3×10^{-4}
Methanoic acid	1.8×10^{-4}
Ethanoic acid	1.8×10^{-5}
Hypochlorous acid	4.0×10^{-8}

Use the table above to answer question 7.

7. The ionization constant of a weak acid is a calculation of the number of ionized molecules (products) in a dilute aqueous solution divided by the number of un-ionized molecules (reactants) in the solution. Which acid is the weakest?

a) hydrofluoric acid

b) methanoic acid

c) ethanoic acid

d) hypochlorous acid

8. A Brønsted-Lowry acid is defined as a(n)

a) acid that donates a hydrogen ion during a chemical reaction.

b) acid that accepts a hydrogen ion during a chemical reaction.

c) strong acid that donates the maximum number of hydrogen ions.

d) weak acid that donates a limited number of hydrogen ions.

9. A solution of 0.600M HCl is used to titrate 15.00 mL of KOH solution. The endpoint of the titration is reached after the addition of 27.13 mL of HCl. What is the concentration of the KOH solution?

a) 9.000M

b) 1.09M

c) 0.332M

d) 0.0163M

Test Taking Tip

Tables If a test question involves a table, skim the table before reading the question. Read the title, column heads, and row heads. Then read the question and interpret the information in the table.

CHAPTER
16 Oxidation-Reduction Reactions

Chapter Preview

Sections

16.1 The Nature of Oxidation-Reduction Reactions

MiniLab 16.1   Corrosion of Iron

ChemLab   Copper Atoms and Ions: Oxidation and Reduction

16.2 Applications of Oxidation-Reduction Reactions

MiniLab 16.2   Testing for Alcohol by Redox

Why Do Things Rust?

When iron corrodes, iron metal reacts with oxygen from the air and water to form iron(III) oxide—rust. Rust is the result of an oxidation-reduction reaction in which iron metal loses electrons to oxygen. Given time, and oxygen from the air and water, all of the drums in this photo will rust away completely.

*L*aunch Lab

Observing an Oxidation–Reduction Reaction

Rust is the result of a reaction of iron and oxygen. Iron nails can also react with substances other than oxygen, as you will find out in this experiment.

Safety Precautions

Always wear safety goggles and an apron in the laboratory.

Materials

- test tube
- iron nail
- steel wool or sandpaper
- 1M copper(II) sulfate ($CuSO_4$)

Procedure

1. Use a piece of steel wool to polish the end of an iron nail.
2. Add about 3 mL 1.0M $CuSO_4$ to a test tube. Place the polished end of the nail into the $CuSO_4$ solution. Let stand and observe for about 10 minutes. Record your observations.

Analysis

What is the substance found clinging to the nail? What happened to the color of the copper(II) sulfate solution? Write the balanced chemical equation for the reaction you observed.

What I Already Know

Review the following concepts before studying this chapter.
Chapter 3: patterns of valence electrons
Chapter 5: predicting oxidation number from the periodic table
Chapter 6: types of reactions

Reading Chemistry

Look through the Section Previews for this chapter, jotting down some key ideas. As you read through the chapter, make an outline using the key ideas you wrote down. For each topic, review any new vocabulary words.

Preview this chapter's content and activities at **chemistryca.com**

The Nature of Oxidation-Reduction Reactions

Oxygen undergoes many reactions when it encounters other substances. One of these reactions is responsible for the browning of fruits. Another forms the rust that eats away at the metal parts of bikes and cars. In both of these cases, a type of reaction called oxidation is taking place. You can probably guess how this reaction got its name; oxygen is a reactant. But you will learn that not all oxidation reactions involve oxygen. And oxidation reactions are never lonely because they always have partners— reduction reactions. You will see what the characteristics of these reactions are and why they always take place together.

What is oxidation-reduction?

Oxygen is the most abundant element in Earth's crust. It is very reactive and can combine with almost every other element. An element that bonds to oxygen to form a new compound, called an oxide, usually loses electrons because oxygen is more electronegative. You will recall that an electronegative element has a strong attraction for electrons. Because of this strong attraction, oxygen is able to pull electrons away from other atoms. The reactions in which elements combine with oxygen to form oxides were among the first to be studied by early chemists, who grouped them together and called them *oxidation reactions*. Later, chemists realized that some other nonmetal elements can combine with substances in the same way as oxygen and that these reactions are similar to oxidation reactions. Modern chemists use the term *oxidation* to refer to any chemical reaction in which an element or compound loses electrons to another substance.

A common oxidation reaction occurs when iron metal loses electrons to oxygen. Each year in the United States, corrosion of metals—especially the iron in steel—costs billions of dollars as automobiles, ships, and bridges and other structures are slowly eaten away. **Figure 16.1** shows some of this damage and how it can be prevented.

Figure 16.1

Corrosion of Iron

When iron corrodes, iron metal reacts with oxygen to form iron(III) oxide—rust. Corrosion of iron can be prevented by covering the surface of exposed steel with paint or other coatings such as plastic. If the protective coating is damaged or cracked, rust forms quickly. ▶

◀ Steel can be protected from oxidation if it is coated with a more active metal such as zinc. Zinc loses electrons to oxygen more readily than iron does, so the zinc is oxidized preferentially, forming a tough protective layer of zinc oxide. The coating of zinc and zinc oxide prevents the formation of rust by keeping oxygen from reaching the iron. Steel that has been coated with zinc is called galvanized steel. The bucket on the left has been galvanized.

Redox

What happens to the zinc in galvanized steel? It reacts with oxygen to form zinc oxide in the following reaction.

$$2Zn(s) + O_2(g) \rightarrow 2Zn^{2+}(s) + 2O^{2-}(s)$$

Does this type of reaction look familiar? You learned in Chapter 6 that this is classified as a synthesis reaction. You also know that early chemists called it an oxidation reaction because oxygen is a reactant. The formation of zinc oxide falls into another, broader class of reactions characterized by the transfer of electrons from one atom or ion to another. This type of reaction is called an **oxidation-reduction reaction,** commonly known as a redox reaction. Many important chemical reactions are redox reactions. Formation of rust is one example; combustion of fuels is another. In each redox reaction, one element loses electrons, and another element takes them.

How do atoms or ions lose electrons in a redox reaction? If you examine the equation for the reaction between zinc and oxygen more closely, you can see which atoms are gaining electrons and which are losing them. You also can determine where the electrons go during a redox reaction by comparing the oxidation number of each type of atom or ion before and after the reaction takes place. Recall from Chapter 5 that the oxidation number of an ion is equal to its charge. All elements, when in their free form, have a charge of zero and are assigned an oxidation number of zero. In the formation of zinc oxide, the zinc atom and the diatomic oxygen molecule that react each has an oxidation number of zero. In the ionic compound formed, each oxide ion has a 2− charge and an oxidation number of 2−. Because the compound must be neutral, the total positive charge must be 4+; thus, each zinc ion must have a charge and an oxidation number of 2+.

Oxidation

You have learned that a reaction in which an element loses electrons is called an **oxidation** reaction. The element that loses the electrons becomes more positively charged; that is, its oxidation number increases. That element is said to be oxidized during the reaction. Zinc is oxidized during the formation of zinc oxide because metallic zinc atoms each lose two electrons. The oxidation reaction can be written by itself to show how zinc changes during the redox reaction. Here's what happens to each atom of zinc.

$$Zn\!:\; \longrightarrow [Zn]^{2+} + 2e^- \qquad \text{(loss of electrons)}$$

Reduction

What happens to the electrons that are lost by the zinc atom? Electrons do not wander around by themselves; they must be transferred to another atom or ion. This is why oxidation reactions never occur alone. They are always paired with reduction reactions. A **reduction** reaction is one in which an element gains one or more electrons. The element that picks up the electrons and becomes more negatively charged during the reaction is said to be reduced. Its oxidation number decreases, or is reduced. Because oxidation and reduction reactions occur together, each is referred to as a half-reaction.

In every redox reaction, at least one element undergoes reduction while another undergoes oxidation. Just as a successful pass in football requires a quarterback to throw the ball and a receiver to catch it, a redox reaction must have one element that gives up electrons and one that accepts them. The electronic structure of both reactants changes during a redox reaction.

Figure 16.2 shows the movement of electrons in the formation of zinc oxide. Oxygen accepts the electrons that zinc loses. Oxygen is reduced during the reaction between zinc and oxygen because each oxygen atom gains two electrons. Like the oxidation reaction, the reduction reaction can be written by itself. Here's what happens to each atom of oxygen.

$$\cdot\ddot{O}\!: + 2e^- \longrightarrow :\ddot{O}\!:^{2-} \qquad \text{(gain of electrons)}$$

WORD ORIGIN

reduction:
  re (L) back
  ducere (L) to lead

In a reduction reaction, the addition of electrons results in a decrease in oxidation number of an atom or ion.

Figure 16.2

Formation of Zinc Oxide

In the formation of zinc oxide, the zinc atom loses two electrons during the reaction, becoming a zinc ion. Its oxidation number increases from zero to 2+. The oxygen atom gains the two electrons from zinc, becoming an oxide ion. Its oxidation number decreases from zero to 2−.

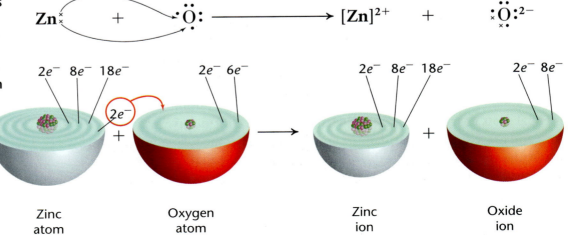

| Zinc atom | Oxygen atom | Zinc ion | Oxide ion |

Corrosion of Iron

Corrosion is the term generally used to describe the oxidation of a metal during its interaction with the environment. In this MiniLab, you will study the corrosion of a nail and determine the factors that affect this process.

Procedure

1. Dissolve a package of clear, unflavored gelatin in about 200 mL of warm water. Stir in 2 mL of phenolphthalein solution and 2 mL of potassium hexacyanoferrate(III) solution. Pour the prepared solution into a widemouth glass jar or petri dish to a depth of about 1 cm.

2. In the liquid gelatin, place a plain iron nail, an aluminum nail, a galvanized iron nail, and a painted iron nail of the type used for paneling. Space the nails far apart.

3. Label the jar or petri dish with your name, and leave it for several hours or overnight. Handle the jar or dish carefully until the gelatin has set.

4. Record your observations regarding any interactions of the nails with the substances in the gelatin.

Analysis

1. Which of the nails have reacted with the substances in the gelatin? What is the evidence of corrosion?

2. If any of the nails have not corroded in the solution, can you suggest a reason why they haven't? What methods are commonly used to prevent or minimize corrosion?

3. Any blue color in the gelatin is due to the formation of iron(II) ions and their interaction with the hexacyanoferrate(III) ion. Any pink or red color in the gelatin is due to the gaining of electrons by oxygen and water molecules, forming basic hydroxide ions that turn the phenolphthalein pink. Which of these reactions is oxidation, and where does it occur on the reacting nail?

Combining the Half-Reactions

The equation for the reduction half-reaction shows one atom of oxygen reacting. However, oxygen is not found in nature as single atoms; two atoms combine to form a diatomic molecule of O_2. The reduction equation must be multiplied by 2 to reflect this. Thus, the balanced equation for the reduction reaction is written as follows.

$$O_2 + 4e^- \rightarrow 2\ddot{\underset{..}{O}}\!:^{2-}$$

Note that four electrons are gained by the oxygen molecule. To produce those four electrons, two atoms of zinc must take part in the reaction. Therefore, the balanced equation for the oxidation reaction must be written as follows.

$$2Zn\!: \rightarrow 2[Zn]^{2+} + 4e^-$$

Figure 16.3

Summarizing the Reaction

Two zinc atoms combine with one diatomic oxygen molecule to form two formula units of zinc oxide. Because zinc loses four electrons in the oxidation reaction and oxygen gains four electrons in the reduction reaction, all electrons are accounted for; the two half-reactions are balanced.

Each atom of oxygen accepts $2e^-$ from a zinc atom and is reduced.

$$2Zn^0 + O_2^0 \rightarrow 2Zn^{2+} + 2O^{2-}$$

$+4e^-$ from Zn^0

$-4e^-$ to O_2

Each atom of zinc donates $2e^-$ to an oxygen atom and is oxidized.

The balanced overall equation for the reaction now can be written as shown in **Figure 16.3.** This equation is the same as the equation for the formation of zinc oxide that you read at the beginning of the discussion on redox reactions. Now you know that it represents the net oxidation-reduction reaction and is the sum of an oxidation half-reaction and a reduction half-reaction.

If an element is gaining electrons, why is this called a reduction reaction? After all, you don't gain weight when you reduce. You have learned that the reason is because there is a reduction in the charge or oxidation number of an atom of the substance that is reduced. An older, historic reason for the use of the term *reduction* is that the name was first applied to processes in furnaces in which metals are isolated from their ores at high temperatures, **Figure 16.4.** During these processes, oxygen is removed from the ores in which it is combined with the metal, so the ore is reduced to the free metal. There is a reduction in the amount of solid material and a considerable decrease in volume.

Figure 16.4

Furnaces: Old and Modern

For thousands of years, metals have been used by many different cultures for making jewelry, cookware, and weapons. Because metals are normally found combined with other elements as ores, furnaces operating at high temperatures are used to separate the free metal from other elements. The positively charged metal ions in the ore are reduced to the elemental state, while oxygen and other negatively charged elements in the ore are oxidized. Iron smelting in medieval England is shown here. ▼

▲ A modern industrial iron blast furnace is shown here.

Identifying a Redox Reaction

The oxidation of zinc is a redox reaction in which oxygen is a reactant. You have learned that elements other than oxygen can accept electrons and become reduced during redox reactions. You are already familiar with the explosive reaction in which sodium and chlorine combine to form table salt.

$$2Na(s) + Cl_2(g) \rightarrow 2NaCl(s)$$

Are electrons transferred during this reaction? Yes, because each sodium atom loses one electron to become a sodium ion with a charge of 1+. The oxidation number of sodium increases from 0 to 1+. Each chlorine atom gains one electron to form a chloride ion. The oxidation number of chlorine decreases from 0 to 1−. Therefore, this is another example of a redox reaction.

$$\overset{\displaystyle \lceil \overset{2e^-}{} \rceil}{2Na^0 + Cl_2^0 \rightarrow 2Na^+ + 2Cl^-}$$

Oxidizing and Reducing Agents

Another redox reaction that doesn't involve oxygen occurs when a strip of zinc metal is placed in a solution of copper(II) sulfate. The progress of this reaction can be followed easily because a readily observable change takes place. As shown in **Figure 16.5**, copper metal quickly begins to form on the zinc strip.

$$Cu^{2+} \xrightarrow[\text{reduced to}]{2e^-} Cu^0 \qquad Zn^0 \xrightarrow[\text{oxidized to}]{2e^-} Zn^{2+}$$

Cu^{2+} is the oxidizing agent, and Zn^0 is the reducing agent. ▶

$$\underset{\substack{\text{oxidizing} \\ \text{agent}}}{Cu^{2+}} + \underset{\substack{\text{reducing} \\ \text{agent}}}{Zn^0} \xrightarrow{\text{reduced to}} Cu^0 + Zn^{2+} \quad \text{oxidized to}$$

ChemLab

Copper Atoms and Ions: Oxidation and Reduction

Copper atoms and ions often take part in reactions by losing or gaining electrons, which are oxidation and reduction, respectively. If copper atoms lose electrons to form positive ions, copper is oxidized. Other atoms or ions must gain the electrons that copper atoms lose. These atoms or ions are reduced and are called oxidizing agents. In this ChemLab, you will observe two reactions that involve the oxidation or reduction of copper.

Problem

What are some typical reactions that involve the oxidation or reduction of copper?

Objectives

- **Observe** reactions that involve the oxidation or reduction of copper.
- **Classify** the reactants as substance oxidized, reducing agent, substance reduced, and oxidizing agent.

PREPARATION

Materials

copper(II) oxide
powdered charcoal (carbon)
weighing paper
balance
large Pyrex or Kimax test tubes (2)
1-hole rubber stopper fitted with glass tube with bend as shown
150-mL beakers (2)
graduated cylinder, small
glass stirring rod
Bunsen or Tirrill burner
ring stand
test-tube clamp

thermal glove
limewater (calcium hydroxide solution)

Safety Precautions

Care should be taken in handling hot objects and when working around open flames. Do not breathe in the fumes that are produced during the teacher demonstration in step 1.

PROCEDURE

1. **Teacher Demonstration** Your teacher will perform this reaction as a demonstration either in the fume hood or outside the building. **CAUTION:** *Do not perform this procedure by yourself.* A 1-cm square of copper foil will be placed in a porcelain evaporating dish. First, 5 mL of water, then 5 mL of concentrated HNO_3 will be added. Note the color of the evolved gas and the color of the resulting solution. Record your observations in a table similar to the one under Data and Observations.

2. On a piece of weighing paper, thoroughly mix approximately 1 g of copper(II) oxide with twice its volume of powdered charcoal. Place the mixture in a clean, dry Pyrex or Kimax test tube. Add about 10 mL of limewater to a second test tube, and stand it in a 150-mL beaker. Assemble the apparatus as shown here, with the copper-oxide test tube sloped slightly downward and the delivery tube extending into the limewater.

3. Heat the mixture in the test tube, gently at first and then strongly. As soon as you notice a change in the limewater, carefully remove the stopper and delivery tube from the reaction test tube. **CAUTION:** *Do not stop heating as long as the tube is in the limewater.* Record your observations of the limewater in your table.

4. Continue heating the reaction test tube until a glow spreads throughout the reactant mixture. Turn off the burner.

5. After the reaction test tube has cooled to nearly room temperature, empty the contents into a beaker that is about half full of water. In a sink, slowly stir the mixture while running water into the beaker until all the unreacted charcoal has washed away. Observe the product that remains in the beaker, and record your observations.

ANALYZE AND CONCLUDE

1. **Interpreting Data** What evidence of chemical change did you observe in each reaction?

2. **Interpreting Data** In the first reaction, a blue-colored solution indicates the presence of Cu^{2+} ions, a brown gas is NO_2, and a colorless gas is NO. In the second reaction, if limewater becomes cloudy and white, carbon dioxide gas has reacted with the calcium hydroxide to form insoluble calcium carbonate. Using this information, analyze your data and observations. Determine which reactants (Cu and HNO_3 for the first reaction, CuO and C for the second reaction) were oxidized and which were reduced in each reaction.

3. **Classifying** Classify each of the four reactants as an oxidizing agent or a reducing agent.

APPLY AND ASSESS

1. The mass of copper produced in the second reaction is less than the mass of the reacting copper(II) oxide. Why, then, is the gain of electrons known as reduction?

2. What mass of copper may be produced from the reduction of 1.000 metric ton of copper(II) oxide? Hint: Determine the formula mass of copper(II) oxide.

3. If the chlorine gas used at a water-treatment plant reacts with organic materials in the water to yield chloride ions, how would you classify the chlorine gas in terms of oxidation and reduction?

DATA AND OBSERVATIONS

	Observations
Step 1: Gas	
Solution	
Step 3: Limewater	
Step 5: Product	

What role do the copper ions play in the redox reaction? Each copper ion is reduced to uncharged copper metal when it accepts electrons from the zinc metal. Because the copper ion is the agent that oxidizes zinc metal to the zinc ion, Cu^{2+} is called an oxidizing agent. An **oxidizing agent** is the substance that gains electrons in a redox reaction. The oxidizing agent is the material that's reduced. Because oxidation and reduction go hand in hand, a reducing agent must be present. Zinc metal is the agent that supplies electrons and reduces the copper ion to copper metal; therefore, zinc is called the reducing agent. A **reducing agent** is the substance that loses electrons in a redox reaction. The reducing agent is the material that's oxidized. **Figure 16.6** summarizes the roles of oxidizing and reducing agents in redox reactions.

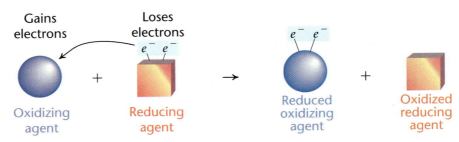

Figure 16.6

Oxidizing and Reducing Agents

When electrons are transferred from one element to another, a combination of an electron-gaining—or reduction—reaction and an electron-losing—or oxidation—reaction takes place. This combination is called a redox reaction. The element that is reduced oxidizes another element by attracting electrons from it, so it is called an oxidizing agent. The element that is oxidized reduces the first element by transferring electrons to it, so it is called a reducing agent.

SECTION REVIEW

Understanding Concepts

1. Name and define the two half-reactions that make up a redox reaction.

2. Identify which reactant is reduced and which is oxidized in each of the following reactions.

   a) $C_5H_{12}(l) + 8O_2(g) \rightarrow 5CO_2(g) + 6H_2O(g)$
   b) $2Al(s) + 3Cu^{2+}(aq) \rightarrow 2Al^{3+}(aq) + 3Cu(s)$
   c) $2Cr^{3+}(aq) + 3Zn(s) \rightarrow 2Cr(s) + 3Zn^{2+}(aq)$
   d) $2Au^{3+}(aq) + 3Cd(s) \rightarrow 2Au(s) + 3Cd^{2+}(aq)$

3. What is the oxidizing agent when iron rusts? What is the reducing agent?

Thinking Critically

4. **Applying Concepts** The following equation represents the reaction between an acid and a base to form a salt and water. Determine the oxidation number for each element. Is this a redox reaction? Explain.

$$2KOH(aq) + H_2SO_4(aq) \rightarrow K_2SO_4(aq) + 2H_2O(l)$$

Applying Chemistry

5. **Antioxidants** Compounds that are easily oxidized can act as antioxidants to prevent other compounds from being oxidized. Vitamins C and E protect living cells from oxidative damage by acting as antioxidants. Why does adding lemon juice to fruit salad prevent browning of the fruit?

chemistryca.com/self_check_quiz

Applications of Oxidation-Reduction Reactions

Natural redox reactions are going on around you every day, everywhere. This is partly due to the abundance of oxygen, which acts as the oxidizing agent as it is reduced in some redox reactions. Other oxidizing agents take part in different redox reactions, especially in environments where not much oxygen gas is found. Near the vents of volcanoes, where sulfur compounds explode out from deep within Earth, enormous deposits of solid yellow sulfur are found. The element sulfur acts both as an oxidizing agent and as a reducing agent in the reaction that forms the sulfur deposits. Can you tell which sulfur compound serves each function in this reaction?

$$2H_2S(g) + SO_2(g) \rightarrow 3S(s) + 2H_2(g) + O_2(g)$$

Note that more than one element in a reaction can be oxidized or reduced. The sulfur in hydrogen sulfide and the oxygen in sulfur dioxide both are oxidized. Sulfur in sulfur dioxide and hydrogen in hydrogen sulfide both are reduced. Each reactant acts as both a reducing agent and an oxidizing agent.

SECTION PREVIEW

Objectives
✓ **Analyze** common redox processes to identify the oxidizing and reducing agents.

✓ **Identify** some redox reactions that take place in living cells.

Review Vocabulary
Oxidation: reaction in which an element loses electrons.

Say Cheese: Redox in Photography

Understanding natural redox reactions such as the one that occurs in sulfur volcanoes has allowed chemists to develop many processes that make use of oxidation and reduction reactions. Without them, photographs or steel wouldn't exist, and stains would be much harder to remove from clothing.

Figure 16.7

Early Photos
In a daguerreotype, a redox reaction between silver and iodine fumes produced a layer of light-sensitive silver iodide on the surface of the polished photographic plate. Exposure to light caused decomposition of the silver iodide into elemental silver, which was then treated with the fumes of heated mercury to form bright amalgam areas. The image of Paris shown here was made by Daguerre himself.

WORD ORIGIN

photograph:
photos (GK) light
graphein (GK) to write
Light is used to record the image of an object in a photograph.

Leonardo da Vinci described a primitive "camera" before 1519, in which someone had to trace images focused on a glass plate inside a box. However, it wasn't until 1838 that the French inventor L.J.M. Daguerre successfully fixed the images in a camera on highly polished, silver-plated copper to make the first photographs. These early photographs were called daguerreotypes in his honor, **Figure 16.7**.

Modern photographic film is made of a plastic backing covered with a layer of gelatin, in which millions of grains of silver bromide are embedded. When light strikes a grain, silver and bromide ions are converted into their elemental forms through a redox reaction. The equation for this redox reaction is as follows.

$$2Ag^+ + 2Br^- \rightarrow 2Ag + Br_2$$

The reaction begins when the shutter on a camera is opened. Light from the scene being photographed passes through the camera's lens and shutter and strikes the light-sensitive silver bromide on the film. The light energy causes electrons to be ejected from a few of the bromide ions, oxidizing them to elemental bromine. The electrons are transferred to silver ions, reducing them to metallic silver atoms. These grains are now activated. The developing chemicals continue the redox reaction by causing the activated grains to be converted to metallic silver. In areas where the light is brightest, more grains are activated, and after developing, they become the darker areas. No silver atoms form in areas of the film that are not struck by light, and that part of the film remains transparent. The exposed film is then developed into a negative, during which time the remaining AgBr and Br_2 is washed away. **Figure 16.8** describes the developing and printing processes.

Figure 16.8

Developing and Printing Pictures

Making photographic negatives by developing exposed film involves several steps. The process describes how black-and-white pictures are made. For color photos, light-sensitive dyes are combined with the silver bromide in layers on the film.

◀ 1. The exposed film is transferred to a canister, where it is developed using a solution of a reducing agent, or developer. The organic compound hydroquinone is usually used for this purpose. The developer reduces all the silver ions to silver atoms in any grain of silver bromide that was hit by light, but it does not react with silver ions in grains that were not exposed to light. Because metallic silver is dark and silver bromide is light, an image having light and dark areas is produced.

◀ 2. After the film has been developed, a solution of a fixer containing thiosulfate ions is added. Thiosulfate ions react with unreduced silver ions to form a soluble complex, which is washed away. This prevents unreduced silver ions from becoming reduced and darkening slowly over time. The reaction follows.

$$AgBr(s) + 2S_2O_3^{2-}(aq) \rightarrow [Ag(S_2O_3)_2]^{3-}(aq) + Br^-(aq)$$

◀ 3. The fixed film is washed to remove any remaining developer or fixer solution. The photographic negative is the reverse of the image photographed; that is, light areas in the scene are dark on the film, and vice versa.

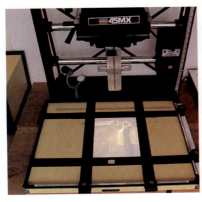

◀ 4. When light is shone through the negative onto light-sensitive photographic paper, a photographic print is made. The print is positive; light and dark areas are identical to those in the scene.

Solid Rocket Booster Engines

If you have ever built and launched a model rocket, you probably noticed that the rocket engine was made of a solid, highly combustible material packed into a cardboard tube. After ignition, the expansion and expulsion of the gases produced enough downward force to launch the lightweight rocket quickly into the air. Space shuttles use a similar type of technology, but on a much larger scale.

Engine systems The space shuttle has two different engine systems. The three main engines attached directly to the shuttle operate on liquid hydrogen and liquid oxygen reservoirs carried in the large, centrally located disposable fuel tank. The two smaller, reusable, strap-on booster rockets on either side of the main fuel tank are loaded with a solid fuel, which undergoes a powerful, thrust-producing, oxidation-reduction reaction that helps boost the shuttle into orbit.

Solid rocket fuel The solid rocket fuel is a mixture containing 12 percent aluminum powder, 74 percent ammonium perchlorate, and 12 percent polymer binder. Once ignited, the engine cannot be extinguished. The extremely reactive ammonium perchlorate supplies oxygen to the easily oxidized aluminum powder, providing a greatly exothermic and fast reaction. The purpose of the polymer binder is to hold the mixture together and to help it burn evenly. The overall redox reaction is shown here.

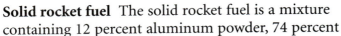

$$8Al + 3NH_4ClO_4 \rightarrow 4Al_2O_3 + 3NH_4Cl$$

(above equation: $24e^-$ transferred)

Shuttle forces Each solid rocket booster weighs 591 000 kg at liftoff, produces 11.5 million N of force, and operates for about two minutes into the flight. For comparison, a 1000-kg car accelerating from 0 to 26.8 m/s (60 mph) in 7 seconds would require a force of only 3830 N. The tremendous release of chemical energy and expansion of hot gases due to the oxidation-reduction reaction through the engine of the solid rocket booster produces the tremendous thrust needed to get the 2 million-kg shuttle from 0 to almost 700 m/s (1500 mph) in just 132 seconds.

Connecting to Chemistry

1. **Applying** Powdered aluminum is used in another greatly exothermic reaction, the thermite reaction, which is used for welding metals. The reaction is as shown.

$$2Al + Fe_2O_3 \rightarrow Al_2O_3 + 2Fe$$

What role does the powdered aluminum play in this reaction?

2. **Acquiring Information** Investigate the lives and research of Robert Goddard and Werner Von Braun, who both experimented with rockets in the 1930s and helped guide the United States into the space age. Write a short report about these men.

Having a Blast: Redox in a Blast Furnace

Iron is seldom found in the elemental form needed to make steel. Metallic iron must be separated and purified from iron ore—usually hematite, Fe_2O_3. This process takes place in a blast furnace in a series of redox reactions. The major reaction in which iron ore is reduced to iron metal uses carbon monoxide gas as a reducing agent.

First, a blast of hot air causes coke, a form of carbon, to burn, producing CO_2 and heat. Limestone, $CaCO_3$, which is mixed with the iron ore in the furnace, decomposes to form lime (CaO) and more carbon dioxide. The carbon dioxide then oxidizes the coke in a redox reaction to form carbon monoxide, which is used to reduce the iron ore to iron. The process is outlined here and illustrated in **Figure 16.9**.

$$CaCO_3(s) \rightarrow CaO(s) + CO_2(g)$$

$$CO_2(g) + C(s) \rightarrow 2CO(g)$$

$$2Fe^{3+}(s) + 3O^{2-}(s) + 3CO(g) \xrightarrow{\quad 6e^- \quad} 2Fe^0(l) + 3CO_2(g)$$

Redox in Bleaching Processes

Bleaches can be used to remove stains from clothing. Where do the stains go? Bleach does not actually remove the chemicals in stains from the fabric; it reacts with them to form colorless compounds. In chlorine bleaches, an ionic chlorine compound in the bleach reacts with the compounds responsible for the stain. This ionic compound is sodium hypochlorite (NaOCl). The hypochlorite ions oxidize the molecules that cause dark stains.

$$OCl^-(aq) + \text{stain molecule(s)} \rightarrow Cl^-(aq) + \text{oxidized stain molecule(s)}$$
$$\text{(colored)} \qquad\qquad\qquad\qquad\qquad \text{(colorless)}$$

▲ Molten iron is drawn off at the bottom of the furnace. A combination of by-products known as slag is also removed at the bottom.

Figure 16.9

Blast Furnace

Iron ore (Fe_2O_3), coke (C), and limestone ($CaCO_3$) are added at the top of the furnace. Hot air at about 900°C, blasted into the bottom of the furnace, burns the coke in an exothermic reaction. This reaction causes temperatures in a blast furnace to reach about 2000°C. ▶

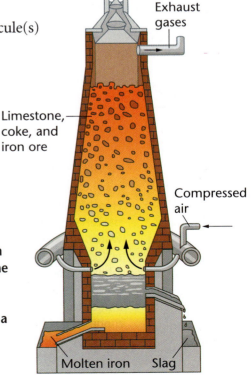

Exhaust gases

Limestone, coke, and iron ore

Compressed air

Molten iron   Slag

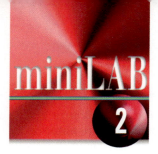

miniLAB 2

Testing for Alcohol by Redox

Organic alcohols react with orange dichromate ions, producing blue-green chromium(III) ions. This reaction is used in a Breathalyzer test to test for the presence of alcohol in a person's breath. In this MiniLab, you will use this reaction to test for the presence of alcohol in a number of household hygiene, cosmetic, and cleaning products.

Procedure

1. Label five small test tubes with the names of the products to be tested.

2. Place approximately 1 mL of each product in the appropriate tube.

3. Wearing apron and goggles, add three drops of dichromate reagent to each tube, and stir to mix the solutions.

CAUTION: *Do not allow dichromate reagent to come into contact with skin. Wash with large volumes of water if it does.*

4. Observe and record any color changes that occur within one minute.

Analysis

1. Which of the products that you tested contain alcohol? Was the presence of alcohol noted on the label of the products?

2. If the orange $Cr_2O_7^{2-}$ ion reacts with alcohol to produce the blue-green Cr^{3+} ion, what substance is the reducing agent in the reaction?

See page 870 in Appendix F for

Testing the Oxidation Power of Bleach

Bleaches containing hypochlorite should be used carefully because hypochlorite is a powerful oxidizing agent that can damage delicate fabrics. These bleaches usually have a warning label telling the user to test an inconspicuous part of the fabric before using the product. In addition to acting as a bleaching agent, hypochlorite ions are also used as disinfectants, as **Figure 16.10** shows.

Figure 16.10

Hypochlorite as a Disinfectant
Hypochlorite is used in disinfectants to kill bacteria in swimming pools and in drinking water. In both cases, the hypochlorite ions act as oxidizing agents. Bacteria are killed when important compounds in them are destroyed by oxidation. In this photo, the amount of chlorine in the water is being monitored. Chlorine reacts with the water to form hypochlorite ions.

How it Works

Breathalyzer Test

The alcohol in beverages, hair spray, and mouthwashes is ethanol. Ethanol is a volatile liquid that evaporates rapidly at room temperature. Because of this volatility, drinking an alcoholic beverage results in a level of gaseous ethanol in the breath that is proportional to the level of alcohol in the bloodstream. About 50 percent of all automobile accidents that result in a fatality are caused by intoxicated drivers. Law officers can determine quickly whether a person is legally intoxicated by using an instrument called a breath analyzer, or Breathalyzer.

1. A simple Breathalyzer device has an inflatable plastic bag attached to a tube containing an orange solution of potassium dichromate and sulfuric acid.

2. During a Breathalyzer test, a person blows into the mouthpiece of the bag.

3. If alcohol vapors are present in the person's breath, ethanol undergoes a redox reaction with the dichromate. As ethanol is oxidized, the orange Cr^{6+} ions are reduced to blue-green Cr^{3+} ions.

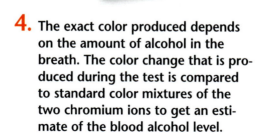

4. The exact color produced depends on the amount of alcohol in the breath. The color change that is produced during the test is compared to standard color mixtures of the two chromium ions to get an estimate of the blood alcohol level.

Thinking Critically

1. Suppose a person used mouthwash shortly before taking a Breathalyzer test. What might be the result?

2. How would the color produced in a Breathalyzer test change as the ethanol content of the blood increases?

Figure 16.11
The Green Lady
The green color of the Statue of Liberty in New York Harbor is due to a layer of patina, or protective coating, that covers the copper sheets making up the statue. The presence of the patina helps keep the statue from corroding further because oxygen cannot get through the patina to reach the copper layers underneath.

Corrosion of Metals

Did you know that the Statue of Liberty is made of copper sheets attached to a steel skeleton? Why does it appear green rather than the reddish-brown color of copper? When copper is exposed to humid air that contains sulfur compounds, it undergoes a slow oxidation process. Under these conditions, the copper metal atoms each lose two electrons to produce Cu^{2+} ions, which form the compounds $CuSO_4 \cdot 3Cu(OH)_2$ and $Cu_2(OH)_2CO_3$. These compounds are responsible for the green coat or patina found on the surface of copper objects that have been exposed to air for long periods of time, **Figure 16.11**.

You have learned that iron is oxidized by oxygen in the air to form rust. Aluminum is a more active metal than iron. As a result of its greater activity, aluminum is oxidized more quickly than iron. If this is true, why does an aluminum can degrade much more slowly than a tin can, which is made of iron-containing steel that is coated with a thin layer of tin? The reason is that, like copper, aluminum is oxidized to form a compound that coats the metal and protects it from further corrosion, as shown in **Figure 16.12**. Aluminum reacts with oxygen to form aluminum oxide in a redox reaction.

$$4Al(s) + 3O_2(g) \rightarrow 2Al_2O_3(s)$$

A coating of aluminum oxide is tough and does not flake off easily, as iron oxide rust does. When rust flakes fall off a surface, additional metal is exposed to air and becomes corroded.

Figure 16.12
Corrosion of Iron and Aluminum
Because iron rust is porous and flaky, it does not form a good protective coating for itself. ▶

A tin coating offers some protection to the iron. However, if a hole or crack develops in the thin tin coating, the underlying iron corrodes rapidly. A tin-coated steel can will degrade completely in about 100 years. The aluminum oxide coating on an aluminum can is tough and closely packed. It protects the underlying aluminum from further corrosion so that the can will take about 400 years to degrade. ▶

Corrosive solution

Iron metal

Steel can

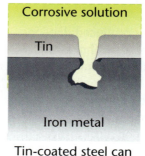

Corrosive solution

Tin

Iron metal

Tin-coated steel can

Corrosive solution

Aluminum oxide

Aluminum metal

Aluminum can

Lightning-Produced Fertilizer

Did you know that one of the main nutrients plants need is nitrogen? Although the air surrounding Earth is almost 80 percent nitrogen, the nitrogen is in the form of N_2 molecules, a form that most plants and animals cannot use. Nitrogen from the air is converted to a form that plants can use by a process called nitrogen fixation. Plants can best use nitrogen when it is in the form of the ammonium ion, NH_4^+, where the nitrogen has an oxidation number of $3-$, but they can also use the nitrate ion, NO_3^-, with nitrogen having an oxidation number of $5+$.

Nitrogen fixation   Nitrogen can be fixed for plants in three ways: by lightning, by nitrogen-fixing bacteria living in the roots of plants or in the soil, and by commercial synthesis reactions such as the Haber ammonia process.

Nitrogen is a fairly inert gas because the triple bond of N_2 is strong and resists breaking. However, the exceptionally high energy and temperatures of lightning can easily break bonds and allow for recombination of gases in the atmosphere.

Lightning-driven reactions   In the process of lightning-driven nitrogen fixation, nitrogen and oxygen combine to form nitrogen monoxide. Nitrogen monoxide then combines with more oxygen to form nitrogen dioxide. This nitrogen dioxide mixes with water in the air to form nitric acid and more nitrogen monoxide, which is available to continue the cycle.

$$N_2 + O_2 \rightarrow 2NO$$
$$2NO + O_2 \rightarrow 2NO_2$$
$$3NO_2 + H_2O \rightarrow 2HNO_3 + NO$$

Fertilizer production   The pH of rainwater is naturally slightly acidic, and some of this acidity is

due to the dissolved nitric acid, HNO_3, from nitrogen fixation. As the rain soaks into the soil, bacteria convert the nitrate ions into ammonium ions.

How does nature's manufacturing of fixed nitrogen compare with commercial production of fixed nitrogen? Lightning may seem uncommon, but it is estimated that there are approximately 10 000 lightning storms every day over the surface of Earth. Stated another way, lightning strikes 100 times a second on the planet as a whole. Approximately 10 billion kg of nitrogen are fixed yearly in the atmosphere. Biological agents such as bacteria fix about 100 billion kg of nitrogen yearly, and an amount equal to that is fixed through the manufacture of fertilizer and other industrial processes.

Exploring Further

1. **Classifying**   Nitrogen fixation in the soil is accomplished by bacteria living in the roots of certain plants. Name some of these plants.

2. **Applying**   In each of the three equations shown, what is oxidized and what is reduced?

3. **Acquiring Information**   The process by which nitrogen is put back into the air is called denitrification. Find out what conditions are necessary for this process and what reaction occurs.

To learn more about the nitrogen cycle, visit the Chemistry Web site at chemistryca.com

Silver Tarnish: A Redox Reaction

WORD ORIGIN

tarnish:
terne (OF) dull, wan

The shiny surfaces of many metal objects lose luster and become dull and tarnished as the metal atoms undergo oxidation.

Imagine if, along with your usual chores of taking out the trash, washing dishes, feeding your pets, and taking care of your younger siblings, you also had to polish the silver—as people did back in your great-grandparents' days. How would you find time for any fun? Fortunately, other materials such as stainless steel have replaced most "silverware." Why do silver utensils have to be polished, but those made of stainless steel or aluminum don't? Silver becomes tarnished through a redox reaction that is a form of corrosion, as rusting is. Tarnish is formed on the surface of a silver object when silver reacts with H_2S in air. The product, black silver sulfide, forms the coating of tarnish on the silver.

$$O_2(g) + 4Ag(s) + 2H_2S(g) \rightarrow 2Ag_2S(s) + 2H_2O(l)$$

Many commercial silver polishes contain abrasives that help to remove tarnish. Unfortunately, they also remove some of the silver. A more gentle way to remove tarnish from the surface of a silver object involves another redox reaction. In this reaction, aluminum foil scraps act as a reducing agent.

$$\overset{\displaystyle \overbrace{}^{6e^-}}{2Al^0(s)} + 6Ag^+(s) + 3S^{2-}(s) + 6H_2O(l) \rightarrow 6Ag^0(s) + 2Al^{3+}(aq) + 6OH^-(aq) + 3H_2S(g)$$

This reaction is essentially the reverse of the reaction that forms tarnish. Here, silver ions in the Ag_2S tarnish are reduced to silver atoms, while aluminum atoms in the foil are oxidized to aluminum ions. The tarnish-removing solution usually includes baking soda (sodium hydrogen carbonate) to help remove any aluminum oxide coating that forms and to make the cleaning solution more conductive. **Figure 16.13** shows how this method of silver cleaning is done.

Figure 16.13
Removing Silver Tarnish

Even though corrosion is an unwanted redox reaction, removing the tarnish makes use of another redox reaction. A nest of crumpled aluminum foil scraps is made at the bottom of a large pot. The tarnished silver object is added, making sure the silver is in contact with the foil scraps. Baking soda is added, and the silver is covered with water. When the pot is heated on a stove, the silver sulfide tarnish is reduced to silver atoms, and the silver object becomes shiny and bright.

Forensic Blood Detection

The gas station at the corner was robbed, and the cashier was shot. On television, police announce that Suspect A has been taken into custody. They have confiscated a jacket, allegedly worn by the suspect. After preliminary examination by the police department, the jacket is sent to a forensic laboratory for scientific investigation. One of the first tests a technician at the laboratory will carry out determines whether or not there are blood stains on the jacket.

The Luminol Test

The technician may choose from several chemical tests for blood, all based on the fact that the hemoglobin in blood catalyzes the oxidation of a number of organic indicators to produce a colored product that emits light, or luminesces.

The technician on this case chooses the luminol test. Luminol has an organic double-ring structure, shown below. In 1928, German chemists first observed the blue-green luminescence when the compound was oxidized in alkaline solution. It was soon found that a number of oxidizing agents, such as hydrogen peroxide, bring about the luminescence. Later, workers noted that the luminescence was greatly enhanced by the presence of blood, which led to its current use in forensic investigations.

Luminol

The technician carefully mixes an alkaline solution of luminol with aqueous sodium peroxide and, in a darkened workplace, sprays the solution onto suspected spots on the jacket. Bingo! An intense, blue-green chemiluminescence is emitted from several spots. Because the glow will last for a few minutes, the technician photographs the spots and their telltale light.

Ruling Out with Luminol

You may wonder if this relatively simple procedure will serve to convict Suspect A. Certainly not. However, if the test had been negative, Suspect A might have been cleared from suspicion. A negative result ensures that a stain is *not* blood. But, because this is not the case with the stains on the jacket, the luminol test is preliminary and will be used with other tests.

The luminol test is especially useful because it works well with both fresh and dried blood. Luminol has one particularly useful feature. The same stains can be made luminescent over and over again if the spray is allowed to dry and the stains are resprayed.

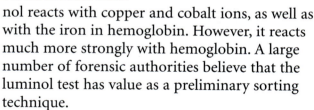

A positive test should not be taken as absolute proof of blood because luminol reacts with copper and cobalt ions, as well as with the iron in hemoglobin. However, it reacts much more strongly with hemoglobin. A large number of forensic authorities believe that the luminol test has value as a preliminary sorting technique.

DISCUSSING THE TECHNOLOGY

1. **Applying** If a luminol test yields a positive reaction, what is the next logical step?

2. **Hypothesizing** Why can it almost never be assumed that stains are uncontaminated, although stain evidence is important in a criminal investigation?

Figure 16.14
Chemiluminescence

◀ When lightning is produced by an electrical discharge in the atmosphere, electrons in molecules of O_2 and N_2 gases are excited to higher energy levels. Energy from the electricity breaks the molecules into atoms. When the atoms recombine to form molecules and the electrons return to lower energy levels, light energy is released through chemiluminescence.

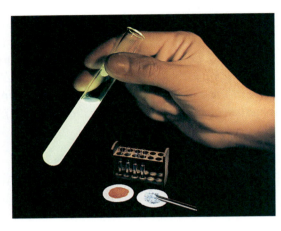

When luminol is oxidized and is observed in the dark, an eerie blue-green glow is produced through chemiluminescence. ▶

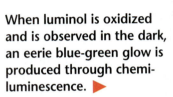

Nitrous oxide is produced by a redox reaction between oxygen and nitrogen during lightning storms. It was discovered and studied in the late 1700s by Joseph Priestley, who found that inhaling it resulted in unusual side effects including laughing, singing, and fighting. For this reason, it was called laughing gas. Its anesthetic properties were discovered by accident in Connecticut in 1844 at a public demonstration given for amusement when a man who inhaled nitrous oxide cut his leg badly in a scuffle but felt no pain until the gas wore off.

Chemiluminescence: It's Cool

Some redox reactions can release light energy at room temperature. The production of this kind of cool light by a chemical reaction is called chemiluminescence. The light from chemiluminescent reactions can be used in emergency light sticks that work without an external energy source. You may recall learning in Chapter 6 how these light sticks work. Now you know that the reaction that takes place when the two solutions in the light sticks are mixed involves an oxidation and a reduction.

Some chemiluminescent redox reactions occur naturally in the atmosphere as a result of lightning, **Figure 16.14.** Other chemiluminescent reactions involve luminol, an organic compound that emits cool light when it is oxidized. Luminol reactions are utilized by forensic chemists to analyze evidence in crime investigations. They spray luminol onto a location where the presence of blood is suspected. If blood is present, the iron(II) ions in the blood oxidize the luminol to form a chemiluminescent compound that glows in the dark. The iron is reduced by the luminol. **Figure 16.14** shows the glow from the oxidized form of luminol.

Biochemical Redox Processes

How are bears able to stay warm enough to keep from freezing during their winter hibernation? How do marathon runners get the energy to finish a race without stopping to eat? In both cases, fats stored in the body are oxidized. Oxygen molecules from the air are reduced as they gain electrons to form water. In a series of redox reactions called respiration,

Figure 16.15
Keeping Warm
Although it is common to think that only mammals keep warm, in truth, all plants and animals maintain a temperature at which their enzymes function best. Plants keep from freezing because heat is produced as a by-product of respiration and photosynthesis. One of the first plants to poke through the snow in early spring is the heat-producing skunk cabbage. The heat it releases allows it to get a head start on other plants and also contributes to the unpleasant odor that gives it its name.

energy is released. **Figure 16.15** shows one effect of this heat in plants. Respiration will be discussed in Chapter 19. Many other redox reactions take place in living things. Electrons are transferred between molecules in redox reactions during photosynthesis and in the reactions that fireflies use to flash light signals to potential mates. You will study photosynthesis in Chapter 20.

Some organisms can use the energy released during redox reactions to convert chemical energy into light energy, a process called bioluminescence. You are probably familiar with the flashing lights given off by fireflies during courtship, but did you know that many different organisms—including some fish, at least one type of mushroom, and a caterpillar known as a glowworm—also are bioluminescent? **Figure 16.16** shows bioluminescence in fireflies.

Now that you have learned what redox reactions are and have read about some of the processes of which they are part, you can reexamine the redox reaction that makes cut fruit turn brown. The color is due to brown pigments that are formed by the oxidation of colorless compounds normally present in the cells of the fruit. Oxygen in the air is the oxidizing agent that reacts with the colorless compounds to produce the brown pigments. The oxygen is reduced when it accepts electrons from the pigments, so the pigments function as reducing agents. This combination of oxidation and reduction goes hand in hand in a redox reaction because electrons that are lost by one element must be gained by another.

WORD ORIGIN

bioluminescent:
bios (GK) life
lumen (L) light
escentis (L) beginning to be, have, or do

A bioluminescent substance undergoes a chemical reaction in living things in which potential energy in chemical bonds is converted into light energy.

Figure 16.16
Firefly Signals
Fireflies use flashes of light to attract mates. Light energy is released during an enzyme-catalyzed redox reaction. Luciferase is the name given to the enzyme that speeds up the reaction in which the organic molecule luciferin is oxidized.

Myoglobin, found in muscle tissue, is an iron-containing protein that stores oxygen. Myoglobin in living muscle tissue is bound to oxygen and is a red color. It becomes pale purple after death when the oxygen is lost. Heating meat results in oxidation of the iron in myoglobin, which then has the brown color that tells you the meat is cooked.

Figure 16.17

Antioxidants

Vitamin C owes its antioxidant properties to the fact that it reacts so readily with oxygen. When added to a food product, oxygen reacts preferentially with vitamin C, thereby sparing the food product from oxidation. Other anti-oxidant food additives include the synthetic compounds BHA and BHT and the natural antioxidant, vitamin E.

The skin of a fruit keeps oxygen out, which is why unbroken fruit does not turn brown. Coating cut fruit with an antioxidant can prevent browning and keep a fruit salad looking fresh longer. The vitamin C in lemons is a good antioxidant. If lemon juice is squirted onto cut banana or apple slices, they will not brown as quickly because the vitamin C reacts with oxygen more readily than do the fruit-browning compounds, **Figure 16.17.**

Connecting Ideas

Most reactions involve electron transfer and thus are redox reactions. You have learned to identify which element is reduced and which is oxidized when you are given the equation for a redox reaction. You might wonder why one element accepts electrons from another and whether you can predict which element will be oxidized and which will be reduced. Learning to make those predictions is the next step in your study of electron-transfer processes in compounds and will help you understand how redox reactions in batteries produce electricity.

Supplemental Problems

For more practice with solving problems, see Supplemental Practice Problems, Appendix B.

SECTION REVIEW

Understanding Concepts

1. What role does the reducing agent hydroquinone play in the production of a photographic negative?

2. How is most of the iron that is used for making steel purified from iron ores?

3. Why do aluminum cans degrade more slowly than cans made of iron?

Thinking Critically

4. **Applying Concepts** Oxygen is required for the production of light by fireflies. What role does the oxygen play in the reaction?

Applying Chemistry

5. **Bleaching** Why can't rust stains be removed with bleach?

REVIEWING MAIN IDEAS

16.1 The Nature of Oxidation-Reduction Reactions

- Oxidation occurs when an atom or ion loses one or more electrons and attains a more positive oxidation number. Reduction takes place when an atom or ion gains electrons and attains a more negative oxidation number.

- Oxidation and reduction reactions always occur together in a net process called a redox reaction.

- An oxidizing agent is the substance that gains electrons and is reduced during a redox reaction. A reducing agent is the substance that loses electrons and is oxidized during a redox reaction.

16.2 Applications of Oxidation-Reduction Reactions

- In photography, light triggers the reduction of silver ions to silver metal on photographic film.

- Bleach removes stains from clothing by oxidizing colored molecules to form colorless molecules.

- Metals such as copper and aluminum are resistant to corrosion even though they are easily oxidized because the products of their reactions with oxygen form protective coatings on the surface of the metal.

- Chemiluminescent reactions in emergency light sticks, lightning, and the luminol reaction convert the energy of chemical bonds into light energy.

- Some organisms use redox reactions to produce light, which they use in communication. This light production is called bioluminescence.

- Cut fruits turn brown because compounds in the fruit cells react with oxygen in a redox reaction to produce brown pigments. Coating the fruits with antioxidants can prevent this browning.

Vocabulary

For each of the following terms, write a sentence that shows your understanding of its meaning.

oxidation
oxidation-reduction reaction
reduction
oxidizing agent
reducing agent

UNDERSTANDING CONCEPTS

1. What is the difference between an oxidizing agent and a reducing agent?

2. Which of the changes indicated are oxidations and which are reductions?
   a) Cu becomes Cu^{2+}
   b) Sn^{4+} becomes Sn^{2+}
   c) Cr^{3+} becomes Cr^{6+}
   d) Ag becomes Ag^+

3. Identify the oxidizing agent in each of the following reactions.
   a) $Cu^{2+}(aq) + Mg(s) \rightarrow Cu(s) + Mg^{2+}(aq)$
   b) $Fe_2O_3(s) + 3CO(g) \rightarrow 2Fe(l) + 3CO_2(g)$

4. What is the oxidizing agent in household bleach?

5. Why does a photographic negative need to be fixed?

6. In which direction do electrons move during a redox reaction: from oxidizing agent to reducing agent or vice versa?

7. Why is aluminum metal used to remove tarnish from silver?

8. What chemical process do hibernating animals use to stay warm?

9. Write the equation for the redox reaction that occurs when a piece of iron metal is dipped in a solution of copper(II) sulfate.

10. Identify the following as an oxidation reaction or a reduction reaction.

$$Fe^{2+} \rightarrow Fe^{3+} + e^-$$

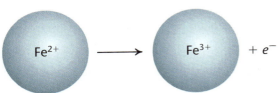

APPLYING CONCEPTS

11. If galvanized nails, which have been coated with zinc, are placed in a brown solution containing I_2, the solution slowly turns colorless. Adding a few drops of bleach to the colorless solution results in a return of the brown color. Explain what makes these changes occur.

12. List several ways in which a steel chain-link fence could be treated to prevent corrosion.

13. When hydrogen peroxide is added to a colorless solution of potassium iodide, a red-brown color appears. What substance is responsible for the color?

14. Write the equation for a reaction that is not a redox reaction. Are electrons transferred in this reaction?

15. Indigo is one of the oldest known dyes. It has been detected in cloth used to wrap mummies that are more than 5000 years old. When cotton jeans are dyed with indigo, they are dipped into a yellow solution of indigo and sodium hydrosulfite, which is a good reducing agent. Within minutes after being taken out of the solution, the jeans turn blue. How can you explain this?

16. A shiny copper mirror can be formed on the inside of a test tube in which the following reaction takes place.

$$H_2C{=}O(aq) + Cu^{2+}(aq) + 2OH^-(aq) \rightarrow$$
formaldehyde
$$Cu(s) + HCOOH(aq) + H_2O(l)$$
formic acid

   a) Identify the substance that is reduced during this reaction.

   b) Identify the substance that is oxidized during this reaction.
   c) What is the oxidizing agent in this reaction?
   d) What is the reducing agent in this reaction?

17. Is oxygen a necessary reactant for an oxidation reaction? Explain.

18. Sodium nitrite is often added to meat to inhibit the growth of microorganisms and to keep the meat from spoiling. Under the acidic conditions in our stomachs, nitrites can be converted into potentially cancer-causing substances. Vitamin C can convert nitrite ions into nitrogen monoxide gas and may help protect us from the effects of these ions.

$$NO_2^-(aq) \rightarrow NO(g)$$

   a) Is the nitrite ion oxidized or reduced in this reaction?

   b) Does vitamin C act as an oxidizing agent or a reducing agent?

19. Why is gold rather than copper used to coat electrical connections in expensive electronic equipment?

Everyday Chemistry

20. Just before World War I, a German chemist named Fritz Haber developed a process for fixing atmospheric nitrogen into ammonia. The ammonia produced this way can be converted into ammonium nitrate, an important fertilizer and explosive.

$$3H_2 + N_2 \rightarrow 2NH_3$$

   a) What element is oxidized during this reaction? What is reduced?

   b) What is the oxidizing agent? What is the reducing agent?

Physics Connection

21. By passing an electric current through water, the water can be separated into its component elements in the reverse of the reaction used to power the main stage of the space shuttle.

$$2H_2O(l) + energy \rightarrow 2H_2(g) + O_2(g)$$

   a) Is this a redox reaction? If so, what element is oxidized?

b) Where does the energy for this endothermic reaction come from?

How It Works

22. If ethanol were less volatile, how might the usefulness of a Breathalyzer test be affected? Explain.

Chemistry and Technology

23. Why should a positive result from the luminol test not be taken as proof of the presence of blood?

THINKING CRITICALLY

Using a Table

24. The table below lists some of the most common compounds that are used as oxidizing agents.

Common Oxidizing Agents	
O_2	$K_2Cr_2O_7$
H_2O_2	HNO_3
$KMnO_4$	$NaClO$
Cl_2	$KClO_3$

a) Name each of the compounds in the table.
b) List at least one practical application, mentioned in this chapter or from a reference book, of each of these oxidizing agents.
c) Make a similar table for common reducing agents. Should any compounds be listed in both tables?

Making Predictions

25. Hydrogen peroxide (H_2O_2) can be used to restore white areas of paintings that have darkened from the reaction of lead paint pigments with polluted air containing hydrogen sulfide gas.

$$PbS + 4H_2O_2 \rightarrow PbSO_4 + 4H_2O$$
(black)   (white)

Could hydrogen peroxide be used to remove tarnish from silver objects? Would the reaction have any undesirable effects?

Interpreting Data

26. ChemLab Write the balanced equation for the reaction that caused the limewater to become cloudy. Is this a redox reaction? Explain.

27. MiniLab 1 Why do you think corrosion seems to occur mostly at the head and point of a nail?

Making Inferences

28. MiniLab 2 When a pile of orange ammonium dichromate is ignited, it decomposes in an exothermic reaction in which the green product and flames shoot upward like an erupting volcano. (**CAUTION:** *Do NOT perform this reaction.*)

$$(NH_4)_2Cr_2O_7(s) \rightarrow$$
$$Cr_2O_3(s) + N_2(g) + 4H_2O(g)$$

a) What is the reducing agent in this reaction? The oxidizing agent?
b) How is this reaction similar to the Breathalyzer reaction?

CUMULATIVE REVIEW

29. Identify each of the following as a pure substance or a mixture. (Chapter 1)

a) petroleum	**d)** diamond
b) fruit juice	**e)** milk
c) smog	**f)** iron ore

30. List some characteristic properties of metals. (Chapter 3)

31. Name each of the following ionic compounds. (Chapter 5)

a) NaF	**d)** $Na_2Cr_2O_7$
b) CaS	**e)** KCN
c) $Al(OH)_3$	**f)** NH_4Cl

32. How many grams of nitrogen are needed to react completely with 346 g of hydrogen to form ammonia by the Haber process? (Chapter 12)

$$N_2 + 3H_2 \rightarrow 2NH_3$$

33. Draw Lewis electron dot diagrams for each of the following covalent molecules. (Chapter 9)

  a) $CHCl_3$ **c)** CH_3CH_3
  b) CH_3CH_2OH

🔴 SKILL REVIEW

34. Designing an Experiment Do you think silver will tarnish more quickly in clean air or in polluted air? Design an experiment to test your hypothesis.

🔴 WRITING IN CHEMISTRY

35. Research the evidence that suggests that the antioxidant properties of vitamin C may help prevent cancer in people who take large doses of this vitamin. Write a summary of your findings in which you propose how you would do more tests to determine whether or not vitamin C has anticarcinogenic properties.

🔴 PROBLEM SOLVING

36. A flask filled with acid-washed steel wool is fitted with a long, thin glass tube in a rubber stopper. When the flask is inverted so the tube opening is in a beaker of colored water, the water slowly begins to rise in the tube. Write a summary of this experiment, as if you had performed it. Explain what makes the water rise. Predict what portion of the flask will be filled with water at the end of the experiment.

37. The patina coating on the Statue of Liberty has preserved most of the copper metal in the statue. Some damage does occur wherever steel rivets are in contact with copper and exposed to water. Do library research to determine why those sites are more susceptible to corrosion than the rest of the statue. Write up your findings in a short report. Include a diagram or make a poster showing the movement of electrons in the process.

38. Metallic lithium reacts vigorously with fluorine gas to form lithium fluoride.

  a) Write an equation for this process.
  b) Is this an oxidation-reduction reaction?
  c) If it is an oxidation-reduction, which element is oxidized? Which is reduced?
  d) If 2.0 g of lithium are reacted with 0.1 L fluorine at STP, which reactant is limiting?
  e) If 0.04 g of lithium fluoride is formed in reaction in part **d.**, what is the percent yield?

39. Identify the oxidizing reagent in each of the following reactions.

  a) $C_2H_5OH(l) + 3O_2(g) \rightarrow 2CO_2(g) + 3H_2O(l)$
  b) $CuO(s) + H_2(g) \rightarrow Cu(s) + H_2O(l)$
  c) $2FeO(s) + C(s) \rightarrow 2Fe(s) + CO_2(g)$
  d) $2Fe^{2+}(aq) + Br_2(l) \rightarrow 2Fe^{3+}(aq) + 2\ Br^-(aq)$

40. When coal and other fossil fuels containing sulfur are burned, sulfur is converted to sulfur dioxide: $S(s) + O_2(g) \rightarrow SO_2(g)$

  a) Is this an oxidation-reduction reaction?
  b) If it is an oxidation-reduction, which element is oxidized? Which is reduced?
  c) If 7.0×10^3 kg of fuels containing 3.5 percent sulfur are burned in a city on a given day, how much SO_2 will be emitted? Assume that the sulfur reacts completely.

41. Sodium nitrite is formed when sodium nitrate reacts with lead:
    $NaNO_3(s) + Pb(s) \rightarrow NaNO_2(s) + PbO(s)$

  a) What is the oxidizing reagent in this reaction? What is the reducing reagent?
  b) If 5.00 g of sodium nitrate is reacted with an excess of lead, what mass of sodium nitrite will form if the yield is 100 percent?

1. The term used to describe a chemical reaction in which a substance loses electrons to another substance is

   a) oxidation.
   b) redox.
   c) reduction.
   d) corrosion.

2. The oxidation numbers of the elements in $CuSO_4$ are

   a) $Cu = +2, S = +6, O = -2$
   b) $Cu = +3, S = +5, O = -2$
   c) $Cu = +2, S = +2, O = -1$
   d) $Cu = +2, S = 0, O = -2$

3. For the reaction $X + Y \rightarrow XY$, the element that will be reduced is the one that is

   a) more reactive.
   b) more massive.
   c) more electronegative.
   d) more radioactive.

4. The reaction between sodium iodine and chlorine is:

   $$2NaI(aq) + Cl_2(aq) \rightarrow 2NaCl(aq) + I_2(aq)$$

   The oxidation state of Na remains unchanged because

   a) Na^+ is a spectator ion.
   b) Na^+ cannot be reduced.
   c) Na^+ is an uncombined element.
   d) Na^+ is a monatomic ion.

5. The reaction between nickel and copper(II) chloride is:

   $$Ni(s) + CuCl_2(aq) \rightarrow Cu(s) + NiCl_2(aq)$$

   The half reactions for this redox reaction are

   a) $Ni \rightarrow Ni^{2+} + 2e^-; Cl_2 \rightarrow 2Cl^- + 2e^-$
   b) $Ni \rightarrow Ni^+ + e^-; Cu+ + e^- \rightarrow Cu$
   c) $Ni \rightarrow Ni^{2+} + 2e^-; Cu^{2+} + 2e^- \rightarrow Cu$
   d) $Ni \rightarrow Ni^{2+} + 2e^-; 2C^+ + 2e^- \rightarrow Cu$

Interpreting Tables: Use the table in the next column to answer questions 6–8.

6. Which of the following elements forms a monatomic ion that is a spectator in the redox reaction?

   a) Zn
   b) O
   c) N
   d) H

Data for Elements in the Redox Reaction $Zn + HNO_3 \rightarrow Zn(NO_3)_2 + NO_2 + H_2O$		
Element	Oxidation Number	Complex ion of which element is a part
Zn	0	none
Zn in $Zn(NO_3)_2$	+2	none
H in HNO_3	+1	none
H in H_2O	?	none
N in HNO_3	?	NO_3^-
N in NO_2	+4	none
N in $Zn(NO_3)_2$	?	NO_3^-
O in HNO_3	-2	NO_3^-
O in NO_2	?	none
O in $Zn(NO_3)_2$	?	NO_3^-
O in H_2O	-2	none

7. The oxidation number of N in $Zn(NO_3)2$ is

   a) +3.
   b) +5.
   a) +1.
   d) +6.

8. The element that is oxidized in this reaction is

   a) Zn.
   b) O.
   c) N.
   d) H.

9. Why are redox reactions so common in everyday situations?

   a) Nitrogen is an abundant reducing agent.
   b) Nitrogen is an abundant oxidizing agent.
   c) Oxygen is an abundant reducing agent.
   d) Oxygen is an abundant oxidizing agent.

Test Taking Tip

Write It Down! Most tests ask you a large number of questions in a small amount of time. Write down your work whenever possible. Write out the half-reactions for a redox problem, and make sure they add up. Do math on paper, not in your head. Underline and reread important facts in passages and diagrams—don't try to memorize them.

Chapter Preview

Sections

17.1 Electrolysis: Chemistry from Electricity

MiniLab 17.1 Electrolysis

17.2 Galvanic Cells: Electricity from Chemistry

MiniLab 17.2 The Lemon with Potential

ChemLab Oxidation-Reduction and Electrochemical Cells

Lightning in a Beaker?

You are familiar with this spectacular site of electrical current. Did you know that electrical current also can flow through solutions in a beaker? Silver plated tableware, chrome trim on cars, and batteries are examples of processes that take advantage of current flowing in solutions.

Launch Lab

A Lemon Battery?

You can purchase a handy package of portable power at any convenience store—a battery. You also can craft a working battery from a lemon. How are these power sources alike?

Safety Precautions

Use caution with electricity.

Materials

- lemon pieces
- zinc metal strip
- copper metal strip
- voltmeter with leads

Procedure

1. Insert the zinc and copper strips into the lemon, about 2 cm apart from each other.
2. Attach the black lead from a voltmeter to the zinc and the red lead to the copper. Read and record the potential difference (voltage) from the voltmeter.
3. Remove one of the metals from the lemon and observe what happens to the potential difference on the voltmeter.

Analysis

What is the purpose of the zinc and copper metals? What is the purpose of the lemon?

What I Already Know

Review the following concepts before studying this chapter.

Chapter 4: electrolytes

Chapter 7: role of electrons in chemical bonding; atomic structure

Chapter 9: how metals conduct electricity

Chapter 16: oxidation; reduction; redox reactions

Reading Chemistry

As you look through the chapter, pay attention to some of the graphics and photos. Read the captions that accompany them. Make a list of some of the ways the chemistry of electricity is used in the world of technology or business. Keep this list in mind while reading the chapter.

Preview this chapter's content and activities at chemistryca.com

Electrolysis: Chemistry from Electricity

SECTION PREVIEW

Objectives

✓ **Explain** how a non-spontaneous redox reaction can be driven forward during electrolysis.

✓ **Relate** the movement of charge through an electrolytic cell to the chemical reactions that occur.

✓ **Apply** the principles of electrolysis to its applications such as chemical synthesis, refining, plating, and cleaning.

Review Vocabulary

Reduction: reaction in which an element gains one or more electrons.

New Vocabulary

electrical current
electrolysis
cathode
anode
electrolytic cell
cation
anion

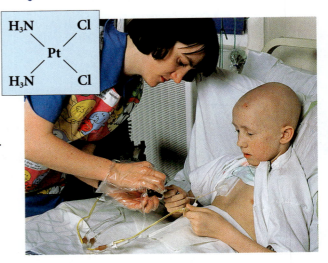

Sometimes, unexpected results in scientific research, such as Galvani's frog, lead to important discoveries. In 1964, a group of researchers studying how electricity affects the growth of bacteria found that bacterial cells stopped dividing if they were subjected to an electric current. This was an important finding because agents that inhibit cell division have the potential to work as cancer treatments. Cancer is a disease in which body cells divide uncontrollably. Upon doing more careful studies, the researchers found that it wasn't the electricity that was preventing the cells from dividing. Rather, a compound was made when platinum from the electrodes used to provide the current took part in a reaction caused by electrical energy. This compound was named cisplatin, and today it is a major medicine used in chemotherapy.

Redox Revisited

You have learned how oxidation and reduction always occur simultaneously. Think about the chemistry of corrosion you studied in Chapter 16. When iron metal reacts with oxygen, a redox reaction creates rust, iron oxide. Electrons are always transferred when a redox reaction occurs. In the rust reaction, electrons are transferred from the reducing agent, iron, to the oxidizing agent, oxygen.

Suppose you could separate the oxidation and reduction parts of a redox reaction and cause the electrons to flow through a wire. The flow of electrons in a particular direction is called an **electrical current.** In other words, you are using a redox reaction to produce an electrical current. This is what occurs in a battery—one form of an electrochemical cell in which chemical energy is converted to electrical energy. You can reverse the process and use a current to cause a redox reaction to occur.

In this section, you will examine **electrolysis,** the process in which electrical energy is used to cause a non-spontaneous chemical reaction to occur. In the second part of the chapter, you will learn about the process that is the reverse of electrolysis—electrochemical reactions that can be used to produce electricity.

*L*aunch Lab

A Lemon Battery?

You can purchase a handy package of portable power at any convenience store—a battery. You also can craft a working battery from a lemon. How are these power sources alike?

Safety Precautions

Use caution with electricity.

Materials

- lemon pieces
- zinc metal strip
- copper metal strip
- voltmeter with leads

Procedure

1. Insert the zinc and copper strips into the lemon, about 2 cm apart from each other.
2. Attach the black lead from a voltmeter to the zinc and the red lead to the copper. Read and record the potential difference (voltage) from the voltmeter.
3. Remove one of the metals from the lemon and observe what happens to the potential difference on the voltmeter.

Analysis

What is the purpose of the zinc and copper metals? What is the purpose of the lemon?

What I Already Know

Review the following concepts before studying this chapter.

Chapter 4: electrolytes

Chapter 7: role of electrons in chemical bonding; atomic structure

Chapter 9: how metals conduct electricity

Chapter 16: oxidation; reduction; redox reactions

Reading Chemistry

As you look through the chapter, pay attention to some of the graphics and photos. Read the captions that accompany them. Make a list of some of the ways the chemistry of electricity is used in the world of technology or business. Keep this list in mind while reading the chapter.

Preview this chapter's content and activities at **chemistryca.com**

Electrolysis: Chemistry from Electricity

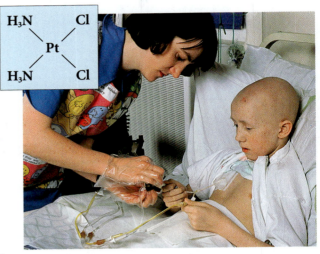

Sometimes, unexpected results in scientific research, such as Galvani's frog, lead to important discoveries. In 1964, a group of researchers studying how electricity affects the growth of bacteria found that bacterial cells stopped dividing if they were subjected to an electric current. This was an important finding because agents that inhibit cell division have the potential to work as cancer treatments. Cancer is a disease in which body cells divide uncontrollably. Upon doing more careful studies, the researchers found that it wasn't the electricity that was preventing the cells from dividing. Rather, a compound was made when platinum from the electrodes used to provide the current took part in a reaction caused by electrical energy. This compound was named cisplatin, and today it is a major medicine used in chemotherapy.

Redox Revisited

You have learned how oxidation and reduction always occur simultaneously. Think about the chemistry of corrosion you studied in Chapter 16. When iron metal reacts with oxygen, a redox reaction creates rust, iron oxide. Electrons are always transferred when a redox reaction occurs. In the rust reaction, electrons are transferred from the reducing agent, iron, to the oxidizing agent, oxygen.

Suppose you could separate the oxidation and reduction parts of a redox reaction and cause the electrons to flow through a wire. The flow of electrons in a particular direction is called an **electrical current.** In other words, you are using a redox reaction to produce an electrical current. This is what occurs in a battery—one form of an electrochemical cell in which chemical energy is converted to electrical energy. You can reverse the process and use a current to cause a redox reaction to occur.

In this section, you will examine **electrolysis,** the process in which electrical energy is used to cause a non-spontaneous chemical reaction to occur. In the second part of the chapter, you will learn about the process that is the reverse of electrolysis—electrochemical reactions that can be used to produce electricity.

Launch Lab

A Lemon Battery?

You can purchase a handy package of portable power at any convenience store—a battery. You also can craft a working battery from a lemon. How are these power sources alike?

Safety Precautions

Use caution with electricity.

Materials

- lemon pieces
- zinc metal strip
- copper metal strip
- voltmeter with leads

Procedure

1. Insert the zinc and copper strips into the lemon, about 2 cm apart from each other.
2. Attach the black lead from a voltmeter to the zinc and the red lead to the copper. Read and record the potential difference (voltage) from the voltmeter.
3. Remove one of the metals from the lemon and observe what happens to the potential difference on the voltmeter.

Analysis

What is the purpose of the zinc and copper metals? What is the purpose of the lemon?

What I Already Know

Review the following concepts before studying this chapter.
Chapter 4: electrolytes
Chapter 7: role of electrons in chemical bonding; atomic structure
Chapter 9: how metals conduct electricity
Chapter 16: oxidation; reduction; redox reactions

Reading Chemistry

As you look through the chapter, pay attention to some of the graphics and photos. Read the captions that accompany them. Make a list of some of the ways the chemistry of electricity is used in the world of technology or business. Keep this list in mind while reading the chapter.

Preview this chapter's content and activities at chemistryca.com

Electrolysis: Chemistry from Electricity

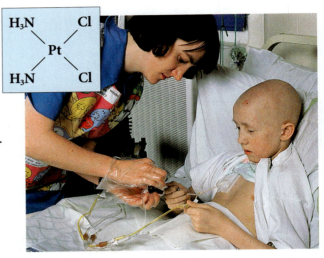

Sometimes, unexpected results in scientific research, such as Galvani's frog, lead to important discoveries. In 1964, a group of researchers studying how electricity affects the growth of bacteria found that bacterial cells stopped dividing if they were subjected to an electric current. This was an important finding because agents that inhibit cell division have the potential to work as cancer treatments. Cancer is a disease in which body cells divide uncontrollably. Upon doing more careful studies, the researchers found that it wasn't the electricity that was preventing the cells from dividing. Rather, a compound was made when platinum from the electrodes used to provide the current took part in a reaction caused by electrical energy. This compound was named cisplatin, and today it is a major medicine used in chemotherapy.

Redox Revisited

You have learned how oxidation and reduction always occur simultaneously. Think about the chemistry of corrosion you studied in Chapter 16. When iron metal reacts with oxygen, a redox reaction creates rust, iron oxide. Electrons are always transferred when a redox reaction occurs. In the rust reaction, electrons are transferred from the reducing agent, iron, to the oxidizing agent, oxygen.

Suppose you could separate the oxidation and reduction parts of a redox reaction and cause the electrons to flow through a wire. The flow of electrons in a particular direction is called an **electrical current.** In other words, you are using a redox reaction to produce an electrical current. This is what occurs in a battery—one form of an electrochemical cell in which chemical energy is converted to electrical energy. You can reverse the process and use a current to cause a redox reaction to occur.

In this section, you will examine **electrolysis,** the process in which electrical energy is used to cause a non-spontaneous chemical reaction to occur. In the second part of the chapter, you will learn about the process that is the reverse of electrolysis—electrochemical reactions that can be used to produce electricity.

Electrolysis

Not too long after Volta's invention of the first electrochemical cell, the British chemist Humphry Davy built a cell of his own and used it to pass electricity through molten salts. An electrochemical cell consists of two electrodes and a liquid electrolyte. One electrode, the **cathode,** brings electrons to the chemically reacting ions or atoms in the liquid; the other electrode, the **anode,** takes electrons away, **Figure 17.1.** The electrons act as chemical reagents at the electrode surface. The liquid electrolyte acts as the chemical reaction medium.

In Davy's electrolysis of molten NaCl, sodium ions were reduced to metallic sodium at the cathode. The oxidation of chloride ions to chlorine gas occurred at the second electrode, the anode. Each half-reaction in the electrolysis of molten sodium chloride is shown.

$$2Na^+(l) + 2e^- \rightarrow 2Na(l) \qquad 2Cl^-(l) \rightarrow Cl_2(g) + 2e^-$$

When the equations for the two half-reactions are combined, the equation for the overall reaction can be written as follows.

$$2Na^+(l) + 2Cl^-(l) \rightarrow 2Na(l) + Cl_2(g)$$

Davy discovered several elements in this way, beginning in 1807. After releasing purified potassium metal from potassium hydroxide, it took him only a year to produce magnesium, strontium, barium, and calcium. Fewer than 30 elements had been isolated by 1800, but by 1850, more than 50 were known. Most of these new elements were isolated using electrolysis. **Figure 17.2** shows the modern commercial electrolysis of molten rock salt. Rock salt is sodium chloride, NaCl. In this process, pure sodium metal and chlorine gas are produced.

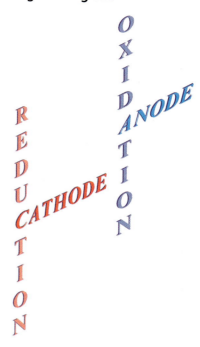

Figure 17.1

Oxidation and Reduction
You can remember that reduction always occurs at the cathode and oxidation always occurs at the anode by studying this diagram.

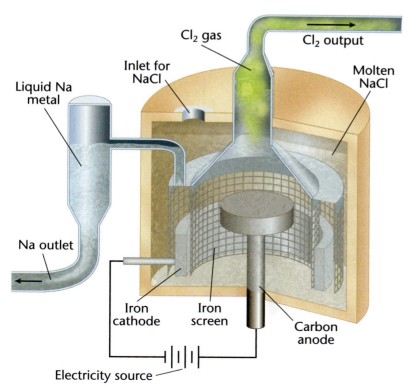

Figure 17.2

Production of Sodium and Chlorine
A device called a Downs cell is used for the electrolysis of molten sodium chloride. As an electrical current is passed through the cell, liquid sodium forms at the circular iron cathode. Because liquid sodium is less dense than molten sodium chloride, the sodium floats to the surface and is collected. Chlorine gas forms at the carbon anode and is collected at the top. An iron screen separates the two electrodes.

The Electrolysis Process

Electrolysis takes place in a type of electrochemical cell called an **electrolytic cell,** in which a source of electricity, such as a battery, is added to an external circuit connecting the electrodes. The electrolysis process occurs when the electrons are transferred between the electronic conductors—the metal electrodes—and the ions or atoms at the electrode surfaces. In the liquid, the charge is conducted by ions such as the Na^+ and Cl^- in the molten rock salt. Of course, ions contain electrons too, but these electrons are held tightly to individual ions. Ions can conduct current through a liquid only when they move through the liquid. This type of conduction is called electrolytic conduction. A positive ion is known as a **cation.** A negative ion is called an **anion.** Notice in **Figure 17.3** that the cations, Na^+ ions, move toward the cathode. The negative anions, Cl^- ions, move toward the anode.

What happens when the moving ions reach an electrode surface? If the electrodes are inert, which means that they don't react chemically with the ions in the solution, then only electron transfer will take place at the electrodes. Electrons are being pumped from the battery toward the cathode, where reduction will occur. At the cathode, the ion that reacts is the one that most readily reacts with electrons. In **Figure 17.3,** Na^+ and Cl^- are both present in the liquid near the surface of the cathode. Na^+ accepts electrons more readily, so each Na^+ cation gains one electron, being reduced to the metal.

At the anode, electrons are transferred from the ion that most easily gives them up to the anode. In this case, Cl^- holds onto its electrons more loosely, so each Cl^- anion loses an electron and is oxidized to a chlorine atom. Chlorine atoms then combine to form Cl_2 molecules. The electrons released by the chloride ions flow through the external circuit to the battery and are recycled to the cathode, where they continue the reduction reaction. Because only as many electrons are available at the cathode as are removed at the anode, the reduction process at the cathode must

Figure 17.3

Electrolytic Cell

Electrolysis, the splitting of compounds by electricity, occurs when two electrodes, an anode and a cathode, are inserted into a liquid electrolyte such as molten sodium chloride and connected to a source of electrical energy such as a battery. When electrical current flows into the electrolytic cell, chemical reactions occur. Anions and cations conduct the current by moving freely through the liquid. In the external circuit, electrons move out of the anode, through the battery, and into the cathode.

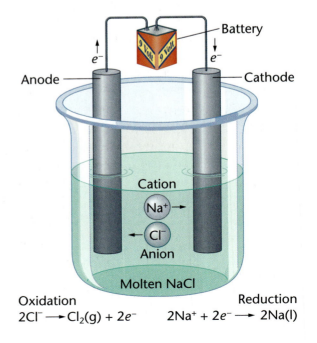

Battery

Anode

Cathode

Cation
$Na^+ \rightarrow$

$\leftarrow Cl^-$
Anion

Molten NaCl

Oxidation
$2Cl^- \longrightarrow Cl_2(g) + 2e^-$

Reduction
$2Na^+ + 2e^- \longrightarrow 2Na(l)$

Electrolysis

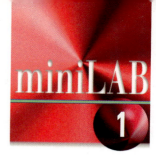
Electrolysis is the process in which an electric current is used to produce a chemical change. In this MiniLab, you will use the current from a 9-V battery to produce chemical changes in a simple electrolytic cell.

Procedure

1. Pour about 200 mL of 0.5M copper(II) sulfate into a 250-mL beaker.

2. Use wires with alligator clips to attach two 4-inch pencil leads (actually graphite, not lead) to the terminals of a 9-V battery.

3. Put the pencil leads into the copper(II) sulfate solution, keeping them as far from each other as possible.

4. Observe the reactions that occur at the two pencil leads for five minutes. Record your observations, including the polarities (+ and −) of the leads.

Analysis

1. Describe the reaction that occurs at the positive electrode, the anode, of the electrolytic cell. Write the equation for the oxidation reaction.

2. Describe the reaction that occurs at the negative electrode, the cathode. Write the equation for the reduction reaction.

3. Describe how you might use an electrolytic cell to silver plate an iron spoon.

always occur together with the oxidation process at the anode. Charge transfer at the two electrodes must exactly balance because, just as in the redox reaction, the liquid and its contents must always remain electrically neutral. Therefore, in an electrolytic cell, the overall result of the two electrolysis processes is to carry out a balanced redox reaction, even though the two half-reactions take place at different locations.

Producing Chemicals by Electrolysis

In the electrolysis of molten sodium chloride, a redox reaction taking place in an electrolytic cell can be used to generate chemicals that are important commercially. Because electrolysis consumes large amounts of energy when it is carried out on a commercial scale, many companies that use this process have made their homes in locations where electric power is inexpensive. The abundant hydroelectric power available from Niagara Falls has made that area of New York State a prime location for companies that use electrolysis. One commercial electrolytic process that is carried out in this area is the electrolysis of rock salt solutions to produce chlorine, hydrogen, and sodium hydroxide. The overall change taking place in this process is a redox as well as a substitution reaction.

$$2NaCl(aq) + 2H_2O(l) \rightarrow Cl_2(g) + H_2(g) + 2NaOH(aq)$$

How is electrolysis of a sodium chloride *solution* different from electrolysis of *molten* sodium chloride? In molten NaCl, the only ions present are Na^+ and Cl^-. What ions are present in an aqueous solution of rock salt? Recall that water dissociates slightly to form H^+ and OH^- ions.

Therefore, a rock salt solution contains Na^+, H^+, Cl^-, and OH^- ions. At the anode, the chloride ions lose electrons more easily than the other ions present, just as they did in the electrolysis of molten rock salt. This oxidation forms chlorine gas that can be used for making PVC plastic and other consumer products. At the cathode, the reaction that occurs in molten salt doesn't occur. Instead, H^+ ions are easier to reduce than Na^+, Cl^-, or OH^- ions, so hydrogen ions pick up electrons from the cathode and are reduced to form hydrogen gas. Hydrogen is used in industrial processes such as the catalytic hydrogenation of vegetable oils to form margarine. The Na^+ and OH^- ions are left dissolved in water after the electrolysis process has removed H^+ and Cl^- ions. This is a solution of the base sodium hydroxide, NaOH. Sodium hydroxide is an important industrial and household chemical.

As you have seen, electrochemical transformation of a simple salt solution has produced three valuable products. Each can be sold to pay for the electrical energy that must be invested to make them, with a little left over for a profit to the manufacturer.

Applications of Electrolysis

Electrolysis has numerous useful applications in addition to the generation of chemical substances. The process can also be used to purify metals from ores, coat surfaces with metal, and purify contaminated water. The applications range from the world of art to the world of heavy industry.

Figure 17.4
Hall-Héroult Method of Producing Aluminum
A Hall-Héroult electrolytic cell is used to produce aluminum metal. It is made of a steel shell lined with carbon that forms the cathode. Anodes of carbon hang down into the solution of aluminum oxide dissolved in cryolite. ▶

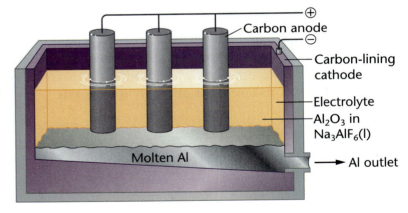

Carbon anode
Carbon-lining cathode
Electrolyte
Al_2O_3 in $Na_3AlF_6(l)$
Molten Al
Al outlet

Molten aluminum is drawn off from the bottom of the electrolytic cell, where it accumulates during the electrolysis process. ▶

Figure 17.5

Recycling Aluminum

◀ Recycling of aluminum provides a relatively cheap source of this metal. Currently, 60 percent of aluminum beverage cans are recycled in the United States.

▲ Scrap aluminum is melted in a furnace and then re-formed into other products.

Refining Ores

Just as sodium can be produced from melted NaCl by electrolysis, many metals are separated from their ores, or refined, using electrolysis. Today, the metal produced in largest quantity by electrolysis is aluminum from aluminum oxide. First, bauxite ore is heated, driving off the water and leaving aluminum oxide, Al_2O_3. Pure aluminum oxide melts at about 2000°C, so cryolite, Na_3AlF_6, is added to lower the melting point to about 1000°C. The molten salt solution is placed in a large electrolytic cell lined with carbon, which acts as the cathode during electrolysis. Large carbon anodes are dipped into the molten salt to complete the cell, as shown in **Figure 17.4.**

The electron transfers cause complex chemical reactions during electrolysis of molten Al_2O_3 / Na_3AlF_6, but the net reactions are simple. At the cathode, electrons reduce aluminum ions to aluminum metal, which is molten at this temperature. At the anodes, oxide ions lose electrons to form oxygen. The oxygen then combines with the carbon anodes to produce carbon dioxide. The carbon anodes must be replaced periodically because they are gradually used up. The overall redox reaction follows.

$$2Al_2O_3(s) + 3C(s) \rightarrow 4Al(l) + 3CO_2(g)$$

This electrolytic method for producing aluminum is called the Hall-Héroult process because it was developed simultaneously by Charles Martin Hall of the United States and Paul Héroult of France in 1886. Aluminum was so rare before this process was developed that it was more expensive than silver or gold. Today, about 10 million metric tons of aluminum are produced per year worldwide using this process.

The Hall-Héroult process is expensive and consumes large amounts of electrical energy. Recycling aluminum metal prevents some of the expense of producing new aluminum by electrolysis. A lot of waste aluminum is available from discarded aluminum containers, and the energy invested in the original refining of aluminum can be saved if metallic aluminum can be melted down, **Figure 17.5.** It takes only about seven percent as much energy to make new aluminum cans from old ones as it does to make new cans from aluminum ore.

CHEMISTRY & TECHNOLOGY

Copper Ore to Wire

It would be hard to imagine life without many of the common metals that are used today. Copper, for example, is used in many of the pots and pans in your kitchen, the cooling coils in the air-conditioning system, and some of the pennies you carry in your pocket. More importantly, copper is the metal of choice for most of the electrical wiring in your appliances, homes, and cars because of its good conductivity and low cost. How does copper get from rocks to the finished wire?

Copper can be found in the ground as the free metal and was used about 5000 B.C. In Roman times, much of the copper was obtained from the island of Cyprus, whose name means "copper." Coppersmiths soon learned that copper could be shaped by exposing it to a slow, softening heat in a process called annealing. Annealing soon led to the development of the smelting process. By 3000 B.C., smiths were adept at the metallurgical processes of hammering, annealing, oxidation and reduction, melting, alloying, and removing impurities. Unfortunately, these processes could produce only small quantities of copper. Large-scale production was not mastered until modern furnaces and rolling techniques were developed.

1. Mining

Copper can be found as the free metal in many parts of the world. However, most copper is found as $CuFeS_2$ (chalcopyrite), Cu_2S (chalcocite), and CuS (covellite). The ores are typically surface mined and ground into powders.

2. Ore Enrichment

Because each ore contains only from one percent to ten percent copper, the copper ore must be concentrated by the flotation process. A frothing agent such as pine oil is mixed with the powdered copper ore, and air is blown in to froth the mixture. Because copper ores are hydrophobic (not wetted by water), the copper and iron sulfides cling to the oil and float to the top, where they can be continuously removed.

Copper mine, Arizona

Ore enrichment

Roasting

3. Roasting

The preparation of copper metal involves roasting the ore with oxygen to convert the metallic sulfides to metallic oxides. Usually, both iron and copper are present in the mix.

$$2Cu_2S(s) + 3O_2(g) \rightarrow 2Cu_2O(s) + 2SO_2(g)$$
$$2FeS(s) + 3O_2(g) \rightarrow 2FeO(s) + 2SO_2(g)$$

4. Smelting

The copper and iron oxides are smelted by mixing and heating them with SiO_2, air, and limestone. The result is dense blister copper, lightweight iron(II) calcium silicate slag, and gaseous

SO_2. The blistered surface of the copper is due to the escaping gas. The blister copper can be drawn off the bottom and cast into large blocks.

5. Purification

The copper can be purified by electrolysis to 99.95 percent purity. The large blocks of blister copper are used as anodes suspended in a solution of aqueous copper(II) sulfate. Pure copper is used as the cathode. During electrolysis, copper is oxidized at the anode, moves through the solution as Cu^{2+} ions, and is deposited on the cathode. Waste products left after the dissolution of the anode produce a sludge on the bottom of the electrolysis vessel. The sludge, which is often rich in silver and gold, can be recovered for profit.

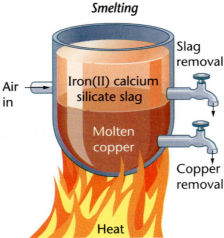

Smelting

Air in

Iron(II) calcium silicate slag

Molten copper

Slag removal

Copper removal

Heat

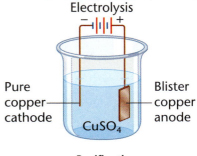

Electrolysis
− +

Pure copper cathode

Blister copper anode

$CuSO_4$

Purification

6. Wire Bar Production

The electrolytic copper is cast into wire bars ranging from 60 to 227 kg. The wire bars are heated to 700-850°C and rolled without further reheating into rods approximately 1 cm in diameter.

Wire bar production

7. Drawing Wire

The 1-cm rods of copper are drawn through successively smaller dies until the desired size of wire is reached. The dies must be made of exceptionally hard materials because of the tremendous amount of wear from drawing the wire. The dies typically are made of tungsten carbide or diamond.

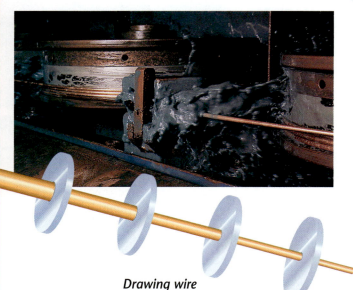

Drawing wire

8. Coatings

The finished wire may be coated with plastic, enamel, or another metal to help protect it from moisture and oxidation, or it may be left bare.

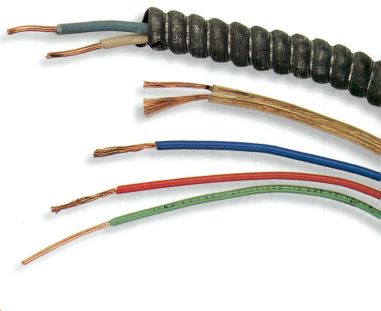

DISCUSSING THE TECHNOLOGY

1. **Thinking Critically** Why is it important that the slag formed during smelting be lightweight?

2. **Acquiring Information** Copper can be made into many different forms and alloys. Research and list several of the forms and uses for each.

3. **Hypothesizing** How might the gold and silver be removed from the sludge formed beneath the anode?

Electroplating

You have learned that many metals can be protected from corrosion by plating them with other metals. Zinc coatings are often used to keep iron from rusting. Metal garbage cans are galvanized by dipping them into molten zinc. This process produces an uneven, lumpy coating both inside and outside of the can. That's OK for a garbage can, but the lumpy surface wouldn't look good under the snazzy red paint job on a new sports car. That's why automobile manufacturers electroplate zinc onto the steel used for car bodies. This process involves dragging a sheet of steel across the surface of an electrolyte in an electrolytic cell. The process produces a thin (cost-saving), uniform (smooth and clean) coating of zinc on only one side of the sheet of steel, saving half the coating cost. Because only one side of the car body is exposed to the corrosive effects of water and salt, it is not worth the cost of coating the inside. **Figure 17.6** shows other uses of electroplating.

In zinc electroplating, zinc ions are reduced to zinc atoms at the surface of the metal object to be coated, which becomes the cathode in an electrolytic cell. At the anode, which is made of zinc, atoms of zinc are oxidized to ions. The electrolyte solution contains dissolved zinc salt.

Try at Home Lab

See page 871 in Appendix F for **Removing Electroplating**

Figure 17.6
Electroplating
Chromium is often electroplated onto a softer metal to improve its hardness, stability, and appearance. Chrome bumpers and trim can be found on many vintage cars. ▼

Reduction of silver ions onto cheaper metals forms silver-plate. The object to be plated is made the cathode. At the pure silver anode, oxidation of silver metal to silver ions replaces the silver ions removed from the solution by plating at the cathode. ▶

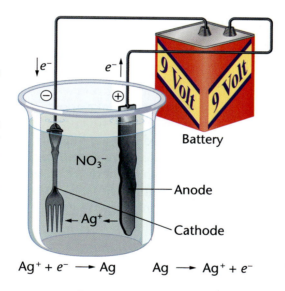

NO$_3^-$

Battery

Anode

← Ag$^+$ →

Cathode

$$Ag^+ + e^- \longrightarrow Ag \qquad Ag \longrightarrow Ag^+ + e^-$$

◀ In the early 1980s, because of high inflation of the U.S. currency and a worldwide shortage of copper, the cost of producing a copper penny became nearly equal to the value of the copper metal itself. The U.S. Mint was instructed to begin making pennies from a cheaper metal, such as zinc. Knowing that the American public would not accept dusty-gray zinc pennies, the Mint began to produce pennies by electroplating zinc disks with a copper coat. Many people think that pennies are still made of pure copper.

Everyday Chemistry

Manufacturing a Hit CD

The sound from your new compact disc (CD) is so clear and crisp that it seems as if the musicians are in the same room. The clarity of the sound is possible thanks to the chemical process of metallic depositing and electroplating used in the manufacturing of CDs.

Data pickup A CD is a collection of binary data in a long, continuous spiral that runs from the inside edge to the outside. Nothing touches the information stored on the CD except laser light, which is reflected from the CD and read by the CD player computer as a binary signal. This signal is then transformed by a tiny computer into audio signals, which are amplified to produce the sound heard from your speakers.

Master copy production Your CDs are stamped from a master copy called a stamper master. But how is this stamper made? The musical data are first transferred onto a glass disc using a high-powered laser that etches small pits in the glass as shown here. This glass master copy then contains all of the musical information in binary data form.

The master is coated with a dilute solution of silver diammine complex—$[Ag(NH_3)_2]^+$—followed

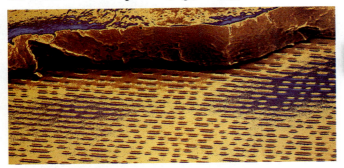

Etched pits, magnified 1200×

by a solution of formaldehyde that acts as a reducing agent for the silver. The result is a redox reaction that deposits a thin silver mirror that plates the etched disc. This mirrored master disc is produced in the following reaction.

$$2Ag^+ + HCHO + H_2O \rightarrow 2Ag + HCOOH + 2H^+$$

Electrodeposition The silver coating forms the surface onto which a thin layer of nickel is electrodeposited to make a nickel-coated disc called the mother disc.

$$Ni^{2+}(aq) + 2e^- \rightarrow Ni(s)$$

A second coating of nickel is then electrodeposited onto the mother disc. This nickel layer, the stamper master, is peeled off and used to stamp the data impression onto melted polycarbonate plastic discs. The polycarbonate disc now has all of the pits found on the original glass master disc etched by the laser. Because polycarbonate is clear, it is vacuum coated with a thin aluminum film to produce the reflective layer required by the laser. This delicate aluminum layer is covered with a protective layer of polycarbonate to prevent aluminum oxidation and marring of the data surface. The back of the CD can now be covered with information in the form of art and lettering.

Stamping characteristics Each nickel stamper master can make about 20 000 copies before it wears out. At this point, the nickel stamper can be recycled into an aqueous nickel solution and used to make more nickel stampers. A hit CD may go through as many as 50 nickel stampers plated onto the mother disc. CD-ROMs for your computer are made in the same way.

Exploring Further

1. **Hypothesizing** Why do you think nickel is used for stamping the polycarbonate discs?

2. **Acquiring Information** Polycarbonate is the base material used for CDs. Research its properties and some of its other major uses.

To find out more about the technology behind the production of compact discs, visit the Chemistry Web site at chemistryca.com

The thickness of the zinc metal coating can be controlled exactly by controlling the total charge (number of electrons) used to plate the object. Only that portion or side of the object that is immersed in the cell electrolyte receives a zinc coat.

Because coatings adhere best to a chemically clean foundation, objects to be coated are usually degreased, cleaned with soap, and then treated with a corrosive fluid to remove any dirt on their surface before electroplating. Then, the object is immersed in an electrolyte containing the salt of the metal to be deposited. Because the object acts as the cathode of the electrolytic cell, it must be a conductor. Metal objects are electroplated most often because metals usually are excellent conductors. The anode is made of the same metal that is being plated so that it will replenish the metal ions in the electrolyte that are removed as the plating proceeds. The net effect is that when an electric current is passed through the electroplating bath, metal is transferred from the anode and distributed over the cathode. Eventually, the entire object becomes coated with a thin film of the desired metal.

WORD ORIGIN

corrosion:
com (L)
thoroughly
rodere (L) to
gnaw
Metals wear away
gradually during
corrosion, as if
they were being
eaten.

Electrolytic Cleaning

Electrolysis can be used to clean objects by pulling ionic dirt away from them. The process has been used to restore some of the many metal artifacts taken from the shipwrecked cruise ship *Titanic*, which sank in the northern Atlantic Ocean in 1912, **Figure 17.7.** Coatings of salts containing chloride ions, which came from the seawater, were removed by electrolysis. The electrolysis cell for this cleaning process includes a cathode that is the object itself, a stainless steel anode, and an alkaline electrolyte. When an electric current is run through the cell, the chloride ions are drawn out. Hydrogen gas forms and bubbles out, helping to loosen corrosion products. Among the objects that have been recovered are a porthole, a chandelier, and buttons from the uniforms of crew members.

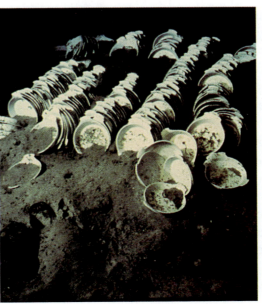

Figure 17.7
When That Great Ship Went Down
Chemistry played a key role in restoring many items taken from the wreck of the *Titanic*. Electrolysis was used to clean and stabilize many metal artifacts, and electrophoresis was used to remove corrosion from bank notes, leather, and objects such as these casserole dishes. Chemicals that attract and hold metal atoms or ions were used to remove iron stains from delicate objects made of organic materials such as newspapers, textiles, and letters. The study of objects from the ship may help scientists compile information for long-term storage and containment under seawater.

PEOPLE in CHEMISTRY

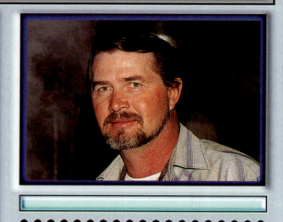

Meet Harvey Morser, Metal Plater

A seasonal highlight in Reno, Nevada, is a celebration called Hot August Nights. On display are thousands of beautifully restored antique cars. "Muscle cars"—like a 1934 Ford sedan with a $12 000 custom paint job and immaculately plated chrome—draw admiring crowds. Harvey Morser watches with pride as the cars sporting his shop's work on grilles and door handles parade by.

On the Job

 Mr. Morser, will you tell us what you do on the job?

Although I'm now the owner of Western Metal Finishing, I still go out on the shop floor and do plating. All iron metal needs some type of protective coating to keep it from rusting. There are different types of applications that you can use. Cadmium plating is probably the best, but it's very toxic. Hardchrome plating also releases fumes into the atmosphere. So, because of environmental concerns, electroless nickel is the favored process. It's a chemically applied nickel plating done without electricity. In my opinion, this process is twice as effective because it goes on easily and consistently, unlike hard plating, which you have to apply and then grind back down. Chemists have developed an electroless nickel that has the same Rockwell factors (the hardness factors) as hardchroming. I know that a major heavy-equipment manufacturer converted from hard-chrome over to electroless nickel and saved something like $3 million the first year.

 What kinds of metal products do you plate?

One of my major accounts is for the metal straps on hearing protectors that the Navy uses on aircraft carriers. At this plant, we also plate things like computer chassis and covers for stereo systems.

 Why does the electroless nickel process produce a more even coating?

Visualize a flat, square plate. When you put a hook in one edge of it and hang it in the tank to hardplate it, the electrical current will reach the corners first, then travel down the side edges, and finally spread into the center of the plate. The plating will go on in the same way, building up probably twice as fast on the edges as in the center. So the edges might have eight ten-thousandths of an inch of plating, whereas the center might only have four ten-thousandths. With electroless nickel, a metal part will plate perfectly evenly because it's put on chemically. Electroless

nickel has the highest corrosion resistance next to cadmium plating, so these plated parts resist corrosion as well as being uniform.

Early Influences

 What training did you have in metal plating?

All my training came on the job. I started out polishing the metal prior to plating. That's a tough and dirty job, but an important one. Plating duplicates a surface, so it has to be polished like a mirror. Otherwise, plating will magnify even a tiny pit. In those early days, I had to work a side job in a bowling alley as a pin chaser, unsticking the balls and unjamming the pins. Along the way, I picked up carpentry, welding, and electrical skills, which have come in very handy in doing the maintenance at the metal plant.

How did you work your way up to owning the plant?

It all boiled down to learning quickly, not whining, and working hard. Plating is tough and heavy work. The heat on the lines is terrible, with the humidity and steam off the tanks. All the tanks are running about 160°, so in the summertime it will be 104° on the line. The joke here is that we don't charge extra for the steam bath. Nine years ago, I became the owner of the business. To me, that's the Great American Dream.

Personal Insights

  If someone came to you just out of high school, would you give him or her a chance on the job?

In a heartbeat. I believe people need to get their schooling, but they also need to have common sense. Every employee here gets on-the-job training, just like I did. What I look for in a prospective employee is honesty and dependability, plus an ability and desire to learn.

 Is this a stressful business to be in?

Absolutely! I carry a pager all the time and even take a cellular phone out on the lake when I go fishing. There's always the possibility of an industrial accident. Earthquakes don't announce that they are on their way. If I see the numbers 1-5 on my pager, I know it's all clear. Seeing five zeros is what I dread.

 What appeals to you about the plating business?

I was always turned on by the fact that I could take something that looked terrible, like an old car part, and make it look gorgeous. I want even the most modest plating job to look good, even when it's on a part that probably won't be visible after it's installed.

CAREER ▸ CONNECTION

These jobs also involve working with metals.

Metallurgical Technician Two-year training program
Mining Engineer Bachelor's degree in engineering
Scrap Metal Processing Worker On-the-job training after high school

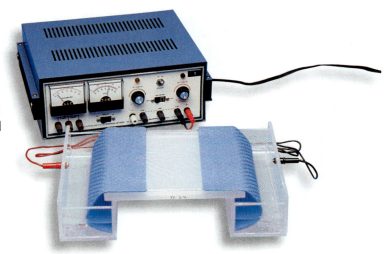

Figure 17.8
Electrophoresis
Electrophoresis is a valuable laboratory tool used to separate and identify large charged particles such as DNA and proteins. Negatively charged particles move toward the anode. Positively charged particles move toward the cathode.

Electrophoresis is another electrochemical process that was used to restore some of the ceramic and organic artifacts from the *Titanic*. Electrophoresis involves placing an artifact in an electrolyte solution between positive and negative electrodes and applying a current. The current breaks up salts, dirt, and other particles as their charged components migrate to the electrodes. Electrophoresis is also used in laboratories to separate and identify large molecules, **Figure 17.8.**

Electrolysis of Toxic Wastes

Supplemental Problems

For more practice with solving problems, see Supplemental Practice Problems, Appendix B.

The plating baths used in the various applications of electrolysis often contain toxic materials or produce toxic by-products. After bath solutions have been used for a period of time, they must be changed and the toxic contents disposed of in a safe manner. Remarkably, electrolysis offers one of the safest and most thorough means of cleaning up toxic metal-containing wastes. When the bath solution is subjected to electrolysis, the toxic metal ions are reduced to free metal at the cathode. The metal can then be recycled or disposed of safely.

SECTION REVIEW

Understanding Concepts

1. Draw and label the parts of an electrolytic cell.
2. Of what value is electroplating? Why is the process used?
3. How is electrolysis used for cleaning objects?

Thinking Critically

4. **Applying Concepts** What effect would electroplating steel jewelry with gold have on the rate of corrosion of the jewelry?

Applying Chemistry

5. **Magnesium from the Sea** Magnesium in seawater is found mostly as $Mg(OH)_2$, which can be converted to $MgCl_2$ by reacting it with HCl. Magnesium metal can then be purified by electrolysis of molten $MgCl_2$.

   a) What reaction takes place at the cathode during electrolysis?
   b) What reaction takes place at the anode during electrolysis?
   c) Write an equation for the net reaction that occurs.

Galvanic Cells: Electricity from Chemistry

Suppose it's nearing the end of half-time during the seventh game of the NBA championship finals, and your favorite team is leading by just two points. Suddenly . . . the electricity goes out. You might find that battery-powered radio you haven't used for months, but its batteries are dead. You can't go buy more batteries because the electricity is out all over town. Must you resign yourself to missing the end of this game?

SECTION PREVIEW

Objectives

✓ **Relate** the construction of a galvanic cell to how it functions to produce a voltage and an electrical current.

✓ **Trace** the movement of electrons in a galvanic cell.

✓ **Relate** chemistry in a redox reaction to separate reactions occurring at electrodes in a galvanic cell.

Review Vocabulary

Electrolysis: process in which electrical energy is used to cause a non-spontaneous chemical reaction to occur.

New Vocabulary

potential difference
voltage
galvanic cell

Electrochemical Cells

You don't have to miss the end of the NBA game because you have at hand all the ingredients for making a battery that will power your radio. All you need are several lemons or pieces of fruit (or even glasses of fruit juice), two different kinds of metal (a penny and a steel nail will do), and some pieces of wire for connecting everything together. Once you have those items, you need only a little knowledge of electrochemistry, and you'll soon be listening to your team go all the way.

How can it be that simple? The energy in oxidation-reduction reactions can be harnessed to do useful work, if listening to an NBA game can be called work. A battery is the tool that makes this possible.

When Luigi Galvani used two dissimilar metals to produce an electrical current that stimulated the nerve in the frog leg, he didn't know that he had invented the first battery. Batteries are electrochemical cells. The battery made out of fruit, which allowed you to hear the end of the basketball game, is such a cell. In a battery, the two halves of a spontaneous redox reaction are separated and made to transfer electrons through a wire.

The Lemon Battery

How does the lemon battery produce electrical energy? The lemon itself is a container for a solution of electrolyte—the lemon juice. As you know, lemon juice is sour; that is, it is acidic. The hydrogen ions from partially dissociated citric acid give it a sour taste and also provide the ions for conduction of charge through the lemon battery. The two dissimilar metal strips are the electrodes at which an oxidation reaction and a reduction reaction take place to provide the battery's power source.

miniLAB
2

The Lemon with Potential

Lemons are good for more than just making lemonade. By adding some metal strips, lemons have other "potential" uses. In this MiniLab, you will investigate the interactions of zinc and aluminum with lead when the metals are placed in a lemon.

Procedure

1. Gently knead a lemon without breaking the skin. Make two slits about 1 cm in depth on opposite sides of the lemon.

2. Insert a strip of zinc in one of the slits and a strip of lead in the other slit.

3. Connect an alligator clip wire to each of the metal strips, touching or connecting the other end of each wire to the poles of a voltmeter. If the voltmeter gives no reading, reverse the wires.

4. Read and record the voltage.

5. Repeat steps 1 through 4 with strips of lead and aluminum, making a new slit for the

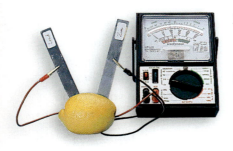

aluminum and lightly buffing the metal with fine steel wool *immediately* before inserting to remove the oxide coating. **CAUTION:** *Discard the lemon. Do not use for food.*

Analysis

1. What causes the potential difference between the zinc and lead strips?

2. Why is the potential difference greater when aluminum is substituted for zinc?

3. If strips of zinc and magnesium, rather than zinc and lead, were used in the MiniLab, would the reaction that occurs at the zinc strip be the same? Explain.

In the lemon battery shown in **Figure 17.9,** a different chemical reaction occurs at each of the metal-strip electrodes. The electrode made of the metal that is more easily oxidized becomes the anode—the electrode at which the oxidation reaction occurs. The second electrode becomes the cathode, and a reduction reaction proceeds at its surface. The substance in a lemon that is most easily reduced is the abundant hydrogen ion of the electrolyte. When these two reactions occur together, in the same cell, they combine to produce a spontaneous redox reaction. This type of reaction is represented by the equation below, where M is the metal that is oxidized.

$$M + 2H^+ \rightarrow M^{2+} + H_2$$

This spontaneous reaction generates the cell voltage of the battery by producing a different electrical potential at each electrode.

Figure 17.9

Lemon Battery

A battery can be made by inserting iron and copper strips into a lemon and connecting them with a conducting wire in an external circuit. Electrons travel through the wire by metallic conduction and through the lemon by electrolytic conduction.

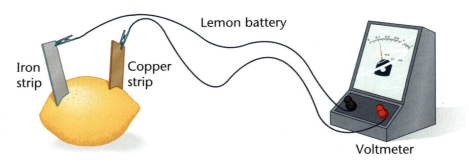

Potential Difference

Electrons in the metal electrodes of the lemon battery move through the external circuit as a current and can do useful work. The portable radio you connected to a lemon battery to listen to the game needs power to work, and it gets this power from the electrons flowing through the wires from the anode to the cathode of the battery. The chemical reaction at the anode gives off electrons, which enter the metal and then flow through the external part of the circuit connecting the anode to the cathode. At the cathode, the electrons are used up in a reduction reaction. Just as adding water to a container raises the level of the water, adding electrons builds up a negative potential at the anode. This electrical potential is often described as a force, or a pressure of electrons produced by raising the level of the electron sea, **Figure 17.10**.

Why do the electrons travel in one direction and not in the reverse? The electron pressure at the cathode is kept low by the reduction reaction, and the electrons flow from a region of high pressure (negative potential at the anode) to a region of low pressure (positive potential at the cathode). This **potential difference** between the electrodes in the lemon battery causes an electrical current to flow. If there is no potential difference between the electrodes, no current will flow. The size of the current depends upon the size of the potential difference. As the electrons move from a region of more negative potential to a region of more positive potential, they lose energy, so the discharge of a lemon battery is a spontaneous process. Potential energy stored in chemical bonds is released as electrical energy and, finally, as heat. An electrical potential difference is called **voltage** and is expressed in units of volts in honor of Alessandro Volta.

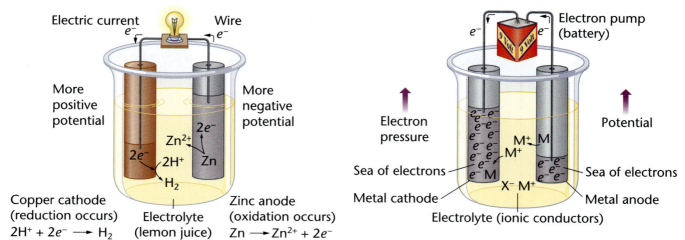

Electric current | Wire | Electron pump (battery)

More positive potential | More negative potential

Zn^{2+} | $2e^-$ | Zn | $2e^-$ | $2H^+$ | H_2

Electron pressure | Potential

Sea of electrons | Sea of electrons

Metal cathode | M^+ M | M^+ | M | X^- M^+ | Metal anode

Copper cathode (reduction occurs) | Electrolyte (lemon juice) | Zinc anode (oxidation occurs)

$2H^+ + 2e^- \longrightarrow H_2$ | $Zn \longrightarrow Zn^{2+} + 2e^-$

Electrolyte (ionic conductors)

Figure 17.10

Potential Difference

▲ In this model of a lemon battery, the level of the electron sea is raised or lowered by the chemical reactions at the electrode surfaces, creating a potential difference across the battery. A spontaneous oxidation reaction raises the electron pressure (potential) at the anode, and a spontaneous reduction reaction reduces the pressure at the cathode. The "sea level" in the lemon juice is uniform throughout and is intermediate between the levels at the two electrodes.

▲ Because the redox reactions that take place during electrolysis are not spontaneous, a battery is needed to pump electrons from an area of low potential to one of high potential.

Table 17.1 Ease of Oxidation of Common Metals

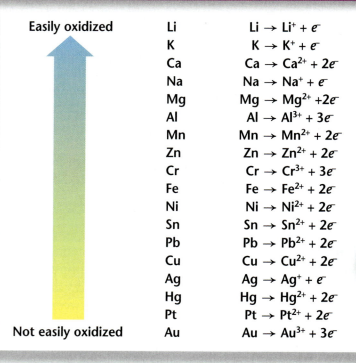

Easily oxidized ↑	
Li	$Li \rightarrow Li^+ + e^-$
K	$K \rightarrow K^+ + e^-$
Ca	$Ca \rightarrow Ca^{2+} + 2e^-$
Na	$Na \rightarrow Na^+ + e^-$
Mg	$Mg \rightarrow Mg^{2+} + 2e^-$
Al	$Al \rightarrow Al^{3+} + 3e^-$
Mn	$Mn \rightarrow Mn^{2+} + 2e^-$
Zn	$Zn \rightarrow Zn^{2+} + 2e^-$
Cr	$Cr \rightarrow Cr^{3+} + 3e^-$
Fe	$Fe \rightarrow Fe^{2+} + 2e^-$
Ni	$Ni \rightarrow Ni^{2+} + 2e^-$
Sn	$Sn \rightarrow Sn^{2+} + 2e^-$
Pb	$Pb \rightarrow Pb^{2+} + 2e^-$
Cu	$Cu \rightarrow Cu^{2+} + 2e^-$
Ag	$Ag \rightarrow Ag^+ + e^-$
Hg	$Hg \rightarrow Hg^{2+} + 2e^-$
Pt	$Pt \rightarrow Pt^{2+} + 2e^-$
Not easily oxidized Au	$Au \rightarrow Au^{3+} + 3e^-$

Gold

Chemists at the University of California at Irvine have made the world's smallest galvanic cell. It is too small to be seen without an electron microscope and much smaller than most human cells. The galvanic cell consists of two mounds each of copper and silver attached to a graphite surface. Although it probably will never be used as a practical battery, it may allow scientists to study redox reactions at the atomic level.

Iron is readily oxidized partly because the transfer of electrons from iron to an oxidizing agent releases a large amount of energy. You learned that other metals also are oxidized in corrosion reactions. However, different substances release different amounts of energy when they become oxidized, and this fact may be used to construct a table such as **Table 17.1.** It may be used as a general guide to the ease with which a substance will lose electrons. By examining this table, you can see why copper, gold, and silver are the metals most commonly used in jewelry. All three are hard to oxidize and are, thus, resistant to corrosion.

Galvanic Cells

In the lemon battery, a redox reaction occurs spontaneously to produce a separation of charge at the two electrodes. The reaction begins as soon as the two electrodes are connected by a conductor so that current can flow. An electrochemical cell in which an oxidation-reduction reaction occurs spontaneously to produce a potential difference is called a **galvanic cell.** In a galvanic cell, chemical energy is converted into electrical energy. *Galvanic cells* are sometimes called *voltaic cells;* both terms refer to the same device. A galvanic cell that has been packaged as a portable power source is often called a battery.

Sometimes, the chemical change taking place in a galvanic cell can be seen easily, such as in the simple magnesium-copper galvanic cell shown in **Figure 17.11.** Because magnesium is more easily oxidized than copper, the magnesium loses electrons and becomes oxidized, forming Mg^{2+} ions. The potential of the magnesium anode becomes more negative because of the increased electrical pressure from the released electrons. At the same time, the Cu^{2+} ions pick up electrons from the copper electrode and are reduced to copper metal. The potential of the copper electrode becomes more positive because electrical pressure is lowered as electrons are removed from the cathode. If a wire is connected between the electrodes, current flows from the magnesium electrode to the copper electrode, and the voltmeter in the external circuit reads a voltage of 2.696 V. The energy released during discharge of the cell can be used to power a device such as a radio by connecting the wire from the electrodes through the radio. The overall reaction in the copper-magnesium cell is a redox reaction.

$$Mg(s) \rightarrow Mg^{2+}(aq) + 2e^- \qquad Cu^{2+}(aq) + 2e^- \rightarrow Cu(s)$$

Oxidation half-reaction   **Reduction half-reaction**

$$Mg(s) + Cu^{2+}(aq) \rightarrow Mg^{2+}(aq) + Cu(s)$$

Net redox reaction

Figure 17.11

Magnesium-Copper Galvanic Cell

A piece of magnesium metal is placed in a beaker containing a solution of magnesium sulfate, and a piece of copper metal is placed in a beaker containing a solution of copper(II) sulfate. The two beakers are connected via a salt bridge, which is a porous barrier containing a salt solution; this prevents the two solutions from mixing but permits the movement of ions from one side of the cell to the other. An external circuit containing a voltmeter connects the two metal electrodes. ▶

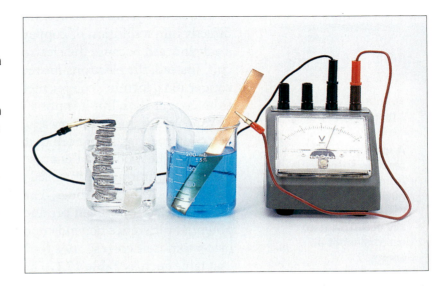

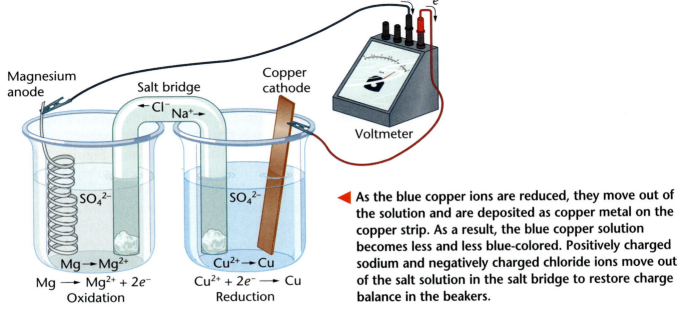

As the blue copper ions are reduced, they move out of the solution and are deposited as copper metal on the copper strip. As a result, the blue copper solution becomes less and less blue-colored. Positively charged sodium and negatively charged chloride ions move out of the salt solution in the salt bridge to restore charge balance in the beakers.

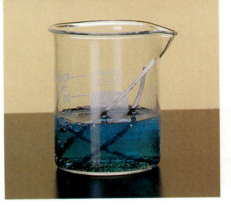

Figure 17.12

Magnesium-Copper Redox Reaction

When magnesium metal is added to a blue solution of $CuSO_4$, both the magnesium metal and the blue color disappear. ▶

The more easily oxidized magnesium forms colorless Mg^{2+} ions, which dissolve in the solution. The blue copper(II) ions are reduced to the red-brown copper metal that can be seen at the bottom of the beaker. ▶

WORD ORIGIN

spontaneous:
sponte (L) of free will

A spontaneous reaction arises from the inherent qualities of the reactants and usually occurs with no external input of energy.

The same overall redox reaction occurs if the magnesium metal is placed directly into a solution of copper sulfate, **Figure 17.12.** However, this is not a galvanic cell because the electrons do not flow through an external circuit. Instead, the electrons move directly from the magnesium metal to the copper ions, forming copper metal. This is a way to make copper metal from copper ions, but it is not a way to make electrical power.

You can see that for every spontaneous redox reaction, you theoretically can construct a galvanic cell that can capture the energy released by the reaction. The amount of energy released depends upon two properties of the cell: the amount of material that is present and the potential difference between the electrodes. The more material there is in the electrode, the more electrons it can produce during the course of the reaction. The potential difference depends upon the nature of the reaction that takes place; that is, it corresponds to the relative positions of the two substances in a table such as **Table 17.1.** The farther apart the two substances are in the table, the greater the potential difference between the electrodes, and the greater the energy delivered by each electron that flows through the external wire.

How do you know which substance will be oxidized and which reduced in any cell? Look back at **Table 17.1.** Experimental chemists such as Humphry Davy and his student Michael Faraday did many experiments from which this type of table could be made. The table is used today to predict the outcome of new experiments. For example, in a Zn-Cu galvanic cell, zinc will be oxidized and copper reduced. Because zinc is more easily oxidized than copper, electrons will flow from zinc to copper.

A cell voltage should register on the voltmeter shown in **Figure 17.11** because a potential difference exists between the magnesium and copper electrodes. What function does the salt bridge serve? As the half-reactions continue, magnesium ions are released into the solution at the anode, and copper ions are removed at the cathode. Ions must be free to move between the electrodes to neutralize positive charge (Mg^{2+} cations) created at the anode and negative charge (anions) left over at the cathode. The solution of ions in the salt bridge allows ionic conduction to complete the electrical circuit and prevent a buildup of excess charge at the electrodes.

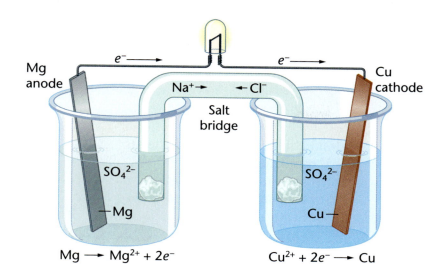

Mg
anode

e^- →

e^- →

Cu
cathode

Na^+ → ← Cl^-

Salt
bridge

SO_4^{2-}

SO_4^{2-}

— Mg

Cu —

$Mg \longrightarrow Mg^{2+} + 2e^-$

$Cu^{2+} + 2e^- \longrightarrow Cu$

Figure 17.13

Batteries Perform Work
When a simple galvanic cell does useful work, it is called a battery. If the external circuit is connected with a wire, electrons flow from the site of oxidation at the magnesium strip and through the LED to the surface of the copper strip, where reduction of Cu^{2+} ions takes place. The voltage pushes electrons through the LED, causing it to light up.

Useful work may be done if the voltmeter is replaced by wires connecting the galvanic cell to a lightbulb. Then, electrical energy will be converted into light energy, a useful process in a dark room. In **Figure 17.13,** wires connect the cell to a light source with a low-voltage requirement, called a light-emitting diode or LED. If the circuit to the cell is complete, the LED lights up, showing that the cell is doing useful work. With time, the light intensity will fade. Why doesn't it stay lit indefinitely? Eventually, all of the magnesium in the anode becomes oxidized. The capacity of the battery has been exceeded, the magnesium is gone and, if there is no electrode, there can be no cell.

Better and Better Batteries

Although the galvanic cell made from magnesium and copper can do useful work, it isn't something you'd want to bring along on a camping trip. The wet solutions could be sloppy, the glass could break easily, and the capacity is limited. Fortunately, scientists have developed much better batteries that are smaller, lighter, provide higher voltages, and last longer. **Figure 17.14** shows an assortment of commonly used batteries. Experimental batteries no thicker than a sheet of paper have already been developed. And, although you might think batteries always have to be made of metal and acids, some batteries of the future may be made of microorganisms that use the energy in sugar to make electricity. A living fuel cell has been developed that someday could be used to power an automobile for up to 15 miles on two pounds of sugar.

How are batteries designed? The farther apart two metals are in **Table 17.1,** the larger the voltage of a battery that can be constructed from them. If you wanted to make a high-voltage battery to power your radio, you would choose metals that are far apart in the table. A copper penny with an iron nail will yield a larger voltage than a penny with a piece of nickel because copper is farther away from iron in the table than it is from nickel.

Figure 17.14

Modern Batteries
Modern batteries come in a wide variety of sizes, shapes, and strengths. Each type of battery serves a different purpose.

ChemLab

Oxidation-Reduction and Electrochemical Cells

Redox reactions involve the loss and gain of electrons. By separating the oxidation process from the reduction process and connecting them electrically through an external circuit, many spontaneous redox reactions can be utilized to produce an electrical potential and an electrical current. Devices that perform these functions are called electrochemical cells. In this ChemLab, you will investigate a redox reaction and use it to construct an electrochemical cell.

Problem

How may a spontaneous redox reaction be used to construct an electrochemical cell?

Objectives

- **Observe** a simple oxidation-reduction reaction.
- **Relate** the reaction to the oxidation tendencies of the reactants.
- **Utilize** the reaction to construct an electrochemical cell that can operate electrical devices.

PREPARATION

Materials

craft-stick support with V-cut and slit cut
25-mm (flat diameter) dialysis tubing (15 cm in length)
magnesium ribbon (10 cm in length)
magnesium ribbon (1 cm in length)
copper foil (10 cm × 1 cm strip)
copper foil (1 cm × 2 mm piece)
metric ruler
250-mL beaker
10 × 100 mm test tubes (2)
wire leads with alligator clips (2)
DC voltmeter with a 2-V or 3-V scale

flashlight bulb for 2 AAA batteries
9-V transistor radio
0.5M sodium chloride solution
0.5M copper(II) chloride solution
0.1M magnesium chloride

Safety Precautions

Wear an apron and safety goggles. Rinse the solutions down the drain with large amounts of tap water. Wash your hands after performing the lab.

PROCEDURE

1. Soak the dialysis tubing in tap water for about ten minutes while you complete steps 2 and 3. Tie two knots near one end of the tubing, and open the other end by sliding the material between your fingers.

2. Pour a small amount of the copper(II) chloride solution into a 10 × 100 mm test tube, and drop a 1-cm length of magnesium ribbon into the solution. Observe the system for about one minute, and record your observations in a data table similar to the table in Data and Observations. Pour the solution down the drain and discard the piece of magnesium in a wastebasket.

3. Repeat step 2 using magnesium chloride solution and the small piece of copper foil.

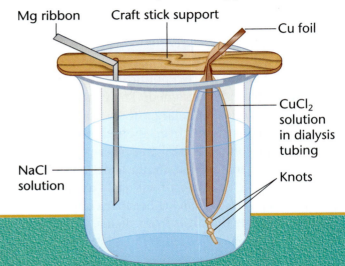

Mg ribbon   Craft stick support   Cu foil
CuCl₂ solution in dialysis tubing
NaCl solution
Knots

4. Pour copper(II) chloride solution into the open end of the tubing to a depth of 6 cm to 8 cm, and insert the strip of copper foil. Slide the top of the tubing and copper strip into the V-cut in the stick as shown below. Suspend the tubing in the beaker, as shown.

5. Slide the length of magnesium ribbon into the slit cut in the craft stick, as shown.

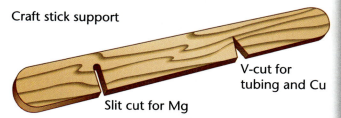

Craft stick support

Slit cut for Mg

V-cut for tubing and Cu

6. Pour about 200 mL of sodium chloride solution into the beaker.

7. Connect leads to the pieces of copper and magnesium, and touch the leads to a DC voltmeter—the lead from the Cu electrode to the + terminal and the lead from the Mg electrode to the − terminal. Read and record the potential difference, or voltage.

8. Cooperate with other lab groups in the following way to light the bulb and to operate the transistor radio. The flashlight bulb requires a voltage of about 3V, and the radio requires a voltage of about 9V. Connect your electrochemical cells in series (copper to magnesium) to provide the desired voltages. **CAUTION:** *Be sure to connect your cells to the battery terminals of the radio in the correct polarity.* Connected in series, the voltages are additive; for example, five 2-V cells in series yield a voltage of 10V. Such combinations of electrochemical cells are called batteries.

9. Disassemble your cell, observing the pieces of copper and magnesium and recording your observations. Rinse the pieces of copper and

magnesium, and dispose of them according to the instructions of your teacher.

DATA AND OBSERVATIONS

	Data and Observations
Mg − Cu^{2+}	
Cu − Mg^{2+}	
Voltage	
Pieces of Cu and Mg	

ANALYZE AND CONCLUDE

1. **Interpreting Data** Write the balanced equation for the single-replacement reaction between magnesium and copper(II) chloride that occurred in step 2. Which metallic element, Cu or Mg, has the greater tendency (or oxidation potential) to lose electrons?

2. **Relating Concepts** In an electrochemical cell, oxidation occurs at the anode, and reduction occurs at the cathode. Which metal was the anode and which was the cathode? Write the equations for the half-reactions.

APPLY AND ASSESS

1. When an electrochemical cell is used to operate an electrical device, in which direction do the electrons move in the external circuit?

2. Is it possible to construct an electrochemical cell in which lead is the anode and lithium is the cathode? Explain.

Although the term *battery* usually refers to a series of galvanic cells connected together, some batteries have only one such cell. Other batteries may have a dozen or more cells. When you put a battery into a flashlight, radio, or CD player, you complete the electrical circuit of the galvanic cell(s), providing a path for the electrons to flow through as they move from the reducing agent (the site of oxidation) to the oxidizing agent (the site of reduction). The most powerful batteries combine strong oxidizing agents and strong reducing agents to give the largest possible potential difference. But those agents aren't necessarily safe, convenient, or economical to use. To get a higher voltage from a cell type with a relatively small potential difference, several of the cells can be connected in series, as **Figure 17.15** shows.

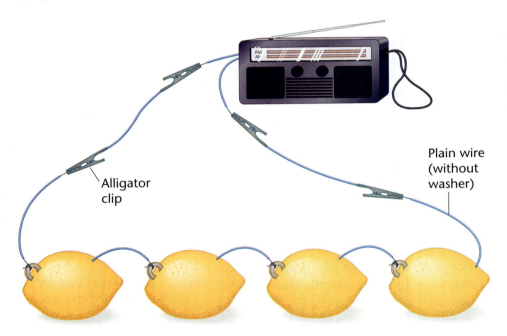

Alligator clip

Plain wire (without washer)

Figure 17.15

A More Powerful Lemon Battery
One lemon cell wouldn't provide enough voltage to power a transistor radio, but several connected together in series would. This means that the positive terminal of one cell is connected to the negative terminal of the next, and so on. The electrodes in this battery are steel washers and copper wire. The total voltage of the battery is the sum of all the voltages of the individual cells.

Carbon-Zinc Dry Cell

Whenever you put two or more common D batteries into a flashlight, you are connecting them in series. They have to be placed in the correct order so that electrons flow through both cells. These relatively inexpensive batteries are carbon-zinc galvanic cells, and they come in several types, including standard, heavy-duty, and alkaline. This type of battery is often called a dry cell because there is no aqueous electrolyte solution; a semisolid paste serves that role. Examine the cutaway view of the carbon-zinc battery in **Figure 17.16** to see if you can locate the parts of the galvanic cell it contains.

Figure 17.16

Carbon-Zinc Dry Cell

A standard D battery is shown both whole and cut in half to reveal the structure of the carbon-zinc dry cell. Beneath the outside paper cover of the battery is a cylinder casing made of zinc. The zinc serves as the anode and will be oxidized in the redox reaction. ▼

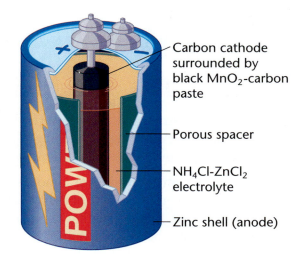

Carbon cathode surrounded by black MnO_2-carbon paste

Porous spacer

NH_4Cl-$ZnCl_2$ electrolyte

Zinc shell (anode)

▲ The carbon rod in the center of the cylinder—surrounded by a moist, black paste of manganese(IV) oxide (MnO_2) and carbon black—acts as a cathode. Ammonium chloride (NH_4Cl) and zinc chloride ($ZnCl_2$) serve as electrolytes. Alkaline batteries contain potassium hydroxide (KOH) in place of the ammonium chloride electrolyte, and they maintain a high voltage for a longer period of time.

What is missing in this galvanic cell? Notice that the circuit is not complete, so the electrons that are produced at the zinc cylinder have no external conductor through which to travel to the carbon. This is by design and is not a defect in the battery. The circuit will be complete when the battery is placed in something designed to be powered by it, such as a flashlight. When the flashlight is turned on, the redox reaction starts. Electrons travel out of the zinc casing into a piece of metal built into the flashlight. There, they travel through a bulb, causing it to light up. The electrons then reenter the battery at the top and move down through the carbon rod and into the black paste, where they take part in the reduction reaction.

The flow of electrons from the zinc cylinder through the electrical circuits of an appliance and back into the battery provides the electricity needed to power a flashlight, radio, CD player, toy, clock, or other item. When electrons leave the casing, zinc metal is oxidized.

$$Zn \rightarrow Zn^{2+} + 2e^-$$

The reactions in the carbon rod and the paste are much more complex, but one major reduction that takes place is that of manganese in manganese(IV) oxide. In this reaction, the oxidation number of manganese is reduced from 4+ to 3+.

$$2MnO_2 + H_2O + 2e^- \rightarrow Mn_2O_3 + 2OH^-$$

Adding the two half-reactions together gives the major redox reaction taking place in a carbon-zinc dry cell.

$$Zn + 2MnO_2 + H_2O \rightarrow Zn^{2+} + Mn_2O_3 + 2OH^-$$

Lithium Batteries in Pacemakers

It's always frustrating to have batteries go dead just when it seems you need them most. However, imagine needing a battery upon which your life depends. What materials could it be made from? Would it last long enough?

Heart stimulation Consider for a moment that, over a period of time, you began to feel light-headed, dizzy, weak, or fatigued. It could be that the chambers of your heart are not beating rhythmically or fast enough. You might be a candidate for a heart pacemaker. This device, which is inserted inside the body, monitors the heart's activity. When necessary, the pacemaker supplies the electrical impulses needed to stimulate the heart. In order to be most effective, the batteries in a pacemaker need to be fully powered for long periods of time and survive in the hostile, saline environment of the human body without breaking down.

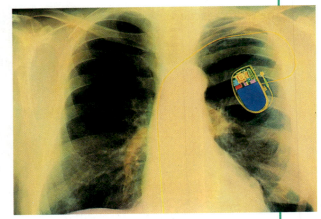

Powerful electrodes One type of battery used in pacemakers is a cell based on lithium and sulfuryl chloride. Lithium is a popular choice for battery anodes because of its strong tendency to be oxidized. Lithium is oxidized during the reaction, and the sulfur in sulfuryl chloride is reduced. The unbalanced half-reactions are given below.

$$Li(s) \rightarrow Li^+(aq) + e^- \qquad \text{oxidation}$$
$$SOCl_2(l) + 4e^- \rightarrow Cl^-(aq) + S(s) + SO_2(g) \quad \text{reduction}$$

Battery characteristics Lithium is the least dense of all nongaseous elements, with a density of only 0.534 g/mL. The lightweight lithium contributes little weight to the small, detachable battery pack, which is a circular disk only about 5 cm by 1 cm in size. The pacemaker battery pack is implanted under the patient's collarbone and has a life expectancy of about seven to ten years, depending upon how often it is needed to stimulate the heart.

Lithium hazards Lithium also presents some potential problems when used in a wet environment. The tremendous activity of lithium makes it dangerously reactive with a variety of compounds, including water. If the battery pack is not adequately sealed against the body's corrosive fluids, the resulting exothermic reaction of lithium in water accompanied by the production of H_2 gas could have serious consequences.

As battery technology continues to advance, medical applications requiring portable, long-lasting power supplies will become more and more common.

Connecting to Chemistry

1. **Interpreting** Write the balanced overall reaction of lithium with sulfuryl chloride.

2. **Interpreting** Write the balanced overall reaction of lithium with water. What is the oxidizing agent? The reducing agent?

Automobile Lead Storage Battery

The most common type of battery used in cars is a lead-acid, 12-volt storage battery. It contains six 2-volt cells connected in series. Although much larger than carbon-zinc batteries and relatively heavy, this type of battery is durable, supplies a large current, and can be recharged. When you turn your key in the ignition, it is the battery that supplies electricity to start the car. It also provides energy for any demands not met by the car's alternator, such as running the radio or using the lights when the engine is off. Leaving on the lights or radio for too long with the engine off can make the battery go dead because it is the engine that recharges the battery as the car runs.

Each galvanic cell in a lead-acid battery has two electrodes—one made of a lead(IV) oxide (PbO_2) plate and the other of spongy lead metal, as **Figure 17.17** shows. In each cell, lead metal is oxidized as lead(IV) oxide is reduced. The lead metal is oxidized to Pb^{2+} ions as it releases two electrons at the anode. The Pb^{4+} ions in lead oxide gain two electrons, forming Pb^{2+} ions at the cathode. The Pb^{2+} ions combine with SO_4^{2-} ions from the dissociated sulfuric acid in the electrolyte solution to form lead(II) sulfate at each electrode. Thus, the net reaction that takes place when a lead-acid battery is discharged results in the formation of lead sulfate at both of the electrodes.

$$PbO_2 + Pb + 2H_2SO_4 \rightarrow 2PbSO_4 + 2H_2O$$

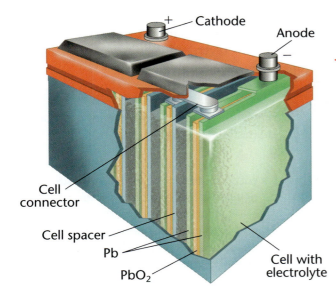

Cathode
Anode
Cell connector
Cell spacer
Pb
PbO_2
Cell with electrolyte

Figure 17.17

Lead Storage Battery

◀ A lead storage battery is not a dry cell because it contains several connected cells filled with an aqueous solution of sulfuric acid, which serves as the electrolyte. The electrodes are alternating plates of lead metal and lead(IV) oxide. The case surrounding the battery is hard plastic. It holds the cells in place and acts as an insulator because it does not conduct electricity itself. This helps keep the electricity inside the battery.

A car with a dead battery can still be started. Electricity from a second car is used to jump-start the car, bypassing the dead battery. ▶

Nicad Rechargeable Batteries

The nickel-cadmium, or nicad, cell is a common storage battery that can usually be discharged and recharged more than 500 times. These batteries are used in calculators, cordless power tools and vacuum cleaners, and rechargeable electric toothbrushes and shavers. Once nicad batteries have been spent, disposal presents a problem because cadmium is toxic. Nicads can be recycled, but the process is expensive. Although rechargeable batteries containing less toxic metals are being developed, none have been found that can sustain a constant rate of discharge as well as the nicad.

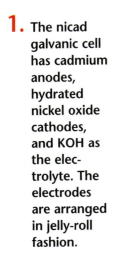

1. The nicad galvanic cell has cadmium anodes, hydrated nickel oxide cathodes, and KOH as the electrolyte. The electrodes are arranged in jelly-roll fashion.

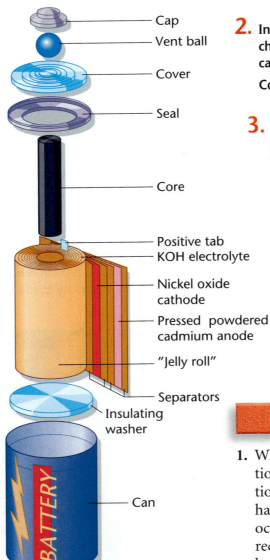

- Cap
- Vent ball
- Cover
- Seal
- Core
- Positive tab
- KOH electrolyte
- Nickel oxide cathode
- Pressed powdered cadmium anode
- "Jelly roll"
- Separators
- Insulating washer
- Can

2. In the redox reaction that takes place during discharge, nickel oxide is reduced at the cathode, and cadmium is oxidized at the anode.

$$Cd + NiO \rightarrow CdO + Ni$$

3. The electrolysis reaction that takes place when an external source of electricity is used to recharge the cell is the reverse of the discharge reaction.

$$CdO + Ni \rightarrow Cd + NiO$$

4. Newly purchased nicad batteries must be charged before use.

5. Nicad batteries are not suitable for devices that are left idle for long stretches—such as smoke detectors, cameras, and flashlights—because they will lose about one percent of their charge daily even when not being used.

Thinking Critically

1. What are the equations for the oxidation and reduction half-reactions that occur during recharging of a nicad battery?

2. What might be an environmental advantage to using nicad batteries?

The reaction that occurs during discharge of a lead-acid battery is spontaneous and requires no energy input. The reverse reaction, which recharges the battery, is not spontaneous and requires an input of electricity from the car's alternator. Current enters the battery and provides energy for the reaction in which lead sulfate and water are converted into lead(IV) oxide, lead metal, and sulfuric acid.

$$2PbSO_4 + 2H_2O \rightarrow PbO_2 + Pb + 2H_2SO_4$$

Sulfuric acid is corrosive. It is important to be careful when working around a car battery, as well as disposing of it properly when it finally goes dead for good. These batteries can usually be discharged and recharged numerous times and last about three to five years.

Better Batteries for Electric Cars

At the end of the 19th century, most cars were powered by steam or by electric batteries, although today most cars have internal combustion engines that are powered by gasoline. Electric cars, **Figure 17.18,** could help reduce our dependence on fossil fuels, cause less pollution, and be more economical in the long run, but they have several disadvantages, such as high initial cost, limited driving range, low speed, and long recharge time. They also present a disposal problem because cadmium is a toxic metal.

These disadvantages would disappear if a battery that is cheap enough, powerful enough, and safe enough for running an electric car could be developed. Two new experimental types of batteries for use in electric cars show early promise as candidates. One is a rechargeable, nickel-metal hydride or NiMH battery. This type of battery is less toxic and has a higher storage capacity than the batteries now used in electric cars. Another experimental battery is a lithium battery with a water-based electrolyte. Lithium is more easily oxidized than any other metal but has a drawback that has limited its use in batteries: it explodes violently when it comes into contact with water. Lithium is used in some batteries

Figure 17.18

Electric Cars
Most of the nickel-cadmium batteries that are used in electric cars today power the car for only 50 to 100 miles before they run down and need to be recharged, a process that takes many hours. In Randers, Denmark, special parking spaces with electric hookups are available for battery-operated cars.

Hydrogen-Oxygen Fuel Cell

Recall that the combustion of a fuel is a redox reaction in which the fuel molecules are oxidized and oxygen is reduced to form an oxide. For years, scientists have worked to find a way to separate the oxidation and reduction reactions to make them produce an electric current. The simplest fuel cell involves the oxidation of the fuel hydrogen gas to form water. Today, hydrogen-oxygen fuel cells are used to supply electricity to the space shuttle orbiters. The fuel cells have a weight advantage over storage batteries, and the water produced during their operation can be used for drinking.

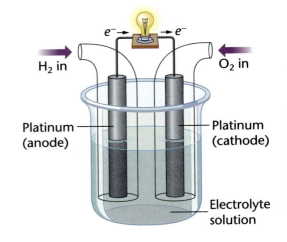

1. A simple hydrogen-oxygen fuel cell differs in two major ways from a galvanic cell: the electrodes are made of an inert material that doesn't react during the process, and hydrogen and oxygen gas are fed in continuously.

2. Hydrogen is fed onto an electrode on one side of the fuel cell, and oxygen is fed onto an electrode on the other side.

3. Concentrated KOH serves as the electrolyte in the fuel cell.

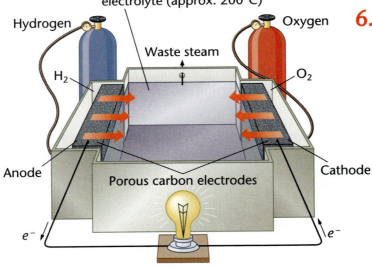

4. The electrons lost by hydrogen molecules, which are oxidized at the anode, flow out of the fuel cell, through a circuit, and then back into the fuel cell at the cathode, where oxygen is reduced.

5. Water vapor—steam—is produced in the fuel cell, as up to 75 percent of the chemical energy is converted into electricity. The steam can be condensed and used for drinking water.

$$2H_2(g) + O_2(g) \rightarrow 2H_2O(g) + energy$$

6. If more inexpensive and longer-lasting fuel cells can be developed, they may someday produce electricity in power plants.

Thinking Critically

1. What causes electrons to flow from hydrogen to oxygen in a fuel cell?

2. If fuel cells are about 75 percent efficient, what happens to the rest of the potential energy?

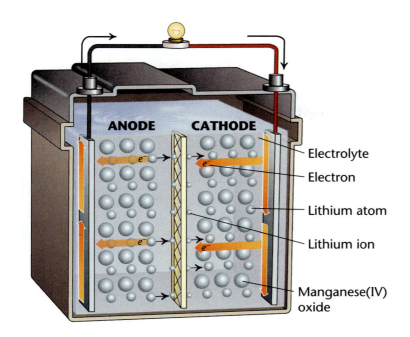

ANODE   CATHODE

Electrolyte

Electron

Lithium atom

Lithium ion

Manganese(IV) oxide

Figure 17.19
Aqueous Lithium Battery

Figure 17.19
Aqueous Lithium Battery
How can a lithium battery have an aqueous electrolyte? Two facets of the construction of this new battery keep the lithium metal from reacting with water. First, the lithium is in the form of individual atoms embedded in a material such as manganese(IV) oxide, rather than as a solid metal. Second, the electrolyte is full of dissolved lithium salts, so the lithium ions that are produced travel to the site of reduction without reacting with water.

to power camcorders, but they require an expensive, nonaqueous electrolyte. **Figure 17.19** shows the construction of the experimental aqueous lithium battery. This battery is less toxic and will probably be cheaper to manufacture than the nickel-cadmium batteries used in most electric cars in operation today.

Connecting Ideas

In your study of electrochemistry, you have seen how chemical reactions in batteries can be used to generate electricity. However, most of the electricity you use comes from another chemical source—the fossil fuels petroleum, natural gas, and coal. Although inorganic chemicals are usually used to fuel batteries, fossil fuels are a major source of a large group of chemicals, the carbon-containing organic compounds. In the next chapter, you'll learn that organic chemicals also provide us with most medicines, dyes, plastics, and textiles.

Supplemental Problems

For more practice with solving problems, see Supplemental Practice Problems, Appendix B.

SECTION REVIEW

Understanding Concepts

1. Describe the movement of electrons in a galvanic cell.

2. Draw a diagram of a simple galvanic cell.

3. How are zinc-carbon and lead-acid batteries different?

Thinking Critically

4. **Using a Table** A piece of copper metal is placed in a $1M$ solution of silver nitrate ($AgNO_3$).

a) Use **Table 17.1** to predict which metal will be reduced and which will be oxidized.
b) Write an equation for the net redox reaction that occurs. HINT: Cu^{2+} is formed.
c) Is this system a galvanic cell? Explain.

Applying Chemistry

5. **Dry Cells** A dry cell cannot really be dry. Explain why.

REVIEWING MAIN IDEAS

17.1 Electrolysis: Chemistry from Electricity

- An electrolytic cell is a chemical system that uses an electric current to drive a non-spontaneous redox reaction. Electrolysis is the process that takes place in such a cell.

- An electrical current is the flow of charged particles such as electrons.

- Reduction takes place at the cathode in an electrolytic cell.

- Oxidation takes place at the anode in an electrolytic cell.

- Electrolysis can be used to produce compounds, separate metals from ores, clean metal objects, and plate metal coatings onto objects.

17.2 Galvanic Cells: Electricity from Chemistry

- A potential difference between two substances is a measure of the tendency of electrons to flow from one to the other.

- A galvanic cell is a chemical system that produces an electric current through a spontaneous redox reaction.

- Batteries contain one or more galvanic cells.

Vocabulary

For each of the following terms, write a sentence that shows your understanding of its meaning.

anion
anode
cathode
cation
electryca current

electrolysis
electrolytic cell
galvanic cell
potential difference
voltage

UNDERSTANDING CONCEPTS

1. What is the function of the salt bridge in a galvanic cell?

2. What is the difference between an electrolytic cell and a galvanic cell?

3. What is a galvanic cell?

4. What happens to the case of a carbon-zinc dry cell as the cell is used to produce an electric current?

5. Why is the electrolyte necessary in both galvanic and electrolytic cells?

6. What products are formed from the electrolysis of an aqueous solution of rock salt?

7. What is the function of the acid in the lead-acid storage battery used in cars?

8. By what process can chlorine gas be prepared commercially?

APPLYING CONCEPTS

9. What will happen if a rod made of aluminum is used to stir a solution of iron(II) nitrate?

10. Can a solution of copper(II) sulfate be stored in a container made of nickel metal? Explain.

11. How would gold-electroplated jewelry compare to jewelry made of solid gold in terms of price, appearance, and durability?

12. Tests conducted on different types of common commercial batteries involved measuring the voltage drop over time during simulated non-stop use of a motorized toy. Based on the following graph of data obtained from testing rechargeable, alkaline, and heavy-duty batteries, which battery type would be best to use if you wanted to run the toy for a long period of time? Which battery type goes dead abruptly?

chemistryca.com/vocabulary_puzzlemaker

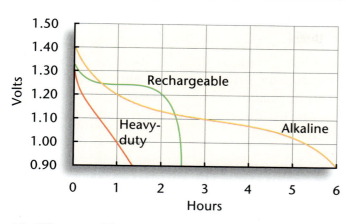

13. What would you expect to see if you placed

   a) a strip of copper metal into a solution of zinc sulfate?

   b) a strip of zinc metal into a solution of copper sulfate?

14. How could lead be removed from drinking water by electrolysis?

Everyday Chemistry

15. The information on a CD stamper master is the reversal of that on the original glass master cut by the recording laser. Explain why this reversal is necessary.

Chemistry and Technology

16. Should the anode or the cathode be made of pure copper in an electrolytic cell designed for refining copper metal? Explain.

Health Connection

17. What happens to lithium metal if it comes into contact with water?

How it Works

18. What are the advantages and disadvantages of using rechargeable batteries instead of conventional types?

How it Works

19. Write the equations for the two half-reactions that take place in a hydrogen-oxygen fuel cell.

THINKING CRITICALLY

Making Predictions

20. **MiniLab 1** Would it be possible to plate a silver spoon or a gold spoon with copper?

Drawing Conclusions

21. **MiniLab 2** Why should the metal pieces used as electrodes in the lemon battery be cleaned with steel wool?

Relating Cause and Effect

22. **ChemLab** How would your result in step 7 of the ChemLab have been different if

   a) a piece of zinc were used instead of the piece of magnesium?

   b) silver were used instead of copper?

Making Decisions

23. What factors must be considered in designing or selecting batteries for the following applications?

   a) flashlight   c) pacemaker
   b) hearing aid   d) toy car

Making Predictions

24. If a strong tendency to be oxidized were the only consideration, what metals other than lithium might be used to power a cardiac pacemaker?

Forming a Hypothesis

25. Why was electrophoresis rather than electrolysis used to restore ceramic and organic artifacts from the *Titanic?*

CUMULATIVE REVIEW

26. Draw Lewis electron dot structures for the ions listed. (Chapter 2)

   a) Ca^{2+} b) Cl^- c) OH^- d) O^{2-}

27. List the names and symbols of all of the noble gases. (Chapter 8)

28. Which requires more energy: boiling 100 g of water or melting 100 g of ice? Explain. (Chapter 10)

29. A 0.543-g piece of magnesium reacts with excess oxygen to form magnesium oxide in the reaction

$$2Mg(s) + O_2(g) \rightarrow 2MgO(s).$$

How much oxygen reacts? What is the mass of magnesium oxide, MgO, produced? (Chapter 12)

30. What are the mass percents of magnesium and oxygen in magnesium oxide in the problem above? (Chapter 12)

31. Compare the hydronium ion concentrations in two aqueous solutions that have pH values of 9 and 11. (Chapter 14)

SKILL REVIEW

32. Interpreting Scientific Illustrations A process called cathodic protection is sometimes used to protect a buried steel pipeline from corrosion. In this process, the pipeline is connected to a more active metal such as magnesium, which is corroded preferentially before the iron. The diagram below illustrates how the two metals are connected and shows the reactions that take place.

   a) What acts as the cathode in this process? What acts as the anode?
   b) What is the oxidizing agent?
   c) Write a short summary describing how the magnesium is preferentially corroded.

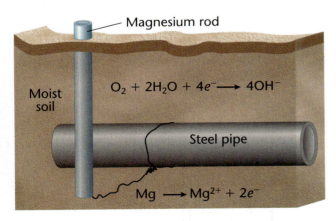

Magnesium rod

Moist soil

$$O_2 + 2H_2O + 4e^- \longrightarrow 4OH^-$$

Steel pipe

$$Mg \longrightarrow Mg^{2+} + 2e^-$$

33. Making Scientific Illustrations Draw a diagram of a galvanic cell in which the reaction is $Ni(s) + 2Ag^+(aq) \rightarrow Ni^{2+}(aq) + 2Ag(s)$. Label the cathode and the anode. Show the ions present in both compartments, and indicate the direction of electron flow in the external circuit.

WRITING IN CHEMISTRY

34. Write an article about the development and uses for the Daniell cell, an early battery made in 1836 by John Frederic Daniell of Great Britain. Find out how it improved on the Volta cell and whether or not this type of battery is used much today.

PROBLEM SOLVING

35. How many grams of aluminum oxide are needed to produce 10 million metric tons (1 metric ton = 1000 kg) of aluminum each year using the Hall-Héroult process? Assume 100 percent yield.

36. In one type of fuel cell, methane gas (CH_4) is "burned" electrochemically to produce electricity:
$$CH_4(g) + H_2O(g) \rightarrow CO(g) + 3H_2(g) + energy$$

   a) Is this a galvanic or an electrolytic cell?
   b) What acts as the oxidizing reagent?
   c) What acts as the reducing reagent?
   d) If 224 L of natural gas are burned in this fuel cell at STP, how many moles of carbon monoxide and hydrogen gases are produced?

37. What will happen to your gold ring if you leave it sitting in a solution of iron(II) chloride ($FeCl_2$) at room temperature?

38. What will happen to a copper bracelet that remains in contact with a solution of silver nitrate ($AgNO_3$) for several hours at room temperature?

1. Electrolysis is
   a) the process of using electrical currents to speed up redox reactions.
   b) the process of using electrical currents to speed up any reaction.
   c) the process of using electrical currents to start reactions that do not occur.
   d) the process of using electrical currents to stop reaction from occurring.

2. The part of an electrochemical cell that carries electrons to a reacting ion is a(n)
   a) cathode.   c) electrode.
   b) anode.   d) electrolyte.

3. Which of the following processes is an application of electrolysis?
   a) electroplating
   b) cleaning toxic wastes
   c) separating metals from ores
   d) all of the above

4. Why does electric current flow between the two electrodes of a lemon battery?
   a) The two electrodes have a potential difference maintaining a flow of electrons from an anode to a cathode.
   b) The two electrodes have a potential difference maintaining a flow of electrons from a cathode to an anode.
   c) A battery attached to the two electrodes creates a potential equivalence between the two different metals.
   d) A battery attached to the two electrodes creates a potential difference between the two different metals.

5. An electrical potential difference is also called
   a) oxidation.   c) corrosion.
   b) reduction.   d) voltage.

Use the table to answer questions 6 and 7.

6. The table of standard reduction potential of substances lists common metals in order of increasing reduction potential and decreas-

Standard Reduction Potentials	
Metal	Standard Reduction Potential
Li	−3.0401
Al	−1.662
Cu	0.521
Ag	0.7996
Au	1.498

ing oxidation potential. Which metal will most easily corrode?
   a) Li   c) Cu
   b) Al   d) Au

7. What would happen if a lemon battery were constructed using a copper strip and a gold strip?
   a) Electrons would flow from the copper strip, where reduction would occur, to the gold strip, where oxidation would occur.
   b) Electrons would flow from the copper strip, where oxidation would occur, to the gold strip, where reduction would occur.
   c) Electrons would flow from the gold strip, where reduction would occur, to the copper strip, where reduction would occur.
   d) Electrons would flow from the gold strip, where oxidation would occur, to the copperstrip, where reduction would occur.

8. A Galvanic cell packaged and sold as a portable power source is called a(n)
   a) voltaic cell.
   b) electrode.
   c) battery.
   d) electrochemical cell.

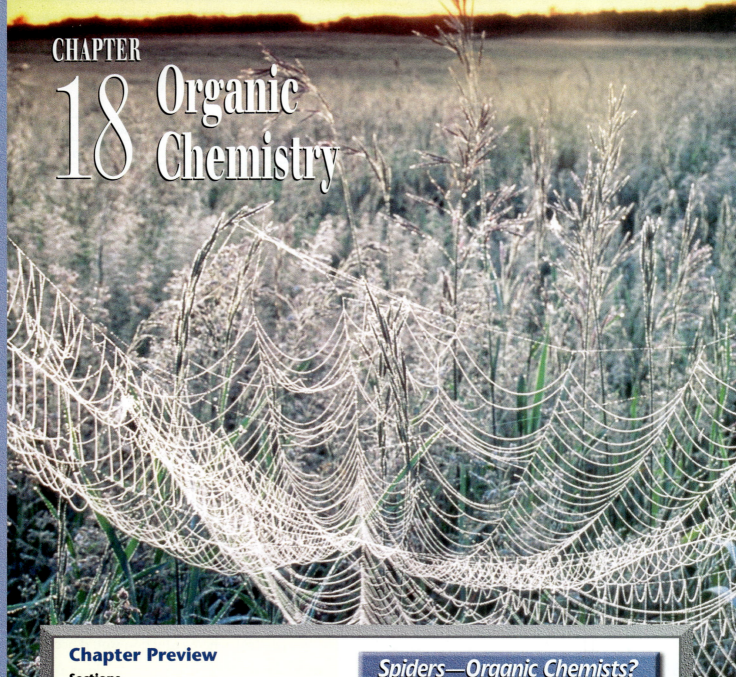

Chapter Preview

Sections

18.1 Hydrocarbons

MiniLab 18.1 How Unsaturated is
Your Oil?

18.2 Substituted Hydrocarbons

MiniLab 18.2 A Synthetic Aroma

18.3 Plastics and Other Polymers

MiniLab 18.3 When Polymers Meet
Water

ChemLab Identification of
Textile Polymers

Spiders—Organic Chemists?

Spiders produce a silk composed of a protein called fibroin. Spider silk is an organic compound that is composed of amino acids. This compound is the same as the one produced by silkworms. The silk is very strong, elastic, and lightweight.

Silk is one of over 10 million different organic compounds that contain carbon. Organic compounds are found in many products you use every day.

Start-up Activities

*L*aunch Lab

Making Slime

In addition to carbon and hydrogen, most organic substances contain other elements that give the substances unique properties. In this lab, you will work with an organic substance consisting of long carbon chains to which many —OH groups are bonded. How will the properties of this substance change when these groups react to form bonds called crosslinks between the chains?

Safety Precautions

Do not allow solutions or product to contact eyes or exposed skin.

Materials

- 4% sodium tetraborate
- (borax) solution
- 4% polyvinyl alcohol solution
- disposable plastic cup
- stirring rod

Procedure

1. Pour 20 mL of 4% polyvinyl alcohol solution into a small disposable plastic cup. Note the viscosity of the solution as you stir it.
2. While stirring, add 6 mL of 4% sodium tetraborate solution to the polyvinyl alcohol solution. Continue to stir until there is no further change in the consistency of the product.
3. Use your gloved hand to scoop the material out of the cup. Knead the polymer into a ball.

Analysis

What physical property of the product differs markedly from those of the reactants?

What I Already Know

Review the following concepts before studying this chapter.

Chapter 4: bonding in covalent compounds

Chapter 5: naming of hydrocarbons

Chapter 6: combustion of hydrocarbons

Chapter 9: bonding in hydrocarbons

Reading Chemistry

Observe the chemical structures that appear throughout the chapter. Note which elements are present in each. Next, look at the key terms for each section. As you read, match up the key terms with the chemical structure examples.

Preview this chapter's content and activities at **chemistryca.com**

Hydrocarbons

Have you ever seen a spectrum of colors on the top of an otherwise-drab puddle of water? What causes those bright colors to appear? They form as a result of pollution, namely small amounts of gasoline or oil that have leaked out of automobiles and formed a miniature oil spill on the puddle.

Gasoline is composed mostly of hydrocarbons, the organic compounds with the names and structures you studied in Chapters 5 and 9. Hydrocarbons have properties that are different from those of water. They are insoluble in water, which is why they form a distinct layer on a puddle of water. Hydrocarbons are less dense than water; that's why the layer floats on top of the puddle. What causes the colorful effect? The spill forms an

extremely thin layer of hydrocarbon molecules on the water, which reflects sunlight. In this section, you will learn more about the structures and names of hydrocarbons, as well as the sources of these useful compounds.

Millions and Millions of Organic Compounds

Carbon is unique among elements in that it can bond to other carbon atoms to form chains containing as many as several thousand atoms. Because a carbon atom can bond to as many as four other atoms at once, these chains can have branches and form closed-ring structures that make possible an almost endless variety of compounds. In addition, carbon can bond strongly to elements such as oxygen and nitrogen, and it can form double and triple bonds. Thus, carbon forms an enormous number of compounds with chains and rings of various sizes, each with a variety of bond types and atoms of other elements bonded to them. Fortunately, you don't need to study each of these millions of compounds to understand organic chemistry because they can be classified into groups of compounds that have similar structures and properties.

Saturated Hydrocarbons

Gasoline is a mixture of organic compounds that is derived from petroleum. Most of the compounds in gasoline are hydrocarbons. You will recall learning that hydrocarbons are organic compounds containing only hydrogen and carbon atoms. A hydrocarbon in which all the carbon atoms are connected to each other by single bonds is called a **saturated hydrocarbon.** Another name for a saturated hydrocarbon is an **alkane.** Although burning alkanes for fuel is their most common use, they are also used as solvents in paint removers, glues, and other products.

Alkanes

Alkanes are the simplest hydrocarbons. The carbons in an alkane can be arranged in a chain or a ring, and both chains and rings can have branches of other carbon chains attached to them. Alkanes that have no branches are called straight-chain alkanes. Methane, CH_4; ethane, C_2H_6; propane, C_3H_8; and butane, C_4H_{10} are all common fuels. Their structural formulas show that each differs from the next by an increment of $-CH_2-$.

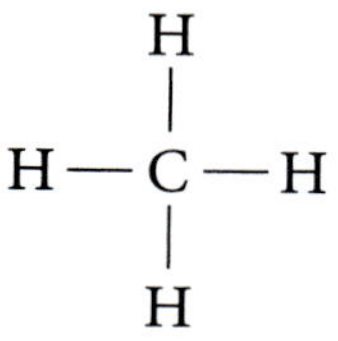

Methane 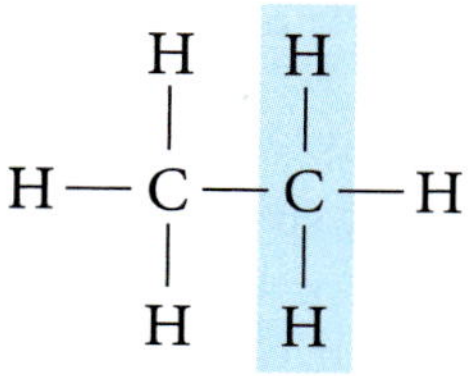   Ethane 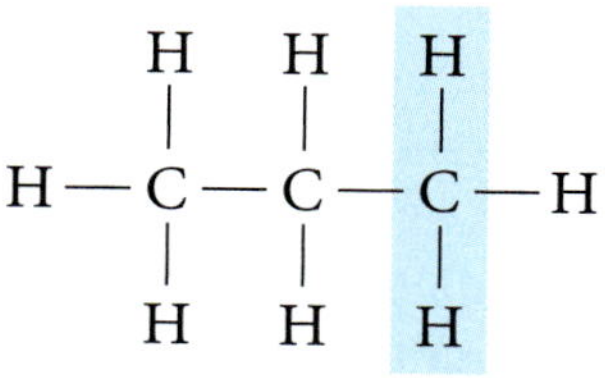   Propane 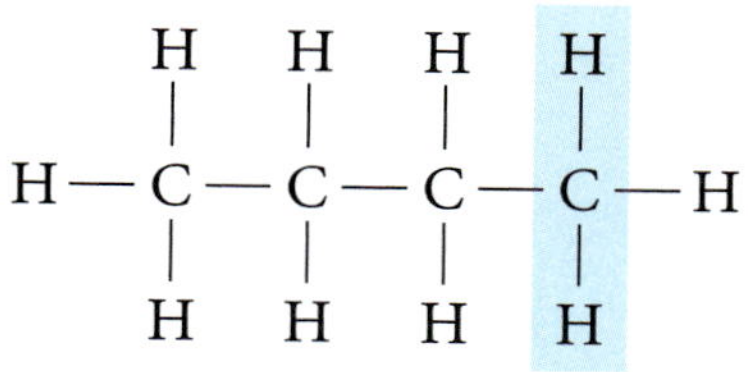   Butane

Some alkanes have a branched structure. In these compounds, a chain of one or more carbons is attached to a carbon in the longest continuous chain, which is called the parent chain. If a chain containing a single carbon is branched off the second carbon of a propane parent chain, a branched alkane with the following structure results.

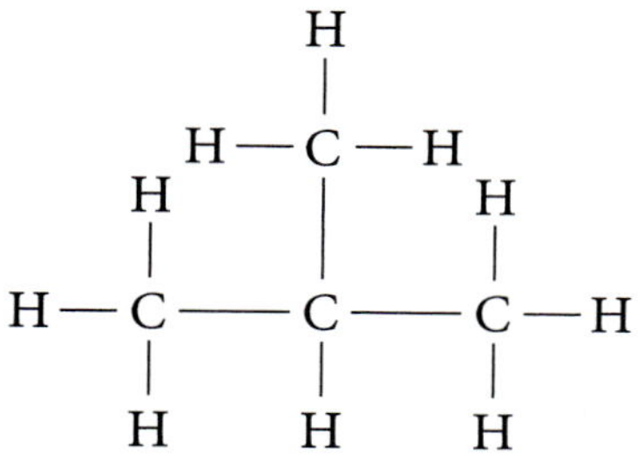

2-methylpropane

The carbon atoms in alkanes can also link up to form closed rings. The most common rings contain five or six carbons. The structures of these compounds can be drawn showing all carbon and hydrogen atoms.

Cyclopentane   Cyclohexane

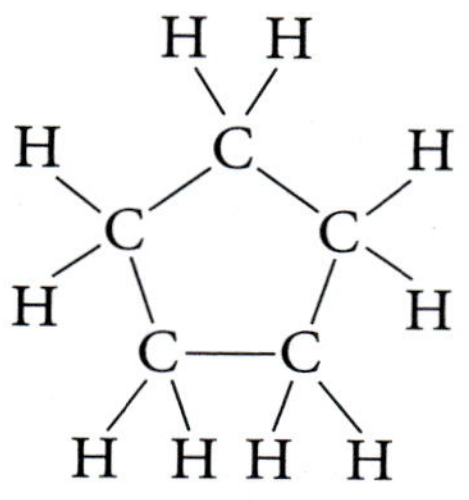

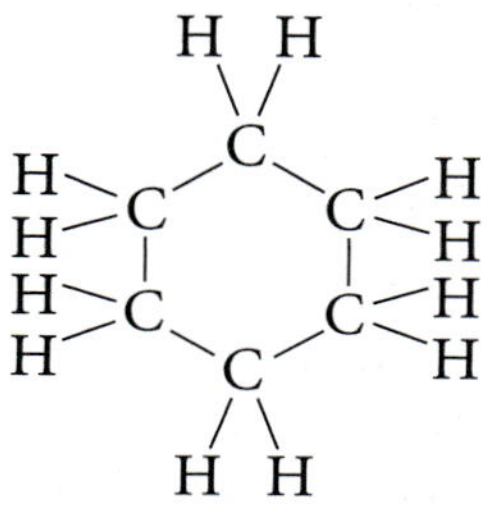

Structural diagrams can be simplified by using straight lines to represent the bonds between atoms in the rings. In these ring diagrams, each corner represents a carbon atom. Because carbon usually forms four bonds, it is understood that enough hydrogen atoms are bonded to each carbon to give it four bonds. Rings containing five and six carbons are drawn as a pentagon and a hexagon, respectively.

Cyclopentane

Cyclohexane

Table 18.1 The First Ten Alkanes	
Formula	**Name**
CH_4	methane
C_2H_6	ethane
C_3H_8	propane
C_4H_{10}	butane
C_5H_{12}	pentane
C_6H_{14}	hexane
C_7H_{16}	heptane
C_8H_{18}	octane
C_9H_{20}	nonane
$C_{10}H_{22}$	decane

Structural diagrams of straight- and branched-chain hydrocarbons also can be written in a simplified way by leaving out some of the bonds. For example, the formula for propane can be written as $CH_3—CH_2—CH_3$. In this type of shorthand structure, the bonds between C and H are understood. In an even more simplified type of shorthand, the condensed structural formula for propane can be written as $CH_3CH_2CH_3$. Here, the bonds both between C and C and between C and H are understood. These condensed structural formulas will be used throughout this chapter.

Figure 18.1 shows the names and structures of some saturated hydrocarbons found in typical gasoline mixtures. The properties of a gasoline mixture—such as how well it burns in an engine—are determined by the relative amounts of various components present in the mixture. Hydrocarbons that contain rings and many branches burn at a more uniform rate than do straight-chain alkanes, which tend to explode prematurely, causing engine knock. The octane number on a gasoline pump is an index that indicates the relative amounts of branched and ring structures in that blend of gasoline.

Figure 18.1

Components of Gasoline and Octane Rating

Gasolines are rated on a scale known as octane rating, which is based on the way they burn in an engine. The higher the octane rating, the greater the percentage of complex-structured hydrocarbons that are present in the mixture, the more uniformly the gasoline burns, and the less knocking there is in the automobile engine. Thus, a gasoline rated 92 octane will burn more smoothly than one rated 87 octane.

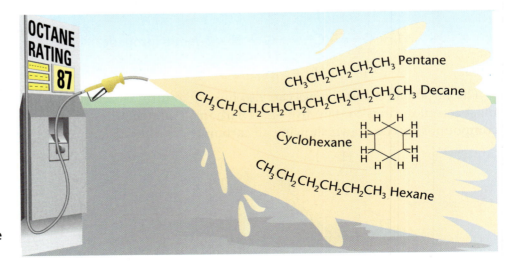

Naming Alkanes

The names of the first ten straight-chain alkanes, shown in **Table 18.1**, are used as the basis for naming most organic compounds. To name a branched alkane, you must be able to answer three questions about its structure.

1. How many carbons are in the longest continuous chain of the molecule?
2. How many branches are on the longest chain and what is their size?
3. To which carbons in the longest chain are the branches attached?

For convenience, the carbon atoms in organic compounds are given position numbers. In straight-chain hydrocarbons, the numbering can begin at either end. It makes no difference. In branched hydrocarbons, the numbering begins at the end closest to the branch.

Examine the structure of this branched alkane.

$$CH_3$$
$$|$$
$$^1CH_3^2CH^3CH_2^4CH_3$$

Four carbons are in the longest continuous chain, so butane is the parent chain and will be part of the compound's name. There is only one branch, and it contains one carbon. Instead of calling this a methane branch, change the *-ane* in methane to *-yl*. Thus, this is a methyl branch. Because the methyl branch is attached to the second carbon of the butane chain, this compound has the name 2-methylbutane.

Now, examine the structure of a different hydrocarbon.

$$CH_3$$
$$|$$
$$^1CH_3^2C^3CH_3$$
$$|$$
$$CH_3$$

Propane will be part of this compound's name because the longest continuous chain has three carbons. Two methyl branches are present, both on the second carbon. To indicate the presence of more than one branch of the same kind, use the same Greek prefixes presented in Chapter 5 for naming hydrates and molecules. The prefix to use when two of anything are present is *di-*. Thus, the name of this compound is 2,2-dimethylpropane. Note that the positions where the methyl groups are attached to the parent chain are written, separated by a comma.

Follow the three-step process to name the following alkane.

$$CH_3$$
$$|$$
$$CH_3—CH_2—CH—CH—CH_2—CH_3$$
$$|$$
$$CH_2$$
$$|$$
$$CH_2$$
$$|$$
$$CH_3$$

Fact of the MATTER

The terms *saturated* and *unsaturated* originated before chemists understood the structures of organic substances. They knew that some hydrocarbons would take up hydrogen in the presence of a catalyst. When the substance would react with no more hydrogen, it was said to be saturated. Hydrocarbons that would take up additional hydrogen were said to be unsaturated. Today, we know that unsaturated hydrocarbons contain double and triple bonds that will react with hydrogen to form single bonds.

1. Count the number of carbons in the longest continuous chain, which may not always be written in a straight line. What looks like a branch may be part of the parent chain. Because this compound has a chain of seven carbons, heptane will be part of the name.

2. Note the size and number of branches present. The compound shown has one branch that is one carbon long and one branch that is two carbons long. *Ethyl* is the name given to a branch that has two carbons, and you know that the name for a one-carbon branch is *methyl*. If a compound has two different branches, they should be named in alphabetical order. Thus, part of the name will be ethylmethylheptane.

3. Number the carbons in the chain starting at the end nearest a branch. In the compound shown, position numbers will start on the right side.

$$CH_3 - CH_2 - \overset{4}{C}H - \overset{3}{C}H - \overset{2}{C}H_2 - \overset{1}{C}H_3$$

with CH₃ above the 4-carbon, and below the 3-carbon: ⁵CH₂ — ⁶CH₂ — ⁷CH₃

The position numbers of the carbons that have branches are 3 and 4. Now, put the name together. The position numbers of the branches come first, followed by a hyphen and then the name of the branch. The complete name for this compound is 4-ethyl-3-methylheptane.

Alkanes containing rings are named using the same rules, but the prefix *cyclo-* is placed before the name. Cyclohexane has six carbons connected in a ring, and cyclononane has a nine-carbon ring.

PRACTICE PROBLEMS

For more practice with solving problems, see Supplemental Practice Problems, Appendix B.

1. Write the structural formulas for the following branched alkanes.

   a) 2-methylbutane
   b) 1,3-dimethylcyclohexane (Hint: Begin numbering at any carbon in the ring, then attach the methyl groups.)
   c) 4-propyldecane
   d) 2,3,4-trimethylheptane

2. Name each of the following alkanes.

   a)

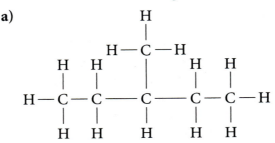

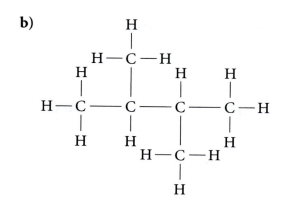

b)

```
                H
                |
            H — C — H
     H          |        H        H
     |          |        |        |
 H — C ————— C ————— C ————— C — H
     |          |        |        |
     H          H        H        H
                         |
                     H — C — H
                         |
                         H
```

c)

```
                                    H
                                    |
                                 H—C—H
   H    H    H    H                 |      H         H
   |    |    |    |                 |      |         |
 H—C——C——C——C———————————C—————————C————C—H
   |    |    |    |                 |      |         |
   H    H    H    H                 H    H—C—H       H
                                           |
                                           H
```

d)

```
   H    H         H         H    H    H
   |    |         |         |    |    |
 H—C——C———————C—————————C——C——C—H
   |    |         |         |    |    |
   H    H     H — C — H     H    H    H
                 |
             H — C — H
                 |
                 H
```

Isomers

Butane and 2-methylpropane are two alkanes with different names and different structures.

```
   H   H   H   H                          H
   |   |   |   |                          |
                                      H — C — H
 H—C——C——C——C—H              H         |        H
   |   |   |   |             |         |        |
   H   H   H   H          H—¹C —————²C —————³C—H
                             |         |        |
                             H         H        H

      Butane                     2-methylpropane
```

Both of these compounds are familiar fuels that burn to give off heat in handheld lighters, **Figure 18.2.** Is there any relationship between these two compounds? If you count the number of carbon and hydrogen atoms in each, you will find that they both have four carbons and ten hydrogens, which gives them the same molecular formula, C_4H_{10}. Compounds that

Figure 18.2
Disposable Lighter Fuel
Although butane and 2-methylpropane are gases at room temperature and atmospheric pressure, they can be liquefied under higher pressures in closed containers. Most disposable lighters contain one or both of these compounds, which are flammable enough to be ignited by a spark.

Figure 18.3

Structure and properties are closely related, as you can see by examining the isomers of pentane. Although all three compounds have the formula C_5H_{12}, differences in the amount of branching affect their properties. Note the differences in the shapes of the molecules.

have the same formula but different structures are called **isomers.** Butane and 2-methylpropane are known as structural isomers. Each has the molecular formula C_4H_{10}, but they have different structural formulas because the carbon chains have different shapes.

Despite their identical molecular formulas, isomers have different properties. The boiling and melting points of 2-methylpropane and butane are different, as are their densities and solubilities in water. In addition, their chemical reactivity is different. **Figure 18.3** shows some property differences in the isomers of pentane.

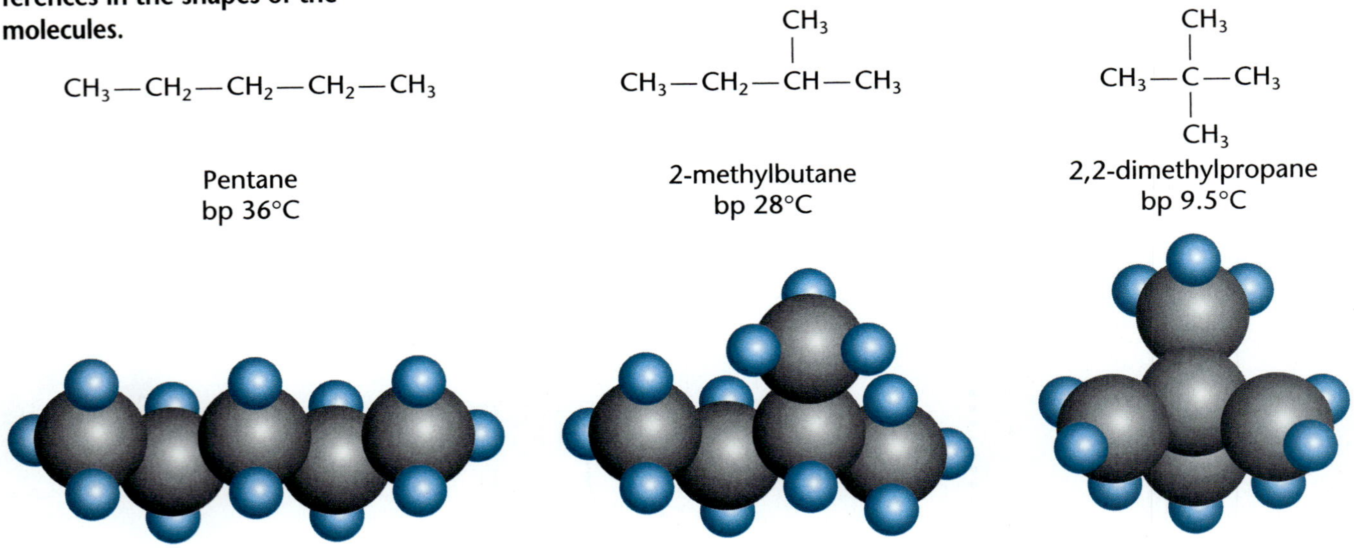

$CH_3{-}CH_2{-}CH_2{-}CH_2{-}CH_3$

$$CH_3{-}CH_2{-}\overset{\displaystyle CH_3}{\overset{\displaystyle |}{CH}}{-}CH_3$$

$$CH_3{-}\overset{\displaystyle CH_3}{\underset{\displaystyle CH_3}{\overset{\displaystyle |}{\underset{\displaystyle |}{C}}}}{-}CH_3$$

Pentane
bp 36°C

2-methylbutane
bp 28°C

2,2-dimethylpropane
bp 9.5°C

Although only two alkane isomers have four carbons, the number of possible isomers increases rapidly as carbon atoms are added to the parent chain. This is because longer chains provide more locations for branches to attach. A methyl branch on a six-carbon parent chain can be attached to either the second or the third carbon from the end of the chain. To make sure that these two compounds are really isomers, count the number of carbon and hydrogen atoms in the structures formed by placing the methyl group at those two positions, and write the molecular formulas.

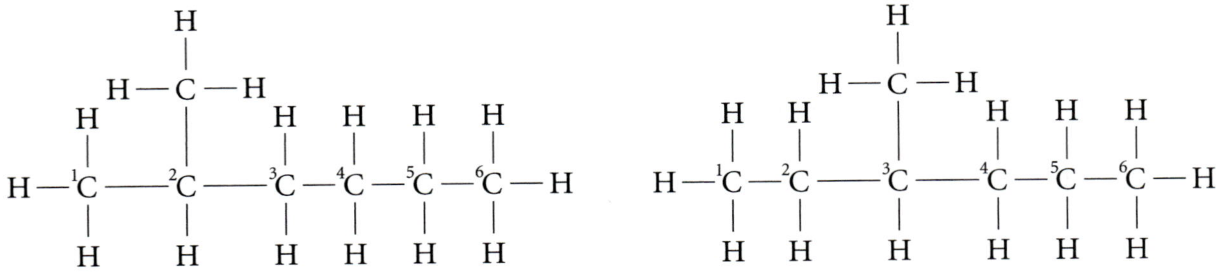

2-methylhexane
C_7H_{16}

3-methylhexane
C_7H_{16}

There are three isomers of pentane, five of hexane, and more than 4 billion isomers of the alkane with the formula $C_{30}H_{62}$.

Properties of Alkanes

Properties are affected by the structure or arrangement of atoms present in a molecule. Another factor that affects properties of alkanes is chain length. In general, the more carbons present in a straight-chain alkane, the higher its melting and boiling points. At room temperature, straight-chain alkanes that have from one to four carbon atoms are gases, those with from five to 16 carbon atoms are liquids, and those with more than 16 carbon atoms are solids.

A property shared by all alkanes is their relative unreactivity. You will recall from Chapter 9 that the carbon-carbon and carbon-hydrogen bonds found in alkanes are nonpolar. Because alkanes don't have any polar bonds, they undergo only a small number of reactions and will dissolve only those organic compounds that are nonpolar or that have low polarity, such as oils and waxes. The nonpolar and low-reactive nature of alkanes makes them good organic solvents. Paints, paint removers, and cleaning solutions often contain hexane or cyclohexane as solvents.

Four ways to represent molecules are shown here. At the top are ball-and-stick models, with springs representing double and triple bonds. At the bottom are space-filling models, which show the actual shape of the molecule. ▶

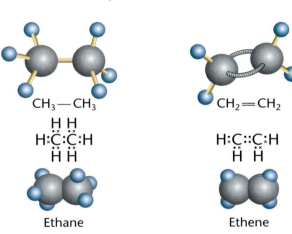

Ethane — Ethene — Ethyne

Figure 18.4
Single, Double, and Triple Bonds

Two carbon atoms can share one, two, or three pairs of electrons. In saturated hydrocarbons, carbon atoms share only one pair, whereas in unsaturated hydrocarbons, the carbon atoms participating in double or triple bonds share two or three pairs. Because a carbon atom forms four bonds, unsaturated hydrocarbons contain fewer than the maximum number of hydrogen atoms.

Unsaturated Hydrocarbons

You learned in Chapter 9 that carbon atoms in organic compounds can be connected by single, double, or triple bonds. A hydrocarbon that has one or more double or triple bonds between carbons is called an **unsaturated hydrocarbon.** Molecules with single, double, and triple bonds are compared in **Figure 18.4.**

Alkenes

Gasoline contains several hydrocarbons with double bonds. A hydrocarbon in which one or more double bonds link carbon atoms together is called an **alkene.** The most common alkenes found in gasoline are pictured in **Figure 18.5.**

Figure 18.5
Alkenes in Gasoline

Gasoline contains some alkenes as well as alkanes. Alkenes burn more uniformly, and their presence raises the octane rating of a gasoline.

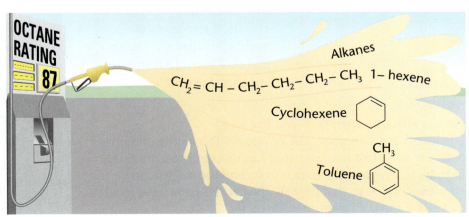

miniLAB 1

How unsaturated is your oil?

Most animal fats are saturated hydrocarbons and are solids at room temperature, whereas most vegetable fats are unsaturated and are liquids at room temperature. Both types of fats are essential in our diets, but medical research has shown that eating high levels of saturated fats can contribute to health problems such as heart disease. Different amounts of unsaturation in fats can be compared by testing how quickly a red-brown iodine solution added to each fat is decolorized. Iodine adds to carbons that take part in multiple bonds, forming colorless organic halogen compounds when the double or triple bond is broken.

Procedure

1. Place 20 mL of peanut oil in one small flask and 20 mL of canola oil in another flask. Label the flasks.

2. Add five drops of tincture of iodine to each oil and swirl to mix well. Note the color of each solution.

3. Heat both flasks on a hot plate set on low.

4. Note which oil returns to its original color first. This oil is more unsaturated than the other.

5. Now read the food label on each bottle of oil, and determine whether your test results agree with how much unsaturated fat the labels say are in each oil.

Analysis

1. What happens to the red-brown iodine when it is added to a relatively unsaturated oil?

2. Examine food labels on bottles of the oils listed below, and predict which of each pair will decolorize faster.

   a) canola or corn oil
   b) coconut or sunflower oil

Alkenes are named using the root names of the alkanes, with the -*ane* ending changed to -*ene*. The simplest alkene is ethene, $CH_2 \!=\! CH_2$, which contains two carbons linked in a chain. Ethene, a gas at room temperature, is the most important organic compound used in the chemical industry. Almost half of the ethene used is converted into plastics. It is also used to make automobile antifreeze, ethylene glycol. **Figure 18.6** shows another use of ethene.

Figure 18.6

Ethene and Ethylene: One and the Same

Ethylene is the common name of ethene. This compound occurs naturally as a plant hormone that functions to speed up ripening of fruits and vegetables. Unripe fruits and vegetables can be treated with ethylene so that they all ripen at the same time, making harvesting more efficient.

The alkene with three carbons in a chain is called propene, $CH_2=CHCH_3$. When four or more carbons are present in a chain, the double bond can be located at more than one possible position. When naming these alkenes, a number must be added at the beginning of the name to indicate where the double bond is located. You should use the following steps in naming alkenes with long carbon chains.

1. Count the number of carbons in the longest continuous chain that contains the double bond, and assign the appropriate alkene name.

2. Number the carbons consecutively in the longest chain, starting at the end of the chain that will result in the lowest possible number for the first carbon to which the double bond is attached.

3. Write the number corresponding to the first carbon in the double bond, followed by a hyphen and then the alkene name.

What is the name of the compound that has the following structure?

$$CH_2=CHCH_2CH_3$$

This compound has four carbons in a chain with one double bond, so butene will be part of its name. Numbering the carbons starting on the left side of the compound gives the first carbon that is part of the double bond the position number of one. Thus, this compound is 1-butene.

$$\overset{1}{C}H_2=\overset{2}{C}H\overset{3}{C}H_2\overset{4}{C}H_3 \quad \text{1-butene}$$

An isomer of 1-butene is 2-butene, $CH_3CH=CHCH_3$. These two compounds have the same molecular formulas but different structures. They are called positional isomers because they differ only by the position of the double bond. Positional isomers have different properties, just as structural isomers do.

The formation of a double bond prevents the carbons on each side of the bond from rotating with respect to each other. If the two groups attached to either carbon are different, the alkene can have two different geometric structures. These structures are geometric isomers. Study the two geometric isomers of 2-butene that are modeled in **Figure 18.7.** In the isomer called *cis*-2-butene, the hydrogen atoms and —CH_3 groups are on the same side of the double bond. In *trans*-2-butene, the hydrogen atoms and —CH_3 groups are on opposite sides of the double bond.

The properties of a pair of geometric isomers vary. *Trans* isomers are more symmetrical than *cis* isomers, and they can pack together more closely when in the solid state. This close packing makes the molecules harder to pull apart. As a result, *trans* isomers generally have higher melting points than do *cis* isomers. ▼

Figure 18.7

Geometric Isomers
Cis-2-butene and *trans*-2-butene are geometric isomers. Note their shapes in both ball-and-stick models and space-filling models.

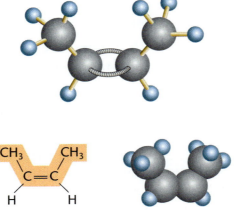

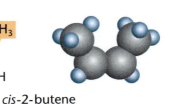

cis-2-butene

trans-2-butene

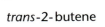

Vision and Vitamin A

Vision is a process usually studied in biology class. That's where you may have learned that light rays pass through the eye to reach the retina, where the rods and cones are located. Rods and cones are nerve receptors that are excited by light. More than 120 million rods in each eye detect white light and provide sharpness of visual images. The 7 million cones in each eye detect color. The pigment molecules responsible for vision are attached to the ends of the rods and cones. One of these pigments is called rhodopsin. Rhodopsin has two parts—a protein called opsin and a small molecule called retinal.

Chemistry of vision Beta-carotene, the natural orange pigment found in carrots and other yellow or green vegetables, breaks down in your body to form vitamin A, which is then converted to 11-*cis*-retinal. In the retina of the eye, 11-*cis*-retinal attaches to the protein opsin to form rhodopsin. When light is absorbed by the 11-*cis* isomer portion of rhodopsin, the energy causes the pigment molecule to undergo a change in shape. The right end of the molecule rotates about a double bond to form all-*trans*-retinal, in which all the groups are in the *trans* position.

11-*trans*-retinal

11-*cis*-retinal

When the retinal portion of rhodopsin is isomerized to the all-*trans* form, it separates from the opsin. Thus, light energy causes the rhodopsin to break down into the substances from which it was formed. As the rhodopsin molecule splits, the rods become excited, probably due to ionic charges that develop on the splitting surfaces. These charges last only for a second, but they generate nerve signals that are transmitted to the optic nerve and then to the brain. After the rhodopsin is activated, the *trans*-retinal returns to the *cis* form and recombines with opsin. This process is relatively slow, which is why your eyes need time to adjust to dim light.

Night vision When large quantities of light energy strike the rods, large amounts of rhodopsin are broken down. As a consequence, the concentration of rhodopsin in the rods falls to a low level. If you leave a bright area and enter a darkened room, the quantity of rhodopsin in the rods at first is small. As a result, you experience temporary blindness. The concentration of rhodopsin gradually builds up until it becomes high enough for even a small amount of light to stimulate the rods. During dark adaptation, the sensitivity of the retina can increase as much as a thousandfold in only a few minutes and as much as 100 000 times in an hour or more.

Connecting to Chemistry

1. **Hypothesizing** Owls and bats avoid daylight and are at home in the dark. Their eyes contain only rods. Hypothesize as to the relative amount of rhodopsin in their rods compared with that in human eyes, and explain.

2. **Analyzing** What is the significance of the change from 11-*cis*-retinal to all-*trans*-retinal for the vision process?

When corn oil is hydrogenated, it becomes solid. This is how margarine is produced and explains why margarine is solid at room temperature, whereas the vegetable oil from which it is derived is liquid.

Figure 18.8

Saturated and Unsaturated Fats

▲ Most vegetable oils contain unsaturated alkenes, whereas butter and lard—both animal products—contain mostly saturated alkanes. Because alkanes have higher melting points than alkenes, fats from animals are solid at room temperature. Fats from plants generally are liquids.

Alkenes are more reactive than alkanes because the two extra electrons in the double bond are not held as tightly to the carbons as are the electrons in a single bond. Alkenes readily undergo synthesis reactions in which smaller molecules or ions bond to the atoms on either side of the double bond. An unsaturated alkene can be converted into a saturated alkane by adding hydrogen to the double bond. This reaction is called hydrogenation. Its effects are shown in **Figure 18.8.**

$$CH_2 {=} CH_2 + H_2 \rightarrow CH_3CH_3$$

Ethene	Ethane
(unsaturated)	(saturated)

Alkynes

Another type of unsaturated hydrocarbon, called an **alkyne,** contains a triple bond between two carbon atoms. Alkynes are named using the alkane root name for a given carbon chain length and changing the *-ane* ending to *-yne.* Ethyne, known more commonly as acetylene, is the most important commercial alkyne. Most acetylene produced in the United States is used to make vinyl and acrylic materials, although about ten percent is burned in oxyacetylene torches. These torches are used to cut and weld metals. Few alkynes are known to occur naturally because they are very reactive. However, they can be synthesized from other organic compounds. The names and structures of some small alkyne molecules are shown in **Table 18.2.**

Table 18.2   Simple Alkynes

Chemical Name	Common Name	Structure
ethyne	acetylene	$HC{\equiv}CH$
propyne	methylacetylene	$HC{\equiv}CCH_3$
1-butyne	—	$HC{\equiv}CCH_2CH_3$
2-butyne	—	$CH_3C{\equiv}CCH_3$

PEOPLE in CHEMISTRY

Meet John Garcia, Pharmacist

"A treasure house going up in smoke"—that's how pharmacist John Garcia describes the fires set to clear land in the Amazon. He's concerned about the potential medicines that humans might be destroying as they make inroads into formerly wild areas. "However," he continues, "endangered sources for new medicines aren't only in jungles. Lincomycin, an antibiotic, was discovered in soil in Lincoln, Nebraska." In this interview, Mr. Garcia describes some of the changes he has seen in his four decades as a pharmacist.

On the Job

 Mr. Garcia, can you tell us what you'll be doing first today in your pharmacy?

 This morning, I have to put together 100 suppositories for a patient with migraine headaches. At least that won't be as difficult as the order for a suppository a colleague of mine once had. That was for an elephant! The medication was made with an aluminum baseball bat mold. Just one dose cost $300.

 Do you prepare prescriptions for other animals?

 I've made them for rabbits. I added raspberry flavor because rabbits enjoy the flavor.

 Do your human patients get a choice of flavors, too?

I've got about 40 flavors—piña colada, bubble gum, peppermint. Tutti-frutti is my personal favorite. These flavorings aren't just frills. For instance, I prepare chloroquine, an antimalarial drug for children, and put it in a chocolate base. Because the molecules of chocolate are fairly large, they block the taste buds and cover up the bitterness of the medicine.

 You own and operate your own independent pharmacy. How does it differ from most pharmacies?

This one is different because I specialize in compounding prescriptions. That means that I can prepare customized formulas in precise doses to meet patients' individual needs. In most pharmacies, the pharmacists do mostly what I call "count and pour." That is, they prepare standardized doses of drugs. A lot of times, I work in partnership with a patient's doctor. He or she will ask my advice about medication. Besides suggesting alternative drugs or doses, I can talk with the doctor about methods of administering a drug.

 How has the practice of pharmacy changed in the 40 years you've been a pharmacist?

 Were you, as a child, aware of pharmacies and medications?

 New laws have given pharmacists the responsibility of explaining carefully to a patient the workings of a drug and its possible side effects. At the same time, changes in the insurance industry have added lots of paperwork and reduced our fees. So that means that pharmacists have to work more quickly while maintaining careful standards.

Seriously ill patients are living longer today. How has that change affected your practice?

I work with several hospices, which care for terminally ill patients. By the time patients are admitted to a hospice, they typically are in the last stages of an illness and have about 40 days to live. Our priority is to make the patient comfortable by using a variety of medications to relieve pain and discomfort.

Early Influences

How did you come to be a pharmacist?

I got an after-school job in a small pharmacy when I was in high school. I started out making deliveries and eventually helped fill prescriptions. I liked working with the public, and the pharmacist, Nathan Fisher, encouraged me to apply to pharmacy school. I feel honored that he was my mentor.

Not really. I don't think I even saw a doctor until I needed a physical exam to play high school sports. My parents are from Mexico, and my mother knew many home remedies. She gave me things like charcoal or mint for an upset stomach.

Personal Insights

 What qualities do you think it takes to be a good pharmacist?

It's obvious that you have to like people and be accurate in mathematical and chemical calculations. I also like to say that pharmacy is a lot like cooking—a little of this, a little of that. Just like a cook, I want my preparations to contain the right proportion of ingredients, to taste good, and also to look elegant.

 What changes do you see coming in the field of pharmacy during the 21st century?

I think people will be able to insert a prescription in something like an automated teller machine. They'll also insert medical cards encoded with their personal medical history and their age, height, and weight. The machine will dispense their medication.

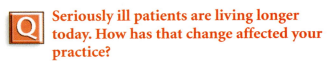

CAREER ▶ CONNECTION

These careers are all related to the field of pharmacy.

Pharmacologist Four to six years post-college study at a medical school or school of pharmacy
Pharmaceutical Technician Two years in a training program after high school
Pharmaceutical Production Worker High school diploma and on-the-job training

Melting and boiling points of alkynes increase with increasing chain length, just as they did for alkanes and alkenes. Alkynes have physical and chemical properties similar to those of alkenes; their melting points are higher than those of alkanes, and they undergo synthesis reactions. For example, hydrogen molecules can be added to an alkyne in a stepwise fashion to form an alkene and then an alkane.

$$CH \equiv CH + H_2 \rightarrow CH_2 = CH_2$$
$$CH_2 = CH_2 + H_2 \rightarrow CH_3CH_3$$

Aromatic Hydrocarbons

Another group of unsaturated hydrocarbons has distinctive, six-carbon ring structures. The simplest compound in this group is benzene, with the molecular formula C_6H_6. Benzene contains six carbons joined together in a flat ring. Its structure was originally thought to contain alternating single and double bonds between the carbons, as seen in **Figure 18.9,** but this structure is now known to be incorrect.

Although it contains double bonds, benzene does not share most of the properties of alkenes. It is unusually resistant to hydrogenation, whereas most alkenes readily become hydrogenated. To account for this inertness, chemists have suggested that the extra electrons are shared equally by all six carbons in the ring rather than being located between specific carbon atoms. The currently accepted structure for the benzene molecule is shown in **Figure 18.9.**

This sharing of electrons among so many carbons gives benzene and similar compounds unique properties. The name **aromatic hydrocarbon** is used to describe a compound that has a benzene ring or the type of bonding exhibited by benzene. Aromatic hydrocarbons were originally named because most of them have distinctive aromas. Naphthalene, formerly used in mothballs to prevent moth damage to woolen clothes,

Figure 18.9

The Structure of Benzene
The structure of benzene can be represented in different ways.

The flat benzene molecule is shown with clouds of shared electrons above and below the plane of the ring. ▼

The hexagon diagram is a more accurate shorthand representation. In this hexagon, each corner represents a carbon atom. The circle in the middle of the structure represents the cloud of six electrons that are shared equally by the six carbon atoms in the molecule. ▼

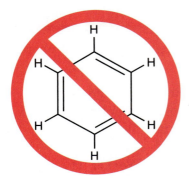

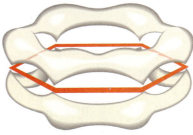

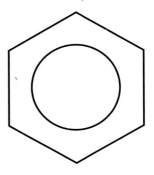

▲ This structure of benzene, showing alternating single and double bonds, is now known to be incorrect because it does not account for benzene's inertness.

consists of two benzene rings attached side-by-side. Aromatic hydrocarbons are unusually stable because the carbons are bonded tightly together by so many electrons.

Sources of Organic Compounds

You have examined structures and learned how to name basic organic compounds, but have you stopped to think about where these compounds come from? Most hydrocarbons come from fossil fuels, especially petroleum, but also natural gas and coal, **Figure 18.10.** Other important sources include wood and the fermentation products of plant materials.

Natural gas contains large quantities of methane, along with smaller amounts of alkanes up to about five carbons in length. Natural gas passes easily through pipes and is used mostly as a fuel, but it also serves as a raw material for making many small organic compounds.

Petroleum is a complex mixture, mostly of alkanes and cyclic alkanes. Products we get from petroleum include gasoline, jet fuel, kerosene, diesel fuel, fuel oil, asphalt, and lubricating oil. To use these organic products, they have to be separated from one another. What properties of hydrocarbons might allow them to be separated?

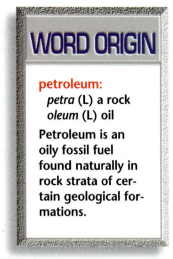

WORD ORIGIN

petroleum:
petra (L) a rock
oleum (L) oil

Petroleum is an oily fossil fuel found naturally in rock strata of certain geological formations.

Figure 18.10

Sources of Aromatic Hydrocarbons

Soot is full of aromatic hydrocarbons produced by combustion of organic materials such as wood or coal. In 1775, a British physician investigating the high incidence of cancer among chimney sweeps theorized that soot in the chimneys was the cause. In the 1930s, it was shown that a number of large hydrocarbons with ring structures found in the coal tar in soot were carcinogenic, or cancer-causing. ▼

▲ Valuable deposits of petroleum and natural gas often are found by offshore oil drilling. These deposits are common under the oceans because microscopic marine organisms—including algae, bacteria, and plankton—died and settled on the seafloor where their remains were altered by high temperatures and pressure. Petroleum deposits also are found under dry land because at one time, this land was under a sea.

Figure 18.11

Fractional Distillation of Petroleum

To separate important components of petroleum using fractional distillation, petroleum or hot crude oil is heated in a furnace. The liquid alkanes are vaporized and allowed to rise in the fractionating tower. Gases that were dissolved in the petroleum are removed at the top of the tower and condensed into liquids that are sold in cylinders. Gasoline is part of the next-lower group of materials drawn off. Below the gasoline fraction come mixtures of heavier hydrocarbons, such as kerosene, fuel and lubricating oils, and asphalt.

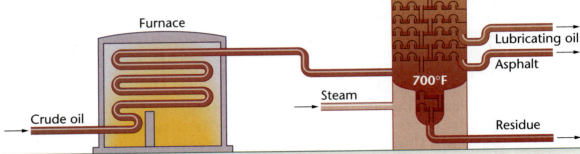

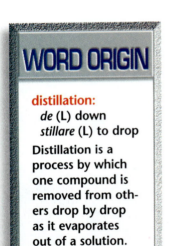

WORD ORIGIN

distillation:
de (L) down
stillare (L) to drop

Distillation is a process by which one compound is removed from others drop by drop as it evaporates out of a solution.

One property is boiling point. As you learned in Chapter 5, distillation is a technique used to separate substances that have different boiling points. In the petroleum industry, huge towers are used to distill petroleum into its component liquids. Inside the tower, many plates provide multiple surfaces on which repeated vaporization-condensation cycles take place. Repeated cycles provide for more efficient separations and allow fractions containing only one or a few different compounds to be isolated. This method of separation is called **fractional distillation.**

Figure 18.11 shows where different products are drawn off according to boiling points at different levels of the tower. The temperature of the tower is controlled so that it is hotter at the bottom and cooler at the top. Because boiling point increases along with molecular weight, the heavier, higher-boiling components condense near the bottom, and the lighter, lower-boiling molecules condense toward the top. Pipes connected at various points allow chemical engineers to draw off each fraction.

The process of fractional distillation does not produce quantities of hydrocarbons in the proportions needed in industry. For example, gasoline is usually the fraction of petroleum most in demand, but this fraction makes up less than half of petroleum. How can the quantities of useful hydrocarbons be increased? The hydrocarbons in the fractions coming off the lower parts of the tower are mostly large alkanes. These can be converted into smaller, more useful alkanes and alkenes by a process called **cracking.** Cracking uses catalysts or high temperatures in the absence of air to break down or rearrange large hydrocarbons. Cracking is also used to increase the yield of natural gas by producing small alkenes from larger molecules. The cracking of propane produces some methane and ethene, in addition to propene and hydrogen.

$$2CH_3CH_2CH_3 \xrightarrow[500-700°C]{} CH_4 + CH_2{=}CH_2 + CH_3CH{=}CH_2 + H_2$$

Underground coal mining can be difficult, dirty, and dangerous work. The coal seams are often found deep underground, and tunnels must be dug so that miners can reach them. A mining machine breaks up the coal and delivers it to a conveyor belt, which carries the coal chunks out of the mine. ▼

Figure 18.12
Sources of Coal

▲ Coal is our most plentiful fossil fuel. The most inexpensive way to obtain coal that is not deeply buried is by strip-mining, in which large areas of land are stripped bare of all vegetation. This causes environmental problems when the exposed soil washes away after the coal is removed. Laws in the United States now require that most stripped areas be restored.

Another process that uses heat, pressure, and catalysts to convert large alkanes into other compounds is called **reforming.** It is used to form aromatic hydrocarbons.

Like petroleum and natural gas, coal also is a fossil fuel. It is formed from the remains of plants that became buried underwater and were subjected to increasing pressures as layers of mud built up. Coal is composed primarily of carbon but also contains many mineral impurities. It is used mostly as a fuel and as a source of aromatic hydrocarbons. Coal must be obtained from underground or surface mines, as shown in **Figure 18.12.**

SECTION REVIEW

Understanding Concepts

1. Name the organic compounds shown.

   a) $CH_3CH = CHCH_2CH_2CH_3$

   b) $CH_3CHCH_2CH_2CH_3$
   $\quad\quad\; |$
   $\quad\quad CH_3$

   c) $CH_3CH_2CH_2CH_2CH_2CH_2CH_3$

   d) $CH_3CH_2CHCH_2CHCH_3$
   $\quad\quad\quad\; |\quad\quad\; |$
   $\quad\quad\quad CH_3\quad CH_3$

2. Draw structures that correspond to the names of the hydrocarbons given.

   a) hexane
   b) 3-ethyloctane
   c) *trans*-2-pentene

   d) ethyne
   e) 1,2-dimethylcyclopropane
   f) 1-butyne

3. What are the major sources of hydrocarbons? Where are they found?

Thinking Critically

4. **Interpreting Chemical Structures** Halogen molecules such as Br_2 can be added to double bonds in a reaction similar to hydrogenation. Draw the structure of the product that forms when Br_2 is added to propene.

Applying Chemistry

5. **Storage of Hydrocarbons** Would it be more important to store octane or pentane in a tightly sealed bottle at a low temperature? Why?

Substituted Hydrocarbons

SECTION PREVIEW

Objectives

✓ **Compare and contrast** the structures of the major classes of substituted hydrocarbons.

✓ **Summarize** properties and uses of each class of substituted hydrocarbon.

Review Vocabulary

Alkane: a saturated hydrocarbon that contains only single bonds between carbon atoms.

New Vocabulary

substituted
  hydrocarbon
functional group

Doesn't that apple pie look good, fresh out of the oven? Looking at the photo, you can almost smell the apples, cinnamon, nutmeg, and perhaps vanilla. What causes these ingredients to have aromas? Apples, cinnamon, nutmeg, vanilla, and many other fruits and spices all contain molecules with a distinctive group of atoms located at the end of the molecule. This atomic arrangement imparts a pleasant odor to the molecule.

Functional Groups

When chlorine and methane gases are mixed in the presence of heat or ultraviolet light, an explosive reaction takes place in which a mixture of the following four products is produced. Because structures of these compounds are the same as hydrocarbons except for the substitution of atoms of another element for part of the hydrocarbon, they are called **substituted hydrocarbons.** The part of a molecule having a specific arrangement of atoms that is largely responsible for the chemical behavior of the parent molecule is called a **functional group.** Functional groups can be atoms, groups of atoms, or bond arrangements. Notice that in the structures shown, one or more of the hydrogen atoms in methane is replaced by a chlorine atom.

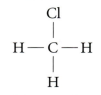

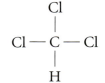
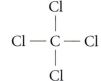

Chloromethane
(methyl chloride)

Dichloromethane
(methylene chloride)

Trichloromethane
(chloroform)

Tetrachloromethane
(carbon tetrachloride)

Replacing part of a hydrocarbon with functional groups changes the structure, properties, and uses that we have for the compounds. For example, chloromethane, a gas, is used as a refrigerant, and dichloromethane, a liquid, is used as a solvent to decaffeinate coffee. You may know trichloromethane by its common name, chloroform. It was an early anesthetic, used to put people to sleep during surgery. Tetrachloromethane, commonly called carbon tetrachloride, is a solvent that was used in dry cleaning and in fire extinguishers.

Some functional groups are complex in structure and consist of a group of atoms rather than a single atom. These groups of atoms often contain oxygen or nitrogen, and some contain sulfur or phosphorus. Double and triple bonds also are considered to be functional groups. Many organic compounds contain more than one type of functional group. Simple organic compounds are grouped into categories depending on the functional group they contain. You will study structures and examples of the most important functional groups.

Functional Groups: Structure and Function

As you study these functional groups, notice that the similarity in properties among molecules containing a given functional group leads to the use of members of that group for similar purposes. In the structures shown below, R and R' each represent a hydrocarbon part of the molecule. R and R' may be the same, or they may be different. For example, in the formula for an ether, $R—O—R'$, where R represents $CH_3CH_2—$ and R' represents CH_3, the ether has the formula $CH_3CH_2OCH_3$.

① Halogenated Compounds

structure: $R—X$, where X = F, Cl, Br, or I

functional group: halogen atoms
properties: high density
uses: refrigerants, solvents, pesticides, moth repellents, some plastics
biological functions: thyroid hormones
examples: chloroform, dichloromethane, thyroxin, Freon, DDT, PCBs, PVC

$$Cl—\overset{\displaystyle F}{\underset{\displaystyle F}{C}}—Cl$$

Freon

Chlorofluorocarbons (CFCs) are substituted compounds containing chlorine and fluorine atoms bonded to carbon. The most common CFC, Freon, has the formula CCl_2F_2. CFCs were once widely used as propellants in aerosol cans; as solvents; as the foaming agent in the manufacture of plastic foam materials; and as refrigerants in air conditioners, refrigerators, and freezers. However, in 1987, the major industrial nations of the world agreed to a gradual reduction in the use of CFCs because they cause a depletion of the valuable ozone in Earth's upper atmosphere. CFCs are being replaced by other halogenated compounds that are not as damaging to the atmosphere.

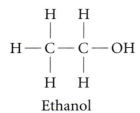

H   H
| |
H—C—C—OH
| |
H   H

Ethanol

Try at Home Lab

See page 871 in
Appendix F for
**Comparing Water
and Alcohol**

2 Alcohols

structure: *R*—O—H

functional group: hydroxyl group (—OH); hydrogen atom bonded to an
  oxygen atom, which is bonded to the hydrocarbon part of the molecule
properties: polar, so water molecules are attracted to it; high boiling point;
  alcohols with low molecular weights are water-soluble
uses: solvents, disinfectants, mouthwash and hair-spray ingredient,
  antifreeze
biological functions: reactive groups in carbohydrates, product of fermentation
examples: methanol, ethanol, isopropanol (one type of rubbing alcohol),
  cholesterol, sugars

An organic compound that contains at least
one hydroxyl group is called an alcohol. The
name of an alcohol ends in *-ol*. Alcohols have
many different uses. One of the most important is as a disinfectant for killing bacteria
and other potentially harmful microorganisms. For this reason, ethanol is put into
mouthwashes, and rubbing alcohol is used as
a disinfectant. Antifreeze also is an alcohol.

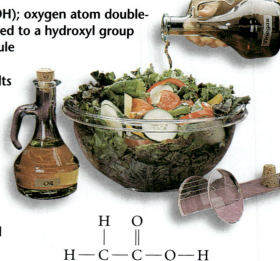

3 Carboxylic Acids

structure: *R*—C—O—H
(with O double-bonded to C)

functional group: carboxyl group (—COOH); oxygen atom double-
  bonded to a carbon, which is also bonded to a hydroxyl group
  and the hydrocarbon part of the molecule
properties: acidic, usually water-soluble,
  strong unpleasant odors, form metal salts
  in acid-base reactions
uses: vinegar, tart flavoring, skin-care
  products, production of soaps and
  detergents
biological functions: pheromones, ant-
  sting toxin; causes rancid-butter and
  smelly-feet odors
examples: acetic acid (in vinegar), formic
  acid, citric acid (in lemons), salicylic acid

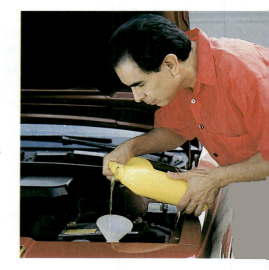

H   O
| ||
H—C—C—O—H
|
H

Acetic acid

A compound containing a carboxyl group is known as a carboxylic acid,
or organic acid. Many pheromones contain carboxyl functional groups.
Pheromones are organic compounds used by animals to communicate
with each other. When an ant finds food, it leaves behind a pheromone
trail that other ants in its colony can follow to get to the food source.

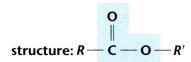 Esters

structure:

$$R - \overset{\overset{\displaystyle O}{\parallel}}{C} - O - R'$$

derived from carboxylic acids in which the —OH of the carboxyl group has
been replaced with an —O*R* from an alcohol
properties: strong aromas, volatile
uses: artificial flavorings and fragrances, polyester fabric
biological functions: fat storage in cells, DNA phosphate-sugar backbone,
natural flavors and fragrances, beeswax
examples: banana oil, oil of wintergreen, triglycerides (fats)

A compound formed from the reaction of an organic acid and an alcohol is called an ester. Some esters are used in the production of polyester fabrics. Many esters are used as flavorings in food products. Natural flavors are often complex mixtures of esters and other compounds, whereas artificial flavors usually contain fewer compounds and may not taste exactly the same as the natural flavor.

$$H - \overset{\overset{\displaystyle H}{|}}{\underset{\underset{\displaystyle H}{|}}{C}} - \overset{\overset{\displaystyle H}{|}}{\underset{\underset{\displaystyle H}{|}}{C}} - \overset{\overset{\displaystyle H}{|}}{\underset{\underset{\displaystyle H}{|}}{C}} - \overset{\overset{\displaystyle O}{\parallel}}{C} - O - \overset{\overset{\displaystyle H}{|}}{\underset{\underset{\displaystyle H}{|}}{C}} - \overset{\overset{\displaystyle H}{|}}{\underset{\underset{\displaystyle H}{|}}{C}} - H$$

Ethyl butyrate
(pineapple flavoring)

5 Ethers

structure: *R*—O—*R'*; oxygen atom bonded to two hydrocarbon groups

properties: mostly unreactive, insoluble in water, volatile
uses: anesthetics, solvents for fats and waxes
examples: diethyl ether

Ethyl ether

An ether is an organic compound in which an oxygen atom is bonded to two hydrocarbon parts of the molecule. Diethyl ether, often simply called ether, is an effective anesthetic. Because it is insoluble in water, it passes readily through the membranes surrounding cells. Ether is rarely used as an anesthetic today because it is highly flammable and because it causes nausea.

6 Ketones and Aldehydes

structures: ketones: $R—C—R'$ aldehydes: $R—C—H$

functional group: carbonyl group (—CO); carbon atom double
 bonded to an oxygen atom
properties: very reactive, distinctive odors
uses: solvents, flavorings, manufacture of plastics and
 adhesives, embalming agent
examples: acetone; formaldehyde; cinnamon,
 vanilla, and almond flavorings

Vanillin

Acetone

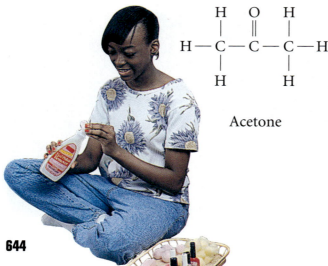

Aldehydes and ketones are compounds that contain a carbonyl group. If the carbonyl group is on the end of the carbon chain, the compound is called an aldehyde. If the carbonyl group is not at the end of the chain, the compound is a ketone. Acetone is a ketone solvent commonly used in nail-polish remover. Nail polish is not soluble in water—if it were, it would come off when you wash your hands. It is soluble in many organic compounds such as acetone, which is used to remove it.

structure: amine: R—NH$_2$ amide: R—$\overset{\overset{\displaystyle O}{\|}}{C}$—NH$_2$

functional group:
amine: amino group (—NH$_2$); two hydrogen atoms bonded to a nitrogen atom, which is bonded to the hydrocarbon part of the molecule
amide: amino group bonded to a carbonyl group (—CONH$_2$)
properties: amines: basic, ammonia-like odor
amides: neutral, most are solids
uses: solvents, synthetic peptide hormones, fertilizer, nylon synthesis
biological functions: in amino acids, peptide hormones, and proteins; distinctive odor of some cheeses
examples: urea, putrescine, cadaverine, Nutrasweet

Putrescine

An organic compound containing an amino group is called an amine. Amines and amides are important biological molecules because they are part of proteins. When an organism dies, its proteins are broken down into many different compounds containing amino functional groups. Given that two of those compounds are named putrescine and cadaverine, what type of odor do you think they have? These compounds have a distinctive, unpleasant smell that specially trained sniffer dogs can use to locate human remains and to help in forensic investigations. Cadaverine also contributes to bad breath.

miniLAB 2

A Synthetic Aroma

When brought together under the proper conditions, an organic acid and an alcohol will react to form an ester. Esters are generally volatile compounds having distinctive odors. In this MiniLab, you will combine methanol and salicylic acid to produce the ester, methyl salicylate.

Procedure

1. Put on an apron and goggles.

2. Set up a hot-water bath for the reaction by filling a 250- or 400-mL beaker about halfway with tap water and setting it on a hot plate or on a wire gauze, iron ring, and ring stand over a burner flame. Allow the water to warm but not boil.

3. Place 3 mL of methyl alcohol and 1 g of salicylic acid in a large (20 mm × 150 mm) test tube. Mix the reactants with a glass stirring rod.

4. Add about 0.5 mL (about ten drops from a dropper) of concentrated sulfuric acid to the reaction mixture and stir. Do not draw the acid into the rubber bulb of the dropper. (**CAUTION:** *Concentrated sulfuric acid is corrosive to skin, eyes, body tissues, and most other organic materials. If any contact occurs, notify your teacher immediately.*)

5. Place the test tube and contents in the hot-water bath and allow it to warm for five or six minutes.

6. Pour the contents of the test tube into about 50 mL of cold, distilled water in a small beaker. Cover the beaker with a watch glass and allow it to stand for a minute or two.

7. Remove the watch glass and waft the methyl salicylate odor toward your nose. Record your observations.

Analysis

1. What familiar odor is caused by methyl salicylate?

2. Look up the structural formulas for the reactants in a reference book, and write the equation for the reaction using structural formulas for all organic reactants and products.

3. What organic reactants would be required to produce the ester, ethyl butyrate?

Plants are a common source of esters.

The Sources of Functional Groups

How can different compounds containing functional groups be made? Alkanes are not very reactive, so it can be difficult to react them directly to form substituted molecules. You read earlier of one reaction used to substitute chlorine atoms for hydrogen atoms in methane. However, that reaction is not practical because a mixture of four different products is formed. Each must be separated and purified before it can be used.

Alcohols are a better choice than alkanes as the raw materials for synthesizing organic molecules. Alcohols can be converted into compounds containing almost every other type of functional group. For example, an acid and an alcohol react together to form an ester and water.

Figure 18.13
Production of Ethanol
Ethanol can be produced industrially by the fermentation of sugars and starches by yeast cells. These sugars and starches come from plant material such as corn, molasses from sugarcane, and the grapes grown in vineyards like this one.

Where do all the alcohols needed by industry come from? Cracking of petroleum products produces alkenes, which can readily be converted into alcohols by reaction with water. This process is used to synthesize ethanol, the two-carbon compound that contains an alcohol functional group.

$$
\begin{array}{cccc}
\text{H}\,\text{C=C}\,\text{H} & + & \text{H}_2\text{O} & \rightarrow \quad \text{H}-\overset{\text{H}}{\underset{\text{H}}{\text{C}}}-\overset{\text{OH}}{\underset{\text{H}}{\text{C}}}-\text{H} \\
\text{Ethene} & + & \text{Water} & \rightarrow \quad \text{Ethanol}
\end{array}
$$

A second important natural source of ethanol provides a convenient supply of this reactive molecule. Ethanol is produced by yeast cells when they ferment the sugars and starches in plant materials, as discussed in **Figure 18.13.** Fermentation produces much smaller quantities of ethanol than does the reaction between ethene and water; it is used in industry mainly to produce the ethanol in alcoholic beverages.

Supplemental Problems

For more practice with solving problems, see Supplemental Practice Problems, Appendix B.

SECTION REVIEW

Understanding Concepts
1. Explain why organic compounds with similar structures often have similar uses.

2. Distinguish between an aldehyde and a ketone.

3. Redraw the following structure of thyroxine, a thyroid hormone. Circle and name each type of functional group present.

$$\text{HO}-\bigcirc\!\!-\text{O}-\bigcirc\!\!-\text{CH}_2\text{CHCOOH}$$
(with I substituents and NH₂)

Thinking Critically
4. **Inferring** Explain why alkenes and alkynes contain functional groups but are not substituted hydrocarbons.

Applying Chemistry
5. **Antifreeze** Ethylene glycol, used as an antifreeze in car radiators, has two hydroxyl groups, as shown below. What can you infer about the boiling point, freezing point, and solubility in water of this compound from its structure?

$$
\text{H}-\overset{\text{OH}}{\underset{\text{H}}{\text{C}}}-\overset{\text{OH}}{\underset{\text{H}}{\text{C}}}-\text{H}
$$

Plastics and Other Polymers

On April 6, 1938, a young chemist working on the preparation of a compound used in refrigeration opened the valve on a tank of the reactant he planned to use in this process, tetrafluoroethene gas. Roy J. Plunkett was puzzled when no gas came out after he opened the valve because the tank was heavy enough to indicate that it was full of gas. He poked a wire through the opening of the valve on the tank and found that it wasn't clogged.

What do you think you would have done if you had been in Plunkett's shoes? Would you have discarded the tank in favor of a new one? Or would you have seen this problem as an intriguing puzzle to be investigated? Plunkett was curious, and when he cut open the tank, he discovered a white, waxy solid in place of the gas he expected. Today, this solid is known as Teflon, a large molecule with remarkable properties. You are probably familiar with Teflon as a nonstick coating on pots and pans, but it is also used for dentures, artificial joints and heart valves, space suits, and fuel tanks on space vehicles.

Monomers and Polymers

How did a white, waxy solid form from the colorless gas tetrafluoroethene (also called tetrafluoroethylene)? One clue to what type of reaction occurred can be found in the properties of those two substances. Most nonpolar molecules that are relatively small tend to be gases at room temperature, whereas larger molecules usually are solids. When the structure of Teflon was determined, the molecule was found to consist of chains of carbons with fluorine atoms attached. Somehow, the tetrafluoroethylene molecules in the gas had reacted with each other to form these long-chain molecules. A large molecule that is made up of many smaller, repeating

Tetrafluoroethylene

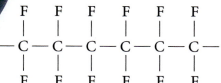

Figure 18.14
Teflon
Teflon provides a good coating for this skillet because it is unreactive and food will not stick to it.

Teflon polymer

units is called a **polymer.** A polymer forms when hundreds or thousands of these small individual units, which are called **monomers,** bond together in chains. The monomers that bond together to form a polymer may all be alike, or they may be different. When Teflon formed in the tank, many small tetrafluoroethylene monomers combined to form long polymers of polytetrafluoroethylene. Examine the structures in **Figure 18.14** to help you visualize this reaction.

The properties of a polymer are different from those of the monomers that formed it. For example, the polyethylene plastic in milk jugs is made when molecules of gaseous ethylene react to form long chains. The unique properties of a given polymer, such as tensile strength, water-repellency, or flexibility, are related to the polymer's enormous size and the way its monomers join together.

Synthetic Polymers

Polymers are everywhere, making fabrics such as nylon and polyester, plastic wrap and bottles, rubber bands, and many more products you see every day. How many polymers can you identify in **Figure 18.15?** What do the polymers that make up such different substances have in common? All are large molecules made of smaller, repeating units.

Chemists have been synthesizing polymers in laboratories for only about 100 years. Can you imagine what life was like before people had these synthetic polymers? You might have gotten wet on the way to school without a nylon raincoat, eaten a stale sandwich for lunch without plastic wrap or a container to put it in, and worn a heavier cotton uniform in your sports activity instead of a lightweight, synthetic fabric. Many polymers have added conveniences that we take for granted in our lives.

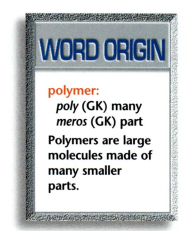

WORD ORIGIN

polymer:
poly (GK) many
meros (GK) part
Polymers are large molecules made of many smaller parts.

Figure 18.15
Synthetic Polymers
Sports activities would be different today without synthetic polymers. Balls, uniforms, artificial turf, bandages used to wrap sprains, and nets used in hoops and goals are usually made of synthetic polymers.

ChemLab

Identification of Textile Polymers

For centuries, polymers have been used for clothing in the form of cotton, wool, and silk. Chemists trying to improve these natural polymers to fit specific purposes have designed many synthetic polymers that also are used to make fabrics. Among these are nylon, rayon, acetate, and the polyesters.

Many synthetic fabrics are designed to look and feel like natural polymers but also have the superior properties for which they were designed, such as being wrinkle-resistant, water-repellent, or quick-drying. However, these imitators can't fool a chemist; simple tests that can be done in minutes will distinguish the real thing from a pretender. This is because differences in their structures lead to differences in properties. In this experiment, you will identify fabric samples by testing them for characteristic properties.

Problem

What tests can be performed that will differentiate among the polymers used to make fabrics?

Objectives

- **Analyze** changes in fabric samples that are subjected to flame and chemical tests.
- **Classify** fabrics by polymer type based on test results.

PREPARATION

Materials

fabric samples A-G
   (7 types; six 0.5 × 0.5
   cm squares of each
   type)
Bunsen burner

beaker containing
   water
forceps
test-tube holder
test-tube rack

medium test tubes (4)
balance
watch glass
red litmus paper
stirring rod
100-mL beakers (2)
10-mL pipet
25-mL graduated cylinder

$Ca(OH)_2$
$1M$ $BaCl_2$
concentrated H_2SO_4
iodine solution
$0.05M$ $CuSO_4$
$3M$ NaOH
acetone

Safety Precautions

Use care when working with an open flame and handling concentrated acids and bases. Do not inhale odors from plastics.

PROCEDURE

Flame Tests

1. Use forceps to hold a square of fabric A in a Bunsen burner flame for 2 seconds.

2. Remove the fabric from the flame, and blow out the flame if the fabric keeps burning.

3. Observe the odor by wafting smoke from the smoldering sample toward your nose. Be sure the fabric is no longer burning by immersing it in a beaker of water.

4. Make a data table similar to Table 1 in Data and Observations on page 652. Record your observations about the way the fabric burns in the flame, odor observed, and characteristics of the residue left after burning.

5. Repeat steps 1 to 4 six times, using fabric samples B-G.

6. Use the following table to make a preliminary identification of your samples.

Polymer Type	Type of Burning	Burning Odor	Residue Type
Silk or wool	Burns and chars	Hair	Crushable bead
Cotton	Burns and chars	Paper	Ash
Nylon, polyester, acetate, or acrylic	Burns and melts	Chemical	Plastic bead

Chemical Tests

Use your preliminary identifications to determine which tests are necessary to identify each fabric sample. You should not have to carry out every test on every sample. Make a data table similar to Table 2 on page 652, and record all your results.

7. **Nitrogen Test** Place a fabric square in a test tube and add 1 g of $Ca(OH)_2$. Using a test-tube holder, heat the tube gently in a Bunsen burner flame while holding a piece of red litmus paper with forceps over the mouth of the tube. If the litmus paper turns blue, nitrogen is present. Only silk, wool, nylon, and acrylic contain nitrogen.

8. **Sulfur Test** Add a fabric square to 10 mL of $3M$ NaOH in a test tube, and gently heat to boiling by holding the tube in a Bunsen burner flame. Be careful that the tube is not pointing toward anyone. Cool the solution, add 30 drops of $BaCl_2$ solution, and observe whether or not a precipitate forms. Only wool contains enough sulfur to give a barium sulfide precipitate.

9. **Cellulose Test** Place a fabric square in a beaker, cover with approximately 2 mL of concentrated H_2SO_4, and then carefully transfer the contents to another beaker containing ten drops of iodine solution in 25 mL of water. Rinse the empty beaker with large quantities of water. Cotton gives a dark blue color within 1 to 2 minutes, and acetate gives this color after 1 to 2 hours. (**CAUTION:** *Handle the beaker containing sulfuric acid with care.*)

10. **Protein Test** Place a fabric square on a watch glass, and add ten drops of $0.05M$ $CuSO_4$. Wait 5 minutes, and then use forceps to dip the fabric into $3M$ NaOH in a test tube for 5 seconds. Silk and wool are protein polymers, and a dark violet color will appear on those fabrics after doing this test.

ChemLab

11. **Formic Acid Test** Place any sample to be subjected to the formic acid test in a test tube and bring it to your teacher, who will perform this test in the fume hood. The teacher will add 1 mL of formic acid to the test tube and stir with a glass rod. Note whether or not the fabric dissolves in the solution. Silk, acetate, and nylon dissolve in formic acid.

12. **Acetone Test** Add a fabric square to 1 mL of acetone in a test tube, stir with a glass rod, and note whether or not the fabric dissolves. Only acetate dissolves in acetone. (**CAUTION:** *Be careful to do this test away from any open flames.*)

13. Dispose of all products of your tests as directed by your teacher.

ANALYZE AND CONCLUDE

1. **Thinking Critically** Why do you think it was necessary to add NaOH and heat before adding $BaCl_2$ in the sulfur test?

2. **Comparing and Contrasting** Can you make any conclusions about the way that synthetic and natural polymers burn?

3. **Classifying** Use your data from the flame and chemical tests to classify each fabric sample you tested as a silk, cotton, wool, nylon, acrylic, acetate, or polyester polymer.

APPLY AND ASSESS

1. What does the plastic, beadlike residue left by some polymers after burning tell about the polymers' structures?

2. If burning silk and wool smell like burning hair, what does this tell about the structure of hair?

3. What results of the flame and chemical tests would you expect to see if you were testing a fabric sample that was a polyester-cotton blend?

4. Cellulose is a major component of wood as well as of cotton. Would you expect wood to also give a dark blue color in the cellulose test?

DATA AND OBSERVATIONS

Table 1   Flame Test Observations

Sample	Type of Burning	Burning Odor	Residue Type
A			

Table 2   Chemical Test Observations

Sample	Nitrogen	Sulfur	Cellulose	Protein	Formic Acid	Acetone
A						

When Polymers Meet Water

Sometimes, the ability of certain polymers to repel water is useful—this property is what keeps you dry when you wear a raincoat. Other times, it is desirable for polymers to absorb water. This is why wool socks keep your feet warm by wicking water away from your skin and why a diaper helps keep a baby dry. Cloth diapers are made of cotton, a natural polymer that absorbs water well. Why do disposable diapers hold so much more water than cloth diapers? They contain a superabsorbent polymer that can hold hundreds of times its weight in water. In this MiniLab, you will determine how much water the polymers in two different brands of diapers can hold.

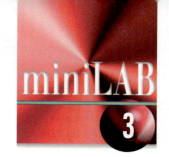
Procedure

1. Obtain two diapers of approximately the same size but of different brands.

2. Fill a 100-mL graduated cylinder with tap water, and slowly pour the water into the center of one of the diapers. Stop when the water begins leaking out of the diaper.

3. Record the volume of water the diaper holds.

4. Repeat steps 2 and 3 with the other diaper.

Analysis

1. What could be different about the two diapers that makes one of them hold more water than the other?

2. How might placing a superabsorbent polymer around the roots of a houseplant help it to grow?

Natural Polymers

Laboratories are not the only place where polymers are synthesized. Living cells are efficient polymer factories. Proteins, DNA, the chitin exoskeletons of insects, wool, silky spiderwebs and moth cocoons, and the jellylike sacs that surround salamander eggs are polymers that are synthesized naturally. The strong cellulose fibers that give tree trunks enough strength and rigidity to grow hundreds of feet tall are formed from monomers of glucose, which is a sweet crystalline solid. Many synthetic polymers were developed by scientists trying to improve on nature. For example, nylon was developed as a possible silk substitute. The idea for the process by which synthetic threads are formed in factories was borrowed from spiders. Observe in **Figure 18.16** the similarities between the spinneret of a spider and an industrial spinneret.

Figure 18.16
Thread Spinners
Long, fine threads are spun when polymer molecules are forced through tiny holes in spinnerets, both natural and industrial.

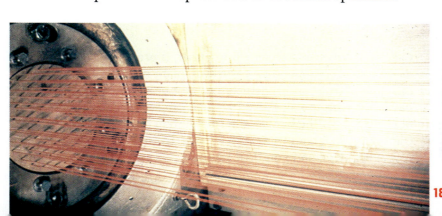

Structure of Polymers

If you examine the structure of a polymer, you can identify the repeating monomers that formed it. Because polymer molecules are large, they are commonly represented by showing just a piece of the chain. The piece shown must include at least one complete repeating unit. Look carefully at the structure of a segment of a cellulose molecule shown in **Figure 18.17.** Cellulose is a polymer found in the cell walls of plant cells such as those of wood, cotton, and leaves. It is responsible for giving plants their structural strength. Can you find the portions of the cellulose structure that repeat? Notice that the ring parts of the molecule are all identical. These are the monomer units that combine to form the polymer. Glucose is the name of the monomer found in cellulose. In **Figures 18.17** and **18.18,** the glucose units are shown in simplified form without carbon and hydrogen atoms. The complete structure of glucose is shown below.

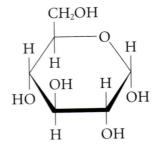

Glucose

Cellulose

Figure 18.17

Cellulose

Cellulose, the main component of plants, is probably the most abundant organic compound on Earth. This plant material, found in the cell walls of fruits and vegetables, cannot be digested by humans. It passes through the digestive tract unchanged, serving as dietary fiber that keeps the human digestive system healthy. Cellulose molecules range in length from several hundred to several thousand glucose units, depending on their source.

Figure 18.18

Starch

Starch is digested by humans and serves as an important nutrient. During digestion, starch is broken down into glucose molecules that can be absorbed in the digestive tract. When starchy potatoes and cereal grains are cooked, they form a paste that thickens soups and stews. Starch molecules are usually several hundred glucose units in length. In addition to forming linear chains, glucose molecules form a branched polymer. Starches are a mixture of linear and branched molecules, both consisting only of glucose monomers.

Starch

Another natural plant polymer that is formed from glucose monomers is starch. Examine the structure of starch in **Figure 18.18** to see if you can find the difference between starch and cellulose. Both cellulose and starch are made from only glucose monomers. The difference between them is the way that these monomers are bonded to each other. In starch, the oxygen atom joining each pair of monomers is pointed downward; in cellulose, the oxygen atom is pointed upward. This appears to be only a minor difference, but it changes the properties of the two polymers dramatically.

Polymerization Reactions

How do you think the many different types of polymers are formed? Polymerization is a type of chemical reaction in which monomers are linked together one after another to make large chains. The two main types of polymerization reactions are addition polymerization and condensation polymerization. The type of reaction that a monomer undergoes depends upon its structure.

Ethylene monomers   Polyethylene

Monomers   Addition   Polymer   +   More   Larger
                                           monomers   polymer

Figure 18.19

Addition Reactions

Ethylene monomers undergo an addition reaction to form the polyethylene that is used in plastic bags, food wrap, and bottles. The extra pair of electrons from the double bond of each ethylene monomer is used to form a new bond to another monomer.

Addition Reactions

The reaction by which Teflon is made from its monomer, tetra-fluoroethylene, is called an **addition reaction.** In this type of reaction, monomers that contain double bonds add onto each other, one after another, to form long chains. The product of an addition polymerization reaction contains all of the atoms of the starting monomers. Notice in **Figure 18.19** that the ethylene monomer contains a double bond, whereas there are none in polyethylene. When monomers are added onto each other in addition polymerization, the double bonds are broken. Thus, all of the carbons in the main chain of an addition polymer are connected by single bonds. Other polymers made by addition polymerization are illustrated in **Figure 18.20.**

Figure 18.20

Polymers Made by Addition Polymerization

Molten low-density polyethylene can be formed into a film. This tough plastic forms a barrier to food odors, which makes it useful for wrapping and storing foods. ▼

▲ Polyvinylacetate is another plastic made by addition polymerization. When mixed with sugar, flavoring, glycerol (for softening), and other ingredients, it becomes chewing gum.

Everyday Chemistry

Chemistry and Permanent Waves

A series of bad-hair days may drive you to the hair stylist in search of a permanent wave. Your hair endures the chemical changes caused by a perm.

What happens in a permanent wave? Permanent-wave lotions are designed to penetrate the scales of the cuticle, the outer layer of the hair shaft. The lotions work because they affect the structure of the proteins that make up the hair. The amino acid cysteine, which contains an atom of sulfur, is found in human hair protein. Sulfur atoms in neighboring cysteine molecules within the hair protein form strong, covalent, disulfide (S—S) bonds. This cross-linking between cysteine molecules holds the strands of hair protein in place and affects its shape and strength.

When you have a permanent wave, the waving lotion that is applied breaks the disulfide bonds. Then, after the hair is set in a new style, it is treated with a neutralizing lotion, which creates new cross-links between cysteine molecules to hold the hair in the new style. Thus, two chemical reactions occur. In the first, the waving lotion reduces each disulfide bond to two —SH groups. In the second, the neutralizing lotion oxidizes the —SH groups to form new disulfide bonds.

Kinds of permanent waves Two kinds of permanent-wave lotions are available. Alkaline lotions have as the reducing agent ammonium thioglycolic acid. The alkalinity causes the scales of the cuticle to swell and open, allowing the solution to penetrate rapidly. The advantages of the alkaline permanent are that it forms stronger, longer-lasting curls; takes a shorter time (usually 5 to 20 minutes); and occurs at room temperature. The alkaline permanent is used on hair capable of resisting damage.

Acid-balanced permanent-wave lotions contain monothioglycolic acid. Normally, acidic lotions penetrate the cuticle only slightly. That's why the process takes a longer time and requires heat in order for curls to develop. The advantages of the acid-balanced wave are that it forms softer waves, is more easily controlled, and can be used for delicate hair or hair that has been colored.

$$\begin{array}{ccc}
\text{NH} & & \text{NH} \\
| & & | \\
\text{O}=\text{C} & & \text{O}=\text{C} \\
| & & | \\
\text{CH}-\text{CH}_2-\text{S}-\text{S}-\text{CH}_2-\text{CH} \\
| & & | \\
\text{NH} & & \text{NH} \\
| & & | \\
\text{O}=\text{C} & & \text{O}=\text{C} \\
| & & |
\end{array}$$

Linkage between two cysteine molecules

Exploring Further

1. **Investigating** Find out more about the protein in hair. Where else is this protein found?

2. **Applying** Why might hair that has been colored or recently permed absorb lotion more rapidly than untreated hair?

To learn more about the chemistry that goes into hair styling, visit the Chemistry Web site at chemistryca.com

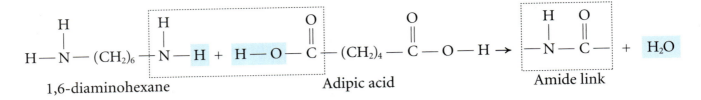

$$H-\overset{\overset{\displaystyle H}{|}}{N}-(CH_2)_6 \overset{\overset{\displaystyle H}{|}}{\underset{}{N}}-H \; + \; H-O-\overset{\overset{\displaystyle O}{\|}}{C}-(CH_2)_4-\overset{\overset{\displaystyle O}{\|}}{C}-O-H \; \rightarrow \; \overset{\overset{\displaystyle H}{|}}{\underset{}{N}}-\overset{\overset{\displaystyle O}{\|}}{C}- \; + \; H_2O$$

1,6-diaminohexane   Adipic acid   Amide link

Figure 18.21

Condensation Reactions

The condensation of two different monomers—1,6-diaminohexane and adipic acid—is used to make the most common type of nylon. Nylons are named according to the number of carbon atoms in each monomer unit. There are six carbons in each monomer, so this type of nylon is called nylon 66.

Condensation Reactions

In the second type of polymerization reactions, monomers add on one after another to form chains, as they do in an addition reaction. However, with every new bond that is formed, a small molecule—usually water—is also formed from atoms of the monomers. In such a reaction, each monomer must have two functional groups so it can add on at each end to another unit of the chain. This type of polymerization is called a **condensation reaction** because a portion of the monomer is not incorporated into the polymer but is split out—usually as water—as the monomers combine. A hydrogen atom from one end of a monomer combines with the —OH group from the other end of another monomer to form water. The condensation reaction used to make one type of nylon is shown in **Figure 18.21.** Other plastics made by condensation polymerization include Bakelite—which is the hard, ovenproof plastic used for handles of toasters and cooking utensils—and Dacron, which is used as fiber for clothing and carpets, backing on audiotapes and videotapes, and plastic wrap.

Figure 18.22

Natural and Synthetic Rubber

Latex or natural rubber harvested from rubber trees is soft and tacky when hot. ▼

Rubber

Another process that often occurs in combination with addition or condensation reactions is the linking together of many polymer chains. This is called **cross-linking,** and it gives additional strength to a polymer, **Figure 18.22.** In 1844, Charles Goodyear discovered that heating the latex from rubber trees with sulfur can cross-link the hydrocarbon chains in the liquid latex. The solid rubber that is formed can be used in tires and rubber balls. The process is called vulcanizing, named for Vulcan, the Roman god of fire and metalworking.

Rubber has a white to brownish-yellow color. Automobile tires are black because the carbon allotrope carbon black is added to strengthen the polymer. ▶

When a piece of vulcanized, cross-linked rubber such as a rubber band is stretched and then released, the cross-links pull the polymer chains back to their original shape. Without vulcanization, the chains would slide past one another. ▼

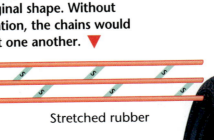

Sulfur links

Unstretched rubber   Stretched rubber

Recycling Plastics

The growing mass of throwaway materials has caused more and more landfills to reach their capacity. In response, recycling has caught on in many parts of the country. People sort their trash into categories: garbage, paper, glass, and plastics. Because garbage and paper are biodegradable and glass can be reused, the focus is on plastics. Thirty percent of the volume of waste in the United States is composed of plastics. Unfortunately, recycling of plastics is more complex than for most other materials. Five plastics are commonly found in landfills. They are polyethylene—both high- and low-density, polyethylene terephthalate, polystyrene, polyvinyl chloride, and polypropylene.

How are plastics used? Polyethylene is the most widely used plastic. High-density polyethylene (HDPE) is used in rigid containers such as milk and water jugs and in household and motor-oil bottles. Low-density polyethylene (LDPE) is usually used for plastic film and bags. Polyethylene terephthalate (PET) is found in rigid containers, particularly carbonated-beverage bottles. Polystyrene (PS) is best known as a foam in the form of plates, cups, and food containers, although in its rigid form, it is used for plastic knives, forks, and spoons. Polyvinyl chloride (PVC) is a tough plastic that is used in plumbing and construction. It is also found as shampoo, oil, and household-product containers. Finally, polypropylene (PP) has a variety of uses, from snack-food packaging to battery cases to disposable-diaper linings.

The Society of the Plastics Industry has designated code numbers and acronyms to help people distinguish among plastics and to make communication about them uniform. The codes are useful in sorting plastics and in deciding what method will be used for recycling. In addition to having a distinct chemical composition, each kind of plastic has distinct physical properties that determine how it can be used.

Code		Material
△ 1	PET	Polyethylene terephthalate
△ 2	HDPE	High-density polyethylene
△ 3	PVC	Polyvinyl chloride
△ 4	LDPE	Low-density polyethylene
△ 5	PP	Polypropylene
△ 6	PS EPS	Polystyrene Foamed polystyrene

Recycling plastics PET beverage bottles and HDPE milk and water bottles receive the most attention because they are the easiest to collect and sort. PET bottles require a drawn-out recycling process because they are made of several materials. Only the body of the bottle is PET. The base is HDPE, the cap is another kind of plastic or aluminum, and the label has adhesives. The bottles are shredded and ground into chips for processing. The adhesives can be removed with a strong detergent. The lighter HDPE is separated from PET in water because one sinks and the other floats. The aluminum is removed electrostatically. What is left are plastic chips that may be sold to manufacturers who can use the chips in making other plastics. However, the FDA prohibits the use of recycled plastics in food containers, making one major market out of bounds for recycled plastics.

Analyzing the Issue

1. **Debating** Debate whether industries should be held responsible for recycling the packaging in which their products are sold.

2. **Writing** Prepare a letter or article for a newspaper in which you support the use of less packaging for food products.

When World War II resulted in the cutting off of the Allies' supply of natural rubber, the polymer industry grew rapidly as chemists searched for rubber substitutes. Some of the most successful substitutes developed were gas- and oil-resistant neoprene, now used to make hoses for gas pumps, and styrene-butadiene rubber (SBR), which is now used along with natural rubber to make most automobile tires. Although synthetic substitutes for rubber have many desirable properties, no one synthetic has all the desirable properties of natural rubber.

Plastics

Although the terms *plastic* and *polymer* are often used synonymously, not all polymers are plastics. Plastics are polymers that can be molded into different shapes. What physical state has a fixed volume that you can pour into a mold and that will take the shape of its mold? Only liquids will. After a polymer has formed, it must be heated enough to become liquefied if it is to be poured into a mold. After pouring, the plastic will harden if it is allowed to cool, **Figure 18.23.**

Some plastics will soften and harden repeatedly as they are heated and cooled. This property is described as being **thermoplastic.** Thermoplastic materials are easy to recycle because each time they are heated, they can be poured into different molds to make new products. Polyethylene and polyvinylchloride are examples of this type of polymer.

Other plastics harden permanently when molded. Because they are set permanently in the shape they first form, they are called **thermosetting** polymers. Thermosetting plastics usually are rigid because they have many cross-links. No matter how much they are reheated, they won't soften enough to be remolded; instead, they get harder when heated because the heat causes more cross-links to form. Bakelite is this type of molecule. Even though thermosetting polymers are more difficult to reuse than thermoplastics, they are durable. **Figure 18.24** shows the relative amounts of various plastics produced for packaging in 1987.

Figure 18.23

Molded Plastics
Molded plastics are used for making objects with strength and durability qualities that are superior to those of earlier materials. Whereas a wood picnic table eventually will decay, plastic is so difficult to break down that it can create disposal problems as plastic materials pile up in landfills. Many picnic tables are already made of recycled plastic.

Figure 18.24

Recycling Plastics

The relative amounts of various plastics produced for packaging in 1987 are shown here. 87 percent of the plastics sold for packaging are thermoplastic. Polyethylenes and PVC are the most recyclable forms of plastic because they are so easy to remelt and reprocess.

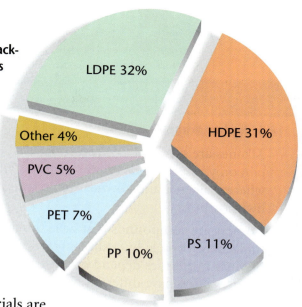

Connecting Ideas

Although polymers are large, complex molecules, their structures are easier to understand when you realize they are made from chains of simpler building blocks. Living things consist mostly of organic compounds, many of them extremely complex polymer structures. Keep in mind, however, that the chemical reactions through which living things get energy and building materials are exquisitely organized processes whereby a relatively small number of building blocks are used to make the many thousands of biochemicals that cells need. You will study these biomolecules in Chapter 19.

SECTION REVIEW

Understanding Concepts

1. Identify the repeating unit that appears in each of the following polymers.

   a) polystyrene

   $\sim CH_2CHCH_2CHCH_2CHCH_2CH \sim$

   b) polyvinylchloride

   $\sim CH_2CHCH_2CHCH_2CHCH_2CH \sim$
   $\quad\quad | \quad\quad\, | \quad\quad\, | \quad\quad\, |$
   $\quad\quad Cl \quad\, Cl \quad\, Cl \quad\, Cl$

   c) Saran

   $\quad\quad Cl \quad Cl \quad Cl \quad Cl$
   $\quad\quad | \quad\, | \quad\, | \quad\, |$
   $\sim CH_2CCH_2CCH_2CCH_2C \sim$
   $\quad\quad | \quad\, | \quad\, | \quad\, |$
   $\quad\quad Cl \quad Cl \quad Cl \quad Cl$

   d) rubber

2. Draw the structure of the polymer that will be formed from each of the monomers shown.

   a) $CH_2{=}CHCl$
   b) NH_2CH_2COOH
   c) $CH_2{=}CHOH$

3. Compare and contrast addition and condensation polymerization reactions.

Thinking Critically

4. **Making Predictions** Can a polymer be made from alkane monomers by addition polymerization? Explain.

Applying Chemistry

5. **Paint Polymers** Paints usually contain three components: a binder that hardens to form a continuous film, a colored pigment, and a volatile solvent that evaporates. In latex paints, one of these components is a polymer. Use what you have learned about polymer properties to decide which of the three most likely is a polymer. Explain.

REVIEWING MAIN IDEAS

18.1 Hydrocarbons

- Alkanes are saturated hydrocarbons that contain only single bonds between carbon atoms, whereas alkenes and alkynes are unsaturated hydrocarbons.

- Isomers are compounds that have the same molecular formula but different structures. Because properties depend upon structure, isomers often have different properties.

- Benzene is one of a group of unusually stable cyclic hydrocarbons. This stability is due to the sharing of six electrons by all the carbon atoms in the molecule.

- Fossil fuels are our major sources of hydrocarbons. The many components in petroleum are separated by fractional distillation.

18.2 Substituted Hydrocarbons

- Special bond organizations, atoms, or groups of atoms that give predictable characteristics to molecules are called functional groups.

- Atoms and groups of atoms are substituted in carbon rings and chains to form substituted hydrocarbons.

- Compounds containing a specific functional group share characteristic properties.

18.3 Plastics and Other Polymers

- Polymers are large molecules made from many smaller units called monomers that repeat over and over again. Many synthetic polymers can be made from smaller monomers. Living cells also make many polymers.

- Addition and condensation are the main reactions by which polymers are made. In addition reactions, all atoms of monomers add end-to-end to form chains. In condensation reactions, a molecule of water is released as each new bond forms between two monomer units.

- Thermoplastic polymers can be recycled for different purposes because they can be repeatedly softened by heating and hardened by cooling. Thermosetting plastics become hardened permanently.

Vocabulary

For each of the following terms, write a sentence that shows your understanding of its meaning.

addition reaction	monomer
alkane	polymer
alkene	reforming
alkyne	saturated
aromatic hydrocarbon	hydrocarbon
condensation reaction	substituted
cracking	hydrocarbon
cross-linking	thermoplastic
fractional distillation	thermosetting
functional group	unsaturated
isomer	hydrocarbon

UNDERSTANDING CONCEPTS

1. Draw structures of the following hydrocarbons.

   a) pentane   d) nonane
   b) propene   e) 2-methylpentane
   c) 1-butyne   f) methylpropene

2. List four natural and four synthetic polymers.

3. What molecule is usually released when monomers combine in a condensation reaction?

4. Name the branched alkanes shown.

   a)

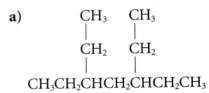

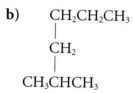
b) $CH_2CH_2CH_3$
                |
            CH_2
                |
        CH_3CHCH_3

5. Describe how the boiling points of alkanes change as chain length increases.

6. Why is there no compound named 4-methyl-hexane? What is the correct name for this compound?

7. Distinguish between thermoplastic and thermosetting materials.

8. List two biological functions of esters.

🔴 APPLYING CONCEPTS

9. Copy the following structure of penicillin G, an antibiotic. Circle all the functional groups in the molecule.

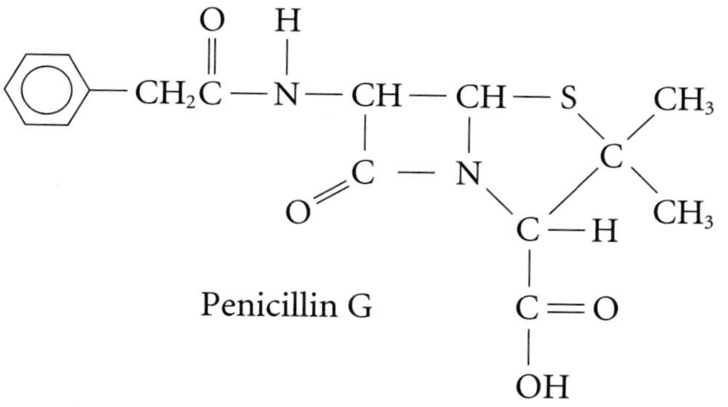

Penicillin G

10. Draw the structure of an ether, where R is —CH_3 and R' is —CH_2CH_3.

11. Predict whether an amine will be more soluble in an acidic or in a basic aqueous solution. Explain.

12. Compare the structures of the following two polymers, and decide which would produce a better fabric to use for making umbrellas. Explain.

a) 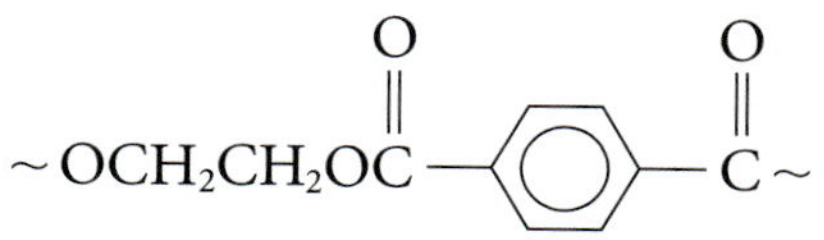
~OCH_2CH_2OC—⬡—C~

b) 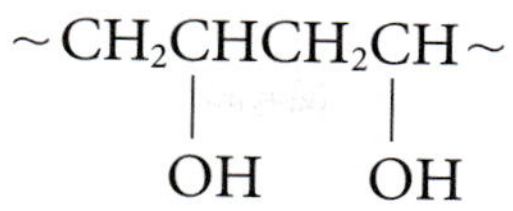 ~CH_2CHCH_2CH~
              | |
            OH   OH

Biology Connection

13. How many moles of H_2 molecules would be required to fully hydrogenate one mole of vitamin A, the structure shown here?

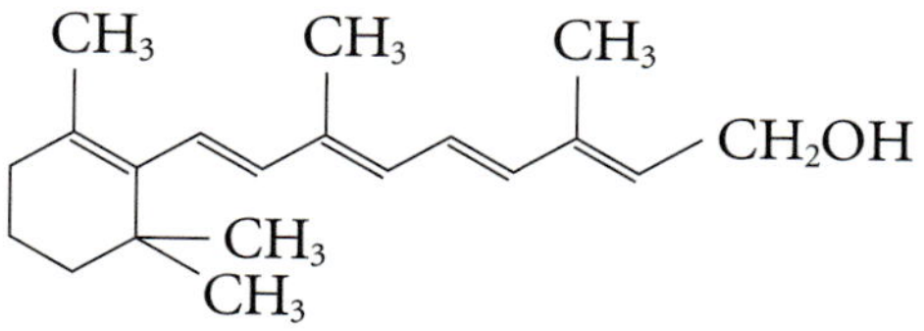

Everyday Chemistry

14. Compare and contrast the two different kinds of permanent-wave lotions.

Chemistry and Society

15. What are the full names and the corresponding codes of the two most commonly recycled plastics?

🔴 THINKING CRITICALLY

Identifying Patterns

16. The molecular formulas of the noncyclic alkanes follow the pattern C_nH_{2n+2}, where n is the number of carbon atoms. What pattern is followed by the noncyclic alkenes with one double bond? What pattern is followed by the noncyclic alkynes with one triple bond?

Designing an Experiment

17. ChemLab Design an experiment that shows how you could distinguish between two fabric samples if one is polyester and the other is a cotton-polyester blend.

Making Predictions

18. MiniLab 1 Predict whether iodine will be decolorized more quickly by melted butter or by melted margarine, assuming that both are at the same temperature.

Applying

19. MiniLab 2 What is the name of the ester formed by reacting ethanol and acetic acid?

Making Predictions

20. MiniLab 3 Disposable diapers contain two different polymers. Examine the structures of the polymers shown, and determine which would best function on the inside and which on the outside of a diaper.

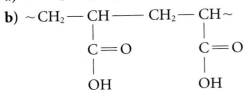

a) $\sim CH_2CH_2CH_2CH_2 \sim$

b)

SKILL REVIEW

21. Interpreting a Graph Examine the following graph and answer the questions.

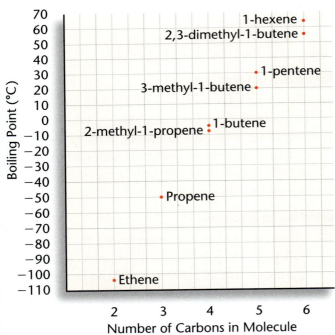

Boiling Point As a Function of Alkene Branching and Molecule Size

a) How does the number of carbon atoms in an alkene affect its boiling point?

b) Is there a relationship between the amount of branching and the boiling point of an alkene? If so, what is the relationship?

22. Hypothesizing The gasoline blend sold in hot climates consists of hydrocarbons of larger molecular mass than the gasoline blend sold in cold climates. Write a report suggesting a reason why refiners might vary the blends in this way.

23. Designing an Experiment Imagine you are a chemist at a textile company and are in charge of a research project aimed at developing a synthetic cotton substitute. What kind of information would you have to collect or questions would you have to ask to solve the problem? Design an experiment for beginning this research.

24. Compare the structures of the following two polymers, and decide which would produce a better fabric to use for making umbrellas. Explain.

a)
$$\sim OCH_2CH_2OC \overset{O}{\underset{\|}{}} - \bigcirc - \overset{O}{\underset{\|}{C}} \sim$$

b) $\sim CH_2CHCH_2CH \sim$ with OH and OH groups

WRITING IN CHEMISTRY

25. Do library research to find information about government regulations controlling levels of sulfur in coal-burning emissions. Write a report giving arguments for or against these regulations. If you can think of a better way to reduce sulfur emissions, give your ideas in your report.

PROBLEM SOLVING

26. a) Write the balanced reaction for the synthesis of ethanol from ethene and water.

b) If 448 L of ethene gas are reacted with excess water at STP, how many grams of ethanol will be produced?

27. a) Write a balanced equation for the complete combustion of hexane.

b) Use the equation you wrote in part **a.** to determine how many moles of CO_2 will be produced from burning 4.25 moles of hexane, assuming 100 percent yield.

1. How many hydrogen atoms will an alkane with 11 carbon atoms have?

a) 11
b) 18
c) 22
d) 24

2. Isomers are

a) compounds with the same chemical formula but different structures.
b) compounds with the same structure but different chemical formula.
c) compounds with the same number of carbon atoms but different types of bonds.
d) compounds with a different number of carbon atoms but the same structure.

3. Hydrocarbons containing a triple bond are called

a) isomers.
b) alkynes.
c) alkenes.
d) alkanes.

4. The most reactive type of hydrocarbon is an

a) isomer.
b) alkyne.
c) alkene.
d) alkane.

5. A hydrocarbon with the formula C_6H_6 will be a(n)

a) carbon ring.
b) alkane.
c) straight chain of carbon atoms.
d) bent chain of carbon atoms.

6. The compound pictured below is a(n)

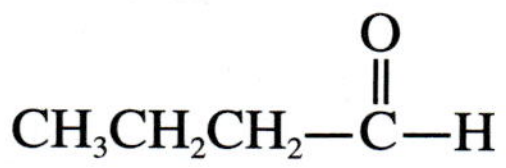

$$CH_3CH_2CH_2-\overset{\overset{\textstyle O}{\|}}{C}-H$$

a) ether.
b) aldehyde.
c) ketone.
d) alcohol.

7. What type of compound does this molecule represent?

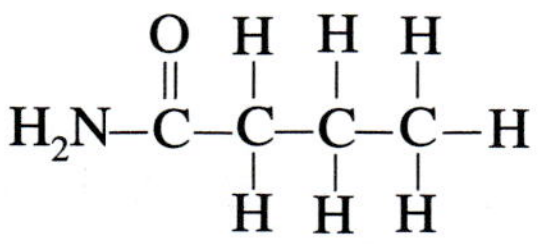

$$H_2N-\overset{\overset{\textstyle O}{\|}}{C}-\overset{\overset{\textstyle H}{|}}{\underset{\underset{\textstyle H}{|}}{C}}-\overset{\overset{\textstyle H}{|}}{\underset{\underset{\textstyle H}{|}}{C}}-\overset{\overset{\textstyle H}{|}}{\underset{\underset{\textstyle H}{|}}{C}}-H$$

a) amine
b) amide
c) ester
d) ether

8. A plastic is a

a) hardened polymer.
b) moldable polymer.
c) heated polymer.
d) heat resistant polymer.

Test Taking Tip

Wear A Watch If you are taking a timed test, you should make sure that you pace yourself and do not spend too much time on any one question, but you shouldn't spend time staring at the clock. When each section of the test begins, set your watch for noon. This will make it very easy for you to figure out how many minutes have passed. After all, it is much easier to know that you started at 12:00 (according to your watch) and you'll be done at 12:30 than it is to figure out that you started at 10:42 and that time will run out at 11:12.

19 The Chemistry of Life

Chapter Preview

Sections

19.1 Molecules of Life

MiniLab 19.1 DNA—The Thread of Life

ChemLab Catalytic Decomposition— It's in the Cells

19.2 Reactions of Life

MiniLab 19.2 Yeast Plus Sugar— Let it Rise

Oxygen Anyone?

Red blood cells like these flow throughout your body. These cells carry oxygen from your lungs to every part of your body and then return with waste gases to be expelled by your lungs. This is just one of the chemical processes that occurs in your body.

Start-up Activities

Launch Lab

Testing for Simple Sugars

Your body constantly uses energy. Many different food sources can supply that energy, which is stored in the bonds of molecules called simple sugars. What foods contain simple sugars?

Safety Precautions

Materials

- 400-mL beaker
- hot plate
- 10-mL graduated cylinder
- Benedict's solution
- 10% glucose solution
- other food solutions such as 10% starch or honey
- test tube
- tongs
- boiling chip
- stirring rod

Procedure

1. Fill a 400-mL beaker one-third full of water. Place this water bath on a hot plate and begin to heat it to boiling.
2. Place 5.0 mL 10% glucose solution in a test tube.
3. Add 3.0 mL Benedict's solution to the test tube. Mix the two solutions using a stirring rod. Add a boiling chip to the test tube.
4. Using tongs, place the test tube in the boiling water bath and heat for five minutes.
5. Record a color change from blue to yellow or orange as a positive test for a simple sugar.
6. Repeat the procedure using food samples such as a 10% starch solution, a 10% gelatin (protein) suspension, or a few drops of honey suspended in water.

Analysis

Was a color change observed? Which foods tested positive for the presence of a simple sugar?

What I Already Know

Review the following concepts before studying this chapter.
Chapter 9: polar molecules; the geometry of hydrocarbons
Chapter 13: hydrogen bonding
Chapter 16: oxidation-reduction reactions
Chapter 18: functional groups, natural polymers

Reading Chemistry

Look through the key terms listed for both sections. As you go through the text, note any words that are used differently in the context of chemistry. After reading the chapter, compare your understanding of the vocabulary with your previous impressions.

Preview this chapter's content and activities at chemistryca.com

Molecules of Life

Many of the most important molecules in your body are polymers. Proteins, carbohydrates, and nucleic acids, all extremely large molecules, are formed from small monomer subunits. Although lipids are usually not considered to be polymers, they, too, are formed from smaller molecules that have been linked together.

You need relatively large amounts of proteins, carbohydrates, and lipids in your diet. Complex reactions in your cells use some of these molecules and a few others to make a fourth group of biomolecules, the nucleic acids, which are needed in smaller amounts. Vitamins and minerals, too, are required by your body. Cells need all of these compounds to form the many structural materials of which they are made. They also use and store some of these compounds as a source of energy.

SECTION PREVIEW

Objectives

✓ **Compare and contrast** the structures and functions of proteins, carbohydrates, lipids, and nucleic acids.

✓ **Analyze** the relationship between the three-dimensional shape of a protein and its function.

Review Vocabulary

Cross-linking: the linking together of many polymer chains, giving the polymer increased strength.

New Vocabulary

biochemistry
protein
amino acid
denaturation
substrate
active site
carbohydrate
lipid
fatty acid
steroid
nucleic acid
DNA
RNA
nucleotide
vitamin
coenzyme

Biochemistry

All living things, no matter how different they appear, are composed of a relatively few kinds of chemicals. Some of these chemicals are molecular and some are ionic. They are arranged into complex and highly organized materials that provide support, carry out energy transformations, or duplicate themselves. The study of the chemistry of living things is called **biochemistry.** This science explores the substances involved in life processes and the reactions they undergo. Other than water, which can account for 80 percent or more of the weight of an organism, most of the molecules of life—the biomolecules—are organic.

Are these biomolecules any different from the other organic chemicals that you have studied? Scientists once thought that these molecules possessed a vital force and did not follow the same chemical and physical principles that govern all other matter. Until the early 1800s, no biomolecule had been synthesized outside of a living cell, although many had been identified. But in 1828, a German physician and chemist named Friedrich Wöhler reported that he could "make urea without needing a kidney."

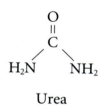

Urea

The organic molecule urea is normally made in your kidneys and excreted in urine to dispose of excess nitrogen. Wöhler had produced urea in the laboratory from ammonia and cyanic acid. He showed that it is possible to take lifeless molecules and produce one of the molecules of life in the laboratory. Today, it is possible to synthesize artificially many thousands of complex biomolecules. However, living cells still are the most efficient laboratories, and it can take months or years for chemists to synthesize a large molecule that a cell can make in seconds or minutes.

The elemental composition of living things is different from the relative abundance of elements in Earth's crust. As shown in **Figure 19.1**, oxygen, silicon, aluminum, and iron are the most abundant atoms in Earth's crust. However, more than 95 percent of the atoms in your body are hydrogen, oxygen, carbon, and nitrogen. All four of these elements can form the strong covalent bonds found in organic molecules. Along with two other elements, sulfur and phosphorus, they are the only elements needed to make most of the proteins, carbohydrates, lipids, and nucleic acids found in every cell.

Figure 19.1

Composition of Earth's Crust and the Human Body
The composition of Earth's crust and that of the human body are significantly different. The numbers represent percentages by mass in each. The elements oxygen and hydrogen are found in both, but carbon is concentrated in living things.

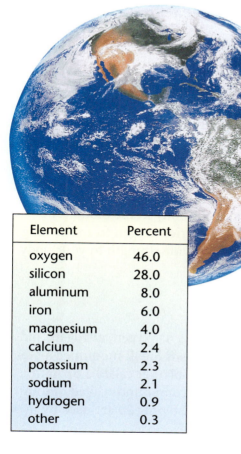

Element	Percent
oxygen	46.0
silicon	28.0
aluminum	8.0
iron	6.0
magnesium	4.0
calcium	2.4
potassium	2.3
sodium	2.1
hydrogen	0.9
other	0.3

Element	Percent
oxygen	65.0
carbon	18.5
hydrogen	9.5
nitrogen	3.3
calcium	1.5
phosphorus	1.0
sulfur	0.3
other	0.9

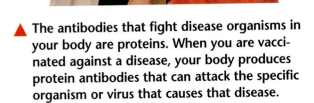

▲ The antibodies that fight disease organisms in your body are proteins. When you are vaccinated against a disease, your body produces protein antibodies that can attack the specific organism or virus that causes that disease.

Figure 19.2

The Functions of Proteins

▲ Keratin, the protein in hair and nails, is also the protein in the soft fur and tough hooves and horns of animals such as these bighorn sheep.

Proteins

What makes foods like chicken, beans, and fish so nutritious? They are composed largely of proteins. Proteins are one of the most important classes of biomolecules needed by living cells. When proteins were first discovered and named in the 19th century, their name was taken from the Greek word *proteios,* meaning "first" or "primary," because they were believed to be the chemical essence of life.

The Role of Proteins

Although about half of the proteins in your body function as catalysts in cell reactions, proteins play other important roles in living systems, as well. **Figure 19.2** shows some of these many functions. Structural proteins include collagen and keratin. Collagen is found in ligaments, tendons, and cartilage, and it provides the structural glue that holds your cells and tissues together. Keratin is the protein in hair, and it is also found in fur, hooves, skin, and fingernails. Other proteins make up your muscles and function to transport substances through your body in the blood. The hemoglobin that makes your blood red and carries oxygen from the lungs to the cells is a protein. Structural proteins provide strength to your body and give it shape.

Structure of Proteins

Proteins contain the elements carbon, hydrogen, oxygen, nitrogen, and sulfur. How are atoms of these elements arranged? Small monomer molecules link together by amide groups to form a polymer called a **protein.** The monomers that form proteins are organic compounds called **amino acids.** Although many different amino acids exist, only 20 are used by human cells to make proteins.

Fact *of the* **MATTER**

Some arctic fishes make unique antifreeze proteins that prevent ice crystals from forming in their blood. Scientists have transferred the genes for these proteins to tomato plants to make the plants tolerant of cold weather.

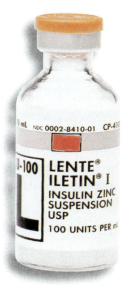

▶ Some proteins help cells communicate with each other. Insulin is a protein made in the pancreas. After a meal, insulin signals the body's cells to take in and use glucose from the blood. It does this by binding to receptor proteins embedded in the membranes surrounding the cells.

All amino acids have certain parts of their structure in common. Look at this general structure of an amino acid and see if you can recognize the two functional groups present. The name *amino acid* should give you a clue.

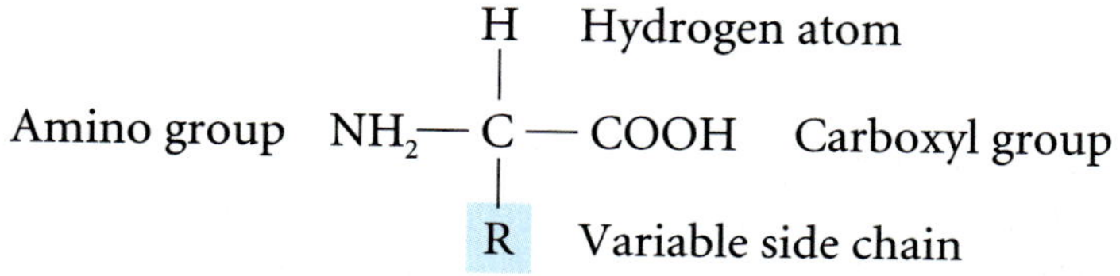

$$\text{Amino group} \quad NH_2 - \underset{\underset{R}{|}}{\overset{\overset{H}{|}}{C}} - COOH \quad \text{Carboxyl group}$$

H   Hydrogen atom

R   Variable side chain

On the left side of the structure is an amino group, —NH_2, and on the right side is a carboxyl group, —COOH. Both of these groups are attached to a central carbon atom that also has a hydrogen atom and a side chain attached. The side chain, represented in the structure by the letter *R*, may consist of a number of different atoms or groups of atoms and is what makes each of the 20 amino acids unique with its own particular properties. The structures of several different amino acids are shown here. Glycine is the simplest amino acid. Its side chain consists simply of a hydrogen atom. Alanine is the next simplest with a methyl-group side chain. Phenylalanine contains a benzene ring as part of its side chain, and cysteine contains a polar —SH group.

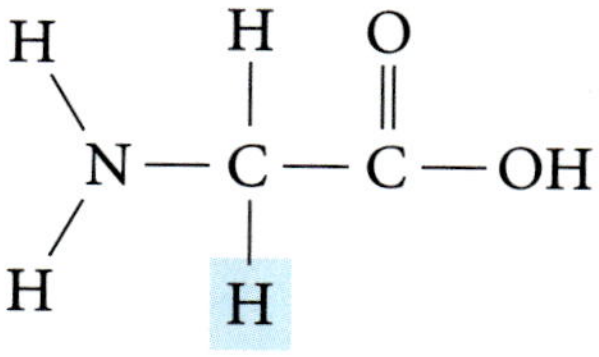

Glycine

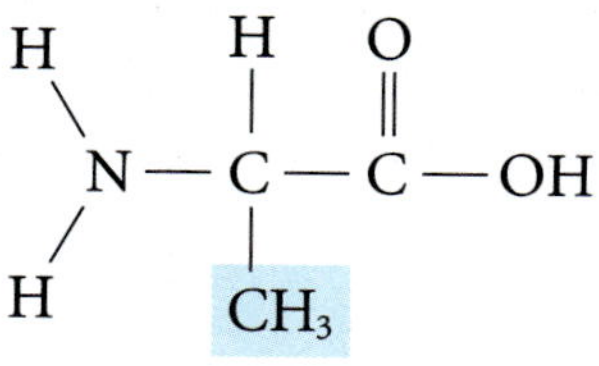

Alanine

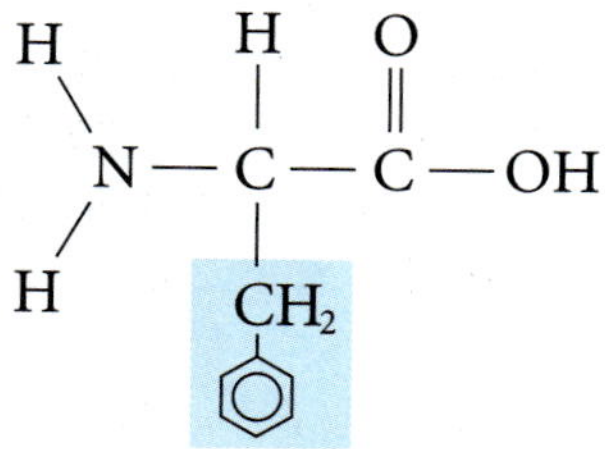

Phenylalanine

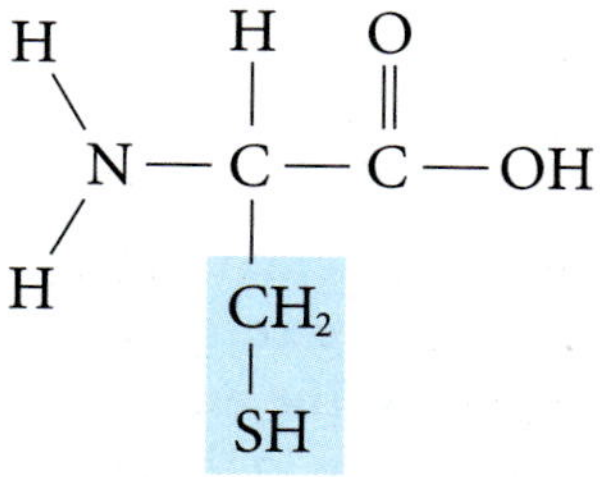

Cysteine

The polymerization reactions that form proteins are condensation reactions, similar to those that are used to make some of the plastics you studied in Chapter 18. When two amino acids bond together, a hydrogen (—H) from the amino group of one amino acid combines with the hydroxyl (—OH) part of the carboxyl group of the other amino acid to form a molecule of water (H_2O). When the water molecule is released, an amide group has formed, linking the two amino acids. Recall that an amide group has the following structure.

$$\begin{matrix} & O & H \\ & \| & | \\ -C & - & N- \end{matrix}$$

Biochemists call an amide group by the name *peptide bond* when it occurs in a protein. Although the names are different, the functional groups are identical. When two amino acids are linked by a peptide bond, the resulting chain of two amino acids is called a dipeptide.

Peptide bond

Glycine + Lysine → A dipeptide + Water

Additional amino acids can be added to a dipeptide by the same reaction to form a long chain. The chain is called a polypeptide because it is a polymer of amino acids held together by peptide bonds. A protein can consist of a single polypeptide chain, although most proteins contain two or more different polypeptide chains.

Three-Dimensional Protein Structure

Proteins have a three-dimensional structure because the polypeptide chains of which they are composed fold up like those shown in **Figure 19.3**. The polypeptide chains are held together by hydrogen bonds, ionic bonds, and disulfide cross-links between the side chains of neighboring amino acids.

Figure 19.3

Three-Dimensional Structure of Proteins

Proteins can fold into either round, globular structures or long, fibrous structures. The amino acid chains are held in place in three-dimensional structures by attractive forces between the side chains of different amino acids, which have been brought close together by the bending and folding of the polypeptide chains.

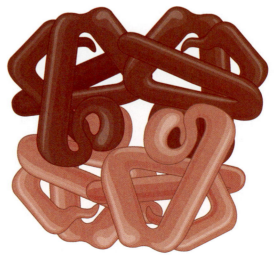

Hemoglobin

Collagen

The amino acids in a folded protein may be twisted into a helix or held tightly in flat sheets by hydrogen bonds that form between the $>$N—H and $>$C$=$O groups of neighboring amino acids. You may remember from Chapter 18 that disulfide bonds in the protein keratin give hair its strength and shape. These bonds form between sulfur atoms in the side chains of two molecules of the amino acid cysteine.

A molecule's geometrical shape is important in determining how chemical interactions take place. For example, the antibodies in your blood are folded into shapes so specific that each will bond to only one type of molecule specific to an invader. These antibodies protect your body from invasion by foreign organisms such as bacteria by bonding to molecules on the surface of the invader. Once bound, the invaders are destroyed by your immune system. Because bonding with invaders is so specific, your body must make a different kind of antibody for each kind of invader.

What happens when the three-dimensional structure of a protein is disrupted? Think of the difference between the consistency of a raw egg white and that of a hard-boiled egg. When the forces holding a polypeptide chain in its three-dimensional shape are broken, the protein is unfolded in a process called **denaturation.** High temperatures can denature proteins, which is why cooking foods that contain proteins results in the proteins' denaturation. Denatured proteins form the solid white of a hard-boiled egg. In addition, proteins can be denatured by extremes in pH, mechanical agitation, and chemical treatments. When egg whites are beaten, the proteins are denatured, as shown in **Figure 19.4.** Because the folded shape of a protein is essential for its function, denaturation of a protein results in loss of its function. This is one reason why organisms can live only in a narrow temperature and pH range.

The proteins of organisms that live in extremely cold or hot surroundings have three-dimensional structures held together by covalent bonds such as disulfide linkages. These bonds are stronger than the non-covalent hydrogen bonds that hold together the proteins of organisms living in more moderate locations. This strong bonding prevents denaturation and loss of protein function. The proteins of bacteria that live in hot springs such as this one in Yellowstone National Park are resistant to heat denaturation. ▼

Figure 19.4

Protein Denaturation

▲ The meringue on a lemon meringue pie owes its structure to denaturation of the proteins in egg whites. Beating the whites incorporates air and denatures some of the proteins, which provide the firm structure of the meringue.

Catalytic Decomposition— It's in the Cells

You may have used hydrogen peroxide, H_2O_2, as an antiseptic when you cut or scratched your skin. If you have, you will have noticed that it seems to fizz and bubble when it contacts your skin. Hydrogen peroxide decomposes to form water and oxygen. The decomposition reaction is more rapid when the hydrogen peroxide comes into contact with your skin. Catalase, an enzyme found in the cells of your skin as well as in most other cells, is responsible for this rapid decomposition.

The activity of an enzyme is affected by environmental factors such as pH and temperature. Every enzyme has optimum conditions at which its reaction rate is fastest. In this ChemLab, you will study the decomposition of hydrogen peroxide as catalyzed by the catalase in carrot cells, and you will determine the optimum temperatures under which this enzyme works.

Problem

How does temperature affect the catalytic decomposition of hydrogen peroxide by catalase from carrot cells?

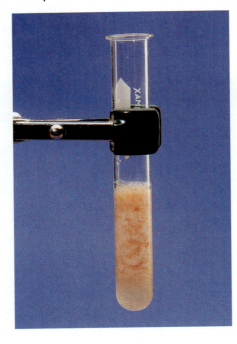

Objectives

- **Observe** the action of catalase on the decomposition of hydrogen peroxide.
- **Compare** the rate of reaction at various temperatures.
- **Make and use graphs** to interpret results.

PREPARATION

Materials

hot plate
small (13 mm × 100 mm), clean, unscratched test tubes (8)
test-tube rack
small beakers (4)
carrot-cell slurry
3% hydrogen peroxide solution

10-mL graduated cylinder
crushed ice
Celsius thermometer
glass stirring rod
timer or clock
metric ruler
marking pencil
thermal glove

Safety Precautions

Hydrogen peroxide can harm the eyes. Wear safety goggles when using it.

PROCEDURE

1. Place carrot-cell slurry into each of four test tubes to a depth of about 2 cm.

2. Carefully measure 3.0 mL of 3% H_2O_2 solution into each of another four test tubes. If large quantities of bubbles are produced, obtain a new tube or thoroughly clean the original one.

3. Set one carrot tube and one hydrogen peroxide tube in each of four small beakers. Label the beakers *A, B, C,* and *D.* Add crushed ice and tap water to one beaker, hot tap water to a second beaker, and room-temperature tap water to the remaining two beakers.

4. Set the hot-water beaker and one of the room-temperature-water beakers on the hot plate, and heat until the water temperatures reach 60-65°C and 37-38°C, respectively.

5. Make a data table similar to the one shown under Data and Observations. Take the temperature of the water in each of the four beakers, and record the temperature and label of each beaker. Let the test tubes sit in the beakers of water at 0°C, room temperature (probably about 18-25°C), 37-38°C, and 60-65°C for five to ten minutes.

6. Pour the tubes of H_2O_2 into the tubes of carrot-cell slurry at the same temperature. Stir each tube quickly with a glass stirring rod, and start the timer at the same moment. Measure the height of the foam, from the top of the liquid mixture to the top of the foam, at one-minute intervals for four minutes or until the foam reaches the top of the test tube. Record your data in the table.

ANALYZE AND CONCLUDE

1. **Interpreting Data** Use your data to construct and label a graph for each of your temperature trials. Plot foam height on the vertical axis and time on the horizontal axis.

2. **Interpreting Data** Construct another graph, plotting foam height at three minutes on the vertical axis and temperature on the horizontal axis.

3. **Comparing and Contrasting** Which of the temperatures appears to be optimum for the catalyzed decomposition of hydrogen peroxide by catalase? How does this temperature compare with human body temperature?

APPLY AND ASSESS

1. Peroxide ions are produced in plants and animals as a result of cellular reactions. Given that these peroxides can oxidize and damage cell structures, why is the presence of catalase in cells beneficial?

2. Suggest a way in which the gas in the foam might have been tested to verify that it was oxygen.

DATA AND OBSERVATIONS

Beaker Number	Temperature of Reactants	Height of Foam After 1 Minute	Height of Foam After 2 Minutes	Height of Foam After 3 Minutes	Height of Foam After 4 Minutes
A					
B					
C					
D					

The Role of Proteins as Enzymes

Although proteins are important for building strong muscles, that is only one of their many functions in living things. One of the most important roles they play is as the biological catalysts called enzymes. You may recall that enzymes help speed up reactions without being changed themselves during the reaction. A spontaneous chemical reaction may occur so slowly that it seems not to occur at all. Enzymes greatly speed up a reaction that would occur eventually, but they cannot make a reaction go if it would never occur spontaneously. How are enzymes able to speed up reactions? Enzymes provide specific sites, usually a pocket or groove in their three-dimensional structure, where each reactant, called a **substrate,** can bind. The pocket that can bind a substrate taking part in the reaction is called the **active site.** After a specific substrate binds to its active site, the active site changes shape to fit the substrate. This process of recognition is called induced fit. The shape of the substrate matches the shape of the active site, much as the shape of a key matches the hole in the lock it opens. Thus, only substrates that have a shape allowing them to bind to the active site can take part in a particular enzyme-catalyzed reaction. **Figure 19.5** shows how enzymes work.

Almost all reactions that take place in the body are catalyzed by enzymes. During digestion, enzymes speed up the breakdown of foods into molecules small enough to be absorbed by cells. Enzymes also make possible the reactions required for cells to extract energy from these nutrients. Enzymes are even involved in the production of other enzymes in cells.

Enzymes are now used in medicine to treat disorders that result from their deficiency. For example, lactase is an enzyme that breaks down milk sugar, which is called lactose, so that the two smaller sugar molecules of which it is composed can be absorbed by the digestive tract. Certain people who suffer from lactose intolerance gradually lose the ability to make the enzyme lactase as they mature. When this enzyme isn't present in the body, eating dairy products causes lactose to accumulate in the digestive system, resulting in gas,

Figure 19.5
Enzyme Action
Substrates are brought close together in the active sites of an enzyme, which lowers the activation energy of the reaction by facilitating the bonding together of the substrates to form a product. After the substrates have reacted, the product is released. The enzyme is then able to bind more substrate molecules and continue catalyzing the reaction. ▼

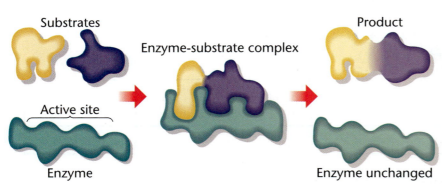

Substrates with shapes that don't match the shape of the active site won't undergo the catalyzed reaction, just as a house key can't be used to start a car. ▶

Figure 19.6

Uses of Enzymes
A group of enzymes found in many common household products is the proteases. These enzymes hydrolyze proteins into free amino acids. The proteases in meat tenderizer make steak more tender by breaking down the proteins. ▼

▲ Proteases in contact lens-cleaning solutions help remove the grimy buildup on lenses that results from proteins secreted by cells around the eyes.

bloating, and diarrhea. If lactase pills are taken along with the dairy products, the enzyme in the medicine breaks down the lactose just as the naturally produced enzyme would. **Figure 19.6** shows other everyday applications of enzymes.

Carbohydrates

Why do marathon runners eat large amounts of pasta the day before a big race? Foods such as pasta, bread, and fruit are rich in carbohydrates. A **carbohydrate** is an organic molecule that contains the elements carbon, hydrogen, and oxygen in a ratio of about two hydrogen atoms and one oxygen atom for every carbon atom. Early chemists thought that carbohydrates were chains of carbon with water attached—hydrated carbon—which is how carbohydrates were named. Chemists now know that carbohydrates are not hydrated carbon chains, but the name persists.

The Role of Carbohydrates

When the carbohydrates in pasta and other foods are broken down in the body, the sugar glucose is formed. Glucose is also called blood sugar because it is the sugar that is used in most body processes. The oxidation of glucose in cells provides most of the energy needed for life. When excess glucose is produced, as when athletes eat large quantities of pasta, the energy that can't be used right away is stored in the cells. Animals store glucose in their livers and muscles as glycogen, a polymer of glucose. In plants, glucose is stored as starch, the polymer you studied in Chapter 18.

PEOPLE in CHEMISTRY

Meet Dr. Lynda Jordan, Biochemist

The door to Dr. Lynda Jordan's office at the North Carolina Agricultural and Technical State University is nearly always open. Students and younger faculty members continually stop by for help with lab problems and personal problems. Dr. Jordan comments, "I know what my life would have been like if there hadn't been people saying, 'You can do it.' That's why I decided to 'put my business on the street.' Students need to know that being a good, hard-working person does pay off."

On the Job

 A Public Broadcasting Service TV special has honored you and your work with enzymes. Can you tell us about this project?

 I'm studying an enzyme involved in diseases such as asthma, arthritis, bronchial disorders, gastrointestinal disorders, preterm labor, and diabetes. I want to find out about this enzyme's structural and functional relationship and how it behaves in human cells. This enzyme is isolated from cells of the human placenta. This is not an easy task. My colleagues and I have classified it by its molecular weight, pH, and calcium dependencies.

Why is this research so important to you, personally?

I first started working on the enzyme, which is called phospholipase A, as part of my post-doctoral research at the Institute Pasteur in Paris, France, in 1985. During that time, we discovered that phospholipase was present in human placentas, and we were one of the first laboratory groups to isolate this calcium-independent protein. It was a very exciting time for me. I know many people that have the diseases associated with this enzyme. If I could understand how this enzyme functions, then perhaps I could make a contribution to understanding what happens in the diseased state. This information would help us in the scientific community to devise strategies that could circumvent these biomedical problems.

Which are the diseases that most interest you?

 Recent scientific evidence has shown a link between the enzyme phospholipase A_2 and diabetes. I am most interested in investigating the relationship between PLA_2 and glucose metabolism. Many of my family members, including myself, have diabetes. Understanding this disease on a molecular level would be rewarding for me on a personal as well as scientific level.

Early Influences

Q **What were you like as a high school student?**

A I was a kid from an inner-city high school, with no hope for going to college.
Although I was a good student, I used to skip classes. One day, when I was 15 years old, I was skipping class with a girlfriend in the rest room. The hall monitor came in, and I ran into the first room I came to, which happened to be an auditorium where an Upward Bound orientation was taking place. The speaker really got my attention. He asked me, 'What are you going to do the rest of your life—stand on the street corner and watch the world go by?' That's about all I was doing, and I was lucky not to end up a sad statistic. I signed up for the program right away.

Q **Can you describe the program?**

A It was a rigorous six-week, college-level program at Brandeis University. We lived in college dormitories and were not allowed to have contact with the inner-city environment. The objective was to change our environment, so we were scheduled from 6 A.M. to 11 P.M. every day. We were expected to give our very best, and we were pushed and pushed.

Personal Insights

Q **Mentors were crucial to your development. Are you yourself a mentor now?**

A Of all the things I do, the main thing is developing people. It's the most taxing, especially when I'm dealing with people who are first-generation college students, as I was. But it's also the most rewarding. I try to help students and younger faculty members with both their scientific problems and their personal problems, so they'll have the confidence to survive in

the big, bad world. Character has a lot to do with success. Honesty, trustworthiness, dependability, and acceptance of responsibility sometimes have to be learned.

Q **Can you draw a parallel between solving problems in the lab and solving them in everyday life?**

A Both are step-by-step processes. First, you have to collect information. Then, you devise a strategy for reaching a goal. In the meantime, you need to think of backup strategies to utilize in case things don't work out as you anticipate. While you're progressing, you need to monitor how far you've come. If you're investigating careers, for example, and are considering being a doctor, the first thing to do is visit a hospital and maybe an emergency room. All along the way to solving any problem, you need to take little baby steps and monitor your progress.

CAREER CONNECTION

These fields are associated with biochemistry.

Medical Technologist Three years of college followed by a one-year program in medical technology

Medical Laboratory Technician High school plus a two-year training program

Medical Records Technician High school plus a two-year training program

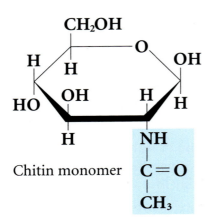

Chitin monomer

Figure 19.7
Chitin
Chitin is a structural carbohydrate that makes up the protective exoskeleton of arthropods such as beetles and lobsters. Chitin is one of the most abundant natural polymers on Earth due to the enormous number of insects on our planet. The monomer in chitin is similar to glucose, but it has a side chain containing an amide group attached to one carbon in the ring.

Carbohydrates also play a structural role in living things. You have learned that cellulose, also a glucose polymer, provides the strength and support needed in plants. Chitin, a natural polymer made from a sugar similar to glucose, forms the hard exteriors of insects and other arthropods, as shown in **Figure 19.7.** Carbohydrates are also found in cotton and rayon fabrics, wood, and paper.

Structures of Carbohydrates

Carbohydrates are a diverse group of molecules that include many simple sugar molecules, as well as larger molecules made by linking together different combinations of these sugars. Most simple sugars contain five, six, or seven carbon atoms arranged in a ring structure. An oxygen atom forms one corner of the ring, and hydroxyl groups are attached to each carbon. The most common simple sugars are glucose, fructose, and ribose.

Glucose   Fructose   Ribose

Simple sugar molecules are called monosaccharides. Two monosaccharides can link together in a condensation reaction to form a disaccharide, such as the most common carbohydrate in cakes and candies, sucrose. The structure of sucrose, which is also called table sugar, is made up of a glucose and a fructose monomer. **Figure 19.8** shows two sources of simple and complex carbohydrates in the diet.

$$\text{Sucrose} + H_2O \rightarrow \text{Glucose} + \text{Fructose}$$

Sucrose   + Water →   Glucose   +   Fructose

In order to be classified as a nutrient, a molecule must be able to enter cells. Disaccharides are too large to pass through cell membranes. However, enzymes in the digestive system catalyze the above reaction, breaking sucrose into glucose and fructose molecules that are small enough to be absorbed and used by cells as nutrients. Other common disaccharides in foods are milk sugar (lactose) and malt sugar (maltose).

Polysaccharides

What happens when you toast a slice of bread? Have you ever noticed that toast has a slightly sweet taste? The sweetness results from maltose, a disaccharide formed from two glucose molecules. Untoasted bread does not seem to taste sweet, and it contains no maltose. Where does the maltose come from when bread is toasted? It is the product of a reaction in which the large carbohydrates in bread—starch and cellulose—are broken down by heat. As you recall, these large carbohydrates are polymers that are formed from many glucose molecules linked together. Another common polymer consisting only of glucose is glycogen, also called animal starch. Starch, cellulose, and glycogen are examples of polysaccharides. Polysaccharides can contain hundreds or even thousands of monosaccharide molecules bonded together. A polysaccharide may contain only one type of monosaccharide, or it may contain more than one type.

Figure 19.8

Carbohydrates

Bread and jelly are both rich in carbohydrates. The sweet taste of jelly is a result of the disaccharide sucrose, whereas the structure and starchy taste of bread come from the polysaccharides cellulose and starch. Wood also contains cellulose. Check the ingredient label on your bread; sometimes, finely ground wood is added to increase the amount of fiber in the bread. This will be listed as *cellulose*.

Figure 19.9

Starch, Cellulose, and Glycogen
The starch in bread, the cellulose in cotton, and the glycogen in meat are all polymers of glucose. The glucose units in cellulose are bonded together somewhat like a chain-link fence. Starch molecules can be either branched or unbranched, and glycogen is highly branched.

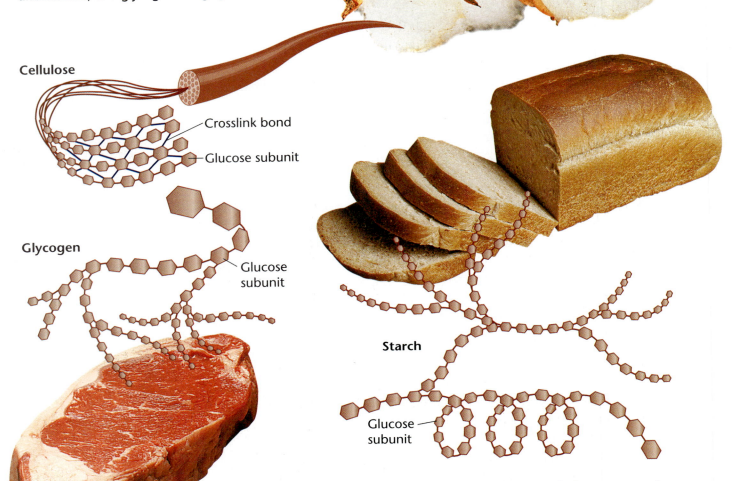

Cellulose

Crosslink bond

Glucose subunit

Glycogen

Glucose subunit

Starch

Glucose subunit

The repeating glucose units in starch, cellulose, and glycogen are the same in the starch of a potato as in the cellulose in the wood of a pencil. So, how can the structures of these three polymers be different? The major differences are the way the glucose molecules are bonded to each other, the number of glucose units in the polymer, and the amount of branching. These small differences in structure lead to large differences in properties and function. Their structures are compared in **Figure 19.9.**

Lipids

A biological compound that contains a large proportion of C—H bonds and less oxygen than in carbohydrates is called a **lipid.** Fats, oils, and waxes are all lipids. Lipids are insoluble in water and soluble in nonpolar organic solvents. In general, lipids that are derived from animals are called fats, and plant lipids are called oils. Lipid molecules are found in

Clues to Sweetness

What makes ripe strawberries taste so deliciously sweet? It has something to do with the sweetness message delivered to your brain when certain molecules from the strawberry fit into receptor sites in the taste buds on your tongue. The molecules lock onto the receptors in a specific way, determined by their structure. When this happens, chemical processes that produce the sweet response are stimulated. Even molecules of artificial sweeteners, many of which are not carbohydrates and are not metabolized by the body, are capable of matching the sweet receptor sites of the taste buds.

Breaking the code for sweetness

Hydrogen bonding is the most common interaction between molecules. When a hydrogen atom is bonded to a highly electronegative atom, its electron is pulled toward the more electronegative atom, and the hydrogen develops a partial positive charge. Consequently, it is capable of attracting other atoms that have partial negative charges. Nitrogen, oxygen, fluorine, and other nonmetals are all highly electronegative atoms that are attracted to hydrogen.

What does hydrogen bonding have to do with sweetness? Decades ago, chemists hypothesized that sweet-tasting molecules have two hydrogen-bonding sites—A and B—located close to each other. They postulated that the sites are separated by a distance of 0.3 nm and that site A has a hydrogen atom covalently bonded to atom A. They called these two sites AH and B, and named the structure the AH,B system. The AH,B system interacts with a similar system, which consists of an $>$NH group and a $>$C$=$O group, located on the sweet receptors of the tongue.

Protein molecules are ideally suited for forming

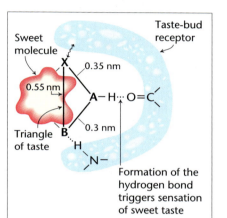

Sweet molecule

0.55 nm

Triangle of taste

X

0.35 nm

A—H···O=C

0.3 nm

B

H

N—

Taste-bud receptor

Formation of the hydrogen bond triggers sensation of sweet taste

pairs of hydrogen bonds because they have both $>$C$=$O and $>$NH sites. The more electronegative oxygen in the $>$C$=$O group seeks a hydrogen atom on another molecule, and the hydrogen in the $>$NH group seeks a more electronegative atom such as nitrogen or oxygen. Interestingly, the $>$C$=$O and $>$NH sites are strategically placed about 0.3 nm apart. A molecule that has complementary sites, also 0.3 nm apart, can bind to the protein. The AH,B system of a sweet molecule, with sites AH and B, fits this description. If the sweet molecule bonds to the protein that forms part of the sweet taste receptors at the tip of the tongue, the brain receives a signal of sweetness.

Refining the sweetness model

Chemists also noticed that sweet molecules shared a third common factor. Part of a sweet molecule is hydrophobic, which means it repels water. This part was called the X site. They found that X must be further from site B than from site AH. The whole molecular arrangement forms a triangle of sweetness.

Exploring Further

1. **Analyzing** What causes something to taste sweet?

2. **Acquiring Information** In a reference book, find a table comparing the relative sweetness of various sugars and artificial sweeteners. How do the following artificial sweeteners compare in sweetness with sucrose (table sugar): sucralose, aspartame, saccharin, and acesulfame-K?

For more on how our taste buds respond to sweeteners, visit the Chemistry Web site at
chemistryca.com

Figure 19.10
Lipids

Honeybees make lipids that they use to form the honey-comb structure of their hive. The walls of the honeycomb are made of a mixture of lipids known as beeswax. ▼

▲ Lipstick is a mixture mostly of oils and waxes. These lipids dissolve the dyes and cause the lipstick to flow smoothly over the lips. They also help keep the skin of the lips soft and moist.

The avocados in this guacamole contain large amounts of oils. ▼

the fatty foods you eat such as butter, margarine, peanut butter, and the oil your french fries are cooked in. The waxes you use to polish the family car or to grease the bottom of a pair of skis also are lipids. Waxes are produced by both plants and animals. **Figure 19.10** shows some other common lipids.

The Structure of Lipids

Most of the oils and fats in your diet consist of long-chain carboxylic acids called **fatty acids** that are bonded to a glycerol molecule. Glycerol is a small carbon chain with three hydroxyl functional groups. Three fatty acid molecules combine with a single glycerol in a condensation reaction to form three molecules of water and a lipid molecule with three ester functional groups. Each fatty acid contributes the hydroxyl part of its carboxyl (—COOH) group, and each hydroxyl group on the glycerol contributes the hydrogen atom to form the water molecules. The lipid formed is called a triglyceride.

$$
\begin{array}{llll}
CH_2OH & HO\overset{\overset{O}{\|}}{C}(CH_2)_{14}CH_3 & CH_2-O-\overset{\overset{O}{\|}}{C}-(CH_2)_{14}-CH_3 & \\
| & & | & \\
CHOH & + \quad HO\overset{\overset{O}{\|}}{C}(CH_2)_{16}CH_3 & CH-O-\overset{\overset{O}{\|}}{C}-(CH_2)_{16}-CH_3 & + \ 3\ H_2O \\
| & & | & \\
CH_2OH & HO\overset{\overset{O}{\|}}{C}(CH_2)_{18}CH_3 & CH_2-O-\overset{\overset{O}{\|}}{C}-(CH_2)_{18}-CH_3 &
\end{array}
$$

Glycerol   +   3 fatty acids   →   Triglyceride   +   Water

Fake Fats and Designer Fats

Fats have a reputation for causing health problems. At the same time, fats are an essential nutrient and are why many foods taste, look, and feel good as you eat them. Food scientists around the world are working on how to provide food that promotes good health, but also keeps the pleasing taste, appearance, and texture of fats.

Carbohydrates as fake fats Two kinds of carbohydrates—starches and cellulose—are being used to replace fats in foods. From the structures of these molecules in the chapter, you might wonder how starch and cellulose molecules can mimic fat's properties. When a starch is mixed with water, it forms gels that have the texture and bulk of fat. The gels can replace the fat in some foods, but they can't be used for frying.

Usually, carbohydrates contribute only four Calories per gram as opposed to the nine Calories per gram that come from fats. But cellulose, a carbohydrate in the cell walls of plants, contributes no Calories at all because the body is not able to metabolize it. Avicel is a form of natural cellulose that, when mixed with water, produces a texture similar to fat. It can replace fats in frozen desserts and bakery products.

Proteins for fats In order for a protein to mimic a fat, it has to be cut into tiny particles that measure only from 0.1 to 3.0 μm in size. Simplesse is the only protein-based fat replacer available in the United States. In one version, egg white and milk protein are subjected to high heat and stress to form small, spherical particles. The small size of these protein particles causes them to be perceived by the mouth as smooth and creamy. Like carbohydrates, protein replacers provide only four Calories per gram. However, they are not suitable for frying, and they lack the fat flavor.

Chemically altered fats Searching for the perfect no-Calorie, health-promoting fat, food chemists have come up with chemically altered fats. They alter the size, shape, or structure of real fat molecules so that the human body digests and uses them to a lesser extent or not at all. One chemically altered fat is Olestra, a sucrose polyester. Food technologists claim that Olestra will satisfy your taste buds without providing saturated fats and Calories. It can be substituted for butter and grease, and can even be used for frying. A fat consists of three fatty acids attached to glycerol. The altered fat Olestra has six to eight fatty acids, all derived from vegetable oils, attached to sucrose. Being larger than a typical fat molecule, it is not absorbed by the cells in the digestive system. Enzymes fail to break the sucrose-fatty acid bond, so the molecule passes through the body undigested.

Exploring Further

1. **Analyzing** What are the dangers of eating only foods in which all the fats have been replaced?

2. **Thinking Critically** Products made with non-digestible fake fats could trap fat-soluble vitamins or medicines in the intestines. What effect would this have on the body? How could this problem be solved?

To find out more about how scientists are working to make fat-free foods taste good, visit the Chemistry Web site at **chemistryca.com**

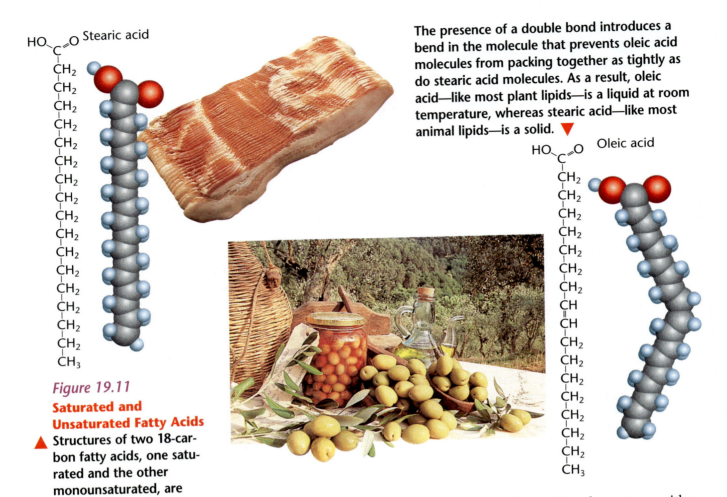

Stearic acid

The presence of a double bond introduces a bend in the molecule that prevents oleic acid molecules from packing together as tightly as do stearic acid molecules. As a result, oleic acid—like most plant lipids—is a liquid at room temperature, whereas stearic acid—like most animal lipids—is a solid. ▼

Oleic acid

Figure 19.11

Saturated and Unsaturated Fatty Acids

▲ Structures of two 18-carbon fatty acids, one saturated and the other monounsaturated, are shown. Stearic acid is found in pork and beef tissue; oleic acid is a major component of olive oil.

The most common fatty acids are chains of 12 to 26 carbon atoms with a carboxylic acid group at one end. They usually have an even number of carbon atoms because they are made from smaller molecules with two carbons. Saturated fatty acids, like saturated hydrocarbons, have only single bonds connecting the carbon atoms. Monounsaturated fatty acids have one double bond between two of the carbon atoms, as you can see in **Figure 19.11.** Fatty acids that are polyunsaturated have two or more double bonds. In general, animal lipids are more saturated than are plant lipids.

Steroids

Another class of lipids is the steroids. A **steroid** is a lipid with a distinctive four-ring structure. Important steroids include cholesterol, some sex hormones, vitamin D, and the bile salts that are produced in the liver to aid in digestion of fats.

Steroid ring structure

Cholesterol

Eating a diet high in saturated fats has been linked to the development of cardiovascular problems such as heart disease. The reason for this is not fully understood, but it may be due partly to the liver's ability to break down fatty acids into small pieces used to make cholesterol. High levels of cholesterol in the blood are associated with a stiffening and thickening of the artery walls known as atherosclerosis, illustrated in **Figure 19.12**. This condition can result in high blood pressure and heart disease. Reducing the amount of saturated fats and cholesterol in the diet—especially animal fats found in eggs, cheese, and red meats—is one way to lower blood cholesterol levels. However, changing the diet alone does not work for everyone because genetics, exercise, stress, and other factors also affect cholesterol level.

Try at Home Lab

See page 872 in Appendix F for
Counting Nutrients

The Functions of Lipids

Lipids have two major biochemical roles in the body. When an organism takes in and processes more food than it needs, excess energy is produced. The organism stores this excess energy for future use by using it to bond atoms together in lipid molecules. Later, when energy is needed, enzymes break these same bonds, releasing the energy used to form them. You have learned that carbohydrates also store energy; however, the process is not as efficient as in lipids. Therefore, long-term storage of energy is usually in the form of lipids.

Lipids also form the membranes that surround cells and many of their parts. The lipids found in membranes include cholesterol and phospholipids. Phospholipids are molecules in which a phosphate group and two fatty acids, rather than three fatty acids, are bonded to the three carbon atoms of glycerol.

$$CH_2O - PO_3^{2-}$$
$$|$$
$$CHO - fatty\ acid$$
$$|$$
$$CH_2O - fatty\ acid$$

Phospholipid

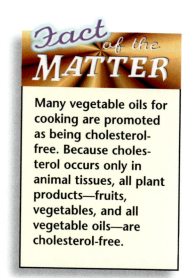

Fact of the **MATTER**

Many vegetable oils for cooking are promoted as being cholesterol-free. Because cholesterol occurs only in animal tissues, all plant products—fruits, vegetables, and all vegetable oils—are cholesterol-free.

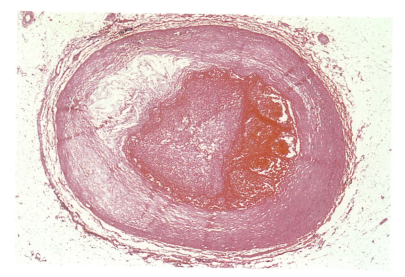

Figure 19.12

Atherosclerosis
Cholesterol forms part of the plaque material clogging this human artery. People with high levels of cholesterol in the blood have an increased risk of developing heart disease due to blocked arteries. Although cholesterol has a bad reputation, cells require it for making membranes, steroid hormones, and bile salts. The human body synthesizes about 1 g of cholesterol each day.

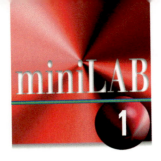
DNA—The Thread of Life

You have learned that DNA is a two-stranded molecule that consists of many polymerized nucleotides. DNA acts as the master blueprint for the cell's activity because it contains the coded instructions for every protein made by the cell. In addition, DNA enables a cell to pass on these instructions to the next generation because it is duplicated before the cell divides. Newly formed cells receive exact copies of the parent cell's DNA. In this MiniLab, you will extract and examine the DNA from wheat-germ cells.

Procedure

1. Use a clean mortar and pestle to grind approximately 5 g of wheat germ in 50 mL of cell lysis solution for one minute. Your teacher will supply the cell lysis solution. It contains chemicals that break open the wheat cells and remove unwanted cellular material.

2. Pour the mixture through a piece of cheesecloth or a kitchen strainer into a 250-mL or 400-mL beaker. Discard the solid.

3. Add 100 mL of 91 percent isopropanol to the liquid, and stir briefly.

4. Carefully wind the strands of DNA onto a glass rod.

5. Spread a thin piece of the DNA on a microscope slide, and add two drops of methylene blue dye. Examine the DNA under a microscope.

Analysis

1. Does the large amount of DNA that you extracted suggest that it is loosely arranged or compacted inside the cells?

2. Does the physical appearance of the DNA enable you to describe its physical structure?

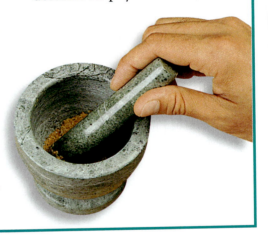

Nucleic Acids

The fourth major class of biomolecules—nucleic acids—is not listed on food-product labels, although members of this group are found in every plant or animal cell that humans use for food. A **nucleic acid** is a large polymer containing carbon, hydrogen, and oxygen, as well as nitrogen and phosphorus. Nucleic acids are present in cells only in tiny amounts and are not essential in the human diet because they can be made by the body from amino acids and carbohydrates. Nucleic acids contain the coded genetic information that cells need to reproduce themselves, and they regulate the cell by controlling synthesis of the proteins that carry out so many functions in cells. The two kinds of nucleic acids found in cells are **DNA** (deoxyribonucleic acid) and **RNA** (ribonucleic acid). Nucleic acids are so named because they were first discovered in the nucleus of cells.

Figure 19.13

Phosphate group

Nitrogen-containing base

Simple sugar

The Structure of Nucleic Acids

Nucleic acids are polymers made from building blocks that look complex at first glance. Understanding the structure of these building blocks, called **nucleotides,** is easier if you break them down into three parts, as shown in **Figure 19.13.**

Nucleotides can contain either of two similar sugars. RNA nucleotides contain a five-carbon sugar called ribose, whereas DNA nucleotides contain deoxyribose, which has the same general structure as ribose but with a single hydrogen atom in place of one of the hydroxyl groups.

Ribose

Deoxyribose

Five different nitrogen-containing bases are found in nucleotides. Thus, there are five different nucleotide building blocks. The names of these bases are often abbreviated by a single letter—A for adenine, C for cytosine, G for guanine, T for thymine, and U for uracil. DNA contains A, C, G, and T, but never U. In RNA, U is found instead of T. A nucleic acid polymer is made of chains in which the sugar of one nucleotide is linked to the phosphate of another. The chain of alternating sugars and phosphates is often called the backbone of the polymer. Attached to each sugar is one of the five bases.

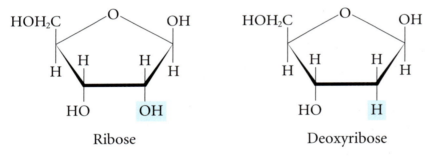

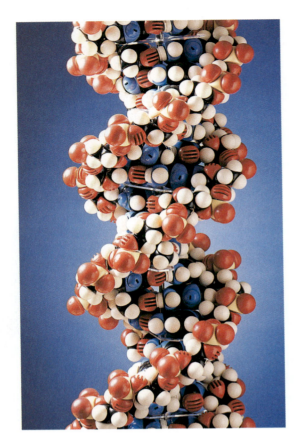

Figure 19.14

The Structure of DNA
This model of a portion of a DNA molecule clearly shows its complexity. A single DNA molecule contains many thousands of nucleotides.

DNA

In the three-dimensional structure of DNA, two sugar-phosphate backbone chains are held together by hydrogen bonds that form between the bases attached to each chain. Each base has a specific shape that allows it to hydrogen-bond to only one other base. In DNA, adenine bonds only to thymine, and cytosine bonds only to guanine. When the bases attached to the two DNA backbones bond together, the DNA forms a structure similar to a ladder. The bases form the rungs of the ladder, and the backbone chains form its sides. Now, imagine winding this ladder around a pole. The DNA ladder twists into a spiral structure known as a double helix, as **Figure 19.14** shows.

The specific sequence of the bases in an organism's DNA forms its genetic code. This master code controls all the characteristics of the organism because it contains the instructions for the structure of every protein made by that organism. The code is passed from one generation to the next because offspring receive copies of their parents' DNA. Francis Crick and James Watson were awarded the 1963 Nobel Prize in Chemistry for determining the structure of DNA. They correctly predicted how this structure permits DNA to be readily duplicated by cells so the genetic information can be transmitted to the next generation of cells. This discovery opened the way to the modern science of genetics and its genetic engineering applications.

Supplemental Problems

For more practice with solving problems, see Supplemental Practice Problems, Appendix B.

RNA

RNA also is a polymer of nucleotides, but some important differences can be found when its structure is compared with that of DNA. You already learned that the sugars found in the two nucleic acids are different and that the base uracil replaces the thymine found in DNA. Uracil binds to adenine in the same way that thymine does. The three-dimensional structure of RNA is different from that of DNA, also. RNA has only a single chain of nucleotides, which is twisted into a helix. RNA functions in the cell to carry the genetic information from the DNA to the site of protein synthesis, where it directs the order of amino acids in the proteins.

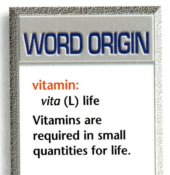

WORD ORIGIN

vitamin:
vita (L) life
Vitamins are required in small quantities for life.

Vitamins

Have you ever noticed that the labels on containers of milk all indicate that vitamin D has been added? Why do you think dairies do this? Cells require one more major group of organic molecules to carry out their functions. **Vitamins** are organic molecules that are required in small amounts in the diet. Vitamin D aids in the absorption of calcium and phosphorus through the wall of the intestine and into the bloodstream. These minerals are essential for the formation of healthy bones. If children do not get enough vitamin D in their diets, they develop rickets, a disease that causes bowed legs and other severe bone malformations. Adding vitamin D to milk assures that children will get enough vitamin D, as well as the calcium and phosphorus that are plentiful in milk.

Vitamins fall into two major classes: the water-soluble vitamins that dissolve in water and the fat-soluble vitamins that dissolve in nonpolar organic solvents. Vitamin D is a fat-soluble vitamin. When added to milk, it dissolves in the milk's fats.

Unlike the functions of proteins, carbohydrates, and lipids, vitamins are not used directly for energy or as building blocks for making structural materials in the body. Many serve as coenzymes in cell reactions. A **coenzyme** is an organic molecule that assists enzymes in catalyzing reactions. Vitamin C is a coenzyme for the reaction that modifies the protein collagen to make its structure stable enough to hold your tissues together.

Deficiencies of vitamins result in specific diseases. Recall the devastating effects of scurvy among British sailors caused by a deficiency of vitamin C. Too much of a fat-soluble vitamin also can result in disease because, unlike water-soluble vitamins, excess fat-soluble vitamins cannot be dissolved in urine and excreted. **Figure 19.15** describes vitamin A, another fat-soluble vitamin.

Figure 19.15

Vitamin A

Vitamin A is needed to maintain healthy eyes, skin, and mucous membranes. It is stored in the body's fat cells, especially in the liver. Polar bears store huge quantities of vitamin A in their livers. In the 19th century, many Arctic explorers died from vitamin A toxicity after eating large amounts of polar bear liver.

SECTION REVIEW

Understanding Concepts

1. What groups of atoms are common to all amino acids?

2. Describe the general structure of a triglyceride.

3. How do the structures of DNA and RNA differ?

Thinking Critically

4. **Applying Concepts** Why is it possible to form two different dipeptides from two given amino acids? Using the structural formulas for glycine and alanine in this section, draw the two possible dipeptides.

Applying Chemistry

5. **Polyunsaturated Fats** Nutritionists believe that polyunsaturated fats in the diet are more healthful than saturated fats. What are polyunsaturated fats, and what is their source in the human diet?

Reactions of Life

Have you ever bitten into a banana that looked ready to eat but wasn't really ripe yet? How did it taste? Probably not very sweet because the sweet taste of a banana results from sugars that are produced when starch is broken down as the banana ripens. The reaction that breaks starch down into sugar is just one of many that occur in cells. Living cells are dynamic arenas in which molecules and elements are rarely on the sidelines. To uncover the action going on, start with the processes cells use to harness energy.

SECTION PREVIEW

Objectives

✓ **Distinguish** between the reactions that cells use in the presence and in the absence of oxygen to extract energy from fuel molecules.

✓ **Explain** how a small number of biochemical building blocks can be used to make the extraordinary variety of molecules needed to perform life's chemical functions.

Review Vocabulary

Substrate: the name given to a reactant in an enzyme-catalyzed reaction.

New Vocabulary

metabolism
hormone
respiration
aerobic
ATP
electron transport
  chain
anaerobic
fermentation

Metabolism

What happens to the sugars and starch in a banana after you eat it? The sugars are small enough to be transported into the cells lining your digestive system. From there, they are moved into your bloodstream and sent to other cells. The starch molecules are too large to be transported into cells. As soon as you bite into the fruit, enzymes in your saliva begin to hydrolyze the starch into glucose. In the same way, the amazing variety of large and complex molecules in foods must be broken down into smaller units before they can be absorbed into cells.

Enzymes—which have been coded by DNA and synthesized with the help of RNA—catalyze the reactions in which proteins, carbohydrates, and lipids are broken down; this process is called digestion. Only small building blocks are able to enter cells to be used in the many reactions that are involved in metabolism. **Metabolism** is the sum of all the chemical reactions necessary for the life of an organism. These numerous intertwined cellular reactions transform the chemical energy stored in the bonds of nutrients into other forms of energy and synthesize the biomolecules needed to provide structure and carry out the functions of living things. Where does the chemical energy that is stored in nutrient molecules come from? This energy was converted from light energy by plants during photosynthesis. Ultimately, all the energy we need to live comes from sunlight. You will learn about photosynthesis and energy conversion in Chapter 20.

Function of Hemoglobin

In 1864, a British physicist discovered that hemoglobin, the pigment of the blood, binds and releases oxygen. As it does so, it changes color from red to bluish-red. Hemoglobin binds oxygen to its iron atoms in the alveoli of the lungs and transports it to the tissues of the body. The shape of the hemoglobin molecule changes as it moves from the lungs through the blood vessels to the capillaries near the cells, and that change in shape is what causes it to drop off the oxygen where it is needed.

Structure and function of hemoglobin The physiological properties of hemoglobin can be explained by studying the structure of the molecule. The hemoglobin molecule (Hb) consists of two copies each of two slightly different polypeptide chains—alpha and beta. Each of the four polypeptide chains has a heme group near its center. Notice that a heme group consists of an iron atom associated with four nitrogen atoms, each of which is part of a ring structure. Thus, the structure of heme is a circle of nitrogenous rings with iron in the center.

Each iron atom in the hemoglobin molecule can bind with one molecule of diatomic oxygen, which contains two oxygen atoms. Thus, every hemoglobin molecule can hold eight oxygen atoms when saturated.

Carbon monoxide poisoning Because it is similar in size to oxygen, the poisonous gas carbon monoxide combines with hemoglobin in almost the same way that oxygen does. Unfortunately, carbon monoxide has an affinity for hemoglobin that is about 210 times that of oxygen. To make matters worse, both oxygen and carbon monoxide combine with hemoglobin at the same point on the molecule. Thus, they can't both be bound to hemoglobin at the same time. Even at concentrations of 0.1 percent, carbon monoxide is dangerous. At that concentration, half of the hemoglobin in the blood will combine with the carbon monoxide, leaving only half of the hemoglobin to combine with oxygen. When the concentration of carbon monoxide rises to about 0.2 percent, the quantity of hemoglobin free to transport oxygen is too small to support life, and death occurs.

Connecting to Chemistry

1. **Hypothesizing** Anemia occurs when the number of red blood cells falls below normal. The patient feels weak and tired. Hypothesize the cause of these symptoms.

2. **Analyzing** When a person is in danger of dying from carbon monoxide poisoning, pure oxygen is administered at six times the normal alveolar oxygen pressure. How might this help?

Look at the metabolic map in **Figure 19.17,** which gives you an idea of how the varied metabolic reactions are interconnected. Some important reactions involve the trapping of energy and its release and use. Other reactions involve the formation and processing of essential molecules, such as protein enzymes, antibodies, and hemoglobin. Cells perform only the reactions they need at any given time, and this allows them to conserve both energy and building block molecules. How do cells switch the reactions off and on? They use an intricate series of control processes. Many of these can be triggered by signal molecules called **hormones** that are made in specific organs of the body and travel through the bloodstream to communicate with cells in other locations. Insulin is one example of a hormone. After a meal, insulin is released by the pancreas to signal to cells that glucose is available. One of the metabolic processes that insulin triggers in cells is the energy-releasing process called respiration.

Respiration

When cells need energy, they oxidize fuels such as carbohydrates and fats. This process results in the formation of carbon dioxide and water, the same products that are formed when a fuel such as gasoline is burned in an engine. Energy is released in this reaction.

fuel molecule + oxygen → carbon dioxide + water + energy

Most of the energy used by cells comes from the oxidation of carbohydrates. In oxidation, as in most chemical reactions, energy is needed initially to break bonds. This energy usually is supplied in the form of heat. In an automobile engine, hydrocarbons undergo combustion after the spark plugs heat the gasoline-oxygen mixture to the point where the molecules can react. Only then can the explosive reaction occur that drives the pistons in the engine. Heat speeds up a reaction, but high temperatures kill living cells. How can oxidation of carbohydrates take place at normal body temperatures? The answer can be found in the catalytic power of enzymes. The complex series of enzyme-catalyzed reactions used to extract the chemical energy from glucose is called **respiration.** Respiration is an **aerobic** process, which means that it takes place only in the presence of oxygen. The athlete shown in **Figure 19.16** could not perform without oxygen.

To harness the energy from glucose, cells must break the bonds in which the energy is stored. The net exothermic reaction that takes place when the bonds in glucose are broken is similar to the combustion of hydrocarbons.

$$C_6H_{12}O_6 + 6O_2 \rightarrow 6CO_2 + 6H_2O + energy$$

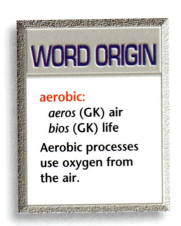

WORD ORIGIN

aerobic:
aeros (GK) air
bios (GK) life
Aerobic processes use oxygen from the air.

Figure 19.16

Respiration and Breathing
Although many people might describe respiration as the heavy breathing that occurs when they compete in an athletic event, a biochemist's definition of respiration explains what happens to molecules in the cells when oxygen is available as a result of breathing. Glucose reacts with oxygen to form water and carbon dioxide, and energy is released.

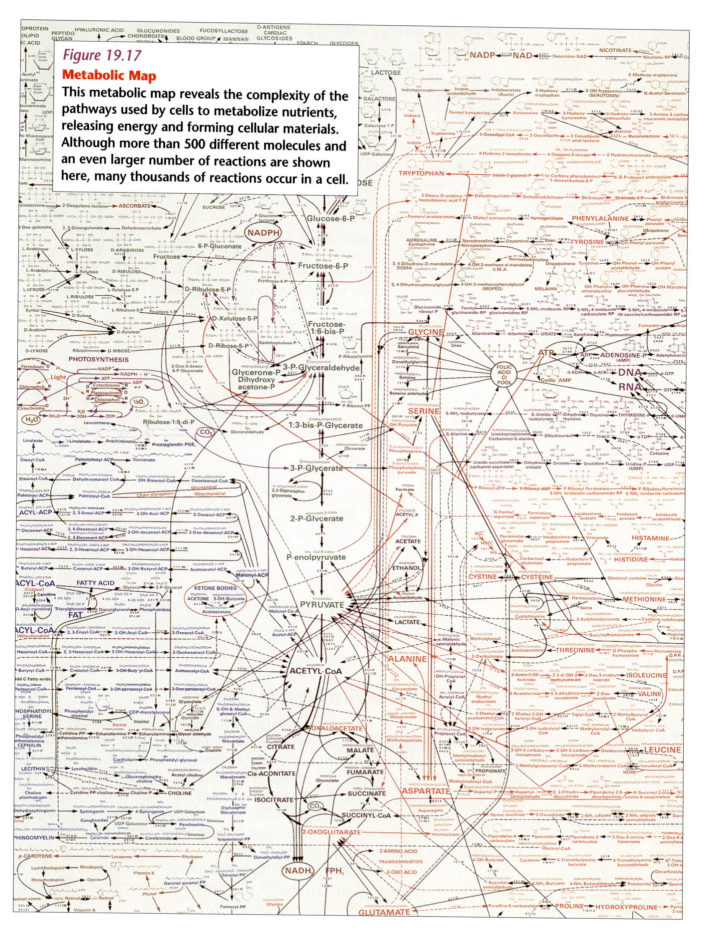

Figure 19.17
Metabolic Map
This metabolic map reveals the complexity of the pathways used by cells to metabolize nutrients, releasing energy and forming cellular materials. Although more than 500 different molecules and an even larger number of reactions are shown here, many thousands of reactions occur in a cell.

ATP and Energy Storage

When gasoline burns in the engine cylinders of a car, large amounts of energy are released as heat in a single explosive reaction. If the energy from the oxidation of glucose and other nutrients were to be released in the same way, the cells would not be able to use it all. Metabolic reactions require small amounts of energy. How can the energy in food be "packaged" so the cell can use it as needed? When nutrients are broken down, the energy is transferred from the broken bonds to energy-storage molecules called adenosine diphosphate, or *ADP* for short. The structure of ADP is identical to one of the building-block nucleotides of nucleic acids, except that ADP has two phosphate groups attached to the ribose molecule, as shown in **Figure 19.18.** Cells store energy by bonding a third phosphate group to ADP to form adenosine triphosphate—**ATP.** When the cell needs energy, the phosphate-phosphate bond in ATP is broken, producing ADP and a phosphate group and releasing stored energy. In cells, the conversion of ADP to ATP and vice versa occurs over and over as metabolic reactions release and use energy. Many of these reactions take place during respiration.

Figure 19.18

ATP and Energy
Energy is stored when ATP is made from ADP and phosphate. Energy is released when ATP breaks down into ADP and phosphate.

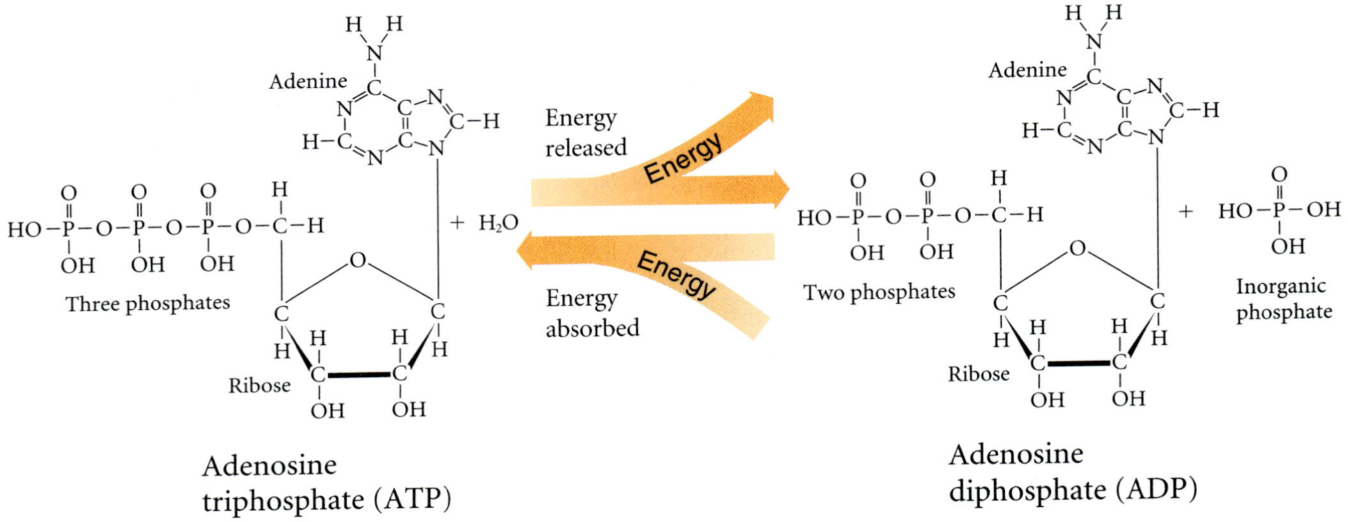

Adenosine triphosphate (ATP)

Adenosine diphosphate (ADP)

Glycolysis—The First Stage of Respiration

The process of respiration consists of many reactions that can be grouped into three stages. **Figure 19.19** summarizes the process, simplifying the reactions by showing only the carbon atoms. In the first stage of respiration, a series of nine reactions breaks down glucose into a pair of three-carbon compounds. Most of the energy from the broken bonds is transferred to ATP. A net yield of two molecules of ATP per molecule of glucose is produced during this stage. This series of reactions is called *glycolysis,* which means "splitting glucose."

Why do cells need ATP? Energy in the bonds of this molecule can be accessed much more quickly and in more manageable amounts than energy in triglycerides, starch, or glycogen.

Figure 19.19

Respiration

In the first stage of respiration, a six-carbon glucose molecule is split into a pair of three-carbon molecules. Hydrogen ions and electrons also are produced in glycolysis. These combine with electron carrier ions called nicotinamide adenine dinucleotide (NAD^+) to form NADH. NADH is a coenzyme that is made from vitamin B_4, which is also called niacin or nicotinic acid. ATP and NADH serve as temporary storage sites for energy and electrons, respectively, during respiration. Two molecules of ATP are used, and four molecules are produced. ▼

Supplemental Problems

For more practice with solving problems, see Supplemental Practice Problems, Appendix B.

1. Glycolysis

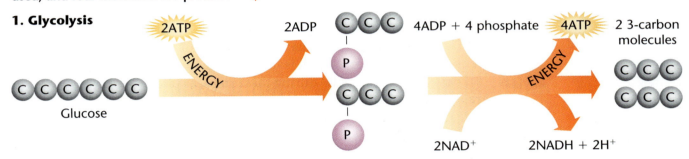

2ATP   2ADP   C C C   4ADP + 4 phosphate   4ATP   2 3-carbon molecules

ENERGY

C C C C C C
Glucose

P

C C C

P

ENERGY

C C C
C C C

2NAD$^+$ 2NADH + 2H$^+$

2. Tricarboxylic Acid Cycle

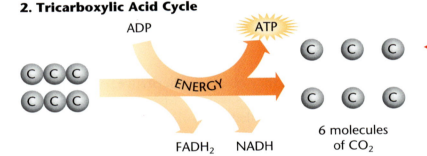

ADP   ATP

C C C
C C C

ENERGY

C   C   C

C   C   C

6 molecules of CO_2

FADH$_2$ NADH

◀ In the second stage, the pair of three-carbon molecules that was made from glucose is converted into six molecules of carbon dioxide. More ATP and NADH are produced, along with another coenzyme molecule called flavin adenine dinucleotide ($FADH_2$).

3. Electron Transport Chain

In the third stage, NADH and $FADH_2$ bring the electrons and hydrogen atoms from glucose to a series of carrier molecules, the electron transport chain. A chain-like series of redox reactions takes place. In the final reaction, the electrons and hydrogen are transferred to oxygen to produce water. At the top of the chain, the electrons have high energy. As they pass down the chain, the energy given off is captured in molecules of ATP. The energy available from the breakdown of each glucose molecule can be used to make as many as 38 ATP molecules, a net of 2 during glycolysis and 36 from the electron transport chain. ▶

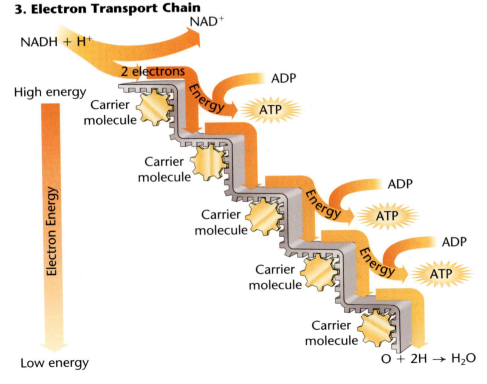

NADH + H$^+$ NAD$^+$

2 electrons   ADP

High energy

Carrier molecule   Energy   ATP

Electron Energy

Carrier molecule

Carrier molecule   Energy   ADP   ATP

Carrier molecule   Energy   ADP   ATP

Carrier molecule

O + 2H → H_2O

Low energy

The Tricarboxylic Acid Cycle—The Second Stage

In the second stage of respiration, carbon dioxide is produced in a series of reactions. These reactions are called the tricarboxylic acid cycle because some of the molecules formed in the intermediate reactions have three carboxyl groups.

The Electron Transport Chain—The Third Stage

Most of the energy from glucose is released in the final stage of respiration, referred to as an **electron transport chain.** The electrons move step-by-step to lower energy levels, allowing for the controlled release of energy, **Figure 19.20.** This is similar to the release of potential energy that happens at every bounce if a ball is dropped down a staircase.

Energy is stored as ATP during these steps, as phosphate groups are transferred to ADP. A final redox reaction transfers the electrons to the oxygen you breathe, forming water. You can see that the electron transport chain can't operate without oxygen because oxygen is the final acceptor of the electrons that came from glucose. If oxygen is not present, the entire transport chain stops because the electrons have nowhere to go.

Figure 19.20
Energy Release
When a marshmallow is burned, the combustion reaction releases carbon dioxide, water, and a lot of energy all at once. In respiration, the same overall reaction occurs, but the energy is released step-by-step.

Fermentation

If cells are deprived of oxygen, they must use alternative routes for extracting the energy of glucose, or they will die. Metabolic processes that occur in the absence of oxygen are called **anaerobic** processes. Cells can generate energy from glucose anaerobically by the process of **fermentation.** Because the first stage of respiration can take place without oxygen, glucose is broken down through the reactions of glycolysis into the two smaller, three-carbon units before fermentation reactions proceed. There are two major types of fermentation. In one, ethanol and carbon dioxide are produced. In the other, the product is lactic acid.

Alcoholic Fermentation

Alcoholic fermentation occurs in some bacteria and in yeast cells. As shown in **Figure 19.21,** the process yields much less energy than respiration; much of the energy from glucose remains in the bonds of the ethanol. Although it is relatively inefficient, fermentation provides enough energy for cells to carry on their basic functions. Fermentation by yeast cells is used in the brewing and wine-making industries, and also in the baking industry. The carbon dioxide produced during alcoholic fermentation causes bread dough to rise, making the dough light and spongy. The ethanol that is also produced evaporates during baking.

Lactic Acid Fermentation

Rapidly contracting muscle cells sometimes use up oxygen faster than it can be supplied by the blood. When the cells run out of oxygen, respiration cannot continue. Instead, a type of fermentation called lactic acid

In the 1600s, distilleries distilled the results of fermentation to concentrate the ethanol. The distillates were tested for flammability by burning them on small piles of gunpowder. If the distillate contained at least 50 percent alcohol, the gunpowder would ignite after the alcohol had burned. This result was proof of the alcohol's quality and is the reason why 50 percent alcohol is called 100 proof.

Yeast Plus Sugar—Let It Rise

You may have used yeast to make bread dough or pizza dough rise. Dry yeast is the dormant form of a single-celled fungus that, when given favorable living conditions and food in the form of a carbohydrate, begins to break down the carbohydrate. One of the products of respiration is carbon dioxide. In this MiniLab, you will mix yeast with a disaccharide, sucrose, and with a polysaccharide, the starch in flour, and compare the rates at which carbon dioxide is produced.

Procedure

1. Obtain two sandwich-size, zipper-close bags, and label them as *sucrose* or *flour*.

2. Fill a water trough or plastic dishpan about ⅔ full of hot water, and adjust the temperature to between 40°C and 50°C.

3. Put one package of dry yeast and one tablespoon of the appropriate carbohydrate into each of the bags. Mix well.

4. Working quickly, measure and add 50 mL (¼ cup) of the warm water from the trough to each of the bags, and thoroughly mix the contents. Remove all the air you can and seal the bags. Start timing when both bags are sealed.

5. Put both bags into the trough of warm water. Determine and record the time required for sufficient carbon dioxide to fill each bag completely. If either of the bags is not completely filled in 30 minutes, estimate the fraction of the bag that is filled.

Analysis

1. Based upon your data, rank the rates at which yeast breaks down each type of carbohydrate.

2. Why might the carbohydrates be broken down at different rates?

Figure 19.21

Alcoholic Fermentation

In alcoholic fermentation, each three-carbon molecule that is produced during glycolysis is split to form a two-carbon molecule—the alcohol ethanol—and a one-carbon molecule, carbon dioxide. Two molecules of ATP are produced during glycolysis. ▼

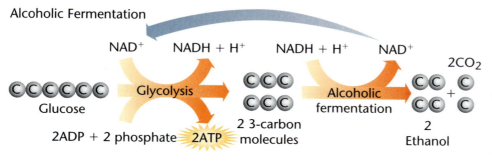

Alcoholic Fermentation

Yeast cells ferment the sugars in fruits to produce ethanol and carbon dioxide. The anaerobic process must be carried out in tightly sealed containers such as these large tanks. One-way valves allow carbon dioxide to escape from the fermenting liquid without letting air in. ▶

Figure 19.22

Lactic Acid Fermentation
Strenuous exercise has caused lactic acid to build up inside this racer's muscle cells, resulting in muscle fatigue. When oxygen becomes available, the body will break down the lactic acid. ▼

fermentation begins. **Figure 19.22** shows the process of lactic acid fermentation. If the anaerobic state continues for too long, lactic acid can build up in the muscles, causing a stiffness called muscle fatigue. Because some of the energy of glucose remains trapped in lactic acid, which cannot be broken down anaerobically, this kind of fermentation does not release as much energy as respiration does.

Lactic Acid Fermentation

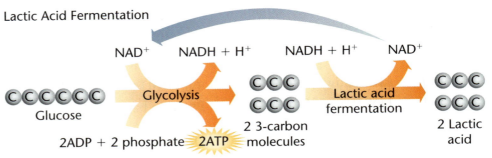

▲ In an anaerobic state, glucose is converted into a pair of three-carbon molecules called lactic acid through glycolysis followed by lactic acid fermentation. Two molecules of ATP are produced during glycolysis. Lactic acid fermentation occurs in some bacteria, in fungi, and in most animals including humans. The dairy industry uses bacteria to make yogurt and buttermilk by this process.

Connecting Ideas

Respiration involves the oxidation of fuel molecules such as glucose in the cells of the body. Animals must take in fuel molecules in the form of food, but how do plants get them? In Chapter 20, you will study photosynthesis, the process in which plants trap energy from sunlight and use this trapped energy to make food molecules. Energy in the form of heat and light commonly is involved in chemical processes. It is important to be able to recognize, as well as measure, the energy changes that accompany chemical reactions. Gaining those skills is the focus of the next chapter.

SECTION REVIEW

Understanding Concepts

1. Explain why respiration provides cells with more energy per molecule of glucose than does fermentation.

2. How do alcoholic fermentation and lactic acid fermentation differ?

3. How many moles of CO_2 are produced from one mole of glucose during respiration?

Thinking Critically

4. **Relating Cause and Effect** Marathon runners go through a crisis called hitting the wall when they use up all their stored glycogen and start using stored lipids as fuel. What would be the advantage to carbohydrate loading, which means eating lots of complex carbohydrates the day before running a marathon?

Applying Chemistry

5. **Aerobic Exercise** Explain what aerobic exercise has in common with aerobic processes in cells.

chemistryca.com/self_check_quiz

REVIEWING MAIN IDEAS

19.1 Molecules of Life

- Proteins, carbohydrates, lipids, and nucleic acids are the four major classes of biomolecules.

- Proteins are polymers of amino acids. These polymers are chains, each containing hundreds or thousands of amino acids, that fold up into a three-dimensional structure.

- Many proteins function as enzymes that speed up reactions. Substrates bind to pockets called active sites on the enzymes, where they are converted into products. The substrate fits tightly and specifically in the active site, much like a key fits in a lock.

- Carbohydrates contain carbon, hydrogen, and oxygen. Simple sugar units combine to form carbohydrate polymers. Simple sugars are used by cells to produce energy, whereas polymers are used to store energy.

- Lipids are molecules that are insoluble in water and soluble in nonpolar solvents. Lipids function to store energy and form cell membranes.

- Nucleic acids are polymers of repeating units called nucleotides, which consist of a sugar, one of five different bases, and a phosphate group. RNA functions in making proteins, and DNA transmits genetic information from one generation to the next.

- Vitamins are complex organic molecules necessary in small amounts for good health. Many vitamins function as coenzymes.

19.2 Reactions of Life

- Metabolism consists of a complex series of interconnected pathways through which fuel molecules are used to trap energy and to make all the different molecules and cell parts needed for the functioning of living cells.

- In the process of respiration, energy is extracted from fuel molecules through a series of enzymatic reactions in which oxygen is used and carbon dioxide and water are produced.

- Fermentation is a process in which energy is extracted from fuel molecules in the absence of oxygen, resulting in the production of ethanol or lactic acid. Less energy is obtained through fermentation than through respiration because some chemical energy is left behind in the products.

Vocabulary

For each of the following terms, write a sentence that shows your understanding of its meaning.

active site	fermentation
aerobic	hormone
amino acid	lipid
anaerobic	metabolism
ATP	nucleic acid
biochemistry	nucleotide
carbohydrate	protein
coenzyme	respiration
denaturation	RNA
DNA	steroid
electron transport chain	substrate
fatty acid	vitamin

UNDERSTANDING CONCEPTS

1. Describe the three-dimensional structure of the DNA double helix.

2. Why do unsaturated fats tend to be liquids rather than solids at room temperature?

3. Distinguish between a disaccharide and a polysaccharide.

4. What kind of functional group is formed when fatty acids combine with glycerol?

5. Describe the process by which a disaccharide forms from two simple sugar molecules.

6. What is the function of an enzyme?

7. What three structural units make up a nucleotide?

8. Compare the functions of DNA and RNA.

9. Where are hydrogen bonds found in the structure of DNA? What function do they have?

10. Lipids are defined by their solubility properties. Why are lipids usually not soluble in water?

APPLYING CONCEPTS

11. Examine the structures of the molecules shown and decide whether each is a carbohydrate, a lipid, or an amino acid.

a) $CH_3-(CH_2)_7-CH=CH-(CH_2)_7-\overset{\overset{\text{O}}{\|}}{C}-OH$

b)

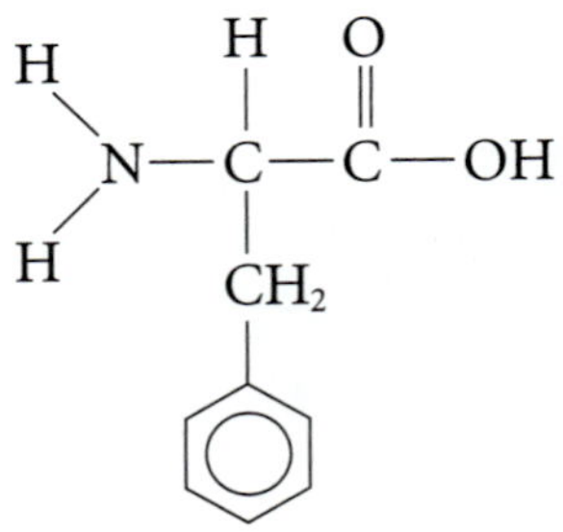

c)

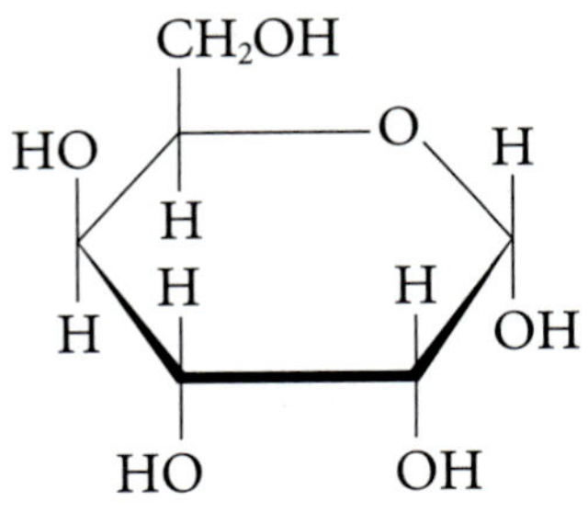

d)

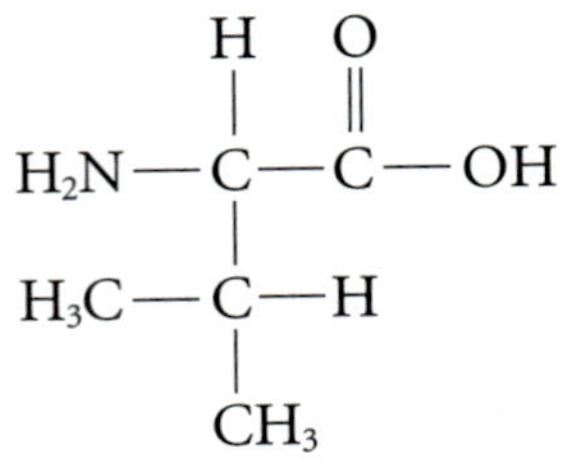

12. Explain how changes in temperature and pH of a cell affect protein structure.

13. Vitamins are classified by their solubility properties into two groups: water-soluble and fat-soluble. Use the structures of the vitamins shown to predict which group each falls into.

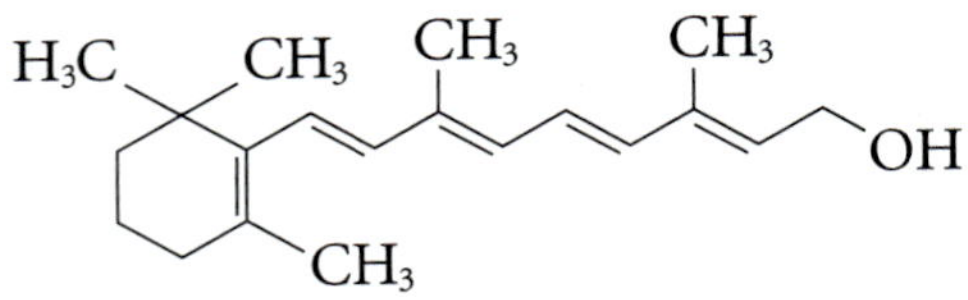

Vitamin A

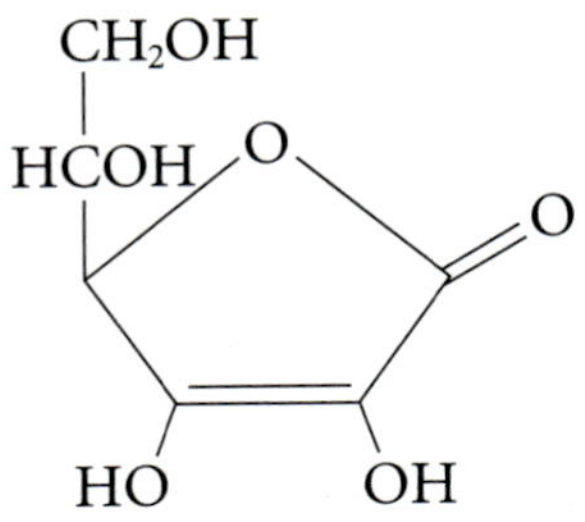

Vitamin C

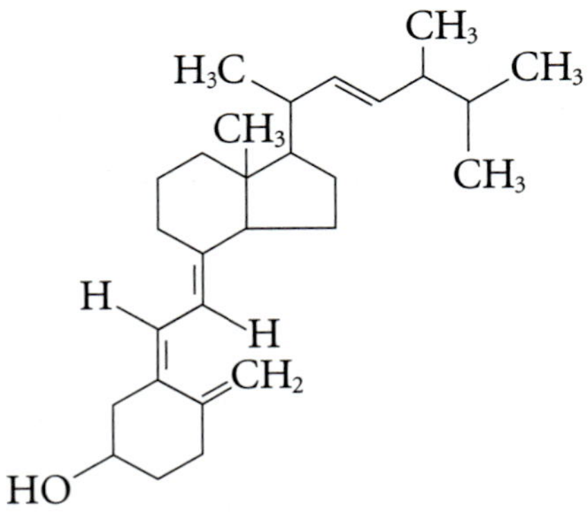

Vitamin D

chemistryca.com/chapter_test

Health Connection

14. Crocodiles can hold their breath under water for up to 90 minutes, which allows them time to kill their prey by drowning. When an animal holds its breath, hydrogen carbonate ions (HCO_3^-) build up in the blood. Normally, only a small portion of the oxygen bonded to hemoglobin is released to cells for use in respiration, but when hydrogen carbonate ions bind to a crocodile's hemoglobin, the hemoglobin delivers most of its oxygen to cells. Suggest a mechanism by which HCO_3^- ions might have this effect on crocodile hemoglobin.

Everyday Chemistry

15. Why are protein molecules ideally suited for forming pairs of hydrogen bonds?

Everyday Chemistry

16. How is the structure of a fat molecule different from the structure of the chemically altered fat Olestra?

● THINKING CRITICALLY

Making Inferences

17. Burning hair has a characteristic odor due to the presence of large amounts of the amino acid cysteine in keratin, the major protein in hair. What element do you think is responsible for this smell?

Forming a Hypothesis

18. When fresh pineapple is added to a gelatin solution, the gelatin fails to gel, or solidify. When cooked or canned pineapple is used, the gelatin gels. Explain.

Interpreting Data

19. **ChemLab** The following table contains data obtained by measuring the activity of the enzyme salivary amylase in solutions buffered at different pH levels.

    a) At which pH is the enzyme most active?
    b) Can you identify any trend in activity as pH moves lower than the optimum pH? Higher?
    c) Why does pH affect enzyme activity?

pH	Time for Color to Disappear
4	10 minutes
5	8 minutes
6	1 minute
7	20 seconds
8	40 seconds
9	4 minutes

Designing an Experiment

20. **MiniLab 1** Design an experiment that would show whether ethanol or isopropanol is more effective at precipitating DNA.

Relating Cause and Effect

21. **MiniLab 2** Why was the experiment conducted in a 40-50°C water bath?

Identifying Patterns

22. Explain why the polymerization reactions that form proteins, polysaccharides, and nucleic acids are condensation and not addition reactions.

● CUMULATIVE REVIEW

23. Explain what a buffer is and why buffers are found in body fluids. (Chapter 15)

24. Name the hydrocarbons shown. (Chapter 18)

    a) $CH_3CH_2CH_2CH_3$

    b)
    $$CH_3CH_2\underset{\underset{\displaystyle CH_3}{|}}{CH}CH_2CH_2CH_3$$

    c)
    $$\begin{array}{c} CH_2 \\ \diagup \quad \diagdown \\ CH_2 \quad\quad CH_2 \\ | \quad\quad\quad | \\ CH_2 - CH_2 \end{array}$$

    d)
    $$CH_3-\underset{\underset{\displaystyle CH_3}{|}}{\overset{\overset{\displaystyle CH_3}{|}}{C}}-CH_3$$

e) $CH_3(CH_2)_5CH_3$

f)
$$\begin{array}{c} H_3C \quad CH_3 \\ | \quad\quad | \\ CH_3CHC-CH_2CH_3 \\ | \\ CH_3 \end{array}$$

SKILL REVIEW

25. **Interpreting Chemical Structures** Use library references to make a poster showing the structures of the 20 common amino acids. Use one color to highlight the side chains that are different in each amino acid. Use other colors to highlight the functional groups that are common to all amino acids.

26. **Making and Using Graphs** Every enzyme has an optimum pH at which it is most active. The following graph shows activity ranges for two digestive enzymes, pepsin and trypsin. Pepsin is found in the stomach, and trypsin is found in the intestine. From the graph, determine the optimum pH for each enzyme. In a paragraph, describe the pH in the organ where each enzyme is found. What would happen in the stomach if pepsin had an optimum pH of 8?

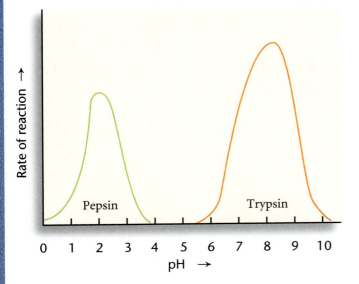

WRITING IN CHEMISTRY

27. Read *In Search of the Double Helix* by John Gribbin, and write a report discussing whether Watson and Crick or Rosalind Franklin really made the discovery of the DNA structure. Include a discussion of how serendipity played a part in the discovery of the structure.

PROBLEM SOLVING

28. Examine the structure of oxygenated hemoglobin on page 693 of this chapter and list the functional groups in this organic molecule.

29. One serving of the cereal represented by the food label shown here provides 26 g of carbohydrates, which is nine percent of the recommended daily value for someone consuming 2000 Calories per day. Calculate how many total grams of carbohydrate that person should eat per day.

Nutrition Facts		
Serving Size: 1 Cup (30g)		
Servings Per Package: About 14		

Amount Per Serving	1 Cup Cereal	Cereal With 1/2 Cup Skim Milk
Calories	110	160
Calories from Fat	10	10

	% Daily Value*	
Total Fat 1g	1%	2%
Saturated Fat 0g	0%	2%
Cholesterol 0mg	0%	1%
Sodium 130mg	5%	8%
Potassium 35mg	1%	7%
Total Carbohydrate 26g	9%	11%
Sugars 14g		
Protein 2g		

*Percent Daily Values are based on a 2000 calorie diet. Your daily values may be higher or lower depending on your calorie needs.

30. How many grams of water will you produce from completely extracting the chemical energy in one can of orange soda that contains 52.0 g of sugar using aerobic respiration?

31. Write a ablanced equation for the hydrogenation of oleic acid. How many moles of H_2 are needed to hydrogenate one mole of oleic acid? What is the name of the product?

1. Which of the following statements about enzymes are NOT true?

   a) Enzymes speed up chemical reactions during which their structure and chemical composition are altered

   b) Enzymes provide pocket or grove sites in their three dimensional structures where substrates can bind.

   c) Active sites alter their initial shape to match the structural shape of a substrate after it binds to the site.

   d) Enzymes dramatically increase the rate of a chemical reaction, but they cannot cause a spontaneous reaction.

2. The general formula for a carbohydrate is

   a) $C_6H_{12}O_6$
   b) $C_{12}H_{24}O_{12}$
   c) CH_2O
   d) $C_2H_3O_2$

3. Which of the following is a polysaccharide?

   a) glucose
   b) glycogen
   c) sucrose
   d) maltose

4. Triglyceride is considered a lipid because

   a) triglyceride molecules are complex, organic molecules derived from animals or animal byproducts.

   b) triglyceride molecules have a carboxylic acid group bonded in a carbon chain with hydrogen atoms.

   c) triglyceride molecules are alkenes composed of many carbon atoms bonded to an oxygen functional group.

   d) triglyceride molecules are biological compounds containing few oxygen bonds and large numbers of C—H bonds.

5. Which of the following factors effect a person's cholesterol level?

   a) diet
   b) genetic make up
   c) stress
   d) all of the above

6. Which of the following statements about RNA nucleotides is true?

   a) RNA contains the five basic nucleotide building blocks including adenine, cytosine, guanine, thymine, and uracil.

   b) RNA nucleotides form bonded pairs in which the sugar of one nucleotide is paired with the sugar of another nucleotide.

   c) RNA chains have replaced the thymine nucleotide with the uracil nucleotide, but they retain the other three nucleotide building blocks.

   d) RNA nucleotides form bonded pairs in which the phosphate of one nucleotide is paired with the phosphate of another nucleotide.

7. Which of the following statements is NOT true about aerobic respiration?

   a) The chemical equation for aerobic respiration is the opposite of the photosynthesis chemical equation.

   b) When glucose bonds are broken during aerobic respiration, the overall reaction resembles the burning of gasoline.

   c) When glucose bonds are broken during aerobic respiration, the overall endothermic reaction absorbs ATP energy.

   d) To break the bonds of one glucose molecule, six molecules of oxygen are required and six molecules of carbon dioxide are required.

8. Which of the following molecules releases energy during respiration?

   a) adenosine triphosphate
   b) adenosine diphosphate
   c) adenosine phosphate
   d) tricarboxylic acid

Test Taking Tip

Skip Around If You Can The questions on some tests start easy and get progressively harder, while other tests mix easy and hard questions. You may want to come back to them later, after you've answered all the easier questions. This will guarantee more points toward your final score. In fact, other questions may help you answer the ones you skipped. Just be sure you fill in the correct ovals on your answer sheet.

20 Chemical Reactions and Energy

Chapter Preview

Sections

20.1 Energy Changes in Chemical Reactions

MiniLab 20.1 Dissolving—Exothermic or Endothermic?

20.2 Measuring Energy Changes

MiniLab 20.2 Heat In, Heat Out
ChemLab Energy Content of Some Common Foods

20.3 Photosynthesis

Wow, That's Hot Stuff!

This chemical reaction produces a lot of thermal energy. In fact, this reaction is used to fuse metals together by creating a metal seam between the two pieces. Two common gases used for this process are acetylene and oxygen.

Start-up Activities

Launch Lab

Speeding Reactions

Many chemical reactions occur so slowly that you don't even know they are happening. For some reactions, it is possible to alter the reaction speed using another substance.

Safety Precautions

Materials

- hydrogen peroxide
- beaker or cup
- baker's yeast
- toothpicks

Procedure

1. Create a "before and after" table and record your observations.
2. Pour about 10 mL of hydrogen peroxide into a small beaker or cup. Observe the hydrogen peroxide.
3. Add a "pinch" (1/8 tsp) of yeast to the hydrogen peroxide. Stir gently with a toothpick and observe the mixture again.

Analysis

Into what two products does hydrogen peroxide decompose? Why aren't bubbles produced in step 1? What is the function of the yeast?

What I Already Know

Review the following concepts before studying this chapter.

Chapter 1: energy in chemical changes

Chapter 6: reasons why reactions go forward; equilibrium; and reaction rates

Chapter 10: temperature and particle motion

Reading Chemistry

Analyze some of the charts, graphs and tables that appear throughout the chapter. Write down any questions or new vocabulary words that are used in the figures. Note the page number of figures you have questions about, and try to find the answers in the text as you read the chapters.

Preview this chapter's content and activities at chemistryca.com

Energy Changes in Chemical Reactions

"Smile," says the photographer, as she pushes a button. The camera shutter opens and an electric current from a small lithium battery sparks across a gap in the flash unit. This spark ionizes xenon gas, creating a bright flash of light. The energy from the chemical reaction in the lithium battery has been successfully put to use.

SECTION PREVIEW

Objectives

✓ **Compare** and **contrast** exothermic and endothermic chemical reactions.

✓ **Analyze** the energetics of typical chemical reactions.

✓ **Illustrate** the meaning of *entropy,* and **trace** its role in various processes.

Review Vocabulary

Electron transport chain: the controlled release of energy from glucose by the step-by-step movement of electrons to lower energy levels.

New Vocabulary

heat
law of conservation
  of energy
fossil fuel
entropy

Exothermic and Endothermic Reactions

As you learned in Chapter 1, chemical reactions can be exothermic or endothermic. Recall that an exothermic reaction releases heat, and an endothermic reaction absorbs heat. **Figure 20.1** pictures an exothermic process as a reaction that's going downhill energetically and an endothermic process as a reaction that's going uphill.

Exothermic Reactions

If you have ever started a campfire or built a fire in a fireplace, you know that the burning of wood is an example of an exothermic process. Once you have ignited the wood, the reaction generates enough heat to keep itself going. A net release of heat occurs, which is what makes the reaction exothermic.

Figure 20.1
Exothermic and Endothermic Reactions
The exothermic reaction (left) gives off heat because the products are at a lower energy level than the reactants. The endothermic reaction (right) absorbs heat because the products are at a higher energy level than the reactants.

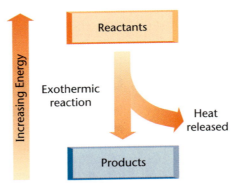

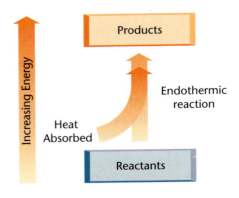

Figure 20.2

The Exothermic Formation of Water

It takes only a small amount of energy to start the reaction between hydrogen and oxygen to form water. The energy released by the reaction is much greater, so the reaction is exothermic. The energy from this reaction has been used to power car and truck engines.

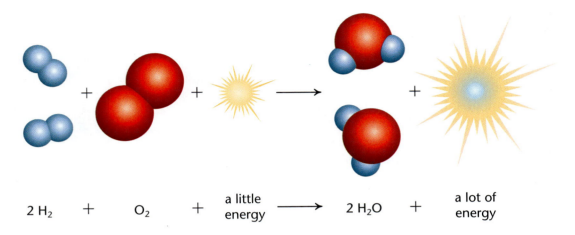

$$2\ H_2\quad +\quad O_2\quad +\quad \text{a little energy}\quad \longrightarrow\quad 2\ H_2O\quad +\quad \text{a lot of energy}$$

The reaction between hydrogen gas and oxygen gas to form water, shown in **Figure 20.2,** is another example of an exothermic reaction. Once a small amount of energy—often just a spark—is added to the mixture of gases, the reaction continues to completion, usually explosively. No additional input of energy from outside is needed to keep it going. Once energy has been supplied to break the covalent bonds in the first few molecules of hydrogen and oxygen, the atoms combine to form water and release enough energy to break the bonds in additional hydrogen and oxygen molecules. The net energy is released as heat.

Endothermic Reactions

Consider the reverse of the reaction just discussed. Just as water can be formed from hydrogen and oxygen, it can also be decomposed to re-form hydrogen and oxygen. In the process of electrolysis, electrical energy is used to break the covalent bonds that unite the hydrogen atoms and the oxygen atoms in the water molecules. The hydrogen atoms pair up to form hydrogen molecules, and the oxygen atoms pair up to form oxygen molecules. The formation of the new bonds releases energy, but not as much as the amount required during the bond breaking. Additional energy must be added continuously during the electrolysis. The reaction absorbs heat energy and is, therefore, endothermic.

All endothermic reactions are characterized by a net absorption of energy. In the History Connection on page 58, you read about another example of an endothermic reaction: the decomposition of orange mercuric oxide into the elements mercury and oxygen. As long as heat is applied, the compound continues to decompose, but if the heat source is removed, the reaction stops. The net absorption of heat energy that is required is what makes the reaction endothermic.

How it Works

Hot and Cold Packs

Instant hot and cold packs create aqueous solutions that form exothermically or endothermically and therefore release or absorb heat. A hot pack generates heat when a salt such as calcium chloride dissolves in water that is stored in the pack. The calcium chloride dissolves exothermically. A cold pack absorbs heat when a salt such as ammonium nitrate dissolves in water. The ammonium nitrate dissolves endothermically. In both cases, the salt and water are separated by a thin membrane. All you have to do is squeeze the pack to mix the components and you have instant heat or cold at your fingertips.

1. The outer casing is strong and flexible. It resists puncture and can be shaped to fit the area that you want to heat or cool.

2. Water is stored in an inner compartment separate from the solid salt.

3. The inner membrane breaks easily when you knead or squeeze the pack or strike it sharply.

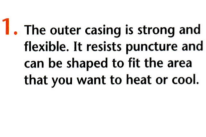

Outer casing

Water

Soluble salt

Membrane of water pack

4. Salt is stored in the outer compartment. When the inner membrane breaks, the salt and water mix. The salt dissolves in the water and releases or absorbs energy.

Thinking Critically

1. Another type of hand warmer contains fine iron powder and chemicals that cause the iron to rust. The rusting of iron lets the hand warmer maintain temperatures above 60°C for several hours. Explain how that is possible.

2. When a solid in a cold pack dissolves in water, the process takes place spontaneously. What causes the solution process to be spontaneous in spite of the fact that it is endothermic?

Heat

The energy that is involved in exothermic and endothermic reactions is usually in the form of heat. **Heat** is defined as the energy transferred from an object at high temperature to an object at lower temperature. Recall that energy is measured in joules; the symbol for joules is *J*. The symbol for a kilojoule, which is equal to 1000 J, is *kJ*.

Using Symbols to Show Energy Changes

Energy changes are frequently included in the equation for a chemical reaction. The amount of heat absorbed or evolved during a reaction is a measure of the energy change that accompanies the reaction. When 1 mol (18.0 g) of liquid water is produced from hydrogen and oxygen gas, 286 kJ of energy are given off. This means that the energy of the uncombined hydrogen gas and oxygen is greater than the energy of the water. When 1 mol (18.0 g) of liquid water decomposes to form hydrogen gas and oxygen gas, 286 kJ of energy are absorbed. This also shows that the energy of the uncombined hydrogen and oxygen is greater than the energy of the water. The graphs in **Figure 20.3** illustrate this relationship.

Scientists have observed that the energy released in the formation of a compound from its elements is always identical to the energy required to decompose that compound into its elements. This observation is an illustration of an important scientific principle known as the **law of conservation of energy.** That law states that energy is neither created nor destroyed in a chemical change, but is simply changed from one form to another. In an exothermic reaction, the heat released comes from the change from reactants at higher energy to products at lower energy. In an endothermic reaction, the heat absorbed comes from the opposite change.

Figure 20.3

Formation and Decomposition of Water
As the graphs show, the energy produced when 1 mol of liquid water forms from the elements hydrogen and oxygen is equal in magnitude to the energy absorbed when the water decomposes.

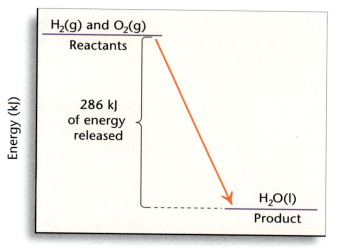

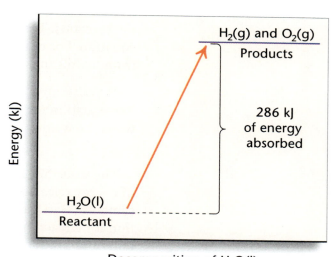

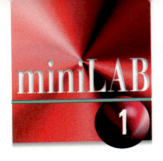

miniLAB

Dissolving—Exothermic or Endothermic?

The dissolving of a solid in water, like most other processes, may liberate energy or absorb energy. If the dissolving process is exothermic, the liberated energy raises the temperature of the solution. If the process is endothermic, it absorbs energy from the solution, lowering the temperature. Examine the dissolution of several common solids in water.

Procedure

1. Measure 100 mL of water into a 250-mL beaker. Set a Celsius thermometer in the water and allow it to come to the water's temperature. Record that temperature. Remove the thermometer from the water.

2. Add to the water approximately 1 tablespoon of the solid to be tested, and stir with a stirring rod for 20 seconds. Put the thermometer back into the solution. Record the temperature.

3. Pour the solution down the drain with large amounts of tap water.

4. Repeat steps 1 through 3 for each of the solids to be tested.

Analysis

1. Which of the solids that you tested dissolve exothermically? Which dissolve endothermically?

2. Which of the solids that you tested could be mixed with water inside a flexible plastic container to produce a cold pack that might be used by medical personnel?

The difference in energy between products and reactants in a chemical change is symbolized ΔH (delta H), where the symbol Δ means a difference or change and the letter H represents the energy. The energy absorbed or released in a reaction ($\Delta H_{reaction}$) is related to the energy of the products and the reactants by the following equation.

$$\Delta H_{reaction} = H_{products} - H_{reactants}$$

For exothermic reactions, ΔH is negative because the energy stored in the products is less than that in the reactants. For endothermic reactions, ΔH is positive because the energy of the products is greater than that of the reactants. The value of ΔH is often shown at the end of a chemical equation. For example, the exothermic formation of 2 mol of liquid water from hydrogen and oxygen gas would be written like this.

$$2H_2(g) + O_2(g) \rightarrow 2H_2O(l) \quad \Delta H = -572 \text{ kJ}$$

The equation for the endothermic decomposition of 2 mol of liquid water would be written like this.

$$2H_2O(l) \rightarrow 2H_2(g) + O_2(g) \quad \Delta H = +572 \text{ kJ}$$

The value 572 kJ in these equations is 2×286 kJ, which is the amount of energy released when 1 mol of liquid water forms. Note the use of the symbols (s), (l), and (g). When energy values are included with an equation for a reaction, it is especially important to show the states of reactants and products because the energy change in a reaction can depend greatly upon physical states.

Activation Energy

The hydrocarbons found in petroleum and natural gas are the remains of plants and other organisms that lived millions of years ago. Oil and natural gas are called **fossil fuels** for this reason. Fossil fuels are a rich source of energy because when they react with oxygen to produce carbon dioxide and water, a great deal of energy is released in the form of heat.

However, fossil fuels do not burn automatically. Energy, usually in the form of heat or light, is needed to get the chemical reaction started. The combustion of hydrocarbon fuels requires this input of energy, called activation energy, to begin the reaction. For example, the butane gas in a disposable lighter requires a spark to start the combustion of the gas.

In Chapter 6, you learned that activation energy is needed to cause particles to collide with enough force to make them react. Activation energy is required in both exothermic and endothermic reactions. The fact that a fuel requires an input of energy—such as from a spark—to begin burning does not mean that the combustion reaction is endothermic. The reaction releases a net amount of heat, and so it is exothermic.

Look more closely at activation energy and heat of reaction, using as an example the burning of methane, which is the main component in natural gas, to yield carbon dioxide and water vapor. The equation for this reaction is as follows.

$$CH_4(g) + 2O_2(g) \rightarrow CO_2(g) + 2H_2O(g) \quad \Delta H = -802 \text{ kJ}$$

A graph of the energy change during the progress of this reaction is shown in **Figure 20.4.** Notice how the energy curve rises, then falls. The rise represents the activation energy, which is the energy difference between the reactants and the maximum energy stage in the reaction. The fall represents the energy liberated by the formation of new chemical substances. When 1 mol of methane burns, 802 kJ of heat are given off. The reaction is exothermic, as is shown by the negative sign of ΔH. The energy stored in the products is less than that stored in the reactants, so a net amount of energy is released. Some of the released energy provides the activation energy needed to keep the reaction going.

WORD ORIGIN

combustion:
combustus (L)
burned

The internal combustion engine in a car burns the hydrocarbons in gasoline.

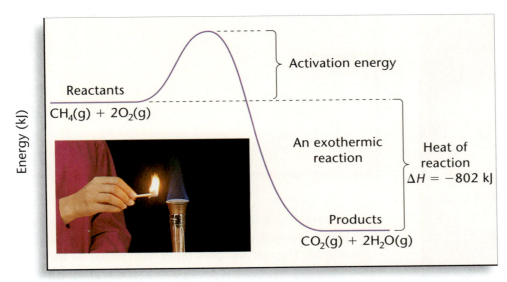

Figure 20.4

Energy in an Exothermic Reaction

In order to occur, the combustion of methane, illustrated in the photo at the left, requires an input of activation energy—in this case, provided by a match or sparking device in the stove. Overall, the reaction releases 802 kJ of energy per mole of methane. Notice from the graph that the products are in a lower energy state than the reactants. The negative value of ΔH reflects this fact.

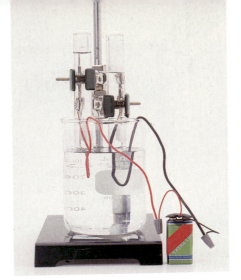

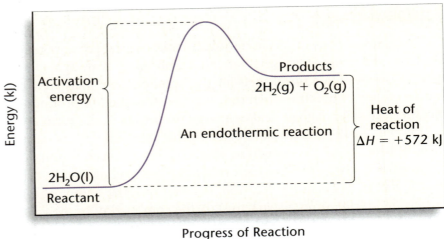

Energy (kJ)

Activation energy

2H₂O(l)
Reactant

$2H_2(g) + O_2(g)$
Products

An endothermic reaction

Heat of reaction
$\Delta H = +572$ kJ

Progress of Reaction

Figure 20.5

Energy in an Endothermic Reaction

The decomposition of water requires a continuous input of energy, such as electrical energy from a battery. Overall, the reaction absorbs 572 kJ of energy for every 2 mol of liquid water. The products are in a higher energy state than the reactant. The positive value of ΔH reflects this fact.

Figure 20.6

Effect of a Catalyst

It is easier for cars to go through the tunnel than for them to climb the mountain. ▼

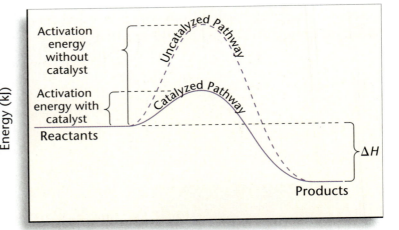

Now, consider an example of an endothermic reaction—one that you have already read about—the decomposition of water. The equation for the reaction is as follows.

$$2H_2O(l) \rightarrow 2H_2(g) + O_2(g) \quad \Delta H = +572 \text{ kJ}$$

A graph of the energy changes during the progress of this reaction is shown in **Figure 20.5.** Notice how the energy curve rises above the level of the reactant, then falls only slightly when the products form. Therefore, the products have more stored energy than the reactants, and the graph clearly shows a net gain of energy. Because of this net gain, ΔH is positive. Energy must be added continuously to keep the reaction going.

Recall from Chapter 6 that a catalyst can be used to speed up a reaction. The catalyst provides a different reaction path—one for which the activation energy is lower. The catalyst thus creates a shortcut or tunnel through the energy hill between the reactants and products. More of the collisions will now be effective because less energy is required. The energy diagram in **Figure 20.6** shows the effect of a catalyst on activation energy.

Energy (kJ)

Activation energy without catalyst

Activation energy with catalyst

Reactants

Uncatalyzed Pathway

Catalyzed Pathway

ΔH

Products

Progress of Reaction

▲ As the graph of a catalyzed reaction shows, the activation energy is lowered, providing an easier reaction pathway. As a result, the reaction goes faster than would be the case without the catalyst. Note that the heat of reaction, ΔH, is the same in both cases.

Everyday Chemistry

Catalytic Converters

Since 1975, every new car sold in the United States has a catalytic converter installed in the exhaust system. This device contains porous, heat-resistant material coated with a catalyst. The purpose of the catalytic converter is to reduce air pollution.

How a catalytic converter works A typical catalytic converter consists of particles of platinum and rhodium deposited on a ceramic structure that is like a honeycomb. The platinum and rhodium catalyze reactions that remove pollutants such as nitrogen monoxide (NO), carbon monoxide (CO), and unburned hydrocarbons. When nitrogen monoxide binds to the rhodium surface, it breaks down to oxygen and nitrogen. The bound oxygen reacts with carbon monoxide, which has also become bound to the rhodium surface. The reaction produces carbon dioxide. The oxidation of unburned hydrocarbons produces carbon dioxide and water.

The honeycomb arrangement of catalytic materials in the converter provides a large surface area for the reactions to take place. Such an arrangement increases the rate at which the pollutants are removed. Catalytic converters have reduced the pollutants released into the air by cars by as much as 90 percent.

Improving the catalytic converter The operating temperature of a catalytic converter is between 316° and 649°C, which is in the range of standard exhaust-gas temperatures for a vehicle being driven on the road. Below 316°C, the device does nothing. The pollutants emitted during the beginning of the driving period, when the converter is just heating up, slip through the catalytic converter without being changed.

In an effort to stop air pollution during the warm-up period, some catalytic converters heat up to 400°C within 5 seconds. The result is that almost as soon as the car is started, air pollutants in the exhaust are broken down. Heated catalytic converters are proving to be highly efficient.

Experimental model of a heated catalytic converter

Someday, perhaps even those 5 seconds of dirty emissions before the converter is heated will be eliminated.

Exploring Further

1. **Acquiring Information** Find out why cars with catalytic converters cannot use leaded gasoline.

2. **Applying** Instead of the honeycomb structure of the catalytic converter described above, some converters contain pellets coated with a platinum-palladium mixture. Why is the use of pellets effective?

Chemistry Online

To learn more about how catalytic converters cut down pollution, visit the Chemistry Web site at chemistryca.com

Forces That Drive Chemical Reactions

When pure aluminum metal is exposed to chlorine gas, aluminum chloride is formed. This chemical change is spontaneous; that is, it just happens on its own. The reaction is highly exothermic. The following equation summarizes the process.

$$2Al(s) + 3Cl_2(g) \rightarrow 2AlCl_3(s) + 1408 \text{ kJ}$$

What allows such reactions to occur spontaneously? The answer has to do with the general forces that drive all reactions.

Order, Disorder, and Entropy

Scientists have observed two tendencies in nature that explain why chemical reactions occur. The first tendency is for systems to go from a state of high energy to a state of low energy. The state of low energy is more stable. Thus, for example, exothermic reactions are more likely to occur than endothermic ones, all other things being equal. The second tendency is for systems to become more disordered. Spontaneous reactions tend to occur if energy decreases and if disorder increases.

Figure 20.7 shows an everyday spontaneous change that increases disorder. The first photo shows a full glass of milk on the breakfast table. The second photo shows the same glass of milk after someone has bumped into it and caused it to fall to the floor. Notice the disorder in the second picture. The glass is broken into seemingly random shards, and the milk is scattered into pools and droplets.

You have already learned a good deal about energy changes, but the concept of disorder in chemical changes may not be as familiar. Scientists use the term **entropy** to describe and measure the degree of disorder. Unlike energy, which is conserved in chemical changes and in the universe as a whole, entropy is not conserved. The natural tendency of entropy is to increase. The fallen glass illustrates how entropy increases in natural, spontaneous changes because when disorder increases, entropy increases. In most spontaneous, naturally occurring processes, entropy increases.

Figure 20.7

Disorder and Spontaneity
When the glass full of milk falls off a table, it breaks and the contents spill. The broken glass is in a state of greater disorder. This breaking and scattering is a spontaneous change. The reverse change—the glass reassembling itself and refilling with the milk—is extremely unlikely to occur. Changes that are spontaneous in one direction are not spontaneous in the other direction under the same conditions.

Some kinds of changes are apt to increase entropy, **Figure 20.8.** These changes include melting and evaporation. Increases in numbers of molecules also tend to increase entropy. Entropy increases during reactions that result in increases in the number of molecules. Such changes result in greater disorder at the submicroscopic level.

Figure 20.8

Increase in Entropy
Humpty-Dumpty's problem was too great an increase in entropy. An input of work can sometimes counter entropy increases. You know that living things require a constant supply of energy for life functions. One of the ways they use that energy is to maintain the strict organization of molecules required for life. The energy is used to do work to overcome entropy increases at the molecular level.

The Direction of a Chemical Reaction

At room temperature, most exothermic reactions tend to proceed spontaneously forward. In other words, they favor the formation of products. In an exothermic reaction, the released energy, usually in the form of heat, raises the temperature of the products and many more atoms and molecules in the surroundings. The energy is distributed more randomly than it was before the reaction. The motion of a greater number of atoms and molecules increases. Therefore, disorder increases.

You have read that both heat and entropy play a role in determining spontaneity. In general, the direction of a chemical reaction is determined by the magnitude and direction of the heat energy and entropy changes. For example, a reaction will proceed in the forward direction, toward formation of products, if that direction results in both a release of heat and an increase in entropy. As an example, consider the combustion of butane, C_4H_{10}.

$$2C_4H_{10}(g) + 13O_2(g) \rightarrow 8CO_2(g) + 10H_2O(g) + \text{heat}$$

This reaction occurs spontaneously in the forward direction because both energy and entropy changes are favorable. The energy decreases because the reaction is exothermic—a favorable change. The entropy increases partially because the total number of molecules increases, from 15 to 18—also a favorable change. The release of heat and increase in entropy combine to drive this reaction forward.

Look at another example of a reaction favored by both energy and entropy changes—the one between calcium carbonate and hydrochloric acid.

Supplemental Problems

For more practice with solving problems, see Supplemental Practice Problems, Appendix B.

$$CaCO_3(s) + 2HCl(aq) \rightarrow CaCl_2(aq) + H_2O(l) + CO_2(g) + heat$$

This reaction favors formation of products because it produces gases and liquids. They are more disordered than the solid $CaCO_3$, so entropy increases. Heat is given off, so the products are at a lower energy state than the reactants. That is also favorable. Once again, the decreased energy and increased entropy both drive the reaction in the forward direction.

What happens if only one of the changes—either energy or entropy—is favorable? If the favorable change is great enough to outweigh the unfavorable one, the reaction will still be spontaneous. Thus, even some endothermic reactions are spontaneous if disorder increases greatly. Also, some reactions that increase order are spontaneous if they are exothermic enough. Spontaneity depends on the balance between energy and entropy factors.

Try at Home Lab

See page 872 in Appendix F for **Observing Entropy**

Table 20.1 summarizes these factors. Note that temperature plays a role in determining spontaneity when one factor is favorable and the other is not.

Table 20.1 Predicting Whether a Reaction Is Spontaneous in the Forward Direction

Energy Change	Entropy Change	Spontaneous?
decrease (exothermic)	increase	yes
decrease (exothermic)	decrease	yes at low temperature; no at high temperature
increase (endothermic)	increase	no at low temperature; yes at high temperature
increase (endothermic)	decrease	no

SECTION REVIEW

Understanding Concepts

1. Are reactions that occur spontaneously at room temperature generally exothermic or endothermic? Explain.

2. Describe the energy changes that occur when a match lights.

3. ΔH for a reaction is negative; compare the energy of the products and the reactants. Is the reaction endothermic or exothermic?

Thinking Critically

4. **Relating Cause and Effect** Describe each of the following processes as involving an increase or decrease in entropy.

   a) Water in an ice-cube tray freezes.

   b) You pick up scattered trash along a highway and pack it into a bag.

   c) Your campfire burns, leaving gray ashes.

Applying Chemistry

5. **Fermentation** Yeast can spontaneously ferment the sugar in grapes or apples and form ethanol. When the process occurs in grapes, it results in wine. In apples, it produces a beverage called hard cider. The first stage in the fermentation process is exothermic, and the balanced equation for the reaction is as follows.

   $$C_6H_{12}O_6(aq) \rightarrow 2C_2H_5OH(aq) + 2CO_2(g)$$

   Determine whether entropy increases in this reaction, and relate the change in heat energy and entropy to the spontaneity of the process.

Measuring Energy Changes

These days, many people are trying to avoid fattening foods and looking for things to eat that taste good and provide adequate nutrition without increasing body weight. Have you ever been on a diet and had to count Calories? If so, you may have wondered how food manufacturers determine the amount of energy available in a certain amount of food.

SECTION PREVIEW

Objectives

✓ **Sequence** the technique of calorimetry and **illustrate** its use.

✓ **Compare** the heat generated by some common fuels and by some foods.

✓ **Analyze** the efficiency of industrial processes and the need to conserve resources.

Review Vocabulary

Entropy: term used to describe and measure the degree of disorder in progress.

New Vocabulary

calorie
kilocalorie
Calorie

Calorimetry

The heat generated in chemical reactions can be measured by a technique called calorimetry and a device called a calorimeter. The device and its use are illustrated in **Figure 20.9,** which deals with an exothermic reaction. A calorimeter can also be used to study the heat absorbed in endothermic reactions. In the case of endothermic reactions, the surrounding water in the calorimeter supplies the heat and decreases in temperature.

In measurements involving a calorimeter, you calculate the heat lost or gained by the surrounding water. This is done by means of the following equation.

$$q_w = (m)(\Delta T)(C_w)$$

In this equation, the symbol q_w stands for heat absorbed by water, m is the mass of the water, ΔT is the temperature change of the water, and C_w is the specific heat of water, which equals 4.184 J/g•°C.

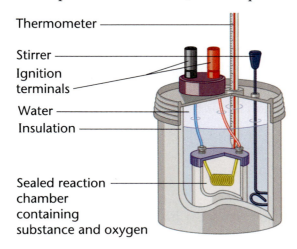

Thermometer
Stirrer
Ignition terminals
Water
Insulation
Sealed reaction chamber containing substance and oxygen

Figure 20.9

Calorimetry

A weighed sample of a substance is burned in pure oxygen inside a container called a reaction chamber. The heat that is released flows into the surrounding water and raises its temperature. Given the temperature change and mass of the water, you can calculate the heat given off in the reaction.

By the law of conservation of energy, there is no net creation or destruction of energy. Any heat absorbed or released by the water has been released or absorbed by the reaction being studied. The symbol $q_{reaction}$ stands for the heat change of the reaction. Heat loss by the reaction means heat gain for the water, and heat gain for the reaction means heat loss for the water. Therefore, the heat of reaction is equal to the negative of the heat change of the water.

$$q_{reaction} = -q_w$$

The following sample problem shows how calorimetry can be used to measure energy change in a chemical reaction.

SAMPLE PROBLEM **1** **Calculating Heat of Reaction for Combustion**

The burning of 1.60 g of methane in oxygen, to yield carbon dioxide gas and liquid water, causes the surrounding 1.52 kg of water in a calorimeter to change in temperature from 20.0°C to 34.0°C. What is the heat of reaction for this combustion of 1 mol of methane?

Analyze • You have all the information you need to find the heat change for the water, q_w. This can be set equal to $-q_{reaction}$ for the burning of the 1.60 g of methane. What you then need to do is convert that value to find $q_{reaction}$ for 1.0 mol of methane. That will equal ΔH.

Set Up • Find the temperature change of the water, using the equation $\Delta T = T_{final} - T_{initial}$. Then, use the equation $q_w = (m)(\Delta T)(C_w)$. The negative of the value of q_w will equal $q_{reaction}$ for the burning of 1.60 g of methane. Use the molar mass of methane, 16.0 g methane/1.0 mol methane, as a conversion factor to find the $q_{reaction}$ for 1.0 mol of methane.

Solve • First, calculate ΔT.

$$\Delta T = T_{final} - T_{initial}$$
$$= 34.0°C - 20.0°C$$
$$= 14.0°C$$

Then, calculate q_w.
$$q_w = (m)(\Delta T)(C_w)$$

> **Problem-Solving HINT**
>
> Be careful with signs in the calculation of ΔT. If the reaction is endothermic, the sign of ΔT will be negative.

$$= \frac{1.52 \times 10^3 \, g \mid 14.0°C \mid 4.184 \, J}{\mid \mid g \cdot °C} = 8.90 \times 10^4 \, J = 89.0 \, kJ$$

Now, find $q_{reaction}$ for the 1.60 g of methane.
$$q_{reaction} = -q_w = -89.0 \, kJ = \text{heat released in burning 1.60 g methane}$$
Now, use a molar-mass conversion factor to find $q_{reaction}$, or ΔH, for 1 mol of methane.

$$q_{reaction} = \frac{-89.0 \, kJ \mid 16.0 \, g \, methane}{1.60 \, g \, methane \mid 1 \, mol \, methane} = -890 \, kJ/mol \, methane$$

$$q_{reaction} = \Delta H = -890 \, kJ/mol \, methane$$

Check • Check to be sure that all units were correctly handled, and review the calculation to be sure that it was sound. The result does check out.

Supplemental Problems

For more practice with solving problems, see Supplemental Practice Problems, Appendix B.

1. How much heat is absorbed by a reaction that lowers the temperature of 500.0 g of water in a calorimeter by 1.10°C?

2. Aluminum reacts with iron(III) oxide to yield aluminum oxide and iron. Calculate the heat given off in the reaction if the temperature of the 1.00 kg of water in the calorimeter increases by 3.00°C.

3. When 1.00 g of a certain fuel gas is burned in a calorimeter, the temperature of the surrounding 1.000 kg of water increases from 20.00°C to 28.05°C. All products and reactants in the process are gases. Calculate the heat given off in this reaction. How much heat would 1.00 mol of the fuel give off, assuming a molar mass of 65.8 g/mol?

Energy Value of Food

Many years ago, chemists measured heat in calories instead of joules. A **calorie** is the heat required to raise the temperature of 1 g of liquid water by 1°C. A **kilocalorie** is a unit equal to 1000 calories. One calorie is equal to 4.184 J. One joule is equal to 0.239 calorie. Some nutritionists and dietitians still use Calories but are switching to kilojoules. The energy value of foods is measured in units called Calories. Note the capital *C*. One food **Calorie** is the same as 1 kilocalorie. One food Calorie is also equal to 4.184 kJ.

Chemical compounds in the food you eat provide you with energy. The compounds undergo slow combustion, combining with oxygen to produce the waste products carbon dioxide and water, along with compounds needed for body growth and development. Fats provide about 9 Calories per gram. In contrast, 1 g of carbohydrate or protein provides about 4 Calories. **Figure 20.10** illustrates some types of nutrients. **Table 20.2** shows how many Calories you get from typical servings of some foods.

Figure 20.10

Types of Nutrients
Foods such as those shown here contain a variety of chemical compounds. They can be grouped into three general types—protein, carbohydrate, and fat—that have different energy contents. Your body processes transform some of the energy released when these foods are broken down into work and heat. However, when you take in more food than you need, your body stores the extra energy by producing fat for later use.

Table 20.2   The Caloric Value of Some Foods

Food	Quantity	Kilojoules	Calories
butter	1 tbsp = 14 g	418	100
peanut butter	1 tbsp = 16 g	418	100
spaghetti	0.5 cup = 55 g	836	200
apple	1	283	70
chicken (broiled)	3 oz = 84 g	502	120
beef (broiled)	3 oz = 84 g	1000	241

Energy Content of Some Common Foods

Foods supply the energy and nutrients you require to build and sustain your body and maintain your levels of activity. The energy is released within cells during the process of respiration. In that process, oxygen combines with energy-storage substances, such as glucose, to produce carbon dioxide, water, and heat. The reaction is essentially a slow combustion process. In this ChemLab, you will compare the amounts of energy liberated from the combustion of three food items.

Problem

How much energy is released during the combustion of some common food items?

Objectives

- **Interpret data** to calculate the energy released during the combustion of pecans, marshmallows, and a food item of your choice.
- **Compare** the energies obtained from the food items.
- **Infer,** based upon the chemical compositions of the foods and upon the amounts of energy obtained, which types of foods contain the greatest amount of energy.

PREPARATION

Materials

oven mitt
bottle opener with can-piercing end
empty, clean soft-drink can with a tab closure
empty clean can, minus the top and bottom lids, of approximately the same diameter as the soft-drink can
ring stand

small iron ring
beaker tongs
Celsius thermometer
100-mL graduated cylinder
glass stirring rod
balance
large paper clip
matches
pecan half
2 small marshmallows
food sample of your choice

Safety Precautions

Be careful in using the match and in handling the cans after each experiment because they may be hot. Perform the experiment in a well-ventilated room because acrid fumes may be produced during the combustion process. Wear an apron and goggles.

PROCEDURE

1. Use a can opener to open holes near the bottom of the open-ended can. Bend a large paper clip to fashion the food support as shown in the figure.

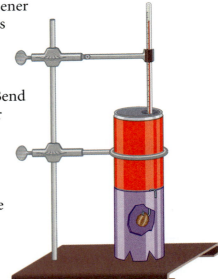

2. Use the graduated cylinder to measure 100 mL of tap water, and pour the water into the soft-drink can. Support this can by the ring stand as shown in the figure. Put the thermometer through the opening in the top of the can and allow it to reach the temperature of the water. Record the initial water temperature to the nearest 0.1°C.

3. Weigh a pecan half, record its mass, and impale it on the paper-clip food support. Hang the support on the edge of the open-ended can with the food sample inside, and rest the can on the ring stand.

4. Use a match to light the pecan. Quickly swing the water-filled soft-drink can assembly directly over the combustion can, allowing a small separation between the two cans.

5. Allow the pecan to burn as completely as possible. If the flame goes out during the process and the pecan is still mostly unburned, you must empty the water in the can and start over with fresh water and a new pecan half. After the pecan has burned out, allow the thermometer to reach the highest temperature. Record this temperature.

6. Carefully disassemble the apparatus, empty the water from the soft-drink can, and dispose of the burned food item.

7. Repeat steps 2-6 with two small marshmallows.

8. Repeat steps 2-6 with a small food item of your choice.

ANALYZE AND CONCLUDE

1. **Interpreting Data** Use your data to calculate the number of kilojoules of energy liberated per gram of each food item. Assume that the water in the soft-drink can has a density of 1.00 g/mL and that the specific heat of the water is 4.19 J/g°C.

2. **Comparing and Contrasting** Compare and rank the food items according to the amounts of energy liberated per gram.

3. **Inferring** Based upon the chemical compositions of the foods you tested (protein, carbohydrate, fat, etc.), what type of food provides the greatest amount of energy per gram?

APPLY AND ASSESS

1. Would it be desirable to eat mainly the type of food that provides the greatest amount of energy per unit mass? Explain.

2. What aspects of your procedure may have caused error in your calculated results? How would each of these sources of error affect your result?

DATA AND OBSERVATIONS

Food Item	Mass (g)	Initial H$_2$O Temperature (°C)	Final H$_2$O Temperature (°C)
Pecan			
Marshmallow			
Other food			

You can use calorimetry to measure the energy content of food. The food is burned rapidly in oxygen, rather than slowly, as in the body, but the amount of energy released is the same. The following sample problem shows you how to calculate food Calories from calorimetry data.

SAMPLE PROBLEM 2 Measuring Food Calories

A 1.00-g sample of nuts reacts with excess oxygen in a calorimeter. The calorimeter contains 1.00 kg of water that has an initial temperature of 15.40°C and a final temperature of 20.20°C. Find the energy content of the nuts. Express your answer in kJ/g and Calories/g.

Analyze
- You first need to find the heat change for the water, q_w, which is equal to $-q_{reaction}$ for the burning of the 1.00 g of nuts.

Set Up
- Find the temperature change of the water. Then, use the equation $q_w = (m)(\Delta T)(C_w)$ after finding the temperature change of the water. The negative of the value of q_w will equal $q_{reaction}$ in kilojoules. A conversion factor can then be used to convert to Calories.

Solve
- $\Delta T = T_{final} - T_{initial}$
$= 20.20°C - 15.40°C = 4.80°C$

$q_w = (m)(\Delta T)(C_w)$

$= \dfrac{1.00 \times 10^3 \, g \mid 4.80°C \mid 4.184 \, J}{\mid g \cdot °C}$

$= 2.01 \times 10^4 \, J = 20.1 \, kJ$

Now, apply a conversion factor to convert to Calories.

$= \dfrac{20.1 \, kJ \mid 1 \, Calorie}{\mid 4.184 \, kJ} = 4.80 \, Calories$

Check
- Check to be sure that all units were correctly handled, and repeat the calculation to be sure that it was sound. The result does check out.

PRACTICE PROBLEMS

Supplemental Problems

For more practice with solving problems, see Supplemental Practice Problems, Appendix B.

4. A group of students decides to measure the energy content of certain foods. They heat 50.0 g of water in an aluminum can by burning a sample of the food beneath the can. When they use 1.00 g of popcorn as their test food, the temperature of the water rises by 24°C. Calculate the heat released by the popcorn, and express your answer in both kilojoules and Calories per gram of popcorn.

5. Another student comes along and tells the group in problem 4 that she has read the label on a popcorn bag that states that 30 g of popcorn yields 110 Calories. What is that value in Calories/gram? How can you account for the difference?

6. A 3.00-g sample of a new snack food is burned in a calorimeter. The 2.00 kg of surrounding water change in temperature from 25.0°C to 32.4°C. What is the food value in Calories per gram?

Energy Economics

Have you ever wondered why recycling has become so important? Part of the answer has to do with energy. For many materials, the energy required for recycling used objects is less than the energy of simply throwing the used objects away and making new ones from fresh raw materials.

Aluminum Production and Recycling

Aluminum is the metal of choice for soft-drink cans, as shown in **Figure 20.11.** It is lightweight, has a low specific heat (0.902 J/g°C), and conducts heat rapidly. As a result of its low specific heat, the aluminum itself absorbs little heat, so the beverage inside the container can be cooled quickly. In Chapter 17, you learned that aluminum is produced by electrolysis of its principal ore, bauxite. The production of aluminum requires large amounts of energy, particularly electrical energy. Therefore, one reason to recycle aluminum is to conserve that energy and reduce the cost of producing aluminum products.

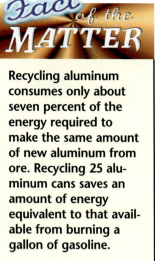

Figure 20.11
Aluminum: A Useful Metal
The soft-drink cans and other objects shown here are all made from aluminum.

A comparison of the energy required to produce new aluminum cans and the energy required to produce cans from recycled aluminum reveals that 15-20 cans can be made from recycled aluminum for every one that can be made from new aluminum ore. Thus, the cost of using recycled aluminum is much less than that of using new aluminum. In addition to saving money on energy costs, recycling aluminum conserves valuable raw materials and uses less fossil fuel, which is the source of most of the energy required to produce the aluminum. To understand where the energy savings come from in reducing use of fossil fuels, it is important to look at the process used to change chemical energy into electrical energy.

miniLAB 2

Heat In, Heat Out

If a chemical reaction is endothermic, the reverse reaction is exothermic. Examine the reaction between sulfite ion (SO_3^{2-}) and hypochlorite ion (OCl^-), which yields sulfate ion (SO_4^{2-}) and chloride ion (Cl^-).

Procedure

1. Pour approximately 40 mL of a commercial chlorine bleach solution into a 150-mL beaker. Put a thermometer into the solution and record the temperature. Be careful handling the bleach and avoid inhaling the fumes. Work in a well-ventilated room.

2. Pour approximately 40 mL of a 0.5M sodium sulfite solution into the beaker that contains the bleach, and stir gently for a few seconds.

3. Record the final temperature of the mixture.

Analysis

1. Based on the temperature change, was the reaction that occurred exothermic or endothermic? Would the reverse reaction be exothermic or endothermic? Refer to the graph for help.

2. Write the balanced equation for the reaction. Identify the oxidizing agent and the reducing agent in the reaction.

[Graph: Energy (kJ) vs Progress of Reaction, showing $SO_3^{2-} + OCl^-$ at higher energy level, activation energy peak, ΔH, and $SO_4^{2-} + Cl^-$ at lower energy level]

Converting Chemical Energy to Electricity

The most convenient form of energy available at the present time is electricity. It provides light, heat, hot water, and the ability to run machines of all kinds. Televisions, air conditioners, home appliances, and computers are but a few of the conveniences you get to enjoy because of electricity. In the United States, most electricity must be produced by burning fossil fuels, typically coal. **Figure 20.12** illustrates the production of electricity in a coal-fired power plant.

Figure 20.12

Energy Processes in a Coal-Fired Power Plant
In a typical electrical power plant, coal burns, producing heat that boils liquid water into steam. The kinetic energy of the moving steam drives a turbine, which drives a generator that produces the electrical energy.

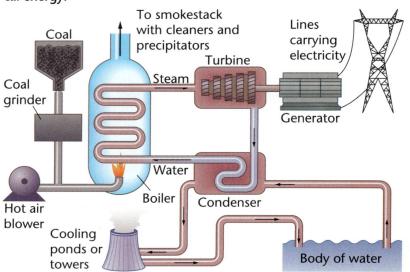

Bacterial Refining of Ores

Roughly 2000 years ago, Roman miners began to investigate a blue liquid that they often noticed around piles of discarded, low-grade copper ore. They associated the blue color with copper, probably based on their experience with blue minerals, such as turquoise, which were often found in the vicinity of that metal. When they heated those minerals in a charcoal fire, they obtained copper.

The blue color of the liquid made them suspect that copper was present in it. They heated the liquid to see whether it really did contain copper. Their efforts yielded success. Pure copper was produced from copper salts in the liquid. What the Romans did not realize was that the copper in the liquid had been leached from the low-grade copper ore by bacteria.

Why bacteria are used to extract metals
After crude ores are mined, they are crushed and then the metals are extracted by chemical treatment or by heating to high temperatures. Another way to extract metals from ore is to use bacteria to do part of the job. This method is less damaging to the environment, costs less in terms of energy input, and offers a way to use low-grade ores productively. Today, 24 percent of the copper produced is the result of bioprocessing by bacteria.

The bacterium *Thiobacillus ferrooxidans* obtains its energy by means of reactions involving various minerals. As a result, it produces acid and an oxidizing solution of Fe^{3+} ions, which can react with metals in crude ore.

The process is simple and requires little energy. Low-grade ore is first treated with sulfuric acid to stimulate the growth of bacteria. The microbes process the ore and release copper ions into solution. The metal is then extracted from the solution.

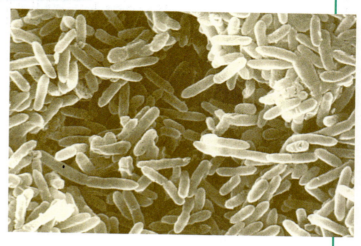

There's gold in them thar' bacteria Gold is another prospect for biorefining. As the sources of high-grade gold ore are disappearing, miners are being forced to mine low-grade ores. The gold in these lesser ores is present in the form of sulfides. To burn off the sulfur, the ores are traditionally treated by roasting or pressure oxidation. Only then can the gold be extracted with cyanide ions.

In one bio-oxidation process, low-grade gold ore is mixed with a brew of bacteria in huge, stirred reactors. A newer process places open piles of gold ore on an impermeable base. Bacterial cultures and the fertilizers needed to nurture them are poured onto the gold ore. In both methods, *T. ferrooxidans* treats the gold sulfides at a lower cost and with an efficiency of recovery increased from 70 to 95 percent.

Connecting to Chemistry

1. **Acquiring Information** Thermophilic bacteria, which thrive at temperatures around 100°C or higher, are being considered as possible candidates for biorefining. Find out what chemical advantage these bacteria have over other bacteria.

2. **Thinking Critically** What are the energy advantages of biorefining?

Alternative Energy Sources

Energy can be neither created nor destroyed. Does that mean that people can keep using the same energy sources at the present rate forever? Fossil fuels are a finite source of energy that should be conserved. Using alternative, renewable sources of energy can prolong the supply of fossil fuels.

Solar Energy

One alternative source of energy is solar energy. The sun will be dispensing energy for the next 5 billion years. More than 300 000 homes in the United States are using solar energy to heat their living quarters. The solar home in the illustration was built to capture sunlight and convert it to heat that warms the rooms. Adobe walls, clay tile floors, triple windows, and heavily insulated walls and ceilings store the sun's warmth in winter. An overhang keeps the high-angle solar rays from entering the house in summer but does not block the lower-angle sun rays in winter.

Photovoltaic cells use solar energy to produce electricity. These solar cells consist of layers of silicon with trace amounts of gallium or phosphorus and are capable of emitting electrons when struck by sunlight. Two main problems must be overcome before solar cells are a reliable source of electricity for everyday use. The first is the efficiency of conversion of solar energy to electrical energy. The second problem is finding efficient ways to store the electricity for use during the dark hours or during periods of cloudy weather.

Geothermal Energy

Magma, which is molten rock, can heat solid rock surrounding the magma chamber. When this occurs, water in the porous rock above the heated solid rock turns to steam. If there are cracks in the solid rock above, the steam escapes to the surface in the form of a geyser or hot spring. This is a natural source of geothermal energy. Many natural geothermal regions lie in the earthquake and volcano belts along Earth's crustal plate edges.

Geysers in Yellowstone National Park are an indication of the heat that is stored beneath the ground. This energy can be tapped and used to run power plants to produce electricity. The largest geothermal power plant in the world is at The Geysers in California. It generates 1000 megawatts of electric power—enough to serve the needs of 1 million people.

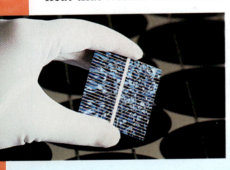

A single photovoltaic cell

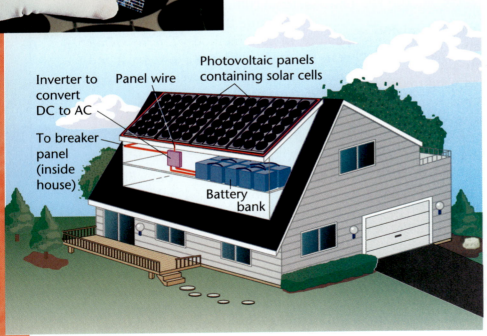

Inverter to convert DC to AC

Panel wire

Photovoltaic panels containing solar cells

To breaker panel (inside house)

Battery bank

Natural geothermal energy is available in only a few spots around the world. To create new sites requires deep drilling. Although a geothermal power plant generates electricity at one-fourth the cost of power from a new nuclear plant and one-half the cost of power from a new coal plant, it is expensive to drill down to a magma chamber. Another disadvantage of these power plants is that they produce air pollution in the form of hydrogen sulfide, ammonia, and radioactive materials released from deep within Earth.

Power plant at The Geysers

Wind Energy

In some parts of the country, wind power is an almost unlimited source of energy. Wind can turn turbines that produce mechanical energy. A generator converts the mechanical energy into electrical energy. Usually, hundreds of wind turbines are needed to run a power plant.

Wind power emits no air pollutants. It requires no water for cooling. However, wind power can be used only in areas that have reliable winds. When the winds die down, a backup system for producing electricity must also be available. At present, wind energy is sometimes more expensive than other energy sources. Cheaper wind turbines could change this.

DISCUSSING THE TECHNOLOGY

1. **Hypothesizing** Which form of alternative energy might be feasible in your state? Explain.

2. **Acquiring Information** Investigate and describe how a solar furnace works.

3. **Thinking Critically** The three kinds of alternative energy sources discussed in this feature are not effective in all places. Despite that fact, how can they help solve the nation's energy problems?

Saving Energy with Catalysts

You have learned in Chapter 18 that polyethylene is a plastic used in many household products. It is made from ethylene, a simple hydrocarbon whose formula is C_2H_4. Ethylene is produced industrially by removing hydrogen from ethane (C_2H_6), another hydrocarbon, according to the following equation.

$$C_2H_6(g) + heat \rightarrow C_2H_4(g) + H_2(g)$$

In theory, 1 mol of ethylene could be made with the addition of only 137 kJ of heat. In practice, the reaction above uses nearly four times that amount of energy to produce 1 mol of ethylene because of inefficiency in transferring energy. Over many years, chemists have introduced refinements into the production process. The development of better catalysts has lowered the energy requirements for this and other chemical processes. Research on energy-saving catalysts continues as energy becomes more expensive.

Entropy: One of the Costs of Using Energy

Any time energy is produced or is converted from one form to another, as it is in power plants and industrial processes, some of the energy is lost. This may seem surprising because of the law of conservation of energy. However, in this case, the energy is not destroyed but is simply not usable to do work. What happens to it? It is wasted as heat. In many processes, the amount of energy lost as heat exceeds the amount of energy that can be harnessed to do useful work.

The waste heat generated in systems that use or produce energy cannot be reclaimed, reused, or recycled because it has increased the random motion, and therefore the disorder, of molecules in the environment. As a result of this increased disorder, entropy increases as shown in **Figure 20.13.** The environment cannot spontaneously reorganize itself to give back the energy that increased its entropy. This increased disorganization of the environment is part of the price of using energy.

Figure 20.13

Using Energy Increases Entropy

Every time energy is used to do useful work such as spinning turbines to generate electricity, some energy is lost to the environment, usually as heat. This heat increases the entropy of the environment and cannot be reclaimed. ▼

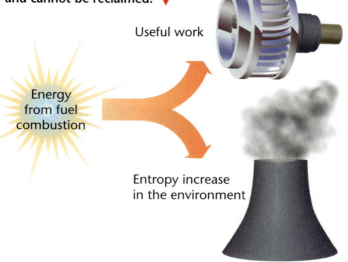

Useful work

Energy from fuel combustion

Entropy increase in the environment

In some cases, the heat is useful in situations where an increase in temperature is desirable, such as heating greenhouses in cold climates. ▼

A process, such as recycling, that saves energy by conserving electricity does more than reduce the total consumption of electricity. It also saves the chemical energy of original fossil fuel that would otherwise be lost forever as waste heat and increased entropy. This sometimes amounts to a savings of 70 percent of the fuel, making its energy available for some other use.

Energy and Efficiency

Because of entropy increases, no power or industrial plant, no matter how well designed, can completely convert heat from chemical reactions into useful work. Waste heat is inevitable. This kind of loss is easiest to understand in terms of efficiency, which is the amount of work that can be obtained from a process compared to the amount that must go into it.

For example, a modern fossil-fuel-burning plant has a maximum theoretical efficiency of only 63 percent. This means that, at most, 63 percent of the chemical energy from the original fuel can be converted to useful work. The remainder is lost as waste heat.

Other inefficiencies further reduce the percentage of energy that is converted to useful work. **Table 20.3** illustrates some of these factors in regard to a modern electrical power plant. The data in the table allow you to compute the plant's overall efficiency, which is simply the product of the efficiencies of all the individual steps multiplied by the maximum theoretical efficiency. If the maximum theoretical efficiency is 63 percent, then the efficiency of generating electricity by the process detailed in **Table 20.3** can be calculated as follows.

$$(0.63)(0.90)(0.75)(0.95)(0.90) = 0.36 = 36\%$$

This low value is typical of the energy efficiency of power and industrial plants.

At one time, scientists thought that heat was a colorless, odorless, weightless fluid. This fluid, which they called *caloric*, was believed to cause other substances to expand when it was added to them.

Table 20.3   Typical Efficiencies in Power Production

Maximum theoretical efficiency	63%
Efficiency of boiler	90%
Mechanical efficiency of turbine	75%
Efficiency of electrical generator	95%
Efficiency of power transmission	90%
Overall actual efficiency	36%

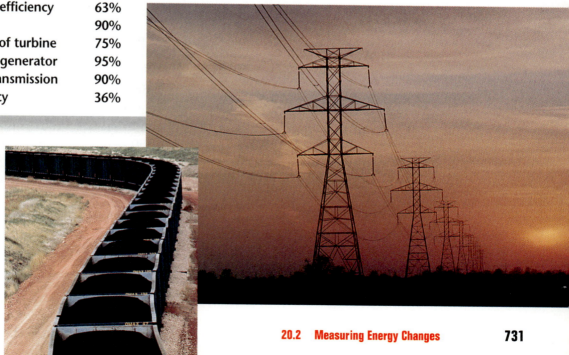

In a coal-fired power plant, a maximum of 36 percent of the energy available from coal is converted to electrical energy. ▶

Figure 20.14
Cooling Tower
A cooling tower is a common way to get rid of waste heat. Inside the tower, hot water is sprayed into the air while large fans draw air through the droplets. Although this process only transfers the heat to the atmosphere, it is an improvement over pumping the hot water back into streams, an action that killed organisms in the stream.

Improving the Efficiency of Industrial Processes

The efficiency of industrial processes can be increased and valuable energy resources can be conserved if plants that use improved energy-saving devices are designed and built. Also, if uses are found for waste heat—such as heating buildings and growing food in cold climates—energy can be conserved. Cooling towers, **Figure 20.14,** are one way of transferring waste heat. The development of more energy-efficient refrigerators, air conditioners, and lightbulbs can also improve the efficiency of electricity use.

Finally, the energy conversion processes of organisms can be studied and perhaps eventually applied to industrial methods. Biological processes convert energy from one form to another and at the same time create highly ordered—that is, low-entropy—systems. These natural processes are far more efficient than most industrial processes.

SECTION REVIEW

Understanding Concepts

1. Explain why it is impossible to convert 100 percent of the chemical energy in a fossil fuel to electrical energy.

2. The specific heat of aluminum is 0.902 J/g°C. The specific heat of copper is 0.389 J/g°C. If the same amount of heat is applied to equal masses of the two metals, which metal will increase more in temperature? Explain.

3. How much heat, in kilojoules, is given off by a chemical reaction that raises the temperature of 700 g of water in a calorimeter by 1.40°C?

Thinking Critically

4. **Using a Table** Using **Table 20.3** as a guide, determine the overall efficiency of a power plant that has a maximum theoretical efficiency of 50 percent, given that the other efficiencies are the same as those in the table.

Applying Chemistry

5. **Calorie Counting** Using **Table 20.2** as a guide, compare the Caloric content of 1 g of butter to 1 g of spaghetti. Which food gives you more Calories per gram?

chemistryca.com/self_check_quiz

Photosynthesis

All organisms need energy to survive, and the main source of energy for Earth is the sun. Somehow, the sun's energy must be captured as chemical energy so that living things can use it. This process is called photosynthesis.

The Basis of Photosynthesis

Plants and other photosynthetic organisms have at least one obvious characteristic in common: they are green, the color of chlorophyll. The chlorophyll in cells of plants and algae is found in organelles called chloroplasts, as shown in **Figure 20.15.** The chlorophyll in photosynthetic bacteria is found on membranes spread throughout the cells. However, in both cases, the process works essentially the same way. Chlorophyll absorbs light and changes it to chemical energy, triggering a complex series of changes that together make up photosynthesis. Let's examine the chemical process in more detail. Note, as you read, that most of the changes absorb energy.

SECTION PREVIEW

Objectives
✓ **Analyze** the process and importance of photosynthesis.
✓ **Compare** the energy efficiency of photosynthesis and processes that produce electricity.
✓ **Trace** how energy from the sun passes through a food web.

Review Vocabulary
Calorie: heat required to raise the temperature of 1 gram of water by 1°C.

New Vocabulary
photosynthesis

Figure 20.15

Chloroplasts

The photo (left) shows some plant cells as seen through a microscope. The green structures inside are chloroplasts, which contain the light-absorbing green pigment chlorophyll. The diagram (right) shows a single chloroplast. Notice the grana, which are stacks of membranes on which the chlorophyll is located. The liquid surrounding the membranes is called stroma and is the place where sugar molecules are synthesized.

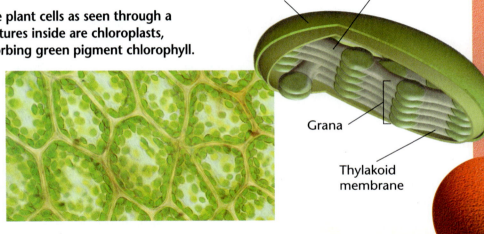

Chloroplast   Stroma

Grana

Thylakoid membrane

The Chemistry of Photosynthesis

The process of **photosynthesis** involves a series of reactions in which green plants and some other organisms manufacture carbohydrates from carbon dioxide and water using energy from sunlight. The net reaction that is carried out in photosynthesis is one that produces simple sugars and oxygen gas from carbon dioxide and water. The balanced equation is generally written as follows.

$$6CO_2 + 6H_2O \rightarrow C_6H_{12}O_6 + 6O_2$$

The formula $C_6H_{12}O_6$ represents glucose, a simple sugar.

Photosynthesis is divided into two basic series of reactions—the light reactions and the Calvin cycle.

The Light Reactions

When light reaches a chlorophyll molecule, the energy of the light is absorbed. The energy excites electrons in the molecule, as described in **Figure 20.16.** A high-energy electron is released from the chlorophyll and transfers some of its energy to a molecule of ADP, adenosine diphosphate. That causes the ADP to bond to a third phosphate to form ATP, adenosine triphosphate. As you may recall from Chapter 19, ATP is the main energy storehouse in cells and also plays a critical role in respiration.

As you can see by following the process shown in **Figure 20.16,** electrons lost by the chlorophyll in the first step are restored to it by a reaction in which water is split into elemental oxygen and hydrogen ions. The oxygen is released into the atmosphere and can later be used by organisms, including you, for respiration.

Figure 20.16
The Light Reactions

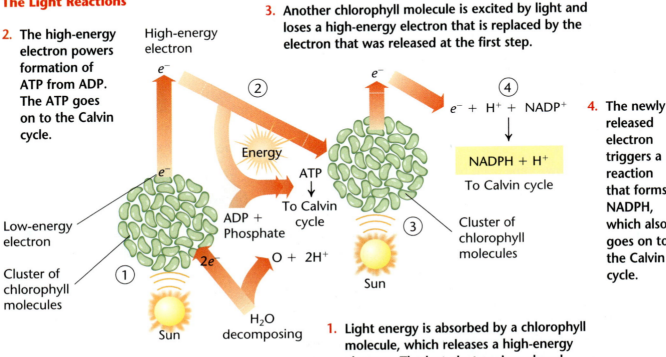

2. **The high-energy electron powers formation of ATP from ADP. The ATP goes on to the Calvin cycle.**

3. **Another chlorophyll molecule is excited by light and loses a high-energy electron that is replaced by the electron that was released at the first step.**

4. **The newly released electron triggers a reaction that forms NADPH, which also goes on to the Calvin cycle.**

1. **Light energy is absorbed by a chlorophyll molecule, which releases a high-energy electron. The lost electron is replaced through the breakdown of water.**

High-energy electron

Low-energy electron

Cluster of chlorophyll molecules

Energy

ATP

To Calvin cycle

ADP + Phosphate

$O + 2H^+$

H_2O decomposing

Sun

$e^- + H^+ + NADP^+$

NADPH + H^+

To Calvin cycle

Cluster of chlorophyll molecules

Sun

Figure 20.17
The Calvin Cycle

1. In the Calvin cycle, atmospheric carbon dioxide reacts with a five-carbon molecule to form an unstable six-carbon molecule. That six-carbon molecule breaks down into two three-carbon molecules.

2. The two three-carbon molecules interact with ATP, NADPH, and H^+ from the light reactions, which provide energy and hydrogen atoms. As a result, the two three-carbon molecules are converted to two PGAL molecules, which are three-carbon sugars.

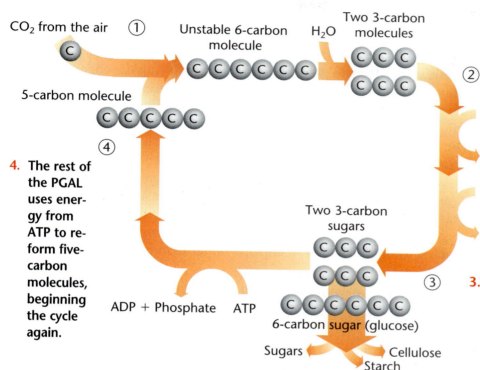

4. The rest of the PGAL uses energy from ATP to reform five-carbon molecules, beginning the cycle again.

3. Some of the PGAL that is made in this way can combine to form the six-carbon sugar glucose. It can also be used directly as an energy source or to make fats or nucleic acids.

The electron from the original chlorophyll has now lost its energy and moves on to join a different chlorophyll molecule—one that also absorbs light energy but is involved in a different reaction. The transferred electron replaces an electron that is energized by light absorption and released. This newly released electron is involved in another reaction, which forms an important coenzyme called NADPH. The NADPH is formed from NADP and from hydrogen ions that were freed in the earlier breakdown of water. Both the NADPH and the ATP formed as a result of the first absorption of energy continue to the second series of photosynthesis reactions, the Calvin cycle.

The Calvin Cycle

Once the light reactions have produced ATP and NADPH, the second series of photosynthesis reactions can occur. As shown in **Figure 20.17**, the Calvin cycle, which takes place in the stroma of chloroplasts, makes use of NADPH from the light reactions, as well as carbon dioxide taken in from the surroundings. Simple sugars, such as glucose, result.

In the Calvin cycle, ATP turns back into ADP and provides energy for a series of reactions. In the first of these, NADPH loses a hydrogen ion and becomes NADP. The hydrogen ion from the NADPH, together with other H^+ ions produced during the light reactions, provides hydrogen for the synthesis of the sugars. The carbon and oxygen for the sugars are provided mainly by the carbon dioxide.

Energy and the Role of Photosynthesis

As you read earlier, most of the reactions that occur during photosynthesis are endothermic. The absorbed energy is stored by using it to synthesize high-energy molecules. Only chlorophyll-containing organisms can carry out photosynthesis and make these high-energy molecules that they need in order to survive. The animals that eat those organisms get the energy they need from them, as shown in **Figure 20.18.** So, photosynthesis is the ultimate process for chemically storing the energy that all organisms in the food web need. You could thus describe photosynthesis as the entry point of energy for life on Earth. See **Figure 20.19** on the next page.

The organization and maintenance of your body, like that of every organism, depend upon energy that is used to power chemical changes. However, energy is not the only factor involved when systems undergo changes. There is also entropy. Natural systems have a tendency toward increased disorder. However, living things are highly ordered, and many of the processes that take place in them, such as building more complex molecules, involve decreases in entropy rather than increases.

When meat-eating animals, such as some kinds of birds, consume the plant eaters, they make use of the energy that was stored in their prey. ▼

Figure 20.18

Energy and the Food Web
Ultimately, the sun is the source of the energy, and photosynthesis is the process that captures it and makes all life possible.

The oak tree absorbs energy from the sun and stores it in the molecules it makes during photosynthesis. The stored energy allows it to carry out essential functions. ▶

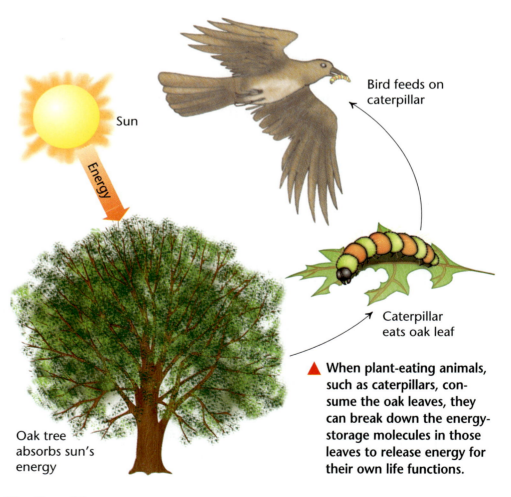

Sun

Energy

Bird feeds on caterpillar

Caterpillar eats oak leaf

Oak tree absorbs sun's energy

▲ When plant-eating animals, such as caterpillars, consume the oak leaves, they can break down the energy-storage molecules in those leaves to release energy for their own life functions.

How is it possible, then, for organisms to overcome the tendency to disorder and maintain their complex structures and life functions? The secret of their success is their ability to absorb energy from outside sources, such as light and food. These outside sources continue pumping in the energy needed to keep the otherwise-nonspontaneous processes going.

When the body converts stored chemical energy to other uses during respiration, heat is produced. The heat, if it built up, could lead to destructive entropy increases and a breakdown of structures and functions. Some of the heat produced is radiated to the environment. The heat given off raises the entropy of the surroundings rather than that of the body. The low entropy of living things is thus maintained partially at the expense of the entropy of the surroundings.

In warm-blooded organisms, such as humans, some of the heat produced during respiration is retained by the body, which uses it to maintain a constant temperature. This temperature allows various biochemical processes to take place at the proper rates.

Supplemental Problems

For more practice with solving problems, see Supplemental Practice Problems, Appendix B.

Figure 20.19

Photosynthesis and Respiration in Balance
This British chemist spent 15 days in a sealed chamber where his oxygen was entirely supplied by 30 000 wheat plants carrying out photosynthesis. All carbon dioxide for photosynthesis came from the man's respiration. This 1995 experiment was the first in a series to determine whether humans could live in a similar but larger chamber on the moon or another planet.

SECTION REVIEW

Understanding Concepts

1. Plants appear green because their leaves reflect green and yellow light. Explain what happens to the energy in the wavelengths of light that are not reflected.

2. How does chlorophyll function in photosynthesis?

3. Explain why a meat-eating animal such as an eagle depends on photosynthesis for its energy.

Thinking Critically

4. **Inferring** What would be the effect on Earth's atmosphere if all plants were temporarily unable to carry out the process of photosynthesis?

Applying Chemistry

5. **Opposite Processes** Compare the process of photosynthesis with the process of respiration in Chapter 19. Explain why they are often considered opposite processes.

CHAPTER 20 ASSESSMENT

REVIEWING MAIN IDEAS

20.1 Energy Changes in Chemical Reactions

- Exothermic reactions give off energy; endothermic reactions absorb energy.

- Energy can be converted from one form to another, but cannot be created or destroyed.

- Activation energy is the energy needed to get a reaction started. A catalyst lowers the activation energy for a given reaction.

- Entropy is a measure of disorder. Spontaneous processes tend to proceed from a state of high energy to a state of low energy. They also tend to proceed from a state of less disorder to a state of greater disorder.

20.2 Measuring Energy Changes

- Chemical reactions are a source of energy because the energy of chemical bonds can be converted to heat, light, or electrical energy. Heat changes can be measured by calorimetry.

- Whenever energy is converted from one form to another, some energy is lost as heat. The lost energy generally cannot be used again or converted back to a useful form of energy.

20.3 Photosynthesis

- Photosynthesis is a process by which plants use light energy to manufacture carbohydrates from carbon dioxide and water. Because such substances are the primary source of energy for all organisms, the results of photosynthesis provide energy for life on Earth.

- The reactions that occur during photosynthesis include the light reactions and the Calvin cycle. Most of the reactions that occur during this process are endothermic.

Vocabulary

For each of the following terms, write a sentence that shows your understanding of its meaning.

calorie
Calorie
entropy
fossil fuel
heat

kilocalorie
law of conservation
   of energy
photosynthesis

UNDERSTANDING CONCEPTS

1. Explain how your car converts energy from one form to another.

2. How does the process by which plants obtain sugars differ from that by which most other organisms obtain sugars?

3. The efficiency of an automobile engine is improved by heating the interior of the car. Why is this the case?

4. Why is energy needed to sustain an endothermic reaction?

5. What is meant by *entropy*?

6. Suppose that a coal-fired power plant converts energy from coal into electrical energy with 36 percent efficiency. How many kilojoules of energy from coal are needed to produce 1.0 kJ of electrical energy?

7. Hydrogen is used as a fuel for the space shuttle because it provides more energy per gram than many other fuels. The combustion of hydrogen is described by the following equation.

$$2H_2(g) + O_2(g) \rightarrow 2H_2O(g)$$
$$\Delta H = -484 \text{ kJ}$$

a) Is the reaction exothermic or endothermic?

b) How much energy does complete combustion of 1.00 g of hydrogen provide?

8. What is the difference between a positive ΔH and a negative ΔH in terms of energy and the kind of reaction involved?

9. Why is outside energy needed to start most exothermic reactions but not to sustain them?

10. Why does it take more energy to decompose water than it does to boil water?

11. Describe the general trends of spontaneous reactions in terms of energy and entropy change.

● APPLYING CONCEPTS

Chemistry and Technology

12. Explain why garbage could be an excellent fuel for the production of electricity.

How It Works

13. Another type of pack that can be used to supply heat to injuries is reusable. It contains a supersaturated solution of a salt and a disc of metal. When the metal disc is bent, the solute begins to crystallize and releases heat. The pack can be reset by heating it in boiling water, which causes the salt to dissolve again. How can you account for the heat given off by this kind of pack?

Everyday Chemistry

14. Write balanced equations for the following reactions, which occur in catalytic converters: a) the breakdown of nitrogen monoxide into oxygen and nitrogen, and b) the reaction of carbon monoxide with oxygen to produce carbon dioxide.

Earth Science Connection

15. The specific heat of lead is 0.128 J/g • °C. How much heat is required to raise the temperature of 10.0 g of lead from 20.0°C to 30.0°C? Compare this amount of heat to the amount needed to raise the temperature of 10.0 g of water by the same amount.

16. List two factors that would make an endothermic reaction more likely to occur.

17. Why will the butane gas in a disposable lighter fail to ignite if the flint is worn out and does not generate a spark?

18. Cite three common examples that illustrate the natural tendency for entropy to increase.

19. Heat from a burning match is necessary to ignite a candle. Why is it incorrect to say that the burning of a candle is, therefore, an endothermic reaction?

● THINKING CRITICALLY

Interpreting Data

20. Account for the fact that there is an energy difference between the following reactions.

$$CH_4(g) + 2O_2(g) \rightarrow CO_2(g) + 2H_2O(g)$$
$$\Delta H = -802 \text{ kJ}$$

$$CH_4(g) + 2O_2(g) \rightarrow CO_2(g) + 2H_2O(l)$$
$$\Delta H = -890 \text{ kJ}$$

Applying Concepts

21. The kinetic model of matter and the concept of entropy both explain the process of diffusion, which is the spontaneous spreading of liquid or gas particles throughout a volume. Explain, in terms of both kinetics and disorder, why a drop of a dye placed into a beaker of water will eventually color the entire volume of water.

Making Predictions

22. The heats of reaction for the formation of the hydrogen halides from their elements are listed in the table below. On the basis of this information, predict which compound is most stable, in terms of not breaking down into its elements. Which is least stable? Explain.

Compound	ΔH (kJ/mol)
HF	−268
HCl	−92
HBr	−36
HI	+25

Interpreting Data

23. **ChemLab** Two different foods are burned in a calorimeter. Sample 1 has a mass of 6.0 g and releases 25 Calories of heat. Sample 2 has a

mass of 2.1 g and releases 9.0 Calories of heat. Which food releases more heat per gram?

Relating Cause and Effect

24. **MiniLab 1** Which factor—entropy, added heat, or both—promotes the dissolution in water of a solid that spontaneously dissolves endothermically? Which factor promotes dissolution in water of a solid that spontaneously dissolves exothermically?

Inferring

25. **MiniLab 2** A certain reaction proceeds in the forward direction with a ΔH of -16 kJ. Will the reverse reaction be endothermic or exothermic? What will be the value of ΔH for the reverse reaction?

● CUMULATIVE REVIEW

26. Distinguish between atomic number and mass number. How do each of these two numbers compare for isotopes of an element? (Chapter 2)

27. Why does the second period of the periodic table contain eight elements? (Chapter 7)

● SKILL REVIEW

28. **Using a Graph** The graph below projects the world production of coal (solid line) until reserves are depleted and also projects the demand for coal (dashed line). Use the graph to answer the following questions.

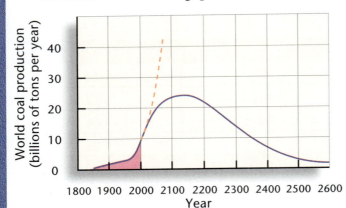

a) How much longer can the present growth rate in coal consumption continue?

b) In what year will maximum coal production be reached, and how much coal will be produced that year?

c) Compare the amount of coal produced in 1900 to the production in the peak year.

29. **Using a Data Table** The chemical formulas, heats of combustion, and formula masses of three hydrocarbons are given in the following table.

Hydrocarbon	Heat of Combustion (kJ/mol)	Formula Mass (g/mol)
Methane, CH_4	-890	16
Butane, C_4H_{10}	-2859	58
Hexane, C_6H_{14}	-4163	86

a) According to the data, which hydrocarbon yields the greatest amount of energy per gram?

b) Using the data table, draw a graph of heat of combustion in kilojoules per mole versus formula weight. Use your graph to estimate the heat of combustion of propane (C_3H_8).

c) Conduct library research to find uses for these four hydrocarbons. Search out advertisements or magazine photos illustrating the uses. Prepare a display or bulletin-board presentation to illustrate your findings.

● PROBLEM SOLVING

30. The compound B_5H_9 was once proposed as a rocket fuel because it has a high heat of combustion. In a combustion experiment, a scientist observes that 1.00 g of B_5H_9 burned in excess oxygen in a calorimeter raises the temperature of 800 g of water from 24.0°C to 44.3°C. How much heat is generated in the reaction?

31. An organic fuel is burned in a calorimeter. The reaction releases 194 kJ of heat. If the initial temperature of the water was 21.0°C, and the final temperature was 51.9°C, how much water was in the calorimeter?

Chemical Reaction	Heat of Reaction (kJ/mol)
$4Fe(s) + 3O_2(g) \rightarrow 2FeO_3(s)$	−1625
$NH_4NO_3(s) \rightarrow NH_4+(aq) + NO_3^-(aq)$	+27
$C_6H_{12}O_6(s) + 6O_2(g) \rightarrow 6CO_2(g) + 6H_2O(l)$	−2808
$H_2O(l) \rightarrow H_2O(g)$	+40.7

Use the table above to answer questions 1–3.

1. The rusting of an iron nail is a(n)

   a) decomposition reaction.
   b) displacement reaction.
   c) endothermic reaction.
   d) exothermic reaction.

2. How much energy is required to boil 0.5 L water into steam?

   a) 1.1 kJ
   b) 20.35 kJ
   c) 732.6 kJ
   d) 1130.6 kJ

3. What is the heat of reaction for the production of 360 g of glucose during the process of photosynthesis ?

   a) 2808 kJ
   b) −2808 kJ
   c) 5616 kJ
   d) −5616 kJ

4. During a catalyzed reaction, the catalyst

   a) provides additional energy to increase the rate of the reaction.
   b) reduces the required activation energy to increase the rate of the reaction.
   c) transforms an endothermic reaction into an exothermic reaction.
   d) transforms an exothermic reaction into an endothermic reaction.

5. Which of the following is NOT an example of entropy?

   a) A perfume bottle is opened releasing evaporated alcohol molecules that carry the perfume scents to every corner of a room.
   b) Solar energy bathes the Earth with light and heat providing energy for photosynthesizing organisms to convert water and carbon dioxide into glucose.
   c) Over thousands of years, ocean waves erode rock and coral into small particles that are carried and deposited to form beaches on shorelines.
   d) A large pile of snow resting on an asphalt surface experiences a drastic increase in temperature and melts into water which runs off the surface into the soil.

6. The specific heat of ethanol is 2.44 J/g °C. How many kilojoules of energy are required to heat 50.0 g of ethanol from −20.0 C to 68.0 °C?

   a) 10.7 kJ   c) 2.44 kJ
   b) 8.30 kJ   d) 1.22 kJ

Unknown Metal Characteristics
Mass of metal = 4.68 g
Quantity of heat absorbed (q) = 256 J
Change in Temperature = 182 °C

Use the information in the table above to answer question 7.

7. Calculate the specific heat for the unknown metal.

   a) 0.15 J/g °C   c) 3.3 J/g °C
   b) 0.301 J/g °C   d) 6.58 J/g °C

Test Taking Tip

Your Mistakes Can Teach You The mistakes you make before the test are helpful because they show you the areas in which you need more work.

CHAPTER
21 Nuclear Chemistry

Chapter Preview

Sections

21.1 Types of Radioactivity
   ChemLab   The Radioactive
               Decay of "Pennium"

21.2 Nuclear Reactions and Energy
   MiniLab 21.1 A Nuclear Fission
                  Chain Reaction

21.3 Photosynthesis
   MiniLab 21.2 Radon—A Problem
                  in Your Home?

How Old is This Tree?

This piece of wood is old enough that it is now composed of rock, and is known as petrified wood. Scientists can determine the ages of old things by measuring the amounts of certain radioactive substances (isotopes) remaining in the objects—this process is called radiochemical dating.

Launch Lab

Chain Reactions

When the products of one nuclear reaction cause additional nuclear reactions to occur, the resulting chain reaction can release large amounts of energy in a short period of time. Explore escalating chain reactions by modeling them with dominoes.

Materials

- 28 domino tiles (1 set)
- stopwatch

Procedure

1. Obtain a set of domino tiles.
2. Stand the individual dominoes on end and arrange them so that when the first domino falls, it causes the other dominoes to fall in series.
3. Practice using different arrangements until you determine how to cause the dominoes to fall in the least amount of time.
4. Time your domino chain reaction. Compare your time with those of your classmates.

Analysis

What arrangement caused the dominoes to fall in the least amount of time? Do the dominoes fall at a steady rate or an escalating rate? What happens to the domino chain reaction if a tile does not contact the next tile in the sequence?

What I Already Know

Review the following concepts before studying this chapter.

Chapter 2: structure of the nucleus

Chapter 6: how to balance a chemical equation

Chapter 8: the actinides

Chapter 19: DNA structure

Reading Chemistry

Before reading the chapter, observe some of the key terms that appear, such as "nuclear" and "radioactivity." Write down the ideas that come to mind upon hearing these words. Where have you heard them used before? As you read the chapter, compare your impression of these words to their scientific use and meaning.

Preview this chapter's content and activities at chemistryca.com

Types of Radioactivity

Some elements have nuclei that are naturally unstable. Such nuclei disintegrate and form more stable nuclei. Marie Curie was one of the scientists who first studied this phenomenon and helped move the field of chemistry into the nuclear age. She and her husband, Pierre, carried out research together in a laboratory in France. They studied the changes that occur in atomic nuclei and discovered several elements that have unstable nuclei. In the days of the Curies, no one understood the dangers of dealing with unstable elements. Today, scientists know that they must be used with extreme care.

SECTION PREVIEW

Objectives

✓ **Analyze** common sources of background radiation.

✓ **Compare and contrast** alpha, beta, and gamma radiation.

✓ **Apply** the concept of half-life of a radioactive element.

Review Vocabulary

Photosynthesis: the process used by certain organisms to capture energy from the Sun.

New Vocabulary

radioactivity
alpha particle
beta particle
gamma ray
half-life

Discovery of Radioactivity

You are probably familiar with novelty items that glow in the dark. Maybe you even have one of those ceiling constellation maps with stars that give off an eerie, green glow long after the room lights are turned out. Have you ever wondered what causes the stars in such maps to glow? The material in them is a phosphorescent form of the compound zinc sulfide, which means that after it is exposed to light, it continues to give off the light it has absorbed, as shown in **Figure 21.1.**

Figure 21.1

Phosphorescence
When light shines on phosphorescent materials, such as these glow-in-the-dark decorations, the absorbed energy raises the electrons in the atoms to higher energy levels. ▼

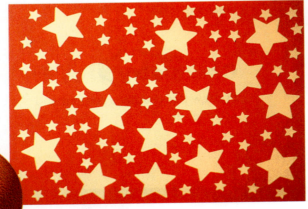

▲ When the electrons move back to lower energy levels, the extra energy is given off as light, as shown by the eerie, green glow of these stars. The glow disappears gradually, as the electrons return to their more stable configurations.

A French scientist named Henri Becquerel was studying a phosphorescent uranium compound in 1896 when he made an accidental discovery that changed the course of history. Becquerel was testing the compound to see whether the phosphorescence was the cause of a recently discovered type of electromagnetic radiation called X rays. After exposing the sample to sunlight, he placed paper-wrapped photographic film near it. X rays would be able to pass through the paper and expose the film, producing an image of the sample on it, just as they make an image of your teeth on film when you have an X ray taken. Becquerel found the image he expected. He believed that X rays were the cause and that the uranium gave off these rays because of absorbed sunlight. However, later he developed a piece of paper-wrapped film that had been in a dark drawer along with a crystal of the uranium. He did not expect to see an image on the film because the uranium sample had not been exposed to sunlight or any other source of energy. To his surprise, the film bore the image of the crystal, as shown in **Figure 21.2.**

What made the uranium give off radiation? It hadn't been exposed to light, so the radiation wasn't caused by phosphorescence. Was a chemical reaction taking place? No known chemical reactions had ever produced this effect. Becquerel suspected that the uranium was spontaneously emitting some type of radiation. Was the radiation the same as X rays? No one could tell for sure. If so, it was odd that the rays could be produced without an input of energy, which was generally needed to produce X rays.

Figure 21.2
Becquerel's Uranium Experiment
Becquerel discovered that uranium compounds spontaneously give off radiation that can pass through paper wrapping and produce an image on film. He also found that a coin prevents the radiation from reaching the film, producing a dark, unexposed circle in the image.

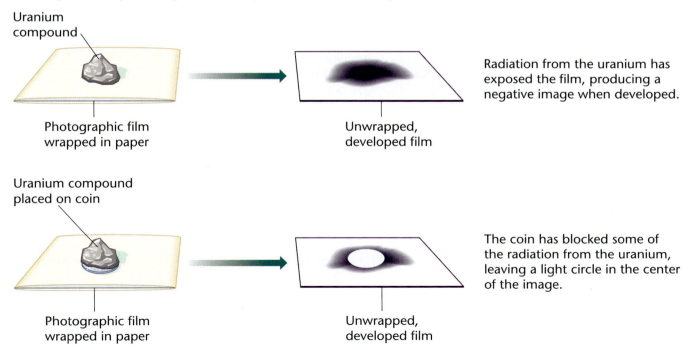

Uranium compound

Photographic film wrapped in paper

Unwrapped, developed film

Radiation from the uranium has exposed the film, producing a negative image when developed.

Uranium compound placed on coin

Photographic film wrapped in paper

Unwrapped, developed film

The coin has blocked some of the radiation from the uranium, leaving a light circle in the center of the image.

Fact of the MATTER

Marie and Pierre Curie's daughter, Irene, also became a scientist and Nobel prize winner. She and her husband, Frederick Joliot-Curie, discovered the first artificially created radioisotope and conducted nuclear fission with uranium. They also hampered Germany's use of nuclear installations during World War II by smuggling "heavy" water (D_2O) out of France after that country was invaded.

Becquerel presented the problem to Marie and Pierre Curie for further study. Their conclusion was that a nuclear reaction was taking place within the uranium atoms. Marie Curie named this spontaneous emission of radiation by an unstable atomic nucleus **radioactivity.**

For the first time, the effects of changes in the nucleus of an atom had been observed and correctly interpreted. The Curies went on to study the properties of radioactivity and discovered elements other than uranium that also give off radiation. Henri Becquerel and the Curies were awarded the 1903 Nobel Prize in Physics for their important work.

Nuclear Notation

Whereas chemical reactions involve changes in the number or configuration of electrons, nuclear reactions involve the protons and neutrons found in the nucleus. During nuclear reactions, a nucleus can lose or gain protons and neutrons. Recall that the number of protons is the same as the atomic number, which identifies the kind of element. Therefore, adding or taking away protons changes the identity of an element. In nuclear reactions, the dream of the alchemists becomes reality as atoms of one element are converted into atoms of another element.

You can represent nuclear changes using equations, just as you do for chemical reactions. The equations will look similar in most ways to those you already know how to write and balance. The reactants will be on the left and products on the right, following an arrow. However, now you will be considering the nuclei of elements as reactants and products.

When you write nuclear equations, it is important to indicate which isotopes of the given elements you are dealing with. For example, the element carbon has natural isotopes with three different atomic masses: carbon-12, carbon-13, and carbon-14. Two of these, carbon-12 and carbon-13, are stable isotopes and are not radioactive. However, carbon-14 is unstable and radioactive. There must be some way to distinguish between these isotopes in nuclear equations. To do this, you write the mass number as a superscript and atomic number as a subscript in the symbol for an isotope. These numbers are both placed to the left of the symbol for the element. So, the three isotopes of carbon are represented as ${}^{12}_{6}C$, ${}^{13}_{6}C$, and ${}^{14}_{6}C$, respectively. Each has six protons, but carbon-12 has a total of 12 protons and neutrons, carbon-13 has a total of 13, and carbon-14 has 14. **Figure 21.3** illustrates the isotopes of carbon and hydrogen.

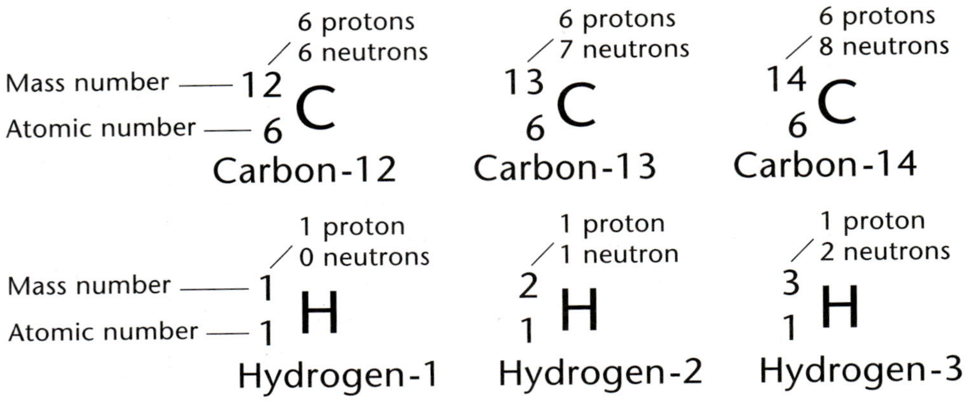

Figure 21.3
Isotope Notation
Numerical superscripts and subscripts are used to show the mass numbers and atomic numbers of isotopes. The isotopes of carbon and of hydrogen are represented here.

Radioactive Decay

The release of radiation by radioactive isotopes—radioisotopes, for short—is called decay. The nuclei of such radioisotopes are unstable. However, not all unstable nuclei decay in the same way. Some give off more powerful radiation than others or different kinds of radiation. Between 1896 and 1903, scientists had discovered three types of nuclear radiation. Each type changes the nucleus in its own way. These three types were named after the first three letters of the Greek alphabet: alpha (α), beta (β), and gamma (γ).

Alpha Decay

Alpha radiation consists of streams of **alpha particles,** which are helium nuclei consisting of two protons and two neutrons. Alpha particles can be represented as $_2^4\text{He}^{2+}$ or simply as α. Their relatively large size and charge, compared with other forms of radiation, cause them to collide frequently with atoms in air and objects they encounter. Alpha radiation does not deeply penetrate into matter and is easily stopped by a thin layer of material, such as paper and clothing, or even by air. Whenever alpha decay occurs, the decaying nucleus loses $2p^+$ and $2n^0$, and turns into the nucleus of an element with an atomic number that is 2 less and a mass number that is 4 less than that of the original.

Radioactive uranium-238 decays spontaneously in a long series of steps, the first of which is an alpha decay. When an alpha particle is emitted from a nucleus of uranium-238, the uranium atom loses two protons and two neutrons. The loss of protons makes the product another element, thorium, as shown in the following equation and in **Figure 21.4.**

$$_{92}^{238}\text{U} \rightarrow \ _{90}^{234}\text{Th} + _2^4\text{He}$$

$_{92}^{238}\text{U}$ $_{90}^{234}\text{Th}$ $_2^4\text{He}$

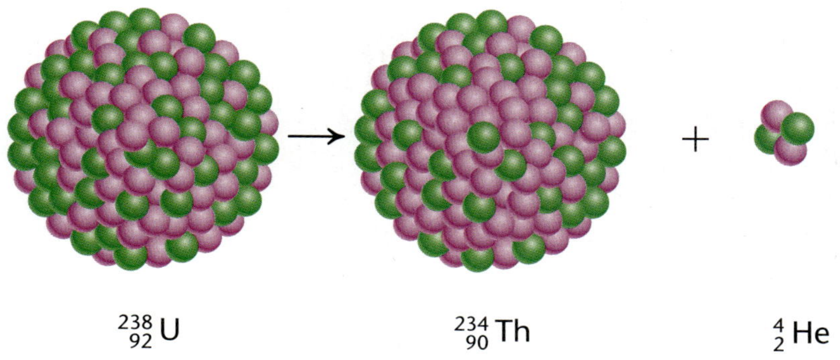

Figure 21.4

Alpha Decay of Uranium-238
When a nucleus of uranium-238 undergoes alpha decay, a nucleus of thorium-234 and a nucleus of helium-4 are produced. The atomic numbers balance on the two sides of the equation, as do the mass numbers.

Is this equation balanced? Check to see whether the sum of the mass numbers on the right side of the equation is 238, which is the mass number of the uranium isotope on the left side. It is, because the mass number of the thorium atom is 234 and that of the alpha particle is 4. Also make sure that the sums of the atomic numbers on both sides of the equation are the same. Both sums equal 92, so the atomic number balances, and the nuclear equation as a whole is balanced. Note that a balanced nuclear equation differs from balanced chemical equations because the kinds of atoms involved need not remain the same.

How it Works

Smoke Detectors

A smoke detector is a device that sounds an alarm in the presence of smoke particles from a smoldering or burning object. An ionizing smoke detector, like the one illustrated here, contains an electrical sensor that detects smoke particles that interrupt an electrical current. Ionizing detectors usually contain the unstable isotope americium-241 ($^{241}_{95}Am$), which decays to form $^{237}_{93}Np$ and $^{4}_{2}He$.

1. A battery or other power source provides electricity.

4. A microchip monitors the flow of current between the electrodes.

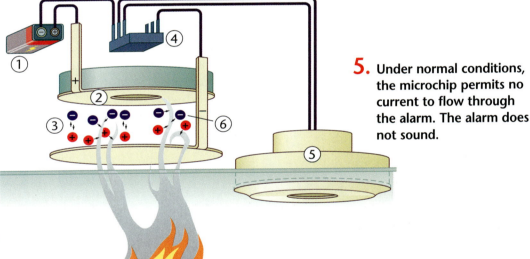

2. Americium-241 provides alpha particles that ionize air molecules.

5. Under normal conditions, the microchip permits no current to flow through the alarm. The alarm does not sound.

3. In the absence of smoke, the ions carry current between a positive and a negative electrode.

6. If a fire occurs, smoke particles rise up into the detector and interfere with the flow of ions between the electrodes. The microchip senses the drop in current and allows electricity to flow through the alarm circuit, sounding the alarm.

Thinking Critically

1. What kind of radioactive decay does americium-241 undergo?

2. Explain why the air that has been exposed to the americium-241 is able to carry a current between charged electrodes.

748   **Chapter 21   Nuclear Chemistry**

Beta Decay

The second type of radiation from unstable atomic nuclei results in beta decay and produces beta radiation. Beta radiation is made up of particles that are much smaller and lighter than alpha particles, so they move much faster and have greater penetrating power. Each **beta particle** is a high-energy electron with a 1− charge and is written as $_{-1}^{0}e$ or β^-. Beta particles pass through matter more easily than do alpha particles. They can be stopped only by thick materials such as stacked sheets of metal, blocks of wood, or heavy clothing.

The electron produced in beta decay is not one of the original electrons from outside the nucleus. It is produced by the change of a neutron into a proton and an electron. Whenever beta decay occurs, transmutation of elements occurs. The decaying nucleus turns into the nucleus of an element with an atomic number that is 1 greater than that of the original and a mass number that is the same.

Carbon-14 is an example of an isotope that decays by beta emission.

$$_{6}^{14}C \rightarrow {}_{7}^{14}N + {}_{-1}^{0}e$$

Notice that transmutation of elements has occurred. An atom of carbon-14 is converted into an atom of nitrogen-14. Because the beta particle has only the mass of an electron, the mass number remains 14. Because the beta decay changes a neutron to a proton, the atomic number of the nucleus increases by one unit, from 6 to 7. The atomic number on the left side of the equation, 6, is the same as the sum on the right side of the equation ($6 = 7 + (-1)$). The nuclear equation as a whole is balanced.

Gamma Decay

In the third type of decay, gamma radiation is produced. A **gamma ray** is a high-energy form of electromagnetic radiation without mass or charge. Because of its high energy and penetrating ability, gamma radiation can cause great harm to living cells. Gamma rays are written as γ. Such rays are much harder to stop than alpha or beta particles, as shown in **Figure 21.5.** They can pass easily through most types of material, and stopping them requires thick blocks of lead or even thicker blocks of concrete.

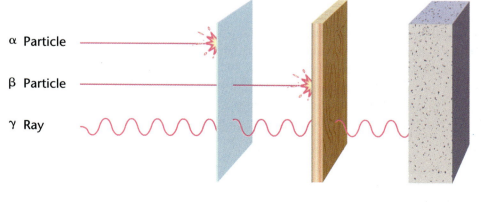

α Particle

β Particle

γ Ray

Sheet of paper   Dense wood   Thick blocks of lead or concrete

Figure 21.5
Blocking Radiation
The three types of radiation vary in their penetrating power. Alpha particles are stopped by even a thin sheet of paper, but beta particles require thicker shields, such as a dense piece of wood. The energetic gamma rays are stopped only by thick blocks of lead or concrete.

During gamma decay, only energy is given off. Gamma radiation is often omitted from nuclear equations because it does not affect mass number or atomic number. Gamma decay generally does not occur alone but accompanies other modes of decay. For example, the alpha decay of uranium to thorium involves simultaneous gamma decay.

$$^{238}_{92}\text{U} \rightarrow\ ^{234}_{90}\text{Th} +\ ^{4}_{2}\text{He} + \gamma$$

Once you know what kind of nucleus is produced by a simple nuclear reaction, you can determine what type of decay has taken place. If both the atomic number and mass number decrease, alpha decay has occurred. If the mass number stays the same but the atomic number increases, beta decay has occurred. If neither atomic number nor mass number changes, only gamma radiation has been emitted.

In more complex reactions, more than one type of decay can take place at the same time. In some cases, one kind of radioactive nucleus decays to form another, which decays to form a third kind, and so on. The decay series ends with the production of a stable nucleus. For example, uranium-238 undergoes a 14-step process involving alpha, beta, and gamma decay, and eventually forms stable lead-206.

Supplemental Problems

For more practice with solving problems, see Supplemental Practice Problems, Appendix B.

SAMPLE PROBLEM **1** **Writing and Balancing a Nuclear Equation**

Potassium-40 can decay to form calcium-40. Write a balanced equation for this nuclear reaction and determine the type of decay.

Analyze
- Write the symbols for the two isotopes involved. Each symbol must include the abbreviation of the element represented, the mass number as a superscript, and the atomic number as a subscript.

    Potassium-40 is $^{40}_{19}\text{K}$ and calcium-40 is $^{40}_{20}\text{Ca}$.

Set Up
- Set up the equation for the reaction. Potassium is the reactant and goes on the left. The product, calcium, goes on the right.

$$^{40}_{19}\text{K} \rightarrow\ ^{40}_{20}\text{Ca} + ?$$

Now you must balance the nuclear equation. Balance the sums of both the mass numbers and the atomic numbers.

> **Problem-Solving HINT**
>
> Do not be tempted to change mass number or atomic numbers in solving this kind of problem. Consider only the mass number and atomic number of the product to produce a balanced equation.

Solve
- Because the mass numbers are already balanced ($40 = 40$), you have to balance only the atomic numbers. Potassium has an atomic number of 19, and the atomic number of calcium is 20. Because a beta particle ($^{0}_{-1}e$) has a negative charge and almost no mass, adding such a particle to the right side will balance the equation for atomic number.

$$^{40}_{19}\text{K} \rightarrow\ ^{40}_{20}\text{Ca} +\ ^{0}_{-1}e$$

Check
- The mass number on the left equals the sum ($40 + 0$) on the right. The atomic number on the left is 19, which equals the sum ($20 + (-1)$) on the right. The equation is balanced. Because an electron (a beta particle) is given off, beta decay has taken place.

1. Write the balanced nuclear equation for the radioactive decay of radium-226 to give radon-222, and determine the type of decay.
2. Write a balanced equation for the nuclear reaction in which neon-23 decays to form sodium-23, and determine the type of decay.

Detecting Radioactivity

Radioactivity cannot be seen, heard, or touched, and it has no taste or smell. Other means of detection must be used to tell whether radioactive materials are nearby. Photographic film can serve as a simple detection device. However, for film to be effective as a detector, the source of radioactivity has to be close. Instruments that measure radiation are shown in **Figure 21.6.**

Figure 21.6

Uses of Radiation Detectors
People who work around radiation sources are required to wear film badges that show their levels of radiation exposure. The film in these badges is developed regularly to show how much exposure to radiation each person has received. ▶

A scintillation counter registers the intensity of radiation by detecting light. Radioactive samples to be measured are mixed with compounds that emit a flash of light, called scintillation, when exposed to radiation. The level of radiation is measured by the number of flashes of light recorded by the device. ▼

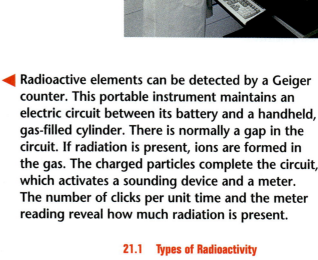

◀ Radioactive elements can be detected by a Geiger counter. This portable instrument maintains an electric circuit between its battery and a handheld, gas-filled cylinder. There is normally a gap in the circuit. If radiation is present, ions are formed in the gas. The charged particles complete the circuit, which activates a sounding device and a meter. The number of clicks per unit time and the meter reading reveal how much radiation is present.

ChemLab

The Radioactive Decay of "Pennium"

The nuclei of unstable atoms disintegrate or decay spontaneously, emitting alpha or beta particles and gamma radiation. Types of atoms that undergo this process are called radioactive isotopes. A decaying reactant isotope is referred to as a parent atom, and the atom produced is a daughter atom. In this ChemLab, heads-up pennies represent individual parent atoms of the fictitious element pennium, and tails-up pennies represent the daughter atoms of the decay. You will study the decay characteristics of pennium and will determine its half-life, which is the time required for one-half of the atoms to decay.

Problem

What is the half-life of the fictitious radio-isotope pennium?

Objectives

- **Infer** the decay characteristics of pennium.
- **Analyze** your data to determine the half-life of pennium.
- **Make** and **use graphs** to interpret data.

PREPARATION

Materials

1 shoe box
120 pennies
stopwatch or watch with a second hand

PROCEDURE

1. Count 120 pennies and lay them heads-up in the bottom of the shoe box. In a table like the one in Data and Observations, record this as 0 daughter atoms and 120 parent pennium atoms at time 0.

2. Put the lid on the shoe box and, holding the lid on tightly, shake the box with moderate force up and down 20 times while your lab partner times the decay process to the nearest second. Assume that all the subsequent decay trial times are identical.

3. Open the box and count the tails-up pennies, which represent the daughter atoms, and remove them from the box. Subtract this number from the original number of heads-up pennies to determine the number of

remaining parent pennium atoms. Record the time taken for the shaking, the number of tails-up atoms removed, and the number of remaining parent atoms.

4. Repeat steps 2 and 3 four more times, recording your data after each trial. Simply add on the original shaking time repeatedly to arrive at the total elapsed time.

ANALYZE AND CONCLUDE

1. **Making a Graph** Construct a graph of your data, plotting the number of remaining parent pennies on the vertical axis (*y*-axis) and the total elapsed time on the horizontal axis (*x*-axis).

2. **Interpreting Data** What is the half-life of pennium in your experiment? Explain how you arrived at this number.

APPLY AND ASSESS

1. Does exactly the same fraction of pennium atoms decay during each half-life? What does this suggest about half-life? Why are such variations not likely to be obvious when actual atoms are involved?

2. If you had started with one mole of pennies (assuming you could actually take the time to count them!), how many would remain after ten half-lives? Is this a large number?

3. If you took a longer time to shake the box in this case, how would half-life be affected? Are the half-lives of atoms also controllable by a change in conditions?

DATA AND OBSERVATIONS

Total Elapsed Time	Number of Tails-Up Daughter Atoms Removed	Number of Heads-Up Parent Atoms Remaining

Archaeological Radiochemistry

What do a piece of wood from an Egyptian pyramid, a bone tool from an archaeological dig, and the Dead Sea Scrolls have in common? They all contain carbon from dead organisms. Thus, archaeologists can determine the objects' ages by the carbon-14 method.

Development of Carbon-14 Dating

Archaeologists need accurate dating techniques for their studies of the artifacts and other remains of early humans. Until about 40 years ago, archaeological dating had to use indirect, time-consuming, and inaccurate methods. In 1946, Willard Libby, working at the University of Chicago, developed the carbon-14 method for dating objects that contain carbon. The first artifact to be dated using carbon-14 was a beam of cypress wood taken from a pharaoh's tomb. Since that time, carbon-14 dating has been widely used to date plant and animal fossils.

Carbon-14 is formed in the upper atmosphere. When cosmic rays collide with atoms in the upper atmosphere, the atoms break up and release subatomic particles. The carbon-14 is made when a neutron collides with a nitrogen atom, causing it to lose a proton.

$$^{14}_{7}N + ^{1}_{0}n \rightarrow ^{14}_{6}C + ^{1}_{1}p$$

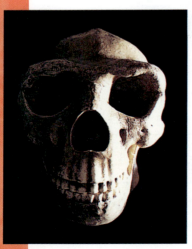

When the ratio of radioactive carbon-14 to radioactive carbon-12 in a once-living thing is determined and compared with the ratio that must have existed while the organism was living, the age of the material can be determined.

The Ice Man

In 1991, a hiker found a man who had been frozen in a glacier for many centuries. Using carbon-14 dating, scientists established that the man lived about 5300 years ago, making the human remains the oldest ever found intact. The ancient ice man had many secrets to reveal about himself and his way of life, for he was completely outfitted and carried various tools and devices with him. Scientists discovered that people at that time had developed wooden-handled daggers. Also, the people cut their hair, had tattoos, and had knowledge of healthful properties of plants. Parts of Neolithic history could now be rewritten, and what made it all possible was being able to find out the age of the remains.

TAMS: An Advanced Dating Method

In the standard lab method of carbon-14 dating, a piece of the object being tested has to be burned. The carbon-14 atoms are then

Tandem Accelerator Mass Spectrometer

depends upon what radioactive substances are present in the material. The methods include rubidium-strontium dating and potassium-argon dating.

Archaeologists are also using a new method to date pottery, called the thermoluminescence (TL) method. As small amounts of radioactive materials such as uranium and thorium decay in clay, they excite other atoms in the clay to a higher energy state. When the clay is heated above 400°C, light called thermoluminescent glow is emitted as electrons fall back to their stable levels. Measuring this glow reveals roughly how long ago the artifact was made.

counted using a scintillator that counts the light flashes from beta radiation. This method requires a fairly large sample because of the small number of carbon-14 atoms, and is effective only for objects up to 60 000 years old. The counting also takes days or weeks because of the slow decay of carbon-14. With the recent development of a device called a tandem accelerator mass spectrometer (TAMS), carbon-14 dating may be effective for objects up to 100 000 years old. A TAMS needs only a few milligrams of sample, and counting can be done in less than an hour.

Other Radiodating Systems Used by Archaeologists

Sometimes an artifact contains little or no carbon or is too old to be dated with carbon-14. So an alternative method has to be used. The method

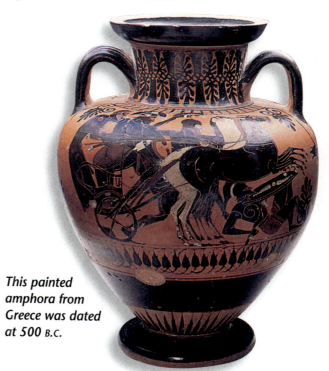

This painted amphora from Greece was dated at 500 B.C.

DISCUSSING THE TECHNOLOGY

1. **Acquiring Information** Carry out library research to obtain information on a radiodating technique not explained in this feature.

2. **Inferring** Iron is not a once-living material. How, then, can iron tools be dated using the carbon-14 method?

3. **Applying** The thermoluminescence method is not highly accurate. However, it is useful in exposing new forgeries in pottery. Explain how this could be so.

Half-Life and Radioisotope Dating

Unlike chemical reaction rates, which are sensitive to factors such as temperature, pressure, and concentration, the rate of spontaneous nuclear decay cannot be changed. Because the decay of an individual nucleus is a random event, it is impossible to predict when a specific nucleus in a sample of a radioactive material will undergo decay. However, the overall rate of decay is constant, which allows you to predict when a given fraction of the sample will have decayed.

Rate of Decay

The **half-life** is the time it takes for half of a given amount of a radioactive isotope to undergo decay. Half-life, which is symbolized $t_{1/2}$, is easy to measure and has been determined for many different radioisotopes, some of which have important uses. Some half-lives are only fractions of a second, whereas others are billions of years. Half-lives for some of the most commonly used radioisotopes are listed in **Table 21.1.** The concept of half-life is illustrated in the graph in **Figure 21.7.**

With its predictable and unchanging rates, radioactive decay has provided scientists with a technique for determining the age of fossils, geological formations, and human artifacts. Using a knowledge of the half-life of a given radioisotope, one can estimate the age of an object in which the isotope is found. Four different isotopes are commonly used for dating objects: carbon-14, uranium-238, rubidium-87, and potassium-40. Now look at one of these techniques in more detail.

Table 21.1	Half-Life Values for Commonly Used Radioactive Isotopes
Isotope	**Half-Life**
$_{1}^{3}\text{H}$	12.26 years
$_{6}^{14}\text{C}$	5730 years
$_{15}^{32}\text{P}$	14.282 days
$_{19}^{40}\text{K}$	1.25 billion years
$_{27}^{60}\text{Co}$	5.271 years
$_{36}^{85}\text{Kr}$	10.76 years
$_{36}^{93}\text{Kr}$	1.3 seconds
$_{37}^{87}\text{Rb}$	48 billion years
$_{43}^{99m}\text{Tc}$	6.0 hours*
$_{53}^{131}\text{I}$	8.07 days
$_{56}^{131}\text{Ba}$	12 days
$_{64}^{153}\text{Gd}$	242 days
$_{81}^{201}\text{Tl}$	73 hours
$_{88}^{226}\text{Ra}$	1600 years
$_{92}^{235}\text{U}$	710 million years
$_{92}^{238}\text{U}$	4.51 billion years
$_{95}^{241}\text{Am}$	432.7 years

*The *m* in the symbol tells you that this is a metastable element, which is one form of an unstable isotope. Technetium 99m gives off a gamma ray to become a more stable form of the same isotope, with no change in either atomic or mass number.

Figure 21.7

Graph Illustrating Half-Life
During each half-life period, half of the radioactive nuclei in a sample decay. After one half-life, 50 percent of the sample remains. After two half-lives, 25 percent of the sample remains. After three half-lives, 12.5 percent remains, and so on.

Carbon-14 Dating

Carbon-14 dating has been widely used to determine the ages of fossils. All organisms take in carbon during their lifetime—plants from carbon dioxide gas and animals from the plants and animals they eat. Most carbon taken in by organisms is in the form of the stable isotopes $^{12}_{6}C$ and $^{13}_{6}C$, but about one carbon atom in every million is the radioisotope $^{14}_{6}C$. As long as an organism is alive, the $^{14}_{6}C$ in its cells remains relatively constant at about one in every million carbon atoms. As $^{14}_{6}C$ atoms decay to form nitrogen and a beta particle, new ones are taken in from the surroundings. After death, the organism no longer takes in carbon, so the proportion of $^{14}_{6}C$ slowly decreases.

In carbon-14 dating, the fraction of $^{14}_{6}C$ that remains in material such as bone or skin is measured and compared with how much was in the material when it was alive. In this way, the age of the object can be estimated. For example, if half of the $^{14}_{6}C$ remains, one half-life period (5730 years) has passed, and the object is about 5730 years old. If only one-fourth of the $^{14}_{6}C$ remains, two half-life periods (11 460 years) have passed. **Figure 21.8** shows some examples in which carbon-14 dating proved useful.

The half-life of carbon-14, 5730 years, is short compared with the age of many fossils and geological formations. As a result, objects more than about 60 000 years old cannot be dated using this technique because the amount of carbon-14 left in them is too small to be measured accurately.

Figure 21.8

Determining the Age of Artifacts
A cypress beam from the cliff dwellings of Mesa Verde in southern Colorado was determined by carbon-14 dating to be about 700 years old. This date was confirmed by counting the annual tree rings in samples of old trees that came from the same area. The Anasazi Indians had moved to Mesa Verde and lived there from around 1100 to 1300 A.D. ▶

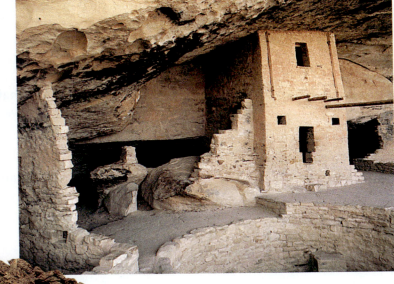

◀ This sandal, found in a cave in Oregon, was analyzed by carbon-14 dating and found to be about 9000 years old.

Supplemental Problems

For more practice with solving problems, see Supplemental Practice Problems, Appendix B.

A fossilized tree killed by a volcano was studied. It had 6.25 percent of the amount of carbon-14 found in a sample of the same size from a tree that is alive today. When did the volcanic eruption take place?

Analyze • The half-life of carbon-14 is 5730 years, and the fraction of carbon-14 remaining is 6.25 percent (0.0625).

Set Up • Calculate the number of half-lives that have passed for the carbon-14 in the sample. During each half-life of a radioactive isotope, one-half of the nuclei decay. The fossil sample has 6.25 percent (0.0625) of its original amount of carbon-14 left, so you need to

> **Problem-Solving HINT**
>
> Repeatedly multiply ½ by itself until the number of factors produces the proper result.

find out how many times 1/2 (0.5) must be used as a factor to produce 0.0625 as a result. The answer is *four times* because 1/2 × 1/2 × 1/2 × 1/2 = 0.5 × 0.5 × 0.5 × 0.5 = 0.0625 (or, carrying out the calculation another way, 100 percent × 1/2 × 1/2 × 1/2 × 1/2 = 6.25 percent). Therefore, four half-lives have gone by.

Solve • Because four half-lives have gone by and each is 5730 years, multiply 5730 by 4.

$$5730 \times 4 = 22\ 920$$

The tree from which the sample was taken must have been killed by the volcano about 22 920 years ago.

Check • Check to be sure that the calculation can be repeated with the same result. All aspects of setup and calculation appear to be correct.

PRACTICE PROBLEMS

3. A rock was analyzed using potassium-40. The half-life of potassium-40 is 1.25 billion years. If the rock had only 25 percent of the potassium-40 that would be found in a similar rock formed today, calculate how long ago the rock was formed.

4. Ash from an early fire pit was found to have 12.5 percent as much carbon-14 as would be found in a similar sample of ash today. How long ago was the ash formed?

Art Forger, van Meegeren–Villain or Hero?

Hans van Meegeren, a skilled Dutch artist, painted forgeries in the style of the great 17th-century Dutch painter Vermeer. In 1937, van Meegeren sold his best forged Vermeer, *Christ and His Disciples at Emmaus,* to a Dutch museum for $280,520. He continued to forge and sell Vermeers and other Dutch masters until 1945.

Wartime forgery During the Nazi occupation of Holland in the early 1940s, van Meegeren sold his forged Vermeer paintings to members of the Nazi government. Part of the price Meegeren exacted for his paintings was the return of 200 works of Dutch art that had been looted earlier in the war by the Nazis. At the end of the war, in 1945, a forged Vermeer was found with the Nazis' treasures and traced to van Meegeren. When he was charged as a Nazi collaborator for selling Dutch national treasures, he confessed to the forgeries. He pointed out that by exchanging the forgeries, he had saved many Dutch art treasures from the Nazis. However, the authorities did not believe that he was a forger. He was sentenced to a year in prison, but died before serving it.

Radioactive dating resolves the problem It was not until 1968 that an American scientist named Keisch presented irrefutable evidence that van Meegeren was an art forger. Keisch's work made use of uranium-lead dating. White lead, a combination of $PbCO_3$ and $Pb(OH)_2$, is an important pigment in oil painting. The parent source of the stable lead-206 in white lead is uranium-238. The uranium-238 has a half-life of 4.5 billion years, decaying into a series of unstable isotopes, including radium-226 and polonium-210, until it finally forms lead-206. During the processing of white lead, most of the isotopes preceding polonium-210 in the uranium-decay series are removed.

However, minute amounts of radium-226 remain. Because the radioactive elements that would be needed to make the polonium-210 are mostly removed, the amount of polonium-210 decreases as the pigment ages. The ratio between the amounts of the radium-226 and the polonium-210 is used to establish the age of paintings. A low ratio of radium to polonium indicates a recent painting.

Proof of a fake Keisch tested a white lead sample from the alleged fake "*Emmaus*" painting, shown in the photo below, and found a low Ra:Po ratio. The ratio indicated that the painting was only about 30 years old, not 260, and therefore a forgery.

Connecting to Chemistry

1. **Acquiring Information** Belgian chemist Coremans also exposed the fraudulent Vermeers. Find out how.

2. **Interpreting** You have two paintings. One is several hundred years old, the other one year old. Painting A has a Po concentration of 12 and a Ra concentration of 0.3. Painting B has a Po concentration of 5 and a Ra concentration of 6. Which painting is which?

3. **Acquiring Information** What are some modern methods of analysis for revealing art forgeries? Write a brief paper tracing this history.

To date objects that are more than 60 000 years old, techniques involving other isotopes must be used. Rocks and minerals that are up to billions of years old can be dated on the basis of the decay of radioisotopes with long half-lives, such as potassium-40 ($t_{1/2}$ = 1.25 billion years), uranium-238 ($t_{1/2}$ = 4.5 billion years), and rubidium-87 ($t_{1/2}$ = 48 billion years). Such techniques are illustrated in **Figure 21.9.**

Figure 21.9

Using Radioactive Dating
Radioactive dating techniques were used to determine that the cyanobacteria that left imprints in these rocks found in a shallow bay in Australia lived about 3.5 billion years ago. ▼

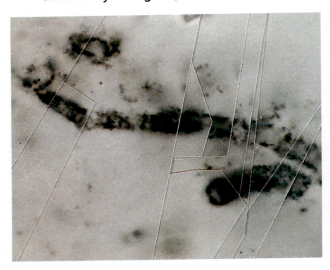

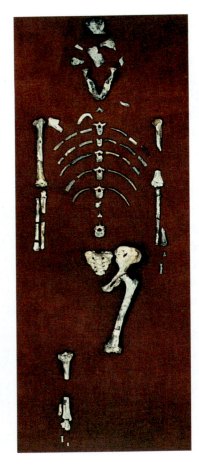

◄ Dating techniques established the age of a fossilized skeleton of *Australopithecus afarensis* found in Ethiopia to be 3.5 million years. The paleontologists who found it named this skeleton of an early humanlike primate "Lucy" after a song by the popular British singing group *The Beatles.* The scientists played the song around a campfire while celebrating their discovery.

SECTION REVIEW

Understanding Concepts

1. Explain what is meant by *radioactivity,* and provide an example of a radioactive isotope.

2. What is a Geiger counter used for? Describe its operation.

3. If half the atoms in a sample decay in one year, why is it incorrect to conclude that the other half will decay in one more year?

Thinking Critically

4. **Inferring** The oldest rocks dated so far on Earth are about 3.8 billion years old. Do you think that this fact was determined by means of carbon-14 dating? Explain.

Applying Chemistry

5. **Using Carbon Dating** Suppose you were given an ancient wooden box. If you analyze the box for carbon-14 activity and find that it is 50 percent of that of a new piece of wood of the same size, how old is the wood in the box?

chemistryca.com/self_check_quiz

Nuclear Reactions and Energy

Both the United States and the former Soviet Union have sent nuclear reactors into space to power satellites. However, another giant nuclear reactor was there first—the sun, our local star. In the giant nuclear furnace of a star, nuclei react with one another and release enormous amounts of energy. You count on this energy to warm your air and water and to power your home, calculators, and other devices that run on solar power. Without this energy, life on Earth as you know it would not be possible. Is it possible to copy the reaction that takes place in stars and make a powerful generator that will provide electricity?

SECTION PREVIEW

Objectives

✓ **Compare and contrast** nuclear fission and nuclear fusion.

✓ **Demonstrate** equations that represent the changes that occur during radioactive decay.

✓ **Trace** the operation and structure of a fission reactor.

Review Vocabulary

Half-life: the time it takes for half of a given radioactive isotope to decay (into a different isotope or element).

New Vocabulary

nuclear fission
nuclear reactor
nuclear fusion
deuterium
tritium

The Power of the Nucleus

Compared to chemical reactions, nuclear reactions involve enormous energy changes. The energy change is due in part to the conversion of small amounts of the mass of nuclear particles into energy. Albert Einstein was the first scientist to realize the enormous amount of potential energy available in matter. He realized that mass and energy are equivalent and related by the following equation.

$$E = mc^2$$

In this equation, E represents energy, m is mass, and c is the speed of light (3.00×10^8 m/s). Because the value of the speed of light is so high, you can see that a small amount of mass can be converted into an enormous amount of energy. That explains why a nuclear weapon, such as an atomic bomb, is so much more powerful than a weapon that involves only chemical reactions, such as a bomb made with dynamite.

Nuclear Fission

Until 1938, all nuclear reactions observed involved the movement of radiation, such as alpha or beta particles, into or out of a decaying nucleus. Shortly before World War II, Otto Hahn, a German chemist, discovered that bombarding uranium with neutrons resulted in the formation of an isotope of barium that is only about 60 percent of the mass of the uranium isotope. This result was so unusual that Hahn asked another physicist, Lise Meitner, shown in **Figure 21.10,** to help him interpret his findings. She determined that the bombardment of uranium with neutrons had resulted in the splitting of the nucleus into two pieces. She called this transformation nuclear fission. The word *fission* means "splitting." When **nuclear fission** takes place, an atomic nucleus is split into two or more large fragments.

The Fission Process

Why did bombarding uranium atoms with neutrons result in fission? Remember that changing the ratio of neutrons to protons can make a nucleus less stable. Adding a neutron to a uranium-235 nucleus causes the nucleus to split. Atoms of two different elements are created along with more neutrons. Here is a typical reaction for uranium fission.

$$^{235}_{92}\text{U} + ^{1}_{0}n \rightarrow ^{140}_{56}\text{Ba} + ^{93}_{36}\text{Kr} + 3^{1}_{0}n$$

Notice that the sums of the mass numbers on the left and right sides are equal, and so are the sums of the atomic numbers. Therefore, the nuclear equation is balanced.

As World War II started, the race was on to find a way to sustain fission in a chain reaction. A chain reaction is a continuing series of reactions in which each produces a product that can react again. Each neutron produced during the fission of one atom of uranium has the potential to cause the fission of another nucleus. Notice that each fission reaction produces three neutrons—more than enough to keep the process going. The fission reactions can continue in a chain, as shown in **Figure 21.11.** For a chain reaction to take place, there must be enough fissionable material present for some of the neutrons produced to be able to collide with other nuclei. However, if the chain reaction takes place too quickly, an explosion will result and release an enormous amount of energy all at once. This is what happens when an atomic bomb explodes. If the rate of the fission chain can be controlled to release energy slowly, the energy can be used to heat materials such as water and then do useful work.

Figure 21.10
Lise Meitner
Lise Meitner was a physicist who was born in Austria but had to flee after the Nazis gained control of her country. She first described nuclear fission. Fission can release amounts of energy that are far greater than those released during radioactive decay.

A Nuclear Fission Chain Reaction

Nuclear fission provides the energy in nuclear power plants. During the fission process, a heavy atomic nucleus generally absorbs a neutron, then splits into two smaller atomic nuclei and several more neutrons. If sufficient fissionable material, known as the critical mass, is present, the neutrons produced from one fission initiate several more fissions, and a chain reaction is sustained. Your group will simulate a nuclear chain reaction with dominoes.

Procedure

1. Obtain two sets of domino tiles.

2. Set the individual tiles on end in patterns such that if you first topple a single tile, it will strike and topple two more tiles, each of which will strike and topple two others, and so on.

3. Practice with possible arrangements until you decide on the best one; that is, one that will topple all the tiles in the quickest, most efficient way. When you are ready, call the teacher to observe your domino chain reaction. You may perform a second trial if your first attempt is unsuccessful. The teacher will compare your best result with those of the class.

Analysis

1. How many tiles were involved in your best chain reaction? In the best of the entire class?

2. If you were able to arrange 1 million dominoes in this manner, so that each falling tile topples exactly two tiles, and assuming 1 second of time required for each of these double topples, how much time would be required to topple all the tiles?

3. Explain how this domino exercise simulates a nuclear fission chain reaction.

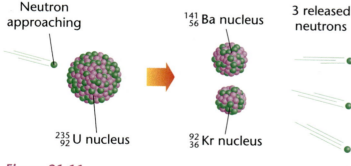

Figure 21.11

A Fission Chain Reaction

During a chain reaction, each step produces a reactant for the next step. In the fission of uranium, a neutron enters a uranium-235 nucleus. As a result, the nucleus breaks into two smaller nuclei. At the same time, it releases more neutrons, which are absorbed by other uranium-235 nuclei, and the reaction continues.

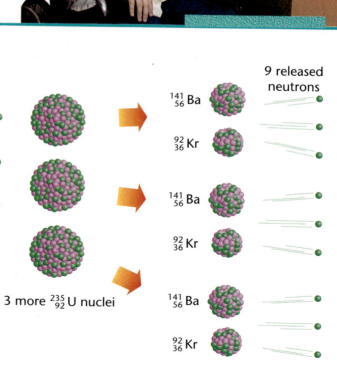

Fission Reactors

The first major step in developing usable, controllable nuclear power took place on December 2, 1942, in a squash court underneath an unused football field at the University of Chicago. It was there that Enrico Fermi, an Italian physicist, successfully carried out a sustained and controlled fission chain reaction.

Today, through a similar procedure, electrical energy is generated in nuclear power plants through the controlled fission of uranium. The device that is used to extract energy from a radioactive fuel is called a **nuclear reactor.** In the most common type of reactor used today, the fission of a sample of uranium enriched in the isotope uranium-235 releases energy that is used to generate electricity. It is, therefore, desirable to concentrate the uranium-235. Refer to the diagram of the nuclear reactor shown in **Figure 21.12** to learn how a fission reactor works.

Figure 21.12

Operation of a Fission Nuclear Power Plant

1. Uranium-235 in fuel rods produces fast-moving neutrons and heat in a fission chain reaction. The neutrons are slowed down by a moderator such as water or graphite so that they are not moving too quickly to be absorbed by other uranium-235 nuclei. The rate of the reaction is maintained using control rods that absorb some of the neutrons. These rods can be raised or lowered in the reaction chamber to slow or speed the reaction, respectively.

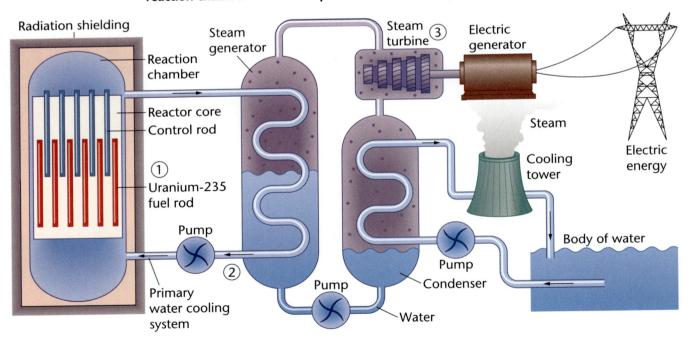

2. The neutrons then enter other fuel rods, where they initiate more fission reactions. The energy produced during these reactions heats water that is pumped among the fuel rods. The water functions as a coolant that keeps the rods from melting and also as a medium to transfer heat from the reaction chamber to a separate basin of water. That water is converted to steam.

3. The steam turns a turbine, which rotates the shaft of an electric generator. When the shaft of the generator turns, electricity is produced. The steam is then condensed back into water, using still other water from an outside source as a coolant. The newly condensed water is cycled back into the process.

More than 100 nuclear power plants are now in operation in the United States, providing almost 20 percent of all the electricity used. Thirty-three states have nuclear plants. Some of these states, such as Vermont, get most of their electricity from nuclear power. Many more plants are operating in other countries. The country that gets the highest percentage of its electrical power from nuclear reactors is France. There, more than 70 percent of the electricity produced comes from nuclear fission reactors.

Plutonium can also be used to power a nuclear fission reactor. Plutonium-239 undergoes a chain reaction that eventually produces more plutonium-239 from uranium-238, as well as heat that is used to generate electricity. This type of fission reactor, which produces fissionable material as it operates, is called a breeder reactor. A breeder reactor can actually produce more fissionable material than it uses. The construction of this type of reactor has been discouraged in many countries because of the health hazard that plutonium presents, as well as the fact that the plutonium-239 generated can be used to make powerful fission bombs.

Unlike burning fossil fuels, nuclear reactions do not produce pollutants such as carbon dioxide and acidic sulfur and nitrogen compounds. However, the nuclear reactions do form highly radioactive waste that is hard to dispose of safely. Other serious problems include the potential release of radioactive materials into the environment when fires or explosions take place, and also the limited supply of fissionable fuel and the higher cost of producing electricity using nuclear fuels rather than fossil fuels. Nuclear reactors that have experienced serious accidents are shown in **Figure 21.13.**

Figure 21.13

Nuclear Accident Sites

What you see coming from the towers of the Three Mile Island nuclear power plant in Pennsylvania is all water. Few chemical pollutants are released during the normal operation of a nuclear plant. Both equipment failure and human error resulted in overheating of the reaction chamber and a partial meltdown of fuel rods at this power plant in 1979. As a result, the building surrounding the reactor became flooded with water contaminated with radioactive material, and radioactive gas was released into the atmosphere. ▶

◀ Both human and mechanical errors led to overheating of the reaction chamber at the Chernobyl nuclear plant in the former Soviet Union in 1986. Water used to cool the chamber decomposed into hydrogen and oxygen gas, which exploded and blew the roof off the building that housed the reactor. A large amount of radioactive debris was released and traveled as far away as Scandinavia and England. Even now, almost 1000 square miles around the plant are considered too radioactive for permanent habitation.

Nuclear Fusion

Energy is also released in a type of nuclear reaction that is the opposite of fission. During **nuclear fusion,** two or more nuclei combine to form a larger nucleus. Fusion is the nuclear process that produces energy in stars like our sun. In the most typical fusion reaction in the sun, hydrogen nuclei fuse to form helium. The enormous amount of energy that is generated sustains all life on Earth.

The Fusion Process

Some of the fusion processes in which hydrogen forms helium have been carried out and studied in laboratories. In one common fusion reaction, two different isotopes of hydrogen combine to form helium and a neutron, as shown in **Figure 21.14.**

$$\ _{1}^{2}\text{H} + \ _{1}^{3}\text{H} \rightarrow \ _{2}^{4}\text{He} + \ _{0}^{1}n$$

Notice that the nuclear equation is balanced. The isotope of hydrogen with a mass number of 2 that is one of the reactants in the equation above is called **deuterium** (D). The hydrogen isotope with a mass number of 3 is called **tritium** (T). Nuclei such as these, which have a small mass, can combine to produce heavier, more stable nuclei. The helium produced has a stable neutron:proton ratio of 1:1. Two deuterium nuclei can also fuse to form either tritium or helium, as shown in the following two equations.

$$\ _{1}^{2}\text{H} + \ _{1}^{2}\text{H} \rightarrow \ _{1}^{3}\text{H} + \ _{1}^{1}p$$

$$\ _{1}^{2}\text{H} + \ _{1}^{2}\text{H} \rightarrow \ _{2}^{3}\text{He} + \ _{0}^{1}n$$

In the first case, tritium and a proton are produced; in the second, $_{2}^{3}$He and a neutron are produced. The fusion of hydrogen produces 20 times the energy produced by the fission of an equal mass of uranium. That large quantity of energy makes the idea of fusion reactors an appealing one. Also, deuterium, the required fuel in fusion, is abundant on Earth. Energy production through nuclear fusion has some other potential advantages over production through fission. No radioactive products are produced, so waste disposal would not be as difficult. A fusion reaction is also easier to control, which would help prevent fires and explosions.

However, there is one major problem with creating a fusion reactor. Compared to uranium nuclei, hydrogen nuclei have much less tendency to react. A large initial input of energy must be provided to start the fusion. In the sun, enormous pressures and temperatures trigger fusion. To initiate a fusion reaction on Earth, a temperature of 200 million kelvins would be required. Any material used to contain the reaction

Try at Home Lab

See page 873 in Appendix F for
Modeling Fusion

Figure 21.14
An Example of Nuclear Fusion
When a nucleus of deuterium (hydrogen-2) and a nucleus of tritium (hydrogen-3) undergo nuclear fusion, a nucleus of helium-4 and a neutron are produced.

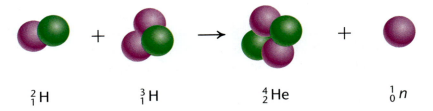

$_{1}^{2}$H $_{1}^{3}$H $_{2}^{4}$He $_{0}^{1}n$

would melt at that temperature. The difficulty in initiating and containing nuclear fusion has prevented its use as a practical energy source.

Scientists in a number of countries are researching ways to make commercially feasible fusion reactors. One promising reactor type is called a tokamak, which is shown in **Figure 21.15.** In this reactor, hydrogen nuclei are trapped by powerful magnetic fields produced by huge electromagnets. The nuclei are heated with radio waves to initiate fusion. Because a magnetic field confines the nuclei, no other container is needed. Such a reactor has been operated at the break-even point, where it produces as much energy as is required to run it, but because it produces no net energy, it is not useful. Better electromagnets and structural materials must be developed to correct the engineering problems encountered in the tokamak.

Figure 21.15

A Tokamak Fusion Reactor
The walls inside the vacuum chamber of the experimental DIII tokamak fusion reactor in San Diego are covered with carbon tiles. They can withstand the radiation and particles produced by the hydrogen undergoing fusion.

SECTION REVIEW

Understanding Concepts

1. What is the difference between nuclear fission and nuclear fusion? List one current use for fission.

2. How is fission controlled and maintained in a nuclear reactor?

3. What advantages would a fusion reactor have over the fission-based reactors currently in use?

Thinking Critically

4. **Comparing and Contrasting** What are the advantages and disadvantages of using energy generated by nuclear fission reactors instead of energy from fossil fuels such as petroleum, natural gas, and coal?

Applying Chemistry

5. **Stellar Fusion** The sun's energy source was a mystery to people for a long time. One 19th-century astronomer calculated that if the sun were made entirely of coal, it would burn for only about 10 000 years. Today, it is known that the sun's fuel is hydrogen, consumed in fusion reactions in which helium is formed. After the sun has used up nearly all of its hydrogen, it is expected to begin a fusion reaction in which helium-4 nuclei combine to form carbon-12. Write a nuclear equation for this process, and verify that it is balanced.

Nuclear Tools

Medical practice, as envisioned in many science fiction works, will differ from what it is today. Surgery, so common today, will supposedly become obsolete, replaced by non-invasive techniques. Only time will tell whether such changes take place. However, thanks to nuclear chemistry, the first steps in the new direction are already being taken. Although traditional surgery is still necessary, diagnosis has become less invasive. As you will see, the inside of the human body can even be explored in detail without the use of a scalpel.

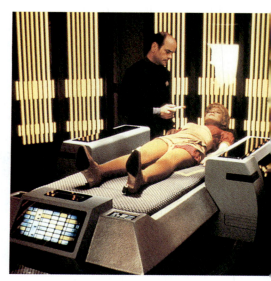
Sick bay on Star Trek's Voyager

Medical Uses of Radioisotopes

You are much more likely to be saved by some form of nuclear medicine than to be killed by the effects of radiation. Radioisotopes are widely used today in diagnosis to generate images of organs and glands, and in treatments for conditions such as cancer. Radioisotopes are also used as tracers to find out where certain chemicals move in the body and to identify abnormalities, as shown in **Figure 21.16**.

Figure 21.16

Use of Tracers in Medicine

A radioactive isotope of any element will undergo the same chemical reactions as a stable isotope. For example, if a molecule of glucose that contains a radioactive atom is injected into a patient, the radiolabeled glucose will be metabolized in a cell in the same way as nonradioactive glucose. A detector set to pick up only the type of radiation given off by the radioisotope in the glucose can be used to trace the molecule—that is, to tell where it moves or concentrates in the body. Because cancer cells take in and use fuels at a much higher rate than do more slowly dividing, normal cells, the radioisotope can be used to locate and identify a tumor.

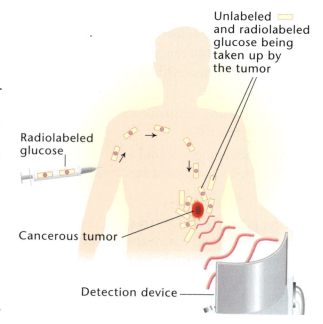

Unlabeled and radiolabeled glucose being taken up by the tumor

Radiolabeled glucose

Cancerous tumor

Detection device

Medical Uses of Radioisotopes in Diagnosis and Treatment

Radioisotopes have become extremely useful tools in medical diagnosis and treatment. Using them allows doctors to detect many diseases early and to treat them more successfully than ever before.

A Positron emission tomography (PET) scans make use of a short-lived positron-emitting radioisotope. A positron is a particle that has the mass of an electron but that is positively charged. When a positron collides with an electron, two gamma rays that move apart at a 180° angle are produced. In a PET scan, rings of highly sensitive detectors surround the patient in the cylindrical enclosure and allow measurement of the distribution of radioactivity in the body. With the aid of a computer, a PET scan can produce a series of two-dimensional imaging slices through an organ. Different radioisotope-containing compounds can be used to measure rates of various processes in the body, such as glucose metabolism or blood flow. The patient shown here is undergoing a PET scan of the brain.

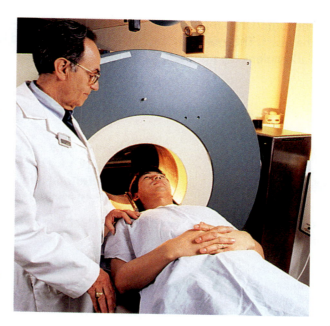

B Most of the iodine that enters your body ends up in the thyroid gland, where it is incorporated into hormones that regulate your growth and metabolism. If you ingest radioactive iodine-131, an image showing the size, shape, and activity of your thyroid gland can be obtained. The image is useful in diagnosing metabolic problems, such as hyperthyroidism. The photo shows a normal thyroid gland.

C Gadolinium-153 is widely used in medicine to detect osteoporosis, the reduction in bone mass that often accompanies aging. Bone density is determined by a scanning device that compares how much of the X rays and gamma rays produced by the decay of gadolinium-153 is absorbed by a bone. The scans shown compare a dense bone with one that has lost a great deal of mineral matter.

D Technetium-99m is a metastable isotope that gives off gamma rays to become a more stable version of the same isotope, with no change in either atomic or mass number. Technetium-99m is one of the most commonly used isotopes in medicine because it produces no alpha or beta particles that could cause unnecessary damage to cells and because it has a half-life of only about six hours. It can be incorporated into different compounds for imaging bones or blood-flow patterns in the heart. The image on the right shows a damaged heart; the dark regions are not getting enough blood flow.

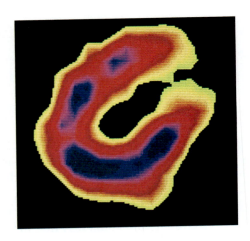

Fact of the MATTER

The first radioactive isotope used in medicine was phosphorus-32, which was administered in 1936 to a woman who had chronic leukemia. The phosphorus-32 was artificially produced in a machine called a cyclotron, in which elements become radioactive after they are bombarded with neutrons and deuterium nuclei.

E Scintigraphy is a bone-scanning technique used to search for stress fractures in racehorses. A radioisotope is injected into bone tissue, and the bone is scanned using a gamma-ray detector.

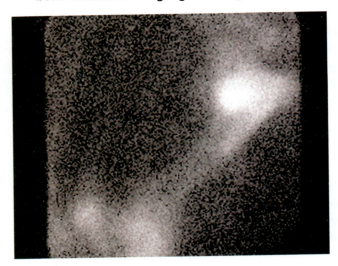

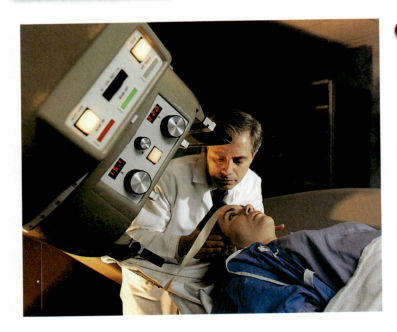

F The goal of radiation therapy is to destroy cancerous cells while minimizing damage to normal cells. Radiation produced by cobalt-60 is carefully aimed at tumor cells, which are more sensitive to the damaging effects of the radiation on DNA because they divide more rapidly than do normal cells. The patient shown in the photo is receiving radiation therapy for cancer. To minimize destruction of healthy cells, the cobalt source is moved around the outside of the patient's body in a circle, with the beam of rays sharply focused on the tumor at all times. This allows any given area of normal tissue to receive only a small dose of radiation while the tumor gets a large total dose.

Nonmedical Uses of Radioisotopes

Hospitals are not the only locations where you'll find radioisotopes in use. You have already read about several nonmedical uses for radioactivity: nuclear reactors, nuclear weapons, and radiodating techniques. As you will see, radioisotopes are also used in research and in the food industry.

Practical Uses of Tracers

If a radioisotope is substituted for a nonradioactive isotope of the same element in a chemical reaction, all compounds formed from that element in a series of steps will also be radioactive. That makes it possible to follow the reaction pathway using instruments that can detect radiation. In this way, the series of steps involved in many important reactions has been studied, as shown in **Figure 21.17.** Tracers containing radioactive phosphorus-32 have also been used in biochemical research to help clarify complicated metabolic pathways. Tracers are also used to test structural weaknesses in mechanical equipment and to follow the pathways taken by pollutants.

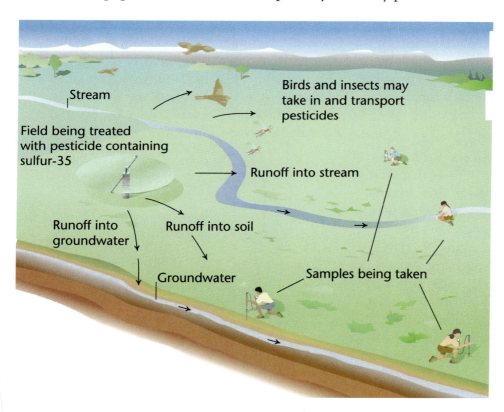

Stream

Field being treated with pesticide containing sulfur-35

Birds and insects may take in and transport pesticides

Runoff into stream

Runoff into groundwater

Runoff into soil

Groundwater

Samples being taken

Figure 21.17

Using a Tracer to Study Pesticide Movement
A pesticide that is sprayed onto a field may be transported to other areas by runoff into soil, streams, and groundwater or by movement of animals that have taken in the pesticide. If a pesticide is "labeled" with a radioisotope, such as sulfur-35, its movement can be traced. Samples of soil or water can be taken at a number of points, and the amount of radiolabeled pesticide at each point can be measured.

Food Irradiation

Gamma radiation disrupts metabolism in cells, sometimes enough to kill them or at least keep them from multiplying. This property makes it useful for sterilization of food and surgical instruments. Exposing food to gamma radiation produced by the decay of cobalt-60 nuclei can keep the food from spoiling. The radiation destroys microorganisms and larger organisms such as insects. The food itself does not become radioactive.

A Biological Mystery Solved with Tracers

For many years, biologists searched for the identity of the genetic material of life. Some scientists thought that the genetic material was protein. Others believed it was nucleic acids. By 1944, strong evidence pointed to DNA. In 1952, Alfred Day Hershey and Martha Chase published the results of their experiments, which confirmed that DNA, not protein, determines heredity.

Radioisotopes used as tracers solve scientific mysteries Radioactive and nonradioactive isotopes of the same element act the same way in a chemical reaction. When scientists want to put tags on a compound, they substitute a radioactive isotope for a nonradioactive one in the compound. Then they can use radiation detectors to track and locate the radioisotope tracer.

Hershey and Chase use tracers Bacteriophages (phages) are simple viruses that attack bacteria. They are composed of two parts: a protein coat and a DNA core, as shown here. The protein is composed of carbon, hydrogen, oxygen, nitrogen, and sulfur. The DNA is composed of carbon, hydrogen, oxygen, nitrogen, and phosphorus. Because sulfur is found only in protein and phosphorus only in DNA, Hershey and Chase chose radioactive ^{35}S and ^{32}P as the tracers for these substances to compare the inheritance of protein and DNA.

The question Hershey and Chase asked was: "When a phage infects a bacterial cell, does the phage inject its DNA core, its protein coat, or both?" They designed experiments using tracers to track the protein coat and DNA core separately. They grew bacteria in a culture containing ^{35}S. When they added phages to this ^{35}S-labeled culture, the new phages produced had protein coats containing

^{35}S. When they infected bacteria having no radioactive materials with the phages containing ^{35}S, they found the ^{35}S only in protein coats of phages outside the new bacteria. The phages made inside the new bacteria had no ^{35}S in their coats. This showed that the protein coat was not injected.

DNA is the genetic material Hershey and Chase did a similar experiment with bacteria grown in a culture containing ^{32}P, to which they added phages. The new phages produced had DNA containing ^{32}P. This time, the DNA with ^{32}P entered unlabeled bacterial cells and produced new phages that contained a significant amount of ^{32}P. The results showed that the phage's DNA core was injected. The phage's DNA alone was able to direct production of an entire phage, with coat. Therefore, Hershey and Chase concluded that DNA, not the protein, was the genetic material. In 1969, Hershey received the Nobel Prize in Medicine for his work.

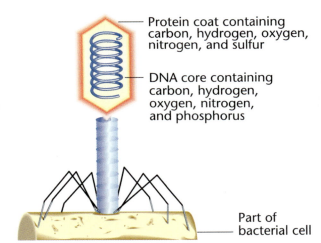

Protein coat containing carbon, hydrogen, oxygen, nitrogen, and sulfur

DNA core containing carbon, hydrogen, oxygen, nitrogen, and phosphorus

Part of bacterial cell

Connecting to Chemistry

1. **Comparing** Compare the structure of a protein with that of DNA.

2. **Interpreting** Explain how Hershey and Chase came to their conclusion.

Irradiation can extend the shelf life of food so that it can be stored for long periods of time without refrigeration, as shown in **Figure 21.18.** However, people who oppose food irradiation are concerned about what are called unique radiolytic products—URPs, for short. These products result from chemical changes caused by the ionizing effects of radiation. Whether URPs present a hazard has not yet been determined for sure. The U.S. Food and Drug Administration (FDA) has approved the use of irradiation for most fruits and vegetables.

Figure 21.18
Food Irradiation
Exposure to gamma radiation can extend the shelf life of food by preventing spoilage. Which group of mushrooms do you think was irradiated before storing?

Sources of Radioisotopes

Using nuclear chemistry, scientists today can change one element into another and even produce elements artificially. How are elements made artificially? Some are produced as by-products in nuclear reactors. However, most are made by bombarding nuclei with small particles that have been accelerated to high speeds. This is done mainly in three instruments, shown in **Figure 21.19.**

Figure 21.19

Accelerators and Particle Generators
◀ The large, expensive, and heavily shielded biomedical cyclotron is used to produce most radioisotopes used in PET scans.

A technetium-99m generator contains radioactive molybdenum-99 that decays to form technetium-99m. Hospitals can use the generator to produce and extract the short-lived technetium-99m just before it is needed in diagnostic techniques. ▶

◀ The Tandem Cascade Accelerator (TCA) is smaller and cheaper to purchase and run than a cyclotron.

Problems Associated with Radioactivity

The understanding of radioactivity has grown rapidly in the 100 years since its discovery. When the Curies worked with radioisotopes, they did not realize how harmful such materials could be. Marie Curie died of leukemia that was probably caused by her years of contact with radio-isotopes. Many more radioactive isotopes exist than the few studied by the Curies. In fact, most of the roughly 2000 known isotopes of all elements are unstable and undergo nuclear decay. Fortunately, most of those do not occur naturally, but are produced synthetically. Your surroundings contain mostly stable isotopes of the common elements, so you are not normally exposed to enough radiation to do you much harm.

However, you may be surprised to learn that you are constantly being bombarded with low levels of radiation. This radiation comes from many sources and is referred to as background radiation. Some of it is in the form of cosmic rays, which are particles that reach Earth from outer space. Small amounts of radioactive elements are found almost every-where on Earth, as well—in wood and bricks used to make buildings, in the fabrics used in clothing, in the foods you eat, and even inside your body. Traces of uranium in rock layers beneath houses may also produce radioactive radon gas that enters the houses and can present health risks. Various sources of radiation are illustrated in **Figure 21.20.**

Exposure to radioactive elements can be hazardous to your health. That's because the radiation they give off is powerful enough to knock electrons loose from atoms and generate ions when it collides with neu-tral matter. Because it can result in the formation of ions, nuclear radia-tion is also known as ionizing radiation. In contrast, most electromagnetic radiation—ordinary visible light, for example—does not have enough energy to knock electrons out of neutral atoms and is, therefore, nonion-izing. Only radiation such as cosmic rays and X rays has sufficient energy to generate ions. It is mainly ionizing power that makes radioactive ele-ments dangerous.

Figure 21.20

Sources of Radiation

As this pie diagram shows, natural radioactivity accounts for about 82 percent of the radiation to which you are exposed. Most of the natural radiation is from radon pro-duced by the decay of urani-um present in rocks and soil. The remaining 18 percent of the radiation comes from sources produced by humans in the last century, including nuclear medical tools such as X rays and nuclear fuel used in reactors.

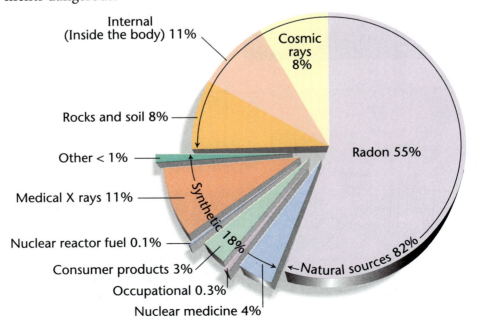

Radon—A problem in your home?

A high level of radon in a home is correlated to a greater risk of lung cancer for the occupants. Radon-222 is an eventual product of the decay of uranium-238, which occurs naturally in many rocks and soils. In this MiniLab, you will study the results of radon tests in several homes.

Procedure

1. Your teacher will supply the class with several commercial radon-detection kits—one for each group of students. The four-day exposure canisters will probably be the most practical to use; however, the 30-day units will usually give more accurate results.

2. Familiarize yourself with the types of homes, geological and geographic features, and factors such as industries in your area. As a class, decide where each group will place its detector. Obtain permission to test for radon from those who own or live in the home you wish to test.

3. Follow the directions on the radon-testing kits, and expose the detectors in the selected home. If there are a sufficient number of kits for the class, some groups can expose their detectors in upper levels of homes, and other groups can do so in basements. Send the detectors to be evaluated according to the instructions.

4. Examine your data and collect and tabulate data from the rest of the class.

5. Look for any safety precautions in your radon kit. Be sure to follow them carefully.

Analysis

1. Did any of the test results exceed the U.S. Environmental Protection Agency's recommended maximum, which is 4 picocuries per liter? How many did so, and by how much?

2. Carefully analyze the class data, looking for correlations between the radon levels and factors such as geological and geographic features, proximity to industrial sites, style of home, and materials used in home construction.

3. Suggest several ways in which radon levels in homes might be reduced.

Because of the danger, elaborate and expensive precautions must be taken to protect people who work with radioisotopes. Highly radioactive waste products that can take many thousands of years to decay must be stored carefully. Choosing storage sites is difficult, and so is transport of the wastes to the storage sites. Many people are concerned about nuclear processes as energy sources. The threat of nuclear war and concern over radioactive fallout have also caused some people to be reluctant to use any form of radioactivity.

Radiation Exposure

As you have read, radiation can do severe damage to parts of a single cell and can cause its death. Damage to DNA can be especially harmful because that substance is the blueprint from which genetic material for future generations is made. When a cell with damaged DNA divides, all cells made from it also have damaged DNA. When this damage occurs in egg or sperm cells of individuals, damaged or mutated DNA can be passed to offspring.

The more energy radiation has, the more dangerous it tends to be. Several different units can be used to measure the energy in a given amount of radiation. A unit used to measure the received dose of radiation is called the **gray.** One gray is equivalent to the transfer of 1 joule of energy in the form of radiation to 1 kg of living cells. However, not all of the radiation that reaches living tissue gets absorbed by the tissue. That fact is not taken into account when radiation is measured in grays. The biological damage caused by radiation is indicated best in terms of how much is actually absorbed, which is measured by a unit called a **sievert** (Sv). One sievert is equal to 1 gray, multiplied by a factor that takes into account how much of the radiation hitting the tissue is absorbed by it. Likely harmful biological effects of single doses of radiation of various strengths are listed in **Table 21.2.**

Table 21.2 Likely Physiological Effects on a Human from a Single Dose of Radiation

Dose (Sv)	Effect
0–0.25	No immediate effect
0.25–0.50	Small temporary decrease in white blood cell count
0.50–1.0	Large decrease in white blood cell count, lesions
1.0–2.0	Nausea, hair loss
2.0–5.0	Hemorrhaging, possible death
>5.0	50 percent chance of death within 30 days of exposure

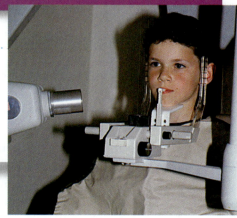

Background ionizing radiation from natural sources results in a dose of about 0.003 Sv per year for the average person. The U.S. government recommends that your total exposure to sources other than background radiation should be limited to 0.005 Sv per year. Workers in nuclear plants are permitted exposure to 0.05 Sv per year. Both these doses are far below the 2- to 5-Sv single dose that can be lethal. Besides the natural background level, the only significant source of radiation for most people who do not work around radioactive materials is medical radiation, mainly X rays. The dose equivalent of a chest X ray is about 0.0005 Sv, and that of a dental X ray is about 0.0002 Sv, presenting risks that most people would consider slight when compared to the potential diagnostic benefits of the X rays.

Radon–An Invisible Killer

If you do not smoke or breathe in cigarette smoke, do you still have a chance of getting lung cancer? The answer is yes. Each year, as many as 15 000 to 20 000 deaths due to lung cancer are the result of radon pollution. Radon, element 86, is the densest noble gas. All of its isotopes are radioactive. Radon-222 has the longest half-life, 3.823 days. Radon is formed in uranium deposits in Earth's crust. Because it is a gas, it can seep through the rocks and soil to the surface.

How does radon cause lung cancer? Even though radon is radioactive, it is not highly dangerous. Most of it is inhaled and rapidly exhaled, like any other gas. However, radon has a short half-life and quickly changes into radioactive isotopes of polonium and lead.

$$^{222}_{86}\text{Rn} \rightarrow ^{218}_{84}\text{Po} + ^{4}_{2}\text{He}$$
$$^{218}_{84}\text{Po} \rightarrow ^{214}_{82}\text{Pb} + ^{4}_{2}\text{He}$$

These radioactive isotopes are solids that can collect on dust. When the dust is inhaled, the radioactive solids remain in the lungs. High-energy alpha particles from the polonium and lead damage the DNA of lung cells, sometimes causing cancer.

Can radon get into your home? Radon can enter homes and other buildings through cracks in concrete foundations and slab floors; porous cinder-block walls; unsealed, poorly ventilated crawl spaces; and small areas around water and sewer pipes. In the early 1970s, radon started to cause more extensive problems. People made their homes airtight to save energy and money. That helped to seal in the unwanted radon.

Is your home safe? Because radon is a dense gas, it accumulates near the bottoms of structures. It is simple to test for its presence using detection kits purchased at a hardware store. The Environmental Protection Agency (EPA) considers a result of more than 4 picocuries per liter (pCi/L) to be unsafe.

How can radon pollution be corrected? Fixing radon pollution problems usually requires only minor repairs or alterations. Installing an exhaust fan in the polluted area or within a foundation might solve the problem. Sealing large cracks or spaces around pipes can also help prevent the entrance of radon.

Exploring Further

1. **Comparing and Contrasting** Compare the use, operation, and effectiveness of short- and long-term radon test devices.

2. **Thinking Critically** Some of the atoms of radon gas in the lungs can turn out to be dangerous. Explain.

Chemistry Online

To find out more about how to prevent radon pollution in your home, visit the Chemistry Web site at chemistryca.com

Lifestyle and environment can result in more than the normal background exposure to radiation as shown in **Figure 21.21.** Cigarette smoke contains significant amounts of radioactive material that can contribute to lung cancer. Smoking two packs of cigarettes a day results in exposure to 0.1 Sv per year. Living at a high altitude or taking frequent trips in airplanes increases your exposure to cosmic rays from outer space. That's because the higher up you are, the less atmosphere there is to block the incoming radiation.

Figure 21.21

Two Sources of Radiation

Medical and dental X rays present only a slight risk due to radiation and are of great benefit in diagnosis. ▶

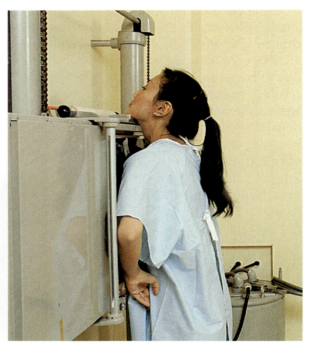

Other sources such as cigarettes present more serious radiation hazards and provide no benefits. ▼

Waste Disposal

You probably associate radioactive waste with nuclear reactors, but more than 80 percent of all such waste is generated in hospitals. How can this hazardous waste be dealt with so that it cannot harm living things? Most of the radioactive waste material produced at hospitals contains isotopes with short half-lives. That kind of waste can simply be stored until the isotopes have decayed to a safe level.

Although waste from nuclear reactors may not be produced in quantities that are large compared to radioactive hospital waste, it is far more dangerous. For example, spent fuel rods from nuclear reactors contain both short-lived and long-lived radioisotopes produced by collisions of fast-moving particles with atoms in the nuclear fuel and in the walls of the reactor itself. These rods are usually stored at the reactor site for several decades until the isotopes with short half-lives have decayed.

What can be done after that point to isolate the radioisotopes with long half-lives? The plan that currently holds the most promise is to incorporate the unstable nuclei into stable material such as glass, which is then surrounded by canisters made of layers of steel and concrete. The canisters can then be buried deep underground in stable rock formations, as shown in **Figure 21.22.** The storage sites would be located in a dry, remote area.

Figure 21.22

Underground Storage of Radioactive Waste
Tunnels are being dug 400 m (about 1300 feet) beneath Yucca Mountain, Nevada, at a proposed site for storage of radioactive waste. It will be the largest radioactive storage facility in the country, capable of holding up to 63 500 tons of waste.

An alternative proposal being tested for nuclear waste disposal is burial in clay sediments deep in the ocean. Recycling of uranium for reuse in reactors is also being considered. The uranium can be extracted from used rods that have been chemically dissolved. This uranium can then be made into new rods. However, this recycling process is currently too expensive to be viable. Whether or not recycling and disposal technologies can keep pace with the use of nuclear materials will influence the development of more technologies using these materials.

Decisions regarding nuclear materials are not simply the business of government officials and scientists. You, too, will have to make decisions about such materials as you help decide how your community supplies energy, whether your country makes nuclear weapons, and, on a more immediate level, whether you undergo medical procedures that employ radioisotopes. How well you understand nuclear chemistry will influence your decisions.

Supplemental Problems

For more practice with solving problems, see Supplemental Practice Problems, Appendix B.

SECTION REVIEW

Understanding Concepts

1. What is the effect of radiation on living cells?

2. What disease is radiation therapy most often used to treat?

3. Explain what a PET scan does.

Thinking Critically

4. **Illustrating** How could a radioactive tracer be used to study the way glucose is metabolized in cells?

Applying Chemistry

5. **Bone Scans** What radioactive isotopes might be useful in scanning techniques to examine bones for fractures and other abnormalities? Include at least two elements not discussed in this chapter, and explain your answer.

CHAPTER 21 ASSESSMENT

REVIEWING MAIN IDEAS

21.1 Types of Radioactivity

- Nuclear reactions involve protons and neutrons rather than electrons, and therefore differ from chemical reactions.

- Radioactivity involves the spontaneous emission of radiation by an unstable nucleus.

- Radioactivity can be detected and measured using film and instruments, including Geiger and scintillation counters.

- Unstable isotopes are called radioisotopes. Symbols for isotopes show the mass number as a superscript and the atomic number as a subscript to the left of the symbol for the element.

- When nuclei decay, they usually emit one or more of the three common types of radiation: alpha particles, beta particles, and gamma rays. During alpha decay, a $^{4}_{2}He$ particle is emitted. During beta decay, an electron is emitted. During gamma decay, high-energy electromagnetic gamma radiation is emitted.

- A nuclear decay process can be represented by an equation in which the sum of the mass numbers and the sum of the atomic numbers must each be equal on both sides of the arrow.

- The half-life of a radioactive element is the time it takes for one-half of the nuclei in a sample to decay. The rate of nuclear decay can be used to date fossils and artifacts.

21.2 Nuclear Reactions and Energy

- A small mass is converted into a large amount of energy during a nuclear reaction, according to Einstein's equation $E = mc^2$.

- Nuclear fission is the breaking apart of a nucleus into two or more smaller nuclei that are similar in size.

- Uranium-235 undergoes a fission chain reaction in a nuclear reactor, which converts some of the released energy into electricity.

- Nuclear fusion is the joining together of two smaller nuclei to form a larger one. Fusion in the sun provides energy for life on Earth. Fusion reactors have the potential to provide energy more safely than fission reactors.

21.3 Nuclear Tools

- Radioactive tracers are used for tracking compounds in an organism and for charting the movement of pollutants. Many different isotopes function as medical tracers for imaging organs and processes inside an organism.

- Radiation therapy involves selectively killing rapidly dividing cancer cells by targeting them with radiation generated by the decay of cobalt-60.

- Gamma radiation from cobalt-60 can be used to irradiate food to keep it from spoiling.

- Mostly natural sources contribute to the ionizing radiation exposure that people undergo all the time. The received dose of radiation is measured in grays. The absorbed dose of radiation that results in damage done to organisms is measured in sieverts.

Vocabulary

For each of the following terms, write a sentence that shows your understanding of its meaning.

alpha particle	nuclear fission
beta particle	nuclear fusion
deuterium	nuclear reactor
gamma ray	radioactivity
gray	sievert
half-life	tritium

chemistryca.com/vocabulary_puzzlemaker

CHAPTER 21 ASSESSMENT

UNDERSTANDING CONCEPTS

1. What part of the atom produces radioactivity?

2. What is ionizing radiation?

3. What is given off by a nucleus undergoing each of the following types of decay?

   a) alpha   **b)** beta   **c)** gamma

4. What types of material will block alpha particles? Beta particles? Gamma rays?

5. What starts the chain reaction that occurs in nuclear fission reactors used today? What keeps it going? How is its rate controlled?

6. What type of nuclear reaction takes place in stars?

7. How is cobalt-60 used to treat cancer?

8. Why are foods irradiated? Does this make them radioactive?

9. Why is nuclear waste so difficult to dispose of safely?

10. Write the symbols for deuterium and tritium.

11. When each of the following is ejected from a nucleus, what happens to the atomic number and mass number of the atom?

    a) an alpha particle
    b) a beta particle
    c) gamma radiation

12. Where does the energy that is produced during nuclear reactions come from?

APPLYING CONCEPTS

13. What radioisotope is generally used in smoke detectors? What function does it serve?

14. Tritium has a half-life of about 12.3 years, which makes it useful for dating objects up to about 100 years old. Tritium dating is sometimes used to verify the dates of aged alcoholic beverages like wines. How old is a bottle of wine if its tritium activity is 12.5 percent as high as that of new wine?

15. What new element is formed if magnesium-24 is bombarded with a neutron and then ejects a proton? Write a balanced nuclear equation for this transmutation.

Biology Connection

16. What results would Hershey and Chase have observed if they had

    a) radiolabeled only protein?
    b) radiolabeled only DNA?
    c) labeled both the protein and DNA with oxygen-18?

Chemistry and Technology

17. Carbon-14 formation in the atmosphere also results in formation of hydrogen. Explain.

Art Connection

18. If the white lead in a sample of paint taken from an alleged forgery of a famous old painting turns out to have a high ratio of radium-226 to polonium-210, what can be concluded?

THINKING CRITICALLY

Relating Concepts

19. Potassium-40 dating techniques can be tricky because of the properties of the decay product argon. What about argon makes it hard to measure?

Designing an Experiment

20. **ChemLab** Design an experiment to determine how long the half-life of pennium is, given standardized shaking of the box.

Drawing Conclusions

21. **MiniLab 1** Why do most chain reactions stop before all of the reactant elements have been used up?

Forming a Hypothesis

22. **MiniLab 2** Would you expect to find higher levels of radon in basements of homes that have floors of dirt or floors of thick concrete? Explain.

CUMULATIVE REVIEW

23. Suppose that a neutral atom of gadolinium has an atomic number of 64 and a mass number of 153. How many electrons, protons, and neutrons does it contain? (Chapter 2)

24. What are four factors that influence the rate of a chemical reaction? (Chapter 6)

25. What happens during oxidation? During reduction? (Chapter 16)

26. What are allotropes? Give an example. (Chapter 5)

27. List two properties of bases that are different from those of acids. (Chapter 14)

SKILL REVIEW

28. Using a Table Examine the table below, which lists information about nuclear power use in selected countries at the end of 1989. Then write a one-paragraph summary of the information, incorporating answers to the questions that follow the table.

Country	Percent Electricity from Nuclear Reactors
Argentina	11.4
Belgium	60.8
Brazil	0.7
Canada	15.6
France	74.6
Hungary	49.8
Japan	27.8
Netherlands	5.4
South Korea	50.2
Former Soviet Union	12.3
Spain	38.4
Sweden	45.1
Switzerland	41.6
United Kingdom	21.7
United States	19.1
Former West Germany	34.3

a) Which country listed gets the highest percentage of its electricity from nuclear power?

b) How many countries listed get more than 40 percent of their electricity from nuclear power?

c) Is the percentage of electricity from nuclear power in the United States greater or less than the average percentage for all countries listed?

WRITING IN CHEMISTRY

29. Write a history of the development of the Tokamak fusion reactor. Research the progress being made in getting experimental tokamak fusion reactors to operate efficiently and economically. When, if ever, do you expect them to exceed the break-even point in terms of energy produced compared to energy input? Are reactors of different designs likely to replace the Tokamak reactor before it is ever used commercially?

PROBLEM SOLVING

30. Why is there more concern about radon levels now than there was 30 years ago?

31. What percentage of C-14 would you expect a piece of 34 000 year-old fossilized bone from a mastodon to have when compared to a similar piece of bone from a modern elephant?

32. How old is an Egyptian scroll made of papyrus that contains 75 percent of the amount of C-14 that would be found in a piece of paper today?

33. Mercury-190 has a half-life of 20 minutes. If you obtain a 36.0-mg sample, how much mercury-190 will remain after one hour?

34. The most common isotope of thorium, $^{232}_{90}$Th, is a radioactive alpha-particle emitter. What product results when thorium-232 decays by emitting an alpha particle? Write an equation for the process.

1. Which of the following is true about beta decay?

   a) Beta particles can be stopped by heavy sheets of paper.

   b) During beta decay, an element will be transformed into a different element.

   c) A beta particle has a similar size and weight as an alpha particle.

   d) After beta decay, an atom will acquire a -1 charge.

2. Geologists use the decay of potassium-40 in volcanic rocks to determine their age. Potassium-40 has a half life of 1.26×10^9 years, so it can be used to date very old rocks. If a sample of rock 3.15×10^8 years old contains 2.73×10^{-7} g of potassium-40 today, how much potassium-40 was originally present in the rock?

   a) 2.3×10^{-7} g

   b) 1.71×10^{-8} g

   c) 3.25×10^{-7} g

   d) 4.37×10^{-6} g

3. A human skeleton is discovered by archeologists that is dated back to the year 9,400 B.C. Approximately what percent of carbon-14 is still present in the skeleton compared with the amount of carbon-14 present in a living person?

   a) 12.5%   c) 45%

   b) 25%   d) 61%

The Radioactive Decay of Stronium-90		
Number of Half Lives	Elapsed Time (Years)	Mass of Stronium-90 Present
0	0	10.0 g
1	29	5.0 g
2	58	?
3	?	1.25 g
4	?	?

Use the table above to answer questions 4–8.

4. What is the half life of strontium-90?

   a) 29 years

   b) 58 years

   c) 87 years

   d) There is insufficient information given to determine the half life of Sr-90.

5. How much stronitum-90 will remain after 174 years?

   a) 0.15625 g   c) 0.625 g

   b) 0.3125 g   d) 1.25 g

6. Approximately what percentage of stronium-90 will remain after 1.5 half lives?

   a) 15%

   b) 25%

   c) 35%

   d) 65%

7. Albert Einstein's equation $E = mc^2$ says

   a) mass and energy are the two basic entities in the universe.

   b) matter and energy are connected and interchangeable at the speed of light.

   c) only small amounts of energy are needed to create large quantities of matter.

   d) small quantities of matter can be converted into enormous amounts of energy.

8. The immense energy released by the sun is due to which of the following reactions occurring within its core?

   a) nuclear fission

   b) gamma decay

   c) nuclear fusion

   d) alpha decay

Test Taking Tip

Slow Down! Check to make sure you're answering the question that each problem is posing. Read the questions and every answer choice very carefully. Remember that doing most of the problems and getting them right is alwyas preferable to doing all the problems and getting lots of them wrong.

APPENDICES CONTENTS

Appendix A   **Chemistry Skill Handbook**   **785**

Measurement in Science   785
    The International System of Units   785
    Other Useful Measurements   786
    SI Prefixes   786
    Relating SI, Metric, and English Measurements   788
Making and Interpreting Measurements   791
    Expressing the Accuracy of Measurements   793
    Expressing Quantities with Scientific Notation   795
Computations with the Calculator   799
Using the Factor Label Method   801
Organizing Information   804
    Making and Using Tables   804
    Making and Using Graphs   805

Appendix B   **Supplemental Practice Problems**   **809**

Appendix C   **Safety Handbook**   **839**

Safety Guidelines in the Chemistry Laboratory   839
First Aid in the Laboratory   839
Safety Symbols   840

Appendix D   **Chemistry Data Handbook**   **841**

Table D.1   Symbols and Abbreviations   841
Table D.2   The Modern Periodic Table   842
Table D.3   Alphabetical Table of the Elements   844
Table D.4   Properties of Elements   845
Table D.5   Electron Configurations of the Elements   848
Table D.6   Useful Physical Constants   850
Table D.7   Names and Charges of Polyatomic Ions   850
Table D.8   Solubility Guidelines   851
Table D.9   Solubility Product Constants   851
Table D.10   Acid-Base Indicators   852

Appendix E   **Answers to In-Chapter Practice Problems**   **853**

Appendix F   **Try at Home Labs**   **863**

APPENDIX A
Chemistry Skill Handbook

Measurement in Science

It's easier to determine if a runner wins a race than to determine if the runner broke a world's record for the race. The first determination requires that you sequence the runners passing the finish line—first, second, third. . . . The second determination requires that you carefully measure and compare the amount of time that passed between the start and finish of the race for each contestant. Because time can be expressed as an amount made by measuring, it is called a quantity. One second, three minutes, and two hours are quantities of time. Other familiar quantities include length, volume, and mass.

The International System of Units

In 1960, the metric system was standardized in the form of Le Système International d'Unités (SI), which is French for the "International System of Units." These SI units were accepted by the international scientific community as the system for measuring all quantities.

SI Base Units   The foundation of SI is seven independent quantities and their SI base units, which are listed in **Table A.1.**

Table A.1   SI Base Units

Quantity	Unit	Unit Symbol
Length	meter	m
Mass	kilogram	kg
Time	second	s
Temperature	kelvin	K
Amount of substance	mole	mol
Electric current	ampere	A
Luminous intensity	candela	cd

SI Derived Units   You can see that quantities such as area and volume are missing from the table. The quantities are omitted because they are derived—that is, computed—from one or more of the SI base units. For example, the unit of area is computed from the product of two perpendicular length units. Because the SI base unit of length is the meter, the SI derived unit of area is the square meter, m^2. Similarly, the unit of volume is derived from three mutually perpendicular length units, each represented by the meter. Therefore, the SI derived unit of volume is the cubic meter, m^3. The SI derived units used in this text are listed in **Table A.2.**

Table A.2   SI Derived Units

Quantity	Unit	Unit Symbol
Area	square meter	m^2
Volume	cubic meter	m^3
Mass density	kilogram per cubic meter	kg/m^3
Energy	joule	J
Heat of fusion	joule per kilogram	J/kg
Heat of vaporization	joule per kilogram	J/kg
Specific heat	joule per kilogram-kelvin	$J/kg \cdot K$
Pressure	pascal	Pa
Electric potential	volt	V
Amount of radiation	gray	Gy
Absorbed dose of radiation	sievert	Sv

Other Useful Measurements

Metric Units   As previously noted, the metric system is a forerunner of SI. In the metric system, as in SI, units of the same quantity are related to each other by orders of magnitude. However, some derived quantities in the metric system have units that differ from those in SI. Because these units are familiar and equipment is often calibrated in these units, they are still used today. **Table A.3** lists several metric units that you might use.

Table A.3   Metric Units

Quantity	Unit	Unit Symbol
Volume	liter (0.001 m^3)	L
Temperature	Celsius degree	°C
Specific heat	joule per kilogram-degree Celsius	$J/kg \cdot °C$
Pressure	millimeter of mercury	mm Hg
Energy	calorie	cal

SI Prefixes

When you express a quantity, such as ten meters, you are comparing the distance to the length of one meter. Ten meters indicates that the distance is a length ten times as great as the length of one meter. Even though you can express any quantity in terms of the base unit, it may not be convenient. For example, the distance between two towns might be 25 000 m. Here, the meter seems too small to describe that distance. Just as you would use 16 miles, not 82 000 feet, to express that distance, you would use a larger unit of length, the kilometer, km. Because the kilometer represents a length of 1000 m, the distance between the towns is 25 km.

In SI, units that represent the same quantity are related to each other by some factor of ten such as 10, 100, 1000, 1/10, 1/100, and 1/1000. In the example above, the kilometer is related to the meter by a factor of 1000; namely, 1 km = 1000 m. As you see, 25 000 m and 25 km differ only in zeros and the units.

To change the size of the SI unit used to measure a quantity, you add a prefix to the base unit or derived unit of that quantity. For example, the prefix *centi-* designates one one-hundredth (0.01). Therefore, a centimeter, cm, is a unit one one-hundredth the length of a meter and a centijoule, cJ, is a unit of energy one one-hundredth that of a joule. The exception to the rule is in the measurement of mass in which the base unit, kg, already has a prefix. To express a different size mass unit, you replace the prefix to the gram unit. Thus, a centigram, cg, represents a unit having one one-hundredth the mass of a gram. **Table A.4** lists the most commonly used SI prefixes.

Table A.4   SI Prefixes

Prefix	Symbol	Meaning	Multiplier	
			Numerical value	Expressed as scientific notation
Greater than 1				
giga-	G	billion	1 000 000 000	1×10^9
mega-	M	million	1 000 000	1×10^6
kilo-	k	thousand	1 000	1×10^3
Less than 1				
deci-	d	tenth	0.1	1×10^{-1}
centi-	c	hundredth	0.01	1×10^{-2}
milli-	m	thousandth	0.001	1×10^{-3}
micro-	μ	millionth	0.000 001	1×10^{-6}
nano-	n	billionth	0.000 000 001	1×10^{-9}
pico-	p	trillionth	0.000 000 000 001	1×10^{-12}

Practice Problems

Use Tables A.1, A.2, and A.3 to answer the following questions.

1. Name the following quantities using SI prefixes. Then write the symbols for each.
   a) 0.1 m
   b) 1 000 000 000 J
   c) 10^{-12} m
   d) 0.000 000 001 m
   e) 10^{-3} g
   f) 10^6 J

2. For each of the following, identify the quantity being expressed and rank the units in increasing order of size.
   a) cm, μm, dm
   b) Pa, MPa, kPa
   c) kV, cV, V
   d) pg, cg, mg
   e) mA, MA, μA
   f) dGy, mGy, nGy

Relating SI, Metric, and English Measurements

Any measurement tells you how much because it is a statement of quantity. However, how you express the measurement depends on the purpose for which you are going to use the quantity. For example, if you look at a room and say its dimensions are 9 ft by 12 ft, you are estimating these measurements from past experience. As you become more familiar with SI, you will be able to estimate the room size as 3 m by 4 m. On the other hand, if you are going to buy carpeting for that room, you are going to make sure of its dimensions by measuring it with a tape measure no matter which system you are using.

Estimating and using any system of measurement requires familiarity with the names and sizes of the units and practice. As you will see in **Figure 1,** you are already familiar with the names and sizes of some common units, which will help you become familiar with SI and metric units.

Figure 1
Relating Measurements

Length
A paper clip and your hand are useful approximations for SI units. For approximating, a meter and a yard are similar lengths. To convert length measurements from one system to another, you can use the relationships shown here.

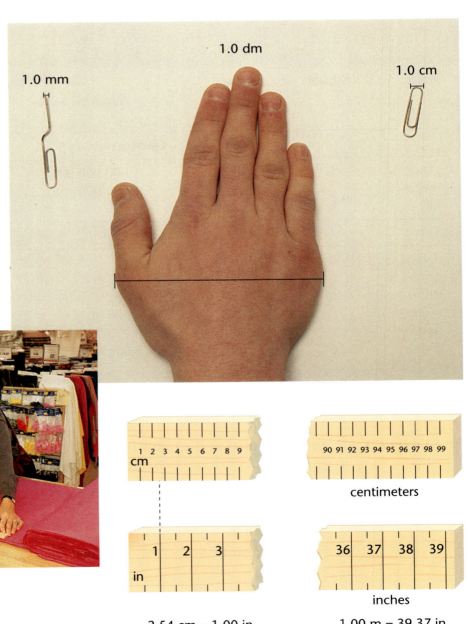

1.0 mm

1.0 dm

1.0 cm

2.54 cm = 1.00 in.

centimeters

inches

1.00 m = 39.37 in.

Volume

For approximating, a liter and a quart are similar volumes. Kitchen measuring spoons are used in estimating small volumes.

1.00 dm³ = 1.0 liter = 1.06 qt. = 1000 mL

Temperature

Having a body temperature of 37° or 310 sounds unhealthy if you don't include the proper units. In fact, 37°C and 310 K are normal body temperatures in the metric system and SI, respectively.

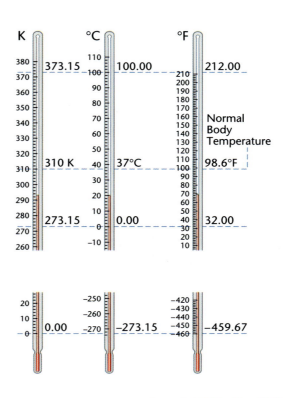

Mass and Weight   One of the most useful ways of describing an amount of stuff is to state its mass. You measure the mass of an object on a balance. Even though you are measuring mass, most people still refer to it as *weighing*. You might think that mass and weight are the same quantity. They are not. The mass of an object is a measure of its inertia; that is, its resistance to changes in motion. The inertia of an object is determined by its quantity of mass. A pint of sand has more matter than a pint of water; therefore, it has more mass.

The weight of an object is the amount of gravitational force acting on the mass of the object. On Earth, you sense the weight of an object by holding it and feeling the pull of gravity on it.

An important aspect of weight is that it is directly proportional to the mass of the object. This relationship means that the weight of a 2-kg object is twice the weight of a 1-kg object. Therefore, the pull of gravity on a 2-kg object is twice as great as the pull on a 1-kg object. Because it's easier to measure the effect of the pull of gravity rather than the resistance of an object to changes in its motion, mass can be determined by a weighing process.

Two instruments used to determine mass are the double-pan balance and a triple-beam balance. How each functions is illustrated in **Figure 2.** You can read how an electronic balance functions in *How it Works* in Chapter 12.

Figure 2
Measuring Mass
In a double-pan balance, the pull of gravity on the mass of an object placed in one pan causes the pan to rotate downward. To balance the rotation, an equal downward pull must be exerted on the opposite pan, as in a seesaw. You can produce this pull by placing calibrated masses on the opposite pan. When the two pulls are balanced, the masses in each pan are equal. ▶

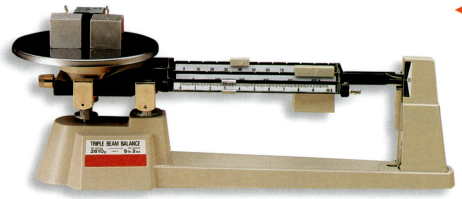

◀ In a triple-beam balance, weights of different sizes are placed at notched locations along each of two beams, and a third is slid along an arm to balance the object being weighed. To determine the mass of the object, you add the numbered positions of the three weights. Using a triple-beam balance is usually a much quicker way to determine the mass of an object than using a double-pan balance.

Figure 3 illustrates several familiar objects and their masses.

Figure 3
Mass
Because weight and mass are so closely related, the contents of cans and boxes are labeled both in pounds and ounces, the British/American weight units, and grams or kilograms, the metric mass units.

Making and Interpreting Measurements

Using measurements in science is different from manipulating numbers in math class. The important difference is that numbers in science are almost always measurements, which are made with instruments of varying accuracy. As you will see, the degree of accuracy of measured quantities must always be taken into account when expressing, multiplying, dividing, adding, or subtracting them. In making or interpreting a measurement, you should consider two points. The first is how well the instrument measures the quantity you're interested in. This point is illustrated in **Figure 4.**

Figure 4

Determining Length

Using the calibrations on the top ruler, you know that the length of the strip is between 4 cm and 5 cm. Because there are no finer calibrations between 4 cm and 5 cm, you have to estimate the length beyond the last calibration. An estimate would be 0.3 cm. Of course, someone might estimate it as 0.2 cm; others as 0.4 cm. Even though you record the measurement of the strip's length as 4.3 cm, you and others reading the measurement should interpret the measurement as 4.3 ± 0.1 cm.

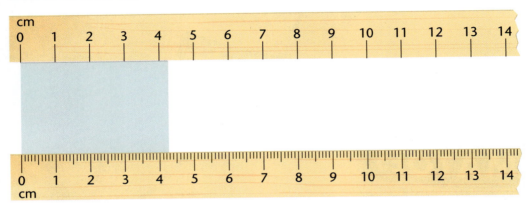

In the bottom ruler, you can see that the edge of the strip lies between 4.2 cm and 4.3 cm. Because there are no finer calibrations between 4.2 cm and 4.3 cm, you estimate the length beyond the last calibration. In this case, you might estimate it as 0.07 cm. You would record this measurement as 4.27 cm. You and others should interpret the measurement as 4.27 ± 0.01 cm.

Because the bottom ruler in **Figure 4** is calibrated to smaller divisions than the top ruler, the lower ruler has more precision than the upper. Any measurement you make with it will be more precise than one made on the top ruler because it will contain a smaller estimated value.

The second point to consider in making or interpreting a measurement is how well the measurement represents the quantity you're interested in. Looking at **Figure 4,** you can see how well the edge of the paper strip aligns with the value 4.27 cm. From sight, you know that 4.27 cm is a better representation of the strip's length than 4.3 cm. Because 4.27 cm better represents the length of the strip (the quantity you're interested in) than does 4.3 cm, it is a more accurate measurement of the strip's length.

In **Figure 5,** you will see that a more precise measurement might not be a more accurate measurement of a quantity.

Figure 5

Precise and Accurate Measurements

Describing the width of the index card as 10.16 cm indicates that the ruler has 0.1-cm calibrations and the 0.06 cm is an estimation. Similarly, the uniform alignment of the edge of the index card with the ruler indicates that the measurement 10.16 cm is also an accurate measurement of width. ▶

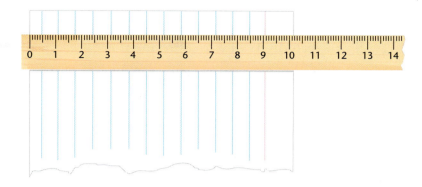

Describing the width of a brick as 10.16 cm indicates that this measurement is as precise as the measurement of the index card. However, 10.16 cm isn't a good representation of the width of the ragged- and jagged-edged brick. ▶

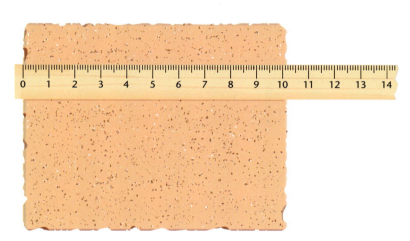

A better representation of the width of the brick is made using a ruler with less precision. As you can see, 10.2 cm is a better representation, and therefore a more accurate measurement, of the brick's width than 10.16 cm. ▶

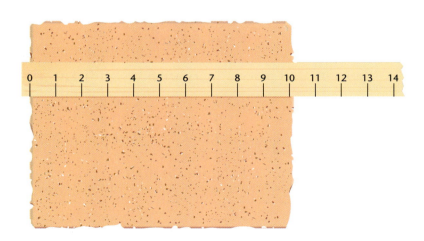

Expressing the Accuracy of Measurements

In measuring the length of the strip as 4.3 cm and 4.27 cm in **Figure 4,** you were aware of the difference in the calibration of the two rulers. This difference appeared in the way each measurement was recorded. In one measurement, the digits 4 and 3 were meaningful. In the second, the digits 4, 2, and 7 were meaningful. In any measurement, meaningful digits are called *significant digits.* The significant digits in a measured quantity include all the digits you know for sure, plus the final estimated digit.

Significant Digits   Several rules can help you express or interpret which digits of a measurement are significant. Notice how those rules apply to readings from a digital balance (left) and a graduated cylinder (right).

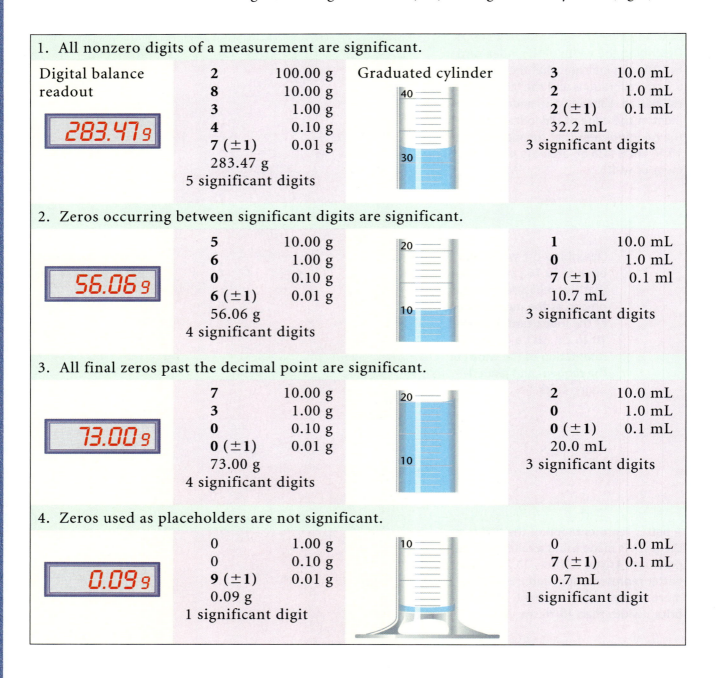

1. All nonzero digits of a measurement are significant.

Digital balance readout

2	100.00 g
8	10.00 g
3	1.00 g
4	0.10 g
7 (±1)	0.01 g
283.47 g	
5 significant digits	

Graduated cylinder

3	10.0 mL
2	1.0 mL
2 (±1)	0.1 mL
32.2 mL	
3 significant digits	

2. Zeros occurring between significant digits are significant.

5	10.00 g
6	1.00 g
0	0.10 g
6 (±1)	0.01 g
56.06 g	
4 significant digits	

1	10.0 mL
0	1.0 mL
7 (±1)	0.1 ml
10.7 mL	
3 significant digits	

3. All final zeros past the decimal point are significant.

7	10.00 g
3	1.00 g
0	0.10 g
0 (±1)	0.01 g
73.00 g	
4 significant digits	

2	10.0 mL
0	1.0 mL
0 (±1)	0.1 mL
20.0 mL	
3 significant digits	

4. Zeros used as placeholders are not significant.

0	1.00 g
0	0.10 g
9 (±1)	0.01 g
0.09 g	
1 significant digit	

0	1.0 mL
7 (±1)	0.1 mL
0.7 mL	
1 significant digit	

The fourth rule sometimes causes difficulty in expressing such measurements as 20 L. Because the zero is a placeholder in the measurement, it is not significant and 20 L has one significant digit. You should interpret the measurement 20 L as 20 L plus or minus the value of the least significant digit, which is 10 L. Thus, a volume measurement of 20 L indicates 20 L ± 10 L, a range of 10-30 L. However, suppose you made the measurement with a device that is accurate to the nearest 1 L. You would want to indicate that both the 2 and the 0 are significant. How would you do this? You can't just add a decimal point after the 20 because it could be mistaken for a period. Adding a decimal point followed by a zero would indicate that the final zero past the decimal is significant (Rule 3), and the measurement would then be 20.0 $\pm$ 0.1 L. To solve the dilemma, you have to express 20 L as 2.0×10^1 L. Now the zero is a significant digit because it is a final zero past the decimal (Rule 3). The measurement has two significant digits and signifies $(2.0 \pm 0.1) \times 10^1$ L.

Expressing Quantities with Scientific Notation

The most common use of scientific notation is in expressing measurements of very large and very small dimensions. Using scientific notation is sometimes referred to as using *powers of ten* because it expresses quantities by using a small number between one and ten, which is then multiplied by ten to a power to give the quantity its proper magnitude. Suppose you went on a trip of 9000 km. You know that $10^3 = 1000$, so you could express the distance of your trip as 9×10^3 km. In this example, it may seem that scientific notation wouldn't be terribly useful. However, consider that an often-used quantity in chemistry is 602 000 000 000 000 000 000 000, the number of atoms or molecules in a mole of a substance. Recall that the mole is the SI unit of amount of a substance. Rather than writing out this huge number every time it is used, it's much easier to express it in scientific notation.

Determining Powers of 10   To determine the exponent of ten, count as you move the decimal point left until it falls just after the first nonzero digit—in this case, 6. If you try this on the number above, you'll find that you've moved the decimal point 23 places. Therefore, the number expressed in scientific notation is 6.02×10^{23}.

Expressing small measurements in scientific notation is done in a similar way. The diameter of a carbon atom is 0.000 000 000 000 154 m. In this case, you move the decimal point right until it is just past the first nonzero digit—in this case, 1. The number of places you move the decimal point right is expressed as a *negative* exponent of ten. The diameter of a carbon atom is 1.54×10^{-13} m. You always move the decimal point until the coefficient of ten is between one and less than ten. Thus, scientific notation always has the form, $M \times 10^n$ where $1 \leq M < 10$.

Notice how the following examples are converted to scientific notation.

Quantities greater than 1

17.16 g   17 . 16 → 1.716×10^1 g   Decimal point moved 1 place left.

152.6 L   152 . 6 → 1.526×10^2 L   Decimal point moved 2 places left.

73 621 kg   73 621. → 7.3621×10^4 kg   Decimal point moved 4 places left.

Quantities between 0 and 1

0.29 mL   0 . 29 → 2.9×10^{-1} mL   Decimal point moved 1 place right.

0.0672 m   0 . 0672 → 6.72×10^{-2} m   Decimal point moved 2 places right.

0.0008 g   0 . 0008 → 8×10^{-4} g   Decimal point moved 4 places right.

Calculations with Measurements   You often must use measurements to calculate other quantities. Remember that the accuracy of measurement depends on the instrument used and that accuracy is expressed as a certain number of significant digits. Therefore, you must indicate which digits in the result of any mathematical operation with measurements are significant. The rule of thumb is that no result can be more accurate than the least accurate quantity used to calculate that result. Therefore, the quantity with the least number of significant digits determines the number of significant digits in the result.

The method used to indicate significant digits depends on the mathematical operation.

Addition and Subtraction
The answer has only as many decimal places as the measurement having the least number of decimal places.

190.2 g

65.291 g

12.38 g

$$
\begin{array}{r}
190.2 \ \ \text{g} \\
65.291 \ \text{g} \\
12.38 \ \ \text{g} \\
\hline
267.871 \ \text{g}
\end{array}
$$

The answer is rounded to the nearest tenth, which is the accuracy of the least accurate measurement.
267.9 g

Because the masses were measured on balances differing in accuracy, the least accurate measurement limits the number of digits past the decimal point.

Multiplication and Division
The answer has only as many significant digits as the measurement with the least number of significant digits.

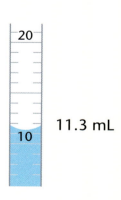

11.3 mL

13.78 g

$$density = \frac{mass}{volume}$$

$$D = \frac{m}{V} =$$

$$\frac{13.78 \text{ g}}{11.3 \text{ mL}} = 1.219469 \text{ g/mL}$$

The answer is rounded to three significant digits, 1.22 g/mL, because the least accurate measurement, 11.3 mL, has three significant digits.

Multiplying or dividing measured quantities results in a derived quantity. For example, the mass of a substance divided by its volume is its density. But mass and volume are measured with different tools, which may have different accuracies. Therefore, the derived quantity can have no more significant digits than the least accurate measurement used to compute it.

Practice Problems

3. Determine the number of significant digits in each of the following measurements.

a) 64 mL   e) 47 080 km

b) 0.650 g   f) 0.072 040 g

c) 30 cg   g) 1.03 mm

d) 724.56 mm   h) 0.001 mm

4. Write each of the following measurements in scientific notation.

a) 76.0°C   e) 0.076 12 m

b) 212 mm   f) 763.01 g

c) 56.021 g   g) 10 301 980 nm

d) 0.78 L   h) 0.001 mm

5. Write each of the following measurements in scientific notation.

a) 73 000 $\pm$ 1 mL   c) 100 $\pm$ 10 cm

b) 4000 $\pm$ 1000 kg   d) 100 000 $\pm$ 1000 km

6. Solve the following problems and express the answer in the correct number of significant digits.

a)   45.761 g   c)   0.340 cg

    $-$ 42.65   g   1.20   cg

                                   $+$ 1.018 cg

b)   1.6   km   d)   6 000 μm

    $+$ 0.62 km $-$ 202 μm

7. Solve the following problems and express the answer in the correct number of significant digits.

a) 5.761 cm $\times$ 6.20 cm

b) $\dfrac{23.5 \text{ kg}}{4.615 \text{ m}^3}$

c) $\dfrac{0.2 \text{ km}}{5.4 \text{ s}}$

d) 11.00 m $\times$ 12.10 m $\times$ 3.53 m

e) $\dfrac{4.500 \text{ kg}}{1.500 \text{ m}^2}$

f) $\dfrac{18.21 \text{ g}}{4.4 \text{ cm}^3}$

Adding and Subtracting Measurements in Scientific Notation   Adding and subtracting measurements in scientific notation requires that for any problem, the measurements must be expressed as the same power of ten. For example, in the following problem, the three length measurements must be expressed in the same power of ten.

$$1.1012 \times 10^4 \text{ mm}$$
$$2.31 \times 10^3 \text{ mm}$$
$$+\ 4.573 \times 10^2 \text{ mm}$$

In adding and subtracting measurements in scientific notation, all measurements are expressed in the same order of magnitude as the measurement with the greatest power of ten. When converting a quantity, the decimal point is moved one place to the left for each increase in power of ten.

$$2.31 \times 10^3 \quad 2.31 \times 10^3 \rightarrow 0.231 \times 10^4$$

$$4.573 \times 10^2 \quad 4.573 \times 10^2 \rightarrow 0.4573 \times 10^3 \rightarrow 0.04573 \times 10^4$$

$$
\begin{array}{r}
1.1012 \times 10^4 \text{ mm} \\
0.231 \times 10^4 \text{ mm} \\
+ \ 0.04573 \times 10^4 \text{ mm} \\
\hline
1.37793 \times 10^4 \text{ mm} = 1.378 \times 10^4 \text{ mm (rounded)}
\end{array}
$$

Multiplying and Dividing Measurements in Scientific Notation   Multiplying and dividing measurements in scientific notation requires that similar operations are done to the numerical values, the powers of ten, and the units of the measurements.

a) The numerical coefficients are multiplied or divided and the resulting value is expressed in the same number of significant digits as the measurement with the least number of significant digits.
b) The exponents of ten are algebraically added in multiplication and subtracted in division.
c) The units are multiplied or divided.

The following problems illustrate these procedures.

Sample Problem 1

$$(3.6 \times 10^3 \text{ m})(9.4 \times 10^3 \text{ m})(5.35 \times 10^{-1} \text{ m})$$
$$= (3.6 \times 9.4 \times 5.35) \times (10^3 \times 10^3 \times 10^{-1}) \ (\text{m} \times \text{m} \times \text{m})$$
$$= (3.6 \times 9.4 \times 5.35) \times 10^{(3+3+(-1))} \ (\text{m} \times \text{m} \times \text{m})$$
$$= (181.044) \times 10^5 \text{ m}^3$$
$$= 1.8 \times 10^2 \times 10^5 \text{ m}^3$$
$$= 1.8 \times 10^7 \text{ m}^3$$

Sample Problem 2

$$\frac{6.762 \times 10^2 \text{ m}^3}{(1.231 \times 10^1 \text{ m})(2.80 \times 10^{-2} \text{ m})}$$

$$= \frac{6.762}{1.231 \times 2.80} \times \frac{10^2}{10^1 \times 10^{-2}} \times \frac{\text{m}^3}{\text{m} \times \text{m}}$$

$$= 1.961819659 \times 10^{(2-(+1-2))} \text{ m}^{(3-(+2))}$$
$$= 1.96 \times 10^{(2-(-1))} \text{ m}^{(1)}$$
$$= 1.96 \times 10^3 \text{ m}$$

Practice Problems

8. Solve the following addition and subtraction problems.
   a) 1.013×10^3 g $+ 8.62 \times 10^2$ g $+ 1.1 \times 10^1$ g
   b) 2.82×10^6 m $- 4.9 \times 10^4$ m
9. Solve the following multiplication and division problems.
   a) 1.18×10^{-3} m $\times 4.00 \times 10^2$ m $\times 6.22 \times 10^2$ m
   b) 3.2×10^2 g $\div 1.04 \times 10^2$ cm^2 $\div 6.22 \times 10^{-1}$ cm

Computations with the Calculator

Working problems in chemistry will require you to have a good understanding of some of the advanced functions of your calculator. When using a calculator to solve a problem involving measured quantities, you should remember that the calculator does not take significant digits into account. It is up to you to round off the answer to the correct number of significant digits at the end of a calculation. In a multistep calculation, you should not round off after each step. Instead, you should complete the calculation and then round off. **Figure 6** shows how to use a calculator to solve a subtraction problem involving quantities in scientific notation.

Figure 6

Subtracting Numbers in Scientific Notation with a Calculator

A quantity in scientific notation is entered by keying in the coefficient and then striking the [EXP] or [EE] key followed by the value of the exponent of ten. At the end of the calculation, the calculator readout must be corrected to the appropriate number of decimal places. The answer would be rounded to the second decimal place and expressed as 2.56×10^4 kg.

To solve the problem

$$2.61 \times 10^4$$
$$- 5.2 \times 10^2$$

Calculator display

Keystrokes

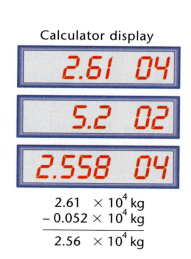

$$
\begin{array}{r}
2.61 \times 10^4 \text{ kg} \\
- 0.052 \times 10^4 \text{ kg} \\
\hline
2.56 \times 10^4 \text{ kg}
\end{array}
$$

Look at **Figure 7** to see how to solve multiplication and division problems involving scientific notation. The problems are the same as the two previous sample problems.

Figure 7
Multiplying and Dividing Measurements Expressed in Scientific Notation

A negative power of ten is usually entered by striking the [EXP] or [EE], entering the positive value of the exponent, and then striking the [±] key. ▼

Keystrokes

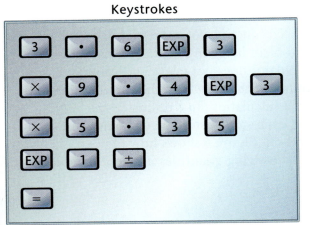

Calculator display

3.6	03
9.4	03
5.35	-01
1.8104	07

Rounded off to 2 significant digits: 1.8×10^7

The numerical value of the answer can have no more significant digits than the measurement that has the least number of significant digits. ▼

Keystrokes

Calculator display

6.762	02
1.231	01
2.80	-02
1.9618	03

Rounded off to 3 significant digits: 1.96×10^3

Practice Problems

10. Solve the following problems and express the answers in scientific notation with the proper number of significant digits.

    a) $\quad 2.01 \;\;\times 10^2$ mL

    $\quad\quad\; 3.1 \;\;\;\times 10^1$ mL

    $\quad + \; 2.712 \times 10^3$ mL

    b) $\quad 7.40 \times 10^2$ mm

    $\quad - \; 4.0 \;\;\times 10^1$ mm

    c) $\quad 2.10 \times 10^1$ g

    $\quad - \; 1.6 \;\;\times 10^{-1}$ g

    d) $\quad 5.131 \times 10^2$ J

    $\quad\quad\; 2.341 \times 10^1$ J

    $\quad + \; 3.781 \times 10^3$ J

11. Solve the following problems and express the answers in scientific notation with the proper number of significant digits.

    a) $(2.00 \times 10^1 \text{ cm})(2.05 \times 10^1 \text{ cm})$

    b) $\dfrac{5.6 \times 10^3 \text{ kg}}{1.20 \times 10^4 \text{ m}^3}$

    c) $(2.51 \times 10^1 \text{ m})(3.52 \times 10^1 \text{ m})(1.2 \times 10^{-1} \text{ m})$

    d) $\dfrac{1.692 \times 10^4 \text{ dm}^3}{(2.7 \times 10^{-2} \text{ dm})(4.201 \times 10^1 \text{ dm})}$

Using the Factor Label Method

The factor label method is used to express a physical quantity such as the length of a pen in any other unit that measures that quantity. For example, you can measure the pen with a metric ruler calibrated in centimeters and then express the length in meters.

If the length of a pen is measured as 14.90 cm, you can express that length in meters by using the numerical relationship between a centimeter and a meter. This relationship is given by the following equation.

$$100 \text{ cm} = 1 \text{ m}$$

If both sides of the equation are divided by 100 cm, the following relationship is obtained.

$$1 = \frac{1 \text{ m}}{100 \text{ cm}}$$

To express 14.90 cm as a measurement in meters, you multiply the quantity by the relationship, which eliminates the cm unit.

$$14.90 \; \cancel{\text{cm}} \times \frac{1 \text{ m}}{100 \; \cancel{\text{cm}}} = \frac{14.90 \; \cancel{\text{cm}}}{1} \times \frac{1 \text{ m}}{100 \; \cancel{\text{cm}}} = \frac{14.90}{100} \text{ m} = 0.1490 \text{ m}$$

Copper weathervane

The factor label method doesn't change the value of the physical quantity because you are multiplying that value by a factor that equals 1. You choose the factor so that when the unit you want to eliminate is multiplied by the factor, that unit and the similar unit in the factor cancel. If the unit you want to eliminate is in the numerator, choose the factor which has that unit in the denominator. Conversely, if the unit you want to eliminate is in the denominator, choose the factor which has that unit in the numerator. For example, in a chemistry lab activity, a student measured the mass and volume of a chunk of copper and calculated its density as 8.80 g/cm^3. Knowing that 1000 g = 1 kg and 100 cm = 1 m, the student could then use the following factor label method to express this value in the SI unit of density, kg/m^3.

$$8.80 \ \frac{g}{cm^3} \ \times \ \frac{1 \ kg}{1000 \ g} \ \times \ \left[\frac{100 \ cm}{1 \ m}\right]^3 =$$

$$8.80 \ \frac{g}{cm \times cm \times cm} \times \frac{1 \ kg}{1000 \ g} \times \frac{100 \ cm}{1 \ m} \times \frac{100 \ cm}{1 \ m} \times \frac{100 \ cm}{1 \ m} =$$

$$\frac{8.80 \times (10^2 \times 10^2 \times 10^2)}{1000} \ \frac{kg}{m^3} = \frac{8.80 \times 10^{(2+2+2)}}{10^3} \ \frac{kg}{m^3} =$$

$$8.80 \times 10^{(6-3)} \ kg/m^3 = 8.80 \times 10^3 \ kg/m^3$$

The factor label method can be extended to other types of calculations in chemistry. To use this method, you first examine the data that you have. Next, you determine the quantity you want to find and look at the units you will need. Finally, you apply a series of factors to the data in order to convert it to the units you need.

Sample Problem 1

The density of silver sulfide (Ag_2S) is 7.234 g/mL. What is the volume of a lump of silver sulfide that has a mass of 6.84 kg?

First, you must apply a factor that will convert kg of Ag_2S to g Ag_2S.

$$\frac{6.84 \text{ kg } Ag_2S}{} \left| \frac{1000 \text{ g } Ag_2S}{1 \text{ kg } Ag_2S} \right. \cdots$$

Next, you use the density of Ag_2S to convert mass to volume.

$$\frac{6.84 \text{ kg } Ag_2S}{} \left| \frac{1000 \text{ g } Ag_2S}{1 \text{ kg } Ag_2S} \right| \frac{1 \text{ mL } Ag_2S}{7.234 \text{ g } Ag_2S} = 946 \text{ mL } Ag_2S$$

Notice that the new factor must have grams in the denominator so that grams will cancel out, leaving mL.

Sample Problem 2

What mass of lead can be obtained from 47.2 g of $Pb(NO_3)_2$?

Because mass is involved, you will need to know the molar mass of $Pb(NO_3)_2$.

$$\begin{aligned} Pb &= 207.2 \text{ g} \\ 2N &= 28.014 \text{ g} \\ 6O &= 95.994 \text{ g} \end{aligned}$$

Molar mass of $Pb(NO_3)_2$ = 331.208 g.
Rounding off according to the rules for significant digits, the molar mass of $Pb(NO_3)_2$ = 331.2 g.

You can see that the mass of lead in $Pb(NO_3)_2$ is 207.2/331.2 of the total mass of $Pb(NO_3)_2$.

Now you can set up a relationship to determine the mass of lead in the 47.2-g sample.

$$\frac{47.2 \text{ g } Pb(NO_3)_2}{} \left| \frac{207.2 \text{ g } Pb}{331.2 \text{ g } Pb(NO_3)_2} \right. = 29.52850242 \text{ g } Pb$$
$$= 29.5 \text{ g } Pb \text{ rounded to}$$
$$\text{3 significant digits.}$$

Notice that the equation is arranged so that the unit *g $Pb(NO_3)_2$* cancels out, leaving only *g Pb*, the quantity asked for in the problem.

Practice Problems in the Factor Label Method

12. Express each quantity in the unit listed to its right.
    a) 3.01 g   cg   e) 0.2 L dm^3
    b) 6200 m   km   f) 0.13 cal/g   J/kg
    c) 6.24×10^{-7} g   μg   g) 5 ft, 1 in.   m
    d) 3.21 L   mL   h) 1.2 qt   L

Organizing Information

It is often necessary to compare and sequence observations and measurements. Two of the most useful ways are to organize the observations and measurements as tables and graphs. If you browse through your textbook, you'll see many tables and graphs. They arrange information in a way that makes it easier to understand.

Making and Using Tables

Most tables have a title telling you what information is being presented. The table itself is divided into columns and rows. The column titles list items to be compared. The row headings list the specific characteristics being compared among those items. Within the grid of the table, the information is recorded. Any table you prepare to organize data taken in a laboratory activity should have these characteristics. Consider, for example, that in a laboratory experiment, you are going to perform a flame test on various solutions. In the test, you place drops of the solution containing a metal ion in a flame, and the color of the flame is observed, as shown in **Figure 8.** Before doing the experiment, you might set up a data table like the one below.

Flame Test Results		
Solution	Metal ion	Color of Flame
KNO_3	K^+	violet-pink

Figure 8

◀ **Flame Test**
A drop of solution containing the potassium ion, K^+, causes the flame to burn with violet color.

While performing the experiment, you would record the name of the solution and then the observation of the flame color. If you weren't sure of the metal ion as you were doing the experiment, you could enter it into the table by checking oxidation numbers afterward. Not only does the table organize your observations, it could also be used as a reference to determine whether a solution of some unknown composition contains one of the metal ions listed in the table.

Making and Using Graphs

After organizing data in tables, scientists usually want to display the data in a more visual way. Using graphs is a common way to accomplish that. There are three common types of graphs—bar graphs, pie graphs, and line graphs.

Bar Graphs   Bar graphs are useful when you want to compare or display data that do not continuously change. Suppose you measure the rate of electrolysis of water by determining the volume of hydrogen gas formed. In addition, you decide to test how the number of batteries affects the rate of electrolysis. You could graph the results using a bar graph as shown in **Figure 9.** Note that you could construct a line graph, but the bar graph is better because there is no way you could use 0.4 or 2.6 batteries.

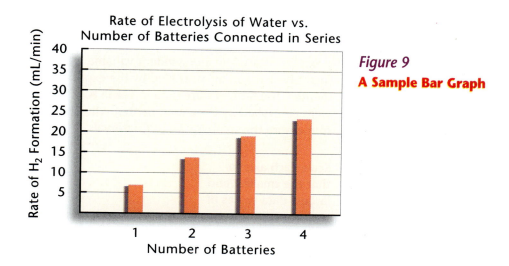

Figure 9

A Sample Bar Graph

Pie Graphs   Pie graphs are especially useful in comparing the parts of a whole. You could use a pie graph to display the percent composition of a compound such as sodium dihydrogen phosphate, NaH_2PO_4, as shown in **Figure 10.**

In constructing a pie graph, recall that a circle has 360°. Therefore, each fraction of the whole is that fraction of 360°. Suppose you did a census of your school and determined that 252 students out of a total of 845 were 17 years old. You would compute the angle of that section of the graph by multiplying $(252/845) \times 360° = 107°$.

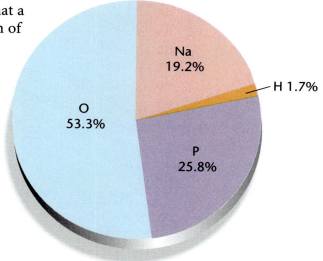

Figure 10

A Sample Pie Graph

Composition of NaH_2PO_4

Line Graphs Line graphs have the ability to show a trend in one variable as another one changes. In addition, they can suggest possible mathematical relationships between variables.

Table A.5 shows the data collected during an experiment to determine whether temperature affects the mass of potassium bromide that dissolves in 100 g of water. If you read the data as you slowly run your fingers down both columns of the table, you will see that solubility increases as the temperature increases. This is the first clue that the two quantities may be related.

To see that the two quantities are related, you should construct a line graph as shown in **Figure 11**.

Table A.5	Effect of Temperature on Solubility of KBr	
Temperature (°C)	**Solubility (g of KBr/100 g H$_2$O)**
10.0	60.2
20.0	64.3
30.0	67.7
40.0	71.6
50.0	75.3
60.0	80.1
70.0	82.6
80.0	86.8
90.0	90.2

Figure 11
Constructing a Line Graph

1. Plot the independent variable on the *x*-axis (horizontal axis) and the dependent variable on the *y*-axis (vertical). The independent variable is the quantity changed or controlled by the experimenter. The temperature data in **Table A.5** were controlled by the experimenter, who chose to measure the solubility at 10°C intervals.

2. Scale each axis so that the smallest and largest data values of each quantity can be plotted. Use divisions such as ones, fives, or tens or decimal values such as hundredths or thousandths.

3. Label each axis with the appropriate quantity and unit.

4. Plot each pair of data from the table as follows.
   - Place a straightedge vertically at the value of the independent variable on the *x*-axis.
   - Place a straightedge horizontally at the value of the dependent variable on the *y*-axis.
   - Mark the point at which the straightedges intersect.

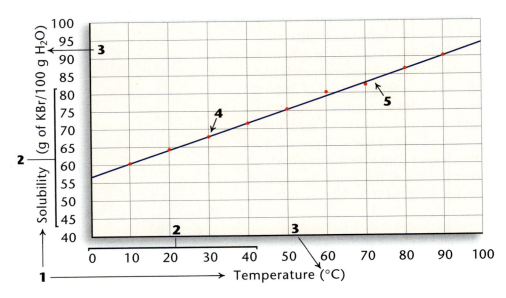

5. Fit the best straight line or curved line through the data points.

One use of a line graph is to predict values of the independent or dependent variables. For example, from **Figure 11** you can predict the solubility of KBr at a temperature of 65°C by the following method:

- Place a straightedge vertically at the approximate value of 65°C on the *x*-axis.

- Mark the point at which the straightedge intersects the line of the graph.

- Place a straightedge horizontally at this point and approximate the value of the dependent variable on the *y*-axis as 82 g/100 g H_2O.

To predict the temperature for a given solubility, you would reverse the above procedure.

Practice Problems

Use Figure 11 to answer these questions.

13. Predict the solubility of KBr at each of the following temperatures.
   a) 25.0°C   **c)** 6.0°C
   b) 52.0°C   **d)** 96.0°C

14. Predict the temperature at which KBr has each of the following solubilities.
   a) 70.0 g/100 g H_2O
   b) 88.0 g/100 g H_2O

Graphs of Direct and Inverse Relationships Graphs can be used to determine quantitative relationships between the independent and dependent variables. Two of the most useful relationships are relationships in which the two quantities are directly proportional or inversely proportional.

When two quantities are directly proportional, an increase in one quantity produces a proportionate increase in the other. A graph of two quantities that are directly proportional is a straight line, as shown in **Figure 12.**

Figure 12

Graph of Quantities That Are Directly Proportional
As you can see, doubling the mass of the carbon burned from 2.00 g to 4.00 g doubles the amount of energy released from 66 kJ to 132 kJ. Such a relationship indicates that mass of carbon burned and the amount of energy released are directly proportional.

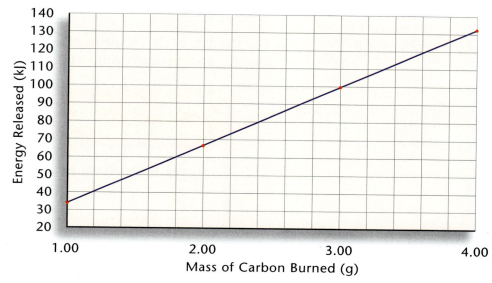

Mass of Carbon Burned (g)

When two quantities are inversely proportional, an increase in one quantity produces a proportionate decrease in the other. A graph of two quantities that are inversely proportional is shown in **Figure 13.**

Figure 13

Graph of Quantities That Are Inversely Proportional

As you can see, doubling the pressure of the gas reduces the volume of the gas by one-half. Such a relationship indicates that the volume and pressure of a gas are inversely proportional.

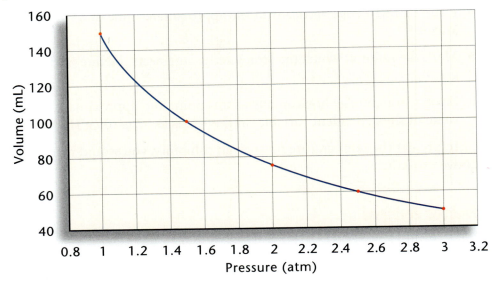

Practice Problems

15. Plot the data in the following table and determine whether the two quantities are directly proportional.

Effect of Temperature on Gas Pressure	
Temperature (K)	**Pressure (kPa)**
300.0	195
320.0	208
340.0	221
360.0	234
380.0	247
400.0	261

16. Plot the data in the following table and determine whether the two quantities are inversely proportional.

Effect of Number of Mini-Lightbulbs on Electrical Current in a Circuit	
Number of mini-lightbulbs	**Current (mA)**
2	3.94
4	1.98
6	1.31
9	0.88

APPENDIX B
Supplemental Practice Problems

This appendix includes *supplemental* practice problems designed to reinforce your knowledge and application skills. See your teacher for the complete set of solutions.

CHAPTER 1

Section 1.1

1. What is matter? Tell whether or not each of the following is matter.

   a. microwaves

   b. helium gas inside a balloon

   c. heat from the sun

   d. velocity

   e. a speck of dust

   f. the color blue

2. Identify each of the following as an element, compound, homogeneous mixture, or heterogeneous mixture.

   a. air   e. ammonia

   b. blood   f. mustard

   c. antimony   g. water

   d. brass   h. tin

3. What is an alloy? Identify each of the following as either a pure metal or an alloy.

   a. zinc   d. copper

   b. steel   e. bronze

   c. sterling silver

4. Which compounds, listed in **Table 1.3,** contain the element nitrogen? Which compounds contain oxygen but do not contain carbon?

Section 1.2

5. The element rubidium melts at 39.5°C and boils at 697°C. What is the physical state of rubidium at room temperature?

6. Which of these substances has the highest melting point: ethanol, helium, or baking soda?

Table 1.3   Some Common Compounds

Compound Name	Formula
Acetaminophen	$C_8H_9NO_2$
Acetic acid	$C_2H_4O_2$
Ammonia	NH_3
Ascorbic acid	$C_6H_8O_6$
Aspartame	$C_{14}H_{18}N_2O_5$
Aspirin	$C_9H_8O_4$
Baking soda	$NaHCO_3$
Butane	C_4H_{10}
Caffeine	$C_8H_{10}N_4O_2$
Calcium carbonate	$CaCO_3$
Carbon dioxide	CO_2
Ethanol	C_2H_6O
Ethylene glycol	$C_2H_6O_2$
Hydrochloric acid	HCl
Magnesium hydroxide	$Mg(OH)_2$
Methane	CH_4
Phosphoric acid	H_3PO_4
Potassium tartrate	$K_2C_4H_4O_6$
Propane	C_3H_8
Salt	NaCl
Sodium carbonate	Na_2CO_3
Sodium hydroxide	NaOH
Sucrose	$C_{12}H_{22}O_{11}$
Sulfuric acid	H_2SO_4
Water	H_2O

7. Your friend says that a certain substance has a freezing point of $-22°C$ and a melting point of $-10°C$. What do you think of your friend's statement?

8. Suppose you have a cube of an unknown metal with a mass of 102 g. You place the cube in a graduated cylinder filled with water to the 40-mL mark. The cube sinks to the bottom of the cylinder and the water level rises to the 54-mL mark. What is the density of the cube?

9. Identify each of the following as either a chemical or a physical property of the substance.

   a. Krypton is an inert gas.
   b. Ethanol is a clear, colorless liquid.
   c. Hydrochloric acid reacts with many metals to form hydrogen gas.
   d. Barium sulfate is practically insoluble in water.
   e. The element tungsten has a very high density.

10. Identify each of the following as either a chemical or a physical change.

   a. An iron bar expands slightly when it is heated.
   b. A banana turns brown when it is left on the counter.
   c. Hydrogen burns in air to form water vapor.
   d. The surface of a pond freezes in winter.
   e. A balloon pops when it is pricked by a pin.

CHAPTER 2

Section 2.1

1. Describe the observations of the chemist Joseph Proust. What scientific law resulted from his studies?

2. Distinguish among a hypothesis, a theory, and a scientific law.

3. How did the discovery of isotopes help to lead to a revised model of atomic structure?

4. What is the atomic number of vanadium? What does that number tell you about an atom of vanadium?

5. A radioactive isotope of iodine is used to diagnose and treat disorders of the thyroid gland. This isotope has 53 protons and 78 neutrons. What are the atomic number and mass number of this isotope? How many electrons does it have?

6. One atom has 45 neutrons and a mass number of 80. Another atom has 36 protons and 44 neutrons. What is the identity of each of these atoms? Which atom has a greater mass number?

7. The element boron has two naturally occurring isotopes. One has atomic mass 10.01, whereas the other has atomic mass 11.01. The average atomic mass of boron is 10.811. Which of the two isotopes is more abundant?

Section 2.2

8. A satellite can orbit Earth at almost any altitude, depending on the amount of energy that is used to launch it. How is this different from the way that an electron orbits the atomic nucleus?

9. Electromagnetic wave A has a wavelength of 10^4 m, whereas electromagnetic wave B has a wavelength of 10^{-2} m. Compare the frequencies and energies of the two waves.

10. For each of the following elements, tell how many electrons are in each energy level and draw the Lewis dot diagram for each atom.

   a. phosphorus, which has 15 electrons
   b. beryllium, which has four electrons
   c. carbon, which has six electrons
   d. helium, which has two electrons

CHAPTER 3

Section 3.1

1. Use the periodic table to separate these ten elements into five pairs of elements having similar properties.
S, Ne, Li, O, Mg, Ag, Na, Sr, Kr, Cu

Section 3.2

2. For each of the following elements, write its group number and period number, its physical state at room temperature, and whether it is a metal, nonmetal, or metalloid.

   a. argon   **d.** fluorine

   b. nickel   **e.** barium

   c. antimony

3. Write the symbol and name of the element that fits each description.

   a. the second-lightest of the halogens

   b. the metalloid with the lowest period number

   c. the only Group 16 element that is a gas at room temperature

   d. the heaviest of the noble gases

   e. the Group 15 nonmetal that is a solid at room temperature

4. Element Q is in the third period. Its outer energy level has six electrons. How does the number of outer-level electrons of element Q compare with that of element Z, which is in the second period, Group 14? Write the name and symbol of each element.

5. Why is the chemistry of the transition elements and the inner transition elements less predictable than the chemistry of the main group metals?

6. Write the symbol of the element that has valence electrons that fit each description.

   a. three electrons in the fourth energy level

   b. one electron in the second energy level

   c. eight electrons in the third energy level

   d. two electrons in the first energy level

   e. five electrons in the sixth energy level

7. Draw the electron dot structure for each of the following elements. What is the group number of each element? Is the element a metal, nonmetal, or metalloid?

   a. As   **e.** Br

   b. Ca   **f.** Al

   c. S   **g.** Si

   d. H   **h.** Xe

8. The chemical formula for sodium oxide is Na_2O. Use the periodic table to predict the formulas of the following similar compounds.

   a. potassium oxide   **c.** potassium sulfide

   b. sodium sulfide   **d.** lithium oxide

9. How can the electrical conductivity of a semiconductor such as silicon be increased?

10. Why are transistors, diodes, and other semiconductor devices useful?

CHAPTER 4

Section 4.1

1. Use carbon dioxide as an example to contrast the properties of a compound with those of the elements of which it is composed.

Section 4.2

2. How is a calcium ion different from a calcium atom? From a potassium ion?

3. Would you expect to find MgO_2 as a stable compound? Explain.

4. Describe the submicroscopic structure of a grain of salt. What is this structure called?

5. Aluminum metal reacts with fluorine to form an ionic compound. Use the periodic table to determine the number of valence electrons for each element. Draw Lewis dot structures to show how they would combine to form ions. What is the formula for the resulting compound?

6. Ethane is a compound with the formula C_2H_6. What kind of compound is ethane? Describe the formation of ethane from carbon and hydrogen atoms, and draw the electron dot structure of ethane.

7. Classify each of the following compounds as ionic or covalent. Use the periodic table to determine whether each component element of the compound is a metal or nonmetal. Make a general statement about the type of elements that make up covalent and ionic compounds.

   a. hydrogen iodide   **d.** calcium sulfide

   b. strontium oxide   **e.** sulfur dioxide

   c. rubidium chloride

8. You are given two clear, colorless solutions, and you are told that one solution consists of an ionic compound dissolved in water and the other consists of a covalent compound dissolved in water. How could you determine which is an ionic solution and which is a covalent solution?

9. You are given a white crystalline substance to identify. You find that it melts at about 90°C and is insoluble in water. Is this substance more likely to be ionic or covalent? Explain.

Sections 4.1 and 4.2

10. Sucrose, or table sugar, has the formula $C_{12}H_{22}O_{11}$. How are the properties of this compound different from those of its elements? Do you think sucrose is more likely to be an ionic or a covalent compound? Explain.

CHAPTER 5

Section 5.1

1. Write the symbols for two ions and one atom that have the same electron configuration as each of the following ions.

   a. Br^- **b.** Ba^{2+} **c.** Na^+ **d.** P^{3-}

2. Label the following as common names or formal names.

   a. saltpeter

   b. sodium hydrogen sulfate

   c. slaked lime

   d. soda ash

   e. potassium nitrite

3. What is indicated by subscripts in chemical formulas?

4. Of the formulas $MgCl_2$ and Mg_2Cl_4, which is the correct formula for magnesium chloride? Explain your choice.

5. What is oxidation number? What determines an element's oxidation number? Write the most common oxidation numbers for each of the following elements.

   a. I   **e.** Se

   b. P   **f.** Al

   c. Cs   **g.** Cu

   d. Ba   **h.** Pb

6. Write the formula for each of the following binary ionic compounds.

   a. strontium chloride

   b. rubidium oxide

   c. sodium fluoride

   d. magnesium sulfide

7. Write the formula for the compound formed from each of the following pairs of elements.

   a. magnesium and bromine

   b. potassium and sulfur

   c. strontium and oxygen

   d. aluminum and phosphorus

8. Examine the list of common polyatomic ions in **Table 5.2**. Which element appears most often in these ions? Write the symbols of the ions that do not contain this element.

9. How many atoms of each element are present in four formula units of ammonium oxalate?

10. The metals in the following compounds can have various oxidation numbers. Predict the charge on each metal ion, and write the name of each compound.

   a. Au_2S **c.** $Pb(C_2H_3O_2)_4$

   b. FeC_2O_4 **d.** Hg_2SO_4

11. Write the formula for the compound made from each of the following pairs of ions.

   a. potassium and dichromate ions

   b. sodium and nitrite ions

   c. ammonium and hydroxide ions

   d. calcium and phosphate ions

12. Write the formula for each of the following compounds containing polyatomic ions.

   a. ammonium sulfate

   b. barium hydroxide

   c. sodium hydrogen sulfate

   d. calcium acetate

13. Write the names of the following compounds containing chromium.

   a. $CrBr_2$ **c.** CrO_3

   b. $Cr_2(SO_4)_3$ **d.** $CrPO_4$

14. Write the formula for the compound made from each of the following pairs of ions.

   a. lead(II) and sulfite

   b. manganese(III) and fluoride

   c. nickel(II) and cyanide

   d. chromium(III) and acetate

Table 5.2 Common Polyatomic Ions

Name of Ion	Formula	Charge
ammonium	NH_4^+	1+
hydronium	H_3O^+	1+
hydrogen carbonate	HCO_3^-	1-
hydrogen sulfate	HSO_4^-	1-
acetate	$C_2H_3O_2^-$	1-
nitrite	NO_2^-	1-
nitrate	NO_3^-	1-
cyanide	CN^-	1-
hydroxide	OH^-	1-
dihydrogen phosphate	$H_2PO_4^-$	1-
permanganate	MnO_4^-	1-
carbonate	CO_3^{2-}	2-
sulfate	SO_4^{2-}	2-
sulfite	SO_3^{2-}	2-
oxalate	$C_2O_4^{2-}$	2-
monohydrogen phosphate	HPO_4^{2-}	2-
dichromate	$Cr_2O_7^{2-}$	2-
phosphate	PO_4^{3-}	3-

Table 5.4 Names of Common Ions of Selected Transition Elements

Element	Ion	Chemical Name
Chromium	Cr^{2+}	chromium(II)
	Cr^{3+}	chromium(III)
	Cr^{6+}	chromium(VI)
Cobalt	Co^{2+}	cobalt(II)
	Co^{3+}	cobalt(III)
Copper	Cu^{+}	copper(I)
	Cu^{2+}	copper(II)
Gold	Au^{+}	gold(I)
	Au^{3+}	gold(III)
Iron	Fe^{2+}	iron(II)
	Fe^{3+}	iron(III)
Manganese	Mn^{2+}	manganese(II)
	Mn^{3+}	manganese(III)
	Mn^{7+}	manganese(VII)
Mercury	Hg^{+}	mercury(I)
	Hg^{2+}	mercury(II)
Nickel	Ni^{2+}	nickel(II)
	Ni^{3+}	nickel(III)
	Ni^{4+}	nickel(IV)

CrO

Cr_2O_3

CrO_3

15. Write the names of the following compounds that contain transition elements not listed in **Table 5.4.**

    a. $RhCl_3$ **b.** WF_6 **c.** Nb_2O_5 **d.** OsF_8

16. Write the formulas for the following compounds. What do the formulas have in common?

    a. hydrogen iodide

    b. calcium selenide

    c. cobalt(II) oxide

    d. gallium phosphide

    e. barium selenide

17. Compare the properties of hygroscopic substances and deliquescent substances.

18. Write the names of the following hydrates.

    a. $MgSO_3 \cdot 6H_2O$

    b. $Hg(NO_3)_2 \cdot H_2O$

    c. $NaMnO_4 \cdot 3H_2O$

    d. $Ni_3(PO_4)_2 \cdot 7H_2O$

19. Write the formula for each of the following hydrates.

    a. nickel(II) cyanide tetrahydrate

    b. lead(II) acetate trihydrate

    c. strontium oxalate monohydrate

    d. palladium(II) chloride dihydrate

20. What is a desiccant? Describe the properties of a compound that would make a good desiccant.

Section 5.2

21. How do the macroscopic properties of molecular substances reflect their submicroscopic structure?

22. What is a molecular element? Name the seven nonmetals that exist naturally as diatomic molecular elements.

23. How are electrons shared differently in H_2, O_2, and N_2 molecules?

24. What is ozone? How is ozone harmful? How is it helpful?

25. Name the following molecular compounds.

   a. SF_4 **b.** CSe_2 **c.** IF_5 **d.** P_2O_5

26. Write the formulas for the following molecular compounds.

   a. nitrogen trichloride
   b. diiodine pentoxide
   c. diphosphorus triselenide
   d. dichlorine heptoxide

27. Name the following molecular compounds, all of which contain carbon.

   a. C_5H_{12} **b.** SiC **c.** CF_4 **d.** C_9H_{20}

Sections 5.1 and 5.2

28. Write formulas for a nitrogen atom, ion, and molecule.

29. Which of the following compounds are ionic and which are molecular?

   a. dihydrogen sulfide
   b. strontium oxide
   c. lithium carbonate
   d. decane
   e. rubidium hydroxide

30. How is $KC_2H_3O_2$ a substance with two different types of bonding?

 CHAPTER 6

Section 6.1

1. If bread is stored at room temperature for a long period of time, it undergoes chemical changes that make it inedible. How can you tell that these chemical changes have occurred?

2. List one piece of evidence you expect to see that indicates that a chemical reaction is taking place in each of the following situations.

   a. An apple that has been peeled begins to oxidize.
   b. A flashlight is turned on.
   c. An egg is fried.
   d. Acidic and basic substances in an antacid tablet react when the tablet is placed in water.

3. Use the equation below to answer the following questions.

   $$3Zn(s) + 2FeCl_3(aq) \rightarrow 2Fe(s) + 3ZnCl_2(aq)$$

   a. What is the physical state of iron(III) chloride? Of iron metal?
   b. What is the coefficient of zinc?
   c. What is the subscript of chlorine in zinc chloride?

4. Balance the following equations.

   a. $Al(s) + HCl(aq) \rightarrow AlCl_3(aq) + H_2(g)$
   b. $H_2O_2(aq) \rightarrow H_2O(l) + O_2(g)$
   c. $HC_2H_3O_2(aq) + CaCO_3(s) \rightarrow$
   $Ca(C_2H_3O_2)_2(aq) + CO_2(g) + H_2O(l)$
   d. $C_2H_6O(l) + O_2(g) \rightarrow CO_2(g) + H_2O(g)$

5. Write balanced chemical equations for the reactions described.

   a. sulfuric acid + sodium hydroxide $\rightarrow$ sodium sulfate solution + water
   b. pentane liquid + oxygen $\rightarrow$ carbon dioxide + water vapor + energy
   c. iron metal + copper(II) sulfate solution $\rightarrow$ copper metal + iron(II) sulfate solution

6. Magnesium metal will burn in air to form magnesium oxide. In the reaction, two magnesium atoms react with one oxygen molecule to form two formula units of magnesium oxide. If you react 80 billion atoms of magnesium with 30 billion molecules of oxygen, will all the reactants be used up? What do you think the results will be?

Section 6.2

7. Which of the five general types of reactions has only one product? Which of the five types always has oxygen as a reactant?

8. Classify each of the following reactions as one of the five general types.

   a. $Zn(s) + 2AgNO_3(aq) \rightarrow$
      $Zn(NO_3)_2(aq) + 2Ag(s)$

   b. $Fe(s) + S(s) \rightarrow FeS(s)$

   c. $2KClO_3(s) \rightarrow 2KCl(s) + 3O_2(g)$

   d. $CH_4(g) + 2O_2(g) \rightarrow$
      $CO_2(g) + 2H_2O(g) + energy$

   e. $Na_2CO_3(aq) + MgSO_4(aq) \rightarrow$
      $MgCO_3(s) + Na_2SO_4(aq)$

9. Use word equations to describe the chemical equations given, and classify each reaction as one of the five major types.

   a. $C_9H_{20}(l) + 14O_2(g) \rightarrow 9CO_2(g) + 10H_2O(g)$

   b. $H_2SO_4(aq) + 2KOH(aq) \rightarrow$
      $K_2SO_4(aq) + 2H_2O(l)$

   c. $2KNO_3(s) + energy \rightarrow 2KNO_2(s) + O_2(g)$

10. Write a balanced equation for the combustion of ethene gas, C_2H_4. How many oxygen molecules will react with 15 trillion ethene molecules?

Sections 6.1 and 6.2

11. Nitrogen oxyfluoride gas, NOF, is formed by a reaction between nitrogen monoxide and fluorine gases. Write a balanced chemical equation for this reaction, and classify the reaction as one of the five major types.

12. When an aqueous solution of copper(II) sulfate is combined with an aqueous solution of sodium hydroxide, a blue precipitate of copper(II) hydroxide forms.

    a. Write word and balanced chemical equations for this reaction.

    b. Classify this reaction as one of the five major types.

13. The unbalanced equation for the reaction between aqueous solutions of NaOH and $MgSO_4$ is as follows.

    $NaOH(aq) + MgSO_4(aq) \rightarrow$
    $Na_2SO_4(aq) + Mg(OH)_2(s)$

    a. Balance the chemical equation.

    b. Write a word equation for the reaction.

    c. Classify this reaction as one of the five major types.

    d. If you observed this reaction, how would you know that a chemical reaction had occurred?

Section 6.3

14. What is a reversible reaction? How can a chemical equation show that a reaction is reversible?

15. Distinguish between the terms *equilibrium* and *dynamic equilibrium.*

16. Describe Le Chatelier's principle. Why is this principle important to chemical engineers?

17. Will an endothermic reaction that is at equilibrium shift to the left or to the right to readjust after each of the following procedures is followed?

    a. More heat is added.

    b. Heat is removed.

    c. More products are added.

    d. More reactants are added.

18. In a closed container, the reversible reaction represented by the equation $N_2O_4(aq) \rightleftarrows 2NO_2(g)$ has come to equilibrium. The volume of the container is then decreased, causing the pressure in the container to increase. Will the reaction be shifted to the left or the right? Explain.

19. A certain exothermic reaction has a very high activation energy. Do you expect this reaction to take place spontaneously under normal conditions? Explain.

20. Why are many valuable historic documents stored in sealed cases with most of the air removed?

21. Road salt on a car speeds the process of rusting. If a car is coated with road salt, would it be more beneficial to wash the car before a period of cold weather or before a period of warm weather? Explain.

22. Hydrogen gas can be produced by reacting aluminum and sulfuric acid, as shown by the equation for the reaction.

$$2Al(s) + 3H_2SO_4(aq) \rightarrow$$
$$Al_2(SO_4)_3(aq) + 3H_2(g)$$

In a particular reaction, 12 billion molecules of H_2SO_4 were mixed with 6 billion atoms of Al.

   a. Which reactant is limiting?

   b. How many molecules of H_2 are formed when the reaction is complete?

23. Fritz Haber carried out his process for the synthesis of ammonia at a temperature of about 600°C. High temperature increases the rate of reaction. Why didn't Haber use a higher temperature for his process?

24. If a catalyst is added to a reversible endothermic reaction that is at equilibrium, will the reaction shift to the left or to the right?

25. Why are catalysts often used in the form of a powder?

26. What are enzymes? Provide some examples of how your body uses enzymes, and list two common products that contain enzymes.

27. Examine the ingredients of some food products in your home, and list the inhibitors—usually called preservatives—that you find in these products.

28. The graph shown represents the concentrations of three compounds, A, B, and C, as they take part in a reaction that reaches equilibrium.

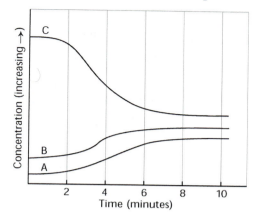

   a. Which compound(s) represent the reactant(s) in this reaction? The product(s)?

   b. How long did it take for the reaction to reach equilibrium?

   c. Explain how the graph will change if more compound C is added one minute after the reaction reaches equilibrium.

Sections 6.1 and 6.3

29. Sulfur dioxide gas and oxygen gas combine in a reversible reaction to form sulfur trioxide gas.

   a. Write a word equation for this reaction.

   b. Write a balanced chemical equation for this reaction.

   c. If more sulfur trioxide gas is added after the reaction reaches equilibrium, will the reaction be shifted to the left or to the right?

30. In one of a series of reactions used to produce nitric acid, ammonia gas and oxygen gas react to form nitrogen monoxide gas and water vapor. Write a balanced equation for this reaction. If 30 trillion ammonia molecules and 35 trillion oxygen molecules are available to react, which is the limiting reactant?

CHAPTER 7

Section 7.1

1. Given the colors yellow, red, and violet, which color has the lowest frequency? The shortest wavelength? The least energy?

2. How many f orbitals can there be in an energy level? What is the maximum number of f electrons in an energy level? What is the lowest energy level that can have f orbitals?

3. What is the Heisenberg uncertainty principle? This principle led to the development of what model to describe electrons in atoms?

4. What is the shape of an s orbital? How do s orbitals at different energy levels differ from one another?

Section 7.2

5. Identify the elements that have the following electron configurations.

   a. $[Ar]3d^{10}4s^24p^6$ d. $1s^22s^1$

   b. $[Ne]3s^23p^1$ e. $[Xe]6s^24f^7$

   c. $[Ar]4s^13d^5$

6. Use the periodic table to help you write electron configurations for the following atoms. Use the appropriate noble gas, inner-core abbreviations.

   a. Br   c. Te   e. Ba

   b. N   d. Ni

7. What are the valence electrons in the electron configuration of germanium, $[Ar]3d^{10}4s^24p^2$?

8. What is the highest occupied sublevel in the structure of each of the following elements?
   a. Ca   b. B   c. In   d. He   e. Bi

9. Refer to the periodic table and compare the similarities and differences in the electron configurations of the following pairs of elements: Ar and Kr, K and Ar, K and Sc.

10. Why do most of the transition elements have multiple oxidation numbers? What other category of elements exhibits multiple oxidation numbers?

CHAPTER 8

Section 8.1

1. From each of the following pairs of atoms, select the one with the larger atomic radius.

   a. Li, Be   c. P, As   e. In, Ba

   b. Ca, Ga   d. Br, O

2. From each of the following pairs of ions, choose the one with the larger ionic radius and explain your choice.

   a. Se^{2-}, S^{2-} d. B^{3+}, N^{3-}

   b. Te^{2-}, Cs^+ e. O^{2-}, P^{3-}

   c. Mg^{2+}, Al^{3+}

3. What is the electron configuration of the potassium ion? What noble gas has that configuration? Name two other ions with the same configuration.

4. Your friend says that the most active metal has a large atomic radius and a large number of valence electrons. Assess the accuracy of your friend's statement.

5. Compare and contrast the properties of the alkali metals and the alkaline earth metals.

6. Both potassium and strontium react with water. Predict the products of the reactions of potassium and strontium with water. Write balanced equations for both reactions.

7. Explain how and why the crystal structures of the ionic compounds CsCl and NaCl differ.

8. Boron is classified as a metalloid. What does this mean in terms of the way that boron reacts with other elements?

9. Describe some uses of the Group 15 elements.

Sections 8.1 and 8.2

10. Give one reason why each of the following metals is essential to human life.

    a. potassium   **c.** calcium

    b. magnesium   **d.** iron

CHAPTER 9

Section 9.1

1. In general, how do electronegativity values change as you move down a column of the periodic table? Across a row of the table? Explain these trends.

2. Using only a periodic table, rank these atoms from the least to the most electronegative. As, Ba, N, Mg, Cs, O

3. Classify the following bonds as ionic, covalent, or polar covalent.

    a. C—O   **c.** Ca—O   **e.** Mn—O

    b. Cu—Cl   **d.** Si—H

4. Using only a periodic table, rank these bonds from the most ionic to the least ionic.

    a. Li—F   **b.** Li—Br   **c.** K—F   **d.** Li—Cl

5. Do you expect KBr or KF to have a higher melting point? Explain.

6. Rank these bonds from the least to the most polar.

    a. S—O   **b.** C—Cl   **c.** Si—Cl   **d.** H—Br

7. What factors determine whether a molecule is polar or nonpolar?

8. In metallic bonding, why are the valence electrons of the metal sometimes described as a "sea of electrons"?

9. Nitric oxide, NO, is a colorless gas used in the manufacture of nitric acid. Is nitric oxide a polar or nonpolar molecule? Explain.

Section 9.2

10. Draw the electron dot diagrams for each of the following molecules.

    a. CS_2 **b.** HI   **c.** CH_3Cl **d.** AsH_3

11. Describe the shape of each molecule in Question 10.

12. Dibromomethane, CH_2Br_2, is a molecule similar in structure to methane. Is dibromomethane a polar or nonpolar molecule? Explain.

13. Ethanol, carbon dioxide, and vitamin C are all covalent compounds. How do these compounds demonstrate the variety of types of covalent compounds?

14. Carbon tetrafluoride (CF_4), also called Freon-14, is used as a low temperature refrigerant and as a gaseous insulator. Are the bonds in CF_4 polar or nonpolar? Is CF_4 a polar molecule? Explain.

15. Compare the molecules phosphorus trichloride (PCl_3) and dichloromonoxide (Cl_2O). How many pairs of electrons surround the central atom? How many of these pairs are bonding? Nonbonding? What are the shapes of the molecules?

16. Tetrachloroethylene, C_2Cl_4, is a derivative of ethene in which the hydrogen atoms have been replaced by chlorine atoms. Draw the electron dot structure of C_2Cl_4. Is the molecule polar? Explain.

17. Formaldehyde, CH_2O, is a colorless gas that, when dissolved in water, is used in embalming fluids. Draw its electron dot structure and describe its geometry.

18. Diethyl ether, $CH_3CH_2OCH_2CH_3$, can be viewed as a derivative of water, in which the hydrogens in water are replaced by CH_3CH_2 (ethyl) groups. Is diethyl ether a polar molecule? Explain.

19. In paper chromatography, what is the stationary phase? What is the mobile phase? Compare the distance migrated for a component with a strong attraction to the paper to the distance migrated for a component with a weak attraction to the paper.

20. Predict the relative boiling points of ammonia(NH_3), phosphine (PH_3), and arsine (AsH_3). Explain your prediction.

CHAPTER 10

Section 10.1

1. Why does a pollen grain suspended in a water droplet trace a random and erratic path?

2. Contrast the behavior of an air hockey puck when it collides with the wall of the game board to the behavior of an ideal gas particle when it collides with the wall of its container.

3. What is the numerical value of atmospheric pressure at sea level? Do humans notice the pressure of the atmosphere? Explain.

4. Compare solids and liquids in terms of particle spacing and particle motion.

Section 10.2

5. Write the Fahrenheit, Celsius, and Kelvin temperatures that correspond to

   a. the freezing point of water.

   b. the boiling point of water.

   c. absolute zero.

6. Rank the following temperature readings in increasing order.

   $110°C$, $212 K$, $212°F$, $273°C$, $273 K$

7. Complete the following table.

Temperature	Celsius, °C	Kelvin, K
Boiling point of helium, He	-268.94	
Hot summer day		305
Melting point of zinc	419	
Boiling point of butane, C_4H_{10}		272.7

8. The boiling points of five gases are provided in either Celsius or Kelvin temperatures. List the gases in order from the one with the lowest boiling point to the one with the highest boiling point.

Chlorine(Cl_2)	239 K
Krypton (Kr)	$-153°C$
Dimethyl ether (C_2H_6O)	$-24°C$
Dihydrogen sulfide (H_2S)	213 K
Sulfur dioxide (SO_2)	$-10°C$

9. Suppose that the outside temperature increases by 1.00 Celsius degree. What is the temperature increase in Fahrenheit degrees? In kelvins?

10. The particles of which of the following gases have the highest average speed? The lowest average speed?

   a. argon at 20°C   **c.** nitrogen at 20°C

   b. nitrogen at 100°C   **d.** helium at 100°C

11. Which gas particles in Question 10 have the highest average kinetic energy? Which have the lowest average kinetic energy?

12. Dihydrogen sulfide (H_2S) and sulfur dioxide (SO_2) are both colorless gases with unpleasant odors. Which of these gases has a higher rate of diffusion in air at the same temperature? Explain your answer.

13. Explain why liquids become cooler as they evaporate.

14. Compare and contrast sublimation and evaporation.

15. Can a liquid in an open container reach equilibrium with its vapor? Explain.

16. Describe how boiling point is affected by increasing pressure.

17. Predict whether the boiling point of water is greater or less than 100°C at your school.

18. Explain the relationship among the terms *freezing point*, *heat of fusion*, and *crystal lattice*.

19. What is the ratio of the amount of energy needed to boil 1 kg of water at 100°C to the amount of energy needed to melt 1 kg of ice at 0°C?

20. The compound cyclohexane, C_6H_{12}, boils at 80.7°C and melts at 6.5°C. A few grams of cyclohexane are cooled from +120°C to −20°C. Graph the cooling curve for cyclohexane. Show time on the horizontal axis and temperature on the vertical axis.

CHAPTER 11

Section 11.1

1. Using **Table 11.1** and the equation 1.00 in = 25.4 mm to convert the following pressure measurements.

   a. 147 mm Hg to psi

   b. 232 psi to kPa

   c. 67.2 kPa to mm Hg

   d. 214 in. Hg to atm

Table 11.1   Equivalent Pressures

1.00 atm	760 mm Hg	14.7 psi	101.3 kPa

2. Convert the pressure measurement 3.50 atm to the following units.

   a. mm Hg   **c.** kPa

   b. psi   **d.** in. Hg

3. What is the difference between gauge pressure and absolute pressure?

4. A tire gauge at a gas station indicates a pressure of 29.0 psi in your tires. What is the absolute pressure in your tires in inches of mercury?

5. Suppose that a certain planet's atmosphere consists primarily of CO_2 gas. Compare the atmospheric pressure on the planet's surface to the pressure that would be exerted if the atmosphere were the same thickness, but consisted primarily of methane gas.

6. Propose a reason why Torricelli used mercury in his barometer rather than some other liquid such as water.

Section 11.2

7. What factors affect gas volume?

8. A cylinder containing 48 g of nitrogen gas at room temperature is placed on a scale. The valve is opened and 12 g of gas are allowed to escape. Assuming that the temperature of the cylinder remains constant, how does the pressure change?

9. A cylinder contains 22 g of air. If 77 g of air are pumped into the cylinder at constant temperature, how does the pressure in the cylinder change?

10. A cylinder contains 36.5 g of argon gas at a pressure of 8.20 atm. The valve is opened and gas is allowed to escape until the pressure is reduced to 4.75 atm at constant temperature. How many grams of argon escaped?

11. If the gas pressure in an aerosol can is 182 kPa at 20.0°C, what is the pressure inside the can if it is heated to 251°C?

12. A 5.3-L sample of argon is at standard atmospheric pressure and 294°C. The sample is cooled in dry ice to −79°C at constant volume. What is its new pressure in mm Hg?

13. A tank for compressed gas can safely withstand a maximum pressure of 955 kPa. The pressure in the tank is 689 kPa at a temperature of 22°C. What is the highest temperature the tank can safely withstand?

14. You use a pressure gauge to measure the air pressure in your bicycle tires on a cold morning when the temperature is −5°C. The gauge reads 53 psi. The next afternoon, the temperature has warmed up to 11°C. If you measure the pressure in your tires again, what would you expect the reading on the pressure gauge to be? Assume the volume of air in the tires is constant.

15. A cylinder contains 98 g of air at 295 K. The pressure in the cylinder is 174 psi. If 32 g of air are allowed to escape and the cylinder is heated to 335 K, what is the new pressure?

16. A cylinder of compressed gas has a volume of 14.5 L and a pressure of 769 kPa. What volume would the gas occupy if it were allowed to escape into a balloon at a pressure of 117 kPa? Assume constant temperature.

17. At a sewage-treatment plant, bacterial cultures produce 2400 L of methane gas per day at 1.0 atm pressure. If one day's production of methane is stored in a 310-L tank, what is the pressure in the tank?

18. A 0.400-L balloon is filled with air at 1.10 atm. If the balloon is squeezed into a 250-mL beaker and doesn't burst, what is the pressure of the air?

19. A sample of argon gas is compressed, causing its pressure to increase by 25 percent at constant temperature. What is the percentage change in the volume of the sample?

20. Compare qualitatively the volume of a sample of air at STP to the volume of the same sample of air under normal conditions of temperature and pressure in your classroom.

21. A balloon filled with 2.5 L of air at 23°C and 1 atm. It is then placed outdoors on a cold winter day when the temperature is −12°C. If the pressure is constant, what is the new volume of the balloon?

22. The volume of a sample of nitrogen is 75 mL at 25°C and 1 atm. What is its volume at STP?

23. At atmospheric pressure, a balloon contains 1.50 L of helium gas. How would the volume change if the Kelvin temperature were only 60 percent of its original value?

24. The volume of a sample of helium is 2.45 L at −225°C and 1 atm. Predict the volume of the sample at +225°C and 1 atm.

25. A cylinder contains 5.70 L of a gas at a temperature of 24°C. The cylinder is heated, and a piston moves in the cylinder so that constant pressure is maintained. If the final volume of the gas in the cylinder is 6.55 L, what is the final temperature?

26. A 3.50-L sample of nitrogen at 10°C is heated at constant pressure until it occupies a volume of 5.10 L. What is the new temperature of the sample?

27. A sample of argon gas has a volume of 425 mL at 463 K. The sample is cooled at constant pressure to a volume of 315 mL. What is the new temperature?

28. Use Charles's law to predict the change in density of a sample of air if its temperature is increased at constant pressure.

29. Suppose you have a 1-L sample of neon gas and a 1-L sample of nitrogen gas; both samples are at STP. Compare the numbers of gas particles in each sample. Compare the masses of the samples.

30. An 825-mL sample of oxygen is collected at 101 kPa and 315 K. If the pressure increases to 135 kPa and the temperature drops to 275 K, what volume will the oxygen occupy?

31. At 37.4 kPa and −35°C, the volume of a sample of krypton gas is 346 mL. What is the volume at STP?

32. A 44-g sample of CO_2 has a volume of 22.4 L at STP. The sample is heated to 71°C and compressed to a volume of 18.0 L. What is the resulting pressure?

33. A gas has a volume of 2940 mL at 58°C and 95.9 kPa. What pressure will cause the gas to have a volume of 3210 mL at 25°C?

34. A balloon is filled with 4.50 L of helium at 15°C and 763 mm Hg. The balloon is released and it rises through the atmosphere. When it reaches an altitude of 2500 m, the temperature is 0°C and the pressure has dropped to 542 mm Hg. What is the volume of the balloon?

35. Compare the number of particles in 1.0 L of nitrogen gas at room temperature and 1.0 atm pressure to the number of particles in 2.0 L of oxygen gas at room temperature and 3.0 atm pressure.

36. How many liters oxygen will be needed to react completely with 15 L of carbon monoxide at the same temperature and pressure to form carbon dioxide?

37. How many liters of nitrogen gas and hydrogen gas must combine to produce 28.6 L of ammonia if all gases are at the same temperature and pressure?

38. A cylinder contains 14.2 g of nitrogen at 23°C and 5030 mm Hg. The cylinder is heated to 45°C, and nitrogen gas is released until the pressure is lowered to 2250 mm Hg. How many grams of nitrogen are left in the cylinder?

39. A 75.0-mL sample of air is at standard pressure and $-25°C$. The air is compressed to a volume of 45.0 mL, and the temperature is adjusted until the pressure of the air doubles to 2.00 atm. What is the final temperature?

40. A gas has a volume of 146 mL at STP. What Celsius temperature will cause the gas to have a volume of 217 mL at a pressure of 615 mm Hg?

CHAPTER 12

Section 12.1

1. Your grandmother gives you a bag of pennies she has saved. If the pennies have a mass of 4.24 kg, and 50 pennies have a mass of 144 g, how many pennies did you receive?

2. What are the molar masses of the following substances?

   a. palladium   **c.** calcium hydroxide

   b. hydrogen   **d.** diphosphorus pentoxide

3. Determine the number of atoms in each sample below.

   a. 12.7 g silver, Ag

   b. 56.1 g aluminum, Al

   c. 162 g calcium, Ca

4. Without calculating, decide whether 10.0 g of zinc or 10.0 g of silicon represents the greater number of atoms. Verify your answer by calculating.

5. Determine the number of moles in each sample below.

   a. 7.62 g cesium chloride, CsCl

   b. 42.5 g propanol, C_3H_8O

   c. 694 g ammonium dichromate, $(NH_4)_2Cr_2O_7$

6. Determine the mass of the following molar quantities.

   a. 0.172 mol ozone, O_3

   b. 2.50 mol heptane, C_7H_{16}

   c. 0.661 mol iron(II) phosphate, $Fe_3(PO_4)_2$

7. Which has the largest mass?

   a. ten atoms of carbon

   b. three molecules of chlorine gas

   c. one molecule of fructose, $C_6H_{12}O_6$

8. Which has the largest mass?

   a. 10.00 mol of carbon

   b. 3.00 mol of chlorine gas

   c. 1.000 mol of fructose, $C_6H_{12}O_6$

9. Determine the number of molecules in 0.127 mol of formic acid, CH_2O_2. What is the mass of this quantity of formic acid?

10. Determine the number of molecules or formula units in each sample below. Identify each as a formula unit or a molecule.

   a. 85.3 g water, H_2O

   b. 100.0 g chlorine, Cl_2

   c. 0.453 g potassium chloride, KCl

   d. 14.6 g acetaminophen, $C_8H_9NO_2$

11. The average atomic mass of oxygen is 4.0 times greater than the average atomic mass of helium. Would you expect the molar mass of oxygen gas to be 4.0 times greater than the molar mass of helium gas? Explain.

12. Vitamin B_2, also called riboflavin, has the chemical formula $C_{17}H_{20}N_4O_6$. What is the molecular mass of vitamin B_2? What is its molar mass?

13. The molecular formula of cholesterol is $C_{27}H_{46}O$. What is the molar mass of cholesterol? What is the mass in grams of a single molecule of cholesterol?

14. What mass of aluminum contains the same number of atoms as 125 g of silver?

Section 12.2

15. The combustion of methanol produces carbon dioxide gas and water vapor.
$$2CH_3OH(l) + 3O_2(g) \rightarrow 2CO_2(g) + 4H_2O(g)$$
What mass of water vapor forms when 52.4 g of methanol burn?

16. Disulfur dichloride is prepared by passing chlorine gas into molten sulfur.
$Cl_2(g) + 2S(l) \rightarrow S_2Cl_2(l)$
How many grams of chlorine will react to produce 20.0 g of S_2Cl_2?

17. Using the reaction in Question 16, how many grams of sulfur will react to produce 20.0 g of S_2Cl_2?

18. Calculate the mass of precipitate that forms when 250 mL of an aqueous solution containing 35.0 g of lead (II) nitrate reacts with excess sodium iodide solution by the following reaction.
$Pb(NO_3)_2(aq) + 2NaI(aq) \rightarrow$
$2NaNO_3(aq) + PbI_2(s)$

19. What mass of carbon must burn to produce 4.56 L of CO_2 gas at STP? The reaction is
$C(s) + O_2(g) \rightarrow CO_2(g)$

20. What volume of hydrogen gas can be produced by reacting 5.40 g of zinc in excess hydrochloric acid at 25.0°C and 105 kPa? The reaction is
$Zn(s) + 2HCl(aq) \rightarrow ZnCl_2(aq) + H_2(g)$.

21. Explain how the ideal gas law confirms the laws of Boyle and Charles.

22. How many moles of argon are contained in a 2.50-L canister at 125 kPa and 15.0°C?

23. What is the volume of 0.300 mol of oxygen at 46.5 kPa and 215°C?

24. What is the pressure of a 2.00-mol sample of nitrogen gas that occupies 10.5 L at −25°C?

25. Determine the value of the ideal gas constant R in units of atm·L/mol·K.

26. Using the value of the ideal gas constant determined in Question 25, calculate the number of moles in a 7.56-L sample of neon gas at −45°C and 4.12 atm. How many grams of neon does the sample contain?

27. Dodecane, $C_{12}H_{26}$, is one of the components of kerosene. Write the balanced chemical equation for the combustion of dodecane to form carbon dioxide gas and water vapor. If 3.00 moles of dodecane burn, how many moles of oxygen are consumed? How many moles of carbon dioxide and water vapor are produced?

28. If 60.0 g of dodecane burn as in Question 27, how many grams of water vapor are produced? How many grams of carbon dioxide?

29. Green plants produce oxygen by the reaction below. If a rosebush produces 50.5 L of oxygen at 27.0°C and 101.3 kPa, how many grams of glucose, $C_6H_{12}O_6$, did the plant also produce?
$6CO_2(g) + 6H_2O(l) \xrightarrow{\text{sunlight}} C_6H_{12}O_6(aq) + 6O_2(g)$

30. Calculate the mass of each product formed when 10.5 g of sodium hydrogen carbonate, $NaHCO_3$, react with excess hydrochloric acid.

31. When a solution of 25.0 g of silver nitrate in 100 g of water is mixed with a solution of 10.0 g of magnesium chloride in 100 g of water, a double displacement reaction occurs. The balanced chemical equation is shown below. Which is the limiting reactant?
$2AgNO_3(aq) + MgCl_2(aq) \rightarrow$
$Mg(NO_3)_2(aq) + 2AgCl(s)$

32. What is the empirical formula of each of the following compounds?

   a. $K_2C_2O_4$ d. $Pb_3(PO_4)_2$
   b. $Na_2S_2O_3$ e. $C_{15}H_{21}N_3O_{15}$
   c. $C_{20}H_{20}O_4$ f. $C_6H_{12}O_7$

33. What is the molecular formula of each of the following compounds?

   a. empirical formula: C_4H_4O;
      molar mass: 136 g/mol
   b. empirical formula: CH_2;
      molar mass: 154 g/mol
   c. empirical formula: As_2S_5;
      molar mass: 310 g/mol

34. Without calculating, which of the following compounds has the greater percentage of nitrogen: $Ca(NO_3)_2$ or $Ca(NO_2)_2$? How do you know?

35. An oxide of chromium is 68.4 percent chromium by mass. The molar mass of the oxide is 152 g/mol. What is the formula of the compound?

36. Determine the percent oxygen in each of the following oxides of nitrogen.

   a. nitrogen monoxide, NO
   b. dinitrogen monoxide, N_2O
   c. dinitrogen pentoxide, N_2O_5

37. A mixture of sodium chloride and potassium chloride is analyzed and is found to contain 22 percent potassium. What percent of the mixture is sodium chloride?

38. Vanillin is a compound that is used as a flavoring agent in many food products. Vanillin is 63.2 percent C, 5.3 percent H, and 31.5 percent O. The molar mass is approximately 152 g/mol. What is the molecular formula of vanillin?

39. What is the formula for a hydrate that consists of 80.15 percent $ZnSO_3$ and 19.85 percent H_2O?

40. A 15.0-g sample of sodium carbonate reacts with excess sulfuric acid to form sodium sulfate by the reaction below. When the reaction is complete, 16.9 g of sodium sulfate are recovered. What is the percent yield?
$Na_2CO_3(s) + H_2SO_4(aq) \rightarrow$
$Na_2SO_4(aq) + CO_2(g) + H_2O(l)$

CHAPTER 13

Section 13.1

1. Give four examples of molecular compounds, other than water, that will form hydrogen bonds.

2. Use **Figure 13.6** to answer the following questions.

   a. What is the density of water at 10.0°C?

   b. At what temperature is the density of water 0.9998 g/mL?

   c. The density of water at 1.0°C is 0.9999 g/mL. At what other temperature does water have the same density?

3. The surface tension of water is about three times greater than that of ethanol. How would a drop of water placed on a flat tabletop behave differently from a drop of ethanol placed on the same surface?

4. Suppose you have a liquid in which both the interparticle attractive forces and the attractive forces between the liquid particles and silicon dioxide are moderately weak. You pour some of the liquid into a 25-mL glass graduated cylinder. How do you expect the liquid surface to appear?

5. Suppose you have 150 mL of water at 20°C in a 250-mL beaker, and 150 g of ethanol at 20°C in an identical beaker. You heat both liquids to 70°C. Which liquid absorbs more heat?

6. Suppose the specific heat of water were 2.0 J/g·°C. How would the climate on Earth be different?

Section 13.2

7. Write equations for the dissociation of the following ionic compounds when they dissolve in water.

   a. $ZnCl_2$ c. $Mg(NO_3)_2$
   b. Rb_2CO_3 d. $(NH_4)_2SO_4$

8. When predicting solubility, scientists often use the phrase *like dissolves like*. Explain how water, a covalent compound, can be "like" an ionic compound.

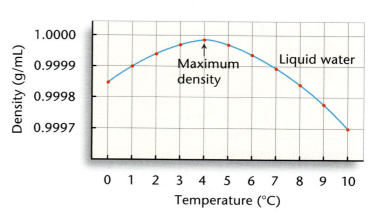

Figure 13.6

Density of Water at Various Temperatures
Water achieves its minimum volume, and therefore its maximum density, at 3.98°C.

9. Do you expect that the liquid compound hexane (C_6H_{14}) will mix with water to form a solution? Explain.

10. Compare and contrast the structures of detergent molecules and soap molecules. Which of the two is more effective in hard water? Explain.

11. Why don't chemists use the words *strong* and *weak* to describe solution concentrations? What terms do they use?

12. You dissolve 5.0 g of $KClO_3$ in 50.0 mL of water at 40°C, then you slowly cool the solution to 20°C. There is no visible change in the solution. Use **Figure 13.20** to determine whether the solution at 20°C is saturated, unsaturated, or supersaturated.

13. Use **Figure 13.20** to compare the solubilities in water of potassium chlorate and cerium(III) sulfate over the temperature range from 10°C to 30°C.

14. Is the process of dissolving in water exothermic or endothermic

    a. for most solid solutes?

    b. for a solute that is used in a hot pack?

15. How would you prepare 3.00 L of a 0.500M solution of calcium nitrate, $Ca(NO_3)_2$?

16. How would you prepare 1.50 L of a 1.75M solution of zinc chloride, $ZnCl_2$?

17. What mass of ammonium sulfate, $(NH_4)_2SO_4$, must be dissolved to make 720.0 mL of a 0.200M solution?

18. What mass of acetone, C_3H_6O, must be dissolved to make 2.50 L of a 1.15M solution?

19. What is the molarity of a solution that contains 2.50 g of potassium carbonate, K_2CO_3, in 1.42 L of solution?

20. Calculate the molarity of a solution that contains 185 g of methanol, CH_3OH, in 1150 mL of solution.

21. If you add 170.0 g of $NaNO_3$ to 1.00 L of water, have you created a 2.00M $NaNO_3$ solution? Explain.

Figure 13.20
Solubility Versus Temperature
The amount of solute required to achieve a saturated solution in water depends upon the temperature, as this graph shows. Most solutes increase in solubility as temperature increases.

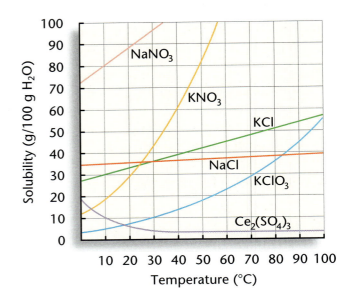

22. For pure water, there is a 100°C difference between the boiling point and the freezing point. Is the corresponding temperature difference for a 1.5M sucrose solution greater or less than 100°C? Explain.

23. Rank these solutions from the one with the lowest to the one with the highest freezing point, and explain your answer.
    a. 1.2 mol of $ZnSO_4$ in 0.900 L of solution
    b. 0.40 mol of KCl in 550 mL of solution
    c. 3.4 mol of $NaNO_3$ in 1.80 L of solution

24. You have two aqueous solutions; one contains 151.6 g of KNO_3 dissolved in 1.00 L of solution, and the other contains 1026 g of sucrose dissolved in 2.00 L of solution. Calculate the molarity of each solution, and determine which solution has a higher boiling point.

25. Ethylenediamine is a molecule with the formula $C_2H_4(NH_2)_2$. Each molecule has two NH_2 groups. Ethylamine, $C_2H_5NH_2$, has one NH_2 group. How would you expect the boiling points of ethylamine and ethylenediamine to compare?

26. A selectively permeable membrane separates two aqueous solutions of sodium chloride. On the left side of the membrane is a solution composed of 74 g of NaCl dissolved in 420 g of water. On the right side of the membrane is a solution composed of 27 g of NaCl dissolved in 125 g of water. In which direction is the net solvent flow?

27. Describe two factors that affect the solubility of a gas in water.

28. Explain the cause of the dangerous condition known as the bends, which is sometimes experienced by divers.

29. What is a liquid aerosol? A foam? Give two examples of each.

30. Compare and contrast colloids and true solutions.

CHAPTER 14

Section 14.1

1. Write the names and formulas of the acids and bases that were among the top ten industrial chemicals produced in the United States in 1994.

2. Write a chemical equation to show how aqueous perchloric acid fits the definition of an acid.

3. Write the balanced equation for the reaction of sulfuric acid with aluminum.

4. Write the formula for each of the following acids and identify each as a monoprotic, a diprotic, or triprotic acid.

   a. nitric acid

   b. hydrobromic acid

   c. citric acid

   d. carbonic acid

   e. benzoic acid

5. Identify the first compound in the following reactions as an acid or a base.

   a. $HBrO + H_2O \rightarrow H_3O^+ + BrO^-$

   b. $N_2H_4 + H_2O \rightarrow N_2H_5^+ + OH^-$

   c. $C_{10}H_{14}N_2 + H_2O \rightarrow C_{10}H_{14}N_2H^+ + OH^-$

   d. $C_6H_5OH + H_2O \rightarrow H_3O^+ + C_6H_5O^-$

6. Classify each of the following as an acid, a base, or neither when mixed with water.

   a. $HC_3H_5O_2$ **c.** HI   **e.** $Fe(OH)_3$

   b. Li_2O **d.** C_2H_6

7. Write the balanced equation for the reaction of calcium hydroxide with formic acid.

8. You want to prepare barium nitrate by an acid-base reaction. Write a reaction that will do this.

9. Consider the oxides SrO and SO_2. For each oxide, tell whether it is an acidic anhydride or a basic anhydride. Write an equation for each to demonstrate its acid-base chemistry.

Section 14.2

10. Give the formula for each of the following hydroxides, and identify each as a strong base or a weak base.

   a. potassium hydroxide

   b. aluminum hydroxide

   c. strontium hydroxide

   d. iron(III) hydroxide

   e. rubidium hydroxide

11. Hydrogen cyanide, HCN, is an extremely poisonous liquid that reacts very slightly with water to form relatively few hydronium ions and cyanide, CN^-, ions. Classify HCN as a strong acid, a weak acid, a strong base, or a weak base.

12. You test several solutions and find that they have pHs of 12.2, 3.5, 8.0, 5.7, 1.2, and 10.0. Which solution has the highest concentration of hydronium ions? Of hydroxide ions? Which solution is the closest to being neutral?

13. Find the pH values of the solutions with the following hydronium ion concentrations.

    a. $10^{-9}M$ **b.** $1M$ **c.** $10^{-3}M$

14. Find the pH values of the solutions with the following hydroxide ion concentrations.

    a. $10^{-6}M$ **b.** $10^{-7}M$ **c.** $10^{-14}M$

15. A solution of sodium carbonate is tested and found to have a pH of 11. Compare the hydronium ion and hydroxide ion concentrations to those of a neutral solution.

16. In an aqueous solution of formic acid, compare the concentrations of hydroxide ions, hydronium ions, formate ions (CHO_2^-), and formic acid molecules.

17. Estimate the molarity of NH_3, NH_4^+, and OH^- in $0.25M\ NH_3$.

18. What are the molarities of HNO_3, H_3O^+, NO_3^-, and OH^- in a $0.010M$ solution of HNO_3? What is the pH of the solution?

19. What are the molarities of NaOH, Na^+, OH^-, and H_3O^+ in a $1.0M$ solution of NaOH? What is the pH of the solution?

Sections 14.1 and 14.2

20. Which of the following solutions is the best conductor of electricity? Which is the weakest conductor?

    a. $1.0M\ HC_2H_3O_2$
    b. $0.1M\ HC_2H_3O_2$
    c. $0.5M\ H_2SO_4$

CHAPTER 15

Section 15.1

Write overall, ionic, and net ionic equations for each reaction in Questions 1 through 6.

1. nitric acid, HNO_3, and magnesium hydroxide, $Mg(OH)_2$

2. formic acid, $HCHO_2$, and lithium hydroxide, LiOH

3. sulfuric acid, H_2SO_4, and sodium hydroxide, NaOH

4. hydrobromic acid, HBr, and ammonia, NH_3

5. lactic acid, $HC_3H_5O_3$, and strontium hydroxide, $Sr(OH)_2$

6. hydrochloric acid, HCl, and iron(III) hydroxide, $Fe(OH)_3$

7. For each of the reactions in Questions 1 through 6, predict whether the pH of the product solution is acidic, basic, or neutral, and explain your prediction.

8. Identify the spectator ions in each reaction in Questions 1 through 6.

9. A reaction occurs in which perchloric acid neutralizes an unknown strong base. Write the net ionic equation for the reaction.

10. Benzoic acid, $HC_7H_5O_2$, is found in small amounts in many types of berries. Write the overall, ionic, and net ionic equations for the reaction of benzoic acid with lithium hydroxide. Will the pH of the product solution be greater than 7, exactly 7, or less than 7? Explain.

11. When each of the following salts is dissolved in water, will its solution be acidic, basic, or neutral? Explain.

    a. NaBr   **b.** $NaC_2H_3O_2$ **c.** NH_4NO_3

12. In terms of hydrogen ion transfer, how are acids and bases defined?

13. Identify the acid and the base in each of the following reactions.

    a. $HC_2H_3O_2(aq) + NH_3(aq) \rightarrow$
        $NH_4^+(aq) + C_2H_3O_2^-(aq)$

    b. $HNO_2(aq) + H_2O(l) \rightarrow$
        $H_3O^+(aq) + NO_2^-(aq)$

    c. $H_2BO_3^-(aq) + H_2O(l) \rightarrow$
        $H_3BO_3(aq) + OH^-(aq)$

14. An amphoteric substance is one that may act as either an acid or a base. The hydrogen sulfite ion is amphoteric. Write reactions that demonstrate this property of HSO_3^-.

15. Provide an example of an acid-base reaction that does not require the presence of water. Explain how you know which reactant is the acid and which is the base.

16. Which type of acid-base reaction does not always go to completion? Explain why.

17. Complete and balance the overall equations for the following acid-base reactions.

    a. $HNO_2(aq) + LiOH(aq) \rightarrow$
    b. $H_2SO_4(aq) \rightarrow Al(OH)_3(s) \rightarrow$
    c. $H_3C_6H_5O_7(aq) + Mg(OH)_2(aq) \rightarrow$
    d. $HC_7H_5O_2(aq) + NH_3(aq) \rightarrow$

18. For each reaction in Question 17, identify the type of acid-base reaction represented by the equation.

19. Write an overall equation for the acid-base reaction that would be required to produce each of the following salts.

    a. $Mg(NO_3)_2$ **c.** K_2SO_4
    b. NH_4I **d.** $Sr(C_2H_3O_2)_2$

Section 15.2

20. If the OH^- concentration in blood increases, what then happens to the concentrations of H_2CO_3, HCO_3^-, and OH^-?

21. If a person's blood pH becomes too high, what can he or she do to cause the pH of the blood to drop to its normal level?

22. Consider an antacid that contains milk of magnesia. Write the overall equation that shows how this antacid reduces the acidity of stomach acid.

23. What is meant by the endpoint of a titration? At the endpoint, what happens to the pH of the solution that is being titrated?

24. How does the endpoint pH of a weak acid-strong base titration compare with that of a weak base-strong acid titration?

25. A $0.200M$ NaOH solution was used to titrate an HCl solution of unknown concentration. At the endpoint, 27.8 mL of NaOH solution had neutralized 10.0 mL of HCl. What is the molarity of of the HCl solution?

26. A student finds that 42.7 mL of $0.500M$ sodium hydroxide are required to neutralize 30.0 mL of a sulfuric acid solution. What is the molarity of the sulfuric acid?

27. A student neutralizes 20.0 L of a potassium hydroxide solution with 11.6 mL of $0.500M$ HCl. What is the molarity of the potassium hydroxide?

28. An NaOH solution of unknown concentration was used to titrate 30.0 mL of a $0.100M$ solution of citric acid, $H_3C_6H_5O_7$. If 16.4 mL of NaOH are used to reach the endpoint, what is the concentration of the NaOH solution?

29. A 35.0-mL sample of a solution of formic acid, $HCHO_2$, is titrated to the endpoint with 68.3 mL of $0.100M$ $Ca(OH)_2$. What is the molarity of the formic acid?

30. A student finds that 31.5 mL of a $0.175M$ solution of LiOH was required to titrate 15.0 mL of a phosphoric acid solution to the endpoint. What is the molarity of the H_3PO_4?

31. How many milliliters of a $0.120M$ $Ca(OH)_2$ are needed to neutralize 40.0 mL of $0.185M$ $HClO_4$?

32. What volume of $0.100M$ HNO_3 is needed to neutralize 56.0 mL of $6.00M$ KOH?

33. A 14.5-mL sample of an unknown triprotic acid is titrated to the endpoint with 35.2 mL of 0.0800M Sr(OH)$_2$. What is the molarity of the acid solution?

34. The following are the endpoint pHs for three titrations. From each endpoint pH, indicate whether the titration involves a weak acid-strong base, a weak base-strong acid, or a strong acid-strong base reaction. Use **Figure 15.17** to select the best indicator for each reaction.

a. pH = 9.38 **b.** pH = 7.00 **c.** pH = 5.32

35. A 25.0-mL sample of an H$_2$SO$_3$ solution of unknown molarity is titrated to the endpoint with 76.2 mL of 0.150M KOH. What is the molarity of the H$_2$SO$_3$? Use **Figure 15.17** to select the best indicator for the reaction.

Figure 15.17

Indicators and Titration

The graphs show some of the more popular acid-base indicators used in the chemistry laboratory and the pH at which they change color.

The reaction between a strong acid and a strong base results in a completely neutral solution. Bromothymol blue is an effective indicator for such reactions because it changes color at a pH of 7. ▶

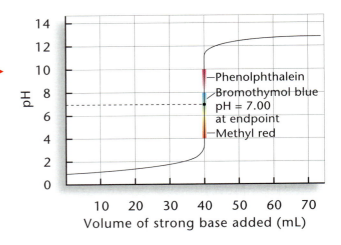

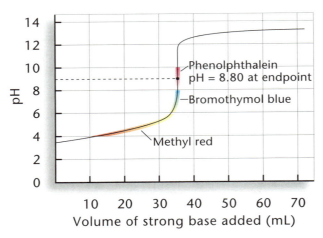

▲ Because the reaction between a weak acid and a strong base results in a slightly basic solution, the endpoint pH for a weak acid-strong base titration is greater than 7. For such a titration, phenolphthalein changes color at the endpoint.

The titration of a weak base with a strong acid has an endpoint pH that is less than 7. For this titration, methyl red is a good indicator because its pH range matches closely the endpoint pH. ▶

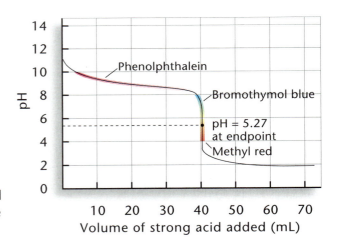

36. An antacid tablet containing $KHCO_3$ is titrated with $0.100M$ HCl. If 0.300 g of the tablet requires 26.5 mL of HCl to reach the endpoint, what is the mass percent of $KHCO_3$ in the tablet?

37. Vitamin C is also known as ascorbic acid, $HC_6H_7O_6$. A solution made by dissolving a tablet containing 0.500 g of vitamin C in 100 mL of distilled water is titrated to the endpoint with $0.125M$ NaOH. Assuming that vitamin C is the only acid present in the tablet, how many milliliters of the NaOH solution are needed to neutralize the vitamin C?

38. A student mixes 112 mL of a $0.150M$ HCl solution and 112 mL of an NaOH solution of unknown molarity. The final solution is found to be acidic. The student then titrates this solution to the endpoint with 16.7 mL of $0.100M$ NaOH. What was the molarity of the original NaOH solution?

Sections 15.1 and 15.2

39. Propanoic acid, $HC_3H_5O_2$, is a weak acid. When propanoic acid completely reacts with sodium hydroxide, the final solution has a pH slightly greater than 7. Why are the products of this "neutralization" reaction not neutral? Use the net ionic equation to help in your explanation.

40. How does a solution that contains dissolved acetic acid and sodium acetate act as a buffer? Use net ionic equations to show how this buffer responds to added H^+ and OH^-.

 CHAPTER 16

Section 16.1

1. What is the difference between an oxidation-reduction reaction and a reaction that is not oxidation-reduction?

2. Which of the following changes are oxidations and which are reductions?

   a. Mg^{2+} becomes Mg   **c.** Fe^{2+} becomes Fe^{3+}

   b. K becomes K^+ **d.** Cl_2 becomes $2Cl^-$

3. Identify the reducing agent in each of the following reactions.

   a. $Zn(s) + 2Ag^+(aq) \rightarrow Zn^{2+}(aq) + 2Ag(s)$

   b. $Cl_2(g) + 2Na(s) \rightarrow 2NaCl(s)$

   c. $2SO_2(g) + O_2(g) \rightarrow 2SO_3(g)$

4. Write the equation for the redox reaction that occurs when magnesium metal reacts with nitric acid. Name the oxidizing agent and the reducing agent.

5. The following equation represents the neutralization reaction between hydrochloric acid and calcium hydroxide. Determine the oxidation number for each element. Is this a redox reaction? Explain.

$$2HCl(aq) + Ca(OH)_2(aq) \rightarrow$$
$$CaCl_2(aq) + 2H_2O(l)$$

6. For each of the following reaction, determine whether or not it is a redox reaction. If it is, identify which reactant is reduced and which is oxidized.

   a. $Fe^{2+}(aq) + Mg(s) \rightarrow Fe(s) + Mg^{2+}(aq)$

   b. $2NaI(aq) + Pb(NO_3)_2(aq) \rightarrow$
      $PbI_2(s) + 2NaNO_3(aq)$

   c. $4Al(s) + 3O_2(g) \rightarrow 2Al_2O_3(s)$

7. When a mixture of iron filings and sulfur crystals is heated, the elements combine to form iron(II) sulfide.

$$Fe(s) + S(s) \rightarrow FeS(s)$$

Is this a redox reaction? If so, which element is oxidized and which is reduced?

8. Mercury(II) oxide is a red crystalline powder that decomposes into mercury and oxygen on exposure to light. Write a balanced equation for this reaction. Is it a redox reaction? If so, which element is oxidized and which is reduced?

Section 16.2

9. Name the compounds that form the green protective coating, or patina, on the Statue of Liberty. What is the oxidation number of copper in these compounds? What useful purpose does the patina serve?

10. What is chemiluminescence? Name two practical uses of chemiluminescence.

CHAPTER 17

Section 17.1

1. Answer the following questions pertaining to an electrolytic cell.

   a. What are positive ions called? What are negative ions called?

   b. Which electrode attracts positive ions? Which attracts negative ions?

   c. Where does oxidation occur? Where does reduction occur?

2. Describe the movement of electrons during electrolysis.

3. Name four uses of electrolysis.

4. What is produced by the Hall-Héroult process? Why was the development of this process so important? What are some drawbacks of the process, and what can people do to circumvent these drawbacks?

Section 17.2

5. A piece of aluminum metal is placed in a 0.50M solution of lead(II) nitrate, $Pb(NO_3)_2$.

   a. Use **Table 17.1** to predict which metal will be reduced and which will be oxidized.

   b. Write an equation for the net redox reaction that occurs.

   c. Is this system a galvanic cell? Explain.

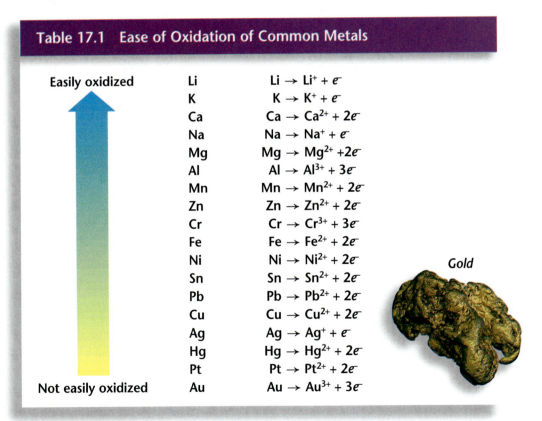

Table 17.1 Ease of Oxidation of Common Metals

Easily oxidized		
	Li	$Li \rightarrow Li^+ + e^-$
	K	$K \rightarrow K^+ + e^-$
	Ca	$Ca \rightarrow Ca^{2+} + 2e^-$
	Na	$Na \rightarrow Na^+ + e^-$
	Mg	$Mg \rightarrow Mg^{2+} + 2e^-$
	Al	$Al \rightarrow Al^{3+} + 3e^-$
	Mn	$Mn \rightarrow Mn^{2+} + 2e^-$
	Zn	$Zn \rightarrow Zn^{2+} + 2e^-$
	Cr	$Cr \rightarrow Cr^{3+} + 3e^-$
	Fe	$Fe \rightarrow Fe^{2+} + 2e^-$
	Ni	$Ni \rightarrow Ni^{2+} + 2e^-$
	Sn	$Sn \rightarrow Sn^{2+} + 2e^-$
	Pb	$Pb \rightarrow Pb^{2+} + 2e^-$
	Cu	$Cu \rightarrow Cu^{2+} + 2e^-$
	Ag	$Ag \rightarrow Ag^+ + e^-$
	Hg	$Hg \rightarrow Hg^{2+} + 2e^-$
	Pt	$Pt \rightarrow Pt^{2+} + 2e^-$
Not easily oxidized	Au	$Au \rightarrow Au^{3+} + 3e^-$

Gold

6. Can a solution of magnesium chloride be stored in a container plate with chromium metal? Explain.

7. Which of the following galvanic cells would you expect to have the highest voltage? The lowest voltage? Explain your answers.

   a. a zinc-copper cell

   b. an iron-copper cell

   c. a magnesium-copper cell

8. What change would occur if you placed

   a. a strip of zinc into a solution of silver nitrate?

   b. a strip of zinc into a solution of sodium sulfate?

9. Draw a diagram of a galvanic cell in which the reaction is

$$Zn(s) + Sn^{2+}(aq) \rightarrow Zn^{2+}(aq) + Sn(s).$$

Label the cathode and the anode. Show the ions present in both compartments, and indicate the direction of electron flow in the external circuit.

10. What are some potential advantages of electric cars? Why are electric cars uncommon in the United States?

CHAPTER 18

Section 18.1

1. Name the following hydrocarbons.

   a. CH_3CH_3

   b. $CH_3CH_2C \equiv CCH_3$

   c. $CH_3CH = CHCH_2CH_2CH_2CH_3$

   d.

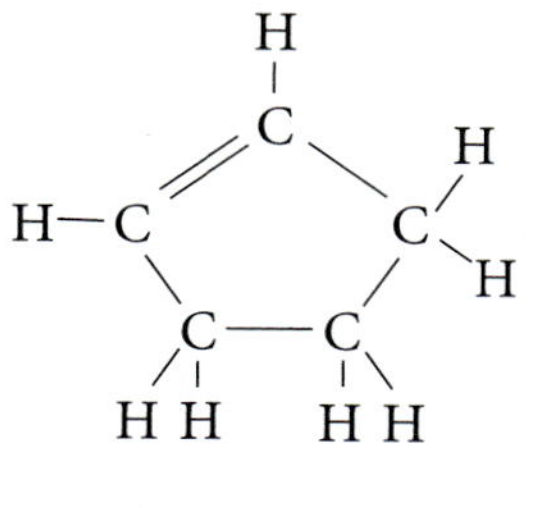

2. Name the branched alkanes shown.

   a. $CH_3CH_2CHCH_2CHCH_2CH_3$
with branches CH_3 and CH_2—CH_3

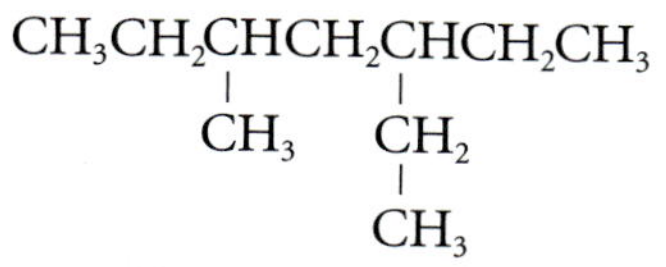

   b. $CH_3CHCH_2CHCH_2CH_2CHCH_3$
with branches CH_3, CH_2—CH_3, and CH_3

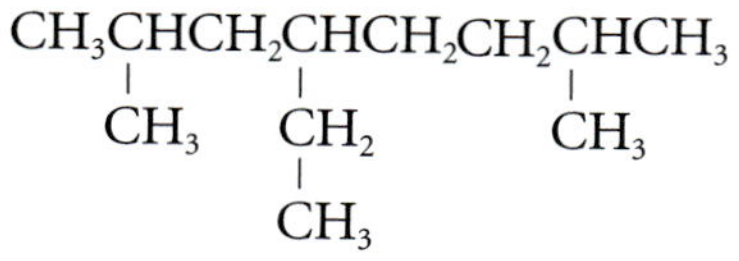

   c. $CH_3CHCH_2CH_3$
with branch CH_2—CH_3

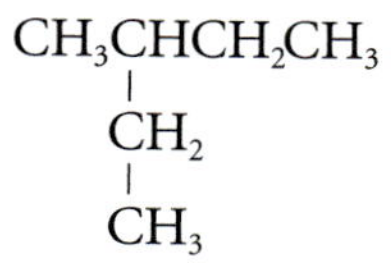

3. Why is there no compound named 3,6-dimethylheptane? What is the correct name for this compound?

4. Compare the reactivities of alkanes and alkenes. Explain the difference.

5. The molecular formula of the cycloalkanes follow the pattern C_nH_{2n}, where n is the number of carbon atoms. What pattern is followed by the cycloalkenes with one double bond?

6. Draw structures of the following hydrocarbons, and identify each as an alkane, alkene, or alkyne.

   a. heptane

   b. *cis*-2-pentene

   c. propyne

   d. 1,2-dimethylcyclopentane

   e. 3-ethyl-2-methylhexane

   f. 2-methyl-2-butene

7. Write the names and structures of one positional isomer, one geometric isomer, and two structural isomers of *cis*-2-butene.

8. How many H_2 molecules would be required to fully hydrogenate one molecule of vitamin D_2, the structure shown here?

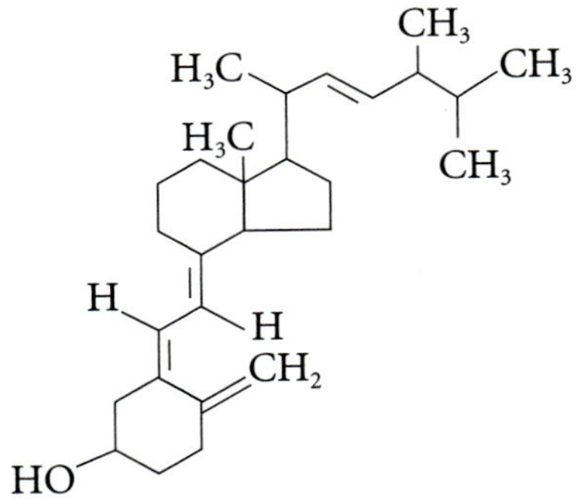

9. Halogen molecules such as Cl_2 can be added to double bonds in a reaction similar to hydrogenation. Draw the structure of the product that forms when Cl_2 is added to 2-butene.

10. Show the structure of the product of each of the following reactions.

   a. $CH_3CH = CHCH_3 + H_2 \rightarrow$?
   b. $CH \equiv CH + 2H_2 \rightarrow$?
   c. $CH_3C \equiv CH + H_2 \rightarrow$?

11. How are the properties of benzene unlike those of most alkenes? How do chemists account for benzene's behavior?

12. What is petroleum? Where is it found? Name six useful products that are derived from petroleum.

13. What technique is commonly used to separate petroleum into its component liquids? What two processes are used to convert large alkanes into smaller hydrocarbons?

Section 18.2

14. Classify each of the following molecules as belonging to one of the categories of substituted hydrocarbons described in the chapter.

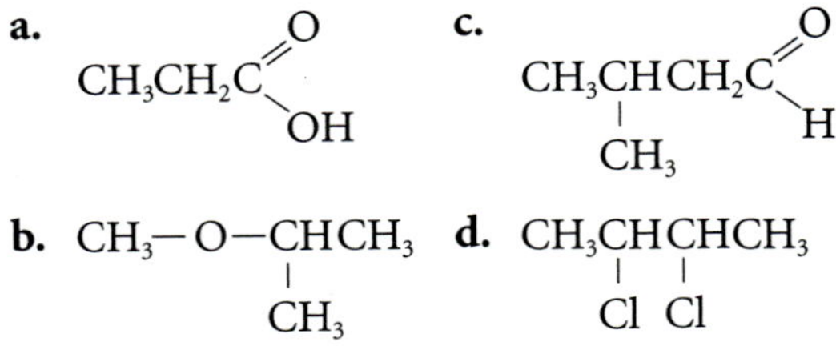

15. Distinguish between

   a. an amide and an amine.
   b. an ether and an ester.

16. Draw the following structures.

   a. an ester, where R is $-CH_2CH_3$ and R' is $-CH_3$
   b. a ketone, where R is $-CHCH_3$ with CH_3 and R' is $-CH_2CH_3$

17. The structure of tyrosine, an amino acid, is shown. Name the functional groups in the molecule.

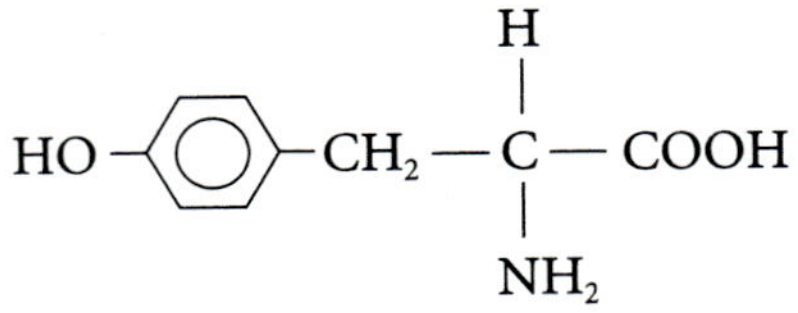

Sections 18.1 and 18.2

18. Name a common example of each of the following classes of organic compounds.

   a. ester
   b. ketone
   c. cycloalkane
   d. aromatic hydrocarbon

Section 18.3

19. Draw the structure of the polymer that will be formed from each of the monomers shown.

   a. $CH_2 = CH_2$

   b. $CH_2 = CHCH_3$

   c. $CH_2 = CCH_3$ with $C = O$, O, CH_3

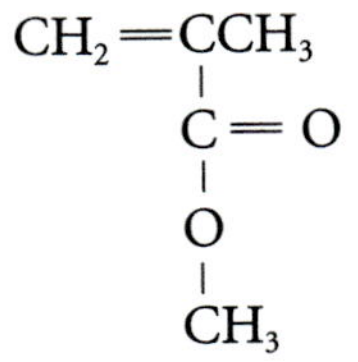

20. Provide one example each of a thermoplastic polymer and a thermosetting polymer. Which type of polymer is easier to recycle? Which type is more durable? Explain.

CHAPTER 19

Section 19.1

1. What elements make up proteins? Name some of the roles that proteins play in living systems.

2. Describe briefly what enzymes are and how they work.

3. Name two common disaccharides found in foods. Are these molecules classified as nutrients? Explain.

4. Describe the functions of lipids and their solubility properties. Are animal lipids or plant lipids generally more saturated?

5. Examine the structure of each of the following molecules shown and decide whether each is a carbohydrate, a lipid, or an amino acid.

   a.

   $$H_2N-\overset{\overset{\displaystyle H}{|}}{C}-\overset{\overset{\displaystyle O}{||}}{C}-OH$$

   with side chain $-CH_2-CH_2-S-CH_3$

   b.

   $$CH_3-(CH_2)_{14}-\overset{\overset{\displaystyle O}{||}}{C}-OH$$

   c.

   HOH_2C ... O ... OH (ring structure with H, H, H, H, HO, HO)

   d.

   $$H_2N-\overset{\overset{\displaystyle H}{|}}{C}-\overset{\overset{\displaystyle O}{||}}{C}-OH$$

   with side chain $-CH_2-C(=O)-OH$

6. What monomers form each of the following types of biomolecule?

   a. protein   **c.** polysaccharide

   b. nucleic acid

7. What are nucleic acids? What functions do they perform in cells? Why are nucleic acids not listed on food-product labels? What are the two kinds of nucleic acids found in cells?

8. Name two diseases that can result from vitamin deficiencies in the diet. Can too much of a vitamin cause disease? Explain.

9. What is a coenzyme? Give an example in which a vitamin acts as a coenzyme.

Section 19.2

10. What are hormones? Give an example of a hormone and describe its function.

CHAPTER 20

Section 20.1

1. Nitric oxide gas (NO) can be prepared by passing air through an electric arc. The heat of reaction is +90.4 kJ/mol NO. Write a balanced chemical equation that includes the reaction heat. Is the reaction exothermic or endothermic?

2. Sketch a graph of the energy changes during the progress of an endothermic reaction with a high activation energy. On the same axes, sketch a graph of energy change versus progress of the reaction if a catalyst is used to lower the activation energy.

3. Describe each of the following processes as involving an increase or decrease in entropy. Explain your answers.

   a. Water boils, forming water vapor.

   b. Baking soda reacts with vinegar to form sodium acetate solution, water, and carbon dioxide gas.

   c. Two moles of nitrogen gas combine with six moles of hydrogen gas to form four moles of ammonia gas.

Table 20.2 The Caloric Value of Some Foods			
Food	**Quantity**	**Kilojoules**	**Calories**
butter	1 tbsp = 14 g	418	100
peanut butter	1 tbsp = 16 g	418	100
spaghetti	0.5 cup = 55 g	836	200
apple	1	283	70
chicken (broiled)	3 oz = 84 g	502	120
beef (broiled)	3 oz = 84 g	1000	241

Section 20.2

4. How much heat is given off by a reaction that raises the temperature of 756 g of water in a calorimeter from 23.2°C to 37.6°C?

5. When 5.00 g of a certain liquid organic compound are burned in a calorimeter, the temperature of the surrounding 2.00 kg of water increases from 24.5°C to 40.5°C. All products of the reaction are gases. Calculate the heat given off in this reaction. How much heat would 1.00 mol of the compound give off, assuming a molar mass of 46.1 g/mol?

6. Explain the relationship among a calorie, a Calorie, and a kilocalorie. If you eat a candy bar that contains 180 Calories, how many calories have you consumed? How many kilocalories?

7. Using **Table 20.2** as a guide, compare the Caloric contents of 1 g of peanut butter and 1 g of broiled chicken. How might you explain the difference?

8. A 10.0-g sample of a certain brand of cereal is burned in a calorimeter. The 4.00 kg of surrounding water increase in temperature from 22.00°C to 30.20°C. What is the food value in Calories per gram?

Sections 20.1 and 20.2

9. List two factors that would make an exothermic reaction less likely to occur.

Section 20.3

10. In photosynthesis, which three products of the light reactions take part in the reactions of the Calvin cycle?

CHAPTER 21

Section 21.1

1. Write the balanced nuclear equation for each of the following nuclear reactions, and determine the type of decay.

    a. Uranium-234 decays to form thorium-230.

    b. Bismuth-214 decays to form polonium-214.

    c. Thallium-210 decays to form lead-210.

2. A compound containing 200 mg of technetium-99m is injected into a patient to get a bone scan, but equipment problems make it impossible to scan the patient before 24 hours have passed. Will the nuclear medical technician be able to get a good scan after that period of time, or should more technetium-99m be injected? The half-life of technetium-99m is 6.0 hours.

3. A fossilized piece of wood is found to have 3.12 percent of the amount of carbon-14 that would be found in a new piece of similar wood. How old is the fossilized wood?

4. A scientist analyzing a rock sample finds that it contains 75 percent of the potassium-40 that would be found in a similar rock formed today. Use **Figure 21.7** and **Table 21.1** to estimate the age of the rock.

Section 21.2

5. In the early 1900s, Albert Einstein developed a famous equation relating mass and energy in nuclear reactions. Use this equation to explain why nuclear reactions release very large amounts of energy.

6. Distinguish between deuterium and tritium. Why is deuterium potentially important as a source of energy?

Sections 21.1 and 21.2

7. How do nuclear reactions differ from chemical reactions?

Table 21.1	Half-Life Values for Commonly Used Radioactive Isotopes
Isotope	**Half-Life**
$^{3}_{1}H$	12.26 years
$^{14}_{6}C$	5730 years
$^{32}_{15}P$	14.282 days
$^{40}_{19}K$	1.25 billion years
$^{60}_{27}Co$	5.271 years
$^{85}_{36}Kr$	10.76 years
$^{93}_{36}Kr$	1.3 seconds
$^{87}_{37}Rb$	48 billion years
$^{99m}_{43}Tc$	6.0 hours*
$^{131}_{53}I$	8.07 days
$^{131}_{56}Ba$	12 days
$^{153}_{64}Gd$	242 days
$^{201}_{81}Tl$	73 hours
$^{226}_{88}Ra$	1600 years
$^{235}_{92}U$	710 million years
$^{238}_{92}U$	4.51 billion years
$^{241}_{95}Am$	432.7 years

*The *m* in the symbol tells you that this is a metastable element, which is one form of an unstable isotope. Technetium 99m gives off a gamma ray to become a more stable form of the same isotope, with no change in either atomic or mass number.

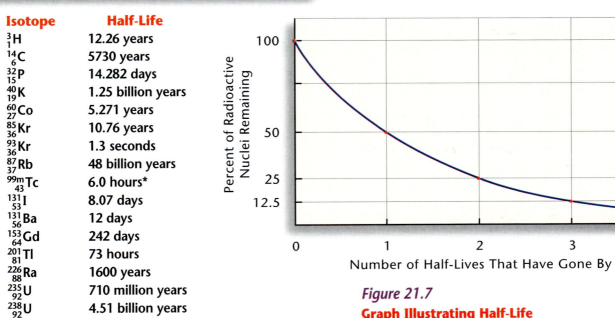

Figure 21.7

Graph Illustrating Half-Life
During each half-life period, half of the radioactive nuclei in a sample decay. After one half-life, 50 percent of the sample remains. After two half-lives, 25 percent of the sample remains. After three half-lives, 12.5 percent remains, and so on.

Section 21.3

8. What is a radioactive tracer? Name three non-medical uses for radioactive tracers.

9. The quantity of nuclear waste generated in hospitals is much greater than the quantity generated in nuclear reactors. Why, then, is nuclear waste from nuclear reactors more dangerous than nuclear waste from hospitals?

10. Using information on pages 776 and 778 of your text, compare the radiation exposure from a chest X ray to the exposure from smoking two packs of cigarettes a day for a year.

Safety Handbook

Safety Guidelines in the Chemistry Laboratory

The chemistry laboratory is a safe place to work if you are aware of important safety rules and if you are careful. You must be responsible for your own safety and for the safety of others. The safety rules given here will protect you and others from harm in the lab. While carrying out procedures in any of the **ChemLabs, Launch Labs,** or **Try at Home Labs,** notice the safety symbols and caution statements. The safety symbols are explained in the chart on the next page.

1. Always obtain your teacher's permission to begin a lab.
2. Study the procedure. If you have questions, ask your teacher. Be sure you understand all safety symbols shown.
3. Use the safety equipment provided for you. Goggles and a safety apron should be worn when any lab calls for using chemicals.
4. When you are heating a test tube, always slant it so the mouth points away from you and others.
5. Never eat or drink in the lab. Never inhale chemicals. Do not taste any substance or draw any material into your mouth.
6. If you spill any chemical, wash it off immediately with water. Report the spill immediately to your teacher.
7. Know the location and proper use of the fire extinguisher, safety shower, fire blanket, first aid kit, and fire alarm.
8. Keep all materials away from open flames. Tie back long hair.
9. If a fire should break out in the classroom, or if your clothing should catch fire, smother it with the fire blanket or a coat, or get under a safety shower. **NEVER RUN.**
10. Report any accident or injury, no matter how small, to your teacher.

Follow these procedures as you clean up your work area.

1. Turn off the water and gas. Disconnect electrical devices.
2. Return materials to their places.
3. Dispose of chemicals and other materials as directed by your teacher. Place broken glass and solid substances in the proper containers. Never discard materials in the sink.
4. Clean your work area.
5. Wash your hands thoroughly after working in the laboratory.

First Aid in the Laboratory

Injury	Safe Response
Burns	Apply cold water. Call your teacher immediately.
Cuts and bruises	Stop any bleeding by applying direct pressure. Cover cuts with a clean dressing. Apply cold compresses to bruises. Call your teacher immediately.
Fainting	Leave the person lying down. Loosen any tight clothing and keep crowds away. Call your teacher immediately.
Foreign matter in eye	Flush with plenty of water. Use eyewash bottle or fountain.
Poisoning	Note the suspected poisoning agent and call your teacher immediately.
Any spills on skin	Flush with large amounts of water or use safety shower. Call your teacher immediately.

Safety Symbols

SAFETY SYMBOLS	HAZARD	EXAMPLES	PRECAUTION	REMEDY
DISPOSAL	Special disposal procedures need to be followed.	certain chemicals, living organisms	Do not dispose of these materials in the sink or trash can.	Dispose of wastes as directed by your teacher.
BIOLOGICAL	Organisms or other biological materials that might be harmful to humans	bacteria, fungi, blood, unpreserved tissues, plant materials	Avoid skin contact with these materials. Wear mask or gloves.	Notify your teacher if you suspect contact with material. Wash hands thoroughly.
EXTREME TEMPERATURE	Objects that can burn skin by being too cold or too hot	boiling liquids, hot plates, dry ice, liquid nitrogen	Use proper protection when handling.	Go to your teacher for first aid.
SHARP OBJECT	Use of tools or glassware that can easily puncture or slice skin	razor blades, pins, scalpels, pointed tools, dissecting probes, broken glass	Practice common-sense behavior and follow guidelines for use of the tool.	Go to your teacher for first aid.
FUME	Possible danger to respiratory tract from fumes	ammonia, acetone, nail polish remover, heated sulfur, moth balls	Make sure there is good ventilation. Never smell fumes directly. Wear a mask.	Leave foul area and notify your teacher immediately.
ELECTRICAL	Possible danger from electrical shock or burn	improper grounding, liquid spills, short circuits, exposed wires	Double-check setup with teacher. Check condition of wires and apparatus.	Do not attempt to fix electrical problems. Notify your teacher immediately.
IRRITANT	Substances that can irritate the skin or mucous membranes of the respiratory tract	pollen, moth balls, steel wool, fiberglass, potassium permanganate	Wear dust mask and gloves. Practice extra care when handling these materials.	Go to your teacher for first aid.
CHEMICAL	Chemicals can react with and destroy tissue and other materials	bleaches such as hydrogen peroxide; acids such as sulfuric acid, hydrochloric acid; bases such as ammonia, sodium hydroxide	Wear goggles, gloves, and an apron.	Immediately flush the affected area with water and notify your teacher.
TOXIC	Substance may be poisonous if touched, inhaled, or swallowed.	mercury, many metal compounds, iodine, poinsettia plant parts	Follow your teacher's instructions.	Always wash hands thoroughly after use. Go to your teacher for first aid.
FLAMMABLE	Flammable chemicals may be ignited by open flame, spark, or exposed heat.	alcohol, kerosene, potassium permanganate	Avoid open flames and heat when using flammable chemicals.	Notify your teacher immediately. Use fire safety equipment if applicable.
OPEN FLAME	Open flame in use, may cause fire.	hair, clothing, paper, synthetic materials	Tie back hair and loose clothing. Follow teacher's instruction on lighting and extinguishing flames.	Notify your teacher immediately. Use fire safety equipment if applicable.

 Eye Safety Proper eye protection should be worn at all times by anyone performing or observing science activities.

 Clothing Protection This symbol appears when substances could stain or burn clothing.

 Animal Safety This symbol appears when safety of animals and students must be ensured.

 Handwashing After the lab, wash hands with soap and water before removing goggles.

APPENDIX D
Chemistry Data Handbook

Table D.1 Symbols and Abbreviations

α	=	particles from radioactive materials, helium nuclei
β	=	particles from radioactive materials, electrons
γ	=	rays from radioactive materials, high-energy quanta
Δ	=	change in
λ	=	wavelength
ν	=	frequency
Π	=	osmotic pressure
A	=	ampere (*electric current*)
Bq	=	becquerel (*nuclear disintegration*)
°C	=	Celsius degree (*temperature*)
C	=	coulomb (*quantity of electricity*)
c	=	speed of light
cd	=	candela (*luminous intensity*)
C_p	=	specific heat
D	=	density
E	=	energy, electromotive force
F	=	force, Faraday
G	=	free energy
g	=	gram (*mass*)
Gy	=	gray (*radiation*)
H	=	enthalpy
Hz	=	hertz (*frequency*)
h	=	Planck's constant
h	=	hour (*time*)
J	=	joule (*energy*)
K	=	kelvin (*temperature*)
K_a	=	ionization constant (acid)
K_b	=	ionization constant (base)
K_{eq}	=	equilibrium constant
K_{sp}	=	solubility product constant
kg	=	kilogram
M	=	molarity
m	=	mass, molality
m	=	meter (*length*)
mol	=	mole (*amount*)
min	=	minute (*time*)
N	=	newton (*force*)
N_A	=	Avogadro's number
n	=	number of moles

P	=	pressure, power
Pa	=	pascal (*pressure*)
$\boldsymbol{p}$	=	momentum
q	=	heat
R	=	gas constant
S	=	entropy
s	=	second (*time*)
Sv	=	sievert (*absorbed radiation*)
T	=	temperature
U	=	internal energy
u	=	atomic mass unit
V	=	volume
V	=	volt (*electromotive force*)
v	=	velocity
W	=	watt (*power*)
w	=	work
x	=	mole fraction

PERIODIC TABLE OF THE ELEMENTS

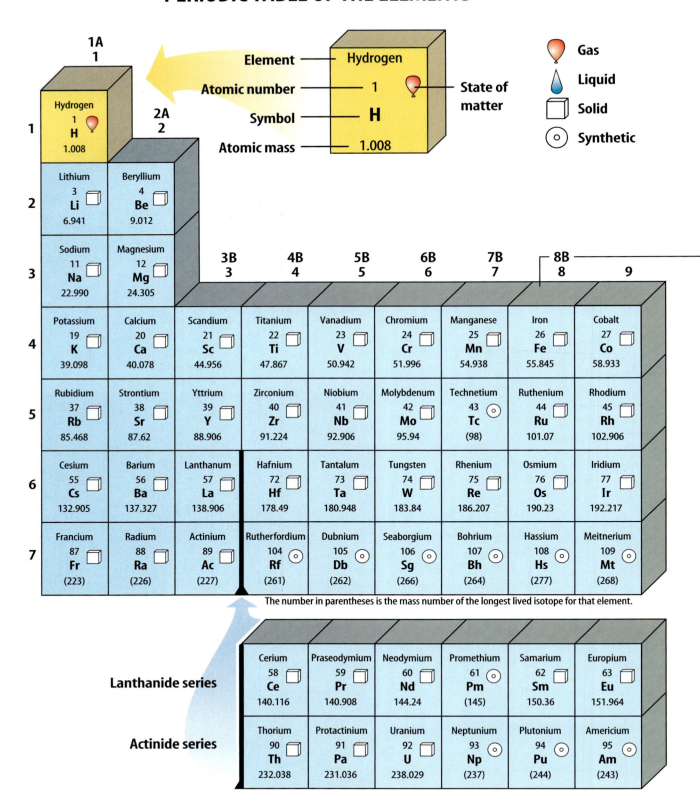

The number in parentheses is the mass number of the longest lived isotope for that element.

Lanthanide series

Cerium	Praseodymium	Neodymium	Promethium	Samarium	Europium
58	59	60	61	62	63
Ce	Pr	Nd	Pm	Sm	Eu
140.116	140.908	144.24	(145)	150.36	151.964

Actinide series

Thorium	Protactinium	Uranium	Neptunium	Plutonium	Americium
90	91	92	93	94	95
Th	Pa	U	Np	Pu	Am
232.038	231.036	238.029	(237)	(244)	(243)

Metal
Metalloid
Nonmetal
Recently discovered

						8A 18
3A 13	**4A** 14	**5A** 15	**6A** 16	**7A** 17		Helium 2 **He** 4.003

10	**1B** 11	**2B** 12	Boron 5 **B** 10.811	Carbon 6 **C** 12.011	Nitrogen 7 **N** 14.007	Oxygen 8 **O** 15.999	Fluorine 9 **F** 18.998	Neon 10 **Ne** 20.180
			Aluminum 13 **Al** 26.982	Silicon 14 **Si** 28.086	Phosphorus 15 **P** 30.974	Sulfur 16 **S** 32.065	Chlorine 17 **Cl** 35.453	Argon 18 **Ar** 39.948
Nickel 28 **Ni** 58.693	Copper 29 **Cu** 63.546	Zinc 30 **Zn** 65.39	Gallium 31 **Ga** 69.723	Germanium 32 **Ge** 72.64	Arsenic 33 **As** 74.922	Selenium 34 **Se** 78.96	Bromine 35 **Br** 79.904	Krypton 36 **Kr** 83.80
Palladium 46 **Pd** 106.42	Silver 47 **Ag** 107.868	Cadmium 48 **Cd** 112.411	Indium 49 **In** 114.818	Tin 50 **Sn** 118.710	Antimony 51 **Sb** 121.760	Tellurium 52 **Te** 127.60	Iodine 53 **I** 126.904	Xenon 54 **Xe** 131.293
Platinum 78 **Pt** 195.078	Gold 79 **Au** 196.967	Mercury 80 **Hg** 200.59	Thallium 81 **Tl** 204.383	Lead 82 **Pb** 207.2	Bismuth 83 **Bi** 208.980	Polonium 84 **Po** (209)	Astatine 85 **At** (210)	Radon 86 **Rn** (222)
Darmstadtium 110 **Ds** (281)	* 111 Unununium **Uuu** (272)	* 112 Ununbium **Uub** (285)		Ununquadium * 114 **Uuq** (289)				

* Names not officially assigned. Discovery of element 114 recently reported. Further information not yet available.

Gadolinium 64 **Gd** 157.25	Terbium 65 **Tb** 158.925	Dysprosium 66 **Dy** 162.50	Holmium 67 **Ho** 164.930	Erbium 68 **Er** 167.259	Thulium 69 **Tm** 168.934	Ytterbium 70 **Yb** 173.04	Lutetium 71 **Lu** 174.967
Curium 96 **Cm** (247)	Berkelium 97 **Bk** (247)	Californium 98 **Cf** (251)	Einsteinium 99 **Es** (252)	Fermium 100 **Fm** (257)	Mendelevium 101 **Md** (258)	Nobelium 102 **No** (259)	Lawrencium 103 **Lr** (262)

Table D.3   Alphabetical Table of the Elements

Element	Symbol	Atomic number	Atomic mass
Actinium	Ac	89	227.027 8*
Aluminum	Al	13	26.981 539
Americium	Am	95	243.061 4*
Antimony	Sb	51	121.757
Argon	Ar	18	39.948
Arsenic	As	33	74.921 59
Astatine	At	85	209.987 1*
Barium	Ba	56	137.327
Berkelium	Bk	97	247.070 3*
Beryllium	Be	4	9.012 182
Bismuth	Bi	83	208.980 37
Bohrium	Bh	107	262*
Boron	B	5	10.811
Bromine	Br	35	79.904
Cadmium	Cd	48	112.411
Calcium	Ca	20	40.078
Californium	Cf	98	251.079 6*
Carbon	C	6	12.011
Cerium	Ce	58	140.115
Cesium	Cs	55	132.905 43
Chlorine	Cl	17	35.452 7
Chromium	Cr	24	51.996 1
Cobalt	Co	27	58.933 20
Copper	Cu	29	63.546
Curium	Cm	96	247.070 3*
Darmstadtium	Ds	110	271*
Dubnium	Db	105	262*
Dysprosium	Dy	66	162.50
Einsteinium	Es	99	252.082 8*
Erbium	Er	68	167.26
Europium	Eu	63	151.965
Fermium	Fm	100	257.095 1*
Fluorine	F	9	18.998 403 2
Francium	Fr	87	223.019 7*
Gadolinium	Gd	64	157.25
Gallium	Ga	31	69.723
Germanium	Ge	32	72.61
Gold	Au	79	196.966 54
Hafnium	Hf	72	178.49
Hassium	Hs	108	265*
Helium	He	2	4.002 602
Holmium	Ho	67	164.930 32
Hydrogen	H	1	1.007 94
Indium	In	49	114.82
Iodine	I	53	126.904 47
Iridium	Ir	77	192.22
Iron	Fe	26	55.847
Krypton	Kr	36	83.80
Lanthanum	La	57	138.905 5
Lawrencium	Lr	103	260.105 4*
Lead	Pb	82	207.2
Lithium	Li	3	6.941
Lutetium	Lu	71	174.967
Magnesium	Mg	12	24.305 0
Manganese	Mn	25	54.938 05
Meitnerium	Mt	109	266*

* The mass of the isotope with the longest known half-life.

Element	Symbol	Atomic number	Atomic mass
Mendelevium	Md	101	258.098 6*
Mercury	Hg	80	200.59
Molybdenum	Mo	42	95.94
Neodymium	Nd	60	144.24
Neon	Ne	10	20.179 7
Neptunium	Np	93	237.048 2
Nickel	Ni	28	58.6934
Niobium	Nb	41	92 .906 38
Nitrogen	N	7	14. 006 74
Nobelium	No	102	259.100 9*
Osmium	Os	76	190.2
Oxygen	O	8	15.999 4
Palladium	Pd	46	106.42
Phosphorus	P	15	30.973 762
Platinum	Pt	78	195.08
Plutonium	Pu	94	244.064 2*
Polonium	Po	84	208.982 4*
Potassium	K	19	39.098 3
Praseodymium	Pr	59	140.907 65
Promethium	Pm	61	144.912 8*
Protactinium	Pa	91	231.035 88
Radium	Ra	88	226.025 4
Radon	Rn	86	222.017 6*
Rhenium	Re	75	186.207
Rhodium	Rh	45	102.905 50
Rubidium	Rb	37	85.467 8
Ruthenium	Ru	44	101.07
Rutherfordium	Rf	104	261*
Samarium	Sm	62	150.36
Scandium	Sc	21	44.955 910
Seaborgium	Sg	106	263*
Selenium	Se	34	78.96
Silicon	Si	14	28.085 5
Silver	Ag	47	107.868 2
Sodium	Na	11	22.989 768
Strontium	Sr	38	87.62
Sulfur	S	16	32.066
Tantalum	Ta	73	180.947 9
Technetium	Tc	43	97.907 2*
Tellurium	Te	52	127.60
Terbium	Tb	65	158.925 34
Thallium	Tl	81	204.383 3
Thorium	Th	90	232.038 1
Thulium	Tm	69	168.934 21
Tin	Sn	50	118.710
Titanium	Ti	22	47.88
Tungsten	W	74	183.85
Ununbium	Uub	112	285*
Ununquadium	Uuq	114	289*
Unununium	Uuu	111	272*
Uranium	U	92	238.028 9
Vanadium	V	23	50.941 5
Xenon	Xe	54	131.29
Ytterbium	Yb	70	173.04
Yttrium	Y	39	88.905 85
Zinc	Zn	30	65.39
Zirconium	Zr	40	91.224

Table D.4 Properties of Elements

Element	Symbol	Atomic Number (Z)	Atomic Mass* (u)	Melting Point (°C)	Boiling Point (°C)	Density (g/cm³) (gases measured at STP)	Atomic Radius (pm)	First Ionization Energy (kJ/mol)	Standard Reduction Potential (V) (for elements from or to oxidation state indicated)	Enthalpy of Fusion (kJ/mol)	Specific Heat (J/g·°C)	Enthalpy of Vaporization (kJ/mol)	Abundance in Earth's Crust (%)	Major Oxidation States
Actinium	Ac	89	[227.0278]	1050	3300	10.07	203	666	(3+) −2.13	14.3	0.120	293	trace	3+
Aluminum	Al	13	26.981539	660.37	2517.6	2.699	143	577.5	(3+) −1.67	10.71	0.9025	290.8	8.1	3+
Americium	Am	95	[243.0614]	994	2600	13.67	183	579	(3+) −2.07	10	—	238.5	—	2+, 3+, 4+
Antimony	Sb	51	121.757	630.7	1635	6.697	161	834	(3+) +0.15	19.5	0.2072	193	2×10^{-5}	3+, 5+
Argon	Ar	18	39.948	−189.37	−185.86	0.001784	98	1521	—	1.18	0.52033	6.52	4×10^{-6}	—
Arsenic	As	33	74.92159	816 (2840 kPa)	615 (sublimes)	5.778	121	947	(3+) +0.24	27.7	0.3289	—	1.9×10^{-4}	3+, 5+
Astatine	At	85	[209.98037]	300	350	—	—	916	(1−) +0.2	23.8	—	90.3	trace	1−, 5+
Barium	Ba	56	137.327	726.9	1845	3.62	222	502.9	(2+) −2.92	8.012	0.2044	140	0.039	2+
Berkelium	Bk	97	[247.0703]	986	2600	14.78	170	601	(3+) −2.01	—	—	—	—	3+, 4+
Beryllium	Be	4	9.012182	1287	2468	1.848	112	899.5	(2+) −1.97	7.895	1.824	297.6	2×10^{-4}	2+
Bismuth	Bi	83	208.98037	271.4	1564	9.808	151	703	(3+) +0.317	10.9	0.1221	179	8×10^{-7}	3+, 5+
Bohrium	Bh	107	[262]	—	—	—	—	—	—	—	—	—	—	—
Boron	B	5	10.811	2080	3865	2.46	85	800.6	(3+) −0.89	50.2	1.026	504.5	9×10^{-4}	3+
Bromine	Br	35	79.904	−7.25	59.35	3.1028	119	1139.9	(1−) +1.065	10.571	0.47362	29.56	2.5×10^{-4}	1−, 1+, 3+, 5+
Cadmium	Cd	48	112.411	320.8	770	8.65	151	867.7	(2+) −0.4025	6.19	0.2311	100	1.6×10^{-5}	2+
Calcium	Ca	20	40.078	841.5	1500.5	1.55	197	589.8	(2+) −2.84	8.54	0.6315	155	4.66	2+
Californium	Cf	98	[251.0796]	900	—	—	186	608	(3+) −2	—	—	—	—	3+, 4+
Carbon	C	6	12.011	3620	4200	2.266	77	1086.5	(4−) +0.132	104.6	0.7099	711	0.018	4−, 2+, 4+
Cerium	Ce	58	140.115	804	3470	6.773	181.8	541	(3+) −2.34	5.2	0.1923	313	0.007	3+, 4+
Cesium	Cs	55	132.90543	28.4	674.8	1.9	262	375.7	(1+) −2.923	2.087	0.2421	67	2.6×10^{-4}	1+
Chlorine	Cl	17	35.4527	−101	−34	0.003214	91	1255.5	(1−) +1.3583	6.41	0.47820	20.41	0.013	1−, 1+, 3+, 5+
Chromium	Cr	24	51.9961	1860	2679	7.2	128	652.8	(3+) −0.74	20.5	0.4491	339	0.01	2+, 3+, 6+
Cobalt	Co	27	58.9332	1495	2912	8.9	125	758.8	(2+) −0.277	16.192	0.4210	382	0.0028	2+, 3+
Copper	Cu	29	63.546	1085	2570	8.92	128	745.5	(2+) +0.34	13.38	0.38452	304	0.0058	1+, 2+
Curium	Cm	96	[247.0703]	1340	3540	13.51	174	581	(3+) −2.06	—	—	—	—	3+, 4+
Darmstadtium	Ds	110	[271]	—	—	—	—	—	—	—	—	—	—	—
Dubnium	Db	105	[262]	—	—	—	—	—	—	—	—	—	—	—
Dysprosium	Dy	66	162.5	1407	2600	8.536	178.1	572	(3+) −2.29	10.4	0.1733	250	6×10^{-4}	2+, 3+
Einsteinium	Es	99	[252.0828]	860	—	—	186	619	(3+) −2	—	—	—	—	3+
Erbium	Er	68	167.26	1497	2900	9.045	176.1	589	(3+) −2.32	17.2	0.1681	293	3.5×10^{-4}	3+
Europium	Eu	63	151.965	826	1439	5.245	208.4	547	(3+) −1.99	10.5	0.1820	176	2.1×10^{-3}	2+, 3+
Fermium	Fm	100	[257.0951]	—	—	—	—	627	(3+) −1.96	—	—	—	—	2+, 3+
Fluorine	F	9	18.9984032	−219.7	−188.2	0.001696	69	1681	(1−) +2.87	0.51	0.8238	6.54	0.0544	1−
Francium	Fr	87	[223.0197]	27	650	—	280	375	—	2	—	63.6	trace	1+
Gadolinium	Gd	64	157.25	1312	3000	7.886	180.4	592	(3+) −2.29	15.5	0.2355	311.7	6.3×10^{-4}	3+
Gallium	Ga	31	69.723	29.77	2203	5.904	134	578.8	(3+) −0.529	5.59	0.3709	256	0.0018	1+, 3+
Germanium	Ge	32	72.61	945	2850	5.323	123	761.2	(4+) +0.124	31.8	0.3215	334.3	1.5×10^{-4}	2+, 4+

* [ ] indicates mass of longest-lived isotope.

Table D.4 Properties of Elements (continued)

Element	Symbol	Atomic Number (Z)	Atomic Mass* (u)	Melting Point (°C)	Boiling Point (°C)	Density (g/cm³) (gases measured at STP)	Atomic Radius (pm)	First Ionization Energy (kJ/mol)	Standard Reduction Potential (V) (for elements from or to oxidation state indicated)	Enthalpy of Fusion (kJ/mol)	Specific Heat (J/g·°C)	Enthalpy of Vaporization (kJ/mol)	Abundance in Earth's Crust (%)	Major Oxidation States
Gold	Au	79	196.96654	1064	2808	19.32	144	889.9	(3+)+1.52	12.4	0.12905	324.4	3×10^{-7}	1+,3+
Hafnium	Hf	72	178.49	2227	4691	13.28	159	654.4	(4+)−1.56	29.288	0.1442	661	3×10^{-4}	4+
Hassium	Hs	108	[265]	—	—	—	—	—	—	—	—	—	—	—
Helium	He	2	4.002602	−269.7 (2536 kPa)	−268.93	0.00017847	31	2372	—	0.02	5.1931	0.084	—	—
Holmium	Ho	67	164.9032	1461	2600	8.78	176.2	581	(3+)−2.33	17.1	0.1646	251	1.5×10^{-4}	3+
Hydrogen	H	1	1.00794	−259.19	−252.76	0.0000899	78	1312	(1+) 0.0000	0.117	14.298	0.904	—	1−,1+
Indium	In	49	114.82	156.61	2080	7.29	167	558.2	(3+)−0.3382	3.26	0.2407	231.8	2×10^{-5}	1+,3+
Iodine	I	53	126.90447	113.6	184.5	4.93	138	1008.4	(1−)+0.5355	15.517	0.21448	41.95	4.6×10^{-5}	1−,1+,5+,7+
Iridium	Ir	77	192.22	2447	4550	22.65	135.5	880	(4+)+0.926	26.4	0.1306	563.6	1×10^{-7}	3+,4+,5+
Iron	Fe	26	55.847	1536	2860	7.874	126	759.4	(3+)−0.4	13.807	0.4494	350	5.8	2+,3+
Krypton	Kr	36	83.8	−157.2	−153.35	0.0037493	112	1351	—	1.64	0.2480	9.03	—	—
Lanthanum	La	57	138.9055	920	3420	6.17	187	538	(3+)−2.37	8.5	0.1952	402	0.0035	3+
Lawrencium	Lr	103	[260.1054]	—	—	—	—	—	(3+)−2.06	—	—	—	—	3+
Lead	Pb	82	207.2	327	1746	11.342	175	715.6	(2+)−0.1251	4.77	0.1276	178	0.0013	2+,4+
Lithium	Li	3	6.941	180.5	1347	0.534	156	520.2	(1+)−3.045	3	3.569	148	0.002	1+
Lutetium	Lu	71	174.967	1652	3327	9.84	173.8	524	(3+)−2.3	11.9	0.1535	414	8×10^{-5}	3+
Magnesium	Mg	12	24.305	650	1105	1.738	160	737.8	(2+)−2.356	8.477	1.024	127.4	2.76	2+
Manganese	Mn	25	54.93805	1246	2061	7.43	127	717.5	(2+)−1.18	12.058	0.4791	219.7	0.1	2+,3+,4+,6+,7+
Meitnerium	Mt	109	[266]	—	—	—	—	—	—	—	—	—	—	—
Mendelevium	Md	101	[258.0986]	—	—	—	—	635	—	—	—	—	—	2+,3+
Mercury	Hg	80	200.59	−38.9	357	13.534	151	1007	(2+)+0.8535	2.2953	0.13950	59.1	2×10^{-6}	1+,2+
Molybdenum	Mo	42	95.94	2623	4679	10.28	139	685	(6+) 0.114	36	0.2508	590	1.2×10^{-4}	4+,5+,6+
Neodymium	Nd	60	144.24	1024	3111	7.003	181.4	530	(3+)−2.32	7.13	0.1903	283.7	0.004	2+,3+
Neon	Ne	10	20.1797	−248.61	−246.05	0.0008999	71	2081	—	0.34	1.0301	1.77	—	—
Neptunium	Np	93	237.0482	640	3900	20.45	155	597	(5+)−0.91	9.46	—	336	—	2+,3+,4+,5+,6+
Nickel	Ni	28	58.6934	1455	2883	8.908	124	736.7	(2+)−0.257	17.15	0.4442	375	0.0075	2+,3+,4+
Niobium	Nb	41	92.90638	2477	4858	8.57	146	664.1	(5+)−0.65	26.9	0.2648	690	0.002	4+,5+
Nitrogen	N	7	14.00674	−210	−195.8	0.0012409	71	1402	(3−)−0.092	0.72	1.0397	5.58	0.002	3−,2−,1−,1+,2+,3+,4+,5+
Nobelium	No	102	[259.1009]	—	—	—	—	642	(2+)−2.5	—	—	—	0.002	2+,3+
Osmium	Os	76	190.2	3045	5025	22.57	135	840	(4+)+0.687	31.7	0.130	627.6	2×10^{-7}	4+,6+,8+
Oxygen	O	8	15.9994	−218.8	−183	0.001429	60	1313.9	(2−) 0.815	0.44	0.91738	6.82	45.5	2−,1−
Palladium	Pd	46	106.42	1552	2940	11.99	137	805	(2+) 0.915	17.6	0.2441	362	3×10^{-7}	2+,4+
Phosphorus	P	15	30.973762	44.2	280.5	1.823	109	1012	(3−)−0.063	0.659	0.76968	49.8	0.11	3−,3+,5+
Platinum	Pt	78	195.08	1769	3824	21.41	138.5	868	(4+)+1.15	19.7	0.1326	510.4	1×10^{-6}	2+,4+
Plutonium	Pu	94	[244.0642]	640	3230	19.86	162	585	(4+)−1.25	2.8	0.138	343.5	—	3+,4+,5+,6+

* [ ] indicates mass of longest-lived isotope.

Table D.4 Properties of Elements (continued)

Element	Symbol	Atomic Number (Z)	Atomic Mass* (u)	Melting Point (°C)	Boiling Point (°C)	Density (g/cm³) (gases measured at STP)	Atomic Radius (pm)	First Ionization Energy (kJ/mol)	Standard Reduction Potential (V) (for elements from or to oxidation state indicated)	Enthalpy of Fusion (kJ/mol)	Specific Heat (J/(g·°C))	Enthalpy of Vaporization (kJ/mol)	Abundance in Earth's Crust (%)	Major Oxidation States
Polonium	Po	84	[208.9824]	254	962	9.4	164	813	$(4+)+0.73$	3.81	0.125	103	—	2−, 2+, 4+, 6+
Potassium	K	19	39.0983	63.2	766.4	0.862	231	418.8	$(1+)-2.925$	2.334	0.7566	76.9	1.84	1+
Praseodymium	Pr	59	140.90765	935	3343	6.782	182.4	522	$(3+)-2.35$	11.3	0.1930	332.6	9.1×10^{-4}	3+, 4+
Promethium	Pm	61	[144.9128]	1168	2460	7.2	183.4	536	$(3+)-2.29$	8.17	—	293	—	3+
Protactinium	Pa	91	231.03588	1552	4227	15.37	163	568	$(5+)-1.19$	14.6	—	481	trace	3+, 4+, 5+
Radium	Ra	88	226.0254	700	1630	5	228	509.1	$(2+)-2.916$	8.36	—	136.8	—	2+
Radon	Rn	86	[222.0176]	−71	−62	0.00973	140	1037	—	16.4	—	16.4	—	—
Rhenium	Re	75	186.207	3180	5650	21.232	137	760	$(7+)+0.34$	33.4	0.1368	707	1×10^{-7}	3+, 4+, 6+, 7+
Rhodium	Rh	45	102.9055	1960	3727	12.39	134	720	$(3+)+0.76$	21.6	0.2427	494	1×10^{-7}	3+, 4+, 5+
Rubidium	Rb	37	85.4678	39.5	697	1.532	248	403	$(1+)-2.925$	2.19	0.36344	69.2	0.0078	1+
Ruthenium	Ru	44	101.07	2310	4119	12.41	134	711	$(4+)+0.68$	25.5	0.2381	567.8	—	2+, 3+, 4+, 5+
Rutherfordium	Rf	104	[261]	—	—	—	—	—	—	—	—	—	—	—
Samarium	Sm	62	150.36	1072	1800	7.536	180.4	542	$(3+)-2.3$	8.9	0.1965	191	7×10^{-4}	2+, 3+
Scandium	Sc	21	44.95591	1539	2831	3	162	631	$(3+)-2.03$	15.77	0.5677	304.8	0.0022	3+
Seaborgium	Sg	106	[263]	—	—	—	—	—	—	—	—	—	—	—
Selenium	Se	34	78.96	221	685	4.79	117	940.7	$(2-)-0.67$	5.43	0.3212	26.3	5×10^{-6}	2−, 2+, 4+, 6+
Silicon	Si	14	28.0855	1411	3231	2.336	118	786.5	$(4-)-0.143$	50.2	0.7121	359	27.2	2+, 4+
Silver	Ag	47	107.8682	961	2195	10.49	144	730.8	$(1+)+0.7991$	11.65	0.23502	255	8×10^{-6}	1+
Sodium	Na	11	22.989768	97.83	897.4	0.968	186	495.9	$(1+)-2.714$	2.602	1.228	97.4	2.27	1+
Strontium	Sr	38	87.62	776.9	1412	2.6	215	549.5	$(2+)-2.89$	7.4308	0.301	137	0.0384	2+
Sulfur	S	16	32.066	115.2	444.7	2.08	103	999.6	$(2-)-0.45$	1.7272	0.7060	9.62	0.03	2−, 4+, 6+
Tantalum	Ta	73	180.9479	2980	5505	16.65	146	760.8	$(5+)-0.81$	36.57	0.1402	737	2×10^{-4}	4+, 5+
Technetium	Tc	43	97.9072	2200	4567	11.5	136	702	$(6+)+0.83$	23.0	—	577	—	2+, 4+, 6+, 7+
Tellurium	Te	52	127.6	450	990	6.25	138	869	$(2-)-1.14$	17.4	0.2016	50.6	2×10^{-7}	2−, 2+, 4+, 6+
Terbium	Tb	65	158.92534	1356	2800	8.272	177.3	564	$(3+)-2.31$	10.3	0.1819	293	1×10^{-4}	3+, 4+
Thallium	Tl	81	204.3833	303.5	1457	11.85	170	589.1	$(1+)-0.3363$	4.27	0.1288	162	7×10^{-5}	1+, 3+
Thorium	Th	90	232.0381	1750	4787	11.78	179	587	$(4+)-1.83$	16.11	0.1177	543.9	8.1×10^{-4}	4+
Thulium	Tm	69	168.93421	1545	1727	9.318	175.9	596	$(3+)-2.32$	18.4	0.1600	213	5×10^{-5}	2+, 3+
Tin	Sn	50	118.71	232	2623	7.265	141	708.4	$(4+)+0.064$	7.07	0.2274	296	2.1×10^{-4}	2+, 4+
Titanium	Ti	22	47.88	1666	3358	4.5	147	658.1	$(4+)-0.86$	14.146	0.5226	425	0.63	2+, 3+, 4+
Tungsten	W	74	183.85	3680	6000	19.3	139	770.4	$(6+)-0.09$	35.4	0.1320	806	1.2×10^{-4}	4+, 5+, 6+
Uranium	U	92	238.0289	1130	3930	19.05	156	584	$(6+)-0.83$	12.6	0.11618	423	2.3×10^{-4}	3+, 4+, 5+, 6+
Ununbium	Uub	112	[285]	—	—	—	—	—	—	—	—	—	—	—
Ununquadium	Uuq	114	[289]	—	—	—	—	—	—	—	—	—	—	—
Unununium	Uuu	111	[272]	—	—	—	—	—	—	—	—	—	—	—
Vanadium	V	23	50.9415	1917	3417	6.11	134	650.3	$(4+)-0.54$	22.84	0.4886	459.7	0.0136	2+, 3+, 4+, 5+
Xenon	Xe	54	131.29	−111.8	−108.09	0.0058971	131	1170	—	2.29	0.15832	12.64	—	—
Ytterbium	Yb	70	173.04	824	1427	6.973	193.3	603	$(3+)-2.22$	7.66	0.1545	155	3.4×10^{-4}	2+, 3+
Yttrium	Y	39	88.90585	1530	3264	4.5	180	616	$(3+)-2.37$	17.15	0.2984	393	0.0035	3+
Zinc	Zn	30	65.39	419.6	907	7.14	134	906.4	$(2+)-0.7626$	7.322	0.3884	115	0.0076	2+
Zirconium	Zr	40	91.224	1852	4400	6.51	160	659.7	$(4+)-1.7$	20.92	0.2780	590.5	0.0162	4+

* [ ] indicates mass of longest-lived isotope.

Table D.5   Electron Configurations of the Elements

	Elements	1s	2s	2p	3s	3p	3d	4s	4p	4d	4f	5s	5p	5d	5f	6s	6p	6d	6f	7s
1	Hydrogen	1																		
2	Helium	2																		
3	Lithium	2	1																	
4	Beryllium	2	2																	
5	Boron	2	2	1																
6	Carbon	2	2	2																
7	Nitrogen	2	2	3																
8	Oxygen	2	2	4																
9	Fluorine	2	2	5																
10	Neon	2	2	6																
11	Sodium	2	2	6	1															
12	Magnesium	2	2	6	2															
13	Aluminum	2	2	6	2	1														
14	Silicon	2	2	6	2	2														
15	Phosphorus	2	2	6	2	3														
16	Sulfur	2	2	6	2	4														
17	Chlorine	2	2	6	2	5														
18	Argon	2	2	6	2	6														
19	Potassium	2	2	6	2	6		1												
20	Calcium	2	2	6	2	6		2												
21	Scandium	2	2	6	2	6	1	2												
22	Titanium	2	2	6	2	6	2	2												
23	Vanadium	2	2	6	2	6	3	2												
24	Chromium	2	2	6	2	6	5	1												
25	Manganese	2	2	6	2	6	5	2												
26	Iron	2	2	6	2	6	6	2												
27	Cobalt	2	2	6	2	6	7	2												
28	Nickel	2	2	6	2	6	8	2												
29	Copper	2	2	6	2	6	10	1												
30	Zinc	2	2	6	2	6	10	2												
31	Gallium	2	2	6	2	6	10	2	1											
32	Germanium	2	2	6	2	6	10	2	2											
33	Arsenic	2	2	6	2	6	10	2	3											
34	Selenium	2	2	6	2	6	10	2	4											
35	Bromine	2	2	6	2	6	10	2	5											
36	Krypton	2	2	6	2	6	10	2	6											
37	Rubidium	2	2	6	2	6	10	2	6			1								
38	Strontium	2	2	6	2	6	10	2	6			2								
39	Yttrium	2	2	6	2	6	10	2	6	1		2								
40	Zirconium	2	2	6	2	6	10	2	6	2		2								
41	Niobium	2	2	6	2	6	10	2	6	4		1								
42	Molybdenum	2	2	6	2	6	10	2	6	5		1								
43	Technetium	2	2	6	2	6	10	2	6	5		2								
44	Ruthenium	2	2	6	2	6	10	2	6	7		1								
45	Rhodium	2	2	6	2	6	10	2	6	8		1								
46	Palladium	2	2	6	2	6	10	2	6	10										
47	Silver	2	2	6	2	6	10	2	6	10		1								
48	Cadmium	2	2	6	2	6	10	2	6	10		2								
49	Indium	2	2	6	2	6	10	2	6	10		2	1							
50	Tin	2	2	6	2	6	10	2	6	10		2	2							
51	Antimony	2	2	6	2	6	10	2	6	10		2	3							
52	Tellurium	2	2	6	2	6	10	2	6	10		2	4							
53	Iodine	2	2	6	2	6	10	2	6	10		2	5							
54	Xenon	2	2	6	2	6	10	2	6	10		2	6							

Elements	1s	2s	2p	3s	3p	3d	4s	4p	4d	4f	5s	5p	5d	5f	6s	6p	6d	6f	7s
															Sublevels				
55 Cesium	2	2	6	2	6	10	2	6	10		2	6			1				
56 Barium	2	2	6	2	6	10	2	6	10		2	6			2				
57 Lanthanum	2	2	6	2	6	10	2	6	10		2	6	1		2				
58 Cerium	2	2	6	2	6	10	2	6	10	2	2	6			2				
59 Praseodymium	2	2	6	2	6	10	2	6	10	3	2	6			2				
60 Neodymium	2	2	6	2	6	10	2	6	10	4	2	6			2				
61 Promethium	2	2	6	2	6	10	2	6	10	5	2	6			2				
62 Samarium	2	2	6	2	6	10	2	6	10	6	2	6			2				
63 Europium	2	2	6	2	6	10	2	6	10	7	2	6			2				
64 Gadolinium	2	2	6	2	6	10	2	6	10	7	2	6	1		2				
65 Terbium	2	2	6	2	6	10	2	6	10	9	2	6			2				
66 Dysprosium	2	2	6	2	6	10	2	6	10	10	2	6			2				
67 Holmium	2	2	6	2	6	10	2	6	10	11	2	6			2				
68 Erbium	2	2	6	2	6	10	2	6	10	12	2	6			2				
69 Thulium	2	2	6	2	6	10	2	6	10	13	2	6			2				
70 Ytterbium	2	2	6	2	6	10	2	6	10	14	2	6			2				
71 Lutetium	2	2	6	2	6	10	2	6	10	14	2	6	1		2				
72 Hafnium	2	2	6	2	6	10	2	6	10	14	2	6	2		2				
73 Tantalum	2	2	6	2	6	10	2	6	10	14	2	6	3		2				
74 Tungsten	2	2	6	2	6	10	2	6	10	14	2	6	4		2				
75 Rhenium	2	2	6	2	6	10	2	6	10	14	2	6	5		2				
76 Osmium	2	2	6	2	6	10	2	6	10	14	2	6	6		2				
77 Iridium	2	2	6	2	6	10	2	6	10	14	2	6	7		2				
78 Platinum	2	2	6	2	6	10	2	6	10	14	2	6	9		1				
79 Gold	2	2	6	2	6	10	2	6	10	14	2	6	10		1				
80 Mercury	2	2	6	2	6	10	2	6	10	14	2	6	10		2				
81 Thallium	2	2	6	2	6	10	2	6	10	14	2	6	10		2	1			
82 Lead	2	2	6	2	6	10	2	6	10	14	2	6	10		2	2			
83 Bismuth	2	2	6	2	6	10	2	6	10	14	2	6	10		2	3			
84 Polonium	2	2	6	2	6	10	2	6	10	14	2	6	10		2	4			
85 Astatine	2	2	6	2	6	10	2	6	10	14	2	6	10		2	5			
86 Radon	2	2	6	2	6	10	2	6	10	14	2	6	10		2	6			
87 Francium	2	2	6	2	6	10	2	6	10	14	2	6	10		2	6			1
88 Radium	2	2	6	2	6	10	2	6	10	14	2	6	10		2	6			2
89 Actinium	2	2	8	2	6	10	2	6	10	14	2	6	10		2	6	1		2
90 Thorium	2	2	6	2	6	10	2	6	10	14	2	6	10		2	6	2		2
91 Protactinium	2	2	6	2	6	10	2	6	10	14	2	6	10	2	2	6	1		2
92 Uranium	2	2	6	2	6	10	2	6	10	14	2	6	10	3	2	6	1		2
93 Neptunium	2	2	6	2	6	10	2	6	10	14	2	6	10	4	2	6	1		2
94 Plutonium	2	2	6	2	6	10	2	6	10	14	2	6	10	6	2	6			2
95 Americium	2	2	6	2	6	10	2	6	10	14	2	6	10	7	2	6			2
96 Curium	2	2	6	2	6	10	2	6	10	14	2	6	10	7	2	6	1		2
97 Berkelium	2	2	6	2	6	10	2	6	10	14	2	6	10	9	2	6			2
98 Californium	2	2	6	2	6	10	2	6	10	14	2	6	10	10	2	6			2
99 Einsteinium	2	2	6	2	6	10	2	6	10	14	2	6	10	11	2	6			2
100 Fermium	2	2	6	2	6	10	2	6	10	14	2	6	10	12	2	6			2
101 Mendelevium	2	2	6	2	6	10	2	6	10	14	2	6	10	13	2	6			2
102 Nobelium	2	2	6	2	6	10	2	6	10	14	2	6	10	14	2	6			2
103 Lawrencium	2	2	6	2	6	10	2	6	10	14	2	6	10	14	2	6	1		2
104 Rutherfordium	2	2	6	2	6	10	2	6	10	14	2	6	10	14	2	6	2		2?
105 Dubnium	2	2	6	2	6	10	2	6	10	14	2	6	10	14	2	6	3		2?
106 Seaborgium	2	2	6	2	6	10	2	6	10	14	2	6	10	14	2	6	4		2?
107 Bohrium	2	2	6	2	6	10	2	6	10	14	2	6	10	14	2	6	5		2?
108 Hassium	2	2	6	2	6	10	2	6	10	14	2	6	10	14	2	6	6		2?
109 Meitnerium	2	2	6	2	6	10	2	6	10	14	2	6	10	14	2	6	7		2?
110 Darmstadtium	2	2	6	2	6	10	2	6	10	14	2	6	10	14	2	6	8		2?
111 Unununium	2	2	6	2	6	10	2	6	10	14	2	6	10	14	2	6	9		2?
112 Unumbium	2	2	6	2	6	10	2	6	10	14	2	6	10	14	2	6	10		2?
114 Ununquadium	2	2	6	2	6	10	2	6	10	14	2	6	10	14	2	6	10	2?	2?

Table D.6 Useful Physical Constants

1 ampere is the constant current which, if maintained in two straight parallel conductors of infinite length, of negligible circular cross-section, and placed 1 meter apart in a vacuum, would produce a force of 2×10^{-7} newtons per meter of length between these conductors.

1 candela is the luminous intensity, in the perpendicular direction, of a surface of $1/600\,000$ m^2 of a blackbody at the temperature of freezing platinum at a pressure of 101 325 pascals.

1 cubic decimeter is equal to 1 liter.

1 kelvin is $1/273.16$ of the thermodynamic temperature of the triple point of water.

1 kilogram is the mass of the international prototype kilogram.

1 meter is the distance light travels in $1/299\,792\,458$ of a second.

1 mole is the amount of substance containing as many elementary entities as there are atoms in 0.012 kilogram of carbon-12.

1 second is equal to 9 192 631 770 periods of the natural electromagnetic oscillation during that transition of ground state $^2S_{1/2}$ of cesium-133, which is designated $(F = 4, M = 0) \leftrightarrow (F = 3, M = 0)$.

Avogadro constant $= 6.022\,136\,7 \times 10^{23}$

1 electronvolt $= 1.602\,177\,33 \times 10^{-19}$ J

Faraday constant $= 96\,485.309$ C/mole e^-

Ideal gas constant $= 8.314\,471$ J/mol $\cdot$ K $= 8.314\,471$ $dm^3 \cdot$ kPa/mol $\cdot$ K

Molar gas volume at STP $= 22.414\,10$ dm^3

Planck's constant $= 6.626\,075 \times 10^{-34}$ J $\cdot$ s

Speed of light $= 2.997\,924\,58 \times 10^8$ m/s

Table D.7 Names and Charges of Polyatomic Ions

1−	2−	3−	4−
Acetate, CH_3COO^-	Carbonate, CO_3^{2-}	Arsenate, AsO_4^{3-}	Hexacyanoferrate(II), $Fe(CN)_6^{4-}$
Amide, NH_2^-	Chromate, CrO_4^{2-}	Arsenite, AsO_3^{3-}	Orthosilicate, SiO_4^{4-}
Astatate, AtO_3^-	Dichromate, $Cr_2O_7^{2-}$	Borate, BO_3^{3-}	Diphosphate, $P_2O_7^{4-}$
Azide, N_3^-	Hexachloroplatinate, $PtCl_6^{2-}$	Citrate, $C_6H_5O_7^{3-}$	
Benzoate, $C_6H_5COO^-$	Hexafluorosilicate, SiF_6^{2-}	Hexacyanoferrate(III), $Fe(CN)_6^{3-}$	
Bismuthate, BiO_3^-	Molybdate, MoO_4^{2-}	Phosphate, PO_4^{3-}	
Bromate, BrO_3^-	Oxalate, $C_2O_4^{2-}$		

1−	2−	1+	2+
Chlorate, ClO_3^-	Peroxide, O_2^{2-}	Ammonium, NH_4^+	Mercury(I), Hg_2^{2+}
Chlorite, ClO_2^-	Peroxydisulfate, $S_2O_8^{2-}$	Neptunyl(V), NpO_2^+	Neptunyl(VI), NpO_2^{2+}
Cyanide, CN^-	Phosphite, HPO_3^{2-}	Plutonyl(V), PuO_2^+	Plutonyl(VI), PuO_2^{2+}
Formate, $HCOO^-$	Ruthenate, RuO_4^{2-}	Uranyl(V), UO_2^+	Uranyl(VI), UO_2^{2+}
Hydroxide, OH^-	Selenate, SeO_4^{2-}	Vanadyl(V), VO_2^+	Vanadyl(IV), VO^{2+}
Hypobromite, BrO^-	Selenite, SeO_3^{2-}		
Hypochlorite, ClO^-	Silicate, SiO_3^{2-}		
Hypophosphite, $H_2PO_2^-$	Sulfate, SO_4^{2-}		
Iodate, IO_3^-	Sulfite, SO_3^{2-}		
Nitrate, NO_3^-	Tartrate, $C_4H_4O_6^{2-}$		
Nitrite, NO_2^-	Tellurate, TeO_4^{2-}		
Perbromate, BrO_4^-	Tellurite, TeO_3^{2-}		
Perchlorate, ClO_4^-	Tetraborate, $B_4O_7^{2-}$		
Periodate, IO_4^-	Thiosulfate, $S_2O_3^{2-}$		
Permanganate, MnO_4^-	Tungstate, WO_4^{2-}		
Perrhenate, ReO_4^-			
Thiocyanate, SCN^-			
Vanadate, VO_3^-			

Table D.8   Solubility Guidelines

You will be working with water solutions, and it is helpful to have a few guidelines concerning what substances are soluble in water. A substance is considered soluble if more than 3 grams of the substance dissolve in 100 mL of water. The more common rules are listed below.

1. All common salts of the Group 1(IA) elements and ammonium ions are soluble.
2. All common acetates and nitrates are soluble.
3. All binary compounds of Group 17(VIIA) elements (other than F) with metals are soluble except those of silver, mercury(I), and lead.
4. All sulfates are soluble except those of barium, strontium, lead, calcium, silver, and mercury(I).
5. Except for those in Rule 1, carbonates, hydroxides, oxides, sulfides, and phosphates are insoluble.

Table D.9   Solubility Product Constants (at 25°C)

Substance	K_{sp}	Substance	K_{sp}	Substance	K_{sp}
AgBr	5.01×10^{-13}	$BaSO_4$	1.10×10^{-10}	Li_2CO_3	2.51×10^{-2}
$AgBrO_3$	5.25×10^{-5}	$CaCO_3$	2.88×10^{-9}	$MgCO_3$	3.47×10^{-8}
Ag_2CO_3	8.13×10^{-12}	$CaSO_4$	9.12×10^{-6}	$MnCO_3$	1.82×10^{-11}
AgCl	1.78×10^{-10}	CdS	7.94×10^{-27}	$NiCO_3$	6.61×10^{-9}
Ag_2CrO_4	1.12×10^{-12}	$Cu(IO_3)_2$	7.41×10^{-8}	$PbCl_2$	1.62×10^{-5}
$Ag_2Cr_2O_7$	2.00×10^{-7}	CuC_2O_4	2.29×10^{-8}	PbI_2	7.08×10^{-9}
AgI	8.32×10^{-17}	$Cu(OH)_2$	2.19×10^{-20}	$Pb(IO_3)_2$	3.24×10^{-13}
AgSCN	1.00×10^{-12}	CuS	6.31×10^{-36}	$SrCO_3$	1.10×10^{-10}
$Al(OH)_3$	1.26×10^{-33}	FeC_2O_4	3.16×10^{-7}	$SrSO_4$	3.24×10^{-7}
Al_2S_3	2.00×10^{-7}	$Fe(OH)_3$	3.98×10^{-38}	TlBr	3.39×10^{-6}
$BaCO_3$	5.13×10^{-9}	FeS	6.31×10^{-18}	$ZnCO_3$	1.45×10^{-11}
$BaCrO_4$	1.17×10^{-10}	Hg_2SO_4	7.41×10^{-7}	ZnS	1.58×10^{-24}

Table D.10 Acid-Base Indicators

Indicator	Lower Color	Range	Upper Color
Methyl violet	yellow-green	0.0–2.5	violet
Malachite green HCl	yellow	0.5–2.0	blue
Thymol blue	red	1.0–2.8	yellow
Naphthol yellow S	colorless	1.5–2.6	yellow
p-Phenylazoaniline	orange	2.1–2.8	yellow
Methyl orange	red	2.5–4.4	yellow
Bromophenol blue	orange-yellow	3.0–4.7	violet
Gallein	orange	3.5–6.3	red
2,5-Dinitrophenol	colorless	4.0–5.8	yellow
Ethyl orange	salmon	4.2–4.6	orange
Propyl red	pink	5.1–6.5	yellow
Bromocresol purple	green-yellow	5.4–6.8	violet
Bromoxylenol blue	orange-yellow	6.0–7.6	blue
Phenol red	yellow	6.4–8.2	red-violet
Cresol red	yellow	7.1–8.8	violet
m-Cresol purple	yellow	7.5–9.0	violet
Thymol blue	yellow	8.1–9.5	blue
Phenolphthalein	colorless	8.3–10.0	dark pink
o-Cresolphthalein	colorless	8.6–9.8	pink
Thymolphthalein	colorless	9.5–10.4	blue
Alizarin yellow R	yellow	9.9–11.8	dark orange
Methyl blue	blue	10.6–13.4	pale violet
Acid fuchsin	red	11.1–12.8	colorless
2,4,6-Trinitrotoluene	colorless	11.7–12.8	orange

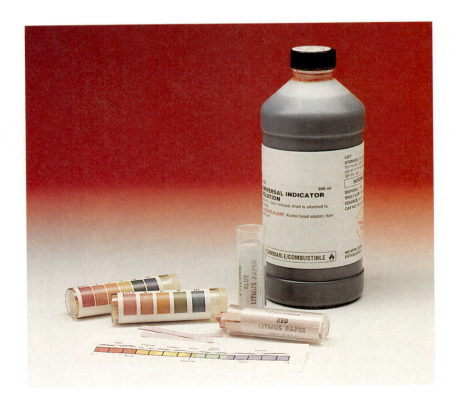

Answers to In-Chapter Practice Problems

CHAPTER 5

1. Write the formula for each of the following compounds.

   a. lithium oxide
   Li_2O

   b. calcium bromide
   $CaBr_2$

   c. sodium oxide
   Na_2O

   d. aluminum sulfide
   Al_2S_3

2. Write the formula for the compound formed from each of the following pairs of elements.

   a. barium and oxygen
   BaO

   b. strontium and iodine
   SrI_2

   c. lithium and chlorine
   $LiCl$

   d. radium and chlorine
   $RaCl_2$

3. Write the formula for the compound made from each of the following ions.

   a. ammonium and sulfite ions
   $(NH_4)_2SO_3$

   b. calcium and monohydrogen phosphate ions
   $CaHPO_4$

   c. ammonium and dichromate ions
   $(NH_4)_2Cr_2O_7$

   d. barium and nitrate ions
   $Ba(NO_3)_2$

4. Write the formula for each of the following compounds.

   a. sodium phosphate
   Na_3PO_4

   b. magnesium hydroxide
   $Mg(OH)_2$

   c. ammonium phosphate
   $(NH_4)_3PO_4$

   d. potassium dichromate
   $K_2Cr_2O_7$

5. Write the formula for the compound made from each of the following pairs of ions.

   a. copper(I) and sulfite
   Cu_2SO_3

   b. tin(IV) and fluoride
   SnF_4

   c. gold(III) and cyanide
   $Au(CN)_3$

   d. lead(II) and sulfide
   PbS

6. Write the names of the following compounds.

   a. $Pb(NO_3)_2$
   lead(II) nitrate

   b. Mn_2O_3
   manganese(III) oxide

   c. $Ni(C_2H_3O_2)_2$
   nickel(II) acetate

   d. HgF_2
   mercury(II) fluoride

7. Name the following molecular compounds.

   a. S_2Cl_2
   disulfur dichloride

   b. CS_2
   carbon disulfide

   c. SO_3
   sulfur trioxide

   d. P_4O_{10}
   tetraphosphorus decoxide

8. Write the formulas for the following molecular compounds.

   a. carbon tetrachloride
   CCl_4

   b. iodine heptafluoride
   IF_7

   c. dinitrogen monoxide
   N_2O

   d. sulfur dioxide
   SO_2

CHAPTER 6

Write word equations and chemical equations for the following reactions.

1. **Magnesium metal and water combine to form solid magnesium hydroxide and hydrogen gas.**
   magnesium + water
   $\rightarrow$ magnesium hydroxide + hydrogen
   $Mg(s) + 2H_2O(1) \rightarrow Mg(OH)_2(s) + H_2(g)$

2. **An aqueous solution of hydrogen peroxide (dihydrogen dioxide) and solid lead(II) sulfide combine to form solid lead(II) sulfate and liquid water.**
   hydrogen peroxide + lead(II) sulfide
   $\rightarrow$ lead(II) sulfate + water
   $4H_2O_2(aq) + PbS(s) \rightarrow PbSO_4(s) + 4H_2O(1)$

3. **When energy is added to solid manganese(II) sulfate heptahydrate crystals, they break down to form liquid water and solid manganese(II) sulfate monohydrate.**
   manganese(II) sulfate heptahydrate + energy
   $\rightarrow$ water + manganese(II) sulfate monohydrate
   $MnSO_4 \cdot 7H_2O(s) + energy$
   $\rightarrow 6H_2O(1) + MnSO_4 \cdot H_2O(s)$

4. **Solid potassium reacts with liquid water to produce aqueous potassium hydroxide and hydrogen gas.**
   potassium + water
   $\rightarrow$ potassium hydroxide + hydrogen gas
   $2K(s) + 2H_2O(l) \rightarrow 2 KOH(aq) + H_2(g)$

1. **Calculate ΔEN for the pairs of atoms in the following bonds.**

   a. Ca—S
   1.5

   b. Ba—O
   2.6

   c. C—Br
   0.3

   d. Ca—F
   3.0

   e. H—Br
   0.7

2. **Use ΔEN to classify the bonds in problem 1 as covalent, polar covalent, or ionic.**

   a. polar covalent

   b. ionic

   c. covalent

   d. ionic

   e. polar covalent

CHAPTER 10

Complete the following table.

1. **Melting point of iron**
   1535; $T_C = 1808 - 273 = 1535°C$

2. **Household oven**
   448 - 478; $T_K = 175 + 273 = 448$; $205 + 273 = 478$

3. **Food freezer**
   -18; $T_C = 255 - 273 = -18°C$

4. **Sublimation point of dry ice, $CO_2(s)$**
   194.5; $T_K = -78.5 + 273 = 194.5$ K

5. **Boiling point of nitrogen, N_2**
   -195.6; $T_C = 77.4 - 273 = -195.6$

Use Table 11.1 and the equation 1.00 in. = 25.4 mm to convert the following measurements.

1. **59.8 in. Hg to psi**

59.8 in. Hg	25.4 mm Hg	14.7 psi
1	1.00 in. Hg	760 mm Hg

   $= \dfrac{59.8 \text{ in. Hg}}{1} \Big| \dfrac{25.4 \text{ mm Hg}}{1.00 \text{ in. Hg}} \Big| \dfrac{14.7 \text{ psi}}{760 \text{ mm Hg}}$

   $= \dfrac{59.8 \times 25.4 \times 14.7 \text{ psi}}{760} = 29.4 \text{ psi}$

2. **7.35 psi to mm Hg**

7.35 psi	1 atm	760 mm Hg
1	14.7 psi	1 atm

   $= \dfrac{7.35 \text{ psi}}{1} \Big| \dfrac{1 \text{ atm}}{14.7 \text{ psi}} \Big| \dfrac{760 \text{ mm Hg}}{1 \text{ atm}}$

   $= \dfrac{7.35 \times 760 \text{ mm Hg}}{14.7} = 3.80 \times 10^2 \text{ mm Hg}$

3. **1140 mm Hg to kPa**

1140 mm Hg	1.00 atm	101.3 kPa
1	760 mm Hg	1.00 atm

   $= \dfrac{1140 \text{ mm Hg}}{1} \Big| \dfrac{1.00 \text{ atm}}{760 \text{ mm Hg}} \Big| \dfrac{101.3 \text{ kPa}}{1.00 \text{ atm}}$

   $= \dfrac{1140 \times 101.3 \text{ kPa}}{760} = 152 \text{ kPa}$

4. **19.0 psi to kPa**

19.0 psi	1.00 atm	101.3 kPa
1	14.7 psi	1.00 atm

   $= \dfrac{19.0 \text{ psi}}{1} \Big| \dfrac{1.00 \text{ atm}}{14.7 \text{ psi}} \Big| \dfrac{101.3 \text{ kPa}}{1.00 \text{ atm}}$

   $= \dfrac{19.0 \times 101.3 \text{ kPa}}{14.7} = 131 \text{ kPa}$

5. 202 kPa to psi

$$\frac{202 \text{ kPa}}{1} \left| \frac{1.00 \text{ atm}}{101.3 \text{ kPa}} \right| \frac{14.7 \text{ psi}}{1.00 \text{ atm}}$$

$$= \frac{202 \text{ kPa}}{1} \left| \frac{1.00 \text{ atm}}{101.3 \text{ kPa}} \right| \frac{14.7 \text{ psi}}{1.00 \text{ atm}}$$

$$= \frac{202 \times 14.7 \text{ psi}}{101.3} = 29.3 \text{ psi}$$

Assume that the temperature remains constant in the following problems.

6. Bacteria produce methane gas in sewage-treatment plants. This gas is often captured or burned. If a bacterial culture produces 60.0 mL of methane gas at 700.0 mm Hg, what volume would be produced at 760.0 mm Hg?

$$\frac{60.0 \text{ mL}}{1} \left| \frac{700.0 \text{ mm Hg}}{760.0 \text{ mm Hg}} \right.$$

$$= \frac{60.0 \text{ mL}}{1} \left| \frac{700.0 \text{ mm Hg}}{760.0 \text{ mm Hg}} \right.$$

$$= \frac{60.0 \text{ mL} \times 700.0}{760.0} = 55.3 \text{ mL}$$

7. At one sewage-treatment plant, bacteria cultures produce 1000 L of methane gas per day at 1.0 atm pressure. What volume tank would be needed to store one day's production at 5.0 atm?

$$\frac{1000 \text{ L}}{1} \left| \frac{1 \text{ atm}}{5 \text{ atm}} = \frac{1000 \text{ L}}{1} \right| \frac{1 \text{ atm}}{5 \text{ atm}}$$

$$= \frac{1000 \text{ L}}{5} = 200 \text{ L}$$

8. Hospitals buy 400-L cylinders of oxygen gas compressed at 150 atm. They administer oxygen to patients at 3.0 atm in a hyperbaric oxygen chamber. What volume of oxygen can a cylinder supply at this pressure?

$$\frac{400 \text{ L}}{1} \left| \frac{150 \text{ atm}}{3 \text{ atm}} = \frac{400 \text{ L}}{1} \right| \frac{150 \text{ atm}}{3 \text{ atm}}$$

$$= \frac{400 \text{ L} \times 150}{3} = 20\,000 \text{ L}$$

9. If the valve in a tire pump with a volume of 0.78 L fails at a pressure of 9.0 atm, what would be the volume of air in the cylinder just before the valve fails?

$$\frac{0.78 \text{ L}}{1} \left| \frac{1.0 \text{ atm}}{9.0 \text{ atm}} = \frac{0.78 \text{ L}}{1} \right| \frac{1.0 \text{ atm}}{9.0 \text{ atm}}$$

$$= \frac{0.78 \text{ L}}{9.0} = 0.087 \text{ L}$$

10. The volume of a scuba tank is 10.0 L. It contains a mixture of nitrogen and oxygen at 290.0 atm. What volume of this mixture could the tank supply to a diver at 2.40 atm?

$$\frac{10.0 \text{ L}}{1} \left| \frac{290.0 \text{ atm}}{2.40 \text{ atm}} = \frac{10.0 \text{ L}}{1} \right| \frac{290.0 \text{ atm}}{2.40 \text{ atm}}$$

$$= \frac{10.0 \text{ L} \times 290.0}{2.40} = 1210 \text{ L}$$

11. A 1.00-L balloon is filled with helium at 1.20 atm. If the balloon is squeezed into a 0.500-L beaker and doesn't burst, what is the pressure of the helium?

$$\frac{1.20 \text{ atm}}{1} \left| \frac{1.00 \text{ L}}{0.500 \text{ L}} = \frac{1.20 \text{ atm}}{1} \right| \frac{1.00 \text{ L}}{0.500 \text{ L}}$$

$$= \frac{1.20 \text{ atm}}{0.500} = 2.40 \text{ atm}$$

Assume that the pressure remains constant in problems 12 to 16.

12. A balloon is filled with 3.0 L of helium at 310 K and 1 atm. The balloon is placed in an oven where the temperature reaches 340 K. What is the new volume of the balloon?

$$\frac{3.0 \text{ L}}{1} \left| \frac{340 \text{ K}}{310 \text{ K}} = \frac{3.0 \text{ L}}{1} \right| \frac{340 \text{ K}}{310 \text{ K}}$$

$$= \frac{3.0 \text{ L} \times 340}{310} = 3.3 \text{ L}$$

13. A 4.0-L sample of methane gas is collected at 30.0°C. Predict the volume of the sample at 0°C.

$$\frac{4.0 \text{ L}}{1} \left| \frac{(0.0 + 273) \text{K}}{(30.0 + 273) \text{K}} \right.$$

$$= \frac{4.0 \text{ L}}{1} \left| \frac{273 \text{ K}}{303 \text{ K}} \right.$$

$$= \frac{4.0 \text{ L} \times 273}{303} = 3.6 \text{ L}$$

14. A 25-L sample of nitrogen is heated from 110°C to 260°C. What volume will the sample occupy at the higher temperature?

$$\frac{25 \text{ L}}{1} \left| \frac{(260 + 273) \text{K}}{(110 + 273) \text{K}} \right.$$

$$= \frac{25 \text{ L}}{1} \left| \frac{533 \text{ K}}{383 \text{ K}} \right.$$

$$= \frac{25 \text{ L} \times 533}{383} = 35 \text{ L}$$

15. The volume of a 16-g sample of oxygen is 11.2 L at 273 K and 1.00 atm. Predict the volume of the sample at 409 K.

$$\frac{11.2 \text{ L}}{1} \left| \frac{409 \text{ K}}{273 \text{ K}} = \frac{11.2 \text{ L}}{1} \right| \frac{409 \text{ K}}{273 \text{ K}}$$

$$= \frac{11.2 \text{ L} \times 409}{273} = 16.8 \text{ L}$$

16. The volume of a sample of argon is 8.5 mL at 15°C and 101 kPa. What will its volume be at 0.00°C and 101 kPa?

$$\frac{8.5 \text{ mL}}{1} \left| \frac{273 \text{ K}}{(15 + 273) \text{K}} \right.$$

$$= \frac{8.5 \text{ mL}}{1} \left| \frac{273 \text{ K}}{288 \text{ K}} \right.$$

$$= \frac{8.5 \text{ mL} \times 273}{288} = 8.1 \text{ mL}$$

17. A 2.7-L sample of nitrogen is collected at 121 kPa and 288 K. If the pressure increases to 202 kPa and the temperature rises to 303 K, what volume will the nitrogen occupy?

$$\frac{2.7\ \text{L}}{1} \left| \frac{121\ \text{kPa}}{202\ \text{kPa}} \right| \frac{303\ \text{K}}{288\ \text{K}}$$

$$= \frac{2.7\ \text{L}}{1} \left| \frac{121\ \cancel{\text{kPa}}}{202\ \cancel{\text{kPa}}} \right| \frac{303\ \cancel{\text{K}}}{288\ \cancel{\text{K}}}$$

$$= \frac{2.7\ \text{L} \times 121 \times 303}{202 \times 288} = 1.7\ \text{L}$$

18. A chunk of subliming carbon dioxide (dry ice) generates a 0.80-L sample of gaseous CO_2 at 22°C and 720 mm Hg. What volume will the carbon dioxide gas have at STP?

$$\frac{0.80\ \text{L}}{1} \left| \frac{720\ \text{mm Hg}}{760\ \text{mm Hg}} \right| \frac{273\ \text{K}}{(22 + 273)\text{K}}$$

$$= \frac{0.80\ \text{L}}{1} \left| \frac{720\ \cancel{\text{mm Hg}}}{760\ \cancel{\text{mm Hg}}} \right| \frac{273\ \cancel{\text{K}}}{295\ \cancel{\text{K}}}$$

$$= \frac{0.80\ \text{L} \times 720 \times 273}{760 \times 295} = 0.70\ \text{L}$$

CHAPTER 12

1. Without calculating, decide whether 50.0 g of sulfur or 50.0 g of tin represents the greater number of atoms. Verify your answer by calculating.

Because sulfur has a smaller molar mass, there will be a greater number of atoms in 50.0 g of sulfur than in 50.0 g of tin.

$$\frac{50.0\ \cancel{\text{g S}}}{1} \left| \frac{1\ \cancel{\text{mol S}}}{32.1\ \cancel{\text{g S}}} \right| \frac{6.02 \times 10^{23}\ \text{S atoms}}{1\ \cancel{\text{mol S}}}$$

$$= \frac{50.0}{1} \left| \frac{1}{32.1} \right| \frac{6.02 \times 10^{23}\ \text{S atoms}}{1}$$

$$= \frac{50.0 \times 6.02 \times 10^{23}\ \text{S atoms}}{32.1}$$

$$= 9.38 \times 10^{23}\ \text{S atoms}$$

$$\frac{50.0\ \cancel{\text{g Sn}}}{1} \left| \frac{1\ \cancel{\text{mol Sn}}}{119\ \cancel{\text{g Sn}}} \right| \frac{6.02 \times 10^{23}\ \text{Sn atoms}}{1\ \cancel{\text{mol Sn}}}$$

$$= \frac{50.0}{1} \left| \frac{1}{119} \right| \frac{6.02 \times 10^{23}\ \text{Sn atoms}}{1}$$

$$= \frac{50.0 \times 6.02 \times 10^{23}\ \text{Sn atoms}}{119}$$

$$= 2.53 \times 10^{23}\ \text{Sn atoms}$$

2. Determine the number of atoms in each sample below.

a. 98.3 g mercury, Hg

$$\frac{98.3\ \cancel{\text{g Hg}}}{1} \left| \frac{\cancel{\text{mol Hg}}}{200.59\ \cancel{\text{g Hg}}} \right| \frac{6.02 \times 10^{23}\ \text{Hg atoms}}{\cancel{\text{mol Hg}}}$$

$$= 2.95 \times 10^{23}\ \text{Hg atoms}$$

b. 45.6 g gold, Au

$$\frac{45.6\ \cancel{\text{g Au}}}{1} \left| \frac{\cancel{\text{mol Au}}}{196.967\ \cancel{\text{g Au}}} \right| \frac{6.02 \times 10^{23}\ \text{Au atoms}}{\cancel{\text{mol Au}}}$$

$$= 1.39 \times 10^{23}\ \text{Au atoms}$$

c. 10.7 g lithium, Li

$$\frac{10.7\ \cancel{\text{g Li}}}{1} \left| \frac{\cancel{\text{mol Li}}}{6.941\ \cancel{\text{g Li}}} \right| \frac{6.02 \times 10^{23}\ \text{Li atoms}}{\cancel{\text{mol Li}}}$$

$$= 9.28 \times 10^{23}\ \text{Li atoms}$$

d. 144.6 g tungsten, W

$$\frac{144.6\ \cancel{\text{g W}}}{1} \left| \frac{\cancel{\text{mol W}}}{183.85\ \cancel{\text{g W}}} \right| \frac{6.02 \times 10^{23}\ \text{W atoms}}{\cancel{\text{mol W}}}$$

$$= 4.73 \times 10^{23}\ \text{W atoms}$$

3. Determine the number of moles in each sample below.

a) 6.84 g sucrose, $C_{12}H_{22}O_{11}$

12 C atoms	$12 \times 12.0\ \text{u} =$	144.0 u
22 H atoms	$22 \times 1.0\ \text{u} =$	22.0 u
11 O atoms	$11 \times 16.0\ \text{u} =$	+176 u
molecular mass $C_{12}H_{22}O_{11}$		342 u
molar mass $C_{12}H_{22}O_{11}$		342 g/mol

$$\frac{6.84\ \text{g }C_{12}H_{22}O_{11}}{1} \left| \frac{1\ \text{mol }C_{12}H_{22}O_{11}}{342\ \text{g }C_{12}H_{22}O_{11}} \right.$$

$$= \frac{6.84\ \cancel{\text{g }C_{12}H_{22}O_{11}}}{1} \left| \frac{1\ \text{mol }C_{12}H_{22}O_{11}}{342\ \cancel{\text{g }C_{12}H_{22}O_{11}}} \right.$$

$$= \frac{6.84}{1} \left| \frac{1\ \text{mol }C_{12}H_{22}O_{11}}{342} \right.$$

$$= 2.00 \times 10^{-2}\ \text{mol }C_{12}H_{22}O_{11}$$

b) 16.0 g sulfur dioxide, SO_2

SO_2

1 S atom	$1 \times 32.1\ \text{u} =$	32.1 u
2 O atoms	$2 \times 16.0\ \text{u} =$	+32.0 u
molecular mass SO_2		64.1 u
molar mass SO_2		64.1 g/mol

$$\frac{16.0\ \text{g }SO_2}{1} \left| \frac{1\ \text{mol }SO_2}{64.1\ \text{g }SO_2} \right.$$

$$= \frac{16.0\ \cancel{\text{g }SO_2}}{1} \left| \frac{1\ \text{mol }SO_2}{64.1\ \cancel{\text{g }SO_2}} \right.$$

$$= \frac{16.0}{1} \left| \frac{1\ \text{mol }SO_2}{64.1} \right.$$

$$= 0.250\ \text{mol }SO_2$$

c) 68.0 g ammonia, NH_3

NH_3

1 N atom	$1 \times 14.0\ \text{u} =$	14.0 u
3 H atoms	$3 \times 1.0\ \text{u} =$	+ 3.0 u
molecular mass NH_3		17.0 u
molar mass NH_3		17.0 g/mol

$$\frac{68.0 \text{ g NH}_3}{1} \, \bigg| \, \frac{1 \text{ mol NH}_3}{17.0 \text{ g NH}_3}$$

$$= \frac{68.0 \text{ g NH}_3}{1} \, \bigg| \, \frac{1 \text{ mol NH}_3}{17.0 \text{ g NH}_3}$$

$$= \frac{68.0}{1} \, \bigg| \, \frac{1 \text{ mol NH}_3}{17.0}$$

$$= 4.00 \text{ mol NH}_3$$

d) 17.5 g copper(II) oxide, CuO

1 Cu atom	1×63.5 u $=$	63.5 u
1 O atom	1×16.0 u $=$	$+16.0$ u
formula mass CuO		79.5 u
molar mass CuO		79.5 g/mol

$$\frac{17.5 \text{ g CuO}}{1} \, \bigg| \, \frac{1 \text{ mol CuO}}{79.5 \text{ g CuO}}$$

$$= \frac{17.5 \text{ g CuO}}{1} \, \bigg| \, \frac{1 \text{ mol CuO}}{79.5 \text{ g CuO}}$$

$$= \frac{17.5}{1} \, \bigg| \, \frac{1 \text{ mol CuO}}{79.5}$$

$$= 0.220 \text{ mol CuO}$$

4. Determine the mass of the following molar quantities.

a) 3.52 mol Si

$$\frac{3.52 \text{ mol Si}}{1} \, \bigg| \, \frac{28.1 \text{ g Si}}{1 \text{ mol Si}}$$

$$= \frac{3.52 \text{ mol Si}}{1} \, \bigg| \, \frac{28.1 \text{ g Si}}{1 \text{ mol Si}}$$

$$= 3.52 \times 28.1 \text{ g Si}$$

$$= 98.9 \text{ g Si}$$

b) 1.25 mol aspirin, $C_9H_8O_4$

$C_9H_8O_4$

9 C atoms	9×12.0 u $=$	108 u
8 H atoms	8×1.0 u $=$	$+8.0$ u
4 O atoms	4×16.0 u $=$	$+64$ u
molecular mass $C_9H_8O_4$		180.0 u
molar mass $C_9H_8O_4$		180.0 g/mol

$$\frac{1.25 \text{ mol } C_9H_8O_4}{1} \, \bigg| \, \frac{180.0 \text{ g } C_9H_8O_4}{1 \text{ mol } C_9H_8O_4}$$

$$= \frac{1.25 \text{ mol } C_9H_8O_4}{1} \, \bigg| \, \frac{180.0 \text{ g } C_9H_8O_4}{1 \text{ mol } C_9H_8O_4}$$

$$= 1.25 \times 180 \text{ g } C_9H_8O_4$$

$$= 225 \text{ g } C_9H_8O_4$$

c) 0.550 mol F_2

F_2

2 F atoms	2×19.0 u $=$	38.0 u
molecular mass F_2		38.0 u
molar mass F_2		38.0 g/mol

$$(0.550 \text{ mol } F_2)(38.0 \text{ g } F_2/\text{mol } F_2)$$

$$= \frac{0.550 \text{ mol } F_2}{1} \, \bigg| \, \frac{38.0 \text{ g } F_2}{1 \text{ mol } F_2}$$

$$= 0.550 \times 38.0 \text{ g } F_2 = 20.9 \text{ g } F_2$$

d) 2.35 mol barium iodide, BaI_2

BaI_2

1 Ba atom	1×137 u $=$	137 u
2 I atoms	2×127 u $=$	$+254$ u
formula mass BaI_2		391 u
molar mass BaI_2		391 g/mol

$$\frac{2.35 \text{ mol } BaI_2}{1} \, \bigg| \, \frac{391 \text{ g } BaI_2}{1 \text{ mol } BaI_2}$$

$$= \frac{2.35 \text{ mol } BaI_2}{1} \, \bigg| \, \frac{391 \text{ g } BaI_2}{1 \text{ mol } BaI_2}$$

$$= 2.35 \times 391 \text{ g } BaI_2 = 919 \text{ g } BaI_2$$

5. The combustion of propane, C_3H_8, a fuel used in backyard grills and camp stoves, produces carbon dioxide and water vapor.

$$C_3H_8(g) + 5O_2(g) \rightarrow 3CO_2(g) + 4H_2O(g)$$

What mass of carbon dioxide forms when 95.6 g of propane burns?

$$\frac{95.6 \text{ g } C_3H_8}{1} \, \bigg| \, \frac{1 \text{ mol } C_3H_8}{44.0 \text{ g } C_3H_8} \cdots$$

$$\bigg| \, \frac{3 \text{ mol } CO_2}{1 \text{ mol } C_3H_8} \, \bigg| \, \frac{44.0 \text{ g } CO_2}{1 \text{ mol } CO_3}$$

$$= \frac{95.6 \text{ g } C_3H_8}{1} \, \bigg| \, \frac{1 \text{ mol } C_3H_8}{44.0 \text{ g } C_3H_8} \cdots$$

$$\bigg| \, \frac{3 \text{ mol } CO_2}{1 \text{ mol } C_3H_8} \, \bigg| \, \frac{44.0 \text{ g } CO_2}{1 \text{ mol } CO_2}$$

$$= 95.6 \times 3 \times \frac{44.0 \text{ g } CO_2}{44.0} = 287 \text{ g } CO_2$$

6. Solid xenon hexafluoride is prepared by allowing xenon gas and fluorine gas to react.

$$Xe(g) + 3F_2(g) \rightarrow XeF_6(s)$$

How many grams of fluorine are required to produce 10.0 g of XeF_6?

$$\frac{10 \text{ g } XeF_6}{1} \, \bigg| \, \frac{1 \text{ mol } XeF_6}{245.3 \text{ g } XeF_6} \cdots$$

$$\bigg| \, \frac{3 \text{ mol } F_2}{1 \text{ mol } XeF_6} \, \bigg| \, \frac{38.0 \text{ g } F_2}{1 \text{ mol } F_2}$$

$$= \frac{10.0 \text{ g } XeF_6}{1} \, \bigg| \, \frac{1 \text{ mol } XeF_6}{245.3 \text{ g } XeF_6} \cdots$$

$$\bigg| \, \frac{3 \text{ mol } F_2}{1 \text{ mol } XeF_6} \, \bigg| \, \frac{38.0 \text{ g } F_2}{1 \text{ mol } F_2}$$

$$= \frac{10.0 \times 3 \times 38.0 \text{ g } F_2}{245.3} = 4.65 \text{ g } F_2$$

7. Using the reaction in Practice Problem 6, how many grams of xenon are required to produce 10.0 g of XeF_6?

$$\frac{10.0 \text{ g } XeF_2}{1} \left| \frac{1 \text{ mol } XeF_6}{245.3 \text{ g } XeF_6} \right. \cdots$$

$$\left| \frac{1 \text{ mol } Xe}{1 \text{ mol } XeF_6} \right| \frac{131.3 \text{ g } Xe}{1 \text{ mol } Xe}$$

$$= \frac{10.0 \text{ g } \cancel{XeF_6}}{1} \left| \frac{1 \text{ mol } \cancel{XeF_6}}{245.3 \text{ g } \cancel{XeF_6}} \right. \cdots$$

$$\left| \frac{1 \text{ mol } \cancel{Xe}}{1 \text{ mol } \cancel{XeF_6}} \right| \frac{131.3 \text{ g } Xe}{1 \text{ mol } \cancel{Xe}}$$

$$= 10.0 \times 1 \times \frac{131.3 \text{ g } Xe}{245.3} = 5.35 \text{ g } Xe$$

8. What mass of sulfur must burn to produce 3.42 L of SO_2 at 273°C and 101 kPa? The reaction is $S(s) + O_2(g) \rightarrow SO_2(g)$.

Note that the volume must be corrected to standard temperature (0°C) by applying the factor

$$\frac{273 \text{ K}}{(273 + 273)\text{K}} = \frac{273 \text{ K}}{546 \text{ K}}$$

$$\frac{3.42 \text{ L } SO_2}{1} \left| \frac{273 \text{ K}}{546 \text{ K}} \right| \frac{1 \text{ mol } SO_2}{22.4 \text{ L } SO_2} \cdots$$

$$\left| \frac{1 \text{ mol } S}{1 \text{ mol } SO_2} \right| \frac{32.1 \text{ g } S}{1 \text{ mol } S}$$

$$= \frac{3.42 \text{ } \cancel{L SO_2}}{1} \left| \frac{273 \text{ } \cancel{K}}{546 \text{ } \cancel{K}} \right| \frac{1 \text{ mol } \cancel{SO_2}}{22.4 \text{ } \cancel{L SO_2}} \cdots$$

$$\left| \frac{1 \text{ mol } \cancel{S}}{1 \text{ mol } \cancel{SO_2}} \right| \frac{32.1 \text{ g } S}{1 \text{ mol } \cancel{S}}$$

$$= 3.42 \times 273 \times \frac{32.1 \text{ g } S}{546 \times 22.4} = 2.45 \text{ g } S$$

9. What volume of hydrogen gas can be produced by reacting 4.20 g sodium in excess water at 50.0°C and 106 kPa? The reaction is $2Na + 2H_2O \rightarrow 2NaOH + H_2$

$$\frac{4.20 \text{ g } Na}{1} \left| \frac{1 \text{ mol } Na}{23.0 \text{ g } Na} \right| \frac{1 \text{ mol } H_2}{2 \text{ mol } Na} \cdots$$

$$\left| \frac{22.4 \text{ L } H_2}{1 \text{ mol } H_2} \right| \frac{323 \text{ K}}{273 \text{ K}} \left| \frac{101 \text{ kPa}}{106 \text{ kPa}} \right.$$

$$= \frac{4.20 \text{ g } \cancel{Na}}{1} \left| \frac{1 \text{ mol } \cancel{Na}}{23.0 \text{ g } \cancel{Na}} \right| \frac{1 \text{ mol } \cancel{H_2}}{2 \text{ mol } \cancel{Na}} \cdots$$

$$\left| \frac{22.4 \text{ L } H_2}{1 \text{ mol } \cancel{H_2}} \right| \frac{323 \text{ } \cancel{K}}{273 \text{ } \cancel{K}} \left| \frac{101 \text{ } \cancel{kPa}}{106 \text{ } \cancel{kPa}} \right.$$

$$= \frac{4.20 \times 22.4 \text{ L } H_2 \times 323 \times 101}{23.0 \times 2 \times 273 \times 106} = 2.31 \text{ L } H_2$$

10. How many moles of helium are contained in a 5.00-L canister at 101 kPa and 30.0°C?

$$n = \frac{VP}{RT} = \frac{(5.00 \text{ L})(101 \text{ kPa})}{\left(\dfrac{8.31 \text{ kPa} \cdot \text{L}}{1 \text{ mol} \cdot \text{K}}\right) 303 \text{ K}}$$

$$= \frac{5.00 \text{ } \cancel{L} \times 101 \text{ } \cancel{kPa} \times \text{mol} \cdot \cancel{K}}{8.31 \text{ } \cancel{kPa} \cdot \cancel{L} \times 303 \text{ } \cancel{K}}$$

$$= \frac{5.00 \times 101 \times 1 \text{ mol}}{8.31 \times 303} = 0.201 \text{ mol}$$

11. What is the volume of 0.020 mol Ne at 0.505 kPa and 27.0°C?

$$V = \frac{nRT}{P} = \frac{(0.020 \text{ mol Ne})\left(\dfrac{8.31 \text{ kPa} \cdot \text{L}}{1 \text{ mol} \cdot \text{K}}\right)(300 \text{ K})}{0.505 \text{ kPa}}$$

$$= 99 \text{ L}$$

12. How much zinc must react in order to form 15.5 L of hydrogen, $H_2(g)$, at 32.0°C and 115 kPa?
$Zn(s) + H_2SO_4(aq) \rightarrow ZnSO_4(aq) + H_2(g)$

First, use the ideal gas law to determine moles of H_2.

$$PV = nRT \qquad n = \frac{PV}{RT}$$

$$n = \frac{(115 \text{ kPa})(15.5 \text{ L})}{\left(\dfrac{8.31 \text{ kPa} \cdot \text{L}}{1 \text{ mol} \cdot \text{K}}\right)(305 \text{ K})}$$

$$n = 0.703 \text{ mol } H_2$$

Now, determine the mass of zinc.

$$\frac{0.703 \text{ mol } H_2}{1} \left| \frac{1 \text{ mol } Zn}{1 \text{ mol } H_2} \right| \frac{65.39 \text{ g } Zn}{1 \text{ mol } Zn}$$

$$= 46.0 \text{ g } Zn$$

CHAPTER 13

1. How would you prepare 1.00 L of a 0.400M solution of copper(II) sulfate, $CuSO_4$?

Dissolve 63.8 g of $CuSO_4$ in 1.00 L of solution;

Molar mass $CuSO_4$ = 159.61 g/mol

$$\frac{1.00 \text{ L soln}}{1} \left| \frac{0.400 \text{ mol } CuSO_4}{1 \text{ L soln}} \right| \frac{159.61 \text{ g } CuSO_4}{1 \text{ mol } CuSO_4}$$

$$= 63.8 \text{ g } CuSO_4$$

2. How would you prepare 2.50 L of a 0.800M solution of potassium nitrate, KNO_3?

Dissolve 202 g of KNO_3 in 2.50 L of solution;

Molar mass KNO_3 = 101.10 g/mol

$$\frac{2.50 \text{ L soln}}{1} \left| \frac{0.800 \text{ mol } KNO_3}{1 \text{ L soln}} \right| \frac{101.10 \text{ g}}{1 \text{ mol}}$$

$$= 202 \text{ g } KNO_3$$

3. What mass of sucrose, $C_{12}H_{22}O_{11}$, must be dissolved to make 460 mL of a 1.10M solution?

173 g of sucrose;

Molar mass $C_{12}H_{22}O_{11}$ = 342.30 g/mol

$$\frac{460 \text{ mL soln}}{1} \left| \frac{1 \text{ L}}{10^3 \text{ mL}} \right| \frac{1.10 \text{ mol } C_{12}H_{22}O_{11}}{1 \text{ L soln}} \cdots$$

$$\left| \frac{342.30 \text{ g}}{1 \text{ mol}} \right| = 170 \text{ g } C_{12}H_{22}O_{11}$$

4. What mass of lithium chloride, LiCl, must be dissolved to make a 0.194M solution that has a volume of 1.00 L?

8.23 g of LiCl;

Molar mass LiCl = 42.40 g/mol

$$\frac{1.00 \text{ L soln}}{1} \left| \frac{0.194 \text{ mol LiCl}}{1 \text{ L soln}} \right| \frac{42.40 \text{ g}}{1 \text{ mol}}$$

= 8.23 g LiCl

5. What is the molarity of a solution that contains 14 g of sodium sulfate, Na_2SO_4, dissolved in 1.6 L of solution?

0.062M Na_2SO_4;

Molar mass Na_2SO_4 = 142.06 g/mol

$$\frac{14 \text{ g } Na_2SO_4}{1.6 \text{ L soln}} \left| \frac{1 \text{ mol}}{142.06 \text{ g}} \right.$$

= 0.062 mol Na_2SO_4/L soln

= 0.062M Na_2SO_4

6. Calculate the molarity of a solution, given that its volume is 820 mL and that it contains 7.4 g of ammonium chloride, NH_4Cl.

0.17M NH_4Cl;

Molar mass NH_4Cl = 53.50 g/mol

$$\frac{7.4 \text{ g } NH_4Cl}{820 \text{ mL}} \left| \frac{1 \text{ mol}}{53.50 \text{ g}} \right| \frac{10^3 \text{ mL}}{1 \text{ L}}$$

= 0.17 mol NH_4Cl/L soln

= 0.17M NH_4Cl

Find the pH of each of the following solutions.

1. The hydronium ion concentration equals:

a) $10^{-5}M$
    5

b) $10^{-12}M$
    12

c) $10^{-2}M$
    2

2. The hydroxide ion concentration equals:

a) $10^{-4}M$
    10

b) $10^{-11}M$
    3

c) $10^{-8}M$
    6

Write overall, ionic, and net ionic equations for each of the following reactions.

1. hydroiodic acid, HI, and calcium hydroxide, $Ca(OH)_2$

$2HI(aq) + Ca(OH)_2(aq) \rightarrow CaI_2(aq) + 2H_2O(l)$
$2H^+(aq) + 2I^-(aq) + Ca^{2+}(aq) + 2OH^-(aq)$
$\quad \rightarrow Ca^{2+}(aq) + 2I^-(aq) + 2H_2O(l)$
$H^+(aq) + OH^-(aq) \rightarrow H_2O(l)$

2. hydrobromic acid, HBr, and lithium hydroxide, LiOH

$HBr(aq) + LiOH(aq) \rightarrow LiBr(aq) + H_2O(l)$
$H^+(aq) + Br^-(aq) + Li^+(aq) + OH^-(aq)$
$\quad \rightarrow Li^+(aq) + Br^-(aq) + H_2O(l)$
$H^+(aq) + OH^-(aq) \rightarrow H_2O(l)$

3. sulfuric acid, H_2SO_4, and strontium hydroxide, $Sr(OH)_2$

$H_2SO_4(aq) + Sr(OH)_2(aq) \rightarrow SrSO_4(aq) + 2H_2O(l)$
$2H^+(aq) + SO_4^{2-}(aq) + Sr^{2+}(aq) + 2OH^-(aq)$
$\quad \rightarrow Sr^{2+}(aq) + SO_4^{2-}(aq) + 2H_2O(l)$
$H^+(aq) + OH^-(aq) \rightarrow H_2O(l)$

4. perchloric acid, $HClO_4$, and barium hydroxide, $Ba(OH)_2$

$2HClO_4(aq) + Ba(OH)_2(aq) \rightarrow 2H_2O(l) + Ba(ClO_4)_2(aq)$
$2H^+(aq) + 2ClO_4^-(aq) + Ba^{2+}(aq) + 2OH^-(aq)$
$\quad \rightarrow 2H_2O(l) + Ba^{2+}(aq) + 2ClO_4^-(aq)$
$H^+(aq) + OH^-(aq) \rightarrow H_2O(l)$

5. perchloric acid, $HClO_4$, and ammonia, NH_3

$HClO_4(aq) + NH_3(aq) \rightarrow NH_4ClO_4(aq)$
$H^+(aq) + ClO_4^-(aq) + NH_3(aq)$
$\quad \rightarrow NH_4^+(aq) + ClO_4^-(aq)$
$H^+(aq) + NH_3(aq) \rightarrow NH_4^+(aq)$

6. hydrochloric acid, HCl, and aluminum hydroxide, $Al(OH)_3$

$3HCl(aq) + Al(OH)_3(aq) \rightarrow AlCl_3(aq) + 3H_2O(l)$
$3H^+(aq) + 3Cl^-(aq) + Al(OH)_3(aq)$
$\quad \rightarrow Al^{3+}(aq) + 3Cl^-(aq) + 3H_2O(l)$
$3H^+(aq) + Al(OH)_3(aq) \rightarrow Al^{3+}(aq) + 3H_2O(l)$

7. sulfuric acid, H_2SO_4, and iron(III) hydroxide, $Fe(OH)_3$

$3H_2SO_4(aq) + 2Fe(OH)_3(aq)$
$\quad \rightarrow Fe_2(SO_4)_3(aq) + 6H_2O(l)$
$6H^+(aq) + 3SO_4^{2-}(aq) + 2Fe(OH)_3(aq)$
$\quad \rightarrow 2Fe^{3+}(aq) + 3SO_4^{2-}(aq) + 6H_2O(l)$
$3H^+(aq) + Fe(OH)_3(aq) \rightarrow Fe^{3+}(aq) + 3H_2O(l)$

8. carbonic acid, H_2CO_3, and sodium hydroxide, NaOH

$$H_2CO_3(aq) + 2NaOH(aq) \rightarrow Na_2CO_3(aq) + 2H_2O(1)$$

$$H_2CO_3(aq) + 2Na^+(aq) + 2OH^-(aq)$$
$$\rightarrow 2Na^+(aq) + CO_3{}^{2-}(aq) + 2H_2O(1)$$

$$H_2CO_3(aq) + 2OH^-(aq) \rightarrow CO_3{}^{2-}(aq) + 2H_2O(1)$$

9. boric acid, H_3BO_3, and potassium hydroxide, KOH

$$H_3BO_3(aq) + 3KOH(aq) \rightarrow K_3BO_3(aq) + 3H_2O(1)$$

$$H_3BO_3(aq) + 3K^+(aq) + 3OH^-(aq)$$
$$\rightarrow 3K^+(aq) + BO_3{}^{3-}(aq) + 3H_2O(1)$$

$$H_3BO_3(aq) + 3OH^-(aq) \rightarrow BO_3{}^{3-}(aq) + 3H_2O(1)$$

10. acetic acid, $HC_2H_3O_2$, and calcium hydroxide, $Ca(OH)_2$

$$2HC_2H_3O_2(aq) + Ca(OH)_2(aq)$$
$$\rightarrow Ca(C_2H_3O_2)_2(aq) + 2H_2O(1)$$

$$2HC_2H_3O_2(aq) + Ca^{2+}(aq) + 2OH^-(aq)$$
$$\rightarrow Ca^{2+}(aq) + 2C_2H_3O_2{}^-(aq) + 2H_2O(1)$$

$$HC_2H_3O_2(aq) + OH^-(aq) \rightarrow C_2H_3O_2{}^-(aq) + H_2O(1)$$

Finding Molarity

11. A $0.100M$ LiOH solution was used to titrate an HBr solution of unknown concentration. At the endpoint, 21.0 mL of LiOH solution had neutralized 10.0 mL of HBr. What is the molarity of the HBr solution?

$0.210M$;

$$\frac{21.0 \text{ mL LiOH}}{1} \left| \frac{1 \text{ L}}{10^3 \text{ mL LiOH}} \right| \frac{0.100 \text{ mol LiOH}}{1 \text{ L}} \cdots$$

$$\left| \frac{1 \text{ mol HBr}}{1 \text{ mol LiOH}} \right| \frac{1}{10.0 \text{ mL HBr}} \left| \frac{10^3 \text{ mL HBr}}{1 \text{ L}} \right.$$

$$= 0.210M \text{ HBr}$$

12. A $0.150M$ KOH solution fills a burette to the 0 mark. The solution was used to titrate 25.0 mL of an HNO_3 solution of unknown concentration. At the endpoint, the burette reading was 34.6 mL. What was the molarity of the HNO_3 solution?

$0.208M$;

$$\frac{34.6 \text{ mL KOH}}{1} \left| \frac{1 \text{ L}}{10^3 \text{ mL KOH}} \right. \cdots$$

$$\left| \frac{0.150 \text{ mol KOH}}{1 \text{ L}} \right| \frac{1 \text{ mol HNO}_3}{1 \text{ mol KOH}} \cdots$$

$$\left| \frac{1}{25.0 \text{ mL HNO}_3} \right| \frac{10^3 \text{ mL HNO}_3}{1 \text{ L}}$$

$$= 0.208M \text{ HNO}_3$$

13. A $Ca(OH)_2$ solution of unknown concentration was used to titrate 15.0 mL of a $0.125M$ H_3PO_4 solution. If 12.4 mL of $Ca(OH)_2$ are used to reach the endpoint, what is the concentration of the $Ca(OH)_2$ solution?

$0.227M$;

$$\frac{15.0 \text{ mL H}_3\text{PO}_4}{1} \left| \frac{1 \text{ L}}{10^3 \text{ mL H}_3\text{PO}_4} \right| \frac{0.125 \text{ mol H}_3\text{PO}_4}{1 \text{ L}} \cdots$$

$$\left| \frac{3 \text{ mol Ca(OH)}_2}{2 \text{ mol H}_3\text{PO}_4} \right| \frac{1}{12.4 \text{ mL}} \left| \frac{10^3 \text{ mL}}{1 \text{ L}} \right.$$

$$= 0.227M \text{ Ca(OH)}_2$$

CHAPTER 18

1. Write the structural formulas for the following branched alkanes.

a) 2-methylbutane

$$\begin{array}{c} CH_3 \\ | \\ CH_3CHCH_2CH_3 \end{array}$$

b) 1, 3-dimethylcyclohexane (Hint: Begin numbering at any carbon in the ring, then attach the methyl groups.)

c) 4-propyldecane

$$\begin{array}{c} CH_3CH_2CH_2CHCH_2CH_2CH_2CH_2CH_2CH_3 \\ | \\ CH_2CH_2CH_3 \end{array}$$

d) 2, 3, 4-trimethylheptane

$$\begin{array}{c} CH_3 \quad CH_3 \quad CH_3 \\ | \qquad | \qquad | \\ CH_3CH\!-\!CH\!-\!CHCH_2CH_2CH_3 \end{array}$$

2. Name each of the following alkanes.

a)

3-methylpentane

b)

2, 3-dimethylbutane

c)

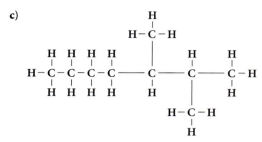

2, 3-dimethylheptane

d)

H H   H   H H H
| | | | | |
H-C-C————C————C-C-C-H
| | | | | |
H H | H H H
      H-C-H
      |
      H-C-H
      |
      H

3-ethylhexane

CHAPTER 20

1. How much heat is absorbed by a reaction that lowers the temperature of 500.0 g of water in a calorimeter by 1.10°C?

$q_w = (m)(\Delta T)(C_w)$
$= (500.0 \text{ g})(1.10°C)(4.184 \text{ J/g°C})$
$= 2300 \text{ J}$
$= 2.30 \text{ kJ}$

2. Aluminum reacts with iron(III) oxide to yield aluminum oxide and iron. Calculate the heat given off in the reaction if the temperature of the 1.00 kg of water in the calorimeter increases by 3.00°C.

$q_w = (m)(\Delta T)(C_w)$
$= (1.00 \times 10^3 \text{ g})(3.00°C)(4.184 \text{ J/g°C})$
$= 12600 \text{ J}$
$= 12.6 \text{ kJ}$

3. When 1.00 g of a certain fuel gas is burned in a calorimeter, the temperature of the surrounding 1.000 kg of water increases from 20.00°C to 28.05°C. All products and reactants in the process are gases. Calculate the heat given off in this reaction. How much heat would 1.00 mol of the fuel give off, assuming a molar mass of 65.8 g/mol?

$\Delta T = 28.05°C - 20.00°C = 8.05°C$

For 1 g:
$q_w = (m)(\Delta T)(C_w)$
$= (1.000 \times 10^3 \text{ g})(8.05°C)(4.184/g°C)$
$= 33700 \text{ J}$
$= 33.7 \text{ kJ}$

For 1 mol:

$q_w = \dfrac{33.7 \text{ kJ}}{1 \text{ g}} \bigg| \dfrac{65.8 \text{ g}}{1 \text{ mol}}$

$= 2220 \text{ kJ/mol}$

4. A group of students decides to measure the energy content of certain foods. They heat 50.0 g of water in an aluminum can by burning a sample of the food beneath the can. When they use 1.00 g of popcorn as their test food, the temperature of the water rises by 24°C. Calculate the heat released by the popcorn, and express your answer in both kilojoules and Calories per gram of popcorn.

$q_w = (m)(\Delta T)(C_w)$
$= (50.0 \text{ g})(24°C)(4.184 \text{ J/g°C})$ for 1 g
$= 5000 \text{ J/g}$
$= 5.0 \text{ kJ/g}$

$= \dfrac{5.0 \text{ kJ}}{1 \text{ g}} \bigg| \dfrac{1 \text{ Cal}}{4.184 \text{ kJ}} = 1.2 \text{ Cal/g}$

5. Another student comes along and tells the group in problem 4 that she has read the label on a popcorn bag that states that 30 g of popcorn yields 110 Calories. What is that value in Calories/gram? How can you account for the difference?

$\dfrac{110 \text{ Cal}}{30.0 \text{ g}} = 3.7 \text{ Cal/g}$

1.2 Cal/g, the experimental value, is much lower than the value given on the popcorn bag because of loss of heat to the air in the student experiment. Also, the combustion of popcorn in the student experiment may be incomplete.

6. A 3.00-g sample of a new snack food is burned in a calorimeter. The 2.00 kg of surrounding water change in temperature from 25.0°C to 32.4°C. What is the food value in Calories per gram?

$\Delta T = 32.4°C - 25.0°C = 7.4°C$
$q_w = (m)(\Delta T)(C_w)$

$= \dfrac{(2.00 \times 10^3 \text{ g})(7.4°C)(4.184 \text{ J/g°C})}{3.00 \text{ g}}$ for 3 g

$= 20.6 \text{ kJ/g}$

$= \dfrac{20.6 \text{ kJ}}{1 \text{ g}} \bigg| \dfrac{1 \text{ Cal}}{4.184 \text{ kJ}} = 4.93 \text{ Cal/g}$

CHAPTER 21

1. Write the balanced nuclear equation for the radioactive decay of radium-226 to give radon-222, and determine the type of decay.

$^{226}_{88}\text{Ra} \rightarrow {}^{222}_{86}\text{Rn} + {}^{4}_{2}\text{He}$; alpha decay

2. **Write a balanced equation for the nuclear reaction in which neon-23 decays to form sodium-23, and determine the type of decay.**

$^{23}_{10}\text{Ne} \rightarrow \ ^{23}_{11}\text{Na} + \ ^{0}_{-1}e$; beta decay

3. **A rock was analyzed using potassium-40. The half-life of potassium-40 is 1.25 billion years. If the rock had only 25 percent of the potassium-40 that would be found in a similar rock formed today, calculate how long ago the rock was formed.**

Because $(0.5)(0.5) = 0.25$, two half-lives have gone by.

2 half-lives $\times$ 1.25 billion years/half-life = 2.50 billion years

4. **Ash from an early fire pit was found to have 12.5 percent as much carbon-14 as would be found in a similar sample of ash today. How long ago was the ash formed?**

Because $(0.5)(0.5)(0.5) = 0.125$, three half-lives have gone by.

3 half-lives $\times$ 5730 years/half-life = 17 190 years

APPENDIX F
Try at Home Labs

 Comparing Frozen Liquids

Real-World Question How do different kitchen liquids react when placed in a freezer?

Possible Materials
- five identical, narrow-necked plastic bottles or photographic film canisters
- large cutting board or cookie sheet
- water
- orange juice
- vinegar
- soft drink
- cooking oil
- freezer

Procedure
1. Obtain permission to use the freezer before beginning this activity.
2. Fill one of the containers with water. The water should come to the top brim of the container.
3. Fill the other four containers in the same manner with the other four liquids.
4. Place the cutting board or cookie sheet in a freezer so that it is level and place the five containers on the board or sheet.
5. Leave the containers in the freezer overnight and observe the effect of the freezer's temperature on each liquid the following day.

Conclude and Apply
1. Describe the effect of the colder temperature on each liquid.
2. Infer why the water behaved as it did.
3. Infer why some liquids froze but others did not.

 Comparing Atom Sizes

Real-World Question How do the sizes of different atoms and subatomic particles compare?

Possible Materials
- metric ruler
- meterstick
- white sheet of paper
- fine-tipped black marker
- masking tape or transparent tape
- three plastic milk containers

Procedure
1. Using a black marker, draw a 0.1-mm-wide dot on one end of a white sheet of paper. This dot represents the diameter of an electron.
2. Measure a distance of 10 cm from the dot and draw a second dot. The distance between the two dots represents the diameter of a proton or a neutron.
3. Securely tape the paper to the top of a plastic milk container.
4. Measure a distance of 6.2 m from the milk container and place a second milk container. This distance represents the diameter of the smallest atom, a helium atom.
5. Measure a distance of 59.6 m from the first milk container and place a third milk container. This distance represents the diameter of a larger atom, a cesium atom.

Conclude and Apply
1. Considering the comparative sizes of protons and neutrons with the sizes of atoms, and infer what makes up most of an atom.

APPENDIX F
Try at Home Labs

Try at Home — Element Hunt

Real-World Question How many elements can be found in your home?

Possible Materials
- periodic table of the elements
- chemistry books or resources

Procedure
1. Create a chart to record the elements that are metals, metalloids, and nonmetals.
2. Use your chemistry textbook and resources to research common uses for elements.
3. Search your home for items made entirely or primarily of metal elements. List the items in your chart.
4. Search your home for items made primarily or entirely of metalloids and nonmetals. List the items in your data chart.

Conclude and Apply
1. List the two most common nonmetals found in your home and identify their locations.
2. List the most common metals found in your home.
3. Infer what group of metals is most commonly found in a home.

Try at Home — Mixing Ionic and Covalent Liquids

Real-World Question Do common ionic and covalent liquids found in a kitchen mix together?

Possible Materials
- clear-glass container
- water
- rubbing alcohol
- soft drink
- vegetable oil
- spoon or stirring rod
- measuring cup with SI units
- dish soap

Procedure
1. Create a data table to record your observations.
2. Measure 100 mL of water and pour it into the container. Measure 100 mL of rubbing alcohol, pour it into the container, and stir the mixture. Observe and record what happens to the two liquids.
3. Thoroughly wash and rinse out your container using dish soap.
4. Repeat steps #2 and #3 using a soft drink and vegetable oil.
5. Measure 100 mL of vegetable oil and pour it into the container. Measure 100 mL of rubbing alcohol, pour it into the container, and stir the mixture. Observe and record what happens to the two liquids.
6. Thoroughly wash and rinse out your container using dish soap.
7. Repeat steps #5 and #6 using a soft drink and rubbing alcohol.

Conclude and Apply
1. Research what type of compound water is.
2. Infer why the liquids mixed or did not mix the way they did.

Try at Home Labs

Try at Home
  Iron Ink

Real-World Question   How can common household items be used to make nineteenth century ink?

Possible Materials

- ceramic mug
- tea bag
- iron sulfate tablets
- coffee filter
- measuring cup
- paintbrush
- glass
- water
- funnel
- oven mitt
- white paper
- microwave oven

Procedure

1. Pour 30 mL of water into the mug.
2. Lay a tea bag in the water and drop five iron sulfate tablets onto the bag.
3. Microwave the mug for 2 min. Remove the tea bag from the mug.
4. Place a coffee filter into a funnel, insert the funnel into a glass, and pour the mixture from the mug into the filter.
5. Place the glass of filtered solution in the refrigerator for 15 min.
6. Remove the funnel and use a paintbrush to write letters with the ink on a white sheet of paper. Set the paper aside to dry.

Conclude and Apply

1. Identify the molecular formula of the compound in the iron tablets that creates the dark color of the ink.
2. Infer the type of compounds involved in this chemical reaction.
3. Research the precipitate that forms and gives the ink its black color.

Try at Home **Preventing a Chemical Reaction**

Real-World Question   How can the chemical reaction that turns apples brown be prevented?

Possible Materials

- Seven identical glasses
- SI measuring cup
- 100-mg vitamin C tablets
- rolling pin
- apple
- bottled water
- wax paper
- paper towels
- kitchen knife
- permanent black marker
- masking tape

Procedure

1. Pour 200 mL of water into each glass.
2. Label glass #1 *no vitamin C*, glass #2 *100 mg*, glass #3 *200 mg*, glass #4 *500 mg*, glass #5 *1000 mg*, glass #6 *2000 mg*, and glass #7 *3000 mg*.
3. Place a 100-mg vitamin C tablet between two sheets of wax paper and use a rolling pin to grind the tablet into powder.
4. Place the powder into glass #2 and stir the mixture vigorously.
5. Repeat steps 3 and 4 for glasses #3–#7 using the appropriate mass of vitamin C.
6. Cut seven equal-sized wedges of apple and immediately place one into each glass.
7. After 5 min, lay the wedges on paper towels in front of the glasses in which they were soaking. Observe the apples every 5 min for 45 min.

Conclude and Apply

1. Describe the results of your experiment.
2. Infer why vitamin C prevents apples from turning brown.

APPENDIX F
Try at Home Labs

Comparing Orbital Sizes

Real-World Question How do the sizes of different electron orbitals compare?

Possible Materials
- metric ruler or meterstick
- 3-m-long piece of white paper (or tape several smaller pieces together)
- markers
- masking tape

Procedure
1. Lay the white paper out on a flat surface.
2. Use a marker to draw a thin, 3-cm-long line at one end of the paper.
3. Measure a 1-cm distance and draw a second line to represent the radius of an atom's nucleus.
4. From the first line, measure a distance of 37 cm to represent the radius of the first electron orbital, 156 cm to represent the radius of the second orbital, 186 cm for the third orbital, 227 cm for the fourth orbital, 248 cm to represent the radius of the fifth orbital, and 265 cm for the radius of the sixth orbital. Mark each length with a new line.
5. On your paper, label the radii with the appropriate orbital number.

Conclude and Apply
1. The average radius of an atom's nucleus is 1×10^{-12}. Calculate the scale you used for your drawing.
2. Infer from your drawing what comprises most of an atom.

Periodic Properties of the Elements

Real-World Question How does the solubility of iodine differ in different liquids?

Possible Materials
- glass jar with lid (baby food jars work well)
- dropper
- mineral oil
- tincture of iodine
- water
- measuring cup
- tablespoon

Procedure
1. Pour water into a glass jar until it is half-full.
2. Add 4 drops of iodine to the water.
3. Tighten the lid and shake the jar vigorously until the solution turns a light brown color.
4. Measure 15 mL of mineral oil and pour it into the jar.
5. Shake the jar vigorously again and allow the mixture to sit for 5 min. Observe the mixture.

Conclude and Apply
1. Describe the mixture in the jar after you added the mineral oil and allowed the jar's contents to sit for 5 min.
2. Infer why the iodine behaved the way it did when the mineral oil was added.
3. Infer other elements or compounds that might behave similarly to the iodine. Explain your answer.

APPENDIX F
Try at Home Labs

Breaking Covalent Bonds

Real-World Question What liquids will break the covalent bonds of polystyrene?

Possible Materials
- polystyrene packing peanuts or polystyrene cups
- large glass
- measuring cup
- nail polish remover
- shallow dish
- rubbing alcohol
- water
- cooking oil

Procedure

1. Pour 200 mL of water into a glass.

2. Drop a polystyrene packing peanut into the water and observe how the polystyrene and water react.

3. Thoroughly wash out the glass and repeat steps 1 and 2 using cooking oil and nail polish remover.

4. Drop several peanuts into any of the liquids that cause a chemical reaction with the polystyrene peanuts and observe what happens to them.

Conclude and Apply

1. Describe the reaction between the polystyrene peanuts and each of the four liquids.

2. Infer why the polystyrene reacted as it did with each of the liquids.

Estimating Metric Temperatures

Real-World Question How can Americans estimate daily metric temperatures?

Possible Materials
- thermometer with only Fahrenheit temperature scale units (try not to use a mercury-filled thermometer)
- calculator

Procedure

1. Create a data table to record your temperature observations and calculations.

2. Place a Fahrenheit scale thermometer outside your home in a location where you can easily view it from a window or door.

3. Observe the air temperature in degrees Fahrenheit each day for two weeks. Observe the temperature at different times of the day.

4. After reading the temperature in degrees Fahrenheit each day, estimate the temperature in degrees Celsius and Kelvins.

5. Calculate the actual temperature in degrees Celsius.

6. Calculate the actual temperature in Kelvins.

Conclude and Apply

1. Compare how well you were able to estimate metric temperatures at the start of the activity with your ability at the end of the lab.

2. Infer why many Americans cannot estimate temperatures in degrees Celsius.

APPENDIX F
Try at Home Labs

 Crushing Cans

Real-World Question How can an aluminum soft-drink can be used to demonstrate the gas laws?

Possible Materials
- aluminum soft-drink can (355 mL)
- stovetop or hot plate
- pan
- oven mitt
- large tongs
- water
- large, plastic container
- large bowl
- measuring cup with SI units
- watch with second hand or stopwatch
- ice

Procedure
1. Fill a pan with water, place the pan on a stovetop, and bring the water to a boil.
2. Fill a large, plastic container with cold water and ice.

3. Measure 25 mL of water and pour it into an aluminum soft-drink can.
4. Using an oven mitt, hold the can right side up in the boiling water for one minute.
5. Quickly invert the aluminum can and submerge it beneath the cold water in the container. Observe the result.

Conclude and Apply
1. Describe what happened to the can when you submerged it beneath the cold water.
2. Explain why the can behaved the way it did using both Charles's Law and Boyle's Law in your explanation.
3. Compare the amount of kinetic energy in the air molecules while the can was held in the boiling water to when the can was submerged in the cold water.

 Measuring Moles of Sugar

Real-World Question How can you measure and calculate the number of moles in a sample of sugar?

Possible Materials
- sugar
- kitchen scale
- measuring cup
- small bowl
- calculator

Procedure
1. Measure the mass of a small bowl.

2. Measure approximately 100 mL of sugar and pour the sugar into the bowl.
3. Measure the mass of the sugar and bowl and calculate the mass of the sugar sample.

Conclude and Apply
1. Research the chemical formula of sucrose (table sugar).
2. Calculate the molar mass of sucrose.
3. Calculate the number of moles in your sugar sample.

Try at Home Labs

 ## Measuring Capillarity

Real-World Question How does the height of a meniscus differ in containers of different widths and shapes?

Possible Materials
- 30-mL medicine cup
- 750-mL bottle
- narrow-necked bottle
- 1-mL, 2-mL, and 5-mL droppers
- glass
- bowl
- water
- metric ruler

Procedure
1. Create a data table to record your data.
2. Measure the diameter of all the containers. Measure the width of the bottle's neck.
3. Suction up water into a 5-mL dropper until it is half-full.
4. Hold the dropper level and measure the height of the meniscus formed by the water.
5. Repeat steps 3 and 4 using 2-mL and 1-mL droppers.
6. Measure the heights of the meniscuses formed by water in a 30-mL medicine cup and the neck of a 750-mL bottle.

Conclude and Apply
1. Explain the relationship between container width and meniscus height.
2. Infer other factors, aside from a container's width, that will determine the height of a meniscus.
3. Infer the maximum width of a container in which water will still form a measurable meniscus.

 ## Testing for Acid Rain

Real-World Question Does your area have acid rain?

Possible Materials
- aquarium pH test kit
- small test tubes (from aquarium kit)
- clean plastic containers

Procedure
1. Create a data table to record your data.
2. Place a container with a wide mouth opening outside in a location where it will capture rain. Set the container away from trees or other objects that might drop debris into the container.
3. Use an aquarium pH test kit to test the pH of your rainwater sample.
4. Collect water samples from streams or rivers in your area and test the pH.

Conclude and Apply
1. Research the pH value of normal rainwater.
2. Research the pH value of acid rain.
3. Test the pH of bottled water and compare the value to the pH of your water samples.
4. Infer whether or not your region has an acid rain problem.

APPENDIX F
Try at Home Labs

Testing for Ammonia

Real-World Question What substances elevate ammonia levels in natural waterways?

Possible Materials

- four glass jars with lids
- measuring cup
- raw chicken (6 ounces)
- water
- ammonia test kit
- kitchen scale
- kitchen knife
- masking tape
- black marker

Procedure

1. Use a kitchen scale to measure 28-g (1 ounce), 56-g (2-ounce), and 84-g (3-ounce) pieces of raw chicken.
2. Fill four jars with 500 mL of water.
3. Add nothing to the first jar. Place 28 g of raw chicken into the second jar, 56 g into the third jar, and 84 g into the fourth jar. Label your jars.
4. Create a data table to record your data.
5. Measure the amount of ammonia in each water sample each day for five days. Also observe the clarity of each sample.

Conclude and Apply

1. Identify possible procedural errors in your experiment.
2. Summarize your experiment results.
3. Infer the common cause for elevated ammonia levels in natural waterways.

Testing the Oxidation Power of Bleach

Real-World Question How well does bleach oxidize the compounds in different stains?

Possible Materials

- cotton swabs
- bleach
- mustard
- ketchup
- grass
- grape juice
- cranberry juice
- dropper or straw
- grape jam
- access to clothes dryer
- white cotton cloth
- rubber gloves
- bowl
- barbeque brush
- permanent black marker
- watch with second hand or stopwatch

Procedure

1. Take a peanut-sized dab of ketchup and place it on one end of a white cloth. Brush the ketchup into the cloth to form a stain that is about the size of a quarter and label the stain with a permanent marker.
2. Thoroughly wash out the brush and repeat step 1 using mustard and grape jam. Place the stains side by side.
3. Use a dropper to make quarter-sized stains of grape and cranberry juices next to the other three stains. Label each stain.
4. Place the cloth in a dryer until the stains dry.
5. Wearing rubber gloves, carefully pour bleach into a bowl.
6. Dip a cotton swab into the bleach and press the swab tip on the ketchup stain for 15 s.
7. Repeat step 6 for the other stains.

Conclude and Apply

1. Explain how bleach removes stains.
2. Describe the effectiveness of the bleach at oxidizing the compounds in the various stains you tested.
3. Infer why the bleach oxidized some stains better than others.

APPENDIX F
Try at Home Labs

 Removing Electroplating

Real-World Question   How can the electroplating on aluminum pans be removed?

Possible Materials
- aluminum pan turned dark on the bottom
- water
- lemon
- kitchen knife
- measuring cup
- stovetop burner or hotplate

Procedure

1. Measure 500 mL of water and pour it into the pan.

2. Place the pan on a stovetop burner and bring the water to a boil.

3. Slice a lemon into several wedges and drop the wedges into the pan of boiling water.

4. Observe the chemical reaction that occurs at the bottom of the pan.

Conclude and Apply

1. Describe the results of your experiment.

2. Infer why the bottom of the pan changed in appearance.

Comparing Water and Alcohol

Real-World Question   How do the properties of water and an alcohol compare?

Possible Materials
- water
- rubbing alcohol
- vegetable oil
- three measuring cups with SI units
- three clear glasses
- two ice cubes
- stirring rod
- spoon
- kitchen scale
- stopwatch or watch with second hand

Procedure

1. Create a data table for comparing the physical properties of water and rubbing alcohol.

2. Compare the color and odor of each liquid.

3. Measure the masses of equal volumes of each liquid and calculate the density of each. Drop an ice cube into each glass and observe any differences in densities of the liquids.

4. Mix 20 mL of water with oil and stir the mixture. Do the same with the alcohol and oil.

Conclude and Apply

1. Describe what happened when an ice cube was dropped into each liquid.

2. Infer why the ice cube behaved as it did in each liquid.

3. Infer what would happen if water and the alcohol were mixed together.

4. Summarize your comparisons of the two liquids.

Try at Home Labs

Counting Nutrients

Real-World Question What percentage of the daily allowance of fats, carbohydrates, and proteins do you consume each day?

Possible Materials
- nutrition facts chart
- packages of foods and drinks consumed during a week
- nutrition guide book
- kitchen scale

Procedure
1. Create a data table to record the mass and percentage of the daily allowance of fats, carbohydrates, and proteins that you consume each day for a week.
2. After each meal or snack, check the nutrition facts charts on your food packaging or a nutrition guide to count up the number of grams of fat, carbohydrates, and proteins you consumed. Be certain you consider how many servings of each food or drink you consumed.
3. Count up the percentage of the daily allowance of fats, carbohydrates, and proteins that you consume each day.
4. Count the grams of fat, carbohydrates, and proteins you consume each day for a week and calculate a daily average for each nutrient.
5. Calculate a daily average for the percentage of the daily allowance of fats, carbohydrates, and proteins you consume.

Conclude and Apply
1. Compare your fat, carbohydrate, and protein intake with the amounts recommended on the bottom table of a nutrition facts chart.
2. Infer if this lab can determine whether or not a person has a healthy diet.
3. Infer possible dietary changes you may want to consider based upon the results of this lab.

Observing Entropy

Real-World Question How quickly do common household liquids enter into a state of entropy?

Possible Materials
- seven identical glass containers
- stopwatch or watch with second hand
- water
- corn syrup
- rubbing alcohol
- measuring cup
- clear soft drink
- vinegar
- milk
- cooking oil
- food coloring

Procedure
1. Fill identical glass containers with equal volumes of the seven different liquids. Label each container.
2. Create a data table to record your observations and measurements.
3. Quickly place one drop of food coloring into each container while a partner simultaneously starts a stopwatch.
4. Observe how the food coloring behaves in each liquid. Time how quickly the coloring and each liquid reach total entropy.

Conclude and Apply
1. Infer the relationship between the rate at which a liquid and the dye achieves a state of total entropy and the time measurement from your data table.
2. Infer why the entropy rates for the different liquids varied.

Try at Home Labs

Modeling Fusion

Real-World Question How can fusion reactions be modeled?

Possible Materials
- 14 red gumdrops
- 12 green gumdrops
- toothpicks
- white paper (9 sheets)
- black marker

Procedure

1. Draw a large black arrow on 3 separate sheets of white paper. Draw a large plus sign on 6 separate sheets of paper.

2. While making your models in the steps below, use red gumdrops to represent protons and green gumdrops to represent neutrons.

3. Construct a deuterium atom, tritium atom, and helium atom. Arrange your three models and a single neutron into a common fusion reaction.

4. Construct two deuterium atoms and a tritium atom. Arrange your models and a single proton into a second fusion reaction.

5. Construct two deuterium atoms and a helium atom. Arrange your three models and a single neutron into a third fusion reaction.

Conclude and Apply

1. Describe what all three fusion reactions have in common.

2. Research the natural abundance of H-1 atoms and deuterium.

3. Infer from your models why nuclear fusion reactions create no radioactive waste.

Glossary/Glosario

English

absolute zero: the temperature below which a substance would have zero kinetic energy. (Chap. 10, p. 349)

acid: a substance that produces hydronium ions when dissolved in water. (Chap. 14, p. 483)

acidic anhydride: a nonmetal oxide that reacts with water to form an acid. (Chap. 14, p. 492)

acidic hydrogen: in an acid, any hydrogen that can be transferred to water. (Chap. 14, p. 485)

actinide: any of the second series of inner transition elements with atomic numbers from 90 to 103; all are radioactive. (Chap. 3, p. 104)

activation energy: the amount of energy the particles in a reaction must have when they collide for the reaction to occur. (Chap. 6, p. 218)

active site: on an enzyme, the pocket or groove that can bind a substrate taking part in a reaction. (Chap. 19, p. 676)

addition reaction: a reaction where monomers that contain double bonds add onto each other to form long chains; the product contains all the atoms of the starting monomers. (Chap. 18, p. 656)

aerobic: a metabolic process that takes place only in the presence of oxygen. (Chap. 19, p. 694)

alkali metal: any element from Group 1: lithium, sodium, potassium, rubidium, cesium, francium. (Chap. 8, p. 263)

alkaline earth metal: any element from Group 2: beryllium, magnesium, calcium, strontium, barium. (Chap. 8, p. 265)

alkane: a saturated hydrocarbon that consists of only carbon and hydrogen atoms with single bonds between all the atoms. (Chap. 18, p. 623)

alkene: a hydrocarbon in which one or more double bonds link carbon atoms. (Chap. 18, p. 629)

alkyne: an unsaturated hydrocarbon that contains a triple bond between two carbon atoms. (Chap. 18, p. 633)

allotrope: any of two or more molecules of a single element that have different crystalline or molecular structures. (Chap. 5, p. 175)

alloy: a solid solution containing different metals, and sometimes nonmetallic substances. (Chap. 1, p. 23)

alpha particle: a helium nucleus consisting of two protons and two neutrons. (Chap. 21, p. 747)

Spanish

absolute zero / cero absoluto: la temperatura debajo de la cual una sustancia carecería de energía cinética. (Cap. 10, pág. 349)

acid / ácido: sustancia que produce iones de hidronio cuando se disuelve en agua. (Cap. 14, pág. 483)

acidic anhydride / anhídrido acídico: óxido no metálico que reacciona con agua para formar un ácido. (Cap. 14, pág. 492)

acidic hydrogen / hidrógeno acídico: en un ácido, es cualquier hidrógeno que puede transferirse al agua. (Cap. 14, pág. 485)

actinide / actínido: cualquiera de la segunda serie de elementos de transición interna con números atómicos de 90 a 103; todos son radioactivos. (Cap. 3, pág. 104)

activation energy / energía de activación: cantidad de energía que deben tener las partículas cuando chocan en una reacción para que ésta ocurra. (Cap. 6, pág. 218)

active site / sitio activo: en una enzima, es el abolsamiento o la ranura que puede unirse a un sustrato que participa en una reacción. (Cap. 19, pág. 676)

addition reaction / reacción de adición: reacción donde monómeros que contienen dobles enlaces se unen unos con otros para formar cadenas largas; el producto contiene todos los átomos de los monómeros iniciadores. (Cap. 18, pág. 656)

aerobic / aeróbico: proceso metabólico que sólo se lleva a cabo en la presencia de oxígeno. (Cap. 19, pág. 694)

alkali metal / metal alcalino: cualquier elemento del Grupo 1: litio, sodio, potasio, rubidio, cesio, francio. (Cap. 8, pág. 263)

alkaline earth metal / metal alcalinotérreo: cualquier elemento del Grupo 2: berilio, magnesio, calcio, estroncio, bario. (Cap. 8, pág. 265)

alkane / alcano: hidrocarburo saturado formado únicamente por átomos de carbono e hidrógeno con enlaces sencillos entre todos los átomos. (Cap. 18, pág. 623)

alkene / alqueno: hidrocarburo en que uno o más enlaces dobles unen átomos de carbono. (Cap. 18, pág. 629)

alkyne / alquino: hidrocarburo insaturado que contiene un enlace triple entre dos átomos de carbono. (Cap. 18, pág. 633)

allotrope / alótropo: cualquiera de dos o más moléculas de un solo elemento que tienen estructuras cristalinas o moleculares diferentes. (Cap. 5, pág. 175)

alloy / aleación: solución sólida que contiene metales diferentes y, algunas veces, sustancias no metálicas. (Cap. 1, pág. 23)

alpha particle / partícula alfa: núcleo de helio que consiste en dos protones y dos neutrones. (Cap. 21, pág. 747)

amino acid: an organic compound; a monomer that forms proteins. (Chap. 19, p. 670)

amorphous material: a substance with a haphazard, disjointed, and incomplete crystal lattice. (Chap. 10, p. 346)

anaerobic: a metabolic process that takes place in the absence of oxygen. (Chap. 19, p. 698)

anhydrous: a compound in which all water has been removed, usually by heating. (Chap. 5, p. 168)

anion: a negative ion. (Chap. 17, p. 586)

anode: the electrode that takes electrons away from the reacting ions or atoms in solution. (Chap. 17, p. 585)

aqueous solution: a solution in which the solvent is water. (Chap. 1, p. 23)

aromatic hydrocarbon: a compound that has a benzene ring or the type of bonding exhibited by benzene; most have distinct odors. (Chap. 18, p. 636)

atom: the smallest particle of a given type of matter. (Chap. 2, p. 53)

atomic number: the number of protons in the nucleus of an atom of an element. (Chap. 2, p. 66)

atomic theory: the idea that matter is made up of fundamental particles called atoms. (Chap. 2, p. 53)

ATP: adenosine triphosphate, the energy storage molecule in cells. (Chap. 19, p. 696)

Avogadro constant: the number of things in one mole of a substance, specifically 6.02×10^{23}. (Chap. 12, p. 405)

Avogadro's principle: statement that at the same temperature and pressure, equal volumes of gases contain equal numbers of particles. (Chap. 11, p. 398)

barometer: an instrument that measures the pressure exerted by the atmosphere. (Chap. 11, p. 376)

base: a substance that produces hydroxide ions when it dissolves in water. (Chap. 14, p. 488)

amino acid / aminoácido: compuesto orgánico; monómero que forma proteínas. (Cap. 19, pág. 670)

amorphous material / material amorfo: sustancia con una estructura cristalina aleatoria, desorganizada e incompleta. (Cap. 10, pág. 346)

anaerobic / anaeróbico: proceso metabólico que se lleva a cabo en ausencia de oxígeno. (Cap. 19, pág. 698)

anhydrous / anhidro: compuesto al que se le ha extraido toda el agua, generalmente por calentamiento. (Cap. 5, pág. 168)

anion / anión: ion negativo. (Cap. 17, pág. 586)

anode / ánodo: electrodo que quita electrones de iones o átomos reactivos en solución. (Cap. 17, pág. 585)

aqueous solution / solución acuosa: solución en la cual el disolvente es agua. (Cap. 1, pág. 23)

aromatic hydrocarbon / hidrocarburo aromático: compuesto que tiene un anillo de benceno o el tipo de enlace exhibido por el benceno; la mayoría posee olores distintivos. (Cap. 18, pág. 636)

atom / átomo: la partícula más pequeña de un cierto tipo de materia. (Cap. 2, pág. 53)

atomic number / número atómico: número de protones en el núcleo del átomo de un elemento. (Cap. 2, pág. 66)

atomic theory / teoría atómica: la idea de que la materia está compuesta de partículas fundamentales llamadas átomos. (Cap. 2, pág. 53)

ATP / ATP: trifosfato de adenosina, la molécula de almacenamiento de energía en las células. (Cap. 19, pág. 696)

Avogadro constant / constante de Avogadro: el número de cosas en un mol de una cierta sustancia, específicamente 6.02×10^{23}. (Cap. 12, pág. 405)

Avogadro's principle / principio de Avogadro: afirmación de que a la misma temperatura y presión, volúmenes iguales de gases contienen números iguales de partículas. (Cap. 11, pág. 398)

barometer / barómetro: instrumento que mide la presión ejercida por la atmósfera. (Cap. 11, pág. 376)

base / base: sustancia que produce iones hidróxido cuando se disuelve en agua. (Cap. 14, pág. 488)

Glossary/Glosario

basic anhydride: a metal oxide that reacts with water to form a base. (Chap. 14, p. 492)

beta particle: a high-energy electron with a 1− charge. (Chap. 21, p. 749)

binary compound: a compound that contains only two elements. (Chap. 5, p. 155)

biochemistry: the study of the chemistry of living things. (Chap. 19, p. 668)

boiling point: the temperature of a liquid where its vapor pressure equals the pressure exerted on its surface. (Chap. 10, p. 358)

Boyle's law: at a constant temperature, the volume and pressure of a gas are inversely proportional. (Chap. 11, p. 383)

Brownian motion: the constant, random motion of tiny chunks of matter. (Chap. 10, p. 342)

buffer: a solution that resists changes in pH when moderate amounts of acids or bases are added to it. (Chap. 15, p. 531)

basic anhydride / anhídrido básico: óxido metálico que reacciona con agua para formar una base. (Cap. 14, pág. 492)

beta particle / partícula beta: electrón de alta energía con una carga de 1−. (Cap. 21, pág. 749)

binary compound / compuesto binario: compuesto que contiene sólo dos elementos. (Cap. 5, pág. 155)

biochemistry / bioquímica: el estudio de la química de los seres vivos. (Cap. 19, pág. 668)

boiling point / punto de ebullición: temperatura de un líquido cuando su presión de vapor iguala a la presión ejercida sobre su superficie. (Cap. 10, pág. 358)

Boyle's law / ley de Boyle: a temperatura constante, el volumen y la presión de un gas son inversamente proporcionales. (Cap. 11, pág. 383)

Brownian motion / movimiento browniano: movimiento aleatorio constante de trozos diminutos de materia. (Cap. 10, pág. 342)

buffer / amortiguador: solución que resiste cambios de pH cuando se le añaden cantidades moderadas de ácidos o bases. (Cap. 15, pág. 531)

calorie: the heat required to raise the temperature of 1 gram of liquid water by 1°C. (Chap. 20, p. 721)

Calorie: a food Calorie, equal to 1 kilocalorie, used to measure the energy value of foods. (Chap. 20, p. 721)

capillarity: the rising of a liquid in a narrow tube, sometimes called capillary action. (Chap. 13, p. 443)

carbohydrate: an organic molecule that contains the elements carbon, hydrogen, and oxygen in a ratio of about two hydrogen atoms and one oxygen atom for each carbon atom. (Chap. 19, p. 677)

catalyst: a substance that speeds up the rate of a reaction without being used up itself or permanently changed. (Chap. 6, p. 222)

cathode: the electrode that brings electrons to the reacting ions or atoms in solution. (Chap. 17, p. 585)

cation: a positive ion. (Chap. 17, p. 586)

calorie / caloría: el calor requerido para incrementar la temperatura de 1 gramo de agua líquida en un 1°C. (Cap. 20, pág. 721)

Calorie / Caloría: una Caloría alimenticia (igual a 1 kilocaloría) se usa para medir el valor energético de los alimentos. (Cap. 20, pág. 721)

capillarity / capilaridad: elevación de un líquido por un tubo estrecho, llamado en ocasiones acción capilar. (Cap. 13, pág. 443)

carbohydrate / carbohidrato: una molécula orgánica que contiene los elementos carbono, hidrógeno y oxígeno en una proporción de aproximadamente dos átomos de hidrógeno y uno de oxígeno por cada átomo de carbono. (Cap. 19, pág. 677)

catalyst / catalizador: sustancia que acelera la velocidad de reacción sin que se la consuma o se cambie permanentemente. (Cap. 6, pág. 222)

cathode / cátodo: electrodo que lleva electrones a los iones o átomos reactivos en solución. (Cap. 17, pág. 585)

cation / catión: ion positivo. (Cap. 17, pág. 586)

Charles's law: at constant pressure, the volume of a gas is directly proportional to its Kelvin temperature. (Chap. 11, p. 392)

chemical change: the change of one or more substances into other substances. (Chap. 1, p. 40)

chemical property: a property that can be observed only when there is a change in the composition of a substance. (Chap. 1, p. 40)

chemical reaction: another term for chemical change. (Chap. 1, p. 40)

chemistry: the science that investigates and explains the structure and properties of matter. (Chap. 1, p. 4)

coefficient: a number placed in front of the parts of a chemical equation to indicate how many are involved; always a positive whole number. (Chap. 6, p. 199)

coenzyme: an organic molecule that assists an enzyme in catalyzing a reaction. (Chap. 19, p. 691)

colloid: a mixture that contains particles that are evenly distributed through a dispersing medium and do not settle out over time. (Chap. 13, p. 472)

combined gas law: the combination of Boyle's law and Charles's law. (Chap. 11, p. 395)

combustion: term for a reaction in which a substance rapidly combines with oxygen to form one or more oxides. (Chap. 6, p. 209)

compound: a chemical combination of two or more different elements joined together in a fixed proportion. (Chap. 1, p. 30)

concentration: the amount of a substance present in a unit volume. (Chap. 6, p. 219)

condensation: the process where gaseous particles come together, that is, condense, to form a liquid or sometimes a solid. (Chap. 10, p. 356)

condensation reaction: reaction to form a polymer where a small molecule, usually water, is given off as each new bond is formed. (Chap. 18, p. 658)

conductivity: a measure of how easily electrons can flow through a material to produce an electrical current. (Chap. 9, p. 313)

covalent bond: the attraction of two atoms for a shared pair of electrons. (Chap. 4, p. 140)

covalent compound: a compound whose atoms are held together by covalent bonds. (Chap. 4, p. 140)

Charles's law / ley de Charles: a presión constante, el volumen de un gas es directamente proporcional a su temperatura en Kelvin. (Cap. 11, pág. 392)

chemical change / cambio químico: cambio de una o más sustancias en otras sustancias. (Cap. 1, pág. 40)

chemical property / propiedad química: propiedad que puede observarse únicamente cuando hay un cambio en la composición de una sustancia. (Cap. 1, pág. 40)

chemical reaction / reacción química: otra forma de nombrar un cambio químico. (Cap. 1, pág. 40)

chemistry / química: ciencia que investiga y explica la estructura y las propiedades de la materia. (Cap. 1, pág. 4)

coefficient / coeficiente: número que se coloca antes de las partes de una ecuación química para indicar cuántas de ellas están involucradas; siempre es un número entero positivo. (Cap. 6, pág. 199)

coenzyme / coenzima: molécula orgánica que ayuda a una enzima a catalizar una reacción. (Cap. 19, pág. 691)

colloid / coloide: mezcla que contiene partículas uniformemente distribuidas a través de un medio dispersante y que no se estabiliza con el tiempo. (Cap. 13, pág. 472)

combined gas law / ley combinada de los gases: combinación de las leyes de Boyle y Charles. (Cap. 11, pág. 395)

combustion / combustión: término para una reacción en la cual una sustancia se combina rápidamente con oxígeno para formar uno o más óxidos. (Cap. 6, pág. 209)

compound compuesto: combinación química de dos o más elementos diferentes unidos en una proporción fija. (Cap. 1, pág. 30)

concentration / concentración: cantidad de sustancia presente en una unidad de volumen. (Cap. 6, pág. 219)

condensation / condensación: proceso en que las partículas gaseosas se unen o se condensan formando un líquido o en algunas ocasiones un sólido. (Cap. 10, pág. 356)

condensation reaction / reacción de condensación: reacción para formar un polímero donde una molécula pequeña, generalmente agua, es eliminada al formarse un nuevo enlace. (Cap. 18, pág. 658)

conductivity / conductividad: medida de la facilidad con que pueden fluir los electrones a través de un material para producir una corriente eléctrica. (Cap. 9, pág. 313)

covalent bond / enlace covalente: atracción de dos átomos por un par compartido de electrones. (Cap. 4, pág. 140)

covalent compound / compueso covalente: compuesto cuyos átomos se mantienen unidos por enlaces covalentes. (Cap. 4, pág. 140)

Glossary/Glosario

cracking: the use of a catalyst or high temperatures in the absence of air to break down or rearrange large hydrocarbons. (Chap. 18, p. 638)

cross-linking: the linking together of many polymer chains, giving the polymer increased strength. (Chap. 18, p. 658)

crystal: a regular, repeating arrangement of atoms, ions, or molecules in three dimensions. (Chap. 4, p. 134)

crystal lattice: the three-dimensional arrangement repeated throughout a solid. (Chap. 10, p. 345)

cracking / cracking: uso de un catalizador o de altas temperaturas en ausencia de aire para romper o rearreglar hidrocarburos grandes. (Cap. 18, pág. 638)

cross-linking / entrecruzamiento: unión de varias cadenas de polímeros, produciendo un polímero con fuerza incrementada. (Cap. 18, pág. 658)

crystal / cristal: arreglo tridimensional, repetitivo y regular de átomos, iones, o moléculas. (Cap. 4, pág. 134)

crystal lattice / red cristalina: arreglo tridimensional repetido a través de un sólido. (Cap. 10, pág. 345)

decomposition: the name applied to a reaction where a compound breaks down into two or more simpler substances. (Chap. 6, p. 204)

deliquescent: a substance that takes up enough water from the air that it dissolves completely to a liquid solution. (Chap. 5, p. 166)

denaturation: the name given to the process of unfolding of a protein when the forces holding the polypeptide chain in shape are broken. (Chap. 19, p. 673)

density: the amount of matter (mass) in a given unit volume. (Chap. 1, p. 36)

deuterium: the hydrogen isotope with a mass number of 2. (Chap. 21, p. 766)

diffusion: the process by which a gas enters a container and fills it, or when the particles of two gases or liquids mix together. (Chap. 10, p. 352)

dissociation: the process by which the charged particles in an ionic solid separate from one another, primarily when going into solution. (Chap. 13, p. 453)

distillation: the method of separating substances in a mixture by evaporation of a liquid and subsequent condensation of its vapor. (Chap. 5, p. 171)

DNA: deoxyribonucleic acid. (Chap. 19, p. 688)

double bond: a bond formed by the sharing of two pairs of electrons between two atoms. (Chap. 9, p. 321)

double displacement: a type of reaction where the positive and negative portions of two ionic compounds are interchanged; at least one product must be water or a precipitate. (Chap. 6, p. 208)

decomposition / descomposición: nombre que se aplica a una reacción donde un compuesto se descompone en dos o más sustancias más simples. (Cap. 6, pág. 204)

deliquescent / delicuescente: sustancia que absorbe suficiente agua del aire como para disolverse completamente en una solución líquida. (Cap. 5, pág. 166)

denaturation / desnaturalización: nombre que se le da al proceso de desdoblamiento de una proteína que ocurre cuando se rompen las fuerzas que mantienen formada la cadena del polipéptido. (Cap. 19, pág. 673)

density / densidad: cantidad de materia (masa) en una unidad dada de volumen. (Cap. 1, pág. 36)

deuterium / deuterio: isótopo de hidrógeno con un número de masa de 2. (Cap. 21, pág. 766)

diffusion / difusión: proceso por el cual un gas entra a un contenedor y lo llena, o cuando se entremezclan las partículas de dos gases o líquidos. (Cap. 10, pág. 352)

dissociation / disociación: proceso por el cual las partículas cargadas en un sólido iónico se separan una de la otra, principalmente cuando entran en solución. (Cap. 13, pág. 453)

distillation / destilación: método de separación de sustancias en una mezcla por evaporación de un líquido y la subsecuente condensación de su vapor. (Cap. 5, pág. 171)

DNA / DNA: ácido desoxirribonucleico. (Cap. 19, pág. 688)

double bond / enlace doble: enlace formado al compartir dos pares de electrones entre dos átomos. (Cap. 9, pág. 321)

double displacement / desplazamiento doble: tipo de reacción en la cual se intercambian las partes positivas y negativas de dos compuestos iónicos. Por lo menos, uno de los productos debe ser agua o un precipitado. (Cap. 6, pág. 208)

ductile: property of a metal that means it can easily be drawn into a wire. (Chap. 9, p. 313)

dynamic equilibrium: term describing a system in which opposite reactions are taking place at the same rate. (Chap. 6, p. 211)

electrical current: the flow of electrons in a particular direction. (Chap. 17, p. 584)

electrolysis: the process in which electrical energy causes a non-spontaneous chemical reaction to occur. (Chap. 17, p. 584)

electrolyte: any compound that conducts electricity when melted or dissolved in water. (Chap. 4, p. 144)

electrolytic cell: the electrochemical cell in which electrolysis takes place. (Chap. 17, p. 586)

electromagnetic spectrum: the whole range of electromagnetic radiation. (Chap. 2, p. 71)

electron: negatively-charged particle. (Chap. 2, p. 61)

electron cloud: the space around the nucleus of an atom where the atom's electrons are found. (Chap. 2, p. 77)

electron configuration: the most stable arrangement of electrons in sublevels and orbitals. (Chap. 7, p. 242)

electron transport chain: the controlled release of energy from glucose by the step-by-step movement of electrons to lower energy levels. (Chap. 19, p. 698)

electronegativity: the measure of the ability of an atom in a bond to attract electrons. (Chap. 9, p. 303)

element: a substance that cannot be broken down into simpler substances. (Chap. 1, p. 24)

emission spectrum: the spectrum of light released from excited atoms of an element. (Chap. 2, p. 74)

empirical formula: the formula of a compound having the smallest whole-number ratio of atoms in the compound. (Chap. 12, p. 428)

ductile / dúctil: propiedad de un metal que permite que se pueda estirar fácilmente formando un alambre. (Cap. 9, pág. 313)

dynamic equilibrium / equilibrio dinámico: término que describe un sistema en que las reacciones opuestas se llevan a cabo a la misma velocidad. (Cap. 6, pág. 211)

electrical current / corriente eléctrica: flujo de electrones en cierta dirección. (Cap. 17, pág. 584)

electrolysis / electrólisis: proceso por el cual la energía eléctrica hace que ocurra una reacción química no espontánea. (Cap. 17, pág. 584)

electrolyte / electrolito: cualquier compuesto que conduce electricidad cuando se funde o disuelve en agua. (Cap. 4, pág. 144)

electrolytic cell / celda electrolítica: celda electroquímica en la cual se lleva a cabo la electrólisis. (Cap. 17, pág. 586)

electromagnetic spectrum / espectro electromagnético: rango completo de la radiación electromagnética. (Cap. 2, pág. 71)

electron / electrón: partícula cargada negativamente. (Cap. 2, pág. 61)

electron cloud / nube electrónica: espacio alrededor del núcleo de un átomo donde se encuentran los electrones del átomo. (Cap. 2, pág. 77)

electron configuration / configuración electrónica: arreglo de electrones más estable en subniveles y orbitales. (Cap. 7, pág. 242)

electron transport chain / cadena de transporte de electrones: liberación controlada de energía a partir de la glucosa por el movimiento gradual de electrones para disminuir los niveles energéticos. (Cap. 19, pág. 698)

electronegativity / electronegatividad: medida de la capacidad de un átomo en un enlace para atraer electrones. (Cap. 9, pág. 303)

element / elemento: sustancia que no puede separarse en sustancias más simples. (Cap. 1, pág. 24)

emission spectrum / espectro de emisión: espectro de luz liberada por átomos excitados de un elemento. (Cap. 2, pág. 74)

empirical formula / fórmula empírica: fórmula de un compuesto que tiene la proporción de átomos en números enteros menor en el compuesto. (Cap. 12, pág. 428)

Glossary/Glosario

endothermic: chemical reaction that absorbs energy. (Chap. 1, p. 43)

energy: the capacity to do work. (Chap. 1, p. 42)

energy level: the regions of space in which electrons can move about the nucleus of an atom. (Chap. 2, p. 75)

entropy: term used to describe and measure the degree of disorder in a process. (Chap. 20, p. 716)

enzyme: a biological catalyst. (Chap. 6, p. 222)

equilibrium: term for a system where no net change occurs in the amount of reactants or products. (Chap. 6, p. 211)

evaporation: the process by which particles of a liquid form a gas by escaping from the liquid surface. (Chap. 10, p. 352)

exothermic: chemical reaction that gives off energy. (Chap. 1, p. 42)

endothermic / endotérmica: reacción química que absorbe energía. (Cap. 1, pág. 43)

energy / energía: capacidad de hacer trabajo. (Cap. 1, pág. 42)

energy level / nivel energético: regiones del espacio en las cuales los electrones se pueden mover alrededor del núcleo de un átomo. (Cap. 2, pág. 75)

entropy / entropía: término para describir y medir el grado de degradación en un proceso. (Cap. 20, pág. 716)

enzyme / enzima: catalizador biológico. (Cap. 6, pág. 222)

equilibrium / equilibrio: término que describe un sistema donde no ocurre un cambio neto en la cantidad de reactivos ni de productos. (Cap. 6, pág. 211)

evaporation / evaporación: proceso por el cual las partículas de un líquido forman un gas al escaparse de la superficie del líquido. (Cap. 10, pág. 352)

exothermic / exotérmica: reacción química que libera energía. (Cap. 1, pág. 42)

factor label method: The problem-solving method in chemistry that uses mathematical relationships to convert one quantity to another. (Chap. 11, p. 380)

family: see group. (Chap. 3, p. 96)

fatty acid: a long-chain carboxylic acid. (Chap. 19, p. 684)

fermentation: the anaerobic process of generating energy from glucose. (Chap. 19, p. 698)

formula: a combination of chemical symbols that show what elements make up a compound and the number of atoms of each element. (Chap. 1, p. 31)

formula mass: the mass in atomic mass units of one formula unit of an ionic compound. (Chap. 12, p. 409)

formula unit: the simplest ratio of ions in a compound. (Chap. 5, p. 156)

fossil fuel: a fuel such as oil or natural gas, comprised of hydrocarbons that are the remains of plants and other organisms that lived millions of years ago. (Chap. 20, p. 713)

factor label method / método del factor: método de solución de problemas en química que utiliza relaciones matemáticas para convertir una cantidad a otra. (Cap. 11, pág. 380)

family / familia: véase grupo. (Cap. 3, pág. 96)

fatty acid / ácido graso: ácido carboxílico de cadena larga. (Cap. 19, pág. 684)

fermentation / fermentación: proceso anaerobio de generación de energía a partir de glucosa. (Cap. 19, pág. 698)

formula / fórmula: combinación de símbolos químicos que muestra qué elementos forman un compuesto y el número de átomos de cada elemento. (Cap. 1, pág. 31)

formula mass / fórmula-masa: la masa en unidades de masa atómicas de una fórmula unitaria en un compuesto iónico. (Cap. 12, pág. 409)

formula unit / fórmula unitaria: la proporción más sencilla de iones en un compuesto. (Cap. 5, pág. 156)

fossil fuel / combustible fósil: combustible como el petróleo o el gas natural, compuesto de hidrocarburos que son restos de plantas y otros organismos que vivieron hace millones de años. (Cap. 20, pág. 713)

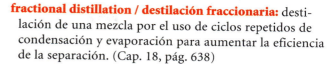

fractional distillation: the distillation of a mixture by the use of repeated vaporization-condensation cycles to increase the efficiency of separation. (Chap. 18, p. 638)

freezing point: the temperature of a liquid when it becomes a crystal lattice. (Chap. 10, p. 364)

functional group: the part of a molecule that is largely responsible for the chemical behavior of the molecule. (Chap. 18, p. 640)

fractional distillation / destilación fraccionaria: destilación de una mezcla por el uso de ciclos repetidos de condensación y evaporación para aumentar la eficiencia de la separación. (Cap. 18, pág. 638)

freezing point / punto de congelación: temperatura de un líquido cuando se convierte en una red cristalina. (Cap. 10, pág. 364)

functional group / grupo funcional: la parte de una molécula que es en gran parte responsable por el comportamiento químico de la molécula. (Cap. 18, pág. 640)

galvanic cell: an electrochemical cell in which an oxidation-reduction reaction occurs spontaneously to produce a potential difference. (Chap. 17, p. 602)

gamma ray: a high-energy form of electromagnetic radiation with no charge and no mass. (Chap. 21, p. 749)

gas: a flowing, compressible substance with no definite volume or shape. (Chap. 10, p. 341)

gray: the unit used to measure a received dose of radiation. (Chap. 21, p. 776)

group: the elements in a vertical column of the periodic table. (Chap. 3, p. 96)

galvanic cell / celda galvánica: celda electroquímica en que se produce espontáneamente una reacción de óxido-reducción para generar una diferencia de potencial. (Cap. 17, pág. 602)

gamma ray / rayo gama: forma de radiación electromagnética de alta energía sin carga ni masa algunas. (Cap. 21, pág. 749)

gas / gas: sustancia comprensible y fluida sin volumen o forma definidos. (Cap. 10, pág. 341)

gray / gray: unidad que se usa para medir una dosis de radiación recibida. (Cap. 21, pág. 776)

group / grupo: elementos en una columna vertical de la tabla periódica. (Cap. 3, pág. 96)

half life: the time it takes for half of a given radioactive isotope to decay (into a different isotope or element). (Chap. 21, p. 756)

halogen: any element from Group 17: fluorine, chlorine, bromine, iodine, astatine. (Chap. 8, p. 278)

heat: the energy transferred from an object at high temperature to an object at lower temperature. (Chap. 20, p. 711)

heat of fusion: the energy released as one kilogram of a substance solidifies at its freezing point. (Chap. 10, p. 364)

half life / media vida: tiempo que tarda la mitad de cierto isótopo radioactivo en descomponerse (en un isótopo o elemento diferentes). (Cap. 21, pág. 756)

halogen / halógeno: cualquier elemento del Grupo 17: flúor, cloro, bromo, yodo, astato. (Cap. 8, pág. 278)

heat / calor: tenergía transferida de un cuerpo a temperatura elevada hacia un cuerpo a una temperatura más baja. (Cap. 20, pág. 711)

heat of fusion / calor de fusión: energía liberada al solidificarse un kilogramo de sustancia en su punto de congelación. (Cap. 10, pág. 364)

Glossary/Glosario

heat of solution: the heat taken in or released in the dissolving process. (Chap. 13, p. 460)

heat of vaporization: the energy absorbed when one kilogram of a liquid vaporizes at its normal boiling point. (Chap. 10, p. 361)

Heisenberg uncertainty principle: the principle that it is impossible to accurately measure both the position and energy of an electron at the same time. (Chap. 7, p. 238)

hormone: a signal molecule that tells cells whether to start or stop a reaction. (Chap. 19, p. 694)

hydrate: a compound in which there is a specific ratio of water to ionic compound. (Chap. 5, p. 165)

hydrocarbon: an organic compound that consists of only hydrogen and carbon. (Chap. 5, p. 183)

hydrogen bonding: a connection between the hydrogen atoms on one molecule and a highly electronegative atom on another molecule, but not a full covalent bond. (Chap. 13, p. 439)

hydronium ion: a hydrogen ion attached to a water molecule. (Chap. 14, p. 483)

hygroscopic: a substance that absorbs water molecules from the air to become a hydrate. (Chap. 5, p. 166)

hypothesis: a prediction that can be tested to explain observations. (Chap. 2, p. 59)

heat of solution / calor de solución: calor consumido o liberado en el proceso de disolución. (Cap. 13, pág. 460)

heat of vaporization / calor de vaporización: energía absorbida cuándo un kilogramo de un líquido se vaporiza en su punto de ebullición normal. (Cap. 10, pág. 361)

Heisenberg uncertainty principle / principio de incertidumbre de Heisenberg: el principio que enuncia que resulta imposible medir exactamente la posición y energía de un electrón al mismo tiempo. (Cap. 7, pág. 238)

hormone / hormona: molécula mensajera que les indica a las células cuando empezar o parar una reacción. (Cap. 19, pág. 694)

hydrate / hidrato: compuesto en que hay una proporción específica de agua y compuesto iónico. (Cap. 5, pág. 165)

hydrocarbon / hidrocarburo: compuesto orgánico cuyos únicos componentes son hidrógeno y carbono. (Cap. 5, pág. 183)

hydrogen bonding / puente de hidrógeno: conexión entre los átomos de hidrógeno en una molécula y un átomo altamente electronegativo en otra molécula, pero no es un enlace covalente completo. (Cap. 13, pág. 439)

hydronium ion / ion hidronio: ion de hidrógeno unido a una molécula de agua. (Cap. 14, pág. 483)

hygroscopic / higroscópico: sustancia que absorbe moléculas de agua del aire para convertirse en un hidrato. (Cap. 5, pág. 166)

hypothesis / hipótesis: predicción que puede comprobarse para explicar observaciones. (Cap. 2, pág. 59)

ideal gas: a gas in which the particles undergo elastic collisions. (Chap. 10, p. 343)

ideal gas law: the equation that expresses exactly how pressure P, volume V, temperature T, and the number of particles n of a gas are related. $PV = nRT$. (Chap. 12, p. 419)

inhibitor: a substance that slows down a reaction. (Chap. 6, p. 223)

inner transition element: one of the elements in the two rows of elements below the main body of the periodic table; the lanthanides and the actinides. (Chap. 7, p. 250)

ideal gas / gas ideal: gas cuyas partículas experimentan choques elásticos. (Cap. 10, pág. 343)

ideal gas law / ley del gas ideal: ecuación que expresa la relación exacta entre la presión P, el volumen V, la temperatura T y el número de partículas n de un gas. $PV = nRT$. (Cap. 12, pág. 419)

inhibitor / inhibidor: sustancia que decelera una reacción. (Cap. 6, pág. 223)

inner transition element / elemento de transición interna: uno de los elementos en dos filas de elementos debajo del cuerpo principal de la tabla periódica; los lantánidos y los actínidos. (Cap. 7, pág. 250)

inorganic compound: a compound that does not contain carbon. (Chap. 5, p. 180)

insoluble: term describing a compound that does not dissolve in a liquid. (Chap. 6, p. 215)

interparticle forces: the forces between the particles that make up a substance. (Chap. 4, p. 144)

ion: an atom or group of combined atoms that has a charge because of the loss or gain of electrons. (Chap. 4, p. 134)

ionic bond: the strong attractive force between ions of opposite charge. (Chap. 4, p. 134)

ionic compound: a compound comprised of ions. (Chap. 4, p. 134)

ionic equation: an equation in which substances that primarily exist as ions in solution are shown as ions. (Chap. 15, p. 517)

ionization: the process where ions form from a covalent compound. (Chap. 14, p. 488)

isomer: a compound with a structure different from another compound with the same formula. (Chap. 18, p. 628)

isotope: any of two or more atoms of an element that are chemically alike but have different masses. (Chap. 2, p. 62)

inorganic compound / compuesto inorgánico: compuesto que no contiene carbono. (Cap. 5, pág. 180)

insoluble / insoluble: término que describe un compuesto que no se disuelve en un líquido. (Cap. 6, pág. 215)

interparticle forces / fuerzas interparticulares: fuerzas entre las partículas que componen una sustancia. (Cap. 4, pág. 144)

ion / ion: átomo o grupo de átomos combinados que tiene(n) una carga debido a la pérdida o la ganancia de electrones. (Cap. 4, pág. 134)

ionic bond / enlace iónico: fuerza de atracción intensa entre iones de carga opuesta. (Cap. 4, pág. 134)

ionic compound / compuesto iónico: compuesto formado por iones. (Cap. 4, pág. 134)

ionic equation / ecuación iónica: ecuación en la cual sustancias que existen principalmente como iones en solución se muestran como iones. (Cap. 15, pág. 517)

ionization / ionización: proceso en que los iones forman un compuesto covalente. (Cap. 14, pág. 488)

isomer / isómero: compuesto con una estructura diferente de otro compuesto con la misma fórmula. (Cap. 18, pág. 628)

isotope / isótopo: cualquiera de dos o más átomos de un elemento que son químicamente semejantes pero que tienen masas diferentes. (Cap. 2, pág. 62)

joule: the SI unit of energy; the energy required to lift a one-newton weight one meter against the force of gravity. (Chap. 10, p. 361)

joule / julio: unidad de energía del SI; energía requerida para levantar un metro el peso de un newton contra la fuerza de gravedad. (Cap. 10, pág. 361)

kelvin (K): a division on the Kelvin scale; the SI unit of temperature. (Chap. 10, p. 350)

Kelvin scale: the temperature scale defined so that temperature of a substance is directly proportional to the average kinetic energy of the particles and so that zero on the scale corresponds to zero kinetic energy. (Chap. 10, p. 349)

kilocalorie: a unit equal to 1000 calories. (Chap. 20, p. 721)

kelvin (K) / kelvin (K): una división en la escala Kelvin; la unidad de temperatura del SI. (Cap. 10, pág. 350)

Kelvin scale / escala Kelvin: escala de temperatura definida de tal forma que la temperatura de una sustancia es directamente proporcional a la energía cinética media de las partículas, de modo que cero en la escala corresponda a la energía cinética cero. (Cap. 10, pág. 349)

kilocalorie / kilocaloría: una unidad igual a 1000 calorías. (Cap. 20, pág. 721)

Glossary/Glosario

kilopascal (kPa): 1000 pascals. (Chap. 11, p. 378)

kinetic theory: the theory that states that submicroscopic particles of all matter are in constant, random motion. (Chap. 10, p. 342)

lanthanide: one of the first series of inner transition elements with atomic numbers 58 to 71. (Chap. 3, p. 104)

law of combining gas volumes: the observation that at the same temperature and pressure, volumes of gases combine or decompose in ratios of small whole numbers. (Chap. 11, p. 396)

law of conservation of energy: statement that energy is neither created nor destroyed in a chemical change, but is simply changed from one form to another. (Chap. 20, p. 711)

law of conservation of mass: in a chemical change, matter is neither created nor destroyed. (Chap. 1, p. 42)

law of definite proportions: the principle that the elements that comprise a compound are always in a certain proportion by mass. (Chap. 2, p. 54)

Lewis dot diagram: a diagram where dots or other small symbols are placed around the chemical symbol of an element to illustrate the valence electrons. (Chap. 2, p. 79)

limiting reactant: the reactant of which there is not enough; when it is used up, the reaction stops and no new product is formed. (Chap. 6, p. 220)

lipid: a biological compound that contains a large proportion of C—H bonds and less oxygen than in a carbohydrate; commonly called fats and oils. (Chap. 19, p. 682)

liquid: a flowing substance with a definite volume but an indefinite shape. (Chap. 10, p. 341)

liquid crystal: a material that loses its rigid organization in only one or two dimensions when it melts. (Chap. 10, p. 345)

kilopascal (kPa) / kilopascal (kPa): 1000 pascales. (Cap. 11, pág. 378)

kinetic theory / teoría cinética: teoría que indica que las partículas submicroscópicas de toda la materia están en movimiento aleatorio constante. (Cap. 10, pág. 342)

lanthanide / lantánido: una de la primeras series de elementos de transición interna con números atómicos 58 al 71. (Cap. 3, pág. 104)

law of combining gas volumes / ley de combinación de volúmenes de gases: la observación de que a la misma temperatura y presión, los volúmenes de gases se combinan o se descomponen en proporciones de números enteros pequeños. (Cap. 11, pág. 396)

law of conservation of energy / ley de conservación de la energía: afima que la energía ni se crea ni se destruye durante un cambio químico, simplemente cambia de una forma a otra. (Cap. 20, pág. 711)

law of conservation of mass / ley de conservación de la masa: durante un cambio químico, la materia ni se crea ni se destruye. (Cap. 1, pág. 42)

law of definite proportions / ley de proporciones definidas: principio que dice que los elementos que forman un compuesto están siempre en una proporción de masa dada. (Cap. 2, pág. 54)

Lewis dot diagram / diagrama de punto de Lewis: esquema en que se colocan puntos u otros símbolos pequeños alrededor del símbolo químico de un elemento para ilustrar los electrones de valencia. (Cap. 2, pág. 79)

limiting reactant / reactivo limitante: el reactivo del que no hay suficiente; cuando se consume completamente, la reacción se detiene y no se forma ningún producto nuevo. (Cap. 6, pág. 220)

lipid / lípido: compuesto biológico que contiene una proporción considerable de enlaces C—H y menor cantidad de oxígeno que en un carbohidrato; llamados comúnmente grasas y aceites. (Cap. 19, pág. 682)

liquid / líquido: sustancia fluida con un volumen definido pero sin forma definida. (Cap. 10, pág. 341)

liquid crystal / cristal líquido: material que cuando se funde pierde su organización rígida en una o dos dimensiones únicamente. (Cap. 10, pág. 345)

malleable: property of a metal meaning it can be pounded or rolled into thin sheets. (Chap. 9, p. 313)

mass: the measure of the amount of matter an object contains. (Chap. 1, p. 4)

mass number: the sum of the neutrons and protons in the nucleus of an atom. (Chap. 2, p. 66)

matter: anything that takes up space and has mass. (Chap. 1, p. 4)

melting point: the temperature of a solid when its crystal lattice disintegrates. (Chap. 10, p. 364)

metabolism: name given to the sum of all the chemical reactions necessary for the life of an organism. (Chap. 19, p. 692)

metal: an element that has luster, conducts heat and electricity, and usually bends without breaking. (Chap. 3, p. 103)

metallic bond: the bond that results when metal atoms release their valence electrons to a pool of electrons shared by all the metal atoms. (Chap. 9, p. 314)

metalloid: an element with some physical and chemical properties of metals and other properties of nonmetals. (Chap. 3, p. 105)

mixture: a combination of two or more substances in which the basic identity of each substance is not changed. (Chap. 1, p. 18)

molar mass: the mass of one mole of a pure substance. (Chap. 12, p. 407)

molar volume: the volume that a mole of gas occupies at a pressure of one atmosphere and a temperature of 0.00°C. (Chap. 12, p. 416)

mole: the unit used to count numbers of atoms, molecules, or formula units of substances. (Chap. 12, p. 405)

molecular element: a molecule formed when atoms of the same element bond together. (Chap. 5, p. 174)

molecular mass: the mass in atomic mass units of one molecule of a covalent compound. (Chap. 12, p. 409)

malleable / maleable: propiedad de un metal que significa que éste puede golpearse o enrollarse en hojas delgadas. (Cap. 9, pág. 313)

mass / masa: medida de la cantidad de materia que contiene un cuerpo. (Cap. 1, pág. 4)

mass number / número de masa: suma de los neutrones y protones en el núcleo de un átomo. (Cap. 2, pág. 66)

matter / materia: todo aquello que ocupa espacio y tiene masa. (Cap. 1, pág. 4)

melting point / punto de fusión: temperatura a la cual se desintegr la estructura cristalina de un sólidoa. (Cap. 10, pág. 364)

metabolism / metabilismo: nombre dado a la suma de todas las reacciones químicas necesarias para la vida de un organismo. (Cap. 19, pág. 692)

metal / metal: elemento que posee lustre, conduce calor y electricidad y que generalmente se dobla sin romperse. (Cap. 3, pág. 103)

metallic bond / enlace metálico: enlace que resulta cuando átomos metálicos liberan sus electrones de valencia a un grupo de electrones compartidos por todos ellos. (Cap. 9, pág. 314)

metalloid / metaloide: elemento con algunas propiedades físicas y químicas de metales y otras propiedades de no metales. (Cap. 3, pág. 105)

mixture / mezcla: combinación de dos o más sustancias en la cual la identidad básica de cada sustancia no se altera. (Cap. 1, pág. 18)

molar mass / masa molar: la masa de un mol de una sustancia pura. (Cap. 12, pág. 407)

molar volume / volumen molar: volumen que ocupa ún mol de gas a una presión de una atmósfera y una temperatura de 0.00°C. (Cap. 12, pág. 416)

mole / mole: unidad que se usa para contar números de átomos, moléculas, o fórmulas unitarias de sustancias. (Cap. 12, pág. 405)

molecular element / elemento molecular: molécula que se forma cuando se unen átomos del mismo elemento. (Cap. 5, pág. 174)

molecular mass / masa molecular: masa en unidades de masa atómica de una molécula de un compuesto covalente. (Cap. 12, pág. 409)

Glossary/Glosario

molecular substance: a substance that has atoms held together by covalent rather than ionic bonds. (Chap. 5, p. 170)

molecule: an uncharged group of two or more atoms held together by covalent bonds. (Chap. 4, p. 140)

monomer: the individual, small units that make up a polymer. (Chap. 18, p. 649)

net ionic equation: the equation that results when ions common to both sides of the equation are removed, usually from an ionic equation. (Chap. 15, p. 520)

neutralization reaction: the reaction of an acid with a base, so called because the properties of both the acid and base are diminished or neutralized. (Chap. 15, p. 516)

neutron: a subatomic particle with a mass equal to a proton but with no electrical charge. (Chap. 2, p. 62)

noble gas: an element from Group 18 that has a full compliment of valence electrons and as such is unreactive. (Chap. 3, p. 98)

noble gas configuration: the state of an atom achieved by having the same valence electron configuration as a noble gas atom; the most stable configuration. (Chap. 4, p. 132)

nonmetal: an element that in general does not conduct electricity, is a poor conductor of heat, and is brittle when solid. Many are gases at room temperature. (Chap. 3, p. 104)

nuclear fission: the process in which an atomic nucleus splits into two or more large fragments. (Chap. 21, p. 762)

nuclear fusion: the process in which two or more nuclei combine to form a larger nucleus. (Chap. 21, p. 766)

nuclear reactor: the device used to extract energy from a radioactive fuel. (Chap. 21, p. 764)

nucleic acid: a large polymer containing carbon, hydrogen, and oxygen, as well as nitrogen and phosphorus; found in all plant and animal cells. (Chap. 19, p. 688)

molecular substance / sustancia molecular: sustancia que tiene átomos unidos por enlaces covalentes en lugar de unirse por enlaces iónicos. (Cap. 5, pág. 170)

molecule / molécula: grupo sin carga de dos o más átomos unidos por un enlace covalente. (Cap. 4, pág. 140)

monomer / monómero: unidades pequeñas e individuales que forman un polímero. (Cap. 18, pág. 649)

net ionic equation / ecuación iónica neta: ecuación que resulta cuando iones comunes a ambos lados de la ecuación se eliminan, generalmente de una ecuación iónica. (Cap. 15, pág. 520)

neutralization reaction / reacción de neutralización: reacción de un ácido con una base, llamada así porque las propiedades tanto del ácido como de la base disminuyen o se neutralizan. (Cap. 15, pág. 516)

neutron / neutrón: partícula subatómica con una masa igual a la de un protón pero sin carga eléctrica. (Cap. 2, pág. 62)

noble gas / gas noble: elemento del Grupo 18 que tiene completos todos sus electrones de valencia y por lo tanto no es reactivo. (Cap. 3, pág. 98)

noble gas configuration / configuración de gas noble: estado que logra un átomo al tener la misma configuración de electrones de valencia que un átomo de gas noble; la configuración más estable. (Cap. 4, pág. 132)

nonmetal / no metal: elemento que por lo general no conduce electricidad, es un mal conductor de calor y es quebradizo en estado sólido. Muchos son gases a temperatura ambiente. (Cap. 3, pág. 104)

nuclear fission / fisión nuclear: proceso en el cual un núcleo atómico se separa en dos o más fragmentos grandes. (Cap. 21, pág. 762)

nuclear fusion / fusión nuclear: proceso en el cual dos o más núcleos se combinan para formar un núcleo más grande (Cap. 21, pág. 766)

nuclear reactor / reactor nuclear: dispositivo que se emplea para extraer energía de un combustible radiactivo. (Cap. 21, pág. 764)

nucleic acid / ácido nucleico: polímero grande que contiene carbono, hidrógeno y oxígeno, así como nitrógeno y fósforo; se halla en toda célula vegetal y animal. (Cap. 19, pág. 688)

nucleotide: the building blocks of nucleic acids; each consists of a simple sugar, a phosphate group, and a nitrogen-containing base. (Chap. 19, p. 689)

nucleus: the small, dense, positively charged central core of an atom. (Chap. 2, p. 65)

octet rule: the model of chemical stability that states that atoms become stable by having eight electrons in their outer energy level except for some of the smallest atoms, which have only two electrons. (Chap. 4, p. 132)

orbital: the space in which there is a high probability of finding an electron. (Chap. 7, p. 239)

organic compound: a compound that contains carbon; a few exceptions exist. (Chap. 5, p. 180)

osmosis: the flow of molecules through a selectively permeable membrane driven by concentration difference. (Chap. 13, p. 467)

oxidation: a reaction in which an element loses electrons. (Chap. 16, p. 556)

oxidation number: the charge on an ion or an element; can be positive or negative. (Chap. 5, p. 157)

oxidation-reduction reaction: a reaction characterized by the transfer of electrons from one atom or ion to another. Also known as a redox reaction. (Chap. 16, p. 555)

oxidizing agent: the substance that gains electrons in a redox reaction. It is the substance that is reduced. (Chap. 16, p. 562)

pascal (Pa): the SI unit for measuring pressure. (Chap. 11, p. 378)

period: a horizontal row in the periodic table. (Chap. 3, p. 96)

nucleotide / nucleótido: constituyentes de ácidos nucleicos; cada uno consiste en un azúcar sencillo, un grupo fosfato y una base nitrogenada. (Cap. 19, pág. 689)

nucleus / núcleo: la parte central y que está cargada positivamente, densa y pequeña de un átomo. (Cap. 2, pág. 65)

octet rule / regla del octeto: modelo de estabilidad química que establece que los átomos llegan a ser estables al tener ocho electrones en su nivel de energía externo, a excepción de los átomos más pequeños, los cuales tienen sólo dos electrones. (Cap. 4, pág. 132)

orbital / orbital: espacio en el cual hay una alta probabilidad de encontrar un electrón. (Cap. 7, pág. 239)

organic compound / compuesto orgánico: compuesto que contiene carbono; existen unas cuantas excepciones. (Cap. 5, pág. 180)

osmosis / ósmosis: flujo de moléculas a través de una membrana permeable selectiva debido a una diferencia de la concentración. (Cap. 13, pág. 467)

oxidation / oxidación: reacción en la cual un elemento pierde electrones. (Cap. 16, pág. 556)

oxidation number / número de oxidación: carga en un ion o elemento; puede ser positivo o negativo. (Cap. 5, pág. 157)

oxidation-reduction reaction / reacción de óxido-reducción: reacción caracterizada por la transferencia de electrones de un átomo o ion a otro. También conocida como reacción redox. (Cap. 16, pág. 555)

oxidizing agent / agente oxidante: sustancia que gana electrones en una reacción de redox. Es la sustancia que se reduce. (Cap. 16, pág. 562)

pascal (Pa) / pascal (Pa): unidad del SI para medir presión. (Cap. 11, pág. 378)

period / período: una fila horizontal en la tabla periódica. (Cap. 3, pág. 96)

Glossary/Glosario

periodic law: the statement that the physical and chemical properties of the elements repeat in a regular pattern when they are arranged in order of increasing atomic number. (Chap. 3, p. 94)

periodicity: the tendency to recur at regular intervals. (Chap. 3, p. 90)

pH: a mathematical scale in which the concentration of hydronium ions in a solution is expressed as a number from 0 to 14. (Chap. 14, p. 502)

photosynthesis: the process used by certain organisms to capture energy from the sun. (Chap. 20, p. 734)

physical change: a change in matter where its identity does not change. (Chap. 1, p. 20)

physical property: a characteristic of matter that is exhibited without a change of identity. (Chap. 1, p. 20)

plasma: an ionized gas. (Chap. 10, p. 347)

polar covalent bond: a bond where the electrons are shared unequally; there is some degree of ionic character to this type of bond. (Chap. 9, p. 310)

polar molecule: a molecule that has a positive pole and a negative pole because of the arrangement of the polar bonds; also called a dipole. (Chap. 9, p. 331)

polyatomic ion: an ion that consists of two or more different elements. (Chap. 5, p. 158)

polymer: a large molecule that is made up of many smaller repeating units. (Chap. 18, p. 649)

potential difference: the difference in electron pressure at the cathode (low) and at the anode (high) in an electrochemical cell. (Chap. 17, p. 601)

pressure: the force acting on a unit area of a surface. (Chap. 10, p. 344)

product: a new substance formed when reactants undergo chemical change. (Chap. 6, p. 192)

property: the characteristics of matter; how it behaves. (Chap. 1, p. 4)

protein: a polymer formed from small monomer molecules linked together by amide groups. (Chap. 19, p. 670)

proton: a positively charged subatomic particle. (Chap. 2, p. 62)

periodic law / ley periódica: establece que las propiedades físicas y químicas de los elementos se repiten en un patrón regular cuando éstos se organizan en orden creciente de número atómico. (Cap. 3, pág. 94)

periodicity / periodicidad: tendencia de volver a ocurrir a intervalos regulares. (Cap. 3, pág. 90)

pH / pH: escala matemática en la cual la concentración de iones hidronio en una solución se expresa como un número entre 0 y 14. (Cap. 14, pág. 502)

photosynthesis / fotosíntesis: proceso que usan ciertos organismos para capturar energía solar. (Cap. 20, pág. 734)

physical change / cambio físico: cambio en la materia donde su identidad no cambia. (Cap. 1, pág. 20)

physical property / propiedad física: característica de la materia que se presenta sin un cambio de identidad. (Cap. 1, pág. 20)

plasma / plasma: gas ionizado. (Cap. 10, pág. 347)

polar covalent bond / enlace covalente polar: enlace en el cual los electrones no se comparten de igual manera; hay algún grado de carácter iónico en este tipo de enlace. (Cap. 9, pág. 310)

polar molecule / molécula polar: molécula con un polo positivo y uno negativo a causa del arreglo de los enlaces polares; también llamada dipolo. (Cap. 9, pág. 331)

polyatomic ion / ion poliatómico: ion que consta de dos o más elementos diferentes. (Cap. 5, pág. 158)

polymer / polímero: molécula grande compuesta de muchas unidades más pequeñas que se repiten. (Cap. 18, pág. 649)

potential difference / diferencia de potencial: diferencia en la presión electrónica en el cátodo (baja) y en el ánodo (alta) en una celda electroquímica. (Cap. 17, pág. 601)

pressure / presión: fuerza que actúa sobre una unidad de área de una superficie. (Cap. 10, pág. 344)

product / producto: sustancia nueva que se forma cuando los reactantes experimentan un cambio químico. (Cap. 6, pág. 192)

property / propiedad: características de la materia; la manera como se comporta. (Cap. 1, pág. 4)

protein / proteína: polímero formado a partir de pequeñas moléculas monoméricas unidas por grupos amida. (Cap. 19, pág. 670)

proton / protón: partícula subatómica cargada positivamente. (Cap. 2, pág. 62)

Glossary/Glosario

qualitative: an observation made without measurement. (Chap. 1, p. 14)

quantitative: an observation made with measurement. (Chap. 1, p. 14)

radioactivity: the spontaneous emission of radiation by an unstable atomic nucleus. (Chap. 21, p. 746)

reactant: a substance that undergoes a reaction. (Chap. 6, p. 192)

reducing agent: the substance that loses electrons in a redox reaction. It is the substance that is oxidized. (Chap. 16, p. 562)

reduction: a reaction in which an element gains one or more electrons. (Chap. 16, p. 556)

reforming: the use of heat, pressure, and catalysts to convert large alkanes into other compounds, often aromatic hydrocarbons. (Chap. 18, p. 639)

respiration: the complex series of enzyme-catalyzed reactions that are used to extract chemical energy from glucose. (Chap. 19, p. 694)

RNA: ribonucleic acid. (Chap. 19, p. 688)

salt: the general term used in chemistry to describe the ionic compound formed from the negative part of an acid and the positive part of a base. (Chap. 15, p. 516)

saturated hydrocarbon: a hydrocarbon with all the carbon atoms connected to each other by single bonds. (Chap. 18, p. 623)

saturated solution: a solution that holds the maximum amount of solute under the given conditions. (Chap. 13, p. 458)

scientific law: a fact of nature that is observed so often that it is accepted as the truth. (Chap. 2, p. 59)

qualitative / cualitativa: observación hecha sin ninguna medición. (Cap. 1, pág. 14)

quantitative / cuantitativa: observación hecha con cierta medición. (Cap. 1, pág. 14)

radioactivity / radiactividad: emisión espontánea de radiación por un núcleo atómico inestable. (Cap. 21, pág. 746)

reactant / reactivo: sustancia que experimenta una reacción. (Cap. 6, pág. 192)

reducing agent / agente reductor: sustancia que pierde electrones en una reacción redox. Es la sustancia que se oxida. (Cap. 16, pág. 562)

reduction / reducción: reacción en la cual un elemento gana uno o más electrones. (Cap. 16, pág. 556)

reforming / reformado: uso de calor, presión y catalizadores para convertir alcanos largos en otros compuestos, a menudo, hidrocarburos aromáticos. (Cap. 18, pág. 639)

respiration / respiración: serie compleja de reacciones catalizadas por enzimas que se usan para extraer energía química de la glucosa. (Cap. 19, pág. 694)

RNA / RNA: ácido ribonucleico. (Cap. 19, pág. 688)

salt / sal: término general que se usa en química para describir el compuesto iónico formado por la parte negativa de un ácido y la parte positiva de una base. (Cap. 15, pág. 516)

saturated hydrocarbon / hidrocarburo saturado: hidrocarburo con todos los átomos de carbono unidos unos a los otros por enlaces sencillos. (Cap. 18, pág. 623)

saturated solution / solución saturada: solución que tiene la cantidad máxima de soluto bajo las condiciones dadas. (Cap. 13, pág. 458)

scientific law / ley científica: un hecho de la naturaleza que se observa con tanta frecuencia que se acepta como cierto. (Cap. 2, pág. 59)

Glossary/Glosario   **889**

Glossary/Glosario

scientific model: a thinking device, built on experimentation, that helps us understand and explain macroscopic observations. (Chap. 1, p. 11)

semiconductor: an element that does not conduct electricity as well as a metal, but that does conduct slightly better than a nonmetal. (Chap. 3, p. 105)

shielding effect: the tendency for the electrons in the inner energy levels to block the attraction of the nucleus for the valence electrons. (Chap. 9, p. 304)

sievert: the unit of radiation equal to one gray multiplied by a factor that assesses how much of the radiation striking tissue is actually absorbed by the tissue and so is a measure of how much biological damage is caused. (Chap. 21, p. 776)

single displacement: a type of reaction where one element takes the place of another in a compound. (Chap. 6, p. 205)

solid: a substance in which the particles occupy fixed positions in a well-defined, three-dimensional arrangement. (Chap. 10, p. 340)

soluble: term describing a substance that dissolves in a liquid. (Chap. 6, p. 215)

solute: the substance that is being dissolved when making a solution. (Chap. 1, p. 23)

solution: a mixture that is the same throughout, or homogeneous. (Chap. 1, p. 22)

solvent: the substance that dissolves the solute when making a solution. (Chap. 1, p. 23)

specific heat: a measure of the amount of heat needed to raise the temperature of 1 gram of a substance 1°C. (Chap. 13, p. 445)

spectator ion: an ion that is present in solution but does not participate in the reaction. (Chap. 15, p. 520)

standard atmosphere (atm): the pressure that supports a column of mercury 760 millimeters in height. (Chap. 11, p. 376)

standard solution: a solution of known molarity used in a titration. (Chap. 15, p. 539)

standard temperature and pressure, STP: the set of conditions 0.00°C and 1 atmosphere. (Chap. 11, p. 395)

steroid: a lipid with a distinctive four-ring structure. (Chap. 19, p. 686)

stoichiometry: the study of relationships between measurable quantities, such as mass and volume, and the number of atoms in chemical reactions. (Chap. 12, p. 404)

scientific model / modelo científico: instrumento de pensamiento, desarrollado sobre la experimentación, que nos ayuda a entender y explicar observaciones macroscópicas. (Cap. 1, pág. 11)

semiconductor / semiconductor: elemento que no conduce la electricidad tan bien como un metal, pero que la conduce ligeramente mejor que un no metal. (Cap. 3, pág. 105)

shielding effect / efecto protector: tendencia de los electrones en los niveles energéticos internos de bloquear la atracción del núcleo hacia los electrones de valencia. (Cap. 9, pág. 304)

sievert / sievert: unidad de radiación igual a un gray multiplicado por un factor que mide la cantidad de radiación que recibe un tejido y es absorbida por el mismo; por lo tanto es una medida del daño biológico causado. (Cap. 21, pág. 776)

single displacement / desplazamiento simple: tipo de reacción en el cual un elemento ocupa el lugar de otro en un compuesto. (Cap. 6, pág. 205)

solid / sólido: sustancia cuyas partículas ocupan posiciones fijas en un arreglo tridimensional bien definido. (Cap. 10, pág. 340)

soluble / soluble: término que describe una sustancia que se disuelve en un líquido. (Cap. 6, pág. 215)

solute / soluto: la sustancia que se disuelve cuando se hace una solución. (Cap. 1, pág. 23)

solution / solución: mezcla uniforme u homogénea. (Cap. 1, pág. 22)

solvent / disolvente: sustancia que disuelve el soluto cuando se hace una solución. (Cap. 1, pág. 23)

specific heat / calor específico: medida de la cantidad de calor requerido para aumentar en 1°C la temperatura de 1 gramo de una sustancia. (Cap. 13, pág. 445)

spectator ion / ion espectador: ion que está presente en la solución pero que no participa en ella. (Cap. 15, pág. 520)

standard atmosphere (atm) / atmósfera estándar (atm): presión que sostiene una columna de mercurio de 760 milímetros de altura. (Cap. 11, pág. 376)

standard solution / solución estándar: solución de molaridad conocida que se usa en una titulación. (Cap. 15, pág. 539)

standard temperature and pressure, STP / temperatura y presión estándares, STP: las condiciones de 0.00°C y 1 atmósfera. (Cap. 11, pág. 395)

steroid / esteroide: lípido con una estructura distintiva de cuatro anillos. (Cap. 19, pág. 686)

stoichiometry / estequiometría: estudio de las relaciones entre cantidades mensurables, tal como masa y volumen y el número de átomos en reacciones químicas. (Cap. 12, pág. 404)

strong acid: an acid that is completely ionized in water; no molecules exist in the water solution. (Chap. 14, p. 498)

strong base: a base that is completely dissociated into separate ions when dissolved in water. (Chap. 14, p. 497)

sublevel: the small energy divisions in a given energy level. (Chap. 7, p. 234)

sublimation: the process by which particles of a solid escape from its surface and form a gas. (Chap. 10, p. 356)

substance: matter with the same fixed composition and properties. (Chap. 1 p. 15)

substituted hydrocarbon: a compound that has the same structure as a hydrocarbon, except that other atoms are substituted for part of the hydrocarbon. (Chap. 18, p. 640)

substrate: the name given to a reactant in an enzyme-catalyzed reaction. (Chap. 19, p. 676)

supersaturated solution: a solution containing more solute than the usual maximum; they are unstable. (Chap. 13, p. 459)

surface tension: the force needed to overcome intermolecular attractions and break through the surface of a liquid or spread the liquid out. (Chap. 13, p. 442)

synthesis: the name applied to a reaction in which two or more substances combine to form a single product. (Chap. 6, p. 203)

strong acid / ácido fuerte: ácido que se ioniza completamente en agua; no existen moléculas en solución acuosa. (Cap. 14, pág. 498)

strong base / base fuerte: base que se disocia completamente en iones separados cuando se disuelve en agua. (Cap. 14, pág. 497)

sublevel / subnivel: pequeñas divisiones energéticas en un nivel dado de energía. (Cap. 7, pág. 234)

sublimation / sublimación: proceso por cual las partículas de un sólido se escapan de su superficie y forman un gas. (Cap. 10, pág. 356)

substance / sustancia: materia con la misma composición fija y propiedades. (Cap. 1 p. 15)

substituted hydrocarbon / hidrocarburo sustituido: compuesto que tiene la misma estructura de un hidrocarburo, excepto que otros átomos reemplazan parte del hidrocarburo. (Cap. 18, pág. 640)

substrate / sustrato: nombre dado a un reactante en una reacción catalizada por enzimas. (Cap. 19, pág. 676)

supersaturated solution / solución supersaturada: solución que contiene más soluto que el máximo normal; son inestables. (Cap. 13, pág. 459)

surface tension / tensión superficial: fuerza necesaria para vencer las atracciones intermoleculares y penetrar la superficie de un líquido o para esparcirlo. (Cap. 13, pág. 442)

synthesis / síntesis: reacción en que dos o más sustancias se combinan para formar un solo producto. (Cap. 6, pág. 203)

temperature: the measure of the average kinetic energy of the particles that make up a material. (Chap. 10, p. 348)

theory: an explanation based on many observations and supported by the results of many experiments. (Chap. 2, p. 59)

thermoplastic: a plastic that will soften and harden repeatedly when heated and cooled. (Chap. 18, p. 660)

thermosetting: a plastic that hardens permanently when first formed. (Chap. 18, p. 660)

titration: the process of determining the molarity of an acid or base by using an acid-base reaction where one reactant is of known molarity. (Chap. 15, p. 539)

temperature / temperatura: medida de la energía cinética promedio de las partículas que componen un material. (Cap. 10, pág. 348)

theory / teoría: explicación basada en muchas observaciones y sostenida por los resultados de muchos experimentos. (Cap. 2, pág. 59)

thermoplastic / termoplástico: plástico que se ablandará y se endurecerá repetidamente cuando se calienta y enfría. (Cap. 18, pág. 660)

thermosetting / fraguado: plástico que se endurece permanentemente cuando se le da forma. (Cap. 18, pág. 660)

titration / titulación: proceso para determinar la molaridad de un ácido o base usando una reacción ácido-base donde un reactivo es de molaridad conocida. (Cap. 15, pág. 539)

Glossary/Glosario

transition element: any of the elements in Groups 3 through 12 of the periodic table, all of which are metals. (Chap. 3, p. 103)

triple bond: a bond formed by sharing three pairs of electrons between two atoms. (Chap. 9, p. 325)

tritium: the hydrogen isotope with a mass number of 3. (Chap. 21, p. 766)

Tyndall effect: the scattering effect caused when light passes through a colloid. (Chap. 13, p. 472)

unsaturated hydrocarbon: a hydrocarbon that has one or more double or triple bonds between carbon atoms. (Chap. 18, p. 629)

unsaturated solution: a solution in which the amount of solute dissolved is less than the maximum that could be dissolved. (Chap. 13, p. 458)

valence electron: an electron in the outermost energy level of an atom. (Chap. 2, p. 78)

vapor pressure: the pressure of a substance in equilibrium with its liquid. (Chap. 10, p. 357)

vitamin: an organic molecule required in small amounts; there are fat-soluble and water-soluble types. (Chap. 19, p. 690)

volatile: description of a substance that easily changes to a gas at room temperature. (Chap. 1, p. 35)

voltage: an electrical potential difference, expressed in units of volts. (Chap. 17, p. 601)

weak acid: an acid in which almost all the molecules remain as molecules when placed into a water solution. (Chap. 14, p. 499)

weak base: a base in which most of the molecules do not react with water to form ions. (Chap. 14, p. 500)

transition element / elemento de transición: cualquiera de los elementos en los Grupos 3 a 12 de la tabla periódica, todos los cuales son metales. (Cap. 3, pág. 103)

triple bond / enlace triple: enlace que se forma cuando dos átomos comparten tres pares de electrones. (Cap. 9, pág. 325)

tritium / tritio: isótopo de hidrógeno con un número de masa de 3. (Cap. 21, pág. 766)

Tyndall effect / efecto Tyndall: efecto de dispersión que ocurre cuando la luz pasa a través de un coloide. (Cap. 13, pág. 472)

unsaturated hydrocarbon / hidrocarburo insaturado: hidrocarburo que tiene uno o más enlaces dobles o triples entre átomos de carbono. (Cap. 18, pág. 629)

unsaturated solution / solución insaturada: solución en que la cantidad de soluto disuelta es menor que el máximo que se puede disolver. (Cap. 13, pág. 458)

valence electron / electrón de valencia: electrón en el nivel energético más externo de un átomo. (Cap. 2, pág. 78)

vapor pressure / presión de vapor: presión de una sustancia en equilibrio con su líquido. (Cap. 10, pág. 357)

vitamin / vitamina: molécula orgánica requerida en cantidades pequeñas; son de tipo liposoluble e hidrosoluble. (Cap. 19, pág. 690)

volatile / volátil: se dice de una sustancia que se convierte fácilmente en un gas a temperatura ambiente. (Cap. 1, pág. 35)

voltage / voltaje: diferencia de potencial eléctrico, expresado en unidades de voltios. (Cap. 17, pág. 601)

weak acid / ácido débil: ácido en que casi todas sus moléculas permanecen como tales cuando se introduce en una solución acuosa. (Cap. 14, pág. 499)

weak base / base débil: base en que la mayor parte de las moléculas no reaccionan con agua para formar iones. (Cap. 14, pág. 500)

This index will help you locate important information in the text quickly and easily. Page numbers given in bold type indicate the location of a definition for that entry.

Absolute zero, 349

Accelerators, 773 *illus.*

Accuracy, 793

Acetaminophen, 33 table, 809 *table*

Acetic acid. *See also* Vinegar; buffer solution with sodium acetate, 531–532, 533 *illus.*; carboxyl group on, 642; dissolution of in water, 485; formula, 33 *table*, 809 *table*; glacial, 528; reaction with carbonates, 483 *illus.*; reaction with strong bases, 528, 529 *illus.*; titration of, 542–543 *lab*; uses of, 33 *table*; as weak acid, 498 *illus.*, 499

Acetone, 644

Acetylene, 324–325, 633

Acetylene welding, 209 *illus.*, 325 *illus.*, 633

Acetylsalicylic acid, 32

Acid-base indicators, 481, 507; bromothymol blue, 518 *lab*; cabbage juice indicator, creating and using, 504–505 *lab*; list of, 852 *table*; pH range of different, 508 *illus.*; requirements for, 545; titration and, 540

Acid-base reactions, 481, 482 *lab*, 516–517, 517 *table*; antacids and, 535–536, 538; blood gas measurements and, 487; buffers and, 531–533, 535; cave formation and, 524; gas-phase, 527, 528 *illus.*; ionic equations for, 517–518, 520, 521 prob.; salt produced by, 516, 518 *lab*; strong acid plus strong base, 517–518, 520, 521 prob., 522; strong acid plus weak base, 522–523, 525 prob., 525–526; weak acid plus strong base, 528–529, 529 prob., 530 prob.; weak acid plus weak bases, 530

Acid-base titration, 539–541, 544; of an acid with a base, 539; concentration from, 541, 542–543

lab, 544, 544 prob., 546 prob.; how to perform, 540; indicators and, 540; standard solutions and, 539

Acid hydrogen, 485

Acidic anhydrides, 492–494

Acid indigestion, 535–536

Acid ionization, 488

Acid lakes, 535

Acid rain, 482, 494, 495, 535

Acids, 483, 485. *See also* Acid-base reactions; Bases; acid-base indicators and, 481, 504–505 *lab*; acidic anhydrides, 492–494; among top ten industrial chemicals, 481 *table*; Brønsted-Lowry definition, 526–527; common names, 182 *table*; as electrolytes, 488; hyrdronium ions, production of in water, 483, 485; ionization, 486, 488, 499; mono- vs. polyprotic, 485 *illus.*, 486 *illus.*; names of common, 182 *table*; pH scale and, 500, 502–503, 506–508; reaction with carbonates, 482, 483 *illus.*; reaction with metals, 236–237, 482, 483 *illus.*; strong, 498, 498 *table*; taste and feel of, 480, 519; titration of. *See* Titration; weak, 498 *illus.*, 499; Word Origin, 485

Actinides, 92–93 *illus.*, 94, **104,** 107, 250; radioactivity of, 295; Word Origin, 295

Activated charcoal, 177

Activation energy, 218, 713–714; catalysts and, 222, 714

Active site, 676

Activities, ChemLabs. *See* ChemLabs; MiniLabs. *See* MiniLabs; safety guidelines, 839; safety symbols, 840

Actual yield, 421

Addition reactions, 656

Adenine (A), 689, 690

Adenosine diphosphate (ADP), 696, 697 *illus.*, 734

Adenosine triphosphate (ATP), 273, 696, 697 *illus.*, 734

Adipic acid, polymerization, 658 *illus.*

ADP. *See* Adenosine diphosphate (ADP)

Aerobic, 694; respiration. *See* Respiration; Word Origin, 694

Aerosols, 470

AFM. *See* Atomic force microscope (AFM)

Air, composition of, 354; density of, 36 *table*; exhaled vs. inhaled, 124 *table*; fractionation, 354–355; as homogeneous mixture, 23; space shuttle, filtering of on, 203, 418 prob.

Air bags, 417

Air pollution, 495; acid rain, 482, 494, 495, 535; catalytic converters and, 715

Alanine, 671

Alchemy, 25 *lab*

Alcoholic fermentation, 698, 699 *illus.*, 699 *lab*

Alcohols, 642; hydrogen bonding and, 439; production of compounds containing, 646–647; testing for presence of, 568 *lab*, 569; uses and properties of, 642; volatility of, 35

Aldehydes, 644

Alkali materials, 498

Alkali metals, 86, 99, **263**–264, 264–265; electron configuration, 263; ionic charge of, 157 *table*; physical properties, 263; reactivity, 263, 264; valence electron configuration, 243–244

Alkaline earth metals, 99, 157 *table*, **265;** ion charges, 266–267 *lab*; properties, 265; reactivity, 97 *lab*, 265, 266–267 *lab*, 268; valence electron configuration, 243–244

Alkanes (saturated hydrocarbons), 323, **623**–626; first ten, 624 *table*; isomers, 627–628; naming, 623–626, 626–627 prob.; production of compounds containing, 646; properties, 629; structural diagrams, 623–624, 626–627 prob.

Index

Alkenes, 324, 629–631, 633; geometric isomers, 631 *illus.*; naming, 630–631; properties, 636

Alkynes, 324, **633**, 633 *table*; naming, 633 *table*; properties, 635

Allotropes, **175**, 179; of carbon, 176–178, 179; of oxygen, 175, 276; of phosphorus, 179, 274

Alloys, **23**, 103 *illus.*, 106, 282; common, 23 *table*; of copper and zinc, 25 *lab*; crystal structure of, 108; misch metal, 294; shape-memory, 108–109; steel. *See* Steel

Alloy steels, 290, 290 *table*

Almond flavor, 644

Alnico steel, 290, 290 *table*

Alpha decay, 747

Alpha particles, 64, 230, **747**

Alpha radiation, 747

Alternative energy sources, 728–729; geothermal energy, 728–729; nuclear energy. *See* Nuclear power plants; solar energy, 728; wind energy, 729

Alumina, **157**

Aluminum, 25 *illus.*, 98 *illus.*, 269; corrosion, 570 *illus.*; density, 36 *table*; in Earth's crust, 669 *illus.*; Hall-Héroult method of producing, 215, 269, 588 *illus.*, 589, 725; importance of, 269; ionic compounds formed by, 135 *lab*; oxidation-reduction (redox) reactions, 570; reaction capacity, 236–237; reaction with chlorine, 716; recycling, 269, 589 *illus.*, 725; uses of, 269, 269–270

Aluminum chloride, 716

Aluminum hydroxide, 269, 489 *illus.*; reaction with hydrobromic acid, 522–523; reaction with weak acid, 530 *illus.*

Aluminum oxide, 157, 269, 570

Aluminum zirconium, 269

Amethyst, 248 *illus.*

Amide groups, 645, 672

Amides, 645

Amines, 645

Amino acids, 500, **670–672**; in hair, 657

Amino group, 645, 671

Ammonia, 274, 809 *table*; acid-base reactions and, 523; aqueous solution of, 493 *illus.*; boiling point, 358; buffer solution with ammonium salt, 531; formula and uses, 33 *table*; Haber process for synthesizing, 216–217, 219; molecular shape of, 322; polarity of, 331; reaction with hydrochloric acid, 527, 528 *illus.*; water vs., 437 *table*; as weak base, 489, 492, 500

Ammonium chloride, 195, 523

Ammonium hydrogen phosphate, 274

Ammonium hydroxide, 493 *illus.*

Ammonium nitrate, 204 *illus.*, 274; in cold packs, 460 *illus.*, 710

Ammonium perchlorate, 566

Ammonium sulfate, 169, 274

Ammonium thiocyanate, 43 *illus.*

Amorphous materials, **346**, 347 *illus.*

Anaerobic, **698**

Anhydrides, **496**; acidic, 492–494; basic, 492, 494, 496

Anhydrous compounds, **168**, 169 *illus.*

Animal fats, 630 *lab*, 633 *illus.*

Animal starch, 681, 682 *illus.*

Anions, **586**

Anodes, **585**, 586

Antacids, 503 *lab*, 535–536, 538; carbonates, 538; compounds used in, 536 *table*; hydroxides, 538

Antibacterials, halogens as, 279

Antibodies, 670 *illus.*, 673

Antifreeze, 466, 642

Antifreeze proteins, 670

Antimony, 273; in alloys, 274; chemical symbol for, 27 *table*; as dopant, 111

Antioxidants, 576

Aqueous solutions, **23**, 450–454. *See also* Colloids; Solutions; boiling-point elevation, 465; concentrated versus dilute, 458; of covalent substances, 453; endothermic/exothermic reactions in, 712 *lab*; freeze-point depression, 465; of gases in water, 469; guidelines for, 851 *table*; identification of, 456–457 *lab*; of ionic substances, 451–453, 452 *lab*; molarity and, 460–461, 462–464 *prob.*; temperature and solubility, 459–460; unsaturated vs. saturated, 458–459

Archaeological radiometry, 754–755, 757–758, 760 *illus.*

Argon, 132 *illus.*, 133, 247, 281

Aromas, synthetic, 646 *lab*

Aromatic hydrocarbons, **636**–637, 639

Arsenic, 273; as dopant, 111; gallium arsenide, 274

Art Connection, Asante brass weights, 411; Chinese porcelain, 163; glass sculptures, 346; radioactive dating of forged art, 759

Art forgeries, 759

Artificial blood, 537

Asante brass weights, 411

Ascorbic acid, 33 *table*, 498 *illus.*, 809 *table*

Aspartame, 5 *illus.*, 33 *table*, 519

Aspirin, 5 *illus.*, 10, 33 *table*, 809 *table*; derivation of from plants, 146; modeling, 11; structure, 10 *illus.*; synthesis, 32

Astatine, 278

Atherosclerosis, 687

Atmosphere, carbon dioxide in, 493–494; composition of, 354; nitrogen in, 273; oxygen in, 276; pollution of, 493–494, 495; pressure of, 344

Atmospheric pressure, 344, 376, 378; boiling and, 358; equivalent units, 379 *table*

Atomic bombs, 232, 761, 764

Atomic collisions, 130, 138–139

Atomic force microscope (AFM), 241

Atomic mass, 67, 68 *illus.*, 86, 92–93 *illus.*

Atomic mass unit, **67**

Atomic models, **10**

Atomic number, **66**, 92–93 *illus.*

Atomic size (radii), atom size-ion size relationship, 260 *illus.*; main group element patterns, 258–259, 258–259, 259 *illus.*, 262 *lab*; transition element patterns, 285

Atomic structure, Bohr's model of, 231; discovery of, 61–65; electron cloud model of, 77, 238–239; nuclear model of, 65; periodic table and, 95–96, 97 *lab*, 98–99, 243–251

Atomic theory of matter, **53**–55, 230–231; conservation of matter and, 53, 55, 56–57 *lab*; Dalton's atomic theory and, 54–55; early Greek ideas about, 52–53; Lavoisier's contributions to, 53; Proust's contributions to, 53–54

Atoms, 7, **53**; chemical changes and, 41–42; conservation of in chemical reactions, 42, 53, 55, 198–199; number in a sample, 408 *prob*.; particles of, 61–62, 66, 67 *table*; size of, 10, 65; Word Origin, 53

ATP. *See* Adenosine triphosphate; Adenosine triphosphate (ATP)

Aurora borealis, 73. *See also* Word Origin

Austenite phase, 108

Automobile air bags, 417

Automobile batteries, lead-acid, 273, 611, 613; rechargeable for electric cars, 613–614

Avicel, 685

Avogadro, Amedeo, 398

Avogadro's constant, **405**–406

Avogadro's principle, **398**, 404

Bacteria, cavities and, 280; refining of ores by, 727

Bakelite, 658

Baking soda, 143 *illus*.; chemical properties of, 41 *illus*.; endothermic reactions of, 43; formula for, 33 *table*, 809 *table*; reaction with vinegar, 41 *illus*., 192–193, 195, 420 *lab*; uses of, 33 *table*

Balances, double-pan, 407 *illus*., 790; electronic, 410; triple-beam, 790

Balancing, of chemical equations, 198–199, 200–201 *prob*.; of nuclear equations, 750–751 *prob*.

Ballooning, 394 *illus*.

Balsa wood, density of, 36 *table*

Bar graphs, 805

Barium, 265; reactivity of, 265, 266–267 *lab*

Barium hydroxide, 498 *table*

Barium hydroxide octahydrate, reaction with ammonium chloride, 195; reaction with ammonium thiocyanate, 43 *illus*.

Barium nitrate, 456

Barium sulfate, 456

Barometers, 283 *illus*., **376**; Word Origin, 376

Barometric pressure units, converting, 379 *prob*.

Bartlett, Neil, 195

Bases, 481, **488**–489, 492. *See also* Acid-base reactions; Acids; acid-base indicators and, 481, 504–505 *lab*; among top ten industrial chemicals, 481 *table*; antacids and, 503 *lab*, 535–536, 538; basic anhydrides, 492, 494, 496; Brønsted-Lowry definition of, 526–527; common names, 182 *table*; as electrolytes, 492; hydroxide ions, production of by, 488–489; ionization of, 489; names of common, 182 *table*; strong, 497–498, 498 *table*; taste and feel of, 480, 519; titration of. *See* Titration; weak, 500

BASF, 217

Basic anhydrides, **492**, 494, 496

Batteries, 601, 602, 605, 608. *See also* Electrochemical cells; carbon-zinc dry cell, 608–609; lead storage, 611, 613; lemon, 599–601, 600 *lab*, 608 *illus*.; lithium, 610, 613, 615; NiCad, 612; rechargeable, 210 *illus*., 612, 613, 615

Bauxite, 215, 269, 589

Becquerel, Henri, 745–746

Beeswax, 684 *illus*.

Behavior of matter, 4, 5 *illus*.

Benz, Karl, 466

Benzene, 636

Benzoic acid, 498 *illus*.

Beryllium, 265; atomic size, 259; electron configuration, 244, 246 *table*; melting point of, 283; reactivity of, 261, 265; uses of, 268

Beta-carotene, 632

Beta decay, 749

Beta particles, **749**

Beta radiation, 749

BHA, 576 *illus*.

BHT, 576 *illus*.

Binary compounds, **155**–156; binary inorganic compounds, 180–182, 181 *prob*.; binary ionic compounds, 155–156; Word Origin (binary), 155

Biochemist, 678–679

Biochemistry, **668**–669

Biological specimens, freeze drying of, 353

Biology Connection, blood gases, measurement of, 487; fluoridation, 280; Hershey and Chase's experiments on DNA with tracers, 772; regulation of air quality in space vehicles, 203; vision and vitamin A, 632

Bioluminescence, 575; Word Origin, 575

Biomolecules, 668–669. *See also* Metabolism; carbohydrates, 677, 680–682; lipids, 682, 684, 685, 686–687; nucleic acids, 688 *lab*, 688–690; proteins, 670–673, 674–675 *lab*, 676–677; vitamins, 690–691

Biorefining, 727

Bismuth, 273

Bitter taste, 519

Black phosphorus, 179

Blast furnaces, 558 *illus*., 567

Bleaching reactions, 194, 567–568

Blood, artificial, 537; buffers and, 531, 533; as complex mixture, 18 *illus*.; forensic detection of, 573, 574; hemoglobin in, 286, 670, 693; pH of, 531, 533

Blood gases, measurement of, 487

Blood sugar. *See* Glucose

Bohr, Neils, 69–70, 74–75, 231, 232

Bohr's atomic model, 69–70, 231 *illus*., 232

Boiling, 356, 358, 360

Boiling point, 35, **358**; bond type and, 311 *illus*.; elevation of in solution, 465; periodic *table* trends, 88

Bonding electron pairs, 318

Bonds, covalent. *See* Covalent bonds; electron sharing model of, 303; electronegativity and, 303–305; electronegativity difference between bonding atoms (?EN), 305–306, 308–311, 312 *prob*.; hydrogen. *See* Hydrogen bonding; ionic. *See* Ionic bonds; metallic, 313–314; peptide, 672

Bone, osteoporosis detection, 769 *illus*.; reaction of with vinegar, 171 *lab*; scintigraphic scanning, 770 *illus*.

Index

Borax, 269

Boric acid, 269

Boron, 269; electron configuration, 246 *table*; reactivity of, 262; uses of, 269

Boyle, Robert, 382–383

Boyle's law, **383**, 388–389 prob.; deduction of, 384–385 *lab*; kinetic explanation of, 386; straws and, 388 *lab*; weather balloons and, 386, 387

Branched hydrocarbons, 623

Brass, 23 *table*, 293

Breakfast cereal, fortified, 30 *lab*

Breathalyzer tests, 568 *lab*, 569

Bromine, 86; as diatomic element, 174–175; in halogen lamps, 107; physical state and color, 87; reactivity, 97 *lab*, 278

Brønsted-Lowry model of acids and bases, 526–527

Bronze, 23 *table*, 106

Bronze Age, 567

Brownian motion, 342

Brown, Robert, 342

Buckminsterfullerene, 178

Buffered lakes, 535

Buffers, 531–533, 532 *lab*; blood, 531, 533; buffered lakes, 535

Butane, 144 *illus.*, 623; formula, 33 *table*, 809 *table*; isomers, 627–628; uses of, 33 *table*

1-Butene, 631

cis-**2-Butene**, 631 *illus.*

trans-**2-Butene**, 631 *illus.*

Butter, 145 *illus.*

Cadaverine, 645

Caffeine, 33 *table*, 809 *table*

Calcium, in Earth's crust, 669 *illus.*; electron configuration, 247; as essential element, 128; in human body, 669 *illus.*; importance of, 265, 268; ionic compounds formed by, 135 *lab*; oxidation number, 157 *illus.*; reaction of bones with vinegar, 171 *lab*; reactivity of, 97 *lab*, 265, 266–267 *lab*, 268; shielding effect in, 305 *illus.*

Calcium-40, 750 prob.

Calcium carbonate. *See* Limestone (calcium carbonate)

Calcium chloride, in hot packs, 460 *illus.*, 710; reaction with potassium hydroxide, 215 *illus.*; use as road salt, 465

Calcium fluoride, 156 *illus.*

Calcium hydroxide, 265, 489 *illus.*, 494, 498 *table*

Calcium oxide, 157 *illus.*, 494

Calcium phosphate, 154 *illus.*

Calcium sulfate dihydrate, 167, 168

Calculators, computations with, 799–801

Californium, 102 *illus.*

Californium-252, 295

Caloric, 731

Calorie (food), **721**; caloric value of some foods, 721 *table*; measurement of, 724 prob.

Calorie (heat), **721**

Calorimeter, 719 *illus.*

Calorimetry, of food, 722–723 *lab*, 724, 724 prob.; heat of reaction and, 719–720, 720 prob., 721 prob.

Calvin cycle, 735

Cancer, detection with tracers, 768 *illus.*; radiation therapy and, 770 *illus.*

Candle, observing burning of, 8–9 *lab*

Candy, chromatography of dyes in, 328–329 *lab*

Capillarity, **443**–444; paper chromatography and, 22 *lab*; Word Origin, 443

Capillary action, 22 *lab*, **443**–444

Capillary tube, 443, 444

Carbohydrates, **677**, 680–682; disaccharides, 680–681; as fake fats, 685; hydrogen bonding in, 439; monosaccharides, 680; oxidation in respiration, 694; polysaccharides, 681–682; role of in living things, 677

Carbon, 107, 263, 270; allotropes of, 176–178, 179; electron configuration, 244, 246 *table*; as essential element, 128; in human body, 669 *illus.*; molar mass of, 407 *illus.*; properties, 125; reactivity, 263

Carbon-14, 749

Carbon-14 dating, 754–755, 757, 758 prob.

Carbon blacks, 176

Carbon dioxide, 124–125, 124–125, 809 *table*; acid rain and, 493–494, 495; as acidic anhydride, 493; air quality in space vehicles and, 203; in atmosphere, 203; blood buffering, 533; dissolved in sodas, 331 *illus.*; electron sharing in, 142 *illus.*; formation of, 199 *illus.*; formula and uses, 33 *table*; molecular modeling of, 321; nonpolarity of, 331; as pollutant, 495; reaction with sodium hydroxide, 199

Carbon disulfide, 180 *illus.*, 309 *illus.*, 311

Carbon monoxide, 495, 693

Carbon monoxide poisoning, 693

Carbon steels, 288 *table*, 288–289

Carbon tetrachloride, 279, 640

Carbon-zinc dry cell batteries, 608–609

Carbonate antacids, 538

Carbonates, reaction with acids, 482, 483 *illus.*; solubility of, 482

Carbonyl group, 644

Carboxyl group, 642

Carboxylic acids, 642

Career Connection. *See also* People in Chemistry; chemical engineer, 317; chemical *lab*oratory technician, 317; cosmetologist, 491; crime *lab* technologist, 13; environmental health inspector, 449; environmental technician, 449; fingerprint classifier, 13; food and drug inspector, 491; horticulturalist, 213; industrial chemical worker, 317; industrial machinery mechanic, 449; landscape architect, 213; manufacturer's sales representative, 491; medical *lab*oratory technician, 679; medical records technician, 679; medical technologist, 679; metallurgical technician, 597; mining engineer, 597; pharmaceutical production worker, 635; pharmaceutical technician, 635; pharmacologist, 635; private investigator, 13; scrap metal processing worker, 597; soil conservation technician, 213

Carothers, Wallace, 654

Cars, air bags, 417; batteries. *See* Automobile batteries; electric, 613–614

Catalase, 674–675 *lab*

Catalysts, 222; activation energy and, 714; commercial energy production and, 730; Haber process (synthesis of ammonia) and, 217; platinum group, 292

Catalytic converters, 715

Catalytic decomposition, 674–675 *lab*

Cathode rays, 61, 62

Cathode-ray tubes, 61, 62

Cathodes, 585, 586

Cations, 586

Caves, formation of, 524

CDs (compact discs), 594

Celluloid, 660

Cellulose, 653, 655, 680, 681–682; as fake fats, 685; structure of, 654 *illus.*

Celsius scale, 349–350, 350 prob., 789

Cement, 167, 445 *table*

Center of gravity of a mass, 26

Cereal, fortified, 30 *lab*

Cerium, 294

Cerium sulfate, 459

Cesium, 263; electronegativity, 304; ionic size, 260 *illus.*; reactivity, 263

Cesium chloride, 260 *illus.*

Challenger **explosion,** 141

Changes of state, 35, 360–361, 362–363 *lab,* 364–365; condensation, 356; evaporation, 352–353; heat of fusion and, 364; heat of vaporization and, 360–361; kinetic energy and, 362–363 *lab*; pressure cookers and, 359; sublimation, 356; vapor pressure and boiling, 356–358, 357 *lab,* 360

Charcoal, 125 *illus.,* 140, 177

Charles, Jacques, 141, 391

Charles's law, 391–**392,** 393 prob.

Chase, Martha, 772

Chemical bonds. *See* Bonds

Chemical change, 40–43; atoms and, 41–42, 198; energy and, 42–43, 191, 195–196, 196 *lab*

Chemical coolants, 466

Chemical engineer, 317

Chemical equations, balancing, 198–201; converting word equations to, 193; energy and, 195–196; writing, 192–193, 195, 201–202 prob.

Chemical formulas. *See* Formulas

Chemical laboratory technician, 317

Chemical properties, 40, 41 *illus.*; of kitchen chemicals, 16–17 *lab*

Chemical reaction rates. *See* Reaction rates

Chemical reactions, 40–43. *See also specific types*; activation energy and, 713–714; atoms and, 41–42; classification of, 202–205, 206–207 *lab,* 208 *table,* 208–209; direction of, 717–718, 718 *table*; endothermic, 43, 709, 712 *lab,* 726 *lab*; energy changes and, 42–43, 195–196, 215, 708–709, 711–712; entropy and, 716–717; equations for. *See* Chemical equations; equilibrium and, 210–211, 214–215; exothermic, 42, 708–709, 712 *lab,* 726 *lab*; heat of reaction from calorimetry, 719–720, 720 prob., 721 prob.; oxidation-reduction (redox). *See* Oxidation-reduction reactions; rate of. *See* Reaction rates; signs of, 190–191, 206–207 *lab*; spontaneity of, 604, 717–718, 718 *table*; synthesis, 203–204; theoretical yield and actual yield, 421, 424–425; word equations for, 193

Chemical symbols, historic origins of, 27 *table*; periodic *table,* 92–93 *illus.*

Chemical worker, industrial, 317

Chemiluminescence, 197, 574 *illus.,* 575 *illus.*

Chemistry, 4

Chemistry and Society, artificial blood, 537; medicines from rain forests, 146; recycling of glass, 60; water-treatment plants, 447

Chemistry and Technology, alternative energy sources, 728–729; archaeological radiochemistry, 754–755; forensic blood detec-

tion, 573; fractionation of air, 354–355; Haber process, 216–217; hyperbaric oxygen (HBO) chambers, 390; microscopes, 240–241; refining of copper ore, 590–592; shape-memory metals, 108–109

Chemistry Data Handbook, 841–852; acid-base indicators, 852 *table*; alphabetical *table* of elements, 844 *table*; aqueous solutions, guidelines for, 851 *table*; electron configuration of each element, 848–849 *table*; periodic *table* of elements, 842–843; physical constants, 850 *table*; polyatomic ions, names and charges of, 850 *table*; solubility product constants, 851 *table*; symbols and abbreviations, 841

Chemistry Skill Handbook, 785–808; calculators, 799–801; factor *lab*el method, 801–803; graphs, making and using, 805–807; measurements, making and interpreting, 791–799; SI units, 785–787; *table*s, making and using, 804

Chemists, 316–317

ChemLabs. *See also* MiniLabs; acid-base indicators, 504–505 *lab*; alkaline earth metals, reactions and ion charges, 266–267 *lab*; aqueous solutions, identification of, 456–457 *lab*; Boyle's law, 384–385 *lab*; burning of a candle, 8–9 *lab*; catalytic decomposition, 674–675 *lab*; change of state and kinetic energy, 362–363 *lab*; chemical reactions, types of, 206–207 *lab*; classification of compounds as ionic or covalent, 172–173 *lab*; composition of a penny, 38–39 *lab*; conservation of matter, 56–57 *lab*; energy content of common foods, 722–723 *lab*; formation and decomposition of zinc iodide, 136–137 *lab*; kitchen chemicals, 16–17 *lab*; mass percents of compounds in a mixture, 422–423 *lab*; oxidation-reduction (redox) reactions in electrochemical cells, 606–607 *lab*; oxidation-reduction (redox)

Index

reactions involving copper, 560–561 *lab*; paper chromatography, 328–329 *lab*; periodic *table* trends, 100–101 *lab*; polymers, differentiating between, 650–652 *lab*; radioactive decay, 752–753 *lab*; reaction capacities of metals and valence electrons, 236–267 *lab*; titration of vinegar, 542–543 *lab*

Chemotherapy, 584

Chernobyl nuclear power plant, 765 *illus.*

Chinese porcelain, 163

Chips, 113

Chitin, 680

Chloride ion, 260 *illus.*

Chlorine, 86, 99; average atomic mass, 68 *table*; compounds of copper and chlorine, 164 *table*; copper compounds containing, 164 *table*; as diatomic element, 174–175; as essential element, 128; physical state and color of, 87 *illus.*; production of by electrolysis, 585 *illus.*, 586 *illus.*; properties, 123; reaction with aluminum, 716; reaction with iron, 129 *illus.*; reaction with sodium, 133, 134 *table*; reactivity, 97 *lab*, 278; uses of, 278, 279

Chlorofluorocarbons (CFCs), 641

Chloroform, 279, 640

Chloromethane, 640

Chlorophyll, 733, 734–735; Word Origin, 734

Chloroplasts, 733

Cholesterol, 642, 686, 687 *illus.*

Chromatography, 326–327; gas, 327; gel, 327; paper, 22 *lab*, 312 *lab*, 326, 328–329 *lab*; thin-layer, 326–327

Chrome plating, 106

Chromium, 293; colored compounds of, 103; electron configuration, 247 *illus.*, 293; electroplating, 593 *illus.*; as essential element, 128; properties of, 293

Cigarettes, radiation exposure from, 778 *illus.*

Cinnamon flavor, 644

cis **configuration**, 631; Word Origin (cis), 631

Cisplatin, 584

Citric acid, 642

Citrine, 248 *illus.*

Clock reactions, 220 *lab*

Clot captor, 109

Clouds, 127 *illus.*

Club soda, 469 *illus.*

Coal, 107, 199 *illus.*, 637, 639

Coal-fired power plants, 726 *illus.*

Cobalt, 292; electron configuration, 247 *illus.*; as essential element, 128; properties of, 292

Cobalt(II) chloride, 166 *lab*

Cobalt fluorides, 247

Codeine, 146

Coefficients, **199**

Coenzymes, **691**

Coffee filters, chromatography with, 312 *lab*

Coinage metals, 86, 110, 292

Cold packs, 460 *illus.*, 710

Cold-working of steel, 290, 291

Collagen, 670, 672 *illus.*

Colloids, 470–471, **472**; emulsions, 470; foams, 471; gels, 471; liquid aerosols, 470; pastes, 471; solid aerosols, 470; sols, 470; Tyndall effect and, 472, 473 *illus.*; Word Origin, 472

Color change, chemical reactions and, 190 *illus.*

Color televisions, 107

Combined gas law, 394–**395**, 395–396 prob.

Combustion reactions, **209**, 713; classification of reactions as, 206–207 *lab*; heat of reaction calculation, 720 prob.; reactants and products of, 8–9 *lab*, 208 *table*, 209; Word Origin, 209, 713

Common names, inorganic compounds, 181–182, 182 *table*

Compact discs (CDs), production of, 594

Composition of matter, 4, 5 *illus.*, 6; classification of matter by, 14

Compounds, **30**–31. *See also* Covalent compounds; Inorganic compounds; Organic compounds; classification of, 172–173 *lab*; common, 33 *table*; decomposition of, 136–137 *lab*; formula of unknown, 427–429; formula units in a sample of, 411, 412 prob.; formulas for, 31, 138;

molar mass of, 409, 412–413 prob.; natural vs. synthetic, 31, 33

Computer chips, 113

Concentrated solutions, 458

Concentration, **219**, 458–460; of acids and bases, 500; concentrated versus dilute solutions, 458; molarity and, 460–461, 462–464 prob.; rate of reaction and, 219–220; from titration, 541, 542–543 *lab*, 544 prob., 546 prob.

Condensation, 35, **356**, 446

Condensation reactions, **658**

Conductivity, **313**, 314, 488 *illus.*

Conservation of energy, **711**–712

Conservation of mass, **42**, 198–199

Conservation of matter, 53 *illus.*, 55, 56–57 *lab*

Contact-lens cleaning solutions, 677 *illus.*

Contact process, production of sulfuric acid by, 424, 425, 484

Cooling towers, 732

Cool light, 197

Copper, 292; alloy of copper and zinc, 25 *lab*; biorefining of, 727; chemical symbol for, 27 *table*; chlorine-containing compounds, 164 *table*; as coinage metal, 86, 110, 292; conductivity of, 313; corrosion, 570 *illus.*; ductility, 313; electron configuration, 247 *illus.*; as essential element, 128; magnesium-copper galvanic cell, 602–603; magnesium-copper redox reaction, 608 *illus.*; melting point, 283; oxidation-reduction (redox) reactions involving, 559, 560–561 *lab*, 562, 570; properties of, 106; refining of by electrolysis, 590–592; specific heat, 445 *table*

Copper sulfate, 143 *illus.*; hydrated, 169 *illus.*; reaction of aqueous with iron, 205 *illus.*

Copper(II) nitrate, 205 *lab*

Copper(II) sulfate, 143 *illus.*, 559

Copper(II) sulfate pentahydrate, 169 *illus.*

Cork, density of, 36 *table*

Corrosion, 554, 570, 572; ease of different metals, 602 *table*; of iron, 555 *illus.*, 557 *lab*, 570; Word Origin, 595

Cosmetic bench chemist, Fe Tayag, 490–491

Cosmetics, pH of, 501

Cosmetologist, 491

Cottrell precipitators, 470

Covalent bonds, 140, 302–303; electronegativity difference between bonding atoms (?*EN*), 308–309; equal electron sharing, 308; nearly equal electron sharing, 309; polar, 310–311

Covalent compounds, 140; classification of compounds as, 172–173 *lab*; electron sharing in, 138–140, 142; electronegativity difference between bonding atoms (?*EN*), 308–309; molecular substances, 170–171, 174; naming of, 179–183; properties of, 144–145, 147, 302 *table*, 332–333; separating from ionic substances, 170–171, 171 *lab*; solubility of, 453

Cracking, 638, 647

Crayons, 170 *illus.*

Crick, Francis, 690

Crime analysis. *See* Forensics

Crime lab technologists, 13

Cross-Curricular Connections. *See* Art Connection; Biology Connection; Earth Science Connection; Health Connection; History Connection; Literature Connection; Physics Connection

Cross-linking, 657, **658**

Crude iron, 286

Crystal lattices, 240 *illus.*, **345**

Crystals, 134; color of, 248; liquid, 345; snowflakes, 360

Curie, Irene, 746

Curie, Marie, 746, 774

Curie, Pierre, 746

Cyclohexane, 623, 624, 626

Cyclononane, 626

Cyclopentane, 623, 624

Cyclotrons, 773 *illus.*

Cysteine, 671

Cytochrome *c*, 174 *illus.*

Cytosine (C), 689, 690

***d* orbitals,** periodic *table* and, 244, 247, 250

***d* sublevels,** 235, 238, 238 *table*

da Vinci, Leonardo, 564

Dacron, 658

Daguerre, L. J. M., 564

Daguerreotypes, 564

Dalton, John, 54–55, 61, 230

Dalton's atomic theory of matter, 54–55, 61, 230, 231 *illus.*

Damascus steel, 288, 289

Davy, Humphrey, 9 *lab*, 585, 604

DDT, 641

Dead Sea, 458

Decomposition reactions, 198–199, **204,** 206–207 *lab*, 208 *table*

Definite proportions, law of, **54**

Deforestation, 146

Degrees, 350

Deicing, 465 *illus.*

Deliquescent substances, 166

Demineralization, 280

Democritus, 53

Denaturation, 673, 674–675 *lab*

Density, 36–37; of common materials, 36 *table*; composition of a penny and, 38–39 *lab*; determination of, 37; periodic *table* trends, 88; of water, 36 *table*, 437, 440–441

Dentrification, 571

Deoxyribonucleic acid. *See* DNA (deoxyribonucleic acid)

Deoxyribose, 689

Dependent variable, 806 *illus.*

Desalination, 465, 468

Desiccants, 168

Designer fats, 685

Detergents, 455

Detergent solutions, 443

Deuterium (D), 766

1,6-Diaminohexane, 658 *illus.*

Diamonds, 24 *illus.*, 107, 177, 179

Diapers, comparing polymers in, 653 *lab*

Diatomic elements, 174–175, 398; covalent bonds between, 308; gumdrop models of, 318

Dichloromethane, 641

Diethyl ether, 644

Diffusion, 352; of gases, 343 *lab*, 352; in liquids, 352

Digestion, enzymes and, 676

Dilute solutions, 458

2,2-Dimethlypropane, 625, 628 *illus.*

Dinitrogen trioxide, 181 *illus.*

Diodes, 112–113

Diprotic acids, 485 *illus.*, 486 *illus.*

Direct relationships, graphs and, 807

Disaccharides, 680–681

Disorder, entropy and, 716–717

Displacement reactions, 205, 206–207 *lab*, 208

Dissociation, 453

Dissolution, 712

Distillation, 171, 638; fractional. *See* Fractional distillation; Word Origin, 638

DNA (deoxyribonucleic acid), 273, **688,** 690; acid-base interactions in, 500; damage to by radiation, 776; double helix structure, 440, 690; extraction of, 688 *lab*; Hershey and Chase's experiments on with tracers, 772; identification of as genetic material, 772; nucleotides in, 689, 690

Dobereiner, J. W., 87

Dobereiner's triads, 87–88

Double bonds, 321, 629 *illus.*, 634

Double-displacement reactions, 206–207 *lab*, **208,** 208 *table*

Double helix, DNA, 440, 690

Double-pan balances, 407 *illus.*, 790

Downs cell, 585 *illus.*

Drinking water, fluoridation of, 279, 280

Drugs, development of new, 32

Dry cells, carbon-zinc, 608–609

Dry ice, 125 *illus.*; sublimation of, 375 *lab*

Drying agents. *See* Desiccants

Ductile metals, 313

Dynamic equilibrium, 211

Dynamite, 104 *illus.*

Dysprosium, 295

Earth, elements in crust of, 126 *illus.*, 272, 286, 669 *illus.*; mass on (weight), 4; surface temperature of and water, 446

Earth Science Connection, cave formation, 524; refining of ores by bacteria, 727; weather balloons, 387

Efficiency of industrial power production, 731 *table*, 731–732

Efficiency of reaction, 421

Eggshells, 154 *illus.*; calcium in, 171 *lab*; reaction with vinegar, 483 *illus.*

Egg whites, protein denaturation in, 673 *illus.*

Einstein, Alfred, 761

Eka-aluminum, 91

Eka-silicon, 91

Elastic collisions, kinetic model of gases and, 343

Electrical conductivity. *See* Conductivity

Electrical current, 584

Electric cars, batteries for, 613, 615

Electricity, 726

Electrochemical cells, 585, 599–615; galvanic cells, 602–605; lemon batteries, 599–601, 600 *lab*; oxidation-reduction (redox) reactions in, 603–604, 606–607 *lab*; potential difference and, 601–602

Electrodeposition, 594

Electrolysis, 584–589, 593, 595, 598; chemicals production and, 587–588; cleaning by, 595, 598; electroplating and, 593, 595; formation and decomposition of zinc iodide by, 136–137 *lab*; formation of metallic sodium from, 264; process of, 586–587, 587 *lab*; refining of ores by, 588 *illus.*, 589, 590–592; of toxic wastes, 598; Word Origin, 586

Electrolytes, 144; acids as, 488; bases as, 492

Electrolytic cells, 585 *illus.*, **586**, 587 *lab*

Electrolytic cleaning, 595 *illus.*, 598

Electrolytic conduction, 586

Electromagnetic radiation, 70–73, 233

Electromagnetic spectrum, **71**–72

Electron cloud model, **77**, 238–239

Electron configurations, 242, 244–246; of each element, 848–849 *table*; first period elements, 244; fourth period elements, 247; inner transition elements, 250; noble gases, 247; periodic *table* trends, 243–247, 250; probability of distribution in 1s orbital, 245 *lab*; second period elements, 244; third period elements, 245–246; transition elements, 247, 250

Electron dot diagram, 133; converting to molecular models, 318–319

Electronegativity, 303–305; covalent compounds and, 308–311; differences in. *See* Electronegativity difference (?EN); hydrogen bonding and, 439; ionic compounds and, 305–306, 308; periodic *table* trends, 304–305; Word Origin, 310

Electronegativity difference (ΔEN), 305–306, 312 prob.; covalent bonding and, 308–310; ionic bonding and, 305–306; polar covalent bonds and, 310–311

Electronic balances, 410

Electrons, 61, 66, 67 *table*. *See also* Valence electrons; Bohr's model of atom and, 69–70; bonding pairs, 318; discovery of, 61–62, 230; electron cloud model of, 77, 238–239; energy levels and. *See* Energy levels; nonbonding, 318; orbits, 69–70, 70 *illus.*; sharing, 138–140, 142; sublevels of, 234–235

Electron transport chain, 697 *illus.*, **698**

Electrophoresis, 598

Electroplating, 593, 594, 595

Elements, 24–25, 27; alphabetical listing of, 844 *table*; cereal fortified with, 30 *lab*; compounds of. *See* Compounds; diatomic, 174–175; in Earth's crust, 126 *illus.*, 669 *illus.*; electron configuration of each, 848–849 *table*; emission spectrum of, 74, 77 *lab*; essential for good health, 128; in human body, 669 *illus.*; ionic charges of some, 157 *table*; mass percent in compounds, 426–427;

molar mass of, 407; number of atoms in a sample of, 407, 408 prob.; Periodic Table of. *See* Periodic *table* of elements; predicting properties of unknown, 89 *lab*; properties of each, 845–847 *table*; relative masses of, 407; symbols for, 27, 27 *table*, 92–93 *illus.*; synthetic, 102

Emergency light sticks, 197

Emission spectra, 74, 77 *lab*, 233–234, 234 *lab*, 235 *illus.*

Empirical formula, 428–429; determining correct molecular formula from, 429; determining mass percents from, 428; Word Origin, 428

Emulsions, 470

Endothermic reactions, 43, 195, 215, 708 *illus.*, 709, 712 *lab*, 726 *lab*; activation energy and, 714; cold packs and, 710; decomposition of orange mercuric oxide, 58, 709; direction of, 718, 718 *table*

Energy, 42, 364; alternative sources of, 728–729; in chemical reactions. *See* Energy changes, chemical reactions and; conservation of, 711–712; in food, 721, 721 *table*, 722–723 *lab*, 724, 724 prob.; in natural systems, 736–737; nuclear reactions and, 761; relationship with mass (E=mc^2), 761; storage of in ATP and ADP, 696; Word Origin, 711

Energy changes, chemical reactions and, 42–43, 191, 196 *lab*, 708–709, 711–714; activation energy and, 713–714; chemical equations and, 195–196; direction of reaction and, 215; dissolution of solids in water and, 712 *lab*; endothermic reactions, 195–196, 708 *illus.*, 709; exothermic reactions, 196, 196 *lab*, 708–709; hot and cold packs and, 710; symbols depicting, 711–712

Energy costs, alternative energy sources and, 728–729; catalysts and, 730; efficiency and, 731–732; electricity generation and, 726; increased entropy,

730–731; production and recycling of aluminum and, 725

Energy levels, 75, 78–79, 214, 233–235; electron distribution in, 235 *illus.*, 238 *table*, 238–242; s*table*, 132–135

Engine coolants, 466

Entropy, 716–717; energy production and, 730–731; living systems and, 736–737

Environmental health inspector, 449

Environmental technician, 449

Enzymes, 222, 676–677; active site model, 676 *illus.*; denaturation of, 673, 674–675 *lab*; metabolism and, 692

Equations, chemical. *See* Chemical equations; nuclear, 746, 750–751 prob.

Equilibrium, 211, 214–215; adding of reactants or energy and, 215; Haber process (synthesis of ammonia) and, 216–217; Le Chatelier's principle, 214; removing products and, 214–215; vapor-liquid, 356–357, 358

Essential elements, 128

Esters, 643, 646 *lab*

Estimated Safe and Adequate Dietary Intake (ESAI), 128

Ethane, 183, 623, 633; dehydrogenation, 730; molecular shape of, 323–324

Ethanol, 140 *illus.*, 642, 809 *table*; Breathalyzer tests for, 569; formula and uses, 33 *table*; production of by fermentation, 698; specific heat, 445 *table*; synthesis of, 647; vapor pressure of, 357; vaporization of, 357 *lab*

Ethene. *See* Ethylene

Ether (diethyl ether), 644

Ethers, 644

Ethnobotany, 146

Ethyl alcohol. *See* Ethanol

Ethyl butyrate, 643

4-Ethyl-3-methylheptane, 626

Ethylene, 630, 630 *illus.*; hydrogenation, 633; molecular modeling of, 324; polymerization reactions, 656 *illus.*; synthesis of, 730

Ethyl ether, 644

Ethylene glycol, 33 *table*, 465 *illus.*, 466, 630, 642, 809 *table*

Ethyne, 633; molecular shape of, 324–325

Europium, 107, 294–295

Evaporation, 352–353, 446

Everyday Chemistry, air bags, 417; bleaching reactions, 194; catalytic converters, 715; chemicals in food, 19; chemicals in the body, 19; coinage metals, 110; compact disc (CD) production, 594; essential elements, 128; fake fats and designer fats, 685; fireworks, 76; flameless ration heaters, 221; freeze drying, 353; gem stones, color of, 248–249; hard water, 160; hiccups, 534; lightning-produced fertilizer, 571; matches, chemistry of, 275; microwave heating and decomposition, 320; permanent waves, 657; popping of popcorn, 397; soaps and detergents, 455; sweetness of foods, 683

Exothermic reactions, 42, 196, 196 *lab*, 708–709, 712 *lab*, 726 *lab*; activation energy and, 218, 713; direction of, 717, 718 *table*; hot packs and, 710

Explosives, 419 *illus.*

Extrusion, 290–291

Eye, vision and vitamin A, 632

f **orbitals,** 247, 250

f **sublevel,** 235, 238, 238 *table*

Fabric, testing for synthetic fibers, 650–652 *lab*

Factor label method, 379 prob., **380,** 380 prob., 801–803

FADH$_2$, 697 *illus.*

Fahrenheit scale, 349–350, 789

Fake fats, 685

Families. *See* Groups

Faraday, Michael, 604; observation of burning candle by, 8 *lab*

Fats, 682. *See also* Lipids; fake, 685; fatty acids in, 684

Fat-soluble vitamins, 691

Fatty acids, 684, 686; monounsaturated, 686; polyunsaturated, 686; saturated, 686

Fermentation, 698, 700; alcoholic, 698, 699 *illus.*, 699 *lab*; lactic acid, 698, 700

Fertilizers, 274, 424, 571

Fillings, 608

Fingerprint classifier, 13

Fingerprints, development of, 12

Fire extinguishers, 125 *illus.*

Fireflies, 575 *illus.*

Fireworks, 76

First aid, 839

Fish, antifreeze proteins in, 670

Fission, nuclear. *See* Nuclear fission

Fission nuclear power plant, 764–765

Fixer, 565 *illus.*

Flame tests, 234 *lab*

Flameless ration heaters, 221

Flavin adenine dinucleotide (FADH$_2$), 697 *illus.*

Flavorings, 643

Flexinol, 109

Fluids. *See* Gases; Liquids

Fluorapatite, 280

Fluorescent lights, 347

Fluoridation, 279, 280

Fluorides, 279, 280

Fluorine, 278; as diatomic element, 174–175; electron configuration, 246 *table*; electronegativity, 304; as essential element, 128; fluoridation and, 279, 280; hydrogen bonding by, 439; reaction with noble gases, 131; reactivity of, 263, 278

Fluorite, 156 *illus.*

Foam, 471

Fog, 470

Food, chemicals found in, 19; chromatography of dyes in, 328–329 *lab*; energy content of, 721, 721 *table*, 722–723 *lab*, 724, 724 prob.; freeze drying of, 353; irradiation of, 771, 773; spoilage of, 191, 223

Food and drug inspector, 491

Food irradiation, 771, 773

Food webs, 736 *illus.*

Fool's gold, 54 *illus.*, 165 *illus.*

Forensic Analytical Specialties, 12–13

Index

Forensics, detection of blood stains with luminol, 573, 574; forensic scientists, 12–13

Forensic scientists, 12–13

Forging, 290

Formaldehyde, 644

Formic acid, 498 *illus.*, 538, 642

Formula mass, **409**, 412 prob.

Formulas, **31**, 138; for binary ionic compounds, 155–156, 158 prob.; determining unknown with mass percents, 427–429; empirical, 428–429; for hydrates, 168; for molecular compounds, 180–181; multiple formula units in, 168–169; for polyatomic ions, 159, 161 prob., 162 prob.; for transition element compounds, 165, 165 prob.

Formula unit, **156**; formula mass and, 409, 412 prob.; representing multiple, 168–169

Fortified cereals, 30 *lab*

Fossil fuels, 637, **713**, 726; coal, 637; natural gas, 637; petroleum, 637–639

Fossils, radioactive dating of, 756–757, 758 prob., 760

Fourth period elements, electron configuration, 247; properties, 283

Fractional distillation, 274, 354–355, **638**

Francium, 263

Frasch, Herman, 276 *illus.*

Frasch process, 276

Freeze drying, 353

Freeze-point depression, 465, 466

Freezing point, 35, **364**

Freon, 641

Frequency, 71 *illus.*, 233

Fructose, 680

Fruit, oxidation of cut surfaces, 575, 576 *illus.*; treatment with ethylene, 630 *illus.*

Fudge, 459 *illus.*

Fuel cells, 614

Fullerenes, 178, 179

Functional groups, **640**–647; amino group, 645; carbonyl group, 644; carboxyl group, 642; halogens, 641; hydroxyl group, 642; notation used for, 641; production of compounds containing, 646–647

Furnaces, 558

Fusion, heat of, 364

Fusion, nuclear, 347, **766**–767; Word Origin, 766

Fusion reactors, 766–767

G

Gadolinium, 295

Gadolinium-153, 769 *illus.*

Galena, 273

Gallium, 91 *illus.*, 269, 270

Gallium arsenide, 274

Galvanic cells, **602**–605; carbon-zinc dry cell batteries, 608–609; lead storage batteries, 611, 613; NiCad rechargeable batteries, 612

Galvani, Luigi, 599

Galvanizing, 293, 555 *illus.*, 593–594

Gamma decay, 749–750

Gamma radiation, 749–750

Gamma rays, 72 *illus.*, **749**–750; food irradiation with, 771

Garcia, John (pharmacologist), 634–635

Gardening, basic anhydrides and, 494

Gas chromatography, 327

Gases, 34–35, **341**; Avogadro's principle, 398; condensation of, 356; diffusion of, 343 *lab*, 352; gas laws governing behavior of. *See* Gas laws; ideal, 343; kinetic model of, 342–343, 343 *lab*, 386 *illus.*, 392 *illus.*; mass-pressure relationship, 373 *illus.*, 373–374; mass-volume relationship, 375 *lab*; molar volume, 416, 418 prob.; particle speed distribution, 348 *illus.*; pressure of. *See* Gas pressure; properties of, 341 *illus.*; release of by chemical reactions, 191; solubility in liquids, 469; temperature change when cooling, 365 *illus.*; temperature-pressure relationship, 374–375; temperature-volume relationship, 391–392, 393 prob.

Gas laws, 382–383, 386, 388–389 prob., 391–398; Boyle's law, 382–383, 384–385 *lab*, 386, 388 *lab*, 388–389 prob.; Charles's law, 391–392, 393 prob.; combined gas law, 394–395, 395–396 prob.; ideal gas law, 419, 419 prob., 420 *lab*, 420 prob.; law of combining gas volumes, 396–398

Gasoline, 23, 145 *illus.*, 622, 623, 624 *illus.*; alkenes in, 629; combustion, 191; octane rating of, 624; production, 638 *illus.*; volatility of, 35

Gas pressure, 344, 372–376, 378–381; devices that measure, 376, 377, 378; number of particles and, 373–374; pressure-volume relationship (Boyle's law), 382–383, 384–385 *lab*, 386, 388 *lab*, 388–389 prob.; temperature and, 374–375, 391–392, 393 prob.; units used for, 378–380, 379 prob., 380 prob.

Gastric ulcers, 536 *illus.*

Geiger counters, 751 *illus.*

Gel chromatography, 327

Gels, 40 *lab*, 471

Gemstones, colors of, 248–249

Genetic code, 690. *See also* DNA (Deoxyribonucleic acid)

Geochemical cycles, conservation of matter and, 55

Geometric isomers, 631

Geothermal energy, 728–729

Geraniol, 426–427

Germanium, 91 *table*, 270

Geysers, 729

Glacial acetic acid, 528

Glass, blowing of soda-lead for sculptures, 346; colored, 249, 294; lanthanides in, 294; recycling of, 60; specific heat, 445 *table*

Glass cullet, 60

Glazes, 163

Glucose, 677, 680, 681; metabolism of, 694, 696–698; production of by photosynthesis, 734; structure, 654 *illus.*

Glycerin, 455

Glycerol, 684

Glycine, 671, 672

Glycogen, 677, 681–682

Glycolysis, 696, 697 *illus.*

Gold. *See also* Fool's gold; Asante weights and, 411; attempts to convert common metals into, 25 *lab*; biorefining of, 727; chemical symbols for, 27 *table*; as coinage metal, 86, 110, 292; density of, 36 *table*; 18-karat, 23 *table*; 14-karat, 23 *table*; malleability of, 313; nuggets of, 24 *illus.*; specific heat, 445 *table*

Gold foil experiment, Rutherford's, 64, 230

Goodyear, Charles, 658

Granite, 21 *illus.*

Graphite, 107, 176, 179

Graphs, 805–807; bar graphs, 805; of direct relationships, 807; of inverse relationships, 808; line graphs, 805–806; pie graphs, 805

Gray (radiation unit), **776**

Greeks, early ideas on matter, 52–53; use of scientific models by, 11

Green chemistry, 425

Green technology, 425

Group 1 elements. *See* Alkali metals

Group 2 elements. *See* Alkaline earth metals

Group 3 elements, 283

Group 5 elements, 283

Group 6 elements, 283

Group 13 elements, 269, 269–270; ionic charge, 157 *table*; properties, 269; uses of, 269, 269–270; valence electron configuration, 243, 244, 269

Group 14 elements, 270, 272; properties, 270; uses of, 272, 272–273; valence electron configuration, 243, 244, 270

Group 15 elements, 273–274; ionic charge of, 157 *table*; valence electron configuration, 243, 244, 273

Group 16 elements, 276–277; ionic charge of, 157 *table*; properties, 276; uses of, 277; valence electron configuration, 234, 244, 276

Group 17 elements, 278, 279; ionic charge of, 157 *table*; reactivity, 97 *lab*, 278; uses of, 279; valence electron configuration, 234, 244, 278

Group 18 elements (noble gases), **98,** 99, 281; electron configuration, 246, 246 *table*, 281; stability of, 131–132, 281; valence electron configuration, 234, 244

Groups, **96**; atomic structure of elements in, 98–99; reactivity within, 97 *lab*

Guanine (G), 689, 690

Gumdrop models, 315, 318; of ammonia, 322; of carbon dioxide, 321; of diatomic elements, 318; of ethane, 323–324; of ethyne, 324–325, 324–325; of methane, 323; of water, 318–319

Haber, Fritz, 216–217

Haber process, 215, 216–217, 571

Hafnium, 89 *illus.*

Hahn, Otto, 762

Hair, chemistry of perms, 657; dissolving of by bases, 480 *illus.*

Half-life, 756; of commonly used isotopes, 756 *table*; modeling, 752–753 *lab*

Half reactions, electrolysis, 585, 587; redox, 556, 557–558

Hall, Charles Martin, 269, 589

Hall-Héroult process, 215, 269, 588 *illus.*, 589, 725

Halogenated compounds, 641

Halogen lamp, 107

Halogens, 86, 99, **278,** 279; halogenated hydrocarbons, 641; properties of, 87, 87 *table*; reactivity, 278; uses, 278, 279; valence electron configuration, 278; Word Origin, 278

Halogen triad, 87 *table*, 87–88

Hard water, 160, 191, 452 *lab*, 455

HBO therapy. *See* Hyperbaric oxygen (HBO) chambers

HDPE, 659

Health Connection, hemoglobin, 693; lithium batteries in pacemakers, 610

Heartburn, antacids and, 503 *lab*

Heart disease, 687

Heart imaging, 770 *illus.*

Heat, 711; absorption of by chemical reactions, 709. *See also* Endothermic reactions; reducing release of waste heat, 731–732; release of by chemical reactions, 709. *See also* Exothermic reactions; units of (joules), 711

Heat of fusion, 364

Heat of reaction, 712; activation energy and, 713–714; measuring by calorimetry, 719–720, 721 prob.

Heat of solution, 460

Heat of vaporization, 361, 446

Heisenberg uncertainty principle, 238–239

Heisenberg, Werner, 238

Helium, 95–96, 98, 281; in airships, 141; in balloons, 94 *illus.*; density, 36 *table*; electron configuration, 132 *illus.*, 244, 246; molar volume, 416 *illus.*

Heme, 286, 693

Hemoglobin, 286, 537, 670, 672 *illus.*, 693

Héroult, Paul, 269, 589

Hershey, Alfred Day, 772

Hertz (Hz), 71 *illus.*

Heterogeneous mixtures, 21–22; Word Origin (heterogeneous), 23

Hexane, isomers, 628; vaporization of, 357 *lab*

Hiccups, 534

High-density polyethylene (HDOE), 659

Hindenburg, 141

History Connection, the *Hindenburg*, 141; Lavoisier, politics and chemistry, 58; lead poisoning in Rome, 271

Holes, semiconductors, 112

Homogeneous mixtures. *See* Solutions

Homogeneous, Word Origin, 24

Honeycombs, 684 *illus.*

Hormones, 694

Horticulturalist, 213

Hot-air balloons, 391

Hot packs, 460 *illus.*, 710

Hot-working of steel, 290–291

Household bleach, 194

Household chemicals, chemical and physical properties of, 16–17 *lab*

Housepaints, as colloids, 470

How It Works, breathalyzer test, 569; cement, 167; electronic balances, 410; emergency light sticks, 197; hot and cold packs, 710; hydrogen-oxygen fuel cell, 614; indicators, 545; lightbulbs,

Index

284; pressure cookers, 359; reverse osmosis units, 468; smoke detectors, 748; taste, 519; tire-pressure gauge, 378

Human body, composition of, 19, 669 *illus.*

Hydrates, 165–166, 168; cement, 167; chemical weather predictors, 166 *lab*; formulas, 168; naming, 168

Hydrides, 311

Hydrobromic acid, 498 *table*, 522–523

Hydrocarbons, 183, 183 *table*, 495. *See also* Alkanes; Alkenes; Alkynes; air pollution and, 495; aromatic, 636–637; branching of, 623; cracking and, 638, 647; fossil fuels, 637–639, 713; functional groups on substituted, 640–647; modeling of, 323–325, 325 *lab*; reforming and, 639; saturated. *See* Alkanes (saturated hydrocarbons); unsaturated, 625, 629–631, 630 *lab*, 633, 636

Hydrochloric acid, 809 *table*; formula and uses, 33 *table*; ionization, 486; polar covalent bonding, 310; production of hydronium when dissolved in water, 485; reaction with ammonia, 527, 528 *illus.*; reaction with calcium carbonate, 717–718; reaction with sodium hydroxide, 516 *illus.*; as strong acid, 498, 498 *table*

Hydrogen, 95, 98; as diatomic element, 174–175; in Earth's crust, 669 *illus.*; electron cloud model of, 239 *illus.*; electron configuration, 244; emission spectrum, 74 *illus.*, 233, 235 *illus.*; energy levels, 75 *illus.*, 78 *illus.*; as essential element, 128; as fuel source, 141; in human body, 669 *illus.*; properties, 127; reaction with oxygen, 127 *illus.*, 200 prob., 218, 709 *illus.*

Hydrogenation, 633

Hydrogen bonding, 439. *See also* Water; sweetness and, 683

Hydrogen carbonate ion, buffers and, 533, 535

Hydrogen chloride. *See* Hydrochloric acid

Hydrogen-oxygen fuel cell, 614

Hydrogen peroxide, catalytic decomposition of, 674–675 *lab*; chemical properties of, 41 *illus.*; inhibitors and, 223; instability, 40, 277; reaction with magnesium dioxide, 41 *illus.*

Hydroiodic acid, 498 *table*

Hydronium ions, 483, 485, 517; pH and, 502, 502 prob., 506–507

Hydroxide antacids, 538

Hydroxide ions, 159 *illus.*, 488–489, 517; pH and, 502, 502 prob., 506–507

Hydroxides, 264

Hydroxyapatite, 280

Hydroxyl group, 642

Hygroscopic substances, 166

Hyperbaric oxygen (HBO) chambers, 390

Hyperventilation, 533

Hypochlorite bleaches, 567–568

Hypothesis, 59

Ice, 127 *illus.*, 436; density, 36 *table*, 437, 440–441; sublimation, 356, 375 *lab*

Icebergs, center of gravity of, 26, 127 *illus.*

Ice man, dating of, 754–755

Ideal gases, 343

Ideal gas law, 419, 419 prob., 420 *lab*, 420 prob.

Incandescent lights, 131 *illus.*

Independent variable, 806 *illus.*

Indicators, acid-base. *See* Acid-base indicators

Indigo, 429

Indium, 269, 270

Industrial chemicals, 481 *table*

Industrial chemical worker, 317

Industrial machinery mechanic, 449

Industrial processes. *See also specific processes*; aluminum production and recycling, 725; catalysts in, 730; efficiency of, 731 *table*, 731–732; energy production and, 726; entropy from, 730–731

Inert gases, 131–132, 281, 284. *See also* Noble gases

Infrared waves, 72 *illus.*

Inhibitors, 223

Inks, chromatography of, 22 *lab*

Inner transition elements, 104, 250, 294–295; electron configuration, 250, 294; radioactivity, 295; uses of, 294–295

Inorganic compounds, 180; naming, 180–182

Insoluble compounds, 215

Insulin, 671 *illus.*, 694

Integrated circuits, 113

International System of Units, 785–786

Interparticle forces, 144–145, 147

Invar, 290 *table*

Inverse relationships, graphs and, 808

Iodine, 86, 107, 278; as antibacterial, 279; biological importance of, 278; as diatomic element, 174–175; as essential element, 128; halogen triad, 87, 87 *table*; ionic compounds formed by, 135 *lab*; properties of, 87 *table*; reaction with starch, 190 *illus.*, 220 *lab*; reactivity, 97 *lab*, 278; size of, 174 *illus.*; sublimation, 356 *illus.*

Iodine-131, 769 *illus.*

Ion exchangers, 160

Ionic attraction, 135

Ionic bonds, 134, 135, 302–303; electronegativity difference between bonding atoms (ΔEN) and, 306, 308

Ionic charges, predicting, 157; of representative elements, 157 *table*

Ionic compounds, 134; binary, 155–156; classification of compounds as, 172–173 *lab*; difference between electronegativities (ΔEN) of bonding atoms and, 305–306, 308; dissolving of by water, 451–453; formation of by electron transfer, 133–134, 135 *lab*, 136–137 *lab*; formulas and names of, 154–162, 164–166, 168–169; hydrates, 165–166, 168; identification of aqueous solutions and, 456–457 *lab*; interparticle forces in, 144 *illus.*; ion charge prediction, 156–157; polyatomic, 158–162; properties, 143–144, 302 *table*, 332–333; separating from covalent sub-

stances, 170–171, 171 *lab*; transition elements, 162, 164 *table*, 164–165
Ionic crystals, 134–135
Ionic equations, 517–518, 520, 521 prob.
Ionic size (radii), 260; atom size-ion size relationship, 260 *illus.*; patterns in main group elements, 260, 261
Ionization, **488**; of acids, 486, 488, 499; of bases, 489
Ionizing radiation, 776
Ionizing smoke detectors, 748
Ions, **134**, 242; charge prediction, 156–158; spectator, 520
Iridium, 292
Iron, 286, 286–287, 292; biological importance, 286; chemical symbol, 27 *table*; in Earth's crust, 669 *illus.*; electron configuration, 247 *illus.*, 285 *lab*; as essential element, 128; molar mass, 407 *illus.*; oxidation states, 283; properties, 6 *illus.*, 40, 41 *illus.*, 106, 282, 292; reaction of with chlorine, 129 *illus.*; reaction with copper sulfate, 205 *illus.*; reaction with HCL, 482, 483 *illus.*; rusting, 41 *illus.*, 122 *lab*, 204 *illus.*, 250, 555 *illus.*, 557 *lab*, 570; separation of from ore, 286; smelting of in blast furnaces, 558 *illus.*, 567; specific heat, 445 *table*
Iron Age, 567
Iron chloride, 140 *illus.*
Iron(III) oxide, 555 *illus.*
Iron pyrite, 54 *illus.*, 165 *illus.*
Iron smelting, 558 *illus.*
Iron(II) sulfate, 162 *illus.*
Iron(III) sulfate, 162 *illus.*
Iron triad, 292
Irradiation, food, 771, 773
Irrigation, 156
Isomers, 627–**628**; geometric, 631; positional, 631; structural, 628
Isopropanol, 642
Isopropyl alcohol, 358, 642
Isotopes, **62**. *See also* Radioisotopes; atomic mass and, 67, 68 *illus.*; half life of common, 756 *table*; nuclear notation and, 746; representing with pennies, 63 *lab*

Jewelry alloys, 23 *table*
Jolliot-Curie, Frederick, 746
Jordan, Lynda, 678–679
Joule (J), **361**, 711

Kekulé, August, 636
Kelvins (K), **350**
Kelvin scale, **349**–350, 350 prob., 789
Keratin, 501, 670
Kerosene, 638 *illus.*
Ketones, 644
Kilocalories, **721**
Kilopascal (kPa), **378**
Kinetic energy, diffusion and, 352; Kelvin scale and, 349–350; mass and speed and particles and, 351; temperature and, 348–350, 362–363 *lab*
Kinetic theory, **342**–345; Boyle's law and, 386; Charles's law and, 392; of gases, 342–344, 343 *lab*, 386 *illus.*; of liquids, 344; of solids, 345
Kitchen chemicals, 16–17 *lab*
Krypton, 131, 132 *illus.*, 281

Labs, ChemLab. *See* ChemLab; MiniLab. *See* MiniLab; safety guidelines, 839; safety symbols, 840; Launch Labs. *See* Launch Labs; Try at Home Labs. *See* Try at Home Labs.
Lactase, 676–677
Lactic acid, 429
Lactic acid fermentation, 698, 700
Lactose, 681
Lactose intolerance, 222, 223 *illus.*, 676–677
Landscape architect, 213
Language Arts. *See* Literature Connection
Lanthanides, 92–93 *illus.*, 94, **104**, 107, 250, 294–295
Lanthanum, 294

Latex paints, 661
Laughing gas, 574
Launch Labs, A Lemon Battery?, 583; Chain Reactions, 743; Defying Density, 339; Elements, Compounds, and Mixtures, 153; How Much is a Mole?, 403; Magnetic Materials, 257; Making Slime, 621; More Than Just Hot Air, 371; Observing a Chemical Reaction, 189; Observing an Oxidation-Reduction Reaction, 553; Observing Electrical Charge, 229; Oil and Vinegar Dressing, 301; Physical or Chemical Change?, 515; Red Liquid to Clear Liquid, 119; Solution Formation, 435; Speeding Reactions, 707; Testing for Simple Sugars, 667; Testing pH Using Natural Indicators, 479; Versatile Metals, 85; What's Inside?, 51; Where is It?, 3
Lavoisier, Antoine, 53, 58, 198
Law of combining gas volumes, **396**–398, 404
Law of conservation of energy, **711**
Law of conservation of matter (mass), **42**, 56–57 *lab*; discovery of, 53, 198; recycling and, 55
Law of definite proportions, **54**
LCDs (liquid crystal displays), 345
LDPE, 659
Le Chatelier, Henri Louis, 214
Le Chatelier's principle, 214
Lead, 270; density, 36 *table*; symbol for, 27 *table*; uses, 273
Lead-acid storage battery, 273, 611, 613
Lead-chamber process, production of sulfuric acid by, 424
Lead(II) iodide, 208 *illus.*
Lead(II) nitrate, 208 *illus.*
Lead poisoning, 271
LEDs, 605 *illus.*
Lemon batteries, 599–601, 600 *lab*, 608 *illus.*
Length, measuring, 788, 792
Levi, Primo, 96
Lewis dot diagrams, **79**; for main groups of elements, 98 *illus.*
Libby, Willard, 754–755
Life, molecules of. *See* Biomolecules
Light, electromagnetic spectrum and, 233

Index

Light-emitting diodes (LEDs), 605 *illus.*

Light reactions, photosynthesis and, 734–735

Lightbulbs, 284

Lighter fuel, 627 *illus.*

Lightning, 571, 574 *illus.*

Light sticks, 197

Lime, 157 *illus.*; formation of, 211, 214; as soil treatment, 494

Limestone (calcium carbonate), acid dissolution, 482 *lab*, 483 *illus.*, 524; cave formation and, 524; damage to sculptures by, 483 *illus.*; decomposition, 211, 214; formula, 33 *table*, 809 *table*; hard water and, 160; neutralization of lakes by, 535; reaction with HCl, 717–718

Limiting reactant, 220

Linear acetylenic carbon, 178

Linear molecular geometry, 318 *illus.*, 321 *illus.*, 325 *illus.*

Line graphs, 805–806

Lipid membranes, 687

Lipids, 682, 684; fake fats and, 685, 686–687; fatty acids and, 684, 686; functions, 687; steroids, 686–687; structure, 684, 686

Lipstick, 684 *illus.*

Liquid aerosols, 470

Liquid crystal displays (LCDs), 345

Liquid crystals, 345

Liquid hydrogen, 141

Liquids, 34–35, 341; boiling, 358, 360; diffusion in, 352; dissolved gases in, 469 *illus.*; evaporation, 352–353; freezing, 364; kinetic model of, 344; properties, 341 *illus.*; surface tension and, 442; volatile, 353, 357

Literature Connection, Jules Verne, icebergs, 26; Primo Levi, language of a chemist, 96

Lithium, 86, 96, 99, 261, 263; atomic size, 259; atomic structure, 66 *illus.*; electron configuration, 244, 245, 246 *table*; ionic compounds formed by, 135 *lab*; ionic radii and, 261; reactivity, 261, 263

Lithium batteries, for electric cars, 613, 615; in pacemakers, 610

Lithium fluoride, 263 *illus.*, 308

Lithium hydroxide, 203, 418 prob., 498 *table*

Litmus paper, 481, 482 *lab*

Lone electron pairs, 318

Low-density polyethylene (LDPE), 656 *illus.*, 659

Luciferase, 575 *illus.*

Luminol, 573, 574

Lung cancer, radon exposure and, 777, 778

Lye, 455

Lysine, 672

Macroscopic world, 7

Magma, 728

Magnesium, 265; in Earth's crust, 669 *illus.*; as essential element, 128; ionic compounds formed by, 135 *lab*; magnesium-copper redox reaction, 604 *illus.*; properties, 268; reactivity, 97 *lab*, 265; in seawater, 598; shielding effect, 304, 305 *illus.*; uses, 268

Magnesium alloys, 268

Magnesium chloride, reaction with silver nitrate, 200–201 prob.

Magnesium-copper galvanic cell, 602–604

Magnesium dioxide, 41 *illus.*

Magnesium hydroxide, 33 *table*, 498 *table*, 809 *table*

Magnesium oxide, 155, 268

Main group elements, 238–274, 258–265, 276–279, 280; alkali metals. *See* Alkali metals; alkaline earth metals. *See* Alkaline earth metals; atomic radii trends, 258–259; halogens. *See* Halogens; ionic radii trends, 260, 261; metal-metalloid-nonmetal-noble gas pattern, 258, 259 *illus.*; metallic property trends, 259 *illus.*; period 2 chemical reactivity, 261–263

Malachite, 249 *illus.*

Malleability, 313

Maltose, 681

Manganese, electron configuration, 247 *illus.*; as essential element, 128

Manganese(IV) oxide, reduction in carbon-zinc dry cells, 609 *illus.*

Manganese steel, 290 *table*

Manhattan project, 232

Manufacturers' sales representative, 491

Marble, acid dissolution of, 482, 483 *illus.*

Margarine, 633 *illus.*

Martensite phase, 108

Mass, 4, 5; atomic. *See* Atomic mass; comparing, 4, 5 *illus.*; conservation of, 42, 198–199; determining number of atoms by, 408 *lab*; measurement of, 410, 790–791; molar. *See* Molar mass; molecular, 409; relationship with energy ($E = mc^2$), 761; relationship with gas volume, 375 *lab*; speed of particles in and kinetic energy, 351

Mass formula, 409

Mass number, 66

Mass percents, determination of, 422–423 *lab*, 426–427; determining chemical formulas from, 427–429

Mass spectrometers, 327

Matches, chemistry of, 275

Matter, 4. *See also* Atoms; Compounds; Elements; States of matter; atomic model of, 11; change in state of, 352–353, 356–358, 360–361, 362–363 *lab*, 364–365; classification of, 14–15, 18, 20–24, 24 *illus.*; composition and behavior of, 4, 5 *illus.*; conservation of, 53 *illus.*, 55, 56–57 *lab*; density of, 36–37; early Greek ideas about, 52–53; kinetic theory of, 342–345; macroscopic view of, 7; mixtures of, 15, 18, 20–24; properties of, 4, 5 *illus.*, 6, 40–43; states of, 34–35, 340–341; submicroscopic view of, 7, 10; substances, 15

Measurements, of length, 789, 792; of mass and weight, 790–791; precision and accuracy of, 793; as a quantitative observation, 14; scientific notation and, 795–799; SI system for, 786–787; significant digits and, 793–795; of temperature, 789; of volume, 789

Medical laboratory technician, 679

Medical records technician, 679

Medical technologist, 679

Medicines from rain forests, 146

Meitner, Lise, 762

Melting point, 35, **364**; periodic *table* trends, 88

Mendeleev, Dmitri, 88–91, 94

Mercury, boiling point, 358; emission spectrum, 235 *illus.*; melting point, 283; physical properties, 25 *illus.*, 106; surface tension, 442, 444 *illus.*

Mercury barometers, 283 *illus.*

Metabolism, **692**, 693. *See also* Biomolecules; fermentation, 698–699, 699 *lab*; map of, 695 *illus.*; photosynthesis, 733–737; respiration, 694, 696, 697 *illus.*, 698

Metal alloys. *See* Alloys

Metal hydroxides, 489

Metallic bonds, **314**

Metalloids, **105**; atomic structure, 105, 106; periodic *table* trends, 100–101 *lab*; position of on periodic *table*, 102; properties, 107; valence electron configuration, 243

Metallurgical technician, 597

Metal plater (Harvey Morser), 596

Metals, 102, **103**–104; atomic structure, 105, 106; bonding in, 313–314; conductivity of, 313, 314; corrosion, 555, 557 *lab*, 570, 572; oxidation, ease of, 602 *table*; oxidation-reduction (redox) reactions involving, 570; periodic *table* trends, 100–101 *lab*; properties, 103, 105 *table*, 106, 107, 313; reaction with acids, 482, 483 *illus.*; shape-memory, 108–109; valence electron configuration, 236–267 *lab*, 243, 313–314

Methane, 182 *illus.*, 183, 309 *illus.*, 623 *illus.*, 809 *table*; combustion, 196, 713 *illus.*; deep-sea deposits of, 181; formula and uses, 33 *table*; molecular shape, 323; nonpolarity, 332; water vs., 437 *table*

Methanol, 406, 439 *illus.*, 642

Methyl chloride, 640

Methyl red, 540 *illus.*

Methyl salicylate, 646 *lab*

2-Methylbutane, 628 *illus.*

2-Methylhexane, 628 *illus.*

3-Methylhexane, 628 *illus.*

2-Methylpropane, 623 *illus.*, 627–628

Metric system, 4, 786

Microscopes, 240–241; atomic force microscope (AFM), 241; scanning probe microscope (SPM), 240; scanning tunneling microscope (STM), 10, 241

Microwave decomposition, 320

Microwave heating, 320

Microwave ovens, 320

Microwave radiation, 71, 72 *illus.*, 320

Milk of Magnesia, 538

Minerals, nutrition and, 128

MiniLabs. *See also* ChemLabs; alcohol, testing for presence of by redox, 568 *lab*; alloy of copper and zinc, 25 *lab*; antacids and acid-base chemistry, 503 *lab*; atomic radii, periodic *table* patterns, 262 *lab*; buffers, 532 *lab*; calcium (bones), reaction with vinegar, 171 *lab*; cereal fortified with elements, 30 *lab*; chemical weather predictors, 166 *lab*; chromatography, 22 *lab*, 312 *lab*; corrosion of iron and oxidation, 557 *lab*; determining number of items by weight, 408 *lab*; diffusion of gases in air, 343 *lab*; DNA extraction, 688 *lab*; electrolysis, 587 *lab*; electron configuration of iron, 285 *lab*; elements, predicting properties of unknown, 89 *lab*; emission spectrum, 77 *lab*, 234 *lab*; esters, aroma of, 646 *lab*; exothermic and endothermic reactions, 196 *lab*, 712 *lab*, 726 *lab*; gel, properties of, 40 *lab*; groups, reactivity within, 97 *lab*; isotopes, representing with pennies, 63 *lab*; lemon batteries, 600 *lab*; line emission spectra of elements, 77 *lab*; mass and volume of a gas, 375 *lab*; mixture of water and alcohol, 21 *lab*; molar volume of a gas, 420 *lab*; molecular shapes, modeling, 325 *lab*; neutralization reactions, 518 *lab*; nuclear fission chain reaction, 763 *lab*; paper chromatography, 22 *lab*, 312 *lab*; polymers, comparing, 653 *lab*; radon levels, 775 *lab*; rusting of iron, 122 *lab*; 1s orbital, probability of electron

distribution in, 245 *lab*; single-displacement reactions, 205 *lab*; starch-iodine clock reaction, 220 *lab*; straws and Boyle's law, 388 *lab*; vaporization rates, 357 *lab*; yeast fermentation, 699 *lab*

Mining engineer, 597

Misch metal, 294

Mixtures, 15, **18**, 20–24; analysis of, 422–423 *lab*; composition of, determining, 16–17 *lab*; examples of everyday, 18 *illus.*; heterogeneous, 21; homogeneous (solutions), 22–23; separation of, 20, 326–327; of water and alcohol, 21 *lab*

Models, 10–11

Molarity, 460–461; calculation of, 463–464 prob.; preparing solutions based on, 462–463 prob.; from titration, 541, 542–543 *lab*, 544, 544 prob.

Molar mass, 406–**407**; of a compound, 409, 412–413 prob.; of an element, 407, 408 prob.; mass of a number of moles of a compound and, 413 prob.; number of atoms in a sample of an element from, 408 prob.; number of formula units in a sample of a compound and, 412 prob.; predicting mass of a product and, 415–416 prob.; predicting the mass of a reactant and, 414–415 prob.

Molar volume, **416**, 418 prob., 420 *lab*

Molded plastics, 660 *illus.*

Mole, **405**; mass of, 413; number of things in (Avogadro's constant), 405–406

Molecular compounds, 179–183, 180 *table*. *See also* Molecular elements

Molecular elements, **174**. *See also* Molecular compounds; allotropes, 175, 179; diatomic elements, 174–175

Molecular formula, from empirical formula, 429

Molecular mass, **409**, 412–413 prob.

Molecular modeling, 314, 318–319, 321–325, 325 *lab*; ammonia, 322; carbon dioxide, 321; diatomic

Index

elements, 318; ethane, 323–324; gumdrop models, 315, 318; methane, 323; water, 318–319

Molecular shape, 315. *See* Polarity

Molecular substances, 170. *See also* Covalent compounds

Molecules, 140

Monomers, 649

Monoprotic acids, 485 *illus.*

Monosaccharides, 680

Monounsaturated fatty acids, 686

Morser, Harvey (metal plater), 596

Mothballs, 35, 636–637

Moving phase, paper chromatography, 312 *lab*

MREs (Meal, Ready-to-Eat), 221

Muscles, 174, 698, 700 *illus.*

Myoglobin, 286, 576

***n*-type semiconductors,** 112

NAD+, 697 *illus.*

NADH, 697 *illus.*

Nagaoka, Hantaro, 63

Nail polish remover, 643

Names, alkanes, 625–626, 626–627 prob.; alkenes, 630–631; alkynes, 633; binary inorganic compounds, 180–182, 181 prob.; binary ionic compounds, 155; organic compounds, 182–183; polyatomic ions, 160; transition element compounds, 164, 164 *table*, 165

Naphthalene, 35, 636

Natural chemicals, vs. synthetic, 32

Natural gas, 637; deposits, 637 *illus.*; processing, 638–639

Natural polymers, 653; comparing with synthetic, 653 *lab*; differentiating from synthetic polymers, 650–652 *lab*

Natural radiation, 774 *illus.*, 776

Natural systems, energy in, 736–737

Neodymium, 107, 294

Neon, 281; electron configuration, 244, 245, 245–246, 246 *table*; emission spectrum, 235 *illus.*; isotopes, 62 *illus.*, 66

Neon lights, 131 *illus.*

Neoprene, 660

Neptunium, 102

Net ionic equations, 520, 521 prob.

Neutralization reaction, 516–517, 518 *lab*. *See also* Acid-base reactions

Neutral solutions, 503

Neutrons, 62, 66; discovery of, 62; properties, 67 *table*

NiCad rechargeable batteries, 612

Nickel, 292; as coinage metal, 110; electron configuration, 247 *illus.*; as essential element, 128; properties, 292

Nickel-cadmium (NiCad) batteries, 612

Nickel-metal hydride batteries, 613

Nicotinimide adenine dinucleotide (NAD+), 697 *illus.*

Night vision, 632

NiMH batteries, 613

Nitinol, 108–109

Nitric acid, 498 *table*

Nitrogen, 273; biological importance of, 273; cycling, 55 *illus.*, 273; as diatomic element, 174; electron configuration, 246 *table*; as essential element, 128; fertilizers from, 274; from fractional distillation, 274, 354–355; in human body, 669 *illus.*; hydrogen bonding by, 439; ionic compounds formed by, 135 *lab*; liquid, 107; reactivity, 263; water vs., 437 *table*

Nitrogen cycle, 55 *illus.*, 273

Nitrogen dioxide, 309 *illus.*

Nitrogen fixation, 55 *illus.*, 273, 571

Nitrogen-fixing bacteria, 216, 273

Nitrogen monoxide, 273, 495

Nitrogen oxides, 494 *illus.*, 495

Nitroglycerin, 42 *illus.*

Nitrous oxide, 574

Noble gas compounds, 195, 281

Noble gas configuration, 132

Noble gases, 98, 99, 281; applications, 131 *illus.*; electron configuration, 132 *illus.*, 246, 246 *table*, 281; periodic *table* and, 94; stability of, 131–132, 281

Nonbonding electron pairs, 318

Nonmetals, 102, **104;** atomic structure, 105, 106; electron configuration, 243; periodic *table* and, 100–101 *lab*, 102; properties, 104, 105 *table*, 107

Normal boiling point, 358

***npn*-junctions,** 112

Nuclear accidents, 765

Nuclear equations, 746; radioactive decay and, 747, 749–750; writing and balancing, 746, 750–751 prob.

Nuclear fission, 232, **762**–765; fission chain reaction, 762, 763 *illus.*, 763 *lab*; fission reactors, 764–765; Word Origin (fission), 762

Nuclear fusion, 347, **766**–767

Nuclear fusion reactors, 767 *illus.*

Nuclear medicine, 768–770

Nuclear model of the atom, 65

Nuclear power plants, 737 *illus.*, 764–765 *illus.*

Nuclear reactions, energy potential of, 761; equations for, 746, 750–751 prob.; fission, 762–765, 763 *illus.*; fusion, 766–767; notation for, 746; radioactive decay, 747–750

Nuclear reactors, 764–765, 767 *illus.*

Nuclear waste disposal, 778–779

Nucleic acids, 688–690; deoxyribonucleic acid (DNA), 688 *lab*, 690; hydrogen bonding between, 439; nucleotides in, 689; ribonucleic acid (RNA), 690

Nucleotides, 689

Nucleus, 65, 230, 231 *illus.*

Nutrient cycles, conservation of matter and, 55

Nutrients, 721 *illus.*

Nutrition, 128

Nylon, 649, 654

Oaktree automatic arm, 109

Observations, 14, 59; qualitative, 14; quantitative, 14

Octane ratings, 624

Octet rule, 132

Odor, changes in and chemical reactions, 191; diffusion of gases in air and, 343 *lab*

Oils, 682. *See also* Lipids; Petroleum; action of soaps and detergents on, 455; fatty acids in, 684; insolubility in water, 454

Oleic acid, 686 *illus.*

Olestra, 685

Operation Trinity, **764**–765

Opsin, 632

Orange juice, 21 *illus.*

Orbitals, 238–**239**; number of electrons in each, 238, 238 *table*; overlapping, 242 *illus.*; periodic *table* trends, 243–247, 250; size, 250–251; Word Origin, 247

***Orbiter* space shuttle**, 203

Ores, refining of by bacteria, 727; separation of iron from, 286, 558 *illus.*

Organic chemistry. *See* Biochemistry; Organic compounds

Organic compounds, **180**, 622. *See also* Saturated hydrocarbons; Unsaturated hydrocarbons; alkanes, 623–626, 626–627 prob., 628–629; alkenes, 629–631, 633; alkynes, 633, 633 *table*, 636; aromatic hydrocarbons, 636–637; functional groups, 640–647; naming, 182–183; polymerization reactions, 655–656, 658, 660; polymers, 649, 650–652 *lab*, 653 *lab*, 653–655; sources, 637–639

Osmium, 292

Osmosis, **467**–468; por*table* reverse osmosis units, 468; Word Origin, 467

Osteoporosis, detection of with radioisotopes, 769 *illus.*

Oxalate ions, as polyatomic ion, 159 *illus.*; reactivity of alkaline earth elements with, 266–267 *lab*

Oxidation number, **157**; multiple in transition elements, 247

Oxidation reactions, 554, **556**. *See also* Oxidation-reduction reactions; combining with reduction half-reactions, 557–558; metals, ease of, 602 *table*; oxidizing and reducing agents and, 559, 562

Oxidation-reduction reactions, **555**; alcohol, testing for presence of by, 568 *lab*; biochemical processes involving, 574–576; blast furnaces and, 567, 568 *illus.*; bleaching processes and, 567–568; chemiluminescent, 574; copper atoms and ions and, 559, 560–561 *lab*, 562; corrosion of iron and oxidation, 555 *illus.*, 557 *lab*; in electrochemical cells, 603–604, 606–607 *lab*; electrolysis and, 584, 585; half-reactions, combining, 557–558; identification of, 559; lightning-produced fertilizer and, 571; oxidizing and reducing agents and, 559, 562; photography and, 563–564, 565 *illus.*; tarnishing of silver, 572

Oxides, 492

Oxidizing agents, 559, **562**

Oxyacetylene torches, 633

Oxygen, 276. *See also* Oxidation reactions; abundance of, 276; allotropes, 175, 276; as diatomic element, 174; dissolved in water, 469 *illus.*; in Earth's crust, 669 *illus.*; electron configuration, 99 *illus.*, 246 *table*; energy levels, 78 *illus.*; as essential element, 128; from fractional distillation, 276, 354–355; in human body, 669 *illus.*; hydrogen bonding by, 439; industrial uses of, 277; ionic compounds formed by, 135 *lab*; properties, 126; reaction with hydrogen, 127 *illus.*, 200 prob., 218, 709 *illus.*; reactions with metals and nonmetals, 276, 276 *table*; reactivity, 263, 276; water vs., 437 *table*

Ozone, 175, 276, 495

p orbitals, 239 *illus.*, 242; periodic *table* and, 243–247, 250

p sublevels, 235, 238

p-type semiconductors, 112

Pacemakers, lithium batteries in, 610

Pacific yew tree, 31, 32

Paints, 470, 661

Palladium, 292

Paper chromatography, 22 *lab*, 312 *lab*, 326, 328–329 *lab*

Parent acids, 518 *lab*

Parent bases, 518 *lab*

Particle accelerators, 102, 773 *illus.*

Particle generators, 773 *illus.*

Pascal, Blaise, 378

Pascal (Pa), **378**

Pastes, 471

Pauling, Linus, 307

PCBs, 641

Pectin, 471

Pennies, composition of, 38–39 *lab*; creation of copper and zinc alloy with, 25 *lab*; electroplating of, 593 *illus.*; modeling radioactive decay with, 752–753 *lab*; representing isotopes with, 63 *lab*; surface tension and, 443 *lab*

Pentane, 628

People in Chemistry. *See also* Career Connection; biochemist, 678–679; chemist, 316–317; cosmetic bench chemist, 490–491; forensic scientist, 12–13; metal plater, 596–597; pharmacist, 634–635; wastewater operator, 448–449

Peptide bonds, 672

Percent yield, 421; improving in chemical synthesis of sulfuric acid, 424–425

Perchloric acid, 498 *table*

Period 2, chemical reactivity in, 261–263

Periodic law, 94

Periodic, Word Origin, 99

Periodicity, 90, 94

Periods, 96; atomic structure of elements in, 98

Periodic table of elements, 27, 28–29 *illus.*, 86–91, 92–93 *illus.*, 842–843. *See also specific groups*; atomic numbers, 66; atomic radius (size) patterns, 258–259, 262 *lab*; atomic structure and, 95–96, 97 *lab*, 98–99, 243–252; classes of elements on, 102–105; development of, 86–91, 89 *lab*; electron configuration trends, 243–247, 246 *table*, 250; electronegativity trends, 304–305; information in each block of, 67 *illus.*; ionic charge trends, 157; ionic size patterns, 260–261; long version of, 104–105 *illus.*; metallic and nonmetallic characteristics and, 100–101 *lab*; modern version of, 94; physical states of

Index

elements and, 102; reactivity of period 2 and, 261–263; synthetic elements and, 102; valence electrons and, 231, 232

Permanent waves, 657

Permeable membranes, 468

Peroxides, 277

Perspiration, 446 *illus.*

Pesticides, tracking with tracers, 771

PET, 659

Petroleum, 145 *illus.*, 637–639; cracking and, 638–639; fractional distillation of, 637–638; reforming and, 639; Word Origin, 637

Pharmaceutical production worker, 635

Pharmaceutical technician, 635

Pharmacist, John Garcia, 634–635

Pharmacologist, 635

Pharmacology, 634–635

Phenolphthalein, 264, 540 *illus.*

Phenylalanine, 671

Pheromones, 642

pH, 502, 502 *prob.*; of blood, 531, 533; buffers and, 531–533, 535; of common materials, 506 *illus.*, 507; of cosmetics, 501; development of pH scale, 500; interpretation of pH scale, 503, 506; measuring with acid-base indicators, 504–505 *lab*, 507–508; protein denaturation and, 673; strong acid plus strong base reactions and, 522; strong acid plus weak base reactions and, 525–526

pH indicators. *See* Acid-base indicators

Phosphate group, 273, 689

Phosphorescence, 744 *illus.*

Phosphoric acid, 33 *table*, 499, 809 *table*

Phosphors, 294

Phosphorus, 273; allotropes, 179, 274; biological importance, 273; as essential element, 128; in human body, 669 *illus.*

Phosphorus pentachloride, 214

Photochemical smog, 495

Photography, developing and printing pictures, 565 *illus.*; oxidation-reduction (redox) reactions and, 563–564; silver bromide in film, 279, 564; Word Origin (photo-

graph), 564

Photosynthesis, 44 *illus.*, 124–125, 733, **734**–737; Calvin cycle, 735; endothermic reactions in, 43, 733, 736; light reactions, 734–735; net equation for, 734; redox reactions in, 575

Photovoltaic cells, 728

pH scale, 500 *illus.*, 503 *illus.*; development of, 500; interpreting, 503, 506

Physical changes, **20**, 35. *See also* Changes of state; separation of mixtures by, 20

Physical processes, 20

Physical properties, **20**, 34–35; composition of a penny and, 38–39 *lab*; kitchen chemicals, 16–17 *lab*; melting and freezing points, 35

Physics Connection, Aurora Borealis, 73; Bohr, Neils, 232; solid rocket booster engines, 566

Pie graphs, 805

Pig iron, 286

Pistons, 374 *illus.*, 375

Plant-care specialists, 212–213

Plants, cellulose in, 680; medicines from, 146; nitrogen fixation by, 273, 571; photosynthesis and. *See* Photosynthesis

Plasmas, 34, **347**

Plaster of paris, 167, 168

Plastics, 660; recycling of, 659, 661 *illus.*; thermoplastic, 660; thermosetting, 660

Platen, 110

Platinum, 292, 715

Platinum group, 292

Plutonium, 102, 295, 765

Plutonium-238, 295

pnp-junctions, 112

Polar covalent bonds, **310**–311; paper chromatography and, 312 *lab*; properties of compounds having, 311; water and, 311

Polarity, 330–333; of ammonia, 331; chromatography and. *See* Chromatography; lack of in carbon dioxide, 331; of water, 311, 330–331

Polar molecules, 331–333; ammonia, 331; attraction to other polar molecules, 332; properties,

332–333; water, 311, 330–331

Polar solvents, 331

Pollution, thermal, 469 *illus.*; tracking with tracers, 771

Polonium, 276

Polyatomic ions, **158**–162; charge of, 159, 850 *table*; common, 159 *table*, 809 *table*; formulas for, 159, 161 *prob.*, 162 *prob.*; naming, 160

Polyester, 643, 649

Polyethylene, 656, 659, 730

Polyethylene terphthalate (PET), 659

Polymerization reactions, 655–656, 658, 660; addition reactions, 656; condensation reactions, 658; cross-linking and rubber, 658, 660; plastics and, 660

Polymers, **649**, 649, 653. *See also* Plastics; comparing natural and synthetic, 653 *lab*; differentiating between natural and synthetic, 650–652 *lab*; polymerization reactions forming, 655–656, 658, 660; structure, 655–656; Word Origin, 649

Polypeptide chains, 672. *See also* Proteins

Polypropylene (PP), 659

Polyprotic acids, 485 *illus.*, 486 *illus.*

Polysaccharides, 681–682

Polyunsaturated fatty acids, 686

Polyvinyl alcohol, 40 *lab*

Polyvinyl chloride (PVC), 659

Polyvinylacetate, 656 *illus.*

Popcorn, popping of, 397

Porcelain, 163

Portable reverse osmosis units, 468

Positional isomers, 631

Positron emission tomography (PET), 769 *illus.*

Potassium, 86, 263; decay of Potassium-40, 750 *prob.*; in Earth's crust, 669 *illus.*; electron configuration, 247; as essential element, 128; symbol, 27 *table*

Potassium-argon dating, 755

Potassium bromide, 308, 518 *lab*

Potassium chloride, 143 *illus.*, 154, 459

Potassium chromate, 293

Potassium dichromate, 293

Potassium hydroxide, 264; reaction

with calcium chloride, 215 *illus.*; as strong base, 498, 498 *table*
Potassium iodide, 154
Potassium perchlorate, 76
Potassium tartrate, 33 *table*, 809 *table*
Potential difference, 600 *lab*, **601**
Power plants, coal-fired, 726 *illus.*; efficiency of, 731 *table*; nuclear, 764 *illus.*, 764–765
Powers of ten. *See* Scientific notation
PP (polypropylene), 659
Practice Problems, alkanes, formulas and names of, 626–627 prob.; answers to, 853–862; Boyle's law, 389 prob.; Charles's law, 393 prob.; converting pressure units, 381 prob.; electronegativity difference between bonding atoms (?*EN*), 312 prob.; food calories, 724 prob.; formulas for binary ionic compounds, 158 prob.; formulas for polyatomic ions, 162 prob.; half-life and rate of decay, 758 prob.; heat of reaction, 721 prob.; hydronium ion and hydroxide ion concentrations, 502 prob.; ideal gas law and, 420 prob.; ionic and net ionic equations, 521 prob., 525 prob., 530 prob.; molar mass, 412 prob., 413 prob.; molar volume, 418 prob.; molarity, 463 prob., 464 prob., 546 prob.; molecular compounds, naming, 181 prob.; nuclear equations, writing and balancing, 751 prob.; predicting the mass of reactants and products, 416 prob.; supplemental problems (Appendix B), 809–838; temperature conversions, 350 prob.; titration, 546 prob.
Praseodymium, 294
Precipitation, chemical reactions and, 191, 215
Precision, 793
Pressure, 344; converting units of, 379 prob., 379–380, 380–381 prob.; gas pressure and volume and. *See* Boyle's law; of gases, 344; Haber process and, 216; ideal gas law and, 419; measurement of, 376, 377, 378; standard atmosphere, 376; STP (standard

temperature and pressure), 395, 395–396 prob.; units of, 378–379; vapor pressure, 357
Pressure cookers, 359
Pressure gauges, 376, 377, 378
Pressure units, 378–380; converting between, 379 prob., 379–380, 380 prob.; equivalent, 379 *table*
Priestly, Joseph, 574
Princeton Review. *See* Standardized Test Practice
Private investigator, 13
Products, 192; predicting mass of, 415–416 prob.; Word Origin, 192
Professional opportunities. *See* Career Connection; People in Chemistry
Promethium, 295
Propane, 33 *table*, 145 *illus.*, 416, 623, 638, 809 *table*
Propene, 631
Properties of matter, 4, 5 *illus.*, 6
Prostratin, 146
Proteases, 222, 677 *illus.*
Proteins, 670–673; denaturation, 673; enzymes, 674–675 *lab*, 676; as fake fats, 685; hydrogen bonding and, 439; monomers of (amino acids), 670–671; role of in human body, 670; structure of, 670–673
Protic, 485 *illus.*
Protons, 62, 66, 230; discovery of, 62; properties, 67 *table*
Proust, Joseph, 53–54
PS EPS, 659
Pure substances, 15
Putrescine, 645
PVC, 641, 659
Pyrite, 165 *illus.*

Qualitative expression of composition, 14, 34
Quantitative expression of composition, 14, 34
Quantities, 785
Quantum theory of chemistry, 307
Quartz, 248 *illus.*
Quinine, 146

R **(ideal gas law constant)**, 419
Radiant heat, 71
Radiation. *See* Electromagnetic radiation
Radiation detectors, 751 *illus.*, 775 *lab*
Radiation exposure, 776
Radiation therapy, 770 *illus.*
Radioactive dating, 756–757; Carbon-14 dating, 754–755, 757, 758 prob.; of forged art, 759
Radioactive decay, 747, 749–750; alpha particles, 747; beta particles, 749; gamma rays, 749–750; modeling, 752–753 *lab*; nuclear equations for, 747, 749–750, 750 prob.
Radioactive isotopes, dating methods using, 754–755, 756–757, 758 prob.; half life of, 756, 756 *table*; modeling decay of, 752–753 *lab*; nuclear equations for, 746, 750–751 prob.; radiation released from, 747, 749–750
Radioactive wastes, 775, 778–779
Radioactivity, 746; detection of, 751, 775 *lab*; discovery of, 744–746; hazards of exposure to, 774–776, 776 *table*, 777, 778; nuclear equations and, 746; radioactive dating and, 754–755, 756–757, 758 prob., 760; radioactive decay and, 747, 749–750, 752–753 *lab*, 756–757, 758 prob.; radon and, 775 *lab*, 777; waste disposal and, 778–779
Radioisotopes, Hershey and Chase's experiments on DNA with, 772; medical applications of, 768–770; nonmedical applications of, 771, 773; release of radiation by, 747, 749–750; sources of, 773
Radio waves, 71, 72 *illus.*
Radium, 265
Radon, 132, 281, 775 *lab*, 777
Radon-testing kits, 775 *lab*
Rain forests, medicines from, 146
Rare earth metals. *See* Lanthanides
Ration heaters, flameless, 221
Reactants, 192; limiting, 220; predicting mass of, 414–415 prob., 420 *lab*

Index

Reaction capacity, metals and valence electrons, 236–267 *lab*

Reaction rates, 218–220, 222–223; activation energy and, 218; catalysts and, 222; concentration and, 219–220; determining speed of, 218–219; inhibitors and, 223; starch-iodine clock reaction, 220 *lab*; temperature and, 219

Recommended Dietary Allowance (RDA), 128

Recycling, 55, 731; aluminum, 589 *illus.,* 725; conservation of matter and, 55; glass, 60; plastics, 659, 661 *illus.*

Redox reactions. *See* Oxidation-reduction reactions

Red phosphorus, 179, 274

Reducing agent, 559, **562**

Reduction reactions, **556.** *See also* Oxidation-reduction reactions; combining with oxidation half-reactions, 557–558; oxidizing and reducing agents and, 559, 562; Word Origin (reduction), 556

Reforming, **638**

Respiration, 124, 574–575, **694,** 696, 697 *illus.,* 698; ATP and energy storage and, 696; electron transport chain, 698; glycolysis, 696; photosynthesis and, 737 *illus.*; tricarboxylic acid cycle, 698

Retinal, 632

Reverse osmosis, 467

Reverse osmosis purification plants, 465

Reverse osmosis units, 468

Reversible reactions, 210–211, 214

Rhodium, 292, 715

Rhodopsin, 632

Ribonucleic acid. *See* RNA (ribonucleic acid)

Ribose, 680, 689

RNA (ribonucleic acid), 273, **688,** 689, 690

Robotic arms, 109

Rocket fuel, 566

Rockets, solid rocket booster engines, 566

Rolling, 291

Roman Empire, lead poisoning and, 271

Roman numerals, in names of transition-element compounds, 164 *table,* 164–165

Rubber, cross-linking in, 658; vulcanized, 658

Rubbing alcohol (isopropyl alcohol), 140 *illus.,* 358, 642

Rubidium, 263

Rubidium-strontium dating, 755

Rubies, 248, 249

Rust, 41 *illus.,* 122 *lab,* 204 *illus.,* 250, 555 *illus.,* 557, 570

Ruthenium, 292

Rutherford, Ernest, 64, 65, 230

***s* orbitals,** 238 *table,* 239; predicting distribution in, 245 *lab*

***s* sublevels,** 235, 238

Saccharin, 5 *illus.*

Safety guidelines, 839

Safety Handbook, 839–840

Safety symbols, 840

Salicylic acid, 32, 642

Salt bridge, 602 *illus.,* 604

Salt (table salt). *See* Sodium chloride

Salts, **516**

Samarium, 295

Sample Problems, atoms, number in a sample of an element, 408 prob.; Boyle's law, 388–389 prob.; Charles's law, 393 prob.; combined gas law, 395–396 prob.; converting pressure units, 379 prob., 380 prob.; food calories, 724 prob.; formula for a binary ionic compound, 158 prob.; formula for a polyatomic ion, 161 prob.; heat of reaction for combustion, 720 prob.; ideal gas law and, 419 prob.; ionic and net ionic equations of strong acid/strong base reactions, 521 prob.; ionic and net ionic equations of strong acid/weak base reactions, 525 prob.; ionic and net ionic equations of weak acid/strong base reactions, 529 prob.; mass of a number of moles of a compound, 413 prob.;

molarity, 462–464 prob.; molarity from titration, 544 prob.; nuclear equations, writing and balancing, 750 prob.; number of formula units in a sample of a compound, 412 prob.; radioactive dating of fossils, 758 prob.; writing chemical equations, 200 prob.

Sapa, 271

Sapphire, 249 *illus.*

Satellites, 70

Saturated fats, 630, 633 *illus.*

Saturated fatty acids, 686

Saturated hydrocarbons (alkanes), **623**–626; isomers, 627–628; properties, 629; structural diagrams and names, 623–626, 626–627 prob.

Saturated solutions, **458**

Scandium, 247

Scanning probe microscope (SPM), 240

Scanning tunneling microscope (STM), 10, 241

Scientific laws, **59**

Scientific method, 59

Scientific models, **11**

Scientific notation, 795–798, 799–800

Scintigraphy, 770 *illus.*

Scintillation counters, 751 *illus.*

Scrap metal processing worker, 597

Scurvy, 691

Seaborgium, 102 *illus.*

Seawater, floatation of icebergs on, 26; as mixture, 18 *illus.*

Second period elements, 244, 246 *table*

Selectively permeable membranes, 467

Selenium, 276, 277; electron configuration, 99 *illus.*; as essential element, 128

Semiconductors, **105,** 111–113; diodes, 112–113; electrical conduction by, 111–112; gallium arsenide in, 274; n-type and p-type, 112; properties of at room temperature, 111; silicon in, 272; Word Origin, 112

Shampoos, pH of, 501

Shape-memory metals, 108–109

Shielding effect, **304,** 305 *illus.*

Sievert (Sv), 776
Significant digits, 793–794
Silica, 272
Silicon, 105, 107, 270; doping, 111, 112 *illus.*; in Earth's crust, 669 *illus.*; as essential element, 128; glass from, 272; uses of, 272; valence electrons, 111 *illus.*
Silicon chips, 113 *illus.*
Silicon dioxide, 272, 347 *illus.*
Silk, 653
Silver, 23, 292; as coinage metal, 86, 110, 292; conductivity, 313; electroplating, 593 *illus.*; symbol for, 27 *table*; tarnishing, 572
Silver bromide, 30 *illus.*, 279, 564, 565 *illus.*
Silver nitrate, 200–201 prob.
Silverplate, 593 *illus.*
Silver sulfide, 572
Simple sugars. *See* Sugars
Single bonds, 629 *illus.*
Single-displacement reactions, 205, 205 *lab*, 206–207 *lab*, 208 *table*
SI units, 785–787; base units, 785; derived units, 785, 786 *table*; English units and, 788; metric units and, 786, 788; prefixes for, 786–787, 787 *table*
Skawindski, William (chemist), 316–317
Skin care products, pH of, 501
Slag, 286, 567 *illus.*
Slime, creation of, 40 *lab*
Smog, 495
Smoke detectors, 748
Snowflakes, 360, 441 *illus.*
Soap scum, 160, 191
Soaps, 455; production of, 455, 494, 496
Soda-lead glass, 346
Sodium, 99, 263; as coinage metal, 86; in Earth's crust, 669 *illus.*; from electrolysis, 585 *illus.*, 586 *illus.*, 587–588; electron configuration, 245; as essential element, 128; properties, 122, 123 *illus.*; reaction with chlorine, 133, 134 *table*, 264; reaction with water, 264; symbol for, 27 *table*
Sodium acetate, 531–532, 533 *illus.*
Sodium azide, air bags and, 417
Sodium bicarbonate, 143 *illus.*
Sodium bromide, 205 *illus.*

Sodium carbonate, 33 *table*, 166, 199, 809 *table*
Sodium chloride, 5 *illus.*, 120–122, 155; crystal structure of, 134–135; electrical conductivity, 121; electrolysis of, 585, 587–588; formation, 120–122, 133, 134 *table*, 516 *illus.*; formula, 33 *table*; ionic bonding, 306, 308; mining, 121 *illus.*; properties, 121, 122; solubility of, 451, 452 *illus.*, 459; uses of, 33 *table*, 121 *illus.*
Sodium fluoride, 154, 279
Sodium hydrogen carbonate, 143 *illus.*
Sodium hydroxide, 809 *table*; as deliquescent substance, 166 *illus.*; formula, 33 *table*; production by electrolysis, 587–588; reaction with acetic acid, 528, 529 *illus.*; reaction with carbon dioxide, 199; reaction with HCl, 516 *illus.*; as strong base, 489, 498, 498 *table*; uses, 33 *table*, 265
Sodium hypochlorite, 194, 567–568
Sodium nitrate, 205 *lab*, 459
Sodium stearate, 160, 455
Sodium sulfate, 456
Soft water, 452 *lab*
Soil, adding lime to, 494 *illus.*; as mixture, 18 *illus.*
Soil conservation technician, 213
Solar energy, 728
Solar stills, 171
Solar wind, 73
Solder, 18 *illus.*, 23 *table*
Solid aerosols, 470
Solid rocket booster engines, 566
Solid rocket fuel, 566
Solids, 34–35, **340**; kinetic model, 345; melting, 364; properties, 340 *illus.*; sublimation, 356, 375 *lab*
Sols, 470
Solubility, of carbonates, 482; of a gas in a liquid, 469; of ionic compounds, 451; "like dissolves like", 454; temperature and, 459–460
Solubility product constants, 851 *table*
Soluble, 215; Word Origin, 214
Soluble compounds, 215
Solutes, 23
Solution, heat of, **460**

Solutions, 22–23. *See also* Aqueous solutions; Colloids; alloys, 23, 23 *table*; boiling-point elevation, 465; concentration, 458–459, 460–461, 462 prob.; freeze-point depression, 465; of gases in liquids, 469; guidelines for, 851 *table*; heat of, 460; identification of, 456–457 *lab*; "like dissolves like", 453; molarity and, 460–461, 462–464 prob.; saturated, 458–459; standard, 539; unsaturated, 458–459; volume when mixed, 21 *lab*
Solvents, 23
Soot, 470, 637 *illus.*
Sørenson, S.P.L., 500
Sour taste, 519
Space shuttle, reaction of hydrogen and oxygen to fuel, 200 prob.; regulation of air quality in, 203, 418 prob.; solid rocket booster engines, 566
Specific heat, 445 *table*, **445**–446
Spectator ions, 520
Spectrum. *See also* Electromagnetic spectrum; Emission spectra; Word Origin, 231
Speed of reaction, 218–219
Sphygmomanometer, 376 *illus.*
Spices, 18 *illus.*
Spinnerets, 653
SPM. *See* Scanning probe microscope (SPM)
Spontaneous reactions, 716–718; predicting, 718 *table*; Word Origin (spontaneous), 604
Stained-glass windows, 472 *illus.*
Stainless steel, 23 *table*
Stalactites, 160, 524
Stalagmites, 160, 524
Standard atmosphere (atm), 376
Standardized Test Practice, 49, 83, 117, 151, 187, 227, 255, 299, 337, 369, 401, 433, 477, 513, 551, 581, 619, 665, 705, 741, 783
Standard solutions, 539
Standard temperature and pressure (STP), 395, 395–396 prob.
Starch, 655, 677, 681–682; as fat replacement, 685; reaction with iodine, 220 *lab*; structure, 655 *illus.*
Starch-iodine clock reaction, 220 *lab*

Index

States of matter, 34–35, 340–347. *See also* Gases; Liquids; Plasmas; Solids; changes in. *See* Changes of state; periodic *table* trends, 92–93 *illus.*

Stationary phase, paper chromatography, 312 *lab*

Statue of Liberty, corrosion of, 570

Steam, 127 *illus.*, 436

Stearic acid, 686 *illus.*

Steel, 23, 106, 282 *illus.*, 288–291; alloy steels, 290; carbon steels, 288 *table*, 289; classification, 288; earliest known objects made of, 288; galvanized, 293, 555 *illus.*, 593–594; production, 287; shaping (working), 290–291

Steelmaking, 287, 288–291; alloy steels, 290, 290 *table*; carbon steels, 288 *table*, 288–289; shaping of steel, 290–291

Sterling silver, 23 *table*

Steroids, 686–687; Word Origin, 687

Stierle, Andrea, 32

Stierle, Donald, 32

STM. *See* Scanning tunneling microscope (STM)

Stoichiometry, 405–409; atoms in a sample of an element, 408 prob.; Avogadro's constant and, 405–406; chemical formulas from, 427–428; formula units in a sample of a compound, 412 prob.; ideal gas law and, 419, 419 prob.; mass of a number of moles of a compound, 413 prob.; mass of a product, predicting, 415–416 prob.; mass of a reactant, predicting, 414–415 prob.; mass percents, 422–423 *lab*, 426–427; molar mass of a compound, 409, 412–413 prob.; molar mass of an element, 407, 408 prob.; molar volumes and, 416, 418 prob., 420 *lab*; molarity from titration, 541, 542–543 *lab*, 544, 544 prob.; percent yield and, 421; theoretical yield and actual yield, 421; Word Origin, 406

Stomach, acidity of and antacids, 535–536, 538

Stone Age, 567

Stoney, George, 61

STP. *See* Standard temperature and pressure (STP)

Straight-chain hydrocarbons, 623

Straws, gas laws and, 388 *lab*

Strong acids, 498, 498 *table*

Strong bases, 497–498, 498 *table*

Strontium, 265; reactivity, 266–267 *lab*; uses, 268

Strontium hydroxide, 498 *table*

Structural isomers, 628

Structural proteins, 670

Structure of matter, 7 *illus.*

Styrene-butadiene rubber, 660

Styrofoam, 36 *illus.*

Sublevels, 234–235

Sublimation, 356, 375 *lab*

Submicroscopic matter, 7, 10

Subscripts, balancing chemical equations and, 199; formulas for binary ionic compounds, 155–156

Substances, 15, 24–25, 27–31; compounds. *See* Compounds; elements. *See* Elements

Substituted hydrocarbons, 640–647. *See also specific groups*; production of, 646–647

Substrates, 676

Sucrose, 5 *illus.*, 33 *table*, 145 *illus.*, 680, 809 *table*; models of, 10; solubility, 453, 454; structure, 10 *illus.*; sweetness of, 519

Sugar of lead, 271

Sugars, 6 *illus.*, 36 *table*, 680–682. *See also* Carbohydrates

Sulfate ion, 159 *illus.*

Sulfonates, 455

Sulfur, 276; electron configuration, 99 *illus.*, 233 *illus.*; in human body, 669 *illus.*; mining of by Frasch process, 276, 276 *illus.*; oxidation-reduction (redox) reactions, 563; uses, 277

Sulfur dioxide, 495

Sulfur oxides, acid rain and, 493–494, 495

Sulfuric acid, 277, 809 *table*; in lead-acid batteries, 273, 611 *illus.*, 613; manufacturing, 424–425, 484; steelmaking and, 486 *illus.*; as strong acid, 498 *table*; uses, 33 *table*, 424, 484

Super HILAC Accelerator, 102 *illus.*

Superplastic steel, 288, 288 *table*, 289

Supersaturated solutions, 459

Supplemental Practice Problems, 809–838

Surface tension, 442, 443 *lab*

Sutliff, Caroline (plant-care specialist), 212–213

Sweetness, 519, 683

Symbols, element, 27 *table*, 92–93 *illus.*

Synthesis reactions, 203–204, 208 *table*; recognizing, 206–207 *lab*; Word Origin, 204

Synthetic aromas, 646 *lab*

Synthetic chemicals, 32

Synthetic elements, 102

Synthetic polymers, 649; comparing with natural, 653 *lab*; differentiating between synthetic and natural, 650–652 *lab*

Synthetic rubber, 658 *illus.*, 660

Tables, making and using, 804

Table salt. *See* Sodium chloride

Table sugar. *See* Sucrose

Tandem accelerator mass spectrometer (TAMS), 754–755

Tandem Cascade Accelerator (TCA), 773 *illus.*

Tarnish, 572; Word Origin, 572

Taste, 519, 683

Taste buds, 519

Taxol, 31, 32, 421

Tayag, Fe (cosmetic bench chemist), 490–491

Technetium, 102

Technetium-99m, 770 *illus.*, 773 *illus.*

Teflon, 648–649

Tellurium, 91, 94, 276, 277

Temperature, 348–349, 789; chemical reaction rate and, 219; gas pressure and, 358, 374–375; gas volume and. *See* Charles's law; Haber process (synthesis of ammonia) and, 216; ideal gas law and, 419; kinetic energy and, 348–350, 362–363 *lab*; measurement of, 789; protein denaturation and, 674, 674–675 *lab*; solubility and, 459–460; standard

temperature and pressure (STP), 395

Temperature scales, 349–350, 789; converting between, 350, 350 prob.

Terbium, 295

Tetrachloromethane, 640

Tetrafluoroethene, 648–649

Tetrahedral molecular geometry, 319 *illus.,* 322 *illus.,* 323 *illus.*

Textiles, polymers in, 650–652 *lab*

Thallium, 269, 270

Theoretical yield, 421

Theories, 59

Thermal pollution, 469 *illus.*

Thermoluminescence (TL) dating, 755

Thermometers, 283 *illus.*

Thermoplastics, 660

Thermosetting, 660

Thin-layer chromatography, 326–327. *See also* Chromatography

Thiobacillus ferroxidans, 727

Third period elements, electron configuration, 245–246, 246 *table*

Thomson, J. J., 61–62, 230

Thomson's atomic model, 63 *illus.,* 230, 231 *illus.*

Thornton, John, 12–13

Three Mile Island, 765 *illus.*

Thymine (T), 689, 690

Thymol blue, 545

Thyroid gland, iodine and, 278, 769 *illus.*

Thyroxin, 641

Tin, 27 *table,* 270, 272, 570 *illus.*

Tin(II) fluoride, 279

Tire-pressure gauges, 377, 378

Titanic, cleaning of by electrolysis, 595

Titanium, 247 *illus.*

Titration, 539–541, 544; concentration from, 541, 542–543 *lab,* 544, 544 prob., 546 prob.; how to perform, 540; indicators and, 540; standard solutions and, 539

TNT, 104 *illus.*

Tokamak Fusion Reactor, 767 *illus.*

Tooth decay, prevention of by fluoridation, 279, 280

Toricelli barometer, 376

Torricelli, Evangelista, 376

Toxic wastes, disposal of radioactive, 778–779; electrolysis of, 598

Tracers, radioisotope, Hershey and Chase's experiments on DNA with, 772; medical applications of, 768 *illus.;* tracking of pollutants by, 771

trans **isomer configuration,** 631; Word Origin (trans), 631

Transistors, 112–113

Transition elements, 94, **103–104,** 106. *See also* Actinides; Lanthanides; atomic size trends, 285; as coinage metals, 292; electron configuration, 247, 250, 285 *lab;* formulas for compounds containing, 164–165, 165 prob.; gemstones and, 248–249; ionic compounds of, 162, 164–165; multiple oxidation states, 283; naming of compounds having, 164, 164 *table;* oxidation number of, 162, 165; properties, 282–283, 285–287; uses, 286–287, 292–293

Transmutation, 749

Triads, Dobereiner's, 87–88

Triangular pyramid molecular geometry, 322 *illus.*

Tricarboxylic acid cycle, 697 *illus.,* 698

Trichloromethane, 640

Triglycerides, 684

Triple-beam balances, 790

Triple bonds, 325, 629 *illus.,* 633

Triprotic acids, 485 *illus.*

Tritium (T), 766

Try at Home Labs, 863–873

Tungsten, chemical symbols, 27 *table;* as light bulb filament, 284; melting point, 283, 284

Tungsten steel, 290 *table*

Twenty Thousand Leagues Under the Sea, **Jules Verne's,** 26

Tyndall effect, 472, 473 *illus.*

Ultrahigh-carbon steel (UHCS), 288 *table,* 289

Ultraviolet waves, 72

Unbranched hydrocarbons, 623

Uncertainty principle, Heisenberg's, **238**

Units of measurement, 785–791

Unsaturated fats, 630, 630 *illus.*

Unsaturated fatty acids, 686 *illus.*

Unsaturated hydrocarbons, 625, **636;** alkenes, 629–631, 633; alkynes, 633, 633 *table,* 636; aromatic hydrocarbons, 636–637; food oils and, 630 *lab*

Unsaturated solutions, 458

Uracil (U), 689, 690

Uranium, 295; alpha decay, 746 *illus.,* 750; Becquerel's radioactivity experiment, 745 *illus.*

Uranium-238, 747

Uranium experiment, Becquerel's, 745–746

Uranium-lead dating, 759

Urea, 669

Vacuum tubes, discovery of electrons and, 61

Valence electrons, 78–79, 98; atomic collisions and, 130; atomic size and, 259; electron configurations and, 243–247; metal-acid reactions and, 236–267 *lab;* in metalloids, 105, 106; in metals, 105, 106; in nonmetals, 105, 106; periodic *table* trends, 231, 233, 243–247, 250; semiconductors and, 111 *illus.;* sharing of, 303 *illus.;* shielding effect, 304, 305 *illus.;* sublevels and, 234–235

Vanadium, electron configuration, 247 *illus.;* melting point, 283

Vanilla flavor, 644

van Meegeren, Hans, 759

Vaporization, 356–358, 360; of different liquids, 357 *lab;* as endothermic process, 446; heat of, 361

Vapor pressure, 357–358, 360

Vegetable fats, 630 *lab,* 633 *illus.*

Vegetable oils, 687

Verne, Jules, 26

Vinegar, 485, 642. *See also* Acetic acid; reaction of bones with, 171 *lab;* reaction with baking soda, 41 *illus.,* 192–193, 195, 420 *lab;* reaction with egg shells, 483

Index

illus.; reaction with strong bases, 528; titration of, 542–543 *lab*; Word Origin, 528

Visible light, 71, 72

Vision, vitamin A and, 632

Vitamin A, 632, 691

Vitamin C, 307, 498 *illus.*, 576 *illus.*, 691

Vitamin D, 691

Vitamin E, 576 *illus.*

Vitamins, 690–691; coenzymes, 691; water- vs. fat-soluble, 691; Word Origin, 690

Volatile substances, 35, 353, 357

Volta, Alessandro, 601

Voltage, 601

Voltaic cells, 602–605

Voltmeters, 604–605

Volume, combined gas law and, 394–395, 395–396 prob.; gas pressure and. *See* Boyle's law; gas pressure and temperature and. *See* Charles's law; ideal gas law and, 419; law of combining gas volumes and, 396–398; measurement of, 789; relationship between volume and mass of a gas, 375 *lab*

Vulcanizing, 658

Wastewater operator (Alice Arellano), 448–449

Water, 5 *illus.*, 126–127, 436–450. *See also* Aqueous solutions; Ice; Steam; bond angle distortion, 319; changes in state of, 35; condensation, 356, 446; density, 36 *table*, 437, 440–441; dissolution of into hydrogen and oxygen, 41–42, 711–712, 712 *lab*; distilled, 452 *lab*; electrolysis, 41 *illus.*; electron sharing in, 138–139; evaporation, 446; formation of, 127 *illus.*, 200 prob., 218, 709 *illus.*, 711 *illus.*; formula, 33 *table*, 809 *table*; freeze-melting, 35 *illus.*, 441 *illus.*; hard, 160, 452 *lab*; heat of fusion, 361; heat of vaporization, 364; hydrates, 165–166, 166 *lab*, 168; hydrogen bonding in, 439, 440, 441, 444, 454; intermolecular

forces in, 438 *illus.*, 438–439; modeling shape of, 318–319; osmosis and, 467; polar nature of, 311, 330–331; properties, 41 *illus.*, 126–127, 332 *table*, 436–437, 467 *table*; regulation of Earth's temperature by, 446; soft, 452 *lab*; as solvent, 23, 449, 450–454; specific heat of, 445; states of, 34, 126, 127 *illus.*, 436, 440–441; surface tension, 442, 443 *lab*; vapor pressure of, 357

Water softeners, 160

Water-soluble vitamins, 691

Water-treatment plants, 447

Watson, James, 690

Wavelength, 71 *illus.*, 233

Waves, 70–71

Waxes, 682. *See also* Lipids

Weak acids, 498 *illus.*, **499**; reaction with strong bases, 528–539, 529–530 prob.; reaction with weak bases, 530

Weak bases, 500; reaction with strong acids, 522–523, 525 prob., 525–526; reaction with weak acids, 530

Weather balloons, 386, 387

Weather predictors, 166 *lab*

Weight, 4, 42, 790–791

White light, spectrum of, 71 *illus.*

White phosphorus, 179, 274

Willow bark, salicylic acid from, 32

Wind energy, 729

Wire, manufacture of copper, 590–592

Wöhler, Friedrich, 669

Wood, specific heat, 445 *table*

Wootz steel, 288

Word equations, 193, 200–201 prob.

Word Origin, acid, 485; actinide, 295; aerobic, 694; allotrope, 175; atom, 53; aurora, 73; barometer, 376; binary, 155; bioluminescent, 575; capillarity, 443; chlorophyll, 734; cis-, 631; colloid, 472; combustion, 209, 713; corrosion, 595; distillation, 638; electronegative, 310; empirical, 428; energy, 711; fission, 762; fusion, 766; halogen, 278; heterogeneous, 23; homogeneous, 24; orbitals, 247; osmosis, 467; periodic, 99; petroleum, 637; photograph, 564; polymer, 649;

product, 192; reduction, 556; semiconductor, 112; soluble, 214; spectrum, 231; spontaneous, 604; steroid, 687; stoichiometry, 406; synthesis, 204; tarnish, 572; trans-, 631; vinegar, 528; vitamin, 690

Work, 42

X rays, 72 *illus.*, 745, 776, 778 *illus.*

Xenon, 131, 132 *illus.*, 281

Xenon hexafluoride, 195, 416 prob.

Xerography, 277

Yawning, 533

Yeasts, fermentation by, 698, 699 *lab*

Yellow dye #5, chromatography of, 328–329 *lab*

Ytterbium, 107

Yttrium, 294

Zinc, 293; alloy of copper and zinc, 25 *lab*; in carbon-zinc dry cells, 608–609; electron configuration, 247; as essential element, 128; oxidation-reduction (redox) reactions and, 555–558, 559

Zinc electroplating, 593, 595

Zinc iodide, electrolysis and, 136–137 *lab*

Zinc oxide, 555–558

Zircon, 292

Zirconium, 292

Cover (l)Peter Steiner/The Stock Market, (r)William Westheimer/The Stock Market.
Chemistry and Society Ken Fisher/Tony Stone Images.

vi (t)Don Mason/The Stock Market, (c)Bill Horsman/Stock Boston, (b)Gary Yeowell/Tony Stone Images; **vii** (t)Phyllis Picardi/Stock Boston, (b)Tim Courlas; **viii** (t)Jean-Loup Charnet/Science Photo Library/Photo Researchers, (c)Zefa/The Stock Market, (b)Tony Freeman/PhotoEdit; **ix** (t)Stephen Frisch/Stock Boston, (c)Matt Meadows, (b)Will Crocker/The Image Bank; **x** (t)David R. Frazier Photolibrary, (c)Pete Salautos/The Stock Market, (b)Mark Steinmetz; **xi** (tl)Matt Meadows, (tr)Bob Mullenix, (b)Dennis Kunkel/PhotoTake NYC; **xii** (t)Kunio Owaki/The Stock Market, (b)Mark Steinmetz; **xiii** (t)John Sims/Tony Stone Images, (b)Stan Osolinski/FPG International; **xiv** (t)John Durhan/Science Photo Library/Photo Researchers, (b)Jean-Perrin/CNRI/Science Photo Library/Photo Researchers; **xv** Skip Comer; **xvi** Zefa/The Stock Market; **xvii, xviii** Matt Meadows; **xix** (t)Al Francekevich/The Stock Market, (bl)David R. Frazier Photolibrary, (br)Doug Martin; **xx** (t)Somatogen, Inc., (b)Phillip Hayson/Photo Researchers; **xxi** (clockwise from top)Jose Luis Banus-March/FPG International, Mark Tuschman, Phillip DeManczuk, Mark Tuschman, Otis Hairston, Daniel Schaefer, Brian Buckley, Ted Corwin/Oman Studio, Mark Tuschman; **xxii** (t)Tom Stewart/The Stock Market, (c)Art Resource, NY, (b)Scala/Art Resource, NY; **1** (t)NASA, (b)Kim Westerskov/Ford Scientific Films/Earth Scenes; **2** Nawrocki Stock Photo; **3** Space Telescope Science Institute/NASA/Science Photo Library/Photo Researchers; **4** Geoff Butler; **5** (tl)Mark Steinmetz, (tc)Douglas Mesney/The Stock Market, (tr)Doug Martin, (c)Matt Meadows, (bl)Matt Meadows, (br)Matt Meadows; **6** (tl)John Henley/The Stock Market, (tc)John Foster/Mastefile, (bl bc br)Chip Clark, (others)Matt Meadows; **7** Roger Ressmeyer/Corbis; **8** The Bettmann Archive; **9** Matt Meadows; **10** (tl)Philippe Plailly/Science Photo Library/Photo Researchers, (tr)courtesy Digital Instrument, Inc.; **11** (t)Dirck Halstead/Gamma Liaison, (b)Bill Horsman/Stock Boston; **12** Mark Tuschman; **13** Mark Tuschman; **14** Doug Martin; **15** Mark Steinmetz; **16** Matt Meadows; **18** (t)Color Box/FPG International, (cl)William S. Nawrocki/Nawrock Stock Photo, (cr)Cindy Charles/PhotoEdit, (bl)Virginia Mason Medical/Custom Medical Stock Photo, (br)Geoff Butler; **19** Geoff Butler; **20** Matt Meadows; **21** (t)Matt Meadows, (bl)Mark Steinmetz, (br)Craig Kramer; **22** Matt Meadows; **23** (t c)Mark Steinmetz, (b)Matt Meadows; **24** (l)Don Mason/The Stock Market, (r)Tino Hammid; **25** (t)Matt Meadows, (bl)Charles D. Winters, (br)Mark Steinmetz; **26** Tom Stewart/The Stock Market; **27** (t)Matt Meadows, (b)Comstock; **28-29** from "The Periodic Systems of the Elements," author P. Menzel. C. by Ernst Klett Schulbuch Verlag GmbH, Stuttgart, Germany. Charts available from Science Import, P.O. Box 465, Sillery, Quebec, Canada G1T 2R8; **30** (t)Mark Burnett, (bl)John Cancalosi/Stock Boston, (br)Charles D. Winters; **31** (l)Matt Meadows, (r)Doug Martin; **32** Tom & Pat Leeson/Photo Researchers; **33** (t)Aaron Haupt, (c b)Mark Steinmetz; **34** David R. Frazier Photolibrary; **35** Terry Qing/FPG International; **36** Geoff Butler; **37** Matt Meadows; **38** (l)Mark Steinmetz, (r)Matt Meadows; **39** Mark Burnett; **40** Matt Meadows; **41** (tl)Matt Meadows, (tr)Richard Megna/Fundamental Photographs, (bl)Stephen Frisch/Stock Boston, (br)Matt Meadows; **42** Ed Lallo/Gamma Liaison; **43** Matt Meadows; **44** Craig Hammell/The Stock Market; **46** Mark Steinmetz; **47** Mark Steinmetz; **50** Dave Nagel/Gamma Liaison; **51** Stephen Marks/The Image Bank; **52** Scala/Art Resource, NY; **53** (l)Charles D. Winter, (c r)Matt Meadows; **54** (l)Charles D. Winters, (r)John Cancalosi/Stock Boston; **56, 57** Matt Meadows; **58** file photo; **60** (t)James L. Amos/Peter Arnold, Inc., (b)Skip Comer; **61** Skip Comer; **63** Matt Meadows; **67** Stephen Simpson/FPG International; **69** NASA/Science Photo Library/Photo Researchers; **71** (t)PhotoEdit, (b)Gary Yeowell/Tony Stone Images; **72** Richard Megna/Fundamental Photographs; **73** Pekka Parviainen/Science Photo Library/Photo Researchers; **74** Rich Treptow/Photo Researchers; **76** (t)Peter Gridley/FPG International, (b)Zambelli Internationale Fireworks; **77** Ted Rice; **78** Morton & White;

81 David McGlynn/FPG International; **84** Jerry Lodriguss/Photo Researchers; **85** (l)Richard Megna/Fundamental Photographs, (r)Matt Meadows; **86** SuperStock; **87** Chip Clark; **88** Stamp from the collection of Prof. C.M. Lang, photograph by Gary Shulfer, University of WI Stevens Point; **89** (t)Charles D. Winters, (b)Matt Meadows; **90** Matt Meadows; **91** Matt Meadows **94** Tim Courlas; **95** Doug Martin; **96** (l)Matt Meadows, (r)Gianni Giansanti/Sygma; **98** Phyllis Picardi/Stock Boston; **101** Matt Meadows; **102** Lawrence Berkeley Laboratories; **103** (l)Chris Bjornberg/Photo Researchers, (r)Nubar Alexanian/Stock Boston; **104** Jim Zipp/Photo Researchers; **106** (tl)Matt Meadows, (tr)National Gallery of Art, Washington, D.C., (cl)KS Studios, (cr)Tom Pantages, (bl)Aaron Haupt, (br)Richard Megna/Fundamental Photographs; **107** (tl)Charlie Westerman/Gamma Liaison, (tc)Ed Wheeler/The Stock Market, (tr)Aaron Haupt, (cl)Paul Silverman/Fundamental Photographs, (c)Dennis O'Clair/Tony Stone Images, (cr)Charles D. Winters, (b)courtesy Texas Instruments Inc.; **108** (b)Tom Pantages, (others)Matt Meadows; **109** (t)Nitinol Medical Technologies, (b)Oaktree Automation, Inc.; **110** (l)Matt Meadows, (r)Gregg Hadel/Tony Stone Images; **112** Mark Steinmetz; **113** (t)Uniphoto, (bl)The Bettmann Archive, (br)Jose L. Pelaez/The Stock Market; **114** Stamp from the collection of Prof. C.M. Lang, photograph by Gary Shulfer, University of WI Stevens Point; **117** Scala/Art Resource, NY; **118-119** Tom Pantages; **119** (r)Matt Meadows; **120** Matt Meadows; **121** (tl)Co Rentmeester/The Image Bank, (c)Craig Newbauer/Peter Arnold, Inc., (others)Matt Meadows; **122** Dick Durrance II/The Stock Market; **123** (l)Richard Megna/Fundamental Photographs, (r)Chip Clark; **124** Matt Meadows; **125** (tl)Walter Kidde Company, (tr)R. Folwell/Science Photo Library/Photo Researchers, (b)Dennis M. Gottlieb Studio Inc./The Stock Market; **127** (t)Joyce Photographics/Photo Researchers, (b)Christopher Swann/Peter Arnold, Inc.; **128** Mark Steinmetz; **129** Charles D. Winters; **131** (t)Peter Steiner/The Stock Market, (b)Osborne Photographic Illustration; **134** Manfred Kage/Peter Arnold, Inc.; **135** (t)Matt Meadows, (b)John Gillmore/The Stock Market; **137, 140** Matt Meadows; **141** (t)The Bettmann Archive, (b)NASA/Peter Arnold, Inc.; **143** Matt Meadows; **145** (tl br)Matt Meadows, (tr)Spencer Swanger/Tom Stack & Associates, (bl)Richard Megna/Fundamental Photographs; **146** Dr. Paul Alan Cox; **150** Matt Meadows; **152** Tony Stone Images; **153** Porterfield-Chickering/Photo Researchers; **154** Mark Burnett; **157** Runk/Shoenberger from Grant Heilman; **160** Matt Meadows; **162** (l, c)Matt Meadows, (r)David Aronson/Stock Boston; **163** Art Resource, NY; **164** Matt Meadows; **165** Tim Courlas; **166** Matt Meadows; **167** (l)Joseph Schuyler/Stock Boston, (r)Wendy Shattil & Bob Rozinski/Tom Stack & Associates; **168** Mark Burnett; **169** Matt Meadows; **170** (t)Mark Burnett, (b)Aaron Haupt; **171** Matt Meadows/Peter Arnold, Inc.; **173** Matt Meadows; **174** (l)Irving Geis/Science Source/Photo Researchers, (r)Gerard Vandystadt/Photo Researchers; **175** Matt Meadows; **176** (t)Matt Meadows, (b)M. Kirk/Peter Arnold, Inc; **177** (t)Phillip Hayson/Photo Researchers, (b)Mark Burnett; **178** Tony Stone Images; **179** Tim Courlas; **180** Chip Clark; **182** Matt Meadows; **185** D.L. Aubrey/The Stock Market; **188, 189** Matt Meadows; **190** (t)Uniphoto, (b)Matt Meadows; **191** (t cl)Matt Meadows, (cr)Aaron Haupt, (b)Chad Slattery/Tony Stone Images; **192** (l)Diamond International, (r)Bob Daemmrich/Uniphoto; **193, 194, 195** Matt Meadows; **196** SuperStock; **197** Matt Meadows; **198** (l)Chip Clark, (r)Jean-Loup Charmet/Science Photo Library/Photo Researchers; **199** Ray Massey/Tony Stone Images; **202** (t)Mark Steinmetz, (bl)Daniel J. Cox/Tony Stone Images, (br)Pat O'Hara/Tony Stone Images; **203** NASA/Mark Marten/Photo Researchers; **204** (t)Michael Cooper/Stock Boston, (b)Charles D. Winters; **205** (b)E.R. Degginger/Color-Pic, (others)Matt Meadows; **207, 208** Matt Meadows; **209** Ron Dorsey/Stock Boston; **210** (t)Mark Steinmetz, (b)Matt Meadows; **211** S.L. Alexander/Uniphoto; **212, 213** Ted Corwin/Oman Studio; **214** Matt Meadows; **215** (t)Comer/Gerard, (b)Matt Meadows; **216** (t)Science & Society Picture Library, (b)Tom Mareschal/The Image Bank; **217** Mitch Kezar/Tony Stone Images; **218** James H. Robinson; **219** (t)Matt Meadows, (b)Matt Meadows; **220** Matt Meadows; **221** U.S. Army Natick Research & Development Center; **222** Matt Meadows; **223** Elaine Shay;

Photo Credits

225 Robert Becker/Custom Medical Stock Photo; 228 Dan McCoy from Rainbow; 229 Stephen Frisch/Stock Boston; 230 Ken Whitmore/Tony Stone Images; 232 The Bettmann Archive; 234 Matt Meadows; 240 (t)The Kobal Collection, (c)IBM Research/Peter Arnold, Inc., (b)Michael Greenlar; 241 (l)IBM Research, (r)IBM Research/Peter Arnold, Inc.; 243 Matt Meadows; 246 Martha Cooper/Peter Arnold, Inc.; 248 (t)Bill Tronca/Tom Stack & Associates, (bl)Zefa/The Stock Market, (br)Michael Dalton/Fundamental Photographs; 249 Michael Dalton/Fundamental Photographs; 250 Bob Daemmrich/Stock Boston; 253 Tony Freeman/PhotoEdit; 254 Matt Meadows; 256 George Hunter/Tony Stone Images; 257 (l)NASA, (lc)Al Francekevich/The Stock Market, (rc)Stefan Merken/Tony Stone Images, (r)Crown Studios; 258, 262 Matt Meadows; 264 (tl)Yoav Levy/PhotoTake NYC, (tr)Chip Clark, (b)Stephen Frisch/Stock Boston; 265 (tl)William Rivelli/The Image Bank, (tr)Charles Gupton/Stock Boston, (b)Matt Meadows; 267 Matt Meadows; 268 (t)Joe Towers/The Stock Market, (c)Richard Megna/Fundamental Photographs, (b)Mark Burnett; 269 Will McIntyre/Photo Researchers; 270 (t)Glencoe photo, (c)David Parker/Science Photo Library/Photo Researchers, (b)Matt Meadows; 271 Scala/Art Resource, NY; 272 (t)H.P. Merten/The Stock Market, (c)Matt Meadows, (b)Ron Chapple/FPG International; 273 Mark Steinmetz; 274 (tl)Matt Meadows, (tr)Grant Heilman, (c)Tim Courlas, (b)GTE Labs/Peter Arnold, Inc.; 275 (t)Matt Meadows, (b)Matt Meadows; 276 Farrell Grehan/Photo Researchers; 277 (t)Yoav Levy/PhotoTake NYC, (cl)John Maher/The Stock Market, (cr)Matt Meadows, (b)SuperStock; 278 (l)Barry Slaven/Medical Images, Inc., (r)Matt Meadows; 279 (t)Skip Comer, (cl)Yoav Levy/PhotoTake NYC, (cr)Gary Hansen/PhotoTake NYC, (b)Matt Meadows; 280 Jerome Tisne/Tony Stone Images; 281 Glencoe photo; 282 (t)Matt Meadows, (b)Jim Cummins/FPG International; 283 (l)Stephen Frisch/Stock Boston, (r)Tom Pantages; 284 Matt Meadows; 285 Keith Ledbury/Uniphoto; 286 (t)Richard Megna/Fundamental Photographs, (c)Gerard Fritz/FPG International, (b)Dawson Jones, Inc./The Stock Market; 287 (t)Alan Schein/The Stock Market, (cl)H.P. Merten/The Stock Market, (cr)David Joel/Tony Stone Images, (b)B. Roland/Image Works; 288 The Robert Elgoord Collection, London; 289 (t)Terry Farmer/Tony Stone Images, (bl)Robert Sherbow/Uniphoto, (br)Matt Meadows; 290 (t)Matt Meadows, (b)Ken Kay/Fundamental Photographs; 291 (l)Jim Pickerell/The Image Works, (r)Aaron Haupt; 293 (t)Matt Meadows, (b)Mark E. Gibson/The Stock Market; 294 Stephen Simpson/FPG International; 300 Matt Meadows, (inset)IBM Research; 301 Matt Meadows; 302 Steven L. Alexander/Uniphoto; 307 Joe McNally/Sygma; 310 Mark Steinmetz; 312 Matt Meadows; 313 (l)Matt Meadows, (r)Mark Burnett; 315 Richard Pasley/Stock Boston; 316, 317 Brian Buckley; 318 Matt Meadows; 319–324 Matt Meadows; 325 (t)Matt Meadows, (bl)Frank Siteman/Stock Boston, (br)Matt Meadows; 326 (c)Matt Meadows, (bl)Mark Steinmetz, (br)Sinclair Stammers/Science Photo Library/Photo Researchers, (others)Matt Meadows; 327 Perkin-Elmer Corporation; 328 Matt Meadows; 329 Matt Meadows; 330 (l)Stan Osolinski/Tony Stone Images, (r)Zefa-N.Y. Gold/The Stock Market; 331 Kristen Brochmann/Fundamental Photographs; 333 (tl)Matt Meadows, (tc)Matt Meadows, (tr)Matt Meadows, (b)Charles D. Winters; 338 SuperStock; 339 Bill Auth/Uniphoto; 340 (t)Peter Menzel/Stock Boston, (cl)Tino Hammid, (cr)Will Crocker/The Image Bank, (b)James L. Amos; 341 (t)Mark Burnett, (bl, br)Matt Meadows; 343, 344 Matt Meadows; 345 Doug Martin; 346 James L. Amos; 348 Uniphoto; 351 Matt Meadows; 353 Mark Steinmetz; 354 Joseph Nettis/Stock Boston; 355 Doug Martin; 356, 357 Matt Meadows; 359 Eric Leigh Simmons/The Image Bank; 360, 362 Matt Meadows; 365 (l)Doug Martin, (r)David R. Frazier; 367 Diane Schiumo/Fundamental Photographs; 368 Matt Meadows; 370 David R. Frazier Photolibrary; 371 Mark Steinmetz; 372 Joseph A. Dichello Jr.; 373, 375 Matt Meadows; 376 Michael Krasowitz/FPG International; 377 David R. Frazier Photolibrary; 378 John Lemker/Earth Scenes; 381 (l)Matt Meadows, (r)Mark Burnett; 382 Matt Meadows; 384 Matt Meadows; 385 The Bettmann Archive; 386 Myrleen Ferguson Cate/PhotoEdit; 387 Kim Westerskov/Ford Scientific Films/Earth Scenes; 388 Bob Mullenix; 389 Mark Steinmetz; 390 Glenn Baglo/The Vancouver Sun; 391 Matt Meadows; 394 Vince Streano/Tony Stone Images; 397 Yoav Levy/PhotoTake NYC; 401 Bob Abraham/The Stock Market; 402 Matt Meadows, (inset)Obremski/The Image Bank; 403 Matt Meadows; 404 John Urban/Stock Boston; 405 Matt Meadows; 406 Mark Burnett; 407–410 Matt Meadows; 411 (t, c)courtesy Rosenthal Art Slides, (b)Lakeview Museum of Arts & Sciences, Richard K. Meyer Collection; 414 Pete Salautos/The Stock Market; 417 Joe Caputo/Gamma Liaison; 419 Jim Cambon/Tony Stone Images; 420 Matt Meadows; 421 Frank Cezus/Tony Stone Images; 422 Matt Meadows; 429 Richard Megna/Fundamental Photos; 430 Mark Steinmetz; 434 Jeanne Drake/Tony Stone Images; 435 Kevin Kelley/Tony Stone Images; 436 Tim Lynch/Stock Boston; 437 (l)Ryan-Beyer/Tony Stone Images, (r)Eric Neurath/Stock Boston; 441 (l)Matt Meadows, (r)Larry West/FPG International; 442 NASA; 443 (t)Geoff Butler, (b)Custom Medical Stock Photo; 444 Richard Megna/Fundamental Photographs; 445 (t)Aaron Haupt, (b)Robert Reiff/FPG International; 446 (t)E. Nagele/FPG International, (b)Bob Daemmrich/The Image Works; 448, 449 Daniel Schaefer; 450 Matt Meadows; 451 Geoff Butler; 452 Matt Meadows; 454 Matt Meadows; 455 Mark Steinmetz; 457 Matt Meadows; 458 Mark Steinmetz; 459 Geoff Butler; 460 Mark Burnett; 461 Matt Meadows; 465 Richard Kavlin/Tony Stone Images; 466 StudiOhio; 469 (t)Stephen Frink/The Stock Market, (bl)Ken Frick, (br)Vince McGuire; 470 (tl)David R. Frazier Photolibrary, (tr)Mark Steinmetz, (bl)PhotoTake NYC, (br)Skip Comer; 471 (t)Peter Johansky/FPG International, (bl)Giraudon/Art Resource, NY, (br)KS Studios; 472 Larry Hamill; 473 (l)Kip & Pat Peticolas/Fundamental Photographs, (r)George Disario/The Stock Market; 476 StudiOhio; 478 Animals Animals/Richard Shiell; 479 Richard Hutchings/PhotoEdit; 480 Mark Steinmetz; 481, 482 Matt Meadows; 483 (tl)Matt Meadows, (tr)Matt Meadows, (b)Martin Rogers/Stock Boston; 484 (t)Julie Houck/Stock Boston, (b)Chuck Holmes; 486 Tony Stone Images; 487 Charles Gupton/Stock Boston; 488, 489 Matt Meadows; 490, 491 Mark Tuschman; 492 Matt Meadows; 494 (t)Kent Knudson/Uniphoto, (b)Stephen R. Swinburne/Stock Boston; 495 (t)Shaun Van Steyn/Uniphoto, (b)Ray Pfortner/Peter Arnold, Inc.; 496 The Bettmann Archive; 497 Mark Burnett; 500 The Bettmann Archive; 501 Mark Steinmetz; 502 (l)Phil Degginger/Color-Pic, (r)Matt Meadows; 503 (t)Mark Steinmetz, (bl)Glencoe photo, (br)Tony Freeman/PhotoEdit; 504 Matt Meadows; 507 Bob Daemmrich/Tony Stone Images; 512 Mark Burnett; 514 Paul Silverman/Fundamental Photographs; 515 Doug Martin; 516 (t)Mark Steinmetz, (b)Matt Meadows; 517, 518 Matt Meadows; 520 Bob Daemmrich/Stock Boston; 522 Matt Meadows; 524 Jeff Gnass/The Stock Market; 526 Matt Meadows; 528 Charles D. Winters; 529, 530 Matt Meadows; 531 James Prince/Photo Researchers; 532 Mark Steinmetz; 533 Matt Meadows; 535 Mark Burnett; 536 (t)CNRI/Science Photo Library/Photo Researchers, (b)Mark Steinmetz; 537 Somatogen, Inc.; 538 Mark Steinmetz; 539–545 Matt Meadows; 547 Dennis Kunkel/PhotoTake NYC; 548 Mark Steinmetz; 549 John Terence Turner/FPG International; 550 Matt Meadows; 552 Charles Krebs/The Stock Market; 553 Hickson-Bender; 554 Kuhn, Inc./The Image Bank; 555 (l)Mark Burnett, (r)Michael Dalton/Fundamental Photographs; 557 Mark Burnett; 558 (l)Ann Ronan Picture Library/Image Select, (r)Nikolay Zurek/FPG International; 559, 560, 561 Matt Meadows; 563 Vulcain/Explorer/Photo Researchers; 564 Gernsheim Collection, Harry Ransom Humanities Research Center, University of TX Austin; 565, (t)Mark Steinmetz, (others)Geoff Butler; 566 NASA; 567 Bryan F. Peterson/The Stock Market; 568 (t)Matt Meadows, (b)C.J. Allen/Stock Boston; 569 (t)Stacy Pick/Stock Boston, (b)Doug Martin; 570 (l)Kunio Owaki/The Stock Market, (r)Bill Bachman/Photo Researchers; 571 Zefa/The Stock Market; 572 Tim Courlas; 573 PhotoEdit; 574 (l)Telegraph Colour Library/FPG International, (r)Hickson Associates; 575 (t)Leonard Lee Rue III/Earth Scenes, (bl)Matt Meadows, (br)Keith Kent/Science Photo Library/Photo Researchers; 576 Geoff Butler; 578 Matt Meadows; 579 Matt Meadows; 581 Matt Meadows; 582 Stephen Dalton/Photo Researchers; 583 Van Bucher/Photo Researchers; 584 Simon Fraser/Royal Victoria Infirmary, Newcastle/Science Photo Library/Photo Researchers; 587 Mark Stein-

metz; **588** ALCOA; **589** (l)ALCOA, (r)M.E. Warren/Photo Researchers; **590** (t)Rick Raymond/Nawrocki Stock Photo, (b)courtesy Kennecott Utah Copper Corporation; **591** (t)John Zoiner/International Stock, (c)Donald L. Miller/International Stock, (b)Peter Arnold, Inc.; **592** (t)Pasquale Caprile/International Stock, (bl)Tom Pantages, (br)Mark Steinmetz; **593** (t)SuperStock, (b)Rick Gayle/The Stock Market; **594** Dr. Jeremy Burgess/Science Photo Library/Photo Researchers; **595** Capital Features/The Image Works; **596, 597** Phillip DeManczuk; **598** Matt Meadows; **599** Jerry Wachter/Photo Researchers; **600** Matt Meadows; **602** Otto Rogge/The Stock Market; **603, 604** Matt Meadows; **605** Mark Steinmetz; **609** StudiOhio; **610** Dept. of Clinical Radiology, Salisbury District Hospital/Science Photo Library/Photo Researchers; **611** Coco McCoy from Rainbow; **612** StudiOhio; **613** Russell D. Curtis/Photo Researchers; **617** Richard Laird/FPG International; **620** Ed Taylor/FPG International; **621** Mark Steinmetz; **622** Anne Sager/Photo Researchers; **627** Matt Meadows; **630** (t)Morton & White, (b)John Lund/PhotoTake NYC; **633** (l)Matt Meadows, (r)Matt Meadows; **634, 635** Mark Tuschman; **637** (l)Saussier-Vanderstockt/Gamma Liaison, (r)Mark A. Leman/Tony Stone Images; **639** (l)Richard W. Brooks/Photo Researchers, (r)Mike Abrahams/Tony Stone Images; **640** Susanne Buckler/Gamma Liaison; **641** (l)Bob Daemmrich/Stock Boston, (r)Mark Steinmetz; **642** (t)Daniel A. Erickson, (c)Matt Meadows, (bl)Roger K. Burnard, (br)Matt Meadows; **643** Matt Meadows; **644** (tl, c)Mark Steinmetz, (tr)The Bettmann Archive, (b)Matt Meadows; **645** (l)Owen Franken/Stock Boston, (r)John Sims/Tony Stone Images; **646** Alvin E. Staffan; **647** Spencer Grant/Photo Researchers; **648** NASA; **649** (t)Stephen Frisch/Stock Boston, (b)Robert J. Hack/International Sports Photo Agency; **651** Matt Meadows; **653** (t)Matt Meadows, (bl)courtesy Amoco Fabric and Fibers Company, (br)Nuridsany et Perennou/Photo Researchers; **654** Glencoe photo; **655** Charles D. Winters/Photo Researchers; **656** (l)Chris Springman/The Stock Market, (r)Mark Steinmetz; **657** Vince Streano/Tony Stone Images; **658** (l)Richard Choy/Peter Arnold, Inc., (r)Doug Martin; **660** Matt Meadows; **663** Mak-1; **666** Giraudon/Art Resource, NY; **667** Michael P. Gadomski/Photo Researchers; **668** Craig Hammell/The Stock Market; **669** (l)Tom Stack & Associates, (r)Matt Meadows; **670** (l)Stan Osolinski/FPG International, (r)Zeva Delbaum/Peter Arnold, Inc.; **671** Matt Meadows; **673** (l)Peter Johanski/FPG International, (r)Stan Osolinski/FPG International; **674** Matt Meadows; **675** Mark Steinmetz; **676** Matt Meadows; **677** (l)Mark Burnett, (r)Matt Meadows; **678, 679** Otis Hairston; **680** David M. Dennis/Tom Stack & Associates; **681** Matt Meadows; **682** (t)Robert Finken/Photo Researchers, (c)Doug Martin, (b)Sheryl McNee/FPG International; **684** (l)Hans Pfletschinger/Peter Arnold, Inc., (c)Lou Jones, (r)Matt Meadows; **685** J. Barry O'Rourke/The Stock Market; **686** (t)Victor Scocozza/FPG International, (b)Tony Craddock/Tony Stone Images; **687** Biophoto Associates/Science Source/Photo Researchers; **688** Matt Meadows; **690** Richard Megna/Fundamental Photos; **691** Art Wolfe/Tony Stone Images; **692** DiMaggio/Kalish/The Stock Market; **694** Walter Iooss/The Image Bank; **695** courtesy D.E. Nicholson, University of Leeds, U.K., and Sigma Chemical Co.; **698** David R. Frazier Photolibrary; **699** (t)Matt Meadows, (b)Kevin Schafer; **700** Bob Daemmrich/Stock Boston; **701** Mark Steinmetz; **702** Dr. Jeremy Burgess/Photo Researchers; **706** Doug Martin; **707** KS Studios; **708** Doug Martin; **710, 712, 713** Matt Meadows; **714** (t)Matt Meadows, (b)David Sailors/The Stock Market; **715** courtesy Corning Inc.; **716** (l)Roy Morsch/The Stock Market, (r)KS Studios; **719** Doug Martin; **721, 725** KS Studios; **726** Grant Heilman/Grant Heilman Photography, Inc.; **727** Newmont Metallurgical Services; **728** Tommaso Guicciardini/Science Photo Library/Photo Researchers; **729** (t)Larry Lee/Westlight, (b)Telegraph Colour Library/FPG International; **730** Inga Spence/Tom Stack & Associates; **731** (l)Brownie Harris/The Stock Market, (r)Rich Brommer; **732** Jeremy Walker/Tony Stone Images; **733** (t)David R. Frazier Photolibrary/Photo Researchers, (b)John Durham/Science Photo Library/Photo Researchers; **737** AP/Wide World Photos; **740** Doug Martin; **742** Eric Lessing/Art Resource, NY, (inset)Scala/Art Resource, NY; **743** Mark Steinmetz; **744** (t)The Bettmann Archive, (bl br)Mark Burnett; **748** Doug Martin; **751** (tl)David R. Frazier Photolibrary, (tr)D.C. Lowe/FPG International, (b)David R. Frazier Photolibrary; **752** Geoff Butler; **753** Skip Comer; **754** (t)Paul Hanny/Gamma Liaison, (bl)Michael Collier/Stock Boston, (br)James King-Holmes/Science Photo Library/Photo Researchers; **755** (t)James King-Holmes/Science Photo Library/Photo Researchers, (b)Smithsonian Institution; **757** (t)Scott Warren, (b)Oregon Historical Society, negative number OrHi 93073; **758** Luis Rosendo/FPG International; **759** Museum Boymans-van Beuningen, Rotterdam; **760** (l)Stephen M. Awramik/University of California/Biological Photo Service, (r)Institute of Human Origins; **761** E. Nagele/FPG International; **762** The Bettmann Archive; **763** Geoff Butler; **765** (t)Ted Clutter/Photo Researchers, (b)Sipa Press; **767** General Atomics; **768** Motion Picture and Television Photo Archive; **769** (t)Michael Tamborrino/The Stock Market, (c)Matt Meadows, (inset)Chris Priest/Science Photo Library/Photo Researchers, (b)Howard Sochurek/Medichrome/Stock Shop; **770** (t)Jean-Perrin/CNRI/Science Photo Library/Photo Researchers, (c)Dr. Bill Hornof, School of Veterinary Medicine, UC Davis, (b)Yoav Levy/PhotoTake NYC; **773** (t)Peticolas/Megna/Fundamental Photographs, (cl)Adam Hart/Science Photo Library/Photo Researchers, (cr)DuPont Pharma, (b)Science Research Labs, Inc.; **775** Mark Steinmetz; **776** J. Taposchaner/FPG International; **777** Geoff Butler; **778** (l)William Morris/Photo Researchers, (r)Ted Rice; **779** US DOE; **781** Francois Gohier/Photo Researchers; **782** Richard Megna/Fundamental Photographs; **788** (t)Mark Burnett, (b)Doug Martin; **789** Mark Burnett; **790** Matt Meadows; **791** Mark Burnett; **800** Mark Steinmetz; **802** Glencoe photo; **804** Doug Martin; **811** Mary Van de Ven/Pacific Stock; **821, 822** Matt Meadows.

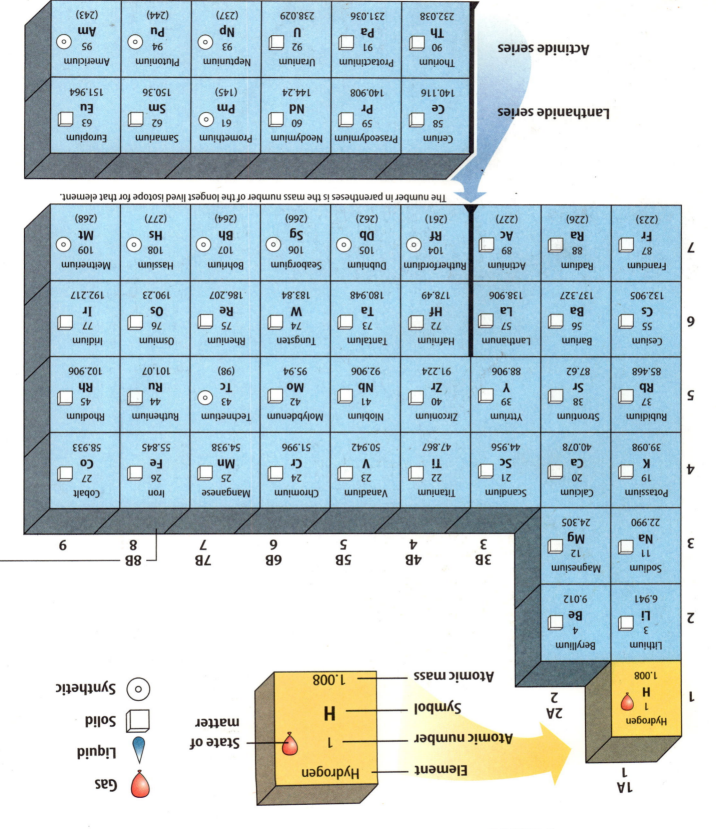

PERIODIC TABLE OF THE ELEMENTS